Mathematics A

Isidore Dressler
Edward P. Keenan

AMSCO

AMSCO SCHOOL PUBLICATIONS, INC.
315 HUDSON STREET, NEW YORK, N.Y. 10013

Reviewers:

Steven J. Balasiano
Assistant Principal, Mathematics
Canarsie High School
Brooklyn, NY

Rosalie David
Assistant Principal, Mathematics
A. Philip Randolph Campus High School
New York, NY

James J. Elliott
Assistant Principal, Mathematics
Port Richmond High School
Staten Island, NY

Lisa Hufnagel
Mathematics Department
Yorktown High School
Yorktown Heights, NY

Mark S. Richman
Mathematics Department
Automotive High School
Brooklyn, NY

Ruby M. Sylvester
Assistant Principal, Mathematics
East New York Transit Tech High School
Brooklyn, NY

James N. Tate
Administrator
City School District of Albany
Albany, NY

Portions of this book were adapted from the following Amsco publications:

Integrated Mathematics, Course I, Third Edition
Revised by Edward P. Keenan and Ann Xavier Gantert
Integrated Mathematics, Course II, Third Edition
Revised by Edward P. Keenan and Ann Xavier Gantert
Preparing for the Regents Examination, Mathematics A
by Richard J. Andres and Joyce Bernstein
Algebra I
by Isidore Dressler

Please visit our Web site at: www.amscopub.com

When ordering this book, please specify:
R 724 P *or* MATHEMATICS A, *Paperback*
or
R 724 H *or* MATHEMATICS A, *Hardbound*

ISBN 1-56765-545-9/NYC Item 56765-545-8 (Softbound edition)
ISBN 1-56765-546-7/NYC Item 56765-546-6 (Hardbound edition)

PRINTED IN THE UNITED STATES OF AMERICA

6 7 8 9 10 05 04 03 02

Preface

Mathematics A is a new compilation of the material that has been a leader in presenting high school mathematics in a contemporary, integrated manner. Over the last decade, this approach has undergone further changes and refinements, which have evolved into Mathematics A.

This Amsco book parallels the integrated approach to the teaching of high school mathematics that is being promoted by the National Council of Teachers of Mathematics (NCTM) in its *Principles and Standards for School Mathematics* and the New York State Core Curriculum for High School Mathematics A. Moreover, the Amsco book implements the range of suggestions set forth in the NCTM Standards and the Core Curriculum, which are the acknowledged guidelines for achieving a higher level of excellence in the study of mathematics.

In this edition:

✔ **The scientific calculator** is introduced and used throughout the book as a routine tool in the study of mathematics, replacing the need for tables and stressing increased attention to estimation skills.

✔ **The real number system** is fully developed in Chapter 1, to help students understand and correctly interpret technological limitations such as calculator displays of rational approximations for irrational numbers.

✔ **Integration** of algebra, geometry, and other branches of mathematics is expanded by the early introduction of selected topics and the inclusion of challenging and non-routine problems.

✔ **Enrichment** is stressed both in the book and in the Teacher's Manual where a host of suggestions is given for teaching strategies and alternative assessment. The Manual includes opportunities for cooperative learning, hands-on activities, extended tasks, and applications with graphing calculators. Reproducible *Enrichment Activities* that lend themselves to student exploration of topics in greater depth are provided in the Teacher's Manual for each chapter.

✔ **Problem solving,** which is a primary goal in all learning standards, is addressed through an introductory chapter where the general technique and alternative strategies are used to solve both standard and non-routine problems, and is further emphasized with fully developed examples in a problem-solving approach throughout the text.

This text has been written to provide effective teaching materials for a unified program appropriate for ninth- and tenth-grade mathematics students, including all topics contained in the Core Curriculum Key Ideas and Performance Indicators for Math A.

Learning standards, whether national in scope such as the NCTM Standards or more regional such as the Core Curriculum of New York State, include goals that are based on clearly defined curriculum content. This text provides that curriculum and gives teachers the opportunity to establish procedures for alternative assessment. As an example, student portfolios may consist of completed copies of the *Enrichment Activities* and suggested chapter test items as well as reports generated by the *Explorations* included in the text. Writing and communication skills are employed frequently when students are asked to explain their reasoning in arriving at solutions.

An intent of the authors was to make this material of greatest service to average students. Since its publication, however, this material has been used successfully with students of varying abilities, and this book has been prepared to address that range of abilities. Specifically:

✔ Concepts are carefully developed using appropriate language and mathematical symbolism.

✔ General principles and procedures are stated clearly and concisely.

✔ Numerous examples are solved as models for students, with detailed step-by-step explanations. In many places, both alternative approaches and calculator solutions are provided.

✔ Varied and carefully graded exercises, in abundance, test student understanding of mathematical skills. Additional enrichment materials challenge the most capable students.

This text is offered so that teachers may effectively continue to help students comprehend, master, and enjoy mathematics from an integrated point of view.

Mathematics A is dedicated to Hilda Dressler and Joan Keenan, in remembrance of Edward P. Keenan and Isidore Dressler.

The Authors

Contents

Chapter 9 First-Degree Equations and Inequalities in One Variable 271

Chapter 10 Angle Measure in Geometry 323

Chapter 11 Congruence and Transformations 355

Chapter 1

The Real Numbers

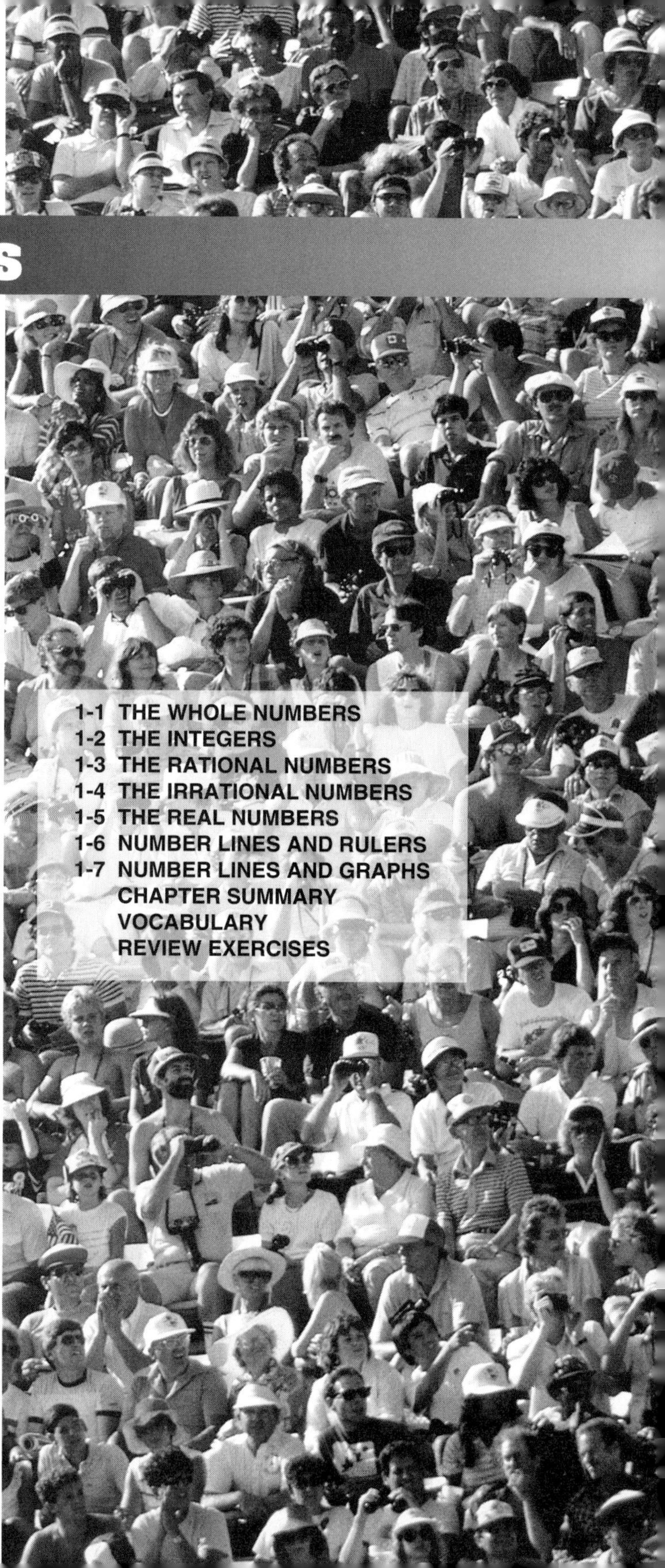

We are surrounded by numbers every day of our lives. Who are *we*?

Every 10 years, the U.S. government takes a census and releases facts discovered.

In 1990, the United States had a population of 250,000,000. Nearly 26% of us were under 18 years of age, and 12% were over 65. About 79% earned high school diplomas, and one-quarter of those people graduated from college.

An average household had 2.62 people, and the average value of a home was \$80,000. The cost of an average home, including mortgage, was \$761 per month, while the average rent was \$462 per month.

How will these numbers change when the data are revealed from the census of the year 2000? Of 2010? Can you make some predictions?

1-1 THE WHOLE NUMBERS

Mathematics is the study of numbers, shapes, arrangements, relationships, and reasoning. Mathematics is both a science and an art that can be used to describe the world in which we live, to solve problems, and to create new ideas.

Numbers, which are only a small part of mathematics, help us to understand algebra, to measure geometric objects, and to make predictions using probability and statistics. In this chapter, we will study numbers such as those shown below:

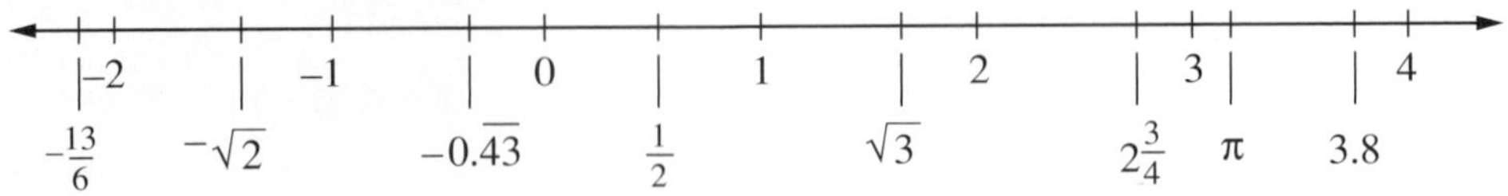

Every point on this number line corresponds to a ***real number***. What are real numbers? What is meant by values such as $\sqrt{3}$ and $-0.\overline{43}$? Let us begin with simpler numbers that we know.

Symbols for Numbers

A ***number*** is really an idea; it is something that we can talk about and think about. We represent numbers in writing by using the symbols 1, 2, 3, 4, and so on. These symbols, called ***numerals*** or ***numerical expressions***, are not numbers but are used to represent numbers.

Counting Numbers or Natural Numbers

The ***counting numbers***, which are also called ***natural numbers***, are represented by the symbols

1, 2, 3, 4, 5, 6, 7, 8, 9, 10, 11, 12, ...

The three dots after the 12 indicate that the numbers continue *in the same pattern* without end. The smallest counting number is 1. Every counting number has a ***successor*** that is 1 more than that number. The successor of 1 is 2, the successor of 2 is 3, and so on. Since this process of counting is endless, there is no last counting number.

On this number line, the points associated with counting numbers are highlighted and an arrow shows that these numbers continue without end.

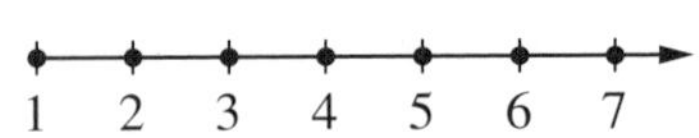

The Set of Whole Numbers

Zero is not a counting number. By combining 0 with all the counting numbers, we form the set of ***whole numbers***. The whole numbers are represented by the symbols

0, 1, 2, 3, 4, 5, 6, 7, 8, 9, 10, 11, 12, ...

The smallest whole number is 0. There is no largest whole number.

Notice that the number line has been extended to include the number 0.

A ***set*** is a collection of distinct objects or elements. A set is usually indicated by enclosing its elements within a pair of braces, { }. For example, the set of whole numbers can be written as {0, 1, 2, 3, 4, . . .}.

Types of Sets

A ***finite set*** is a set whose elements can be counted. For example, the set of ***digits*** consists of only ten symbols, 0 through 9, that are used to write our numerals:

$$\{0, 1, 2, 3, 4, 5, 6, 7, 8, 9\}.$$

An ***infinite set*** is a set whose elements cannot be counted because there is no end to the set. For example, the counting numbers and the whole numbers are both infinite sets.

The ***empty set*** or ***null set*** is a set that has no elements, written as { } or ∅. For example, the set of months beginning with the letter *Q* is empty, and the set of negative counting numbers is also empty, or { }.

Symbols for Arithmetic Operations

In earlier years, you studied the four basic operations in arithmetic: addition, subtraction, multiplication, and division. An ***operation*** is a procedure or a rule that tells you how to combine elements or numbers in a set to obtain other numbers.

Operation	Symbol	Result
Addition	+	Sum
Subtraction	−	Difference
Multiplication	×	Product
Division	÷	Quotient

Sometimes, an operation can be shown by the use of different symbols. For example, ***multiplication*** is shown by ×, by a raised centered dot, or by parentheses.

$$6 \times 2 = 12 \quad \text{or} \quad 6 \cdot 2 = 12 \quad \text{or} \quad 6(2) = 12 \quad \text{or} \quad (6)(2) = 12.$$

Division is shown by the symbol ÷ or by writing a fraction in which the numerator is divided by the denominator.

$$6 \div 2 = 3 \quad \text{or} \quad \frac{6}{2} = 3.$$

Numerical Expressions

A ***numerical expression*** is a way of writing a number in symbols. The expression can be a single numeral, or it can be a collection of numerals with one or more operation symbols. For example:

$$6 + 2 \qquad 18 - 10 \qquad 4 \times 2 \qquad 640 \div 80$$

$$2 \times 2 \times 2 \qquad 2 + 2 + 2 + 2 \qquad 1 \times 7 + 1 \qquad 8$$

Each of these expressions, when simplified, becomes the same value or same number, 8. In general, to ***simplify*** a numerical expression means to find its ***value***.

EXAMPLES

In 1 and 2, each row contains four numerical expressions. Three of the four represent the same whole number. Which expression does *not* represent that number?

1. $8 + 4$ $\qquad 3 \times 4$ $\qquad 12 \div 1$ $\qquad 5 + 6$

Solution In order, the expressions represent 12, 12, 12, and 11. Therefore, $5 + 6$ does not represent the same number that the other expressions represent.
Answer: $5 + 6$

---

2. $0 + 8$ $\qquad 8 - 0$ $\qquad 8 \div 0$ $\qquad 8 \times 1$

Solution The first, second, and fourth expressions represent 8. The expression $8 \div 0$ does not represent any number, since division by 0 is not possible. Remember, however, that $0 \div 8 = 0$.
Answer: $8 \div 0$

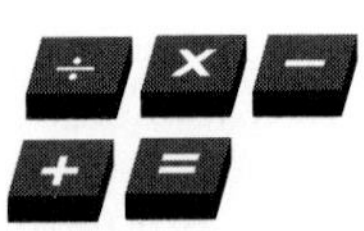

In Examples 3 and 4, a calculator is used. The primary purpose of any calculator is to perform arithmetic operations—in particular, the four basic operations: addition, subtraction, multiplication, and division. The keys pictured at the left are used to perform these operations.

3. Add 6 to the product of 3 and 9.

Solution

Enter: 3 [×] 9 [+] 6 [=]

Display: [33.]

Answer: 33

---

4. From the quotient of 10 and 2, subtract 1.

Solution *Enter:* 10 [÷] 2 [−] 1 [=]

Display: [4.]

Answer: 4

EXERCISES WITH OPEN-RESPONSE PROBLEMS

In 1–16, use the set of whole numbers, {0, 1, 2, 3, . . .}, and the indicated operations to find whole-number answers. If no whole-number answer is possible, write the word *None*.

1. $4 + 8$	**2.** $4 - 8$	**3.** 4×8	**4.** $4 \div 8$
5. $20 + 20$	**6.** $20 - 20$	**7.** 20×20	**8.** $20 \div 20$
9. $6 + 0$	**10.** $6 - 0$	**11.** 6×0	**12.** $6 \div 0$
13. $78 + 97$	**14.** $93 - 19$	**15.** 27×38	**16.** $594 \div 11$

In 17–26, determine, in each case, the whole number that answers the question or is the result of performing the operations.

17. What is the product of 2 and 8?

18. What is the difference of 12 and 3?

19. What is the sum of 18 and 6?

20. What is the quotient of 10 and 2?

21. From the quotient of 20 and 4, subtract 2.

22. From the difference of 20 and 4, subtract 4.

23. To the product of 20 and 4, add 8.

24. To the sum of 20 and 4, add 8.

25. Subtract 3 from the product of 4 and 8.

26. Add 7 to the quotient of 6 and 3.

In 27–31, each row contains four numerical expressions. Three of the four expressions result in the same numerical value. In each case, which expression has a numerical value different from the other three?

27. $3 + 6$	$12 - 3$	$3 \cdot 3$	$3 - 12$
28. $10 + 0$	$10 - 0$	10×0	$0 + 10$
29. $14 - 14$	14×0	$0 \div 14$	$14 \div 14$
30. 75×80	25×200	40×125	8×625
31. $522 \div 29$	$697 \div 41$	$\frac{476}{28}$	$\frac{612}{36}$

In 32–41, write the numeral that is the answer to each instruction. If no answer is possible, explain why.

32. Name the first counting number.

33. Name the first whole number.

34. Name the successor of the whole number 75.

35. Name the successor of the counting number 999.

36. Name the successor of the natural number 1,909.
37. Name a whole number that is *not* a counting number.
38. Name a counting number that is *not* a whole number.
39. Name the largest whole number.
40. Name the product of the smallest whole number and the largest one-digit counting number.
41. Name the product of the smallest counting number and the largest one-digit whole number.

In 42–49, tell whether the set is finite, infinite, or empty.

42. Counting numbers
43. Counting numbers up to 100
44. Digits
45. Whole numbers
46. States beginning with the letter *Z*
47. States beginning with the letter *C*
48. Three-digit numbers
49. U.S. presidents born in Hawaii

1-2 THE INTEGERS

The temperature on a winter day may be below 0 degree, and someone may write a check for an amount that cannot be covered by funds in a checking account. Both situations describe negative numbers.

Just as the number line was extended in Section 1-1 to include 0, we can extend it to include negative numbers. A number that is 1 less than 0 is -1, a number 2 less than 0 is -2, and so on.

Imagine that a mirror is placed at the number 0, and the counting numbers (which are treated as *positives*) are reflected in the mirror to show *negative* numbers. Our new number line extends forever in two directions; it has no beginning and no end.

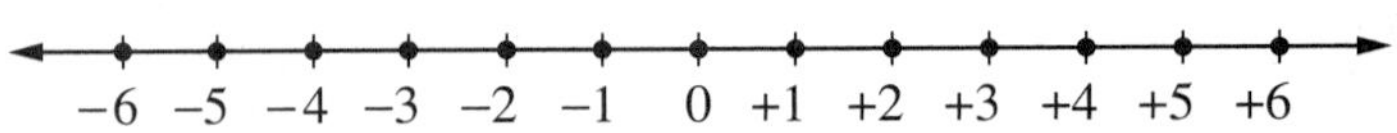

Each positive number can be paired with a negative number that is the same distance away from 0 and on the opposite side of 0. The numbers of each pair are called ***opposites***.

The opposite of $+1$ is -1, and the opposite of -1 is $+1$.

The opposite of $+2$ is -2, and the opposite of -2 is $+2$; and so on.

Notice that 0 is neither positive nor negative.

The Set of Integers

This new set of numbers, called the ***integers***, consists of the positive whole numbers (1, 2, 3, . . .), the opposites or negatives of these whole numbers (-1, -2, -3, . . .), and 0.

Two common ways to write the integers as a set are as follows:

$\{\ldots, -3, -2, -1, 0, +1, +2, +3, \ldots\}$ and $\{0, +1, -1, +2, -2, +3, -3, \ldots\}$

Subsets of the Integers

Set A is called a ***subset*** of set B, written $A \subset B$, if every element of set A is also an element of set B.

Using this definition, we know that the whole numbers and the counting numbers are subsets of the integers. Counting numbers also form a subset of the whole numbers. These subsets can be illustrated in a diagram, as shown below.

Whole numbers $\subset$ Integers

Counting numbers $\subset$ Integers

Counting numbers $\subset$ Whole numbers

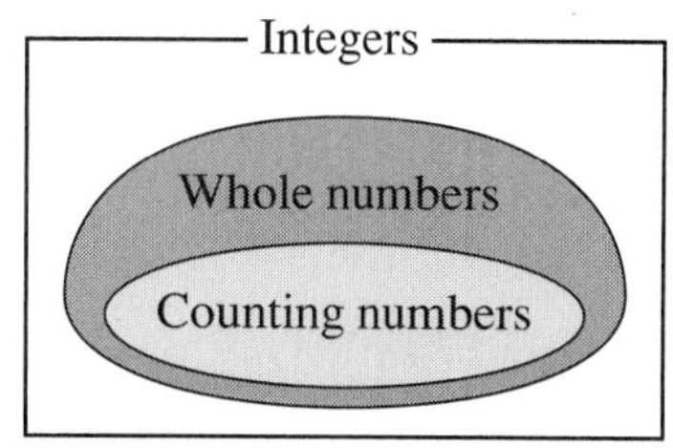

There are many other subsets of the integers, such as:

1. Odd whole numbers $\{1, 3, 5, 7, 9, 11, \ldots\}$
2. Odd integers $\{\ldots, -5, -3, -1, 1, 3, 5, \ldots\}$
3. Even whole numbers $\{0, 2, 4, 6, 8, 10, \ldots\}$
4. Even integers $\{\ldots, -6, -4, -2, 0, 2, 4, 6, \ldots\}$
5. One-digit whole numbers $\{0, 1, 2, 3, 4, \ldots, 9\}$

Ordering the Integers

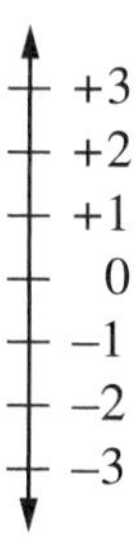

There are two standard forms of the number line that we use. On a *vertical* number line (as pictured left), such as one seen on a thermometer, the higher up we go, the greater will be the number or the higher the temperature. Just as 4 is greater than 3, so is 3 greater than 0, and 0 greater than -1. It follows that -1 is greater than -2, and -2 is greater than -20.

On a *horizontal* number line (as pictured below), positive numbers usually appear to the right, and the greater of any two numbers will be the one to the right. Again, 2 is greater than 0, and -1 is greater than -3.

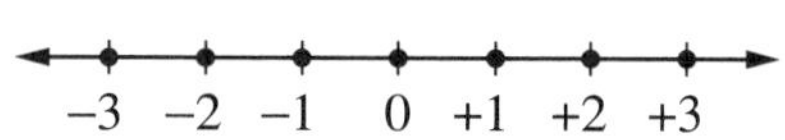

Symbols of Inequality

In our daily lives, we are often asked to compare quantities. Which is cheaper? Which weighs more? Who is taller? Which will last longer? Are two objects the same size? The answers to these questions are given by comparing quantities that are stated in numerical terms.

If two numbers are not equal, the relationship between them can be expressed in several different ways.

Symbol	Example	Read
$>$	$8 > 3$	8 is greater than 3.
$<$	$3 < 8$	3 is less than 8.
$\ngtr$	$3 \ngtr 8$	3 is not greater than 8.
$\nless$	$8 \nless 3$	8 is not less than 3.
$\geq$	$8 \geq 3$	8 is greater than or equal to 3.
$\leq$	$3 \leq 8$	3 is less than or equal to 8.
$\neq$	$8 \neq 3$	8 is not equal to 3.

Notice that in an ***inequality*** the symbols $>$ and $<$ point to the smaller number.

EXAMPLES

In 1–4, tell whether each statement is true or false.

1. $-3 > -5$ *Answer:* True

2. $0 < -4$ *Answer:* False

3. $12 - 5 > 2$ *Answer:* True

4. $(2)(7) \leq 14$ *Answer:* True

---

5. Use the symbol $<$ to order the numbers -4, 2, and -7.

Answer: $-7 < -4$, and $-4 < 2$ or $-7 < -4 < 2$

---

6. If two different numbers are each negative, how can you tell which one is greater?

Answer: The greater of two negative numbers is closer to 0, *or* if two numbers are negative, the smaller is further from 0.

In 7 and 8, write, in each case, at least three true statements to compare the numbers in the order given.

7. 8 and 2 *Answer:* $8 > 2$; $8 \neq 2$; $8 \nless 2$; $8 \geq 2$

8. 12 and 12 *Answer:* $12 = 12$; $12 \ngtr 12$; $12 \nless 12$; $12 \geq 12$; $12 \leq 12$

EXERCISES WITH OPEN-RESPONSE PROBLEMS

In 1–4, state whether each sentence is true or false. Give a reason for each answer.

1. $+5 > +2$ **2.** $-3 < 0$ **3.** $-7 > -1$ **4.** $-2 > -10$

In 5–10, write each inequality using the symbol > or the symbol <.

5. $+8$ is greater than $+6$.
6. -8 is less than 0.
7. -5 is less than -2.
8. -5 is greater than -25.
9. The sum of 16 and 3 is greater than the product of 9 and 2.
10. The product of 6 and 7 is less than the quotient of 100 and 2.

In 11–14, express each inequality in words.

11. $+7 > -7$ **12.** $-20 < -3$ **13.** $-4 < 0$ **14.** $-9 \geq -90$

In 15–25, state whether each sentence is true or false.

15. $+16 > +4$ **16.** $-16 > -4$ **17.** $-8 > +2$ **18.** $-8 > -9$
19. $-2 < +5$ **20.** $+8 \geq +2$ **21.** $-7 \leq -12$ **22.** $-3 \leq -3$
23. $+6 > -1 > -7$ **24.** $-15 > -13 > -2$ **25.** $-65 < -34 < -22$

In 26–29, use the symbol < to order the numbers.

26. $-4, +8$ **27.** $-3, -6$ **28.** $+3, -2, -4$ **29.** $-2, +8, 0$

In 30–33, write, in each case, three true statements to compare the numbers, using the order in which they are given.

30. 8 and 14 **31.** 9 and 3 **32.** 15 and 15 **33.** 6 and -2

34. What is the smallest positive integer?
35. What is the greatest negative integer?
36. **a.** Is there a greatest positive integer? **b.** Why?
37. **a.** Is there a least negative integer? **b.** Why?
38. In Column I, sets of numbers are described in words. In Column II, the sets are listed using patterns and dots. Match the lists in Column II with their correct sets in Column I.

Column I	*Column II*
1. Counting numbers	a. 0, 1, 2, . . . , 9
2. Whole numbers	b. 0, 1, 2, . . .
3. Even whole numbers	c. 0, 2, 4, 6, . . .
4. Odd whole numbers	d. 0, 2, 4, 6, 8
5. Even counting numbers	e. 0, 2, −2, 4, −4, 6, −6, . . .
6. Odd integers	f. 1, 2, 3, 4, . . .
7. Even integers	g. 1, 2, 3, . . . , 9
8. One-digit whole numbers	h. 1, 3, 5, 7, . . .
9. One-digit counting numbers	i. 1, 3, 5, 7, 9
10. Odd whole numbers less than 10	j. 2, 4, 6, 8, . . .
11. Even whole numbers less than 10	k. −2, −1, 0, 1, 2, 3, 4, . . .
12. Integers greater than −3	l. 1, −1, 3, −3, 5, −5, . . .

39. Give three examples in which a negative number can be used in describing a measurement or an event.

1-3 THE RATIONAL NUMBERS

In earlier years, you worked with many numbers other than integers, such as fractions, decimals, and mixed numbers. These numbers from arithmetic, which can be located on the real number line, behave in a special way. Consider the following examples:

$$\frac{3}{5} \qquad 3\frac{1}{2} = \frac{7}{2} \qquad 0.9 = \frac{9}{10} \qquad -\frac{41}{100}$$

$$8.1 = 8\frac{1}{10} = \frac{81}{10} \qquad 0.25 = \frac{25}{100} = \frac{1}{4}$$

Each of the numbers shown here is written in the form of a fraction. In fact, every integer can be written as a fraction by writing the integer with a denominator of 1:

$$5 = \frac{5}{1} \qquad 0 = \frac{0}{1} \qquad -12 = \frac{-12}{1}$$

In general, any integer n can be written as $\frac{n}{1}$, which is a quotient of two integers.

The Set of Rational Numbers

The ***rational numbers*** are all numbers that can be expressed in the form $\frac{a}{b}$ where a and b are integers and $b \neq 0$.

Notice that the first five letters of the word *rational* form the word *ratio*, which means a comparison of two quantities by division.

The counting numbers, the whole numbers, and the integers are all subsets of the set of rational numbers, as illustrated in the diagram.

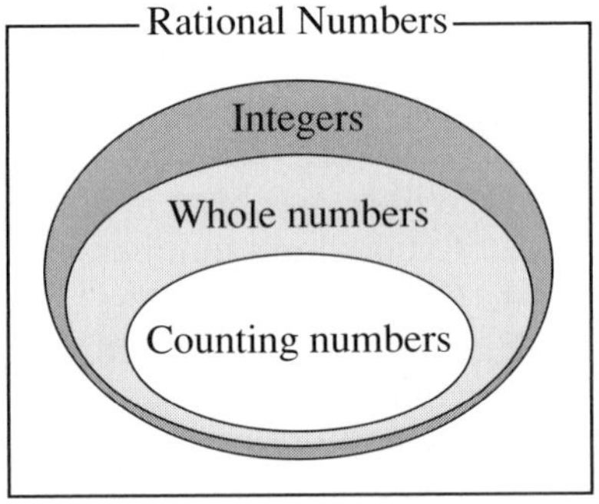

The Rational Number Line

Every rational number can be associated with a point on the real number line. For example, $\frac{1}{2}$ is midway between 0 and 1, and -2.25 $\left(\text{or } -2\frac{1}{4}\right)$ is one-quarter of the way between -2 and -3, closer to -2.

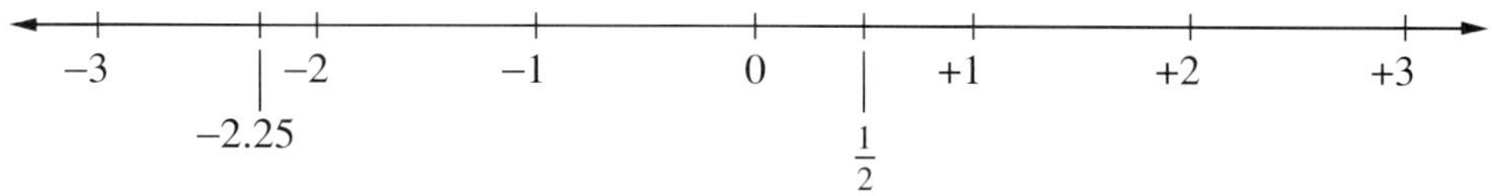

The rational numbers, like other sets studied earlier, can be **ordered.** In other words, given any two unequal rational numbers, we can tell which one is greater. For example, we know that $\frac{1}{2} > -1$ because, on the standard number line, $\frac{1}{2}$ is to the right of -1. There are also other ways to determine which of two rational numbers is greater, as shown in the following example.

EXAMPLE

Which is the greater of the numbers $\frac{7}{9}$ and $\frac{8}{11}$?

Solution METHOD 1. Express the numbers as equivalent fractions with a common denominator, and compare the numerators.

$$\frac{7}{9} = \frac{7}{9} \cdot \frac{11}{11} = \frac{77}{99}$$

$$\frac{8}{11} = \frac{8}{11} \cdot \frac{9}{9} = \frac{72}{99}$$

Since $\frac{77}{99} > \frac{72}{99}$, then $\frac{7}{9} > \frac{8}{11}$.

METHOD 2. Change the fractions to decimals by dividing each numerator by its denominator to see which is greater. The answers here are from a calculator that shows eight places in each display.

Enter: 7 [÷] 9 [=]
Display: [0.7777777]

Enter: 8 [÷] 11 [=]
Display: [0.7272727]

Answer: $\frac{7}{9} > \frac{8}{11}$

A Property of the Rational Numbers

- **The set of rational numbers is everywhere *dense*.** In other words, given two unequal rational numbers, it is always possible to find a rational number that lies between them.

For example, some rational numbers between 1 and 2 are $1\frac{1}{2}$, $1\frac{1}{8}$, $1\frac{2}{3}$, and $1\frac{9}{10}$. In fact, there is an infinite number of rational numbers between two rational numbers.

One way to find a rational number between two rational numbers is to find their *average*, called the ***mean***. To find the mean of two numbers, add the numbers and divide by 2. The mean (or average) of 3 and 5 = (3 + 5) ÷ 2 = 8 ÷ 2 = 4, the number that is exactly between 3 and 5 on a number line. In the same way, a rational number between $\frac{1}{4}$ and $\frac{3}{4}$ is found as follows:

$$\left(\frac{1}{4} + \frac{3}{4}\right) \div 2 = 1 \div 2 = \frac{1}{2}$$

A calculator can be used to do this.

Enter: (1 ÷ 4 + 3 ÷ 4) ÷ 2 =

Display: 0.5

Note that, on a calculator, the rational number $\frac{1}{2}$ is written as 0.5.

Expressing a Rational Number as a Decimal

To express as a decimal a rational number named as a fraction, we simply perform the indicated division.

EXAMPLES

1. Express as a decimal: **a.** $\frac{1}{2}$ **b.** $\frac{3}{4}$ **c.** $\frac{1}{16}$

Solutions **a.** $\frac{1}{2} = 2\overline{)1.000000}$ = 0.500000 *Ans.* **b.** $\frac{3}{4} = 4\overline{)3.000000}$ = 0.750000 *Ans.* **c.** $\frac{1}{16} = 16\overline{)1.000000}$ = 0.062500 *Ans.*

In each of the examples $\frac{1}{2}$, $\frac{3}{4}$, and $\frac{1}{16}$, when we perform the division, we reach a point after which we continually obtain only zeros in the quotient. Decimals that result from such divisions, for example, 0.5, 0.75, and 0.0625, are called ***terminating decimals.***

Not all rational numbers can be expressed as terminating decimals, as shown in the following examples.

EXAMPLES

2. Express as a decimal: **a.** $\frac{1}{3}$ **b.** $\frac{2}{11}$ **c.** $\frac{1}{6}$ **d.** $\frac{3}{55}$

Solutions **a.** $\frac{1}{3} = 3\overline{)1.000000}$ = 0.333333 . . . *Ans.*

b. $\frac{2}{11} = 11\overline{)2.000000}$ = 0.181818 . . . *Ans.*

c. $\frac{1}{6} = 6\overline{)1.000000}$ = 0.166666 . . . *Ans.*

d. $\frac{3}{55} = 55\overline{)3.000000}$ = 0.054545 . . . *Ans.*

In each of the above examples, when we perform the division, we find, in the quotient, that the same group of digits is continually repeated in the same order. Decimals that keep repeating endlessly are called ***repeating decimals*** or ***periodic decimals***.

A repeating decimal may be written in abbreviated form by placing a bar ($\overline{\ }$) over the group of digits that is to be continually repeated. For example:

$$0.333333\ldots = 0.\overline{3} \qquad 0.181818\ldots = 0.\overline{18} \qquad 0.166666\ldots = 0.1\overline{6}$$

The two preceding examples illustrate the truth of the following statement:

- **Every rational number can be expressed as either a terminating decimal or a repeating decimal.**

Note that the equalities $0.5 = 0.5\overline{0}$ and $0.75 = 0.75\overline{0}$ illustrate the fact that every terminating decimal can be expressed as a repeating decimal that, after a point, repeats with all 0's. Therefore, we may say:

- **Every rational number can be expressed as a repeating decimal.**

Since every terminating decimal can be expressed as a repeating decimal, we will henceforth regard terminating decimals as repeating decimals.

Expressing a Decimal as a Rational Number

Recall the steps you used to change a terminating decimal to a fraction:

1. Read it (using place value): 0.8 is read as 8 tenths.

2. Write it (as a fraction, using the same words as in step 1): $\frac{8}{10}$ is also 8 tenths.

3. Reduce it (if possible): $\frac{8}{10} = \frac{4}{5}$

EXAMPLES

1. Express as a fraction: **a.** 0.3 **b.** 0.37 **c.** 0.139 **d.** 0.0777

Answers:

a. $0.3 = \frac{3}{10}$ **b.** $0.37 = \frac{37}{100}$ **c.** $0.139 = \frac{139}{1{,}000}$ **d.** $0.0777 = \frac{777}{10{,}000}$

A few simple operations with numbers can be used to show how a repeating decimal is expressed as a fraction. Study these steps before looking at the examples.

1. Think of any number: Let the number be 5.
2. Be aware that 10 times the number minus the number equals 9 times the number: $10(5) - 5 =$ $50 - 5 = 45$ or $9(5)$
3. Divide the result by 9, and the original number appears: $\frac{45}{9} = 5$

These same steps, or a variation of them, are used to change any repeating decimal to a fraction, as shown in the following examples.

EXAMPLES

2. Find a fraction that names the same rational number as:

a. 0.6666 . . . **b.** 0.4141 . . . .

Solution **a.** Since the pattern repeats in one place, multiply by 10 to shift the decimal point one place.

Let the number = 0.6666 . . . .

10 times the number = 6.6666 . . .
Subtract the number: −0.6666 . . .
9 times the number = 6

Divide by 9 to get the original number.

The number $= \frac{6}{9} = \frac{2}{3}$

b. Since the pattern repeats every two places, multiply by 100 to shift the decimal point two places.

Let the number = 0.4141 . . . .

100 times the number = 41.4141 . . .
Subtract the number: −0.4141 . . .
99 times the number = 41

Divide by 99 to get the original number.

The number $= \frac{41}{99}$

Answers: **a.** $0.6666\ldots = \frac{2}{3}$ **b.** $0.4141\ldots = \frac{41}{99}$

(*Note:* To check these results, enter the fractions on a calculator.)

The preceding examples illustrate the fact that:

- **Every repeating decimal is a rational number.**

Knowing this, and knowing that every rational number can be expressed as a repeating decimal, we can now make the following statement:

- **A number is a rational number if and only if it can be represented by a repeating decimal.**

EXERCISES WITH OPEN-RESPONSE PROBLEMS

In 1–10, write each rational number in the form $\frac{a}{b}$, where a and b are integers, and $b \neq 0$.

1. 0.7 **2.** 0.18 **3.** -0.21 **4.** 9 **5.** -3

6. 0 **7.** $5\frac{1}{2}$ **8.** $-3\frac{1}{3}$ **9.** 0.007 **10.** -2.3

In 11–22, state, in each case, which of the given numbers is the greater.

11. $\frac{5}{2}, \frac{7}{2}$ **12.** $-\frac{9}{3}, -\frac{11}{3}$ **13.** $\frac{5}{6}, -\frac{13}{6}$ **14.** $-\frac{1}{5}, -5$ **15.** $\frac{5}{2}, \frac{7}{4}$

16. $-\frac{10}{3}, -\frac{13}{6}$ **17.** $\frac{13}{6}, \frac{15}{10}$ **18.** $-\frac{5}{8}, -\frac{5}{12}$ **19.** $1.4, 1\frac{3}{5}$ **20.** $-3.4, -3\frac{1}{3}$

21. $0.06, \frac{1}{6}$ **22.** $-\frac{15}{11}, -\frac{11}{15}$

In 23–32, find a rational number midway between each pair of given numbers.

23. 5, 6 **24.** $-4, -3$ **25.** $-1, 0$ **26.** $\frac{1}{4}, \frac{1}{2}$ **27.** $\frac{1}{2}, \frac{7}{8}$

28. $-\frac{3}{4}, -\frac{2}{3}$ **29.** $-2.1, -2.2$ **30.** $2\frac{1}{2}, 2\frac{5}{8}$ **31.** $-1\frac{1}{3}, -1\frac{1}{4}$ **32.** $3.05, 3\frac{1}{10}$

In 33–42, write each rational number as a repeating decimal. (*Hint:* Every terminating decimal has a repeating zero, as in $0.3 = 0.3\overline{0}$.)

33. $\frac{5}{8}$ **34.** $\frac{9}{4}$ **35.** $-5\frac{1}{2}$ **36.** $\frac{13}{8}$ **37.** $-\frac{7}{12}$

38. $\frac{5}{3}$ **39.** $\frac{7}{9}$ **40.** $\frac{2}{11}$ **41.** $\frac{5}{99}$ **42.** $-\frac{5}{6}$

In 43–48, find a fraction that names the same rational number as each decimal.

43. 0.5 **44.** 0.555 . . . **45.** $-0.\overline{2}$ **46.** $0.125\overline{0}$ **47.** 0.2525 . . . **48.** $0.0\overline{7}$

In 49–55, tell whether each statement is true or false, and give a reason for each answer.

49. Every integer is a rational number.
50. Whole numbers can be negative.
51. On a number line, the greater of two numbers is always the number further away from 0.
52. Every rational number can be written as a repeating decimal.
53. Between 0 and 1, there is an infinite number of fractions.
54. There is an infinite number of numbers between -2 and -1.
55. There is a smallest positive rational number.

1-4 THE IRRATIONAL NUMBERS

We have learned that on the real number line, there is one point for every rational number. We also know that there is an infinite number of rational numbers and, in turn, an infinite number of points assigned to these numbers. These points are so dense and crowded together that the line appears to be complete. However, there are still points on the real number line that are not associated with rational numbers.

The Set of Irrational Numbers

Recall that every repeating decimal is a rational number. This includes terminating decimals, where 0 is repeated. There are infinitely many decimals, however, that do not terminate and are nonrepeating. Here is one example of such a decimal:

$$0.030030003000030000003\ldots$$

Observe that, in this number, only the digits 0 and 3 appear. First, there is a zero; to the right of the decimal point, there is a 3 preceded by one 0, then a 3 preceded by two 0's, then a 3 preceded by three 0's, and so on. Since this number is not a repeating decimal, it is not a rational number.

A nonrepeating decimal is called an ***irrational number***. An irrational number cannot be expressed in the form $\frac{a}{b}$, where a and b are integers ($b \neq 0$).

When writing an irrational number, we use three dots (. . .) after a series of digits to indicate that the number does not terminate. The dots do *not* indicate a pattern, and no raised bar can be placed over any digits. In an irrational number, we are never certain what the next digit will be when these dots (. . .) are used.

In this section, we will see more examples of irrational numbers, both positive and negative. First, however, we need to review a few terms you learned in earlier mathematics courses.

A SIDE TRIP:

SQUARES AND SQUARE ROOTS

To ***square*** a number means to multiply the number by itself. For example:

$3^2 = 3 \cdot 3 = 9$ The square of 3 is 9.

$4^2 = 4 \cdot 4 = 16$ The square of 4 is 16.

Many calculators have a special key, [x^2], that will square a number. If your calculator does not have this key, use repeated multiplication. To find 5^2:

Enter:

METHOD 1: 5 [x^2]

METHOD 2: 5 [×] 5 [=]

Display: 25.

To find the ***square root*** of a number means to find the number that, when multiplied by itself, gives the value under the ***radical sign***, $\sqrt{\ }$. For example:

$\sqrt{9} = 3$ The square root of 9 is 3 because $3 \cdot 3 = 9$.

$\sqrt{16} = 4$ The square root of 16 is 4 because $4 \cdot 4 = 16$.

Calculators also have a key, [$\sqrt{x}$] or [$\sqrt{\ }$], that will display the square root of a number. For example:

Enter: 25 [$\sqrt{x}$]

Display: 5.

More Irrational Numbers

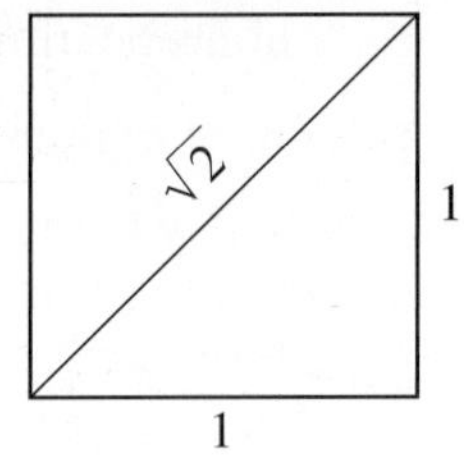

When a square measures 1 unit on every side, its diagonal measures $\sqrt{2}$ units. You can use a ruler to measure the diagonal and then show the placement of $\sqrt{2}$ on a number line.

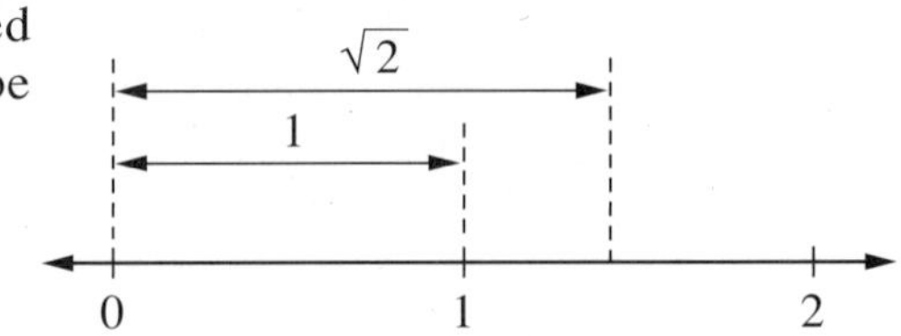

What is the value of $\sqrt{2}$? Can we find a decimal number that, when multiplied by itself, equals 2? We expect $\sqrt{2}$ to be somewhere between 1 and 2.

Use a calculator to find the value.

Enter: 2 [$\sqrt{x}$]

Display: [1.4142136]

To check this answer, multiply: $1.4142136 \times 1.4142136 = 2.0000001$, too large.

Try a smaller value: $1.4142135 \times 1.4142135 = 1.9999998$, too small.

No matter how many digits can be displayed on a calculator, no terminating decimal, nor any repeating decimal, can be found for $\sqrt{2}$ because

$\sqrt{2}$ is an irrational number.

In the same way, an infinite number of square roots are irrational numbers, for example:

$$\sqrt{3} \qquad \sqrt{5} \qquad \sqrt{3.2} \qquad \sqrt{0.1} \qquad -\sqrt{2} \qquad -\sqrt{3}$$

The values displayed on a calculator for these square roots are called rational approximations. A ***rational approximation*** for an irrational number is actually a rational number that is *close to*, but *not equal to*, the value of the irrational number.

The symbol $\approx$ means approximately equal to. Therefore, it is not correct to write $\sqrt{3} = 1.732$, but it is correct to write $\sqrt{3} \approx 1.732$.

Another interesting number that you have encountered in earlier courses is π, read as "pi." Recall that π equals the circumference of a circle divided by its diameter, or $\pi = \frac{C}{d}$.

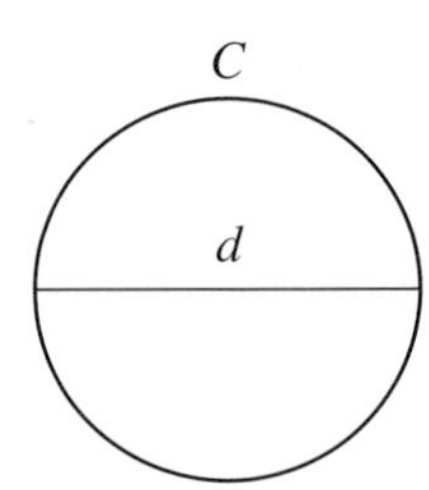

π is an irrational number.

There are many *rational approximations* for π, including these:

$$\pi \approx 3.14 \qquad \pi \approx \frac{22}{7} \qquad \pi \approx 3.1416$$

If π is doubled, or divided in half, or if a rational number is added to or subtracted from π, the result is again an irrational number. There are infinitely many such irrational numbers, for example:

$$2\pi \qquad \frac{\pi}{2} \qquad \pi + 7 \qquad \pi - 3$$

Estimation

Throughout this text, we will make use of a scientific calculator. A ***scientific calculator*** combines all of the capabilities of a simple four-function calculator with many other keys and features that are used in science and mathematics. For example, a scientific calculator has a key that, when pressed, will place in the display a rational approximation for π that is more accurate than the ones given above.

Enter: [π]

Display: [3.1415927]

With a calculator, however, you must be careful how you interpret and use the information given in the display. At times, the value shown is exact, but, more often, displays that fill the screen are rational approximations.

To write a rational approximation to a given number of decimal places, ***round*** the number.

PROCEDURE. To round to a given decimal place:

1. Look at the digit in the place at the immediate right.
2. If the digit being examined is *less than* 5, drop that digit and all digits to the right. (Example: π or 3.1415927 . . . rounded to two decimal places is 3.14.)
3. If the digit being examined is *greater than or equal to* 5, add 1 to the digit in the place to which you are rounding and then drop all digits to the right. (Example: π or 3.1415927 . . . rounded to four decimal places is 3.1416.)

Correct use of a calculator requires good skills in estimation and in the mathematics being performed. We will study estimation many times in this text, beginning with the examples that follow.

EXAMPLES

1. *True* or *False*: $\sqrt{5} + \sqrt{5} = \sqrt{10}$. Explain why.

Solution METHOD 1. *Use estimation.*

(1) Think of perfect squares: $1^2 = 1$, $2^2 = 4$, $3^2 = 9$, $4^2 = 16$, and so on.

Place 10 between the closest perfect squares:	$9 < 10 < 16$
Take the square roots of all these numbers:	$\sqrt{9} < \sqrt{10} < \sqrt{16}$
Simplify the perfect squares:	$3 < \sqrt{10} < 4$

(2) Follow the same procedure to estimate $\sqrt{5}$.

Place 5 between the closest perfect squares:	$4 < 5 < 9$
Take the square roots of all these numbers:	$\sqrt{4} < \sqrt{5} < \sqrt{9}$
Simplify the perfect squares:	$2 < \sqrt{5} < 3$

(3) Apply your knowledge of values.

Since $2 < \sqrt{5}$, you know that $\sqrt{5} > 2$.

Then, $\sqrt{5} + \sqrt{5} > 2 + 2$, or $\sqrt{5} + \sqrt{5} > 4$.

Note that on the number line

$$\sqrt{10} < 4,$$

while $\sqrt{5} + \sqrt{5} > 4$.

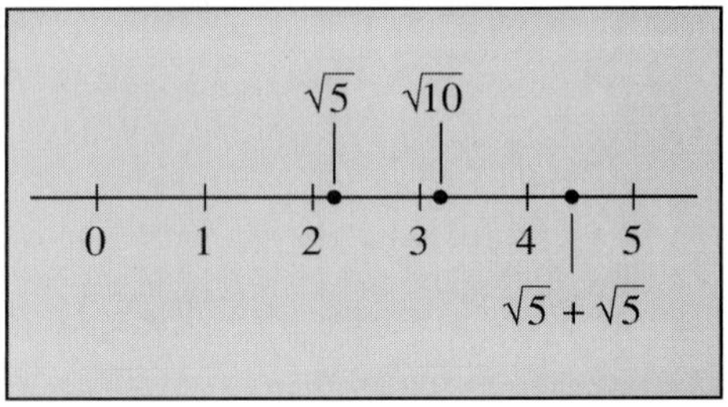

METHOD 2. *Use a calculator.*

Enter: 5 [√x] [+] 5 [√x] [=]

Display: 4.472136

Enter: 10 [√x]

Display: 3.1622777

Use these *rational approximations* to see that the values are not equal.

Answer: False. $\sqrt{5} + \sqrt{5} \neq \sqrt{10}$ because $\sqrt{5} + \sqrt{5}$ is greater than 4, while $\sqrt{10}$ is less than 4.

---

2. Find a rational approximation for each irrational number, to the *nearest hundredth.*

a. $\sqrt{3}$ **b.** $-\sqrt{0.1}$

Solutions **a.** Use a calculator.

Follow the rules for rounding. The digit in the thousandths place, 2, is *less than 5*. Drop this digit and all digits to the right of it.

Enter: 3 [√x]

Display: 1.7320508

↑

1.73

Answer: **a.** $\sqrt{3} \approx 1.73$

b. Use a calculator.

The digit in the thousandths place, 6, is *greater than or equal to 5*. Add 1 to the digit in the hundredths place and drop all digits to the right of it. Then make this a negative number.

Enter: .1 [$\sqrt{x}$]

Display: [0.3162277]

0.31
+ 1
0.32 Write −0.32.

Answer: **b.** $-\sqrt{0.1} \approx -0.32$

---

3. The circumference C of a circle with a diameter d is found by using the formula $C = \pi d$.

a. Find the *exact* circumference of a circle whose diameter is 8.

b. Find, to the nearest thousandth, a *rational approximation* of the circumference of this circle.

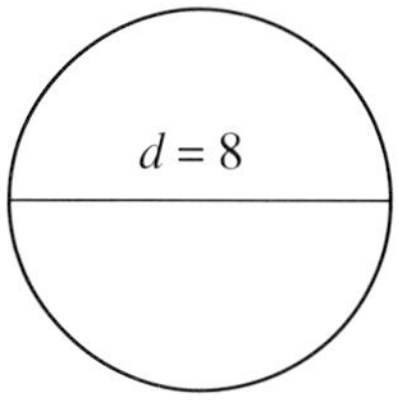

Solutions **a.** $C = \pi d$

$C = \pi \cdot 8$ or 8π

b. Use a calculator.

Enter: [π] [×] 8 [=]

Display: [25.132741]

Round the number in the display to three decimal places: 25.133.

Answers:

a. 8π is the *exact* circumference, shown by an *irrational* number.

b. 25.133 is the *rational approximation* of the circumference, to the *nearest thousandth.*

---

4. Which of the following four numbers is an irrational number?
(1) 0.12 (2) 0.12121212 . . . (3) 0.12111111 . . . (4) 0.12112111211112 . . .

Solution Since each of the first three numbers can be written as a fraction, choices (1), (2), and (3) are rational numbers. Note the repeating patterns in choices (2) and (3).

$$(1)\ 0.12 = \frac{12}{100} = \frac{3}{25} \quad (2)\ 0.12121212\ldots = \frac{12}{99} = \frac{4}{33} \quad (3)\ 0.121111111\ldots = \frac{109}{900}$$

There is no repeating pattern, however, in choice (4).

Answer: (4) 0.12112111211112 . . . , is irrational.

EXERCISES WITH OPEN-RESPONSE PROBLEMS

In 1–20, tell whether each number is rational or irrational.

1. 0.36	**2.** $0.36363636\ldots$	**3.** $0.3\overline{6}$	**4.** $0.363363336\ldots$	**5.** $\sqrt{8}$
6. 10π	**7.** $0.12131415\ldots$	**8.** $\sqrt{16}$	**9.** $0.989989998\ldots$	**10.** 0.725
11. $\sqrt{121}$	**12.** $\pi + 30$	**13.** -5.28	**14.** $0.14141414\ldots$	**15.** $-\sqrt{5}$
16. $-\pi$	**17.** $\sqrt{48}$	**18.** $\sqrt{49}$	**19.** $0.24\overline{682}$	**20.** $\pi - 2$

In 21–30, write the rational approximation of each given number: **a.** as shown on a calculator display **b.** rounded to the *nearest thousandth* (three decimal places) **c.** rounded to the *nearest hundredth* (two decimal places).

21. $\sqrt{5}$	**22.** $\sqrt{7}$	**23.** $\sqrt{19}$	**24.** $\sqrt{75}$	**25.** $\sqrt{63}$
26. $\sqrt{90}$	**27.** $-\sqrt{14}$	**28.** $-\sqrt{22}$	**29.** $\sqrt{0.2}$	**30.** $\sqrt{0.3}$

31. A rational approximation for $\sqrt{3}$ is 1.732. **a.** Multiply 1.732 by 1.732. **b.** What is the difference between 3 and the product found in part **a**?

32. **a.** Find $(3.162)^2$. **b.** Find $(3.163)^2$. **c.** Is 3.162 or 3.163 a better approximation for $\sqrt{10}$? Explain why.

33. A calculator display shows 3.1415927 as a rational approximation for π. Is 3.14 or $\frac{22}{7}$ closer to the real value of π? Explain your answer.

In 34–38, use the formula $C = \pi d$ to find, in each case, the circumference C of a circle when the diameter d is given. **a.** Write the exact value of C by using an irrational number. **b.** Find a rational approximation of C to the nearest hundredth.

34. $d = 7$	**35.** $d = 15$	**36.** $d = 72$	**37.** $d = \frac{1}{2}$	**38.** $d = 3\frac{1}{3}$

39. An irrational number that lies between 0.11 and 0.12 is 0.113111311113. . . . Find an irrational number that lies: **a.** between 0 and 1 **b.** between 7 and 8 **c.** between 0.15 and 0.16 **d.** between 2 and 2.1

40. *True* or *False*: $\sqrt{4} + \sqrt{4} = \sqrt{8}$? Explain why or why not.

41. *True* or *False*: $\sqrt{18} + \sqrt{18} = \sqrt{36}$? Explain why or why not.

42. Is $\sqrt{0.35}$ a rational or an irrational number? Explain your answer.

1-5 THE REAL NUMBERS

Recall that rational numbers can be written as repeating decimals, and that irrational numbers are decimals that do not repeat. Taken together, rational and irrational numbers comprise the set of all numbers that can be written as decimals.

The set of ***real numbers*** is the set that consists of all rational numbers and all irrational numbers.

The accompanying diagram shows that the rational numbers are a subset of the real numbers, and the irrational numbers are also a subset of the real numbers. Notice, however, that the rationals and irrationals take up different spaces in the diagram because they have no numbers in common. Together, these two sets of numbers form the real numbers. The cross-hatched shaded portion in the diagram contains no real numbers. The cross-hatched shading indicates that no other numbers except the rationals and irrationals are real numbers.

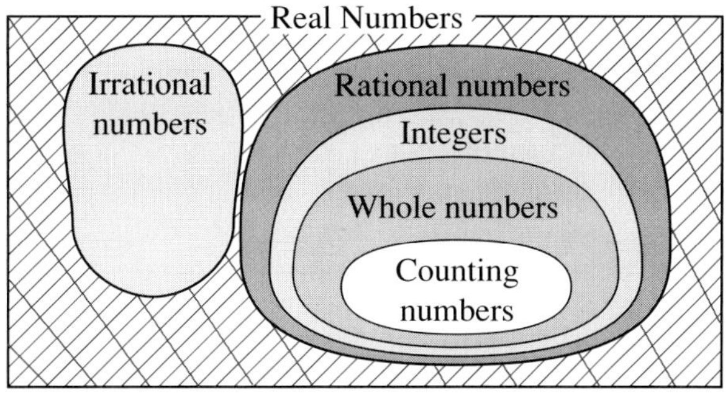

We have seen that there are an infinite number of rational numbers and an infinite number of irrationals. For every rational number, there is a corresponding point on the number line, and, for every irrational, there is a corresponding point on the number line. All of these points, taken together, make up the ***real number line***.

Since there are no more holes in this line, we say that the real number line is now complete. The ***completeness property of real numbers*** may be stated as follows:

- **Every point on the real number line corresponds to a real number, and every real number corresponds to a point on the real number line.**

Ordering Real Numbers

There are two ways in which we can order real numbers:

1. Use a *number line*. On the standard horizontal real number line, the graph of the greater number is always to the right of the graph of the smaller number.
2. Use *decimals*. Given any two real numbers that are not equal, we can express them in decimal form (even using rational approximations) to see which is greater.

EXAMPLES

1. The number line that was first seen in Section 1-1 is repeated below.

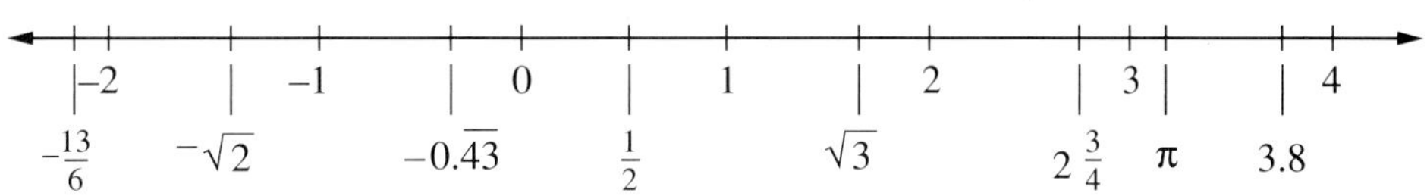

Of the numbers shown here, tell which are: **a.** counting numbers **b.** whole numbers **c.** integers **d.** rational numbers **e.** irrational numbers **f.** real numbers.

Answers:

a. Counting numbers: $1, \ 2, \ 3, \ 4$
b. Whole numbers: $0, \ 1, \ 2, \ 3, \ 4$
c. Integers: $-2, \ -1, \ 0, \ 1, \ 2, \ 3, \ 4$
d. Rational numbers: $-\frac{13}{6}, \ -2, \ -1, \ -0.\overline{43}, \ 0, \ \frac{1}{2}, \ 1, \ 2, \ 2\frac{3}{4}, \ 3, \ 3.8, \ 4$
e. Irrational numbers: $-\sqrt{2}, \ \sqrt{3}, \ \pi$
f. Real numbers All: $-\frac{13}{6}, \ -2, \ -\sqrt{2}, \ -1, \ -0.\overline{43}, \ 0, \ \frac{1}{2}, \ 1, \ \sqrt{3}, \ 2, \ 2\frac{3}{4}, \ 3, \ \pi, \ 3.8, \ 4$

---

2. Order these real numbers from smallest to greatest, using the symbol <.

$$0.3 \qquad \sqrt{0.3} \qquad 0.\overline{3}$$

Solution (1) Write each real number in decimal form:

$0.3 = 0.3000000\ldots$
$\sqrt{0.3} \approx 0.5477225$ (a rational approximation, displayed on a calculator)
$0.\overline{3} = 0.3333333\ldots$

(2) Compare these decimals: $0.3000000\ldots < 0.3333333\ldots < 0.5477225$

(3) Replace each decimal with the number in its original form: $0.3 < 0.\overline{3} < \sqrt{0.3}$

Answer: $0.3 < 0.\overline{3} < \sqrt{0.3}$

---

3. Which is the greater number, -1.7 or -1.9?

Solution (1) Place the numbers on a standard horizontal number line. Note that -1.9 is closer to -2.

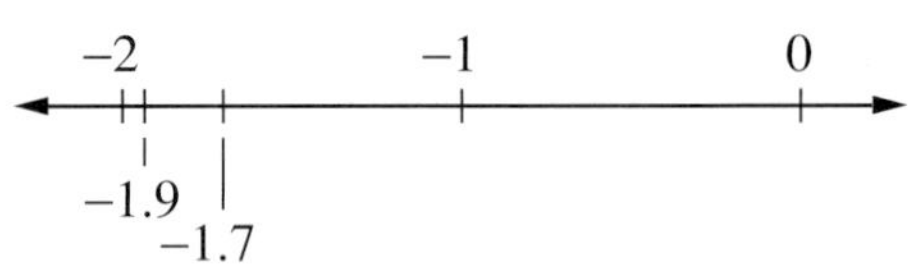

(2) Recall that, on a standard horizontal number line, the greater number is always to the right.

Answer: -1.7 is the greater number.

EXERCISES

1. Twelve numbers have been placed on a number line as shown here.

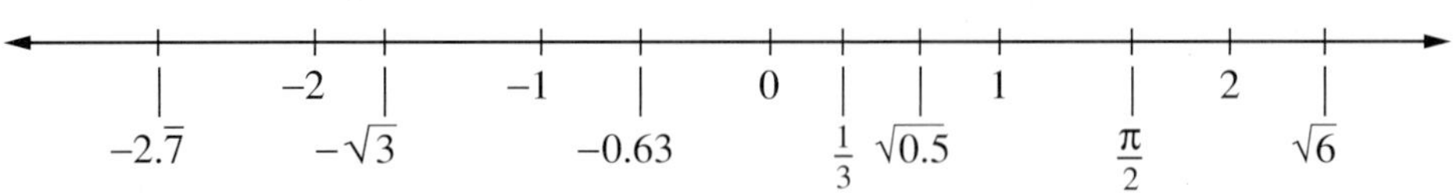

Of these numbers, tell which are: **a.** counting numbers **b.** whole numbers **c.** integers **d.** rational numbers **e.** irrational numbers **f.** real numbers

2. Given the following series of numbers: $\sqrt{0}$, $\sqrt{1}$, $\sqrt{2}$, $\sqrt{3}$, $\sqrt{4}$, $\sqrt{5}$, $\sqrt{6}$, $\sqrt{7}$, $\sqrt{8}$, $\sqrt{9}$. Of these ten numbers, tell which is (are): **a.** rational **b.** irrational **c.** real

3. Given the following series of numbers: π, 2π, 3π, 4π, 5π, 6π, 7π, 8π, 9π. Of these nine numbers, tell which is (are): **a.** rational **b.** irrational **c.** real

In 4–17, determine, for each pair, which is the greater number.

4. 2 or 2.5 **5.** 8 or $\sqrt{8}$ **6.** $0.\overline{2}$ or 0.22 **7.** $0.\overline{2}$ or 0.23

8. 0.7 or $0.\overline{7}$ **9.** -5.6 or -5.9 **10.** -0.43 or -0.431 **11.** -90 or $\sqrt{2}$

12. $0.\overline{21}$ or $0.\overline{2}$ **13.** 3.14 or π **14.** -5 or $-\sqrt{5}$ **15.** 0.5 or $\sqrt{.5}$

16. 0.41414141 . . . or 0.414114111 . . . **17.** 0.57575757 . . . or 0.57577577757777 . . .

In 18–26, order the numbers in each group from smallest to greatest by using the symbol $<$.

18. 0.202, $0.\overline{2}$, 0.2022 **19.** $0.\overline{4}$, 0.45, 0.4499 **20.** $0.\overline{67}$, $0.\overline{6}$, $0.\overline{667}$

21. $-\sqrt{2}$, $-\sqrt{3}$, -1.5 **22.** 0.5, $0.\overline{5}$, $\sqrt{0.3}$ **23.** $-\sqrt{3}$, $-\sqrt{5}$, -2

24. π, $\sqrt{10}$, 3.5 **25.** $\sqrt{16}$, $4.\overline{4}$, $\frac{13}{3}$ **26.** $\sqrt{17}$, 4, $\frac{17}{4}$

In 27–36, tell whether each statement is true or false.

27. Every real number is a rational number.

28. Every rational number is a real number.

29. Every irrational number is a real number.

30. Every real number is an irrational number.

31. Every rational number corresponds to a point on the real number line.

32. Every point on the real number line corresponds to a rational number.

33. Every irrational number corresponds to a point on the real number line.

34. Every point on the real number line corresponds to an irrational number.

35. Some numbers are both rational and irrational.

36. Every repeating decimal corresponds to a point on the real number line.

1-6 NUMBER LINES AND RULERS

Building a Number Line

Two points, taken anywhere, determine a straight line. To build a number line, follow these steps:

1. Start with two points. Label one point 0 and the other point 1.

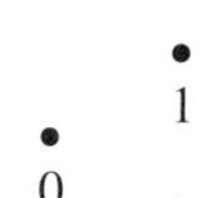

2. Connect the points to form a segment whose length, from 0 to 1, is called the ***unit measure***.

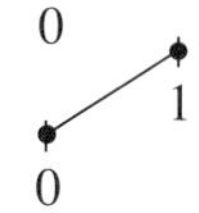

3. Extend this segment to form a straight line, with arrowheads to show that the line is endless in both directions. Use the unit measure to mark off equally spaced points that are integers: 2, 3, 4, . . . in one direction and −1, −2, −3, . . . in the other. All real numbers can be placed on this number line, using the order learned in earlier sections.

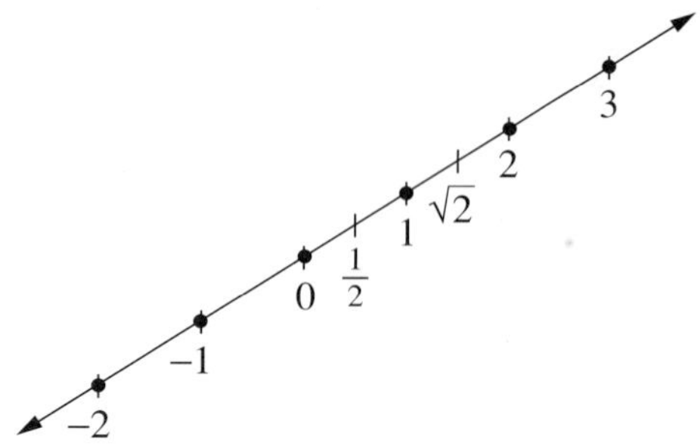

The number that is associated with a point on the number line is called the ***coordinate*** of that point, and the point itself is called the ***graph*** of the number.

Rulers

There are many uses of number lines in mathematics. For example, imagine holding a ruler in your hand. A ruler is simply part of an infinite number line. A ruler can be held to show the standard vertical and horizontal positions of number lines, but it can also be held at an angle or even upside down.

On a typical ruler, 0 may not appear, but this coordinate is assigned to the starting point of the ruler. The length from the start of the ruler to the number 1 shows its unit measure. Unit measures may differ from one ruler to another.

The *English ruler* uses one ***inch*** (1 in.) as its unit measure, and parts of this unit are usually given as fractions: $\frac{1}{2}, \frac{1}{4}, \frac{1}{8}, \frac{1}{16}$, and so on.

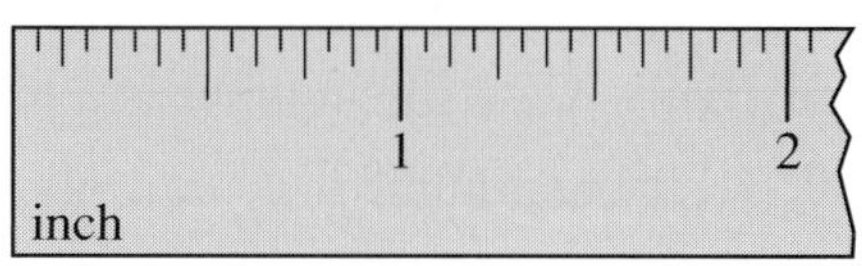

The *metric ruler* use one ***centimeter*** (1 cm) as its unit measure, and parts of this unit are treated as decimals: 0.1, 0.2, and so on. Each one-tenth centimeter equals one *millimeter* (0.1 cm = 1 mm).

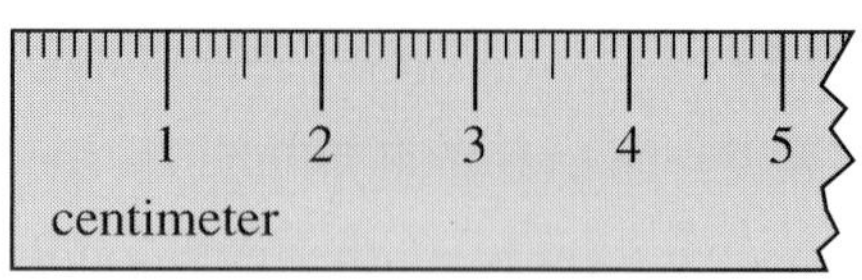

EXAMPLES

1. Use the accompanying number line to answer parts **a–c**.

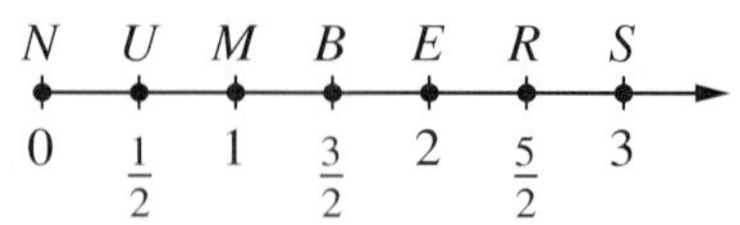

Answers:

a. Name the number assigned to point *S*. 3

b. Name the coordinate of point *M*. 1

c. Name the point that is the graph of $1\frac{1}{2}$. *B*

2. Use the accompanying number line, where the letters shown are equally spaced.

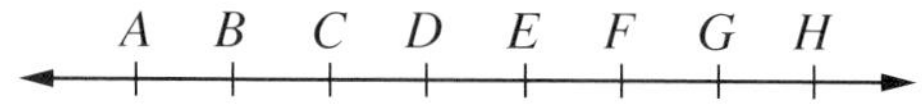

Find the coordinates of all points named on the number line when:
a. $C = 0$ and $D = 1$ **b.** $B = 0$ and $E = 1$

Solutions Copy the number line, place 0 and 1 at the given letters, and use this unit measure to name the coordinates of integers and any fractional parts. Remember that numbers increase in the direction from 0 to 1.

Answers:

a. $C = 0$ and $D = 1$

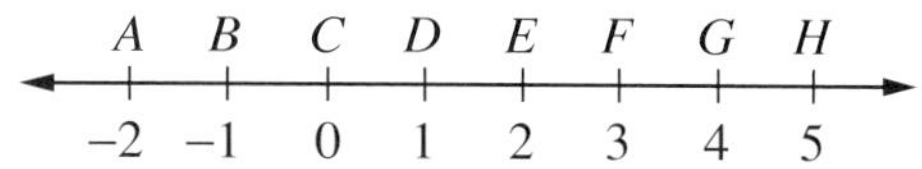

b. $B = 0$ and $E = 1$

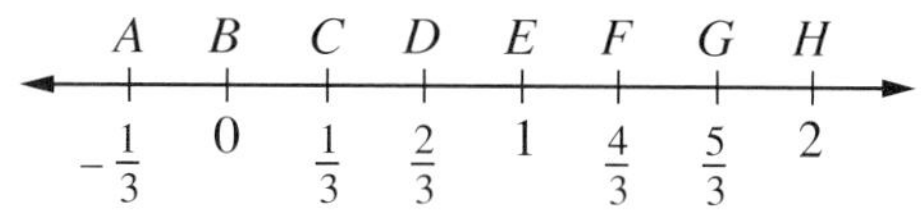

---

3. On the accompanying number line, between what two consecutive given points is the graph of: **a.** 0.25 **b.** $\sqrt{5}$ **c.** $\pi - 2$

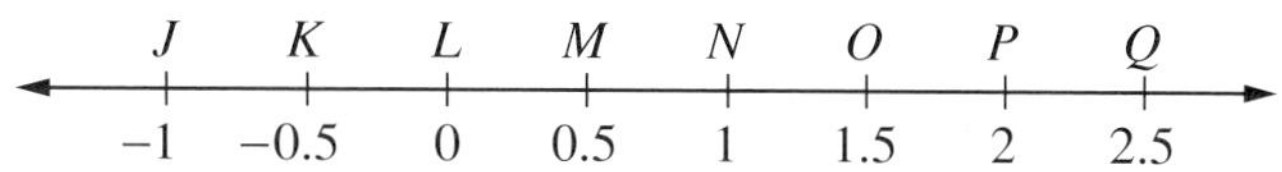

Solutions **a.** Since $0 < 0.25$ and $0.25 < 0.50$, the graph of 0.25 is between 0 and 0.50, or points L and M.

b. Since $\sqrt{5} \approx 2.237$, its graph is between 2 and 2.5, or points P and Q.

c. Since $\pi - 2 \approx 3.14 - 2$, or 1.14, its graph is between 1 and 1.5, or points N and O.

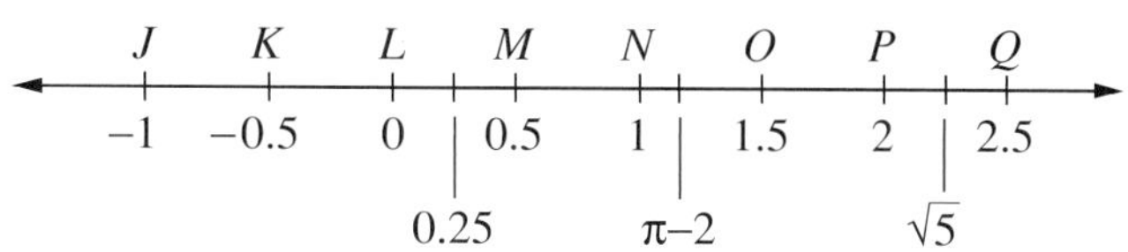

Answers: **a.** L and M **b.** P and Q **c.** N and O

EXERCISES WITH OPEN-RESPONSE PROBLEM

In 1–4, the markings on each number line are equally spaced and the locations of 0 and 1 are shown. Name the coordinate of each labeled point on the number line.

1. 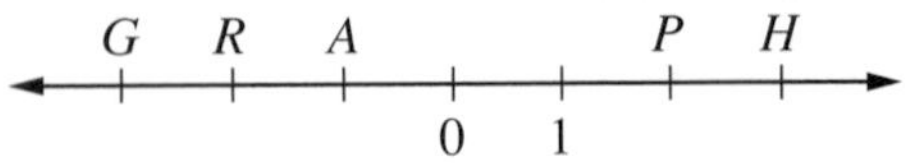

2. 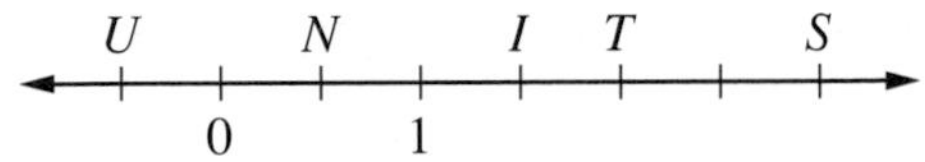

3. 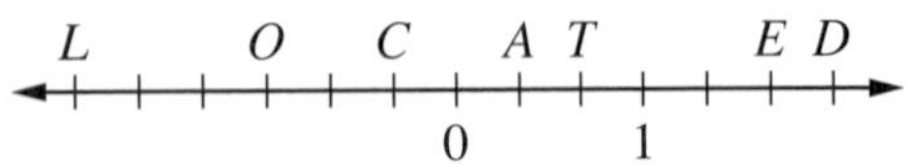

4. 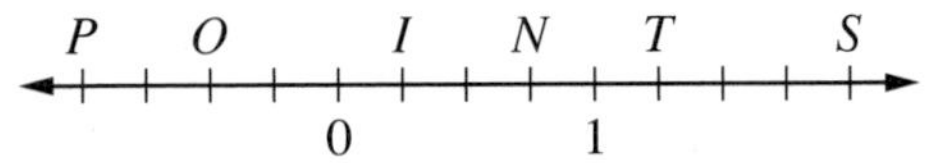

In 5 and 6, use the following number line:

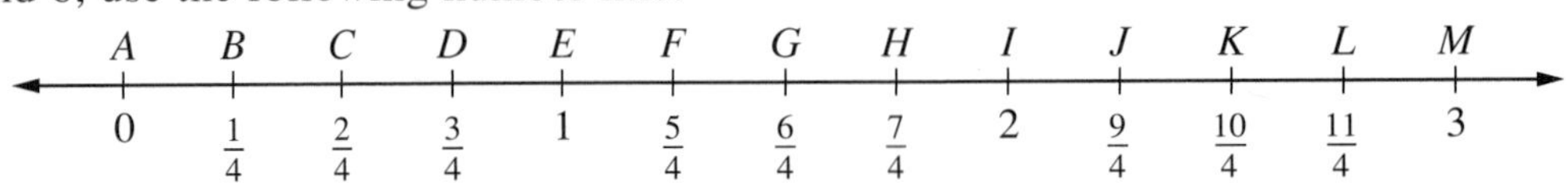

Ex. 5–6

5. Name the coordinate assigned to each given point.

a. A **b.** B **c.** M **d.** D **e.** J **f.** G

6. Name the point that is the graph of each number.

a. 1 **b.** $\frac{8}{4}$ **c.** $1\frac{3}{4}$ **d.** $2\frac{1}{2}$ **e.** 0.5 **f.** 1.25 **g.** 2.75

7. Use the following number line, where the letters shown are equally spaced:

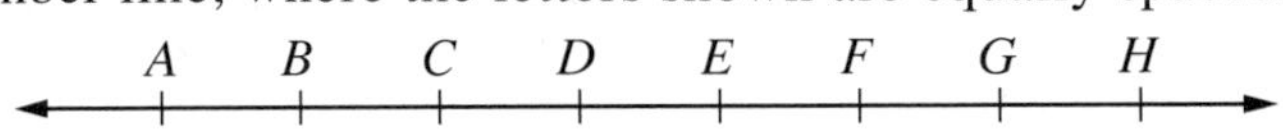

In parts **a–f**, draw each number line and find the coordinates of all points named on this line when:

a. $D = 0$ and $E = 1$ **b.** $C = 0$ and $E = 1$ **c.** $A = 0$ and $F = 1$

d. $F = 0$ and $G = 1$ **e.** $C = 0$ and $G = 1$ **f.** $D = 0$ and $G = 1$

8. Use the following number line, where $K = 0$, $M = 1$, and the letters are equally spaced:

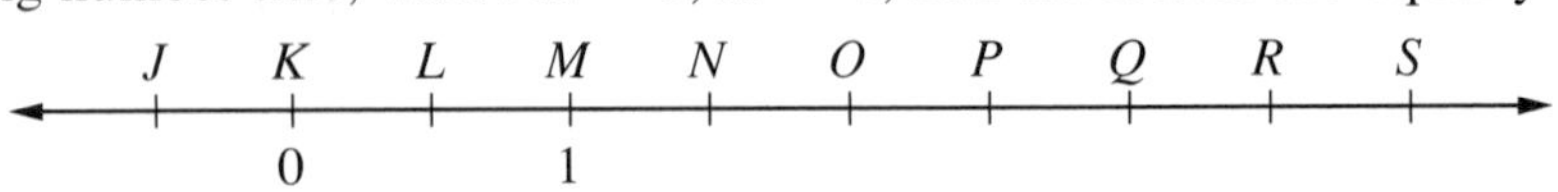

Between what two consecutive points on this number line is the graph of each of the following?

a. 0.75 **b.** π **c.** $\sqrt{2}$ **d.** $\frac{8}{3}$ **e.** $-\frac{1}{4}$

f. $\sqrt{5}$ **g.** $\sqrt{0.36}$ **h.** $\frac{\pi}{2}$ **i.** $3.\overline{82}$ **j.** $\frac{0.382}{4}$

In 9–20: **a.** In each case, draw a standard horizontal number line. Show points for 0 and 1 and any other integers that may help you as guides. Then graph the three given numbers. **b.** Write an inequality to show the order of the three numbers in each group (as in $2 < 7 < 8$).

9. 2, 4, 3 **10.** 2, -4, -3 **11.** -2, 3, -4 **12.** 3, $0.\overline{3}$, 2

13. 2, 0.2, 0.02 **14.** 1.5, 0.15, 0.5 **15.** $\sqrt{8}$, 5, $\sqrt{20}$ **16.** -3, $\sqrt{3}$, 0.3

17. $\frac{1}{3}, \frac{1}{5}, \frac{1}{4}$ **18.** $-\frac{2}{3}, -\frac{3}{2}, -\frac{3}{4}$ **19.** 2, π, $\frac{\pi}{2}$ **20.** -1, π, $\pi - 1$

21. In parts **a**–**h**, for each pair shown, tell which is the larger unit measure.
 a. 1 inch, 1 foot **b.** 1 centimeter, 1 millimeter **c.** 1 yard, 1 mile **d.** 1 in., 1 cm
 e. 1 foot, 1 meter **f.** 1 kilometer, 1 meter **g.** 1 yard, 1 meter **h.** 1 mile, 1 kilometer

22. Measure the length of this page from top to bottom: **a.** in inches **b.** in centimeters.
 c. Are these lengths exact or rational approximations? Explain why.

1-7 NUMBER LINES AND GRAPHS

Number lines play an important role in the presentation of information through graphs. A ***graph*** is a picture of numerical facts, organized in such a way that this picture is easier to understand than a table of numbers.

Number lines are used in bar graphs, picture graphs, and line graphs.

The Bar Graph

A ***bar graph*** is used to compare *different* items, such as the sales records for different people in a company or the heights of different buildings.

In a bar graph, the length of each bar represents a numerical fact. To understand these numbers, a ***scale*** is drawn, beginning at zero and consisting of equally spaced intervals. This scale, which is actually a *number line*, may be drawn vertically or horizontally. The bars are then drawn in the same direction as the number line.

WORLD'S LONGEST SUSPENSION BRIDGES

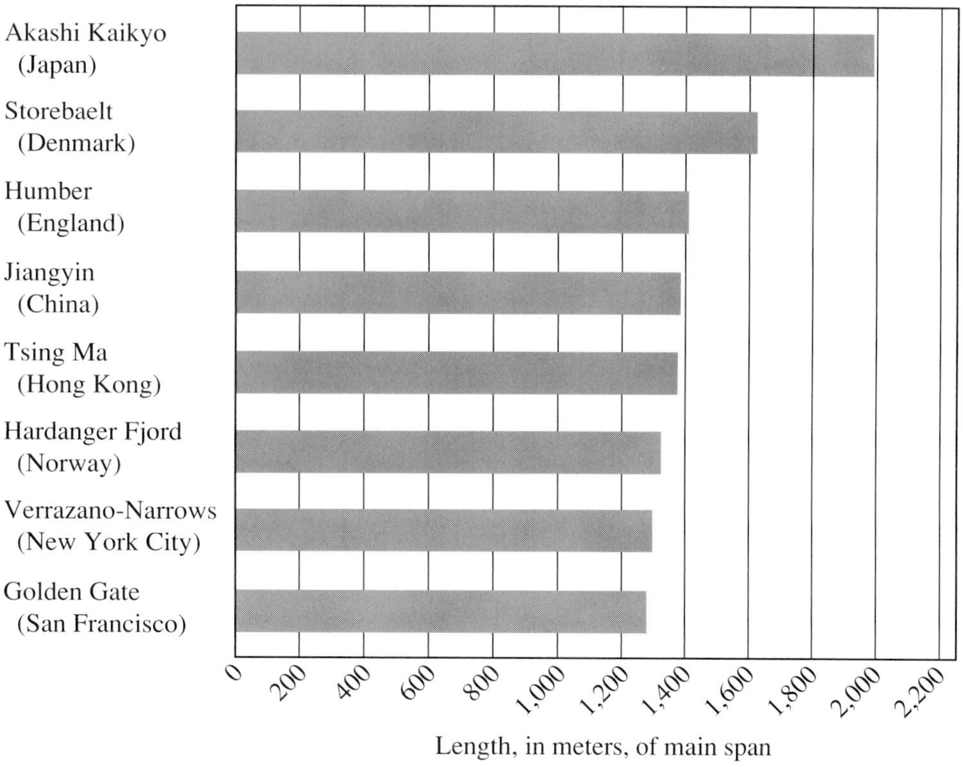

The actual lengths of the main spans of the world's eight longest suspension bridges are not stated in the preceding bar graph. However, the scale at the bottom of the graph allows us to compare the lengths of these eight bridges. By observation we can learn several numerical facts, such as:

1. There are only three suspension bridges in the world whose main spans are longer than 1,400 meters.
2. The span closest to 1,300 meters in length is on the Verrazano Narrows Bridge.
3. The longest span in the world, over 1,900 meters in length, is on the Akashi Kaikyo, completed in Japan in 1998.
4. Each of the world's eight longest suspension bridges has a main span that is longer than 1,200 meters.

The Picture Graph

A ***picture graph***, which, like a bar graph, compares *different* items, presents information by using a *hidden* or understood number line. Instead of showing a scale or number line, a ***symbol*** is used to represent a definite quantity.

1995 MOST POPULATED COUNTRIES

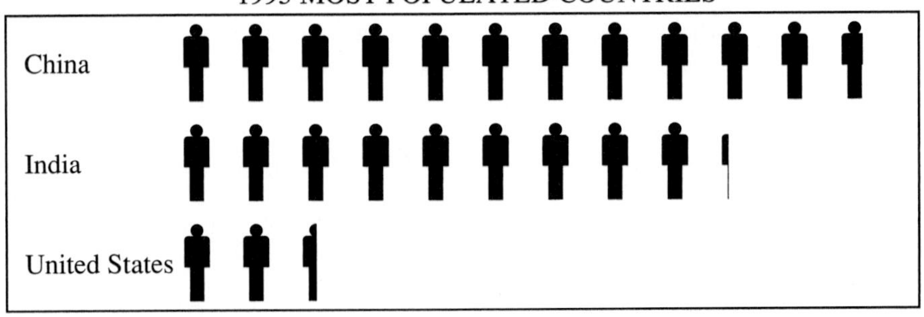

Each symbol, , represents 100,000,000 people.

In the accompanying picture graph, the actual populations of the three countries are not given. However, knowing that each symbol represents 100 million people, we can learn many numerical facts:

1. The 11.9 symbols shown for China tell us that its population is approximately $11.9 \times 100{,}000{,}000$, or 1,190,000,000.
2. India (the second largest population) has approximately $9.1 \times 100{,}000{,}000$, or 910,000,000 people.
3. The United States ranks third with $2.6 \times 100{,}000{,}000$, or 260,000,000 people.

The Line Graph

A ***line graph***, sometimes called a ***broken line graph***, shows *changes in one* particular item. For example, temperature, the price of a stock, or a population can increase or decrease.

In making a line graph, *two* number lines are used, one drawn vertically and the other horizontally. Points are marked on the graph to show numerical facts. These points are then connected in consecutive order to form the line graph.

The accompanying line graph shows the sales record of Mrs. Bixhorn over a 6-month period. Exactly six points are drawn and connected. From this graph, we can observe that:

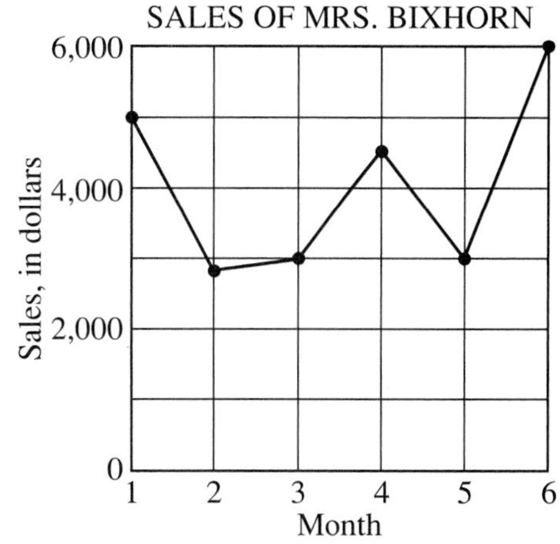

1. Mrs. Bixhorn's highest sales, about \$6,000, were made in month 6.
2. Her lowest sales, about \$2,800, were made in month 2.
3. Her greatest sales increase occurred from month 5 to month 6, with a gain of approximately \$3,000.
4. The two months between which her sales changed the least were months 2 and 3.
5. The two months closest in sales, both near \$3,000, were months 3 and 5.
6. Her monthly sales never fell below \$2,500.

Note that it is not appropriate to use a line graph to compare *different* items, such as the length of several bridges. A line graph, which shows increases and decreases in an single item, would be meaningless when comparing two or more items.

Other graphs, such as the circle graph and histogram, will be studied in later chapters of this book.

EXERCISES WITH OPEN-RESPONSE PROBLEMS

1. Of the tallest buildings in the United States, three are in Chicago: the Sears Tower, the Amoco Building, and the John Hancock Building. Three others are in New York City: the Empire State Building, the World Trade Center, and the Chrysler Building. Use the accompanying graph to answer the following questions:

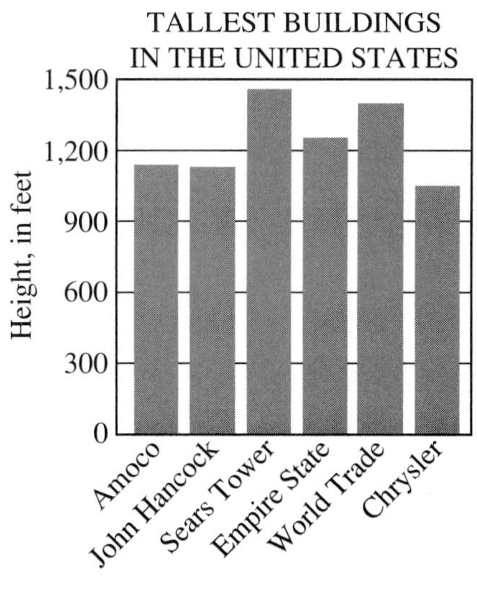

   a. Find the height, to the nearest 50 feet, of each building.
   b. Name the tallest building.
   c. What is the approximate difference in height between the World Trade Center and the Empire State Building?
   d. What building is closest in height to the Amoco Building?

e. What is the approximate difference between the height of the John Hancock Building and the height of the Sears Tower?
f. Which buildings, if any, are less than 1,000 feet tall?

2. Use the accompanying graph to answer the following questions:
a. How many telephones, to the nearest 5,000, are there in each of the towns?
b. Which two towns have about the same number of telephones?
c. Parr City has how many times as many telephones as Tyne?
d. How does the number of telephones in Calcot compare to the total number of telephones in the other three towns?

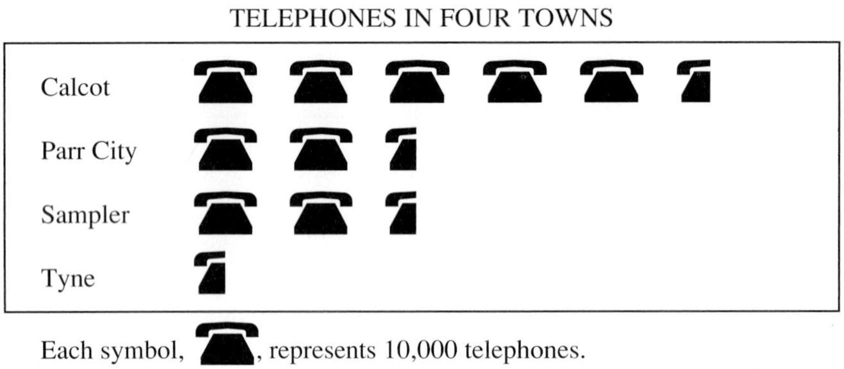

3. The accompanying bar graph shows the number of acres of different vegetables that a farmer planted during one season.

a. How many acres of carrots were planted?

b. For which vegetable was the smallest number of acres planted?

c. What was the total number of acres planted?

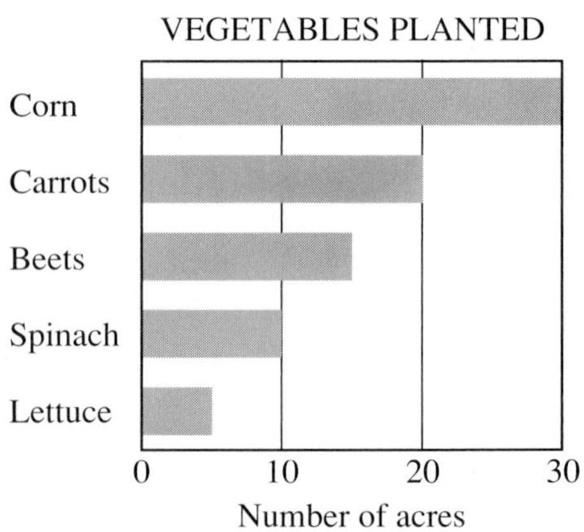

4. In a bar graph, 1 centimeter represents 30 kilometers. Find the length of the bar needed to represent each given distance.

a. 60 km **b.** 300 km
c. 15 km **d.** 75 km

5. In a bar graph, $\frac{1}{4}$ inch represents 100 people. In each case, find the length of the bar needed to represent the given number of people.

a. 200 **b.** 400 **c.** 2,000
d. 500 **e.** 1,500 **f.** 50

6. The accompanying graph shows the closing price per share of a certain stock over a period of 11 days.

a. Approximately how much was the price per share on the tenth day?
b. On what day was the price the lowest?
c. Between what two days did the price decrease most sharply?
d. Between what two days did the price remain constant?
e. For what two days did the closing price equal $1.25 per share?

7. In the scale of the accompanying line graph, a break appears between 0 and 60 to allow for a closer look at the increases in life expectancies for U.S. males over the period from 1940 to 1990, inclusive. If this same scale was used without a break, the graph would extend beyond the length of this page.

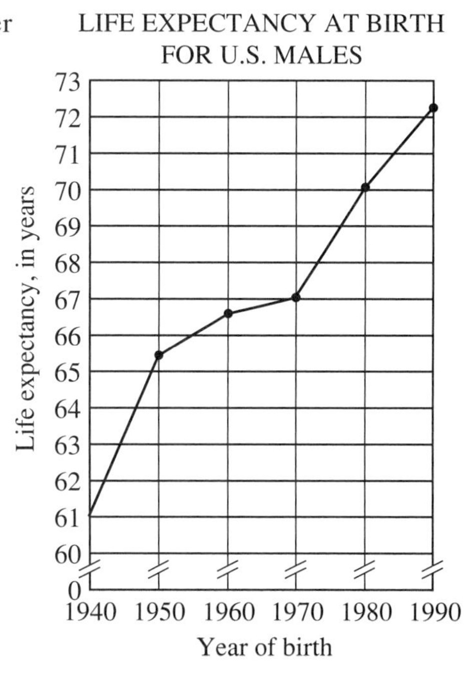

a. In what decade (10-year period) did the life expectancy increase the most?
b. In what decade did the life expectancy increase the least?
c. What was the life expectancy for U.S. males born in 1940, to the nearest year?
d. What was the life expectancy for U.S. males born in 1950, to the nearest year?
e. By how many years did the life expectancy for U.S. males increase from 1980 to 1990?

8. **a.** Draw a line graph to show the life expectancy at birth for U.S. females from 1940 to 1990 based on these figures: 1940, 65.2 years; 1950, 71.1 years; 1960, 73.1 years; 1970, 74.7 years; 1980, 77.5 years; 1990, 78.8 years.

b. Use a different color ink or pencil to copy the line graph from Exercise 7 onto the graph drawn in Part a. Compare these two line graphs and state two conclusions.

9. The unemployment rate in the United States is given in 5-year intervals for the years 1960–1990 in the accompanying table.

U.S. Unemployment Rate (per 100 workers)	
1960	5.4
1965	4.4
1970	4.8
1975	8.3
1980	7.0
1985	7.2
1990	5.5

a. Construct a bar graph to represent these data.

b. Tell why it is appropriate to display these data with a line graph.

c. Construct a line graph.

CHAPTER SUMMARY

A ***set*** is a collection of objects or elements.

The ***counting numbers*** or ***natural numbers*** are {1, 2, 3, 4, . . .}.

The ***whole numbers*** are {0, 1, 2, 3, 4, . . .}.

The ***integers*** are {. . . , −4, −3, −2, −1, 0, 1, 2, 3, 4, . . .}.

These sets of numbers form the basis for a ***number line***, on which the length of a segment from 0 to 1 is called the ***unit measure*** of the line.

The ***rational numbers*** are all numbers that can be expressed in the form $\frac{a}{b}$ where a and b are integers and $b \neq 0$. Every rational number can be expressed as a repeating decimal or as a terminating decimal (which is actually a decimal in which 0 is repeated).

The ***irrational numbers*** are decimal numbers that do not terminate and do not repeat. On calculators and in many operations, *rational approximations* are used to show values that are close to, but not equal to, irrational numbers.

The ***real numbers*** consist of all rational numbers and all irrational numbers taken together. These numbers can be graphed and ordered on a ***real number line***.

Number lines have many uses in mathematics, including rulers. Number lines are also used in constructing bar graphs and picture graphs to compare different items, and line graphs to compare the same item at different times.

VOCABULARY

1-1 Mathematics Number Counting numbers Whole numbers Finite, infinite, and empty sets Numerical expression Simplify

1-2 Opposites Integers Subset Inequality

1-3 Rational numbers Mean (average) Terminating and repeating decimals

1-4 Irrational numbers Squares Square roots Radical sign Pi (π) Rational approximation Scientific calculator Rounding

1-5 Real numbers Completeness property of real numbers

1-6 Unit measure Coordinate Graph

1-7 Bar graph Scale Picture graph Line graph

REVIEW EXERCISES WITH OPEN-RESPONSE PROBLEM

In 1–4, simplify each expression.

1. 18×0 **2.** $18 \div 2$ **3.** $18 \cdot 2$ **4.** 18^2

5. Find: **a.** the product of 9 and 5 **b.** the quotient of 300 and 12
c. the difference of 143 and 98 **d.** the sum of 3 squared and 4

6. Order the numbers -5, 3, and -1 using the symbol $>$.

In 7–10, state whether each sentence is true or false.

7. $7 > -8$ **8.** $-7 > -2$ **9.** $4 < -8$ **10.** $9 \leq 9$

In 11–16, write each rational number in the form $\frac{a}{b}$, where a and b are integers and $b \neq 0$.

11. 0.9 **12.** 0.45 **13.** $8\frac{1}{2}$ **14.** 14 **15.** $0.\overline{3}$ **16.** -63

17. Find a rational number that is midway between 19.9 and 20.

In 18–22, tell whether each number is rational or irrational.

18. $0.\overline{64}$ **19.** $\sqrt{6}$ **20.** $\sqrt{64}$

21. π **22.** 0.040040004 . . .

In 23–27, write a rational approximation of each given number:
a. as shown on a calculator display **b.** rounded to the *nearest hundredth*

23. $\sqrt{11}$ **24.** $\sqrt{0.7}$ **25.** $\sqrt{905}$ **26.** $\sqrt{1599}$ **27.** π

In 28–32, determine which is the greater number in each pair.

28. 5 or $\sqrt{20}$ **29.** -5 or -10 **30.** 3.2 or π **31.** 0.41 or $0.\overline{4}$

32. $0.\overline{12}$ or 0.121

In 33–38, tell whether each statement is true or false.

33. Every integer is a real number.

34. Every rational number is an integer.

35. Every whole number is a counting number.

36. Every irrational number is a real number.

37. Between 0 and 1, there is an infinite number of rational numbers.

38. Whole numbers that are negative are called integers.

39. Draw a number line, showing the graphs of these numbers: 0, 1, 4, -3, -1.5, and π.

In 40–41, use the given number line, where the letters are equally spaced.

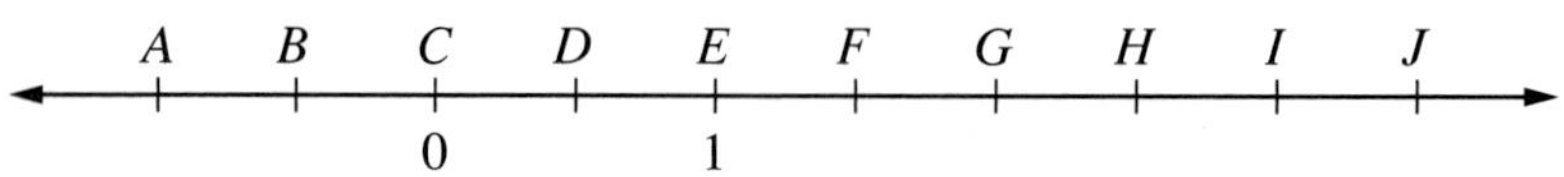

Ex. 40–41

40. Find the coordinates of all letters shown on the number line when $C = 0$ and $E = 1$.

41. Between what two consecutive points on this number line is the graph of:
a. 1.8 **b.** -0.6 **c.** $\sqrt{2}$ **d.** π **e.** $\sqrt{6}$

42. The heights, in feet, of five dams are as follows: Boulder, 730; Kensico, 310; Shasta, 600; Grand Coulee, 450; and Gatun, 120.
a. Make a bar graph to represent these data.
b. Tell why it is not appropriate to display these data with a line graph.

Chapter 2

Operations and Properties

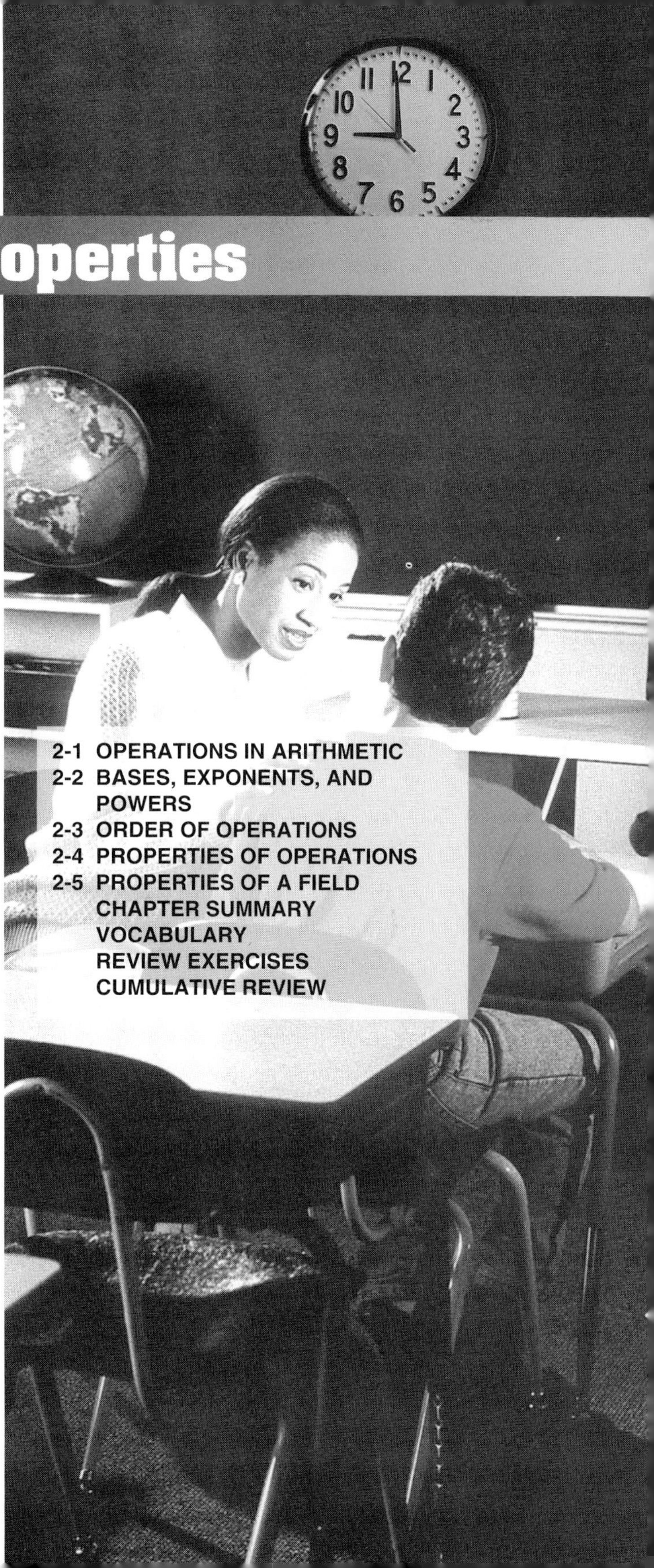

Jesse kept looking at the clock. He was racing to finish a timed test. Here are his answers for the last ten questions on the test:

$$3 + 5 = 8$$
$$6 + 6 = 12$$
$$8 + 5 = 1$$
$$10 + 4 = 2$$
$$6 + 10 = 4$$
$$12 + 9 = 9$$
$$12 + 12 = 12$$
$$11 + 10 = 9$$
$$7 + 6 = 1$$
$$5 + 10 = 3$$

When Jesse handed in his paper, his teacher said that almost all of his answers were wrong. After Jesse had explained how he got his answers, however, the teacher said, "You're right. You have a grade of 100%."

What explanation did Jesse give his teacher to show that all of his answers were correct?

2-1 OPERATIONS IN ARITHMETIC

The Four Basic Operations in Arithmetic

Bicycles have two wheels. Bipeds walk on two feet. Biceps are muscles that have two points of origin. Bilingual people can speak two languages. What do these *bi*-words have in common with the following examples?

$$6.3 + 0.9 = 7.2 \qquad 21.4 \times 3 = 64.2$$

$$11\tfrac{3}{7} - 2\tfrac{1}{7} = 9\tfrac{2}{7} \qquad 9 \div 2 = 4\tfrac{1}{2}$$

The prefix *bi*- means "two." In each example above, an operation or rule was followed to replace *two* rational numbers with a *single* rational number. These familiar operations of addition, subtraction, multiplication, and division are called ***binary operations***. Each of these operations can be performed with any pair of rational numbers, except that division by zero is meaningless and is not allowed.

In every binary operation, two elements from a set are replaced by exactly one element from the same set. There are some important concepts to remember when working with binary operations:

1. A *set* must be identified, such as the set of whole numbers or the set of rational numbers. When no set is identified, use the set of all numbers you know.
2. The *rule* for the binary operation must be clear, such as the rules you know for addition, subtraction, multiplication, and division.
3. The order of the elements is important. Later in this book, we will use the notation (a, b) to indicate an ***ordered pair*** in which a is the first element and b is the second element. For now, be aware that answers may be different depending on which element is first and which is second. Consider subtraction.
   If 8 is the first element and 5 is the second element, then: $8 - 5 = 3$
   But if 5 is the first element and 8 is the second element, then: $5 - 8 = -3$
4. Every problem using a binary operation must have an *answer*, and there must be *only one* answer. We say that each answer is ***unique***, meaning there is *one and only one* answer.

To show a binary operation in general terms, think of a as the first element, b as the second element, $*$ as the operation symbol, and c as the unique answer from the set. Then we can write:

$$a \quad * \quad b \quad = \quad c$$

a	$*$	b	c
First element from the set	**Operation symbol**	**Second element from the set**	**Unique answer from the set**

- ***Definition.*** A ***binary operation*** in a set assigns to every ordered pair of elements from the set a unique answer from the set.

Note that, even when we find the sum of three or more numbers, we still add only two numbers at a time, indicating the binary operation:

$$4 + 9 + 7 = (4 + 9) + 7 = 13 + 7 = 20$$

The Property of Closure

When numbers behave in a certain way for an operation, we describe this behavior as a ***property***. The four basic operations are binary operations for the set of rational numbers. Is this true for other sets of numbers? Let us examine the basic operations with the set of whole numbers {0, 1, 2, 3, . . .}, and discuss a property called ***closure***.

Add any two whole numbers. The sum is *always* a whole number. For example, $23 + 11 = 34$, and $78 + 48 = 126$. This tells us that:

1. addition is a binary operation for the set of whole numbers, *and*

2. the set of whole numbers is ***closed*** under the operation of addition.

Multiply any two whole numbers. The product is *always* a whole number. For example, $(2)(4) = 8$, and $25(92) = 2{,}300$. This tells us that:

1. multiplication is a binary operation for the set of whole numbers, *and*

2. the set of whole numbers is ***closed*** under the operation of multiplication.

Subtract any two whole numbers. Many differences are whole numbers, but wait! Some results are *not* whole numbers, such as $5 - 7 = -2$. We conclude that:

1. subtraction is *not* a binary operation for the set of whole numbers, *and*

2. the set of whole numbers is ***not closed*** under the operation of subtraction.

Divide any whole number by a nonzero whole number. (Remember that division by 0 is not allowed.) The quotients are *not* always whole numbers, such as $9 \div 2 = 4.5$, and $10 \div 3 = 3.3333. . . .$ We conclude that, even after ruling out division by 0:

1. division is *not* a binary operation for the set of whole numbers, *and*

2. the set of whole numbers is ***not closed*** under the operation of division.

These examples with whole numbers lead to a general definition of ***closure***:

A set is said to be ***closed*** under an operation when every pair of elements from the set, under the given operation, yields an element from that set.

In later chapters of this book, we will study operations with signed numbers and operations with irrational numbers. For now, we will simply make these observations:

- **The set of whole numbers is closed under the operations of addition and multiplication.**
- **The set of integers is closed under the operations of addition, subtraction, and multiplication.**

- **The set of rational numbers is closed under the operations of addition, subtraction, and multiplication, and is closed under division by nonzero rational numbers.**

Other Operations in Arithmetic

There are many operations in arithmetic other than addition, subtraction, multiplication, and division. Here we will study two other types of arithmetic operations that are defined not only for the rational numbers but for all real numbers.

An operation: the mean or average (avg) Students are often concerned about their averages in a subject. To find the *mean* or ***average*** of two numbers, add the numbers and divide the sum by 2.

For example, the average of 75 and 87 $= \frac{75 + 87}{2} = \frac{162}{2} = 81$.

This is a binary operation. We can rewrite this problem in the form of the binary operation, $a * b = c$. Replacing $*$, the symbol for the operation, with *avg*, we can write 75 avg 87 = 81.

- ***Definition:*** If a and b are real numbers, then:

$$a \text{ avg } b = \frac{a + b}{2}$$

An operation: the maximum (max) Sometimes we have to discover which of two numbers is the larger. Finding the larger of two numbers can be thought of as a binary operation called ***maximum***. For example, the maximum of 13 and 17 is 17. To rewrite this problem in the general form of a binary operation, $a * b = c$, we replace $*$ with *max*. The result is 13 max 17 = 17.

- ***Definition:*** If a and b are real numbers and $a \neq b$, then:

$$a \text{ max } b = \text{the larger of the numbers } a \text{ and } b$$

(*Note.* If the numbers are equal, we define a max $a = a$. For example, 4 max 4 = 4.)

EXAMPLES

1. The high temperature of the day was 33° Celsius, and the low temperature was 19° Celsius. What is the recorded average temperature for the day as reported by radio and TV weather commentators?

Solution Use the operation of average. Add the given numbers and divide by 2.

$$33 \text{ avg } 19 = \frac{33 + 19}{2} = \frac{52}{2} = 26$$

Alternative Solution Find the average on a calculator.

Enter:

METHOD 1. [( ] 33 [+] 19 [ )] [÷] 2 [=]

METHOD 2. 33 [+] 19 [=] [÷] 2 [=]

Display: 26.

Answer: The recorded average temperature (sometimes called the mid-range) was 26° Celsius.

---

2. Is there more gasoline in the tank of a car when the tank is $\frac{2}{3}$ full or when the tank is $\frac{3}{5}$ full?

Solution Answer this problem by evaluating $\frac{2}{3}$ max $\frac{3}{5}$.

(1) Convert the fractions to decimal form:

Enter: 2 [÷] 3 [=] *Display:* 0.6666666

Enter: 3 [÷] 5 [=] *Display:* 0.6

(2) Write the expression with fractions. Substitute their decimal values, and select the larger decimal:

$$\frac{2}{3} \text{ max } \frac{3}{5}$$
$$= 0.6666666 \text{ max } 0.6000000$$
$$= 0.6666666$$

(3) Rewrite the decimal answer in its original fraction form:

$$= \frac{2}{3}$$

Answer: $\frac{2}{3}$

---

3. In a chemistry lab, a rack can hold 24 test tubes. How many racks are needed for:

a. 48 test tubes? **b.** 57 test tubes? **c.** 10 test tubes? **d.** k test tubes?

Solutions To begin, divide the number of test tubes by 24, the number that a rack can hold. Then, when necessary, round up. In other words, find the smallest whole number greater than or equal to the quotient.

a. $\frac{48}{24} = 2$

b. $\frac{57}{24} = 2\frac{9}{24}$ Round up to 3

c. $\frac{10}{24}$ Round up to 1

d. $\frac{k}{24}$ If this is a whole number, it is the answer. If not, round up.

Answers:

a. 2 **b.** 3 **c.** 1 **d.** The smallest whole number equal to or greater than $\frac{k}{24}$

EXERCISES

In 1–5, tell whether each statement is true or false.

1. The set of whole numbers is closed under addition and multiplication.
2. The set of whole numbers is closed under subtraction.
3. The set of integers is closed under addition, subtraction, and multiplication.
4. When any rational number is divided by any other rational number, there is always an answer.
5. The set of rational numbers is closed under division by nonzero rational numbers.

6. In each part, choose one or more answers from the following symbols that may or may not represent numbers: 0, 1, $\frac{0}{1}$, $\frac{1}{0}$.

a. Which represent counting numbers?
b. Which represent whole numbers?
c. Which represent numbers that can be written as fractions?
d. Which are meaningless?

In 7–26, write the number that each numerical expression represents.

7. $\frac{3}{5} + \frac{7}{5}$ **8.** $0.75 + 1.25$ **9.** $\frac{3}{8} + \frac{5}{8}$ **10.** $\frac{3}{2} + \frac{7}{4}$ **11.** $\frac{12}{7} - \frac{5}{7}$

12. $4.65 - 2.25$ **13.** $\frac{7}{8} - \frac{2}{8}$ **14.** $\frac{15}{3} - \frac{1}{2}$ **15.** $12 \times \frac{1}{3}$ **16.** $\frac{21}{2} \times \frac{1}{7}$

17. $\frac{1}{4} \times 4$ **18.** 2.5×0.64 **19.** $6.5 - 0.35$ **20.** 1.25×4 **21.** $6.28 \div 4$

22. $60 \div 1.25$ **23.** $6 \div 0.75$ **24.** $\frac{1}{8} \times 0.4$ **25.** $\frac{2}{3} \times 3$ **26.** $5 \div \frac{1}{2}$

In 27–61, perform the indicated operations within the set of rational numbers.

27. $\frac{3}{5} + \frac{1}{2}$ **28.** $1\frac{2}{5} + \frac{3}{10}$ **29.** $\frac{5}{9} + 3\frac{1}{2}$ **30.** $\frac{7}{8} + \frac{3}{4}$ **31.** $3\frac{1}{3} - \frac{1}{5}$

32. $7\frac{1}{3} - \frac{1}{2}$ **33.** $5\frac{4}{5} - 2\frac{1}{2}$ **34.** $6\frac{1}{4} - 3\frac{7}{8}$ **35.** $2\frac{1}{4} \times \frac{2}{3}$ **36.** $3\frac{1}{5} \times \frac{1}{4}$

37. $2\frac{2}{5} \times 2\frac{1}{2}$ **38.** $6\frac{2}{3} \times 2\frac{2}{5}$ **39.** $2\frac{1}{2} \div 3\frac{3}{4}$ **40.** $3\frac{1}{8} \div 1\frac{1}{4}$ **41.** $2\frac{1}{3} \div 7$

42. $7 \div 2\frac{1}{3}$ **43.** $3.7 + 0.37$ **44.** $1.9 + 0.09$ **45.** $1.9 - 0.09$ **46.** $3.2 - 0.31$

47. 0.8×0.5 **48.** 0.375×0.8 **49.** $0.125 \div 5$ **50.** $0.008 \div 0.4$ **51.** $3 - 1\frac{1}{5}$

52. $3 - 1.2$ **53.** $12\frac{3}{4} - 8$ **54.** $12.75 - 8$ **55.** $1\frac{1}{2} \times 0.6$ **56.** $\frac{2}{3} \div 1.5$

57. $\frac{4}{5} - 0.45$ **58.** $1.8 + \frac{1}{4}$ **59.** $\frac{7}{3} \div \frac{1}{6}$ **60.** $\frac{5}{8} \times 0.3$ **61.** $\frac{7}{8} - 0.25$

In 62–76, find the unique answer using the set of rational numbers.

62. $1\frac{1}{2}$ avg $4\frac{1}{2}$ **63.** 3.4 avg 6.6 **64.** $1\frac{1}{4}$ avg $\frac{3}{4}$ **65.** 8 avg 3 **66.** 80 avg 85

67. 2.9 avg 4.1 **68.** $\frac{3}{8}$ avg $\frac{7}{8}$ **69.** $\frac{5}{6}$ avg $\frac{3}{6}$ **70.** $\frac{6}{8}$ avg $\frac{1}{2}$ **71.** 0.5 avg 0.2

72. 1.8 avg 0.9 **73.** 0.24 avg 2.4 **74.** 38 avg 3.08 **75.** 0 avg 19.98 **76.** $\frac{3}{8}$ avg 0.12345

In 77–79, the average of two numbers is 20. Find the missing number, n.

77. n avg 17 = 20 **78.** n avg 29 = 20 **79.** 21.375 avg n = 20

In 80–97, find the maximum using the set of real numbers.

80. 8 max 3 **81.** 8 max 11 **82.** 8 max 8 **83.** 1.2 max 0.12 **84.** 0.2 max 0.15

85. 0.8 max 0.30 **86.** 0.4 max 0.40 **87.** $\frac{1}{5}$ max $\frac{2}{5}$ **88.** $\frac{1}{5}$ max $\frac{1}{10}$ **89.** $\frac{8}{5}$ max 1.3

90. $\frac{8}{5}$ max 1.6 **91.** $\frac{3}{5}$ max $\frac{5}{8}$ **92.** 5 max $\sqrt{20}$ **93.** −16 max −4 **94.** 3 max π

95. −12.5 max −12 **96.** $\sqrt{2.89}$ max 1.7 **97.** 7.2 max 7.21

In 98–100, describe all possible values for b that would make the expression true.

98. 4 max b = 7 **99.** 4 max b = 4 **100.** 4 max b = 3

101. Elaine agrees to sell her bicycle to the person who makes the best offer. Mary Rose offers her $35, and Barbara offers her $27.50.
 a. To whom will Elaine sell the bicycle?
 b. What binary operation did you use to answer the question?

102. On Monday, Pam walked 3 miles in the morning and 1.5 miles in the afternoon. Anne walked the same total distance as Pam, but she walked equal distances in the morning and in the afternoon.
 a. How far did Anne walk in the morning?
 b. What binary operation did you use to answer the question?

103. A stock clerk is packaging books in cartons for shipment. A carton can hold 24 books. How many cartons are needed to package:
 a. 40 books? **b.** 75 books? **c.** 18 books? **d.** b books?

104. In a school store, pens cost 19 cents each. How many pens can be bought for:
 a. 50 cents? **b.** 38 cents? **c.** 12 cents? **d.** c cents?

2-2 BASES, EXPONENTS, AND POWERS

Factors

When two or more numbers are multiplied to give a certain product, each number is called a ***factor*** of the product. For example:

Since 1 × 16 = 16, then 1 and 16 are factors of 16.
Since 2 × 8 = 16, then 2 and 8 are factors of 16.
Since 4 × 4 = 16, then 4 is a factor of 16.

The numbers 1, 2, 4, 8, and 16 are all factors of 16.

Bases, Exponents, Powers

When the same number appears as a factor many times, we can rewrite the expression using *exponents*. For example, the exponent 2 indicates that the factor appears twice. In the following examples, the repeated factor is called a *base*.

Squares. $4 \times 4 = 16$ can be written as $4^2 = 16$.

4^2 is read as "4 squared," or "4 raised to the second power," or "the second power of 4."

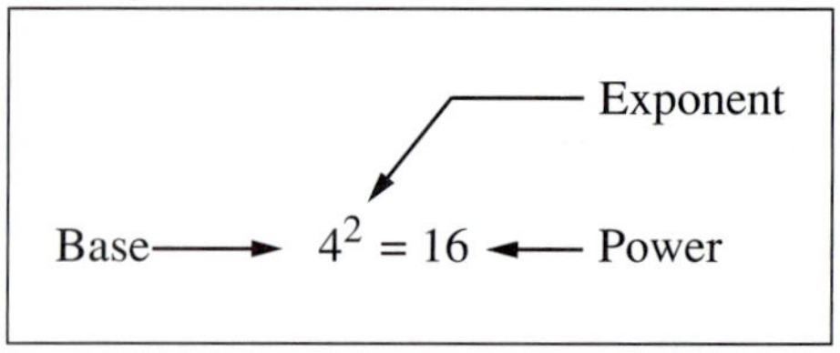

In Chapter 1, we learned that most calculators have a special key, [x^2], to square a number. Repeated multiplication can also be used to find a square.

Enter: METHOD 1. 4 [x^2]

METHOD 2. 4 [×] 4 [=]

Display: 16.

Cubes. $4 \times 4 \times 4 = 64$ can be written as $4^3 = 64$.

4^3 is read as "4 cubed," or "4 raised to the third power," or "the third power of 4."

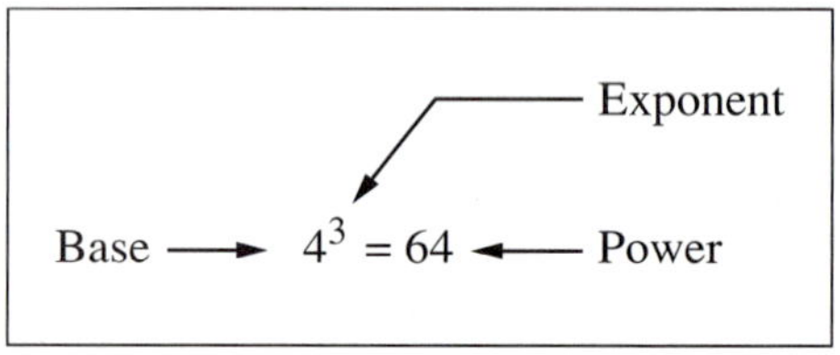

Scientific calculators have an exponent key, usually indicated as [y^x], that evaluates powers when the exponent is greater than 2. Repeated multiplication may also be used.

Enter: METHOD 1. 4 [y^x] 3 [=]

METHOD 2. 4 [×] 4 [×] 4 [=]

Display: 64.

The examples shown above lead to the following definitions:

A ***base*** is a number that is used as a factor in the product.

An ***exponent*** is a number that tells how many times the base is to be used as a factor. The exponent is written, in a smaller size, to the upper right of the base.

A ***power*** is a number that is a product in which all of its factors are equal.

A number raised to the first power is equal to the number itself, as in $6^1 = 6$. Also, when no exponent is shown, the exponent is 1, as in $9 = 9^1$.

Raising a number to a power is another binary operation. To evaluate 3^4 on a calculator, you would enter 3 [y^x] 4 [=] and get 81 in the display. Compare this to the general form of a binary operation, $a * b = c$, to see that the base is the first element of the ordered pair, the exponent is the second, and the power is the result.

The operation symbol on a calculator is [y^x], but, when writing exponents, we do not use the operation symbol. We simply write the exponent in smaller size to the upper right, as in $3^4 = 81$.

Examples like $3^4 = 81$ and $4^3 = 64$ point out again that order is important in a binary operation.

EXAMPLES

1. Compute the value of 4^5.

Solution

$$4 \times 4 \times 4 \times 4 \times 4 = 1{,}024$$

Alternative Solution Use the exponent key on a calculator.

Enter: 4 [y^x] 5 [=]

Display: [1024.]

A Third Solution Some calculators have a repeat function, and some have a constant key [K], which allows you to repeat multiplication without entering the base each time. The following works only for these calculators:

Enter: METHOD 1. 4 [×] [=] [=] [=] [=]

METHOD 2. 4 [×] [K] [=] [=] [=] [=]

Display: [1024.]

Answer: 1,024

---

2. Find $\left(\frac{2}{3}\right)^3$: **a.** as an exact value. **b.** as a rational approximation obtained by using a calculator.

Solution **a.** The exact value is a fraction.

$$\left(\frac{2}{3}\right)^3 = \frac{2}{3} \times \frac{2}{3} \times \frac{2}{3} = \frac{8}{27}$$

b. A calculator gives an approximate answer.

Enter: [(] 2 [÷] 3 [)] [y^x] 3 [=]

Display: [0.2962962]

Answers: **a.** $\frac{8}{27}$

b. 0.2962962 (*Note:* The actual value is 0.296296296 . . . or $0.\overline{296}$.)

---

3. Compute the value of $\left(1\frac{1}{2}\right)^2$

Solution $$\left(1\frac{1}{2}\right)^2 = 1\frac{1}{2} \times 1\frac{1}{2} = \frac{3}{2} \times \frac{3}{2} = \frac{9}{4} = 2\frac{1}{4}$$

Answer: $2\frac{1}{4}$

EXERCISES WITH OPEN-RESPONSE PROBLEM

In 1–24, find the exact value of each expression. In many cases, answers can be obtained by quick mental arithmetic.

1. 9^2 **2.** 10^3 **3.** 5^3 **4.** 15^2 **5.** 10^5 **6.** 3^5

7. $\left(\frac{1}{3}\right)^2$ **8.** $\left(\frac{1}{2}\right)^2$ **9.** $\left(\frac{1}{10}\right)^3$ **10.** $\left(\frac{3}{4}\right)^2$ **11.** $\left(\frac{1}{10}\right)^4$ **12.** $\left(\frac{8}{9}\right)^1$

13. $(0.8)^2$ **14.** $(0.5)^3$ **15.** $(0.3)^2$ **16.** $(0.2)^4$ **17.** $(0.1)^5$ **18.** $(0.25)^2$

19. $(1.1)^3$ **20.** $(3.1)^2$ **21.** $(2.5)^2$ **22.** $\left(2\frac{1}{2}\right)^2$ **23.** $\left(3\frac{1}{10}\right)^2$ **24.** $\left(1\frac{1}{3}\right)^3$

In 25–32: **a.** Find, in each case, the value of each of the three given expressions. **b.** Name, in each case, the expression that has the greatest value.

25. $(5)^2$ $(5)^3$ $(5)^4$

26. $(0.5)^2$ $(0.6)^2$ $(0.7)^2$

27. $(1.1)^2$ $(1.1)^3$ $(1.1)^4$

28. $(1.0)^2$ $(1.2)^2$ $(1.4)^2$

29. $\left(\frac{1}{5}\right)^2$ $\left(\frac{1}{6}\right)^2$ $\left(\frac{1}{7}\right)^2$

30. $(0.1)^3$ $(0.2)^2$ $(0.3)^1$

31. $(0.5)^3$ $(0.4)^2$ $(0.2)^4$

32. $(1.5)^3$ $(1.4)^2$ $(1.2)^4$

In 33–37, find the value of each expression.

33. 94^2 **34.** 17.8^3 **35.** 0.62^3 **36.** 8.5^4 **37.** 5.9^4

In 38–42, find the value of each given power rounded to four decimal places.

38. 0.84^3 **39.** 0.9^8 **40.** 1.3^7 **41.** 3.44^5 **42.** 1.7^{10}

In 43–48: **a.** Find the exact value of each given expression. **b.** Using a calculator, find a rational approximation for each expression.

43. $\left(\frac{2}{7}\right)^2$ **44.** $\left(\frac{4}{9}\right)^2$ **45.** $\left(\frac{1}{6}\right)^3$ **46.** $\left(\frac{5}{12}\right)^2$ **47.** $\left(2\frac{1}{6}\right)^2$ **48.** $\left(1\frac{2}{3}\right)^4$

49. The value of \$1 invested at 6% for 20 years is equal to 1.06^{20}. Find the value of this investment to the nearest cent.

50. a. How many years will it be required for \$1 invested at 8% to double in value? (*Hint:* Guess at values of n to find where 1.08^n is closest to 2.00.)

b. How many years will be required for \$100 invested at 8% to double in value?

2-3 ORDER OF OPERATIONS

When a numerical expression involves two or more different operations, we need to agree on the order in which they are performed. Consider this example:

$$11 - 3 \times 2$$

Suppose one person multiplied first.	Suppose another person subtracted first.
$11 - 3 \times 2 = 11 - 6$	$11 - 3 \times 2 = 8 \times 2$
$= 5$	$= 16$

Who is correct?

To guarantee that there will be one and only one correct answer to problems like this, mathematicians have agreed to follow this ***order of operations***:

1. Simplify powers (terms with exponents).

2. Multiply and divide, from left to right.

3. Add and subtract, from left to right.

Therefore, we multiply before we subtract, and $11 - 3 \times 2 = 11 - 6 = 5$ is correct.

A different problem involving powers is solved in this way:

1. Simplify powers: $5 \cdot 2^3 + 3 = 5 \cdot 8 + 3$

2. Multiply and divide: $= 40 + 3$

3. Add and subtract: $= 43$

Calculators and the Order of Operations

Two students had different calculators. They entered the problem at the top of this page and were surprised to see what happened.

On the first calculator,	On the second calculator,
Enter: 11 [−] 3 [×] 2 [=]	***Enter:*** 11 [−] 3 [×] 2 [=]
Display: [5.] *correct*	***Display:*** [16.] *incorrect*
(*A mathematical order calculator*)	(*An entry order calculator*)

A *mathematical order (MO) calculator* does not perform any operations until the entry key, [=], is pressed. It then follows the correct order of operations. These calculators are sometimes described as having *algebraic logic*.

Every *scientific calculator* is a mathematical order calculator, with the correct order of operations built into the device.

An *entry order (EO) calculator* performs the operations in the order in which they are entered. Most basic four-function calculators use this entry order. To get the correct answer to the given problem, it is necessary to enter the values in a different order, so that the multiplication is performed before the subtraction.

To find the correct answer to $11 - 3 \times 2$ on an entry order calculator, follow these steps:

Enter: 3 [×] 2 [+/–] [+] 11 [=]

The first three entries give the product of 6.
The [+/–] key, called the ***sign-change key***, changes positive 6 to negative 6.
Finally, 11 is added to -6, which is the same as subtracting 6 from 11, giving the correct answer.

Display: 5.

Clearly, it is easier to use a scientific calculator, in which the numerical expression can be directly entered as presented, and the order of operations is followed.

Expressions with Grouping Symbols

In mathematics, ***parentheses*** ( ) act as grouping symbols, giving different meanings to expressions. For example, $(4 \times 6) + 7$ means "add 7 to the product of 4 and 6," while $4 \times (6 + 7)$ means "multiply the sum of 6 and 7 by 4."

When simplifying any numerical expression, always perform the operations within parentheses first.

$$(4 \times 6) + 7 = 24 + 7 = 31 \qquad 4 \times (6 + 7) = 4 \times 13 = 52$$

Besides parentheses, other symbols are used to indicate grouping, such as *brackets*, [ ]. The expressions $2(5 + 9)$ and $2[5 + 9]$ have the same meaning: 2 is multiplied by the sum of 5 and 9. A *bar*, or *fraction line*, also acts as a symbol of grouping, telling us to perform the operations in the numerator and/or denominator first.

$$\frac{20 - 8}{3} = \frac{12}{3} = 4 \qquad \frac{6}{3 + 1} = \frac{6}{4} = \frac{3}{2} = 1\frac{1}{2}$$

When there are *two or more* grouping symbols in an expression, we perform the operations on the numbers in the *innermost symbol first*. For example:

$$\begin{aligned} & 5 + 2[6 + (3 - 1)^3] \\ &= 5 + 2[6 + 2^3] \\ &= 5 + 2[6 + 8] \\ &= 5 + 2[14] \\ &= 5 + 28 \\ &= 33 \end{aligned}$$

PROCEDURE. To simplify a numerical expression, follow the correct order of operations:

1. Simplify any numerical expressions within parentheses or within other grouping symbols, starting with the innermost.
2. Simplify any powers.
3. Do all multiplications and divisions in order from left to right.
4. Do all additions and subtractions in order from left to right.

EXAMPLES

1. Simplify the numerical expression $80 - 4(7 - 5)$.

How to Proceed:	*Solution:*
(1) Write the expression:	$80 - 4(7 - 5)$
(2) Simplify the value within the parentheses:	$80 - 4(2)$
(3) Multiply:	$80 - 8$
(4) Subtract:	72

A Calculator Solution: Parentheses can be entered on scientific calculators. Remember that, in the given expression, $4(7 - 5)$ means 4 *times* the value in the parentheses. When no operation symbol appears between a number and the start of parentheses, multiplication is indicated and the [×] key must be entered on the calculator.

Enter: 80 [−] 4 [×] [(] 7 [−] 5 [)] [=]

Display: 72.

Answer: 72

---

2. Evaluate: $6(4 + 1)^2 + (8 - 5)^3$.

How to Proceed:	*Solution:*
(1) Write the expression:	$6(4 + 1)^2 + (8 - 5)^3$
(2) Simplify expressions within parentheses:	$6(5)^2 + (3)^3$
(3) Evaluate the powers:	$6(25) + 27$
(4) Multiply:	$150 + 27$
(5) Add:	177

A Calculator Solution: Remember that the first set of parentheses must be multiplied by 6, so the [×] key is needed. When squaring, the [x^2] key is used. Also, when raising to the third power, the [y^x] key is followed by the exponent 3.

Enter: 6 [×] [(] 4 [+] 1 [)] [x^2] [+] [(] 8 [−] 5 [)] [y^x] 3 [=]

Display: 177.

Answer: 177

---

3. Evaluate: $28 \div 4 - 2(8 - 7)^2 + 5 \times 3$.

How to Proceed:	*Solution:*
(1) Write the expression:	$28 \div 4 - 2(8 - 7)^2 + 5 \times 3$
(2) Simplify the expression within the parentheses:	$28 \div 4 - 2(1)^2 + 5 \times 3$
(3) Evaluate the power:	$28 \div 4 - 2(1) + 5 \times 3$
(4) Do multiplication and division from left to right:	$7 - 2 + 15$
(5) Do addition and subtraction from left to right:	$5 + 15$ 20

A Calculator Solution:

Enter: 28 [÷] 4 [−] 2 [×] [(] 8 [−] 7 [)] [x^2] [+] 5 [×] 3 [=]

Display: 20.

Answer: 20

EXERCISES WITH OPEN-RESPONSE PROBLEM

In 1–8, state the meaning of each expression in part **a** and the meaning of each expression in part **b**, and simplify the expression in each part.

1. a. $20 + (6 + 1)$ **b.** $20 + 6 + 1$

2. a. $18 - (4 + 3)$ **b.** $18 - 4 + 3$

3. a. $12 - \left(3 - \frac{1}{2}\right)$ **b.** $12 - 3 - \frac{1}{2}$

4. a. $15 \times (2 + 1)$ **b.** $15 \times 2 + 1$

5. a. $(12 + 8) \div 4$ **b.** $12 + 8 \div 4$

6. a. $48 \div (8 - 4)$ **b.** $48 \div 8 - 4$

7. a. $7 + 5^2$ **b.** $(7 + 5)^2$

8. a. 4×3^2 **b.** $(4 \times 3)^2$

In 9–14, use parentheses to express the sentence in symbols.

9. The sum of 10 and 8 is to be found, and then 5 is to be subtracted from this sum.
10. 15 is to be subtracted from 25, and 7 is to be added to the difference.
11. 8 is to be multiplied by the difference of 6 and 2.
12. 12 is to be subtracted from the product of 10 and 5.
13. The difference of 12 and 2 is to be multiplied by the sum of 3 and 4.
14. The quotient of 20 and 5 is to be subtracted from the product of 16 and 3.

Basic Operations

In 15–20, simplify each numerical expression.

15. $6 \times 5 - 8 \times 2$
16. $20 + 20 \div 5 + 5$
17. $36 - 12 \div 4 - 1$
18. $36 + \frac{1}{2} \times 10$
19. $24 - 4 \div \frac{1}{2}$
20. $28 + 0 \div 4 - 10 \times 0.2$

Basic Operations and Powers

In 21–35, simplify each numerical expression.

21. 2×3^2
22. 4×5^2
23. $81 \times \left(\frac{1}{3}\right)^3$
24. $64 \times (0.5)^2$
25. $2^3 \times 1^2$
26. $10^2 \times 3^3$
27. $1^4 \times 9^2$
28. $(2^5)\left(\frac{1}{2}\right)^3$
29. $5^2 + 12^2$
30. $16^2 + 9^2$
31. $13^2 - 5^2$
32. $20^2 - 12$
33. $6 + 4(5)^2$
34. $3(2)^2 + 6$
35. $120 - 6(2)^4$

Grouping Symbols

In 36–59, simplify each numerical expression.

36. $10 + (1 + 4)$
37. $13 - (9 + 1)$
38. $36 - (10 - 8)$
39. $7(5 + 2)$
40. $(6 - 1)10$
41. $20 \div (7 + 3)$
42. $\frac{48}{15 - 3}$
43. $\frac{24 - 8}{2}$
44. $\frac{17 + 13}{25 - 10}$
45. $15 - (15 \div 5)$
46. $3(6 + 3) - 4$
47. $25 + 3(10 - 4)$
48. $26 - 4(7 - 5)$
49. $25 \div (6 - 1) + 3$
50. $3(6 + 4)(6 - 4)$
51. $100(6)^2 - 75$
52. $(4 + 6)^2$
53. $2[14 - (1 + 7)]$
54. $(20 - 15)^2$
55. $[7 - (2 + 3)]^3$
56. $2(4 + 6)^2 - 10$
57. $200 - 3(5 - 1)^3$
58. $12(5^2 - 4^2)$
59. $(7^2 - 6^2)(1^2 + 2^2)$

In 60–63: **a.** Write a numerical expression for each of the following. **b.** Evaluate each numerical expression written in answer to part **a.**

60. The cost of two chocolate chip and three peanut butter cookies if each cookie costs 8 cents.

61. The number of miles traveled by Ms. McCarthy if she drove 30 miles per hour for $\frac{3}{4}$ hour and 55 miles per hour for $1\frac{1}{2}$ hours.

62. The cost of two pens at \$0.38 each and three notebooks at \$0.69 each.
63. The cost of five pens at \$0.29 each and three notebooks at \$0.75 each if ordered from a mail order company that adds \$1.75 in postage and handling charges.
64. Insert operational symbols (+, −, ×, ÷) and parentheses to make each of the following statements true.

a. 3 __ 2 __ 1 = 4 b. 1 __ 3 __ 1 = 4
c. 1 __ 2 __ 3 __ 4 = 5 d. 4 __ 3 __ 2 __ 1 = 5
e. 6 __ 6 __ 6 __ 6 = 5 f. 6 __ 6 __ 6 __ 6 = 6

2-4 PROPERTIES OF OPERATIONS

When numbers behave in a certain way for an operation, we describe this behavior as a *property*. For example, in Section 2-1, we studied the property of closure.

As we examine other properties, we will use examples from the set of rational numbers. We are familiar with these numbers from our study of arithmetic and our review in Chapter 1. No proofs are given here, but the examples help us to see that these properties make sense and that they are true for the set of rational numbers.

Commutative Property of Addition

When we add rational numbers, we assume that we can change the order in which two numbers are added without changing the sum. For example, $4 + 5 = 5 + 4$, and $\frac{1}{2} + \frac{1}{4} = \frac{1}{4} + \frac{1}{2}$. These examples illustrate the ***commutative property of addition***.

In general, we assume that for every number a and every number b:

$$a + b = b + a$$

Commutative Property of Multiplication

In the same way, when we multiply rational numbers, we assume that we can change the order of the factors without changing the product. For example, $5 \times 4 = 4 \times 5$, and $\frac{1}{2} \times \frac{1}{4} = \frac{1}{4} \times \frac{1}{2}$. These examples illustrate the ***commutative property of multiplication***.

In general, we assume that for every number a and every number b:

$$a \cdot b = b \cdot a$$

The General Commutative Property

If we think in terms of the general form of a binary operation, $a * b = c$, and we can change the order of every pair of elements, a and b, without changing the result c, then we can say that $*$ is a commutative operation where:

$$\boxed{a * b = b * a}$$

However, if we find even one case where $a * b$ and $b * a$ produce different answers, then the operation $*$ is not commutative. For example:

The operation *subtraction* is not commutative because $5 - 4 \neq 4 - 5$.

The operation *division* is not commutative because $8 \div 4 \neq 4 \div 8$.

Many other operations seem to be commutative, such as avg and max.

5 avg 8 = 6.5, and 8 avg 5 = 6.5 Thus, 5 avg 8 = 8 avg 5	9.3 max 24 = 24, and 24 max 9.3 = 24 Thus, 9.3 max 24 = 24 max 9.3
In general, for every a and b: $\boldsymbol{a \text{ avg } b = b \text{ avg } a}$	In general, for every a and b: $\boldsymbol{a \text{ max } b = b \text{ max } a}$

Associative Property of Addition

Addition is a binary operation; that is, we add two numbers at a time. If we wish to add three numbers, we find the sum of two and add that sum to the third. For example:

$$\begin{aligned} 2 + 5 + 8 &= (2 + 5) + 8 \\ &= 7 + 8 \\ &= 15 \end{aligned} \quad \text{or} \quad \begin{aligned} 2 + 5 + 8 &= 2 + (5 + 8) \\ &= 2 + 13 \\ &= 15 \end{aligned}$$

Therefore, we see that $(2 + 5) + 8 = 2 + (5 + 8)$. This example illustrates the ***associative property of addition***. In general, we assume that for every number a, every number b, and every number c:

$$(\boldsymbol{a} + \boldsymbol{b}) + \boldsymbol{c} = \boldsymbol{a} + (\boldsymbol{b} + \boldsymbol{c})$$

Associative Property of Multiplication

In a similar way, to find a product that involves three factors, we first multiply any two factors and then multiply this result by the third factor. We assume that we do not change the product when we change the grouping. For example:

$$\begin{aligned} 5 \times 4 \times 2 &= (5 \times 4) \times 2 \\ &= 20 \times 2 \\ &= 40 \end{aligned} \quad \text{or} \quad \begin{aligned} 5 \times 4 \times 2 &= 5 \times (4 \times 2) \\ &= 5 \times 8 \\ &= 40 \end{aligned}$$

Therefore, $(5 \times 4) \times 2 = 5 \times (4 \times 2)$. This example illustrates the ***associative property of multiplication***.

In general, we assume that for every number a, every number b, and every number c:

$$(\boldsymbol{a} \cdot \boldsymbol{b}) \cdot \boldsymbol{c} = \boldsymbol{a} \cdot (\boldsymbol{b} \cdot \boldsymbol{c})$$

The General Associative Property

In a binary operation, symbolized by $*$, we work with two numbers at a time. According to the order of operations, we must simplify operations within parentheses first.

In general, when for every number a, for every number b, and for every number c:

$$\boxed{(a * b) * c = a * (b * c)}$$

then $*$ is an associative operation.

Remember that, with any operation, only one case is needed to show when the property fails. *Subtraction* and *division* are not associative, as shown in these examples:

$(10 - 8) - 2 \ ? \ 10 - (8 - 2)$	$(16 \div 4) \div 2 \ ? \ 16 \div (4 \div 2)$
$2 - 2 \ ? \ 10 - 6$	$4 \div 2 \ ? \ 16 \div 2$
$0 \neq 4$	$2 \neq 8$

Thus: $(a - b) - c \neq a - (b - c)$ and $(a \div b) \div c \neq a \div (b \div c)$

The Distributive Property

We know $4(3 + 2) = 4(5) = 20$ and also $4(3) + 4(2) = 12 + 8 = 20$. Therefore, we see that $4(3 + 2) = 4(3) + 4(2)$.

This result can be illustrated geometrically. Recall that the area of a rectangle is equal to the product of its length and its width.

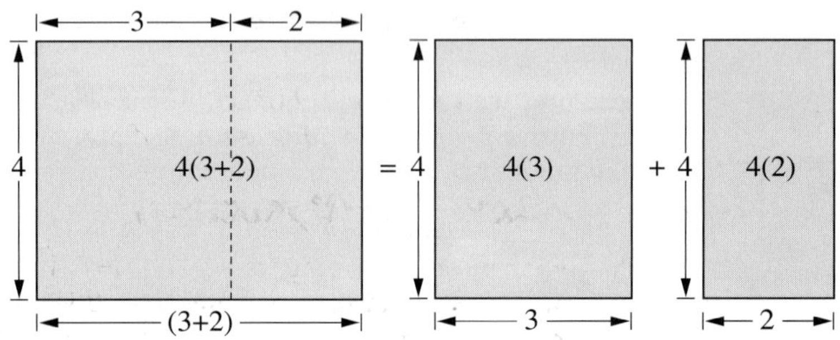

This example illustrates the ***distributive property of multiplication over addition***, also called the ***distributive property***. This means that the product of one number times the sum of a second and a third number equals the product of the first and second numbers plus the product of the first and third numbers.

In general, we assume that for every number a, every number b, and every number c:

$$\boldsymbol{a(b + c) = ab + ac} \quad \text{and} \quad \boldsymbol{ab + ac = a(b + c)}$$

The distributive property is also assumed to be true for subtraction:

$$\boldsymbol{a(b - c) = ab - ac} \quad \text{and} \quad \boldsymbol{ab - ac = a(b - c)}$$

Observe how we can use the distributive property to find the following products:

1. $6 \times 23 = 6(20 + 3) = 6 \times 20 + 6 \times 3 = 120 + 18 = 138$

2. $9 \times 3\frac{1}{3} = 9\left(3 + \frac{1}{3}\right) = 9 \times 3 + 9 \times \frac{1}{3} = 27 + 3 = 30$

3. $6.5 \times 8 = (6 + 0.5)8 = 6 \times 8 + 0.5 \times 8 = 48 + 4 = 52$

Notice how we can use the distributive property to change the form of an expression from a product to a sum or a difference:

1. $5(a + b) = 5a + 5b$ **2.** $9(a - b) = 9a - 9b$

Working backward, we can also use the distributive property to change the form of an expression from a sum or a difference to a product:

1. $3c + 3d = 3(c + d)$ **2.** $9y - 4y = (9 - 4)y = 5y$

Note that the ***substitution principle*** allows us to replace one expression for another expression that represents the same number, such as $(9 - 4)y = 5y$. Here $(9 - 4)$ was replaced by 5.

Addition Property of Zero and the Additive Identity Element

The sentences $5 + 0 = 5$ and $0 + 2.8 = 2.8$ are true. They illustrate that the sum of a rational number and zero is the number itself. These examples lead us to observe that:

1. The ***addition property of zero*** states that for every number a:

$$\boldsymbol{a + 0 = a} \qquad \text{and} \qquad \boldsymbol{0 + a = a}$$

2. The ***identity element of addition***, or the ***additive identity element***, is **0**. Thus, for any number a: if $a + x = a$, or if $x + a = a$, it follows that $x = 0$.

Additive Inverses (Opposites)

When we first studied integers, we learned about *opposites*. For example, the opposite of 4 is -4, and the opposite of -10 is 10.

Every rational number a has an ***opposite***, $-a$, such that their sum is 0, the identity element in addition. The opposite of a number is called the ***additive inverse*** of the number.

In general, for every rational number a and its opposite $-a$:

$$\boldsymbol{a + (-a) = 0}$$

On a calculator, the sign-change key, [+/–], shows the opposite of a number. If any nonzero number is displayed on the calculator, its opposite will appear when the sign-change key is pressed.

For example, to enter -4.25:

Enter: 4.25 [+/–]

Display: [–4.25]

Multiplication Property of One and the Multiplicative Identity Element

The sentences $5 \times 1 = 5$ and $1 \times 4.6 = 4.6$ are true. They illustrate that the product of a rational number and one is the number itself. These examples lead us to observe that:

1. The ***multiplication property of one*** states that for every number a:

$$\mathbf{a \cdot 1 = a} \quad \text{and} \quad \mathbf{1 \cdot a = a}$$

2. The ***identity element of multiplication***, or the ***multiplicative identity element***, is **1**.

Multiplicative Inverses (Reciprocals)

When the product of two numbers is 1 (the identity element in multiplication), then each of these numbers is called the ***multiplicative inverse*** or ***reciprocal*** of the other. Consider these examples:

$4 \cdot \frac{1}{4} = 1$	$\left(2\frac{1}{2}\right)\left(\frac{2}{5}\right) = \left(\frac{5}{2}\right)\left(\frac{2}{5}\right) = 1$
The reciprocal of 4 is $\frac{1}{4}$.	The multiplicative inverse of $2\frac{1}{2}$ or $\frac{5}{2}$ is $\frac{2}{5}$.
The reciprocal of $\frac{1}{4}$ is 4.	The multiplicative inverse of $\frac{2}{5}$ is $\frac{5}{2}$ or $2\frac{1}{2}$.

Since there is no number that, when multiplied by 0, gives 1, the number 0 has no reciprocal, or no multiplicative inverse.

In general, for every nonzero rational number a, there is a unique rational number $\frac{1}{a}$ such that:

$$\mathbf{a \cdot \frac{1}{a} = 1}$$

On many calculators, a special key, **1/x**, displays the reciprocal. For example, if each of the numbers shown above is entered and the ***reciprocal key***, **1/x**, is pressed, the reciprocal appears in decimal form.

Enter:	4 **1/x**	***Enter:***	2.5 **1/x**
Display:	0.25	***Display:***	0.4

For many other numbers, however, the decimal form of the reciprocal is not shown in its entirety in the display. For example, we know that the reciprocal of 6 is an exact value, $\frac{1}{6}$, but what appears is a rational approximation of $\frac{1}{6}$.

Enter: 6 **1/x**

Display: 0.1666666

A calculator may store more decimal places in its operating system than it has in its display. If this is true, the decimal displayed times the original number will equal 1. If the product does not equal 1, the calculator will use only the approximation shown in its display. Test your calculator to see which of the following is true.

Enter: 1/x × 6 =	*Enter:* 6 1/x × 6 =
Display: 1.	*Display:* 0.9999996
This calculator holds more decimals in its operating system than displayed.	This calculator does *not* hold more decimals. It works only with the approximations in the display.

Multiplication Property of Zero

The sentences $7 \times 0 = 0$ and $0 \times \frac{3}{4} = 0$ are true. They illustrate that the product of a rational number and zero is zero. These examples lead us to observe that:

In general, for every number a:

$$\mathbf{a \cdot 0 = 0} \quad \text{and} \quad \mathbf{0 \cdot a = 0}$$

EXAMPLES

Express $6t + t$ as a product and give the reason for each step of the procedure.

Solution

Step	*Reason*
(1) $6t + t = 6t + 1t$	(1) Multiplication property of 1.
(2) $= (6 + 1)t$	(2) Distributive property.
(3) $= 7t$	(3) Substitution principle.

Answer: $7t$

EXERCISES WITH OPEN-RESPONSE PROBLEMS

1. Name the number that is the additive identity element for the rational numbers.

2. Name the identity element of multiplication for the rational numbers.

3. Give the value of each expression.

a. $9 + 0$ **b.** 9×0 **c.** 9×1 **d.** $\frac{2}{3} \times 0$ **e.** $0 + \frac{2}{3}$ **f.** $1 \times \frac{2}{3}$

In 4–21: **a.** Replace each question mark with the number that makes the sentence true. **b.** Name the property illustrated in each sentence that is formed when the replacement is made.

4. $8 + 6 = 6 + ?$

5. $17 \times 5 = ? \times 17$

6. $(3 \times 9) \times 15 = 3 \times (9 \times ?)$

7. $6(5 + 8) = 6(5) + ?(8)$

8. $\left(\frac{1}{3} + \frac{1}{6}\right) + \frac{1}{2} = ? + \left(\frac{1}{6} + \frac{1}{2}\right)$

9. $4 + 0 = ?$

10. $3(x + 5) = 3x + 3(?)$

11. $(3 \times 7) + 5 = (? \times 3) + 5$

12. $(19 \times 2) \times 50 = ? \times (19 \times 2)$

13. $(19 \times ?) + 50 = 19 + 50$

14. $3(8 + 2) = (? + 2)3$

15. $(5 + ?) + 2 = 5 + (6 + 2)$

16. $\frac{1}{2}$ avg 6 = ? avg $\frac{1}{2}$

17. $\frac{1}{2} \times 10 = 10 \times ?$

18. $3(y + ?) = 3y$

19. $?x = x$

20. (7 max 9) max 4 = 7 max (? max 4)

21. $5\,(4 + 3) = 5\,(3 + ?)$

In 22–27, name the additive inverse (opposite) of each number.

22. 7 **23.** 1 **24.** -10 **25.** 2.5 **26.** -1.8 **27.** $\frac{1}{9}$

In 28–33, name the multiplicative inverse (reciprocal) of each number.

28. 7 **29.** 1 **30.** -10 **31.** 2.5 **32.** -1.8 **33.** $\frac{1}{9}$

In 34–39, write the display shown on a calculator for each entry.

34. ***Enter:*** 4 [+/−]

35. ***Enter:*** 4 [1/x]

36. ***Enter:*** 0.8 [1/x]

37. ***Enter:*** 0.8 [+/−]

38. ***Enter:*** 20 [1/x] [×] 20 [=]

39. ***Enter:*** 17.94 [+/−] [+] 17.94 [=]

40. **a.** Write the display shown on a calculator for each of the following entries.
(*1*) ***Enter:*** [(] 3 [+] 9 [)] [+/−] [=]
(2) ***Enter:*** 3 [+/−] [+] 9 [+/−] [=]
b. Are the two displays in part **a** the same? If they are, explain why. If they are not, what change could be made so that the same number is displayed for both entries.

41. **a.** Find the reciprocal of $\frac{5}{8}$. **b.** Find the reciprocal of 0.625.
c. What relationship exists between the answers to parts **a** and **b**? Explain why.

In 42–51, state whether each sentence is a correct application of the distributive property. If you believe that it is not, state your reason.

42. $6(5 + 8) = 6 \times 5 + 6 \times 8$

43. $10\left(\frac{1}{2} + \frac{1}{5}\right) = 10 \times \frac{1}{2} + \frac{1}{5}$

44. $(7 + 9)5 = 7 + 9 \times 5$

45. $3(x + 5) = 3x + 3 \times 5$

46. $2(y + 6) = 2y + 6$

47. $(b + 2)a = ba + 2a$

48. $4a(b + c) = 4ab + 4ac$

49. $4b(c - 2) = 4bc - 2$

50. $8m + 6m = (8 + 6)m$

51. $14x - 4x = (14 - 4)x$

In 52–57: **a.** Tell whether each sentence is true or false.
b. Tell whether the commutative property holds for the given operation.

52. $357 + 19 = 19 + 357$ **53.** $2 \div 1 = 1 \div 2$ **54.** $25 - 7 = 7 - 25$

55. $(18)\left(2\frac{1}{2}\right) = \left(2\frac{1}{2}\right)(18)$ **56.** $\frac{2}{5} + \frac{3}{10} = \frac{3}{10} + \frac{2}{5}$ **57.** $5 - 0 = 0 - 5$

In 58–63: **a.** Tell whether each sentence is true or false.
b. Tell whether the associative property holds for the given operation.

58. $(73 \times 68) \times 92 = 73 \times (68 \times 92)$

59. $(24 \div 6) \div 2 = 24 \div (6 \div 2)$

60. $(19 - 8) - 5 = 19 - (8 - 5)$

61. $(9 + 0.3) + 0.7 = 9 + (0.3 + 0.7)$

62. $(8 \div 4) \div 2 = 8 \div (4 \div 2)$

63. $(40 - 20) - 10 = 40 - (20 - 10)$

64. If $r + s = r$: **a.** What is the numerical value of s? **b.** What is the numerical value of rs?

65. If $xy = x$, what is the numerical value of y?

66. If $xy = 0$ and $x \neq 0$, what is the numerical value of y?

In 67–70, complete each sentence so that it is an application of the distributive property.

67. $4(p + q) =$ ____ **68.** ____ $= 2x - 2y$ **69.** $8t + 13t =$ ____ **70.** ____ $= (15 - 7)m$

71. **a.** Find the unique solution of (10 avg 14) avg 2.
b. Find the unique solution of 10 avg (14 avg 2).
c. Are the solutions in parts **a** and **b** the same or different?
d. Does it appear that the operation of averaging two numbers is associative?

In 72–73, give a reason for each step used to simplify the given expression.

72. (1) $3w - w = 3w - 1w$
(2) $= (3 - 1)w$
(3) $= 2w$

73. (1) $(8x + 4x) - 11x = (8 + 4)x - 11x$
(2) $= 12x - 11x$
(3) $= (12 - 11)x$
(4) $= 1x$
(5) $= x$

74. Insert parentheses to make each statement true.

a. $3 \times 2 + 1 \div 3 = 3$ **b.** $4 \times 3 \div 2 + 2 = 3$ **c.** $8 + 8 \div 8 - 8 \times 8 = 8$
d. $3 \div 3 + 3 \times 3 - 3 = 1$ **e.** $3 \div 3 + 3 \times 3 - 3 = 0$ **f.** $0 \times 12 \times 3 - 16 \div 8 = 0$

2-5 PROPERTIES OF A FIELD

In Section 2-4, we studied many properties of operations using the set of rational numbers. Most of those properties are found in the definition of a field, and you may think of this definition as a compact way to remember those properties.

A ***field*** is simply a set of numbers for which, under the operations of addition and multiplication, the following 11 properties are true:

Property	*In Symbols*
1. The set is closed under addition (or addition is a binary operation).	**1.** $a + b = c$ (c is a unique number in the set.)
2. Addition is commutative.	**2.** $a + b = b + a$

Property	*In Symbols*
3. Addition is associative.	**3.** $(a + b) + c = a + (b + c)$
4. The number 0 is the additive identity.	**4.** $a + 0 = a$ and $0 + a = a$
5. Every number a has an additive inverse $-a$.	**5.** $a + (-a) = 0$
6. The set is closed under multiplication (or multiplication is a binary operation).	**6.** $ab = c$ (c is a unique number in the set.)
7. Multiplication is commutative.	**7.** $ab = ba$
8. Multiplication is associative.	**8.** $(ab)c = a(bc)$
9. The number 1 is the multiplicative identity.	**9.** $a \cdot 1 = a$ and $1 \cdot a = a$
10. Every nonzero number a has a unique multiplicative inverse $\frac{1}{a}$.	**10.** $a \cdot \frac{1}{a} = 1$ $(a \neq 0)$
11. Multiplication is distributive over addition.	**11.** $a(b + c) = ab + ac$

The set of *rational numbers*, under the operations of addition and multiplication, is a field. We have seen how the 11 properties are true for rational numbers.

The set of *real numbers*, under the operations of addition and multiplication, is also a field. In other words, the 11 properties listed above are true for all real numbers. We will study some of these properties later in this book.

EXERCISES WITH OPEN-RESPONSE PROBLEM

1. Which is an illustration of the commutative property of addition?
(1) $ab = ba$ (2) $a + 0 = a$
(3) $a + b = b + a$ (4) $(a + b) + c = a + (b + c)$

2. Which is an illustration of the associative property of multiplication?
(1) $ab = ba$ (2) $a(0) = 0$ (3) $a(1) = a$ (4) $(ab)c = a(bc)$

3. Which is an illustration of the distributive property of multiplication over addition?
(1) $a + b = b + a$ (2) $a(b + c) = ab + ac$
(3) $(a + b) + c = a + (b + c)$ (4) $a(b + c) = ab + c$

4. What is the additive inverse of the real number represented by n?

5. What is the multiplicative inverse of the nonzero real number represented by n?

6. What is the additive identity for the set of real numbers?

7. What is the multiplicative identity for the set of real numbers?

8. **a.** Name the additive inverse or opposite of: (*1*) 8 (*2*) 50 (*3*) 999
b. Each of the numbers in part **a** is a whole number. Are the additive inverses of these numbers also whole numbers?
c. *True* or *False*: The set of whole numbers is not a field because its additive inverses are not whole numbers.

9. **a.** Name the multiplicative inverse or reciprocal of: (*1*) 8 (*2*) 50 (*3*) 999
b. Each of the numbers in part **a** is a whole number. Are the multiplicative inverses of these numbers also whole numbers?
c. *True* or *False*: The set of whole numbers is not a field because its multiplicative inverses are not whole numbers.

10. Explain why the set of integers is not a field by naming one of the 11 properties that is not true for this set.

CHAPTER SUMMARY

A ***binary operation*** in a set assigns to every ordered pair of elements from the set a unique answer from that set. The general form of a binary operation is $a * b = c$, where a, b, and c are elements of the set and $*$ is the operation symbol. Binary operations exist in arithmetic, in geometry, and in sets.

Operations in arithmetic include addition, subtraction, multiplication, division, raising to a power, average (avg), and maximum (max).

Powers are the result of repeated multiplication of the same ***factor***, as in $5^3 = 125$. Here, the ***base*** 5 with an ***exponent*** of 3 equals $5 \times 5 \times 5$ or 125, the power.

Numerical expressions are simplified by following a clear ***order of operations***: (1) simplify parentheses; (2) simplify powers; (3) multiply and divide from left to right; (4) add and subtract from left to right.

Many ***properties*** are used in operations with numbers, including 11 important field properties: ***closure*** under addition and multiplication; ***commutative*** and ***associative*** properties for these operations; the ***distributive*** property; the ***additive identity*** (0); an ***additive inverse*** (opposite), called $-a$ for every element a; the ***multiplicative identity*** (1); and a ***multiplicative inverse*** (reciprocal), $\frac{1}{a}$, for every nonzero element a.

VOCABULARY

2-1 Binary operation Unique Closure Average (avg) Maximum (max)

2-2 Factor Base Exponent Power

2-3 Order of operations — Calculator sign-change key, [+/–]
Parentheses

2-4 Commutative property: $a * b = b * a$
Associative property: $(a * b) * c = a * (b * c)$
Distributive property of multiplication over addition:
$a \cdot (b + c) = a \cdot b + a \cdot c$
Identities and Inverses
Reciprocal key, [1/x]

2-5 Field

REVIEW EXERCISES

In 1–16, perform the operation indicated within the set of rational numbers.

1. $8.5 + 0.64$ **2.** $8.5 - 0.64$ **3.** 8.5×0.64 **4.** $6 \div 0.2$

5. $\frac{2}{3} + \frac{1}{6}$ **6.** $3\frac{1}{2} - \frac{4}{9}$ **7.** $5\frac{1}{3} \times 4\frac{1}{2}$ **8.** $1\frac{1}{2} \div \frac{3}{8}$

9. 7 avg 31 **10.** 1.2 avg 2 **11.** 3 max 18 **12.** 0.3 max 0.18

13. $\frac{3}{8}$ max $\frac{2}{5}$ **14.** $2\frac{1}{3}$ avg $\frac{1}{3}$ **15.** -2 max 2 **16.** 0 max -1

In 17–22, simplify each numerical expression.

17. $20 - 3 \times 4$ **18.** $(20 - 3)4$ **19.** $8 + 16 \div 4 - 2$

20. $(0.16)^2$ **21.** $6^2 + 8^2$ **22.** $7(9 - 7)^3$

In 23–28, find the display shown when each of the following entries is made on a calculator with mathematical order (a calculator that uses the order of operations).

23. ***Enter:*** 1.8 [x^2] **24.** ***Enter:*** 23.00826 [+/–]

25. ***Enter:*** 1.2 [y^x] 3 [=] **26.** ***Enter:*** [(] 5 [÷] 8 [)] [×] 3 [=]

27. ***Enter:*** 0.125 [1/x] [=] **28.** ***Enter:*** 5 [÷] [(] 8 [×] 3 [)] [=]

In 29–34: **a.** Replace each question mark with a number that makes the sentence true. **b.** Name the property illustrated in each sentence that is formed when the replacement is made.

29. $8 + (2 + 9) = 8 + (9 + ?)$ **30.** $8 + (2 + 9) = (8 + ?) + 9$

31. $3(?) = 3$ **32.** $3(?) = 0$

33. $5(7 + 4) = 5(7) + ?(4)$ **34.** $5(7 + 4) = (7 + 4)?$

In 35–42, to each property named in Column I, match the correct application of the property found in Column II.

Column I	*Column II*
35. Associative property of multiplication	**a.** $3 + 4 = 4 + 3$
36. Associative property of addition	**b.** $3 \cdot 1 = 3$
37. Commutative property of addition	**c.** $0 \cdot 4 = 0$
38. Commutative property of multiplication	**d.** $3 + 0 = 3$
39. Identity element of multiplication	**e.** $3 \cdot 4 = 4 \cdot 3$
40. Identity element of addition	**f.** $3(4 + 5) = 3 \cdot 4 + 3 \cdot 5$
41. Distributive property	**g.** $3(4 \cdot 5) = (3 \cdot 4)5$
42. Multiplication property of zero	**h.** $(3 + 4) + 5 = 3 + (4 + 5)$

CUMULATIVE REVIEW WITH OPEN-RESPONSE PROBLEMS

In 1–4, perform each operation within the set of rational numbers.

1. 7(3.5) **2.** 7 max 3.5 **3.** $7 \div 3.5$ **4.** 7 avg 3.5

5. Which of the four numbers listed below is different from the other three? Explain why it is different.

$$-30, \quad 10, \quad 15, \quad 18$$

Exploraton

Interview a person who uses mathematics at work. Ask the person to describe the kinds of numbers that he/she uses. Are the numbers used natural numbers, counting numbers, integers, rational numbers or irrational numbers? Ask the person if he/she uses approximate values, and if so, to how many decimal places are the numbers rounded?

Chapter 3
Problem Solving

The athletic department arranged to transport 125 students, including the basketball team and supporters, to a playoff game. If each bus accommodated 48 students, how many buses were needed for the trip?

The distance from the school to the game was 125 miles. If the bus traveled at an average rate of 48 miles per hour, how long did the trip take?

On the trip, each student drank a can of soda. The 125 cans were collected for recycling and placed in cases of 48 cans each. Only full cases of empties could be returned to the distributor for a deposit refund. How many cases could be returned?

Each of these is a simple problem. How are the three problems alike? How are they different? In solving a problem, we must both select a plan that will provide a solution and determine how to use that plan to answer the question.

3-1 GENERAL TECHNIQUE FOR SOLVING PROBLEMS

We are living in exciting times. Our world is changing at a faster pace than ever before. Since you will probably see, in the years to come, advances in science and technology that today exist only as ideas in the minds of creative persons, you cannot foresee some of the kinds of problems you will have to solve on your job or in your life.

This chapter is about learning to solve problems. The emphasis is not on any particular type of problem. Rather, the focus is on the general technique that you can use to solve any problem. As you will discover, the strategies developed here can help you to become a better problem solver.

PROCEDURE. To solve any problem, it is useful to work through these four basic steps:

1. *Understand the problem.*
   Read and reread the problem. Be sure that you understand what information is given and what the problem asks you to do.
2. *Make a plan.*
   Gather and organize the information given, discarding whatever is not necessary for the solution of the problem.
   Decide on an approach that may lead to a solution. If several steps are involved, list them in order.
   If possible and appropriate, make an estimate of what a reasonable answer might be.
3. *Solve the problem.*
   Carry out the steps in your plan.
4. *Check the solution.*
   Test your result. Is it reasonable? Does it satisfy the given conditions? If your check shows that you have correctly solved the problem, you may write your solution as the answer.

Not every problem is solved in the first attempt. If carrying out your plan does not result in a solution that fulfills all of the conditions given, or if it is impossible to carry out your plan, start again. Reread the problem for information or ideas that you may have overlooked. Try another strategy—a different plan or approach that may lead to a solution.

The following strategies are some of the more commonly used.

1. Guessing and checking
2. Using a simpler related problem
3. Working backward
4. Discovering patterns
5. Drawing pictures and diagrams
6. Making lists and charts

It is impossible to list here all the ways in which a problem may be solved. These six strategies will help you to formulate a plan for solving a problem.

There is no one "right" strategy for a given problem. Often, a problem can be solved by any one of several strategies or by combining strategies. Problems sometimes require that you use familiar ideas in new ways.

As you work through subsequent chapters of this book and develop algebraic skills, you will be able to add to the six strategies listed above a seventh one: using an algebraic equation. Because algebra is important in the development of advanced mathematics, this algebraic strategy is the one that you will use most often in this course.

3-2 GUESSING AND CHECKING

This strategy is often called *trial and error*. It is particularly useful when the answer must be a whole number between limits.

EXAMPLES

1. Keegan celebrated his birthday by inviting a group of friends to join him in having sundaes. The bill amounted to \$27.43. If the cost of each sundae was the same: **a.** how many friends did Keegan invite? **b.** what was the cost of a sundae?

Understand: Think of the amount of the bill as 2,743 cents. The cost, in cents, of each sundae and the number of sundaes must be whole numbers whose product is 2,743. If the product of two whole numbers is 2,743, then 2,743 divided by one of the whole numbers is the other whole number.

Plan: Guess a possible number of sundaes. Use a calculator to divide 2,743 by that number. If the quotient is a whole number, the guess and the quotient are possible solutions. If the quotient is not a whole number, guess another number. Since 2,743 is an odd number, it must be the product of two odd numbers. Also, neither of these odd numbers can end in 5 because the units digit of 2,743 is not 5 or 0.

Solutions **a** and **b.** Divide 2,743 by odd numbers that do not end in 5 (that is, divide by 3, 7, 9, 11, 13, 17, . . .) until a whole-number quotient is found.

Enter: 2,743 [÷] 3 [=]
Display: 914.33333 No

Enter: 2,743 [÷] 7 [=]
Display: 391.85714 No

Enter: 2743 [÷] 9 [=]
Display: 304.77778 No

Enter: 2743 [÷] 11 [=]
Display: 249.36364 No

In each case shown above, the quotient is not a whole number.

Enter: 2743 [÷] 13 [=]

Display: [211.] A whole number

Since 13(211) = 2,743:

or 13 sundaes at \$2.11 each = \$27.43 (reasonable answers)
211 sundaes at \$0.13 each = \$27.43 (211 is not a reasonable number of people, nor is \$0.13 a reasonable cost for a sundae)

Are these the only possible answers? If we try other numbers, we find that 13 and 211 are the only two whole numbers, other than 1 and 2,743, whose product is 2,743. No solution makes sense except the one found.

Check: 13(\$2.11) = \$27.43

Answers: **a.** Keegan invited 12 friends (the 13th sundae was for himself).
b. Each sundae cost \$2.11.

---

2. Ann has eight bills, consisting of \$5 bills and \$10 bills, worth \$50. What bills does she have?

Plan: Try possible combinations of \$5 bills and \$10 bills until you find the combination whose value is \$50.

If all eight bills were \$5 bills, the value would be \$40. Since this total is a little too small, and the problem implies that Ann has at least one \$10 bill, you may estimate that you need a small number of \$10 bills.

Solution

Try seven \$5 bills = \$35
one \$10 bill = \$10
total value = \$45

Since you again need to increase the total value, you must add another \$10 bill.

Try six \$5 bills = \$30
two \$10 bills = \$20
total value = \$50

Check: You checked as part of the solution. Note that the small number of \$10 bills agrees with the estimate.

Answer: Ann has six \$5 bills and two \$10 bills.

EXERCISES WITH OPEN-RESPONSE PROBLEMS

1. Find two whole numbers whose sum is 60 and whose difference is 12.
2. Cynthia is 5 years older than her sister Sylvia. The sum of their ages is 13. How old is each girl?
3. Anthony has 24 coins, all nickels and dimes, worth $2.20. How many of each coin has he?
4. Mr. Strapp makes stools with three legs and with four legs. One week, he used 70 legs to make 18 stools. How many stools of each kind did he make?
5. The sophomores made 50 corsages to present to their mothers at Open House. Each corsage consisted of either two gardenias or three roses. How many of each kind of corsage were made if 130 flowers were used?
6. Two different numbers use the same two digits. The sum of the digits is 7, and the difference of the numbers is 27. Find the numbers.
7. Place 40 checkers in two stacks so that, if five checkers are moved from the taller stack to the shorter one, there will be the same number of checkers in each stack. How many checkers were in the taller stack?
8. If Shelly gives John $5, they will have the same amount of money. If John gives Shelly $5, she will have twice as much money as he will have. How much money does each have?
9. A tour-bus operator sold $1,517 worth of tickets for a trip to Niagara Falls. Each ticket cost the same amount and could be paid for with three bills.
   a. How many tickets were sold?
   b. What was the cost of a ticket?
10. An auditorium seats 1,073 people. Each row has the same number of seats, and there are more than 30 rows.
    a. How many rows are there?
    b. How many seats are in each row?

3-3 USING A SIMPLER RELATED PROBLEM

If a problem uses large numbers or consists of many cases, it is often possible to find the solution by first solving a similar problem with smaller numbers or fewer cases.

Study the following examples that are solved by using a simpler related problem. In Example 1, the problem is changed from finding the sum of the whole numbers from 30 to 39 to finding the sum of the whole numbers from 0 to 9. In Example 2, the problem is changed from forming, under given conditions, three stacks of pennies made with 100 pennies to finding the number of pennies needed for the smallest possible stacks with the given conditions. In each case, the simpler problem leads to a solution of the original problem.

EXAMPLES

1. A man has 100 pennies. He tries to place them all in three stacks so that the second stack has twice as many pennies as the first, and the third stack has twice as many pennies as the second.

a. What is the largest number of pennies that can be placed in each stack?

b. How many pennies are left over?

Plan: Make the smallest possible stacks, and determine how many such stacks can be made with 100 pennies.

Solutions **a** and **b.** To make the smallest possible stacks (1, 2, and 4 pennies), you need 7 pennies.

To find how many such stacks can be made from 100 pennies, divide:

$$100 \div 7 = 14, \text{ with a remainder of } 2$$

Therefore, you can make each of these small stacks 14 times as large and use all but two of the pennies.

$$\begin{aligned} 14(1) &= 14 \quad \text{first stack} \\ 14(2) &= 28 \quad \text{second stack} \\ 14(4) &= \underline{56} \quad \text{third stack} \\ &\ 98 \quad \text{pennies} \end{aligned}$$

There would be 2 pennies left over.

Alternative Solution Start with simple numbers that fit the problem. Then use multiplication to increase these numbers until you find the answer.

	Stack 1	Stack 2	Stack 3	Total
The smallest possible numbers are:	1	2	4	7
Multiply each of these numbers by 10:	10	20	40	70
Multiply the top row by 15:	15	30	60	105 (too large)
Multiply the top row by 14:	14	28	56	98

There would be 100 − 98 or 2 pennies left over.

Check: The second stack has twice as many as the first stack ($2 \cdot 14 = 28$), and the third stack has twice as many as the second stack ($2 \cdot 28 = 56$).

Answers: **a.** The three stacks contain 14, 28, and 56 pennies.
b. Two pennies are left over.

2. Find the sum of the whole numbers from 30 through 39.

Plan: First, find the sum of the whole numbers from 0 through 9. Since each number in the sum required is 30 larger, add 10(30) to the sum of the numbers from 0 through 9.

To estimate an answer, use the fact that, since all but one of the 10 numbers are larger than 30, the sum should be larger than 10(30), or 300. Also, since all 10 numbers are less than 40, the sum should be smaller than 10(40), or 400.

Solution

$$0 + 1 + 2 + 3 + 4 + 5 + 6 + 7 + 8 + 9 = 45$$

$$45 + 10(30) = 45 + 300 = 345$$

Check: You can use a calculator to check the sum. Notice that this answer is within the range that was estimated.

Answer: The sum of the whole numbers from 30 through 39 is 345.

EXERCISES WITH OPEN-RESPONSE PROBLEMS

1. Dora has a bank that contains 4 times as many quarters as dimes. The bank contains more than \$10 and less than \$12. How many dimes and quarters are in the bank?

2. Agnes won \$10,000. She wants to spend part for a trip and 3 times as much to make repairs on her house. She also plans to put twice as much as she spends on her house into a savings account. How much money does she plan to save?

3. A man wishes to divide his herd of horses among his children so that each child gets half as many horses as the next older child. Any extra horses will be sold. The man has four children and 48 horses. **a.** How many horses will each child get? **b.** How many horses will be sold?

4. Find the sum of the whole numbers from 51 through 59.

5. The directions on a bottle of concentrated cleaning liquid recommend that 2 ounces of the concentrate be mixed with 30 ounces of water. How many ounces of the concentrate must be used to make 3 gallons (384 ounces) of cleaning solution?

6. A health-foods store prepares a snack mix by combining 1 pound of nuts, $\frac{1}{2}$ pound of dried apricots, and 2 pounds of dried apples. How many pounds of apricots are needed to prepare 35 pounds of this mix?

7. Khong bought twice as many loaves of bread as he did pounds of meat. Each pound of meat cost 3 times as much as a loaf of bread. If Khong paid \$12 for the meat and bread, how much did he spend for bread?

8. In the town of Tranquility, each of the 5,285 families has no, one, or two children in the local school. The majority of families have one child. Half of the remaining families have two children. How many children from the town attend the local school? (*Hint:* Try small numbers of families, and assign a number at random to the number of families who have one child.)

9. Find the largest three-digit number that has exactly three factors. (*Hint:* Think of small numbers that have exactly three factors. For example, the factors of 4 are 1, 2, and 4. The factors of 9 are 1, 3, and 9.)

10. Find the sum:

$$\frac{1}{2} + \frac{1}{4} + \frac{1}{8} + \frac{1}{16} + \frac{1}{32} + \frac{1}{64} + \frac{1}{128} + \frac{1}{256} + \frac{1}{512} + \frac{1}{1{,}024}$$

$\left(\right.$*Hint:* Find smaller sums such as $\frac{1}{2} + \frac{1}{4}$ and $\frac{1}{2} + \frac{1}{4} + \frac{1}{8}$.$\left.\right)$

3-4 WORKING BACKWARD

This strategy is often useful when you know the result of a series of events and want to find a value present at the beginning of the series.

Inverses are often used when working backward. If a problem involves addition or subtraction, use the additive inverse or opposite. If a problem involves multiplication or division, use the multiplicative inverse or reciprocal.

EXAMPLES

1. Bob runs the elevator in an apartment building. He took Mr. Sloan up six floors from the floor on which he lives. Then Bob went down five floors, where he picked up Mrs. Rice. He took her down ten floors to the first-floor lobby. What is the number of the floor on which Mr. Sloan lives?

Plan: Work backward, reversing the course of the elevator, to find the number of Mr. Sloan's floor.

Solution From the lobby, go up 10 floors to where Mrs. Rice got on. This is the 11th floor. Go up 5 more floors to where Mr. Sloan got off. This is the 16th floor. From the 16th floor, go down 6 floors to where Mr. Sloan got on. This is the 10th floor.

Alternative Solution The elevator went up 6 floors, down 5 floors, then down 10 to get to floor 1.

Use integers: $+6$ -5 and -10 to get 1

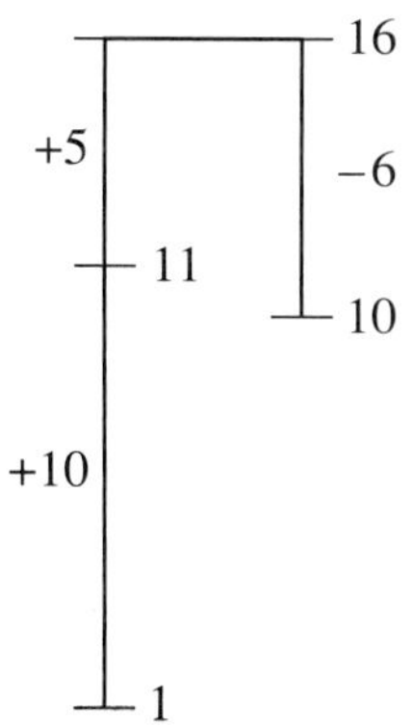

Start at 1 and apply additive inverses in the reverse order of the problem.

The additive inverse of -10 is $+10$, so: $1 + 10 = 11$

The additive inverse of -5 is $+5$, so: $11 + 5 = 16$

The additive inverse of $+6$ is -6, so: $16 - 6 = 10$ (the 10th floor)

Check: Let Mr. Sloan get on at the 10th floor and follow the original course of the elevator. The elevator went from the 10th floor up 6 floors to the 16th. From there, it went down 5 floors, to the 11th, and then down the remaining 10 floors to the 1st floor, or lobby.

Answer: Mr. Sloan lives on the 10th floor.

2. Each weekend Jamaal earns money delivering groceries. He deposits four-fifths of what he earns in his savings account and keeps the rest to spend. Last weekend, he kept \$12. How much did Jamaal earn?

Plan: If Jamaal deposited $\frac{4}{5}$ of his money, he kept $\frac{1}{5}$ to spend. Therefore, \$12 is $\frac{1}{5}$ of what he earned.

Solution When a fractional part of an amount is calculated, multiplication is the operation used. Since \$12 is $\frac{1}{5}$ of Jamaal's earnings, multiply by the inverse, $\frac{5}{1}$ or 5, to find his total earnings.

$$\$12 \times 5 = \$60$$

He earned \$60.

Check: $\frac{4}{5} \times \$60 = \$48$ Therefore, \$48 was saved.
$\$60 - \$48 = \$12$ Therefore, \$12 was left to spend.

Answer: Jamaal earned \$60 last weekend.

EXERCISES WITH OPEN-RESPONSE PROBLEMS

1. Yolanda was standing on the middle step of a staircase, painting a wall. She went up four steps to touch up a spot, then she went down five steps to continue painting. When she finished everything within reach, she went down six steps to the first step.
a. Which step was the middle step?
b. How many steps were there in the staircase?

2. The crosstown bus left the west-end terminal with several passengers. At the first stop, three persons left the bus and five boarded. At the second stop, two persons boarded. At the third stop, five persons left the bus and four boarded. The remaining six passengers left the bus at the east-end terminal, which was the next stop. How many passengers were on the bus when it left the west-end terminal?

3. Shaquille scored twice as many points in the second quarter as in the first quarter of a basketball game. He scored 5 points in the third quarter and 2 points in the fourth quarter. If he scored 19 points in the game, how many did he score in the first quarter?

4. During the month of October, Alisha deposited \$325.00 in her checking account, wrote checks for \$114.37 and \$54.25, and was charged a bank fee of \$2.50. If her balance at the end of October was \$287.15, what was her balance at the end of September?

5. After four pickup stops, every passenger seat in a school bus was taken. Half as many students got on at the second stop as at the first stop, and half as many got on at the third stop as at the second stop. At the fourth stop, five students got on, the same number as at the third stop. How many passenger seats were there on the bus?

6. Dolores bought a box of pastries at the bakery. She gave half of them to friends that she met on her way home. At home, she gave one to her brother, ate one herself, and had two left. How many pastries did she buy?

7. Kim won some money. She spent $25 to have her hair cut, and lent one-fifth of the remaining money to a friend. After she deposited two-thirds of what was left in the bank, she still had $40. How much money did she win?

8. Marsha had some math exercises to do for homework. She did one-half during study period and two-thirds of those remaining while waiting for her friend after school. She had three to finish at home that evening. How many exercises did she have to do for homework?

3-5 DISCOVERING PATTERNS

Many problems about sets of numbers that follow a pattern can be solved by making use of the patterns involved.

EXAMPLES

1. What is the next number in the following sequence?

$$1, 2, 4, 7, 11, \ldots$$

Plan: Since the numbers are increasing whole numbers, look for patterns that add whole numbers or multiply by whole numbers.

Solution Look for a multiplication pattern. The second number is twice the first, and the third number is twice the second. However, the fourth number is not twice the third. Thus, the sequence is not the result of multiplication by the same number.

Look for an addition pattern.

$$\begin{array}{ccccccccc} 1 & & 2 & & 4 & & 7 & & 11 \\ & +1 & & +2 & & +3 & & +4 & \end{array}$$

Each number is obtained by adding a number larger by 1 than the number previously added. If this pattern is continued, the next number is obtained by adding 5 to 11. Using this pattern, you find that the next number is 16.

Check:

$$\begin{array}{ccccccccccc} 1 & & 2 & & 4 & & 7 & & 11 & & 16 \\ & +1 & & +2 & & +3 & & +4 & & +5 & \end{array}$$

Answer: The next number in the sequence is 16.

2. What is the sum of the whole numbers 1 through 80?

Plan: To estimate an answer, consider that, of these 80 numbers, the middle, or "average," number is about 40. If 80 numbers have an average of 40, then 80(40) or 3,200 is a good estimate of the sum of these numbers.

One possible solution is simply to add the 80 numbers or you can look for a pattern. There are many ways to find a pattern. Two methods are shown in the solutions that follow.

Solution Match the smallest and largest numbers. Then match the second smallest with the second largest. Notice that $1 + 80 = 81$, $2 + 79 = 81$, $3 + 78 = 81$, and so on.

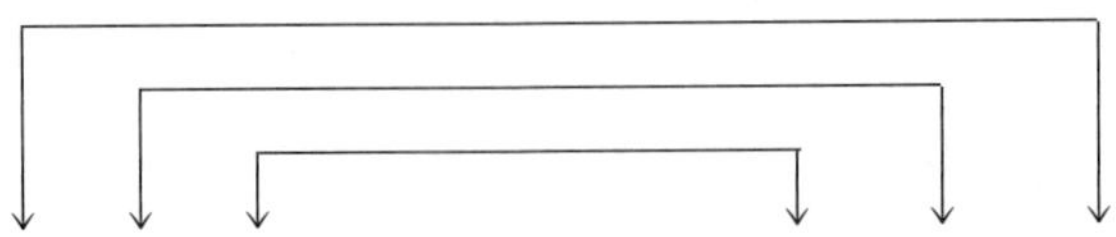

$$1 + 2 + 3 + 4 + \ldots + 77 + 78 + 79 + 80$$

Since there are 80 numbers, there are $\frac{1}{2}(80)$ or 40 pairs. The sum of these 40 pairs is 40(81) or 3,240.

Alternative Solution Write the sum twice: first with the numbers in increasing order as given, and then, on a second line, in decreasing order. Add the two lines as shown below:

$$\begin{array}{r}
1 + 2 + 3 + 4 + 5 + \ldots + 76 + 77 + 78 + 79 + 80 \\
\underline{80 + 79 + 78 + 77 + 76 + \ldots + 5 + 4 + 3 + 2 + 1} \\
81 + 81 + 81 + 81 + 81 + \ldots + 81 + 81 + 81 + 81 + 81
\end{array}$$

There are 80 pairs, each having a sum of 81. This third line, which equals 80(81), is *twice* the sum of the whole numbers from 1 to 80. Therefore, the sum we are looking for is one-half of 80(81) or

$$\frac{80(81)}{2} = 40(81) = 3{,}240$$

Check: You could use a calculator to check this sum. Notice that the result is close to the estimate.

Answer: The sum of the whole numbers 1 through 80 is 3,240.

EXERCISES WITH OPEN-RESPONSE PROBLEMS

1. Find the sum of the counting numbers 1 through 100.

2. Find the sum of the first 50 even counting numbers.

3. Find the sum of the first 10 odd counting numbers.

4. If a clock strikes once on the half hour and strikes the hour on the hour (that is, strikes once at 1 o'clock, twice at 2 o'clock, and so on), how many times does the clock strike from 12:15 A.M. to 12:15 P.M.?

5. Find the next number in the sequence 1, 3, 7, 13, 21, 31, 43, ...

6. Find the next number in the sequence 2, 6, 18, 54, 162, 486, ...

7. Find the next number in the sequence 1, 3, 7, 15, 31, 63, 127, ...

8. If $4 * 3 = 24$, $8 * 2 = 32$, and $1 * 5 = 10$, what is the value of $6 * 7$?

9. If $2 * 3 = 12$, $3 * 5 = 45$, and $4 * 5 = 80$, what is the value of $3 * 4$?

3-6 DRAWING PICTURES AND DIAGRAMS

A picture or diagram can often help you visualize a problem and may suggest a way to solve it.

EXAMPLES

1. Mr. Vroman had a rectangular vegetable garden. He decided to increase the size by making the length twice that of the original garden and the width 3 times that of the original garden. How many times as large as the original garden is the new garden?

Plan: Draw a diagram of the original garden and of the new garden and compare them.

Solution Sketch the original garden.

Double the length.

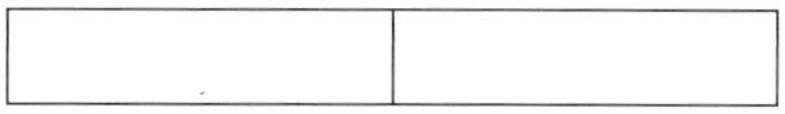

Triple the width.

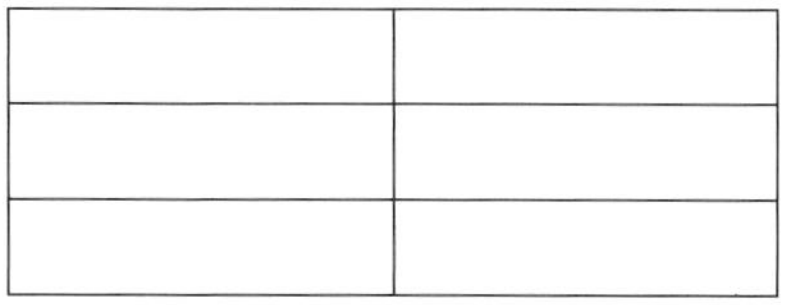

The new garden consists of six gardens each of the same size as the original one.

Check: Assign numbers to the length and width to show that the relationship holds.

Original:	1 by 5	Area = 5	3 by 4	Area = 12
New:	3 by 10	Area = 30	9 by 8	Area = 72
		30 = 6(5)		72 = 6(12)

Answer: The new garden is 6 times as large as the original one.

---

2. Saleem planted some trees alongside his family's driveway. If the distance from the first tree to the last tree was 200 feet and he planted the trees 50 feet apart, how many trees did he plant?

Solution

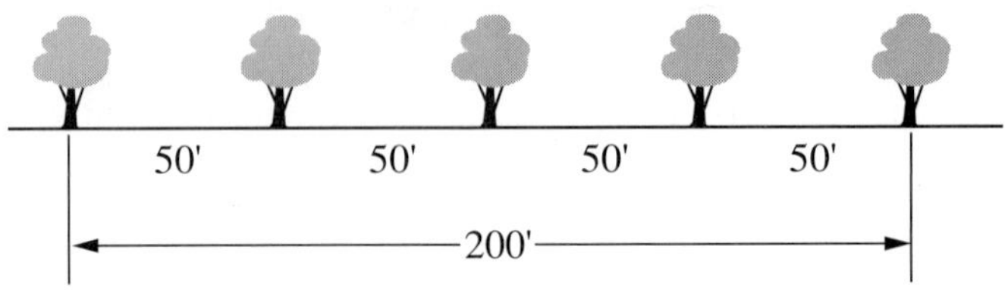

Answer: Saleem planted five trees.

---

3. Boxes measuring 3″ by 2″ by 5″ are to be packed in a carton whose dimensions are 9″ by 15″ by 5″.

a. If all the boxes must be aligned in the same way, how many boxes will fit into the carton and still allow the carton to be closed?

b. If the boxes can be placed in *any* way, how many can fit and still allow the carton to be closed?

Plan: Since both the carton and the boxes to be put into the carton are 5″ high, draw diagrams to show how the 3″-by-2″ boxes will fit in the 9″-by-15″ space.

Solutions **a.**

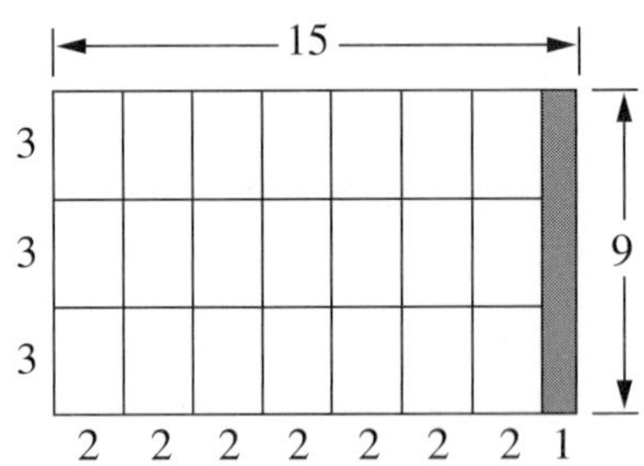

3(7) = 21

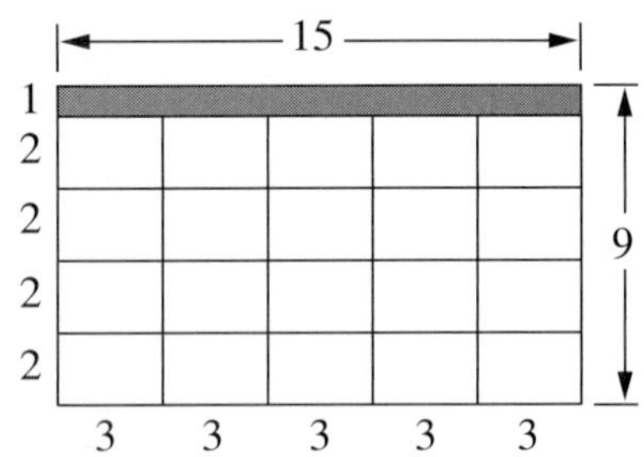

4(5) = 20

The diagrams show that, at most, 21 boxes will fit.

Answer: **a.** 21 boxes aligned in the same way can fit into the carton.

b.

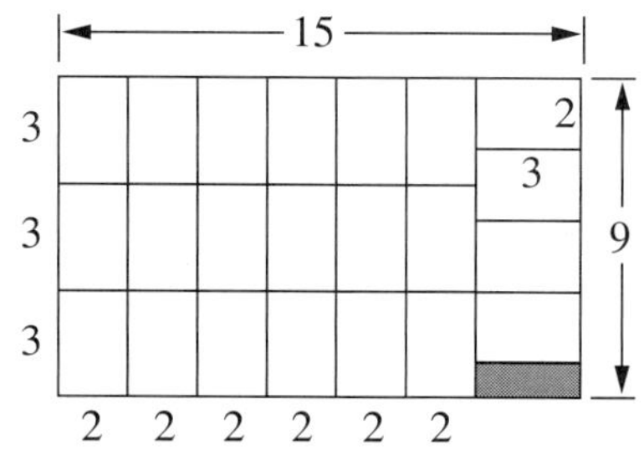

Answer: **b.** 22 boxes placed *any* way can fit into the carton.

EXERCISES WITH OPEN-RESPONSE PROBLEMS

1. Mrs. Rodriguez has a rectangular flower garden that she plans to enlarge by making it 4 times as long and twice as wide. How many times as large as the original garden will the new garden be?

2. One side of a 30-foot walkway is to be fenced using posts placed 6 feet apart. How many posts are needed?

3. The McMahons decided to build a fence along the back of their property. They planned to use a post every 8 feet. Since the posts were more expensive than they had expected, however, they bought 3 fewer posts than planned and placed the posts 10 feet apart. How many posts did they purchase?

4. A parking lot is 100 feet wide. Traffic markers are placed in rows so that they are 3 feet apart and no marker is closer than 20 feet to the sides of the lot. How many markers are there in a row?

5. A man built some pens to house his dogs. If he puts one dog in each pen, he will have two dogs left over. If he puts two dogs in each pen, he will have one empty pen. How many dogs and pens has he?

6. Cora wants to arrange her doll collection on the bookshelves in her room. If she puts two dolls on each shelf, she will have one doll left over. If she puts three dolls on each shelf, she will have two extra shelves. How many dolls and how many shelves has she?

7. A milk crate holds bottles in four rows of five each. How can you place 14 bottles of milk in the crate so that each row and each column has an even number of bottles?

8. A milk crate holds bottles in five rows of five each. How can you place 11 bottles of milk so that each row and each column has an odd number of bottles?

3-7 MAKING LISTS AND CHARTS

Problems in which different possible solutions are to be investigated can often be solved by making organized lists or charts.

EXAMPLES

1. Three friends, Jane, Rose, and Phyllis, study different languages and have different career goals. One wants to be an artist, one a doctor, and the third a lawyer.
   (1) The girl who studies Italian does not plan to be a lawyer.
   (2) Jane studies French and does not plan to be an artist.
   (3) The girl who studies Spanish plans to be a doctor.
   (4) Phyllis does not study Italian.

   Find the language and career goal of each girl.

   Plan: Make a chart and fill in the information, starting with the most definite clues first.

Solution

	Jane	Rose	Phyllis
Language			
Career			

Clue (2) gives the most definite information about language. After using that, use Clue (4) to determine the other languages.

	Jane	Rose	Phyllis
Language	French	Italian	Spanish
Career			

Now that you know which girl studies which language, Clue (3) gives definite information about a career choice. Clues (1) and (2) enable you to determine the other career choices.

	Jane	Rose	Phyllis
Language	French	Italian	Spanish
Career	lawyer	artist	doctor

Check: Compare each clue with the chart.

Rose studies Italian and does not plan to be a lawyer.
Jane studies French and does not plan to be an artist.
Phyllis studies Spanish and plans to be a doctor.
Phyllis does not study Italian.

Answer: The chart displays the language and career goal of each girl.

---

2. When the local newspaper wanted to introduce a new cartoon, subscribers were surveyed to determine which of two possible cartoons they liked. Of the 500 persons who replied, 275 liked cartoon *A*, 172 liked both cartoon *A* and cartoon *B*, and 37 liked neither. Which cartoon was liked by the larger number of people?

Plan: Draw a rectangle to represent all of the persons who responded. Inside the rectangle, draw a circle to represent the persons who liked cartoon *A* and another circle to represent the people who liked cartoon *B*. These circles should overlap since the people who liked both must be included in both circles. The space outside both circles represents the people who liked neither cartoon. (This way of picturing information is called a *Venn diagram.*) Our plan is to place numbers in the four regions of this diagram, and then compare the number in circle *A* with the number in circle *B*.

Solution

1. Write 172 in the intersection of *A* and *B* to show the number who liked both cartoons.

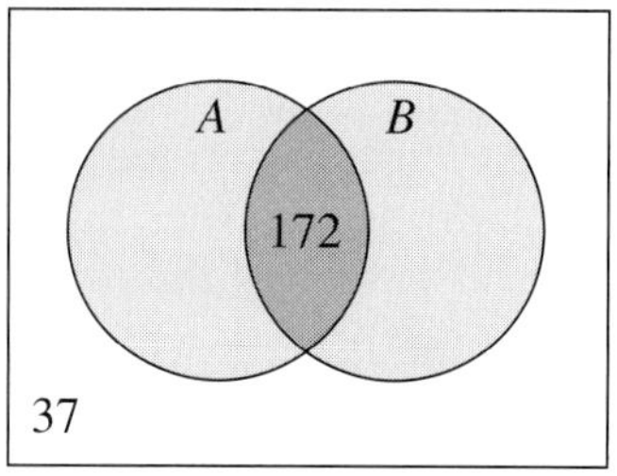

2. Write 37 outside the circles to show the number who liked neither.

3. Find the number who liked *A* only. All who liked *A*, less those who liked both, equals 275 − 172 or 103 who liked *A* only.

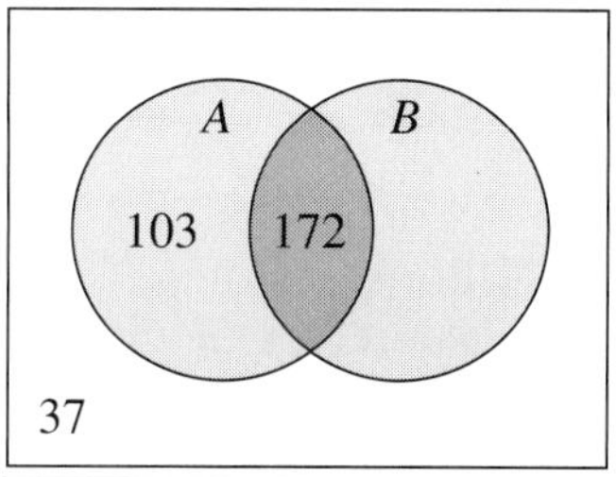

4. Find the number who liked *B* only. Subtract the three numbers found from the 500 people surveyed: 500 − (172 + 103 + 37) or 500 − 312 equals 188 who liked *B* only.

5. Compare the numbers liking *A* and *B*.
   Those who liked *A* = 103 + 172 = 275.
   Those who liked *B* = 188 + 172 = 360.

   Check: Reread the problem to see that the final diagram meets all given conditions.

   Answer: The larger number of people liked cartoon *B*.

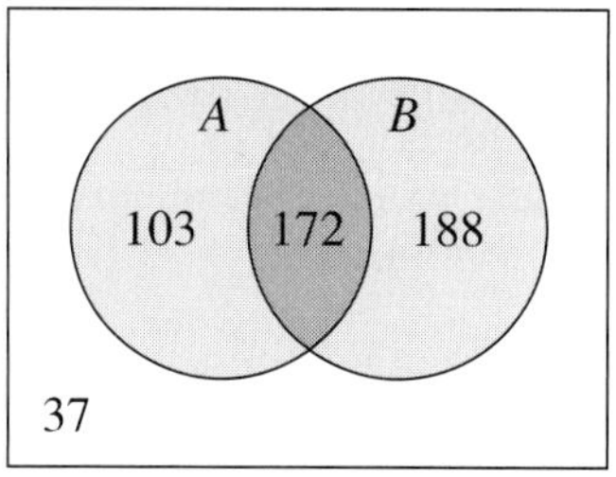

---

3. Magda needs 50 cents for a coin machine that takes nickels, dimes, and quarters. In how many different ways can she have the correct change?

Plan: List the coins that can be used and all possible combinations of these coins that make 50 cents.

Solution

Nickels	10	8	6	5	4	3	2	1	0	0
Dimes	0	1	2	0	3	1	4	2	5	0
Quarters	0	0	0	1	0	1	0	1	0	2

Check: Verify that the value of the set of coins in each column is 50 cents. For example, $10 \times 5 = 50$; $(8 \cdot 5) + 10 = 50$.

Answer: Magda can have the correct change in 10 different ways.

EXERCISES WITH OPEN-RESPONSE PROBLEMS

1. Beta has three boxes of paper. The cover of the first is labeled "white," of the second is labeled "blue," and of the third is labeled "assorted." She knows that the three labels correctly described the contents of the boxes as delivered, but the covers were replaced carelessly so that each box now has an incorrect label. She opens the box marked "assorted" and finds that the top sheet of paper is white. Which incorrect label is on which box?

2. Mark, Jay, and Fred have each invited a sister of one of the other two boys to the senior dance. From the clues given below, determine the name of the girl that each boy has invited to the dance.
(1) No girl is going to the dance with her brother.
(2) Sally is not Mark's sister and is not going to the dance with Joan's brother.
(3) Marion is Jay's sister.

3. A florist charges \$3 for a rose and \$2 for three carnations. Carnations cannot be purchased separately. One customer paid \$20 for flowers. What assortments could have been purchased?

4. In how many ways can change for a \$20 bill be made in \$1, \$5, and/or \$10 bills?

5. The eight bills in Marilyn's purse are worth less than \$30. What bills could she have?

6. The eight bills in Marilyn's purse are worth less than \$30. She makes a \$12 purchase and gives the exact amount in payment. What bills could she have had in her purse? (Consider the combinations you found in answer to Exercise 5.)

7. Find the smallest number that will leave a remainder of 1 when divided by 3, a remainder of 3 when divided by 4, and a remainder of 4 when divided by 5.

8. Find the smallest number that will leave a remainder of 1 when divided by 2, a remainder of 2 when divided by 3, a remainder of 3 when divided by 4, and a remainder of 4 when divided by 5.

9. Of 100 households surveyed, 28 had neither a dog nor a cat, 18 had both a dog and a cat, and 32 had a cat. How many households had dogs?

10. Of the 200 students in the ninth grade at West High School, 70 study French and biology, 10 study French but not biology, and 12 study neither French nor biology. How many students in the ninth grade at West High School study biology?

3-8 CHOOSING AND COMBINING STRATEGIES

Many problems can be solved by using several different strategies or a combination of strategies.

EXAMPLES

1. Brian and Linda work at a restaurant that is open 7 days a week. Brian works for 5 days and then has 2 days off. Linda works for 3 days and then has 1 day off. Neither Brian nor Linda worked on Sunday, April 1, and both worked on April 2. What other days in April will both Linda and Brian have off?

Plan 1: Make a chart in the form of a calendar on which you can mark the days on which each person works.

Solution Start the chart with a blank for April 1, on which neither Brian nor Linda worked.
Mark Brian's schedule, beginning with April 2.
Mark Linda's schedule, beginning with April 2.

S	M	T	W	Th	F	S
1	2 B L	3 B L	4 B L	5 B	6 B L	7 L
8 L	9 B	10 B L	11 B L	12 B L	13 B	14 L
15 L	16 B L	17 B	18 B L	19 B L	20 B L	21
22 L	23 B L	24 B L	25 B	26 B L	27 B L	28 L
29	30 B L					

Answer: Brian and Linda will both be off on April 21 and 29.
There are other strategies that you can use to solve this problem.

Plan 2: Make a list of the days each person has off and find the dates that occur in both lists.

Solution Brian is off every sixth and seventh day after April 1. Add 6 and 7 to the last date that Brian had off.
Brian: 7 8 14 15 21 22 28 29

Linda is off every fourth day after April 1. Add 4 to the last date that Linda had off.
Linda: 5 9 13 17 21 25 29

April 21 and April 29 appear in both lists.

Answer: Brian and Linda will both be off on April 21 and 29.

---

2. Mr. Breiner has a machine that can harvest his corn in 40 hours. His neighbor has a larger machine that can harvest the same number of acres of corn in 30 hours. If they work together, how long will it take to harvest Mr. Breiner's corn using both machines?

Plan: Use trial and error. To determine a starting guess, estimate from the facts. The smaller machine needs 20 hours to do half of the job, and the larger machine needs 15 hours to do half of the job. Working together, the larger machine will do more than half, and the smaller machine will do less than half. A good estimate would be between 15 and 20 hours.

Solution Try 17 hours and test the result. If the smaller machine takes 40 hours to do the job, it can do $\frac{1}{40}$ of the job every hour and $\frac{17}{40}$ in 17 hours. If the larger machine takes 30 hours do the job, it can do $\frac{1}{30}$ of the job every hour and $\frac{17}{30}$ in 17 hours. Together, the two machines must do one whole job.

$$\frac{17}{40} + \frac{17}{30} = \frac{51}{120} + \frac{68}{120} = \frac{119}{120}$$

The result is very close but not exact. The answer is probably a fractional part of an hour, and, therefore, the number of possibilities to try is endless. This is not a good strategy, but what you have done is not useless. It suggests that you need to consider what part of the work can be completed in 1 hour.

Use the strategy of a simpler related problem. Use the same problem but change the question: What part of the work can be done by both machines in 1 hour?

You have already seen that, in 1 hour, the smaller machine can do $\frac{1}{40}$ and the larger machine can do $\frac{1}{30}$ of the job. Together, in 1 hour, the two machines can do $\frac{1}{40} + \frac{1}{30}$ of the job.

$$\frac{1}{40} + \frac{1}{30} = \frac{3}{120} + \frac{4}{120} = \frac{7}{120}$$

Now, use the strategy of working backward.

$$\frac{7}{120} \text{ of what number is 1?}$$

$$1 \div \frac{7}{120} = 1 \cdot \frac{120}{7} = \frac{120}{7} = 17\frac{1}{7}$$

Check: Smaller machine: $\frac{1}{40} \cdot 17\frac{1}{7} = \frac{1}{40} \cdot \frac{120}{7} = \frac{3}{7}$

Larger machine: $\frac{1}{30} \cdot 17\frac{1}{7} = \frac{1}{30} \cdot \frac{120}{7} = \frac{4}{7}$

Together: $\frac{3}{7} + \frac{4}{7} = \frac{7}{7} = 1$

Answer: Working together, the two machines will do the job in $17\frac{1}{7}$ hours.

EXERCISES WITH OPEN-RESPONSE PROBLEMS

1. A nine-inch square is divided into 81 one-inch squares that are colored alternately red and black. How many red and how many black squares are there if the corner squares are red?

2. A kitten in a bucket weighs 3 kilograms. A rabbit in the same bucket weighs 5 kilograms. The kitten and the rabbit together weigh 6 kilograms. What is the weight of the bucket?

3. A group of people enter a room in which there are some benches. If one person sits on each bench, four persons will remain standing. If two people sit on each bench, one bench will be empty. How many people and how many benches are there?

4. At a school play, adults pay $4 and children pay $2.50. The cashier collected $300 for 99 tickets. How many adults and how many children attended the play?

5. Greg agreed to work for 1 year for $10,000 and a car. After 6 months, he quit and was paid $1,000 and the car. What was the value of the car?

6. At the end of an hour, a cook in a diner noted that one-third of the customers had ordered sandwiches, one-half had ordered hamburgers, one-eighth had ordered steak, and the remaining two had ordered just a salad. If no customer ordered more than one of these items, how many customers were served?

7. Draw two squares that intersect in:
a. exactly 1 point **b.** exactly 2 points **c.** exactly 3 points **d.** exactly 4 points
e. exactly 5 points **f.** exactly 6 points **g.** exactly 7 points **h.** exactly 8 points

8. Both Rosa and Tony work Monday through Friday of each week. Rosa is paid $32 per day, including holidays. Tony is paid $680 per month.
a. Who is paid more in a month? **b.** Who is paid more in a year?

9. Four gumdrops cost a penny, and chocolate drops cost 4 cents each. Glenn bought 20 pieces of candy for 20 cents. What did he buy?

10. How many triangles are there in the figure?

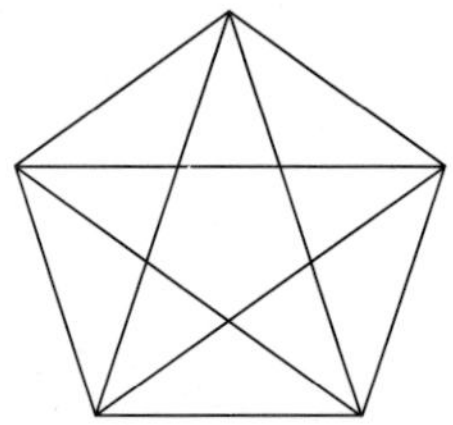

11. Mary takes 20 minutes to wash and dry dishes after dinner. Eddy takes 30 minutes to do the same chore. How long would Mary and Eddy take to wash and dry dishes together?
12. Mary takes 20 minutes to read the daily newspaper. Eddy takes 30 minutes to read the newspaper. If they share sections of the paper, how long would Mary and Eddy take to read the newspaper at the same time?
13. Mary takes 20 minutes to walk home from school. Eddy takes 30 minutes to walk home. How long would Mary and Eddy take to walk home from school together?
14. Compare Exercises 11, 12, and 13. How are they the same, and how are they different?

CHAPTER SUMMARY

The general technique for solving problems includes four basic steps:

1. Understand the problem.
2. Make a plan.
3. Solve the problem.
4. Check the solution.

Some strategies for problem solving are as follows:

1. Guessing and checking
2. Using a simpler related problem
3. Working backward
4. Discovering patterns
5. Drawing pictures and diagrams
6. Making lists and charts

Strategies may also be combined.

REVIEW EXERCISES WITH OPEN-RESPONSE PROBLEMS

1. Solve the three problems on the introductory page of this chapter. Explain why these problems have different answers.
2. The box office attendant at a theater reported that he collected $935 for 155 tickets sold. Adults paid $6.00, and children paid $4.50, for a ticket. What conclusion can you draw?

3. Five students were in the cafeteria line today. Bernie was first in line. Joel was two places behind Edith. Lester was ahead of Pierre, who was fifth. Who was second in line?
4. Carl is 4 years older than Chris. Five years ago, Carl was twice as old as Chris. How old is Carl now?
5. A student wants to give 25 cents to each of several charities but finds that she is 5 cents short. If she gives 20 cents to each, she will have 15 cents left. How much money does she have to start with?
6. Ernestine is 3 times as old as her sister Lucy. In 5 years, Ernestine will be twice as old as Lucy. How old are Ernestine and Lucy now?
7. The 400 voters in the town of Euclid voted on two issues. There were 225 in favor of the first issue and 355 in favor of the second. If 40 persons voted against both issues, how many voted in favor of both issues?
8. If $2 * 4 = 12$, $3 * 5 = 16$, and $1 * 2 = 6$, what is $2 * 3$?
9. How many boxes 3″ by 4″ by 5″ will fit in a carton whose dimensions are 9″ by 15″ by 10″?
10. A group of Americans and Canadians were on a bus tour of Niagara Falls. There were 8 boys, 5 American children, 9 men, 6 Canadian boys, 10 Americans, 3 American males, and 15 Canadian females. How many persons were on the bus tour?
11. On what day of the week must the year start in order that July 4 will fall on a Monday?
12. A scrap dealer pays a trucking company to transport metal to a recycling center. The company has two types of trucks. The larger truck can carry a maximum load of 2,000 pounds, takes 3 hours to make a round trip to the recycling center, and costs a base fee of \$40.00 per load plus \$50.00 per hour. The smaller truck can carry a maximum load of 500 pounds, takes 2 hours to make the round trip to the recycling center, and costs a base fee of \$5.00 per load plus \$25.00 per hour. What is the minimum cost of transporting 9,000 pounds of scrap metal?
13. On Monday, Nicole invested \$7,500 in 100 shares of Gardener and 50 shares of Kent. For the rest of the week, through Friday, the total value of her stock remained unchanged each day, although the cost of a share of each changed. The chart at the right shows the cost of a share of Kent from Monday through Friday. Make a similar chart to show the cost of a share of Gardener during the same period.

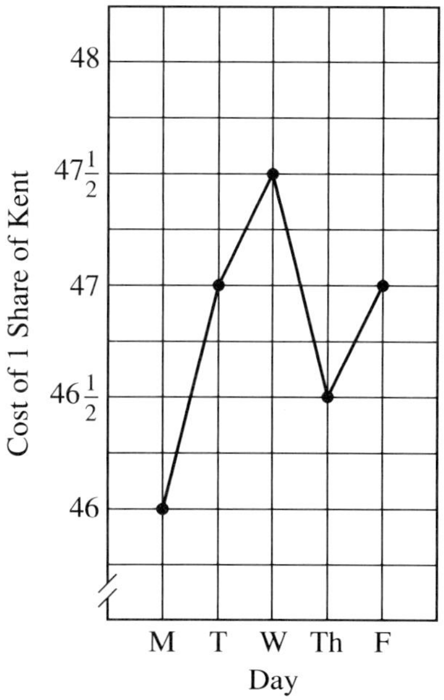

CUMULATIVE REVIEW WITH OPEN-RESPONSE PROBLEMS

1. Points A through J on the number line represent numbers in the set $\left\{-4, 0, 0.5, -2, \pi, \frac{2}{3}, -2.7, -\sqrt{3}, 1, -\frac{\sqrt{5}}{3}\right\}$. Point F represents 0, and point I represents 1.

a. Draw the number line, and label each of the other points with the number that it represents.
b. Name the counting number(s).
c. Name the whole number(s).
d. Name the rational number(s).
e. Name the irrational number(s).
f. Name the real numbers.
g. Which number is the identity for addition?
h. Which number is the identity for multiplication?
i. Which number is the opposite or additive inverse of 2?
j. Which number is the reciprocal or multiplicative inverse of 1.5?

In 2–4, order the numbers in each group from smallest to largest using the symbol $<$.

2. $8.1, -8.1, 0$ **3.** $\sqrt{3}, 1.7, 1.\overline{7}$ **4.** $-\pi, -3.14, -3.\overline{14}$

In 5–10, evaluate each expression.

5. 1.4×9^2 **6.** $\frac{167.7 - 5.2}{13}$ **7.** $\frac{21^3}{63}$

8. $15(3^4 - 12)$ **9.** $74 - 6 \times 8$ **10.** $13 + 4(8)^2$

In 11 and 12, insert parentheses to make the statement true.

11. $3 + 3 \times 3 - 3 = 3$ **12.** $1 + 2 \times 3 - 4 = 5$

13. Match each of line graphs (1) through (4) with one of paragraphs **a** through **d** on the next page.

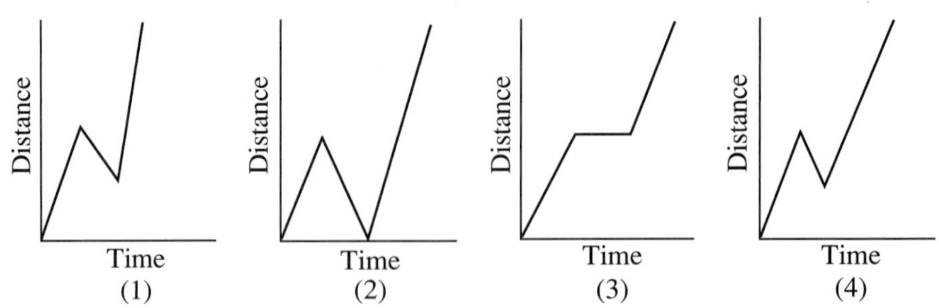

a. When Elfie was halfway to school, he remembered that he had left an assignment that was due that day on the kitchen table. He went back home to get it and then started out again for school.

b. When Elfie was halfway to school, he remembered that he had left an assignment that was due that day on the kitchen table. He started back home to get it; but when he was halfway there, he met his brother, who had noticed the assignment and brought it with him. Elfie took it and then continued on to school.

c. When Elfie was halfway to school, he remembered that he had left an assignment that was due that day on the kitchen table. He started back home to get it; but when he was half way home, his mother met him. She had found the assignment on the table and brought it to him. She drove him the rest of the way to school so that he would not be late.

d. When Elfie was halfway to school, he met a friend who was walking in the opposite direction. They stood and talked for a few minutes before Elfie continued on his way to school.

Exploration

Many people use mental arithmetic for simple computations. Many mental computations are based on the properties of the real numbers. Investigate the processes that are used:

(1) to multiply by 10.
(2) to multiply by 5.
(3) to divide by 10.
(4) to divide by 5.
(5) to multiply by a two-digit number.
(6) to determine whether a number is divisible by 2, by 3, by 4, by 5, by 6, by 9, by 10.

Explain your findings.

Chapter 4

Algebraic Expressions, Geometric Formulas, and Open Sentences

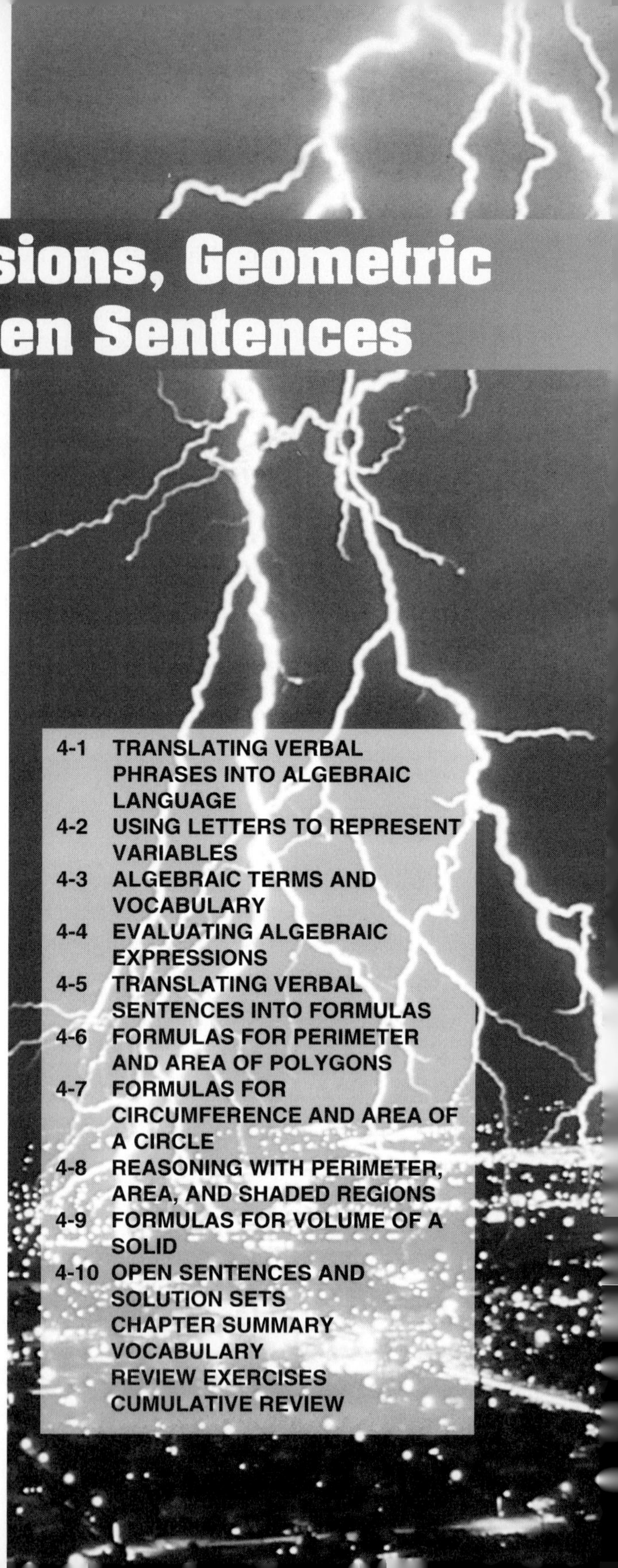

Have you ever experienced seeing a flash of lightning and hearing a crack of thunder at the same time? You may have thought, "Wow, that was close."

Actually, light travels much faster than sound. The speed of light is 186,000 miles per second, but the speed of sound in air is generally measured as only 1,088 *feet* per second. For this reason, when a storm is nearby, you see the lightning almost instantly, but some time passes before the sound of the thunder reaches you.

According to folklore, some people believe that they can approximate the distance from where a lightning bolt hit by counting the seconds, "1, 2, 3, 4, . . . ," that elapse between the time they see the lightning and the time they hear the thunder. This can be written as a rule or formula:

$$D = n$$

where D is the distance in miles and n is the number of seconds counted before hearing the thunder.

Since 1,088 feet is about $\frac{1}{5}$ mile, would a better rule be $D = \frac{1}{5}n$? Test the rules and decide for yourself the next time you experience a thunderstorm.

In this chapter, you will see how algebra is used to write general expressions and formulas that can be applied to many situations.

4-1 TRANSLATING VERBAL PHRASES INTO ALGEBRAIC LANGUAGE

Eggs are usually sold by the dozen, that is, 12 in a carton. Therefore, we know that:

In 1 carton, there are 12×1 or 12 eggs.
In 2 cartons, there are 12×2 or 24 eggs.
In 3 cartons, there are 12×3 or 36 eggs.
In n cartons, there are $12 \times n$ or $12n$ eggs.

Here, n is called a ***variable*** or a ***placeholder*** that can represent different numbers from a set, such as $\{1, 2, 3\}$. The set of numbers that can replace a variable is called the ***domain*** or the ***replacement set*** of that variable.

Recall that a numerical expression contains only numbers. An ***algebraic expression***, such as $12n$, however, is an expression or phrase that contains one or more variables. An algebraic expression is also called an ***open expression*** or an ***open phrase***.

In this section, you will see how verbal phrases are translated into algebraic expressions, using letters as variables and using symbols to represent operations.

Verbal Phrases Involving Addition

The algebraic expression $a + b$ may be used to represent several different verbal phrases, such as:

a plus b	a and b are added	a is increased by b
the sum of a and b	b is added to a	b more than a

The word *exceeds* means "is more than." Thus, *7 exceeds 5 by 2* means "7 *is 2 more than* 5" or $7 = 5 + 2$.

Compare the numerical and algebraic expressions shown below.

A numerical expression: The number that exceeds 5 by 2 is $5 + 2$, or 7.
An algebraic expression: The number that exceeds a by b is $a + b$.

Verbal Phrases Involving Subtraction

The algebraic expression $a - b$ may be used to represent several different verbal phrases, such as:

a minus b	a decreased by b	b less than a
the difference between a and b	a diminished by b	a reduced by b
b subtracted from a		

Verbal Phrases Involving Multiplication

The algebraic expressions $a \times b$, $a \cdot b$, $(a)(b)$, and ab may be used to represent several different verbal phrases, such as:

a times b the product of a and b b multiplied by a

The preferred form to indicate multiplication in algebra is ab. Here, a product is indicated by using *no symbol* between the factors being multiplied.

The multiplication symbol $\times$ is avoided in algebra because it can be confused with the letter or variable x. The raised dot, which is sometimes mistaken for a decimal point, is also avoided.

Parentheses are used to write numerical expressions: (3)(5)(2) or 3(5)(2). Note that all but the first number must be in parentheses. In algebraic expressions, parentheses may be used but they are not needed: $3(b)(h) = 3bh$.

Verbal Phrases Involving Division

The algebraic expressions $a \div b$ and $\frac{a}{b}$ may be used to represent several different verbal phrases, such as:

a divided by b the quotient of a and b

The symbols $a \div 4$ and $\frac{a}{4}$ mean one-fourth of a as well as a divided by 4.

Phrases and Commas

In some verbal phrases, using a comma can prevent misreading. For example, in "the product of x and y, decreased by 2," the comma after y makes it clear that the phrase means $(xy) - 2$ and not $x(y - 2)$.

EXAMPLES

1. Use mathematical symbols to translate the following verbal phrases into algebraic language:

	Answers:
a. w more than 3	$3 + w$
b. w less than 3	$3 - w$
c. r decreased by 2	$r - 2$
d. the product of $5r$ and s	$5rs$
e. twice x, decreased by 10	$2x - 10$
f. 25, diminished by 4 times n	$25 - 4n$
g. the sum of t and u, divided by 6	$\frac{t + u}{6}$
h. 100 decreased by twice $(x + 5)$	$100 - 2(x + 5)$

---

2. Represent in algebraic language:

a. a number that exceeds 5 by m	$5 + m$
b. a number that x exceeds by 5	$x + 5$
c. twice the sum of x and y	$2(x + y)$

EXERCISES

In 1–28, use mathematical symbols to translate the verbal phrases into algebraic language.

1. y plus 8
2. 8 plus y
3. r minus 4
4. 4 minus r
5. 7 times x
6. x times 7
7. x divided by 10
8. 10 divided by x
9. the product of 6 and d
10. c decreased by 6
11. 15 added to b
12. one-tenth of w
13. the sum of b and 8
14. x diminished by y
15. the product of x and y
16. the quotient of s and t
17. 12 increased by a
18. 5 less than d
19. 8 divided by y
20. y multiplied by 10
21. the product of $2c$ and $3d$
22. t more than w
23. one-third of z
24. twice the difference of p and q
25. a number that exceeds m by 4
26. one-half of the sum of L and W
27. 5 times x, increased by 2
28. 10 decreased by twice a

In 29–39, using the letter n to represent "a number," write each verbal phrase in algebraic language.

29. a number increased by 2
30. 20 more than a number
31. 8 increased by a number
32. a number decreased by 6
33. 2 less than a number
34. 3 times a number
35. three-fourths of a number
36. 4 times a number, increased by 3
37. 3 less than twice a number
38. 10 times a number, decreased by 2
39. the product of 5 more than a number, and 4

In 40–45, translate each verbal phrase into algebraic language, representing the two numbers by L and W, with L being the larger.

40. the sum of the two numbers
41. the product of the two numbers
42. the larger number decreased by the smaller number
43. the smaller number divided by the larger number
44. the sum of twice the larger number and twice the smaller number
45. 10 times the smaller number, decreased by 6 times the larger number

46. Let L and W in Exercises 40–45 be natural numbers with $L > W$. Which of Exercises 40–45 illustrate algebraic expressions that:
a. must be natural numbers?
b. must be integers?
c. must be positive numbers?
d. must be rational numbers?

4-2 USING LETTERS TO REPRESENT VARIABLES

A knowledge of arithmetic is important in algebra. Since the variables represent numbers that are familiar to you, it will be helpful to apply the strategy of using a simpler related problem; that is, to relate similar arithmetic problems to the given algebraic ones.

PROCEDURE. To write an algebraic expression involving variables:

1. Think of a similar problem in arithmetic.
2. Write an expression for the arithmetic problem, using numbers.
3. Write a similar expression for the problem, now using letters or variables.

EXAMPLES

Represent by an algebraic expression:

1. a distance that is 20 meters shorter than x meters

How to Proceed:	*Solution:*
(1) Think of a similar problem in arithmetic:	Think of a distance that is 20 meters shorter than 50 meters.
(2) Write an expression for this arithmetic problem:	$50 - 20$
(3) Write a similar expression involving the letter x:	$x - 20$

Answer: $(x - 20)$ m

---

2. a bill for n baseball caps, each costing d dollars

How to Proceed:	*Solution:*
(1) Think of a similar problem in arithmetic:	Think of a bill for 6 caps, each costing 5 dollars.
(2) Write an expression for this arithmetic problem:	$6(5)$
(3) Write a similar expression using the letters n and d:	nd

Answer: nd dollars
(*Note:* After some practice, you will be able to do steps (1) and (2) mentally.)

---

3. a weight that is 40 pounds heavier than p pounds. *Answer:* $(p + 40)$ pounds

4. an amount of money that is twice d dollars *Answer:* $2d$ dollars

EXERCISES WITH OPEN-RESPONSE PROBLEMS

In 1–18, represent each answer in algebraic language, using the variable mentioned in the problem.

1. The number of kilometers traveled by a bus is represented by x. If a train traveled 200 kilometers farther than the bus, represent the number of kilometers traveled by the train.

2. Mr. Gold invested $1,000 in stocks. If he lost d dollars when he sold the stocks, represent the amount he received for them.

3. The cost of a mountain bike is 5 times the cost of a skateboard. If the skateboard costs x dollars, represent the cost of the mountain bike.

4. The length of a rectangle is represented by ℓ. If the width of the rectangle is one-half of its length, represent its width.

5. After 12 centimeters had been cut from a piece of lumber, c centimeters were left. Represent the length of the original piece of lumber.

6. Paul and Martha saved 100 dollars. If the amount saved by Paul is represented by x, represent the amount saved by Martha.

7. The sum of two numbers is s. If one number is represented by x, represent the other number in terms of s and x.

8. A suit costs $150. Represent the cost of n suits.

9. A ballpoint pen sells for 39 cents. Represent the cost of x pens.

10. Represent the cost of t feet of lumber that sells for g cents a foot.

11. If Hilda weighed 45 kilograms, represent her weight after she had lost x kilograms.

12. Ronald, who weighs c pounds, is d pounds overweight. Represent the number of pounds Ronald should weigh.

13. A woman spent $250 for jeans and a ski jacket. If she spent y dollars for the ski jacket, represent the amount she spent for the jeans.

14. A man bought an article for c dollars and sold it at a profit of $25. Represent the amount for which he sold it.

15. The width of a rectangle is represented by w meters. Represent the length of the rectangle if it exceeds the width by 8 meters.

16. The width of a rectangle is x centimeters. Represent the length of the rectangle if it exceeds twice the width by 3 centimeters.

17. If a plane travels 550 kilometers per hour, represent the distance it will travel in h hours.

18. If an automobile traveled for 5 hours at an average rate of r kilometers per hour, represent the distance it traveled.

19. Represent algebraically the number of:

a. centimeters in m meters
b. meters in i centimeters
c. days in w weeks
d. weeks in d days
e. hours in d days
f. days in h hours
g. feet in c inches
h. grams in k kilograms

20. Represent the total number of days in w weeks and d days.

21. An auditorium with m rows can seat a total of c people. If each row in the auditorium has the same number of seats, represent the number of seats in one row.
22. Represent the total number of calories in x peanuts and y potato chips if each peanut contains 15 calories and each potato chip contains 18 calories.
23. The charges for a telephone call are \$0.45 for the first 3 minutes and \$0.09 for each additional minute. Represent the cost of a telephone call that lasts m minutes when m is greater than 3.
24. A printing shop charges a 50-cent minimum for the first 8 photocopies of a flyer. Additional copies cost 6 cents each. Represent the cost of c copies if c is greater than 8.
25. A utilities company measures gas consumption by the hundred cubic feet, CCF. The company has a three-step rate schedule for gas customers. First, there is a minimum charge of \$5.00 per month for up to 3 CCF of gas used. Then, for the next 6 CCF, the charge is \$0.75 per CCF. Finally, after 9 CCF, the charge is \$0.55 per CCF. Represent the cost of g CCF of gas if g is greater than 9.
26. **a.** Represent the number of pounds of grapes you can buy with d dollars if each pound costs b dollars.
    b. Does the algebraic expression in part **a** always represent a whole number? Explain your answer by showing examples with numbers.
27. **a.** If x apples cost c cents, represent the cost of one apple.
    b. If x apples cost c cents, represent the cost of n apples.
    c. Do the algebraic expressions in parts **a** and **b** always represent whole numbers? Explain your answer.

4-3 ALGEBRAIC TERMS AND VOCABULARY

Term

A ***term*** is a number, a variable, or any *product* or *quotient* of numbers and variables. For example: 5, x, $4y$, $8ab$, k^2, and $\frac{x}{y}$ are terms.

An algebraic expression involving a *sum* or a *difference* has more than one term. For example, $4a + 2b - 5c$ has three terms. These terms, $4a$, $2b$, and $5c$, are separated by $+$ and $-$ signs.

Factors of a Term

If a term contains two or more numbers or variables, then each number, each variable, and any product of them are called ***factors*** of the term, or factors of the product. As seen in the table, the factors of $3xy$ are 1, 3, x, y, $3x$, $3y$, xy, and $3xy$. When we factor numbers, we write only factors that are integers.

Factors		Term
$1 \cdot 3xy$	=	$3xy$
$3 \cdot xy$	=	$3xy$
$x \cdot 3y$	=	$3xy$
$y \cdot 3x$	=	$3xy$

Coefficient Any factor of an algebraic term is called the ***coefficient*** of the remaining factor, or product of factors, of that term.

For example, consider the algebraic term $3xy$:

3 is the coefficient of xy	$3x$ is the coefficient of y.
$3y$ is the coefficient of x.	xy is the coefficient of 3.

When an algebraic term consists of a number and one or more variables, the number is called the ***numerical coefficient*** of the term. For example:

In $8y$, the numerical coefficient is 8. In $4abc$, the numerical coefficient is 4.

Some special notation: When the word *coefficient* is used alone, it usually means a numerical coefficient. Also, since x names the same term as $1x$, the coefficient of x is understood to be 1. This is true of all terms that contain only variables. For example:

7 is the coefficient of $7b$.	1 is the coefficient of ax.
2.25 is the coefficient of $2.25gt$	1 is the coefficient of k^2.

Base, exponent, power You learned in Chapter 2 that: a ***power*** is the product of equal factors. A power has a base and an exponent.

The ***base*** is one of the equal factors of the power.

The ***exponent*** is the number of times the base is used as a factor. (If a term is written without an exponent, the exponent is understood to be 1.)

In $4^2 = 4 \cdot 4 = 16$:	base $= 4$,	exponent $= 2$,	power $= 4^2$ or 16
In $x^3 = x \cdot x \cdot x$:	base $= x$,	exponent $= 3$,	power $= x^3$
In $35m$:	base $= m$,	exponent $= 1$,	power $= m^1$ or m
In $5d^2 = 5 \cdot d \cdot d$:	base $= d$,	exponent $= 2$,	power $= d^2$

An exponent refers only to the number or variable that is directly to its left, as seen in the last example, where 2 refers only to the base d. To show the product $5d$ as a base (or to show any sum, difference, product or quotient as a base), we must enclose the base in parentheses.

In $(5d)^2 = 5d \cdot 5d = 25d^2$: base $= 5d$, exponent $= 2$, power $= 25d^2$
In $(a + 4)^2 = (a + 4)(a + 4)$: base $= (a + 4)$, exponent $= 2$, power $= (a + 4)^2$

EXAMPLE

For each term, name the coefficient, base, and exponent.

a. $4x^5$	*Answer:* coefficient $= 4$,	base $= x$,	exponent $= 5$
b. $-w^8$	*Answer:* coefficient $= -1$,	base $= w$,	exponent $= 8$
c. $2\pi r$	*Answer:* coefficient $= 2\pi$,	base $= r$,	exponent $= 1$

(*Note:* Remember that coefficient means *numerical coefficient*, and that 2π is a real number.)

EXERCISES

In 1–6, name the factors (other than 1) of each product.

1. xy **2.** $3a$ **3.** $5n$ **4.** $7mn$ **5.** $13xy$ **6.** $11st$

In 7–12, name, in each case, the numerical coefficient of x.

7. $8x$ **8.** $(5 + 2)x$ **9.** $\frac{1}{2}x$ **10.** x **11.** $-1.4x$ **12.** $2 + 7x$

In 13–18, name, in each case, the base and exponent of the power.

13. m^2 **14.** $-s^3$ **15.** t **16.** 10^6 **17.** $(5y)^4$ **18.** $(x + y)^5$

In 19–30, write each expression, using exponents.

19. $m \cdot m \cdot m$ **20.** $b \cdot b \cdot b \cdot b \cdot b$ **21.** $4 \cdot x \cdot x \cdot x \cdot x \cdot x$

22. $\pi \cdot r \cdot r$ **23.** $a \cdot a \cdot a \cdot a \cdot b \cdot b$ **24.** $7 \cdot r \cdot r \cdot r \cdot s \cdot s$

25. $9 \cdot c \cdot c \cdot c \cdot d$ **26.** $(6a)(6a)(6a)$ **27.** $(x + y)(x + y)$

28. $(a - b)(a - b)(a - b)$ **29.** the square of $(b - 5)$ **30.** the fourth power of $(m + 2n)$

In 31–36, write each term as a product without using exponents.

31. r^6 **32.** $5x^4$ **33.** x^3y^5 **34.** $4a^4b^2$ **35.** $3c^2d^3e$ **36.** $(3y)^5$

In 37–42, name, for each given term, the coefficient, base, and exponent.

37. $-3k$ **38.** $-k^3$ **39.** πr^2 **40.** $(ax)^5$ **41.** $\sqrt{2}y$ **42.** $0.0004t^{12}$

4-4 EVALUATING ALGEBRAIC EXPRESSIONS

Benjamin has 1 more tape than 3 times the number of tapes that Julia has. If Julia has n tapes, then Benjamin has $3n + 1$ tapes.

The algebraic expression $3n + 1$ represents an unspecified number. Only when the variable n is replaced by a specific number does $3n + 1$ become a specific number. For example:

If $n = 10$, then $3n + 1 = 3(10) + 1 = 30 + 1 = 31$.

If $n = 15$, then $3n + 1 = 3(15) + 1 = 45 + 1 = 46$.

When we determine the number that an algebraic expression represents for specific values of its variables, we are ***evaluating the algebraic expression***.

PROCEDURE. *To evaluate an algebraic expression*, replace the variables by their specific values, and then follow the rules for the *order of operations*.

The steps of this procedure, in greater detail, are as follows:

1. Replace the variables by their specific values.
2. Evaluate the expressions within grouping symbols such as parentheses, always simplifying the expressions in the innermost groupings first.
3. Simplify all powers and roots. (Roots will be studied in Chapter 20.)
4. Multiply and divide, from left to right.
5. Add and subtract, from left to right.

EXAMPLES

1. Evaluate $50 - 3x$ when $x = 7$.

How to Proceed:	*Solution:*
(1) Write the expression:	$50 - 3x$
(2) Replace the variable by its given value:	$50 - 3(7)$
(3) Do the multiplication:	$50 - 21$
(4) Do the subtraction:	29

Answer: 29

An Alternative Solution Use a scientific calculator that follows the order of operations.

Enter: 50 [−] 3 [×] 7 [=]

Display: [29.]

Answer: 29

---

2. Evaluate $2x^2 - 5x + 4$ when $x = 7$.

How to Proceed:	*Solution:*
(1) Write the expression:	$2x^2 - 5x + 4$
(2) Replace the variable by its given value:	$2(7)^2 - 5(7) + 4$
(3) Evaluate the power:	$2(49) - 5(7) + 4$
(4) Do the multiplication:	$98 - 35 + 4$
(5) Do the addition and subtraction:	67

Answer: 67

An Alternative Solution Use a scientific calculator that follows the order of operations.

Enter: 2 [×] 7 [x^2] [−] 5 [×] 7 [+] 4 [=]

Display: 67.

Answer: 67

---

3. Evaluate $\frac{2a}{5} + (n - 1)d$ when $a = 40$, $n = 10$, and $d = 3$.

How to Proceed:	*Solution:*
(1) Write the expression:	$\frac{2a}{5} + (n - 1)d$
(2) Replace the variables by their given values:	$\frac{2(40)}{5} + (10 - 1)(3)$
(3) Simplify the expression grouped by parentheses or a fraction bar:	$\frac{80}{5} + (9)(3)$
(4) Do the multiplication and division:	$16 + 27$
(5) Do the addition:	43

Answer: 43

An Alternative Solution Use a scientific calculator that follows the order of operations.

Enter: 2 [×] 40 [÷] 5 [+] [(] 10 [−] 1 [)] [×] 3 [=]

Display: 43.

Answer: 43

---

4. Evaluate $(2x)^3 - 2x^3$ when $x = 4$.

How to Proceed:	*Solution*
(1) Write the expression:	$(2x)^3 - 2x^3$
(2) Replace the variable by its given value:	$(2 \cdot 4)^3 - 2(4)^3$
(3) Simplify the expression within parentheses:	$(8)^3 - 2(4)^3$
(4) Evaluate the powers:	$512 - 2(64)$
(5) Do the multiplication:	$512 - 128$
(6) Do the subtraction:	384

Answer: 384

An Alternative Solution Use a scientific calculator that follows the order of operations.

Enter: [(] 2 [×] 4 [)] [y^x] 3 [−] 2 [×] 4 [y^x] 3 [=]

Display: 384.

Answer: 384

EXERCISES

To truly understand this topic, you should first evaluate the expressions in Exercises 1 to 74 without a calculator. Then, use a calculator to check your work.

Basic Operations

In 1–22, find the numerical value of each expression. Use $a = 8$, $b = 6$, $d = 3$, $x = 4$, $y = 5$, and $z = 1$.

1. $5a$ **2.** $9b$ **3.** $\frac{1}{2}x$ **4.** $0.3y$ **5.** $a + 3$ **6.** $7 + y$

7. $a - 2$ **8.** $5 - y$ **9.** $a - b$ **10.** ax **11.** $3xy$ **12.** $\frac{2b}{3}$

13. $\frac{3bd}{9}$ **14.** $2x + 9$ **15.** $3y - b$ **16.** $20 - 4z$ **17.** $5x + 2y$

18. $ab - dx$ **19.** $a + 5d + 3x$ **20.** $9y + 6b - d$ **21.** $ab - d - xy$ **22.** $0.5a - 0.8y$

Basic Operations and Powers

In 23–51, find the numerical value of each expression. Use $a = 8$, $b = 6$, $d = 3$, $x = 4$, $y = 5$, and $z = 1$.

23. a^2 **24.** x^2 **25.** b^3 **26.** y^3 **27.** d^4 **28.** z^5

29. $2x^2$ **30.** $3b^2$ **31.** $4d^3$ **32.** $6z^5$ **33.** $\frac{b^2}{9}$ **34.** $\frac{1}{2}x^3$

35. $\frac{3}{4}x^3$ **36.** a^2d **37.** xy^2 **38.** $3z^2a$ **39.** $2a^2b^3$

40. $\frac{1}{4}x^2y^2$ **41.** $(2d)^2$ **42.** $a^2 + b^2$ **43.** $b^2 - y^2$ **44.** $a^2 + b^2 - d^2$

45. $x^2 + x$ **46.** $b^2 + 2b$ **47.** $y^2 - 4y$ **48.** $2b^2 + b$ **49.** $9a - a^2$

50. $x^2 + 3x + 5$ **51.** $y^2 + 2y - 7$

Grouping Symbols

In 52–71, find the numerical value of each expression. Use $w = 10$, $x = 8$, $y = 5$, and $z = 2$.

52. $2(x + 5)$ **53.** $x(y - 2)$ **54.** $3(2x + z)$ **55.** $4(2x - 3y)$

56. $\frac{x}{2}(y + z)$ **57.** $\frac{1}{2}x(y + z)$ **58.** $3y - (x - z)$ **59.** $2x + 5(y - 1)$

60. $2(x + z) - 5$ **61.** $3x^2$ **62.** $(3x)^2$ **63.** $y^2 + z^2$

64. $(y + z)^2$ **65.** $w^3 - x^3$ **66.** $(w - x)^3$ **67.** $3w^2 - 2x^2$

68. $(3w - 2x)^2$ **69.** $(3w)^2 - (2x)^2$ **70.** $(xy)^2$ **71.** $(w^2)(x^2)$

72. Find the value of $x^2 - 8y$ when $x = 5$ and $y = \frac{1}{2}$.

73. Find the value of $r^2 + 4s$ when $r = 3$ and $s = 0.5$.

74. Evaluate $\frac{5}{9}(F - 32)$ when the value of F is:

a. 50 **b.** 77 **c.** 86 **d.** 32 **e.** 212

Evaluating Algebraic Expressions with a Calculator

In 75–89, find the numerical value of each expression by using a calculator. Let $b = 0.04$, $c = 3.5$, $x = 6.75$, and $y = 2.96$.

75. bcx **76.** cy^2 **77.** $(cy)^2$ **78.** b^2x **79.** b^2x^3

80. $bx + c$ **81.** $b(x + c)$ **82.** $cx - by$ **83.** $c(x - by)$ **84.** $(cx - b)y$

85. $\frac{y^2}{b}$ **86.** $\left(\frac{y}{b}\right)^2$ **87.** $\frac{y}{b^2}$ **88.** $(b + y)^2$ **89.** $b^3 + b^2 + b$

90. Which expression has the *smallest* numerical value when $k = 4.2$ and $n = 0.03$?
(1) kn^2 (2) k^2n (3) $(kn)^2$ (4) $\frac{k^2}{n}$ (5) $\frac{k}{n}$

91. Which expression has the *greatest* numerical value when $g = 1.05$ and $x = 3.75$?
(1) gx^2 (2) g^2x (3) $(gx)^2$ (4) $(x + g)^2$ (5) $(x - g)^3$

4-5 TRANSLATING VERBAL SENTENCES INTO FORMULAS

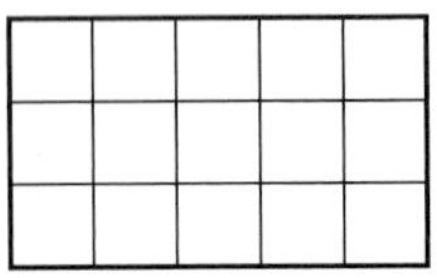

A ***formula*** uses mathematical language to express the relationship between two or more variables. Some formulas can be found by the strategy of looking for patterns. For example, how many square units are shown in the rectangle on the left? This rectangle, measuring 5 units in length and 3 units in width, contains a total of 15 square units of area.

Several such examples lead to the conclusion that the area of a rectangle is equal to the product of its length and width. This relationship can be expressed by the formula $A = \ell w$. Here, A, ℓ, and w are variables that represent, respectively, the area, the length, and the width of a rectangle.

In formulas, the word *is* is translated into the symbol $=$.

EXAMPLES

1. Write a formula for each of the following relationships.

Solution **a.** The perimeter P of a square is equal to 4 times the length of each side s.

Answer: $P = 4s$

b. The total cost C of an article is equal to its price p plus an 8% tax on the price.

8% (or 8 percent) means 8 hundredths, written as 0.08 or $\frac{8}{100}$.

Answer: $C = p + 0.08p$ or $C = p + \frac{8}{100}p$

2. Write a formula that expresses the number of months m that are in y years.

Solution Use the strategy of discovering a pattern.

In 1 year, there are 12 months.
In 2 years, there are 12(2) or 24 months.
In 3 years, there are 12(3) or 36 months.
In y years, there are $12(y)$ or $12y$ months. This equals m months.

Answer: $m = 12y$

EXERCISES

In 1–15, write a formula that expresses each relationship.

1. The total length ℓ of 10 pieces of lumber, each m meters in length, is 10 times the length of each piece of lumber.
2. An article's selling price S equals its cost c plus the margin of profit m.
3. The perimeter P of a rectangle is equal to the sum of twice its length ℓ and twice its width w.
4. The average M of three numbers, a, b, c, is their sum divided by 3.
5. The area A of a triangle is equal to one-half the length of the base b multiplied by the length of the altitude h.
6. The area A of a square is equal to the square of the length of a side s.
7. The volume V of a cube is equal to the cube of the length of an edge e.
8. The surface area S of a cube is equal to 6 times the square of the length of an edge e.
9. The surface area S of a sphere is equal to the product of 4π and the square of the radius r.
10. The average rate of speed r is equal to the distance that is traveled d divided by the time spent on the trip t.
11. The Fahrenheit temperature F is 32° more than nine-fifths of the Celsius temperature C.
12. The Celsius temperature C is equal to five-ninths of the difference between the Fahrenheit temperature F and 32.
13. The dividend D equals the product of the divisor d and the quotient q plus the remainder r.
14. A sales tax T that must be paid when an article is purchased is equal to 8% of the value of the article v.
15. A salesman's weekly earnings E is equal to his weekly salary s increased by 2% of his total volume of sales v.

In 16–20, each formula you write will express one of the variables in terms of the others.

16. Write a formula for finding the number of trees N in an orchard containing r rows of t trees each.
17. Write a formula for the total number of seats N in a school auditorium that has two sections, each with r rows having s seats in each row.
18. A group of n persons in an automobile takes a ferry to cross over a body of water. Write a formula for the total ferry charge C in dollars, if the charge is \$20.00 for the car and driver, plus d dollars for each additional passenger.

19. Write a formula for the cost C in cents of a telephone conversation lasting 9 minutes if the charge for the first 3 minutes is x cents and the charge for each additional minute is y cents.

20. Write a formula for the cost D in dollars of sending a fax of 18 pages if the cost of sending the first page is a dollars and each additional page costs b dollars.

21. A gasoline dealer is allowed a profit of 12 cents a gallon for each gallon she sells. If she sells more than 25,000 gallons in a month, she is given an additional profit of 3 cents for every gallon over that number. Assuming that she always sells more than 25,000 gallons a month, express as a formula the number of dollars D in her monthly income in terms of the number n of gallons sold.

4-6 FORMULAS FOR PERIMETER AND AREA OF POLYGONS

There are many formulas that help us to measure geometric figures. When we use a formula, we are simply evaluating an algebraic expression. We will study geometric figures in greater detail in Chapters 10 and 11. In this chapter, we are simply using these formulas to help strengthen our algebraic skills.

Any figure that can be drawn on a plane or a flat surface is called a ***plane figure***; examples are a square, a triangle, and a circle. A ***polygon*** is a closed plane figure whose sides are line segments.

Evaluating Perimeter Formulas

To find the ***perimeter*** of a polygon, we add the lengths of its sides. For example, to find the perimeter P of a square whose side is s, we can use the formula

$$P = s + s + s + s \qquad \text{or} \qquad P = 4s$$

In each case, we are evaluating an algebraic expression, $s + s + s + s$ or $4s$, to find the perimeter P.

When finding the perimeter of a figure, all lengths should be expressed in the same unit of measure.

EXAMPLE

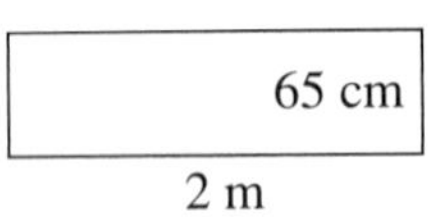

Find the perimeter of a rectangle whose length ℓ is 2 meters and whose width w is 65 centimeters by using the formula $P = 2\ell + 2w$.

How to Proceed:	*Solutions:* *In Centimeters*	*In Meters*
(1) Write the formula:	$P = 2\ell + 2w$	$P = 2\ell + 2w$
(2) Replace the variables ℓ and w by values in meters or in centimeters: (Recall that 100 cm = 1 m.)	$P = 2(200) + 2(65)$	$P = 2(2) + 2(0.65)$

	In Centimeters	*In Meters*
(3) Perform the operations: First multiply, and then add:	$P = 400 \quad + 130$ $P = 530$	$P = 4 \quad + 1.30$ $P = 5.30$

Answer: 530 cm or 5.3 m

(*Note:* The sides of a rectangle may also be called the base b and the height h. The formula for the perimeter is then written as $P = 2b + 2h$.)

Evaluating Area Formulas

The ***area*** of a plane figure is the number of square units contained within the figure.

Recall that the area A of a rectangle is the product of its base b and its height h, written as

$$A = bh$$

Rectangle

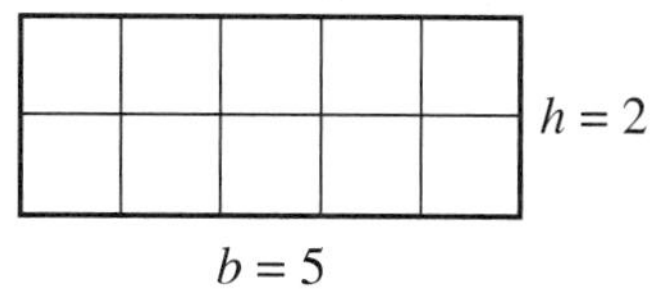

$$A = bh$$
$$= 5(2) = 10 \text{ square units}$$

In a polygon, the ***base*** is a side and the ***height*** (or altitude) is a segment drawn perpendicular to the base from an opposite vertex.

The areas of many four-sided figures are found by using a variation of the formula $A = bh$. In the diagrams that follow, notice how each geometric figure can be transformed or changed back to a basic rectangle. For example, by sliding a small triangular piece either away from or towards the slant edges of a parallelogram, we form a rectangle, so the area of the parallelogram is found by the same formula, $A = bh$.

In a trapezoid, where two parallel sides b and c have unequal lengths, the average length of these sides is multiplied by the height h, giving us the area formula

$$A = h\left(\frac{b + c}{2}\right) \quad \text{or} \quad A = \frac{1}{2}h(b + c)$$

Parallelogram

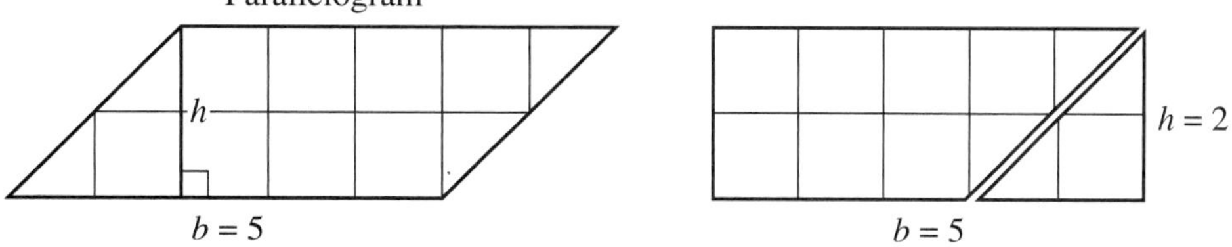

$A = bh$, or $A = 5(2) = 10$ square units

Trapezoid

$c = 3$

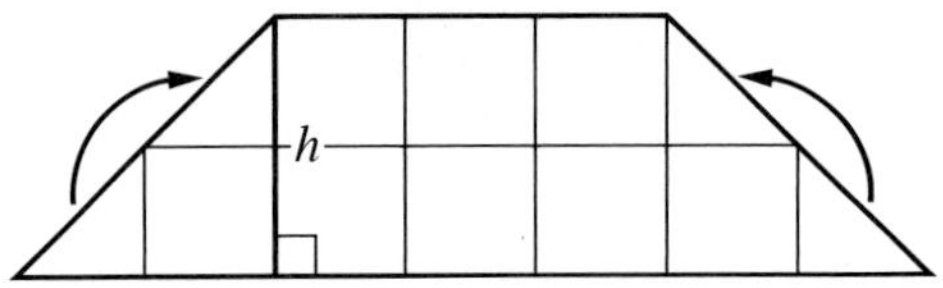

$b = 7$

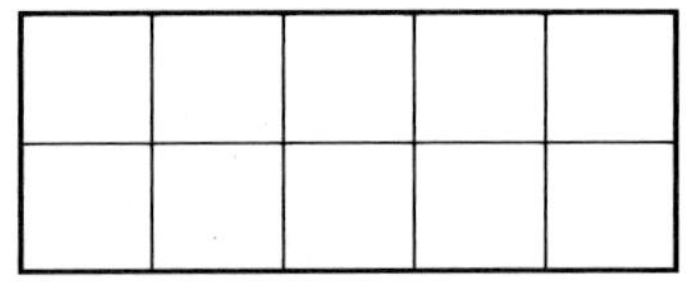

$h = 2$

Average base = 5

$$A = h\left(\frac{b + c}{2}\right), \text{ or } a = 2\left(\frac{7 + 3}{2}\right) = 2(5) = 10 \text{ square units}$$

The area of a triangle, which can be shown to be one-half of a parallelogram or a rectangle, is found by using the formula

$$A = \frac{1}{2}bh.$$

Triangle

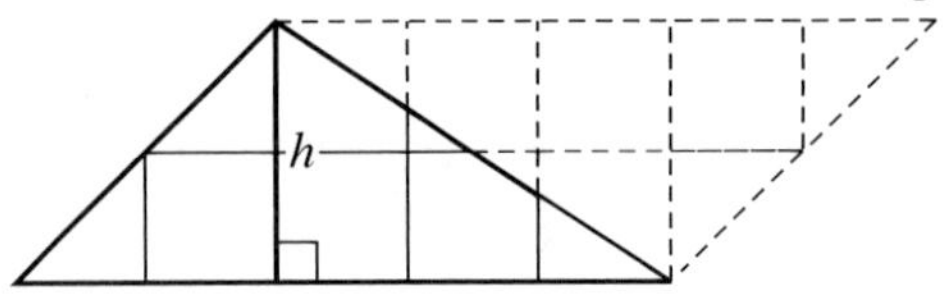

b

h

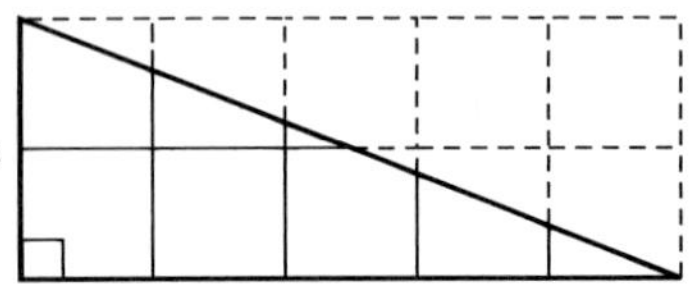

b

$$A = \frac{1}{2}bh, \text{ or } a = \frac{1}{2}(5)(2) = 5 \text{ square miles}$$

In the English system, square units are usually abbreviated and written with periods, as in sq. ft. (square feet) and sq. in. (square inches). In the metric system, area is expressed using powers or square units, as in m^2 (square meters) and cm^2 (square centimeters). Notice that periods are not used in metric abbreviations.

When finding the area of a plane figure, all lengths *must* be expressed in the same unit of measure, and the area is expressed as square units of that measure.

EXAMPLES

1. Find the area of a square whose side s is 3.9 yards by using the formula $A = s^2$.

$s = 3.9$ yd.

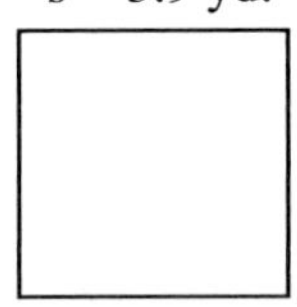

Solution

$$A = s^2$$
$$= (3.9)^2$$
$$= (3.9)(3.9) = 15.21$$

Answer: 15.21 sq. yd.

Alternative Solution:
Enter: 3.9 [x^2]
Display: [15.21]

2. The base b of a parallelogram measures 30 centimeters, and its height h measures 79 millimeters. Find the area of the parallelogram using the formula $A = bh$.

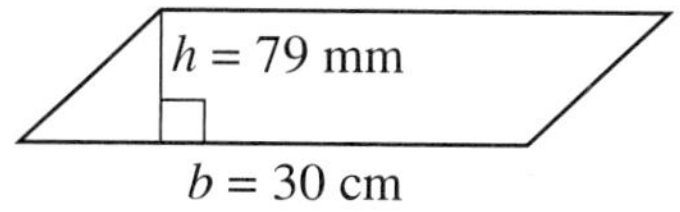

Solution

	In Centimeters	*In Millimeters*
1. Write the formula:	$A = bh$	$A = bh$
2. Replace the variables b and h by values in centimeters or millimeters: (*Note:* 10 mm = 1 cm.)	$= 30\ (7.9)$	$= 300(79)$
3. Perform the multiplication: (*Note:* 100 mm^2 = 1 cm^2)	$= 237.0$	$= 23{,}700$

Answer: 237 cm^2 or 23,700 mm^2

EXERCISES

1. The formula for the perimeter of a triangle is $P = a + b + c$. Find P when:

a. $a = 15$ in., $b = 10$ in., $c = 7$ in.
b. $a = 4.5$ m, $b = 1.7$ m, $c = 3.8$ m
c. $a = 9$ ft, $b = 8$ ft, $c = 18$ in.
d. $a = 7\frac{1}{2}$ ft, $b = 5\frac{3}{4}$ ft, $c = 6\frac{1}{2}$ ft
e. $a = 75$ cm, $b = 1.5$ m, $c = 2$ m

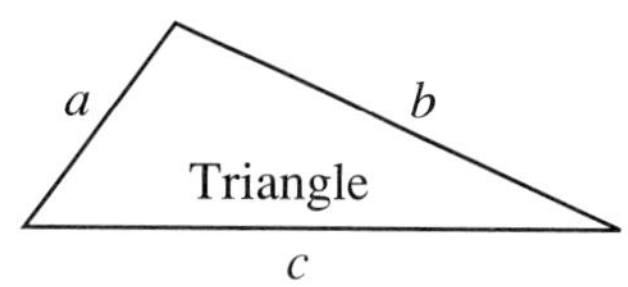

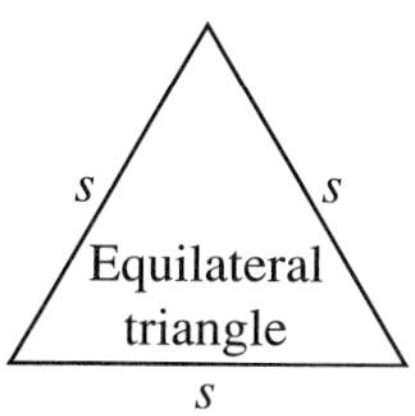

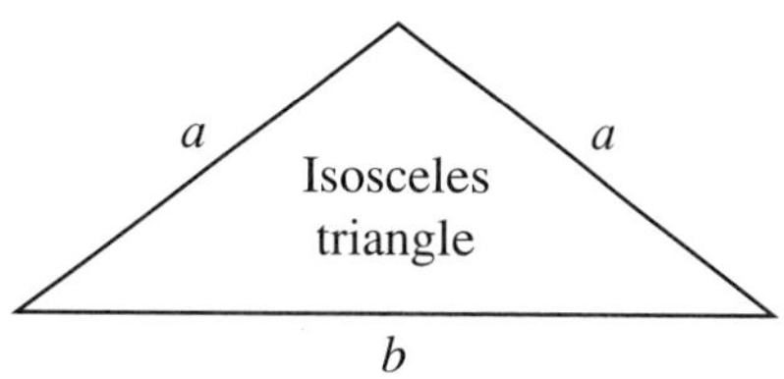

2. The formula for the perimeter of an equilateral triangle is $P = 3s$. Find P when s equals:

a. 12 cm **b.** 4.8 m

c. $9\frac{1}{3}$ ft **d.** 2 ft 8 in.

e. 72.98 m

3. The formula for the perimeter of an isosceles triangle is $P = 2a + b$. Find P when:

a. $a = 6$ m, $b = 4$ m

b. $a = 7.5$ m, $b = 5.4$ m

c. $a = 3\frac{1}{2}$ ft, $b = 5$ ft

d. $a = 9.2$ m, $b = 92$ cm

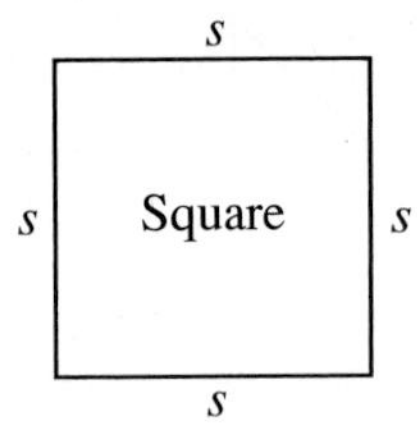

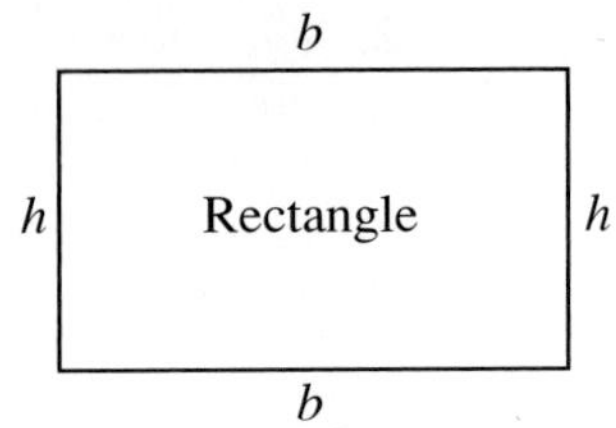

4. The formula for the perimeter of a square is $P = 4s$. Find P when s equals:

a. 4 cm **b.** 3.5 m

c. $8\frac{3}{4}$ in **d.** 7 ft 5 in.

e. 80.9 cm

5. The formula for the perimeter of a rectangle is $P = 2b + 2h$. Find P when:

a. $b = 20$ cm, $h = 9$ cm

b. $b = 7.3$ m, $h = 6.9$ m

c. $b = 5\frac{1}{2}$ in., $h = 5\frac{1}{4}$ in.

d. $b = 4.57$ m, $h = 48$ cm

6. The formula for the area of a rectangle is $A = bh$. Find A when:

a. $b = 15$ ft, $h = 13$ ft **b.** $b = 7.5$ m, $h = 3.4$ m

c. $b = 8\frac{1}{2}$ ft, $h = 6$ ft **d.** $b = 1$ m, $h = 40$ cm

e. $b = 2$ ft, $h = 18$ in. **f.** $b = 2.6$ m, $h = 87$ cm

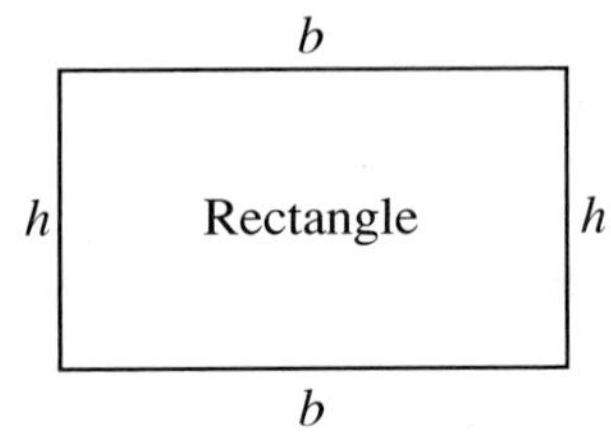

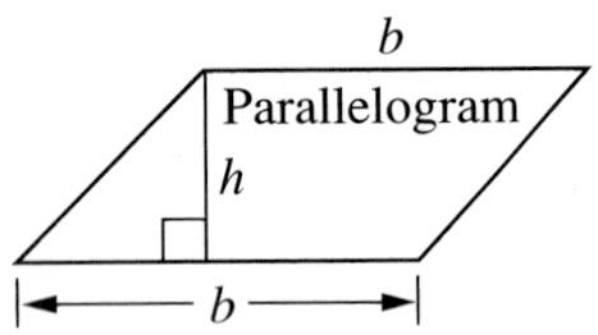

7. The formula for the area of a parallelogram is $A = bh$. Find A when:

a. $b = 8$ ft, $h = 12$ ft **b.** $b = 3.5$ m, $h = 6.4$ m

c. $b = 7\frac{1}{2}$ in., $h = 8$ in. **d.** $b = 1$ m, $h = 10$ cm

e. $b = 0.75$ cm, $h = 6.8$ mm **f.** $b = 54$ in. $h = 3$ ft

8. The formula for the area of a rhombus (a parallelogram all of whose sides have the same length) is $A = bh$. Find A when:

a. $b = 5$ m, $h = 3$ m **b.** $b = 7$ in., $h = 4.5$ in.

c. $b = 10$ ft, $h = 8\frac{1}{2}$ ft **d.** $b = 14.5$ cm, $h = 11.4$ cm

e. $b = 12$ ft, $h = 30$ in. **f.** $b = 1.82$ m, $h = 59$ cm

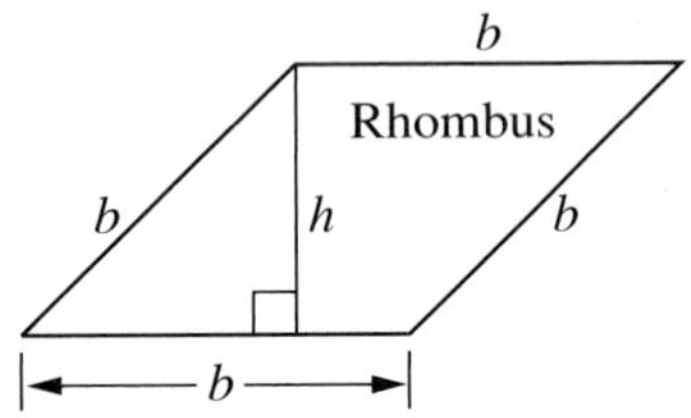

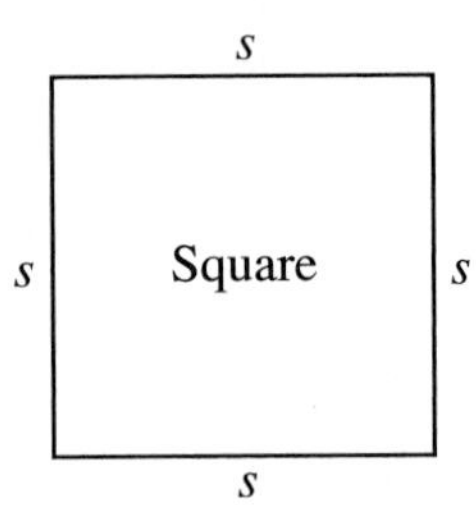

9. The formula for the area of a square is $A = s^2$. Find A when s equals:

a. 25 in. **b.** 9 cm **c.** $2\frac{1}{2}$ ft **d.** 6.1 m **e.** 14.27 km

10. The formula for the area of a triangle is $A = \frac{1}{2}bh$. Find A when:

a. $b = 10$ cm, $h = 6$ cm **b.** $b = 10.5$ m, $h = 7.6$ m

c. $b = 3\frac{1}{2}$ in., $h = 8$ in. **d.** $b = 1$ ft, $h = 5\frac{1}{2}$ in.

e. $b = 145$ cm, $h = 2.5$ m **f.** $b = 9$ yd, $h = 32$ ft

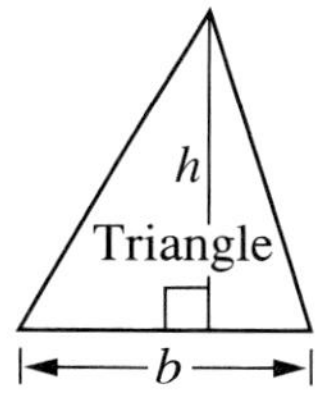

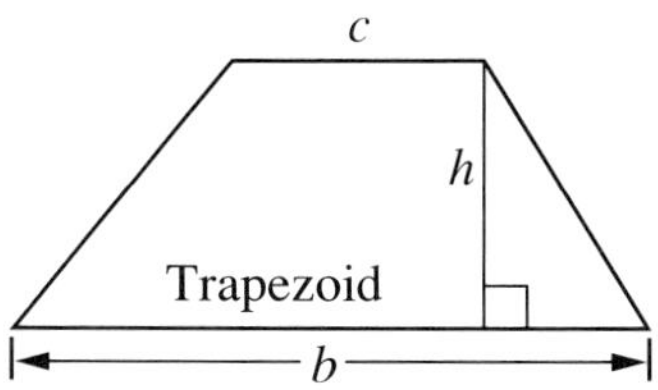

11. The formula for the area of a trapezoid is $A = \frac{1}{2}h(b + c)$. Find A when:

a. $h = 9$ ft, $b = 14$ ft, $c = 8$ ft

b. $h = 5$ in., $b = 3\frac{1}{4}$ in., $c = \frac{3}{4}$ in.

c. $h = 2$ m, $b = 1.8$ m, $c = 1.1$ m

d. $h = 2$ ft, $b = 18$ in., $c = 10$ in.

e. $h = 60$ cm, $b = 1.7$ m, $c = 95$ cm

4-7 FORMULAS FOR CIRCUMFERENCE AND AREA OF A CIRCLE

Terms and Definitions

A ***circle*** is a set of points in a plane such that each point is the same distance from a fixed point called the ***center***. When the center is named by the capital letter O, the circle is called circle O.

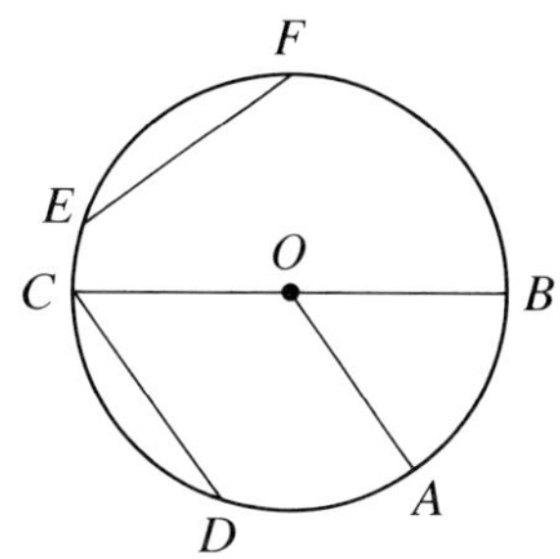

A ***radius*** of a circle (plural, ***radii***) is a line segment from the center of the circle to any point of the circle. In the diagram, $\overline{OA}$, $\overline{OB}$, and $\overline{OC}$ are radii of circle O.

A ***diameter*** of a circle is a line segment that contains the center of the circle and whose endpoints are points of the circle. A diameter of circle O in the diagram is $\overline{CB}$.

A ***chord*** is a line segment whose endpoints are points of the circle. In the diagram, $\overline{CD}$, $\overline{CB}$, and $\overline{EF}$ are chords of circle O. Note that any diameter of a circle is always the largest chord for that circle.

The words *radius* and *diameter* mean both the line segments of a circle and the lengths of these line segments. Thus, it is correct to say either that the length of a radius is 5 or that the radius is 5.

Circumference of a Circle

The distance around a circle, or the perimeter of the circle, is called its ***circumference***. If we measure different circles, no matter how large or small the circle, we discover that the circumference of a circle is always slightly more than 3 times the diameter of the circle. Mathematicians have shown that, for every circle, the ratio of the circumference C to the diameter d is always the same value. This constant value is represented by the Greek letter π (read as "pi"). Therefore:

$$\frac{\text{Circumference}}{\text{diameter}} = \text{a constant} \quad \text{or} \quad \frac{C}{d} = \pi$$

We saw in Chapter 1 that π is an irrational number and an *exact* value. Other values sometimes used for π, including $\frac{22}{7}$, 3.14, 3.1416, and the value of $\boxed{\pi}$ displayed on a scientific calculator, are *rational approximations* of π.

There are two formulas for finding the circumference of a circle.

1. We know $\frac{C}{d} = \pi$, so we multiply both sides of the equation by d to get $\frac{C}{d} \cdot d = \pi \cdot d$, or

$$\boxed{C = \pi d}$$

2. The diameter of a circle equals twice the radius, so $C = \pi d$ becomes $C = \pi \cdot 2r$, or

$$\boxed{C = 2\pi r}$$

Area of a Circle

The ***area of a circle*** means the area of the region enclosed by the circle. Mathematicians have discovered that the area A of a circle is equal to π multiplied by the square of the radius r. The formula for the area of a circle is:

$$\boxed{A = \pi r^2}$$

EXAMPLES

1. In the diagram, two circles are shown with center O. Points A, B, and G are on the smaller circle; points C, D, E, and F are on the larger circle. $\overline{DF}$ is a line segment containing points B, O, and G; $\overline{OC}$ is a line segment containing point A; $\overline{EC}$ is a line segment containing point B.

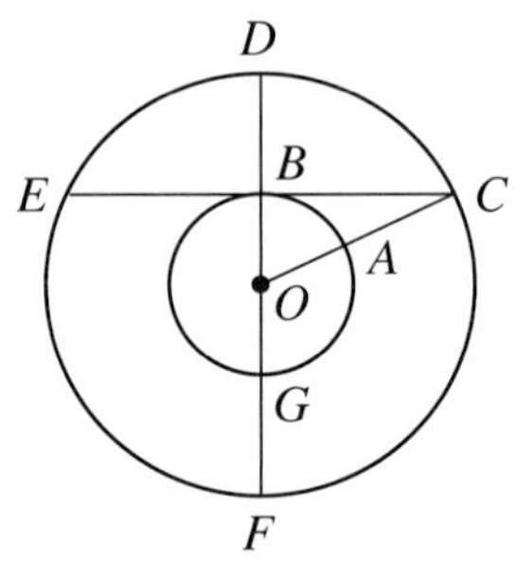

Name the points or segments that describe:

Answers:

a. The radii of the smaller circle — **a.** $\overline{OA}$, $\overline{OB}$, $\overline{OG}$

b. The radii of the larger circle — **b.** $\overline{OC}$, $\overline{OD}$, $\overline{OF}$

c. The chords shown for the larger circle — **c.** $\overline{EC}$, $\overline{DF}$

d. The diameter of the larger circle — **d.** $\overline{DF}$

2. A circle has a radius of 20 centimeters.

a. Find its circumference in terms of π.

b. Find its circumference to the *nearest thousandth* of a centimeter.

Solutions **a.** METHOD 1. Substitute radius $r = 20$ in the formula:

$$\begin{aligned} C &= 2\pi r \\ &= 2\pi(20) \\ &= 40\pi \end{aligned}$$

METHOD 2. If radius $r = 20$, then diameter $d = 2r$ or $d = 40$. Use the formula:

$$\begin{aligned} C &= \pi d \\ &= \pi(40) \\ &= 40\pi \end{aligned}$$

b. To find a value to the nearest thousandths, a four-place decimal number is needed before rounding. (For this reason, approximations of π such as $\frac{22}{7}$ and 3.14 cannot be used here.) On a scientific calculator, $\pi \approx 3.1415927$.

METHOD 1. Use the formula $C = 2\pi r$.

Enter: 2 [×] [π] [×] 20 [=]

Display: [125.66371]

Round to the nearest thousandth: 125.664

METHOD 2. Use the formula $C = \pi d$.

Enter: [π] [×] 40 [=]

Display: [125.66371]

Answers: **a.** 40π cm **b.** 125.664 cm

---

3. Find the area of a circle whose radius is 14 feet:

a. expressed in terms of π **b.** using $\pi = 3.14$ **c.** using $\pi = \frac{22}{7}$

d. using the [π] key on a scientific calculator

Solutions **a.**

$$\begin{aligned} A &= \pi r^2 \\ &= \pi \cdot (14)^2 \\ &= \pi \cdot 14 \cdot 14 \\ &= \pi \cdot 196 \\ &= 196\pi \end{aligned}$$

b.

$$\begin{aligned} A &= \pi r^2 \\ &= (3.14) \cdot (14)^2 \\ &= (3.14) \cdot (196) \\ &= 615.44 \end{aligned}$$

c.

$$\begin{aligned} A &= \pi r^2 \\ &= \frac{22}{7} \cdot (14)^2 \\ &= \frac{22}{\cancel{7}} \cdot \overset{2}{\cancel{14}} \cdot 14 \\ &= 616 \end{aligned}$$

d. Think of the formula $A = \pi r^2$.

Enter: [π] [×] 14 [x^2] [=]

Display: 615.75216

Answers: **a.** 196π sq. ft. **b.** 615.44 sq. ft. **c.** 616 sq. ft.
d. 615.75216 sq. ft.

(*Note:* If a radius is given in feet, the area of the circle is measured in square feet. Also, only the answer to part **a** is exact. All other answers are rational approximations.)

EXERCISES WITH OPEN-RESPONSE PROBLEM

1. Find the diameter of a circle whose radius measures:

a. 2 in. **b.** 13 ft **c.** $2\frac{1}{8}$ in. **d.** 36.5 mm **e.** 496.37 cm **f.** 20.05 m

2. Find the radius of a circle whose diameter measures:

a. 4 ft **b.** 7 in. **c.** $3\frac{1}{2}$ in. **d.** 13.9 cm **e.** 295 mm **f.** 195.45 m

3. Find, in each case, the circumference of a circle, expressed in terms of π, for the given radius r or diameter d.

a. $r = 5$ cm **b.** $d = 9$ ft **c.** $r = 2.4$ mm **d.** $d = \frac{1}{3}$ in.
e. $d = 25$ yd **f.** $r = 4.375$ m **g.** $d = 9$ mm ***h*** $r = 97.5$ cm

4. Using $\pi = 3.14$, find the approximate length of the circumference of a circle when:

a. $r = 8$ cm **b.** $d = 8$ in. **c.** $r = 2.5$ mm **d.** $d = 3\frac{1}{2}$ in.

5. Using $\pi = \frac{22}{7}$, find the approximate length of the circumference of a circle when:

a. $r = 14$ cm **b.** $d = 7$ ft **c.** $r = 3.5$ mm **d.** $d = 10\frac{1}{2}$ in.

6. Use a calculator to find the circumference of each circle, correct to the *nearest thousandth*, given its radius r or diameter d.

a. $r = 10$ mm **b.** $r = 3.8$ cm **c.** $d = 32$ in. **d.** $d = 17$ ft
e. $r = \frac{3}{4}$ yd **f.** $d = 0.6$ km **g.** $r = 4.37$ m **h.** $d = 35$ mi

7. How many inches does Joan's bicycle go in one turn of the wheels if the diameter of each wheel is 28 inches? Express your answer to the nearest inch.

8. Find the area of a circle, expressed in terms of π, when its radius is:

a. 4 **b.** 9 **c.** 0.8 **d.** 0.3 **e.** $\frac{1}{3}$ **f.** $1\frac{1}{2}$

9. Find the area of a circle, expressed in terms of π, for the given radius r or diameter d:

a. $r = 8$ in. **b.** $d = 18$ cm **c.** $r = \frac{1}{4}$ in.

d. $d = 0.6$ mm **e.** $r = 1.5$ m **f.** $d = 1\frac{1}{3}$ ft

10. Using $\pi = 3.14$, find the approximate area of a circle when:

a. $r = 10$ cm **b.** $d = 20$ ft **c.** $r = 0.4$ mm

d. $d = 3$ in. **e.** $r = 0.1$ m **f.** $d = 1\frac{1}{2}$ cm

11. Using $\pi = \frac{22}{7}$, find the approximate area of a circle when:

a. $r = 7$ ft **b.** $d = 140$ mm **c.** $r = 21$ cm

d. $d = 7$ in. **e.** $r = \frac{1}{2}$ in. **f.** $d = 1.5$ mm

12. Use a calculator to find the area of each circle, correct to the *nearest thousandth*, given its radius r or diameter d.

a. $d = 12$ cm **b.** $r = 1$ ft **c.** $d = 2.4$ mm **d.** $r = 32$ in.

e. $r = 0.8$ km **f.** $d = 2\frac{1}{2}$ yd **g.** $r = 3.26$ m **h.** $d = \frac{2}{5}$ mi

13. The radius of a circular flower bed is 3.5 meters. **a.** Express its area in terms of π.
b. Use $\pi = \frac{22}{7}$ to find its approximate area.

14. A circular mirror has a diameter of 18 inches. Find its approximate area to the *nearest hundredth* of a square inch.

15. A circular flower bed has a diameter of 4.2 meters. How many meters of fencing will be needed to enclose this garden? Express your answer to the *nearest tenth* of a meter.

In 16–18, select the best answer from the four choices given.

16. If the diameter of a circle is 7, the circumference of the circle is:
(1) exactly 21.98 (2) exactly 22
(3) between 21 and 22 (4) exactly 44

17. If the diameter of a circle is 20, the area of the circle is:
(1) exactly 314 (2) exactly $314\frac{2}{7}$
(3) between 314 and 315 (4) between 1,256 and 1,257

18. If the radius of a circle is 14, the circumference of the circle is:
(1) between 43 and 44 (2) exactly 43.96
(3) between 87 and 88 (4) exactly 88

19. The distance around a circular track is 440 yards. Using $\pi = \frac{22}{7}$, find the diameter of the circular track.

20. A student reasoned that, if the radius of a circle is 2, the circumference and the area of the circle are equal because they both measure 4π. This reasoning, however, is not correct. Where did the student make an error?

4-8 REASONING WITH PERIMETER, AREA, AND SHADED REGIONS

It is relatively easy to work with formulas to find the perimeter and area of a geometric figure. In some cases, however, additional reasoning is needed to solve a problem involving these figures and the formulas you have learned.

EXAMPLES

1. In the figure on the left, square $ABCD$ is drawn about circle O. The radius of circle O is 5. The region between the square and the circle is shaded. Find:

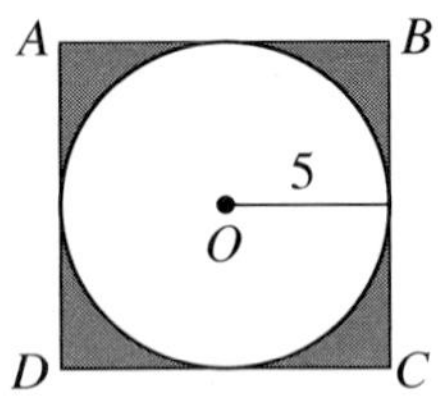

a. the diameter of the circle
b. the length of a side of the square
c. the area of the square
d. the area of the circle
e. the area of the shaded region.
f. the area of the shaded region to the *nearest hundredth*.

Solutions **a.** The length of diameter d is twice the length of radius r, so $d = 2(5) = 10$.
b. Here, the length of side s of the square equals diameter d, so $s = 10$.
c. The area of a square is found by the formula $A = s^2$. Here, $A = 10^2 = 100$.
d. The area of a circle is found by the formula $A = \pi r^2$. Here, $A = \pi(5)^2 = 25\pi$.
e. The area of the shaded region is found by subtracting the area of the circle from the area of the square: $100 - 25\pi$.
f. Evaluate the expression $100 - 25\pi$ on a calculator.

Enter: 100 [−] 25 [×] [π] [=]

Display: [21.460184]

Round the display to two decimal places, namely, to 21.46 square units.

Answers: **a.** 10 **b.** 10 **c.** 100 **d.** 25π **e.** $100 - 25\pi$ **f.** 21.46

$P = 24$ cm

2. The perimeter of a square is 24 centimeters. What is the area of this square?

How to Proceed:	*Solution:*
(1) Write the formula for the perimeter of a square:	$P = 4s$
(2) Substitute the given value for the perimeter:	$24 = 4s$
(3) To find the length of one side s, divide each side of the equation by 4, or multiply each side of the equation by the reciprocal $\frac{1}{4}$:	$\frac{1}{4}(24) = \frac{1}{4}(4s)$ $6 = s$
(4) Write the formula for the area of a square:	$A = s^2$
(5) Substitute the value for the side s:	$= (6)^2$
(6) Perform the computation:	$= 36$

Answer: 36 cm^2

EXERCISES WITH OPEN-RESPONSE PROBLEMS

In 1–8: For each figure, find: **a.** the perimeter **b.** the area. Recall that: (1) For figures such as the parallelogram and the rectangle, the opposite sides are equal in length. (2) For the square and the rhombus, all sides are equal in length.

1.

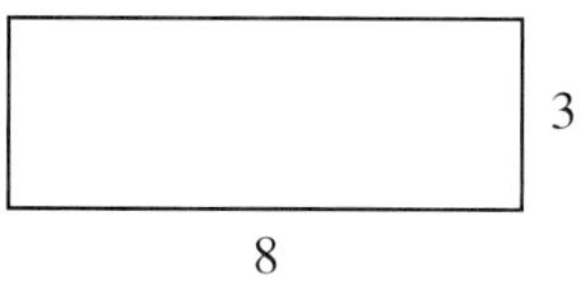

Rectangle

2.

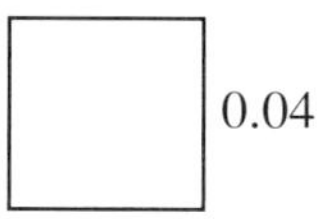

Square

3.

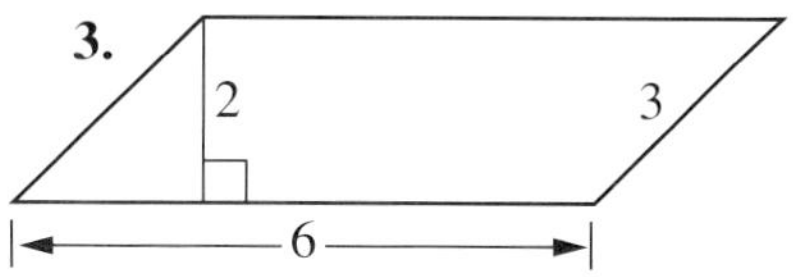

Parallelogram

4.

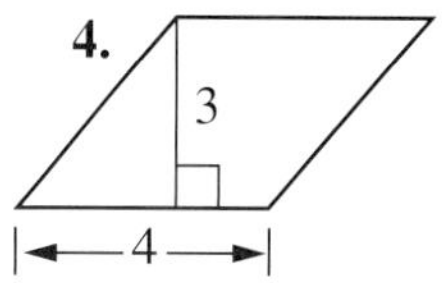

Rhombus

5.

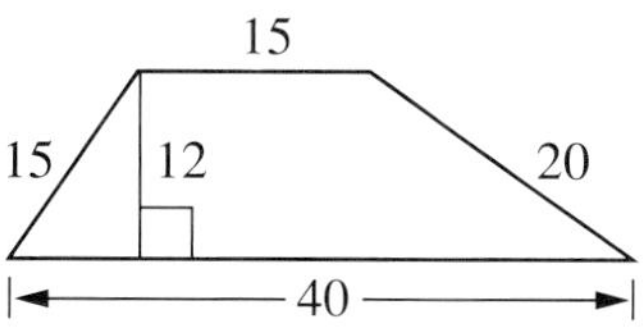

Trapezoid

6.

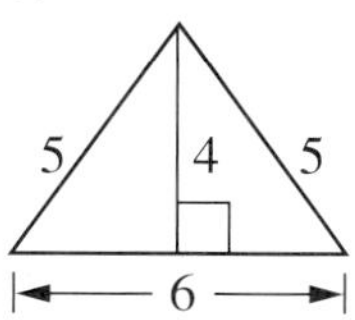

Triangle

7.

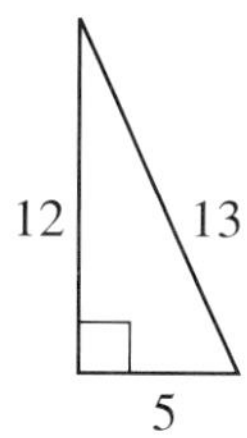

Triangle

8.

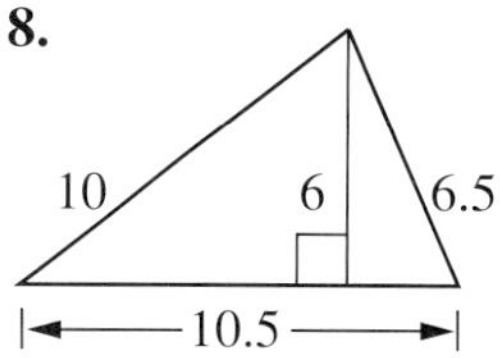

Triangle

9. Find the area of a rectangle in which the base measures 20 centimeters and the height is half the length of the base.

10. Find the area of a parallelogram in which the height measures 14 inches and the base measures 6 inches more than the height.

11. Find the area of a triangle in which the base measures 8.5 centimeters and the height measures 3 centimeters less than the base.

In 12–16, the given measure represents the perimeter of a square. Find: **a.** the length of a side of the square **b.** the area of the square

12. 20 cm **13.** 100 mm **14.** 4 ft. **15.** 2 in. **16.** 2.8 m

In 17–29: **a.** Find the measure of each line segment whose algebraic representation is given, when $x = 5$ and $y = 4$. **b.** Using the results obtained in part **a**, find the area of the geometric figure.

17.

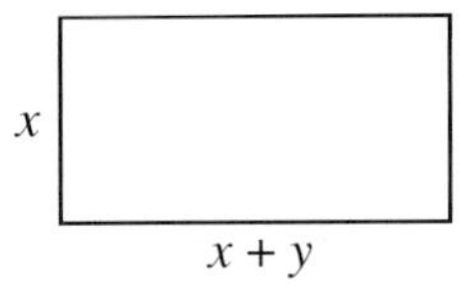

Rectangle

18.

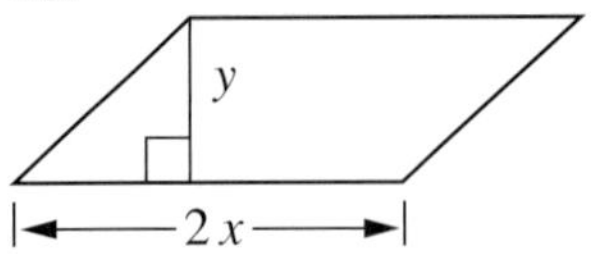

Parallelogram

19.

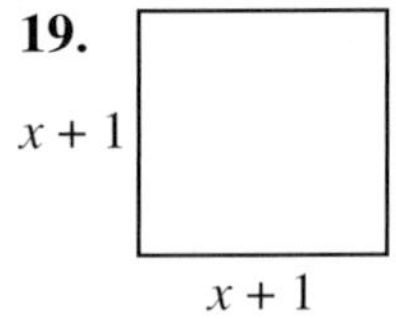

Square

20.

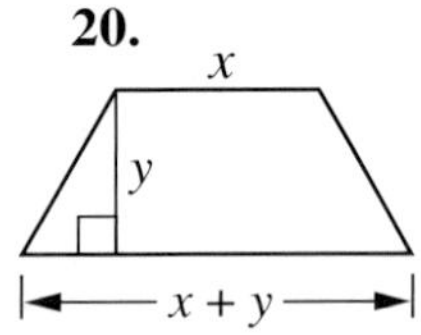

Trapezoid

21.

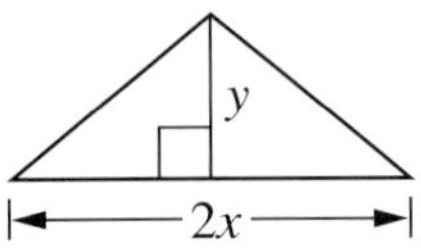

Triangle

22.

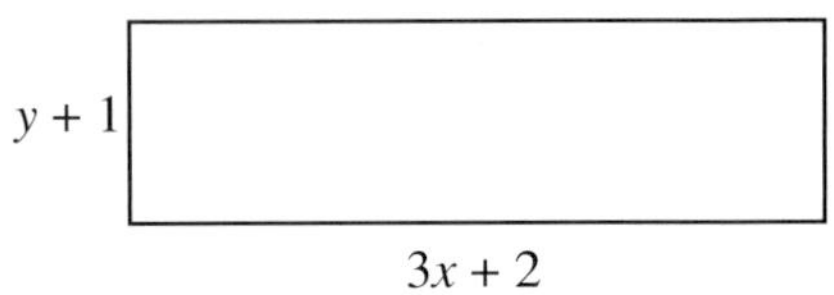

Rectangle

23.

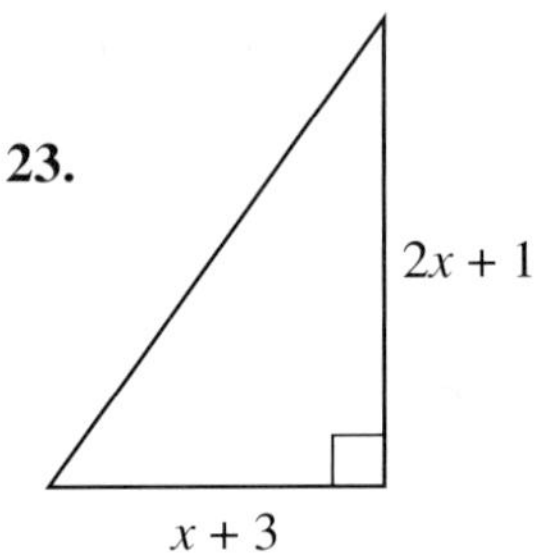

Triangle

24.

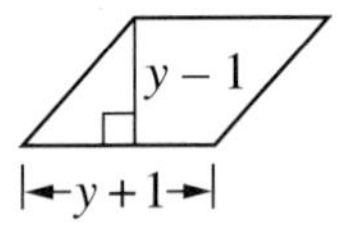

Parallelogram

25. A triangle: measure of the base $= 3x + 1$; height $= 2x$
26. A rectangle: length $= 2x$; width $= y + 2$
27. A square: length of a side $= 2x$
28. A parallelogram: measure of the base $= \frac{y}{2}$; height $= x + 7$
29. A trapezoid: height $= x + 3$; measures of the bases $= y + 3$ and $x + y$

30. A rectangular mall located between two parallel streets is 50 feet wide and 200 yards long. Find the area of the mall.

31. Chow Li is planting a border of 48 seedlings that will completely surround a rectangular garden plot. He wants to plant the seedlings 2 feet apart. If the garden plot is to be 30 feet long, how wide should it be?

32. In the figure at the right, a circle is drawn in a square whose side has a length of 12.

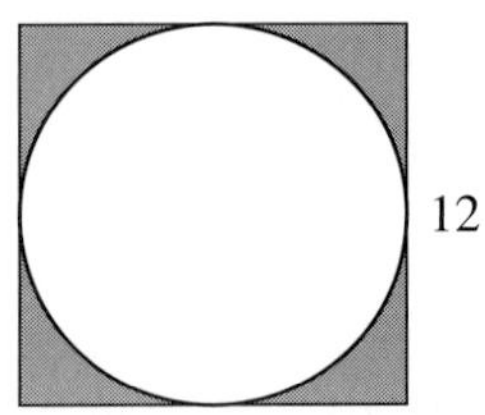

a. Find the diameter of the circle.
b. Find the radius of the circle.
c. Find the area of the square.
d. Find the area of the circle in terms of π.
e. Using the results of parts **c** and **d**, express the area of the shaded portion of the diagram in terms of π.

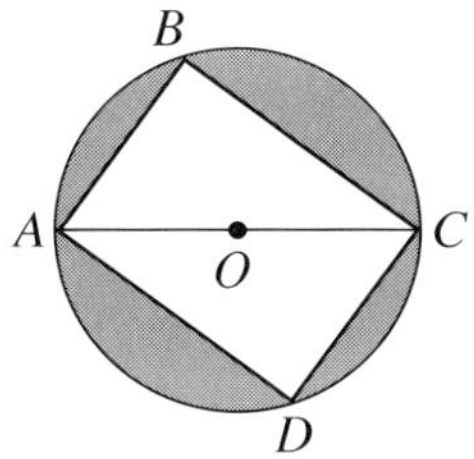

33. In the diagram on the left, rectangle $ABCD$ is drawn in circle O. $AB = 6$, $BC = 8$, and $AC = 10$. The diagonal of the rectangle, $\overline{AC}$, is also the diameter of the circle. [Wherever possible, answers may be left in terms of π.]

a. Find OC.

b. Find the circumference of circle O.

c. Find the perimeter of the rectangle.

d. Find the area of circle O.

e. Find the area of rectangle $ABCD$.

f. Find the area of the shaded region.

34. An 8-inch pizza has a diameter of 8 inches; a 12-inch pizza has a diameter of 12 inches. Two boys purchased an 8-inch pizza for \$3.20 and shared the pie equally. Four girls purchased a 12-inch pizza for \$5.40 and shared the pie equally. Assume that both pies are of the same thickness.

a. Who got a larger piece of pizza, a boy or a girl? Explain why. (*Hint:* Relate the problem to the area of a circle.)

b. How much did each boy pay for his share of the pie?

c. How much did each girl pay for her share of the pie?

d. Which size pie is the better buy? Explain why.

35. A square is folded in half horizontally as shown in the figure at the right. If the perimeter of the resulting figure is 24 inches, what was the area of the original square?

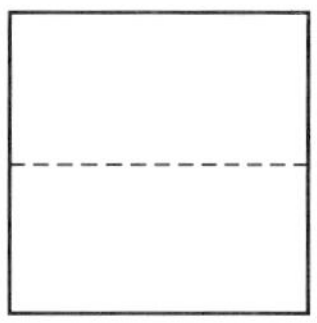

36. An area of a lobby floor measures 15 feet by 18 feet. In one corner, an 8-foot-by-7-foot rectangular portion of the floor space is to be covered with tile. The remainder of the floor is to be carpeted.

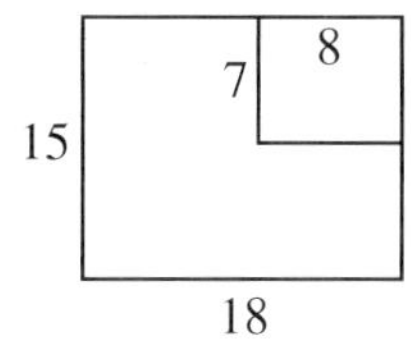

a. How many square feet of carpet are needed?

b. When the carpet is laid, a metal strip is used along each edge of the carpet to hold it in place. How many feet of stripping are needed?

c. If the uncarpeted area were in the center along one side as shown in the diagram, would the amount of carpet change? Would the amount of stripping change?

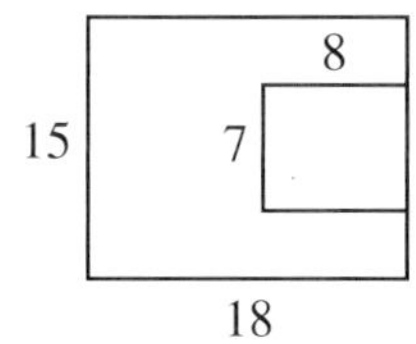

d. If the uncarpeted section were in the center of the area as shown in the diagram, would the amount of carpet change? Would the amount of stripping change?

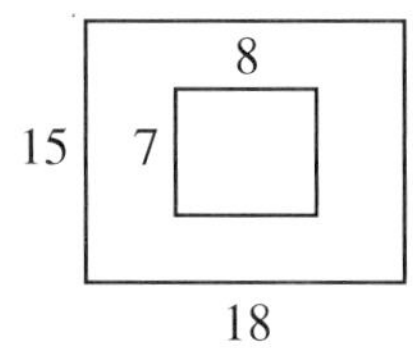

In 37–41, find the area of the shaded region.

37.

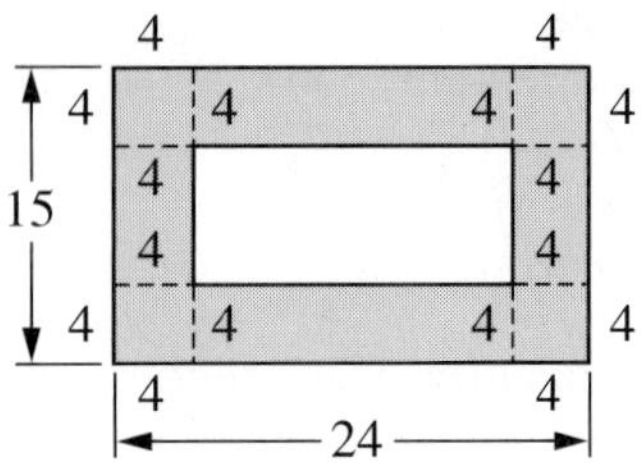

38.

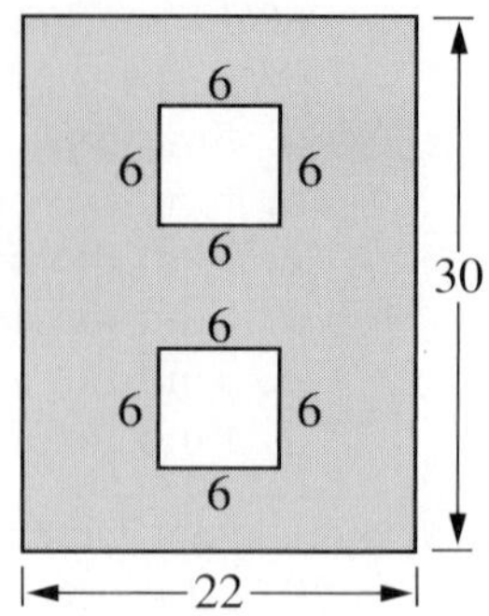

39.

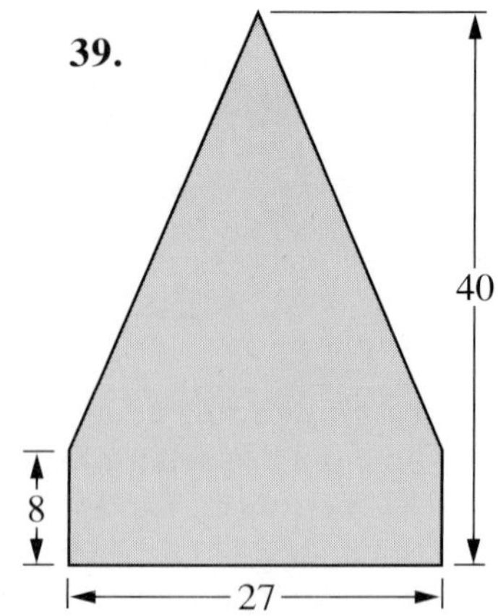

40.

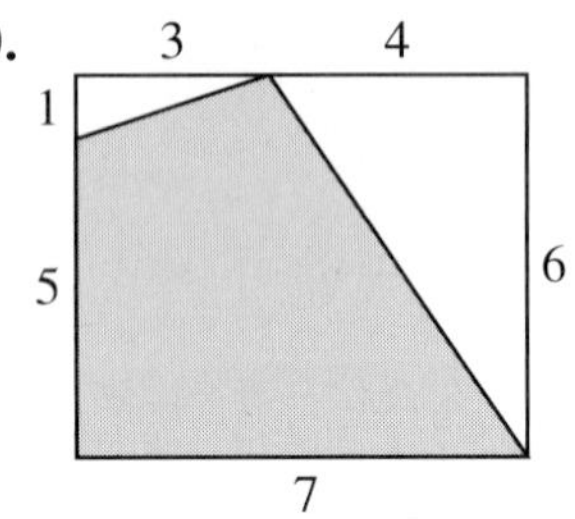

41.

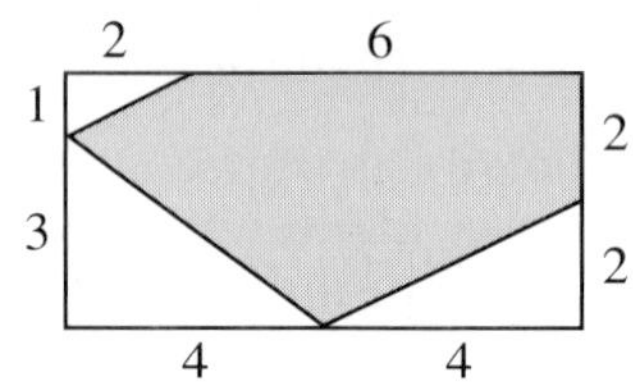

42. Ann wants to make a rectangular pen for her rabbit by using one side of the garage as one long side of the pen. She has 26 feet of fencing to use for the other three sides.

a. If she makes the pen 12 feet long and uses all of the fencing, what are the dimensions of the pen? What is the area?

b. If she makes the pen 10 feet long and uses all of the fencing, what are the dimensions of the pen? Will the rabbit have more room?

c. If she uses the garage wall and all 26 feet of fencing, what are the dimensions of the largest possible pen?

4-9 FORMULAS FOR VOLUME OF A SOLID

The ***volume*** of a solid is the number of unit cubes (or cubic units) that it contains. To find the volume of a solid, all lengths *must* be expressed in the same unit of measure. The volume is then expressed in cubic units of this length.

A ***right prism*** is a solid with bases that have the same size and shape, and with a height that is perpendicular to

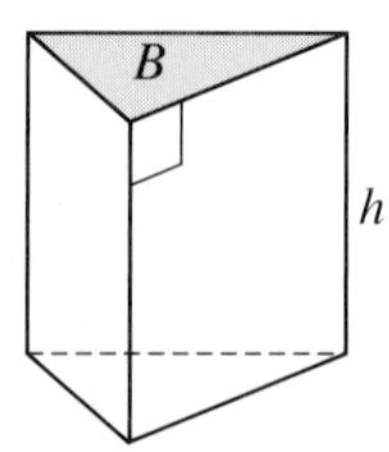

Triangular right prism

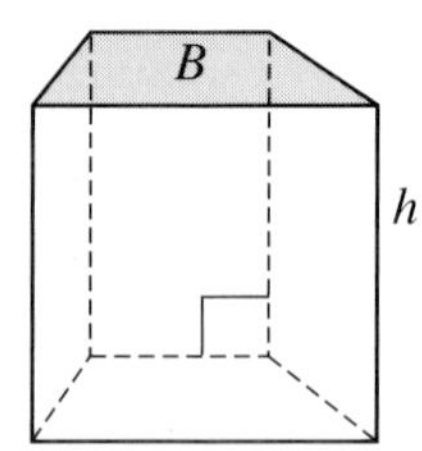

Trapezoidal right prism

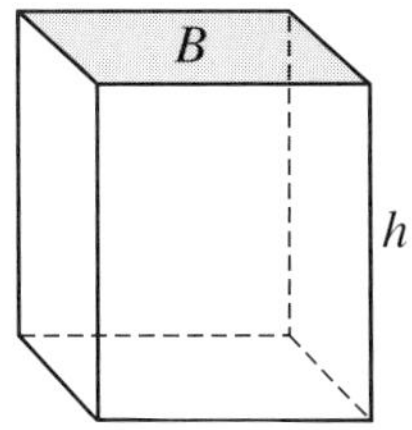

Rectangular solid (a prism)

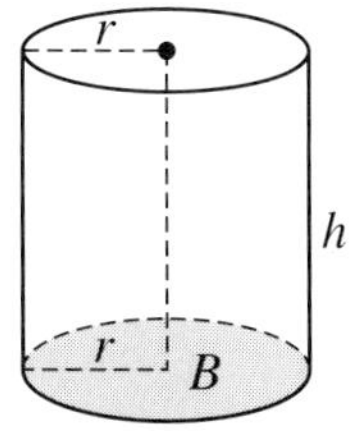

Right circular cylinder

these bases. Some examples of right prisms, as seen in the diagrams, include solids whose bases are triangles, trapezoids, and rectangles. The two bases may be any polygons, as long as they have the same size and shape.

A ***right circular cylinder*** is a solid with two bases that are circles of the same size, and with a height that is perpendicular to these bases.

Both the volume V of a right prism and the volume V of a right circular cylinder can be found by multiplying the area B of the base by the height h. This formula is written as

$$\boxed{V = Bh}$$

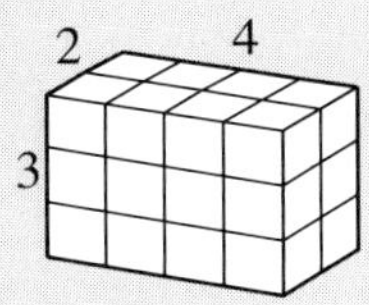

$V = Bh$
$V = (\ell w)\,h$
$= (4 \cdot 2) \cdot 3$
$= 8 \cdot 3$
$= 24 \text{ cm}^3$

For example, in the rectangular solid or right prism shown at left, the base is a rectangle, 4 centimeters by 2 centimeters. The area B of this base $= \ell w = 4(2) = 8 \text{ cm}^2$. Then, using a height h of 3 centimeters, the volume V of the rectangular solid $= Bh = (8 \text{ cm}^2)(3 \text{ cm}) = 24 \text{ cm}^3$.

For a rectangular solid, note that the volume formula can be written in two ways: as $V = Bh$ or as $V = (\ell w)h$.

To understand volume, count the cubes in the diagram at the left. There are 3 layers, each containing 8 cubes, for a total of 24 cubes. Note that 3 corresponds to the height h, that 8 corresponds to the area B of the base, and that 24 corresponds to the volume V in cubic units.

Other formulas for volume will be given and explained in the exercises.

EXAMPLE

A cylindrical can of soup has a radius of 1.5 inches and a height of 5 inches.

Find the volume of this can:

a. in terms of π
b. to the nearest cubic inch

Solutions This can is a right circular cylinder. Use the formula $V = Bh$. Since the base is a circle whose area equals πr^2, the area B of the base can be replaced by πr^2.

a. $V = Bh = (\pi r^2)h = \pi(1.5)^2(5) = \pi(2.25)(5) = \pi(11.25) = 11.25\pi$

b. Use a calculator to evaluate $V = \pi r^2 h$ for the given measures. Then round.

Enter: [π] [$\times$] 1.5 [x^2] [$\times$] 5 [=]

Display: [35.342917]

Round to the nearest whole number, 35.

Answers: **a.** 11.25π cu. in. **b.** 35 cu. in.

EXERCISES WITH OPEN-RESPONSE PROBLEMS

In 1–5, use the formula $V = \ell wh$ and the given dimensions to find the volume of each rectangular solid.

Rectangular solid

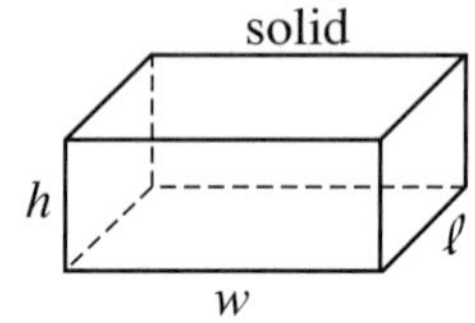

1. $\ell = 5$ ft, $w = 4$ ft, $h = 7$ ft
2. $\ell = 8$ cm, $w = 7$ cm, $h = 5$ cm
3. $\ell = 8.5$ m, $w = 4.2$ m, $h = 6.0$ m
4. $\ell = 2\frac{1}{2}$ in., $w = 8$ in., $h = 5\frac{1}{4}$ in.
5. $\ell = 7.25$ cm, $w = 6.4$ cm, $h = 3.6$ cm

In 6–11, use the formula $V = e^3$ to find the volume of each cube for which edge e is given.

Cube

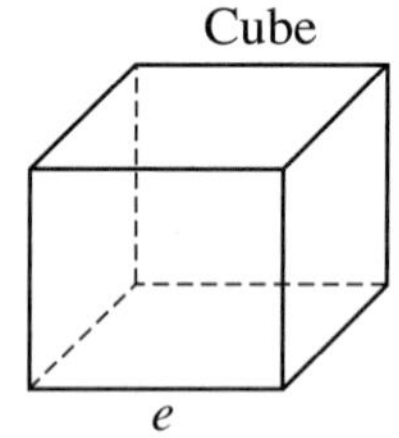

6. 2 yd **7.** 3 m **8.** 8 cm
9. $\frac{1}{3}$ ft **10.** 1.5 in. **11.** 2.76 m

12. Use the formula $V = Bh$ for the volume of a triangular right prism.

a. The base of the right prism is a triangle, with one side measuring 8 centimeters and the altitude to that side measuring 6 centimeters. The height h of the prism is 9.7 centimeters. (1) Find the area B of the base. (2) Find the volume of the triangular right prism.

b. On the triangular base of a right prism, one side = 17 inches, the altitude to that side = 13 inches, and the height of the prism = 30.5 inches. Find the volume of the prism.

Triangular right prism

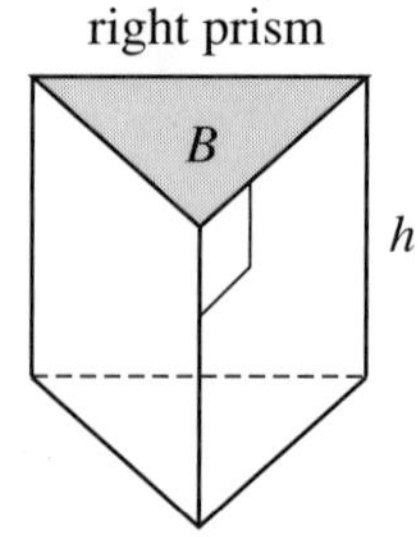

13. Use the formula $V = Bh$ for the volume of a trapezoidal right prism.

a. The base of the right prism is a trapezoid. In this trapezoid, there are two bases, 6 feet and 10 feet long, separated by a height or altitude of 4 feet. The height h of the prism is 12 feet.

(*1*) Find the area B of the trapezoidal base.

(*2*) Find the volume of the trapezoidal right prism.

b. A trapezoidal right prism is 35.7 centimeters high. The base B, which is a trapezoid, has bases of 23 centimeters and 29 centimeters, separated by a height of 18 centimeters. Find the volume of the trapezoidal right prism.

Trapezoidal right prism

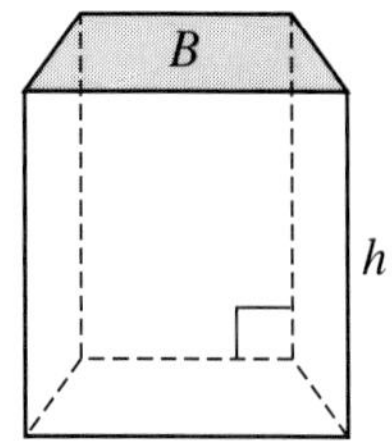

A ***pyramid*** is a solid figure with a base that is a polygon and triangular faces or sides that meet at a common point. The volume of a pyramid is found by using the formula

$$V = \frac{1}{3}Bh$$

In the ***pyramid*** and ***prism*** shown here, the bases are the same in size and shape, and the heights are equal in measure. If the pyramid could be filled with water and poured into the prism, exactly three pyramids of water would be needed to fill the prism. In other words, the volume of a pyramid is one-third the volume of a right prism with the same base and same height as those of the pyramid.

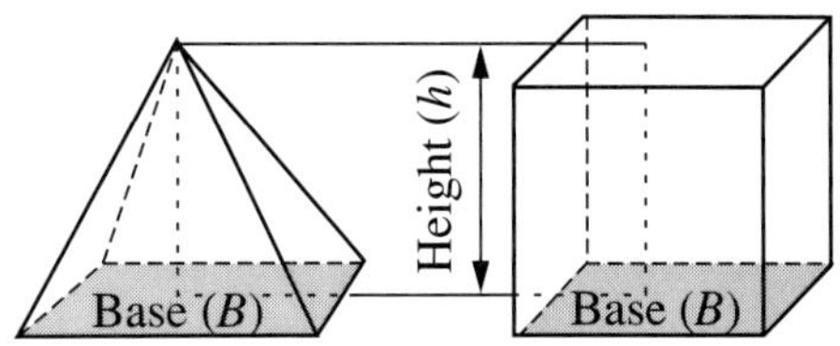

Pyramid = $\frac{1}{3}$ prism

Ex. 14–16

In 14–15, use the formula $V = \frac{1}{3}Bh$ and the given values to find the volume of each pyramid when:

14. The base of the pyramid has an area B of 48 millimeters squared, and a height h of 13 millimeters.

15. The height of the pyramid is 4 inches, and the base is a rectangle 6 inches long and $3\frac{1}{2}$ inches wide.

16. The largest pyramid in the world was built around 2500 B.C. by Khufu, or Cheops, a king of ancient Egypt. The pyramid had a square base, 230 meters (756 ft) on each side, and a height of 147 meters (482 ft). Find the volume of Cheops' pyramid: **a.** in cubic meters **b.** in cubic feet

In 17–21, each algebraic expression represents the length of an edge of a cube. If $x = 3$, find the volume of the cube.

17. x **18.** $x + 2$ **19.** $2x - 5$ **20.** $2(x - 1)$ **21.** $\frac{x}{2}$

n 22–24, the given algebraic expressions represent the dimensions of a rectangular solid. If $x = 4$, l the volume of the rectangular solid.

Length $= x + 2$, width $= x - 1$, height $= 2x$

. Length $= 2x + 1$, width $= \frac{1}{2}x$, height $= \frac{3}{4}x + 1$

4. Length $= x + 0.7$, width $= 7x$, height $= 0.7x$

In 25–28, use the formula $V = \pi r^2h$ to find the volume of each right circular cylinder that has the given dimensions: **a.** in terms of π **b.** to the nearest cubic unit

25. $r = 21$ in., $h = 10$ in. **26.** $r = 28$ ft., $h = \frac{3}{4}$ ft.

27. $r = 10$ cm, $h = 4.2$ cm **28.** $r = 1.4$ m, $h = 6$ m

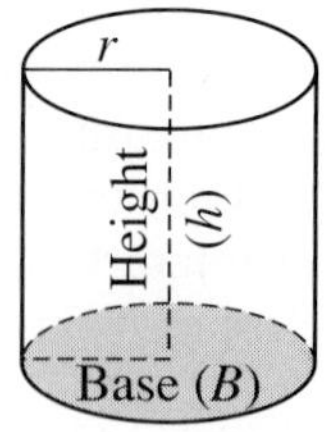

Ex. 25–28

The volume of a ***cone*** is one-third of the volume of a right circular cylinder, when both the cone and the cylinder have circular bases that are equal in measure and heights that are equal in measure.

The formula for the volume of the cone is

$$V = \frac{1}{3}Bh \quad \text{or} \quad V = \frac{1}{3}\pi r^2h$$

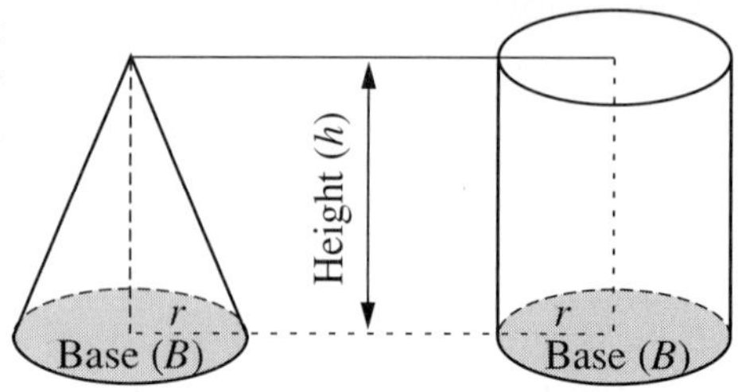

Ex. 29–32

In 29–32, find the volume of each cone that has the given dimensions: **a.** in terms of π **b.** to the nearest cubic unit

29. $r = 10$ in., $h = 12$ in. **30.** $r = 6$ ft., $h = \frac{1}{3}$ ft.

31. $r = 1$ mm, $h = 3$ mm **32.** $r = 2$ cm, $h = 2.4$ cm

A ***sphere*** is a *solid* figure whose points are all equally distant from one fixed point in space called its ***center***.

A sphere is *not* a circle, which is drawn on a flat surface or a *plane*, but similar terminology is used for a sphere:

A line segment, such as $\overline{OC}$, that joins the center O to any point on the sphere is called a ***radius*** of the sphere. A line segment, such as $\overline{AB}$, that joins two points of the sphere and passes through its center is called a ***diameter*** of the sphere.

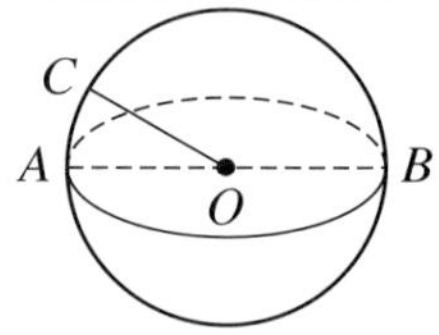

Sphere

Ex. 33–36

The volume of a sphere with radius r is found by using the formula:

$$V = \frac{4}{3}\pi r^3$$

In 33–36, find the volume of each sphere with the given radius r:
a. in terms of π **b.** to the nearest cubic unit

33. $r = 10$ ft. **34.** $r = 0.5$ mm **35.** $r = 9$ in. **36.** $r = 2.6$ cm

37. An official handball has a diameter of 4.8 centimeters (1.875 inches). Find its volume:
a. to the nearest cubic centimeter **b.** to the nearest cubic inch

38. A tank in the form of a right circular cylinder is used for storing water. It has a diameter of 12 feet and a height of 14 feet. How many gallons of water will it hold? Use $\pi = \frac{22}{7}$. (1 cubic foot contains 7.5 gallons.)

39. A cone and a cylinder each have a radius of 3 centimeters and a height of 6 centimeters. A sphere has a radius of 3 centimeters.
a. Find, in terms of π, the volume of the
(*1*) cone (*2*) sphere (*3*) cylinder
b. Note in the diagram that the cone and sphere can fit exactly inside the cylinder. Describe the relationships found for the volumes of these figures.

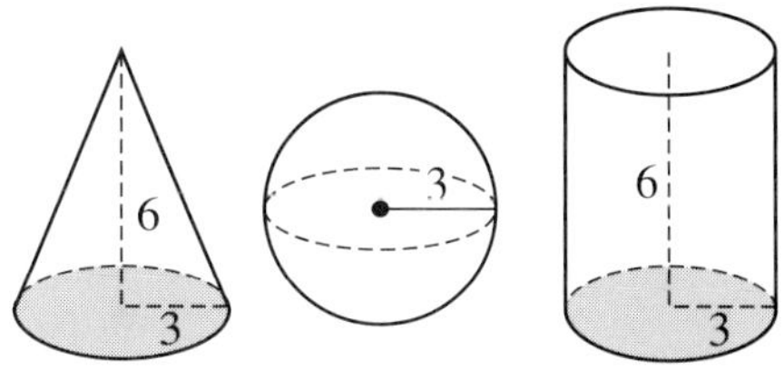

4-10 OPEN SENTENCES AND SOLUTION SETS

In this chapter, you learned how to translate words into formulas and algebraic sentences. These sentences may be equations or inequalities. Every sentence that contains a variable is called an ***open sentence***.

$$x + 6 = 9 \qquad 3y = 12 \qquad 2n > 0 \qquad x + 5 \leq 8$$

An open sentence is neither true nor false.

The sentence will be true or false only when the variables are replaced by numbers from a ***domain*** or a ***replacement set***, such as {0, 1, 2, 3}.

The numbers from the domain that make the sentence *true* are the elements of the ***solution set*** of the open sentence. A solution set, as seen below, can contain one or more numbers or, at times, no numbers at all, from the replacement set.

EXAMPLES

1. Using the domain {0, 1, 2, 3}, find the solution set of each open sentence:

a. $x + 6 = 9$

Procedure: Replace x in the open sentence with numbers from the domain {0, 1, 2, 3}.

b. $2n > 0$

Procedure: Replace n in the open sentence with numbers from the domain {0, 1, 2, 3}.

Solution

$$x + 6 = 9$$

Let $x = 0$. Then $0 + 6 = 9$ is a false sentence.
Let $x = 1$. Then $1 + 6 = 9$ is a false sentence.
Let $x = 2$. Then $2 + 6 = 9$ is a false sentence.
Let $x = 3$. Then $3 + 6 = 9$ is a *true sentence*.
Here, only when $x = 3$ does the open sentence become a true sentence.

Answer: Solution set = {3}

Solution

$$2n > 0$$

If $n = 0$, then $2(0) > 0$, or
$0 > 0$. False

If $n = 1$, then $2(1) > 0$, or
$2 > 0$. *True*
If $n = 2$, then $2(2) > 0$, or
$4 > 0$. *True*
If $n = 3$, then $2(3) > 0$, or
$6 > 0$. *True*
Here, three elements of the domain result in true sentences.

Answer: Solution set = {1, 2, 3}

---

2. Find the solution set for the open sentence $3y = 12$ using:

a. domain = {3, 5, 7}

Procedure: Replace y with 3, 5, and 7.

Solution

$$3y = 12$$

If $y = 3$, then $3(3) = 12$, or
$9 = 12$. False
If $y = 5$, then $3(5) = 12$, or
$15 = 12$. False
If $y = 7$, then $3(7) = 12$, or
$21 = 12$. False
When y is replaced by each of the numbers in this domain, no true sentence is found. The solution set is the ***empty set*** or the ***null set***, written in symbols as { } or ∅.

Answer: Solution set = { } or ∅

b. domain = whole numbers

Procedure: Of course, you cannot replace y with every whole number, but you can use multiplication facts learned previously.

Solution

You know: $3(4) = 12$
Let $y = 4$. Then $3(4) = 12$ is *true*.
No other whole number would make the open sentence $3y = 12$ a true sentence.

Answer: Solution set = {4}

EXERCISES

In 1–6, tell whether each sentence is an open sentence.

1. $2 + 3 = 5 + 0$ **2.** $x + 10 = 14$ **3.** $y - 4 = 12$
4. $3 + 2 < 10 \times 0$ **5.** $n > 7$ **6.** $r < 5 + 2$

In 7–9, name the variable in each equality.

7. $x + 5 = 9$ 8. $4y = 20$ 9. $r - 6 = 12$

In 10–21, use the domain {0, 1, 2, 3, 4, 5} to find all the replacements that will change each open sentence to a true sentence. If no replacement will make a true sentence, write *None*.

10. $n + 3 = 7$ 11. $5 - n = 2$ 12. $5z = 0$ 13. $2m = 7$

14. $x - x = 0$ 15. $n > 2$ 16. $n + 3 > 9$ 17. $2n + 1 < 8$

18. $\frac{n + 1}{2} = 2$ 19. $\frac{2n + 1}{3} = 4$ 20. $\frac{n}{4} > 1$ 21. $\frac{3x}{2} < x$

In 22–29, using the replacement set {1, 2, 3, 4, 5, 6, 7, 8, 9, 10}, find the solution set for each open sentence.

22. $x + 6 = 9$ 23. $8 - x = 5$ 24. $2x + 1 = 24$ 25. $16 = 18 - x$

26. $y > 9$ 27. $4 < m$ 28. $2m > 17$ 29. $2x - 1 > 50$

In 30–37, using the domain $\{2, 2\frac{1}{2}, 3, 3\frac{1}{2}, 4, 4\frac{1}{2}\}$, find the solution set for each open sentence.

30. $x + 2 = 4\frac{1}{2}$ 31. $2x = 7$ 32. $5 - r = \frac{1}{2}$ 33. $\frac{x}{2} = 2.25$

34. $y > 4$ 35. $m < 3$ 36. $2x > 8$ 37. $3a < 4.5$

In 38–41, using the domain {2.1, 2.2, 2.3, 2.4, 2.5}, find the solution set for each open sentence.

38. $x + 0.1 = 2.4$ 39. $3x - 4 = 2.3$ 40. $\frac{y}{2} = 3.6$ 41. $2x + 3 < 6.5$

In 42–49, write the solution set for each open sentence using the domain of whole numbers.

42. $4 + x = 20$ 43. $4x = 20$ 44. $x - 4 = 20$ 45. $x < 4$

46. $4x \leq 20$ 47. $6x = 20$ 48. $6x < 20$ 49. $x - 20 = 6$

50. Find the solution set of $2n = 5$ using: **a.** domain = {1, 2, 3} **b.** domain = {1, 1.5, 2, 2.5, 3} **c.** domain = whole numbers **d.** domain = rational numbers **e.** domain = real numbers

CHAPTER SUMMARY

An ***algebraic expression***, such as $x + 6$, is an expression or a phrase that contains one or more variables, such as x. The ***variable*** is a placeholder for numbers. To evaluate an expression, replace the variable with a number and follow the order of operations.

A ***term*** is a number, a variable, or any product or quotient of numbers and variables. In the term $6by$, the ***coefficient*** of y is $6b$, and 6 by itself is the ***numerical coefficient***. In the term n^3, the ***base*** is n, the ***exponent*** is 3, and the ***power*** is n^3, which is equal to a product in which the base n is a factor 3 times.

A ***formula*** is a sentence that shows the relationship between two or more variables. Many formulas are used to measure geometric figures:

Perimeter is the length around a plane figure. In a polygon, we add the lengths of the sides. In a circle, the perimeter is called the ***circumference***, C, and is found by using the formula $C = \pi d$, or $C = 2\pi r$.

Area is the number of square units contained in a plane figure.

In a rectangle, parallelogram, and rhombus, area $A = bh$. In a square, $A = s^2$.

In a trapezoid, $A = \frac{1}{2}h(b + c)$. In a triangle, $A = \frac{1}{2}bh$. In a circle, $A = \pi r^2$.

Volume is the number of cubic units contained in a solid figure. In a right prism and right circular cylinder, with a base whose area is B with a height h, the volume $V = Bh$.

Thus, in a rectangular solid $V = \ell wh$, and in a right circular cylinder $V = \pi r^2 h$.

In a pyramid, $V = \frac{1}{3}Bh$. In a cone, $V = \frac{1}{3}\pi r^2 h$. In a sphere, $V = \frac{4}{3}\pi r^3$.

An ***open sentence***, which can be an equation or an inequality, contains a variable. When the variable is replaced by numbers from a ***domain***, the numbers that make the open sentence true are the elements of the ***solution set*** of the sentence.

VOCABULARY

4-1 Variable Placeholder Domain Algebraic expression

4-3 Term Factors Coefficient Base Exponent Power

4-4 Evaluating an algebraic expression

4-5 Formula

4-6 Perimeter Area Plane figure Polygon Rectangle Parallelogram Rhombus Square Triangle Trapezoid

4-7 Circle Radius Diameter Chord Circumference

4-9 Volume Right prism Right circular cylinder Rectangular solid Pyramid Cone Sphere

4-10 Open sentence Solution Set

REVIEW EXERCISES WITH OPEN-RESPONSE PROBLEM

In 1–4, use mathematical symbols to translate the verbal phrases into algebraic language.

1. x divided by b

2. 4 less than r

3. q decreased by d

4. 3 more than twice g

5. Nancy and Gary scored a total of 30 points in a game. If Gary scored x points, represent Nancy's score in terms of x.

In 6–13, find the value of each expression when $a = 10$, $b = 8$, $c = 5$, and $d = 24$.

6. $ac - d$

7. $4c^2$

8. $3b + c$

9. $d - b + c$

10. $\frac{bc}{a}$

11. $2a^2 - 2a$

12. $(2a)^2 - 2a$

13. $a(b - c)$

14. If $p = 5$ and $q = 1$, find the value of $(p - q)^3$.

15. Write a formula to express the number of grams g in k kilograms.

16. If P represents the perimeter of an equilateral triangle, represent the length of one of its sides in terms of P.

17. If the base of a triangle is 8 centimeters and its height is 12 centimeters, find the number of square centimeters in the area of the triangle.

18. The lengths, in inches, of the sides of a triangle are represented by y, $y + 3$, and $2y - 2$. Find the perimeter of the triangle when $y = 7$.

19. The length and width, in meters, of a rectangle can be represented by $2x$ and $x - 5$. Find the area of the rectangle when $x = 12$.

In 20–27, using the domain $\{1, 3, 5, 7, 9\}$, find the solution set for each open sentence.

20. $2a < 8$

21. $3b > 21$

22. $\frac{x}{5} < 2$

23. $5 < 12 - y$

24. $3n = 8$

25. $3n = 24$

26. $n + 3 = 8$

27. $n + 3 \geq 8$

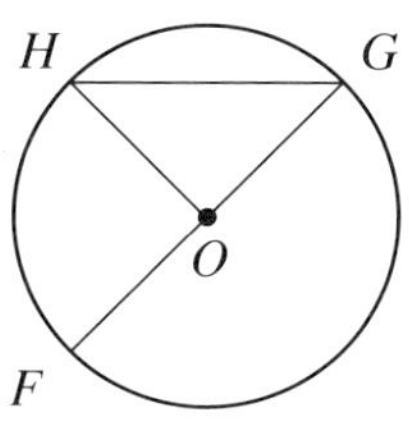

28. In the figure, F, G, and H are points on circle O. Name all segments that are:

a. radii

b. diameters

c. chords

29. If the circumference of a circle is 28π, find the radius of the circle.

30. The diameter of a half-dollar is 30 millimeters. Find the area, in square millimeters, of one face of the coin, expressed:

a. in terms of π **b.** to the nearest whole number

31. The formula for the volume V of a right circular cylinder is $V = \pi r^2 h$. Find the volume, in cubic inches, of a cylinder with a radius $r = 10$ inches and a height $h = 25$ inches:

a. in terms of π **b.** to the nearest whole number of cubic inches

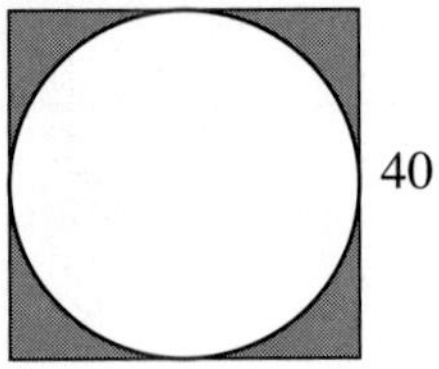

32. A circle is drawn in a square whose side has a length of 40. Expressing each answer in terms of π, find:

a. the circumference of the circle
b. the area of the circle
c. the area of the shaded region

In 33–37, select the numeral preceding the correct answer.

33. What is the total number of cents in n nickels and q quarters?
(1) $n + q$ (2) $n + 5q$ (3) $5n + 25q$ (4) $30nq$

34. If the perimeter of a square is $4y$, then its area is represented as
(1) $4y$ (2) y^2 (3) $4y^2$ (4) $16y^2$

35. In the term $2xy^3$, the base that is used 3 times as a factor is
(1) y (2) xy (3) $2xy$ (4) $2x$

36. One member of the solution set of $19 < y \leq 29$ is
(1) 9 (2) 19 (3) 29 (4) 39

37. What is the radius of a circle whose area is 144π?
(1) 6 (2) 12 (3) 24 (4) 72

CUMULATIVE REVIEW WITH OPEN-RESPONSE PROBLEMS

In 1–6, name the property of real numbers that justifies each statement.

1. $2 + (-3) = -3 + 2$
2. $-3 + 0 = -3$
3. $-5(-1 + 7) = -5(-1) + (-5)(7)$
4. $-3(0) = 0$
5. $(-4 \times -2) \times 8 = -4 \times (-2 \times 8)$
6. $-3 + (-1)$ is a real number

7. A list of numbers that follows a pattern begins with the numbers 2, 5, 8, 11 . . .
a. Find the next number in the list.
b. Write a rule or explain how the next number is determined
c. What is the 25th number in the list?

8. Which of the numbers given below is different from the others? Explain why it is different.

6 3 35 9

9. Three apples cost as much as two oranges. One orange costs the same as a kiwi and an apple. How many apples cost the same as four kiwis?

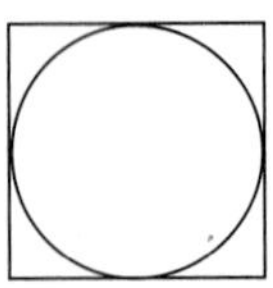

Exploration

The figure at the left shows a circle inscribed in a square. Explain how the figure shows that $\pi r^2 < (2r)^2$.

Chapter 5

Signed Numbers

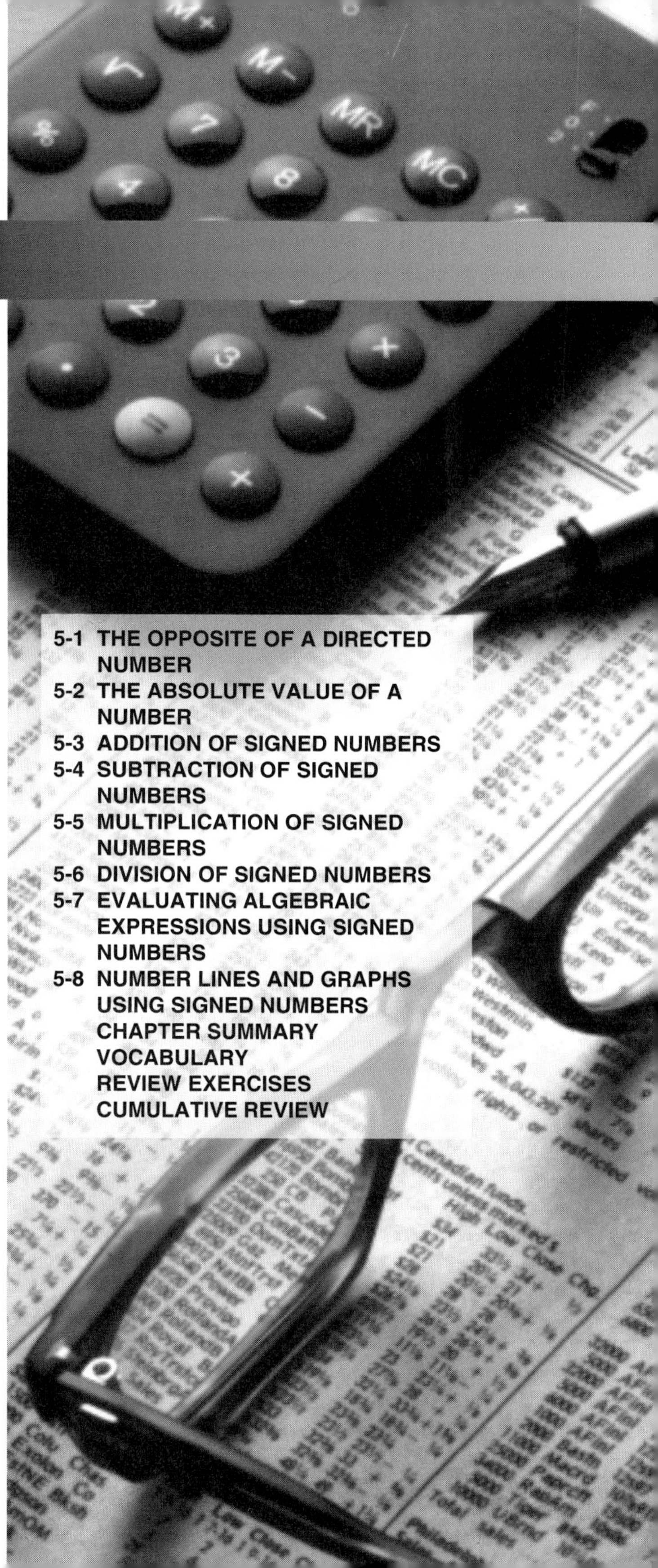

Change is always taking place in our world, and comparisons are constantly being made to monitor that change. Positive and negative numbers are used to represent both the amount of change and the direction in which the change took place.

In a golf tournament, performance is often recorded in terms of number of strokes above or below par.

The stock market records the high and low price of a share of stock for the day, as well as the change in price as an increase or a decrease over the closing price of the preceding day.

Geographers measure the elevation of land or of the ocean floor in terms of distance above or below sea level.

The use of positive and negative numbers to distinguish an increase from a decrease, or a value above the norm from one below it, is an efficient way to display and to compute with the numbers that represent these ideas.

5-1 THE OPPOSITE OF A DIRECTED NUMBER

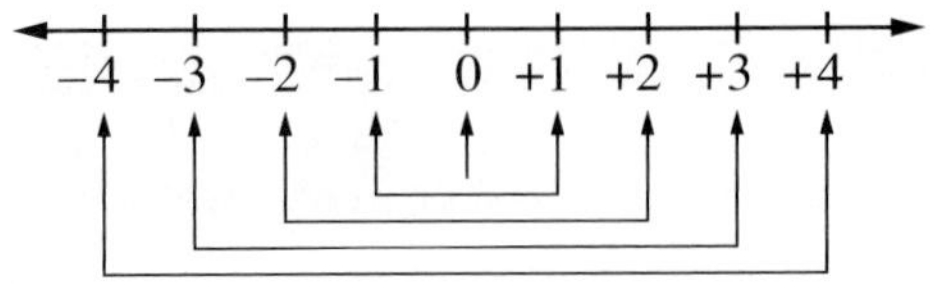

In Chapter 1, we defined the ***opposite*** of a number to be a number that is the same distance from 0 and on the opposite side of 0. The opposite of a positive number is a negative number, and the opposite of a negative number is a positive number. The opposite of 0 is 0.

The opposite of a number or variable is symbolized by a minus sign (−) placed before the number or variable.

$-(+1) = -1$ is read as "the opposite of $+1$ is -1."
$-(-2) = +2$ is read as "the opposite of -2 is $+2$."
$-a$ is read as "the opposite of a."
When a is positive, $-a$ is negative.
When a is negative, $-a$ is positive.

On a scientific calculator, the [+/–] key is used to enter the opposite of a number. A positive number or 0 is entered with no sign. Since a negative number is the opposite of a positive number, to enter -4:

Enter: 4 [+/–]
Display: [–4.]

The opposite of -4 is the opposite of the opposite of 4, which is 4.

Enter: 4 [+/–] [+/–]
Display: [4.]

EXAMPLES

In 1–4: **a.** Show the calculator entry that would be used to find the opposite of each given number. **b.** Write the opposite of the given number in simplest form.

Solutions

	a. Calculator entry	**b.** Opposite of number
1. 15	***Enter:*** 15 [+/–]	-15 *Answer*
2. -10	***Enter:*** 10 [+/–] [+/–]	10 *Answer*
3. $(4 + 8)$	***Enter:*** [( ] 4 [+] 8 [ )] [+/–]	-12 *Answer*
4. $-[-(9 - 3)]$	***Enter:*** [( ] 9 [–] 3 [ )] [+/–] [+/–] [+/–]	-6 *Answer*

EXERCISES

In 1–12, write the simplest symbol that represents the opposite of each number.

1. 8 **2.** -8 **3.** $+3\frac{1}{2}$ **4.** -6.5
5. $(10 + 9)$ **6.** $(24 - 10)$ **7.** $(9 - 9)$ **8.** 8×0
9. $-(-7)$ **10.** $-\left(-\frac{3}{4}\right)$ **11.** $-[-(+5)]$ **12.** $-[-(6 + 8)]$

In 13–16, select, in each case, the greater of the two numbers.

13. $10, -5$ **14.** $-1, 7$ **15.** $-8, -4$ **16.** $-12, 0$

In 17–25, tell whether each statement is true or false.

17. If a is a real number, then $-a$ is always a negative number.
18. If a is a negative number, then $-a$ is always a positive number.
19. The opposite of a number is always a different number.
20. On a standard horizontal number line, the opposite of a positive number is to the left of the number.
21. On a standard horizontal number line, the opposite of any number is always to the left of the number.
22. If x is a positive number, then x is greater than its opposite.
23. The opposite of the opposite of a number is that number itself.
24. If $x > 0$, then $-x < 0$. **25.** If $x < 0$, then $-x > 0$.

5-2 THE ABSOLUTE VALUE OF A NUMBER

In every pair of nonzero opposites, the positive number is the greater. On a standard horizontal number line, the positive number is always to the right of the negative number that is its opposite. For example, 10 is greater than its opposite -10; on a number line, 10 is to the right of -10.

The greater of a nonzero number and its opposite is called the ***absolute value*** of the number. The absolute value of 0 is 0.

The absolute value of a number, a, is symbolized as $|a|$. Since 10 is the greater of the two numbers 10 and -10, the absolute value of 10 is 10 and the absolute value of -10 is 10.

$$|10| = 10$$
$$|-10| = 10$$
$$|10| = |-10|$$

- **The absolute value of a positive number is the number itself; the absolute value of a negative number is the opposite of the number.**

The absolute value of a number can also be thought of as the distance between 0 and the graph of that number on the real number line. For example, $|3| = 3$, the distance between 0 and P, the graph of 3 on the real number line shown below; $|-3| = 3$, the distance between 0 and S, the graph of -3 on the real number line.

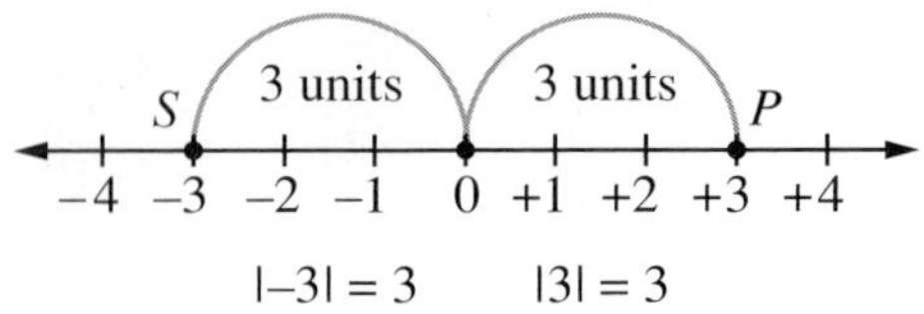

Observe, too, that the absolute value of any real number x is the maximum of the number and its opposite, symbolized as follows:

$$|x| = x \max (-x)$$

EXAMPLES

1. Find the value of the expression $|12| + |-3|$.

Solution Since $|12| = 12$, and $|-3| = 3$, $|12| + |-3| = 12 + 3 = 15$ *Answer*

---

2. Find the value of the expression $|12 - 3|$.

Solution First, evaluate the expression inside the absolute value symbol since this is a grouping symbol. Then, find the absolute value.

$|12 - 3| = |9|$, and $|9| = 9$ *Answer*

EXERCISES

In 1–10: **a.** Give the absolute value of each given number. **b.** Give another number that has the same absolute value.

1. 3 **2.** -5 **3.** $+18$ **4.** -13 **5.** -20

6. $1\frac{1}{2}$ **7.** $-3\frac{3}{4}$ **8.** $-1\frac{1}{2}$ **9.** $+2.7$ **10.** -1.4

In 11–18, state whether each sentence is true or false.

11. $|20| = 20$ **12.** $|-13| = 13$ **13.** $|-15| = -15$ **14.** $|-9| = |9|$

15. $|-7| < |7|$ **16.** $|-10| > |3|$ **17.** $|8| < |-19|$ **18.** $|-21| > 21$

In 19–33, find the value of each expression.

19. $|9| + |3|$ **20.** $|+8| - |+2|$ **21.** $|-6| + |4|$

22. $|-10| - |-5|$ **23.** $|4.5| - |4.5|$ **24.** $|+6| + |-4|$

25. $|8 + 6|$ **26.** $|7 - 2|$ **27.** $|15 - 15|$

28. $|9| + |-3| - |-4|$ **29.** $|-8| - |-2| + |-3|$ **30.** $|10| - |-6| - |4|$

31. $|(8 - 4)| + |-3|$ **32.** $-(|-9| - |7|)$ **33.** $-(|-8| - 2)$

In 34–42, state whether each sentence is true or false.

34. $|+5| - |-5| = 0$ **35.** $|+9| + |-9| = 0$ **36.** $|3| \cdot |-3| = -9$

37. $2 \cdot |-4| = |-2| \cdot |-4|$ **38.** $\frac{|-8|}{|-4|} = -|+2|$ **39.** $|4| \cdot |-2| - \frac{|-16|}{|2|} = 0$

40. $|-6| \cdot |4| > 0$ **41.** $|6| + |-4| < 6 - 4$ **42.** $\frac{|-5|}{5} - \frac{|7|}{7} \leq 0$

5-3 ADDITION OF SIGNED NUMBERS

The number line can be used to find the sum of two number. Start at 0. To add a positive number, move to the right. To add a negative number, move to the left.

EXAMPLES

1. Add: $+3$ and $+2$.

Solution Start at 0 and move 3 units to the right to $+3$; then move 2 more units to the right, arriving at $+5$.

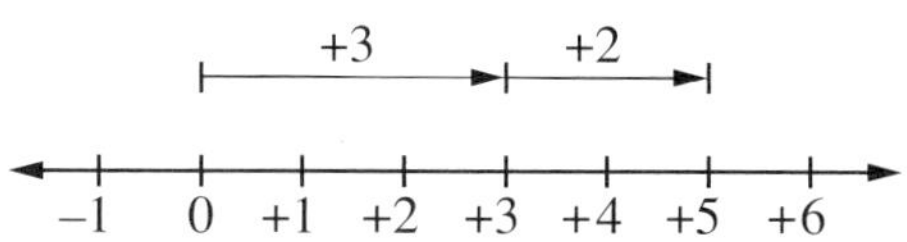

Answer: 5

The sum of two positive integers is the same as the sum of two whole numbers. The sum $+5$ is a number whose absolute value is the sum of the absolute values of $+3$ and $+2$ and whose sign is the same as the sign of $+3$ and $+2$.

Calculator Solution

Enter: 3 [+] 2 [=]

Display: [5.]

Answer: 5

2. Add -3 and -2.

Solution Start at 0 and move 3 units to the left to -3; then move 2 more units to the left, arriving at -5.

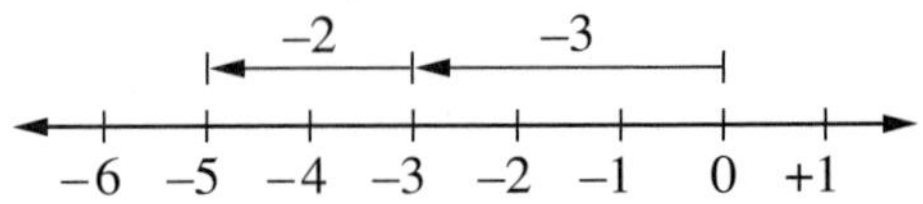

Answer: -5

The sum -5 is a number whose absolute value is the sum of the absolute values of -3 and -2 and whose sign is the same as the sign of -3 and -2.

Calculator Solution

Enter: 3 [+/–] [+] 2 [+/–] [=]

Display: [–5.]

Answer: -5

Examples 1 and 2 illustrate that the sum of two numbers with the same sign is a number whose absolute value is the sum of the absolute values of the numbers and whose sign is the sign of the numbers.

PROCEDURE. To add two numbers that have the same sign:
1. find the sum of the absolute values,
2. give this sum the common sign.

EXAMPLE

3. Add: $+3$ and -3.

Solution Start at 0 and move 3 units to the right to $+3$; then move 3 units to the left, arriving at 0.

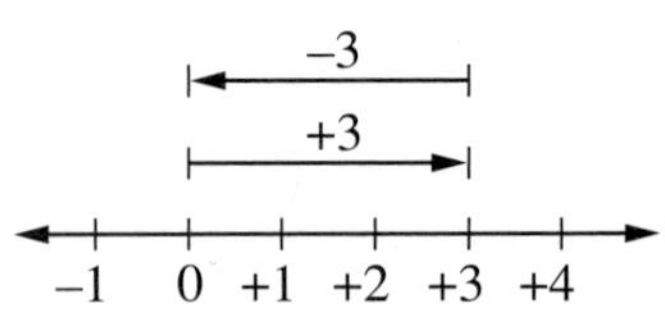

Answer: 0

The sum of a number and its opposite is always 0. The opposite of a number is also called the ***additive inverse*** of the number.

Addition Property of Opposites: For every real number a and its opposite $-a$:

$$\mathbf{a + (-a) = 0}$$

EXAMPLE

4. Add: 0 and -3.

Solution Start at 0 and move neither to the right nor to the left; then move 3 units to the left, arriving at -3.

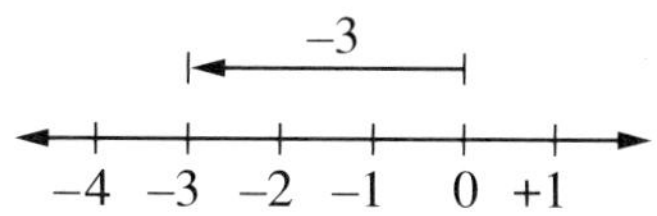

Answer: -3

Addition Property of Zero: For every real number a:

$$\mathbf{a + 0 = 0 + a = a}$$

EXAMPLES

5. Add: $+3$ and -2.

Solution Start at 0 and move 3 units to the right to $+3$; then move 2 units to the left, arriving at $+1$.

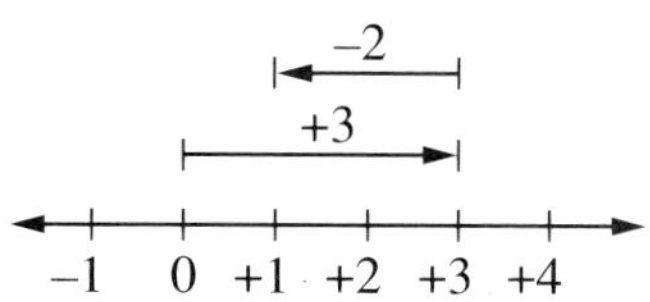

Answer: 1

This sum can also be found by using properties. In the first step, substitution is used, replacing $(+3)$ with the sum $(+1) + (+2)$.

$(+3) + (-2) = [(+1) + (+2)] + (-2)$	Substitution
$= (+1) + [(+2) + (-2)]$	Associative property
$= +1 + 0$	Addition property of opposites
$= +1$	Addition property of zero

The sum $+1$ is a number whose absolute value is the difference of the absolute values of $+3$ and -2 and whose sign is the same as the sign of $+3$, the number with the greater absolute value.

Calculator Solution

Enter: 3 [+] 2 [+/−] [=]

Display: [1.]

Answer: 1

---

6. Add: -3 and $+2$.

Solution Start at 0 and move 3 units to the left to -3; then move 2 units to the right, arriving at -1.

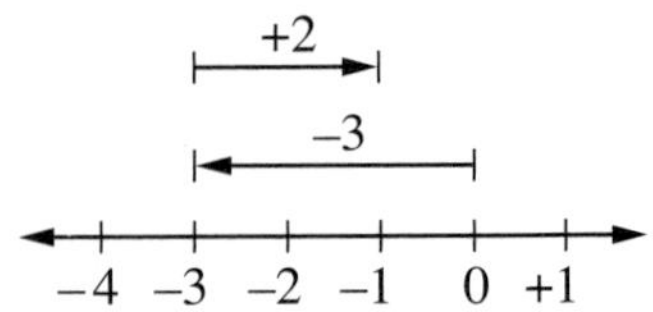

Answer: 1

To find the sum by using properties, replace (-3) with the sum $(-1) + (-2)$. Then:

$(-3) + (+2) = [(-1) + (-2)] + (+2)$	Substitution
$= (-1) + [(-2) + (+2)]$	Associative property
$= -1 + 0$	Addition property of opposites
$= -1$	Addition property of zero

The sum -1 is a number whose absolute value is the difference of the absolute values of -3 and $+2$ and whose sign is the same as the sign of -3, the number with the greater absolute value.

Calculator Solution

Enter: 3 [+/−] [+] 2 [=]

Display: [−1.]

Answer: -1

Examples 5 and 6 illustrate that the sum of a positive number and a negative number is a number whose absolute value is the difference of the absolute values of the numbers and whose sign is the sign of the number having the larger absolute value.

PROCEDURE. To add two numbers that have different signs:

1. find the difference of the absolute values of the numbers.
2. give this difference the sign of the number that has the greater absolute value.
3. The sum is 0 if both numbers have the same absolute value.

EXAMPLE

7. Find the sum of $-3\frac{3}{4}$ and $+1\frac{1}{4}$.

How to Proceed:	*Solution:*
(1) Since the numbers have different signs, find the difference of their absolute values:	*Think:* $3\frac{3}{4} - 1\frac{1}{4} = 2\frac{2}{4}$
(2) Give this difference the sign of the number with the greater absolute value:	*Write:* $-2\frac{2}{4} = -2\frac{1}{2}$

A mixed number such as $3\frac{3}{4}$ is the sum of two numbers: $3\frac{3}{4} = 3 + \frac{3}{4}$

Its opposite, $-3\frac{3}{4}$, is also the sum of two numbers: $-3\frac{3}{4} = (-3) + \left(-\frac{3}{4}\right)$

Calculator Solution When a calculator is used to find the sum, there are two ways to enter $3\frac{3}{4}$.

METHOD 1. Enclose the absolute value of the negative mixed number in parentheses, and then press the [+/–] key.

Enter: [( ] 3 [+] 3 [÷] 4 [ )] [+/–] [+] 1 [+] 1 [÷] 4 [=]

Display: -2.5

METHOD 2. Press the [+/–] key after each part of the negative mixed number.

Enter: 3 [+/–] [+] 3 [÷] 4 [+/–] [+] 1 [+] 1 [÷] 4 [=]

Display: -2.5

Answer: $-2\frac{1}{2}$ or -2.5

The two methods used in Example 7 to enter the negative mixed number $-3\frac{3}{4}$ illustrate that the opposite of a sum of two or more numbers is the sum of the opposites of the numbers.

Property of the Opposite of a Sum: For all real numbers a and b:

$$-(a + b) = (-a) + (-b)$$

When adding more than two signed numbers, the commutative and associative properties allow us to arrange the numbers in any order and group them in any way. It may prove helpful to add positive numbers first, add negative numbers next, and then add the two results.

EXAMPLE

8. Add: $(+6) + (-2) + (+7) + (-4)$.

How to Proceed:	*Solution:*
(1) Write the expression:	$(+6) + (-2) + (+7) + (-4)$
(2) Use the commutative and associative properties and add the positive and negative numbers separately:	$[(+6) + (+7)] + [(-2) + (-4)]$ $\begin{array}{r} +6 \\ \underline{+7} \\ +13 \end{array} \qquad \begin{array}{r} -2 \\ \underline{-4} \\ -6 \end{array}$
(3) Add the positive and negative sums:	$(+13) + (-6) = +7$

Answer: 7

(*Note:* This sum could have been written as $6 - 2 + 7 - 4$. We will agree that this expression will mean the sum of the four signed numbers $+6$, -2, $+7$, and -4.)

EXERCISES WITH OPEN-RESPONSE PROBLEMS

In 1–54, add the numbers in each case. Use a calculator to check your answers.

1. $\begin{array}{r} +6 \\ \underline{+4} \end{array}$ **2.** $\begin{array}{r} +7 \\ \underline{+6} \end{array}$ **3.** $\begin{array}{r} -14 \\ \underline{-23} \end{array}$ **4.** $\begin{array}{r} -17 \\ \underline{-28} \end{array}$ **5.** $\begin{array}{r} +8 \\ \underline{-6} \end{array}$

6. $\begin{array}{r} -9 \\ \underline{+7} \end{array}$ **7.** $\begin{array}{r} +8 \\ \underline{-4} \end{array}$ **8.** $\begin{array}{r} +2 \\ \underline{-9} \end{array}$ **9.** $\begin{array}{r} +23 \\ \underline{-35} \end{array}$ **10.** $\begin{array}{r} +6 \\ \underline{0} \end{array}$

11. $\begin{array}{r} -5 \\ \underline{0} \end{array}$ **12.** $\begin{array}{r} 0 \\ \underline{+4} \end{array}$ **13.** $\begin{array}{r} 0 \\ \underline{-9} \end{array}$ **14.** $\begin{array}{r} +5 \\ \underline{-5} \end{array}$ **15.** $\begin{array}{r} -9 \\ \underline{+9} \end{array}$

16. $\begin{array}{r} +15 \\ \underline{+9} \end{array}$ **17.** $\begin{array}{r} -28 \\ \underline{-38} \end{array}$ **18.** $\begin{array}{r} -15 \\ \underline{-15} \end{array}$ **19.** $\begin{array}{r} +6\frac{2}{3} \\ \underline{+1\frac{1}{3}} \end{array}$ **20.** $\begin{array}{r} -5\frac{1}{2} \\ \underline{-3\frac{1}{2}} \end{array}$

21. $\begin{array}{r} 9\frac{1}{2} \\ \underline{8\frac{3}{4}} \end{array}$ **22.** $\begin{array}{r} -6\frac{5}{6} \\ \underline{-1\frac{2}{3}} \end{array}$ **23.** $\begin{array}{r} -33\frac{1}{3} \\ \underline{+19\frac{2}{3}} \end{array}$ **24.** $\begin{array}{r} -5\frac{3}{4} \\ \underline{8\frac{1}{2}} \end{array}$ **25.** $\begin{array}{r} -5.6 \\ \underline{-2.2} \end{array}$

26. $\begin{array}{r} +5.4 \\ \underline{+2.9} \end{array}$ **27.** $\begin{array}{r} -8.8 \\ \underline{-7.5} \end{array}$ **28.** $\begin{array}{r} 7.9 \\ \underline{-5.6} \end{array}$ **29.** $\begin{array}{r} -6.9 \\ \underline{9.4} \end{array}$ **30.** $\begin{array}{r} +7 \\ \underline{-8\frac{3}{4}} \end{array}$

31. $(+8) + (-14)$ **32.** $(-12) + (+37)$ **33.** $(+40) + (-17)$ **34.** $(-18) + 0$

35. $0 + (-28)$ **36.** $(+15) + (-15)$ **37.** $|-34| + |+20|$ **38.** $-|7| + (-10)$

39. $|-2| + (-2)$ **40.** $|-8| + |+8|$ **41.** $|-9| + (-|9|)$ **42.** $|15| + (-|-15|)$

43. $\begin{array}{r} +27 \\ -9 \\ \underline{-12} \end{array}$ **44.** $\begin{array}{r} -45 \\ +12 \\ \underline{+13} \end{array}$ **45.** $\begin{array}{r} 15 \\ -28 \\ \underline{13} \end{array}$ **46.** $\begin{array}{r} +20 \\ -12 \\ \underline{-8} \end{array}$ **47.** $\begin{array}{r} -1.5 \\ +3.7 \\ \underline{-8.3} \end{array}$ **48.** $\begin{array}{r} 8\frac{1}{2} \\ -4\frac{1}{4} \\ \underline{7\frac{3}{4}} \end{array}$

49. $+18 - 15 + 9$ **50.** $30 - 18 - 12$

51. $-19 + 8 - 15$ **52.** $-17 - 1 + 40$

53. $48 - 32 + 19 - 41$ **54.** $-4\frac{1}{3} + 7 + 8\frac{1}{3} - 11$

In 55–58, find the coordinate of the midpoint of each line segment if points *A* through *E* are points on the number line as shown below.

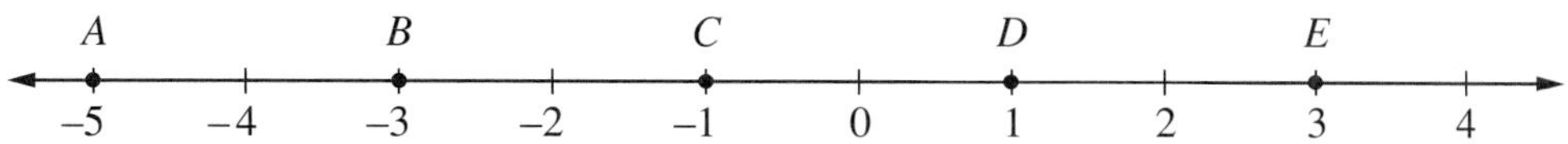

Ex. 55–58

55. $\overline{AC}$ **56.** $\overline{AD}$ **57.** $\overline{BD}$ **58.** $\overline{CE}$

59. The sum of two numbers is positive. One of the numbers is a positive number that is larger than the sum. Is the other number positive or negative? Explain your answer.

60. The sum of two numbers is positive. One of the numbers is negative. Is the other number positive or negative? Explain your answer.

61. The sum of two numbers is negative. One of the numbers is a negative number that is smaller than the sum. Is the other number positive or negative? Explain your answer.

62. The sum of two numbers is negative. One of the numbers is positive. Is the other number positive or negative? Explain your answer.

63. Is it possible for the sum of two numbers to be smaller than either of the numbers? If so, give an example.

In 64–68, use signed numbers to solve the problem.

64. In 1 hour the Celsius temperature rose 4°, and in the next hour it rose 3°. What was the net change in temperature during the 2-hour period?

65. An elevator started on the ground floor and rose 30 floors. Then it came down 12 floors. At which floor was it at that time?

66. A football team gained 7 yards on the first play, lost 2 yards on the second, and lost 8 yards on the third. What was the net result of the three plays?

67. Fay deposited $250 in a bank. During the next month, she made a deposit of $60 and a withdrawal of $80. How much money did Fay have in the bank at the end of that month?

68. During a 4-day period, the dollar value of a share of stock rose $1\frac{1}{2}$ on the first day, dropped $\frac{5}{8}$ on the second day, rose $\frac{1}{8}$ on the third day, and dropped $1\frac{3}{4}$ on the fourth day. What was the net change in the stock during this period?

5-4 SUBTRACTION OF SIGNED NUMBERS

In arithmetic, to subtract 3 from 7, we find the number that, when added to 3, gives 7. We know that $7 - 3 = 4$ because $3 + 4 = 7$. Subtraction is the inverse operation of addition.

- **In general, for every number c and every number b, the expression $c - b$ is the number a such that $b + a = c$.**

We use this definition in order to subtract signed numbers.

To subtract -2 from $+3$, written as $(+3) - (-2)$, we must find a number that, when added to -2, will give $+3$. We write:

$$(-2) + (?) = +3$$

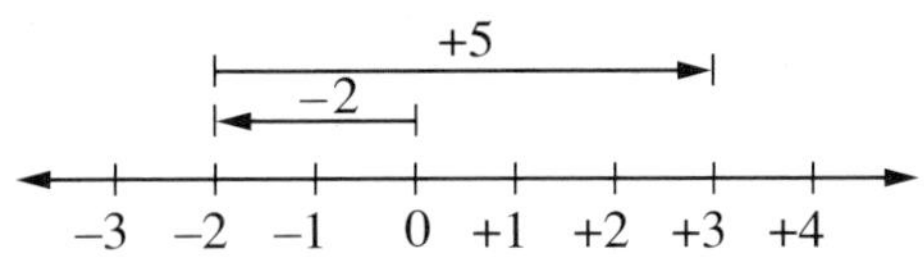

We can use a number line to find the answer to $(-2) + (?) = +3$. We think as follows: From a point 2 units to the left of 0, what motion must be made to arrive at a point 3 units to the right of 0? We must move 5 units to the right. This motion is represented by $+5$.

Therefore, $(+3) - (-2) = +5$ because $(-2) + (+5) = +3$. We can write $(+3) - (-2) = +5$ vertically as follows:

$$\begin{array}{r} (+3) \\ \underline{-(-2)} \\ +5 \end{array} \quad or \quad \textit{Subtract:}\ \begin{array}{rl} (+3) & \text{minuend} \\ \underline{(-2)} & \text{subtrahend} \\ +5 & \text{difference} \end{array}$$

Check each of the following examples by using a number line to answer the related question: subtrahend + (?) = minuend.

$$\textit{Subtract:}\ \begin{array}{r} +9 \\ \underline{+6} \\ +3 \end{array} \quad \begin{array}{r} -7 \\ \underline{-2} \\ -5 \end{array} \quad \begin{array}{r} +5 \\ \underline{-2} \\ +7 \end{array} \quad \begin{array}{r} -3 \\ \underline{+1} \\ -4 \end{array}$$

Now, consider another way in which addition and subtraction are related. In each of the following examples, compare the result obtained when a signed number is subtracted with the result obtained when the opposite of that signed number is added.

Subtract	*Add*	*Subtract*	*Add*	*Subtract*	*Add*	*Subtract*	*Add*
$+9$	$+9$	-7	-7	$+5$	$+5$	-3	-3
$\underline{+6}$	$\underline{-6}$	$\underline{-2}$	$\underline{+2}$	$\underline{-2}$	$\underline{+2}$	$\underline{+1}$	$\underline{-1}$
$+3$	$+3$	-5	-5	$+7$	$+7$	-4	-4

Observe that, in each example, adding the opposite (the additive inverse) of a signed number gives the same result as subtracting that signed number. It therefore seems reasonable to define subtraction as follows:

If a is any signed number and b is any signed number, then:

$$\boldsymbol{a - b = a + (-b)}$$

PROCEDURE. To subtract one signed number from another, add the opposite (additive inverse) of the subtrahend to the minuend.

Uses of the Symbol −

We have used the symbol − in three ways:

- To indicate that a number is negative:

  -2 Negative 2

- To indicate the opposite of a number:

  $-(-4)$ Opposite of negative 4
  $-a$ Opposite of a

- To indicate subtraction:

  $4 - (-3)$ Difference between 4 and -3

EXAMPLES

1. Perform the indicated subtractions.

	Solutions and answers:
a. $(+30) - (+12)$	$(-30) + (-12) = +18$
b. $(-19) - (-7)$	$(-19) + (+7) = -12$
c. $(-4) - (0)$	$(-4) + (0) = -4$
d. $0 - 8$	$0 + (-8) = -8$

2. How much greater than -3 is 9?

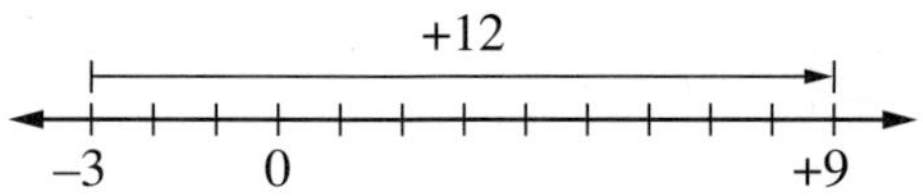

Solution $9 - (-3) = 9 + 3 = 12$

Answer: 9 is 12 greater than -3.

---

3. From the sum of -2 and $+8$, subtract -5.

Solution

$$(-2 + 8) - (-5) = +6 + (+5) = +11$$

Calculator Solution

Enter: 2 [+/–] [+] 8 [–] 5 [+/–] [=]

Display: [11.]

Answer: 11

---

4. Subtract the sum of -7 and $+2$ from -4.

Solution

$$-4 - (-7 + 2) = -4 - (-5) = -4 + (+5) = +1$$

Calculator Solution

Enter: 4 [+/–] [–] [(] 7 [+/–] [+] 2 [)] [=]

Display: [1.]

Answer: 1

Note that parentheses need not be entered in the calculator in Example 3 since the additions and subtractions are to be done in the order in which they occur in the expression. Parentheses are needed, however, in Example 4 since the sum -7 and $+2$ is to be found first, before subtracting this sum from -4.

EXERCISES

In 1–36, subtract, in each case, the lower number from the upper number. Check your answers using a calculator.

1. $\begin{array}{r}+9\\ \underline{+3}\end{array}$	**2.** $\begin{array}{r}+25\\ \underline{+18}\end{array}$	**3.** $\begin{array}{r}+6\\ \underline{+8}\end{array}$	**4.** $\begin{array}{r}+16\\ \underline{+24}\end{array}$	**5.** $\begin{array}{r}-7\\ \underline{-3}\end{array}$	**6.** $\begin{array}{r}-26\\ \underline{-12}\end{array}$
7. $\begin{array}{r}-2\\ \underline{-6}\end{array}$	**8.** $\begin{array}{r}-34\\ \underline{-50}\end{array}$	**9.** $\begin{array}{r}+5\\ \underline{-1}\end{array}$	**10.** $\begin{array}{r}+26\\ \underline{-19}\end{array}$	**11.** $\begin{array}{r}+4\\ \underline{-9}\end{array}$	**12.** $\begin{array}{r}+65\\ \underline{-75}\end{array}$
13. $\begin{array}{r}-6\\ \underline{+2}\end{array}$	**14.** $\begin{array}{r}-63\\ \underline{+29}\end{array}$	**15.** $\begin{array}{r}-5\\ \underline{+7}\end{array}$	**16.** $\begin{array}{r}-36\\ \underline{+50}\end{array}$	**17.** $\begin{array}{r}+10\\ \underline{0}\end{array}$	**18.** $\begin{array}{r}0\\ \underline{+7}\end{array}$
19. $\begin{array}{r}0\\ \underline{-20}\end{array}$	**20.** $\begin{array}{r}-19\\ \underline{-19}\end{array}$	**21.** $\begin{array}{r}+18\\ \underline{+29}\end{array}$	**22.** $\begin{array}{r}+15\\ \underline{+15}\end{array}$	**23.** $\begin{array}{r}+36\\ \underline{-15}\end{array}$	**24.** $\begin{array}{r}-39\\ \underline{+15}\end{array}$
25. $\begin{array}{r}-45\\ \underline{+17}\end{array}$	**26.** $\begin{array}{r}-6\\ \underline{+6}\end{array}$	**27.** $\begin{array}{r}-8\\ \underline{-8}\end{array}$	**28.** $\begin{array}{r}0\\ \underline{-15}\end{array}$	**29.** $\begin{array}{r}+8.7\\ \underline{+6.5}\end{array}$	**30.** $\begin{array}{r}+8.3\\ \underline{-6.2}\end{array}$
31. $\begin{array}{r}-6.9\\ \underline{+3.7}\end{array}$	**32.** $\begin{array}{r}5.9\\ \underline{7.2}\end{array}$	**33.** $\begin{array}{r}+9\frac{1}{2}\\ \underline{+6\frac{1}{2}}\end{array}$	**34.** $\begin{array}{r}-3\frac{1}{4}\\ \underline{-7\frac{3}{4}}\end{array}$	**35.** $\begin{array}{r}7\frac{3}{4}\\ \underline{-2\frac{1}{4}}\end{array}$	**36.** $\begin{array}{r}-6\frac{5}{6}\\ \underline{+3\frac{1}{3}}\end{array}$

In 37–42, perform each indicated subtraction.

37. $(+19) - (+30)$ **38.** $(-12) - (-25)$ **39.** $22 - (-8)$
40. $(+6.4) - (+8.1)$ **41.** $(-3.7) - (-5.2)$ **42.** $(-9.2) - 8.3$

43. How much is 18 decreased by -7?
44. How much greater than -15 is 12?
45. How much greater than -4 is -1?
46. How much less than 6 is -3?
47. What number is 6 less than -6?
48. From the sum of 25 and -10, subtract -4.
49. Subtract 8 from the sum of -6 and -12.
50. From -2 subtract the sum of -8 and 3.

In 51–56, state the number that must be added to each given number to make the result equal to 0.

51. $+5$ **52.** -3 **53.** $+8.5$ **54.** -3.7 **55.** $+1\frac{7}{8}$ **56.** $-\frac{9}{2}$

In 57–62, find the value of each given expression.

57. $(+7) + (+9) - (-4)$ **58.** $(-12) - (+9) + (-20)$ **59.** $32 - 49 - 21 + 10$
60. $-15 + 8 - 5 + 12$ **61.** $6\frac{1}{4} - 5 + 7\frac{3}{4} - 1\frac{1}{2}$ **62.** $-5\frac{1}{3} + 8 + 9\frac{1}{3} - 12$

In 63–71, use signed numbers to do the problems.

63. Find the change when the Celsius temperature changes from:
a. $+5°$ to $+8°$ **b.** $-10°$ to $+18°$ **c.** $-6°$ to $-18°$ **d.** $+12°$ to $-4°$

64. Find the change in altitude when you go from a place that is 15 meters below sea level to a place that is 95 meters above sea level.

65. In a game, Sid was 35 points "in the hole." How many points must he make in order to have a score of 150 points?

66. The record high Fahrenheit temperature in New City is 105°; the record low is $-9°$. Find the difference between these temperatures.

67. At one point, the Pacific Ocean is 0.5 kilometer in depth; at another point it is 0.25 kilometer in depth. Find the difference between these depths.

68. State whether each of the following sentences is true or false:
a. $(+5) - (-3) = (-3) - (+5)$ **b.** $(-7) - (-4) = (-4) - (-7)$

69. If x and y represent numbers:
a. Does $x - y = y - x$ for all replacements of x and y?
b. Does $x - y = y - x$ for any replacements of x and y? For which values of x and y?
c. What is the relation between $x - y$ and $y - x$ for all replacements of x and y?
d. Is the operation of subtraction commutative? In other words, for all signed numbers x and y, does $x - y = y - x$?

70. State whether each of the following sentences is true or false:
a. $(15 - 9) - 6 = 15 - (9 - 6)$
b. $[(-10) - (+4)] - (+8) = (-10) - [(+4) - (+8)]$

71. Is the operation of subtraction associative? In other words, for all signed numbers x, y, and z, does $(x - y) - z - (y - z)$?

In 72–77, find the length of each line segment if A through E are points on the number line as shown below.

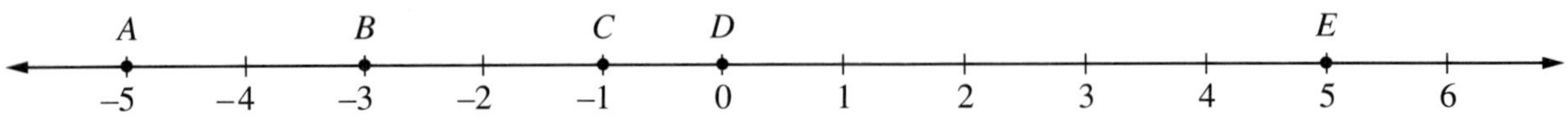

72. $\overline{AC}$ **73.** $\overline{BD}$ **74.** $\overline{CE}$ **75.** $\overline{AE}$ **76.** $\overline{BC}$ **77.** $\overline{DE}$

5-5 MULTIPLICATION OF SIGNED NUMBERS

Four Possible Cases in the Multiplication of Signed Numbers

We will use a common experience to illustrate the various cases that can arise in the multiplication of signed numbers. We will represent a gain in weight by a positive number and a loss of weight by a negative number, and represent a number of weeks in the future by a positive number and a number of weeks in the past by a negative number.

Case 1. Multiplying a Positive Number by a Positive Number
If a girl gains 1 kilogram each week, 4 weeks from now she will be 4 kilograms heavier than she is now. Using signed numbers, we may write:

$$(+4) \bullet (+1) = +4$$

The product of the two positive numbers is a positive number.

Case 2. Multiplying a Negative Number by a Positive Number
If a girl loses 1 kilogram each week, 4 weeks from now she will be 4 kilograms lighter than she is now. Using signed numbers, we may write:

$$(+4) \bullet (-1) = -4$$

The product of the negative number and the positive number is a negative number.

Case 3. Multiplying a Positive Number by a Negative Number

If a girl has gained 1 kilogram each week, 4 weeks ago she was 4 kilograms lighter than she is now. Using signed numbers, we may write:

$$(-4) \cdot (+1) = -4$$

The product of the positive number and the negative number is a negative number.

Case 4. Multiplying a Negative Number by a Negative Number

If a girl has lost 1 kilogram each week, 4 weeks ago she was 4 kilograms heavier than she is now. Using signed numbers, we may write:

$$(-4) \cdot (-1) = +4$$

The product of the two negative numbers is a positive number.

In all four cases, the absolute value of the product, 4, is equal to the product of the absolute values of the factors, 4 and 1.

These examples illustrate the reasonableness of the following rules.

Rules for Multiplying Signed Numbers

- **Rule 1.** The product of two positive numbers or of two negative numbers is a positive number whose absolute value is the product of the absolute values of the numbers.
- **Rule 2.** The product of a positive number and a negative number is a negative number whose absolute value is the product of the absolute values of the numbers.

In general, if a and b are both positive or are both negative, then:

$$ab = |a| \cdot |b|$$

If one of the numbers a and b is positive and the other is negative, then:

$$ab = -(|a| \cdot |b|)$$

In effect, these rules tell us that:

PROCEDURE. To multiply two signed numbers:

1. Find the product of the absolute values.
2. Write a plus sign before this product when the two numbers have the same sign.
3. Write a minus sign before this product when the two numbers have different signs.

EXAMPLES

1. Multiply the two numbers in each set.

Solutions

	a.	**b.**	**c.**	**d.**	**e.**	**f.**
	$+12$	-13	$+18$	-15	$+3.4$	$-7\frac{1}{8}$
	$+4$	-5	-3	6	-3	-3
	$+48$	$+65$	-54	-90	-10.2	$+21\frac{3}{8}$ *Answers*

---

2. Find the value of $(+5)[(-17)(-2)]$.

Solution

$$\begin{aligned}(+5)[(-17)(-2)] &= (+5)[(-2)(-17)] && \text{Commutative property of multiplication}\\ &= [(+5)(-2)](-17) && \text{Associative property of multiplication}\\ &= (-10)(-17)\\ &= +170 \quad \textit{Answer}\end{aligned}$$

---

3. Use the distributive property of multiplication over addition to find the product $8(-72)$.

Solution

$$\begin{aligned}8(-72) &= 8[(-70) + (-2)]\\ &= 8(-70) + 8(-2)\\ &= (-560) + (-16)\\ &= -576 \quad \textit{Answer}\end{aligned}$$

---

4. Find the value of $(-2)^3$.

Solution

$$(-2)^3 = (-2)(-2)(-2) = -8 \quad \textit{Answer}$$

(*Note:* The answer is negative because there is an odd number (3) of negative factors.)

---

5. Find the value of $(-3)^4$.

Solution

$$(-3)^4 = (-3)(-3)(-3)(-3) = +81 \quad \textit{Answer}$$

(*Note:* The answer is positive because there is an even number (4) of negative factors.)

In this example, the value of $(-3)^4$ was found to be $+81$. This is not equal to -3^4, which is the opposite of 3^4 or $-1(3^4)$. To find the value of -3^4, first find the value of 3^4, which is 81, and then write the opposite of 81, -81. Thus, $(-3)^4 = 81$ and $-3^4 = -81$.

On a calculator, evaluate $(-3)^4$.

Enter: 3 [+/−] [y^x] 4 [=]

Display: [81.]

On a calculator, evaluate -3^4.

Enter: METHOD 1: 3 [y^x] 4 [=] [+/−]

METHOD 2: 1 [+/−] [×] 3 [y^x] 4 [=]

Display: [−81.]

EXERCISES

In 1–27, find the product of each set of numbers.

1. $\begin{array}{r} +6 \\ \underline{+4} \end{array}$ **2.** $\begin{array}{r} -7 \\ \underline{-1} \end{array}$ **3.** $\begin{array}{r} -15 \\ \underline{-7} \end{array}$ **4.** $\begin{array}{r} -8 \\ \underline{+4} \end{array}$ **5.** $\begin{array}{r} +15 \\ \underline{-8} \end{array}$ **6.** $\begin{array}{r} 24 \\ \underline{-6} \end{array}$

7. $\begin{array}{r} -12 \\ \underline{0} \end{array}$ **8.** $\begin{array}{r} -25 \\ \underline{-4} \end{array}$ **9.** $\begin{array}{r} -24 \\ \underline{+8} \end{array}$ **10.** $\begin{array}{r} 0 \\ \underline{-5} \end{array}$ **11.** $\begin{array}{r} -75 \\ \underline{-3} \end{array}$ **12.** $\begin{array}{r} 15 \\ \underline{-9} \end{array}$

13. $\begin{array}{r} +1.5 \\ \underline{-2.4} \end{array}$ **14.** $\begin{array}{r} -0.25 \\ \underline{80} \end{array}$ **15.** $\begin{array}{r} +8 \\ \underline{+\frac{1}{2}} \end{array}$ **16.** $\begin{array}{r} -15 \\ \underline{+\frac{3}{5}} \end{array}$ **17.** $\begin{array}{r} -\frac{1}{2} \\ \underline{-\frac{1}{3}} \end{array}$ **18.** $\begin{array}{r} +16 \\ \underline{-2\frac{1}{4}} \end{array}$

19. $(+8)\left(+\frac{1}{4}\right)$ **20.** $\left(-\frac{3}{5}\right)(-20)$ **21.** $(+2)\left(-\frac{1}{2}\right)$

22. $(+4)(+3)(+2)$ **23.** $(-1)(-7)(-8)$ **24.** $(-3)(-5)(+4)(-1)$

25. $(-7)(+2)(0)$ **26.** $|+10| \cdot |-3| \cdot (-4)$ **27.** $(+8)(-9)(0)(-10)$

In 28–39, find the value of each expression.

28. $(+4)^2$ **29.** $(-3)^2$ **30.** $(+5)^3$ **31.** $(-4)^3$

32. $(-5)^3$ **33.** $(-1)^4$ **34.** $\left(+\frac{1}{2}\right)^2$ **35.** $\left(-\frac{1}{2}\right)^2$

36. $\left(+\frac{2}{3}\right)^3$ **37.** $\left(-\frac{3}{5}\right)^3$ **38.** $\left(-\frac{1}{4}\right)^3$ **39.** $\left(-\frac{1}{5}\right)^4$

In 40–45, fill in the blanks so that the resulting sentences illustrate the distributive property.

40. $5(9 + 7) =$ __ **41.** $-4(x + y) =$ __

42. __ $= 6(-3) + 6(-5)$ **43.** __ $= 7a + 7b$

44. 8(__) = (__)(5) + (__)(−3) **45.** __ (−3+3) = 2 (−3) + __ (__)

In 46–49, name, in each case, the multiplication property illustrated.

46. $(-6) \cdot (-5) = (-5) \cdot (-6)$

47. $[(-3) \cdot 4] \cdot 7 = (-3) \cdot [4 \cdot 7]$

48. $-8 \cdot [4 + (-1)] = (-8) \cdot (4) + (-8) \cdot (-1)$

49. $5x + 5 \cdot (-y) = 5 \cdot [x + (-y)]$

50. Name the property that justifies each statement.

a. $2(1 + 3) = 2(1) + 2(3)$

b. $(-2) + (-1) = (-1) + (-2)$

c. $(1 + 2) + 3 = 1 + (2 + 3)$

d. $2 + 0 = 2$

e. $(-5) \times 7 = 7 \times (-5)$

f. $-3x(2 \times 0) = (-3x \times 2) \times 0$

51. State whether each of the following statements is true or false:

a. $5(7 - 3) = 5(7) - 5(3)$

b. $8[(+4) - (-2)] = 8(+4) - 8(-2)$

52. Is the operation of multiplication distributive over subtraction? In other words, for all signed numbers x, y, and z, does $x(y - z) = xy - xz$?

5-6 DIVISION OF SIGNED NUMBERS

Using the Inverse Operation in Dividing Signed Numbers

Division may be defined as the inverse operation of multiplication, just as subtraction is defined as the inverse operation of addition. To divide 6 by 2 means to find a number that, when multiplied by 2, gives 6. That number is 3 because $3(2) = 6$. Thus, $\frac{6}{2} = 3$, or $6 \div 2 = 3$. The number 6 is the ***dividend***, 2 is the ***divisor***, and 3 is the ***quotient***.

It is impossible to divide a signed number by 0; that is, division by 0 is undefined. For example, to solve $(-9) \div 0 = ?$, we would have to find a number that, when multiplied by 0, would give -9. There is no such number since the product of any signed number and 0 is 0.

- In general, for all signed numbers a and b ($b \neq 0$):

$a \div b$, or $\frac{a}{b}$, means to find a unique number c such that $cb = a$.

In dividing nonzero signed numbers, there are four possible cases. Consider the following examples:

Case 1. $\frac{+6}{+3} = ?$ implies $(?)(+3) = +6$. *Answer:* $\frac{+6}{+3} = +2$

Case 2. $\frac{-6}{-3} = ?$ implies $(?)(-3) = -6$. *Answer:* $\frac{-6}{-3} = +2$

Case 3. $\frac{-6}{+3} = ?$ implies $(?)(+3) = -6$. *Answer:* $\frac{-6}{+3} = -2$

Case 4. $\frac{+6}{-3} = ?$ implies $(?)(-3) = +6$. *Answer:* $\frac{+6}{-3} = -2$

In the preceding examples, observe that:

1. When the dividend and divisor are both positive or both negative, the quotient is positive.
2. When the dividend and divisor have opposite signs, the quotient is negative.
3. In all cases, the absolute value of the quotient is the absolute value of the dividend divided by the absolute value of the divisor.

Cases 1–4 illustrate the following rules of division.

Rules for Dividing Signed Numbers

- **Rule 1.** The quotient of two positive numbers, or of two negative numbers, is a positive number whose absolute value is the absolute value of the dividend divided by the absolute value of the divisor.
- **Rule 2.** The quotient of a positive number and a negative number is a negative number whose absolute value is the absolute value of the dividend divided by the absolute value of the divisor.

In general, if a and b are both positive or are both negative, then:

$$\frac{a}{b} = \frac{|a|}{|b|}$$

If one of the numbers a and b is positive and the other number is negative, then:

$$\frac{a}{b} = -\left(\frac{|a|}{|b|}\right)$$

Rule for Dividing Zero by a Nonzero Number

If the expression $\frac{0}{-5} = ?$, then $(?)(-5) = 0$. Since 0 is the only number that can replace ? and result in a true statement, $\frac{0}{-5} = 0$. This illustrates that 0 divided by any nonzero number is 0.

In general, if a is a nonzero number $(a \neq 0)$:

$$\frac{\mathbf{0}}{a} = \mathbf{0}$$

EXAMPLE

Perform each indicated division, if possible.

Solutions and answers:

a. $\frac{+60}{+15}$ $+\left(\frac{60}{15}\right) = +4$

b. $\frac{+10}{-90}$ $-\left(\frac{10}{90}\right) = -\frac{1}{9}$

c. $\frac{-27}{-3}$ $+\left(\frac{27}{3}\right) = +9$

d. $(-45) \div 9$ $-(45 \div 9) = -5$

e. $9 \div 9$ 0

f. $-3 \div (0)$ Undefined

Using the Reciprocal in Dividing Signed Numbers

When the product of two numbers is 1, one number is called the ***reciprocal*** or ***multiplicative inverse*** of the other. For example, since $(+8)\left(+\frac{1}{8}\right) = 1$, we say that $+\frac{1}{8}$ is the reciprocal or multiplicative inverse of +8 and +8 is the reciprocal or multiplicative inverse of $+\frac{1}{8}$.

Since $\left(\frac{3}{5}\right)\left(\frac{5}{3}\right) = 1$, we say that $\frac{5}{3}$ is the reciprocal or multiplicative inverse of $\frac{3}{5}$ and $\frac{3}{5}$ is the reciprocal or multiplicative inverse of $\frac{5}{3}$.

Since $\left(-\frac{1}{2}\right)(-2) = 1$, we say that -2 is the reciprocal or multiplicative inverse of -2.

Since there is no number that, when multiplied by 0, gives 1, the number 0 has no reciprocal.

In general, for every nonzero number a, there is a unique number $\frac{1}{a}$ such that:

$$a \cdot \frac{1}{a} = 1$$

Using the reciprocal of a number, we can define division in terms of multiplication as follows:

For all numbers a and b $(b \neq 0)$:

$$a \div b = \frac{a}{b} = a \cdot \frac{1}{b} \quad (b \neq 0)$$

PROCEDURE. To divide a signed number by a nonzero signed number, multiply the dividend by the reciprocal of the divisor.

Notice that, if we exclude division by 0, the set of signed numbers is closed with respect to division because every nonzero signed number has a unique reciprocal, and multiplication by this reciprocal is always possible. In other words, if division by 0 is excluded, division is a binary operation for the set of signed numbers.

EXAMPLE

Perform each indicated division.

	Solutions and answers:
a. $\frac{+10}{+2}$	$(+10)\left(+\frac{1}{2}\right) = +5$
b. $\frac{-12}{+8}$	$(-12)\left(+\frac{1}{8}\right) = -\frac{3}{2}$
c. $\frac{-28}{-7}$	$(-28)\left(-\frac{1}{7}\right) = +4$
d. $\frac{0}{-3}$	$(0)\left(-\frac{1}{3}\right) = 0$
e. $(+18) \div \left(-\frac{1}{2}\right)$	$(+18)(-2) = -36$

EXERCISES

In 1–12, name the reciprocal (the multiplicative inverse) of each given number.

1. 6 **2.** -5 **3.** 9 **4.** -7

5. 1 **6.** -1 **7.** $\frac{1}{5}$ **8.** $-\frac{1}{10}$

9. $\frac{3}{4}$ **10.** $-\frac{2}{3}$ **11.** $x\ (x \neq 0)$ **12.** $-x\ (x \neq 0)$

In 13–48, find the indicated quotients or write "undefined" if no quotient exists.

13. $\frac{+10}{+2}$ **14.** $\frac{-63}{-9}$ **15.** $\frac{-8}{+4}$ **16.** $\frac{-48}{+16}$ **17.** $\frac{+25}{-1}$ **18.** $\frac{+84}{-14}$

19. $\frac{-1}{-1}$ **20.** $\frac{-10}{+10}$ **21.** $\frac{0}{+6}$ **22.** $\frac{+52}{-4}$ **23.** $\frac{+84}{-12}$ **24.** $\frac{-30}{-6}$

25. $\frac{+100}{-25}$ **26.** $\frac{-108}{0}$ **27.** $\frac{-65}{+5}$ **28.** $\frac{0}{3}$ **29.** $\frac{+4}{-8}$ **30.** $\frac{-6}{-9}$

31. $\frac{-15}{-12}$ **32.** $\frac{+18}{-4}$ **33.** $\frac{16}{+6}$ **34.** $\frac{-34}{4}$ **35.** $\frac{+100}{-8}$ **36.** $\frac{-36}{-8}$

37. $\frac{0}{-4}$ **38.** $\frac{-5}{-9}$ **39.** $\frac{3}{-7}$ **40.** $\frac{8.4}{-4}$ **41.** $\frac{-9.6}{-0.3}$ **42.** $\frac{-3.6}{1.2}$

43. $(+48) \div (-6)$ **44.** $(-75) \div (-15)$ **45.** $(-50) \div (0)$

46. $(+12) \div \left(-\frac{1}{3}\right)$ **47.** $\left(-\frac{3}{4}\right) \div (+6)$ **48.** $\left(-\frac{3}{4}\right) \div \left(-\frac{2}{3}\right)$

49. a. Find the value of x for which the denominator of the fraction $\frac{1}{x-2}$ has a value of 0.

b. State the value of x for which the multiplicative inverse of $(x - 2)$ is not defined.

In 50–53, give the multiplicative inverse of each expression, and state the value of x for which the multiplicative inverse is not defined.

50. $x - 5$ **51.** $x + 3$ **52.** $2x - 1$ **53.** $3x + 1$

54. State whether each of the following sentences is true or false:

a. $(+10) \div (-5) = (-5) \div (+10)$ **b.** $(-16) \div (-2) = (-2) \div (-16)$

55. If x and y represent signed numbers:

a. Does $x \div y = y \div x$ for all replacements of x and y?

b. Does $x \div y = y \div x$ for any replacements of x and y? If your answer is yes, give an example.

c. What is the relation between $x \div y$ and $y \div x$ when $x \neq 0$ and $y \neq 0$?

d. Is division commutative? In other words, does $x \div y = y \div x$ for every nonzero signed number x and every nonzero signed number y?

56. State whether each of the following sentences is true or false:

a. $[(+16) \div (+4)] \div (+2) = (+16) \div [(+4) \div (+2)]$

b. $[(-36) \div (+6)] \div (-2) = (-36) \div [(+6) \div (-2)]$

57. Is division associative? In other words, does $(x \div y) \div z = x \div (y \div z)$ for every signed number x, y, and z, when $y \neq 0$ and $z \neq 0$?

58. State whether each of the following sentences is true or false:

a. $(12 + 6) \div 2 = 12 \div 2 + 6 \div 2$

b. $[(+25) - (-10)] \div (+5) = (+25) \div (+5) - (-10) \div (+5)$

59. Does it appear that the operation of division is distributive over addition? In other words, does $(x + y) \div z = x \div z + y \div z$ for every signed number x, y, and z when $z \neq 0$?

5-7 EVALUATING ALGEBRAIC EXPRESSIONS USING SIGNED NUMBERS

When you evaluate algebraic expressions by replacing the variables with signed numbers, you follow the same procedure that you used when you evaluated algebraic expressions by replacing the variables with positive numbers.

EXAMPLES

1. Find the value of $x^2 - 3x - 54$ when $x = -5$.

How to Proceed:	*Solution:*
(1) Write the expression:	$x^2 - 3x - 54$
(2) Replace the variable by its given value:	$(-5)^2 - 3(-5) - 54$
(3) Evaluate the power:	$25 - 3(-5) - 54$
(4) Do the multiplication:	$25 + 15 - 54$
(5) Do the addition:	-14

Calculator Solution

METHOD 1. Replace x with -5 in the expression $x^2 - 3x - 54$.

Enter: 5 [+/–] [x^2] [–] 3 [×] 5 [+/–] [–] 54 [=]

Display: [–14.]

METHOD 2. Think of the expression as the sum of three terms:
$x^2 + (-3x) + (-54)$.

Enter: 5 [+/–] [x^2] [+] 3 [+/–] [×] 5 [+/–] [+] 54 [+/–] [=]

Display: [–14.]

Answer: -14

---

2. Find the value of $\frac{3a + 7}{5b^2}$ when $a = -4$ and $b = -2$.

The line of the fraction is a grouping symbol, like parentheses. Evaluate the numerator and denominator before dividing.

How to Proceed:	*Solution:*
(1) Write the expression:	$\frac{3a + 7}{5b^2}$
(2) Replace the variables by the given values:	$\frac{3(-4) + 7}{5(-2)^2}$
(3) Evaluate the powers:	$\frac{3(-4) + 7}{5(4)}$
(4) Perform the multiplications:	$\frac{-12 + 7}{20}$
(5) Simplify the numerator:	$\frac{-5}{20}$
(6) Divide the numerator by the denominator:	$-\frac{1}{4} = -0.25$

Calculator Solution Enclose the numerator and denominator in parentheses.

Enter: [( ] 3 [×] 4 [+/–] [+] 7 [ )] [÷] [( ] 5 [×] 2 [+/–] [x^2] [ )] [=]

Display: [–0.25]

Answer: -0.25

EXERCISES

In 1–52, find the numerical value of each expression. Use $a = -8$, $b = +6$, $d = -3$, $x = -4$, $y = 5$, and $z = -1$.

1. $6a$	**2.** $-5b$	**3.** ab	**4.** $2xy$
5. $-4bz$	**6.** $\frac{1}{3}d$	**7.** $-\frac{2}{3}b$	**8.** $\frac{3}{8}a$
9. $\frac{1}{2}xy$	**10.** $-\frac{3}{4}ab$	**11.** a^2	**12.** d^3
13. $-y^2$	**14.** $-d^2$	**15.** $-z^3$	**16.** $2x^2$
17. $-3y^2$	**18.** $-3b^2$	**19.** $4d^2$	**20.** $-2z^3$
21. xy^2	**22.** a^2b	**23.** $2d^2y^2$	**24.** $\frac{1}{2}db^2$
25. $-2d^3z^2$	**26.** $a + b$	**27.** $a - x$	**28.** $2x + z$
29. $3y - b$	**30.** $a - 2d$	**31.** $b - 4d$	**32.** $5x + 2y$
33. $7b - 5x$	**34.** $x^2 + x$	**35.** $2b^2 + b$	**36.** $y^2 - y$
37. $2d^2 - d$	**38.** $2a + 5d + 3x$	**39.** $8y + 5b - 6d$	**40.** $9b - 3z - 2x$
41. $x^2 + 3x + 5$	**42.** $z^2 + 2z - 7$	**43.** $a^2 - 5a - 6$	**44.** $d^2 - 4d + 6$
45. $2x^2 - 3x + 5$	**46.** $15 + 5z - z^2$	**47.** $2(a + b)$	**48.** $3(2x - 1) + 6$
49. $10 - 3(x - 4)$	**50.** $(x + 2)(x - 1)$	**51.** $(a - b)(a + b)$	**52.** $(x + d)(x - 4z)$

53. Find the value of $a^2 - 9b$ when $a = 4$ and $b = \frac{1}{3}$.

54. Find the value of $9x^2 - 4y^2$ when $x = \frac{1}{3}$ and $y = \frac{1}{2}$.

55. If $x = 4$, find the value of: **a.** $2x^2$ **b.** $(2x)^2$.

56. If $y = -2$, find the value of: **a.** $3y^2$ **b.** $(3y)^2$.

57. If $z = -\frac{1}{2}$, find the value of: **a.** $4z^2$ **b.** $(4z)^2$.

In 58–62, find the value of each expression. Use $a = -12$, $b = +6$, and $c = -1$.

58. $\frac{ac}{-3b}$ **59.** $\frac{b^2c}{a}$ **60.** $\frac{3a^2c^3}{b^3}$ **61.** $\frac{a - b^2}{-2c^2}$ **62.** $\frac{b^2 - a^2}{b^2 + a^2}$

5-8 NUMBER LINES AND GRAPHS USING SIGNED NUMBERS

Even though we know that the surface of the earth is approximately the surface of a sphere, we often model the earth by using maps that are plane surfaces. To locate a place on a graph, we choose two reference lines, the equator and the prime meridian. The location of a city is given in terms of east or west ***longitude***, that is, distance from the prime meridian, and north or south ***latitude***, or distance from the equator. Thus the city of Lagos in Nigeria is located at 3° east longitude and 6° north latitude, and the city of Dakar in Senegal is located at 17° west longitude and 15° north latitude.

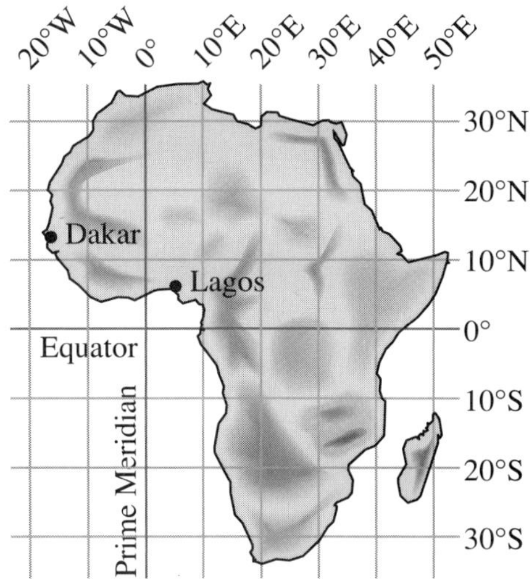

Points on a Plane

The method used to locate cities on a map can be used to locate any point on a plane. The reference lines are a horizontal line called the ***x-axis*** and a vertical line called the ***y-axis***. These two number lines, which have the same scale and are drawn perpendicular to each other, are the ***coordinate axes***.

In a ***coordinate plane***, the intersection of the two lines is called the ***origin*** and is indicated as point O. At this point of intersection, both x and y equal 0.

Moving to the right and moving up are regarded as movements in positive directions. In the coordinate plane, positive numbers are to the right of 0 on the x-axis and above 0 on the y-axis. This is also true for lines parallel to the axes.

Moving to the left and moving down are regarded as movements in negative directions. In the coordinate plane, negative numbers are to the left of 0 on the x-axis and below 0 on the y-axis. This is again true for lines parallel to the axes.

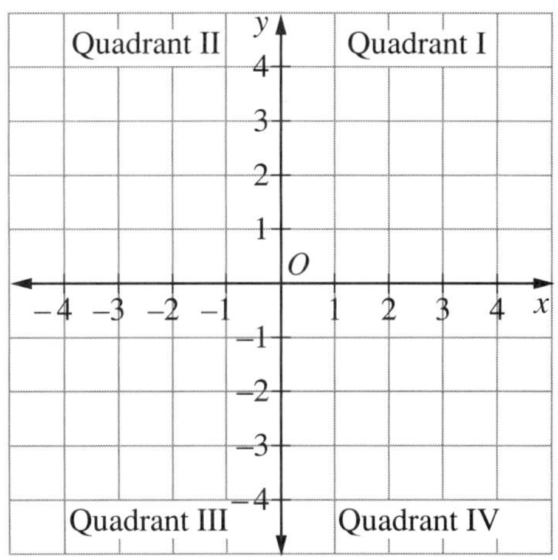

The x-axis and the y-axis separate the plane into four regions called ***quadrants***. These quadrants are numbered I, II, III, and IV in a counterclockwise order, beginning at the upper right, as shown in the accompanying drawing. The points on the axes are not in any quadrant.

Coordinates of a Point

Every point on the plane can be described by two numbers, called the ***coordinates*** of the point, usually written as an ***ordered pair***. The first number in the pair is called the ***x-coordinate*** or the ***abscissa***. The second number is the ***y-coordinate*** or the ***ordinate***. In general, the coordinates of a point are represented as (x, y).

In the graph at the right, point A, which is the ***graph of the ordered pair*** $(+2, +3)$, lies in quadrant I. Here, A lies a distance of 2 units to the right of the origin (in a positive direction along the x-axis) and then up a distance of 3 units (in a positive direction parallel to the y-axis).

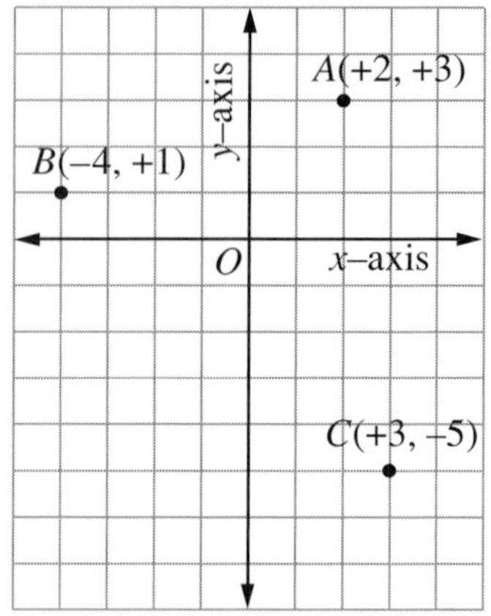

Point B, the graph of $(-4, +1)$ in quadrant II, lies a distance of 4 units to the left of the origin (in a negative direction along the x-axis) and then up 1 unit (in a positive direction parallel to the y-axis).

In quadrant III, every ordered pair (x, y) consists of two negative numbers.

Point C, the graph of $(+3, -5)$ in quadrant IV, lies 3 units to the right of the origin (in a positive direction along the x-axis) and then down 5 units (in a negative direction parallel to the y-axis).

Point O, the origin, has the coordinates (0, 0).

TO GRAPH OR PLOT THE COORDINATES OF A POINT ON A PLANE

On a pair of coordinate axes, graph points $A(-3, -4)$, $B(4, 0)$, and $C(0, -5)$.

Procedure: For each point, move along the x-axis the number of units given by the x-coordinate. Move to the right if the number is positive and to the left if the number is negative. If the x-coordinate is 0, there is no movement along the x-axis. Then, from the point on the x-axis, move parallel to the y-axis the number of units given by the y-coordinate. Move up if the number is positive and down if the number is negative. If the y-coordinate is 0, there is no movement in the y direction.

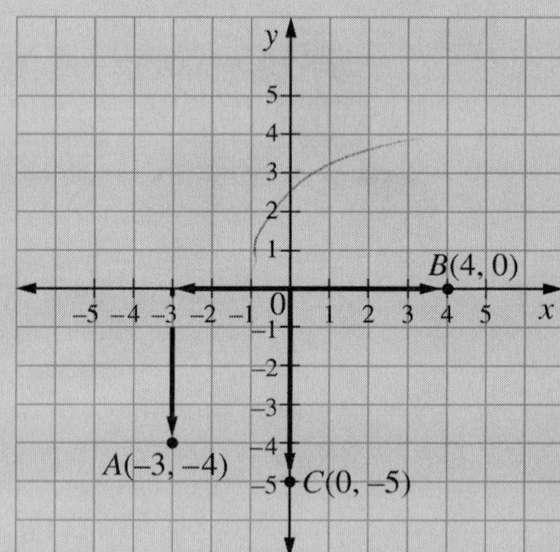

TO READ THE COORDINATES OF A POINT ON A PLANE

Find the coordinates of points *R*, *S*, *T*, and *V*.

Procedure: To find the coordinates of a point:

1. Draw a vertical line from the point to the *x*-axis and a horizontal line from the point to the *y*-axis.

2. The number assigned to the point at which the vertical line intersects the *x*-axis is the *x*-coordinate of the point.

3. The number assigned to the point at which the horizontal line intersects the *y*-axis is the *y*-coordinate of the point.

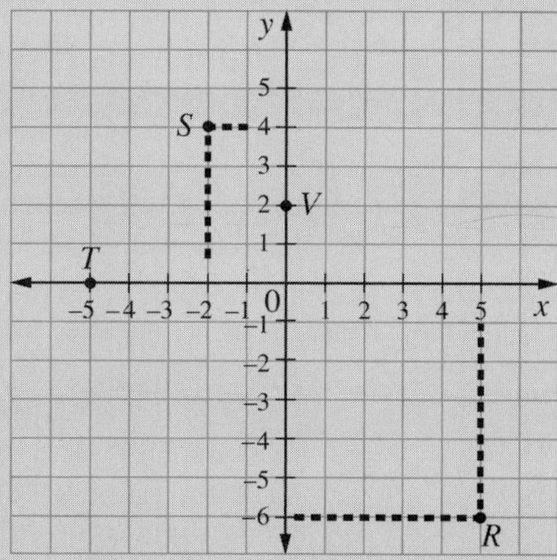

If a point lies on the *x*-axis, its abscissa is the *x* value, its ordinate is 0, and its coordinates are $(x,0)$. If a point lies on the *y*-axis, its abscissa is 0, its ordinate is the *y* value, and its coordinates are $(0, y)$.

Here, the coordinates are $R(5,-6)$, $S(-2, 4)$, $T(-5, 0)$, and $V(0, 2)$.

Graphing Polygons

A quadrilateral (a four-sided figure) can be represented in the coordinate plane by locating its vertices and then drawing the sides by connecting the vertices in order.

The graph at the right shows the rectangle *ABCD*. The vertices are $A(3, 2)$, $B(-3, 2)$, $C(-3,-2)$ and $D(3,-2)$. From the graph, note the following:

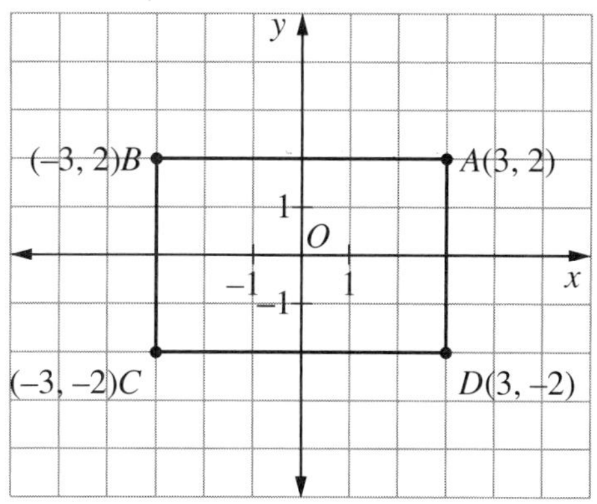

1. Points *A* and *B*, which have the same ordinate, are on a line that is parallel to the *x*-axis.
2. Points *C* and *D*, which have the same ordinate, are on a line that is parallel to the *x*-axis.
3. Lines that are parallel to the *x*-axis are parallel to each other.
4. Lines that are parallel to the *x*-axis are perpendicular to the *y*-axis.
5. Points *B* and *C*, which have the same abscissa, are on a line that is parallel to the *y*-axis.

6. Points A and D, which have the same abscissa, are on a line that is parallel to the y-axis.
7. Lines that are parallel to the y-axis are parallel to each other.
8. Lines that are parallel to the y-axis are perpendicular to the x-axis.

Now, we know that quadrilateral $ABCD$ is a rectangle, because it is a parallelogram with right angles.

From the graph, we can find the dimensions of this rectangle.

To find the length of the rectangle, we can count the number of units from A to B or from C to D. $AB = CD = 6$. Because points on the same horizontal line have the same y-coordinate, we can also find these measures by subtracting their x-coordinates.

$$AB = CD = 3 - (-3) = 3 + 3 = 6$$

To find the width of the rectangle, we can count the number of units from B to C or from D to A. $BC = DA = 4$. Because points on the same vertical line have the same x-coordinate, we can find these measures by subtracting their y-coordinates.

$$BC = DA = 2 - (-2) = 2 + 2 = 4$$

EXAMPLE

Graph the following points: $A(4,1)$, $B(1,5)$, $C(-2,1)$. Then draw $\triangle ABC$ and find its area.

Solution The graph at the right shows $\triangle ABC$.

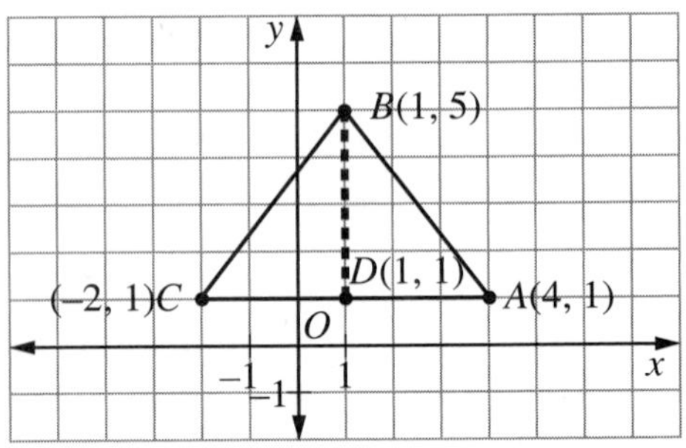

To find the area of the triangle, we need to know the lengths of the base and of the altitude drawn to that base.

The base of $\triangle ABC$ is $\overline{AC}$.

$$AC = 4 - (-2) = 4 + 2 = 6$$

The altitude to the base is the line segment $\overline{BD}$, drawn from B perpendicular to $\overline{AC}$. $BD = 5 - 1 = 4$.

$$\text{Area} = \tfrac{1}{2}(AC)(BD)$$

$$= \tfrac{1}{2}(6)(4) = 12$$

Answer: The area of $\triangle ABC$ is 12 square units.

EXERCISES WITH OPEN-RESPONSE PROBLEMS

1. Write as ordered number pairs the coordinates of points A, B, C, D, E, F, G, H, and O in the graph.

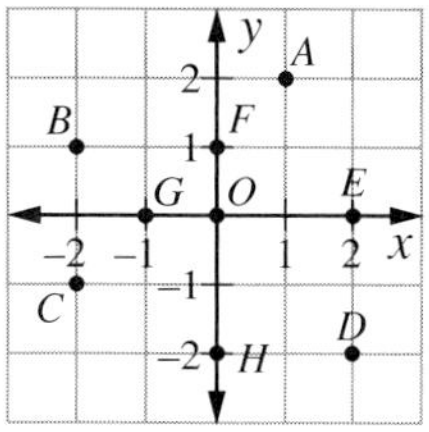

In 2–21, draw a pair of coordinate axes on a sheet of graph paper and graph the point associated with each ordered number pair. Label each point with its coordinates.

2. (5, 4) **3.** (−3, 2) **4.** (2, −6) **5.** (−4, −5)
6. (1, 6) **7.** (−8, 5) **8.** (4, −4) **9.** (−2, −7)
10. (−1.5, −2.5) **11.** (5, 0) **12.** (−3, 0) **13.** (8, 0)
14. (−10, 0) **15.** (0, 4) **16.** (0, −6) **17.** (0, 1)
18. (0, −4) **19.** (0, 0) **20.** $(|2|, |4|)$ **21.** $(|-5|, |3|)$

In 22–26, name the quadrant in which the graph of each point described appears.

22. (5, 7) **23.** (−3, −2) **24.** (−7, 4) **25.** (1, −3) **26.** $(|-2|, |-3|)$

27. Graph several points on the x-axis. What is the value of the ordinate for every point in the set of points on the x-axis?

28. Graph several points on the y-axis. What is the value of the abscissa for every point in the set of points on the y-axis?

29. What are the coordinates of the origin in the coordinate plane?

In 30–39: **a.** Graph the points and connect them with straight lines in order, forming a polygon. **b.** Tell what kind of polygon is drawn. **c.** Find the area of the polygon.

30. $A(1, 1)$, $B(8, 1)$, $C(1, 5)$ **31.** $P(0, 0)$, $Q(5, 0)$, $R(5, 4)$, $S(0, 4)$
32. $C(8, -1)$, $A(9, 3)$, $L(4, 3)$, $F(3, -1)$ **33.** $H(-4, 0)$, $O(0, 0)$, $M(0, 4)$, $E(-4, 4)$
34. $H(5, -3)$, $E(5, 3)$, $N(-2, 0)$ **35.** $F(5, 1)$, $A(5, 5)$, $R(0, 5)$, $M(-2, 1)$
36. $B(-2, -2)$, $A(2, -2)$, $R(2, 2)$, $N(-2, 2)$ **37.** $P(-3, 0)$, $O(0, 0)$, $N(2, 2)$, $D(-1, 2)$
38. $R(-4, 2)$, $A(0, 2)$, $M(0, 7)$ **39.** $M(-1, -1)$, $I(3, -1)$, $L(3, 3)$, $K(-1, 3)$

40. Graph points $A(1, 1)$, $B(5, 1)$, and $C(5, 4)$. What must be the coordinates of point D if $ABCD$ is a rectangle?

41. Graph points $P(-2, -4)$ and $Q(2, -4)$. What are the coordinates of R and S if $PQRS$ is a square? (Two answers are possible.)

42. **a.** Graph points $S(3, 0)$, $T(0, 4)$, $A(-3, 0)$, and $R(0, -4)$, and draw the rhombus $STAR$.
b. Find the area of $STAR$ by adding the areas of the triangles into which the axes separate the rhombus.

43. **a.** Graph points $P(2, 0)$, $L(1, 1)$, $A(-1, 1)$, $N(-2, 0)$, $E(-1, -1)$, and $T(1, -1)$, and draw the hexagon $PLANET$.
b. Find the area of $PLANET$. (*Hint:* Use the x-axis to separate the hexagon into two parts.)

CHAPTER SUMMARY

The ***opposite*** of a number is the number that is the same distance from 0 and on the opposite side of 0. The ***absolute value*** of a number is the larger of a number and its opposite.

To add two numbers that have the same sign, find the sum of the absolute values of the numbers and give this sum the common sign. To add two numbers having different signs, find the difference of the absolute values of the numbers and give this difference the sign of the number that has the greater absolute value. The sum of a number and its opposite is 0.

For every number c and every number b, the expression $c - b$ is the number a such that $b + a = c$. To subtract two signed numbers, add the opposite of the subtrahend to the minuend.

The product of two numbers that have the same sign is the product of their absolute values. The product of two numbers that have different signs is the opposite of the product of their absolute values.

The quotient of two numbers that have the same sign is the quotient of their absolute values. The quotient of two numbers that have different signs is the opposite of the quotient of their absolute values.

The location of a point on a plane is given by an ***ordered pair*** of numbers that indicate the distance of the point from two reference lines, a horizontal line and a vertical line, called ***coordinate axes***. The horizontal line is called the ***x-axis***, the vertical line is the ***y-axis***, and their intersection is the ***origin***. The pair of numbers that are used to locate a point on the plane are called the ***coordinates*** of the point. The first number in the pair is called the ***x-coordinate*** or ***abscissa***, and the second number is the ***y-coordinate*** or ***ordinate***. The coordinates of a point are represented as (x, y).

VOCABULARY

5-1 Opposite

5-2 Absolute value

5-3 Additive inverse Addition property of opposites Addition property of zero Property of the opposite of a sum

5-6 Reciprocal Multiplicative inverse

5-8 x-axis y-axis Coordinate axes Origin Abscissa x-coordinate Ordinate y-coordinate

REVIEW EXERCISES WITH OPEN-RESPONSE PROBLEMS

In 1–3, find each sum.

1. $\begin{array}{r} +9 \\ \underline{-2} \end{array}$ **2.** $\begin{array}{r} -17 \\ \underline{+4} \end{array}$ **3.** $\begin{array}{r} -18 \\ \underline{-36} \end{array}$

In 4–6, find each difference.

4. $\begin{array}{r} -8 \\ \underline{+6} \end{array}$ **5.** $\begin{array}{r} +19 \\ \underline{-43} \end{array}$ **6.** $\begin{array}{r} -8.3 \\ \underline{-9.6} \end{array}$

In 7–9, find each product.

7. $\begin{array}{r} -7 \\ \underline{+6} \end{array}$ 8. $\begin{array}{r} +26 \\ \underline{-4} \end{array}$ 9. $\begin{array}{r} -1.5 \\ \underline{-0.6} \end{array}$

In 10–12, find each quotient.

10. $\frac{-6}{+2}$ 11. $\frac{-56}{-8}$ 12. $\frac{+2}{-8}$

In 13–18, state whether each sentence is true or false.

13. $-7 > -2$
14. $|-7| > |-2|$
15. $|-9| = -9$
16. $|-4| + |+4| = 0$
17. $-3.9 < -3.2$
18. $\frac{|-24|}{|+4|} = -6$

19. How much greater than -8 is $+14$?

20. Name the greatest negative integer.

21. From the sum of 15 and -18, subtract -4.

In 22–30, find the value of each expression.

22. $-7 + (-9)$
23. $(-3)(+17)$
24. $(-75) \div (-5)$
25. $-28 - (+4)$
26. $-1.7 - (-2.3)$
27. $(-3)(-10)(-8)$
28. $(-14)\left(-\frac{3}{7}\right)$
29. $\frac{0}{-12}$
30. $\left(-\frac{1}{3}\right)^2$

In 31–42, find the value of each expression when $a = -3$, $b = +2$, $c = -4$, $x = -1$, and $y = -5$.

31. $-2x$
32. $-2 + x$
33. $4a^2$
34. $(4a)^2$
35. cx^2
36. $y^2 + y$
37. $b^2 + by$
38. $abcx$
39. $a - b + c$
40. $(a + b)(a - b)$
41. $(c - 5)(c + 8)$
42. $|-bcx - y|$

43. What is the opposite of $-(-2)$?

44. **a.** Write the coordinates of each of points A, B, C, and D shown on the graph at the right.

b. What is the area of trapezoid $ABCD$, formed by joining the points in order?

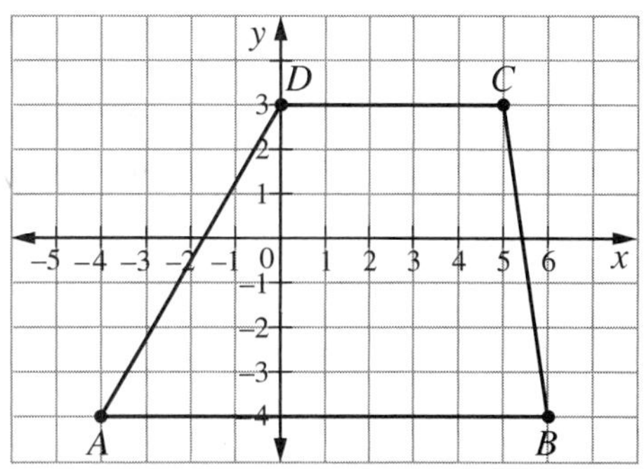

In 45–49, draw a pair of coordinate axes and graph each point for which the ordered pair is given.

45. (2, 5)
46. (−3, 2)
47. (−1, −4)
48. (0, 3)
49. (−5, 0)

50. Maurice answered all of the 60 questions on a multiple-choice test. To penalize for guessing, the test was scored by using the formula $S = R - \frac{W}{4}$, where S is the score on the test, R is the number right, and W is the number wrong.

a. What is the lowest possible score?

b. How many answers did Maurice get right if his score was -5?

c. Is it possible to get a score of -4?

CUMULATIVE REVIEW WITH OPEN-RESPONSE PROBLEMS

1. Is it possible for each boy in a family to have twice as many sisters as brothers, and each girl in the same family to have twice as many brothers as sisters? If so, how many boys and how many girls are there in the family?

In 2–5, write each of the following as an algebraic expression.

2. Four more than twice a number n

3. Twelve divided by 3 times a number x

4. The cost of two loaves of bread at b cents each and one gallon of milk that costs m cents a gallon

5. The perimeter of a rectangle whose length is 5 centimeters and whose width is w centimeters.

6. Tom has started a business, buying cookies for \$4.50 a dozen and selling them on the street for \$0.60 each for 2 hours each day. The average cost for his license, insurance, and other expenses, based on the number of days he operates his stand, is \$2.10 per day. Because he prides himself on selling only fresh cookies, he buys his cookies each morning and gives away at the end of the day any cookies that he has left.

a. Let D represent the number of dozens of cookies Tom buys, and S represent the number of cookies that he sells. Use D and S to write an expression for Tom's net earnings (income − expenses) for 1 day.

b. On Monday he bought 12 dozen cookies and sold all of them. What were his net earnings for Monday?

c. On Tuesday he bought 10 dozen cookies, but because it rained, he sold only 71 cookies. What were his net earnings for Tuesday? Explain the meaning of the negative answer.

Exploration

In each of the following business procedures: sports, and games, negative numbers occur. In (1) to (6), write a word or phrase that is used to express a negative number.

(1) golf (2) card games (3) football
(4) stock market (5) financial reports (6) cartography

Chapter 6
Introduction to Solving Equations

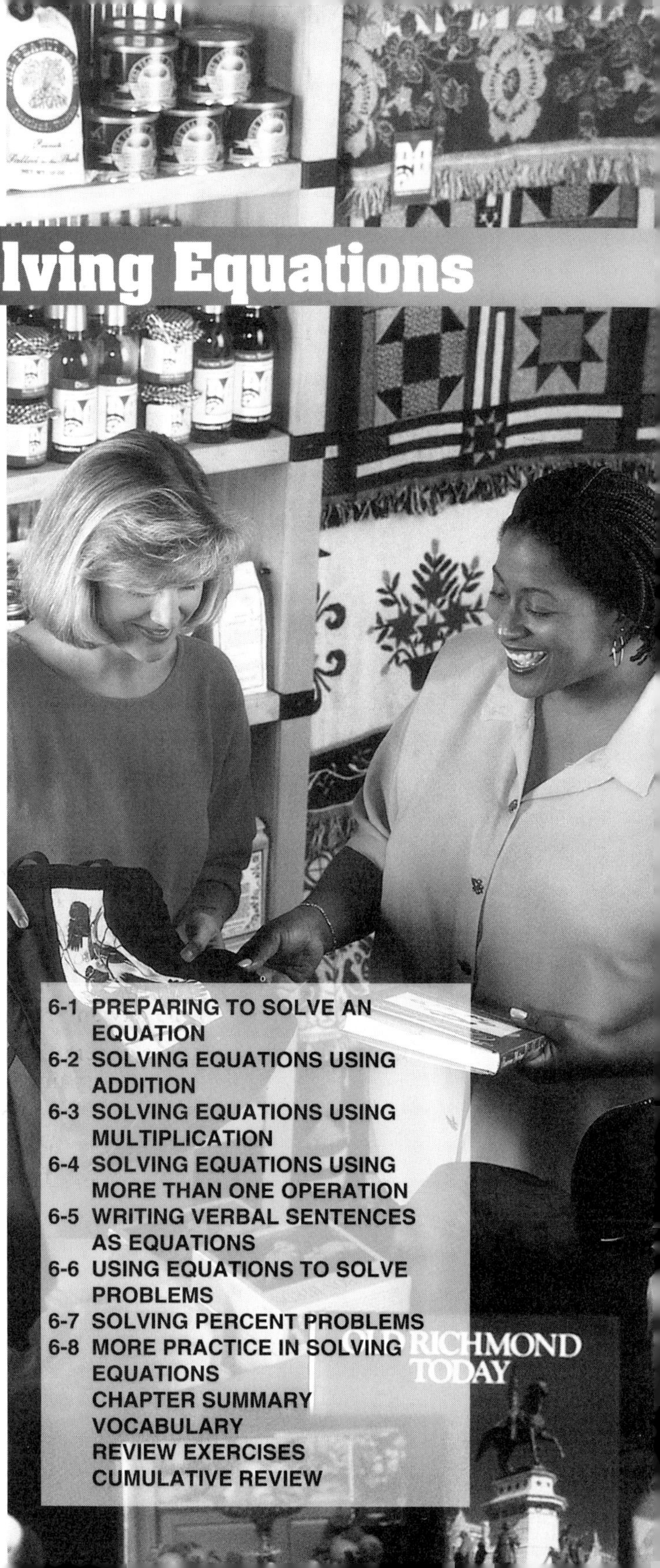

An equation is an important problem-solving tool. A successful business person must make many decisions about business practices. Some of these decisions are based on known facts, but others require the use of information obtained from these facts by solving equations.

The owner of a trucking company knows the weight of his trucks and the weight limits of the roads and bridges over which a truck will travel in making a delivery. When this information is used in an equation, the maximum load that the truck can carry can be determined.

A store owner knows the amount that she paid for merchandise and the cost of doing business. The solution of an equation will determine at what price the merchandise must be sold in order to realize a profit.

Many of these problem-solving equations are complex. Before you can cope with complex equations, you must learn the basic principles involved in solving any equation.

6-1 PREPARING TO SOLVE AN EQUATION

Some Terms and Definitions

An ***equation*** is a sentence that states that two algebraic expressions are equal. For example, $x + 3 = 9$ is an equation in which $x + 3$ is called the ***left side***, or ***left member***, and 9 is the ***right side***, or ***right member***.

An equation may be a *true* sentence such as $5 + 2 = 7$, a *false* sentence such as $6 - 3 = 4$, or an *open* sentence such as $x + 3 = 9$. The number that can replace the variable in an open sentence to make the sentence true is called a ***root***, or a ***solution***, of the equation. For example, 6 is a root of $x + 3 = 9$.

As discussed in Chapter 4, the ***replacement*** set or ***domain*** is the set of possible values that can be used in place of the variable in an open sentence. If no replacement set is given, the replacement set is the set of real numbers. The set consisting of all elements of the replacement set that are solutions of the open sentence is called the ***solution set*** of the open sentence. For example, if the replacement set is the set of real numbers, the solution set of $x + 3 = 9$ is $\{6\}$. If no element of the replacement set makes the open sentence true, the solution set is the empty or null set, $\varnothing$ or $\{\ \}$. ***To solve an equation*** is to find its solution set.

If every element of the domain satisfies an equation, the equation is called an ***identity***. Thus, $5 + x = x - (-5)$ is an identity when the domain is the set of real numbers because every element of the domain makes the sentence true.

If not every element of the domain makes the sentence true, the equation is called a ***conditional equation***, or simply an equation. Therefore, $x + 3 = 9$ is a conditional equation.

EXAMPLES

1. Is 7 a root of the equation $5x - 10 = 25$?

How to Proceed:	*Solution:*
(1) Write the equation:	$5x - 10 = 25$
(2) Replace the variable, x, by 7:	$5(7) - 10 \stackrel{?}{=} 25$
(3) Simplify the left member:	$35 - 10 \stackrel{?}{=} 25$
	$25 = 25$

Answer: Yes

---

2. Find the solution set of $3 - x = 5$ if the domain is $\{-4, -3, -2, -1, 0, +1\}$.

Solution Replace x by each element of the domain in turn.

$3 - (-4) = 5$	False	$3 - (-1) = 5$	False
$3 - (-3) = 5$	False	$3 - (\ 0) = 5$	False
$3 - (-2) = 5$	True	$3 - (+1) = 5$	False

Since -2 is the only element of the domain that makes the equation true, the solution set is $\{-2\}$.

Answer: $\{-2\}$

---

3. Find the solution set of the equation $15 + y = 3$ using the given domain.
a. Domain = positive integers **b.** Domain = negative integers

Solutions

a. The sum of 15 and any positive integer is greater than 15. Since the sum cannot equal 3, the solution is the empty set for this domain.

b. We know $15 - 12 = 3$. (Let $y = -12$, in the given domain.)

$$15 + y = 3$$
$$15 + (-12) = 3$$
$$3 = 3 \quad \text{True}$$

Answers: **a.** $\varnothing$ or $\{\ \}$ **b.** $\{-12\}$

---

4. Write the solution set of each equation using the domain of real numbers.
a. $x - x = 0$ **b.** $x = x + 1$

Solutions

a. Any real number subtracted from itself is 0. This equation, which is true for all real numbers, is an identity. The solution set consists of *all* numbers in the domain.

b. Since the right side of the equation is always 1 greater than the left side, *no* replacement for x ever results in a true statement. The solution is the empty set.

Answers: **a.** {Real numbers} **b.** $\varnothing$ or $\{\ \}$

Postulates of Equality

In mathematics, any statement that we accept as being true without proof is called an ***assumption***, a ***postulate***, or an ***axiom***. At this point, we will state several postulates of equality. These postulates, and others that are stated in later sections, will be used to solve equations in a systematic manner.

- **POSTULATE 1. Reflexive Property of Equality** The *reflexive property of equality* states that for every number a:

$$a = a.$$

- **POSTULATE 2. Symmetric Property of Equality** The *symmetric property of equality* states that for all numbers a and b:

$$\textbf{If } a = b, \textbf{ then } b = a.$$

- **POSTULATE 3. Transitive Property of Equality** The *transitive property of equality* states that for all numbers a, b, and c:

If $a = b$, and $b = c$, then $a = c$.

EXERCISES WITH OPEN-RESPONSE PROBLEMS

In 1–9, tell, in each case, whether the number in parentheses is a root of the given equation.

1. $5x = 10$ (10) **2.** $\frac{1}{2}x = 18$ (36) **3.** $\frac{1}{3}y = 12$ (4)

4. $x - 11 = -5$ (6) **5.** $8 - y = 10$ (−2) **6.** $\frac{2}{3}x = 9$ $\left(\frac{9}{2}\right)$

7. $m - 4\frac{1}{2} = 9$ $\left(4\frac{1}{2}\right)$ **8.** $7 - 2x = 21$ (−7) **9.** $19 = 4x - 1$ (4.5)

In 10–21, using the domain $\{-4, -3, -2, -1, 0, 1, 2, 3, 4\}$, find the solution set of each equation. If the equation has no roots, indicate the solution set as the empty set, $\varnothing$.

10. $x + 5 = 7$ **11.** $y - 3 = 1$ **12.** $2x + 3 = 1$ **13.** $6 = 3x - 1.8$

14. $\frac{1}{2}x + 4 = 3$ **15.** $3 - 2x = 9$ **16.** $|2x| = 4$ **17.** $x^2 = 9$

18. $|x - 1| = -2$ **19.** $|x| - 2 = -1$ **20.** $|x - 2| = 2$ **21.** $x^2 - 1 = 3$

In 22–27, tell whether each equation is a conditional equation or an identity.

22. $x + 3 = 3 + x$ **23.** $x + 3 = 10$ **24.** $y + 3 + 4 = 7 + y$

25. $5a = a \cdot 5$ **26.** $5a = 40$ **27.** $5 \cdot 2 \cdot a = a \cdot 2 \cdot 5$

In 28–33, name the property of equality that each sentence illustrates.

28. $5 + 2 = 5 + 2$
29. If $6 + 2 = 5 + 3$, then $5 + 3 = 7 + 1$, then $6 + 2 = 7 + 1$.
30. If $4 + 3 = 6 + 1$, then $6 + 1 = 4 + 3$.
31. $x + y = x + y$
32. If $x = y$, then $y = x$.
33. If $m + n = r + s$, and $r + s = x + y$, then $m + n = x + y$.

34. **a.** Can the equation $x + 10 = 3$ have a solution that is a positive integer? Explain why or why not.
b. Can the equation $x + 10 = 3$ have a solution that is a negative integer? Explain why or why not.

35. **a.** Can the equation $x + 1 = x$ have a solution that is a positive integer? Explain why or why not.
b. Can the equation $x + 1 = x$ have a solution that is a negative integer? Explain why or why not.
c. Does the equation $x + 1 = x$ have a solution that is a real number? Explain why or why not.

6-2 SOLVING EQUATIONS USING ADDITION

To solve an equation, you need to work backward or "undo" what has been done by using inverses. To undo the addition of a number, add its opposite.

Number	*Addition*	*Addition of the Opposite*		*Result*
8	$8 + 2$	$8 + 2$	$+ (-2)$	8
-12	$-12 + 7$	$-12 + 7$	$+ (-7)$	-12
4	$4 + (-5)$	$4 + (-5)$	$+ 5$	4
x	$x + 3$	$x + 3$	$+ (-3)$	x
x	$x + (-3)$	$x + (-3)$	$+ 3$	x
x	$x - 3$	$x - 3$	$+ 3$	x

- **POSTULATE 4. Addition Property of Equality** The *addition property of equality* states that for all numbers a, b, and c:

If $a = b$, then $a + c = b + c$.

Therefore, if the same number is added to both sides of an equality, the resulting sums are equal.

For example, consider this equation: $x = 7$

Add 5 to both sides: $x = 7$ becomes $x + 5 = 7 + 5$ or $x + 5 = 12$

Add -4 to both sides: $x = 7$ becomes $x + (-4) = 7 + (-4)$ or $x - 4 = 3$

The three equations shown (above right) are called ***equivalent equations*** because they all have the same solution. Using the domain of real numbers, we find that this solution is $\{7\}$.

When we solve an equation, we are actually ***simplifying the equation***. In other words, we are finding the simplest *equivalent* form, such as $x = 7$, so that we can identify the solution set more easily.

In $x - 4 = 3$, one side of the equation contains a variable, x, that is added to a ***constant***, -4. By adding $+4$ (the opposite of the constant) to both sides, we simplify the equation and find the solution set. In Example 1 below, we simplify $x + 5 = 12$.

EXAMPLES

1. Find the solution set for $x + 5 = 12$ when the domain is the set of real numbers. Since 5 is added to the variable, simplify the equation by adding its inverse, -5, to both sides.

How to Proceed:	*Solution:*
(1) Write the equation:	$x + 5 = 12$
(2) Use the addition property of equality:	$(x + 5) + (-5) = 12 + (-5)$
(3) Use the associative property of addition:	$x + [5 + (-5)] = 7$
(4) Use the inverse property of addition:	$x + 0 = 7$
(5) Use the identity property of addition:	$x = 7$

Check the solution *only in the original equation.* Show that a true statement results when x is replaced by 7.

Answer: $\{7\}$

---

2. Solve and check: $x - 4 = 7$.

How to Proceed:	*Solution:*	*Check:*
(1) Write the equation:	$x - 4 = 7$	$x - 4 = 7$
(2) Add $+4$, the additive inverse (opposite) of -4, to both members of the equation, and simplify each member:	$x - 4 + (+4) = 7 + (+4)$ $x + 0 = 11$ $x = 11$	$11 - 4 \stackrel{?}{=} 7$ $7 = 7$ ✔

Answer: $x = 11$ or The solution set is $\{11\}$.

(*Note:* By writing $x = 11$ as the answer, we show that no other number can replace x and result in a true statement. The answer given above shows two acceptable ways to write the solution.)

---

3. Solve and check: $3 = x + 12$.

How to Proceed:	*Solution:*	*Check:*
(1) Write the equation:	$3 = x + 12$	$3 = x + 12$
(2) Add -12, the opposite of 12, to both members of the equation, and simplify each member:	$3 + (-12) = x + 12 + (-12)$ $-9 = x$	$3 \stackrel{?}{=} -9 + 12$ $3 = 3$ ✔

Answer: $\{-9\}$ or $x = -9$

---

4. Write the solution of $x + \sqrt{2} = 5$ to the *nearest tenth.*

How to Proceed:	*Solution:*
(1) Write the equation:	$x + \sqrt{2} = 5$
(2) Add $-\sqrt{2}$ (the opposite of $\sqrt{2}$) to both sides of the equation to obtain an equivalent equation with the variable x by itself:	$x + \sqrt{2} + (-\sqrt{2}) = 5 + (-\sqrt{2})$ $x = 5 + (-\sqrt{2})$

(3) Use a calculator to find the approximate decimal value of x.

Enter: 5 [+] [( ] 2 [√x] [ )] [+/–] [=]

Display: 3.585786438

(4) Round the number in the display to the nearest tenth.

Answer: 3.6

EXERCISES

In 1–36, solve for the variable in each equation and check. The domain is the set of real numbers.

1. $x + 3 = 8$ **2.** $y + 7 = 12$ **3.** $m - 8 = 4$ **4.** $w + 7 = 5$

5. $x - 2 = 23$ **6.** $d + 9 = 3$ **7.** $6 + t = -3$ **8.** $2 = y + 17$

9. $x - 1.5 = 3.2$ **10.** $y + 7 = 5.4$ **11.** $0.4 + x = 0.01$ **12.** $12 = d + 7.2$

13. $x + \frac{1}{6} = \frac{5}{6}$ **14.** $y - \frac{5}{8} = \frac{1}{4}$ **15.** $\frac{2}{9} + x = \frac{1}{3}$ **16.** $y - 1\frac{1}{3} = 4$

17. $4\frac{2}{3} = m - \frac{1}{3}$ **18.** $12\frac{1}{2} = y + 7\frac{3}{4}$ **19.** $1\frac{1}{4} = c + 6\frac{1}{2}$ **20.** $-3 = 1\frac{2}{5} + x$

21. $1\frac{2}{5} = 4 + c$ **22.** $2 + x = 2$ **23.** $y + 52 = 0$ **24.** $12.5 = a + 12.5$

25. $-3\frac{7}{8} = b + 3\frac{7}{8}$ **26.** $12\frac{1}{8} + w = 12$ **27.** $5\frac{7}{10} = k + 18\frac{3}{10}$

28. $x - 2\frac{7}{8} = -1$ **29.** $y - 37\frac{1}{2} = -48$ **30.** $-27\frac{5}{7} = x - 30$

31. $-102 = y - 97.5$ **32.** $-0.01 = w - 0.001$ **33.** $10 = b + 12.7$

34. $-15 = c - 7.5$ **35.** $x - 0.999 = 0.001$ **36.** $0.5 = d + 102.5$

In 37–39, find the solution of each equation to the *nearest tenth.*

37. $x + \sqrt{5} = 12$ **38.** $x - \pi = 7$ **39.** $x + \sqrt{7} = \sqrt{15}$

40. If $g + 9 = 11$, find the value of $7g$.

41. If $t - 0.5 = 2.5$, find the value of $t + 7$.

42. If $8 = 22 + y$, find the value of $-4y$.

43. If $c - 1\frac{1}{4} = 2\frac{1}{2}$, find the value of $8c - 2$.

44. If $2.7 + b = 1.8$, find the value of $\frac{1}{3}b + 0.3$.

6-3 SOLVING EQUATIONS USING MULTIPLICATION

You learned in Section 6-2 that, to solve an equation, you need to work backward or "undo" what has been done by using inverses. To undo multiplication by a number, multiply by its reciprocal or multiplicative inverse.

Number	*Multiplication*	*Multiplication by the Reciprocal*	*Result*
8	2(8)	$\frac{1}{2}(2)(8) = \frac{2(8)}{2}$	8
−12	$\frac{1}{7}(-12)$	$7\left(\frac{1}{7}\right)(-12)$	−12
4	−3(4)	$-\frac{1}{3}(-3)(4)$	4
x	$5x$	$\frac{1}{5}(5x) = \frac{5x}{5}$	x
x	$-\frac{2}{3}x$	$-\frac{3}{2}\left(-\frac{2}{3}x\right)$	x

Note that to multiply by a number is the same as to divide by the inverse of the number. For example, to multiply by $\frac{1}{5}$ is the same as to divide by 5; that is, $\frac{1}{5}(5x) = \frac{5x}{5} = x$.

It is also true that to divide by a number is the same as to multiply by the multiplicative inverse of that number. For example, to divide by $\frac{3}{4}$ is the same as to multiply by $\frac{4}{3}$.

$$\left(\frac{3}{4}x\right) \div \left(\frac{3}{4}\right) = \left(\frac{3}{4}x\right) \cdot \left(\frac{4}{3}\right) = \frac{4}{3}\left(\frac{3}{4}x\right) = x$$

- **POSTULATE 5. Multiplication Property of Equality** The *multiplication property of equality* states that for all numbers a, b, and c:

If $a = b$, then $ac = bc$.

Therefore, if both sides of an equality are multiplied by the same number, the resulting products are equal.

When one side of an equation contains a variable multiplied by a constant, we multiply both sides of the equation by the reciprocal of the constant to simplify the equation and find the solution set. For example, if one side of an equation is $-\frac{1}{2}x$, x has been multiplied by $-\frac{1}{2}$. Multiply both sides of the equation by $-\frac{2}{1}$ or -2, the reciprocal of $\frac{1}{2}$, to "undo" the original multiplication.

EXAMPLES

1. Find the solution set for $5x = 35$ when the domain is the set of real numbers.
METHOD 1. Since the variable is multiplied by 5, multiply both sides of the equation by its multiplicative inverse or reciprocal, $\frac{1}{5}$.

How to Proceed:	*Solution:*
(1) Write the equation:	$5x = 35$
(2) Use the multiplicative property of equality:	$\frac{1}{5}(5x) = 35\left(\frac{1}{5}\right)$
(3) Use the associative property of multiplication:	$\left[\frac{1}{5}(5)\right]x = 7$
(4) Use the inverse property of multiplication:	$1x = 7$
(5) Use the identity property of multiplication:	$x = 7$

METHOD 2. Since multiplying by $\frac{1}{5}$ is equivalent to dividing by 5, divide both sides of the equation by 5.

How to Proceed:	*Solution:*	*Check:*
(1) Write the equation:	$5x = 35$	$5x = 35$
(2) Divide both sides of the equation by 5:	$\frac{5x}{5} = \frac{35}{5}$	$5(7) \stackrel{?}{=} 35$
(3) Simplify each member:	$1x = 7$	$35 = 35$ ✔
	$x = 7$	

Check the solution by showing that, when x is replaced by 7 in the original equation, a true statement results.
Answer: $\{7\}$

---

2. Find the solution set for $\frac{x}{3} = 4$ when the domain is the set of real numbers.

An algebraic expression such as $\frac{x}{3}$ can be thought of as x divided by 3 or as $\frac{1}{3}$ times x. Therefore, an equation such as $\frac{x}{3} = 4$ can be solved by multiplying both sides of the equation by the reciprocal of $\frac{1}{3}$, which is $\frac{3}{1}$ or 3.

How to Proceed:	*Solution:*	*Check:*
(1) Write the equation:	$\frac{x}{3} = 4$	$\frac{x}{3} = 4$
(2) Multiply both sides of the equation by 3:	$3\left(\frac{1}{3}x\right) = 4(3)$	$\frac{12}{3} \stackrel{?}{=} 4$
(3) Simplify each member:	$1x = 12$	$4 = 4$ ✔
	$x = 12$	

Answer: $\{12\}$

---

3. Solve and check: $-\frac{2}{3}y = 18$.

How to Proceed:	*Solution:*	*Check:*
(1) Write the equation:	$-\frac{2}{3}y = 18$	$-\frac{2}{3}y = 18$
(2) Multiply both members of the equation by $-\frac{3}{2}$, the multiplicative inverse (reciprocal) of the coefficient $-\frac{2}{3}$:	$\left(-\frac{3}{2}\right)\left(-\frac{2}{3}y\right) = \left(-\frac{3}{2}\right)(18)$	$\left(-\frac{2}{3}\right)(-27) \stackrel{?}{=} 18$ $18 = 18$ ✔
(3) Simplify each member:	$1 \cdot y = -27$ $y = -27$	

Answer: $y = -27$

---

4. Find to the *nearest tenth* the solution of $2\pi r = 12$.
In the equation, the variable r is multiplied by the constant 2π.

How to Proceed:	*Solution:*
(1) Write the equation:	$2\pi r = 12$
(2) To simplify the equation, multiply both sides by $\frac{1}{2\pi}$ (the reciprocal of the constant):	$\frac{1}{2\pi}(2\pi r) = \frac{1}{2\pi}(12)$ $\left[\frac{1}{2\pi}(2\pi)\right]r = \frac{12}{2\pi}$
(3) Simplify the left member:	$1r = \frac{12}{2\pi}$ $r = \frac{12}{2\pi}$

(4) Use a calculator to find the approximate decimal value of r.

Enter: 12 [÷] [(] 2 [×] [π] [)] [=]

Display: 1.909859317

(5) Round the number in the display to the nearest tenth.

Answer: 1.9

EXERCISES

In 1–36, solve and check each equation. The domain is the set of real numbers.

1. $3x = 15$ **2.** $10c = 90$ **3.** $20 = 5d$ **4.** $84 = 7y$

5. $5c = 5$ **6.** $6s = 0$ **7.** $-5d = 45$ **8.** $-8 = 2w$

9. $8x = 1$ **10.** $1 = 6b$ **11.** $4d = -0.8$ **12.** $0.36 = -6m$

13. $\frac{1}{2}d = 18$ **14.** $20 = \frac{1}{10}r$ **15.** $\frac{2}{3}x = 18$ **16.** $\frac{a}{2} = -3$

17. $3y = 3$ **18.** $2c = 0$ **19.** $81 = -1d$ **20.** $15 = 2y$

21. $6c = 44$ **22.** $90w = 10$ **23.** $20 = -0.1x$ **24.** $2a = -0.6$

25. $0.4x = 3.2$ **26.** $1.4x = -5.6$ **27.** $-0.12 = -0.04d$ **28.** $4 = 0.2y$

29. $\frac{1}{8}x = \frac{1}{4}$ **30.** $\frac{1}{3} = \frac{m}{4}$ **31.** $\frac{2}{3}b = 8$ **32.** $-\frac{5}{3}y = 45$

33. $-\frac{7}{2}b = -28$ **34.** $\frac{x}{0.5} = 3$ **35.** $\frac{3}{4}m = 5\frac{1}{4}$ **36.** $-\frac{5}{8}a = -7\frac{1}{2}$

In 37–40, find the solution to each equation to the *nearest tenth*.

37. $\sqrt{2}x = 10$ **38.** $9.7y = 19.6$ **39.** $\frac{2}{3}x = \pi$ **40.** $\frac{a}{3} = \sqrt{5}$

41. If $9x = 36$, find the value of $2x$.

42. If $2y = -0.7$, find the value of $\frac{2}{5}y$.

43. If $-\frac{t}{3} = 4$, find the value of $-8t - 3$.

44. If $0.08m = 0.96$, find the value of $3 - \frac{1}{2}m$.

6-4 SOLVING EQUATIONS USING MORE THAN ONE OPERATION

In the equation $2x + 3 = 15$, there are two operations in the left member: *multiplication and addition*. In forming the left member of the equation, x was first multiplied by 2, and then 3 was added to the product. To solve this equation, we must undo these operations by using the inverse elements in the reverse order. Since the last operation was to add 3, the first step in solving the equation is to add its opposite, -3, to both sides of the equation.

$$2x + 3 = 15$$
$$2x + 3 + (-3) = 15 + (-3)$$
$$2x = 12$$

Now we have a simpler equivalent equation that has the same solution set as the original and includes only multiplication by 2. To solve this simpler equation, we multiply both sides of the equation by $\frac{1}{2}$, the reciprocal of 2, or divide both sides of the equation by 2.

$$2x = 12 \qquad\qquad 2x = 12$$
$$\frac{1}{2}(2x) = 12\left(\frac{1}{2}\right) \quad \text{or} \quad \frac{2x}{2} = \frac{12}{2}$$
$$x = 6 \qquad\qquad x = 6$$

EXAMPLES

1. Solve and check: $7x + 15 = 71$.

How to Proceed:	*Solution:*
(1) Write the equation:	$7x + 15 = 71$
(2) Add -15 to each member:	$\underline{\quad -15 = -15}$ $7x = 56$
(3) Divide each member by 7:	$\frac{7x}{7} = \frac{56}{7}$ $x = 8$
(4) Check the solution:	$7x + 15 = 71$ $7(8) + 15 \stackrel{?}{=} 71$ $56 + 15 \stackrel{?}{=} 71$ $71 = 71$ ✔

Answer: $x = 8$

---

2. Find the solution set and check: $\frac{3}{5}x - 6 = -18$.

Solution:

$$\frac{3}{5}x - 6 = -18$$
$$\underline{\quad + 6 = +6}$$
$$\frac{3}{5}x = -12$$
$$\frac{5}{3}\left(\frac{3}{5}x\right) = -12\left(\frac{5}{3}\right)$$
$$x = -20$$

Check:

$$\frac{3}{5}x - 6 = -18$$
$$\frac{3}{5}(-20) - 6 \stackrel{?}{=} -18$$
$$-12 - 6 \stackrel{?}{=} -18$$
$$-18 = -18 \checkmark$$

Answer: The solution set is $\{-20\}$.

---

3. Solve and check: $7 - x = 9$

METHOD 1: Think of $7 - x$ as $7 + (-1x)$.

$$7 + (-1x) = 9$$
$$\underline{-7 \qquad = -7}$$
$$-1x = 2$$
$$\frac{-1x}{-1} = \frac{2}{-1}$$
$$x = -2$$

Check:

$$7 - x = 9$$
$$7 - (-2) \stackrel{?}{=} 9$$
$$7 + 2 \stackrel{?}{=} 9$$
$$9 = 9 \checkmark$$

METHOD 2:

How to Proceed:	*Horizontal Format:*	*Vertical Format:*
(1) Write the equation:	$7 - x = 9$	$7 - x = 9$
(2) Add x to each side of the equation so that the variable has a positive coefficient:	$7 - x + x = 9 + x$ $7 = 9 + x$	$\underline{+ x} \quad \underline{+ x}$ $7 \quad = 9 + x$
(3) Solve the equivalent equation $7 = 9 + x$ by adding -9 to both sides:	$-9 + 7 = -9 + 9 + x$ $-2 = x$	$\underline{-9} \quad \underline{-9}$ $-2 \quad = \quad x$
(4) Check in the original equation:	$7 - x = 9$ $7 - (-2) \stackrel{?}{=} 9$ $7 + 2 \stackrel{?}{=} 9$ $9 = 9$ ✔	

Answer: $\{-2\}$ or $x = -2$

EXERCISES

In 1–36, solve and check each equation.

1. $3x + 5 = 35$ **2.** $5a + 17 = 2$ **3.** $4x - 1 = 15$ **4.** $3y - 5 = 16$

5. $55 = 6a + 7$ **6.** $17 = 8c - 7$ **7.** $75 = 11 - 16x$ **8.** $9 - 1x = 7$

9. $11 = 15t + 16$ **10.** $15 - a = 3$ **11.** $11 = -6d - 1$ **12.** $8 - y = 1$

13. $\frac{3a}{8} = 12$ **14.** $\frac{2}{3}x = -8$ **15.** $12 = \frac{3}{4}y$ **16.** $\frac{5t}{4} = \frac{45}{2}$

17. $-\frac{3}{5}m = 30$ **18.** $7.2 = \frac{4m}{5}$ **19.** $\frac{x}{3} + 4 = 13$ **20.** $\frac{a}{4} + 9 = 5$

21. $15 = \frac{b}{7} - 8$ **22.** $-2 = \frac{y}{5} + 3$ **23.** $0.6m + \frac{1}{3} = 18\frac{1}{3}$ **24.** $9d - \frac{1}{2} = 17\frac{1}{2}$

25. $4a + 0.2 = 5$ **26.** $4 = 3t - 0.2$ **27.** $\frac{1}{4}x + 11 = 5$ **28.** $\frac{1}{9}x + 4 = 13$

29. $13 = 5 - \frac{2}{3}y$ **30.** $\frac{4}{5}t + 7 = 47$ **31.** $0.2x + 1.6 = 1$ **32.** $8 - 0.1b = 3$

33. $0.04c + 1.6 = 0$ **34.** $15x + 14 = 19$ **35.** $14 = 12b + 8$ **36.** $8 = 18c - 1$

6-5 WRITING VERBAL SENTENCES AS EQUATIONS

Often a problem can be solved by writing an equation that describes how the numbers in the problem are related and then solving this equation. At the start, this procedure may seem more difficult than solving the problem by other methods, such as guess and check. However, as you practice writing equations, you will find them easier to write; and as problems become more complex, you will find an algebraic strategy to be powerful and efficient.

Verbal Sentence:	Four times a number s			equals	20.
Equation:	$4s$			$=$	20
Verbal Sentence:	A number y	increased by	6	equals	8.
Equation:	y	$+$	6	$=$	8
Verbal Sentence:	A number x	decreased by	3	equals	5.
Equation:	x	$-$	3	$=$	5
Verbal Sentence:	A number n	divided by	2	equals	4.
Equation:	n	$\div$	2	$=$	4

Note that the words of a sentence may need to be replaced by symbols in a different order.

Verbal Sentence:	Eight less than a number	is	10.
Equation:	$x - 8$	$=$	10

PROCEDURE. To write a verbal sentence as an equation:
1. Choose a letter to represent the unknown quantity in the problem.
2. Use this letter, with the symbols for the arithmetic operations, to write the verbal sentence as an equation.

EXAMPLE

Write the following sentence as an equation: "Five times a number decreased by 7 equals 13."

Solution Let x represent the number.

Five times a number	decreased by	7	equals	13.
$5x$	$-$	7	$=$	13

Answer: $5x - 7 = 13$

EXERCISES WITH OPEN-RESPONSE PROBLEMS

In 1–9, select the equation that represents, in terms of the given variable, the numerical relationship in each sentence.

1. Three times Harold's height is 108 inches. Let h = Harold's height.
(1) $h + 3 = 108$ (2) $3h = 108$
(3) $h - 3 = 108$ (4) $\frac{1}{3}h = 108$

2. One-half of Mary's weight is 20 kilograms. Let w = Mary's weight.
(1) $\frac{1}{2}w = 20$ (2) $2w = 20$ (3) $w + \frac{1}{2} = 20$ (4) $w - \frac{1}{2} = 20$

3. When the original members of the Chess Club were joined by seven new members, there were 28 members. Let n = the original number of members.
(1) $7n = 28$ (2) $n + 7 = 28$ (3) $n - 7 = 28$ (4) $\frac{1}{7}n = 28$

4. After Lyle spent 5 dollars of what he earned this week he had 15 dollars left. Let x = the number of dollars Lyle earned this week.
(1) $x + 5 = 15$ (2) $5x = 15$ (3) $\frac{1}{5}x = 15$ (4) $x - 5 = 15$

5. After a carpenter cut a piece 7 inches long from a board, the remaining piece was 17 inches long. Let y = the length of the board before cutting.
(1) $7 - y = 17$ (2) $y - 7 = 17$ (3) $17 - y = 7$ (4) $y + 7 = 17$

6. The owner of a clothing store bought 15 suits and now has 75 suits in stock. Let s = the number of suits he had before the purchase.
(1) $s + 15 = 75$ (2) $s - 15 = 75$ (3) $15s = 75$ (4) $\frac{s}{15} = 75$

7. In a rectangle, 4 times the width is 100 centimeters. Let w = the width of the rectangle.
(1) $w + 4 = 100$ (2) $4w = 100$ (3) $w - 4 = 100$ (4) $\frac{w}{4} = 100$

8. The measure of the area of a parallelogram is 24 square centimeters, and the measure of the base is 6 centimeters. Let h = the height of the parallelogram.
(1) $h + 6 = 24$ (2) $6h = 24$ (3) $\frac{h}{6} = 24$ (4) $24h = 6$

9. The measure of the base of a triangle is 4 centimeters, and the measure of the area is 32 square centimeters. Let h = the height of the triangle.
(1) $4h = 32$ (2) $2(4h) = 32$ (3) $\frac{1}{2}(h + 4) = 32$ (4) $\frac{1}{2}(4h) = 32$

In 10–26, write each sentence as an equation. Use n to represent "a number."

10. Eight more than a number is 15.
11. Four less than a number is 24.
12. Twelve added to a number is 26.
13. A number decreased by 5 equals 25.
14. A number multiplied by 3 equals 39.
15. A number divided by 4 equals 16.
16. The product of 7 and a number equals 70.
17. One-half of a number, decreased by 7, equals 11.
18. Twice a number, increased by 7, equals 27.
19. Twice a number, decreased by 5, equals 25.
20. The sum of 3 times a number and 7 is 22.

21. When 9 is subtracted from 5 times a number, the result is 31.
22. The sum of 100 and a number is equal to 3 times that number.
23. If 3 times a number is increased by 12, the result is the same as twice the number increased by 24.
24. If 8 times a number is decreased by 20, the result is the same as 3 times the number increased by 80.
25. The sum of a number and twice that number equals 45.
26. Three times a number decreased by half of that number equals 40.

In 27–30, write a problem that can be solved using each given equation. In each case, state what the variable represents.

27. $x + 12 = 18$ **28.** $5x = 50$ **29.** $2x + 2(5) = 80$ **30.** $y - 6 = 14$

6-6 USING EQUATIONS TO SOLVE PROBLEMS

Now you are ready to use an algebraic strategy to solve problems.

PROCEDURE. To solve a problem by using an equation involving one variable:
1. Read the problem carefully.
2. Select a variable that can be used to represent the number to be found.
3. Write an equation using a relationship given in the problem, a previously known relationship, or a formula.
4. Solve the equation.
5. Check the answer by testing it in the given problem.

EXAMPLES

1. After Alphie had completed 12 assigned problems, he still had nine left to do. Find the number of problems assigned.

Solution (1) Read the problem. Decide what is to be found: the number of problems assigned.

(2) Choose a variable to represent what is to be found.
Let x = the number of problems assigned.

(3) Write an equation.
The number assigned minus the number completed is 9.

$$x \quad - \quad 12 \quad = 9$$

(4) Solve the equation.

$$\begin{aligned} x - 12 &= 9 \\ +12 &= +12 \\ \hline x &= 21 \end{aligned}$$

(5) Check the answer.
Is 21 minus 12 equal to 9? Yes

Answer: The number of problems assigned is 21.

2. Emma has addressed seven invitations to her birthday party. This is one more than half of the invitations that she intends to send. How many invitations does she intend to send?

Solution (1) Read the problem. Decide what is to be found: the number of invitations to be sent.

(2) Choose a variable.
Let x = the number of invitations to be sent

(3) Write an equation.
Number addressed is one more than half the number to be sent.

$$7 \quad = 1 \quad + \quad \frac{1}{2}x$$

(4) Solve the equation.

$$\begin{aligned} 7 &= 1 + \frac{1}{2}x \\ -1 &= -1 \\ \hline 6 &= \frac{1}{2}x \\ 2(6) &= 2\left(\frac{1}{2}x\right) \\ 12 &= x \end{aligned}$$

(5) Check the answer.
Half of 12 is 6. One more than 6 is 7, the number of invitations Emma has addressed.

Answer: Emma intends to send 12 invitations.

EXERCISES

In 1–49, solve each problem using a variable and an equation.

1. A number decreased by 20 equals 36. Find the number.

2. Seven times a number is 63. Find the number.

3. If 7 is subtracted from a number, the result is 46. Find the number.

4. Five times a number is 50. Find the number.

5. What number increased by 25 equals 40?

6. When a number is doubled, the result is 36. Find the number.

7. If 18 is added to a number, the result is 32. Find the number.

8. A number divided by 5 equals 17. Find the number.

9. Ten less than a number is 42. Find the number.

10. A number divided by 4 is $3\frac{1}{2}$. Find the number.

11. One-half of a number is 12. Find the number.
12. Three-fifths of a number is 30. Find the number.
13. The sum of 42 and a number is 96. Find the number.
14. A number multiplied by 0.3 is 6. Find the number.
15. Four-hundredths of a number is 16. Find the number.
16. After Helen had spent \$0.25, she had \$0.85 left. How much money did she have originally?
17. After he had lost 13 kilograms, Ben weighed 90 kilograms. Find Ben's original weight.
18. After $2\frac{1}{2}$ feet had been cut from a piece of lumber, $9\frac{1}{2}$ feet were left. What was the original length of the piece of lumber?
19. After a car had increased its rate of speed by 24 kilometers per hour, it was traveling 76 kilometers per hour. What was its original rate of speed?
20. During a charity drive, the boys in a class contributed \$3.75 more than the girls. If the boys contributed \$8.25, how much did the girls contribute?
21. The width of a rectangle is 8 feet less than its length. If the width is 9.5 feet, find the length of the rectangle.
22. A high school admitted 1,125 sophomores, a number that was 78 fewer than the number admitted last year. How many sophomores were admitted last year?
23. How many hours do you have to work to earn \$132 if you are paid \$5.50 per hour?
24. A man earned \$1,000 in 4 weeks. What was his average weekly salary?
25. A dealer sold an electric broiler for \$39.98. This amount was \$12.50 more than the broiler had cost her. How much did the broiler cost the dealer?
26. After using his baseball glove for some time, Charles sold it for \$12.25 less than he paid for it. If Charles sold the glove for \$3.50, how much did he pay for it originally?
27. Baseballs cost \$3.50 each. How many baseballs can be bought for 28.00?
28. The width of a rectangle is $\frac{1}{5}$ of its length. If the width of the rectangle is 6 meters, what is its length?
29. Sue wishes to buy a radio that costs \$38. If she has already saved \$26 for this purpose, how much must she still save to buy the radio?
30. A merchant bought 8 dozen shirts. If he has sold all but 18 of them, how many shirts has he sold?
31. Mr. Alvarez withdrew from his savings account \$25 per week for each of 8 weeks. He then had \$1,623 in his account. How much did he have in his account before he made these withdrawals?
32. In a basketball game, the Knicks scored 12 points more than the Celtics. If the Knicks scored 108 points, how many points did the Celtics score?
33. Last week, Juan worked 7 hours more than 3 times the number of hours that George worked. If Juan worked 40 hours, how long did George work?
34. Ten times a number, increased by 9, is 59. Find the number.
35. The sum of 8 times a number and 5 is 37. Find the number.
36. If 6 times a number is decreased by 4, the result is 68. Find the number.
37. The difference between 4 times a number and 3 is 25. Find the number.

38. If a number is multiplied by 7, and the product is increased by 2, the result is 100. Find the number.
39. When 12 is subtracted from 3 times a number, the result is 24. Find the number.
40. If 9 is added to one-half of a number, the result is 29. Find the number.
41. If two-thirds of a number is decreased by 4, the result is 56. Find the number.
42. If 38 is added to $\frac{5}{9}$ of a number, the result is 128. Find the number.
43. The sum of $\frac{3}{5}$ of a number and 2.3 is 14.6. Find the number.
44. The larger of two numbers is 12 more than twice the smaller. The larger number is 36. What is the smaller number?
45. The larger of two numbers exceeds 6 times the smaller by 10. If the larger number is 76, find the smaller number.
46. Find the width of a rectangle if the length is 12 centimeters and the perimeter is 42 centimeters.
47. When Rosalie bought four packages of paper napkins, the tax was $0.22. If the total cost of the napkins and the tax was $3.38, what was the cost of one package of napkins?
48. The bookstore ordered boxes of pencils for resale. The cost of the pencils was $2.25 per box, and the shipping cost of the order was $2.80. If the total cost of the pencils and shipping was $29.80, how many boxes of pencils were ordered?
49. Dr. Cortez spent 30 minutes driving from home to Seaview Hospital. This was 12 minutes less than twice the time it took him to drive home. How long did it take him to drive home?

6-7 SOLVING PERCENT PROBLEMS

A ***percent*** is a way of expressing the quotient of a number divided by 100. When used in computation, a percent is always expressed as an equivalent decimal or common fraction. For example, 12% = 12 hundredths, written as 0.12 or $\frac{12}{100}$. Use *either* the fraction or the decimal to replace the percent.

In Arithmetic	*In Algebra*
1. To find 12% of 420, write 12% as 0.12: $12\% \text{ of } 420 = 0.12(420) = 50.40$	**1.** To express 12% of a number n, write 12% as 0.12: $12\% \text{ of a number } n = 0.12n$
2. To find 7% of $300, write 7% as $\frac{7}{100}$: $7\% \text{ of } \$300 = \frac{7}{100}(300) = \mathbf{\$21}$	**2.** To express 7% of an amount a, write 7% as $\frac{7}{100}$: $7\% \text{ of an amount } a = \frac{7}{100}a$

EXAMPLE

The Hesters save 8% of the family income. What was their income last week if they saved $34?

How to Proceed:	*Solution:*
(1) Represent last week's income by a variable:	Let n = last week's income.
(2) Using the same variable, represent last week's savings:	Then, $0.08n$ = last week's savings.
(3) Write an equation:	$0.08n = 34$
(4) Solve the equation:	$n = 425$
(5) Check:	Is 8% of 425 equal to 34? Yes, $0.08(425) = 34$.

Answer: The Hester family's income last week was $425.

EXERCISES

In 1–18, use an algebraic equation to solve each problem.

1. 4% of a number is 8. Find the number.
2. 25% of a number is 12. Find the number.
3. 120% of a number is 54. Find the number.
4. 15% of what number is 4.5?
5. $33\frac{1}{3}$ % of what number is $3\frac{1}{3}$?
6. Marvin saved 25% of his allowance. If he saved $2.50, how much was his allowance?
7. The Giants won 24 games. If that number was 60% of all the games they played, how many games did they play?
8. Pearl bought a coat at a sale that offered 40% off. If she saved $48, what was the original price of the coat?
9. Patricia answered 90% of the questions on a test correctly. If she answered 45 questions correctly, how many questions were on the test?
10. At a sale, a radio sold for $20. This amount was 80% of the original price. What was the original price?
11. In a restaurant, Mrs. Green left a tip of $1.20, which was 15% of the amount of the bill. What was the amount of the bill?
12. There are 210 freshmen at East High School. The number of freshmen is 30% of the number of students enrolled in the school. How many students are enrolled in East High School?

13. One day 92 students, a number that was 8% of the total enrollment, were absent. Find the total enrollment.

14. Mr. Hendrick's salary this year is 106% of last year's salary. If this year's salary is \$26,341, what was last year's salary?

15. A dealer sold a suit for 150% of the amount he paid for it. If the dealer sold the suit for \$120, how much did it cost him?

16. The selling price of an article is 175% of the dealer's cost price. If the dealer sold the article for \$28, how much did she pay for it?

17. When Irene bought a dress, she had to pay an 8% sales tax. If the sales tax was \$2.40, what was the price of the dress?

18. The Surevalve Company tested 5% of the valves it had produced. If 350 valves were tested, how many valves had the company produced?

6-8 MORE PRACTICE IN SOLVING EQUATIONS

In previous sections you learned some procedures that are useful in solving different types of equations. To solve an equation, it is necessary to select a procedure that is appropriate for the given equation.

EXERCISES

In 1–44, solve and check each equation.

1. $9x = 108$ **2.** $9b + 8 = 8$ **3.** $4y + 2 = 38$ **4.** $0.15x = 0.06$

5. $79 = 5x - 6$ **6.** $19 + 5y = 4$ **7.** $18 = 10 - 4a$ **8.** $14 = n - 5.6$

9. $\frac{5}{7}b = 35$ **10.** $39 = 5x$ **11.** $7x - 2 = 68$ **12.** $50 = 1 - 7m$

13. $48 = 5.7b - 3.3$ **14.** $\frac{2}{5}x + 15 = 7$ **15.** $3x - \frac{1}{3} = 5\frac{2}{3}$ **16.** $87 = 13 - 2x$

17. $-0.11a = -44$ **18.** $x - 3\frac{1}{8} = 7\frac{1}{4}$ **19.** $3y + 94 = 19$ **20.** $\frac{3t}{4} = 84$

21. $6x + 3 = 60$ **22.** $10r + 2 = 0$ **23.** $\frac{s}{4} + 15 = 7$ **24.** $0.8c - 4 = 3.2$

25. $0.01x = 5$ **26.** $0.9b = 0.18$ **27.** $5 - 5n = 30$ **28.** $8m + 3 = 91$

29. $0.3x + 0.2 = 8$ **30.** $\frac{n}{6} + 5\frac{2}{3} = 12\frac{1}{3}$ **31.** $\frac{4y}{5} + 32 = 28$ **32.** $29 = \frac{1}{3}a + 12$

33. $9x - 3 = 91.5$ **34.** $\frac{15}{4} = -\frac{r}{8}$ **35.** $\frac{7}{10}x - 1 = 34$ **36.** $\frac{1}{8}t + \frac{1}{4} = \frac{3}{2}$

37. $\frac{x}{14} = \frac{5}{7}$ **38.** $x + \frac{1}{8} = \frac{7}{8}$ **39.** $1 - \frac{1}{3}c = 14$ **40.** $3m - 6\frac{1}{2} = 8\frac{1}{2}$

41. $\frac{y}{8} + 5 = \frac{9}{2}$ **42.** $18 = 6 - 8c$ **43.** $7 = 2 - \frac{1}{2}x$ **44.** $2y + \frac{1}{2} = \frac{3}{4}$

CHAPTER SUMMARY

An ***equation*** is a sentence that states that two algebraic expressions are equal.

The ***replacement set*** or ***domain*** is the set of possible values for the variable in an open sentence. Any number from the replacement set that makes an open sentence true is a ***root*** or ***solution*** of the equation. The set consisting of all elements of the replacement set that make an equation true is called the ***solution set*** of the equation.

To ***solve*** an equation means to find its solution set. If no element of the replacement set makes an equation true, the solution set is the empty set. If all elements of the replacement set make an equation true, the equation is called an ***identity***.

The ***reflexive property of equality*** states that for every number a: $a = a$.

The ***symmetric property of equality*** states that for all numbers a and b: If $a = b$, then $b = a$.

The ***transitive property of equality*** states that for all numbers a, b, and c: If $a = b$, and $b = c$, then $a = c$.

The ***addition property of equality*** states that for all numbers a, b, and c: If $a = b$, then $a + c = b + c$.

The ***multiplication property of equality*** states that for all numbers a, b, and c: If $a = b$, then $ac = bc$.

To solve a problem by using an equation involving one variable:

1. Read the problem carefully.
2. Select a variable that can be used to represent the number to be found.
3. Write an equation using a relationship given in the problem, a previously known relationship, or a formula.
4. Solve the equation.
5. Check the answer by testing it in the given problem.

VOCABULARY

6-1 Equation   Left side (left member)   Right side (right member)   Root (solution)   Identity   Conditional equation   Assumption   Postulate   Axiom   Reflexive property of equality   Symmetric property of equality   Transitive property of equality

6-2 Addition property of equality   Equivalent equations   Simplifying (solving) the equation   Constant

6-3 Multiplication property of equality

6-7 Percent

REVIEW EXERCISES WITH OPEN-RESPONSE PROBLEMS

In 1–3, using the domain {1, 3, 5, 7, 9}, find the solution set of each equation. The empty set, ∅ or { }, is a possible answer.

1. $18 - 3x = 9$ **2.** $0.2y + 3 = 4$ **3.** $4t - 7 = 17$

4. Twelve less than a number is 18. Find the number.

5. If $6k = 48$, find the value of $k + 5$.

6. If $d + \frac{3}{4} = 2\frac{1}{2}$, what is the value of $12d$?

In 7 and 8, select the numeral preceding the correct response to each question.

7. Which of the numbers is a root of equation $0.2w = 4$?
(1) 0.2 (2) 2 (3) 20 (4) 4

8. Which statement is true if $x = a + b$ and $a + b = 2r$?
(1) $x = r$ (2) $x = 2r$ (3) $x = 4r$ (4) $a = 2r$

In 9–20, solve and check each equation.

9. $t - 9 = 14$ **10.** $23 + y = -1$ **11.** $6m = 3$

12. $w - 0.5 = 9$ **13.** $8 = \frac{1}{4}b$ **14.** $7x + 2 = 1$

15. $5b - 3 = 32$ **16.** $2x + 4 = 0$ **17.** $0.03p = 0.3$

18. $\frac{k}{10} = \frac{5}{2}$ **19.** $3t = \frac{3}{4}$ **20.** $8 - \frac{2}{3}c = 80$

In 21–24: **a.** Write each sentence as an equation in which n is used to represent the number. **b.** Solve for n.

21. When twice a number is decreased by 5, the result is 27. Find the number.

22. If one-sixth of a number is $1\frac{1}{3}$, find the number.

23. Five times a number increased by 4 is 40. Find the number.

24. A store sold 5% of a new shipment of toys. If the store sold 16 of these toys, how many toys were in the shipment?

25. Use an algebraic equation to solve the following problem: On a quiz show, Georgette won \$10 less than 3 times as much as Charlie. If Georgette won \$800, how much did Charlie win?

26. In his will, Mr. Banks left 50% of the value of his property to his wife, 22% to his daughter, 16% to his son, and the remaining \$30,000 to charity. What was the value of his property?

CUMULATIVE REVIEW WITH OPEN-RESPONSE PROBLEMS

1. The measures of the sides of a triangle are 12 centimeters, 14 centimeters, and 18 centimeters. Find the perimeter of the triangle.

2. The measures of two sides of a triangle are 13.7 inches and 12.6 inches, and the perimeter is 32.1 inches.

a. Write an equation that can be used to find the measure of the third side of the triangle.

b. Find the measure of the third side of the triangle by solving the equation written in part **a**.

3. Graph the solution set of $x \geq -2$ on the real number line.

4. a. Graph the rectangle $ABCD$, with vertices $A(-2, 1)$, $B(5, 1)$, $C(5,-7)$, and $D(-2,-7)$, on the coordinate plane.

b. Find the area of the rectangle.

5. a. The perimeter of a rectangle can be found using the formula $P = 2\ell + 2w$. Using this formula, write an equation to find the width of a rectangle if $\ell = 3.5$ and $P = 17.6$.

b. Solve the equation written in part **a** to find the width of the rectangle.

6. An operation $*$ is defined so that the following are true.

$3 * 5 = 16$
$1 * 8 = 18$
$4 * 4 = 16$
$\frac{1}{2} * \frac{1}{2} = 2$

a. Write a rule for $a * b$ or describe in words how the answer is obtained.

b. Find the value of $10 * 2$.

c. If a and b represent the length and width of a rectangle, what does $a * b$ represent?

7. Although Albert usually uses a calculator for computation, he enjoys challenging himself by performing mental computations.

a. Albert multiplies a number by 5 in this way: he first places a 0 at the end of the number (to multiply by 10) and then he takes half of this new number. Explain why this method is correct. What property of the real numbers is Albert using?

b. Albert multiplied 24(15) by first multiplying 24(10), then multiplying 24(5), and finally adding the results. What property of the real numbers was Albert using?

Exploration

In each of the following areas (1) to (6), find an equation that is used in the indicated field of science or business, and explain its use.

(1) biology (2) chemistry (3) physics
(4) earth science (5) retailing (6) banking

Chapter 7

Mathematical Reasoning

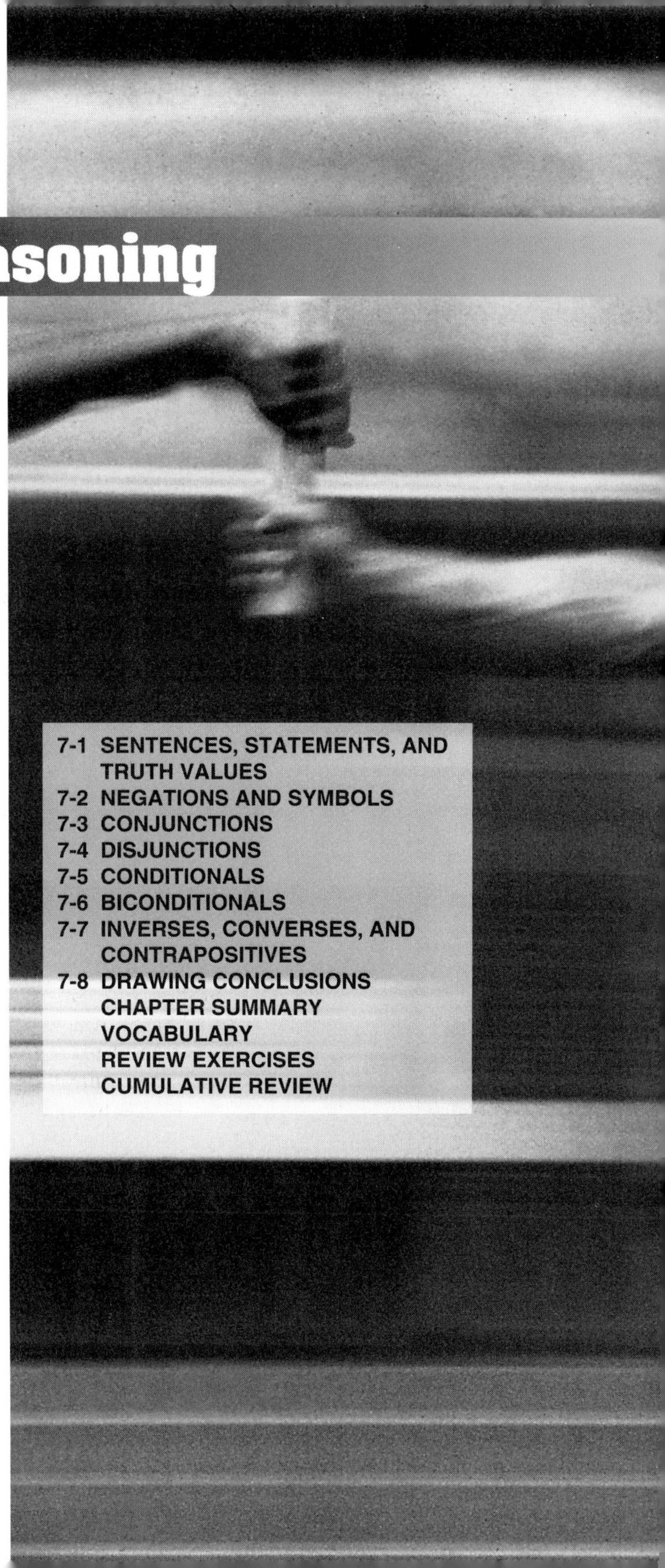

Every team that competes in a 4 × 100-meter relay race consists of four people. Each person runs a separate part of the race, 100 meters in length, called a *leg*. As one runner finishes his leg of the race, he passes a baton to the next runner in a designated exchange zone.

Here are some facts about a team of young men entered in a 4 × 100-meter relay race:

1. The last names of the runners, in no special order, are Cavallaro, Dunn, McCormack, and Jones.
2. Sal runs his leg of the race at some point before Jones races.
3. After Al finishes racing, he passes the baton to McCormack, who runs his leg of the race and then passes the baton to Gary.
4. Dunn passes the baton to Cavallaro.
5. Joe is not the first runner on this relay team.

Use your reasoning powers to try to find the full name, both first and last, of each of the four runners on this team and the number of the leg that each ran, from leg 1 for the first runner to leg 4 for the last runner.

If you can solve this problem, you are using the thinking skill called *logic*.

7-1 SENTENCES, STATEMENTS, AND TRUTH VALUES

Still another effective approach to problem solving is through a branch of mathematics called logic. ***Logic*** is the study of reasoning. All *reasoning*, whether in mathematics or in everyday living, is based on the ways in which we put sentences together.

Sentences That Are True or False

In mathematics, we are concerned with only one type of sentence. A ***mathematical sentence*** must state a fact or contain a complete idea. A mathematical sentence is like a simple declarative sentence in English. It contains a subject and a predicate. Because every mathematical sentence states a fact, we can usually judge such a sentence to be true or false. For example:

1. Every triangle has three sides.	True mathematical sentence
2. $5 + 7 = 12$	True mathematical sentence
3. Chicago is a city.	True mathematical sentence
4. $3 + 5 = 7$	False mathematical sentence
5. Chicago is the capital of Texas.	False mathematical sentence

Nonmathematical Sentences and Phrases

Sentences that ask questions or give commands are not mathematical sentences. We never use these sentences in reasoning because we cannot judge whether they are true or false. For example:

1. Did you do your homework?	This is not a mathematical sentence because it asks a question.
2. Get home early.	This is not a mathematical sentence because it gives a command.

A ***phrase*** is an expression that is only part of a sentence. Because a phrase is not a sentence it does not contain a complete idea, and we cannot judge whether it is true or false. For example:

1. Every triangle	This is a phrase, not a mathematical sentence.
2. $4 + 5$	This is a phrase, not a mathematical sentence.

In reasoning, we try to judge whether sentences are true or false. Some sentences contain complete thoughts that are true for some people and false for others. For example:

1. It's hot in this room.
2. That sound is too loud.
3. Liver tastes really good.

Conclusions based on sentences like these may also be true for some people and false for others.

Open Sentences

Sometimes it is impossible to tell whether a sentence is true or false because more information is needed. Such sentences contain ***variables*** or unknowns. You have seen that a variable may be a symbol such as n or x, but a variable may also be a pronoun such as *he* or *it*. Any sentence that contains a variable is called an ***open sentence***. For example:

1. $x + 3 = 7$	Open sentence; the variable is x.
2. She is my sister.	Open sentence; the variable is *she*.
3. It's on TV at 8 o'clock.	Open sentence; the variable is *it*.

In studying equations, you have seen that a variable is a placeholder for an element from a set of replacements called the ***domain*** or ***replacement set***. The set of all replacements that will change the open sentence into true sentences is called the ***solution set*** or ***truth set***. For example:

Open sentence: $x + 10 = 13$

Variable: x

Domain: $\{1, 2, 3, 4\}$

When $x = 3$, then $3 + 10 = 13$ is a true sentence.

Solution set: $\{3\}$

The concepts that we apply to sentences in algebra are exactly the same as those we apply to sentences that we speak to one another. Of course, we would not use a domain like $\{1, 2, 3, 4\}$ for the open sentence "It is the capital of France." Common sense tells us to use a domain consisting of the names of cities. The following example compares this open sentence with the algebraic sentence used above. Open sentences, variables, domains, and solution sets behave in exactly the same way.

Open sentence: It is the capital of France.

Variable: It

Domain: {Names of cities}

When "It" is replaced by "Paris," then "Paris is the capital of France" is a true sentence.

Solution set: {Paris}

Sometimes a solution set contains more than one element. If Mary has only two sisters, Joanna and Jennifer, then for her the sentence "She is my sister" has the solution set {Joanna, Jennifer}. Here the domain is the set of girls' names.

Some people have no sisters. For them, the solution set for the open sentence "She is my sister" is the empty set, ∅ or { }.

Statements

A sentence that can be judged to be true or false is called a ***statement*** or a ***closed sentence***. In a statement, there are no variables. The following diagram shows how different kinds of sentences are related to one another.

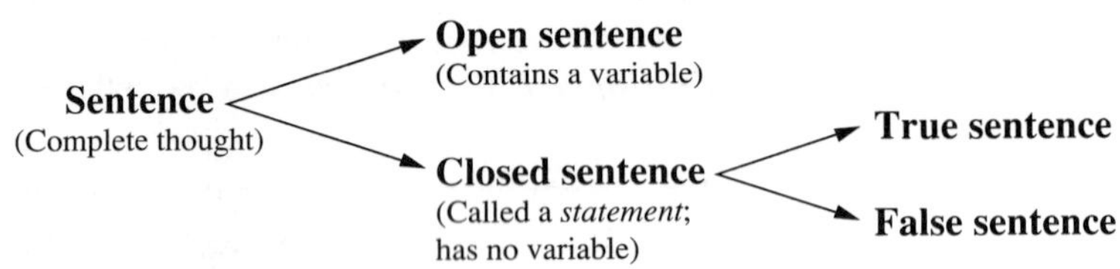

EXAMPLE

Identify each of the following as a true sentence, a false sentence, an open sentence, or not a mathematical sentence at all.

	Answers:
a. Arnold Schwarzenegger is a U.S. President.	False sentence
b. Arnold Schwarzenegger is a movie star.	True sentence
c. He starrred in many action movies.	Open sentence (The variable is *he*.)
d. Do you like action movies?	Not a mathematical sentence (It asks a question.)
e. Read this book.	Not a mathematical sentence (It gives a command.)
f. $3x + 5 < 26$	Open sentence (The variable is x.)
g. $5 + 7 + 8$	Not a mathematical sentence (It is a phrase.)

EXERCISES

In 1–8, tell whether each is or is not a mathematical sentence.

1. The school day ends at 3:15.
2. Take the bus.
3. Are you going?
4. If John goes.
5. Atlanta is a city in Alaska.
6. $x + 3 = 2x + 1$
7. Two teaspoons, 3 times a day.
8. Do the next seven problems.

In 9–16, all of the sentences are open sentences. Find the variable in each sentence.

9. She is smart.
10. We elect a president.
11. It is my favorite color.
12. $4y < 20$
13. This is a great country.
14. It is a counting number.
15. He was the most valuable player in the World Series.
16. $3x - 4 = 10$

In 17–24: **a.** Tell whether each sentence is true, false, or open. **b.** If the sentence is an open sentence, identify the variable.

17. The United States of America declared independence in 1776.
18. They celebrate Independence Day on July 14 every year.
19. San Francisco is a city in New York State.
20. A rectangle is a four-sided polygon.
21. $5x + 2 = 17$
22. $5(10) + 2 = 17$
23. $5(3) + 2 = 17$
24. $2^3 = 3^2$

In 25–29, use the replacement set {New York, Florida, California, Hawaii, Kansas} to find the truth set for each open sentence.

25. Its capital is Sacramento.
26. It does not border on or touch an ocean.
27. It is on the east coast of the United States.
28. It is one of the states of the United States of America.
29. It is one of the last two states admitted to the United States of America.

In 30–41, use the domain of natural numbers to find the solution set for each open sentence. It is possible that no replacements make true sentences.

30. $x + 5 = 17$
31. $x - 5 = 17$
32. $\frac{2x}{3} = 12$
33. $56 = 3x - 4$
34. $2x + x = 6$
35. $2x - x = 1$
36. $x < 3$
37. $x + 1 < 6$
38. $\frac{x}{8} = \frac{1}{2}$
39. $\frac{2}{3} = \frac{x}{6}$
40. $0.2x + 3 = 6$
41. $x + \frac{1}{3} = 3$

In 42–49, use the domain {square, triangle, rectangle, parallelogram, rhombus, trapezoid} to find the truth set for each open sentence.

42. It has three and only three sides.
43. It has exactly six sides.
44. It has fewer than four sides.
45. It must contain only right angles.
46. It has four sides that are all equal in measure.
47. It has two pairs of opposite sides that are parallel.
48. It has exactly one pair of opposite sides that are parallel.
49. It has four angles that are equal in measure and four sides that are equal in measure.

7-2 NEGATIONS AND SYMBOLS

A sentence that has a ***truth value*** is called a statement. There are two truth values: ***true*** and ***false***, indicated by the symbols ***T*** and ***F***. Every statement is either true or false. A statement cannot be both true and false at the same time.

In reasoning, you will learn how to make new statements based upon statements that you already know. One of the simplest examples of this type of reasoning is negating a statement.

The ***negation*** of a statement is usually formed by adding the word *not* to the original or given statement. The negation will always have the opposite truth value of the original statement. For example:

1. Original: John Kennedy was a U.S. president. (True)
   Negation: John Kennedy was *not* a U.S. president. (False)
2. Original: An owl is a fish. (False)
   Negation: An owl is *not* a fish. (True)

There are many ways to insert the word *not* into a statement to form its negation. One method starts the negation with the phrase "It is not true that . . . ." For example:

Original: The post office handles mail. (True)
Negation: It is *not* true that the post office handles mail. (False)
Negation: The post office does *not* handle mail. (False)

Both negations express the same false statement.

The First Symbols in Logic

In logic, we use a single letter to represent a single complete thought. This means that an entire sentence can be replaced by a single letter of the alphabet. Although any letter can represent a simple statement, the letters most often used in logic are p, q, and r. For example, p might represent "John Kennedy was a U.S. president."

To show a negation of a simple statement, we place the symbol $\sim$ before the letter for the original or given statement. Thus $\sim p$ would represent "John Kennedy was *not* a U.S. president." We read the symbol $\sim p$ as "not p."

	Symbol	*Statement in Words*	*Truth Value*
1.	p:	There are 12 months in every year.	(True)
	$\sim p$:	There are *not* 12 months in every year.	(False)
2.	q:	$8 + 9 = 10$	(False)
	$\sim q$:	$8 + 9 \neq 10$	(True)

When p is true, then its negation $\sim p$ is false. When q is false, then its negation $\sim q$ is true.

- **A statement and its negation have opposite truth values.**

However, every sentence is not necessarily a statement with a known truth value. Questions arise in reasoning when we must deal with a sentence having an uncertain truth value, such as "That music is too loud."

Finally, be aware that many negations can be given in a statement. Each time another negation is included, the truth value of the statement will change. For example:

1. k:	A dime is a coin.	(True)
2. $\sim k$:	A dime is not a coin.	(False)
3. $\sim(\sim k)$:	It is not true that a dime is not a coin.	(True)
4. $\sim(\sim(\sim k))$:	It is not the case that it is not true that a dime is not a coin.	(False)

Of course, we don't usually talk like the third and fourth examples because such statements are too confusing. Since $\sim(\sim k)$ always has the same truth value as k, we can use k in place of $\sim(\sim k)$. Therefore, we can negate a sentence that contains the word *not* by omitting that word.

$\sim k$: A dime is not a coin.
$\sim(\sim k)$: A dime is a coin.

EXAMPLES

In 1–6: Let q represent "Oatmeal is a cereal."
Let r represent "She has cereal every morning."

For each given sentence: **a.** Write the sentence in symbolic form. **b.** Tell whether the sentence is true, false, or open.

	Answers:	
1. Oatmeal is a cereal.	**a.** q	**b.** True
2. Oatmeal is not a cereal.	**a.** $\sim q$	**b.** False
3. It is not true that oatmeal is a cereal.	**a.** $\sim q$	**b.** False
4. She has cereal every morning.	**a.** r	**b.** Open
5. She does not have cereal every morning.	**a.** $\sim r$	**b.** Open
6. It is not true that oatmeal is not a cereal.	**a.** $\sim(\sim q)$	**b.** True

In 7 and 8, symbols are used to represent statements and the truth value of each statement is given.

k:	π is an irrational number.	(True)
m:	π is equal to 3.14.	(False)

For each sentence given in symbolic form: **a.** Write a complete sentence in words to show what the symbols represent. **b.** Tell whether the statements are true or false.

Answers:

7. $\sim k$ **a.** π is *not* an irrational number. (Or It is *not* the case that π is an irrational number.)
b. False.

8. $\sim m$ **a.** π is *not* equal to 3.14. (Or $\pi \neq 3.14$) **b.** True.

EXERCISES

In 1–8, write the negation of each sentence.

1. The school has a cafeteria.
2. Georgia is not a city.
3. A school bus is painted yellow.
4. $18 + 20 \div 2 = 28$
5. The measure of a right angle is 90°.
6. $1 + 2 + 3 \neq 4$
7. There are 100 centimeters in a meter.
8. Today is not Saturday.

In 9–18, for each given sentence: **a.** Write the sentence in symbolic form, using the symbols shown below. **b.** Tell whether the sentence is true, false, or open.

Let p represent "A cat is an animal."
Let q represent "A poodle is a cat."
Let r represent "His cat is gray."

9. A cat is an animal.
10. A poodle is a cat.
11. A poodle is not a cat.
12. A cat is not an animal.
13. His cat is gray.
14. His cat is not gray.
15. It is not true that a poodle is a cat.
16. It is not the case that a cat is an animal.
17. It is not the case that a cat is not an animal.
18. It is not the case that a poodle is not a cat.

In 19–26, the symbols represent sentences.

p: Summer follows spring.
q: Baseball is a sport.
r: Baseball is a summer sport.
s: He likes baseball.

For each sentence given in symbolic form: **a.** Write a complete sentence in words to show what the symbols represent. **b.** Tell whether the sentence is true, false, or open.

19. $\sim p$ **20.** $\sim q$ **21.** $\sim r$ **22.** $\sim s$
23. $\sim(\sim q)$ **24.** $\sim(\sim p)$ **25.** $\sim(\sim r)$ **26.** $\sim(\sim s)$

In 27–30, supply a word, a phrase, or a symbol that can be placed in the blank to make the resulting sentence true.

27. When p is true, then $\sim p$ is ______.
28. When p is false, then $\sim p$ is ______.
29. $\sim(\sim p)$ has the same truth value as ______.
30. $\sim(\sim(\sim p))$ has the same truth value as ___.

7-3 CONJUNCTIONS

You have seen that a single letter can be used in logic to represent a single complete thought. Sometimes, however, a sentence contains more than one thought. In English, connectives are used to form compound sentences that express two or more thoughts. One of the simplest connectives is the word *and.*

In logic, a ***conjunction*** is a compound sentence formed by combining two simple sentences, using the word *and.* Each of the simple sentences is called a ***conjunct.*** When p and q represent simple sentences, the conjunction ***p and q*** is written in symbols as $\boldsymbol{p \wedge q}$. For example:

p: There is no school on Saturday.
q: I sleep late.
$p \wedge q$: There is no school on Saturday and I sleep late.

For this sentence to be true, both parts must be true: "There is no school on Saturday" must be true, and "I sleep late" must also be true.

The truth table for a compound sentence contains more than two rows. The first thought, p, can be true or false, and the second thought, q, can be true or false. You must consider every possible combination of these true and false statements.

The diagram shown below is called a ***tree diagram.*** By following its "branches," you can see that there are four possible "true-false" combinations for every two simple statements.

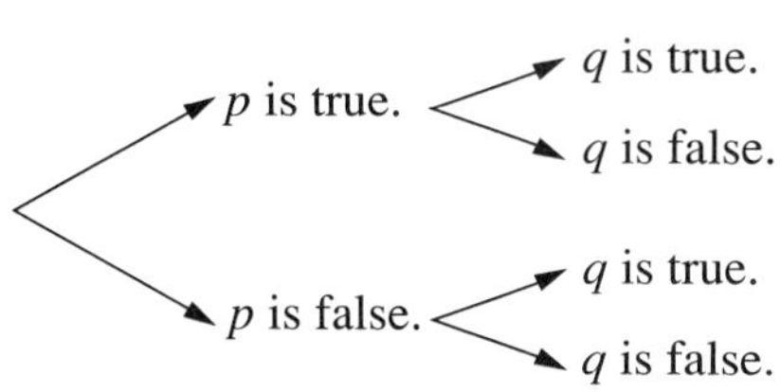

p is true and q is true.
p is true and q is false.
p is false and q is true.
p is false and q is false.

The truth value of every compound sentence depends on the truth values of the simple sentences that form the compound sentence.

- **The conjunction *p and q* is true only when both parts are true: *p* must be true, and *q* must be true.**

If p is false, or if q is false, or if both are false, then the conjunction *p and q* must be false.

Consider these open sentences:

Let p represent "x is an even number."

Let q represent "$x < 8$."

Then, $p \wedge q$ represents "x is an even number and $x < 8$."

1. Let $x = 6$.

p:	6 is an even number.	(True)
q:	$6 < 8$	(True)
$p \wedge q$:	6 is an even number and $6 < 8$.	(True)

When both parts of the conjunction are true, then *p and q* is true.

2. Let $x = 10$.

p:	10 is an even number.	(True)
q:	$10 < 8$	(False)
$p \wedge q$:	10 is an even number and $10 < 8$.	(False)

If any part of the conjunction is false, then *p and q* is false.

3. Let $x = 7$.

p:	7 is an even number.	(False)
q:	$7 < 8$	(True)
$p \wedge q$:	7 is an even number and $7 < 8$.	(False)

If any part of the conjunction is false, then *p and q* is false.

4. Let $x = 13$.

p:	13 is an even number.	(False)
q:	$13 < 8$	(False)
$p \wedge q$:	13 is an even number and $13 < 8$.	(False)

When both parts of the conjunction are false, then *p and q* is false.

A compound sentence may contain both negations and conjunctions at the same time. See Examples 4, 6, 10, and 11, which follow.

EXAMPLES

In 1–6: Let p represent "Coffee is a beverage." (True)
Let q represent "Toast is a beverage." (False)
Let r represent "10 is divisible by 2." (True)
Let s represent "10 is divisible by 3." (False)

For each given sentence: **a.** Write the sentence in symbolic form. **b.** Tell whether the statement is true or false.

1. Coffee is a beverage and 10 is divisible by 2.

Answers: **a.** $p \wedge r$ **b.** True

---

2. Coffee and toast are beverages.

Answers: **a.** $p \wedge q$ **b.** False

---

3. Toast is a beverage and 10 is divisible by 2.

Answers: **a.** $q \wedge r$ **b.** False

---

4. 10 is divisible by 2 and 10 is not divisible by 3.

Answers: **a.** $r \wedge \sim s$ **b.** True

---

5. 10 is divisible by 2 and by 3.

Answers: **a.** $r \wedge s$ **b.** False

---

6. Toast is not a beverage and 10 is not divisible by 3.

Answers: **a.** $\sim q \wedge \sim s$ **b.** True

---

7. Use the domain {1, 2, 3} to find the truth set for the open sentence.

$$(x > 1) \wedge (x < 4)$$

Solution Let $x = 1$. $\quad (1 > 1) \wedge (1 < 4)$

$F \wedge T =$ False statement.

Let $x = 2$. $\quad (2 > 1) \wedge (2 < 4)$

$T \wedge T =$ True statement.

Let $x = 3$. $\quad (3 > 1) \wedge (3 < 4)$

$T \wedge T =$ True statement.

A conjunction is true only when both simple sentences are true. This condition is met here when $x = 2$ and $x = 3$. Thus, the truth set or solution set is {2, 3}.

Answer: {2, 3}.

In 8–11, symbols are used to represent statements about a rectangle with length ℓ and width w. For each statement made, the truth value is noted.

Let p represent "Area $= \ell w$." (True)
Let q represent "Perimeter $= \ell + w$." (False)
Let r represent "Perimeter $= 2\ell + 2w$." (True)

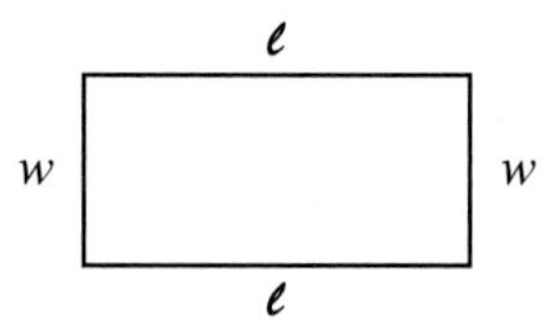

For each sentence given in symbolic form:

a. Write a complete sentence in words to show what the symbols represent.
b. Tell whether the statement is true or false.

---

8. $p \wedge q$ *Answers:* **a.** Area $= \ell w$ *and* Perimeter $= \ell + w$.
b. $T \wedge F =$ False statement.

9. $p \wedge r$ *Answers:* **a.** Area $= \ell w$ *and* Perimeter $= 2\ell + 2w$.
b. $T \wedge T =$ True statement.

10. $\sim q \wedge p$ *Answers:* **a.** Perimeter $\neq \ell + w$ *and* Area $= \ell w$.
b. $\sim F \wedge T = T \wedge T =$ True statement.

11. $\sim(q \wedge p)$ *Answers:* **a.** It is *not* the case that Perimeter $= \ell + w$ *and* Area $= \ell w$.
b. $\sim(F \wedge T) = \sim(F) =$ True Statement.

(*Note:* "It is not the case that . . ." applies to the *entire* statement that follows, as shown by the use of the parentheses in part **b** of Example 11.)

---

12. Three sentences are written below. The truth values are given for the first two sentences. Determine whether the third sentence is true, is false, or has an uncertain truth value.

Trudie likes steak and Phil likes fish.	(False)
Trudie likes steak.	(True)
Phil likes fish.	(?)

Solution (1) Use symbols to represent the sentences. Indicate their truth values.

$p \wedge q$	(False):	Trudie likes steak and Phil likes fish.
p	(True):	Trudie likes steak.
q	(?):	Phil likes fish.

(2) The first sentence, $p \wedge q$, is given as false. The second sentence, p, is given as true. Recall that *p and q* is true only when both p and q are true. You can conclude that the sentence q, Phil likes fish, is false since $p \wedge q$ is false.

Answer: "Phil likes fish" is false.

---

13. Find integral values of x such that "x divides 12 and x does not divide 8" is true.

Solution "x divides 12 is true when x is 1, 2, 3, 4, 6, or 12.
"x does not divide 8 is true when x is 3, 5, 6, 7, 8, or any integer greater than 8.
When $x = 3$ or $x = 5$, the conjunction is true since both conjuncts are true.

---

14. Three sentences are given; the truth values are noted for the first two sentences. Determine whether the third sentence is true, is false, or has an uncertain truth value.

$p \wedge q$:	Jean likes hot weather and Bob likes cold weather.	(False)
p:	Jean likes hot weather.	(False)
q:	Bob likes cold weather.	(?)

Solution The first sentence, $p \wedge q$, is false. The second sentence, p, is false. There are *two* cases where $p \wedge q$ is false and where p is false. Since q could be true or q could be false, conclude that q has an uncertain truth value.

Answer: The truth of "Bob likes cold weather" is uncertain.

EXERCISES

In 1–10, write each sentence in symbolic form, using the given symbols.
Let p represent "It is cold."
Let q represent "It is snowing."
Let r represent "The sun is shining."

1. It is cold and it is snowing.
2. It is cold and the sun is shining.
3. It is not cold.
4. It is not cold and the sun is shining.
5. It is snowing and the sun is not shining.
6. It is not cold and it is not snowing.
7. The sun is not shining and it is not cold.
8. The sun is not shining and it is cold.
9. It is not the case that it is cold and it is snowing.
10. It is not the case that it is snowing and it is not cold.

In 11–20, for each given statement: **a.** Write the statement in symbolic form, using the symbols shown below. **b.** Tell whether the statement is true or false.

Let b represent "Water boils at 100°C." (True)
Let f represent "Water freezes at 0°C." (True)
Let t represent "Normal body temperature is 37°C." (True)
Let r represent "Room temperature is 60°C." (False)

11. Normal body temperature is 37°C and water boils at 100°C.
12. Normal body temperature is 37°C and room temperature is 60°C.
13. Water freezes at 0°C and boils at 100°C.
14. Water freezes at 0°C and room temperature is 60°C.
15. Water does not boil at 100°C and water does not freeze at 0°C.
16. Room temperature is not 60°C and water boils at 100°C.
17. Normal body temperature is not 37°C and room temperature is 60°C.
18. It is not the case that water does not boil at 100°C.
19. It is not true that water boils at 100°C and freezes at 0°C.
20. It is not the case that water boils at 100°C and room temperature is 60°C.

In 21–27, tell whether each sentence is true, false, or open.

21. $x = 28 - 17$ and $x = 11$.
22. 4 [y^x] 3 [=] 64 and 3 [y^x] 4 [=] 64.
23. 4 [y^x] 2 [=] 16 and 2 [y^x] 4 [=] 16.
24. It is a math book and it contains problems.
25. Tuesday follows Monday and $37.5 + 62.5 = 100$.
26. Every square contains four right angles and every triangle contains one right angle.
27. The Surgeon General of the United States has determined that cigarette smoking is dangerous to your health and warning labels are printed on cigarette packs.
28. Use the domain of whole numbers to find the truth set for each compound open sentence.
    a. $(x > 5) \wedge (x < 8)$ **b.** $(x > 4) \wedge (x \leq 6)$ **c.** $(x \geq 3) \wedge (x < 7)$
    d. $(x < 5) \wedge (x < 2)$ **e.** $(x \leq 3) \wedge (x < 4)$ **f.** $(x > 8) \wedge (x > 3)$
    g. $(x \geq 3) \wedge (x + 5 < 7)$ **h.** $(x - 8 \leq 2) \wedge (3x > 24)$ **i.** $(2x < 0) \wedge (x + 5 = 0)$

In 29–40, symbols are assigned to represent sentences.

Let b represent "A banjo is a stringed instrument."
Let d represent "A drum is a stringed instrument."
Let g represent "A guitar is a stringed instrument."
Let s represent "She plays a guitar."

For each sentence given in symbolic form: **a.** Write a complete sentence in words to show what the symbols represent. **b.** Tell whether the sentence is true, false, or open.

29. $b \wedge g$
30. $b \wedge d$
31. $g \wedge s$
32. $b \wedge \sim d$
33. $b \wedge \sim g$
34. $b \wedge \sim s$
35. $g \wedge \sim d$
36. $\sim s \wedge d$
37. $\sim d \wedge \sim b$
38. $\sim(d \wedge b)$
39. $\sim(b \wedge g)$
40. $\sim(g \wedge s)$

In 41–48, supply the word, phrase, or symbol that can be placed in the blank to make the resulting sentence true.

41. When p is true and q is true, then $p \wedge q$ is ______.
42. When p is false, then $p \wedge q$ is ______.
43. If p is true, or q is true, but not both, then $p \wedge q$ is ______.
44. When $p \wedge q$ is true, then p is ______ and q is ______.
45. When $p \wedge \sim q$ is true, then p is ______ and q is ______.
46. When $\sim p \wedge q$ is true, then p is ______ and q is ______.
47. When p is false and q is true, then $\sim(p \wedge q)$ is ______.
48. If both p and q are false, then $\sim p \wedge \sim q$ is ______.

In 49–55, three sentences are written. The truth values are given for the first two sentences. Determine whether the third sentence is true, is false, or has an uncertain truth value.

49. It is raining and I get wet. (True)
It is raining. (True)
I get wet. (?)

50. I have a headache and I take a nap. (False)
I have a headache. (True)
I take a nap. (?)

51. I have a headache and I take a nap. (False)
I have a headache. (False)
I take a nap. (?)

52. Both a potato and a hurricane have many eyes. (False)
A potato has many eyes. (True)
A hurricane has many eyes. (?)

53. Anna loves TV and Anna loves to stay home. (True)
Anna loves TV. (True)
Anna loves to stay home. (?)

54. Joe and Edna like to stay at home. (False)
Joe likes to stay at home. (True)
Edna likes to stay at home. (?)

55. Juan takes the train and the bus to go to work. (False)
Juan takes the bus to go to work. (False)
Juan takes the train to go to work. (?)

In 56 and 57, a compound sentence is given using a conjunction. After examining the truth value of the compound sentence, determine whether each sentence that follows is true or false.

56. Most plants need light and water to grow. (True)
a. Most plants need light to grow.
b. Most plants need water to grow.
c. Most plants do not need light to grow.

57. I do not exercise and I know that I should. (True)
a. I do not exercise.
b. I know that I should exercise.
c. I exercise.

7-4 DISJUNCTIONS

Many different connectives are used in everyday conversation. Some connectives that might be heard at the local diner include the following: bacon and eggs; lettuce and tomato; white bread or rye; tea or coffee; mustard or catsup. In addition to *and*, another common connective in the English language is the word *or*.

In logic, a ***disjunction*** is a compound sentence formed by combining two simple sentences using the word *or*. Each of the simple sentences is called a ***disjunct***. When p and q represent simple sentences, the disjunction ***p or q*** is written in symbols as $\boldsymbol{p \vee q}$. For example:

p: You may use a pencil to answer the test.
q: You may use a pen to answer the test.
$p \vee q$: You may use a pencil or you may use a pen to answer the test.

In this example, the compound sentence p *or* q is true in many cases:

1. You answer in pencil. Here p is true and q is false. The disjunction $p \vee q$ is true.
2. You answer in pen. Here p is false while q is true. The disjunction $p \vee q$ is true.
3. You answer in pen, but soon you run out of ink. You finish the test in pencil. Here p is true and q is true. Since you obeyed the rules and did nothing false, the disjunction $p \vee q$ is true.

- **The disjunction *p or q* is true when any part of the compound sentence is true: *p* is true, or *q* is true, or both *p* and *q* are true.**

In fact, there is only one case where the disjunction p *or* q is false: when both p and q are false.

Two Uses of the Word *Or*

When we use the word *or* to mean that *one or both* of the simple sentences are true, we call this the ***inclusive or***. The truth table we have just developed shows the truth values for the *inclusive or*.

Sometimes, however, the word *or* is used in a different way, as in "He is in grade 9 or he is in grade 10." Here it is not possible for both simple sentences to be true at the same time. When we use the word *or* to mean that *one and only one* of the simple sentences is true, we call this the ***exclusive or***. The truth table for the *exclusive or* will be different from the table shown for disjunction. In the *exclusive or*, the disjunction p *or* q will be true when p is true, or when q is true, but not both.

In everyday conversation, it is often evident from the context which of these uses of *or* is intended. In legal documents or when ambiguity can cause difficulties, the *inclusive or* is sometimes written as *and/or*.

- **We will use only the *inclusive or* in this book. Whenever we speak of disjunction, p *or* q will be true when p is true, or when q is true, or when *both* p and q are true.**

Logic and Sets

Some people believe that logic is not like any other kind of mathematics that they have ever seen. This is not true. Conjunction and disjunction behave in exactly the same way as operations with sets.

The truth sets or solution sets of open sentences that use the connectives *not*, *and*, and *or* can be compared to the *complement*, *intersection*, and *union* of sets.

Logic	*Sets*
Let the domain = {1, 2, 3, 4, 5}.	U = Universe = {1, 2, 3, 4, 5}
Let p represent $x < 4$. The truth set of p is {1, 2, 3}.	$P = \{1, 2, 3\}$
Let q represent $x > 1$. The truth set of q is {2, 3, 4, 5}.	$Q = \{2, 3, 4, 5\}$
The truth set of $\sim p$ is the set of values *not* less than 4, or {4, 5}.	$\overline{P} = \{4, 5\}$ (*complement* of P)
The truth set of $\sim q$ is the set of values *not* greater than 1, or {1}.	$\overline{Q} = \{1\}$ (*complement* of Q)
The truth set of $p \wedge q$ is the set of numbers less than 4 *and* greater than 1, which is {2, 3}.	$P \cap Q = \{2, 3\}$ (*intersection*)
The truth set of $p \vee q$ is the set of numbers less than 4 *or* greater than 1, which is {1, 2, 3, 4, 5}.	$P \cup Q = \{1, 2, 3, 4, 5\}$ (*union*)

EXAMPLES

In 1–6: Let k represent "Kevin won the play-off."
Let a represent "Alexis won the play-off."
Let n represent "Nobody won."

Write each given sentence in symbolic form. *Answers:*

1. Kevin or Alexis won the play-off. $k \vee a$

2. Kevin won the play-off or nobody won. $k \vee n$

3. Alexis won the play-off or Alexis didn't win. $a \vee \sim a$

4. It is not true that Kevin or Alexis won the play-off. $\sim(k \vee a)$

5. Either Kevin did not win the play-off or Alexis did not win. $\sim k \vee \sim a$

6. It's not the case that Alexis and Kevin won the play-off. $\sim(a \wedge k)$

In 7–10, symbols are used to represent three statements as shown. For each statement, the truth value is noted.

Let k represent "Every square is a rhombus." (True)
Let m represent "Every rhombus is a square." (False)
Let q represent "Every square is a parallelogram." (True)

For each sentence given in symbolic form: **a.** Write a complete sentence in words to show what the symbols represent. **b.** Tell whether the statement is true or false.

Answers:

7. $k \vee q$
a. Every square is a rhombus or every square is a parallelogram.
b. $T \vee T$ is a true disjunction.

8. $k \vee m$
a. Every square is a rhombus or every rhombus is a square.
b. $T \vee F$ is a true disjunction.

9. $m \vee \sim q$
a. Every rhombus is a square or every square is not a parallelogram.
b. $F \vee \sim T = F \vee F =$ a false disjunction.

10. $\sim(m \vee q)$
a. It is not the case that every rhombus is a square or every square is a parallelogram.
b. $\sim(F \vee T) = \sim(T) =$ a false statement.

In 11–13, use the domain {0, 1, 2, 3, 4, 5, 6} to find the truth set of each compound sentence.

11. $(x > 4) \vee (x \leq 2)$

$\{5, 6\} \cup \{0, 1, 2\} = \{0, 1, 2, 5, 6\}$ *Answer*

---

12. $(x \not> 4) \wedge (x \not\leq 2)$

$\{0, 1, 2, 3, 4\} \cap \{3, 4, 5, 6\} = \{3, 4\}$ *Answer*

---

13. $(x \geq 5) \wedge (x \leq 3)$

$\{5, 6\} \cap \{0, 1, 2, 3\} = \varnothing$ *Answer*

EXERCISES

In 1–10, write each sentence in symbolic form, using the given symbols.

Let s represent "I will study."
Let p represent "I will pass the test."
Let f represent "I am foolish."

1. I will study or I am foolish.
2. I will study or I will not pass the test.
3. I will study and I will pass the test.
4. I will pass the test or I am foolish.
5. I am not foolish and I will pass the test.
6. I will not study or I am foolish.
7. I will study or I will not study.
8. It is not true that I will study or I am foolish.
9. I will study and I will pass the test, or I am foolish.
10. It is not the case that I will not study or I am not foolish.

In 11–20, for each given statement: **a.** Write the statement in symbolic form, using the symbols given below. **b.** Tell whether the statement is true or false.

Let c represent	"A meter contains 100 centimeters."	(True)
Let m represent	"A meter contains 1,000 millimeters."	(True)
Let k represent	"A kilometer is 1,000 meters."	(True)
Let ℓ represent	"A meter is a liquid measure."	(False)

11. A meter contains 1,000 millimeters or a kilometer is 1,000 meters.

12. A meter contains 100 centimeters or a meter is a liquid measure.

13. A meter contains 100 centimeters or 1,000 millimeters.

14. A kilometer is not 1,000 meters or a meter does not contain 100 centimeters.

15. A meter is a liquid measure or a kilometer is 1,000 meters.

16. A meter is a liquid measure and a meter contains 100 centimeters.

17. It is not the case that a meter contains 100 centimeters or 1,000 millimeters.

18. It is false that a kilometer is not 1,000 meters or a meter is a liquid measure.

19. A meter contains 100 centimeters and 1,000 millimeters or a meter is a liquid measure.

20. It is not true that a meter contains 100 centimeters or a meter is a liquid measure.

21. Use the domain of one-digit whole numbers {0, 1, 2, . . . , 9} to find the truth set of each compound open sentence.

a. $(x < 3) \vee (x < 2)$ **b.** $(x \geq 8) \vee (x < 1)$ **c.** $(x > 9) \vee (x \leq 3)$
d. $(x > 2) \wedge (x < 7)$ **e.** $(x < 5) \vee (x > 9)$ **f.** $(x < 5) \wedge (x > 8)$

In 22–33, symbols are assigned to represent sentences.

Let b represent "Biology is a science."
Let s represent "Spanish is a language."
Let h represent "Homemaking is a language."
Let d represent "It's a difficult course."

For each sentence given in symbolic form: **a.** Write a complete sentence in words to show what the symbols represent. **b.** Tell whether the sentence is true, false, or open.

22. $s \vee h$ **23.** $b \vee s$ **24.** d **25.** $\sim s \vee h$
26. $\sim d$ **27.** $b \vee \sim h$ **28.** $\sim b \vee \sim s$ **29.** $\sim(s \vee h)$
30. $s \wedge h$ **31.** $s \wedge b$ **32.** $\sim(\sim b \vee s)$ **33.** $\sim(\sim s \vee h)$

In 34–40, supply the word, phrase, or symbol that can be placed in the blank to make the resulting sentence true.

34. When p is true, then $p \vee q$ is ______. **35.** When q is true, then $p \vee q$ is ______.
36. When p is false and q is false, then $p \vee q$ is ______.
37. When $p \vee \sim q$ is false, then p is ______ and q is ______.
38. When $\sim p \vee q$ is false, then p is ______ and q is ______.
39. When p is false and q is true, then $\sim(p \vee q)$ is ______.
40. When p is false and q is true, then $\sim p \vee \sim q$ is ______.

In 41–45, three sentences are written. The truth values are given for the first two sentences. Determine whether the third sentence is true, is false, or has an uncertain truth value.

41. She will sink. (False)
She will swim. (True)
She will sink or she will swim. (?)

42. I will work after school or I will study more. (True)
I will work after school. (False)
I will study more. (?)

43. Michael cannot swim or Michael cannot skate. (False)
Michael cannot swim. (False)
Michael cannot skate. (?)

44. Nicolette is my friend or I have two left feet. (True)
Nicolette is my friend. (True)
I have two left feet. (?)

45. Jennifer draws well or plays the cello. (True)
Jennifer does not draw well. (False)
Jennifer plays the cello. (?)

7-5 CONDITIONALS

The connective used most often in reasoning can be found in the following sentence:

If the fever continues, then he should see a doctor.

To find the connective, first write the simple sentences using p and q.

p: The fever continues. q: He should see a doctor.

The connective is the words that remain, *If . . . then*.

In English, such a sentence is called a *complex sentence*. In mathematics, however, all sentences formed by connectives are called *compound sentences*.

In logic, a ***conditional*** is a compound sentence formed by using the words *if . . . then* to combine two simple sentences. When p and q represent simple sentences, the conditional ***if p then q*** is written in symbols as $\boldsymbol{p \rightarrow q}$.

Since a conditional is sometimes called an ***implication***, the symbols for the conditional $\boldsymbol{p \rightarrow q}$ can be read as ***p implies q***. Here is another example:

p: It is snowing.

q: The temperature is below freezing.

$p \rightarrow q$: *If* it is snowing, *then* the temperature is below freezing.

or

$p \rightarrow q$: It is snowing *implies* that the temperature is below freezing.

Certainly we would agree that the compound sentence *if p then q* is true for this example: "If it is snowing, then the temperature must be below freezing." However, if we reverse the order of the simple sentences to form the new conditional *if q then p*, we will get a sentence with a completely different meaning:

$q \rightarrow p$: *If* the temperature is below freezing, *then* it is snowing.

When the temperature is below freezing, it does not necessarily follow that snow is falling. The conditional *if q then p* is not necessarily a true statement. Changing the ***order*** in which we connect two simple sentences can sometimes result in forming two conditionals with *different truth values*.

Parts of a Conditional

The parts of the conditional *if p then q* can be identified by name:

p is called the ***premise***, the ***hypothesis***, or the ***antecedent***. It is an assertion or a sentence that begins an argument. The antecedent usually follows the word *if*.

q is called the ***conclusion*** or the ***consequent***. It is an ending or a sentence that closes an argument. The consequent usually follows the word *then*.

There are different ways to write the conditional *if* p *then* q. Notice that the antecedent p is connected to the word *if* in the examples shown:

$p \rightarrow q$: If $\underbrace{\text{Alice scores one more point}}_{\text{antecedent or hypothesis}}$, then $\underbrace{\text{our team will win}}_{\text{consequent or conclusion}}$.

$p \rightarrow q$: $\underbrace{\text{Our team will win}}_{\text{consequent}}$ if $\underbrace{\text{Alice scores one more point}}_{\text{antecedent}}$.

Both sentences say the same thing: We hypothesize or hope that Alice scores one more point. When that happens, we can conclude that our team will win. Although the word order of a conditional may vary, the antecedent is always written first when using symbols.

When we wish to consider all the truth values that can be assigned to sentences, we can use a device called a truth table. A ***truth table*** is a compact way of listing symbols to show all possible truth values for a set of sentences. It can be used as a tool to organize and clarify the truth values of complex sentences.

Building a Truth Table for the *Conditional* $p \rightarrow q$

1. Let p serve as the symbol for the antecedent, and let q serve as the symbol for the consequent.
   For example: Dr. Cathy Russo told her patient, Bill, "If you take the medicine, then you'll feel better in 24 hours."

   p: You take the medicine.
   q: You'll feel better in 24 hours.

2. List all possible truth values for the antecedent p and the consequent q in the first two columns. Use the same order established for the other connectives.

p	q
T	T
T	F
F	T
F	F

3. In the heading of the third column, write the conditional *if* p *then* q in symbolic form.

p	q	$p \rightarrow q$
T	T	
T	F	
F	T	
F	F	

4. Assign truth values to the conditional by considering the truth values for p and q in each row.

Row 1:

Bill takes the medicine. (p is true.)
He feels better in 24 hours. (q is true.)
Dr. Russo has told Bill the truth.
The conditional statement is true.

	p	q	$p \rightarrow q$
→	T	T	T
	T	F	
	F	T	
	F	F	

Row 2:

Bill takes the medicine, (p is true.)
He does not feel better in 24 hours. (q is false.)
Dr. Russo did not tell Bill the truth.
The conditional statement is false.

	p	q	$p \rightarrow q$
	T	T	T
→	T	F	F
	F	T	
	F	F	

Rows 3 and 4:

Bill does not take the medicine. (p is false.)
It is possible that Bill feels better, (Row 3, q is true.),
or that he does not feel better. (Row 4, q is false.)

	p	q	$p \rightarrow q$
	T	T	T
	T	F	F
→	F	T	?
→	F	F	?

In both cases, the doctor did not tell Bill a lie because she told him only what would happen if he took the medicine. Since the doctor did not make a false statement in these two cases, we will assign T (*true*) to the conditional statement.

5. Assign the truth values to complete the table.

$p \rightarrow q$: "If you take the medicine, then you'll feel better in 24 hours" will be true in all cases except one: when Bill takes the medicine and he does not feel better in 24 hours.

p	q	$p \rightarrow q$
T	T	T
T	F	F
F	T	T
F	F	T

Truth Table for Conditional

- **A conditional is false when a true hypothesis leads to a false conclusion.** In all other cases, the conditional is true.

Hidden Conditionals

We constantly use conditionals in our everyday lives. Often the words *if . . . then* do not appear in a sentence. In such a case, we say that the sentence has a ***hidden conditional***. We can still identify the sentence as a conditional, however, because it contains an antecedent p and a consequent q. We can rewrite the words in the sentence so that the conditional form *if* p *then* q becomes more obvious. For example:

1. "When this assignment has been written, you should hand it in" becomes:

   $p \rightarrow q$: *If* this assignment has been written, *then* you should hand it in.

2. "For good health, exercise regularly" becomes:

   $p \rightarrow q$: *If* you want to have good health, *then* you should exercise regularly.

3. "Vote for me and I'll whip unemployment" becomes:

   $p \rightarrow q$: *If* you vote for me, *then* I'll whip unemployment.

4. "$x + 3 = 10$; therefore $x = 7$" becomes:

   $p \rightarrow q$: *If* $x + 3 = 10$, *then* $x = 7$.

EXAMPLES

In 1–3, for each given sentence: **a.** Identify the hypothesis p. **b.** Identify the conclusion q.

1. If Mrs. Garbowski assigns homework, then you'd better do it.

Answers: **a.** p: Mrs. Garbowski assigns homework.
b. q: You'd better do the homework.

---

2. You can assemble the bicycle if you follow these easy directions.

Answers: **a.** p: You follow these easy directions.
b. q: You can assemble the bicycle.

---

3. I carry an umbrella when the forecast predicts rain and the weather looks threatening.

Answers: **a.** p: The forecast predicts rain and the weather looks threatening.
b. q: I carry an umbrella.

In 4–7, identify the truth value to be assigned to each conditional statement.

4. If $2^2 = 4$, then $2^3 = 8$.

Solution The hypothesis p is "$2^2 = 4$", which is true.
The conclusion q is "$2^3 = 8$," which is true.

The conditional $p \rightarrow q$ is true. *Answer*

---

5. If 9 is an odd number, then 9 is prime.

Solution The hypothesis p is "9 is an odd number," which is true.
The conclusion q is "9 is prime," which is false because 9 is divisible by 3.

The conditional $p \rightarrow q$ is false. *Answer*

---

6. If a square has five sides, then $5 + 5 = 10$.

Solution The antecedent p is "A square has five sides," which is false.
The consequent q is "$5 + 5 = 10$," which is true.

The conditional $p \rightarrow q$ is true. *Answer*

---

7. If time goes backward, then I'll get younger every day.

Solution The antecedent p is "Time goes backward," which is false.
The consequent q is "I'll get younger every day," which is false.

The conditional $p \rightarrow q$ is true. *Answer*

In 8–12, for each given statement: **a.** Write the statement in symbolic form, using the symbols given below. **b.** Tell whether the statement is true or false.

Let m represent "Tuesday follows Monday."	(True)
Let w represent "There are 7 days in 1 week."	(True)
Let h represent "There are 40 hours in a day."	(False)

Answers:

8. If Tuesday follows Monday, then there are 7 days in 1 week.
a. $m \rightarrow w$
b. $T \rightarrow T$ is true.

9. If there are 7 days in 1 week, then there are 40 hours in a day.
a. $w \rightarrow h$
b. $T \rightarrow F$ is false

10. If Tuesday does not follow Monday, then there are not 7 days in 1 week.
a. $\sim m \rightarrow \sim w$
b. $F \rightarrow F$ is true.

11. Tuesday follows Monday if there are not 7 days in 1 week.

a. $\sim w \rightarrow m$
b. $F \rightarrow T$ is true.

12. If Tuesday follows Monday and there are 7 days in 1 week, then there are 40 hours in a day.

a. $(m \wedge w) \rightarrow h$
b. $(T \wedge T) \rightarrow F$
$T \rightarrow F$ is false.

EXERCISES WITH OPEN-RESPONSE PROBLEMS

In 1–8, for each given sentence: **a.** Identify the hypothesis p. **b.** Identify the conclusion q.

1. If it rains, then the game is canceled.
2. If it is 9:05 A.M., then I'm late to class.
3. When it rains, then I do not have to water the lawn.
4. You can get to the stadium if you take the Third Avenue bus.
5. The perimeter of a square is $4x + 8$ if one side of the square is $x + 2$.
6. If the shoe fits, wear it.
7. If a polygon has exactly three sides, it is a triangle.
8. When you have a headache, you should take time out and get some rest.

In 9–14, write each sentence in symbolic form, using the given symbols.

p: The test is easy.
q: Sam studies.
r: Sam passes the test.

9. If the test is easy, then Sam will pass the test.
10. If Sam studies, then Sam will pass the test.
11. If the test is not easy, then Sam will not pass the test.
12. Sam will not pass the test if Sam doesn't study.
13. The test is easy if Sam studies.
14. Sam passes the test if the test is easy.

In 15–22, for each given statement: **a.** Write the statement in symbolic form, using the symbols given below. **b.** Tell whether the conditional statement is true or false, based upon the truth values given.

r: The race is difficult. (True)
p: Karen practices. (False)
w: Karen wins the race. (True)

15. If Karen practices, then Karen will win the race.
16. If Karen wins the race, then Karen has practiced.
17. If Karen wins the race, the race is difficult.
18. Karen wins the race if the race is not difficult.
19. Karen will not win the race if Karen does not practice.
20. Karen practices if the race is difficult.
21. If the race is not difficult and Karen practices, then Karen will win the race.
22. If the race is difficult and Karen does not practice, then Karen will not win the race.

In 23–30, find the truth value to be assigned to each conditional statement.

23. If $5 + 7 = 12$, then $7 + 5 = 12$.

24. If $3 > 10$, then $10 > 13$.

25. If $1 \cdot 1 = 1$, then $1 \cdot 1 \cdot 1 = 1$.

26. $12 \div 3 = 9$ if $12 \div 9 = 3$.

27. $2 + 2 = 2^2$ if $3 + 3 = 3^2$.

28. $2^3 = 3^2$ if $2^4 = 4^2$.

29. If $1 + 2 + 3 = 1 \cdot 2 \cdot 3$, then $1 + 2 = 1 \cdot 2$.

30. If every square is a rectangle, then every rectangle is a square.

In 31–46, symbols are assigned to represent sentences, and truth values are assigned to these sentences.

Let j represent "I jog."	(True)
Let d represent "I diet."	(False)
Let g represent "I feel well."	(True)
Let h represent "I get hungry."	(True)

For each compound sentence in symbolic form: **a.** Write a complete sentence in words to show what the symbols represent. **b.** Tell whether the compound sentence is true or false.

31. $j \rightarrow g$
32. $d \rightarrow h$
33. $h \rightarrow d$
34. $\sim g \rightarrow j$
35. $\sim g \rightarrow \sim j$
36. $g \rightarrow \sim h$
37. $h \rightarrow \sim d$
38. $\sim d \rightarrow \sim h$
39. $\sim j \rightarrow \sim g$
40. $j \rightarrow d$
41. $(j \wedge h) \rightarrow g$
42. $j \rightarrow (h \wedge g)$
43. $(j \vee d) \rightarrow h$
44. $d \rightarrow (h \wedge \sim g)$
45. $\sim j \rightarrow (d \wedge h)$
46. $(j \wedge h) \rightarrow d$

In 47–52, supply the word, phrase, or symbol that can be placed in the blank to make each resulting sentence true.

47. When p and q represent two simple sentences, the conditional *if p then q* is written symbolically as ________.
48. The conditional *if q then p* is written symbolically as ________.
49. The conditional $p \rightarrow q$ is false only when p is ________ and q is ________.
50. When the conclusion q is true, then $p \rightarrow q$ must be ________.
51. When the hypothesis p is false, then $p \rightarrow q$ must be ________.
52. If the hypothesis p is true and conditional $p \rightarrow q$ is true, then the conclusion q must be ________.

In 53–57, three sentences are written in each case. The truth values are given for the first two sentences. Determine whether the third sentence is true, is false, or has an uncertain truth value.

53. If you read in dim light, then you can strain your eyes. (True)
You read in dim light. (True)
You can strain your eyes. (?)

54. If the quadrilateral has four right angles, then the quadrilateral is a square. (False)
The quadrilateral has four right angles. (True)
The quadrilateral is a square. (?)

55. If n is an odd number, then $2 \cdot n$ is an even number. (True)
$2 \cdot n$ is an even number. (True)
n is an odd number. (?)

56. If the report is late, then you will not get an A. (True)
The report is late. (False)
You will not get an A. (?)

57. Area $= \frac{1}{2}bh$, if the polygon is a triangle. (True)
The polygon is a triangle. (True)
Area $= \frac{1}{2}bh$ (?)

7-6 BICONDITIONALS

The truth table for the **conditional** $\boldsymbol{p \to q}$ is shown below. The conditional $p \to q$ is false when the hypothesis p is true and the conclusion q is false, as shown in row 2. In all other cases, the conditional $p \to q$ is true.

Hypothesis (p) ↓ **Conclusion** (q) ↓

p	q	$p \to q$
T	T	T
T	F	F
F	T	T
F	F	T

When we reverse the order of sentences p and q, we form a new **conditional** $\boldsymbol{q \to p}$. The conditional $q \to p$ is false when the hypothesis q is true and the conclusion p is false, as shown in row 3. In all other cases, the conditional $q \to p$ is true.

Conclusion (p) ↓ **Hypothesis** (q) ↓

p	q	$q \to p$
T	T	T
T	F	T
F	T	F
F	F	T

The prefix *bi-* means "two," as in *bicycle*, *binary*, and *bifocals*. The biconditional consists of two conditionals. In logic, the ***biconditional*** is a compound sentence formed by the conjunction of the two conditionals $p \to q$ and $q \to p$. To find the truth value of the biconditional, we can construct a truth table for the sentence $(p \to q) \wedge (q \to p)$.

p	q	$p \to q$	$q \to p$	$(p \to q) \wedge (q \to p)$
T	T	T	T	T
T	F	F	T	F
F	T	T	F	F
F	F	T	T	T

The compound sentence formed in the truth table is lengthy both to read and to write.

$(p \rightarrow q) \wedge (q \rightarrow p)$: **If p then q, and if q then p**

or

p implies q, and q implies p.

We shorten the writing of the biconditional by introducing the symbol $p \leftrightarrow q$ to replace the compound sentence. We shorten the reading of the biconditional by using the words ***p if and only if q***. These abbreviated versions, as well as the lengthy ones, are all acceptable ways of indicating the biconditional. From the truth table, we observe that:

- **The biconditional *p if and only if q* is true when p and q are both true or both false.**

In other words, $p \leftrightarrow q$ is true when p and q have the **same truth value**. When p and q have different truth values, the biconditional is false.

Building the Truth Table for the *Biconditional* $p \leftrightarrow q$

1. Assign two letters p and q, each to serve as a symbol for a different simple sentence.
2. List all possible truth values for p and q in columns 1 and 2.
3. In the third column, write the biconditional in symbolic form.

p	q	$p \leftrightarrow q$
T	T	
T	F	
F	T	
F	F	

4. Finally, assign truth values for the biconditional. When p and q have the same truth value, as in rows 1 and 4, the biconditional is true. When p and q have different truth values, as in rows 2 and 3, the biconditional is false.

p	q	$p \leftrightarrow q$
T	T	T
T	F	F
F	T	F
F	F	T

Truth Table for Biconditional

Applications of the Biconditional

The truth table shows that the biconditional is not true in all cases. For example:

$p \leftrightarrow q$: $x > 5$ *if and only if* $x > 3$.

or

$p \leftrightarrow q$: If $x > 5$, then $x > 3$, and if $x > 3$, then $x > 5$.

This biconditional is *false* because it fails for certain numbers. Let $x = 4$. Then p, $x > 5$, is false while q, $x > 3$, is true. Since the truth values are not the same, the biconditional is not true.

There are many examples, however, where the biconditional is always true. Consider the following cases.

Case 1. Biconditionals are used to write definitions.

Two conditionals are stated.

$p \rightarrow t$: If a polygon has exactly three sides, then it is a triangle.

$t \rightarrow p$: If a polygon is a triangle, then it has exactly three sides.

The conjunction $(p \rightarrow t) \wedge (t \rightarrow p)$ is the biconditional $(p \leftrightarrow t)$. When p is true, t is true. When p is false, t is false. Since the truth values are the same, the biconditional is always true. *We use the biconditional to serve as a definition.*

$p \leftrightarrow t$: A polygon is a triangle *if and only if* it has exactly three sides.

Precise definitions in mathematics are often stated using the words *if and only if*. The *if* tells us what we must include, as in "exactly three sides." The *only if* tells us what to exclude, as in "all polygons that do not have exactly three sides."

Case 2. Biconditionals are used to solve equations.

$$\begin{aligned} 2x + 1 &= 17 \\ 2x &= 16 \\ x &= 8 \end{aligned}$$

In simplifying equations, we follow a process, as shown at the left. Since $x = 8$ will be true in every equation listed here and any other value of x will be false, we can reverse this process.

$a \rightarrow b$: $2x + 1 = 17$ implies $2x = 16$.

$b \rightarrow a$: $2x = 16$ implies $2x + 1 = 17$.

Equations of this type form biconditional statements.

$a \leftrightarrow b$: $2x + 1 = 17$ *if and only if* $2x = 16$.

$b \leftrightarrow c$: $2x = 16$ *if and only if* $x = 8$.

Case 3. Biconditionals are used to state equivalences.

When any two statements always have the same truth value, we can substitute one statement for the other. The statements are said to be ***logically equivalent***. You will see examples of equivalences in the next section.

EXAMPLES

Identify the truth value to be assigned to each biconditional.

a. Cars stop if and only if there is a red light.

Solution Let p represent "Cars stop."
Let q represent "There is a red light."

When p is true, it does not follow that q must be true. Cars stop also at stop signs and railroad crossings, or when parking. Since p and q do not have the same truth value, the biconditional $p \leftrightarrow q$ is false.

Answer: False

b. $x + 2 = 7$ if and only if $x = 5$.

Solution Let p represent "$x + 2 = 7$."
Let q represent "$x = 5$."

When $x = 5$, both p and q are true. When $x \neq 5$, both p and q are false. In any event, p and q have the same truth value. Thus, the biconditional $p \leftrightarrow q$ is true.

Answer: True

EXERCISES

In 1–5, write each biconditional in symbolic form, using the symbols given.

t: The triangle is a right triangle.
r: The triangle contains a right angle.
n: The triangle contains a 90° angle.

1. A triangle is a right triangle if and only if it contains a right angle.
2. A triangle contains a 90° angle if and only if it is a right triangle.
3. A triangle contains a 90° angle if and only if it contains a right angle.
4. If a triangle is a right triangle, then it contains a 90° angle, and if the triangle contains a 90° angle, then it is a right triangle.
5. A triangle is not a right triangle if and only if it does not contain a 90° angle.

In 6–15, identify the truth value to be assigned to each biconditional.

6. $x + 3 = 30$ if and only if $x = 27$.
7. A polygon is a pentagon if and only if it has exactly three sides.
8. $x + 4 = 12$ if and only if $x + 6 = 14$.
9. A rectangle is a square if and only if the rectangle has all sides of equal length.

10. The refrigerator runs if and only if the electricity is on.
11. An angle is right if and only if it contains 90°.
12. A set is empty if and only if it contains no elements.
13. A number is even if and only if it is exactly divisible by 2.
14. Two angles have the same measure if and only if they are right angles.
15. A sentence is a statement if and only if it has truth values.
16. Using the symbols p and q, write each compound sentence in symbolic form.
  a. p if and only if q.
  b. If p then q and if q then p.
  c. p implies q and q implies p.
  d. q if and only if p.
  e. If p then q or if q then p.

17. Of the five responses given in Exercise 16, four name the biconditional, $p \leftrightarrow q$. Which sentence does *not* represent the biconditional?

7-7 INVERSES, CONVERSES, AND CONTRAPOSITIVES

The conditional *if p then q* is the connective most often used in reasoning. Too often, people attempt to win an argument or to make a point by "twisting words around." To help you avoid becoming the victim of this kind of argument, this section presents three new conditionals, each of which is formed by making some changes in an original conditional statement. These new conditionals are called the *inverse*, the *converse*, and the *contrapositive*.

The Inverse

Starting with an original conditional ($p \to q$), we form the ***inverse*** ($\sim p \to \sim q$) by negating the hypothesis and negating the conclusion. The symbols for the inverse may be read as *not p implies not q*, or *if not p, then not q*.

				Conditional	Inverse
p	q	$\sim p$	$\sim q$	$p \to q$	$\sim p \to \sim q$
T	T	F	F	T	T
T	F	F	T	F	T
F	T	T	F	T	F
F	F	T	T	T	T

Note that a conditional and its inverse are *not equivalent* statements because they do not have matching truth values in every case.

Illustration 1. In advertising a product, the manufacturer sometimes would like the reader to assume the truth of the inverse of a true conditional statement. For example, the Spritz Company makes only root beer. We can agree that the conditional "If you have a Spritz, then you have a root beer" is true. Does it follow that the company's slogan, "When you're out of Spritz, then you're out of root beer," is also true? Let us see.

Conditional ($p \rightarrow q$): If you have a Spritz, then you have a root beer.

Inverse ($\sim p \rightarrow \sim q$): If you do *not* have a Spritz, then you do *not* have a root beer.

Notice that the slogan is a "hidden" form of the inverse.

Inverse ($\sim p \rightarrow \sim q$): When you're out of Spritz, then you're out of root beer.

Suppose we do not have a Spritz, but we do have a root beer from another manufacturer. Then p is false and q is true, as seen in row 3 of the truth table on page 216.

Here, the conditional $p \rightarrow q$, with the truth values F $\rightarrow$ T, is true. However, the inverse $\sim p \rightarrow \sim q$, with the truth values T $\rightarrow$ F, is false.

As seen here, the conditional and its inverse do not always have the same truth value. The Spritz slogan is misleading and, as shown in row 3 of the truth table, is actually a false statement.

Illustration 2. Sometimes, when a conditional and its inverse do not have the same truth value, only the inverse is true.

s: It is spring. m: The month is May.

Conditional ($s \rightarrow m$): If it is spring, then the month is May.
Inverse ($\sim s \rightarrow \sim m$): If it is not spring, then the month is not May.
The calendar time for spring begins in March and ends in June.

To compare the truth values of the conditional and its inverse, suppose the month is April. Then s is true and m is false, as seen in row 2 of the truth table.

Here, the conditional $s \rightarrow m$, with the truth values $T \rightarrow F$, is false. However, the inverse $\sim s \rightarrow \sim m$, with the truth values $F \rightarrow T$, is true. Thus we see again, this time using row 2 of the truth table, that a conditional and its inverse do not always have the same truth value.

Illustration 3. In some instances a conditional and its inverse do have the same truth value.

Conditional ($e \rightarrow f$): If $x + 3 = 7$, then $x = 4$.
Inverse ($\sim e \rightarrow \sim f$): If $x + 3 \neq 7$, then $x \neq 4$.

When $x = 4$, e and f are both true, as in row 1 of the truth table.
When $x \neq 4$, e and f are both false, as in row 4 of the truth table.
There are no values of x for which e is true and f is false, as in row 2, or for which e is false and f is true, as in row 3.
Therefore, this conditional and its inverse always have the same truth value.

Hence, we must judge the truth value of each inverse on its own merits.

- **Conclusion: A conditional ($p \rightarrow q$) and its inverse ($\sim p \rightarrow \sim q$) may or may not have the same truth value.**

The Converse

Starting with an original conditional ($p \to q$), we form the ***converse*** **($q \to p$)** by interchanging the hypothesis and the conclusion. The symbols for the converse may be read as *q implies p* or *if q, then p.*

Does the converse have the same truth value as the conditional? To answer this question, we will again construct a truth table. Note that the conditional and its converse are *not equivalent* statements because they do not have matching truth values in every case.

		Conditional	**Converse**
p	q	$p \to q$	$q \to p$
T	T	T	T
T	F	F	T
F	T	T	F
F	F	T	T

Illustration 1. Often a conditional and its converse do not have the same truth value.

p: It rained. q: The ground got wet.

Conditional ($p \to q$): If it rained, then the ground got wet.
Converse ($q \to p$): If the ground got wet, then it rained.

In rows 1, 3, and 4 of the truth table, the conditional ($p \to q$) is true: If it rained, then the ground got wet. However, we notice that in row 3 the converse ($q \to p$) is false. Suppose that the ground got wet, but the conclusion, it rained, is false. A sprinkler or a burst pipe or snow could have caused the ground to become wet.

Since we can find instances where the converse is not true when the conditional is true, we conclude that the conditional and its converse do not always have the same truth value.

Illustration 2. In another example of different truth values, a television commercial shows a series of beautiful women, all of whom use Cleanse soap. Assuming that these models really do use the product, let us also assume that the following conditional is true:

Conditional ($b \to c$): If you are beautiful, then you use Cleanse soap.

Of course, what the advertiser wants you to believe is that the converse is true, but this is not necessarily so.

Converse ($c \to b$): If you use Cleanse soap, then you will be beautiful.

This converse is false because using the soap will not guarantee that you will become beautiful.

Illustration 3. In some instances a conditional and its converse are both true.

Conditional ($h \rightarrow s$): If a polygon is a hexagon, then the polygon has exactly six sides.

Converse ($s \rightarrow h$): If a polygon has exactly six sides, then the polygon is a hexagon.

We must judge the truth value of each converse on its own merits.

- **Conclusion: A conditional ($p \rightarrow q$) and its converse ($q \rightarrow p$) may or may not be true.**

Remember that we normally start with a true conditional when reasoning. Observe that, when we start with a false conditional, as is shown in the second row of the truth table on page 218, the converse is true.

The Contrapositive

Starting with a conditional ($p \rightarrow q$), we form the ***contrapositive*** ($\sim q \rightarrow \sim p$) by negating both the hypothesis and the conclusion, and then interchanging the resulting negations. The symbols for the contrapositive may be read as *not q implies not p* or *if not q, then not p.*

Does the contrapositive have the same truth value as the conditional? To answer this question, we will again construct a truth table.

				Conditional	Contrapositive
p	q	$\sim q$	$\sim p$	$p \rightarrow q$	$\sim q \rightarrow \sim p$
T	T	F	F	T	T
T	F	T	F	F	F
F	T	F	T	T	T
F	F	T	T	T	T

From the table, we see that *the conditional and its contrapositive are logically equivalent statements.*

Illustration 1. Let us compare the truth values of a conditional and its contrapositive.

p: It rained. q: The ground got wet.

Conditional ($p \rightarrow q$): If it rained, then the ground got wet.

Contrapositive ($\sim q \rightarrow \sim p$): If the ground did not get wet, then it did not rain.

Because the conditional and its contrapositive have matching truth values in every case, we can write a *tautology*:

$$(p \rightarrow q) \leftrightarrow (\sim q \rightarrow \sim p)$$

Therefore, this conditional and its contrapositive always have the same truth value.

Illustration 2. Let us consider an example in which we start with a false conditional. We will see that the contrapositive, as in row 2 of the truth table above, must also be false.

Conditional ($o \rightarrow r$): If 15 is an odd number, then 15 is a prime number.

Here, ($o \rightarrow r$) is false because (T $\rightarrow$ F) is false.

Contrapositive ($\sim r \rightarrow \sim o$): If 15 is not a prime number, then 15 is not an odd number.

Here, ($\sim r \rightarrow \sim o$) is false because (T $\rightarrow$ F) is false.

- **Conclusions: A conditional ($p \rightarrow q$) and its contrapositive ($\sim q \rightarrow \sim p$) must have the same truth value.**
  When a conditional is true, its contrapositive must be true.
  When a conditional is false, its contrapositive must be false.

Logical Equivalents

You have seen that a *conditional* and its *contrapositive* are ***logical equivalents*** because they have the same truth value.

Write the converse of a conditional and the inverse of the same conditional. Notice that one is the contrapositive of the other. For example:

Conditional ($p \rightarrow q$): If it snows, then it is cold.

Converse ($q \rightarrow p$): If it is cold, then it snows.

Inverse ($\sim p \rightarrow \sim q$): If it does not snow, then it is not cold.

Since the contrapositive of ($q \rightarrow p$) is ($\sim p \rightarrow \sim q$), we can say that a *converse* of a statement and an *inverse* of the same statement are *logical equivalents*. We may also verify that the converse and inverse are logically equivalent by examining their truth tables.

				Conditional	**Inverse**	**Converse**	**Contrapositive**
p	q	$\sim p$	$\sim q$	$p \rightarrow q$	$\sim p \rightarrow \sim q$	$q \rightarrow p$	$\sim q \rightarrow \sim p$
T	T	F	F	T	T	T	T
T	F	F	T	F	T	T	F
F	T	T	F	T	F	F	T
F	F	T	T	T	T	T	T

- **The truth table shows that the conditional and the contrapositive are logically equivalent and that the inverse and converse are logically equivalent.**

Forming Related Statements

A conditional statement can be formed by using a negated hypothesis or a negated conclusion. Consider, for example, "If it does not rain, then we'll go to the park." In this case:

r: It rains.

$\sim r$: It does *not* rain.

k: We'll go to the park.

$\sim r \to k$: If it does not rain, then we'll go to the park.

Using ($\sim r \to k$) as the original conditional statement, we can form its inverse, its converse, and its contrapositive by following the rules established within each definition.

The General Case and Rules	A Specific Case
Conditional: $p \to q$ The hypothesis is p; the conclusion is q.	**Conditional:** $\sim r \to k$ If it does not rain, then we'll go to the park.
Inverse: $\sim p \to \sim q$ Negate the hypothesis; negate the conclusion.	**Inverse:** $r \to \sim k$ If it rains, then we will not go to the park.
Converse: $q \to p$ Interchange the hypothesis and conclusion.	**Converse:** $k \to \sim r$ If we go to the park, then it does not rain.
Contrapositive: $\sim q \to \sim p$ Negate the hypothesis; negate the conclusion; interchange these negations.	**Contrapositive:** $\sim k \to r$ If we do not go to the park, then it rained.

Since the conditional ($\sim r \to k$) and the contrapositive ($\sim k \to r$) are logically equivalent, we can write the tautology:

$$(\sim r \to k) \leftrightarrow (\sim k \to r)$$

Also, since the inverse ($r \to \sim k$) and the converse ($k \to \sim r$) are logically equivalent, we can write the tautology:

$$(r \to \sim k) \leftrightarrow (k \to \sim r)$$

EXAMPLES

1. Write each required statement in symbolic form.

		Answers:
a.	the inverse of $k \to t$	$\sim k \to \sim t$
b.	the inverse of $\sim m \to r$	$m \to \sim r$
c.	the converse of $v \to y$	$y \to v$
d.	the converse of $s \to \sim t$	$\sim t \to s$
e.	the contrapositive of $l \to m$	$\sim m \to \sim l$
f.	the contrapositive of $\sim c \to d$	$\sim d \to c$

---

2. Given this true statement: If the polygon is a rectangle, then it has four sides. Which statement must also be true?
(1) If the polygon has four sides, then it is a rectangle.
(2) If the polygon is not a rectangle, then it does not have four sides.
(3) If the polygon does not have four sides, then it is not a rectangle.
(4) If the polygon has four sides, then it is not a rectangle.

Solution A conditional and its contrapositive always have the same truth value. When the conditional states "rectangle → four sides," the contrapositive is "not four sides → not rectangle."

Answer: (3)

EXERCISES

In 1–4, write the inverse of each statement in symbolic form.

1. $p \to q$ **2.** $t \to \sim w$ **3.** $\sim m \to p$ **4.** $\sim p \to \sim q$

In 5–8, write the converse of each statement in symbolic form.

5. $p \to q$ **6.** $t \to \sim w$ **7.** $\sim m \to p$ **8.** $q \to p$

In 9–12, write the contrapositive of each statement in symbolic form.

9. $p \to q$ **10.** $t \to \sim w$ **11.** $\sim m \to p$ **12.** $\sim q \to \sim p$

In 13–16, write the inverse of each statement in words.

13. If you use Charm face powder, then you will be beautiful.
14. If you buy Goal toothpaste, then your children will brush longer.
15. When you serve imported sparkling water, you show that you have good taste.
16. The man who wears Cutrite clothes is well dressed.

In 17–20: **a.** Write the inverse of each conditional statement in words. **b.** Give the truth value of the conditional. **c.** Give the truth value of the inverse.

17. If a polygon is a triangle, then the polygon has exactly three sides.
18. If a polygon is a trapezoid, then the polygon has exactly four sides.
19. If $2 \cdot 2 = 4$, then $2 \cdot 3 = 6$.
20. If $2^2 = 4$, then $3^2 = 6$.

In 21–24, write the converse of each statement in words.

21. If you live to an old age, then you eat Nano yogurt.
22. If you take pictures of your family with a Blinko camera, then you care about your family.
23. If you drive a Superb car, then you'll get good mileage.
24. If you use Dust and Roast, you'll make a better chicken dinner.

In 25–28: **a.** Write the converse of each conditional statement in words. **b.** Give the truth value of the conditional. **c.** Give the truth value of the converse.

25. If a number is even, then the number is exactly divisible by 2.
26. If two segments are 5 centimeters each, then the two segments are equal in measure.
27. If $5 = 1 + 4$, then $5^2 = 1^2 + 4^2$.
28. If $2(5) + 3 = 10 + 3$, then $2(5) + 3 = 13$.

In 29–32, write the contrapositive of each statement in words.

29. If you care enough to send the best, then you send Trademark cards.
30. If you use Trickle deodorant, then you won't have body odor.
31. If you brush with Brite, then your teeth will be pearly white.
32. If you want a good job, then you'll get a high school diploma.

In 33–37: **a.** Write the contrapositive of each conditional statement in words. **b.** Give the truth value of the conditional. **c.** Give the truth value of the contrapositive.

33. If opposite sides of a quadrilateral are parallel, then the quadrilateral is a parallelogram.
34. If two segments are 8 centimeters each, then the two segments are equal in measure.
35. If $1 + 2 = 3$, then $2 + 3 = 4$.
36. If all angles of a quadrilateral are equal in measure, then the quadrilateral is a rectangle.
37. If a number is prime, then it is not an even number.

In 38–42, write the numeral preceding the expression that best answers the question.

38. When $p \rightarrow q$ is true, which related conditional must be true?
(1) $q \rightarrow p$ (2) $\sim p \rightarrow \sim q$ (3) $p \rightarrow \sim q$ (4) $\sim q \rightarrow \sim p$

39. Which is the contrapositive of "If winter comes, then spring is not far behind"?
(1) If spring is not far behind, then winter comes.
(2) If spring is far behind, then winter comes.
(3) If spring is not far behind, then winter does not come.
(4) If spring is far behind, then winter does not come.

40. Which is the converse of "If a polygon is a trapezoid, its area $= \frac{1}{2}(b + c) \cdot h$"?
(1) If a polygon is not a trapezoid, then its area $\neq \frac{1}{2}(b + c) \cdot h$.
(2) If a polygon has area $= \frac{1}{2}(b + c) \cdot h$, then it is a trapezoid.
(3) If a polygon has area $\neq \frac{1}{2}(b + c) \cdot h$, then it is not a trapezoid.
(4) If the area of a trapezoid is $\frac{1}{2}(b + c) \cdot h$, then it is a polygon.

41. Which is the inverse of "If $x = 2$, then $x + 3 \neq 9$"?
(1) If $x + 3 \neq 9$, then $x = 2$. (2) If $x + 3 = 9$, then $x \neq 2$.
(3) If $x \neq 2$, then $x + 3 = 9$. (4) If $x \neq 2$, then $x + 3 \neq 9$.

42. Which is the contrapositive of "If $x > 2$, then $3x + 5x \neq 16$"?
(1) If $3x + 5x \neq 16$, then $x > 2$. (2) If $3x + 5x = 16$, then $x \not> 2$.
(3) If $3x + 5x \neq 16$, then $x \not> 2$. (4) If $3x + 5x = 16$, then $x > 2$.

43. For a conditional statement $(p \rightarrow q)$:
a. Write its inverse in symbolic form.
b. Write the converse of this inverse in symbolic form.
c. What is the relationship between the conditional statement $(p \rightarrow q)$ and the converse of the inverse of that statement?
d. What might be another way to define the contrapositive of a conditional statement?

In 44–51, for each conditional statement, write in words: **a.** the inverse **b.** the converse **c.** the contrapositive.

44. If today is Friday, then tomorrow is Saturday.
45. If Douglas does well in college, then he will apply to medical school.
46. Arlette will get a role in the play if she auditions.
47. Dorothea will graduate from law school in January if she takes courses this summer.
48. If John is accepted at the Culinary Institute, then he has a chance to earn a high salary as a chef.
49. If a man is honest, he does not steal.
50. If Julia doesn't water the plants, then the plants will die.
51. Rachel will not get her allowance if she forgets to do her chores.

In 52–54, if each given statement is assumed to be true, which of the four statements that follow must also be true?

52. If a figure is a square, then it is a polygon.
(1) If the figure is a polygon, then it is a square.
(2) If the figure is not a square, then it is not a polygon.
(3) If the figure is not a polygon, then it is not a square.
(4) If the figure is not a square, then it is a polygon.

53. If $a = b$, then $b \neq d$.
(1) If $a \neq b$, then $b \neq d$.
(2) If $b \neq d$, then $a = b$.
(3) If $b = d$, then $a \neq b$.
(4) If $a \neq b$, then $b = d$.

54. I'll get into shape if I exercise.
(1) If I don't exercise, then I won't get into shape.
(2) If I don't get into shape, then I do not exercise.
(3) If I get into shape, then I exercise.
(4) I exercise if I get into shape.

In 55–60, assume that each conditional statement is true. Then:

a. (*1*) Write its converse in words. (*2*) State whether the converse is true, false, or uncertain.
b. (*1*) Write its inverse in words. (*2*) State whether the inverse is true, false, or uncertain.
c. (*1*) Write its contrapositive in words. (*2*) State whether the contrapositive is true, false, or uncertain.

55. If Eddie lives in San Francisco, then he lives in California.

56. If a number is divisible by 12, then it is divisible by 3.

57. If I have the flu, then I am ill.

58. If one pen costs 29 cents, then three pens cost 87 cents.

59. If a polygon is a rhombus, then it has four sides.

60. If Alex loves computers, then he will learn how to write programs.

7-8 DRAWING CONCLUSIONS

In the process of reasoning, we use statements that we believe to be true and the rules of logic in order to draw conclusions, that is, to determine other statements that are true. When people disagree on the truth of the conclusions that have been drawn, often the reason is that they disagree on the truth values of the initial statements to which the laws of logic are applied.

EXAMPLES

1. What conclusion or conclusions can be drawn when the following statements are true?

Today is Monday or I have gym.
Today is not Monday.

Solution (1) Use symbols to represent the sentences that you know to be true.

$p \vee q$: Today is Monday or I have gym.
$\sim p$: Today is not Monday.

(2) Construct a truth table for $p \vee q$.

(a) Since $p \vee q$ is given as true, eliminate the case where $p \vee q$ is false by crossing out row 4.

p	q	$p \vee q$
T	T	T
T	F	T
F	T	T
~~F~~	~~F~~	~~F~~

(b) Since $\sim p$ is given as true, p is false. Eliminate the cases where p is true by crossing out rows 1 and 2.

p	q	$p \vee q$
~~T~~	~~T~~	~~T~~
~~T~~	~~F~~	~~F~~
F	T	T
~~F~~	~~F~~	~~T~~

(3) Note that only one case remains. Row 3 tells you that q is true. Therefore, you may conclude that "I have gym" is a true statement.

Answer: I have gym.

Note: In most cases, there will be many possible conclusions. Any true statement could be considered to be a conclusion. The following statements could all be shown to be true when the given statements are true.

Today is not Monday and I have gym.
If today is not Monday, then I have gym.
If I do not have gym, then today is Monday.

---

2. Draw a conclusion or conclusions based on the following true statements.

If I do not finish my English essay, I will get a *C* in English.
I do not get a *C* in English.

Solution (1) Use symbols to represent the sentences that you know to be true.

$\sim p \rightarrow q$: If I do not finish my English essay, I will get a *C* in English.
$\sim q$: I do not get a *C* in English.

(2) Construct a truth table for $\sim p \rightarrow q$.

(a) Since $\sim p \rightarrow q$ is true, eliminate the case where $p \rightarrow q$ is false by crossing out row 4.

p	q	$\sim p$	$\sim p \rightarrow q$
T	T	F	T
T	F	F	T
F	T	T	T
~~F~~	~~F~~	~~T~~	~~F~~

(b) Since $\sim q$ is true, q is false. Eliminate the cases where q is true by crossing out rows 1 and 3.

p	q	$\sim p$	$\sim p \rightarrow q$
~~T~~	~~T~~	~~F~~	~~T~~
T	F	F	T
~~F~~	~~T~~	~~T~~	~~F~~
~~F~~	~~F~~	~~T~~	~~F~~

(3) Note that only one case remains. Row 2 tells you that p is true. Therefore, you may conclude that "I did finish my English essay" is a true statement.

Answer: I did finish my English essay.

---

3. What conclusions, if any, can be drawn from the following true statements?

If today is Monday, then I have gym.
Today is not Monday.

Solution (1) Use symbols to represent the sentences that you know to be true.

$p \rightarrow q$: If today is Monday, then I have gym.
$\sim p$: Today is not Monday.

(2) Construct a truth table for $p \rightarrow q$.

(a) Since $p \rightarrow q$ is true, eliminate the case where $p \rightarrow q$ is false by crossing out row 2.

p	q	$p \rightarrow q$
T	T	T
~~T~~	~~F~~	~~F~~
F	T	T
F	F	T

(b) Since $\sim p$ is true, p is false. Eliminate the cases where p is true. Row 2 has already been eliminated; cross out row 1.

p	q	$p \rightarrow q$
~~T~~	~~T~~	~~T~~
~~T~~	~~F~~	~~F~~
F	T	T
F	F	T

(3) Two cases remain. Row 3 tells you that q is true. Row 4 tells you that q is false. Therefore, you can draw no conclusion about the truth value of q.

Answer: There is no simple sentence that is a conclusion.

Note: It may, of course, be possible to write some conclusions that do not depend on the truth value of q. For example, the contrapositive of a conditional always has the same truth value as the conditional. Therefore, when "If today is Monday, then I have gym" is true, the contrapositive, "If I do not have gym, then today is not Monday," is also true.

---

4. What conclusion or conclusions can be drawn from the following true statement?

Today is Monday and I have gym.

Solution When a conjunction is true, each of the simple statements from which it is constructed must be true. Therefore, each of the following statements is true.

Answer: Today is Monday.
I have gym.

EXERCISES

In 1–10, assume that the first two sentences in each group are true. Determine whether the third sentence is true, is false, or cannot be found to be true or false.

1. I study hard or I do not pass.
I pass.
I study hard.

2. If Toy is a dog, then Toy is an animal.
Toy is a dog.
Toy is an animal.

3. I like skating or skiing.
I like skating
I like skiing.

4. I like skating or skiing.
I do not like skating.
I like skiing.

5. I live in New York State if I live in Albany.
I do not live in Albany.
I live in New York State.

6. $x > 10$ if $x = 15$.
$x = 15$.
$x > 10$.

7. If I am late for dinner, then my dinner will be cold.
I am late for dinner.
My dinner is cold.

8. If I am late for dinner, then my dinner will be cold.
I am not late for dinner.
My dinner is not cold.

9. I will go to college if and only if I work this summer.
I do not work this summer.
I will go to college.

10. The average of two numbers is 10 when the numbers are 7 and 13.
The average of two numbers is 10.
The two numbers are 7 and 13.

In 11–25, assume, in each case, that the given sentences are true. Write a conclusion that is a simple sentence, if possible. If no conclusion is possible, write "No conclusion."

11. If I do my assignments, I pass the course.
I do my assignments.

12. If x is a prime, then $x \neq 9$.
$x = 9$.

13. On Saturday, we go skiing or we play hockey.
Last Saturday, we did not go skiing.

14. If a parallelogram contains a right angle, that parallelogram is a rectangle.
Parallelogram *ABCD* is not a rectangle.

15. $x \geq 10$ or $x < 10$.
$x \nless 10$.

16. If $2x + 5 = 7$, then $2x = 2$.
If $2x = 2$, then $x = 1$.
$2x + 5 = 7$.

17. x is even and a prime if and only if $x = 2$.
$x = 2$.

18. 3 is a prime if and only if 3 has exactly two factors.
3 is a prime.

19. If x is divisible by 4, then x is divisible by 2.
x is divisible by 2.

20. x is prime or x is even
x is prime.

21. If it is July, then it is summer.
It is not summer.

22. $\sqrt{7}$ is rational or irrational.
$\sqrt{7}$ is not rational.

23. $\sqrt{5} = 2.236068$ if and only if $(2.236068)^2 = 5$.
On a calculator, $(2.236068)^2 = 5.0000001$.

24. I study math, and I study French or Spanish.
I do not study French.

25. It is October or November, and it is not spring.
It is not November.

CHAPTER SUMMARY

Logic is the study of reasoning. In logic, a ***mathematical sentence*** is a sentence that contains a complete thought. Mathematical sentences are judged to be true or false.

A single letter is used in logic to represent a single complete thought. For example:

Let p represent "I sing." and Let q represent "People flee."

These letters are combined with other symbols to form more complex sentences, such as:

	In Symbols	*In Words*
Negation (*not p*)	$\sim p$	"I do *not* sing."
Conjunction (*p and q*)	$p \wedge q$	"I sing *and* people flee."
Disjunction (*p or q*)	$p \vee q$	"I sing *or* people flee."
Conditional (*if p then q*)	$p \rightarrow q$	"*If* I sing, *then* people flee."

Truth tables are used to show all possible truth values that can be assigned to the simple sentences and, in turn, to other more complex sentences."

A compound sentence contains more than one connective. When the truth value of each of its components is known, the compound sentence is called a ***compound statement***. The truth value of a compound statement is found:

1. either by simplifying truth values within the statement, working first within parenthesis, then with negations, and finally with all other connectives;
2. or by using truth tables.

A ***biconditional*** (p if and only if q) is written in symbols as $\boldsymbol{p \leftrightarrow q}$. The biconditional is a compound sentence formed by the conjunction of two conditionals. In other words:

$(p \leftrightarrow q)$ is ***logically equivalent*** to $(p \rightarrow q) \wedge (q \rightarrow p)$. Thus, ($p$ if and only if q) has the same meaning as the compound sentence [(if p then q) and (if q then p)].

p	q	$p \leftrightarrow q$
T	T	T
T	F	F
F	T	F
F	F	T

Given any conditional statement, three other conditionals, called the *inverse*, *converse*, and *contrapositive* of the original conditional statement, can be formed.

	In Symbols	*An Example*
Conditional:	$p \rightarrow q$	If I am 14, then I am a teenager.
Inverse:	$\sim p \rightarrow \sim q$	If I am not 14, then I am not a teenager.
Converse:	$q \rightarrow p$	If I am a teenager, then I am 14.
Contrapositive:	$\sim q \rightarrow \sim p$	If I am not a teenager, then I am not 14.

A conditional $(p \rightarrow q)$ and its contrapositive $(\sim q \rightarrow \sim p)$ always have the same truth value. If one is true, so is the other, as shown in the example above. Also, if a conditional is false, its contrapositive will be false.

When the truth value of a conditional $(p \rightarrow q)$ is known, however, no assumptions can be made about the truth values of its inverse $(\sim p \rightarrow \sim q)$ and its converse $(q \rightarrow p)$, which may be true, false, or uncertain.

VOCABULARY

7-1 Logic Mathematical sentence Phrase Open sentence Solution set (truth set) Statement (closed sentence)

7-2 Truth value Negation $(\sim p)$

7-3 Conjunction $(p \wedge q)$ Conjunct Tree diagram

7-4 Disjunction $(p \vee q)$ Disjunct Inclusive *or* Exclusive *or*

7-5 Conditional $(p \rightarrow q)$ Implication Hidden conditional premise (hypothesis, antecedent) Conclusion (consequent)

7-6 Biconditional $(p \leftrightarrow q)$ Logically equivalent

7-7 Inverse Converse Contrapositive Logical equivalents

REVIEW EXERCISES WITH OPEN-RESPONSE PROBLEMS

In 1–5, for each given statement: **a.** Write the statement in symbolic form, using the symbols given below. **b.** Tell whether the statement is true or false.

Let j represent "July follows June." (True)
Let a represent "August follows July." (True)
Let w represent "July is a winter month." (False)

1. If July follows June, then August follows July.

2. July follows June and July is a winter month.

3. July is a winter month or July follows June.

4. If August follows July, then July does not follow June.

5. August does not follow July and July is not a winter month.

6. Let p represent "I sleep," and let q represent "I'm grouchy." Using p and q, write in symbolic form: "If I don't sleep, I'm grouchy."

7. Let p represent "A parallelogram is a rhombus," and let q represent "Its sides are congruent." Using p and q, write in symbolic form: "A parallelogram is a rhombus if its sides are congruent."

In 8–18, write the numeral preceding the expression that best answers the question or completes the statement.

8. Which whole number, when substituted for y, will make the following sentence true?

$$(y + 3 > 8) \wedge (y < 7)$$

(1) 5 (2) 6 (3) 7 (4) 8

9. What is the converse of the statement "If it snowed, the temperature is below 32°."

(1) If the temperature is not below 32°, it did not snow.
(2) If the temperature is below 32°, it snowed.
(3) If it did not snow, the temperature is not below 32°.
(4) The temperature is below 32° if and only if it snowed.

10. Let p represent "x is prime." Let q represent "$x > 25$." When x is 21, which sentence is true?

(1) p (2) q (3) $p \wedge q$ (4) $p \rightarrow q$

11. Let e represent "x is even." Let d represent "x is divisible by 6." When x is 46, which sentence is true?

(1) $e \wedge d$ (2) $e \vee d$ (3) $e \rightarrow d$ (4) $\sim e \wedge d$

12. The sentence $m \vee r$ is false if and only if

(1) m is false and r is true (2) both m and r are false
(3) m is true and r is false (4) both m and r are true

13. If $p \leftrightarrow q$ is true, which sentence is also true?

(1) $p \rightarrow q$ (2) $p \wedge q$ (3) $p \vee q$ (4) $\sim p \wedge q$

14. What is the inverse of the sentence "If $8x = 24$, then $x = 3$"?
(1) If $x = 3$, then $8x = 24$. (2) If $8x \neq 24$, then $x \neq 3$.
(3) If $8x = 24$, then $x \neq 3$. (4) If $x \neq 3$, then $8x \neq 24$.

15. Which is the converse of the sentence $\sim p \rightarrow q$?
(1) $\sim q \rightarrow p$ (2) $q \rightarrow \sim p$ (3) $p \rightarrow \sim q$ (4) $q \rightarrow p$

16. Which has the same truth value as the sentence "If Will won, then Lon lost"?
(1) If Lon lost, then Will won.
(2) If Will did not win, then Lon did not lose.
(3) If Lon lost, then Will did not win.
(4) If Lon did not lose, then Will did not win.

17. Which sentence is *always* false?
(1) $p \vee \sim p$ (2) $q \wedge \sim q$ (3) $p \wedge \sim q$ (4) $q \vee \sim p$

18. Which of the following sentences is equivalent to $\sim(p \vee q)$?
(1) $\sim p \vee \sim q$ (2) $\sim p \vee q$ (3) $\sim p \wedge \sim q$ (4) $\sim p \rightarrow q$

19. Assume that the given sentences are true. Write a simple sentence that is a conclusion.

If $\triangle ABC$ is isosceles, then $AB = AC$.
$AB \neq AC$

20. At a track meet, Janice, Kay, and Virginia were the first three finishers of a 50-meter dash. Virginia did not come in second. Kay did not come in third. Virginia came in ahead of Janice. In what order did they finish the race?

21. The statement "If an integer, n, is divisible by 4, then it is divisible by 2" is always true.

a. Find a value of n that makes both the hypothesis and the conclusion true.

b. Find a value of n that makes both the hypothesis and the conclusion false.

c. Find a value of n that makes the hypothesis false and the conclusion true.

d. Can you find a value of n that makes the hypothesis true and the conclusion false? Explain why.

CUMULATIVE REVIEW WITH OPEN-RESPONSE PROBLEMS

1. Solve and check: $-3x + 13 = 1$.

2. A list of numbers that follows a pattern begins with the numbers $-16, 8, -4, 2, \ldots$

a. Find the next number in the list.
b. Write a rule or explain how the next number is determined.
c. What is the 10th number in the list?

3. In each of the following, perform the indicated operation and tell whether the number is rational or irrational.

   a. $\frac{11}{6} - \frac{8}{3}$ b. $0 + \sqrt{5}$ c. $\sqrt{4} + \sqrt{9}$

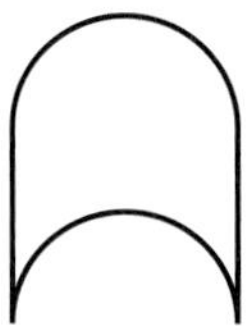

4. a. In the figure at the left, the curved portions are semicircles with a radius of 3 centimeters. The straight edges have a length of 6 centimeters. What is the area of the figure?

   b. Compare the area of this figure with the area of a square whose sides measure 6 centimeters.

5. What is the smallest number of coins (quarters, dimes, nickels, and pennies) that you would need to have the correct change for any amount between \$0.01 and \$0.99?

6. Solve for x: $2\pi x + 3 = 18$.

   a. Express the answer in terms of π.

   b. Express the answer to the nearest *hundredth*.

7. What is the solution set of $3x + 2 \geq 7$ when the domain is $\{0, 1, \sqrt{2}, \sqrt{3}, 2, \sqrt{5}, \sqrt{6}, \sqrt{7}, \sqrt{8}, 3\}$?

8. What is the truth value of "If $\sqrt{9}$ is irrational, then $\sqrt{5}$ is irrational"?

9. The length of a rectangle is 2 centimeters less than three times the width. What is the width of the rectangle if the length is 19 centimeters?

10. A list of numbers that follows a pattern begins with the numbers 1, 2, 4, 8, . . .

    a. Find the next number in the list.

    b. Write a rule or explain how the next number is determined.

    c. What is the 20th number in the list?

11. Peter, Carl, and Ralph play different musical instruments and different sports. The instruments that the boys play are violin, cello, and flute. The sports that the boys play are baseball, tennis, and soccer. From the clues given below, determine what instrument and what sport each boy plays.

    (1) The violinist plays tennis.
    (2) The boy who plays the flute does not play soccer.
    (3) Peter does not play the cello.
    (4) Ralph plays baseball.

Exploration

Collect articles from magazines and newspapers that use logic in presenting an argument. Discuss the use of logic in each article. Were the rules of logic correctly applied? Do you agree or disagree with the conclusions drawn?

Chapter 8
Operations with Algebraic Expressions

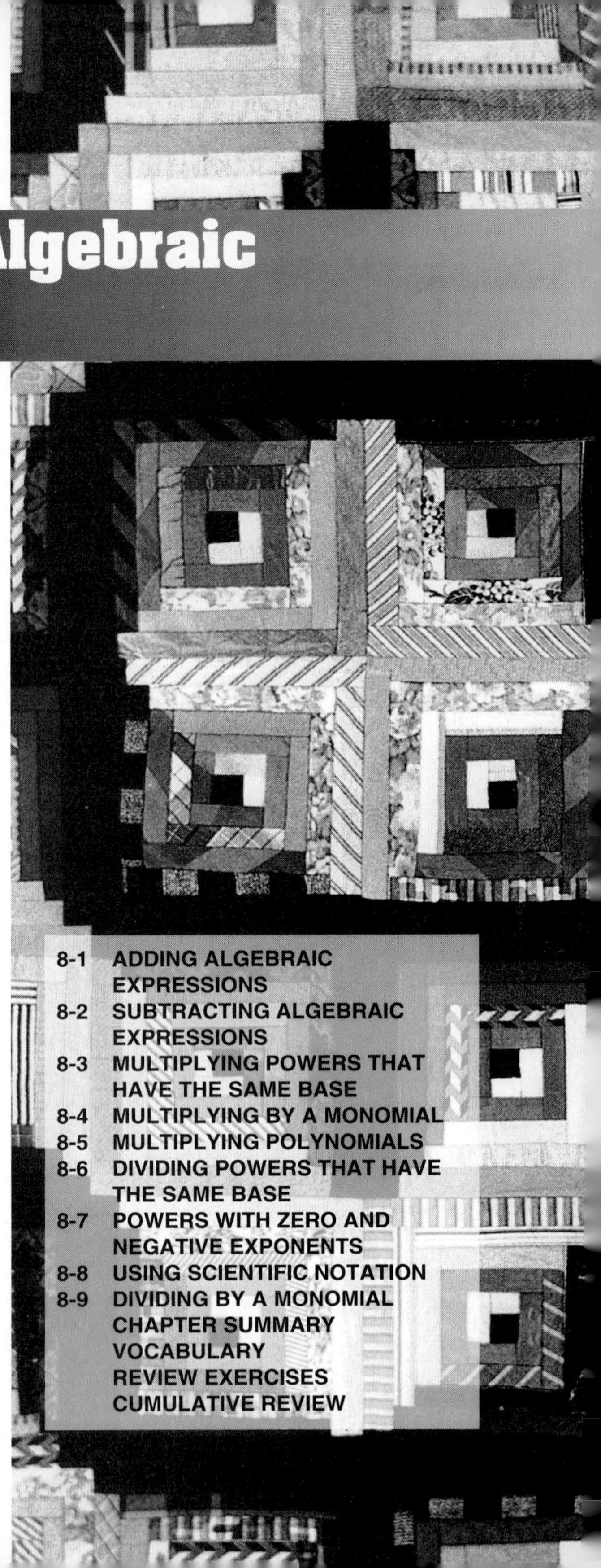

Patchwork is an authentic American craft, developed by our frugal ancestors in a time when nothing was wasted and useful parts of discarded clothing were stitched into warm and decorative quilts.

Quilt patterns, many of which acquired names as they were handed down from one generation to the next, were the products of creative and industrious people. In the Log Cabin pattern, squares are formed by adding pieces of material that are uniform in width and have lengths that are consecutive integers. The creators of this pattern were perhaps more aware of the pleasing effect achieved than of the mathematical relationship

$$(n + 1)^2 = n^2 + n + (n + 1)$$

involved.

8-1 ADDING ALGEBRAIC EXPRESSIONS

An algebraic expression that is a number, a variable, or a product or quotient of numbers and variables is called a ***term***. Examples of terms are:

$$5 \qquad x \qquad 8z \qquad -\frac{4}{7}y^2 \qquad 0.7a^2b^3 \qquad -\frac{5}{w}$$

Like and Unlike Terms

Two or more terms that contain the same variable or variables, with corresponding variables having the same exponents, are called ***like terms*** or ***similar terms***. For example, the following pairs are like terms.

$$6k \text{ and } k \qquad 5x^2 \text{ and } -7x^2 \qquad 9ab \text{ and } \frac{2}{5}ab \qquad \frac{9}{2}x^2y^3 \text{ and } -\frac{11}{3}x^2y^3$$

Two terms are ***unlike terms*** when they contain different variables, or the same variable or variables with different exponents. For example, the following pairs are unlike terms.

$$3x \text{ and } 4y \qquad 5x^2 \text{ and } 5x^3 \qquad 9ab \text{ and } \frac{2}{5}a \qquad \frac{8}{3}x^3y^2 \text{ and } \frac{4}{7}x^2y^3$$

To add like terms, we use the distributive property of multiplication over addition.

$$9x + 2x = (9 + 2)x = 11x$$

$$-16cd + 3cd = (-16 + 3)cd = -13cd$$

$$18y^2 + (-y^2) = [18 + (-1)]y^2 = 17y^2$$

Since the distributive property is true for any number of terms, we can express the sum of any number of like terms as a single term.

$$-3ab^2 + 4ab^2 - 2ab^2 = (-3 + 4 - 2)ab^2 = -1ab^2 = -ab^2$$

$$x^3 + 11x^3 - 8x^3 - 4x^3 = (1 + 11 - 8 - 4)x^3 = 0x^3 = 0$$

Note that in the above examples, when like terms are added:

1. The sum has the same variable or variables as the original terms.
2. The numerical coefficient of the sum is the sum of the numerical coefficients of the terms that were added.

The sum of unlike terms cannot be expressed as a single term. For example, $2x + 3y$ cannot be written as a single term.

We often represent the measures of the sides and the measures of the angles of a geometric figure as algebraic expressions. For example, the length of each side of the square on page 236 is represented by s. The perimeter of the square,

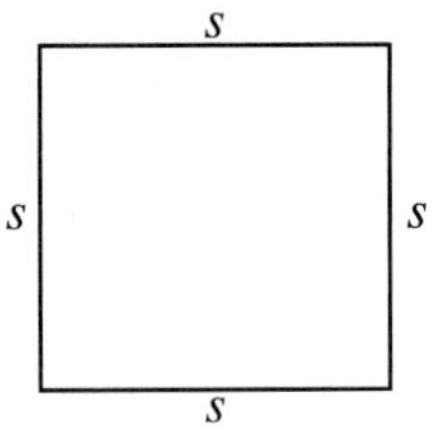

P, is the sum of the lengths of the sides. When like terms are added, the formula $P = s + s + s + s$ becomes $P = 4s$.

$$\begin{aligned} P &= s + s + s + s \\ &= 1s + 1s + 1s + 1s \\ &= (1 + 1 + 1 + 1)s \\ &= 4s \end{aligned}$$

When an algebraic expression involving a variable is used to represent the measure of a line segment or the measure of an angle of a geometric figure, the domain of the variable must be restricted to the values that will yield positive values for the measures. For example, for a square the length of the side, s, cannot be negative or zero.

EXAMPLES

1. Add:

Solutions and answers:

a. $+7x + (-3x)$ $= [+7 + (-3)]x = +4x$

b. $-3y^2 + (-5y^2)$ $= [-3 + (-5)]y^2 = -8y^2$

c. $-15abc + 6abc$ $= [-15 + 6]abc = -9abc$

d. $8x^2y - x^2y$ $= [8 - 1]x^2y = 7x^2y$

e. $-9y + 9y$ $= [-9 + 9]y = 0y = 0$

f. $2(a + b) + 6(a + b)$ $= [2 + 6](a + b) = 8(a + b)$

---

2. The length of each of the two equal sides of an isosceles triangle is twice the length of the base of the triangle. If the length of the base is represented by n, represent in simplest form the perimeter of the triangle.

Solution n represents the length of the base.
$2n$ represents the length of one of the equal sides.
$2n$ represents the length of the other equal side.

$$\text{Perimeter} = n + 2n + 2n = (1 + 2 + 2)n = 5n \quad \textit{Answer}$$

Note that, since the length of a side of a geometric figure is represented by a positive number, the variable n must be positive.

Monomials and Polynomials

A term that has no variable in the denominator is called a ***monomial***. For example, 5, $-5w$, and $\frac{3w^2}{5}$ are monomials, but $\frac{5}{w}$ is *not* a monomial.

A ***polynomial*** is a sum of monomials.

A monomial such as $4x^2$ may be considered to be a polynomial of one term. (*Mono-* means "one"; *poly-* means "many.")

A polynomial of two unlike terms, such as $10a + 12b$, is called a ***binomial.*** (*Bi-* means "two.")

A polynomial of three unlike terms, such as $x^2 + 3x + 2$, is called a ***trinomial.*** (*Tri-* means "three.")

A polynomial such as $5x^2 + (-2x) + (-4)$ is usually written as $5x^2 - 2x - 4$.

A polynomial has been simplified or is in ***simplest form*** when it contains no like terms. For example, $5x^3 + 8x^2 - 5x^3 + 7$, when expressed in simplest form, becomes $8x^2 + 7$.

A polynomial is said to be in ***descending order*** when the exponents of a particular variable decrease as we move from left to right. For example, the polynomial $x^3 + 5x^2 - 4x + 9$ is in a descending order of x.

A polynomial is said to be in ***ascending order*** when the exponents of a particular variable increase as we move from left to right. For example, the polynomial $4 + 5y - y^2$ is in an ascending order of y.

To add two polynomials, we use the commutative, associative, and distributive properties to combine like terms.

EXAMPLES

1. Simplify: $3ab + 5b - ab + 4ab - 2b$.

How to Proceed:	*Solution:*
(1) Write the expression:	$3ab + 5b - ab + 4ab - 2b$
(2) Group like terms together by using the commutative and the associative properties:	$3ab - ab + 4ab + 5b - 2b$ $(3ab - ab + 4ab) + (5b - 2b)$
(3) Use the distributive property.	$(3 - 1 + 4)ab + (5 - 2)b$
(4) Add:	$6ab + 3b$

Answer: $6ab + 3b$

---

2. Find the sum: $(3x^2 + 5) + (6x^2 + 8)$.

How to Proceed:	*Solution:*
(1) Write the expression:	$(3x^2 + 5) + (6x^2 + 8)$
(2) Use the associative property:	$3x^2 + (5 + 6x^2) + 8$
(3) Use the commutative property:	$3x^2 + (6x^2 + 5) + 8$
(4) Use the associative property:	$(3x^2 + 6x^2) + (5 + 8)$
(5) Add like terms:	$9x^2 + 13$

The sum of polynomials can also be arranged vertically, placing like terms under one another. The sum of $3x^2 + 5$ and $6x^2 + 8$ can be arranged as shown at the right.

$$\begin{array}{r} 3x^2 + 5 \\ \underline{6x^2 + 8} \\ 9x^2 + 13 \end{array}$$

Addition can be checked by substituting any convenient value for the variable and evaluating each polynomial and the sum.

Check: Let $x = 4$.

$$\begin{array}{rcrcr} 3x^2 + 5 & = & 3(4)^2 + 5 & = & 53 \\ \underline{6x^2 + 8} & = & 6(4)^2 + 8 & = & \underline{104} \\ 9x^2 + 13 & = & 9(4)^2 + 13 & = & 157 \checkmark \end{array}$$

Answer: $9x^2 + 13$

---

3. Simplify: $6a + [5a + (6 - 3a)]$.

When one grouping symbol appears within another, first simplify the expression within the innermost grouping symbol.

How to Proceed:	*Solution:*
(1) Write the expression:	$6a + [5a + (6 - 3a)]$
(2) Use the commutative property:	$6a + [5a + (-3a + 6)]$
(3) Use the associative property:	$6a + [(5a - 3a) + 6]$
(4) Combine like terms:	$6a + [2a + 6]$
(5) Use the associative property:	$(6a + 2a) + 6$
(6) Combine like terms:	$8a + 6$

Answer: $8a + 6$

---

4. A rectangle has a width of $x - 1$ and a length of $2x - 6$.

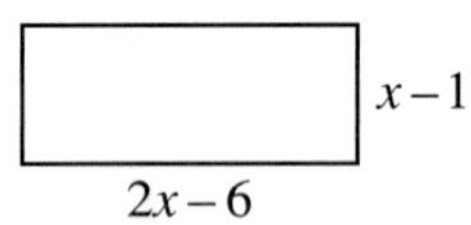

a. Express the perimeter of the rectangle in simplest form.

b. Name one value for x that is *not* possible for this figure.

Solutions **a.** $P = (2x - 6) + (x - 1) + (2x - 6) + (x - 1)$
$= 6x - 14$

b. For the width $(x - 1)$ to be positive, x must be greater than 1.
For the length $(2x - 6)$ to be positive, x must be greater than 3.
If x had a value of 3 or less, the rectangle could not exist because the measures of its sides would not be positive. For example, if $x = 3$, then the width $= x - 1 = (3) - 1 = 2$, and the length $= 2x - 6 = 2(3) - 6 = 0$. Such a figure cannot exist.

Answers: **a.** $P = 6x - 14$ **b.** 3, or any value less than 3

EXERCISES WITH OPEN-RESPONSE PROBLEM

In 1–36, add in each case.

1. $(+8c) + (+7c)$
2. $(+10t) + (-3t)$
3. $(-4a) + (-6a)$
4. $(-20r) + (5r)$
5. $(-7w) + (+7w)$
6. $(5ab) + (-9ab)$

7. $\begin{array}{r} +7c \\ \underline{+8c} \end{array}$
8. $\begin{array}{r} -39r \\ \underline{-22r} \end{array}$
9. $\begin{array}{r} -19t \\ \underline{+6t} \end{array}$
10. $\begin{array}{r} +14c \\ \underline{-c} \end{array}$
11. $\begin{array}{r} -1.5m \\ \underline{+1.2m} \end{array}$

12. $\begin{array}{r} +3e \\ \underline{-3e} \end{array}$
13. $\begin{array}{r} +2x^2 \\ \underline{+9x^2} \end{array}$
14. $\begin{array}{r} -48y^2 \\ \underline{-13y^2} \end{array}$
15. $\begin{array}{r} -d^2 \\ \underline{+7d^2} \end{array}$
16. $\begin{array}{r} 0.5y^3 \\ \underline{0.8y^3} \end{array}$

17. $\begin{array}{r} +\frac{5}{3}c^4 \\ \underline{-\frac{7}{3}c^4} \end{array}$
18. $\begin{array}{r} -10r^3 \\ \underline{10r^3} \end{array}$
19. $\begin{array}{r} 8rs \\ \underline{6rs} \end{array}$
20. $\begin{array}{r} -6mn \\ \underline{-mn} \end{array}$
21. $\begin{array}{r} -4xyz \\ \underline{+5xyz} \end{array}$

22. $\begin{array}{r} +0.4cd \\ \underline{-0.8cd} \end{array}$
23. $\begin{array}{r} -8xy \\ \underline{+8xy} \end{array}$
24. $\begin{array}{r} +3(x+y) \\ \underline{+9(x+y)} \end{array}$
25. $\begin{array}{r} +6a^2b \\ \underline{+7a^2b} \end{array}$
26. $\begin{array}{r} -xy^2 \\ \underline{-3xy^2} \end{array}$

27. $\begin{array}{r} 5x + 3y \\ \underline{6x + 9y} \end{array}$
28. $\begin{array}{r} 4a - 6b \\ \underline{9a + 3b} \end{array}$
29. $\begin{array}{r} -6m + n \\ \underline{-4m - 5n} \end{array}$
30. $\begin{array}{r} -9ab + 8cd \\ \underline{3ab - 8cd} \end{array}$

31. $\begin{array}{r} 15x - 26y + 8z \\ \underline{3x - 14y - 3z} \end{array}$
32. $\begin{array}{r} x^2 - 33x + 15 \\ \underline{-4x^2 + 18x - 36} \end{array}$
33. $\begin{array}{r} -5a^2 - 6ab - 4b^2 \\ \underline{+7a^2 + 6ab - 3b^2} \end{array}$

34. $\begin{array}{r} x^2 + 3x + 5 \\ 2x^2 - 4x - 1 \\ \underline{-5x^2 + 2x + 4} \end{array}$
35. $\begin{array}{r} 5c^2 - 4cd + 6d^2 \\ -c^2 + 3cd + 2d^2 \\ \underline{-3c^2 + cd - 8d^2} \end{array}$
36. $\begin{array}{r} 2.1 + 0.9z + z^2 \\ -0.7z - 0.2z^2 \\ \underline{-0.9 + 0.2z \qquad\quad} \end{array}$

In 37–58, simplify each expression by combining like terms.

37. $(+6x) + (-4x) + (-5x) + (+10x)$
38. $-5y + 6y + 9y - 14y$
39. $(+7c) + (-15c) + (+2c) + (+12c)$
40. $4m + 9m - 12m - m$
41. $(+8x^2) + (-x^2) + (-12x^2) + (+2x^2)$
42. $13y^2 - 15y^2 - y^2 + 8y^2$
43. $4a + (9a + 3)$
44. $7b + (4b - 6)$
45. $8c + (7 - 9c)$
46. $(-6x - 4) + 6x$
47. $r + (s + 2r)$
48. $8d^2 + (6d^2 - 4d)$
49. $(5x + 3) + (6x - 5)$
50. $-6y + [7 + (6y - 7)]$
51. $(5 - 6y) + (-9y + 2)$
52. $5a + [3b + (-2a + 4b)]$
53. $(5x^2 - 4) + (-3x^2 - 4)$
54. $3y^2 + [6y^2 + (3y - 4)]$
55. $(x^3 + 3x^2) + (-2x^2 - 9)$
56. $-d^2 + [9d + (2 - 4d^2)]$
57. $(x^2 + 5x - 24) + (-x^2 - 4x + 9)$
58. $(x^3 + 9x - 5) + (-4x^2 - 12x + 5)$

59. Express the perimeter of each of the following figures, and simplify the result by combining like terms.

a.

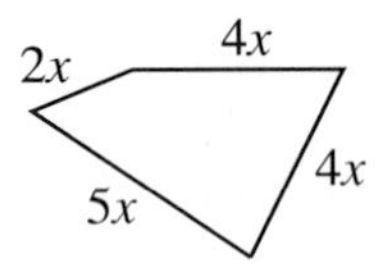

b.

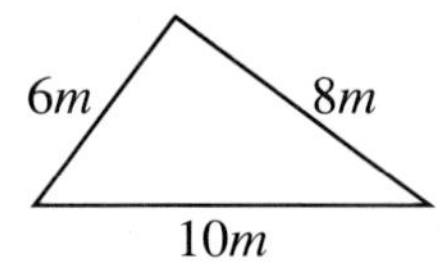

c.

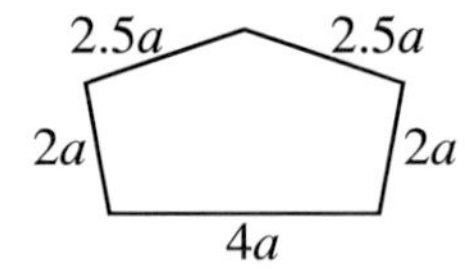

d.

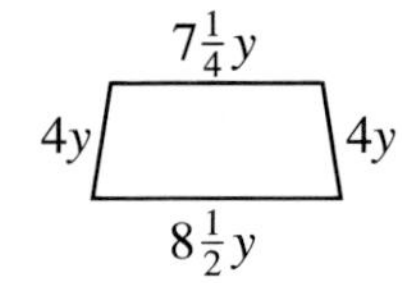

In 60–63: **a.** Express the perimeter of each figure as a polynomial in simplest form. **b.** Name one value for x that is not possible.

60.

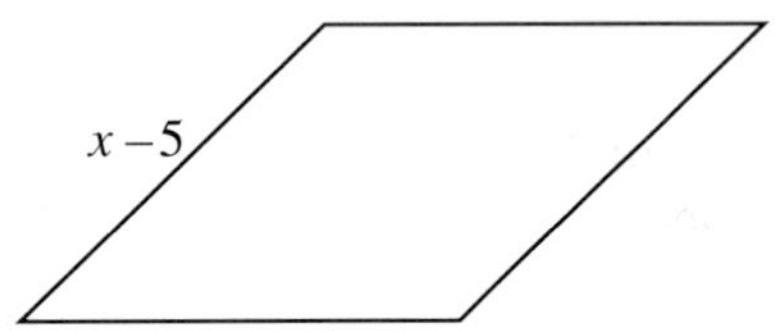

Rhombus

61.

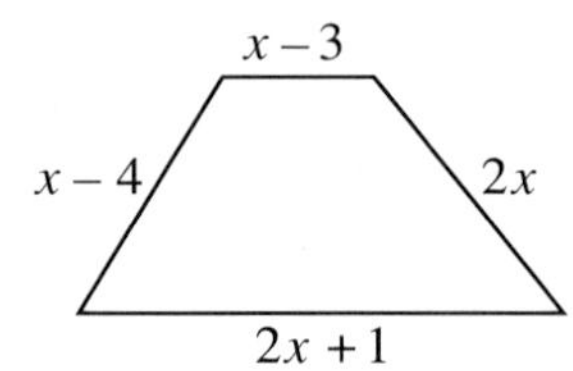

Trapezoid

62.

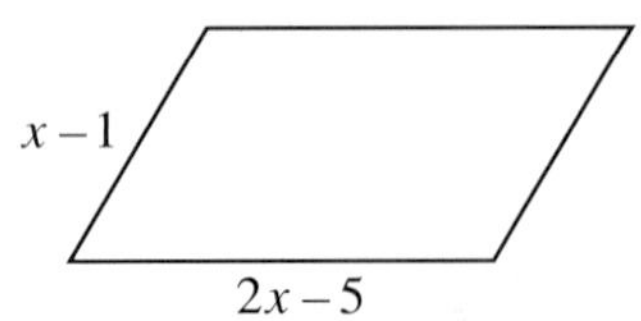

Parallelogram

63.

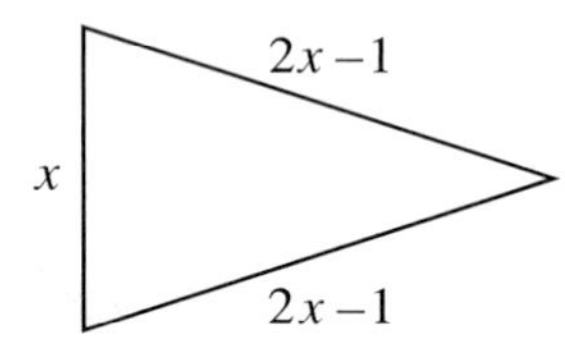

Triangle

In 64–72, write each answer as a polynomial in simplest form.

64. A cheeseburger costs 3 times as much as a soft drink, and an order of fries costs twice as much as a soft drink. If a soft drink costs s cents, express in simplest form the total cost of a cheeseburger, an order of fries, and a soft drink.

65. Jack deposited some money in his savings account in September. In October he deposited twice as much as in September, and in November he deposited one-half as much as in September. If x represents the amount of money deposited in September, represent in simplest form the total amount Jack deposited in the 3 months.

66. On Tuesday Melita read 3 times as many pages as she did on Monday. On Wednesday she read 1.5 times as many pages as she did on Monday, and on Thursday she read half as many pages as she did on Monday. If Melita read p pages on Monday, represent in simplest form the total number of pages she read in the 4 days.

67. The cost of 12 gallons of gas is represented by $12x$, and the cost of a quart of oil is represented by $2x - 30$. Represent the cost of 12 gallons of gas and a quart of oil.

68. In the last basketball game of the season, Tom scored $2x$ points, Tony scored $x + 5$ points, Walt scored $3x + 1$ points, Dick scored $4x - 7$ points, and Dan scored $2x - 2$ points. Represent the total points scored by these five players.

69. The length of a rectangle is 4 more than the width. If x represents the width of the rectangle, represent the perimeter in terms of x.

70. The cost of a chocolate shake is 40 cents less than the cost of a hamburger. If h represents the cost, in cents, of a hamburger, represent in terms of h the cost of a hamburger and a chocolate shake.

71. Ann spent 12 dollars more for fabric for a new dress than she did for buttons, and 1 dollar less for thread than she did for buttons. If b represents the cost, in dollars, of the buttons, represent in terms of b the total cost of the materials needed for the dress.

72. Last week, Greg spent twice as much on bus fare as he did on lunch, and 3 dollars less on entertainment than he did on bus fare. If x represents the amount, in dollars, spent on lunch, express in terms of x the total amount Greg spent on lunch, bus fare, and entertainment.

In 73–76, state whether each expression is a monomial, a binomial, a trinomial, or none of these.

73. $8x + 3$ **74.** $7y$ **75.** $-2a^2 + 3a - 6$ **76.** $x^3 + 2x^2 + x - 7$

77. **a.** Give an example of the sum of two binomials that is a binomial.
b. Give an example of the sum of two binomials that is a monomial.
c. Give an example of the sum of two binomials that is a trinomial.
d. Give an example of the sum of two binomials that has four terms.
e. Can the sum of two binomials have more than four terms?

8-2 SUBTRACTING ALGEBRAIC EXPRESSIONS

Subtracting Like Terms

We can subtract like terms in algebraic expressions by using the same method that we use to subtract signed numbers: we add the opposite of the subtrahend to the minuend.

$$(+7) - (-3) = (+7) + (+3) = +10$$

$$(+7x) - (-3x) = (+7x) + (+3x) = +10x$$

Subtracting a Polynomial from a Polynomial

To subtract one polynomial from another, we again use the same procedure. For example, to subtract $2x^2 - 5x - 3$ from $4x^2 + 2x - 3$, we add the opposite of $2x^2 - 5x - 3$ to $4x^2 + 2x - 3$. This can be done horizontally or vertically.

Horizontal Method

(1) Write the subtraction problem: $(4x^2 + 2x - 3) - (2x^2 - 5x - 3)$

(2) To subtract, add the opposite of the subtrahend (the polynomial being subtracted). Write the opposite of each term: $(4x^2 + 2x - 3) + (-2x^2 + 5x + 3)$

(3) Group like terms together: $(4x^2 - 2x^2) + (2x + 5x) + (-3 + 3)$

(4) Add like terms: $2x^2 + 7x + 0$

$2x^2 + 7x$

Vertical Method

(1) Write the subtraction problem:
(2) Change the sign of each term to be subtracted:
(3) Then add like terms:

$$\begin{array}{rrrr} & 4x^2 & +\ 2x & -\ 3 \\ & - & + & + \\ \text{Write} & \oplus 2x^2 & \ominus\ 5x & \ominus\ 3 \\ \hline & 2x^2 & +\ 7x & \end{array}$$

Answer: $2x^2 + 7x$

Subtraction can be checked by adding the subtrahend and the difference. The result should equal the minuend.

EXAMPLES

1. From $-5a^2$ subtract $+12a^2$.

Solution

$$-5a^2 - (+12a^2) = -5a^2 + (-12a^2)$$
$$= -17a^2$$

Answer: $-17a^2$

---

2. Subtract and check: $(5x^2 - 6x + 3) - (2x^2 - 9x - 6)$.

Solution Change all signs in the subtrahend, and add.

$$\begin{array}{rrrl} 5x^2 & -\ 6x & +\ 3 & \text{Minuend} \\ - & + & + & \\ \oplus 2x^2 & \ominus\ 9x & \ominus\ 6 & \text{Subtrahend} \\ \hline 3x^2 & +\ 3x & +\ 9 & \text{Difference} \end{array}$$

Check: Add. The subtrahend plus the difference should equal the minuend.

$$\begin{array}{rrrl} 2x^2 & -\ 9x & -\ 6 & \text{Subtrahend} \\ 3x^2 & +\ 3x & +\ 9 & \text{Difference} \\ \hline 5x^2 & -\ 6x & +\ 3 & \text{Minuend} \ \checkmark \end{array}$$

Answer: $3x^2 + 3x + 9$

---

3. Simplify the expression $9x - [7 - (4 - 2x)]$.

Solution Perform the subtraction involving the expression within the innermost grouping symbol first.

$$\begin{aligned} 9x - [7 - (4 - 2x)] &= 9x - [7 + (-4 + 2x)] \\ &= 9x - [7 - 4 + 2x] \\ &= 9x - [3 + 2x] \\ &= 9x + [-3 - 2x] \\ &= 9x - 3 - 2x \\ &= 7x - 3 \end{aligned}$$

Answer: $7x - 3$

EXERCISES

In 1–6, write the opposite (additive inverse) of each expression.

1. $9x$ **2.** $-5x$ **3.** $-6x - 6y$

4. $2x^2 - 3x + 2$ **5.** $-y^2 + 5y - 4$ **6.** $7ab - 3bc$

In 7–20, subtract and then check each result.

7. $\begin{array}{l} 10a \\ \underline{4a} \end{array}$ **8.** $\begin{array}{l} 5b \\ \underline{4b} \end{array}$ **9.** $\begin{array}{l} 6d + 6e \\ \underline{9d - 8e} \end{array}$ **10.** $\begin{array}{r} 8x - 3y \\ \underline{-4x + 8y} \end{array}$

11. $\begin{array}{l} 4r - 7s \\ \underline{5r - 7s} \end{array}$ **12.** $\begin{array}{l} 0 \\ \underline{8a - 6b} \end{array}$ **13.** $\begin{array}{l} 6rs - 7bc \\ \underline{9rs - 7bc} \end{array}$ **14.** $\begin{array}{r} 5xy \\ \underline{-3xy + cd} \end{array}$

15. $\begin{array}{r} x^2 - 6x + 5 \\ \underline{3x^2 - 2x - 2} \end{array}$ **16.** $\begin{array}{r} 3y^2 - 2y - 1 \\ \underline{-5y^2 - 2y + 6} \end{array}$ **17.** $\begin{array}{r} 3a^2 - 2ab + 3b^2 \\ \underline{-\ a^2 - 5ab + 3b^2} \end{array}$

18. $\begin{array}{l} 7a + 6b - 9c \\ \underline{3a \qquad\ \ - 6c} \end{array}$ **19.** $\begin{array}{r} x^2 \qquad\quad - 9 \\ \underline{-2x^2 + 5x - 3} \end{array}$ **20.** $\begin{array}{r} 5 - 6d - d^2 \\ \underline{-\ 4d - d^2} \end{array}$

In 21–47, simplify each expression.

21. $(+9r) - (+2r)$ **22.** $(+15s) - (-5s)$ **23.** $(-15t) - (-15t)$

24. $(-17n) - (11n)$ **25.** $(+8x) - (0)$ **26.** $(0) - (+8x)$

27. $5x - (2x + 5)$ **28.** $3y - (5y - 4)$ **29.** $4z - (6z - 2)$

30. $9m - (6 + 6m)$ **31.** $m - (m - n)$ **32.** $4d - (5c + 4d)$

33. $5c - (4c - 6c^2)$ **34.** $8r - (-6s - 8r)$ **35.** $-2x - (5x + 8)$

36. $-9d - (2c - 4d) + 4c$ **37.** $(3y + z) + (z - 5y) - (2z - 2y)$

38. $(a - b) - (a + b) - (-a - b)$ **39.** $(x^2 - 3x) + (5 - 9x) - (5x^2 - 7)$

40. $5c - [8c - (6 - 3c)]$ **41.** $12 - [-3 + (6x - 9)]$

42. $10x + (3x - (5x - 4)]$ **43.** $x^2 - [-3x + (4 - 7x)]$

44. $3x^2 - [7x - (4x - x^2) + 3]$ **45.** $9a - [5a^2 - (7 + 9a - 2a^2)]$

46. $x^2 - [2x^2 - (5x - x^2) + 3x]$ **47.** $4y^2 - \{4y + [3y^2 - (6y + 2) + 6]\}$

48. From $4x + 2y$, subtract $x - 4y$.
49. Subtract $5a - 7b$ from $3a - 9b$.
50. From $5x^2 + 5x - 4$, subtract $x^2 - 3x + 5$.
51. Subtract $7r^2 + 3r - 8$ from $10r^2 - 3r - 7$.
52. From $m^2 + 5m - 7$, subtract $m^2 - 3m - 4$.
53. From $12x - 6y + 9z$, subtract $-x - 3z + 6y$.
54. Subtract $2x^2 - 3x + 7$ from $x^2 + 6x - 12$.
55. Subtract $3x^2 - 9$ from $5x^2 + 2x$.

In 56–59, the perimeter of triangle ABC and the lengths of $\overline{AB}$ and $\overline{BC}$ are expressed in terms of a variable. In each case, express the length of $\overline{AC}$ in terms of that variable.

56. Perimeter $= 18x$, $AB = 6x$, $BC = 4x$

57. Perimeter $= 15y$, $AB = y$, $BC = 7y$

58. Perimeter $= 35k$, $AB = 3k$, $BC = 15k$

59. Perimeter $= x$, $AB = \frac{1}{3}x$, $BC = \frac{1}{3}x$

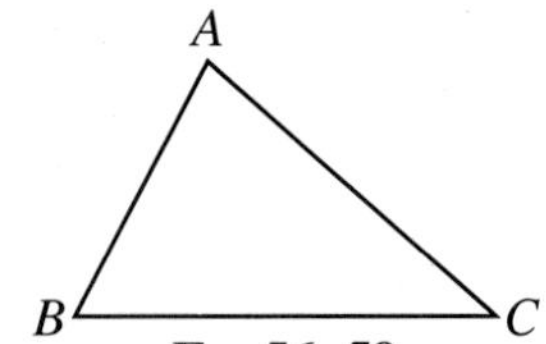

Ex. 56–59

60. Of three sisters, Ellen weighs twice as much as Susan. Lisa weighs 6 times as much as Susan. If Susan weighs p pounds, express in simplest form how much more Lisa weighs than Ellen.
61. Kelly has 5 times as many dimes in her pocket as does Jill. Jill has twice as many dimes as does Leon. If Leon has d dimes, express in simplest form the difference between the numbers of dimes that Kelly and Jill have.
62. **a.** By how much does 13 exceed 10? **b.** By how much does $7x + 5$ exceed $4x - 3$?
63. By how much does $a + b + c$ exceed $a + b - c$?
64. What algebraic expression must be added to $2x^2 + 5x + 7$ to give $8x^2 - 4x - 5$ as the result?
65. What algebraic expression must be added to $4x^2 - 8$ to make the result equal to 0?
66. What algebraic expression must be added to $-3x^2 + 7x - 5$ to give 0 as the result?
67. From the sum of $y^2 + 2y - 7$ and $2y^2 - 4y + 3$, subtract $3y^2 - 8y - 10$.
68. Subtract the sum of $c^2 - 5$ and $-2c^2 + 3c$ from $4c^2 - 6c + 7$.

8-3 MULTIPLYING POWERS THAT HAVE THE SAME BASE

Finding the Product of Powers

We know that y^2 means $y \cdot y$ and y^3 means $y \cdot y \cdot y$. Therefore:

$$y^2 \cdot y^3 = \overbrace{(y \cdot y)}^{2} \cdot \overbrace{(y \cdot y \cdot y)}^{3} = \overbrace{y \cdot y \cdot y \cdot y \cdot y}^{5} = y^5$$

Similarly,

$$c^2 \cdot c^4 = \overbrace{(c \cdot c)}^{2} \cdot \overbrace{(c \cdot c \cdot c \cdot c)}^{4} = \overbrace{c \cdot c \cdot c \cdot c \cdot c \cdot c}^{6} = c^6$$

and

$$x \cdot x^3 = \overbrace{(x)}^{1} \cdot \overbrace{(x \cdot x \cdot x)}^{3} = x^4$$

The exponent in each product is the sum of the exponents in the factors, as shown in these examples.

In general, when x is a signed number and a and b are positive integers:

$$x^a \cdot x^b = x^{a+b}$$

EXAMPLES

Simplify each of the following products:

Solutions and answers

a. $x^5 \cdot x^2 = x^{5+2} = x^7$

b. $a^7 \cdot a = a^{7+1} = a^8$

c. $3^2 \cdot 3^4 = 3^{2+4} = 3^6$

Note: When we multiply powers with like bases, we do not actually perform the operation of multiplication but rather *count up* the number of times that the base is to be used as a factor to find the product. In Example **c** above, the answer does not give the value of the product but indicates only the number of times that 3 must be used as a factor to obtain the product. We can use the power key [y^x] to evaluate the products $3^2 \cdot 3^4$ and 3^6 to show that they are equal.

Enter: 3 [y^x] 2 [×] 3 [y^x] 4 [=]

Display: [729.]

Enter: 3 [y^x] 6 [=]

Display: [729.]

Finding a Power of a Power

Since $(x^3)^4 = x^3 \cdot x^3 \cdot x^3 \cdot x^3$, then $(x^3)^4 = x^{12}$. The exponent 12 can be obtained by addition: $3 + 3 + 3 + 3 = 12$, or by multiplication: $4 \times 3 = 12$.

In general, when x is a signed number and a and c are positive integers:

$$(x^a)^c = x^{ac}$$

An expression such as $(x^5y^2)^3$ can be simplified by using the commutative and associative properties:

$$\begin{aligned}(x^5y^2)^3 &= (x^5y^2)\,(x^5y^2)\,(x^5y^2)\\ &= (x^5 \cdot x^5 \cdot x^5)\,(y^2 \cdot y^2 \cdot y^2)\\ &= x^{15}y^6\end{aligned}$$

When the base is the product of two or more factors, we apply the rule for the power of a power to each factor.

$$(x^5y^2)^3 = (x^5)^3\,(y^2)^3 = x^{5(3)}\,y^{2(3)} = x^{15}\,y^6$$

Thus,

$$(x^a\,y^b)^n = (x^a)^n\,(y^b)^n = x^{an}\,y^{bn}$$

The expression $(5 \cdot 4)^3$ can be evaluated in two ways.

$$(5 \cdot 4)^3 = 5^3 \cdot 4^3 = 125 \cdot 64 = 8{,}000$$
$$(5 \cdot 4)^3 = 20^3 = 8{,}000$$

EXAMPLES

Simplify each expression in two ways.

a. $(a^2)^3 = a^2 \cdot a^2 \cdot a^2 = a^{2+2+2} = a^6$
or $(a^2)^3 = a^{2 \cdot 3} = a^6$

Answer: a^6

b. $(ab^2)^4 = ab^2 \cdot ab^2 \cdot ab^2 \cdot ab^2 = (a \cdot a \cdot a \cdot a)\,(b^2 \cdot b^2 \cdot b^2 \cdot b^2) = a^4b^8$
or $(a \cdot b^2)^4 = a^{1(4)}b^{2(4)} = a^4b^8$

Answer: a^4b^8

c. $(3^2 \cdot 4^2)^3 = (3^2)^3 \cdot (4^2)^3 = 3^6 \cdot 4^6 = (3 \cdot 4)^6 = 12^6$
or $(3^2 \cdot 4^2)^3 = ((3 \cdot 4)^2)^3 = (12^2)^3 = 12^6$

Answer: 12^6

To evaluate the expressions in Example **c**, use a calculator.

Enter: (3 y^x 2 × 4 y^x 2) y^x 3 =

Display: 2985984. Here, $(3^2 \cdot 4^2)^3 = 2{,}985{,}984$.

Enter: 12 y^x 6 =

Display: 2985984. Here, $12^6 = 2{,}985{,}984$.

EXERCISES

In 1–35, multiply in each case.

1. $a^2 \cdot a^3$ **2.** $b^3 \cdot b^4$ **3.** $c^2 \cdot c^5$ **4.** $d^4 \cdot d^6$ **5.** $r^2 \cdot r^4 \cdot r^5$
6. $t^2 \cdot t^2$ **7.** $r^3 \cdot r^3$ **8.** $s^4 \cdot s^4$ **9.** $e^5 \cdot e^5$ **10.** $z^3 \cdot z^3 \cdot z^5$
11. $x^3 \cdot x^2$ **12.** $a^5 \cdot a^2$ **13.** $s^6 \cdot s^3$ **14.** $y^4 \cdot y^2$ **15.** $t^8 \cdot t^4 \cdot t^2$
16. $x \cdot x$ **17.** $a^2 \cdot a$ **18.** $b^4 \cdot b$ **19.** $c \cdot c^5$ **20.** $e^4 \cdot e \cdot e^5$
21. $2^3 \cdot 2^2$ **22.** $3^4 \cdot 3^3$ **23.** $5^2 \cdot 5^4$ **24.** $4^3 \cdot 4$ **25.** $2^4 \cdot 2^5 \cdot 2$

26. $(x^3)^2$ **27.** $(a^4)^2$ **28.** $(y^2)^4$ **29.** $(y^5)^2$ **30.** $(z^3)^2 \cdot (z^4)^2$

31. $(x^2y^3)^2$ **32.** $(ab^2)^4$ **33.** $(rs)^3$ **34.** $(2^2 \cdot 3^2)^3$ **35.** $(5 \cdot 2^3)^4$

In 36–40, multiply in each case. (All exponents are positive integers.)

36. $x^a \cdot x^{2a}$ **37.** $y^c \cdot y^2$ **38.** $c^r \cdot c^2$ **39.** $x^m \cdot x$ **40.** $(3y)^a \cdot (3y)^b$

In 41–49, state whether each sentence is true or false.

41. $10^4 \cdot 10^3 = 10^7$ **42.** $2^4 \cdot 2^2 = 2^8$ **43.** $3^3 \cdot 2^2 = 6^5$

44. $3^3 \cdot 2^2 = 6^6$ **45.** $5^4 \cdot 5 = 5^5$ **46.** $2^2 + 2^2 = 2^3$

47. $(2^2)^3 = 2^5$ **48.** $(2^3)^5 = 2^{15}$ **49.** $2^3 + 3^3 = 5^3$

8-4 MULTIPLYING BY A MONOMIAL

Multiplying a Monomial by a Monomial

We know that the commutative property of multiplication makes it possible to arrange the factors of a product in any order and that the associative property of multiplication makes it possible to group the factors in any combination. For example:

$$(5x)\ (6y) = (5)\ (6)\ (x)\ (y) = (5 \cdot 6)\ (x \cdot y) = 30xy$$

$$(3x)\ (7x) = (3)\ (7)\ (x)(x) = (3 \cdot 7)\ (x \cdot x) = 21x^2$$

$$(-2x^2)\ (+5x^4) = (-2)(x^2)\ (+5)(x^4) = [(-2)(+5)]\ [(x^2)(x^4)] = -10x^6$$

$$\begin{aligned}(-3a^2b^3)\ (-4a^4b) &= (-3)\ (a^2)\ (b^3)(-4)\ (a^4)\ (b)\\ &= [(-3)(-4)]\ [(a^2)(a^4)]\ [(b^3)(b)] = +12a^6b^4\end{aligned}$$

In the preceding examples, the factors may be rearranged and grouped mentally.

PROCEDURE. To multiply a monomial by a monomial:

1. Use the commutative and associative properties to rearrange and group the factors. This may be done mentally.
2. Multiply the numerical coefficients.
3. Multiply powers with the same base by adding exponents.
4. Multiply the products obtained in Steps 2 and 3 and any other variable factors by writing them with no sign between them.

EXAMPLES

1. Multiply:

Solutions and answers:

a. $(+8xy)(+3z)$ $= +24xyz$
b. $(-4a^3)(-5a^5)$ $= +20a^8$
c. $(-6y^3)(y)$ $= -6y^4$
d. $(+3a^2b^3)(+4a^3b^4)$ $= +12a^5b^7$
e. $(-5x^2y^3)(-2xy^2)$ $= +10x^3y^5$
f. $(+6c^2d^3)\left(-\frac{1}{2}d\right)$ $= -3c^2d^4$
g. $(-3x^2)^3$ $= (-3x^2)(-3x^2)(-3x^2) = -27x^6$
or $(-3x^2)^3$ $= (-3)^3(x^2)^3 = -27x^6$

---

2. Represent the area of a rectangle whose length is $3x$ and whose width is $2x$.

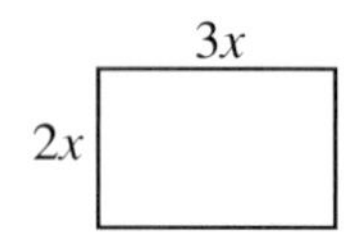

How to Proceed: *Solution:*

(1) Write the area formula: $A = \ell \cdot w$
(2) Substitute the values of ℓ and w: $= (3x) \cdot (2x)$
(3) Perform the multiplication: $= (3 \cdot 2) \cdot (x \cdot x)$
$= 6x^2$

Note that the same answer can be obtained by using the following geometric model:

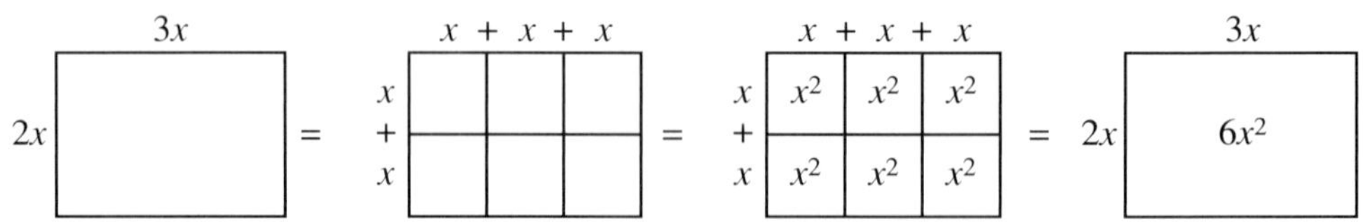

Answer: $6x^2$

Multiplying a Polynomial by a Monomial

The distributive property of multiplication over addition is used to multiply a polynomial by a monomial.

Therefore,

$$a(b + c) = ab + ac$$

$$x(4x + 3) = x(4x) + x(3)$$
$$= 4x^2 + 3x$$

This result can be illustrated geometrically. Let us separate a rectangle whose length is $4x + 3$ and whose width is x into two smaller rectangles such that the length of one rectangle is $4x$ and the length of the other is 3.

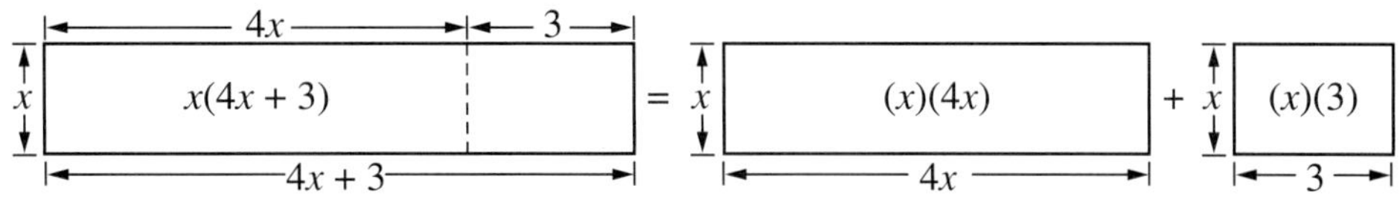

Since the area of the largest rectangle is equal to the sum of the areas of the two smaller rectangles:

$$x(4x + 3) = x(4x) + x(3) = 4x^2 + 3x$$

PROCEDURE. To multiply a polynomial by a monomial, use the distributive property: multiply each term of the polynomial by the monomial and write the result as the sum of these products.

Multiplication and Grouping Symbols

When an algebraic expression involves grouping symbols such as parentheses, we follow the general order of operations and perform operations with algebraic terms.

In the example at the right, we must first	$8y - 2(7y - 4y) + 5$
simplify the expression within parentheses:	$8y - 2(3y) + 5$
Next, we multiply:	$8y - 6y + 5$
As a last step, we combine like terms by addition or subtraction:	$2y + 5$ *Answer*

In many expressions, however, the terms within parentheses cannot be combined because they are unlike terms. When this happens, we use the distributive property to clear parentheses and then follow the order of multiplying before adding.

For example, here we must clear parentheses	$3 + 7(2x + 3)$
by using the distributive property:	$3 + 7(2x) + 7(3)$
Next, we multiply:	$3 + 14x + 21$
As a last step, we combine like terms by addition:	$24 + 14x$ *Answer*

The multiplicative identity states that $a = 1a$. By using this property, we can say that $5 + (2x - 3) = 5 + 1(2x - 3)$ and then follow the procedures shown above.

$$5 + (2x - 3) = 5 + 1(2x - 3) = 5 + 2x - 3 = 2 + 2x$$

Also, since $-a = -1a$, we can use this property to simplify certain expressions:

$$6y - (9 - 7y) = 6y - 1(9 - 7y)$$
$$= 6y - 1(9) - 1(-7y) = 6y - 9 + 7y = 13y - 9$$

EXAMPLES

1. Multiply:

	Answers:
a. $5(r - 7)$	$= 5r - 35$
b. $8(3x - 2y + 4z)$	$= 24x - 16y + 32z$
c. $-5x(x^2 - 2x + 4)$	$= -5x^3 + 10x^2 - 20x$
d. $-3a^2b^2(4ab^2 - 3b^2)$	$= -12a^3b^4 + 9a^2b^4$

---

2. Simplify:

a. $-5x(x^2 - 2) - 7x$

How to Proceed:	*Solution:*
(1) Write the expression:	$-5x(x^2 \; -2) - 7x$
(2) Use the distributive property:	$-5x(x^2) - 5x(-2) - 7x$
(3) Multiply:	$-5x^3 + 10x - 7x$
(4) Add like terms:	$-5x^3 + 3x$

Answer: $-5x^3 + 3x$

b. $3a - (5 - 7a)$

How to Proceed:	*Solution:*
(1) Write the expression:	$3a - (5 - 7a)$
	$3a - 1(5 - 7a)$
(2) Use the distributive property:	$3a - 5 + 7a$
(3) Add like terms:	$10a - 5$

Answer: $10a - 5$

EXERCISES

1. In an algebraic term, how do you show the product of a constant times a variable or the product of different variables?
2. In the expression $2 + 3(7y)$, which operation is performed first?

3. In the expression $(2 + 3)(7y)$, which operation is performed first?
4. In the expression $5y(2y + 3)$, which operation is performed first?
5. Can the sum $x^2 + x^3$ be written in simpler form?
6. Can the product $x^2(x^3)$ be written in simpler form?

In 7–30, find each product.

7. $(+6)(-2a)$
8. $(-4b)(-6b)$
9. $(+5)(-2y)(-3y)$
10. $(4a)(5b)$
11. $(-8r)(-2s)$
12. $(+7x)(-2y)(3z)$
13. $(+6x)\left(-\frac{1}{2}y\right)$
14. $\left(-\frac{3}{4}a\right)(+8b)$
15. $(-6x)\left(\frac{1}{2}y\right)\left(-\frac{1}{3}z\right)$
16. $(+5ab)(-3c)$
17. $(-7r)(5st)$
18. $(-2)(+6cd)(-e)$
19. $(+9xy)(-2cd)$
20. $(3s)(-4m)(5cd)$
21. $(+5a^2)(-4a^2)$
22. $(-6x^4)(-3x^3)$
23. $(20y^3)(-7y^2)$
24. $(18r^5)(-5r^2)$
25. $(+3z^2)(+4z)$
26. $(-8y^5)(+5y)$
27. $(-9z)(8z^4)(z^3)$
28. $(+6x^2y^3)(-4x^4y^2)$
29. $(-7a^3b)(+5a^2b^2)$
30. $(+4ab^2)(-2a^2b^3)$

In 31–48, write each product as a polynomial.

31. $3(6c + 3d)$
32. $-5(4m - 6n)$
33. $-2(8a + 6b)$
34. $10\left(2x - \frac{1}{5}y\right)$
35. $12\left(\frac{2}{3}m - 4n\right)$
36. $-8\left(4r - \frac{1}{4}s\right)$
37. $-16\left(\frac{3}{4}c - \frac{5}{8}d\right)$
38. $4x(5x + 6)$
39. $5d(d^2 - 3d)$
40. $-5c^2(15c - 4c^2)$
41. $mn(m + n)$
42. $-ab(a - b)$
43. $3ab(5a^2 - 7b^2)$
44. $-r^3s^3(6r^4s - 3s^4)$
45. $10d(2a - 3c + 4b)$
46. $-8(2x^2 - 3x - 5)$
47. $3xy(x^2 + xy + y^2)$
48. $5r^2s^2(-2r^2 + 3rs - 4s^2)$

In 49–51, represent the area of each rectangle whose length and width are given.

49. $\ell = 5y$, $w = 3y$
50. $\ell = 3x$, $w = 5y$
51. $\ell = 3c$, $w = 8c - 2$

52. In terms of x:
   a. Express the area of the outer rectangle pictured at the right.
   b. Express the area of the inner rectangle pictured at the right.
   c. Express as a polynomial in simplest form the area of the shaded region.

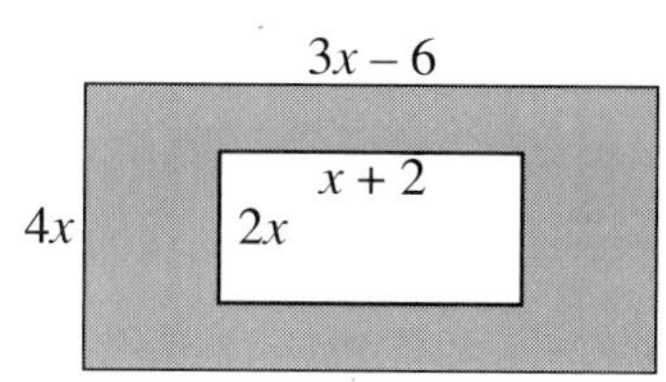

In 53–74, simplify each expression.

53. $5(d + 3) - 10$
54. $3(2 - 3c) + 5c$
55. $7 + 2(7x - 5)$
56. $-2(x - 1) + 6$
57. $-4(3 - 6a) - 7a$
58. $5 - 4(3e - 5)$
59. $8 + (4e - 2)$
60. $a + (b - a)$
61. $(6b + 4) - 2b$
62. $9 - (5t + 6)$
63. $4 - (2 - 8s)$
64. $-(6x - 7) + 14$

65. $5x(2x - 3) + 9x$
66. $12y - 3y(2y - 4)$
67. $7x + 3(2x - 1) - 8$
68. $7c - 4d - 2(4c - 3d)$
69. $3a - 2a(5a - a) + a^2$
70. $(a + 3b) - (a - 3b)$
71. $4(2x + 5) - 3(2 - 7x)$
72. $3(x + y) + 2(x - 3y)$
73. $5x(2 - 3x) - x(3x - 1)$
74. $y(y + 4) - y(y - 3) - 9y$

75. If 1 pound of grass seed costs $25x$ cents, represent in simplest form the cost of 7 pounds of seed.

76. If a bus travels at the rate of $10z$ miles per hour for 4 hours, represent in simplest form the distance traveled.

77. If Lois has $2n$ nickels, represent in simplest form the number of cents she has.

In 78–87, write each answer as a polynomial in simplest form.

78. If the cost of a notebook is $2x - 3$, express the cost of five notebooks.

79. If the length of a rectangle is $5y - 7$ and the width is $3y$, represent the area of the rectangle.

80. If the measure of the base of a triangle is $3b + 2$ and the height is $4b$, represent the area of the triangle.

81. Represent the distance traveled in 3 hours by a car traveling at $3x - 7$ miles per hour.

82. Represent in terms of x and y the amount saved in $3y$ weeks if $x - 2$ dollars are saved each week.

83. The length of a rectangular skating rink is 2 less than 3 times the width. If w represents the width of the rink, represent the area in terms of w.

84. A mail-order catalog lists books for $3x - 5$ dollars each. The cost of postage is 2 dollars for five books or fewer. Represent the total cost of an order for four books.

85. A store advertises skirts for $x - 5$ dollars and allows an additional 10-dollar reduction on the total purchase if three or more skirts are bought. Represent the cost of five skirts.

86. A store advertises skirts for $x - 5$ dollars and allows an additional 2-dollar reduction on each skirt if three or more skirts are purchased. Represent the cost of five skirts.

87. A store advertises skirts for $x - 5$ dollars and tops for $2x - 3$ dollars. Represent the cost of two skirts and three tops.

8-5 MULTIPLYING POLYNOMIALS

As discussed in Section 8-4, to find the product $(x + 4)(a)$, we use the distributive property of multiplication over addition:

$$(x + 4)(a) = x(a) + 4(a)$$

Now, let us use this property to find the product of the two polynomials $(x + 4)(x + 3)$.

$$\begin{aligned}(x + 4)(a) &= x(a) + 4(a)\\ (x + 4)(x + 3) &= x(x + 3) + 4(x + 3)\\ &= x^2 + 3x + 4x + 12\\ &= x^2 + 7x + 12\end{aligned}$$

This result can also be illustrated geometrically.

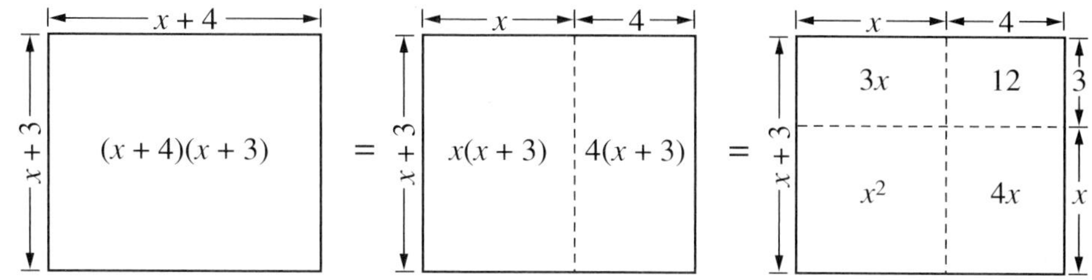

$$(x + 4)(x + 3) \quad = \quad x(x + 3) + 4(x + 3) \quad = \quad x^2 + 3x + 4x + 12$$

In general, for all a, b, c, and d:

$$\begin{aligned}\boldsymbol{(a + b)(c + d)} &= a(c + d) + b(c + d)\\ &= \boldsymbol{ac + ad + bc + bd}\end{aligned}$$

Notice that each term of the first polynomial multiplies each term of the second.

At the right is a convenient vertical arrangement of the preceding multiplication, similar to the arrangement used in arithmetic multiplication. Note that multiplication is done from left to right.

$$\begin{array}{rl} & x + 4 \\ & \underline{x + 3} \\ (x + 4)x \rightarrow & x^2 + 4x \\ (x + 4)3 \rightarrow & \underline{\quad\;\; + 3x + 12} \\ \text{Add like terms:} & x^2 + 7x + 12 \end{array}$$

The word ***FOIL*** serves as a convenient way to remember the steps necessary to multiply two binomials.

$$\begin{array}{rcccccccc} (2x - 5)(x + 4) = & 2x(x) & + & 2x(+4) & - & 5(x) & - & 5(+4) \\ & \textit{First} & & \textit{Outside} & & \textit{Inside} & & \textit{Last} \\ = & 2x^2 & + & 8x & - & 5x & - & 20 \\ = & 2x^2 & & & + & 3x & - & 20 \end{array}$$

PROCEDURE. To multiply a polynomial by a polynomial, first arrange each polynomial in descending or ascending powers of the same variable. Then, use the distributive property: multiply each term of one polynomial by each term of the other.

EXAMPLES

Simplify:

a. $(3x - 4)(4x + 5)$

Solution:

$$\begin{aligned}(3x - 4)(4x + 5) &= 3x(4x + 5) - 4(4x + 5)\\ &= 12x^2 + 15x - 16x - 20\\ &= 12x^2 - x - 20\end{aligned}$$

Alternate Solution:

$$\begin{array}{r} 3x - 4 \\ \underline{4x + 5} \\ 12x^2 - 16x \\ \underline{+ 15x - 20} \\ 12x^2 - x - 20 \end{array}$$

Answer: $12x^2 - x - 20$

b. $(x^2 + 3xy + 9y^2)(x - 3y)$

Solution

$$\begin{aligned}(x^2 + 3xy + 9y^2)(x - 3y) &= x^2(x - 3y) + 3xy(x - 3y) + 9y^2(x - 3y)\\ &= x^3 - 3x^2y + 3x^2y - 9xy^2 + 9xy^2 - 27y^3\\ &= x^3 + 0x^2y + 0xy^2 - 27y^3\end{aligned}$$

Answer: $x^3 - 27y^3$

c. $(2x - 5)^2 - 2(x - 3)$

Solution

$$\begin{aligned}(2x - 5)^2 - 2(x - 3) &= (2x - 5)(2x - 5) - 2(x - 3)\\ &= 2x(2x) + 2x(-5) - 5(2x) - 5(-5) - 2(x) - 2(-3)\\ &= 4x^2 - 10x - 10x + 25 - 2x + 6\\ &= 4x^2 - 22x + 31\end{aligned}$$

Answer: $4x^2 - 22x + 31$

EXERCISES WITH OPEN-RESPONSE PROBLEM

In 1–52, multiply in each case.

1. $(a + 2)(a + 3)$
2. $(c + 6)(c + 1)$
3. $(x - 5)(x - 3)$
4. $(d - 6)(d - 5)$
5. $(d + 9)(d - 3)$
6. $(x - 7)(x + 2)$
7. $(m + 3)(m - 7)$
8. $(z - 5)(z + 8)$
9. $(f + 10)(f - 8)$
10. $(t + 15)(t - 6)$
11. $(b - 8)(b - 10)$
12. $(w - 13)(w + 7)$
13. $(6 + y)(5 + y)$
14. $(8 - e)(6 - e)$
15. $(12 - r)(6 + r)$
16. $(x + 5)(x - 5)$
17. $(y + 7)(y - 7)$
18. $(a + 9)(a - 9)$
19. $(2x + 1)(x - 6)$
20. $(c - 5)(2c - 4)$
21. $(2a + 9)(3a + 1)$
22. $(5y - 2)(3y - 1)$
23. $(2x + 3)(2x - 3)$
24. $(3d + 8)(3d - 8)$
25. $(x + y)(x + y)$
26. $(a - b)(a - b)$
27. $(a + b)(a - b)$
28. $(a + 2b)(a + 3b)$
29. $(2c - d)(3c + d)$
30. $(x - 4y)(x + 4y)$

31. $(2z + 5w)(3z - 4w)$

32. $(9x - 5y)(2x + 3y)$

33. $(5k + 2m)(3r + 4s)$

34. $(3x + 4y)(3x - 4y)$

35. $(r^2 + 5)(r^2 - 2)$

36. $(x^2 - y^2)(x^2 + y^2)$

37. $(x^2 + 3x + 5)(x + 2)$

38. $(2c^2 - 3c - 1)(2c + 1)$

39. $(3 - 2d - d^2)(5 - 2d)$

40. $(c^2 - 2c + 4)(c + 2)$

41. $(2x^2 - 3x + 1)(3x - 2)$

42. $(3x^2 - 4xy + y^2)(4x + 3y)$

43. $(x^3 - 3x^2 + 2x - 4)(3x - 1)$

44. $(2x + 1)(3x - 4)(x + 3)$

45. $(x^2 - 4x + 1)(x^2 + 5x - 2)$

46. $(x + 4)(x + 4)(x + 4)$

47. $(a + 5)^3$

48. $(x - y)^3$

49. $(5 + x^2 - 2x)(2x - 3)$

50. $(5x - 4 + 2x^2)(3 + 4x)$

51. $(2xy + x^2 + y^2)(x + y)$

52. $a(a + b)(a - b)$

In 53–60, simplify each expression.

53. $(x + 7)(x - 2) - x^2$

54. $2(3x + 1)(2x - 3) + 14x$

55. $8x^2 - (4x + 3)(2x - 1)$

56. $(x + 4)(x + 3) - (x - 2)(x - 5)$

57. $(3y + 5)(2y - 3) - (y + 7)(5y - 1)$

58. $(y + 4)^2 - (y - 3)^2$

59. $r(r - 2s) - (r - s)$

60. $a[(a + 2)(a - 2) - 4]$

In 61–63, use grouping symbols to write an algebraic expression that represents the answer. Then, express each answer as a polynomial in simplest form.

61. The length of a rectangle is $2x - 5$ and its width is $x + 7$. Express the area of the rectangle.

62. The dimensions of a rectangle are represented by $11x - 8$ and $3x + 5$. Represent the area of the rectangle.

63. A plane travels at a rate of $(x + 100)$ kilometers per hour. Represent the distance it can travel in $(2x + 3)$ hours.

In 64–69: **a.** Find the area of the rectangle whose length and width are given. **b.** Check the result found in part **a** as follows:

(1) Copy the rectangle shown below.
(2) Represent the length and width of each of the four small rectangles.
(3) Represent the area of each of the four small rectangles.
(4) Add the four areas found in Step (3) to find the area of the original given rectangle.

64. Length $= x + 3$, width $= x + 2$
65. Length $= x + 6$, width $= x + 5$
66. Length $= 2x + 5$, width $= 5x + 3$
67. Length $= 3x + 1$, width $= 3x + 1$
68. Length $= 2x + 3y$, width $= x + y$
69. Length $= 2x + 5y$, width $= 3x + 4y$

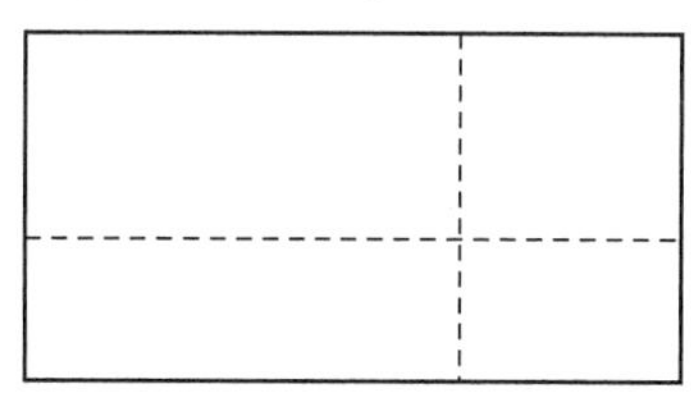

Ex. 64–69

70. The product of two binomials in simplest form can have four terms, three terms, or two terms.
a. When does the product of two binomials have four terms?
b. When is the product of two binomials a trinomial?
c. When is the product of two binomials a binomial?

8-6 DIVIDING POWERS THAT HAVE THE SAME BASE

We know that any nonzero number divided by itself is 1. Therefore, $x \div x = 1$ and $y^3 \div y^3 = 1$.

In general, when $x \neq 0$ and a is a positive integer:

$$\mathbf{x^a \div x^a = 1}$$

Therefore,

$$\frac{x^5}{x^3} = \frac{x^2 \cdot x^3}{x^3} = x^2 \cdot 1 = x^2$$

$$\frac{y^9}{y^4} = \frac{y^5 \cdot y^4}{y^4} = y^5 \cdot 1 = y^5$$

$$\frac{c^5}{c} = \frac{c^4 \cdot c}{c} = c^4 \cdot 1 = c^4$$

These same results can be obtained by using the relationship between division and multiplication:

$$\textbf{If } a \cdot b = c\textbf{, then } c \div b = a.$$

Since $x^2 \cdot x^3 = x^5$, then $x^5 \div x^3 = x^2$.
Since $y^5 \cdot y^4 = y^9$, then $y^9 \div y^4 = y^5$.
Since $c^4 \cdot c = c^5$, then $c^5 \div c = c^4$. (Remember that c means c^1.)

Observe that the exponent in each quotient is the difference between the exponent of the dividend and the exponent of the divisor.

In general, when $x \neq 0$ and a and b are positive integers with $a > b$:

$$\mathbf{x^a \div x^b = x^{a-b}}$$

PROCEDURE. To divide powers of the same base, find the exponent of the quotient by subtracting the exponent of the divisor from the exponent of the dividend. The base of the quotient is the same as the base of the dividend and the base of the divisor.

EXAMPLES

1. Simplify by performing each indicated division.

Solutions and answers:

a. $x^9 \div x^5 \quad = x^{9-5} = x^4$
b. $y^5 \div y \quad = y^{5-1} = y^4$

c. $c^5 \div c^5 \quad = 1$
d. $10^5 \div 10^3 \quad = 10^{5-3} = 10^2$

---

2. Write $\frac{5^7 \cdot 5^4}{5^8}$ in simplest form:
a. by using the rules for multiplying and dividing powers with like bases.
b. by using a calculator.

Solutions

a. First simplify the numerator. Then, apply the rule for division of powers with the same base.

$$\frac{5^7 \cdot 5^4}{5^8} = \frac{5^{7+4}}{5^8} = \frac{5^{11}}{5^8} = 5^{11-8} = 5^3 = 125$$

b. On a calculator:

Enter: 5 [y^x] 7 [×] 5 [y^x] 4 [÷] 5 [y^x] 8 [=]

Display: [125.]

Answer: 125

EXERCISES

In 1–20, divide in each case.

1. $x^8 \div x^2$ **2.** $a^{10} \div a^5$ **3.** $b^7 \div b^3$ **4.** $c^5 \div c^4$ **5.** $d^7 \div d^7$

6. $\frac{d^4}{d^2}$ **7.** $\frac{e^9}{e^3}$ **8.** $\frac{m^{12}}{m^4}$ **9.** $\frac{n^{10}}{n^9}$ **10.** $\frac{r^6}{r^6}$

11. $x^8 \div x$ **12.** $y^7 \div y$ **13.** $z^{10} \div z$ **14.** $t^5 \div t$ **15.** $m \div m$

16. $2^5 \div 2^2$ **17.** $10^6 \div 10^4$ **18.** $3^4 \div 3^2$ **19.** $5^3 \div 5$ **20.** $10^4 \div 10$

In 21–26, divide in each case. (All exponents are positive integers.)

21. $x^{5a} \div x^{2a}$ **22.** $y^{10b} \div y^{2b}$ **23.** $r^c \div r^d \ (c > d)$

24. $s^x \div s^2 \ (x > 2)$ **25.** $a^b \div a^b$ **26.** $2^a \div 2^b \ (a > b)$

In 27–32: **a.** Simplify each expression by using the rules for multiplying and dividing powers with like bases. **b.** Evaluate the expression using a calculator. Compare your answers to parts **a** and **b**.

27. $\frac{2^3 \cdot 2^4}{2^2}$ **28.** $\frac{5^8}{5^4 \cdot 5}$ **29.** $\frac{10^2 \cdot 10^3}{10^4}$

30. $\frac{10^6}{10^2 \cdot 10^4}$ **31.** $\frac{10^8 \cdot 10^2}{(10^5)^2}$ **32.** $\frac{6^4 \cdot 6^9}{6^2 \cdot 6^3}$

In 33–35, tell whether each sentence is true or false.

33. $4^5 \div 2^3 = 2^2$ **34.** $5^6 \div 5^2 = 5^4$ **35.** $3^8 \div 3^4 = 1^4$

8-7 POWERS WITH ZERO AND NEGATIVE EXPONENTS

Nonpositive integers such as 0, −1, and −2 can also be used as exponents. We will define powers having zero and negative integral exponents in such a way that the properties that were valid for positive integral exponents will also be valid for nonpositive integral exponents. In other words, the following properties will be true when the exponents a and b are positive integers, negative integers, or 0:

$$x^a \cdot x^b = x^{a+b} \qquad x^a \div x^b = x^{a-b} \qquad (x^a)^b = x^{a \cdot b}$$

The Zero Exponent

We know that, for $x \neq 0$, $\frac{x^3}{x^3} = 1$.

$$\text{If } \frac{x^3}{x^3} = x^{3-3} = x^0$$

is to be a meaningful statement, we must let $x^0 = 1$ since x^0 and 1 are each equal to $\frac{x^3}{x^3}$. This leads to the following definition:

$x^0 = 1$ if x is a number such that $x \neq 0$.

It can be shown that all the laws of exponents remain valid when x^0 is defined as 1. For example:

Using the definition $10^0 = 1$, we have $10^3 \cdot 10^0 = 10^3 \cdot 1 = 10^3$.

Using the law of exponents, we have $10^3 \cdot 10^0 = 10^{3+0} = 10^3$.

The two procedures result in the same product.

The definition $x^0 = 1$ $(x \neq 0)$ permits us to say that the zero power of any number except 0 equals 1.

$$4^0 = 1 \qquad (-4)^0 = 1 \qquad (4x)^0 = 1 \qquad (-4x)^0 = 1$$

A scientific calculator will return this value. For example, to evaluate 4^0:

Enter: 4 [y^x] 0 [=]

Display: [1.]

Note that $4x^0 = 4^1 \cdot x^0 = 4 \cdot 1 = 4$,

but $(4x)^0 = 4^0 \cdot x^0 = 1 \cdot 1 = 1$.

The Negative Integral Exponent

We know that, for $x \neq 0$, $\frac{x^3}{x^5} = \frac{x \cdot x \cdot x}{x \cdot x \cdot x \cdot x \cdot x} = \frac{1}{x \cdot x} = \frac{1}{x^2}$.

$$\text{If } \frac{x^3}{x^5} = x^{3-5} = x^{-2}$$

is to be a meaningful statement, we must let $x^{-2} = \frac{1}{x^2}$ since x^{-2} and $\frac{1}{x^2}$ are each equal to $\frac{x^3}{x^5}$. This leads to the following definition:

$$\mathbf{x^{-n} = \frac{1}{x^n}} \textbf{ if } \mathbf{x} \textbf{ is a number such that } \mathbf{x \neq 0.}$$

A scientific calculator will return equal values for x^{-2} and $\frac{1}{x^2}$. For example, let $x = 5$.

Evaluate 5^{-2}	Evaluate $\frac{1}{5^2}$.
Enter: 5 [y^x] 2 [+/–] [=]	***Enter:*** 1 [÷] 5 [y^x] 2 [=]
Display: [0.04]	***Display:*** [0.04]

It can be shown that all the laws of exponents remain valid if x^{-n} is defined as $\frac{1}{x^n}$. For example:

Using the definition $2^{-4} = \frac{1}{2^4}$, we have $2^2 \cdot 2^{-4} = 2^2 \cdot \frac{1}{2^4} = \frac{2^2}{2^4} = \frac{1}{2^2} = 2^{-2}$.

Using the law of exponents, we have $2^2 \cdot 2^{-4} = 2^{2+(-4)} = 2^{-2}$.

The two procedures give the same result.

Now we can say that, for all integral values of a and b,

$$\mathbf{\frac{x^a}{x^b} = x^{a-b} \quad (x \neq 0)}$$

EXAMPLES

1. Transform each given expression into an equivalent one with a positive exponent.

Solutions and answers:

a. $4^{-3} \qquad = \frac{1}{4^3}$

b. $10^{-1} \qquad = \frac{1}{10^1}$

Solutions and answers:

c. $\frac{1}{2^{-5}}$ $= 1 \div 2^{-5} = 1 \div \frac{1}{2^5} = 1 \bullet \frac{2^5}{1} = 2^5$

d. $\left(\frac{5}{3}\right)^{-2}$ $= \frac{5^{-2}}{3^{-2}} = \frac{1}{5^2} \div \frac{1}{3^2} = \frac{1}{5^2} \bullet \frac{3^2}{1} = \frac{3^2}{5^2} = \left(\frac{3}{5}\right)^2$

---

2. Compute the value of each expression.

Solutions and answers:

a. 3^0 $= 1$

b. 10^{-2} $= \frac{1}{10^2} = \frac{1}{100}$

c. $(-5)^0 + 2^{-4}$ $= 1 + \frac{1}{2^4} = 1 + \frac{1}{16} = 1\frac{1}{16}$

d. $6(3^{-3})$ $= 6\left(\frac{1}{3^3}\right) = \frac{6}{27} = \frac{2}{9}$

---

3. Use the laws of exponents to perform the indicated operations.

Solutions and answers:

a. $2^7 \cdot 2^{-3}$ $= (2)^{7+(-3)} = 2^4$

b. $3^{-6} \div 3^{-2}$ $= (3)^{-6-(-2)} = 3^{-4}$

c. $(x^4)^{-3}$ $= x^{(4)(-3)} = x^{-12}$

d. $(y^{-2})^{-4}$ $= y^{(-2)(-4)} = y^8$

EXERCISES

In 1–5, transform each given expression into an equivalent expression involving a positive exponent.

1. 10^{-4} **2.** 2^{-1} **3.** $\left(\frac{2}{3}\right)^{-2}$ **4.** m^{-6} **5.** r^{-3}

In 6–21, compute each value using the definitions of zero and negative exponents. Compare your answers with the results obtained using a calculator.

6. 10^0 **7.** $(-4)^0$ **8.** y^0 **9.** $(2K)^0$

10. 3^{-2} **11.** 2^{-4} **12.** $(-6)^{-1}$ **13.** $(-1)^{-5}$

14. 10^{-1} **15.** 10^{-2} **16.** 10^{-3} **17.** 10^{-4}

18. $4(10)^{-2}$ **19.** $1.5(10)^{-3}$ **20.** $7^0 + 6^{-2}$ **21.** $\left(\frac{1}{2}\right)^0 + 3^{-3}$

In 22–33, use the laws of exponents to perform each indicated operation.

22. $10^{-2} \cdot 10^5$	**23.** $3^{-4} \cdot 3^{-2}$	**24.** $10^{-3} \div 10^{-5}$	**25.** $3^4 \div 3^0$
26. $(4^{-1})^2$	**27.** $(3^{-3})^{-2}$	**28.** $a^0 \cdot a^4$	**29.** $x^{-5} \cdot x$
30. $m^2 \div m^7$	**31.** $t^{-6} \div t^2$	**32.** $(a^{-4})^3$	**33.** $(x^{-2})^0$

34. Find the value of $7x^0 - (6x)^0$. **35.** Find the value of $5x^0 + 2x^{-1}$ when $x = 4$.

8-8 USING SCIENTIFIC NOTATION

Scientists and mathematicians often work with numbers that are very large or very small. In order to write and compute with such numbers more easily, these workers use ***scientific notation***. A number is expressed in scientific notation when it is written as the product of two quantities: the first is a number greater than or equal to 1 but less than 10, and the second is a power of 10. In other words, a number is in scientific notation when it is written as $\boldsymbol{a \times 10^n}$, where $1 \leq a < 10$ and n is an integer.

Writing Numbers in Scientific Notation

To write a number in scientific notation, first write it as the product of a number between 1 and 10 times a power of 10. Then express the power of 10 in exponential form.

The table at the right shows some integral powers of 10. When the exponent is a positive integer, the power can be written as 1 followed by the number of 0's equal to the exponent of 10. When the exponent is a negative integer, the power can be written as a decimal value with the number of decimal places equal to the absolute value of the exponent of 10.

Powers of 10
$10^5 = 100{,}000$
$10^4 = 10{,}000$
$10^3 = 1{,}000$
$10^2 = 100$
$10^1 = 10$
$10^0 = 1$
$10^{-1} = \frac{1}{10^1} = 0.1$
$10^{-2} = \frac{1}{10^2} = 0.01$
$10^{-3} = \frac{1}{10^3} = 0.001$
$10^{-4} = \frac{1}{10^4} = 0.0001$
$10^{-5} = \frac{1}{10^5} = 0.00001$

$$3{,}000{,}000 = 3 \times 1{,}000{,}000 = 3 \times 10^6$$
$$780 = 7.8 \times 100 = 7.8 \times 10^2$$
$$3 = 3 \times 1 = 3 \times 10^0$$
$$0.025 = 2.5 \times 0.01 = 2.5 \times 10^{-2}$$
$$0.00003 = 3 \times 0.00001 = 3 \times 10^{-5}$$

When writing a number in scientific notation, keep in mind the following:

- A number equal to or greater than 10 has a positive exponent of 10.
- A number equal to or greater than 1 but less than 10 has a zero exponent of 10.
- A number between 0 and 1 has a negative exponent of 10.

EXAMPLES

1. The distance from the earth to the sun is approximately 93,000,000 miles. Write this number in scientific notation.

How to Proceed:	*Solution:*
(1) Write the number, placing a decimal point after the last digit.	93,000,000.
(2) Place a caret (∧) after the first nonzero digit so that replacing the caret with a decimal point will give a number between 1 and 10.	$9_\wedge 3{,}000{,}000.$
(3) Count the number of digits between the caret and the decimal point. This is the exponent of 10 in scientific notation. The exponent is positive because the given number is greater than 10.	$9_\wedge\underbrace{3{,}000{,}000}_{7}.$
(4) Write the number in the form $a \times 10^n$, where a is found by replacing the caret with a decimal point and n is the exponent found in Step 3.	9.3×10^7 *Answer*

---

2. Express 0.0000029 in scientific notation.

Solution Since the number is between 0 and 1, the exponent will be negative.

$$0.0000029 = 0.\underbrace{000002}_{6}{}_\wedge 9 = 2.9 \times 10^{-6} \text{ Answer}$$

Some scientific calculators and all graphing calculators can be placed in scientific notation mode and will return the results shown in Examples 1 and 2 when the given numbers are entered.

Enter: 0.0000029 [=]

Display: [2.9 E-6]

This display is read as 2.9×10^{-6}, where the integer following E is the exponent to the base 10 used to write the number in scientific notation.

Changing to Ordinary Decimal Notation

We can change a number that is written in scientific notation to ordinary decimal notation by expanding the power of 10 and then multiplying the result by the number between 1 and 10.

EXAMPLES

1. The approximate population of the United States in 1990 was 2.53×10^8. Find the approximate number of people in the United States at that time.

How to Proceed:	*Solution:*
(1) Evaluate the second factor, which is a power of 10:	$2.53 \times 10^8 = 2.53 \times 100{,}000{,}000$
(2) Multiply the factors:	$2.53 \times 10^8 = 253{,}000{,}000$ *Answer*

Note that we could have multiplied 2.53 by 10^8 quickly by moving the decimal point in 2.53 eight places to the right.

---

2. The diameter of a red blood corpuscle is expressed in scientific notation as 7.5×10^{-4} centimeter. Write the number of centimeters in the diameter as a decimal fraction.

How to Proceed:	*Solution:*
(1) Evaluate the second factor, which is a power of 10:	$7.5 \times 10^{-4} = 7.5 \times 0.0001$
(2) Multiply the factors:	$7.5 \times 10^{-4} = 0.00075$ *Answer*

Note that we could have multiplied 7.5 by 10^{-4} quickly by moving the decimal point in 7.5 four places to the left.

---

3. Use a calculator to find the product: $45{,}000 \times 570{,}000$.

Solution

Enter: 45000 [×] 570000 [=]

Display: [2.565 E 10]

A scientific calculator will shift to scientific notation when the number is too large or too small for the display. This display can be changed to decimal notation by using the procedure shown in Examples 1 and 2.

$$2.565 \times 10^{10} = 2.565 \times 10{,}000{,}000{,}000 = 25{,}650{,}000{,}000 \text{ } \textit{Answer}$$

---

4. Use a calculator to find the mass of 2.70×10^{15} hydrogen atoms if the mass of one hydrogen atom is 1.67×10^{-24} gram. Round the answer to three significant digits.

Solution Use a calculator to multiply the mass of one hydrogen atom by the number of hydrogen atoms.

A number that is too small or too large for the display of a calculator can be entered in scientific notation. The **EE** key allows us to do this. This key is used to enter the exponent to the base 10 as shown here.

Enter: 1.67 **EE** 24 **+/–** **×** 2.7 **EE** 15 **=**

Display: 4.509E-9

Round 4.509 to 4.51, which has three significant digits.
$4.51 \times 10^{-9} = 4.51 \times 0.000000001 = 0.00000000451$ gram *Answer*

EXERCISES

In 1–3, write each number as a power of 10 using a positive exponent.

1. 100,000 **2.** 1,000,000,000 **3.** 1,000,000,000,000

In 4–15, find the number that is expressed by each numeral.

4. 10^7 **5.** 10^{10} **6.** 10^{13}
7. 10^{15} **8.** 3×10^5 **9.** 4×10^8
10. 6×10^{14} **11.** 9×10^9 **12.** 1.3×10^4
13. 8.3×10^{12} **14.** 1.27×10^3 **15.** 6.14×10^{10}

In 16–23, find the number that can replace the question mark to make each resulting statement true.

16. $120 = 1.2 \times 10^?$ **17.** $760 = 7.60 \times 10^?$
18. $9{,}300 = 9.3 \times 10^?$ **19.** $52{,}000 = 5.2 \times 10^?$
20. $5{,}280 = 5.28 \times 10^?$ **21.** $375{,}000 = 3.75 \times 10^?$
22. $1{,}610{,}000 = 1.61 \times 10^?$ **23.** $872{,}000{,}000 = 8.72 \times 10^?$

In 24–27, write each number as a power of 10 using a negative exponent.

24. 0.01 **25.** 0.00001 **26.** 0.00000001 **27.** 0.0000000001

In 28–39, express each given number as a decimal fraction.

28. 10^{-6} **29.** 10^{-8} **30.** 10^{-15} **31.** 10^{-18}

32. 4×10^{-3} **33.** $7 \times (10^{-9})$ **34.** 8×10^{-10} **35.** $9 \times (10^{-13})$

36. 1.2×10^{-4} **37.** $3.6 \times (10^{-5})$ **38.** 7.4×10^{-11} **39.** $3.14 \times (10^{-14})$

In 40–45, state the number that can replace the question mark to make each statement true.

40. $0.023 = 2.3 \times 10^{?}$ **41.** $0.000086 = 8.6 \times 10^{?}$

42. $0.000000019 = 1.9 \times 10^{?}$ **43.** $0.000000000041 = 4.1 \times 10^{?}$

44. $0.00156 = 1.56 \times 10^{?}$ **45.** $0.000000873 = 8.73 \times 10^{?}$

In 46–61, express each number in scientific notation.

46. 8,400 **47.** 27,000 **48.** 54,000,000 **49.** 320,000,000

50. 6,750 **51.** 81,600 **52.** 453,000 **53.** 375,000,000

54. 0.0052 **55.** 0.00061 **56.** 0.0000039 **57.** 0.000000014

58. 0.156 **59.** 0.00381 **60.** 0.0000763 **61.** 0.000000917

In 62–65, compute the result of each operation. Then represent the result in: **a.** scientific notation **b.** ordinary decimal notation

62. $(2.9 \times 10^{3})(3.0 \times 10^{-3})$ **63.** $(2.5 \times 10^{-2})(3 \times 10^{-3})$

64. $(7.5 \times 10^{-4}) \div (2.5 \times 10^{3})$ **65.** $(6.8 \times 10^{-5}) \div (3.4 \times 10^{-8})$

In 66–70, express each number in scientific notation.

66. A light-year, which is the distance light travels in 1 year, is approximately 9,500,000,000,000 kilometers.

67. A star that is about 12,000,000,000,000,000,000,000 miles away can be seen by the Palomar telescope.

68. A microampere is 0.000001 of an ampere.

69. The radius of an electron is about 0.0000000000005 centimeter.

70. The diameter of some white blood corpuscles is approximately 0.0008 inch.

In 71–75, express each number in ordinary decimal notation.

71. The diameter of the universe is 2×10^{9} light-years.

72. The distance from the earth to the moon is 2.4×10^{5} miles.

73. In a motion-picture film, the image of each picture remains on the screen approximately 6×10^{-2} second.

74. Light takes about 2×10^{-8} second to cross a room.

75. The mass of the earth is approximately 5.9×10^{24} kilograms.

8-9 DIVIDING BY A MONOMIAL

Dividing a Monomial by a Monomial

We know that

$$\frac{a}{b} \cdot \frac{c}{d} = \frac{ac}{bd}$$

Therefore, by the reflexive property

$$\frac{ac}{bd} = \frac{a}{b} \cdot \frac{c}{d}$$

Using this relationship, we can write:

$$\frac{-30x^6}{2x^4} = \frac{-30}{2} \cdot \frac{x^6}{x^4} = -15x^2$$

$$\frac{-21a^5b^4}{-3a^4b} = \frac{-21}{-3} \cdot \frac{a^5}{a^4} \cdot \frac{b^4}{b} = +7a^1b^3 \text{ or } 7ab^3$$

$$\frac{12y^2z^2}{4y^2z} = \frac{12}{4} \cdot \frac{y^2}{y^2} \cdot \frac{z^2}{z} = 3 \cdot 1 \cdot z^1 \text{ or } 3z$$

PROCEDURE. To divide a monomial by a monomial:

1. Divide the numerical coefficients.
2. When variable factors are powers with the same base, divide by subtracting exponents.
3. Multiply the quotients obtained in Steps 1 and 2.

If the area of a rectangle is 42 and its length is 6, we can find its width by dividing the area, 42, by the length, 6. Thus, $42 \div 6 = 7$, which is the width. Similarly, if the area of a rectangle is represented by $42x^2$ and its length by $6x$, we can find its width by dividing the area, $42x^2$, by the length, $6x$. We get $42x^2 \div 6x = 7x$, which represents the width.

EXAMPLES

1. Divide:

Solutions and answers:

a. $\frac{+24a^5}{-3a^2} = \frac{+24}{-3} \cdot \frac{a^5}{a^2} = -8a^3$

b. $\frac{-18x^3y^2}{-6x^2y} = \frac{-18}{-6} \cdot \frac{x^3}{x^2} \cdot \frac{y^2}{y} = +3xy$

c. $\dfrac{+20a^3c^4d^2}{-5a^3c^3} = \dfrac{+20}{-5} \cdot \dfrac{a^3}{a^3} \cdot \dfrac{c^4}{c^3} \cdot d^2 = -4(1)cd^2 = -4cd^2$

---

2. The area of a rectangle is $24x^4y^3$. Express in terms of x and y the length of the rectangle if the width is $3xy^2$.

Solution The width of a rectangle can be found by dividing the area by the length.

$$\frac{24x^4y^3}{3xy^2} = 8x^3y \quad \textit{Answer}$$

Dividing a Polynomial by a Monomial

We know that to divide by a number is the same as to multiply by its reciprocal. Therefore:

$$\frac{a + c}{b} = \frac{1}{b}(a + c) = \frac{a}{b} + \frac{c}{b}$$

Similarly,

$$\frac{2x + 2y}{2} = \frac{1}{2}(2x + 2y) = \frac{2x}{2} + \frac{2y}{2} = x + y$$

and

$$\frac{21a^2b - 3ab}{3ab} = \frac{1}{3ab}(21a^2b - 3ab) = \frac{21a^2b}{3ab} - \frac{3ab}{3ab} = 7a - 1$$

Usually, the two middle steps are done mentally.

PROCEDURE. To divide a polynomial by a monomial, divide each term of the polynomial by the monomial.

EXAMPLE

Divide:

a. $(8a^5 - 6a^4) \div 2a^2 = 4a^3 - 3a^2$ *Answer*

b. $\dfrac{24x^3y^4 - 18x^2y^2 - 6xy}{-6xy} = -4x^2y^3 + 3xy + 1$ *Answer*

EXERCISES

In 1–31, divide in each case.

1. $18x$ by 2
2. $14x^2y^2$ by -7
3. $-36y^{10}$ by $+6y^2$
4. $\dfrac{18x^6}{2x^2}$
5. $\dfrac{5x^2y^3}{-5y^3}$
6. $\dfrac{-49c^4b^3}{7c^2b^2}$
7. $\dfrac{-24x^2y}{-3xy}$
8. $\dfrac{-56abc}{8abc}$
9. $\dfrac{-27xyz}{9xz}$
10. $(10x + 20y) \div 5$
11. $(18r - 27s) \div 9$
12. $(14x + 7) \div 7$
13. $(cm + cn) \div c$
14. $(tr - r) \div r$
15. $\dfrac{12a - 6b}{-2}$
16. $\dfrac{8c^2 - 12d^2}{-4}$
17. $\dfrac{m^2 + 8m}{m}$
18. $\dfrac{p + prt}{p}$
19. $\dfrac{y^2 - 5y}{-y}$
20. $\dfrac{18d^3 + 12d^2}{6d}$
21. $\dfrac{20x^2 + 15x}{5x}$
22. $\dfrac{18r^5 + 12r^3}{6r^2}$
23. $\dfrac{16t^5 - 8t^4}{4t^2}$
24. $\dfrac{9y^9 - 6y^6}{-3y^3}$
25. $\dfrac{8a^3 - 4a^2}{-4a^2}$
26. $\dfrac{3ab^2 - 4a^2b}{ab}$
27. $\dfrac{4c^2d - 12cd^2}{4cd}$
28. $\dfrac{36a^4b^2 - 18a^2b^2}{-18a^2b^2}$
29. $\dfrac{-5y^5 + 15y - 25}{-5}$
30. $\dfrac{-2a^2 - 3a + 1}{-1}$
31. $\dfrac{2.4y^5 + 1.2y^4 - 0.6y^3}{-0.6y^3}$
32. If five oranges cost $15y$ cents, represent in simplest form the cost of one orange.
33. If the area of a rectangle is $32ab$ and the width is $8a$, represent in simplest form the length.
34. If a train traveled $54r$ miles in 9 hours, represent the distance traveled in 1 hour.
35. If $40ab$ chairs are arranged in $5a$ rows with equal numbers of chairs in each row, represent the number of chairs in one row.

CHAPTER SUMMARY

Two or more terms that contain the same variable, with corresponding variables having the same exponents, are called ***like terms***. The sum of like terms is the sum of the coefficients of the terms times the common variable factor of the terms.

A term that has no variable in the denominator is called a ***monomial***. A ***polynomial*** is the sum of monomials.

To subtract one polynomial from another, add the opposite of the polynomial to be subtracted (the subtrahend) to the polynomial from which it is to be subtracted (the minuend).

When x is a nonzero real number and a and b are integers:

$$x^a \cdot x^b = x^{a+b} \qquad (x^a)^b = x^{ab} \qquad x^a \div x^b = x^{a-b} \qquad x^0 = 1 \qquad x^{-a} = \frac{1}{x^a}$$

A number is in ***scientific notation*** when it is written as $a \times 10^n$, where $1 \leq a < 10$ and n is an integer.

Let $x = a \times 10^n$. Then:

When $x \geq 10$, n is positive.
When $1 \leq x < 10$, n is zero.
When $0 < x < 1$, n is negative.

To multiply a polynomial by a polynomial, multiply each term of one polynomial by each term of the other polynomial and then add similar terms.

To divide a polynomial by a monomial, divide each term of the polynomial by the monomial and write the quotient as the sum of these results.

VOCABULARY

8-1 Like terms (similar terms) Unlike terms Monomial Polynomial Binomial Trinomial Simplest form Descending order Ascending order

8-2 Opposite of a polynomial Minuend Subtrahend Difference

8-3 Power Base Exponent

8-5 FOIL

8-8 Scientific notation

REVIEW EXERCISES

In 1–15, simplify each expression.

1. $3r^2 + 2r^2$ **2.** $5bc - bc$ **3.** $3y^2 - 2y + y^2 - 8y - 2$

4. $5t - (4 - 8t)$ **5.** $8mg(-3g)$ **6.** $3x^2(4x^2 + 2x - 1)$

7. $(4x + 3)(2x - 1)$ **8.** $(-6ab^3)^2$ **9.** $(-6a + b)^2$

10. $(2a + 5)(2a - 5)$ **11.** $(2a - 5)^2$ **12.** $2x - x(2x - 5)$

13. $\frac{40b^3x^6}{-8b^2x}$ **14.** $5y + \frac{6y^4}{-2y^3}$ **15.** $\frac{6w^3 - 8w^2 + 2w}{2w}$

In 16–20, write each expression in an equivalent form, using positive exponents only. Where possible, compute the answer.

16. 8^{-2} **17.** k^{-6} **18.** 10^{-3} **19.** $\left(\frac{2}{5}\right)^{-2}$ **20.** $2x^{-4}$

In 21–24, use the laws of exponents to perform the operations, and simplify.

21. $3^{-5} \cdot 3^4$ **22.** $(7^{-3})^0$ **23.** $(10^{-2})^{-1}$ **24.** $12^{-3} \cdot 12$

25. If the length of one side of a square is $\frac{1}{2}h$, find the representation of
a. the perimeter of the square **b.** the area

26. The perimeter of a triangle is $41px$. If the lengths of two sides are $18px$ and $7px$, represent the length of the third side.

27. If one side of a square is $x + 5$, write the polynomial that represents:
a. the perimeter of the square **b.** its area

28. Find the value of $8x^0 + 4x^{-1}$ when $x = 2$.

29. The cost of a pizza is 20 cents less than 9 times the cost of a soft drink. If x represents the cost, in cents, of a soft drink, express in simplest form the cost of two pizzas and six soft drinks.

In 30–33, express each number in scientific notation.

30. 5,800 **31.** 14,200,000 **32.** 0.00006 **33.** 0.00000277

In 34–37, find the decimal number that is expressed by each given numeral.

34. 4×10^4 **35.** 3×10^{-3} **36.** 3.9×10^8 **37.** 1.03×10^{-4}

CUMULATIVE REVIEW WITH OPEN-RESPONSE PROBLEM

1. The statement "If I am hungry, then I will stop for lunch," is false. Which of the following is true?
(1) I am hungry and I stop for lunch.
(2) I am hungry and I do not stop for lunch.
(3) If I do not stop for lunch, then I am not hungry.
(4) I am not hungry or I stop for lunch.

2. The vertices of $\triangle ABC$ are $A(-1, -2)$, $B(6, -2)$, and $C(6, 5)$. What is the area of $\triangle ABC$?

Exploration

Study the squares of two-digit numbers that end in 5. From what you observe, can you devise a method for finding the square of such a number mentally? Can this method be applied to the square of a three-digit number that ends in 5?

Study the squares of consecutive integers. From what you observe, can you devise a method that uses the square of an integer to find the square of the next consecutive integer?

Chapter 9

First-Degree Equations and Inequalities in One Variable

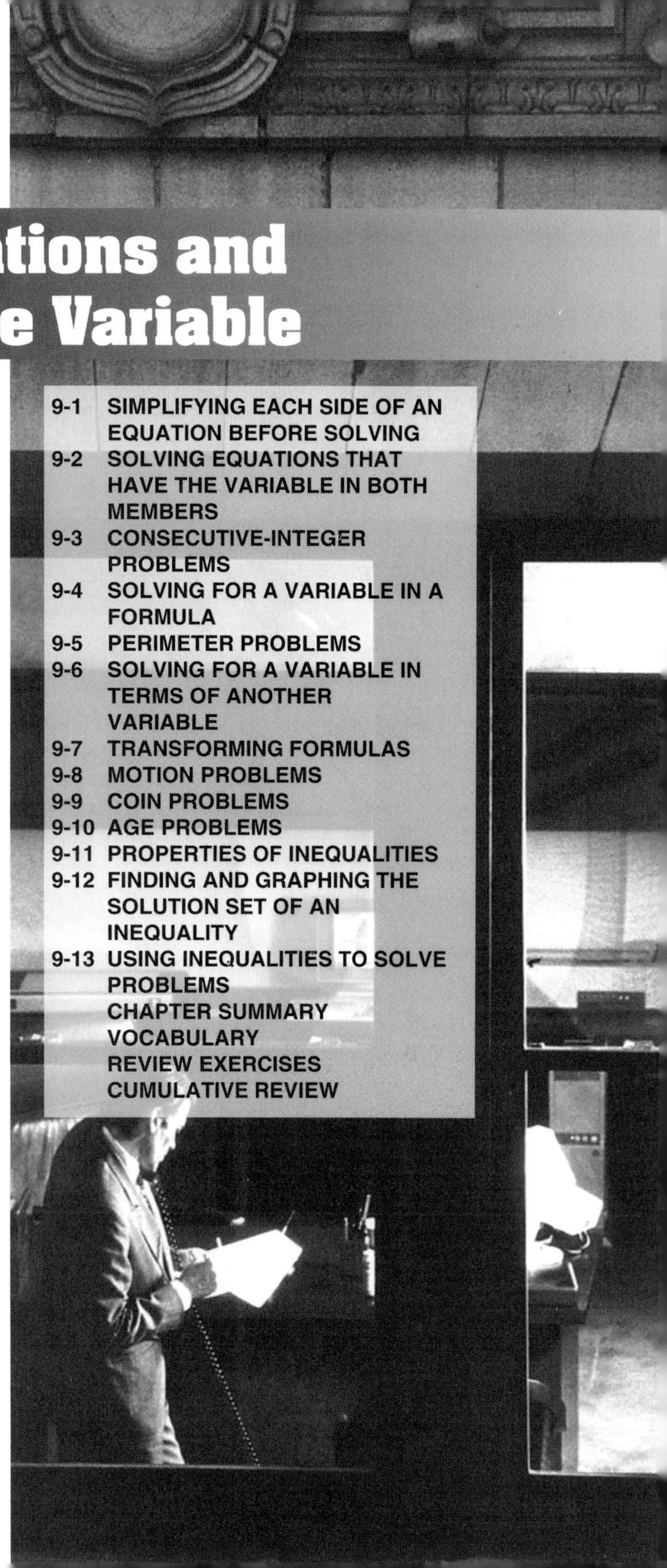

Not all decisions involve quantities that must be equal. Often a decision requires the use of an inequality to determine which of two quantities is larger and which is smaller.

When renting a car, a traveler compares the rates from different car-rental firms in order to determine the best choices for the number of days and number of miles that the car will be used.

When selecting a long-distance telephone company, an office manager uses information from several companies to estimate their monthly charges for the types of services required. On the basis of the results obtained, she can choose the company that will provide the needed services at the lowest cost.

In Chapter 6 you learned to solve simple equations in which the variable appeared in only one term. In this chapter you will learn to solve problems that require the use of equations and inequalities in which the variable occurs more than once.

9-1 SIMPLIFYING EACH SIDE OF AN EQUATION BEFORE SOLVING

An equation is often written in such a way that one or both sides are not in simplest form. Before starting to solve the equation by using additive and multiplicative inverses, you should simplify each side by removing parentheses if necessary and adding like terms.

EXAMPLE

Solve and check:

a. $2x + 3x + 4 = -6$

How to Proceed:	*Solution:*
(1) Write the equation:	$2x + 3x + 4 = -6$
(2) Combine like terms:	$5x + 4 = -6$
(3) Add -4, the additive inverse of $+4$, to each side:	$\underline{\quad -4 \quad -4}$ $5x = -10$
(4) Multiply by $\frac{1}{5}$, the multiplicative inverse of 5:	$\frac{1}{5}(5x) = \frac{1}{5}(-10)$
(5) Simplify each side:	$x = -2$

Algebraic Check:

$$2x + 3x + 4 = -6$$
$$2(-2) + 3(-2) + 4 \stackrel{?}{=} -6$$
$$(-4) + (-6) + 4 \stackrel{?}{=} -6$$
$$-6 = -6 \checkmark$$

Calculator Check:

Evaluate the left-hand member using $x = -2$. The display shows that this equals the right-hand member.

Enter: 2 [×] 2 [+/−] [+] 3 [×] 2 [+/−] [+] 4 [=]

Display: [−6.] ✔

Answer: $x = -2$

b. $27x - 3(x - 6) = 6$

Since $-3(x - 6)$ means that $(x - 6)$ is to be multiplied by -3, we will use the distributive property.

How to Proceed:	*Solution:*
(1) Write the equation:	$27x - 3(x - 6) = 6$
(2) Use the distributive property:	$27x - 3x + 18 = 6$
(3) Combine like terms:	$24x + 18 = 6$
(4) Add -18, the additive inverse of $+18$, to each side:	$\underline{\quad -18 \quad -18}$ $24x = -12$

(5) Divide each side by 24: $\frac{24x}{24} = \frac{-12}{24}$

(6) Simplify each side: $x = -\frac{1}{2}$

The check is left to you.

Answer: $x = -\frac{1}{2}$

Representing Two Numbers with the Same Variable

Problems often involve finding two or more different numbers. It is useful to express these numbers in terms of the same variable. For example, if you know the sum of two numbers, you can express the second in terms of the sum and the first number.

If the sum of two numbers is 12 and one of the numbers is 5, then the other number is 12 − 5 or 7.
If the sum of two numbers is 12 and one of the numbers is 9, then the other number is 12 −9 or 3.
If the sum of two numbers is 12 and one of the numbers is x, then the other number is $12 - x$.

A problem can often be solved algebraically in more than one way, by writing and solving different equations, as shown in the example that follows. The methods used to obtain the solution are different, but both use the facts stated in the problem and arrive at the same solution.

EXAMPLE

The sum of two numbers is 43. The larger number minus the smaller number is 5. Find the numbers.

Note: This problem states two facts:

Fact 1: The sum of the numbers is 43.
Fact 2: The larger number minus the smaller number is 5. In other words, the larger number is 5 more than the smaller.

Solution (1) Represent each number in terms of the same variable using fact 1: the sum of the numbers is 43.

Let x = the larger number.

Then, $43 - x$ = the smaller number.

(2) Write an equation using fact 2:

$$\underbrace{\text{The larger number}}_{x}\quad \underbrace{\text{minus}}_{-}\quad \underbrace{\text{the smaller number}}_{(43 - x)}\quad \underbrace{\text{is}}_{=}\quad \underbrace{5.}_{5}$$

(3) Solve the equation.

(a) Write the equation: $x - (43 - x) = 5$

(b) To subtract $(43 - x)$, add its opposite: $x + (-43 + x) = 5$

(c) Combine like terms: $2x - 43 = 5$

(d) Add the opposite of -43 to each side: $\underline{+43\quad +43}$

$2x = 48$

(e) Divide each side by 2: $\frac{2x}{2} = \frac{48}{2}$

$x = 24$

(4) Find the numbers.

The larger number $= x = 24$.

The smaller number $= 43 - x = 19$.

Check. A word problem *cannot* be checked, either algebraically or with a calculator, by substituting numbers in an equation. The equation formed may not be correct. Check the numbers against the facts stated in the original wording of the problem.

The sum of the numbers is 43: $24 + 19 = 43$.

The larger number minus the smaller number is 5: $24 - 19 = 5$.

Alternative Solution Reverse the way in which the facts are used.

(1) Represent each number in terms of the same variable using fact 2: the larger number is 5 more than the smaller.

Let x = the larger number.

Then, $x - 5$ = the smaller number.

(2) Write an equation using fact 1:

$$\underbrace{\text{The sum of the numbers}}_{x + (x - 5)}\ \underbrace{\text{is}}_{=}\ \underbrace{43.}_{43}$$

(3) Solve the equation.

(a) Write the equation: $x + (x - 5) = 43$

(b) Combine like terms: $2x - 5 = 43$

(c) Add the opposite of -5 to each side: $\underline{+5 \quad +5}$

$2x = 48$

(d) Divide each side by 2: $\frac{2x}{2} = \frac{48}{2}$

$x = 24$

(4) Find the numbers.

The larger number $= x = 24$.

The smaller number $= x - 5 = 19$.

(5) *Check.* (See the first solution.)

Answer: The numbers are 24 and 19.

EXERCISES WITH OPEN-RESPONSE PROBLEMS

In 1–32, solve and check each equation.

1. $x + (x - 6) = 20$
2. $x - (12 - x) = 38$
3. $(15x + 7) - 12 = 4$
4. $(14 - 3c) + 7c = 94$
5. $x + (4x + 32) = 12$
6. $7x - (4x - 39) = 0$
7. $5(x + 2) = 20$
8. $3(y - 9) = 30$
9. $8(2c - 1) = 56$
10. $6(3c - 1) = -42$
11. $30 = 2(10 - y)$
12. $4(c + 1) = 32$
13. $5t - 2(t - 5) = 19$
14. $18 = -6x + 4(2x + 3)$
15. $55 = 4 + 3(m + 2)$
16. $5(x - 3) - 30 = 10$
17. $7(x + 2) = 5(x + 4)$
18. $3(a - 5) - 2(2a + 1) = 0$
19. $3(2b + 1) - 7 = 50$
20. $5(3c - 2) + 8 = 43$
21. $7r - (6r - 5) = 7$
22. $8y - (5y + 2) = 16$
23. $114 = 40 + (7x + 4)$
24. $10z - (3z - 11) = 17$
25. $8b - 4(b - 2) = 24$
26. $5m - 2(m - 5) = 17$
27. $9 + 2(5v + 3) = 135$
28. $28r - 6(3r - 5) = 40$
29. $3a + (2a - 5) + 2(a + 2) = 13$
30. $4(2r + 1) - 3(2r - 5) = 29$
31. $\frac{1}{2}(8x - 6) = 25$
32. $\frac{3}{4}(8 + 4x) - \frac{1}{3}(6x + 3) = 9$

In 33–38, write an algebraic expression to represent each phrase.

33. Twice the number represented by x, increased by 8

34. Three times the number represented by x, decreased by 12

35. Four times the number that is 3 more than x

36. Three times the number that exceeds x by 5

37. Twice the sum of the number represented by x and 5

38. Ten times the number obtained when twice x is decreased by 10

39. If the smaller of two numbers is represented by x, represent the greater when their sum is:
a. 10 **b.** 25 **c.** 36 **d.** 50 **e.** 100 **f.** 3,000

40. If the sum of two numbers is represented by S, and the smaller number is represented by x, represent the greater in terms of S and x.

41. If the sum of two numbers is represented by S, and the greater number is represented by ℓ, represent the smaller in terms of S and ℓ.

42. Sandi bought 6 yards of material. She wants to cut it into two pieces so that the difference between the lengths of the two pieces will be 1.5 yards. What should be the length of each piece?

43. The Tigers won eight games more than they lost, and there were no ties. If the Tigers played 78 games, how many games did they lose?

44. This month Erica saved $20 more than last month. In the 2 months she saved a total of $70. How much did she save each month?

45. On a bus tour, there are 100 passengers on three buses. Two of the buses each carry four fewer passengers than the third bus. How many passengers are on each bus?

46. Rahul has 25 coins, all quarters and dimes.

a. If x is the number of dimes Rahul has, express, in terms of x, the number of quarters he has.

b. Express the value of the dimes in terms of x.

c. Express the value of the quarters in terms of x.

d. If the total value of the dimes and quarters is $4.90, how many dimes and how many quarters does Rahul have?

47. Carlos works Monday through Friday and sometimes on Saturday. Last week Carlos worked 38 hours.

a. If x is the total number of hours Carlos worked Monday through Friday, express, in terms of x, the number of hours he worked on Saturday.

b. Carlos earns $8.50 an hour when he works Monday through Friday. Express, in terms of x, his earnings Monday through Friday.

c. Carlos earns $12.75 an hour when he works on Saturday. Express, in terms of x, his earnings on Saturday.

d. Last week Carlos earned $340. How many hours did he work on Saturday?

9-2 SOLVING EQUATIONS THAT HAVE THE VARIABLE IN BOTH MEMBERS

A variable represents a number. As you know, any number may be added to both members of an equation without changing the solution set. Therefore, the same variable (or the same multiple of the same variable) may be added to or subtracted from both members of an equation without changing the solution set.

To solve $8x = 30 + 5x$, first eliminate $5x$ from the right member of the equation. Then add $-5x$ to both members.

Method 1

$$\begin{array}{rcl} 8x & = & 30 + 5x \\ \underline{-5x} & & \underline{\quad\;\; - 5x} \\ 3x & = & 30 \\ x & = & 10 \end{array}$$

Method 2

$$\begin{aligned} 8x &= 30 + 5x \\ 8x + (-5x) &= 30 + 5x + (-5x) \\ 3x &= 30 \\ x &= 10 \end{aligned}$$

The check is left to you.

Answer: $x = 10$

To solve an equation that has the variable in both members, transform it into an equivalent equation in which the variable appears in only one member. Then, solve the equation.

EXAMPLES

1. Solve and check:

a. $7x = 63 - 2x$

How to Proceed:	*Solution:*	*Check:*
(1) Write the equation:	$7x = 63 - 2x$	$7x = 63 - 2x$
(2) Add $2x$ to each member of the equation:	$\underline{+2x \qquad + 2x}$ $9x = 63$	$7(7) \stackrel{?}{=} 63 - 2(7)$
(3) Divide each member of the equation by 9:	$x = 7$	$49 \stackrel{?}{=} 63 - 14$ $49 = 49$ ✔

Answer: $x = 7$

b. $3y + 7 = 5y - 3$

To solve this equation, write an equivalent equation with only a variable term on one side and only a constant on the other. Then solve the simplified equation. The two possible solutions below show how this can be done.

How to Proceed:	*Solution:* *Method 1*	 *Method 2*
(1) Write the equation.	$3y + 7 = 5y - 3$	$3y + 7 = 5y - 3$
(2) Write an equivalent equation with no variable term on one side by adding the opposite of one variable term to both sides of the equation.	$\underline{-5y} \qquad \underline{-5y}$ $-2y + 7 = \quad - 3$	$\underline{-3y} \qquad \underline{-3y}$ $+ 7 = 2y - 3$
(3) Add the opposite of the constant term that is on the same side as the remaining variable terms to both sides of the equation.	$\underline{\quad -7} \qquad \underline{\quad -7}$ $-2y \quad = \quad -10$	$\underline{\quad +3} \qquad \underline{\quad +3}$ $10 = 2y$
(4) Divide both sides by the coefficient of the variable term.	$\frac{-2y}{-2} = \frac{-10}{-2}$ $y = +5$	$\frac{10}{2} = \frac{2y}{2}$ $5 = y$

Algebraic Check:

$$3y + 7 = 5y - 3$$
$$3(5) + 7 \stackrel{?}{=} 5(5) - 3$$
$$15 + 7 \stackrel{?}{=} 25 - 3$$
$$22 = 22 \checkmark$$

Calculator Check:

Evaluate the left side and evaluate the right side separately. The displays should be equal.

Enter: 3 [×] 5 [+] 7 [=]
Display: 22.

Enter: 5 [×] 5 [−] 3 [=]
Display: 22. ✔

Answer: $y = 5$

---

2. The larger of two numbers is 4 times the smaller. If the larger number exceeds the smaller number by 15, find the number.

> *Note:* When s represents the smaller number and $4s$ represents the larger number, "the larger number ***exceeds*** the smaller by 15" has the following meanings. Use any one of them.
>
> 1. The larger equals 15 more than the smaller, written as $4s = s + 15$.
> 2. The larger decreased by 15 equals the smaller, written as $4s - 15 = s$.
> 3. The larger decreased by the smaller equals 15, written as $4s - s = 15$.

Solution

Let s = the smaller number.

Then, $4s$ = the larger number.

$$\underbrace{\text{The larger}}_{4s}\ \underbrace{\text{is}}_{=}\ \underbrace{\text{15 more than the smaller.}}_{s + 15}$$

$$\begin{aligned} 4s &= s + 15 \\ -s &= -s \\ \hline 3s &= 15 \\ s &= 5 \\ 4s &= 20 \end{aligned}$$

Check: The larger number, 20, is 4 times the smaller number, 5. The larger number, 20, exceeds the smaller number, 5, by 15.

Answer: The larger number is 20; the smaller number is 5.

---

3. In his will, Uncle Clarence left \$5,000 to his two nieces. Emma's share is to be \$500 more than Clara's. How much should each niece receive?

Solution (1) Use the fact that the sum of the two shares is 5,000 to express each share in terms of a variable.

$$\text{Let } x = \text{Clara's share.}$$

$$\text{Then } 5{,}000 - x = \text{Emma's share.}$$

(2) Use the fact that Emma's share is 500 more than Clara's share to write an equation.

$$\underbrace{\text{Emma's share}}_{5{,}000 - x} \quad \underbrace{\text{is}}_{=} \quad \underbrace{\text{\$500 more than Clara's share.}}_{500 + x}$$

(3) Solve the equation to find Clara's share:

$$\begin{aligned} 5{,}000 - x &= 500 + x \\ +x \quad & \quad\; +x \\ 5{,}000 &= 500 + 2x \\ -500 \quad & \quad -500 \\ 4{,}500 &= 2x \\ \frac{4{,}500}{2} &= \frac{2x}{2} \\ 2{,}250 &= x \\ x &= 2{,}250 \end{aligned}$$

(4) Find Emma's share:

$$5{,}000 - x = 2{,}750$$

Alternative Solution (1) Use the fact that Emma's share is 500 more than Clara's share to express each share in terms of a variable.

$$\text{Let } x = \text{Clara's share.}$$

$$\text{Then } x + 500 = \text{Emma's share.}$$

(2) Use the fact that the sum of the two shares is 5,000 to write an equation.

$$\underbrace{\text{Clara's share}}_{x} \quad \underbrace{\text{plus}}_{+} \quad \underbrace{\text{Emma's share}}_{(x + 500)} \quad \underbrace{\text{is}}_{=} \quad \underbrace{\text{\$5,000.}}_{5{,}000}$$

(3) Solve the equation to find Clara's share:

$$\begin{aligned} x + (x + 500) &= 5{,}000 \\ 2x + 500 &= 5{,}000 \\ -\ 500 \quad & \quad -500 \\ 2x \quad &= 4{,}500 \\ \frac{2x}{2} &= \frac{4{,}500}{2} \\ x &= 2{,}250 \end{aligned}$$

(4) Find Emma's share:

$$x + 500 = 2{,}750$$

Check: \$2,750 is \$500 more than \$2,250, and \$2,750 + \$2,250 = \$5,000.

Answer: Clara's share is \$2,250, and Emma's share is \$2,750.

EXERCISES WITH OPEN-RESPONSE PROBLEMS

In 1–38, solve each equation and check.

1. $7x = 10 + 2x$

2. $9x = 44 - 2x$

3. $5c = 28 + c$

4. $y = 4y + 30$

5. $2d = 36 + 5d$

6. $2\frac{1}{4}y = 1\frac{1}{4}y - 8$

7. $0.8m = 0.2m + 24$

8. $8y = 90 - 2y$

9. $2.3x + 36 = 0.3x$

10. $2\frac{3}{4}x + 24 = 3x$

11. $5a - 40 = 3a$

12. $5c = 2c - 81$

13. $x = 9x - 72$

14. $0.5m - 30 = 1.1m$

15. $4\frac{1}{4}c = 9\frac{3}{4}c + 44$

16. $7r + 10 = 3r + 50$

17. $4y + 20 = 5y + 9$

18. $7x + 8 = 6x + 1$

19. $x + 4 = 9x + 4$

20. $9x - 3 = 2x + 46$

21. $y + 30 = 12y - 14$

22. $c + 20 = 55 - 4c$

23. $2d + 36 = -3d - 54$

24. $7y - 5 = 9y + 29$

25. $2m - 1 = 6m + 1$

26. $4x - 3 = 47 - x$

27. $3b - 8 = 14 - 8b$

28. $\frac{2}{3}t - 11 = 64 - 4\frac{1}{3}t$

29. $18 - 4n = 6 - 16n$

30. $-2y - 39 = 5y - 18$

31. $7x - 4 = 5x - x + 35$

32. $10 - x - 3x = 7x - 23$

33. $8a - 15 - 6a = 85 - 3a$

34. $8c + 1 = 7c - 14 - 2c$

35. $12x - 5 = 8x - x + 50$

36. $6d - 12 - d = 9d + 53 + d$

37. $3m - 5m - 12 = 7m - 88 - 5$

38. $5 - 3z - 18 = z - 1 + 8z$

39. Eight times a number equals 35 more than the number. Find the number.

40. Six times a number equals 3 times the number, increased by 24. Find the number.

41. If 3 times a number is increased by 22, the result is 14 less than 7 times the number. Find the number.

42. The greater of two numbers is 1 more than twice the smaller. Three times the greater exceeds 5 times the smaller by 10. Find the numbers.

43. The second of three numbers is 6 more than the first. The third number is twice the first. The sum of the three numbers is 26. Find the three numbers.

44. The second of three numbers is 1 less than the first. The third number is 5 less than the second. If the first number is twice as large as the third, find the three numbers.

45. It took the Gibbons family 2 days to travel 925 miles to their vacation home. They traveled 75 miles more on the first day than on the second. How many miles did they travel each day?

46. During the first 6 month of last year, the interest on an investment was \$130 less than during the second 6 months. The total interest for the year was \$1,450. What was the interest for each 6-month period?

47. Gemma has 7 more five-dollar bills than ten-dollar bills. The value of the five-dollar bills equals the value of the ten-dollar bills. How many five-dollar bills and ten-dollar bills has she?

48. Leonard wants to save \$100 in the next 2 months. He knows that in the second month he will be able to save \$20 more than during the first month. How much should he save each month?

49. To rent a car, the ABC Company charges \$75 a day plus \$0.05 a mile. How many miles did Mrs. Kiley drive if the cost of renting the car was \$92.40?

9-3 CONSECUTIVE-INTEGER PROBLEMS

Preparing to Solve Consecutive-Integer Problems

As you know, an integer is any whole number or its opposite. Examples of integers are 5, -3, and 0.

Consecutive integers are integers that follow one another in order. To obtain a set of consecutive integers, start with any integer and count by 1's. Each number in the set is 1 more than the preceding number. Each of the following is a set of consecutive integers:

1. $\{5, 6, 7, 8\}$

2. $\{-5, -4, -3, -2\}$

3. $\{x, x + 1, x + 2, x + 3\}$ $\quad x \in \{\text{integers}\}$

Consecutive even integers are even integers that follow one another in order. To obtain a set of consecutive even integers, start with any even integer and count by 2's. Each number in the set is 2 more than the preceding number. Each of the following is a set of consecutive even integers:

1. $\{2, 4, 6, 8\}$

2. $\{-12, -10, -8, -6\}$

3. $\{x, x + 2, x + 4, x + 6\}$ $\quad x \in \{\text{even integers}\}$

Consecutive odd integers are odd integers that follow one another in order. To obtain a set of consecutive odd integers, start with any odd integer and count by 2's. Each number in the set is 2 more than the preceding number. Each of the following is a set of consecutive odd integers:

1. $\{3, 5, 7, 9\}$
2. $\{-5, -3, -1, 1\}$
3. $\{x, x + 2, x + 4, x + 6\}$ $\quad x \in \{\text{odd integers}\}$

Solving Consecutive-Integer Problems

EXAMPLES

1. Find two consecutive integers whose sum is 95.

Solution

$$\text{Let } n = \text{the first integer.}$$
$$\text{Then, } n + 1 = \text{the second integer.}$$

$$\underbrace{\text{The sum of the two integers}}\ \underbrace{\text{is}}\ \underbrace{95.}$$
$$n + (n + 1) \qquad = 95$$

$$\begin{aligned} n + n + 1 &= 95 \\ 2n + 1 &= 95 \\ 2n + 1 - 1 &= 95 - 1 \\ 2n &= 94 \\ n &= 47 \\ \text{Then } n + 1 &= 48 \end{aligned}$$

Check: The sum of the consecutive integers, 47 and 48, is 95.

Answer: 47 and 48

---

2. Find three consecutive positive even integers such that 4 times the first decreased by the second is 12 more than twice the third.

Solution

$$\text{Let } n = \text{the first even integer.}$$
$$\text{Then, } n + 2 = \text{the second even integer.}$$
$$\text{Then, } n + 4 = \text{the third even integer.}$$

Four times the first decreased by the second	is	12 more than twice the third.
$4n - (n + 2)$	$=$	$2(n + 4) + 12$

$$\begin{aligned} 4n - n - 2 &= 2n + 8 + 12 \\ 3n - 2 &= 2n + 20 \\ 3n - 2 + 2 &= 2n + 20 + 2 \\ 3n &= 2n + 22 \\ 3n + (-2n) &= 2n + 22 + (-2n) \\ n &= 22 \\ \text{Then } n + 2 &= 24 \\ \text{and } n + 4 &= 26 \end{aligned}$$

Check: The three integers 22, 24, and 26 satisfy the conditions given in the problem: $4(22) - 24$ is 12 more than $2(26)$.

Answer: 22, 24, 26

EXERCISES WITH OPEN-RESPONSE PROBLEMS

1. If $x = 3$, what numbers do x, $x + 1$, $x + 2$, and $x + 3$ represent?

2. If $n = -3$, what numbers do n, $n + 1$, $n + 2$, and $n + 3$ represent?

3. If $x = 8$, what numbers do x, $x + 2$, $x + 4$, and $x + 6$ represent?

4. If $n = -5$, what numbers do n, $n + 2$, $n + 4$, and $n + 6$ represent?

5. Write four consecutive integers beginning with each of the following integers (y is an integer).
a. 15 **b.** -6 **c.** -1 **d.** y **e.** $y - 1$

6. Write four consecutive even integers beginning with each of the following integers (y is an even integer).
a. 12 **b.** -10 **c.** -2 **d.** y **e.** $y - 2$

7. Write four consecutive odd integers beginning with each of the following integers (y is an odd integer).
a. 17 **b.** -7 **c.** -15 **d.** y **e.** $y + 4$

8. Write four consecutive integers whose sum is an even integer.

9. **a.** Is it possible to write four consecutive integers whose sum is an odd integer? **b.** Explain your answer.

10. Find two consecutive integers whose sum is 91.

11. Find two consecutive integers whose sum is -17.

12. Find three consecutive integers whose sum is 93.

13. Find three consecutive integers whose sum is -18.

14. Find four consecutive integers whose sum is 110.

15. Find two consecutive even integers whose sum is 86.

16. Find three consecutive even integers whose sum is -12.

17. Find four consecutive even integers whose sum is 42.

18. Find two consecutive odd integers whose sum is 28.
19. Find three consecutive odd integers whose sum is 255.
20. Find four consecutive odd integers whose sum is 456.
21. Find three consecutive integers such that the sum of the first and the third is 40.
22. Find four consecutive integers such that the sum of the second and the fourth is 132.
23. Find two consecutive odd integers such that 4 times the larger is 29 more than 3 times the smaller.
24. Find two consecutive even integers such that twice the smaller is 26 less than 3 times the larger.
25. Find three consecutive integers such that twice the smallest is 12 more than the largest.
26. Find three consecutive integers such that the sum of the first two is 24 more than the third.
27. Find three consecutive even integers such that the sum of the smallest and twice the second is 20 more than the third.
28. Find two consecutive integers such that 4 times the larger exceeds 3 times the smaller by 23.
29. Find four consecutive odd integers such that the sum of the first three exceeds the fourth by 18.
30. Find three consecutive odd integers such that the largest decreased by 3 times the second is 47 less than the smallest.
31. **a.** Can the sum of three consecutive integers be even?
    b. Give some examples of three consecutive integers whose sum is even.
    c. Can the sum of three consecutive integers be odd?
    d. Give some examples of three consecutive integers whose sum is odd.
    e. When is the sum of three consecutive integers even, and when is it odd?
32. Show that the sum of three consecutive integers is always equal to 3 times the middle integer.

9-4 SOLVING FOR A VARIABLE IN A FORMULA

To solve for the subject of a formula, substitute the known values in the formula and perform the required computation. Thus, to find the area of a triangle when $b = 4.7$ centimeters and $h = 3.2$ centimeters, substitute the given values in the formula for the area of a triangle:

$$\begin{aligned} A &= \frac{1}{2}bh \\ &= \frac{1}{2}(4.7 \text{ cm})(3.2 \text{ cm}) \\ &= 7.52 \text{ cm}^2 \end{aligned}$$

Now that you can solve equations, you will be able to find the value of any variable in a formula when the values of the other variables are known. To do this:

1. Write the formula.
2. Substitute the given values in the formula.
3. Solve the resulting equation.

The values assigned to the variables in a formula often have a unit of measure. It is convenient to solve the equation without writing the unit of measure, but the answer should always be given in terms of the correct unit of measure.

EXAMPLE

The perimeter of a rectangle is 48 centimeters. If the length of the rectangle is 16 centimeters, find its width. (Use $P = 2\ell + 2w$.)

Solution:	*Check:*
$P = 2\ell + 2w$	$P = 2\ell + 2w$
$48 = 2(16) + 2w$	$48 \stackrel{?}{=} 2(16) + 2(8)$
$48 = 32 + 2w$	$48 \stackrel{?}{=} 32 + 16$
$16 = 2w$	$48 = 48$ ✔
$8 = w$	

Answer: 8 centimeters

EXERCISES

1. If $P = a + b + c$, find c when $P = 80$, $a = 20$, and $b = 25$.

2. If $P = 4s$, find s when: **a.** $P = 20$ **b.** $p = 32$ **c.** $P = 6.4$

3. If $A = \ell w$, find w when: **a.** $A = 100$, $\ell = 5$ **b.** $A = 3.6$, $\ell = 0.9$

4. If $A = \frac{1}{2}bh$, find h when: **a.** $A = 24$, $b = 8$ **b.** $A = 12$, $b = 3$

5. If $V = \ell wh$, find w when $V = 72$, $\ell = \frac{3}{4}$, and $h = 12$.

6. If $P = 2\ell + 2w$, find w when: **a.** $P = 20$, $\ell = 7$ **b.** $P = 36$, $\ell = 9\frac{1}{2}$

7. If $P = 2\ell + 2w$, find ℓ when: **a.** $P = 28$, $w = 3$ **b.** $P = 24.8$, $w = 4.7$

8. If $P = 2a + b$, find b when $P = 80$ centimeters and $a = 30$ centimeters.

9. If $P = 2a + b$, find a when $P = 18.6$ centimeters and $b = 5.8$ centimeters.

10. If $A = \frac{1}{2}h(b + c)$, find h when $A = 30$, $b = 4$, and $c = 6$.

11. If $A = \frac{1}{2}h(b + c)$, find b when $A = 50$ square centimeters, $h = 4$ centimeters, and $c = 11$ centimeters.

12. If $D = rt$, find r when: **a.** $D = 120$, $t = 3$ **b.** $D = 40$, $t = \frac{1}{2}$

13. If $I = prt$, find p when $I = \$135$, $r = 6\%$, and $t = 3$ years.

14. If $F = \frac{9}{5}C + 32$, find C when: **a.** $F = 95°$ **b.** $F = 68°$ **c.** $F = 59°$ **d.** $F = 32°$ **e.** $F = 212°$ **f.** $F = -13°$

15. Find the length of a rectangle whose perimeter is 34.6 centimeters and whose width is 5.7 centimeters.

16. The area of a triangle is 36 square centimeters. Find the measure of the altitude drawn to the base when:
a. the base = 8 centimeters **b.** the base = 12 centimeters

In 17–20, the perimeter of a square is given.

a. Find the length of each side of the square. **b.** Find the area of the square.

17. $P = 28$ centimeters

18. $P = 3$ inches

19. $P = 16.8$ centimeters

20. $P = 2$ feet

9-5 PERIMETER PROBLEMS

You know that the ***perimeter*** of a geometric figure is the sum of the lengths of all of its sides. When solving a perimeter problem, it is helpful to draw and label a figure to model the region.

EXAMPLES

1. A garden is in the shape of an isosceles triangle. The length of the base of the triangle is 2 feet greater than the length of each side. Write an algebraic expression for the perimeter of the garden.

Solution Let x = the length of each of the sides.

Then $x + 2$ = the length of the base.

$$\text{Perimeter} = x + x + x + 2$$
$$= 3x + 2$$

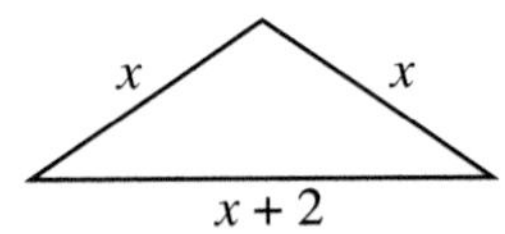

Answer: $3x + 2$

2. The perimeter of a rectangle is 40 feet. The length is 2 feet more than 5 times the width. Find the dimensions of the rectangle.

Solution Let w = the width, in feet, of the rectangle.

Then $5w + 2$ = the length, in feet, of the rectangle.

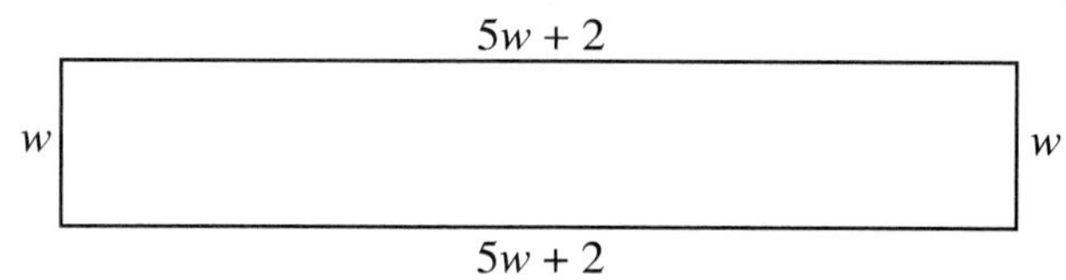

The sum of the lengths of all the sides is 40.

$$\begin{aligned} w + (5w + 2) + w + (5w + 2) &= 40 \\ w + 5w + 2 + w + 5w + 2 &= 40 \\ 12w + 4 &= 40 \\ 12w + 4 - 4 &= 40 - 4 \\ 12w &= 36 \\ w &= 3 \\ 5w + 2 &= 17 \end{aligned}$$

Check:

$$\begin{aligned} 3 + 17 + 3 + 17 &\stackrel{?}{=} 40 \\ 40 &= 40 \checkmark \\ 17 &\stackrel{?}{=} 5(3) + 2 \\ 17 &= 17 \checkmark \end{aligned}$$

Answer: The width is 3 feet; the length is 17 feet.

EXERCISES WITH OPEN-RESPONSE PROBLEMS

1. Represent the perimeter of each of the following figures:

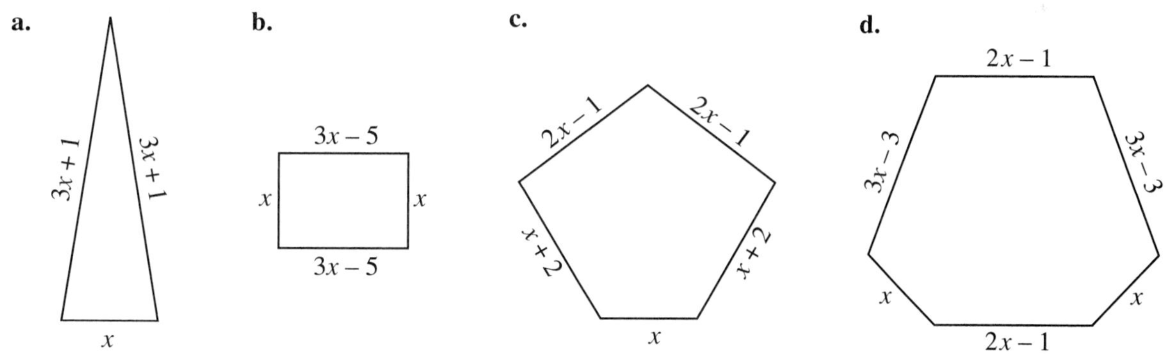

2. Represent the length and the perimeter of a rectangle whose width is represented by x and whose length:

a. is twice its width
b. is 4 more than its width
c. is 5 less than twice its width
d. is 3 more than twice its width

3. The length of a side of an equalateral triangle is represented by $x + 5$. Represent the perimeter of the triangle.
4. If the length of each side of a square is represented by $2x - 1$, represent the perimeter of the square.
5. If the length of each side of an equilateral pentagon (five sides) is represented by $2x + 3$, represent the perimeter of the pentagon.
6. The length of each side of an equilateral hexagon (six sides) is represented by $2x - 3$. Represent the perimeter of the hexagon.
7. The perimeter of an equilateral polygon is $12x - 24$. Express the length of one side if the polygon is:
   a. a triangle **b.** a square **c.** a hexagon
8. The length of a rectangle is 3 times its width. The perimeter of the rectangle is 72 centimeters. Find the dimensions of the rectangle.
9. The length of a rectangle is $2\frac{1}{2}$ times its width. The perimeter of the rectangle is 84 centimeters. Find the dimensions of the rectangle.
10. The length of the second side of a triangle is 2 inches less than the length of the first side. The length of the third side is 12 inches more than the length of the first side. The perimeter of the triangle is 73 inches. Find the length of each side of the triangle.
11. Two sides of a triangle are equal in length. The length of the third side exceeds the length of one of the other sides by 3 centimeters. The perimeter of the triangle is 93 centimeters. Find the length of each of the shorter sides of the triangle.
12. The length of a rectangle is 5 meters more than its width. The perimeter is 66 meters. Find the dimensions of the rectangle.
13. The width of a rectangle is 3 yards less than its length. The perimeter is 130 yards. Find the length and the width of the rectangle.
14. The perimeter of a rectangular parking lot is 146 meters. Find the dimensions of the lot if the length is 7 meters less than 4 times the width.
15. The perimeter of a rectangular tennis court is 228 feet. If the length of the court exceeds twice its width by 6 feet, find the dimensions of the court.
16. The length of the base of an isosceles triangle is 10 meters less than twice the length of one of its legs. If the perimeter of the triangle is 50 meters, find the length of the base of the triangle.
17. The base of an isosceles triangle and one of its legs have lengths that are consecutive integers. The leg is longer than the base. The perimeter of the triangle is 20. Find the length of each side of the triangle.
18. The length of a rectangle is twice the width. If the length is increased by 4 inches and the width is decreased by 1 inch, a new rectangle is formed whose perimeter is 198 inches. Find the dimensions of the original rectangle.
19. The length of a rectangle exceeds its width by 4 feet. If the width is doubled and the length is diminished by 2 feet, a new rectangle is formed whose perimeter is 8 feet more than the perimeter of the original rectangle. Find the dimensions of the original rectangle.
20. A side of a square is 10 meters longer than the side of an equilateral triangle. The perimeter of the square is 3 times the perimeter of the triangle. Find the length of each side of the triangle.
21. The length of each side of a hexagon is 4 inches less than the length of a side of a square. The perimeter of the hexagon is equal to the perimeter of the square. Find the length of a side of the hexagon and the length of a side of the square.

9-6 SOLVING FOR A VARIABLE IN TERMS OF ANOTHER VARIABLE

An equation may contain more than one variable. For example, the equation $ax + b = 3b$ contains the variables a, b, and x. To solve this equation for x means to express x in terms of the other variables.

To plan the steps in the solution, it is helpful to use the strategy of using a simpler related problem, that is, to compare the solution of this equation with the solution of a simpler equation that has only one variable. In example **a** below, the solution of $ax + b = 3b$ is comparable with the solution of $2x + 5 = 15$. The same operations are used in the solution of both equations.

EXAMPLE

Solve for x:

a. $ax + b = 3b$

Compare with $2x + 5 = 15$.

$$\begin{aligned} 2x + 5 &= 15 \\ -5 &= -5 \\ \hline 2x &= 10 \\ \frac{2x}{2} &= \frac{10}{2} \\ x &= 5 \end{aligned}$$

Solution:

$$\begin{aligned} ax + b &= 3b \\ -b &= -b \\ \hline ax &= 2b \\ \frac{ax}{a} &= \frac{2b}{a} \\ x &= \frac{2b}{a} \end{aligned}$$

Check:

$$\begin{aligned} ax + b &= 3b \\ a\left(\frac{2b}{a}\right) + b &= 3b \\ 2b + b &= 3b \\ 3b &= 3b \ ✔ \end{aligned}$$

Answer: $x = \frac{2b}{a}$

b. $x + a = b$

Compare with $x + 5 = 9$.

$$\begin{aligned} x + 5 &= 9 \\ -5 &= -5 \\ \hline x &= 4 \end{aligned}$$

Solution:

$$\begin{aligned} x + a &= b \\ -a &= -a \\ \hline x &= b - a \end{aligned}$$

Check:

$$\begin{aligned} x + a &= b \\ b - a + a &\stackrel{?}{=} b \\ b &= b \ ✔ \end{aligned}$$

Answer: $x = b - a$

c. $2ax = 10a^2 - 3ax \qquad (a \neq 0)$

Compare with $2x = 10 - 3x$.

$$\begin{aligned} 2x &= 10 - 3x \\ +3x &= \quad + 3x \\ \hline 5x &= 10 \\ \frac{5x}{5} &= \frac{10}{5} \\ x &= 2 \end{aligned}$$

Solution:

$$\begin{aligned} 2ax &= 10a^2 - 3ax \\ +3ax &= \quad + 3ax \\ \hline 5ax &= 10a^2 \\ \frac{5ax}{5a} &= \frac{10a^2}{5a} \\ x &= 2a \end{aligned}$$

The check is left for you.

Answer: $x = 2a$

EXERCISES

In 1–32, solve each equation for x or y and check.

1. $5x = b$
2. $sx = 8$
3. $ry = s$
4. $3y = t$
5. $cy = 5$
6. $hy = m$
7. $x + 5 = r$
8. $x + a = 7$
9. $y + c = d$
10. $4 + x = k$
11. $d + y = 9$
12. $3x - q = p$
13. $x - 2 = r$
14. $y - a = 7$
15. $x - c = d$
16. $3x - e = r$
17. $cy - d = 4$
18. $ax + b = c$
19. $rx - s = 0$
20. $r + sy = t$
21. $m = 2(x + n)$
22. $4x - 5c = 3c$
23. $bx = 9b^2$
24. $cx + c^2 = 5c^2 - 3cx$
25. $bx - 5 = c$
26. $a = by + 6$
27. $ry + s = t$
28. $abx - d = 5d$
29. $rsx - rs^2 = 0$
30. $m^2x - 3m^2 = 12m^2$
31. $9x - 24a = 6a + 4x$
32. $8ax - 7a^2 = 19a^2 - 5ax$

9-7 TRANSFORMING FORMULAS

A ***formula*** is an equation that contains more than one variable. Sometimes you want to solve for a variable in the formula that is different from the subject of the formula. For example, the formula for distance D in terms of rate r and time t is $D = rt$. Distance is the subject of the formula, but you might want to rewrite the formula so that it expresses time in terms of distance and rate. You do this solving the equation $D = rt$ for t in terms of D and r.

EXAMPLES

1. a. Solve the formula $D = rt$ for t.

b. Use the answer obtained in part **a** to find the value of t when $D = 200$ miles and $r = 40$ miles per hour.

Solutions **a.**

$$D = rt$$
$$\frac{D}{r} = \frac{rt}{r}$$
$$= t$$

Answer: $t = \frac{D}{r}$

b.

$$t = \frac{D}{r}$$
$$= \frac{200}{40}$$
$$= 5$$

Answer: $t = 5$ hours

Note that the rate is 40 miles per hour, that is, $\frac{40 \text{ miles}}{1 \text{ hour}}$.

Therefore,

$$200 \text{ miles} \div \frac{40 \text{ miles}}{1 \text{ hour}} = 200 \cancel{\text{miles}} \times \frac{1 \text{ hour}}{40 \cancel{\text{miles}}} = 5 \text{ hours}$$

We can think of canceling miles in the numerator and the denominator of the fractions being multiplied.

---

2. Solve the formula
$V = \frac{1}{3}Bh$ for B.

Solution:

$$V = \frac{1}{3}Bh$$

$$3V = 3\left(\frac{1}{3}Bh\right)$$

$$= Bh$$

$$\frac{3V}{h} = \frac{Bh}{h}$$

$$\frac{3V}{h} = B$$

Answer: $B = \frac{3V}{h}$

3. Solve the formula
$P = 2(\ell + w)$ for w.

Solution:

$$P = 2(\ell + w)$$

$$= 2\ell + 2w$$

$$P + (-2\ell) = 2\ell + 2w + (-2\ell)$$

$$P - 2\ell = 2w$$

$$\frac{P - 2\ell}{2} = \frac{2w}{2}$$

$$\frac{P - 2\ell}{2} = w$$

Answer: $w = \frac{P - 2\ell}{2}$

EXERCISES WITH OPEN-RESPONSE PROBLEMS

In 1–19, transform each given formula by solving for the indicated letter.

1. $A = 6\,h$ for h
2. $36 = bh$ for h
3. $P = 4s$ for s
4. $D = rt$ for r
5. $V = \ell wh$ for ℓ
6. $p = br$ for r
7. $A = bh$ for b
8. $A = \ell w$ for ℓ
9. $V = \ell wh$ for h
10. $V = 4bh$ for h
11. $I = prt$ for p
12. $400 = bh$ for b
13. $A = \frac{1}{2}bh$ for h
14. $V = \frac{1}{3}bh$ for h
15. $S = \frac{1}{2}gt^2$ for g
16. $\ell = c - s$ for c
17. $P = 2\ell + 2w$ for ℓ
18. $F = \frac{9}{5}C + 32$ for C
19. $2S = n(a + \ell)$ for a

20. If $A = bh$, express h in terms of A and b. (Solve for h.)

21. If $P = 2a + b$, express b in terms of P and a.

22. If $P = 2a + b + c$, express a in terms of the other variables.

23. The concession stand at a movie theater wants to sell popcorn in containers that are in the shape of a cylinder. The volume of the cylinder is given by the formula $V = \pi r^2 h$, where V is the volume, r is the radius of the base, and h is the height of the container.

a. Solve the formula for h.

b. If the container is to hold 1,400 cubic centimeters of popcorn, find, to the *nearest tenth*, the height of the container if the radius of the base is:

(*1*) 4 centimeters (*2*) 5 centimeters (*3*) 8 centimeters

c. Which do you think would be the best height for the container? Why?

24. A bus travels from Buffalo to Albany, stopping at Rochester and Syracuse. At each city there is a 30-minute stopover to unload and load passengers and baggage. The driving distance from Buffalo to Rochester is 75 miles, from Rochester to Syracuse is 85 miles, and from Syracuse to Albany is 145 miles. The bus travels at an average speed of 50 miles per hour.

a. Solve the formula $D = rt$ for t to find the time needed for each part of the trip.

b. Make a schedule for the times of arrival and departure for each city if the bus leaves Buffalo at 9:00 A.M.

25. The graph at the right shows the relationship between Celsius and Fahrenheit temperature readings. The horizontal axis is the Celsius axis, and the vertical axis is the Fahrenheit axis. Each point on the line represents an ordered pair in which the first element is the Celsius temperature and the second element is the corresponding Fahrenheit temperature.

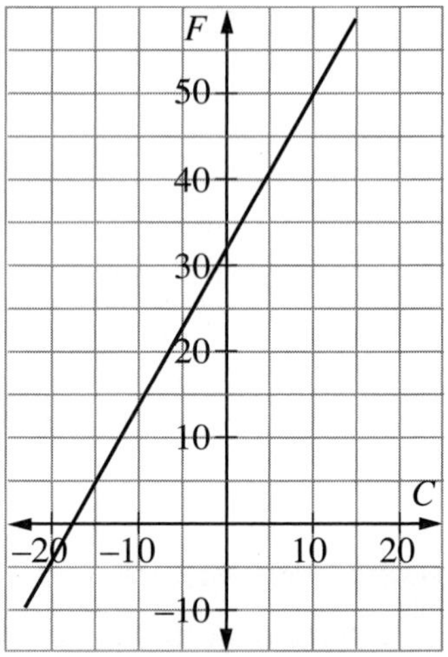

a. Use the graph to estimate:

(1) the Fahrenheit temperature when $C = 10$

(2) the Fahrenheit temperature when $C = 0$

(3) the Fahrenheit temperature when $C = -15$

(4) the Celsius temperature when $F = 14$

(5) the Celsius temperature when $F = 0$

b. Use the formula $F = \frac{9}{5}C + 32$ to check your estimates by finding exact conversions when:

(*1*) $C = 10$ (*2*) $C = 0$ (*3*) $C = -15$ (*4*) $F = 14$ (*5*) $F = 0$

9-8 MOTION PROBLEMS

Preparing to Solve Motion Problems

If a car traveled at the rate of 50 miles per hour, in 2 hours it traveled 2(50), or 100 miles. In this case, the three quantities that are related are:

1. the distance traveled, 100 miles
2. the rate of speed, or rate, 50 miles per hour (mph)
3. the time traveled, 2 hours

The relation involving the distance, D, the rate, R, and the time, T, may be expressed in the following ways:

$$D = RT \qquad T = \frac{D}{R} \qquad R = \frac{D}{T}$$

In our work, rate will represent either of the following:

1. The **uniform rate of speed,** which represents a rate of speed which remains constant (does not change) throughout the trip. Thus, if a car is traveling at a uniform rate of 50 miles per hour on a 2-hour trip, the car is constantly traveling 50 miles per hour during the entire trip.
2. The **average rate of speed,** which represents the total distance traveled divided by the total time traveled. Thus, a car which traveled 165 miles in 3 hours was traveling at an average rate of $165 \div 3$, or 55 miles per hour.

Solving Motion Problems

In solving motion problems, it is helpful to draw a diagram. The facts in the problem may be organized by using verbal sentences or by using a table.

EXAMPLES

1. Two cars start from the same point at the same time and travel in opposite directions. The slow car travels at 56 miles per hour and the fast car travels at 60 miles per hour. In how many hours will the cars be 464 miles apart?

Solution Since both cars start at the same time, they will travel the same number of hours.

Let h = the number of hours traveled by each car.
Then $56h$ = the distance traveled by the slow car.
And $60h$ = the distance traveled by the fast car.

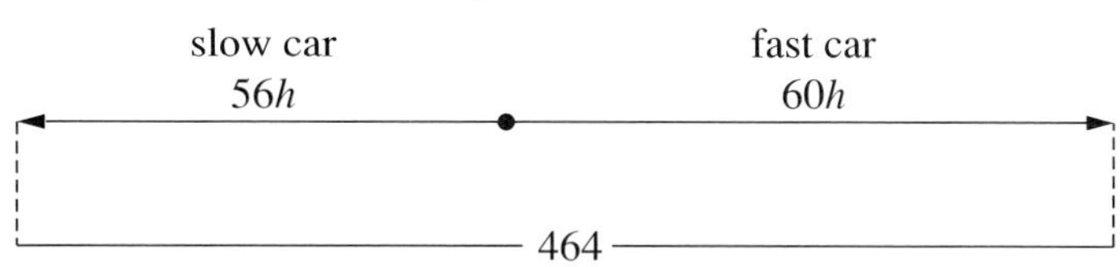

The total distance traveled is 464 miles.

$$\begin{aligned} 56h + 60h &= 464 \\ 116h &= 464 \\ h &= 4 \end{aligned}$$

Check: $56(4) + 60(4) = 224 + 280 = 464$
Answer: 4 hours

Alternative Solution Arrange the information in a table.
Let h = the number of hours traveled by each car.

First fill in the rate and time for each car. Then represent the distance for each car.

	Rate (mph)	Time (hr)	Distance (mi)
Slow car	56	h	$56h$
Fast car	60	h	$60h$

The total distance traveled is 464 miles.

$$56h + 60h = 464$$

Complete the solution as before.

---

2. A passenger train and a freight train start at the same time from stations that are 405 miles apart and travel toward each other. The rate of the passenger train is twice the rate of the freight train. In 3 hours, the trains pass each other. Find the rate of each train.

Solution

Let r = the rate of the freight train.
Then, $2r$ = the rate of the passenger train.
And, $3r$ = the distance traveled by the freight train.
And, $3(2r)$ = the distance traveled by the passenger train.

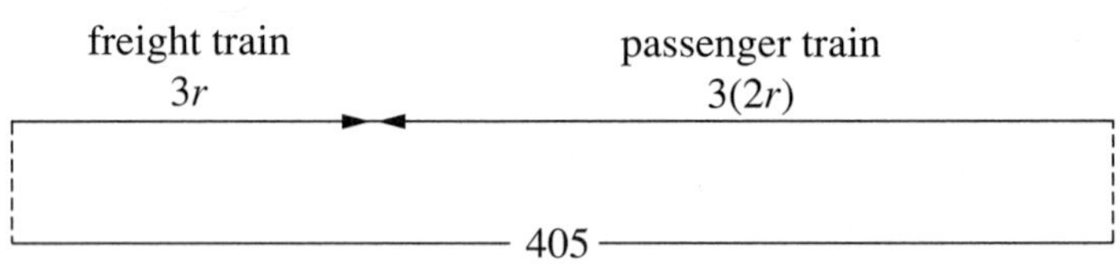

The total distance traveled was 405 miles.

$$\begin{aligned} 3r + 3(2r) &= 405 \\ 3r + 6r &= 405 \\ 9r &= 405 \\ r &= 45 \\ 2r &= 90 \end{aligned}$$

Check: $3(45) + 3(90) = 135 + 270 = 405$

Answer: The rate of the freight train was 45 miles per hour. The rate of the passenger train was 90 miles per hour.

The following table may also be used

Let r = the rate of the freight train.
Then, $2r$ = the rate of the passenger train.

First fill in the time and rate for each train. Then represent the distance for each train.

	Rate (mph)	Time (hr)	Distance (mi)
Freight train	r	3	$3r$
Passenger train	$2r$	3	$3(2r)$

The total distance traveled was 405 miles.

$$3r + 3(2r) = 405$$

Complete the solution as before.

---

3. Two trains started at the same time from stations which were 360 miles apart and traveled toward each other. The rate of the fast train exceeded the rate of the slow train by 10 miles per hour. At the end of 2 hours, the trains were still 120 miles apart. Find the rate of each train.

Solution Let r = the rate of the slow train.
Then, $r + 10$ = the rate of the fast train.

First fill in the time and rate for each train. Then represent the distance for each train.

	Rate (mph)	Time (hr)	Distance (mi)
Slow train	r	2	$2r$
Fast train	$r + 10$	2	$2(r + 10)$

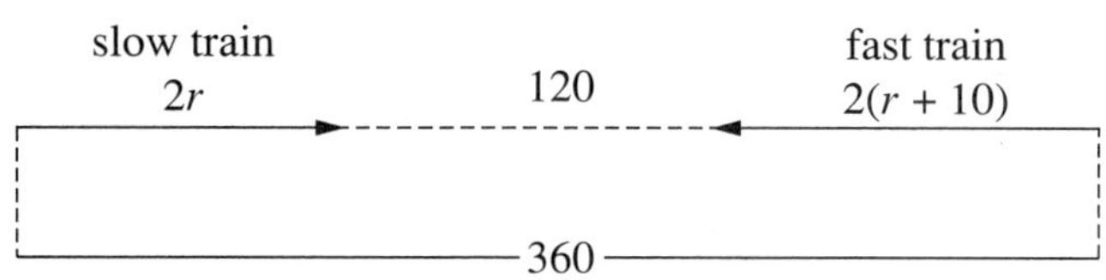

The total distance between the stations was 360 miles.

$$2r + 120 + 2(r + 10) = 360$$
$$2r + 120 + 2r + 20 = 360$$
$$4r + 140 = 360$$
$$4r + 140 - 140 = 360 - 140$$
$$4r = 220$$
$$r = 55$$
$$r + 10 = 65$$

Check:

$$2(55) = 110$$
$$2(65) = \underline{130}$$
$$\text{Total} = 240$$
$$360 - 240 = 120$$

Answer: The rate of the slow train was 55 miles per hour. The rate of the fast train was 65 miles per hour.

---

4. How far can a man drive out into the country at the average rate of 60 miles per hour and return over the same road at the average rate of 45 miles per hour if he travels a total of 7 hours?

Solution Let h = the number of hours he spent driving out.
Then, $7 - h$ = the number of hours he spent driving back.

First fill in the rate and time for each trip. Then represent the distance for each trip.

	Rate (mph)	Time (hr)	Distance (mi)
Trip out	60	h	$60h$
Trip back	45	$7 - h$	$45(7 - h)$

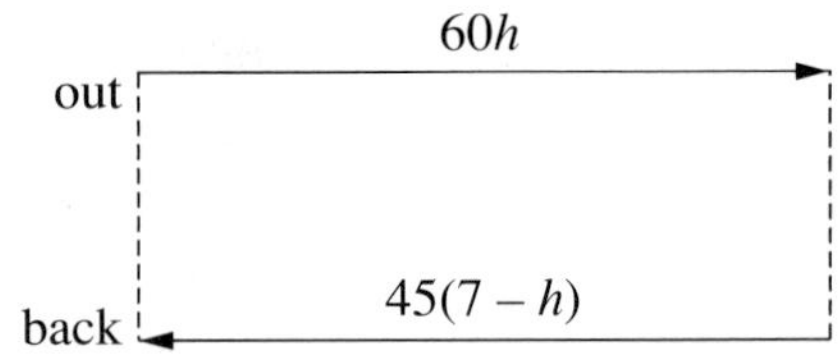

The distance out is the same as the distance back.

$$60h = 45(7 - h)$$
$$60h = 315 - 45h$$
$$60h + 45h = 315 - 45h + 45h$$
$$105h = 315$$
$$h = 3$$
$$60h = 180$$

Check

$7 - 3 = 4$

$3(60) = 180$

$4(45) = 180$

The distances are the same.

Answer: He can travel 180 miles out into the country.

Alternative Solution Since he drives out to a certain point and returns over the same road to his starting point, the distance traveled going out is the same as the distance traveled coming back.

Let d = the number of miles he can travel out into the country.
First fill in the rate and distance for each trip. Then represent the time for each trip.

	Rate (mph)	Time (hr)	Distance (mi)
Trip out	60	$\frac{d}{60}$	d
Trip back	45	$\frac{d}{45}$	d

Total time spent in traveling is 7 hours.

$$\frac{d}{60} + \frac{d}{45} = 7$$

$$180\left(\frac{d}{60} + \frac{d}{45}\right) = 180(7)$$

$$180\left(\frac{d}{60}\right) + 180\left(\frac{d}{45}\right) = 180(7)$$

$$3d + 4d = 1260$$

$$7d = 1260$$

$$d = 180$$

Check

$$\frac{180}{60} = 3$$

$$\frac{180}{45} = 4$$

$$3 + 4 = 7$$

Answer: He can travel 180 miles out into the country.

EXERCISES WITH OPEN-RESPONSE PROBLEMS

1. Find the average rate for the entire trip if a car travels:
 a. 1 hour at 40 mph and 1 hour at 50 mph
 b. 1 hour at 30 mph, 1 hour at 40 mph, and 1 hour at 50 mph
 c. 1 hour at 30 mph and 2 hours at 36 mph
 d. 2 hours at 40 mph and 3 hours at 50 mph

2. Two cars started from the same point and traveled for x hours in opposite directions at rates of 30 mph and 40 mph.
 a. Represent in terms of x the distance traveled by the slow car.
 b. Represent in terms of x the distance traveled by the fast car.
 c. Represent how far apart the two cars were at the end of x hours.
 d. Write an open sentence which would indicate that the two cars were 140 miles apart at the end of x hours.

3. Saratoga and New York are 180 miles apart. A truck traveled from New York toward Saratoga at the rate of 65 miles per hour. Another truck traveled from Saratoga toward New York at the rate of 55 miles per hour. How many miles did each travel before they met?

4. Two planes started at the same time from the same airport and flew in opposite directions. One flew 60 miles per hour faster than the other. In 5 hours, they were 5,300 miles apart. Find the rate of each plane.

5. Two trains started from the same station at the same time and traveled in opposite directions. After traveling 10 hours, they were 1,400 miles apart. The rate of the fast train exceeded the rate of the slow train by 5 miles per hour. Find the rate of each train.

6. Two trains started from the same place at the same time and traveled in opposite directions at rates that differed by 20 miles per hour. In 5 hours, they were 650 miles apart. Find the rate of each train.

7. Two planes left at the same time from two airports which are 4,500 miles apart and flew toward each other. In 5 hours, they passed each other. The rate of the fast plane was twice the rate of the slow plane. Find the rate of each plane.

8. Roland made a trip of 450 miles in 8 hours. Before noon he averaged 60 miles per hour, and after noon he averaged 50 miles per hour. At what time did he begin his trip and when did he end it?
9. Two planes started at the same time from two airports that are 3,600 kilometers apart and flew toward each other. One plane flew at 792 kilometers per hour, and the other flew at 888 kilometers per hour. In how many hours were the planes still 240 kilometers apart?
10. At 3 P.M. two ships started sailing toward each other from ports which were 265 miles apart at average rates of 18 and 23 miles per hour. At what time were the ships still 60 miles apart?
11. At 9:00 A.M. two cars started from the same town and traveled north on the same road. One car averaged 45 miles per hour, and the other car averaged 40 miles per hour. In how many hours were the cars 30 miles apart?
12. At 6:00 A.M. two planes started from the same airport and flew west. One plane averaged 520 miles per hour, and the other plane averaged 600 miles per hour. At what time were the planes 280 miles apart?
13. A ship left a port and sailed east at the rate of 20 miles per hour. One hour later, a second ship left the same port at the rate of 25 miles per hour, also traveling east. In how many hours did the second ship overtake the first ship?
14. Susan left her home at 7 A.M., driving her car at the rate of 45 miles per hour. At 9 A.M. her sister Marion drove after her along the same highway, traveling at the rate of 60 miles per hour. In how many hours did Marion pass Susan?
15. A plane left LaGuardia Airport at 3 P.M., flying west at the rate of 450 miles per hour. At 3:30 P.M. another plane left the same airport, flying west at the rate of 500 miles per hour. At what time did the second plane overtake the first plane?
16. Mr. Stone started from home on a trip, planning to average 48 miles per hour. How fast must his son Carl plan to travel in order to overtake him in 3 hours if Carl started 30 minutes after his father?
17. A cargo plane left an airport at noon and flew toward New York at the average rate of 400 miles per hour. At 2 P.M. a passenger plane left the same airport for New York and flew the same route as the cargo plane at the average rate of 560 miles per hour. How many miles did the passenger plane fly before it overtook the cargo plane?
18. Mr. Fields spent 6 hours hiking along a trail and back. He walked out at the rate of 4 miles per hour and walked back at the rate of 2 miles per hour. How far along the trail did he go?
19. Ronald rode away from home in a car at the rate of 32 miles per hour. He walked back at the rate of 4 miles per hour. The round trip required $2\frac{1}{4}$ hours. How far did Ronald ride?
20. A round trip in a helicopter lasted 4 hours 30 minutes. If the helicopter flew away from the airport at 100 miles per hour and returned at the rate of 50 miles per hour, what was its greatest distance from the airport?

In 21 and 22, select the numeral preceding the expression that best answers the question.

21. Darryl can travel to the store on his skateboard three times faster than he can return carrying his package and skateboard. If the total trip takes 30 minutes, how many minutes does it take him to return from the store?
(1) 7.5 (2) 15 (3) 22.5 (4) 45
22. Ms. Westworth drove her car from her home to Chicago on local streets. She returned from Chicago on the highway at a rate 1.5 times as fast. If her time going exceeded her time returning by 30 minutes, how many hours did it take her to drive to Chicago?
(1) 1 (2) 1.5 (3) 60 (4) 90

9-9 COIN PROBLEMS

Preparing to Solve Coin Problems

In solving problems which deal with coins of different denominations—for example, nickels, dimes, and quarters—it is often helpful to represent the values of the coins in the same unit of money. In the following examples, the unit of money is cents:

The value of 3 nickels in cents is 3(5), or 15 cents.

The value of d dimes in cents is $d(10)$, or $10d$ cents.

The value of q quarters in cents is $q(25)$, or $25q$ cents.

Note: Number of coins × Value of each coin in cents = Total value of the coins in cents

Solving Coin Problems

To solve coin problems, organize the facts in the problem using verbal sentences or a table.

EXAMPLES

1. A boy has \$3.20 in his bank made up of nickels, dimes, and quarters. There are 3 times as many quarters as nickels, and 5 more dimes than nickels. How many coins of each kind are there?

Solution

Let n = the number of nickels.
Then, $3n$ = number of quarters.
And, $n + 5$ = the number of dimes.
Then, $5n$ = the value of the nickels in cents.
And, $25(3n)$ = the value of the quarters in cents.
And, $10(n + 5)$ = the value of the dimes in cents.

The total value of all the coins is 320 cents.

$$
\begin{aligned}
5n + 25(3n) + 10(n + 5) &= 320 \\
5n + 75n + 10n + 50 &= 320 \\
90n + 50 &= 320 \\
90n + 50 + (-50) &= 320 + (-50) \\
90n &= 270 \\
n &= 3 \\
3n &= 9 \\
n + 5 &= 8
\end{aligned}
$$

Check:

9 is 3 times 3

8 is 5 more than 3

Value of 3 nickels	=	\$0.15
Value of 9 quarters	=	2.25
Value of 8 dimes	=	0.80
Total value	=	\$3.20

Answer: There are 3 nickels, 9 quarters, and 8 dimes.

The information may also be arranged in a table.
Let n = the number of nickels.

Kind of coin	Number of coins	Value of each coin in cents	Total value in cents
Nickel	n	5	$5n$
Quarter	$3n$	25	$25(3n)$
Dime	$n + 5$	10	$10(n + 5)$

The total value of all the coins is 320 cents.

$$5n + 25(3n) + 10(n + 5) = 320$$

Complete the solution as before.

---

2. A purse contains \$1.35 in nickels and dimes. In all there are 15 coins. How many coins of each kind are there?

Solution Let d = the number of dimes.
Then $15 - d$ = the number of nickels.

Kind of coin	Number of coins	Value of each coin in cents	Total value in cents
Dime	d	10	$10d$
Nickel	$15 - d$	5	$5(15 - d)$

The total value of all the coins is 135 cents.

$$\begin{aligned} 10d + 5(15 - d) &= 135 \\ 10d + 75 - 5d &= 135 \\ 5d + 75 &= 135 \\ 5d + 75 + (-75) &= 135 + (-75) \\ 5d &= 60 \\ d &= 12 \\ 15 - d &= 3 \end{aligned}$$

Check:
$12 + 3 = 15$
Value of 12 dimes = \$1.20
Value of 3 nickels = .15
Total value = \$1.35

Answer: There are 12 dimes and 3 nickels.

EXERCISES WITH OPEN-RESPONSE PROBLEMS

1. Bill has 4 times as many quarters as dimes. In all he has $2.20. How many coins of each type does he have?
2. Patria has 3 times as many dimes as nickels. In all she has $1.40. How many coins of each type does she have?
3. Paul has twice as many dimes as pennies and 3 times as many nickels as pennies. In all he has $1.80. How many coins of each type does he have?
4. Sally has three times as many dimes as nickels and twice as many quarters as dimes. In all she has $5.55. How many coins of each type does she have?
5. Howard has $2.05 in quarters and dimes. He has 4 more quarters than dimes. Find the number he has of each coin.
6. Helen has $1.35 in her bank in nickels and dimes. There are 9 more nickels than dimes. Find the number she has of each kind.
7. Viki has $2.80 in quarters and dimes. The number of dimes is 7 less than the number of quarters. Find the number she has of each kind.
8. Dylan has $2.30 in dimes and nickels. The number of dimes exceeds the number of nickels by 5. Find the number he has of each kind.
9. Maria has $5.05 in quarters and dimes. The number of quarters exceeds twice the number of dimes by 1. Find the number she has of each kind.
10. Mr. Boyce deposited $170 in his bank. The number of $5 bills was 3 times the number of $10 bills, and the number of $1 bills was 30 more than the number of $5 bills. How many bills of each type did he deposit?
11. Harriet deposited $4.50 in nickels, quarters, and dimes in her coin bank. The number of dimes exceeded the number of nickels by 5, and the number of quarters was 16 less than the number of nickels. Find the number of each kind of coin she deposited.
12. Mildred bought 1-cent stamps, 33-cent stamps, and 34-cent stamps for $22.45. The number of 1-cent stamps exceeded the number of 34-cent stamps by 50. The number of 33-cent stamps was 10 less than twice the number of 34-cent stamps. How many of each kind did she buy?
13. A class contributed $19.00 in quarters and dimes to a local homeless shelter. In all there were 100 coins. How many were there of each kind.
14. A purse containing $3.20 in quarters and dimes has, in all, 20 coins. Find the number of each kind of coin.
15. A purse contains $4.70 in nickels and quarters. There are 30 coins in all. How many of each kind are there?
16. Mr. Perkins cashed a $185 check in his bank. He received $1 bills, $5 bills, and $10 bills. In this order, the numbers of the three types of bills he received were three consecutive integers. How many bills of each type did he receive?
17. Saul paid a bill of $4.60 with nickels, dimes, and quarters. The number of nickels was 3 less than the number of dimes. The number of dimes was 5 more than the number of quarters. How many coins of each type did he use?
18. Roger has $3.10 consisting of quarters, dimes, and nickels. He has twice as many quarters as dimes and 3 more dimes than nickels. Find the number of each kind of coin.
19. In Paul's bank, there is $2.60 in pennies, nickels, and dimes. In all there are 45 coins. If there are twice as many nickels as pennies, find how many of each kind there are.

20. Selma paid a bill of \$2.70 with quarters, dimes, and pennies. She had 5 less quarters than dimes and 4 less quarters than pennies. How many coins of each kind did she use to pay the bill?

21. Sue counted her money and found that her 25 coins which were nickels, dimes, and quarters were worth \$3.20. The number of dimes exceeded the number of nickels by 4. How many coins of each kind did she have?

22. Is it possible to have \$4.50 in dimes and quarters and have twice as many quarters as dimes?

23. Is it possible to spend \$30.00 for 100 stamps consisting of 20-cent stamps and 33-cent stamps?

9-10 AGE PROBLEMS

Preparing to Solve Age Problems

If Ken is 20 years old now, 5 years ago Ken was $20 - 5$, or 15 years old; 5 years from now Ken will be $20 + 5$, or 25 years old.

> *Note:*
> 1. To represent a past age, subtract from the present age.
> 2. To represent a future age, add to the present age.

Solving Age Problems

EXAMPLES

1. Bill is 3 times as old as Pietro. Eighteen years from now, Bill will be twice as old as Pietro will be then. Find the present age of both Bill and Pietro.

Solution

Let x = Pietro's present age.
Then, $3x$ = Bill's present age.
Then, $x + 18$ = Pietro's age 18 years from now.
Then, $3x + 18$ = Bill's age 18 years from now.

Eighteen years from now, Bill will be twice as old as Pietro will be then.

$$\begin{aligned} 3x + 18 &= 2(x + 18) \\ 3x + 18 &= 2x + 36 \\ 3x + 18 + (-18) &= 2x + 36 + (-18) \\ 3x &= 2x + 18 \\ 3x + (-2x) &= 2x + 18 + (-2x) \\ x &= 18 \\ 3x &= 54 \end{aligned}$$

Check:

$$\begin{aligned} 54 &= 3(18) \\ 18 + 18 &= 36 \\ 54 + 18 &= 72 \\ 72 &= 2(36) \end{aligned}$$

Answer: Pietro's age is 18 years. Bill's age is 54 years.

The information may also be arranged in a table.

Person	**Present age in years**	**Future age, 18 years from now**
Pietro	x	$x + 18$
Bill	$3x$	$3x + 18$

Eighteen years from now, Bill will be twice as old as Pietro will be then.

$$3x + 18 = 2(x + 18)$$

Complete the solution as before.

---

2. Helen is now 40 years old and Arlene is 20 years old. How many years ago was Helen three times as old as Arlene was then?

Solution

Let x = the required number of years ago.
Then $40 - x$ = Helen's age x years ago.
And $20 - x$ = Arlene's age x years ago.

Helen was three times as old as Arlene was then.

$$40 - x = 3(20 - x)$$
$$40 - x = 60 - 3x$$
$$40 - x + (+3x) = 60 - 3x + (+3x)$$
$$40 + 2x = 60$$
$$40 + 2x + (-40) = 60 + (-40)$$
$$2x = 20$$
$$x = 10$$

Check:

$40 - 10 = 30$
$20 - 10 = 10$
$30 = 3(10)$

Answer: 10 years ago

The information may also be arranged in a table.

Let x = the required number of years ago.

Person	**Present age in years**	**Past age, *x* years ago**
Helen	40	$40 - x$
Arlene	20	$20 - x$

Helen was three times as old as Arlene was then.

$$40 - x = 3(20 - x)$$

Complete the solution as before.

EXERCISES

1. John is now 3 times as old as his brother Sam. In 5 years, John will be twice as old as Sam will be then. Find their present ages.
2. A man is now 6 times as old as his son. In 6 years, the father will be 3 times as old as the son will be then. Find their present ages.
3. Justin is now 3 times as old as Arthur. In 10 years, Justin will be twice as old as Arthur will be then. Find their present ages.
4. Marie is one-ninth as old as her mother. In 3 years, she will be one-fifth as old as her mother will be then. Find their present ages.
5. A father is now 24 years older than his son. In 8 years, the father will be twice as old as his son will be then. Find their present ages.
6. Mr. Wilson is 15 years older than Mr. Connors. Five years from now, Mr. Wilson will be $1\frac{1}{2}$ times as old as Mr. Connors will be then. How old is each man now?
7. Marion is twice as old as Judy. Six years ago, Marion was three times as old as Judy was then. Find the age of each girl now.
8. A father is 3 times as old as his son. Fifteen years ago, the father was 9 times as old as his son was then. Find their present ages.
9. Selma is now 3 times as old as Joyce. Four years ago, Selma was 4 times as old as Joyce was then. Find their present ages.
10. Robert is one-half as old as his father. Twelve years ago, Robert was one-third as old as his father was then. Find their present ages.
11. Phil is 24 years older than Stanley. Four years ago, he was 7 times as old as Stanley was then. Find their present ages.
12. Mrs. Barry is 20 years older than Mrs. Cook. Sixteen years ago, Mrs. Barry was 3 times as old as Mrs. Cook was then. Find their present ages.
13. Josephine is 22 years old and Ruth is 10 years old. In how many years will Josephine be twice as old as Ruth will be then?
14. A man is 40 years old and his son is 8 years old. In how many years will the man be 3 times as old as his son will be then?
15. Gene is 23 years old and Stanley is 15 years old. How many years ago was Gene 3 times as old as Stanley was then?
16. Mr. Atkins is 33 years old and Mr. Speyer is 27 years old. How many years ago was Mr. Atkins $1\frac{1}{2}$ times as old as Mr. Speyer was then?
17. The sum of a man's age and his daughter's age is 50 years. Eight years from now, the man will be twice as old as his daughter will be then. Find the present age of each.
18. Mrs. Sanford is three times as old as Mrs. Fox. Eight years from now, Mrs. Sanford's age will exceed twice Mrs. Fox's age at that time by 14 years. Find the present age of each.
19. Hal is 10 years older than Gerald. In 6 years, Hal will be $1\frac{1}{2}$ times as old as Gerald will be then. Find their present ages.
20. Mrs. Watson was 25 years old when her daughter Rose was born. Now Mrs. Watson's age exceeds 4 times Rose's age by 10 years. Find their ages now.
21. The sum of Wilbur's age and Fred's age is 20 years. Wilbur's age 1 year from now will be 9 times Fred's age 1 year ago. Find the present age of each.

9-11 PROPERTIES OF INEQUALITIES

The Order Property of the Real Numbers

If two real numbers are graphed on the number line, only one of the following three situations can be true:

x is to the left of y.

$x < y$

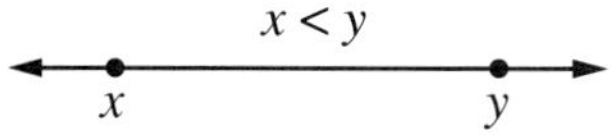

x and y are at the same point.

$x = y$

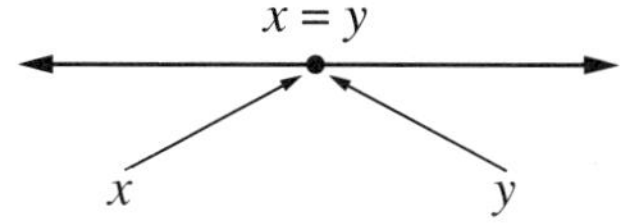

x is to the right of y.

$x > y$

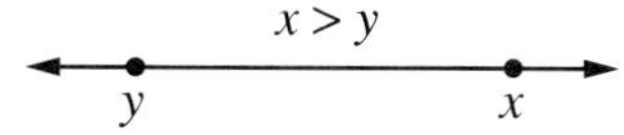

These three cases illustrate the ***order property of the real numbers***.

- If x and y are two real numbers, then one and only one of the following can be true:

$$x < y \qquad x = y \qquad x > y$$

Let y be a fixed point, for example, $y = 3$. Then y separates the real numbers into three sets. For any real number, one of the following must be true: $x < 3$, $x = 3$, $x > 3$.

The real numbers, x, that make $x < 3$ true are to the left of 3 on the number line. The circle at 3 indicates that 3 is the boundary value of the set. The circle is not filled in, indicating that 3 does not belong to the set.

$x < 3$

The real number that makes the corresponding equality, $x = 3$, true is a single point on the number line. This point is also the boundary between the values of x that make $x < 3$ true and the values of x that make $x > 3$ true. The circle is filled in, indicating that 3 belongs to the set. Here, 3 is the only element of the set.

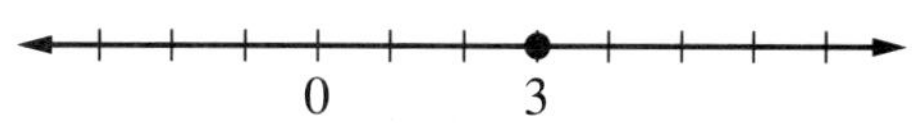

$x = 3$

The real numbers, x, that make $x > 3$ true are to the right of 3 on the number line. Again, the circle at 3 indicates that 3 is the boundary value of the set. The circle is not filled in, indicating that 3 does not belong to the set.

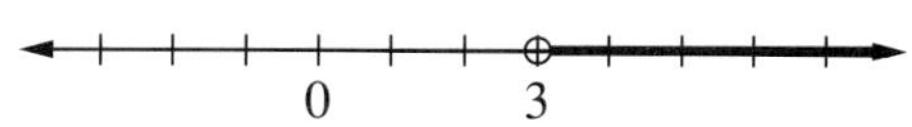

$x > 3$

The Transitive Property of Inequality

From the graph at the right, you can see that, if x lies to the left of y, and y lies to the left of z, then x lies to the left of z.

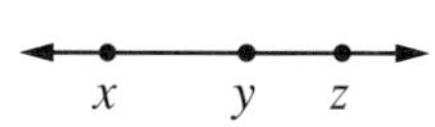

The graph illustrates the ***transitive property of inequality***:

- If x, y, and z are real numbers, then:

  If $x < y$ and $y < z$, then $x < z$;

  and

  if $z > y$ and $y > x$, then $z > x$.

The Addition Property of Inequality

The following table shows the result of adding a number to both members of an inequality.

True Sentence	Number to Add to Both Members	Result
$9 > 2$ Order is $>$.	$9 + 3 \ ? \ 2 + 3$ Add a positive number.	$12 > 5$ (true) Order is unchanged.
$2 < 9$ Order is $<$.	$2 - 3 \ ? \ 9 - 3$ Add a negative number.	$-1 < 6$ (true) Order is unchanged.

The table illustrates the **addition property of inequality**:

- If x, y, and z are real numbers, then:

  If $x < y$, then $x + z < y + z$;

  and

  if $x > y$, then $x + z > y + z$.

Since subtracting a number from both members of an inequality is equivalent to adding the additive inverse to both members of the inequality, the following is true:

- When the same number is added to or subtracted from both members of an inequality, the order of the new inequality is the same as the order of the original one.

EXAMPLE

Use the inequality $5 < 9$ to write a new inequality:

a. by adding 6 to both members
b. by adding -9 to both members

Solutions **a.** $5 + 6 < 9 + 6$
$11 < 15$

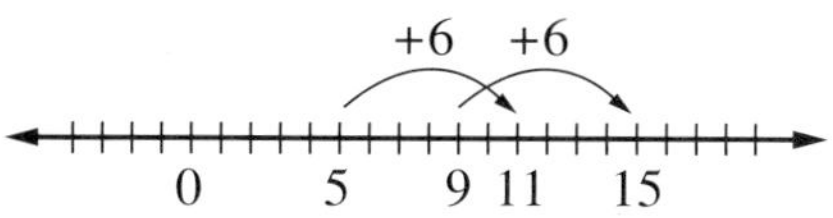

b. $5 + (-9) < 9 + (-9)$
$-4 < 0$

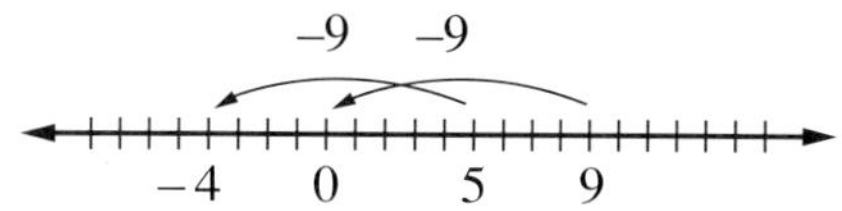

In **a**, since $5 < 9$, note that 5 is to the left of 9 on the number line. When 6 is added to each member, each sum moves 6 units to the right, and the sum of 5 and 6 is to the left of the sum of 9 and 6.

In **b**, when -9 is added to each member, each sum moves 9 units to the left, and the sum of 5 and -9 is to the left of the sum of 9 and -9.

Answers: **a.** $11 < 15$ **b.** $-4 < 0$

The Multiplication Property of Inequality

The following table shows the result of multiplying both members of an inequality by the same number.

True Sentence	Number by Which to Multiply Both Members	Result
$2 < 9$ Order is $<$.	2(3) ? 9(3) Multiply by a positive number.	$6 < 27$ (true) Order is unchanged.
$5 > 2$ Order is $>$.	5(3) ? 2(3) Multiply by a positive number.	$15 > 6$ (true) Order is unchanged.
$2 < 9$ Order is $<$.	$2(-3)$? $9(-3)$ Multiply by a negative number.	$-6 < -27$ (false) $-6 > -27$ (true) Order is changed.
$5 > 2$ Order is $>$.	$5(-3)$? $2(-3)$ Multiply by a negative number.	$-15 > -6$ (false) $-15 < -6$ (true) Order is changed.

The table illustrates that the order does not change when both members are multiplied by a positive number, but does change when both members are multiplied by a negative number.

In general terms, the **multiplication property of inequality** states:

- If x, y, and z are real numbers, then:

  If z is positive ($z > 0$) and $x < y$, then $xz < yz$.
  If z is positive ($z > 0$) and $x > y$, then $xz > yz$.

  If z is negative ($z < 0$) and $x < y$, then $xz > yz$.
  If z is negative ($z < 0$) and $x > y$, then $xz < yz$.

Since dividing both members of an inequality by a number is equivalent to multiplying both members by the multiplicative inverse of the number, the following is true:

- When both members of an inequality are multiplied or divided by the same positive number, the order of the new inequality is the same as the order of the original one. When both members of an inequality are multiplied or divided by the same negative number, the order of the new inequality is the opposite of the order of the original one.

EXAMPLE

Use the inequality $6 < 9$ to write a new inequality:

a. by multiplying both members by 2

b. by multiplying both members by $-\frac{1}{3}$

Solutions **a.** $6(2) \; ? \; 9(2)$
$12 < 18$

b. $6\left(-\frac{1}{3}\right) \; ? \; 9\left(-\frac{1}{3}\right)$
$-2 > -3$

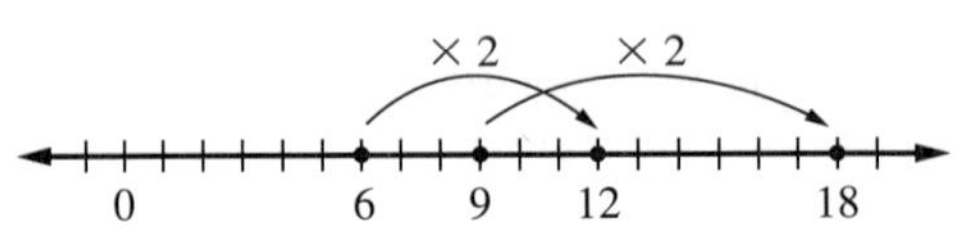

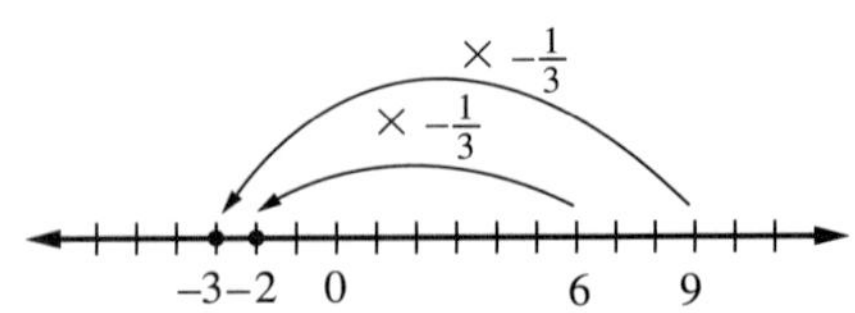

In **a**, since $6 < 9$, note that 6 is to the left of 9 on the number line. When each member is multiplied by 2, the product of 6 and 2 is still to the left of the product of 9 and 2.

In **b**, when each member is multiplied by $-\frac{1}{3}$, the products move to the opposite side of 0 on the number line, and the product of 6 and $-\frac{1}{3}$ is now to the right of the product of 9 and $-\frac{1}{3}$.

Answers: **a.** $12 < 18$ **b.** $-2 > -3$

EXERCISES

In 1–29, replace each question mark with the symbol > or < so that the resulting sentence will be true.

1. Since $8 > 2$, $8 + 1 \; ? \; 2 + 1$.
2. Since $-6 < 2$, $-6 + (-4) \; ? \; 2 + (-4)$.
3. Since $9 > 5$, $9 - 2 \; ? \; 5 - 2$.
4. Since $-2 > -8$, $-2 - \left(\frac{1}{4}\right) \; ? \; -8 - \left(\frac{1}{4}\right)$.
5. Since $7 > 3$, $\frac{2}{3}(7) \; ? \; \frac{2}{3}(3)$.
6. Since $-4 < 1$, $(-2)(-4) \; ? \; (-2)(1)$.
7. Since $-8 < 4$, $(-8) \div (4) \; ? \; (4) \div (4)$.
8. Since $9 > 6$, $(9) \div \left(-\frac{1}{3}\right) \; ? \; (6) \div \left(-\frac{1}{3}\right)$.
9. If $5 > x$, then $5 + 7 \; ? \; x + 7$.
10. If $y < 6$, then $y - 2 \; ? \; 6 - 2$.
11. If $20 > r$, then $4(20) \; ? \; 4(r)$.
12. If $t < 64$, then $t \div 8 \; ? \; 64 \div 8$.
13. If $x > 8$, then $-2x \; ? \; (-2)(8)$.
14. If $y < 8$, then $y \div (-4) \; ? \; 8 \div (-4)$.
15. If $x + 2 > 7$, then $x + 2 + (-2) \; ? \; 7 + (-2)$, or $x \; ? \; 5$.
16. If $y - 3 < 12$, then $y - 3 + 3 \; ? \; 12 + 3$, or $y \; ? \; 15$.
17. If $x + 5 < 14$, then $x + 5 - 5 \; ? \; 14 - 5$, or $x \; ? \; 9$.
18. If $2x > 8$, then $\frac{2x}{3} \; ? \; \frac{8}{2}$, or $x \; ? \; 4$.
19. If $\frac{1}{3}y < 4$, then $3\left(\frac{1}{3}y\right) \; ? \; (3(4)$, or $y \; ? \; 12$.
20. If $-3x < 36$, then $\frac{-3x}{-3} \; ? \; \frac{36}{-3}$, or $x \; ? \; -12$.
21. If $-2x > 6$, then $\left(-\frac{1}{2}\right)(-2x) \; ? \; \left(-\frac{1}{2}\right)(6)$, or $x \; ? \; -3$.
22. If $x < 5$ and $5 < y$, then $x \; ? \; y$.
23. If $m > -7$ and $-7 > a$, then $m \; ? \; a$.
24. If $3 < 7$, then $7 \; ? \; 3$.
25. If $-4 > -12$, then $-12 \; ? \; -4$.
26. If $9 > x$, then $x \; ? \; 9$.
27. If $-7 < a$, then $a \; ? \; -7$.
28. If $x < 10$ and $10 > z$, then $x \; ? \; z$.
29. If $a > b$ and $c < b$, then $a \; ? \; c$.

9-12 FINDING AND GRAPHING THE SOLUTION SET OF AN INEQUALITY

When an inequality contains a variable, the *domain* or *replacement set* of the inequality is the set of all possible numbers that can be used to replace the variable. When an element from the domain is used in place of the variable, the inequality may be true or it may be false. The ***solution set of an inequality*** is the set of numbers from the domain that make the inequality true. Inequalities that have the same solution set are ***equivalent inequalities***.

To find the solution set of an inequality, solve the inequality by methods similar to those used in solving an equation. Use the properties of inequalities to transform the given inequality into a simpler ***equivalent inequality*** whose solution set is evident.

EXAMPLES

In Examples 1–3, the domain is the set of real numbers.

1. Find and graph the solution of the inequality $x - 4 > 1$.

How to Proceed:	*Solution:*
(1) Write the inequality:	$x - 4 > 1$
(2) Use the addition property of inequalities. Add 4 to each side:	$\begin{array}{r} +4 \quad +4 \\ \hline x > 5 \end{array}$

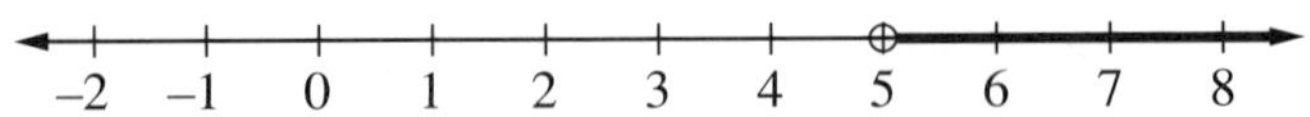

The graph above shows the solution set. The circle at 5 indicates that 5 is the boundary between the numbers to the right, which belong to the solution set, and the numbers to the left, which do not belong. Since 5 is not included in the solution set, the circle is not filled in.

Check:

(1) Check one value from the solution set, for example, 7.
This value will make the inequality true. $7 - 4 > 1$ is true. ✔

(2) Check the boundary value, 5. This value, which separates the values that make the inequality true from the values that make it false, will make the corresponding equality, $x - 4 = 1$, true. $5 - 4 = 1$ is true. ✔

Answer: $x > 5$

---

2. Find and graph the solution of $5x + 4 \leq 11 - 2x$.
Since $5x + 4 \leq 11 - 2x$ means $5x + 4 < 11 - 2x$ or $x + 4 = 11 - 2x$, the solution set of $5x + 4 \leq 11 - 2x$ includes all values of the domain for which $5x + 4 < 11 - 2x$ or $5x + 4 = 11 - 2x$ are true.

How to Proceed:	*Solution:*
(1) Write the inequality:	$5x + 4 \leq 11 - 2x$
(2) Add $2x$ to each side:	$\begin{array}{r} +2x \qquad\quad +2x \\ \hline 7x + 4 \leq 11 \end{array}$
(3) Add -4 to each side:	$\begin{array}{r} -4 \quad -4 \\ \hline 7x \leq 7 \end{array}$
(4) Divide each side by 7:	$\dfrac{7x}{7} \leq \dfrac{7}{7}$
	$x \leq 1$

The graph of the solution set is shown at the right. Since 1 is included in the solution, the circle at 1 is filled in.

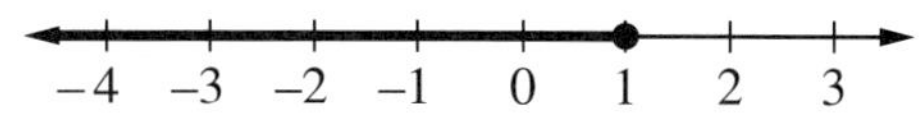

Answer: $x \leq 1$ (The check is left to you.)

---

3. Find and graph the solution set: $2(2x - 8) - 8x \leq 0$.

How to Proceed:	*Solution:*
(1) Write the inequality:	$2(2x - 8) - 8x \leq 0$
(2) Use the distributive property:	$4x - 16 - 8x \leq 0$
(3) Combine like terms in the left member:	$-4x - 16 \leq 0$
(4) Add 16 to each member:	$\underline{\quad + 16 \quad} \quad \underline{+16}$ $-4x \leq 16$
(5) Multiply both members by $-\frac{1}{4}$:	$-\frac{1}{4}(4x) \geq -\frac{1}{4}(16)$
(6) Reverse the order of the inequality:	$x \geq -4$

The graph of the solution set is shown at the right. Note that -4 is included in the graph.

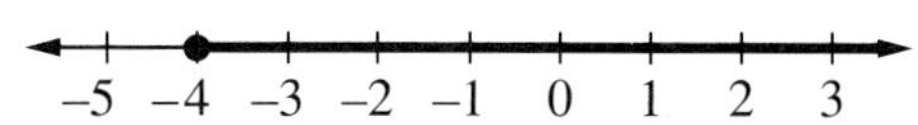

Answer: $x \geq -4$ (The check is left to you.)

Graphing a Conjunction

The inequality $3 < x < 6$ is equivalent to the conjunction $(3 < x) \land (x < 6)$. Since a conjunction is true when both simple statements are true, the solution set of this inequality consists of all of the numbers that are in the solution set of both simple inequalities. The graph of the conjunction can be drawn as shown below.

How to Proceed: *Solution:*

(1) Draw the graph of the solution set of the first inequality a few spaces above the number line:

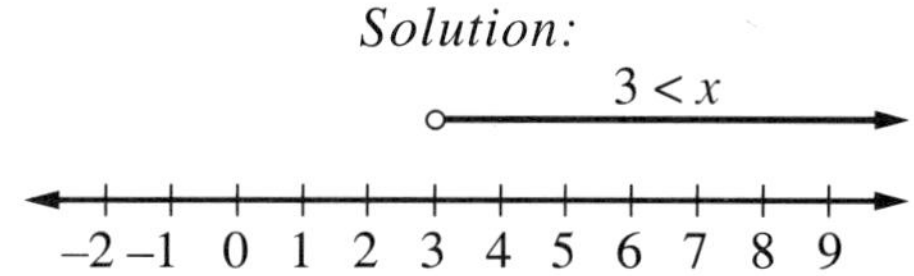

(2) Draw the graph of the solution set of the second inequality above the number line, but below the graph of the first inequality:

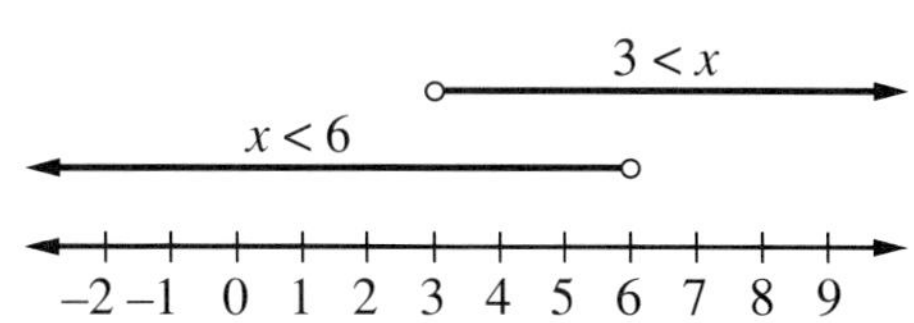

(3) Draw the graph of the conjunction by shading, on the number line, the points that belong to the solution set of both simple inequalities:

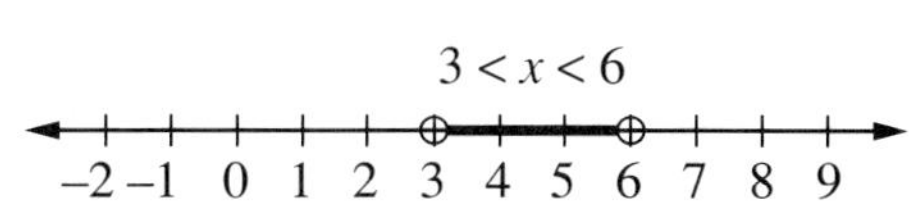

Since 3 is in the solution set of $x < 6$ but not in the solution set of $3 < x$, 3 is not in the solution set of the conjunction. Also, since 6 is in the solution set of $3 < x$ but not in the solution set of $x < 6$, 6 is not in the solution set of the conjunction. Therefore, the circles at 3 and 6 are not filled in, indicating that these boundary values are not elements of the solution set of $3 < x < 6$.

EXAMPLE

Solve the inequality and graph the solution set: $-7 < x - 5 > 0$.

How to Proceed: *Solution:*

(1) First solve the inequalities for x:

Method 1

$$\begin{array}{rcl} -7 < x - 5 & & x - 5 < \;\; 0 \\ \underline{+5 \quad\; + 5} & & \underline{\;\; + 5} \quad \underline{+5} \\ -2 < x \quad\;\; & \text{and} & x \qquad < \;\; 5 \end{array}$$

Method 2

$$\begin{array}{c} -7 < x - 5 < \;\; 0 \\ \underline{+5} \quad \underline{\;\; + 5} \quad \underline{+5} \\ -2 < x \qquad < \;\; 5 \end{array}$$

(2) Draw the graphs of $-2 < x$ and $x < 5$ above the number line:

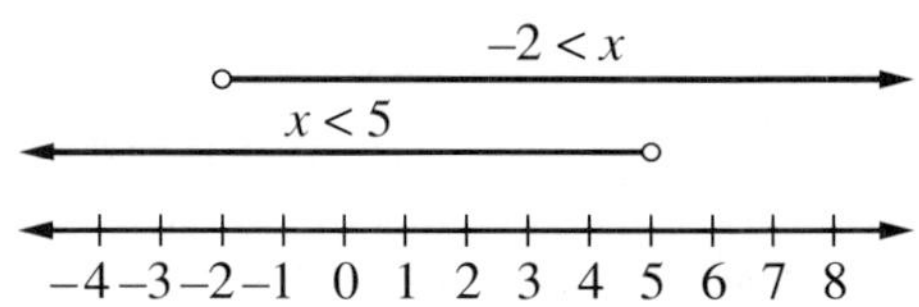

(3) Draw the graph of all points that are common to the graphs of $-2 < x$ and $x < 5$:

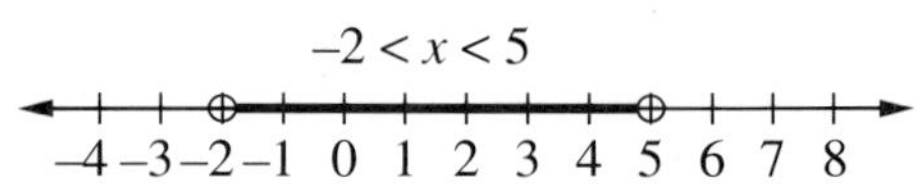

Answer: $-2 < x < 5$

Graphing a Disjunction

The inequality $(x > 3) \vee (x > 6)$ means $(x > 3)$ or $(x > 6)$. Since a disjunction is true when one or both of the simple statements are true, the solution set of the inequality consists of all of the numbers that are in the solution set of at least one of the simple inequalities. The graph of the disjunction can be drawn as shown below.

How to Proceed: *Solution:*

(1) Draw the graph of the solution set of the first inequality a few spaces above the number line:

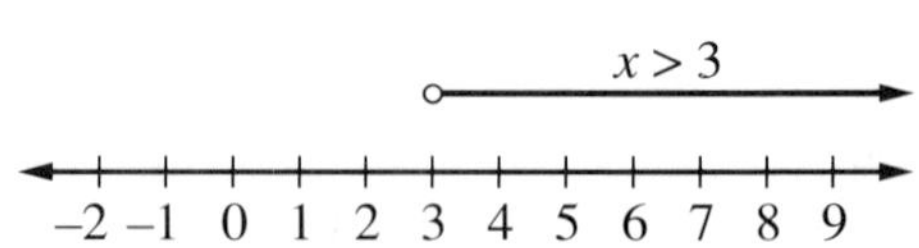

(2) Draw the graph of the solution set of the second inequality above the number line, but below the graph of the first inequality:

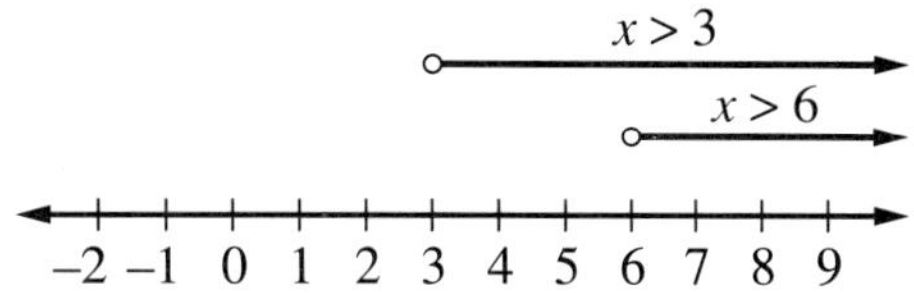

(3) Draw the graph of the disjunction by shading, on the number line, the points that belong to the solution of one or both of the simple inequalities:

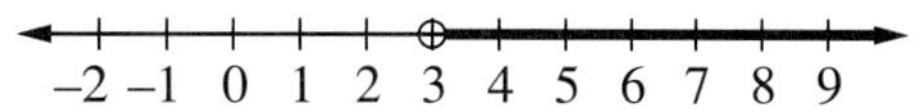

Since 3 is not in the solution set of either inequality, 3 is not in the solution set of the conjunction. Therefore, the circle at 3 is not filled in.

EXAMPLE

Solve the inequality and graph the solution set: $(x + 2 < 0)$ or $(x - 3 > 0)$.

How to Proceed: *Solution:*

(1) Solve each inequality for x:

$$\begin{array}{rcl} x + 2 & < & 0 \\ -\ 2 & & -2 \\ \hline x & < & -2 \end{array} \quad \text{or} \quad \begin{array}{rcl} x - 3 & > & 0 \\ +\ 3 & & +3 \\ \hline x & > & 3 \end{array}$$

(2) Draw the graphs of $x < -2$ and $x > 3$ above the number line:

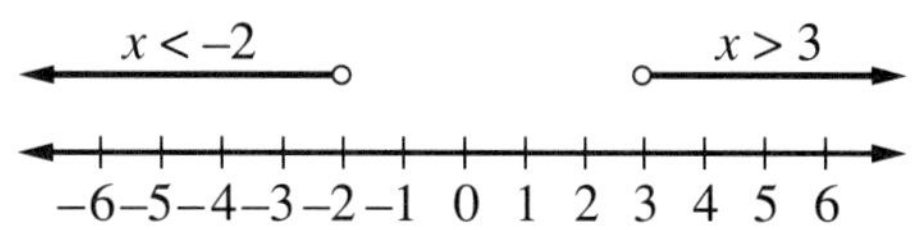

(3) Draw the graph of all points of the graphs of $-2 < x$ or $x > 3$:

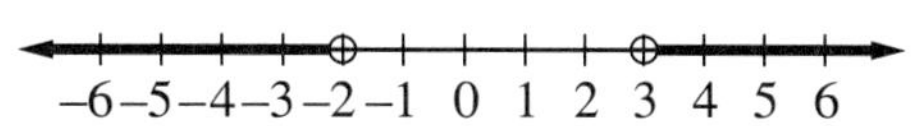

Answer: $x < -2$ or $x > 3$

EXERCISES WITH OPEN-RESPONSE PROBLEMS

In 1–51, find and graph the solution set of each inequality. The domain is the set of real numbers.

1. $x - 2 > 4$ **2.** $z - 6 < 4$ **3.** $y - \frac{1}{2} > 2$

4. $x - 1.5 < 3.5$ **5.** $x + 3 > 6$ **6.** $19 < y + 17$

7. $d + \frac{1}{4} > 3$ **8.** $-3\frac{1}{2} > c + \frac{1}{2}$ **9.** $y - 4 \geq 4$

10. $25 \leq d + 22$
11. $3t > 6$
12. $2x \leq 12$
13. $15 \leq 3y$
14. $-10 \leq 4h$
15. $-6y < 24$
16. $27 > -9y$
17. $-10x > -20$
18. $12 \leq -1.2r$
19. $\frac{1}{3}x > 2$
20. $-\frac{2}{3}z \geq 6$
21. $\frac{x}{2} > 1$
22. $\frac{y}{3} \leq -1$
23. $\frac{1}{2} \leq \frac{z}{4}$
24. $-0.4y \leq 4$
25. $-10 \geq 2.5z$
26. $2x - 1 > 5$
27. $3y - 6 \geq 12$
28. $5x - 1 > -31$
29. $-5 \leq 3y - 2$
30. $3x + 4 > 10$
31. $5y + 3 \geq 13$
32. $6c + 1 > -11$
33. $4d + 3 \leq 17$
34. $5x + 3x - 4 > 4$
35. $8y - 3y - 1 \leq 29$
36. $6x + 2 - 8x < 14$
37. $3x + 1 > 2x + 7$
38. $7y - 4 < 6 + 2y$
39. $4 - 3x \geq 16 + x$
40. $2x - 1 > 4 - \frac{1}{2}x$
41. $2c + 5 \geq 14 + 2\frac{1}{3}c$
42. $\frac{x}{3} - 1 \leq \frac{x}{2} + 3$
43. $4(x - 1) > 16$
44. $8x < 5(2x + 4)$
45. $12\left(\frac{1}{4} + \frac{x}{3}\right) > 15$
46. $8m - 2(2m + 3) \geq 0$
47. $12r - (8r - 20) > 12$
48. $0 < x + 3 < 6$
49. $-5 \leq x + 2 < 7$
50. $(x - 1 > 3) \vee (x + 1 > 9)$
51. $(2x < 2) \vee (x + 5 \geq 10)$

52. To which of the following is $y + 4 \geq 9$ equivalent?
(1) $y > 5$ (2) $y \geq 5$ (3) $y \geq 13$ (4) $y = 13$

53. To which of the following is $5x < 4x + 6$ equivalent?
(1) $x > 6$ (2) $x = 6$ (3) $x = \frac{6}{5}$ (4) $x < 6$

54. Which of the following is the smallest member of the solution set of $3x - 7 \geq 8$?
(1) 3 (2) 4 (3) 5 (4) 6

55. Which of the following is the largest member of the solution set of $4x \leq 3x + 2$?
(1) 1 (2) 2 (3) 3 (4) 4

In 56–61, write an inequality for each graph.

56. 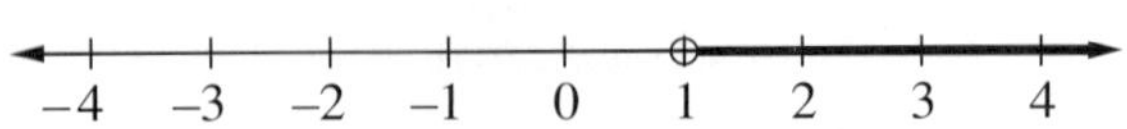

57. 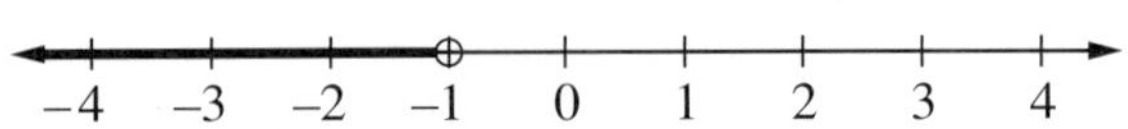

58. 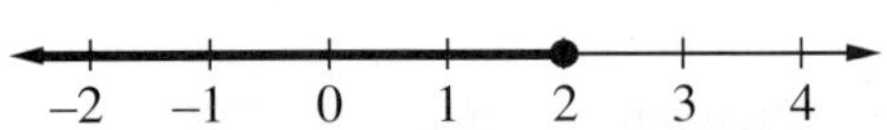

59. 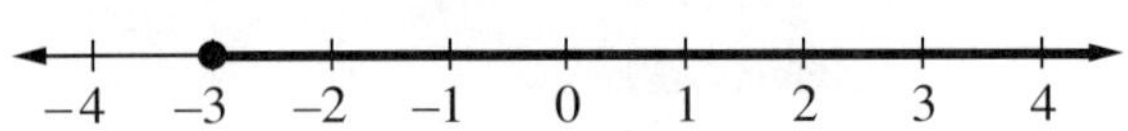

60. 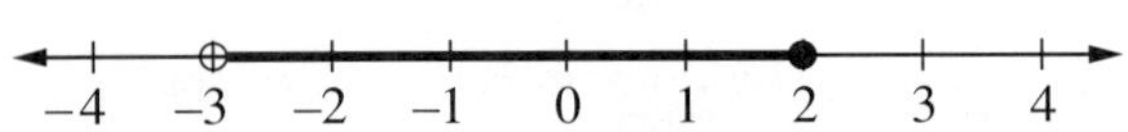

61. 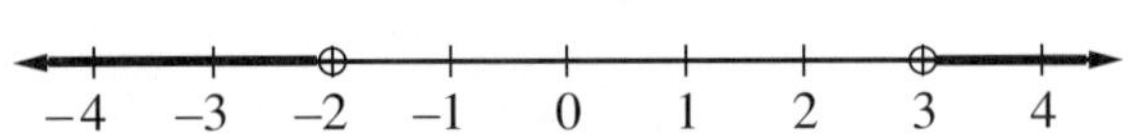

62. **a.** Graph the inequality $(x > 2) \vee (x = 2)$.
b. Write an inequality equivalent to $(x > 2) \vee (x = 2)$.

63. **a.** If $-7x < -21$, is it correct to write $\frac{-7x}{-7} \nless \frac{-21}{-7}$?

b. Explain your answer.

64. **a.** Solve each of the following:

(*1*) $3x < 12 - x$ (*2*) $3x = 12 - x$ (*3*) $3x > 12 - x$

b. What do you observe about the solutions of the equation and inequalities in part **a**?

c. (*1*) If the domain is the set of real numbers, does the solution set of $3x < 12 - x$ have a largest element?

(*2*) Explain your answer.

9-13 USING INEQUALITIES TO SOLVE PROBLEMS

Many problems can be solved by writing an inequality that describes how the numbers in the problem are related and then solving the inequality. An inequality can be expressed in words in different ways. For example:

A number is more than 25. A number exceeds 25. A number is greater than 25. A number is over 25.	$x > 25$
A number is at least 25. A number has a minimum value of 25. A number is not less than 25. A number is not under 25.	$x \geq 25$
A number is less than 25. A number is under 25.	$x < 25$
A number is at most 25. A number has a maximum value of 25. A number is not greater than 25. A number does not exceed 25. A number is not more than 25.	$x \leq 25$

The procedure on the following page can be very helpful in solving problems that involve inequalities.

PROCEDURE. To solve a problem that involves an inequality:

1. Choose a variable to represent one of the unknown quantities in the problem.
2. Express other unknown quantities in terms of the same variable.
3. Write an inequality using a relationship given in the problem, a previously known relationship, or a formula.
4. Solve the inequality.
5. Check the solution using the words of the problem.
6. Use the solution of the inequality to answer the question in the problem.

EXAMPLES

1. Sadie has \$53.50 in her pocket and wants to purchase shirts at a sale price of \$14.95 each. How many shirts can she buy?

Solution The domain is the set of whole numbers.

Let $x =$ the number of shirts that she can buy.

Then, $14.95x =$ the cost of the shirts.

$$\underbrace{\text{The cost of the shirts}}_{14.95x} \quad \underbrace{\text{is less than or equal to}}_{\leq} \quad \underbrace{\$53.50.}_{53.50}$$

$$\frac{14.95x}{14.95} \leq \frac{53.50}{14.95}$$

Use a calculator to complete the computation.

Enter: 53.50 [÷] 14.95 [=]

Display: [3.5785953]

Therefore, $x \leq 3.5785953$. Since the domain is the set of whole numbers, the solution set is $\{0, 1, 2, 3\}$.

Check: 0 shirt costs \$14.95(0) = \$0

1 shirt costs \$14.95(1) = \$14.95

2 shirts cost \$14.95(2) = \$29.90

3 shirts cost \$14.95(3) = \$44.85

4 or more shirts cost more than \$14.95(4) or more than \$59.80

Answer: Sadie can buy 0, 1, 2, or 3 shirts.

---

2. The length of a rectangle is 5 centimeters more than its width. The perimeter of the rectangle is at least 66 centimeters. Find the minimum measures of the length and the width.

Solution If the perimeter is at least 66 centimeters, then the sum of the measures of the four sides is either equal to 66 centimeters or is greater than 66 centimeters.

Let $x =$ the width of the rectangle.

Then, $x + 5 =$ the length of the rectangle.

$$\underbrace{\text{The perimeter of the rectangle}}\ \underbrace{\text{is at least}}\ \underbrace{\text{66 centimeters.}}$$

$$\begin{aligned} x + (x + 5) + x + (x + 5) &\geq 66 \\ x + x + 5 + x + x + 5 &\geq 66 \\ 4x + 10 &\geq 66 \\ 4x &\geq 56 \\ x &\geq 14 \end{aligned}$$

(The width is at least 14 centimeters.)

Since we are looking for the minimum measures:

the smallest possible width is $x = 14$

and

the smallest possible length is $x + 5 = 19$

Answer: The minimum width is 14 centimeters, and the minimum length is 19 centimeters.

---

3. Two sides of a triangle have lengths 2 and 5. Find all possible lengths of the third side.

Solution We know that the shortest distance between two points is a straight line. Therefore, the sum of the lengths of two sides of a triangle is greater than the length of the third side.

(1) Let $s =$ length of third side of triangle.

(2) Of the lengths 2, 5, and s, the longest side is either 5 or s.

(3) Assume the longest side of the triangle is 5.

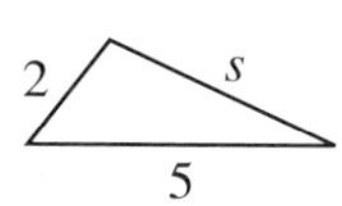

Then, $2 + s > 5$

$s > 5 - 2$

or $s > 3$

(4) Assume the longest side of the triangle is s.

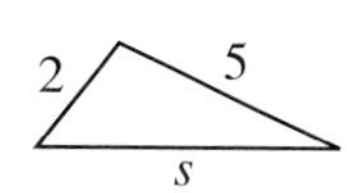

Then, $2 + 5 > s$

$7 > s$

or $s < 7$

Answer: Length s is greater than 3 and less than 7.

Note that the solution $(s > 3) \wedge (s < 7)$ may be rewritten as $(3 < s) \wedge (s < 7)$, which becomes $(3 < s < 7.)$

EXERCISES WITH OPEN-RESPONSE PROBLEMS

In 1–9, represent each sentence as an algebraic inequality.

1. x is less than or equal to 15.
2. y is greater than or equal to 4.
3. x is at most 50.
4. x is more than 50.
5. The greatest possible value of $3y$ is 30.
6. The sum of $5x$ and $2x$ is at least 70.
7. The maximum value of $4x - 6$ is 54.
8. The minimum value of $2x + 1$ is 13.
9. The product of $3x$ and $x + 1$ is less than 35.

In 10–16, in each case write and solve the inequality that represents the given conditions.

10. Six less than a number is greater than 4.
11. Six less than a number is less than 4.
12. Six times a number is less than 72.
13. A number increased by 10 is greater than 50.
14. A number decreased by 15 is less than 35.
15. Twice a number, increased by 6, is less than 48.
16. Five times a number, decreased by 24, is greater than 3 times the number.

In 17 and 18, select the numeral preceding the choice that best completes the sentence.

17. The lengths of two sides of a triangle are 6 and 3. The length of the third side may be
(1) 12 (2) 9 (3) 3 (4) 6

18. The lengths of two sides of a triangle are 10 and 14. The length of the third side may be
(1) 24 (2) 2 (3) 22 (4) 4

In 19–22, using the two given lengths of the sides of a triangle, find all possible lengths for the third side, s.

19. 2 and 4
20. 4 and 7
21. 5 and 6
22. 3 and 9

In 23–33, in each case write an inequality and solve the problem algebraically.

23. Mr. Burke had a sum of money in a bank. After he deposited an additional sum of \$100, he had at least \$550 in the bank. At least how much money did Mr. Burke have in the bank originally?

24. The members of a club agree to buy at least 250 tickets for a theater party. If they expect to buy 80 fewer orchestra tickets than balcony tickets, what is the least number of balcony tickets they will buy?

25. Mrs. Scott decided that she would spend no more than \$120 to buy a jacket and a skirt. If the price of the jacket was \$20 more than 3 times the price of the skirt, find the highest possible price of the skirt.

26. Three times a number increased by 8 is at most 40 more than the number. Find the greatest value of the number.

27. The length of a rectangle is 8 meters less than 5 times its width. If the perimeter of the rectangle is at most 104 meters, find the greatest possible width of the rectangle.

28. The length of a rectangle is 10 centimeters less than 3 times its width. If the perimeter of the rectangle is at most 180 centimeters, find the greatest possible length of the rectangle.

29. Mrs. Diaz wishes to save at least \$1,500 in 12 months. If she saves \$300 during the first 4 months, what is the least possible average amount that she must save in each of the remaining 8 months?

30. Two consecutive even numbers are such that their sum is greater than 98 decreased by twice the larger. Find the smallest possible values for the integers.

31. Minou needed \$29 to buy some compact discs. Her father agreed to pay her \$6 an hour for gardening in addition to her \$5 weekly allowance for helping around the house. What is the minimum number of hours Minou must work at gardening to earn \$29 this week?

32. Fred bought three shirts, each at the same price, and received less than \$2.00 change from a \$20.00 bill. What is the minimum cost of one shirt?

33. Allison has 2 to 3 hours to spend on her homework. She has work in math, English, and social studies. She plans to spend equal amounts of time in studying English and studying social studies, and to spend twice as much time in studying math as in studying English.

a. What is the minimum time she can spend on English homework?

b. What is the maximum time she can spend on social studies?

c. What is the maximum time she can devote to math?

CHAPTER SUMMARY

Before solving an equation, simplify each member if necessary. To solve an equation that has the variable in both members, transform it into an equivalent equation in which the variable appears in only one member. Do this by adding the opposite of one of the variable terms on one side to both sides of the equation.

Any equation or formula containing two or more variables can be transformed so that one variable is expressed in terms of all other variables. To do this, think of solving a simpler but related equation that contains only one variable.

Consecutive integers are integers that follow one another in order. ***Consecutive even integers*** are even integers that follow one another in order. ***Consecutive odd integers*** are odd integers that follow one another in order.

A ***uniform rate of speed*** remains constant. The ***average rate of speed*** is the total distance traveled divided by the time traveled.

Order property of numbers: For real numbers x and y, one and only one of the following can be true: $x < y$, $x = y$, or $x > y$.

Transitive property of inequality: For real numbers x, y, and z, if $x < y$ and $y < z$, then $x < z$, and if $x > y$ and $y > z$, then $x > z$.

Addition property of inequality: When the same number is added to or subtracted from both members of an inequality, the order of the new inequality is the same as the order of the original one.

Multiplication property of inequality: When both members of an inequality are multiplied or divided by the same positive number, the order of the new inequality is the same as the order of the original one. When both members of an inequality are multiplied or divided by the same negative number, the order of the new inequality is the opposite of the order of the original one.

The ***domain or replacement set*** of the inequality is the set of all possible numbers that can be used to replace the variable. The ***solution set*** of an inequality is the set of numbers from the domain that make the inequality true. Inequalities that have the same solution set are ***equivalent inequalities.***

VOCABULARY

9-2 Exceeds

9-3 Consecutive integers Consecutive even integers Consecutive odd integers

9-5 Perimeter

9-7 Formula

9-8 Uniform rate of speed Average rate of speed

9-11 Order property of the real numbers Transitive property of inequality Addition property of inequality Multiplication property of inequality

9-12 Solution set of an inequality Equivalent inequality

REVIEW EXERCISES WITH OPEN-RESPONSE PROBLEMS

In 1–16, in each case solve for the variable and check.

1. $x - 9 = 14$ **2.** $-18 = 3m$ **3.** $8c + 2 = 6$

4. $4 - x = -3$ **5.** $0.3d + 2 = 8$ **6.** $\frac{5}{8}b = -40$

7. $\frac{3}{5}p - 4 = 26$ **8.** $2(z - 6) = 14$ **9.** $8w = 60 - 4w$

10. $8w - 4w = 60$ **11.** $4h + 3 = 23 - h$ **12.** $5y + 3 = 2y$

13. $8z - (6z - 5) = 1$ **14.** $2(b - 4) = 4(2b + 1)$

15. $3(4x - 1) - 2 = 17x + 10$ **16.** $(x + 2) - (3x - 2) = x + 3$

In 17–20, solve each equation for x in terms of a, b, and c.

17. $a + x = b + c$ **18.** $cx + a = b$

19. $bx - a = 5a + c$ **20.** $\frac{a + c}{2}x = b$

21. **a.** Solve $A = \frac{1}{2}bh$ for h in terms of A and b.

b. Find h when $A = 5.4$ and $b = 0.9$.

22. If $P = 2\ell + 2w$, find w when $P = 17$ and $\ell = 5$.

23. If $F = \frac{9}{5}C + 32$, find C when $F = 68$.

24. Syracuse and New York are 240 miles apart. One car left Syracuse for New York, averaging 65 mph. At the same time, another car left New York for Syracuse, averaging 55 mph. How far from Syracuse did the cars pass each other?

25. A pilot plans to make a flight lasting 2 hours and 30 minutes. How far can he fly from the airport at the rate of 600 mph and return over the same route at the rate of 400 mph?
26. James bought 80 postage stamps for which he paid \$18.40. Some were 1-cent stamps and some were 33-cent stamps. How many of each kind did he buy?
27. A postal clerk sold 75 stamps for \$19.90. Some were 20-cent and some were 34-cent stamps. How many of each kind did he sell?
28. The sum of Carl's age and Ellen's age is 40 years. Carl's age 10 years from now will be 1 less than 4 times Ellen's age 6 years ago. Find their present ages.
29. Mark is 10 years younger than Larry. Larry's age 8 years from now will exceed twice Mark's age 3 years ago by 4 years. How old is each now?

In 30–37, find and graph the solution set of each inequality.

30. $\frac{1}{3}x < 1$
31. $-x \geq 4$
32. $6 + x > 3$
33. $2x - 3 \geq -5$
34. $-3 < x \leq 2$
35. $(x \geq 3) \wedge (x < 7)$
36. $(x \leq -2) \vee (x > 0)$
37. $(x - 4 \geq 1) \wedge (-2x > -18)$

In 38–41, tell whether each statement is sometimes, always, or never true.

38. If $x > y$, then $a + x > a + y$.
39. If $x > y$, then $ax > ay$.
40. If $x > y$ and $y > z$, then $x > z$.
41. If $x > y$, then $-x > -y$.

In 42–45, select, in each case, the numeral preceding the inequality that correctly completes the statement or answers the question.

42. An inequality that is equivalent to $4x - 3 > 5$ is
(1) $x > 2$ (2) $x < 2$ (3) $x > \frac{1}{2}$ (4) $x < \frac{1}{2}$

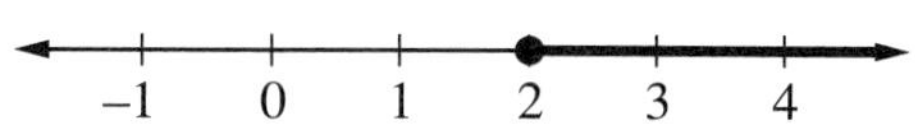

43. The solution set of which inequality is shown in the graph above?
(1) $x - 2 \geq 0$ (2) $x - 2 > 0$ (3) $x - 2 < 0$ (4) $x - 2 \leq 0$

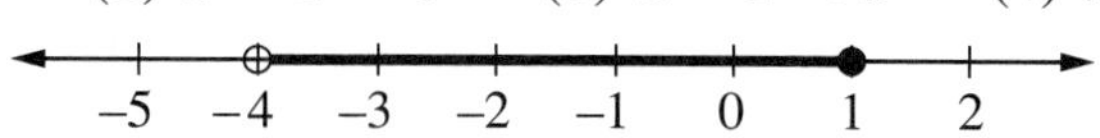

44. The above graph shows the solution set of which inequality?
(1) $-4 < x < 1$ (2) $-4 \leq x < 1$ (3) $-4 < x \leq 1$ (4) $-4 \leq x \leq 1$
45. The lengths of the sides of a triangle may be
(1) 4, 9, 13 (2) 5, 8, 12 (3) 5, 5, 13 (4) 5, 5, 10
46. The length of a rectangular room is 5 feet more than 3 times the width. The perimeter of the room is 62 feet. Find the dimensions of the room.
47. Find three consecutive integers such that 3 times the largest integer is twice the sum of the other two integers.

48. A truck must cross a bridge that can support a maximum weight of 24,000 pounds. The weight of the empty truck is 1,500 pounds, and the driver weighs 190 pounds. What is the maximum load that the truck can carry?

49. In an apartment building there is one elevator, and the maximum load that it can carry is 2,000 pounds. The maintenance supervisor wants to move a replacement part for the air-conditioning unit to the roof. The part weighs 1,600 pounds, and the mechanized cart on which it is being moved weighs 250 pounds. When the maintenance supervisor drives the cart onto the elevator, the alarm sounds to signify that the elevator is overloaded.

a. How much does the maintenance supervisor weigh?

b. How can the replacement part be delivered to the roof if the part cannot be disassembled?

50. A mail-order film developer charges \$1.99 per roll of film plus a \$2.98 handling fee for each order. A local developer charges \$2.50 per roll of film. How many rolls of film must be developed in order that:

a. the local developer offers the lower price?

b. the mail-order developer offers the lower price?

CUMULATIVE REVIEW WITH OPEN-RESPONSE PROBLEMS

1. The figure at the right consists of two squares and two isosceles right triangles. Express the area of the figure in terms of s, the length of one side of a square.

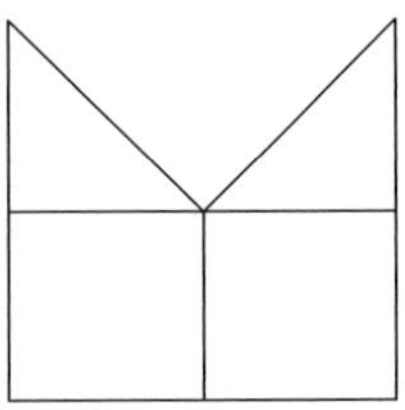

2. Write the quotient $\frac{0.003}{2{,}000}$ in scientific notation.

3. Write as a polynomial in simplest form: $(2x - 3)^2 - 2x^2$.

4. "I am thinking of a number that has exactly three factors or it is divisible by 3." This statement is false if the number I am thinking of is
(1) 4 (2) 9 (3) 16 (4) 25

5. Express in terms of w the number of days in w weeks and 4 days.

Exploration

The program below is written in the computer language BASIC.

```
10 LET X = 0
20 X = X + 1
30 IF X = 25 THEN 60
40 PRINT X * X
50 GO TO 20
60 END
```

a. What does the program do?

b. How does the statement X = X + 1 differ from the algebraic equation $x = x + 1$?

c. How does the statement IF X = 25 THEN 60 differ from a conditional in logic?

Chapter 10

Angle Measure in Geometry

A carpenter is building a deck on the back of a house. He begins by driving some stakes into the ground to locate the corners of the deck. Between each pair of stakes, he stretches a cord to indicate the edges of the deck.

As he works, he is following a plan that he made in the form of a drawing or blueprint. On the blueprint, the positions of the stakes are shown as points and the cords as segments of straight lines. His blueprint is a model of the deck that he is building.

Geometry combines points, lines, and planes to model the world in which we live. An understanding of geometry enables us to understand relationships involving the sizes of physical objects and the magnitudes of the forces that interact in daily life.

10-1 POINTS, LINES, AND PLANES

Undefined Terms

We ordinarily define a word by using a simpler term. The simpler term can be defined by using one or more still simpler terms. But this process cannot go on endlessly; there comes a time when the definition must use a term whose meaning is *assumed* to be clear to all people. Because the meaning is accepted without definition, such a term is called an ***undefined term***.

In geometry, we use such ideas as *point*, *line*, and *plane*. Since we cannot give satisfactory definitions of these words by using simpler defined words, we will consider them to be undefined terms.

Although *point*, *line*, and *plane* are undefined words, we must have a clear understanding of what they mean. Knowing the properties and characteristics they possess helps us to achieve this understanding.

A ***point*** indicates a place or position. It has no length, width, or thickness. A point is usually indicated by a dot and named with a capital letter. For example, point A is shown at the left.

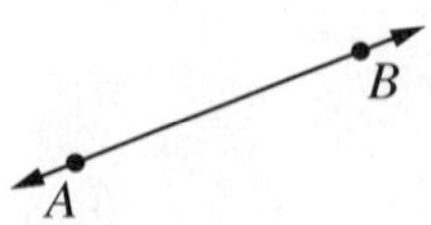

Straight line

Curved line

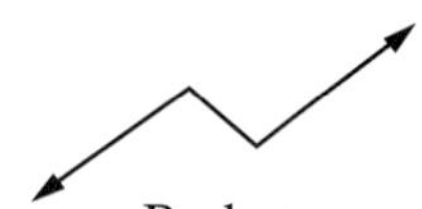
Broken line

A ***line*** is a set of points. The set of points may form a curved line, a broken line, or a straight line. A ***straight line*** is a line that is suggested by a stretched string but that extends without end in both directions. Unless otherwise stated, in this discussion the term *line* will mean *straight* line. A line is named by any two points on the line, for example, the straight line shown at the left is line AB or line BA, usually written as $\overleftrightarrow{AB}$ or $\overleftrightarrow{BA}$. A line can also be named by one point that is written with a lower case letter, for example, line l shown at left. The arrows remind us that the line continues beyond what is drawn in the diagram.

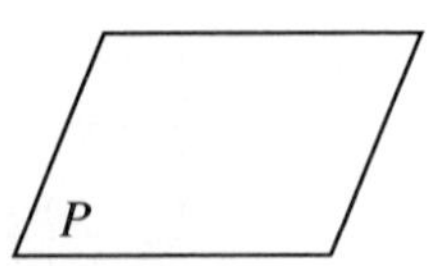

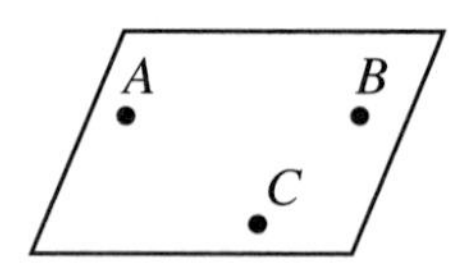

A ***plane*** is a set of points suggested by a flat surface. A plane extends endlessly in all directions. A plane may be named by a single letter, as plane P in the diagram at the left. A plane can also be named by three points of the plane, as plane ABC in the diagram above.

Using these three undefined terms, we can define other geometric terms.

Facts About Straight Lines

A statement that is accepted as true without proof is called an ***axiom*** or a ***postulate***. If we examine the three accompanying figures pictured below, we see that it is reasonable to accept the following three statements as postulates:

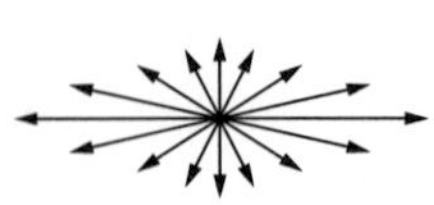

1. **In a plane, an infinite number of straight lines can be drawn through a given point.**

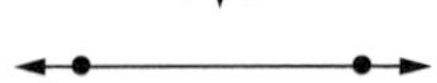

2. **One and only one straight line can be drawn that contains two given points.** (Two points determine a straight line.)

3. **In a plane, two different non-parallel straight lines will intersect at only one point.**

Line Segment

A ***line segment*** or ***segment*** is a part of a line consisting of two points called ***endpoints*** and all points on the line between these endpoints.

At the left is pictured a line segment whose endpoints are points R and S. We use these endpoints to name this segment as segment RS, which may be written as $\overline{RS}$ or $\overline{SR}$.

As we have learned, the *measure of a line segment* or the *length of a line segment* is the distance between its endpoints. We use a number line to associate a number with each endpoint. Since the coordinate of A is 0 and the coordinate of B is 5, the length of $\overline{AB}$, written as AB, is $5 - 0$ or 5.

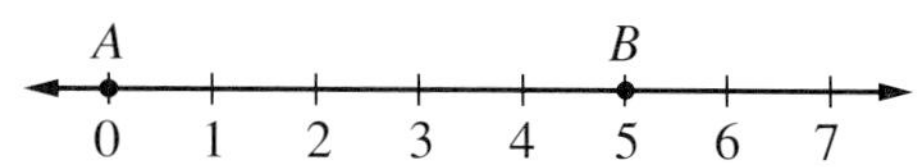

The ***midpoint*** of a line segment is the point that divides the segment into two segments that have the same length. If the length of $\overline{AB}$ is 5, then the length of each of the equal segments, $\overline{AM}$ and $\overline{MB}$, is 2.5.

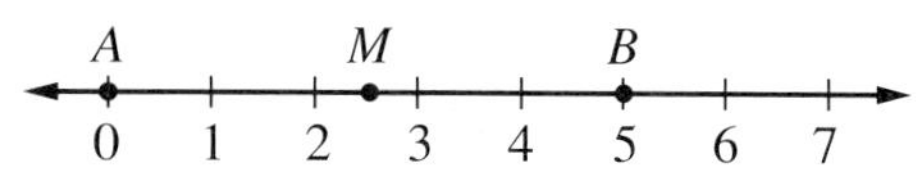

The midpoint of $\overline{AB}$ is M, and $AM = MB$.

EXAMPLE

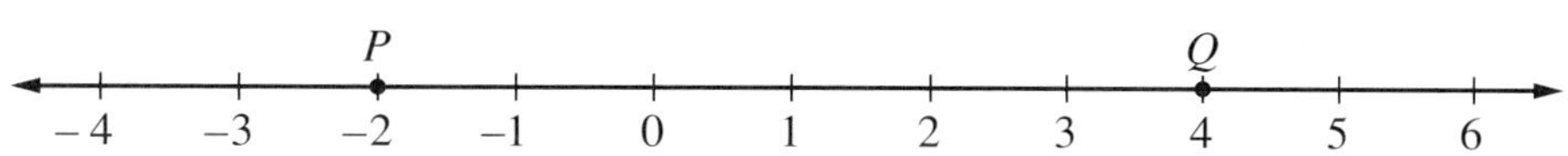

In the diagram, -2 is the coordinate of P and 4 is the coordinate of Q.

a. Find PQ.

b. What is the coordinate of the midpoint of $\overline{PQ}$?

Solutions **a.** The length of $\overline{PQ}$ is the distance between the endpoints. Count the number of spaces.
$PQ = 6$

b. Since the length of $\overline{PQ}$ is 6, the distance from each endpoint to the midpoint M is 3. Therefore the coordinate of M is 1.

Answers: **a.** $PQ = 6$ **b.** The coordinate of the midpoint is 1.

Operations Involving Sets

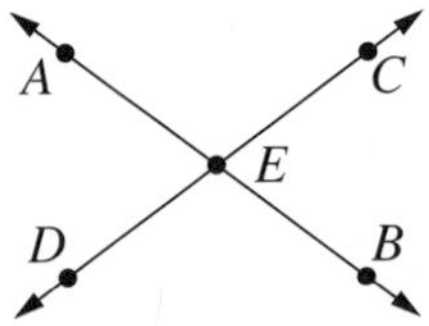

The ***intersection*** of two sets of points is the set of all points that belong to both of those sets of points. For example, the intersection of the set of points contained in line *AB* and the set of points contained in line *CD* is the set that has one element, point *E*. We can write $\overleftrightarrow{AB} \cap \overleftrightarrow{CD} = \{E\}$.

Also, since the intersection of segment *EG* and segment *FH* is segment *FG*, we can write $\overline{EG} \cap \overline{FH} = \overline{FG}$.

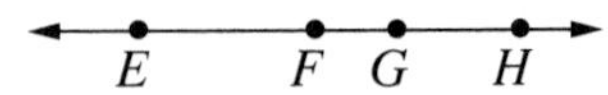

The ***union*** of two sets of points is the set of points that belongs to either or both of those sets of points. For example, since the union of segment *EG* and segment *FH* is segment *EH*, we can write $\overline{EG} \cup \overline{FH} = \overline{EH}$.

EXERCISES

For 1–7, use this number line:

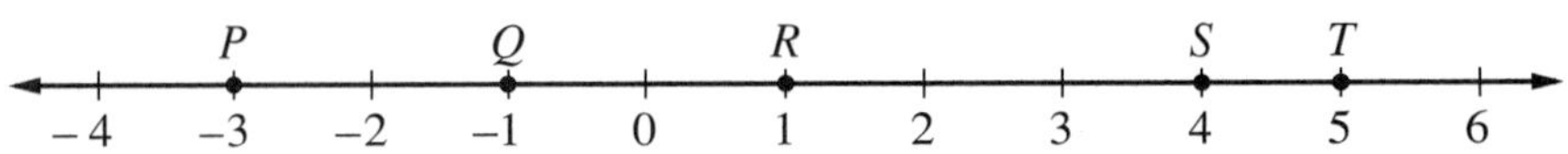

1. Write the meaning of each symbol: **a.** $\overleftrightarrow{PR}$ **b.** *PR* **c.** $\overline{PR}$
2. Name two segments on the line.
3. Name the midpoint of $\overline{PT}$.
4. Is point *T* on $\overleftrightarrow{RS}$?
5. Is point *T* on $\overline{RS}$?
6. Find: **a.** *PT* **b.** *PQ* **c.** *QS* **d.** *RT*
7. Find: **a.** $\overline{QR} \cup \overline{RS}$ **b.** $\overline{QR} \cap \overline{RS}$ **c.** $\overline{PS} \cup \overline{RT}$
   d. $\overline{PS} \cap \overline{RT}$ **e.** $\overline{PQ} \cap \overline{RT}$

10-2 ANGLES, ANGLE MEASURES, AND PERPENDICULARITY

Half-Lines and Rays

When we choose any point *A* on a line, that point separates the line into two sets of points that lie on opposite sides of *A* and are called ***half-lines***. In the diagram at the right, point *A* separates $\overleftrightarrow{CD}$ into two half-lines. The two points *B* and *D* belong to the same half-

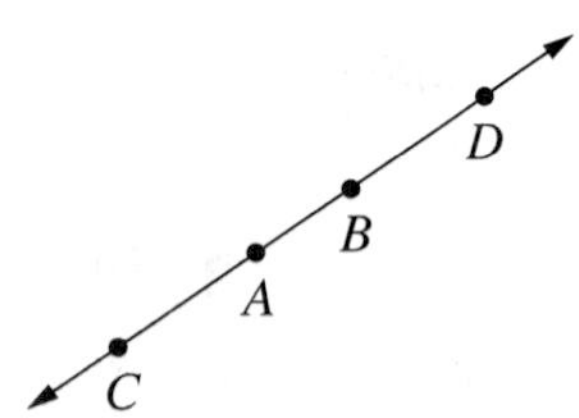

line since A is not a point of $\overline{BD}$. Points B and C, however, do not belong to the same half-line since A is a point of $\overline{CB}$. All points in the same half-line are said to be on the same side of A.

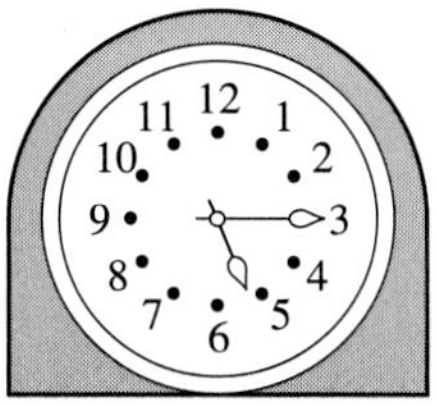

Each hand of a clock can be represented by a point, and a half-line can be determined by that point. A ray of light can be thought of in the same way; here the source of light is a point such as the sun or a bulb in a flashlight, and the path of light in one direction is a half-line.

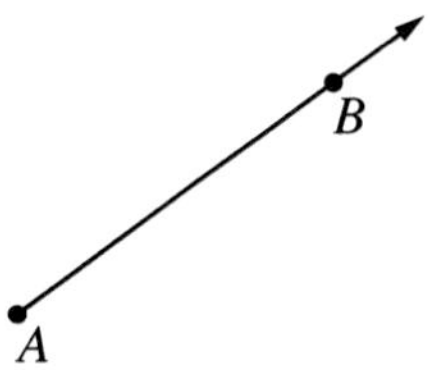

A ***ray*** is a part of a line that consists of a point on the line, called an *endpoint*, and all the points on one side of the endpoint. To name a ray we use two capital letters and an arrow with one arrowhead. The first letter must be the letter that names the endpoint. The second letter is the name of any other point on the ray. The figure at the left shows ray AB, which is represented as $\overrightarrow{AB}$.

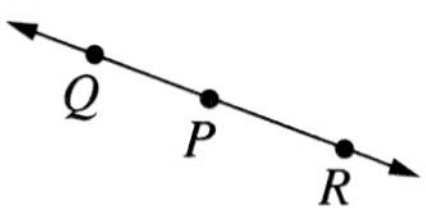

Two rays are called ***opposite rays*** if they are rays of the same line that have a common endpoint but no other points in common. In the diagram at the left, $\overrightarrow{PQ}$ and $\overrightarrow{PR}$ are opposite rays.

Angles

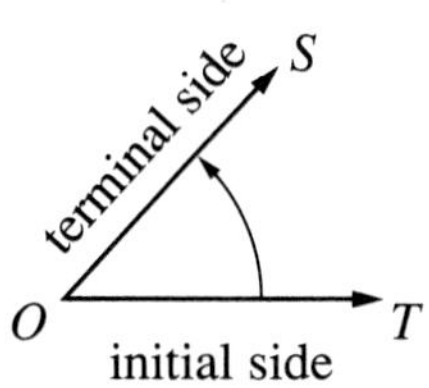

An ***angle*** is a set of points that is the union of two rays having the same endpoint. The common endpoint of the two rays is the ***vertex*** of the angle.

In the diagram at the left, we can think of $\angle TOS$ as having been formed by rotating $\overrightarrow{OT}$. If $\overrightarrow{OT}$ is rotated in a counterclockwise direction about vertex O, it will assume position $\overrightarrow{OS}$, forming $\angle TOS$.

Measuring Angles

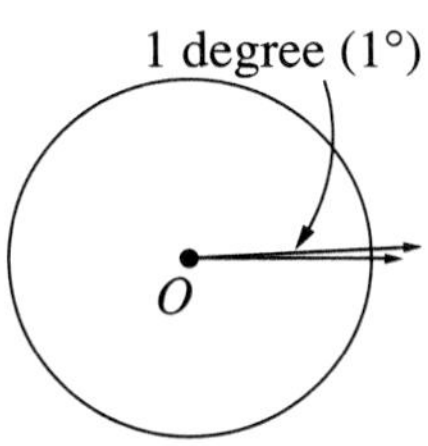

To measure an angle means to determine the number of units of measure it contains. A common standard unit of measure of an angle is the ***degree***; 1 degree is written as 1°. A degree is $\frac{1}{360}$ of a complete rotation of a ray about a point. Thus, a complete rotation contains 360 degrees, written as 360°.

A ***protractor*** is used to measure angles. The diagram shows that the measure of $\angle AOB$ is 45°:

$$m\angle AOB = 45$$

Note: When we use the notation for angle measure (as in $m\angle AOB$), we omit the degree sign.

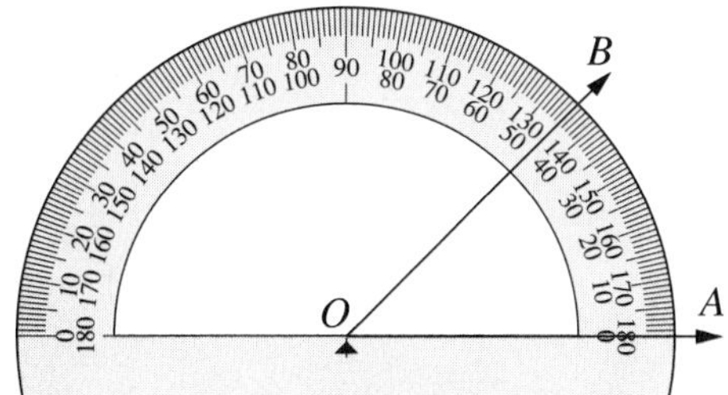

Types of Angles

Angles are classified according to their measures.

An ***acute angle*** is an angle whose measure is greater than 0° and less than 90°; that is, its measure is between 0° and 90°. Angle GHI is an acute angle where $0 < m\angle h < 90$.

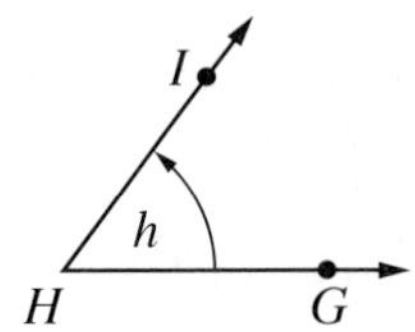

Acute angle

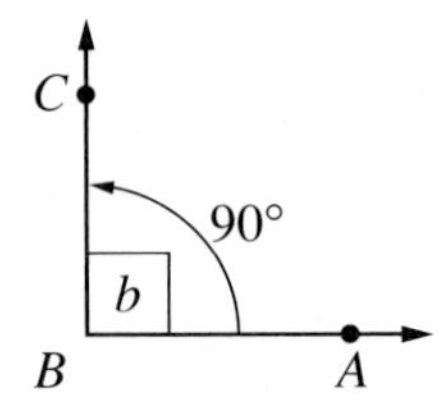

Right angle

A ***right angle*** is an angle whose measure is 90°. Angle ABC is a right angle. We can say that $m\angle ABC = 90$, or $m\angle b = 90$. The symbol ⅂ at B shows that $\angle ABC$ is a right angle.

An ***obtuse angle*** is an angle whose measure is greater than 90° but less than 180°; that is, its measure is between 90° and 180°. Angle LKN is an obtuse angle where $90 < m\angle k < 180$.

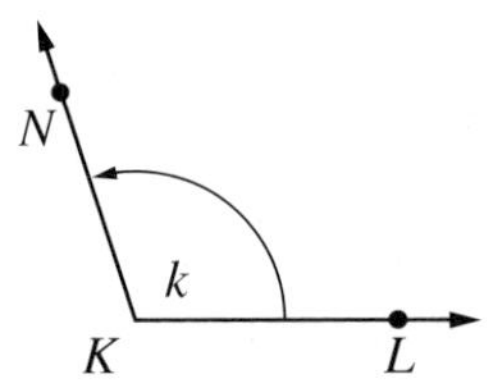

Obtuse angle

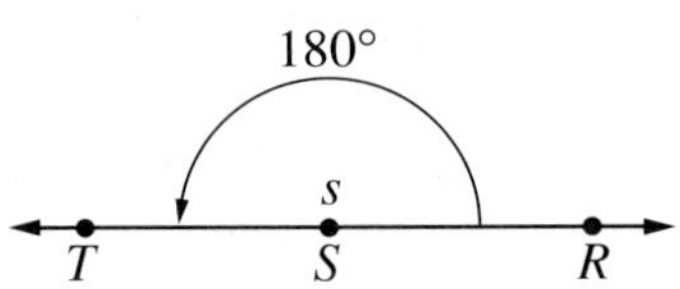

Straight angle

A ***straight angle*** is an angle whose measure is 180°. Angle RST is a straight angle where $m\angle RST = 180$, or $m\angle s = 180$. A straight angle is the union of two opposite rays.

A ***reflex angle*** is an angle whose measure is greater than 180° but less than 360°; that is, its measure is between 180° and 360°. Angle XYZ is a reflex angle where $180 < m\angle y < 360$.

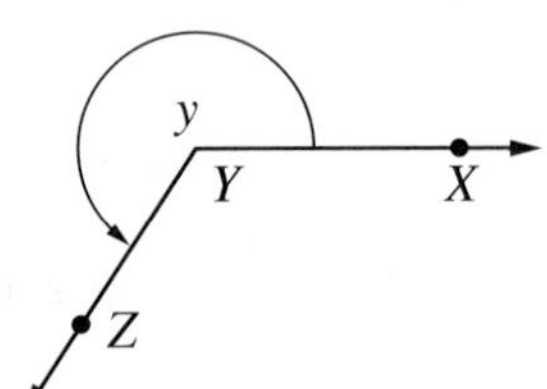

Reflex angle

Here are three important facts about angles:

1. The measure of an angle depends only on the amount of rotation, not on the pictured lengths of the rays forming the angle.
2. Since every right angle measures 90°, all right angles are equal in measure.
3. Since every straight angle measures 180°, all straight angles are equal in measure.

Perpendicularity

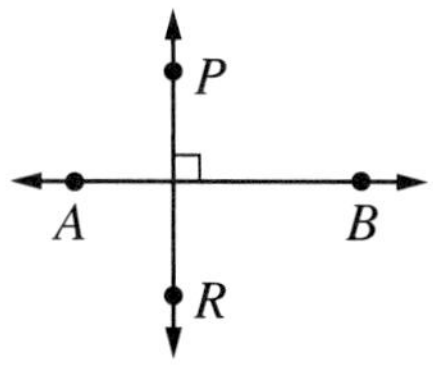

Two lines are ***perpendicular*** if and only if the two lines intersect to form right angles. The symbol for perpendicularity is $\perp$.

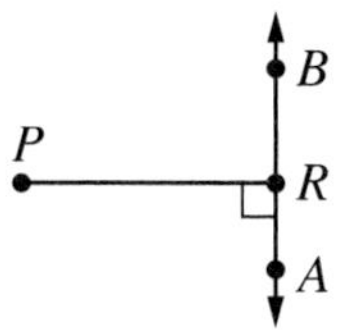

In the diagram, $\overleftrightarrow{PR}$ is perpendicular to $\overleftrightarrow{AB}$, symbolized as $\overleftrightarrow{PR} \perp \overleftrightarrow{AB}$. The symbol ⏋, indicates that a right angle exists where the perpendicular lines intersect.

Segments of perpendicular lines that contain the point of intersection of the lines are also perpendicular. In the diagram, $\overline{PR} \perp \overleftrightarrow{AB}$.

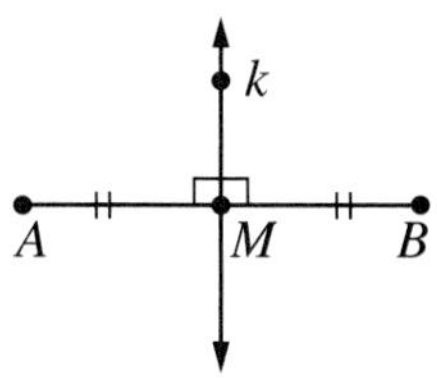

A line that is perpendicular to a line segment at the midpoint of the line segment is called the ***perpendicular bisector*** of the line segment. In the diagram, line k is the perpendicular bisector of $\overline{AB}$. Thus, line $k \perp \overline{AB}$, M is the midpoint of $\overline{AB}$, and $AM = MB$.

EXERCISES

In 1–5: **a.** Use a protractor to measure each angle. **b.** State what kind of angle each is.

1.

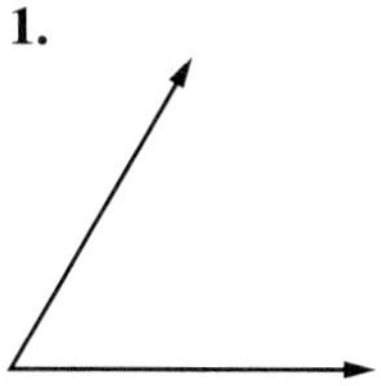

2.

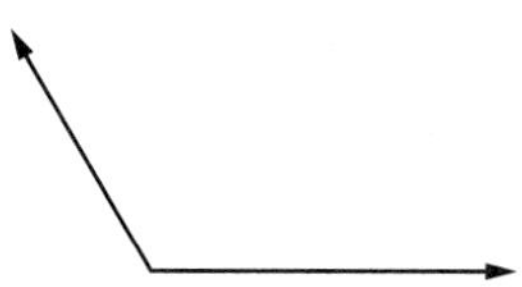

3.

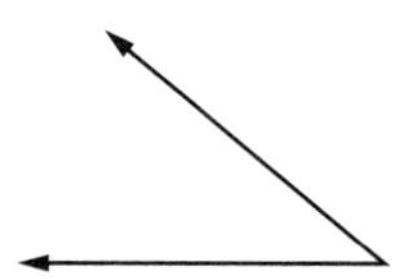

4.

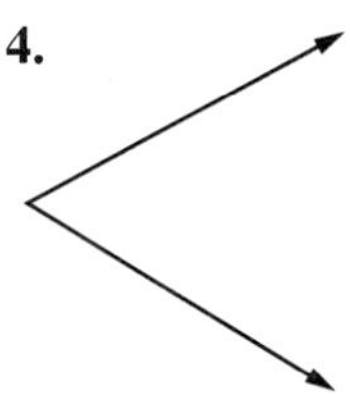

5.

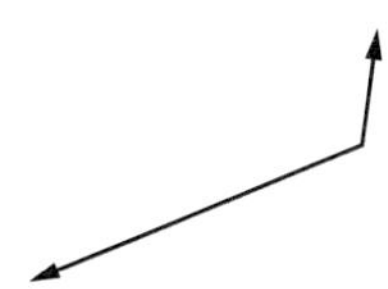

In 6–11, use a protractor to draw an angle whose measure is:

6. 30° **7.** 90° **8.** 48° **9.** 120° **10.** 180° **11.** 138°

12. For the figure at the right:

a. Name $\angle x$ by using three capital letters.

b. Give the shorter name for $\angle COB$.

c. Name one acute angle.

d. Name one obtuse angle.

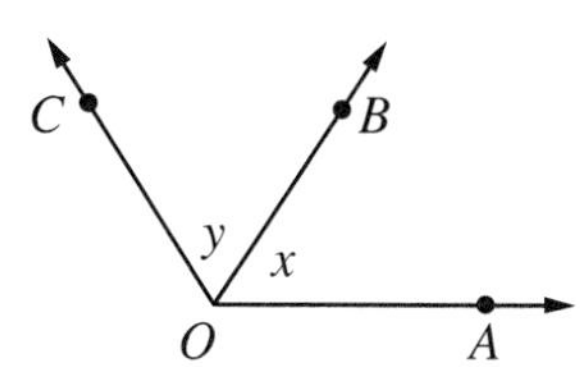

13. Find the number of degrees in:

a. $\frac{1}{2}$ of a complete rotation **b.** $\frac{3}{4}$ of a complete rotation

c. $\frac{1}{3}$ of a right angle **d.** $\frac{5}{6}$ of a straight angle

In 14–17, find the number of degrees in the angle formed by the hands of a clock at each given time.

14. 1 P.M. **15.** 4 P.M. **16.** 6 P.M. **17.** 5:30 P.M.

18. Name a time when the hands of a clock form an angle of 0°.

10-3 PAIRS OF ANGLES

An angle divides the points in a plane that are not points of the angle into two sets of points called *regions*. As illustrated at the right, one region is called the ***interior of the angle***; the other, the ***exterior of the angle***.

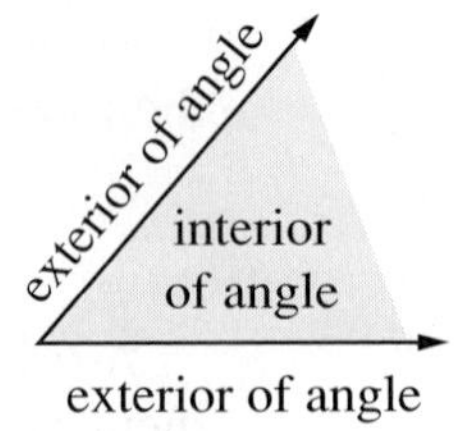

Adjacent Angles

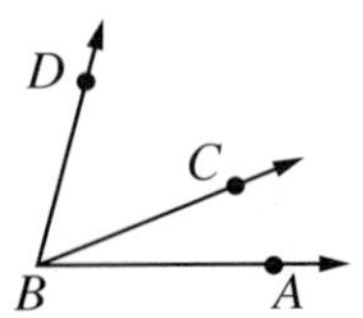

Adjacent angles are two angles in the same plane that have a common vertex and a common side but do not have any interior points in common. In the figure at the left, $\angle ABC$ and $\angle CBD$ are adjacent angles.

Complementary Angles

Two angles are ***complementary angles*** if and only if the sum of their measures is 90°. Each angle is the ***complement*** of the other. In the figures shown below, because $m\angle CAB + m\angle FDE = 25 + 65 = 90$, $\angle CAB$ and $\angle FDE$ are complementary angles. Also, because $m\angle HGI + m\angle IGJ = 53 + 37 = 90$, $\angle HGI$ and $\angle IGJ$ are complementary angles. If an angle contains 50°, its complement contains 90° − 50°, or 40°. If an angle contains x°, its complement contains $(90 - x)$°.

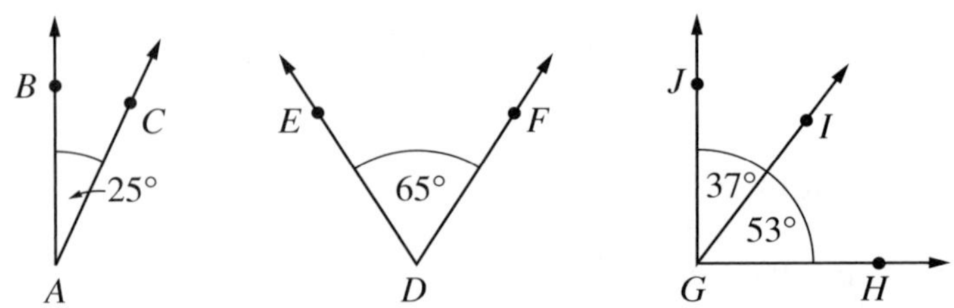

Supplementary Angles

Two angles are ***supplementary angles*** if and only if the sum of their measures is 180°. Each angle is the ***supplement*** of the other. As shown in the figures that follow, because $m\angle LKM + m\angle ONP = 50 + 130 = 180$, $\angle LKM$ and $\angle ONP$ are supplementary angles. Also, $\angle RQS$ and $\angle SQT$ are supplementary angles because

$m\angle RQS + m\angle SQT = 115 + 65 = 180$. If an angle contains 70°, its supplement contains $180° - 70°$, or 110°. If an angle contains $x°$, its supplement contains $(180 - x)°$.

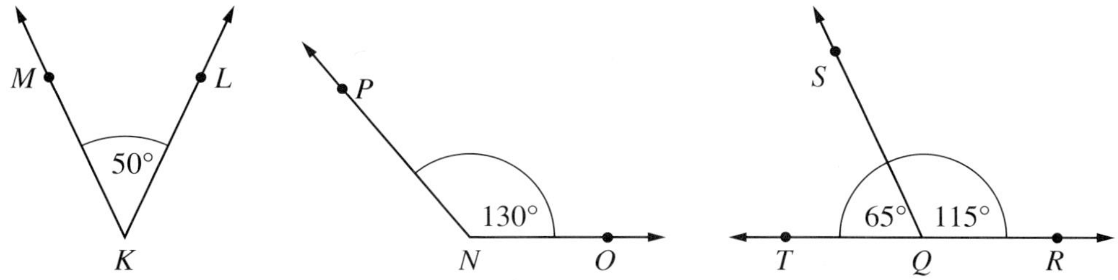

Linear Pair

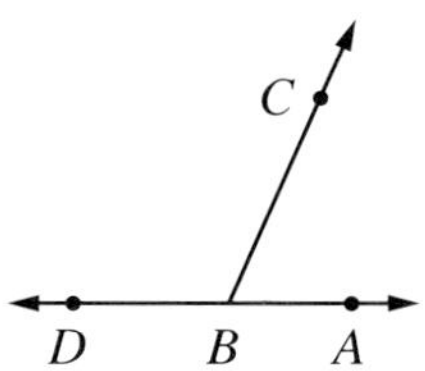

Through two points, D and A, we draw a line. We choose any point B on $\overline{DA}$ and any point C not on $\overleftrightarrow{DA}$, and we draw $\overrightarrow{BC}$. The adjacent angles formed, $\angle DBC$ and $\angle CBA$, are called a ***linear pair***. A linear pair of angles are adjacent angles that are supplementary. The two sides that they do not share in common are opposite rays. The term *linear* tells us that a *line* is used to form this pair. Since the angles are supplementary, if $m\angle DBC = x$, then $m\angle CBA = (180 - x)°$.

Vertical Angles

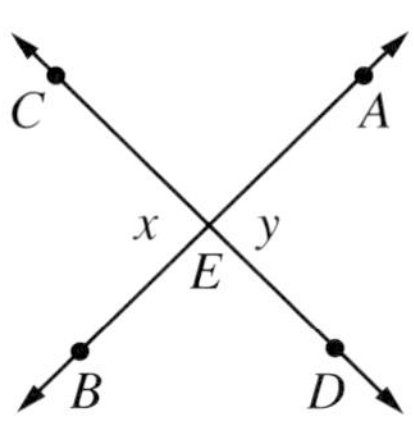

If two straight lines such as $\overleftrightarrow{AB}$ and $\overleftrightarrow{CD}$ intersect at E, $\angle x$ and $\angle y$ share a common vertex at E but do not share a common side. Angles x and y are a pair of ***vertical angles***. Two angles are vertical angles if and only if the sides of one angle are opposite rays of the sides of the second angle.

If two lines intersect, four angles are formed that have no common interior point. In the diagram at the left, $\overleftrightarrow{AB}$ and $\overleftrightarrow{CD}$ intersect at E. There are four linear pairs of angles:

$\angle AED$ and $\angle DEB$ $\quad$ $\angle DEB$ and $\angle BEC$

$\angle BEC$ and $\angle CEA$ $\quad$ $\angle CEA$ and $\angle AED$

The angles of each linear pair are supplementary.

If $m\angle y = 50$, then $m\angle AEC = 180 - 50 = 130$.
If $m\angle AEC = 130$, then $m\angle x = 180 - 130 = 50$.
Therefore, $m\angle x = m\angle y$.

- **When two angles have equal measures, they are *congruent*.**

We use the symbol $\cong$ to represent "is congruent to." Here, we would write $\angle BEC \cong \angle AED$, read as "angle BEC is congruent to angle AED." There are different correct ways to indicate angles with equal measures:

1. *The angle measures are equal:* $m\angle BEC = m\angle AED$ *or* $m\angle x = m\angle y$.

2. *The angles are congruent:* $\angle BEC \cong \angle AED$ *or* $\angle x \cong \angle y$.

It would not be correct to say that the angles are equal, or that the angle measures are congruent.

If we were to draw and measure additional pairs of vertical angles, we would find in each case that the vertical angles would be equal in measure. No matter how many examples of a given situation we consider, however, we cannot *assume* that a conclusion that we draw from these examples will always be true. We must *prove* the conclusion. Statements that we prove are called ***theorems***.

We will use algebraic expressions and properties to write an informal proof of the following statement.

- **If two lines intersect, the vertical angles formed are equal in measure; that is, they are congruent.**

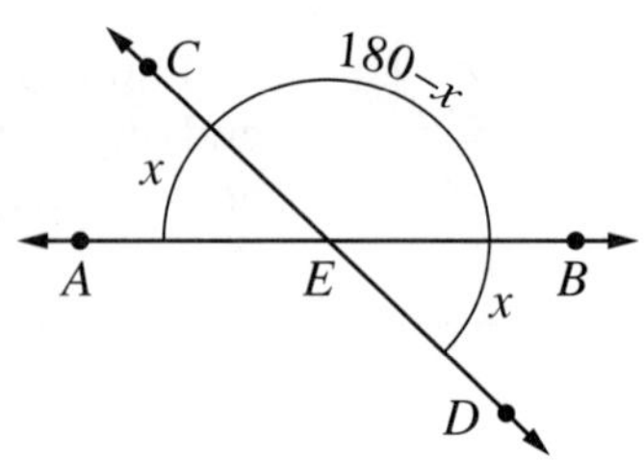

(1) If $\overleftrightarrow{AB}$ and $\overleftrightarrow{CD}$ intersect at E, then $\angle AEB$ is a straight angle whose measure is 180°. Therefore, $m\angle AEC + m\angle CEB = 180$.
(2) $m\angle AEC = x$, then $m\angle CEB = 180 - x$.
(3) Likewise, $\angle CED$ is a straight angle whose measure is 180°. Hence, $m\angle CEB + m\angle BED = 180$.
(4) Since $m\angle CEB = 180 - x$, then $m\angle BED = 180 - (180 - x) = 180 - 180 + x = x$.
(5) Since both $m\angle AEC = x$ and $m\angle BED = x$, then $m\angle AEC = m\angle BED$; that is, $\angle AEC \cong \angle BED$.

EXAMPLES

1. The measure of the complement of an angle is 4 times the measure of the angle. Find the measure of the angle.

Solution

Let x = measure of angle.

Then $4x$ = measure of complement of angle.

The sum of the measures of an angle and its complement is 90°.

$$x + 4x = 90$$
$$5x = 90$$
$$x = 18$$

Check: The measure of the first angle is 18°.
The measure of the second angle is 4(18°) or 72°.

The sum of the measures of the angles is $18° + 72° = 90°$.
Thus, the angles are complementary.

Answer: The measure of the angle is 18°.

Note: The unit of measure is very important in the solution of a problem. While it is not necessary to include the unit of measure in each step of the solution, each term in an equation must represent the same unit and the unit of measure must be included in the answer.

---

2. Find the measure of an angle if its measure is 40° more than the measure of its supplement.

Solution

Let x = measure of supplement of angle.

Then $x + 40$ = measure of angle.

The sum of the measures of an angle and its supplement is 180°.

$$x + x + 40 = 180$$
$$2x + 40 = 180$$
$$2x = 140$$
$$x = 70$$
$$x + 40 = 110$$

Answer: The measure of the angle is 110°.

---

3. The measures in degrees of a pair of vertical angles are represented by $5w - 20$ and $2w + 16$.

a. Find the value of w.

b. Find the measure of each angle.

Solutions *Vertical angles are equal in measure.*

a.

$$5w - 20 = 2w + 16$$
$$3w - 20 = 16$$
$$3w = 36$$
$$w = 12$$

b.

$$5w - 20 = 5(12) - 20 = 60 - 20 = 40$$
$$2w + 16 = 2(12) + 16 = 24 + 16 = 40$$

Check: Since each angle has a measure of 40°, the vertical angles are equal in measure.

Answers: **a.** $w = 12$ **b.** Each angle measures 40°.

EXERCISES WITH OPEN-RESPONSE PROBLEMS

Complementary Angles

In 1–10, write the measure of the complement of each angle whose measure is given.

1. 40° **2.** 25° **3.** 45° **4.** 69.5° **5.** $87\frac{1}{3}°$

6. $m°$ **7.** $d°$ **8.** $(90 - y)°$ **9.** $(x + 10)°$ **10.** $(x - 20)°$

In 11–14, ∠A and ∠B are complementary. Find the measure of each angle if the measures of the two angles are represented by the given expressions. Solve the problem algebraically using an equation.

11. $m\angle A = x$, $m\angle B = 5x$
12. $m\angle B = x$, $m\angle A = x + 50$
13. $m\angle A = x$, $m\angle B = x - 40$
14. $m\angle B = y$, $m\angle A = 2y + 30$

In 15–19, solve each problem algebraically using an equation.

15. Two angles are complementary. One angle is twice as large as the other. Find the number of degrees in each angle.
16. The complement of an angle is 8 times as large as the angle. Find the measure of the complement.
17. The complement of an angle is one-fourth of the measure of the angle. Find the measure of the angle.
18. The complement of an angle measures 20° more than the angle. Find the number of degrees in the angle.
19. Find the number of degrees in an angle that measures 8° less than its complement.

Supplementary Angles

In 20–29, write the measure of the supplement of each angle whose measure is given.

20. 40° **21.** 69° **22.** 90° **23.** 110° **24.** $167\frac{1}{2}°$

25. $m°$ **26.** $c°$ **27.** $(2y)°$ **28.** $(180 - t)°$ **29.** $(x + 40)°$

In 30–33, ∠A and ∠B form a linear pair. (They are supplementary.) Find the measure of each angle if the measures of the two angles are represented by the given expressions. Solve the problem algebraically using an equation.

30. $m\angle A = x$, $m\angle B = 3x$
31. $m\angle B = y$, $m\angle A = \frac{1}{4}y$
32. $m\angle A = w$, $m\angle B = w - 30$
33. $m\angle B = x$, $m\angle A = x + 80$

In 34–37, solve each problem algebraically using an equation.

34. Two angles are supplementary. The measure of one angle is twice as large as the measure of the other. Find the number of degrees in each angle.

35. The measure of the supplement of an angle is 40° more than the measure of the angle. Find the number of degrees in the supplement.

36. Find the number of degrees in the measure of an angle that is 20° less than 4 times the measure of its supplement.

37. The difference between the measures of two supplementary angles is 80°. Find the measure of the larger of the two angles.

38. The supplement of the complement of an acute angle is always:
(1) an acute angle (2) a right angle
(3) an obtuse angle (4) a straight angle

Vertical Angles

In 39–43, $\overleftrightarrow{AB}$ and $\overleftrightarrow{CD}$ intersect at E. Find the measure of $\angle BEC$ when $\angle AED$ measures:

39. 30° **40.** 65° **41.** 90° **42.** 128.4° **43.** $175\frac{1}{2}°$

In 44–46, $\overleftrightarrow{MN}$ and $\overleftrightarrow{RS}$ intersect at T.

44. If $m\angle RTM = 5x$ and $m\angle NTS = 3x + 10$, find $m\angle RTM$.

45. If $m\angle MTS = 4x - 60$ and $m\angle NTR = 2x$, find $m\angle MTS$.

46. If $m\angle RTM = 7x + 16$ and $m\angle NTS = 3x + 48$, find $m\angle NTS$.

Miscellaneous

In 47–54, find the measure of each angle named, based on the given conditions.

47. *Given:* $\overleftrightarrow{EF} \perp \overleftrightarrow{GH}$; $m\angle EGI = 62$.

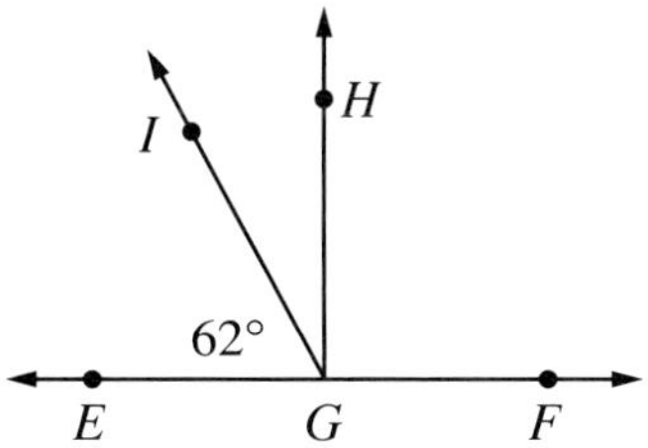

Find: $m\angle FGH$; $m\angle HGI$.

48. *Given:* $\overleftrightarrow{JK} \perp \overrightarrow{LM}$; $\overleftrightarrow{NLO}$ is a line; $m\angle NLM = 48$.

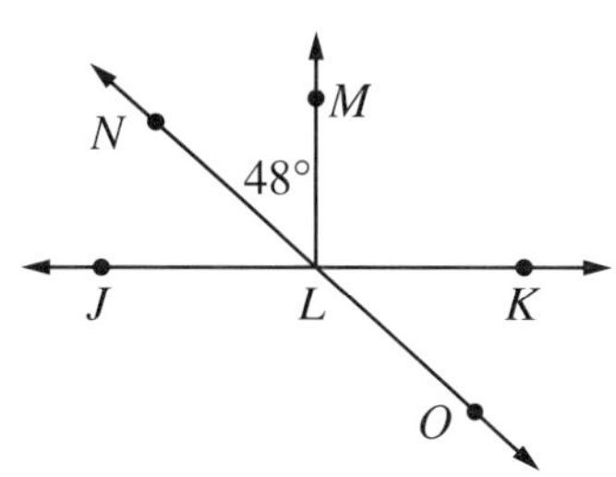

Find: $m\angle JLN$; $m\angle MLK$; $m\angle KLO$; $m\angle JLO$.

49. *Given:* $\angle GKH$ and $\angle HKI$ are a linear pair; $\overrightarrow{KH} \perp \overrightarrow{KJ}$; m$\angle IKJ = 34$.

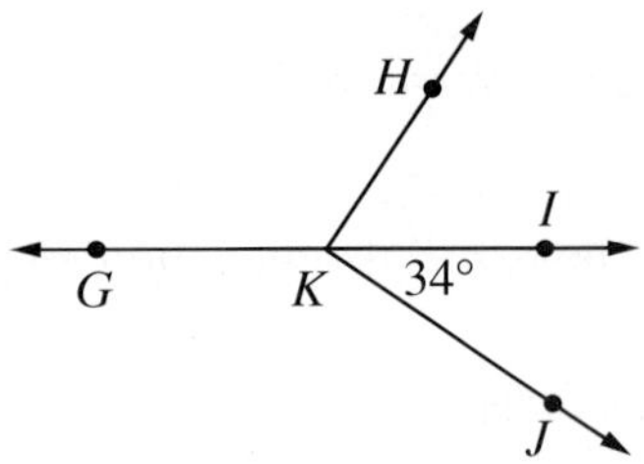

Find: m$\angle HKI$; m$\angle HKG$; m$\angle GKJ$.

50. *Given:* $\overrightarrow{MO} \perp \overrightarrow{MP}$; $\overleftrightarrow{LMN}$ is a line; m$\angle PMN = 40$.

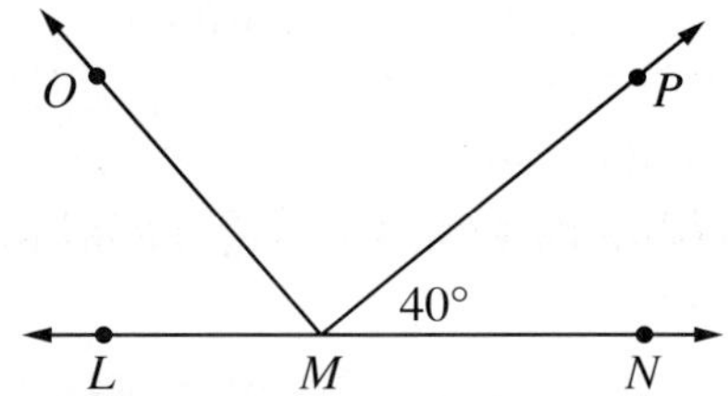

Find: m$\angle PMO$; m$\angle OML$.

51. *Given:* $\overleftrightarrow{RST} \perp \overrightarrow{SQ}$; m$\angle RSU = 89$.

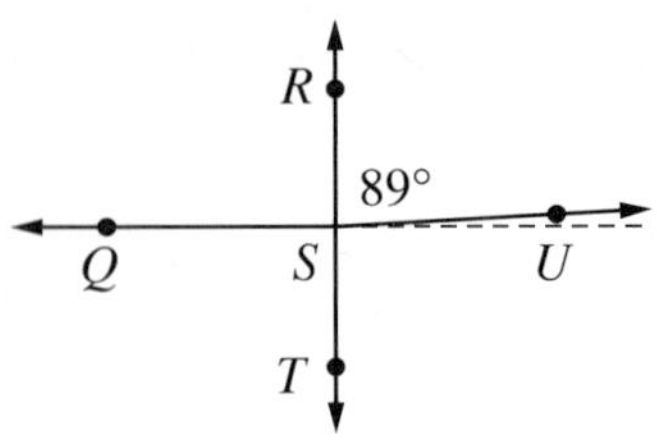

Find: m$\angle RSQ$; m$\angle QST$; m$\angle TSU$.

52. *Given:* lines $\overleftrightarrow{VWX}$ and $\overleftrightarrow{YWZ}$; m$\angle VWZ = 89$.

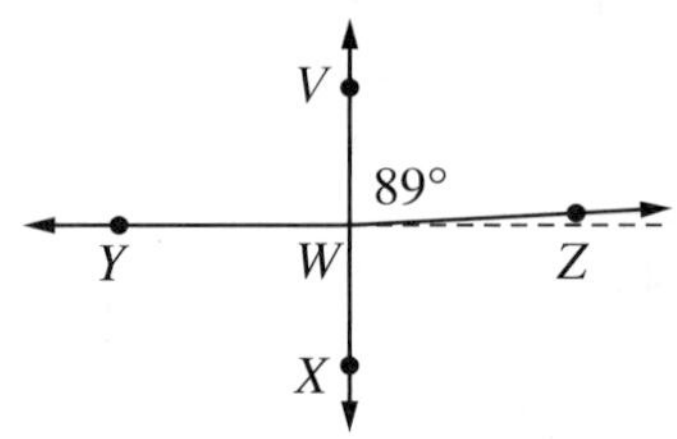

Find: m$\angle VWY$; m$\angle YWX$; m$\angle XWZ$.

53. *Given:* $\angle ABE$ and $\angle EBC$ form a linear pair; m$\angle EBC = 40$; $\angle ABD \cong \angle DBE$.

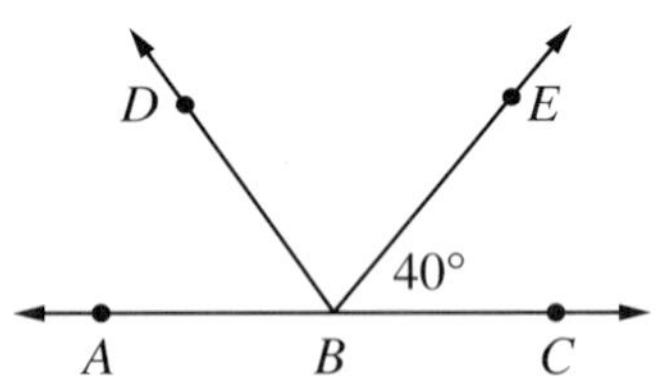

Find: m$\angle ABD$; m$\angle DBE$.

54. *Given:* $\overleftrightarrow{FI}$ intersects $\overleftrightarrow{JH}$ at K; m$\angle HKI = 40$; $\angle FKG \cong \angle FKJ$.

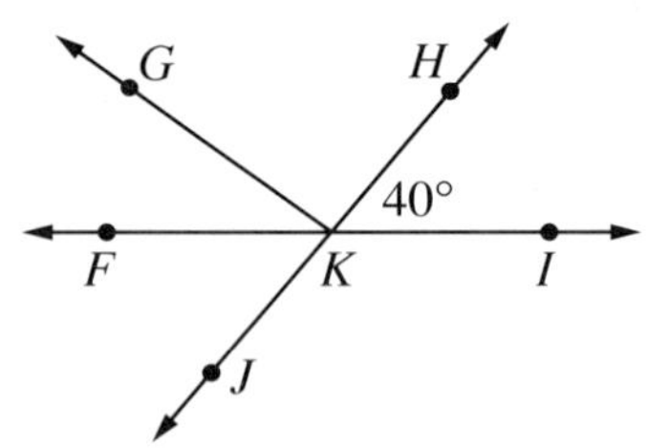

Find: m$\angle FKJ$; m$\angle FKG$; m$\angle GKH$; m$\angle JKI$.

In 55–58, sketch and label a diagram in each case; then find the measure of each angle named.

55. *Given:* $\overleftrightarrow{AB}$ intersects $\overleftrightarrow{CD}$ at E; m$\angle AED = 20$.
Find: m$\angle CEB$; m$\angle BED$; m$\angle CEA$.

56. *Given:* $\angle PQR$ and $\angle RQS$ are complementary; m$\angle PQR = 30$; $\overleftrightarrow{RQT}$ is a line.
Find: m$\angle RQS$; m$\angle SQT$; m$\angle PQT$.

57. *Given:* $\overleftrightarrow{LM}$ intersects $\overleftrightarrow{PQ}$ at R. The number of degrees in $\angle LRQ$ is 80 more than the number of degrees in $\angle LRP$.
Find: m$\angle LRP$; m$\angle LRQ$; m$\angle PRM$.

58. *Given:* $\overleftrightarrow{CD}$ is perpendicular to $\overleftrightarrow{AB}$ at E. Point F is in the interior of $\angle CEB$. The measure of $\angle CEF$ is 8 times the measure of $\angle FEB$.
Find: m$\angle FEB$; m$\angle CEF$; m$\angle AEF$.

59. Two angles, $\angle ABD$ and $\angle DBC$, form a linear pair and are congruent. What must be true about $\overleftrightarrow{AC}$ and $\overleftrightarrow{BD}$?

10-4 ANGLES AND PARALLEL LINES

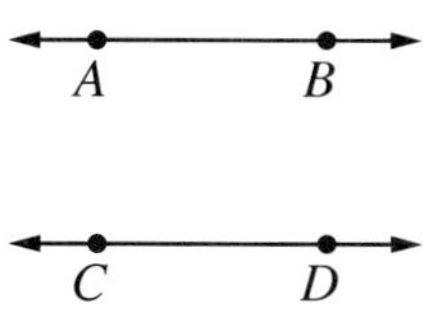

Not all lines in the same plane intersect. Two or more lines are called ***parallel lines*** if and only if the lines lie in the same plane but do not intersect.

In the figure at the left, $\overleftrightarrow{AB}$ and $\overleftrightarrow{CD}$ lie in the same plane but do not intersect. Hence, we say that $\overleftrightarrow{AB}$ is parallel to $\overleftrightarrow{CD}$. Using the symbol $\parallel$ for *is parallel to*, we write $\overleftrightarrow{AB} \parallel \overleftrightarrow{CD}$. When we speak of two parallel lines, we will mean two *distinct* lines. (In more advanced courses, you will see that a line is parallel to itself.)

Line segments and rays are parallel if the lines that contain them are parallel.

If two lines such as $\overleftrightarrow{AB}$ and $\overleftrightarrow{CD}$ lie in the same plane, they must be either intersecting lines or parallel lines, as shown in the following figures.

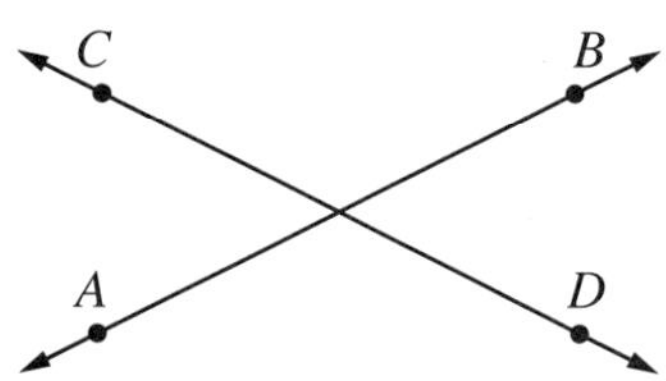

$\overleftrightarrow{AB}$ intersects $\overleftrightarrow{CD}$.

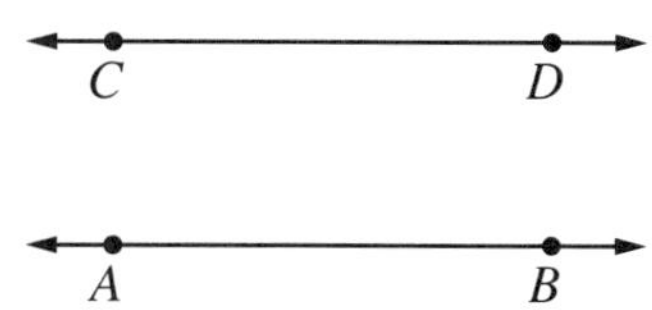

AB is parallel to *CD*.

When two lines such as $\overleftrightarrow{AB}$ and $\overleftrightarrow{CD}$ are parallel, they have no points in common. Hence, the intersection set of $\overleftrightarrow{AB}$ and $\overleftrightarrow{CD}$ is the empty set symbolized as $\overleftrightarrow{AB} \cap \overleftrightarrow{CD} = \varnothing$.

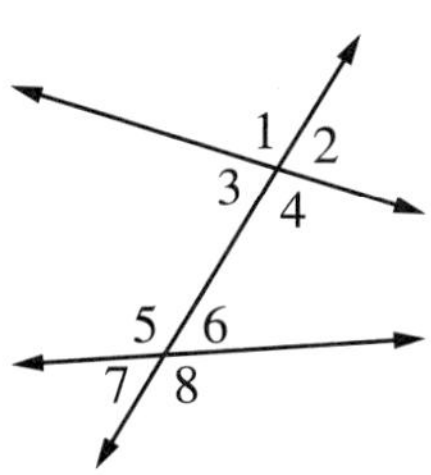

When two lines are cut by a third line, called a ***transversal***, two sets of angles, each containing four angles, are formed. In the figure at the left:

Angles 3, 4, 5, 6 are called ***interior angles***.

Angles 1, 2, 7, 8 are called ***exterior angles***.

Angles 4 and 5 are interior angles on opposite sides of the transversal and do not have the same vertex. They are called ***alternate interior angles***. Angles 3 and 6 are another pair of alternate interior angles.

Angles 1 and 8 are exterior angles on opposite sides of the transversal and do not have the same vertex. They are called ***alternate exterior angles***. Angles 2 and 7 are another pair of alternate exterior angles.

Angles 4 and 6 are ***interior angles on the same side of the transversal***. Angles 3 and 5 are another pair of interior angles on the same side of the transversal.

Angles 1 and 5 are on the same side of the transversal, one interior and one exterior, and at different vertices. They are called ***corresponding angles***. Other pairs of corresponding angles are 2 and 6, 3 and 7, 4 and 8.

Alternate Interior Angles and Parallel Lines

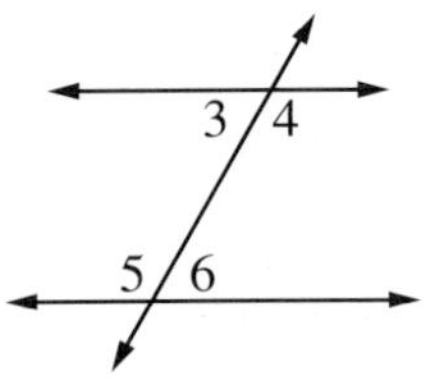

In the figure at the left, a transversal intersects two parallel lines, forming a pair of alternate interior angles, $\angle 3$ and $\angle 6$.

If we measure $\angle 3$ and $\angle 6$ with a protractor, we will find that each angle measures 60°. Here, alternate interior angles 3 and 6 have equal measures, and $\angle 3 \cong \angle 6$. If we draw other pairs of parallel lines intersected by transversals, we will find again that pairs of alternate interior angles have equal measures. Yet, we would be hard pressed to prove that this is *always* true. Therefore, we accept, without proof, the following postulate:

- **If two parallel lines are cut by a transversal, then the alternate interior angles that are formed have equal measures, that is, they are congruent.**

Note that $\angle 4$ and $\angle 5$ are another pair of alternate interior angles formed by a transversal that intersects parallel lines. Therefore, $\angle 4 \cong \angle 5$.

Corresponding Angles and Parallel Lines

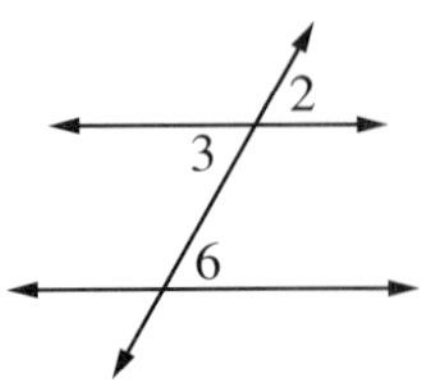

If two lines are cut by a transversal, four pairs of corresponding angles are formed. One such pair of corresponding angles is $\angle 2$ and $\angle 6$, as shown in the figure at the left. If the original two lines are parallel, do these corresponding angles have equal measures? We are ready to prove in an informal manner that they do.

(1) Let $m\angle 2 = x$.
(2) If $m\angle 2 = x$, then $m\angle 3 = x$ (because $\angle 2$ and $\angle 3$ are vertical angles, which we have previously shown must have equal measures).
(3) If $m\angle 3 = x$, then $m\angle 6 = x$ (because $\angle 3$ and $\angle 6$ are alternate interior angles of parallel lines, and we have just accepted the postulate that they have the same measure).
(4) Therefore $m\angle 2 = m\angle 6$ (because the measure of each angle is x).

These four steps serve as a proof of the following theorem:

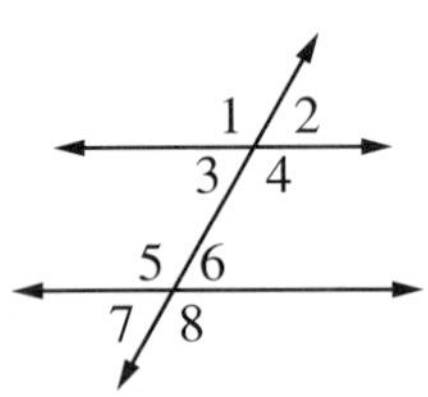

- **If two parallel lines are cut by a transversal, then the corresponding angles formed have equal measures, that is, they are congruent.**

Note that this theorem is true for all pairs of corresponding angles: $\angle 1 \cong \angle 5$; $\angle 2 \cong \angle 6$; $\angle 3 \cong \angle 7$; $\angle 4 \cong \angle 8$.

Alternate Exterior Angles and Parallel Lines

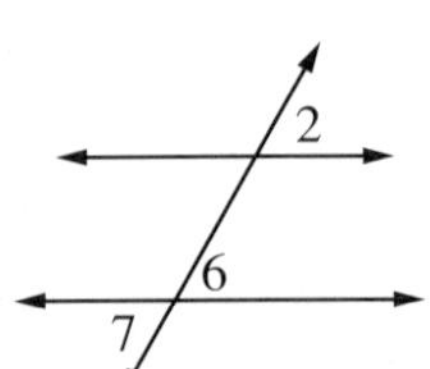

If two parallel lines are cut by a transversal, we can prove informally that the alternate exterior angles formed have equal measures. One such pair of alternate exterior angles consists of $\angle 2$ and $\angle 7$, as shown in the figure at the left.

(1) Let $m\angle 2 = x$.
(2) If $m\angle 2 = x$, then $m\angle 6 = x$ (because $\angle 2$ and $\angle 6$ are corresponding angles of parallel lines, proved to have the same measure).

(3) If m∠6 = x, then m∠7 = x (because ∠6 and ∠7 are vertical angles, previously proved to have the same measure).
(4) Therefore m∠2 = m∠7 (because the measure of each angle is x).

These four steps serve as a proof of the following theorem:

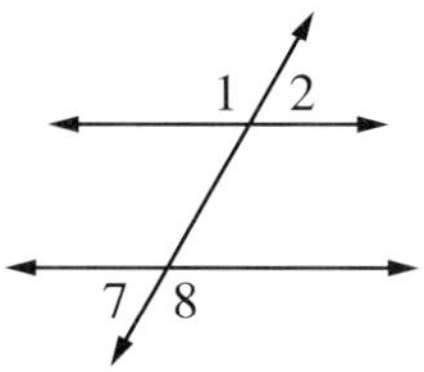

- **If two parallel lines are cut by a transversal, then the alternate exterior angles formed have equal measures, that is, they are congruent.**

Note that this theorem is true for all pairs of alternate exterior angles: ∠1 ≅ ∠8; ∠2 ≅ ∠7.

Interior Angles on the Same Side of the Transversal

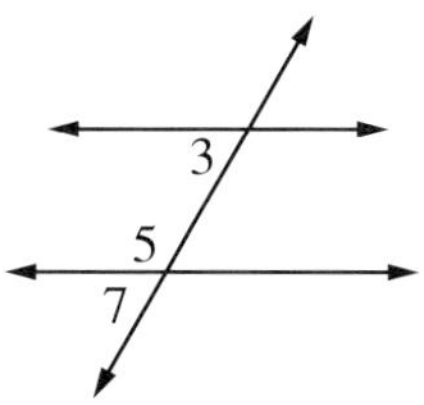

When two parallel lines are cut by a transversal, we can prove informally that the sum of the measures of the interior angles on the same side of the transversal is 180°. One such pair of interior angles on the same side of the transversal consists of ∠3 and ∠5, as shown in the figure at the left.

(1) m∠5 + m∠7 = 180 (∠5 and ∠7 are supplementary angles).
(2) m∠7 = m∠3 (∠7 and ∠3 are corresponding angles).
(3) m∠5 + m∠3 = 180 (by substituting m∠3 for m∠7).

These three steps serve as a proof of the following theorem:

- **If two parallel lines are cut by a transversal, then the sum of the measures of the interior angles on the same side of the transversal is 180°.**

EXAMPLE

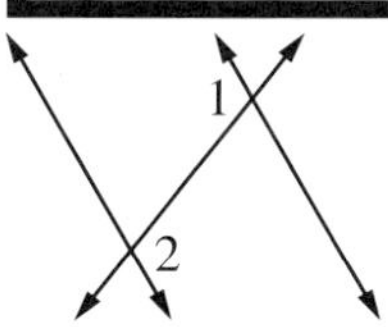

In the figure, the parallel lines are cut by a transversal. If m∠1 = $(5x - 10)$ and m∠2 = $(3x + 60)$, find the measures of ∠1 and ∠2.

Solution Since the lines are parallel, the alternate interior angles, ∠1 and ∠2, have equal measures. Write and solve:

$$5x - 10 = 3x + 60$$
$$5x = 3x + 70$$
$$2x = 70$$
$$x = 35$$
$$5x - 10 = 5(35) - 10 = 175 - 10 = 165$$
$$3x + 60 = 3(35) + 60 = 105 + 60 = 165$$

Answers: m∠1 = 165 and m∠2 = 165.

EXERCISES WITH OPEN-RESPONSE PROBLEM

In 1–4, the figure at the right shows two parallel lines cut by a transversal. In each exercise, given the measure of one angle, find the measures of the other seven angles.

1. m∠3 = 80 **2.** m∠6 = 150

3. m∠5 = 60 **4.** m∠1 = 75

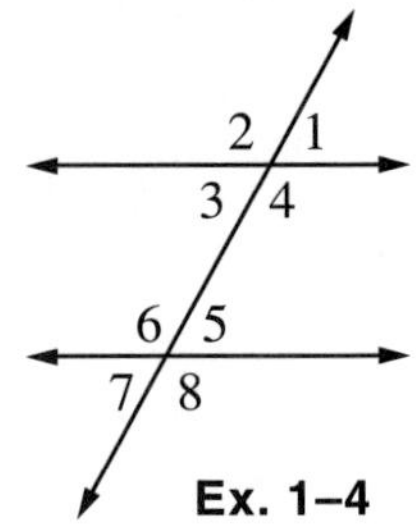

Ex. 1–4

5. If $\overleftrightarrow{AB} \parallel \overleftrightarrow{CD}$, m∠5 = 40, and m∠4 = 30, find the measures of the other angles in the figure.

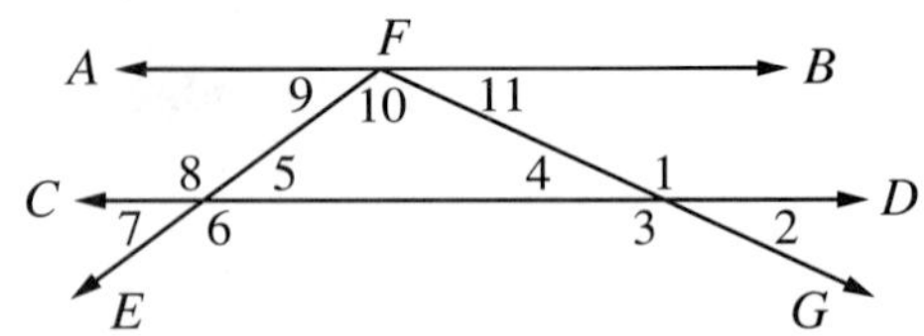

In 6–10, the figure at the right shows two parallel lines cut by a transversal. In each exercise, find the measures of all eight angles under the given conditions.

6. m∠3 = $2x + 40$ and m∠7 = $3x + 27$

7. m∠4 = $4x - 10$ and m∠6 = $x + 80$

8. m∠4 = $3x + 40$ and m∠5 = $2x$

9. m∠4 = $2x - 10$ and m∠2 = $x + 60$

10. ∠8 ≅ ∠3

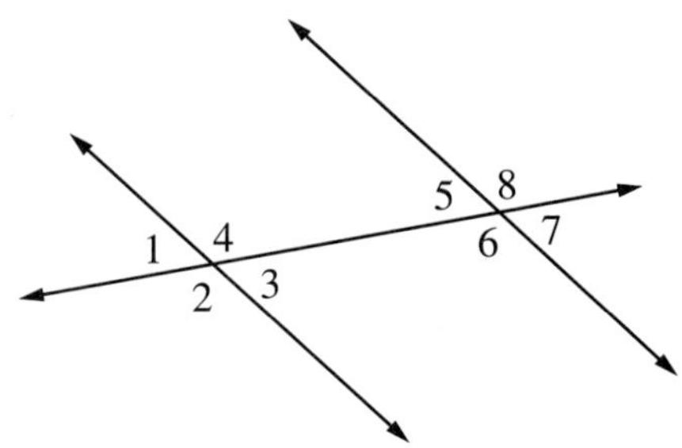

Ex. 6–10

In 11–16, tell whether each statement is always, sometimes, or never true.

11. If two distinct lines intersect, then they are parallel.
12. If two distinct lines do not intersect, then they are parallel.
13. If two angles are alternate interior angles, then they are on opposite sides of the transversal.
14. If two parallel lines are cut by a transversal, then the alternate interior angles are congruent.
15. If two parallel lines are cut by a transversal, then the alternate interior angles are complementary.
16. If two parallel lines are cut by a transversal, then the corresponding angles are supplementary.
17. In the figure at the right, two parallel lines are cut by a transversal. Write an informal proof that ∠1 and ∠2 have equal measures.

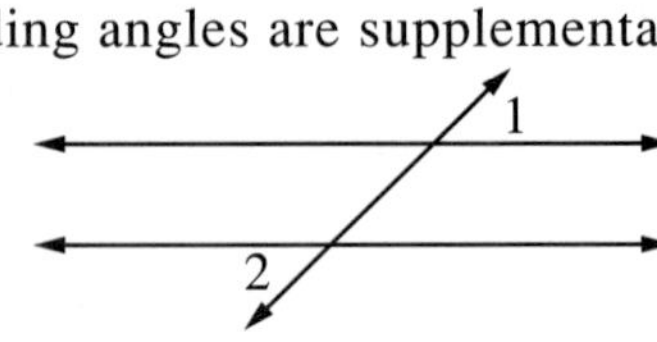

10-5 TRIANGLES AND ANGLES

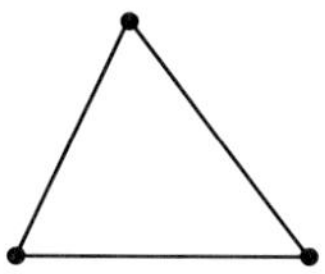

When three points that are not all on the same line are joined in pairs by line segments, the figure formed is a triangle. Each of the given points is the vertex of an angle of the triangle, and each line segment is a side of the triangle.

There are many practical uses of the triangle, especially in construction work such as the building of bridges, radio towers, and airplane wings, because the triangle is a ***rigid figure***. The shape of a triangle cannot be changed without changing the length of at least one of its sides.

We begin our discussion of the triangle by considering triangle *ABC*, shown below, which is symbolized as $\triangle ABC$. In $\triangle ABC$, points *A*, *B*, and *C* are the ***vertices***, and $\overline{AB}$, $\overline{BC}$, and $\overline{CA}$ are the ***sides***. Angles *A*, *B*, and *C* of the triangle are symbolized as $\angle A$, $\angle B$, and $\angle C$.

We make the following observations:

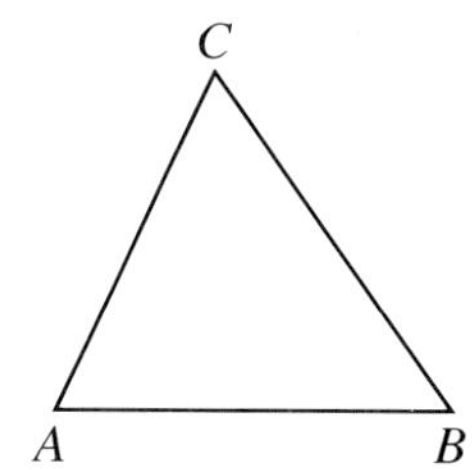

1. Side $\overline{AB}$ is included between $\angle A$ and $\angle B$.
2. Side $\overline{BC}$ is included between $\angle B$ and $\angle C$.
3. Side $\overline{CA}$ is included between $\angle C$ and $\angle A$.
4. Angle *A* is included between sides $\overline{AB}$ and $\overline{AC}$.
5. Angle *B* is included between sides $\overline{BA}$ and $\overline{BC}$.
6. Angle *C* is included between sides $\overline{CA}$ and $\overline{CB}$.

Classifying Triangles According to Angles

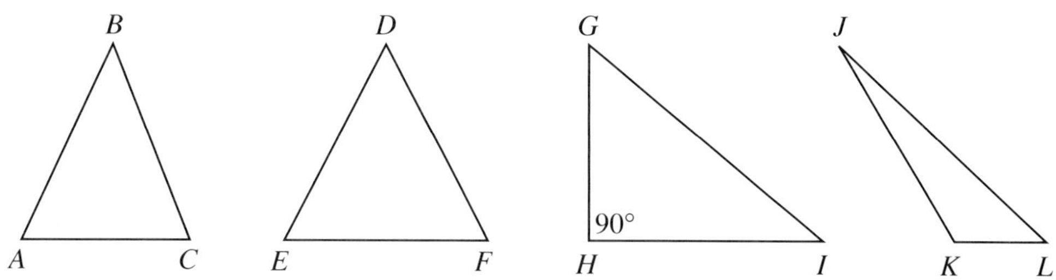

Acute triangle | Equiangular triangle | Right triangle | Obtuse triangle

An ***acute triangle*** has three acute angles.

An ***equiangular triangle*** has three angles equal in measure.

A ***right triangle*** has one right angle.

An ***obtuse triangle*** has one obtuse angle.

In right triangle *GHI* above, the two sides of the triangle that form the right angle, $\overline{GH}$ and $\overline{HI}$, are called the ***legs*** of the right triangle. The side opposite the right angle, $\overline{GI}$, is called the ***hypotenuse***.

Sum of the Measures of the Angles of a Triangle

When we change the shape of a triangle, changes take place also in the measures of its angles. Is there any relationship among the measures of the triangle that does not change? Let us see.

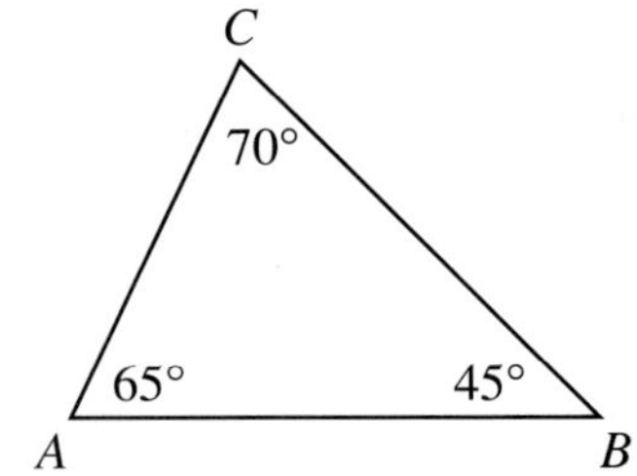

Draw several triangles of different shapes. In each triangle, measure the three angles and find the sum of the three measures. For example, in $\triangle ABC$ at the right, $m\angle A + m\angle B + m\angle C$ is equal to:

$$65 + 45 + 70 = 180.$$

If we measured accurately, we should have found that in each triangle, regardless of shape, the sum of the measures of the three angles is 180°. You can see that this is so by tearing off two angles of any triangle and placing them adjacent to the third angle, as is shown in the figures below.

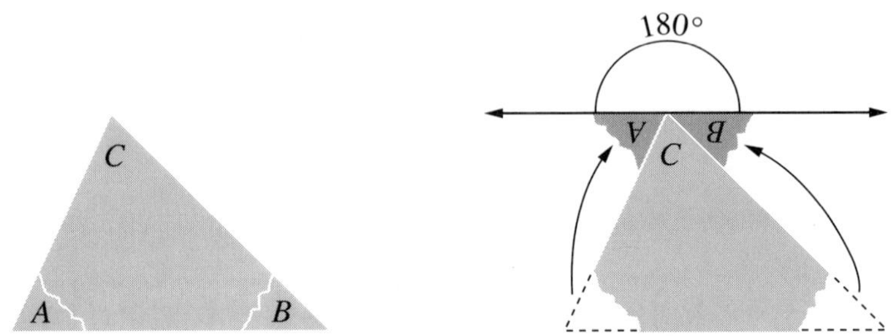

We can write an informal algebraic proof of the following statement:

- **The sum of the measures of the angles of a triangle is 180°.**

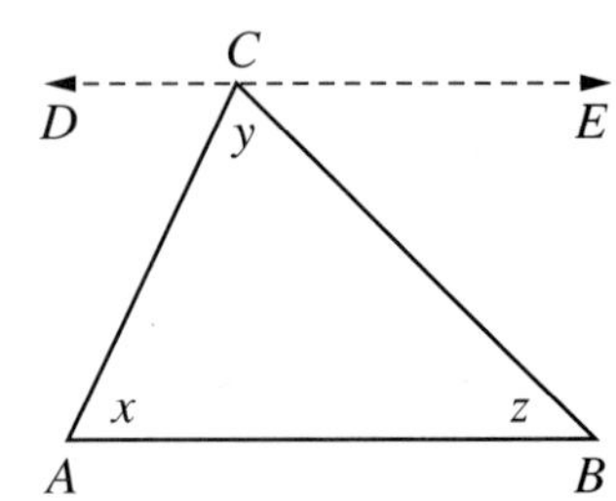

(1) In $\triangle ABC$, let $m\angle A = x$, $m\angle ACB = y$, and $m\angle B = z$.
(2) Let $\overleftrightarrow{DCE}$ be a line parallel to $\overline{AB}$.
(3) Since $\angle DCE$ is a straight angle, $m\angle DCE = 180$.
(4) $m\angle DCE = m\angle DCA + m\angle ACB + m\angle BCE$.
(5) $m\angle DCA + m\angle ACB + m\angle BCE = 180$.
(6) $m\angle DCA = m\angle A = x$.
(7) $m\angle BCE = m\angle B = z$.
(8) Substituting values from lines (6), (1), and (7) into the equation from line (5) gives $x + y + z = 180$.

EXAMPLES

1. Find the measure of the third angle of a triangle if the measures of two of the angles are 72.6° and 84.2°.

Solution Subtract the sum of the known measures from 180:

$$180 - (72.6 + 84.2) = 23.2$$

Calculator Solution

Enter: 180 [−] [( ] 72.6 [+] 84.2 [ )] [=]

Display: [23.2]

Answer: 23.2°

---

2. In $\triangle ABC$, the measure of $\angle B$ is twice the measure of $\angle A$, and the measure of $\angle C$ is 3 times the measure of $\angle A$. Find the number of degrees in each angle of the triangle.

Solution

Let x = number of degrees in $\angle A$.
Then $2x$ = number of degrees in $\angle B$.
Then $3x$ = number of degrees in $\angle C$.

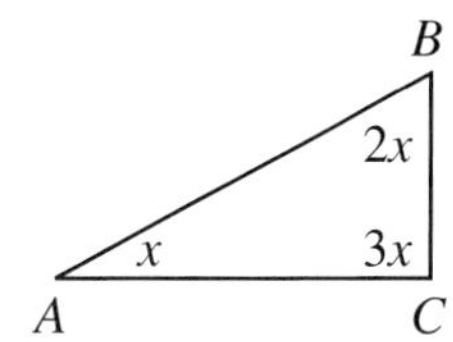

The sum of the measures of the angles of a triangle is 180°.

$$x + 2x + 3x = 180$$

$$6x = 180$$

$$x = 30$$

$$2x = 60$$

$$3x = 90$$

Check:

$$60 = 2 \times 30$$

$$90 = 3 \times 30$$

$$30 + 60 + 90 = 180$$

Answer: $m\angle A = 30$, $m\angle B = 60$, $m\angle C = 90$

EXERCISES WITH OPEN-RESPONSE PROBLEMS

In 1–3, state, in each case, whether the angles given can be the three angles of the same triangle.

1. 30°, 70°, 80° **2.** 70°, 80°, 90° **3.** 30°, 110°, 40°

In 4–7, find, in each case, the measure of the third angle of the triangle if the measures of the first two angles are:

4. 60°, 40° **5.** 100°, 20° **6.** 54.5°, 82.3° **7.** $24\frac{1}{4}°$, $81\frac{3}{4}°$

8. What is the number of degrees in each angle of an equiangular triangle.
9. Can a triangle have: **(a)** two right angles? **(b)** two obtuse angles? **(c)** one right and one obtuse angle? Why?
10. What is the sum of the measures of the two acute angles of a right triangle?
11. If two angles in one triangle contain the same number of degrees as two angles in another triangle, what must be true of the third pair of angles in the two triangles? Why?
12. In a triangle, the measure of the second angle is 3 times the measure of the first angle, and the measure of the third angle is 5 times the measure of the first angle. Find the number of degrees in each angle of the triangle.
13. In a triangle, the measure of the second angle is 4 times the measure of the first angle. The measure of the third angle is equal to the sum of the measures of the first two angles. Find the number of degrees in each angle of the triangle.
14. In a triangle, the measure of the second angle is 30° more than the measure of the first angle, and the measure of the third angle is 45° more than the measure of the first angle. Find the number of degrees in each angle of the triangle.
15. In a triangle, the measure of the second angle is 5° more than twice the measure of the first angle. The measure of the third angle is 35° less than 3 times the measure of the first angle. Find the number of degrees in each angle of the triangle.

16. $\overleftrightarrow{AEFB}$ is a straight line; $m\angle AEG = 130$, $m\angle BFG = 140$.
   a. Find $m\angle x$, $m\angle y$, and $m\angle z$.
   b. What kind of triangle is $\triangle EFG$?

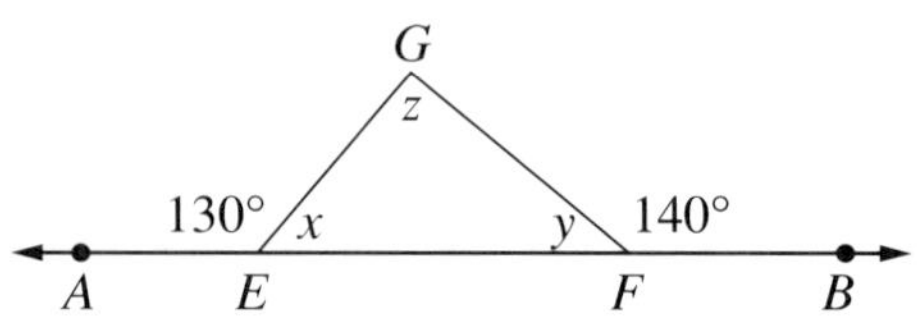

17. In $\triangle RST$, $m\angle R = x$, $m\angle S = x + 30$, $m\angle T = x - 30$.
   a. Find the measures of the three angles.
   b. What kind of triangle is $\triangle RST$?

18. In $\triangle KLM$, $m\angle K = 2x$, $m\angle L = x + 30$, $m\angle M = 3x - 30$.
   a. Find the measures of the three angles.
   b. What kind of triangle is $\triangle KLM$?

The Exterior Angle of a Triangle

In the figure at the right, side $\overline{AC}$ is extended to form $\angle BCE$ at vertex C. Notice $\angle BCE$ is in the *exterior* of $\triangle ABC$, and that $\angle BCE$ and $\angle BCA$ are supplementary, forming a linear pair. We call $\angle BCE$ an ***exterior angle*** of $\triangle ABC$, drawn at vertex C.

If $m\angle BCA = x$, then $m\angle BCE = 180 - x$.

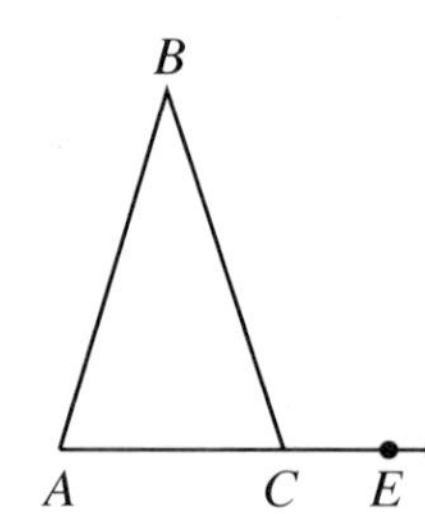

- **An exterior angle of a triangle is an angle that forms a linear pair with one of the angles of the triangle.**

In the figure at the right, $\triangle DEF$ is shown with exterior $\angle EFG$ at vertex F. There are three angles in the *interior* of $\triangle DEF$. We know that $\angle EFD$ is *adjacent* to exterior $\angle EFG$ because these angles form a linear pair. The two angles at the remaining vertices, $\angle D$ and $\angle E$, are called ***remote interior angles*** to the exterior $\angle EFG$. Using this triangle and exterior $\angle EFG$:

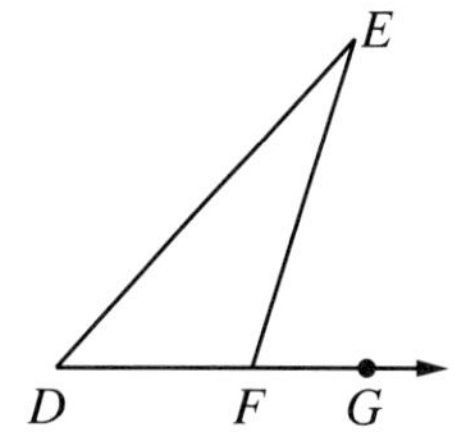

(1) If $m\angle D = 45$ and $m\angle E = 30$, then $m\angle EFD = 180 - (45 + 30) = 105$.
(2) If $m\angle EFD = 105$ and this angle forms a linear pair with exterior $\angle EFG$, then
$m\angle EFG = 180 - 105 = 75$.
(3) Hence, $m\angle D + m\angle E = m\angle EFG$, since $45 + 30 = 75$.

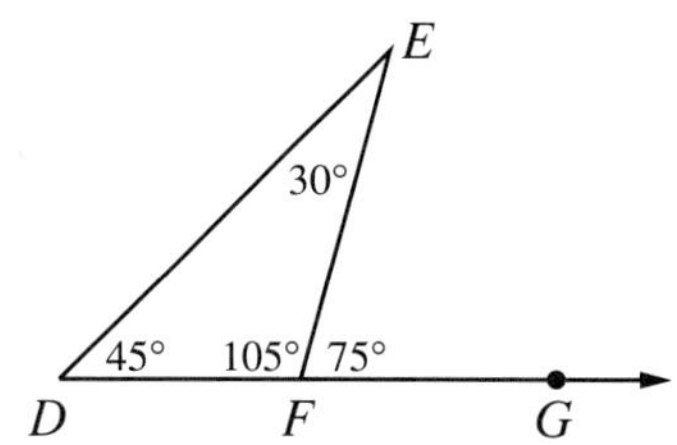

This example illustrates the truth of the following statement, which is proved informally by replacing the given values by variables.

- **The measure of an exterior angle of a triangle is equal to the sum of the measures of the two remote interior angles.**

(1) Let $x = m\angle D$
$y = m\angle E$,
$z = m\angle EFD$,
and $w = m\angle EFG$.
(2) $x + y + z = 180$ or $x + y = 180 - z$
(3) $w + z = 180$ or $w = 180 - z$
(4) Therefore, $x + y = w$.

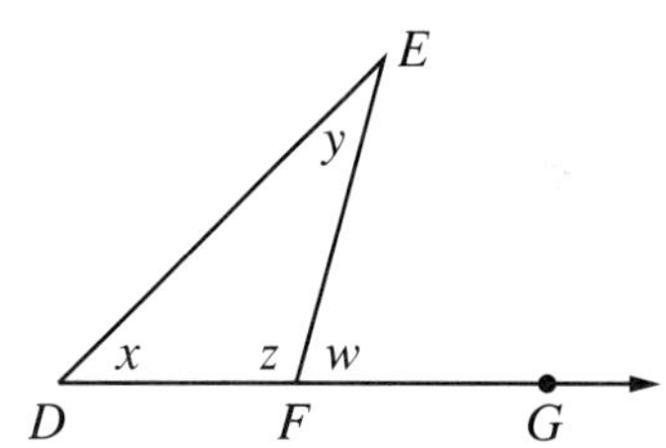

Properties of Triangles

1. The sum of the measures of the angles of a triangle is 180°.
2. The acute angles of a right triangle are complementary.
3. If the measures of two angles of one triangle are equal, respectively, to the measures of two angles of another triangle, then the remaining angles are equal in measure.
4. The measure of an exterior angle of a triangle is equal to the sum of the measures of the two remote interior angles.

EXAMPLE

In $\triangle ABC$, $m\angle A = 37.5$ and the measure of an exterior angle at B is 104.4°. Find the measures of $\angle B$ and $\angle C$.

Solution The measure of an exterior angle is equal to the sum of the measures of the two remote interior angles:

$$\begin{aligned} m\angle A + m\angle C &= 104.4 \\ 37.5 + m\angle C &= 104.4 \\ m\angle C &= 104.4 - 37.5 \\ &= 66.9 \\ m\angle B &= 180 - 104.4 \\ &= 75.6 \end{aligned}$$

Calculator Solution

To find $m\angle C$:	To find $m\angle B$:
Enter: 104.4 [−] 37.5 [=]	***Enter:*** 180 [−] 104.4 [=]
Display: [66.9]	***Display:*** [75.6]

Answer: The measure of $\angle B$ is 75.6° and of $\angle C$ is 66.9°.

EXERCISES WITH OPEN-RESPONSE PROBLEMS

In 1–5: **a.** Name the exterior angle of the triangle that is shown in each diagram. **b.** Name the two remote interior angles to that exterior angle.

1.

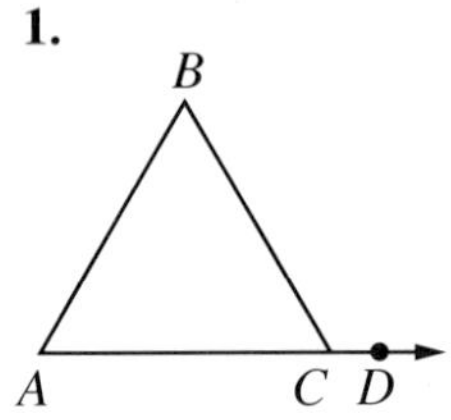

2.

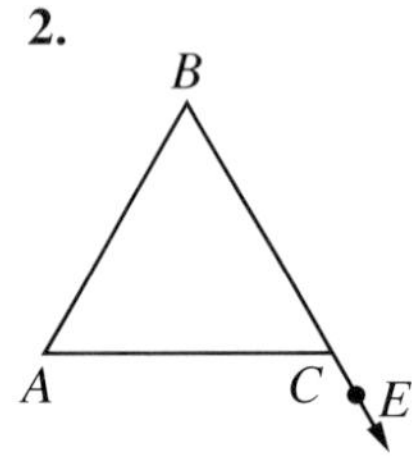

3.

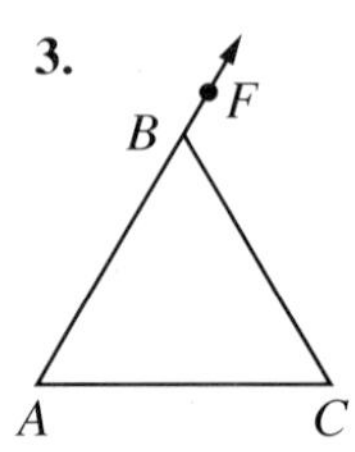

4.

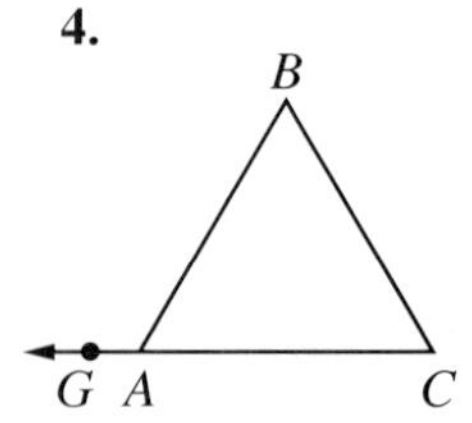

5.

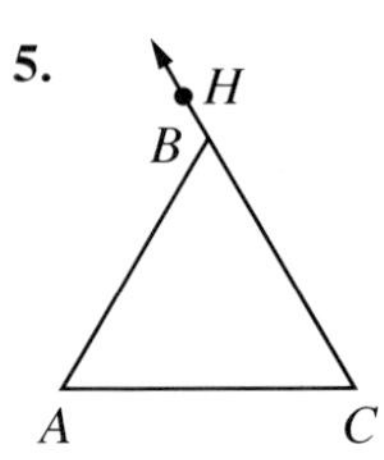

In 6–10, find, in each case, the number of degrees in the value of x.

6.

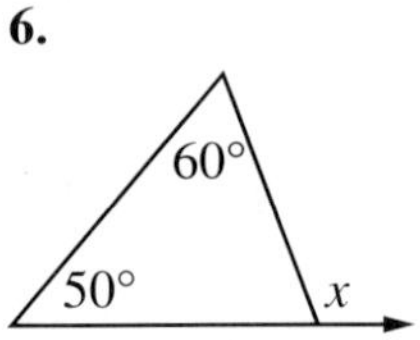

7.

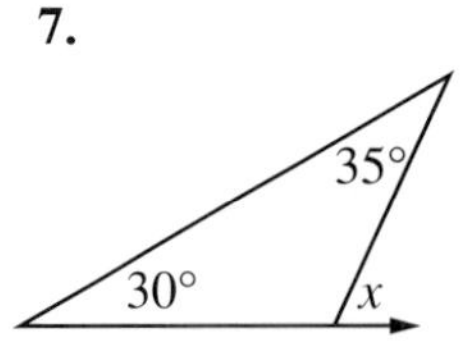

8.

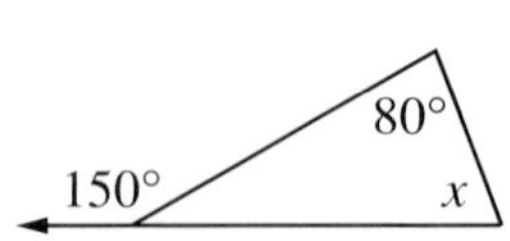

9.

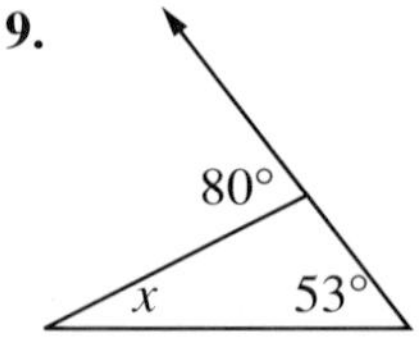

10.

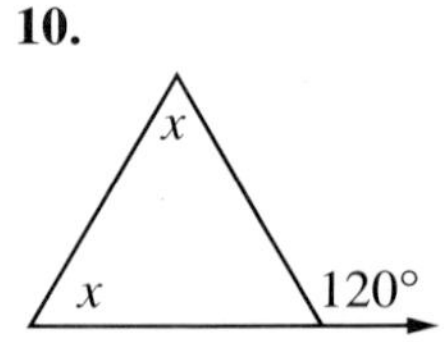

In 11–13, draw and label a diagram to fit each set of conditions, and then find the value of x.

11. In $\triangle ABC$, $m\angle A = x$, $m\angle B = x + 10$, and the measure of an exterior angle at C is 70°.

12. In $\triangle PQR$, $m\angle P = 2x$, $m\angle Q = 3x$, and the measure of an exterior angle at R is 150°.

13. In $\triangle RST$, $m\angle R = 36$, $m\angle S = 2x$, and the measure of an exterior angle at T is $(5x)°$.

In 14 and 15, draw and label a diagram for each case, and then find the specified measure.

14. In $\triangle DEF$, $m\angle D = 80$, and the measure of an exterior angle at E is 3 times the measure of $\angle F$. Find the measure of $\angle F$.

15. In $\triangle KLM$, $m\angle K = 85$, and the measure of an exterior angle at L is 135°. Find the measure of an exterior angle at M.

16. In an equiangular triangle, what is the measure of any one of its exterior angles?

17. In a right triangle, what is the measure of the smallest exterior angle?

18. An exterior angle is drawn to a triangle. If this exterior angle is acute, then the triangle must be: (1) acute (2) right (3) obtuse (4) equilateral.

19. In the figure at the right, three exterior angles are drawn to $\triangle ABC$.

a. What is the sum of the measures of the three exterior angles of $\triangle ABC$?

b. Will the answer to part **a** be true for the sum of the measures of the exterior angles of any triangle?

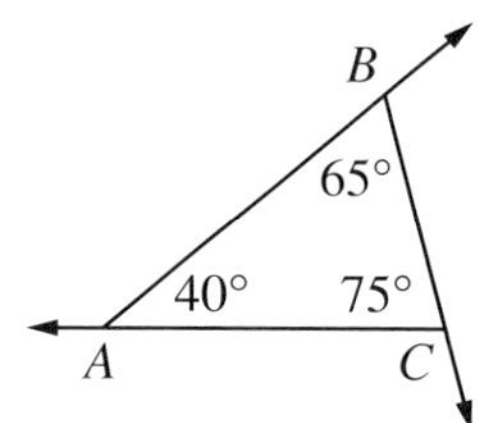

20. In the figure at the right, $\angle ACD$ is an exterior angle at C. Through C, draw a line parallel to $\overline{AB}$. Use alternate interior angles and corresponding angles to show that $m\angle ACD = m\angle A + m\angle B$.

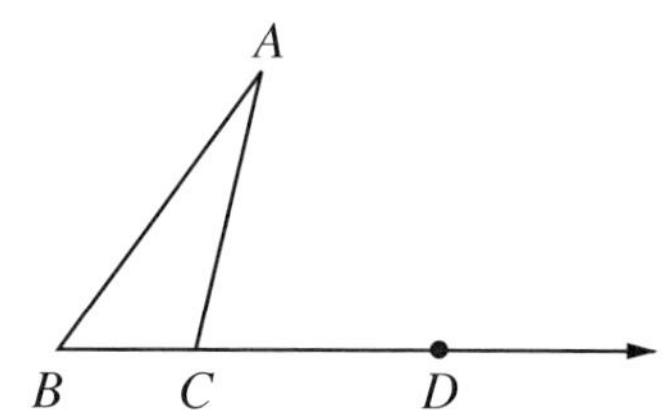

10-6 TRIANGLES WITH CONGRUENT ANGLES

Classifying Triangles According to Sides

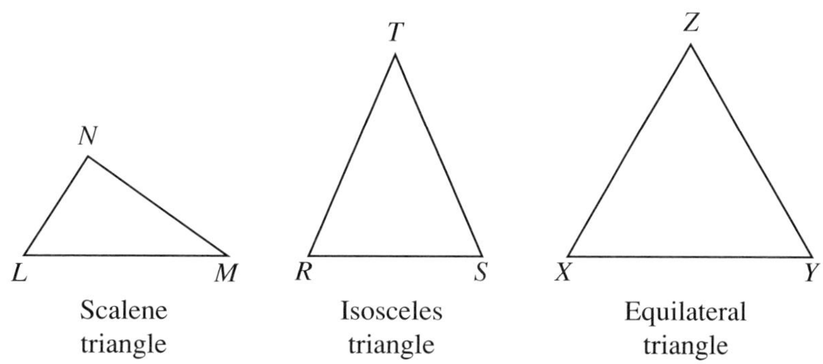

A *scalene triangle* has no sides equal in length.
An *isosceles triangle* has two sides equal in length.
An *equilateral triangle* has three sides equal in length.

The Isosceles Triangle

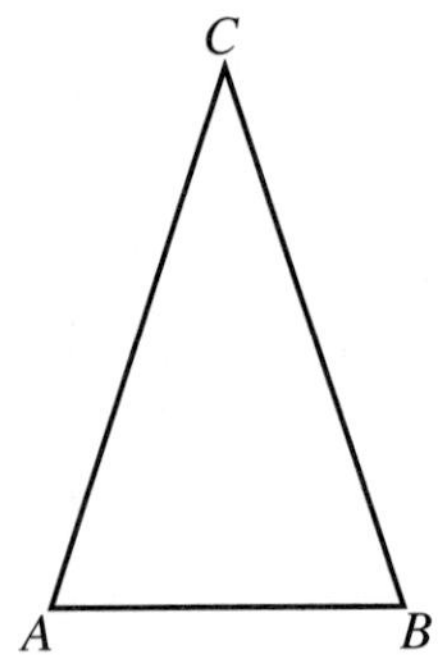

In isosceles triangle ABC, shown at the left, the two sides that are equal in measure, $\overline{AC}$ and $\overline{BC}$, are the called ***legs***. The third side, $\overline{AB}$, is the ***base***.

The angle formed by the two congruent sides, $\angle C$, is called the ***vertex angle***. The two angles at the endpoints of the base, $\angle A$ and $\angle B$, are the ***base angles***.

In isosceles triangle ABC, if we measure the base angles, $\angle A$ and $\angle B$, we find that each angle contains 65°. Therefore, $m\angle A = m\angle B$. Similarly, if we measure the base angles in any other isosceles triangle, we find that they are equal in measure. Thus, we will accept the truth of the following statement.

- **The base angles of an isosceles triangle are equal in measure; that is, they are congruent.**

This statement may be rephrased in a variety of ways. For example:

1. If a triangle is isosceles, then its base angles are equal in measure.
2. If two sides of a triangle are congruent, then the angles opposite these sides are congruent.

The converse of the preceding statement is also true.

- **If two angles of a triangle are equal in measure, then the triangle is an isosceles triangle.**

This statement may be rephrased as follows:

If two angles of a triangle are congruent, then the sides opposite these angles are congruent.

The Equilateral Triangle

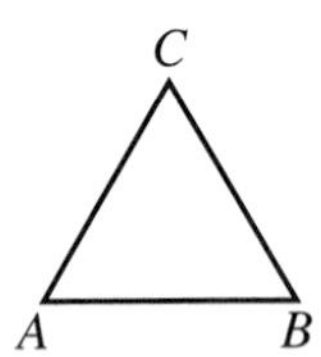

Triangle ABC is an equilateral triangle. Since $AB = BC$, $m\angle C = m\angle A$; also, since $AC = BC$, $m\angle B = m\angle A$. Therefore, $m\angle A = m\angle B = m\angle C$.

In an equilateral triangle, the measures of all of the angles are equal.

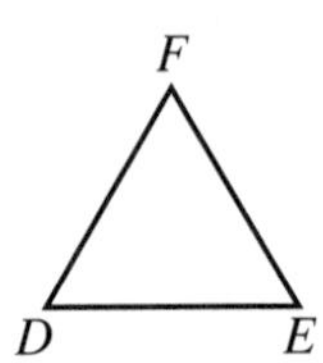

In $\triangle DEF$, all of the angles are equal in measure. Since $m\angle D = m\angle E$, $EF = DF$; also, since $m\angle D = m\angle F$, $EF = DE$. Therefore, $DE = EF = DF$, and $\triangle DEF$ is equilateral.

- **A triangle is equilateral if and only if it is equiangular.**

Properties of Special Triangles

1. If two sides of a triangle are equal in measure, the angles opposite these sides are also equal in measure. (The base angles of an isosceles triangle are equal in measure.)

2. If two angles of a triangle are equal in measure, the sides opposite these angles are also equal in measure.

3. All of the angles of an equilateral triangle are equal in measure. (An equilateral triangle is equiangular.)

4. If three angles of a triangle are equal in measure, the triangle is equilateral. (An equiangular triangle is equilateral.)

EXAMPLE

In isosceles triangle ABC, the measure of vertex angle C is 30° more than the measure of each base angle. Find the number of degrees in each angle of the triangle.

Solution

Let x = number of degrees in one base angle, A.
Then x = number of degrees in the other base angle, B,
and, $x + 30$ = number of degrees in the vertex angle, C.

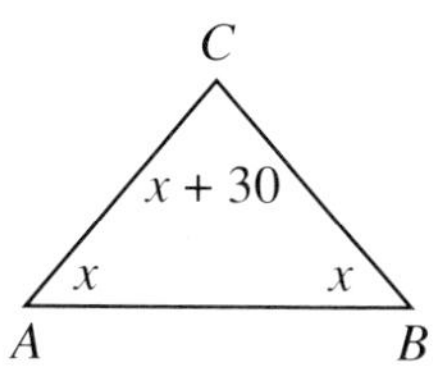

The sum of the measures of the angles of a triangle is 180°.

$$x + x + x + 30 = 180$$
$$3x + 30 = 180$$
$$3x = 150$$
$$x = 50$$
$$x + 30 = 80$$

Check:

$$50 + 50 + 80 = 180$$

Answer: $m\angle A = 50$, $m\angle B = 50$, $m\angle C = 80$

EXERCISES WITH OPEN-RESPONSE PROBLEMS

1. Name the legs, base, vertex angle, and base angles in each of the following isosceles triangles:

a.

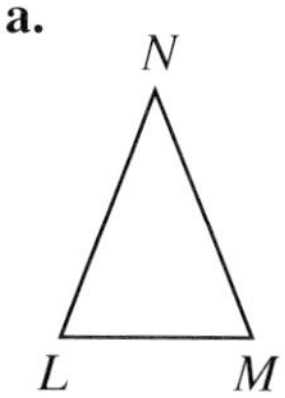

b.

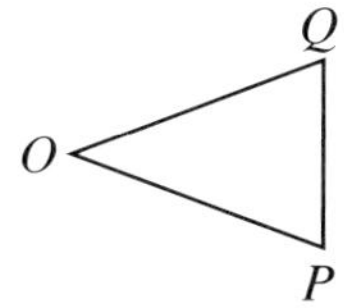

c.

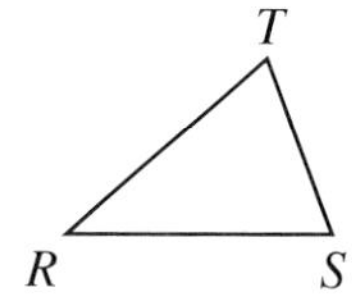

d.

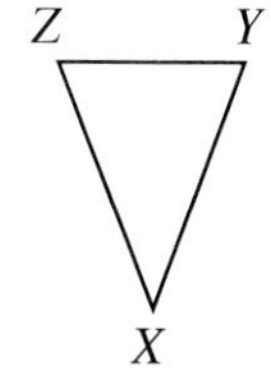

2. In $\triangle ABC$, $AC = 4$ centimeters, $CB = 6$ centimeters, and $AB = 6$ centimeters.
   a. What type of triangle is $\triangle ABC$?
   b. Name two angles in $\triangle ABC$ whose measures are equal.
   c. Why are these angles equal in measure?

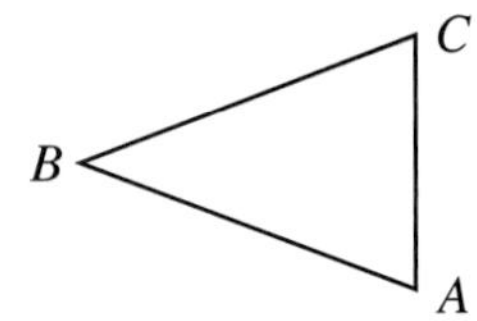

3. In $\triangle RST$, $m\angle R = 70$ and $m\angle T = 40$.
   a. Find the measure of $\angle S$.
   b. Name two sides in $\triangle RST$ that are congruent.
   c. Why are the two sides congruent?
   d. What type of triangle is $\triangle RST$?

4. Draw an isosceles triangle that is: **a.** an acute triangle **b.** a right triangle **c.** an obtuse triangle
5. Can a base angle of an isosceles triangle be: **a.** a right angle? **b.** an obtuse angle? Explain your answer.
6. Find the measure of the vertex angle of an isosceles triangle if the measure of each base angle is:
   a. 80° **b.** 55° **c.** 42° **d.** $22\frac{1}{2}°$ **e.** 51.5°

7. Find the measure of each base angle of an isosceles triangle if the vertex angle measures:
   a. 40° **b.** 50° **c.** 76° **d.** 100° **e.** 65°

8. What is the number of degrees in each acute angle of an isosceles right triangle?
9. The measure of each base angle of an isosceles triangle is 7 times the measure of the vertex angle. Find the measure of each angle of the triangle.
10. The measure of each of the congruent angles of an isosceles triangle is one-half of the measure of the vertex angle. Find the measure of each angle of the triangle.
11. The measure of the vertex angle of an isosceles triangle is 3 times the measure of each base angle. Find the number of degrees in each angle of the triangle.
12. The measure of the vertex angle of an isosceles triangle is 15° more than the measure of each base angle. Find the number of degrees in each angle of the triangle.
13. The measure of each of the congruent angles of an isosceles triangle is 6° less than the measure of the vertex angle. Find the measure of each angle of the triangle.
14. The measure of each of the congruent angles of an isosceles triangle is 9° less than 4 times the vertex angle. Find the measure of each angle of the triangle.
15. In $\triangle ABC$, $m\angle A = x$, $m\angle B = x + 30$, and $m\angle C = 2x - 10$.
    a. Find the measures of the three angles.
    b. What kind of triangle is $\triangle ABC$?
16. If a triangle is equilateral, what is the measure of each angle?
17. The measures of the angles of $\triangle ABC$ can be represented by $(x + 30)°$, $2x°$, and $(4x - 60)°$.
    a. What is the measure of each angle?
    b. What kind of triangle is $\triangle ABC$?

18. **a.** Graph points $A(0, 4)$, $B(0, 0)$, and $C(4, 0)$, and connect the points in order to form a triangle.
b. What kind of triangle is $\triangle ABC$?
c. Find the measure of each angle of $\triangle ABC$.
d. Find the area of $\triangle ABC$.

19. The vertex angle of an isosceles triangle is 80° in measure. What is the measure of an exterior angle to one of the base angles of this triangle?

20. The measure of an exterior angle to a base angle of an isosceles triangle is 115°. What is the measure of the vertex angle of the triangle?

21. The measure of an exterior angle to the vertex angle of an isosceles triangle is 60°. What is the measure of one of the base angles of the triangle?

22. In $\triangle ABC$, $m\angle A = m\angle B$, and the measure of the exterior angle to $\angle C$ is 120°.
a. Find the measure of each angle of $\triangle ABC$.
b. What kind of triangle is $\triangle ABC$?

23. What is the measure of each exterior angle of an equilateral triangle?

24. Given this statement: "If two sides of a triangle are equal in measure, then the angles opposite these sides are equal in measure."
a. What is the truth value of the statement?
b. Write the converse of the statement.
c. What is the truth value of the converse?
d. Using the statement and its converse, write a biconditional statement.
e. What is the truth value of this biconditional?

CHAPTER SUMMARY

Point, ***line***, and ***plane*** are undefined terms that are used to define other terms. A ***line segment*** is a part of a line consisting of two points on the line, called ***endpoints***, and all of the points on the line between the endpoints. The ***midpoint*** of a line segment separates the segment into two congruent parts. The ***perpendicular bisector*** of a line segment is perpendicular to the segment at its midpoint.

An ***angle*** is the union of two rays with a common endpoint. Two angles are ***complementary*** if the sum of their measures is 90°. If the measure of an angle is $x°$, the measure of its complement is $(90 - x)°$.

Two angles are ***supplementary*** if the sum of their measures is 180°. If the measure of an angle is $x°$, the measure of its supplement is $(180 - x)°$.

The angles of a linear pair are supplementary. Vertical angles are ***congruent***.

If two parallel lines are cut by a transversal, then:

- The alternate interior angles are congruent.
- The alternate exterior angles are congruent.
- The corresponding angles are congruent.
- Interior angles on the same side of the transversal are supplementary.

The sum of the measures of the angles of a triangle is 180°.
The base angles of an isosceles triangle are congruent.
An equilateral triangle is equiangular.

VOCABULARY

10-1 Undefined term Point Line Straight line Plane Axiom (postulate) Line segment (Segment) Endpoints Midpoint Intersection Union

10-2 Half-line Ray Opposite rays Angle Vertex Degree Protractor Acute angle Right angle Obtuse angle Straight angle Reflex angle Perpendicular Perpendicular bisector

10-3 Interior of an angle Exterior of an angle Adjacent angles Complementary angles Supplementary angles Linear pair Vertical angles Congruent Theorem

10-4 Parallel lines Transversal Interior angles Alternate interior angles Exterior angles Alternate exterior angles Interior angles on the same side of the transversal Corresponding angles

10-5 Rigid figure Vertices Sides Angles Acute triangle Equiangular triangle Right triangle Obtuse triangle Legs Hypotenuse Exterior angle Remote interior angle

10-6 Scalene triangle Isosceles triangle Equilateral triangle Base of an isosceles triangle Vertex angle Base angle

REVIEW EXERCISES WITH OPEN-RESPONSE PROBLEMS

1. Find the number of degrees in three-eighths of a complete rotation.
2. If $\overleftrightarrow{AB}$ and $\overleftrightarrow{CD}$ intersect at E, $m\angle AEC = x + 10$, and $m\angle DEB = 2x - 30$, find the measure of $\angle AEC$.
3. If two angles of a triangle are complementary, what is the measure of the third angle?
4. The measure of the complement of an angle is 20° less than the measure of the angle. Find the number of degrees in the angle.
5. If each base angle of an isosceles triangle measures 55°, find the measure of the vertex angle of the triangle.

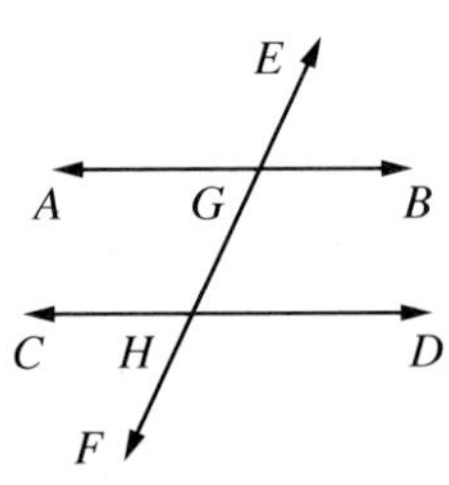

In 6–8, $\overleftrightarrow{AB}$ is parallel to $\overleftrightarrow{CD}$, and these lines are cut by transversal $\overleftrightarrow{EF}$ at points G and H, respectively.

6. If $\angle AGH$ measures 73°, find $m\angle GHD$.
7. If $m\angle EGB = 70$ and $m\angle GHD = 3x - 2$, find x.
8. If $m\angle HGB = 2x + 10$ and $m\angle GHD = x + 20$, find $m\angle GHD$.

9. In $\triangle ART$, $m\angle A = y + 10$, $m\angle R = 2y$, and $m\angle T = 2y - 30$.
   a. Find the measure of each of the three angles.
   b. Choose one: $\triangle ART$ is
      (*1*) right (*2*) isosceles (*3*) equilateral (*4*) scalene

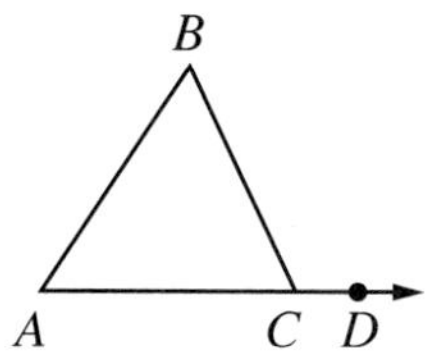

Ex. 10–13

In 10–13, $\angle BCD$ is an exterior angle to $\triangle ABC$ at vertex C.

10. If $m\angle A = 57$ and $m\angle B = 62$, find $m\angle BCD$.
11. If $m\angle BCD = 118$ and $m\angle A = 54$, find $m\angle B$.
12. If $m\angle A = x + 12$, $m\angle B = x + 4$, and $m\angle BCD = 120$, find x.
13. If $m\angle B = y$ and $m\angle BCD = y + 65$, find $m\angle A$.
14. In the diagram, $\triangle ABC$ is isosceles with $AC = BC$, $m\angle ACB = 4x$, and the measure of exterior angle $BAD = 5x$.
    a. Express the measure of $\angle B$ in terms of x.
    b. Find the value of x.
    c. Find the measure of each of the indicated angles:
       (*1*) $\angle ABC$ (*2*) $\angle ACB$
       (*3*) $\angle BAD$

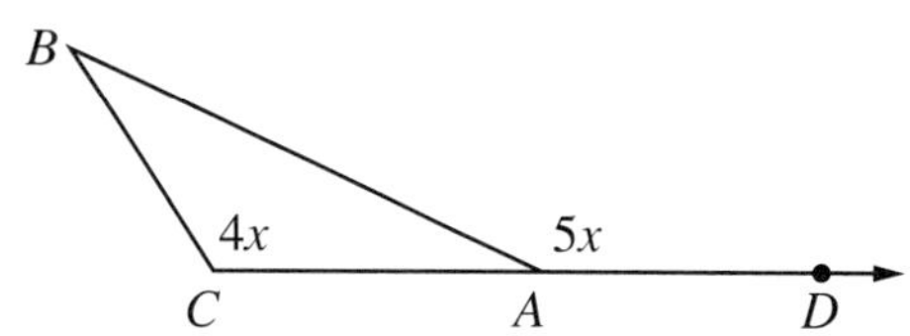

15. The measure of each base angle of an isosceles triangle is 15° more than the measure of the vertex angle. Find the measure of each angle.
16. The measure of an angle is 20° less than 3 times the measure of its supplement. What is the measure of the angle and of its supplement?

In 17–19, select, in each case, the numeral preceding the correct answer.

17. In $\triangle ABC$, the measure of $\angle B$ is $\frac{3}{2}$ the measure of $\angle A$ and the measure of $\angle C$ is $\frac{5}{2}$ the measure of $\angle A$. What is the measure of the smallest angle?
    (1) 9° (2) 18° (3) 36° (4) 40°
18. The measure of one angle is 3 times that of another angle, and the sum of these measures is 120°. Which statement is true?
    (1) One angle must be obtuse.
    (2) Both angles are acute.
    (3) One angle must be a right angle.
    (4) The angles are complementary.
19. The measure of the smaller of two supplementary angles is $\frac{4}{5}$ of the measure of the larger. The measure of the smaller angle is:
    (1) 10° (2) 20° (3) 40° (4) 80°

CUMULATIVE REVIEW WITH OPEN-RESPONSE PROBLEMS

1. Graph the solution set of $x \geq -2$ on the real number line.
2. If p represents "$\frac{3}{4}$ is a rational number," and q represents "$\sqrt{2}$ is a rational number," what is the truth value of $p \wedge \sim q$?

3. When $p \vee q$ is true and $p \wedge q$ is false, which of the following must be false?
(1) $p \vee \sim q$ (2) $\sim p \wedge q$ (3) $p \rightarrow q$ (4) $p \leftrightarrow q$

4. What constant must be added to $x^2 + 8x$ so that the sum is the square of $(x + 4)$?

5. Solve for x and check: $7(x - 2) + 3(x + 2) = 8x$.

6. Solve and check: $1.3x - 1 = 0.5x + 7$.

7. **a.** Solve for x in terms of a and b: $ax + 3b = 7$.
b. Find, to the *nearest hundredth*, the value of x when $a = \sqrt{3}$ and $b = \sqrt{5}$.

8. Evaluate $-3a^2 - a$ when $a = -1.5$.

9. Simplify: $7xy(x^2 - y) + \frac{5x^4y^4}{x\,y^3}$.

10. The measures of the angles of a triangle are consecutive even integers. Find the measure of each angle.

11. Mr. Popowich mailed two packages. The larger package weighed 3 pounds more than the smaller. If the total weight of the packages was 17 pounds, how much did each package weigh?

Exploration

Origami is the Japanese art of paper folding. Investigate the principles and methods of this art and make at least one figure.

Chapter 11

Congruence and Transformations

An employee who works for a company that sells custom clothing is cutting out a blouse, using a paper pattern to determine the exact size and shape of each piece needed. The pieces of cloth that are being cut will be congruent to the pattern pieces.

Some of the pattern pieces, such as the one for the back of the blouse, are placed with one edge on a fold of the material so that, when the material is unfolded, the pieces on the two sides will be mirror images of each other. Because the back of the blouse is symmetric with respect to the center line, the worker can use a transformation called a *line reflection* in order to cut out the pieces more quickly and accurately.

In this chapter, you will study congruence and symmetry, which are common elements in nature, art, and architecture.

11-1 GEOMETRIC FIGURES

Any set of points is a ***geometric figure***. A geometric figure may be a set of one point or a set of many points.

Plane geometric figures are figures all of whose points are in the same plane. Plane geometric figures can be pictured on a flat surface.

Curves

If a picture of a set of points can be drawn without removing the pencil from the paper, the figure is called a ***curve***. A curve that starts and ends at the same point is a ***closed curve***. Of the eight curves pictured at the right, only curves (5), (6), (7), and (8) are closed curves. We see that curves (7) and (8) cross themselves, whereas curves (5) and (6) do not. Closed curves that do not cross themselves are called ***simple closed curves***.

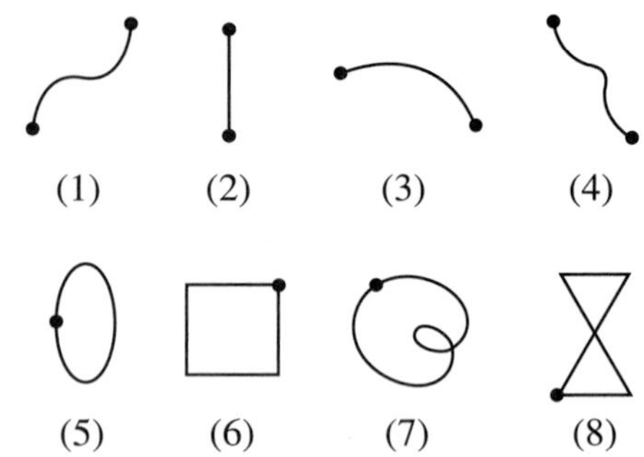

As shown in the figure at the right, a simple closed curve divides the plane into three sets of points:

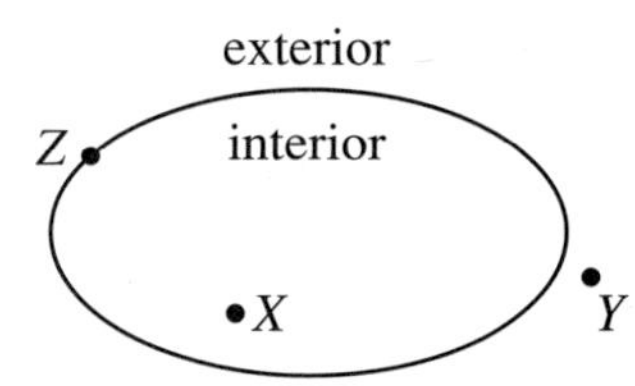

1. The set of points *inside* the curve is called the ***interior region***. For example, point X is in the interior of the curve.
2. The set of points *outside* the curve is called the ***exterior region***. For example, point Y is in the exterior of the curve.
3. The set of points that are on the curve is called the ***boundary*** between the interior and the exterior. For example, point Z is in the boundary of the curve.

Polygons

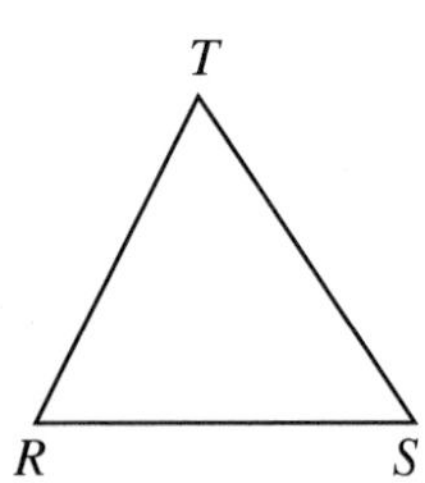

A ***polygon*** is a simple closed plane curve that consists of line segments. In a polygon, each line segment is called a ***side*** of the polygon. A common endpoint of two line segments is a ***vertex*** of the polygon.

We can name a polygon by naming each vertex with a capital letter. The polygon at the right is named RST. Its sides are $\overline{RS}$, $\overline{ST}$, and $\overline{TR}$. Its vertices are R, S, and T.

Classification of polygons As shown in the following figures, a polygon is classified and named according to the number of sides it has:

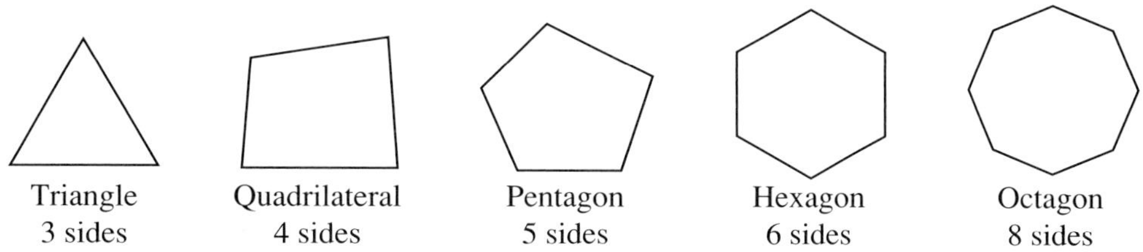

Other polygons with special names are the **heptagon** (7 sides), **nonagon** (9 sides), and **decagon** (10 sides).

A ***regular polygon*** is a polygon in which all of the sides have equal measures, and all of the angles have equal measures. In other words, a regular polygon is equilateral and equiangular. For example, of the figures pictured above, the triangle, the hexagon, and the octagon are regular polygons; the quadrilateral and the pentagon are not. Of course, there are regular pentagons and regular quadrilaterals. For example, a square is a regular quadrilateral.

In Chapter 10, we said that two angles are congruent if their measures are equal. Similarly, we say that two line segments are congruent if their measures are equal. Therefore, in a regular polygon all sides are congruent and all angles are congruent.

There are two commonly accepted ways to indicate that two distinct line segments, such as $\overline{AB}$ and $\overline{CD}$, have equal measures:

1. *The line segments are congruent:* $\overline{AB} \cong \overline{CD}$.

2. *The measures of the segments are equal:* $AB = CD$.

It is correct to say that $m\overline{AB} = m\overline{CD}$, but this symbolism is cumbersome. It is not correct, however, to say that $\overline{AB}$ and $\overline{CD}$ are equal.

Graphing polygons A polygon can be represented in the coordinate plane by plotting its vertices and then drawing the line segments connecting the vertices in order.

The graph at the right shows quadrilateral $ABCD$. The vertices are $A(3, 2)$, $B(-3, 2)$, $C(-3, -2)$, and $D(3, -2)$. From the graph, we note the following:

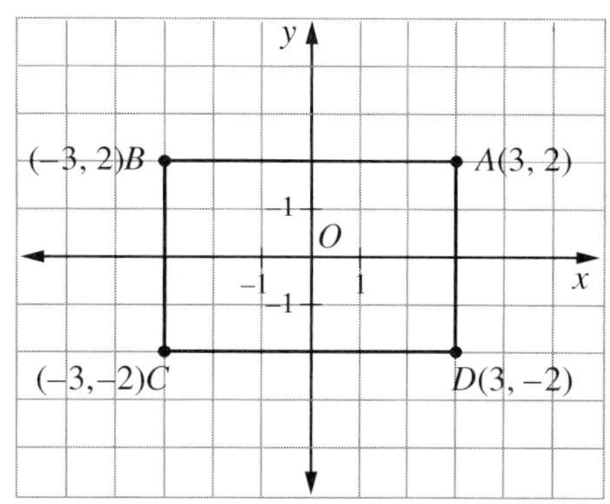

1. Points A and B, which have the same ordinate, are on a line parallel to the x-axis.

2. Points C and D, which have the same ordinate, are on a line parallel to the x-axis.

3. Lines that are parallel to the x-axis are parallel to each other, so $\overline{BA} \parallel \overline{CD}$.

4. Lines that are parallel to the x-axis are perpendicular to the y-axis.

5. Points B and C, which have the same abscissa, are on a line parallel to the y-axis.

6. Points A and D, which have the same abscissa, are on a line parallel to the y-axis.

7. Lines that are parallel to the y-axis are parallel to each other, so $\overline{BC} \parallel \overline{AD}$.

8. Lines that are parallel to the y-axis are perpendicular to the x-axis.

Since each side of this polygon is parallel to an axis, we can find the lengths of the sides by counting the number of units on the graph between the endpoints of the segments:

$$AB = CD = 6 \qquad \text{and} \qquad BC = DA = 4$$

Because points A and B are on the same horizontal line, they have the same y-coordinates, and we can also find the length of $\overline{AB}$ by subtracting the x-coordinates of A and B:

$$AB = 3 - (-3) = 3 + 3 = 6$$

Similarly, points B and C are on the same vertical line, they have the same x-coordinates, and we can also find the length of $\overline{BC}$ by subtracting the y-coordinates of B and C:

$$BC = 2 - (-2) = 2 + 2 = 4$$

EXAMPLE

a. Graph the following points: $A(4, 1)$, $B(1, 5)$, $C(-2, 1)$, and draw $\triangle ABC$.

b. Find its area.

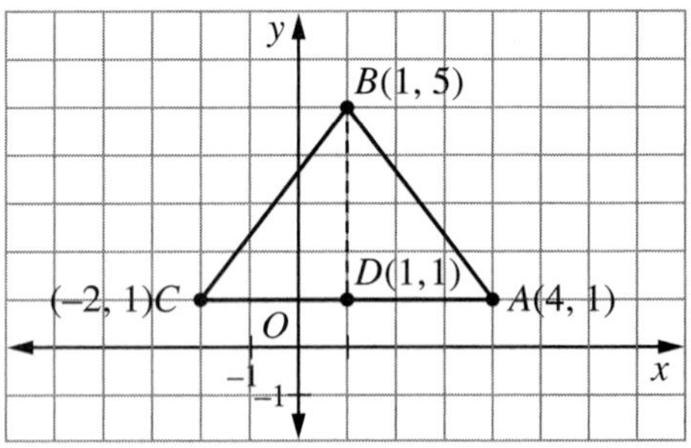

Solutions **a.** The graph at the right shows $\triangle ABC$.

b. To find the area of the triangle, we need to know the lengths of the base and the altitude drawn to that base.

The base of $\triangle ABC$ is $\overline{AC}$.

$$AC = 4 - (-2) = 4 + 2 = 6$$

The altitude to the base is a line segment, $\overline{BD}$, drawn from B perpendicular to $\overline{AC}$.

$$BD = 5 - 1 = 4$$

$$\text{Area} = \frac{1}{2}(AC)(BD)$$

$$= \frac{1}{2}(6)(4) = 12$$

Answer: The area of $\triangle ABC$ is 12 square units.

EXERCISES WITH OPEN-RESPONSE PROBLEMS

1. Which of the figures shown below are: **a.** closed curves? **b.** simple closed curves?

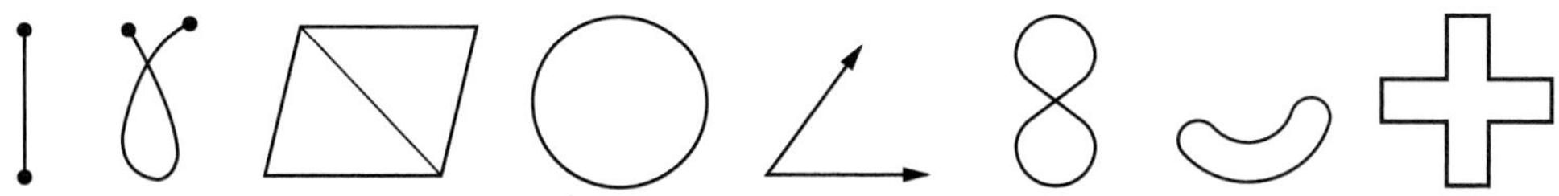

(1) (2) (3) (4) (5) (6) (7) (8)

2. For the figure at the right:
a. Name the points on the curve.
b. Name the points in the interior of the curve.
c. Name the points in the exterior of the curve.

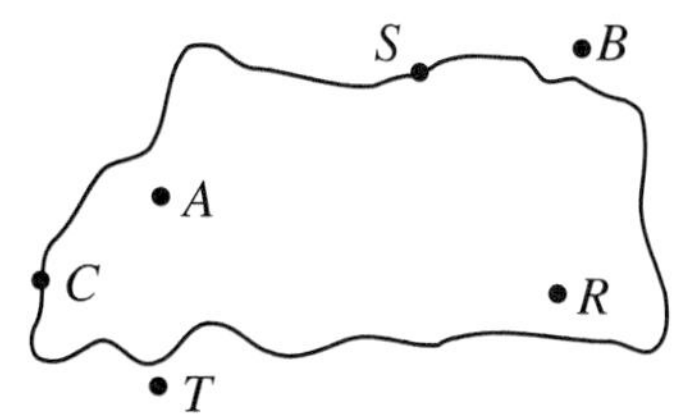

3. Tell the number of sides each of the following polygons has:
a. hexagon **b.** quadrilateral **c.** triangle **d.** octagon **e.** pentagon

4. Which of the figures shown below represent(s): **a.** polygons? **b.** a triangle?
c. a quadrilateral? **d.** a pentagon? **e.** a hexagon? **f.** an octagon?

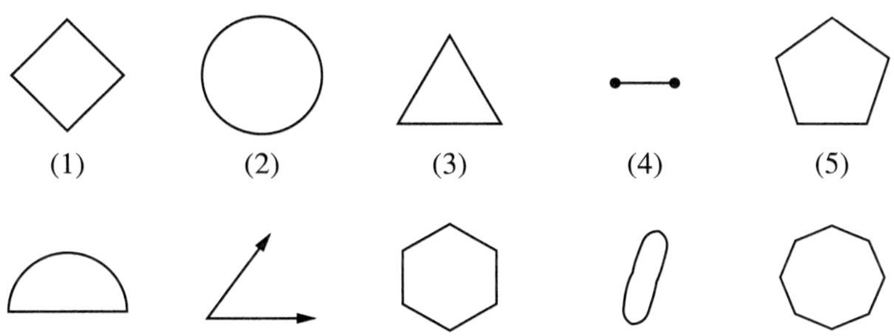

(1) (2) (3) (4) (5)
(6) (7) (8) (9) (10)

5. Each of the following can be represented or modeled by a geometric figure. Name the figure.
a. The front cover of your textbook
b. A baseball diamond
c. The side view of a ladder leaning against a wall
d. A stop sign

In 6–9, in each case: **a.** Graph the given points, and connect them with straight lines in order, forming a polygon. **b.** Tell what kind of polygon was drawn.

6. $A(1, 1)$, $B(6, 1)$, $C(7, 3)$, $D(4, 5)$

7. $D(-4, 4)$, $E(-2, 1)$, $F(3, 1)$, $G(3, 5)$, $H(0, 5)$

8. $P(2, -1)$, $Q(2, 1)$, $R(5, 0)$

9. $S(-2, 0)$, $T(-2, -2)$, $U(0, -3)$, $V(2, -2)$, $W(2, 0)$, $X(0, 1)$

In 10–15, in each case: **a.** Using the given points, graph the polygon. **b.** Find the area of the polygon.

10. $A(-3, 1)$, $B(2, 1)$, $C(4, 0)$

11. $D(-4,-2)$, $E(4,-2)$, $F(4, 3)$, $G(-4, 3)$

12. $R(0, 3)$, $S(2,-3)$, $T(4, 3)$

13. $A(-2,-1)$, $B(-2,-6)$, $C(3,-6)$, $D(3,-1)$

14. $J(4, 4)$, $K(2, 1)$, $L(-3, 1)$, $M(-1, 4)$

15. $A(1, 1)$, $B(4, 1)$, $C(8, 4)$, $D(4, 4)$, $E(1, 4)$

16. In rectangle *ABCD*, the coordinates of three of the vertices are as follows: $A(4, 3)$, $B(-3, 3)$, and $C(-3,-2)$.
a. What are the coordinates of *D*? **b.** What is the area of the rectangle?

17. Polygon *ABCDEF* is a six-sided figure with coordinates $A(1, 4)$, $B(1, 1)$, $C(4, 1)$, $D(4,-1)$, $E(-2,-1)$, and $F(-2, 4)$. **a.** Graph the polygon. **b.** Find its area.

11-2 CONGRUENT TRIANGLES

Congruent Polygons

In modern industry, it is often necessary to make many copies of a part in such a way that the original part and all copies have the same size and shape. For example, a machine can stamp out many duplicates of a piece of metal, each copy having the same size and shape as the original. We say that the original and all its copies are ***congruent***.

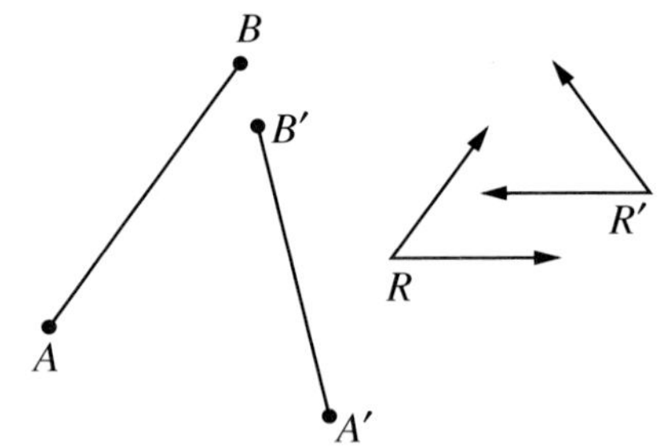

In Section 11-1, we talked about congruent segments, which are segments that are equal in length. In Chapter 10, we also discussed congruent angles, which are angles whose measures are equal. In the figures at the right that show congruent segments and congruent angles, we read the symbol ′ as "prime." We often use this symbol to show a matching element.

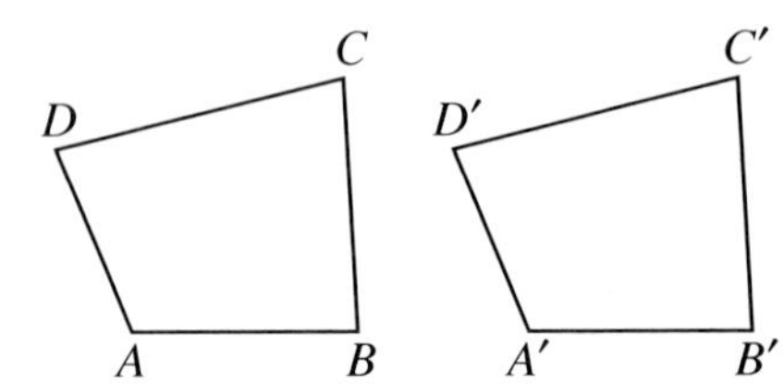

Congruent polygons
$ABCD \cong A'B'C'D'$

Also pictured at the right are two polygons that have the same size and shape. Such polygons are ***congruent polygons***: polygon $ABCD \cong$ polygon $A'B'C'D'$.

One way to discover whether two polygons have the same size and shape is to place one polygon on top of the other. If the figures can be turned in such a way that the sides of one polygon *fit exactly* upon the sides of the other and the angles of one polygon *fit exactly* upon the angles of the other, we say that the polygons *coincide*. The sides that fit one upon the other are called ***corresponding sides***, and the angles that fit one upon the other are called ***corresponding angles***.

In the given polygons $ABCD$ and $A'B'C'D'$, the pairs of corresponding sides are:

$$\overline{AB} \text{ and } \overline{A'B'},\ \overline{BC} \text{ and } \overline{B'C'},\ \overline{CD} \text{ and } \overline{C'D'},\ \overline{DA} \text{ and } \overline{D'A'}$$

angles are:

$$\angle A \text{ and } \angle A' \qquad \angle B \text{ and } \angle B' \qquad \angle C \text{ and } \angle C' \qquad \angle D \text{ and } \angle D'$$

Thus, we can say:

- **If the sides of a first polygon are congruent to the corresponding sides of a second polygon, and if the angles of the first polygon are congruent to the corresponding angles of the second, then the two polygons are congruent.**

Conversely, we can also say:

- **If two polygons are congruent, then their corresponding sides are congruent and their corresponding angles are congruent.**

We can see that, for two triangles to be congruent, three pairs of corresponding sides must be congruent and three pairs of corresponding angles must be congruent. When we know the size of certain sides and angles of a triangle, the sizes of the remaining side and angle are fixed. Let us see whether it is possible to prove two triangles congruent by proving that fewer than three pairs of sides and three pairs of angles are congruent.

Congruent Triangles Involving Two Sides and the Included Angle

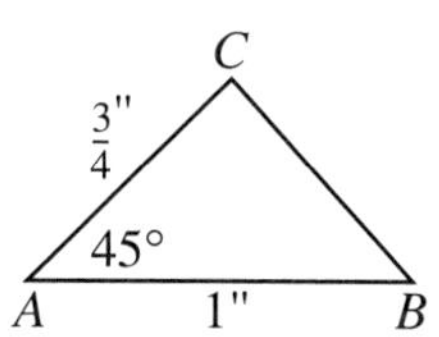

In $\triangle ABC$ at the left, $AB = 1$ inch, $m\angle A = 45$, and $AC = \frac{3}{4}$ inch. Angle A is said to be *included* between side $\overline{AB}$ and side $\overline{AC}$ because these two segments are the sides of the angle.

Perform the following experiment. On a sheet of paper, draw $\triangle A'B'C'$ so that $A'B' = 1$ inch, $A'C' = \frac{3}{4}$ inch, and the measure of included angle A' is 45°.

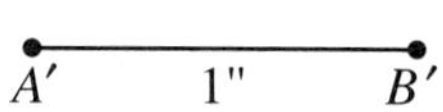

1. Begin by drawing a 1-inch line segment, and label as $\overline{A'B'}$.

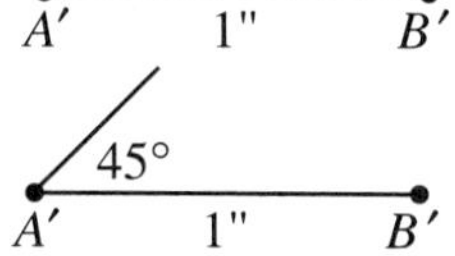

2. With a protractor, draw a 45° angle whose vertex is at point A'.

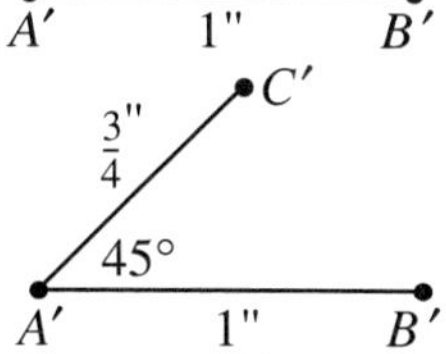

3. On the side of $\angle A'$ that was last drawn, measure off a line segment $\frac{3}{4}$ inch in length, beginning at point A' and ending at point C'.

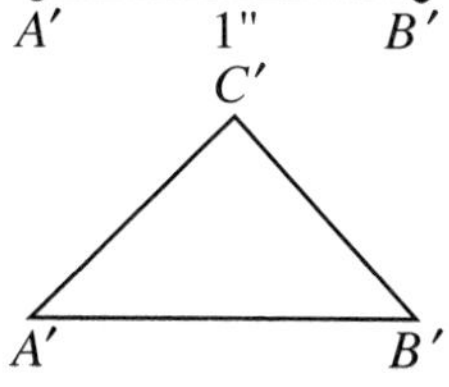

4. Then, draw side $\overline{C'B'}$ to complete the triangle.

If you measure sides $\overline{CB}$ in the given triangle and $\overline{C'B'}$ in the triangle you drew, you will find that their measures are equal: $\overline{CB} \cong \overline{C'B'}$. Now measure $\angle C$ and $\angle C'$; their measures are also equal: $\angle C \cong \angle C'$.

Similarly, $\angle B$ and $\angle B'$ have equal measures: $\angle B \cong \angle B'$. Also, if you cut out $\triangle A'B'C'$, you can make it coincide with $\triangle ABC$. Thus, $\triangle A'B'C'$ appears to be congruent to $\triangle ABC$.

If you repeat the same experiment several times with different sets of measurements for the two sides and the included angle, in each experiment the remaining pairs of corresponding parts of the triangles will appear to be congruent, and the triangles themselves will appear to be congruent. It seems reasonable, therefore, to accept the truth of the following statement:

- **Two triangles are congruent if two sides and the included angle of one triangle are congruent, respectively, to two sides and the included angle of the other [s.a.s. ≅ s.a.s.].**

In $\triangle ABC$ and $\triangle A'B'C'$: If $\overline{AB} \cong \overline{A'B'}$, $\angle A \cong \angle A'$, and $\overline{AC} \cong \overline{A'C'}$, then $\triangle ABC \cong \triangle A'B'C'$ [s.a.s. ≅ s.a.s.].

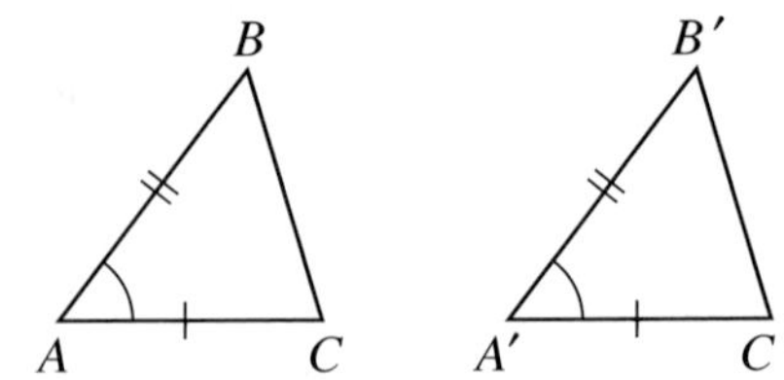

Triangles in which the angle is not included between the sides Now, see what happens when each of two triangles has one side that measures 1 inch, a second side that measures $\frac{3}{4}$ inch, and a 45° angle that is not included between these sides; that is, s.s.a. ≅ s.s.a. Must two such triangles always be congruent?

Using a ruler and a protractor, draw two different triangles that have the given measures, as shown at the right. Even without cutting out $\triangle ABC$ and $\triangle DEF$, you see that they would not coincide. Hence, when the angle is *not included* between the two sides, two triangles need *not* be congruent even though two sides and an angle of one triangle are congruent to two sides and an angle of the other. Therefore, proving that s.s.a. ≅ s.s.a. in two triangles is *not sufficient* to prove the triangles congruent.

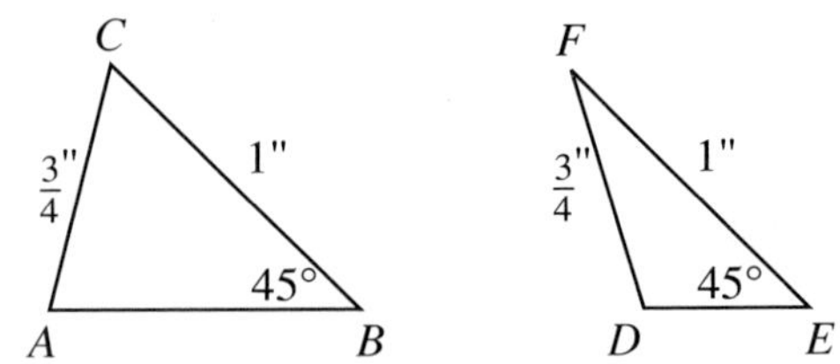

Congruent Triangles Involving Two Angles and the Included Side

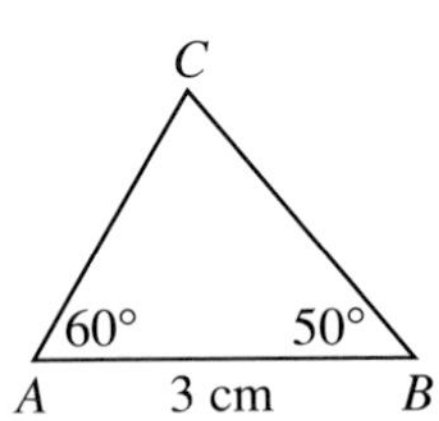

In $\triangle ABC$, m$\angle A = 60$, $AB = 3$ centimeters, and m$\angle B = 50$. Side $\overline{AB}$ *is included* between $\angle A$ and $\angle B$ because $\overline{AB}$ is drawn between vertex A and vertex B.

Perform another experiment. Draw $\triangle A'B'C'$ so that m$\angle A' = 60$, m$\angle B' = 50$, and the included side, $\overline{A'B'}$, measures 3 centimeters.

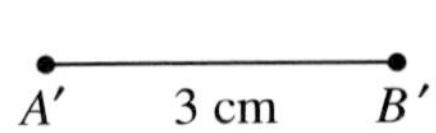

1. Begin by drawing a line segment that measures 3 centimeters. Label it $\overline{A'B'}$.

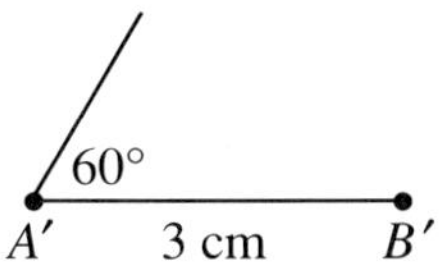

2. With a protractor, draw a 60° angle whose vertex is at point A'.

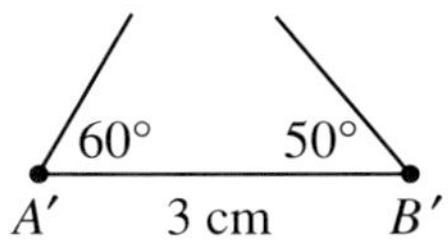

3. Then, draw a 50° angle whose vertex is at point B'.

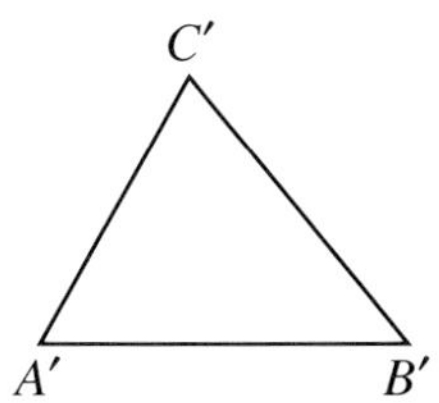

4. To complete the triangle, draw the sides of $\angle A'$ and $\angle B'$ so that they intersect at point C'.

You can see that $\triangle ABC$ and $\triangle A'B'C'$ appear to have the same size and shape, or that $\triangle ABC$ appears to be congruent to $\triangle A'B'C'$.

If you repeat the same experiment several times with different sets of measurements for the two angles and the included side, the triangles in each experiment will appear to be congruent. Therefore, it seems reasonable to accept the truth of the following statement:

- **Two triangles are congruent if two angles and the included side of one triangle are congruent, respectively, to two angles and the included side of the other triangle [a.s.a. ≅ a.s.a.].**

In $\triangle RST$ and $\triangle R'S'T'$:

If $\angle R \cong \angle R'$, $\overline{RS} \cong \overline{R'S'}$, and $\angle S \cong \angle S'$, then $\triangle RST \cong \triangle R'S'T'$ [a.s.a. ≅ a.s.a.].

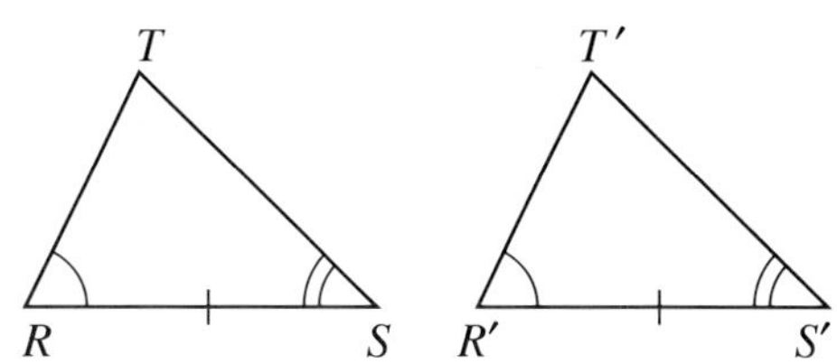

Congruent Triangles Involving Three Sides

The following statement about triangles is also true:

- **Two triangles are congruent if three sides of one triangle are congruent, respectively, to three sides of the other triangle [s.s.s. ≅ s.s.s.].**

In $\triangle RST$ and $\triangle R'S'T'$:

If $\overline{RS} \cong \overline{R'S'}$, $\overline{ST} \cong \overline{S'T'}$, and $\overline{TR} \cong \overline{T'R'}$, then $\triangle RST \cong \triangle R'S'T'$ [s.s.s. ≅ s.s.s.].

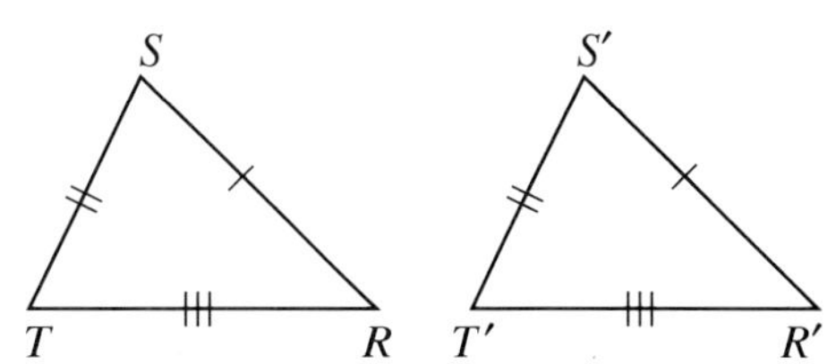

EXAMPLE

In the diagram, $\overline{AB}$ intersects $\overline{CD}$ at E, $\angle C$ and $\angle D$ are right angles, and $\overline{CE} \cong \overline{DE}$.

a. Prove informally that $\triangle AEC \cong \triangle BED$.

b. If $AE = 3x$ and $BE = 2x + 10$, find AE and BE.

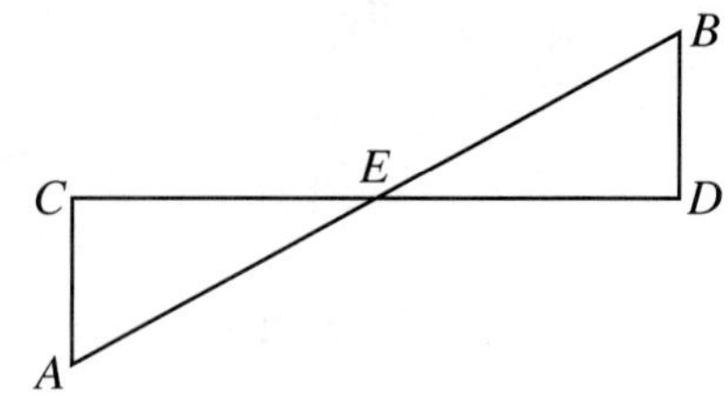

Solutions **a.** (1) It is given that $\overline{CE} \cong \overline{DE}$.

(2) It is given that $\angle C$ and $\angle D$ are right angles. Since all right angles are congruent, $\angle C \cong \angle D$.

(3) Since $\angle CEA$ and $\angle DEB$ are a pair of vertical angles, and vertical angles are congruent, $\angle CEA \cong \angle DEB$.

(4) $\triangle AEC \cong \triangle BED$ because two angles and the included side of one triangle are congruent to two angles and the included side of the other triangle [a.s.a. ≅ a.s.a.].

It may be helpful to mark off the congruent angles and congruent sides as shown in the accompanying diagram.

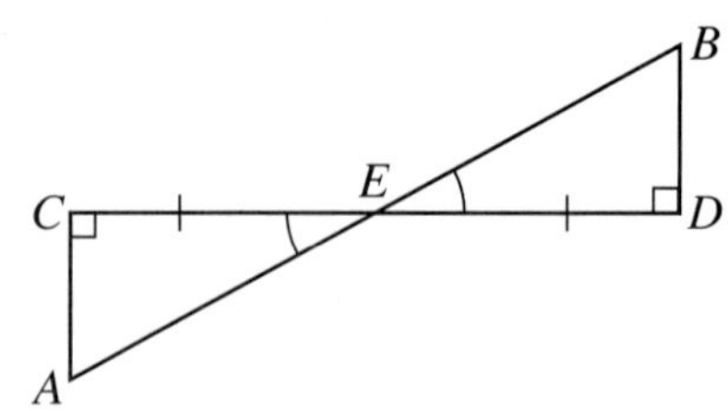

b. Since $\triangle AEC$ and $\triangle BED$ are congruent, their corresponding sides must be congruent. Therefore, $\overline{AE} \cong \overline{BE}$. Thus:

$$AE = BE$$

$$3x = 2x + 10$$

$$x = 10$$

$$3x = 30 \text{ and } 2x + 10 = 2(10) + 10 = 30$$

Answer: $AE = 30$ and $BE = 30$.

EXERCISES WITH OPEN-RESPONSE PROBLEMS

1. In each case, choose among the five polygons the ones that appear to be congruent.

a.

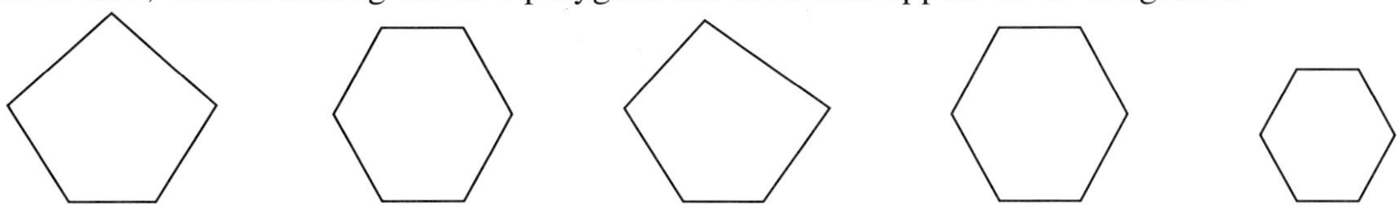

(1) (2) (3) (4) (5)

b.

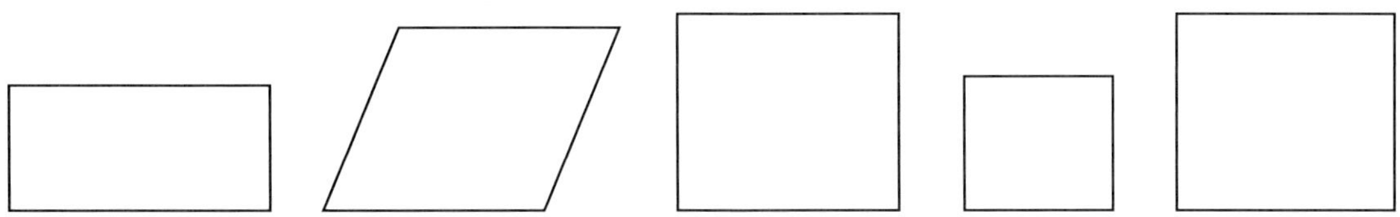

(1) (2) (3) (4) (5)

c.

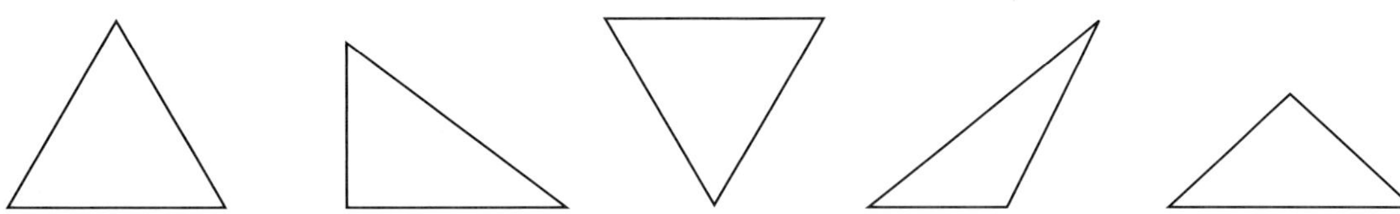

(1) (2) (3) (4) (5)

2. a. If several copies are made of the same photograph, will figures in the original photograph and in the copies be congruent?

b. Why?

3. a. If a photograph is enlarged, what can be said of figures in the original and figures in the enlargement?

b. Why?

In 4–6, two triangles are to be drawn for each problem. **a.** Use a ruler and a protractor to draw $\triangle ABC$ and $\triangle DEF$, starting with the measures given for each triangle. **b.** If the triangles are congruent, state the reason why. If the triangles are not congruent, explain why.

4. In $\triangle ABC$: $AB = 2$ inches, $m\angle B = 60$, and $BC = 1\frac{1}{2}$ inches.
In $\triangle DEF$: $DE = 1\frac{1}{2}$ inches, $m\angle E = 60$, and $EF = 2$ inches.

5. In $\triangle ABC$: $AB = 3$ inches, $m\angle A = 40$, and $m\angle B = 80$.
In $\triangle DEF$: $m\angle E = 80$, $EF = 3$ inches, and $m\angle F = 40$.

6. In $\triangle ABC$: $BC = 2.5$ centimeters, $m\angle C = 90$, and $m\angle B = 60$.
In $\triangle DEF$: $EF = 2.5$ centimeters, $m\angle E = 60$, and $\angle F$ is a right angle.

7. Sam and Rita each drew a triangle in which two sides and an angle measured, respectively, 5 centimeters, 8 centimeters, and 70°. The triangles were not congruent. Tell why this could have happened.

8. **a.** From the following triangles, select pairs that are congruent. Tell why they are congruent.
   b. For each congruent pair, name the corresponding angles and the corresponding sides.

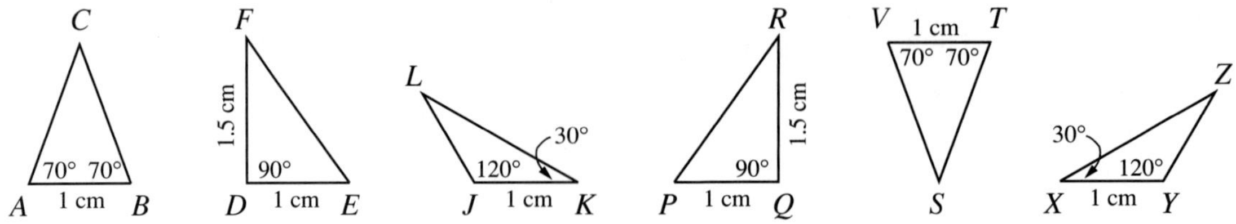

9. **a.** Tell why $\triangle RST \cong \triangle R'S'T'$.
   b. Find $m\angle RTS$ and $m\angle R'T'S'$.

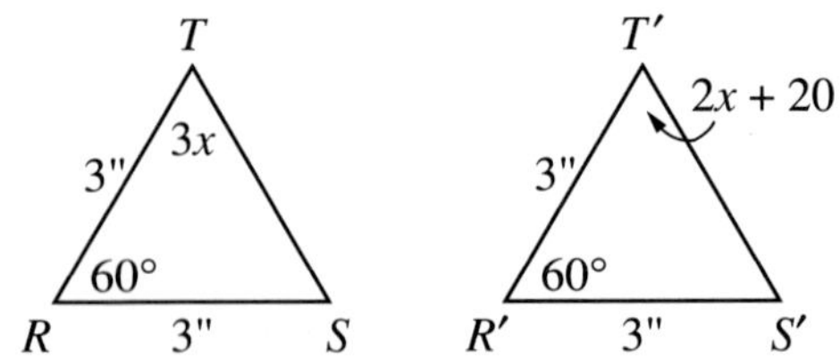

10. **a.** Tell why $\triangle ABC \cong \triangle A'B'C'$.
    b. Find BC and $B'C'$.

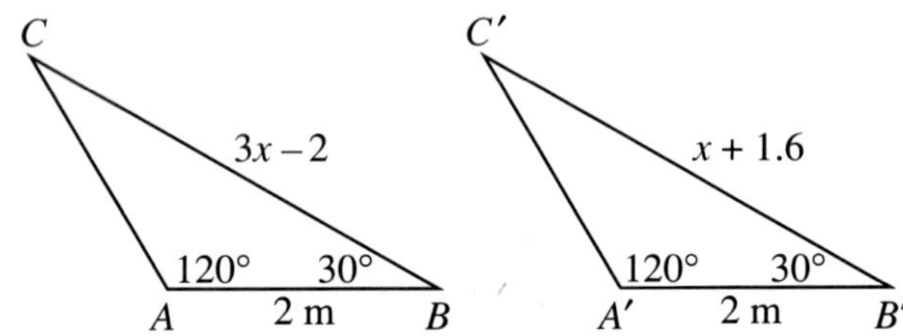

11. **a.** $\overline{BC}$ and $\overline{AD}$ intersect at E. What is the relationship between $m\angle CED$ and $m\angle BEA$?
    b. Why is $\triangle CED \cong \triangle BEA$?
    c. What is the relationship between CD and BA?
    d. Find BA and CD.

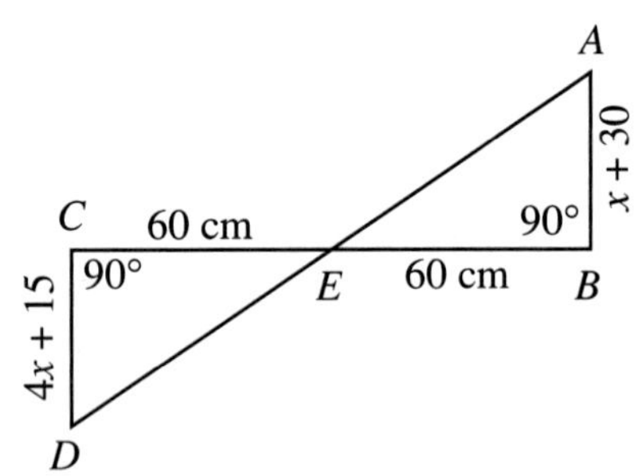

12. **a.** What relationship exists between $\triangle ABE$ and $\triangle CBE$? Why?
    b. Find AE and CE.

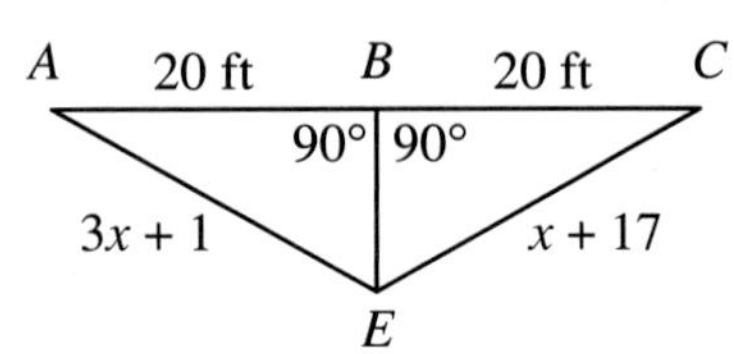

13. **a.** $\triangle ABC$ is isosceles, and D is the midpoint of base $\overline{BC}$. Why is $\triangle ABD \cong \triangle ACD$?
    b. If $m\angle B = 3x + 10$ and $m\angle C = 70$, find the value of x.

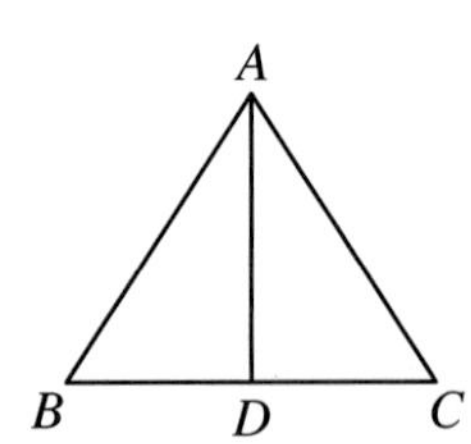

14. On graph paper, draw $\triangle ABC$ if A is point (1, 1), $m\angle A = 90$, B is on the same horizontal line as A, $AB = 6$, and the area of $\triangle ABC$ is 12.

a. How many triangles are possible?

b. What are the possible coordinates of B?

c. What are the possible coordinates of C?

d. Are all possible triangles congruent? Why?

15. On graph paper, draw $\overline{DE}$ if D is point $(-2, -1)$, and E is point $(6, -1)$.

a. Find the coordinates of two points, F and G, such that the areas of $\triangle DEF$ and $\triangle DEG$ are each 20 but $\triangle DEF$ and $\triangle DEG$ are not congruent.

b. Let H be any point such that the area of $\triangle DEH$ is 20. Describe the position of H on the graph.

11-3 QUADRILATERALS

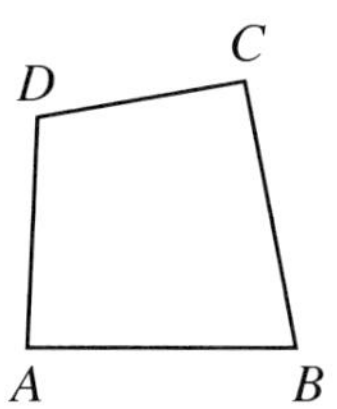

A ***quadrilateral*** is a polygon that has four sides. A point at which any two sides of the quadrilateral meet is called a ***vertex*** of the quadrilateral. At each vertex, the two sides that meet form an angle of the quadrilateral. Thus, $ABCD$ on the left is a quadrilateral whose sides are $\overline{AB}$, $\overline{BC}$, $\overline{CD}$, and $\overline{DA}$. Its vertices are A, B, C, and D. Its angles are $\angle ABC$, $\angle BCD$, $\angle CDA$, and $\angle DAB$.

In a quadrilateral, two angles whose vertices are the endpoints of a side are called ***consecutive angles***. For example, in quadrilateral $ABCD$, $\angle A$ and $\angle B$ are consecutive angles; $\angle B$ and $\angle C$ are consecutive angles; and so on. Two angles that are not consecutive angles are called ***opposite angles***; $\angle A$ and $\angle C$ are opposite angles, as are $\angle B$ and $\angle D$.

When we vary the shape of the quadrilateral by making some of its sides parallel, by making some of its sides equal in length, or by making its angles right angles, we get different members of the family of quadrilaterals, as shown and named below:

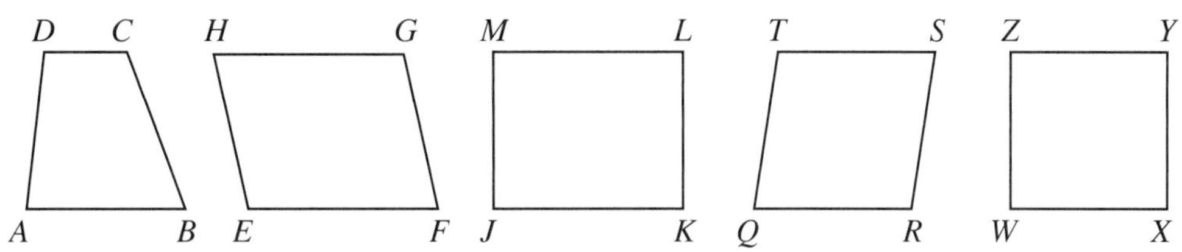

Trapezoid Parallelogram Rectangle Rhombus Square

A ***trapezoid*** is a quadrilateral in which two and only two opposite sides are parallel. In trapezoid $ABCD$, $\overline{AB} \parallel \overline{CD}$. Parallel sides $\overline{AB}$ and $\overline{CD}$ are called the ***bases*** of the trapezoid. In an ***isosceles trapezoid,*** the nonparallel sides are equal.

A ***parallelogram*** is a quadrilateral in which both pairs of opposite sides are parallel. In parallelogram $EFGH$, $\overline{EF} \parallel \overline{GH}$ and $\overline{EH} \parallel \overline{FG}$. The symbol for parallelogram is $\square$.

A ***rectangle*** is a parallelogram in which all four angles are right angles. Rectangle $JKLM$ is a parallelogram in which $\angle J$, $\angle K$, $\angle L$, and $\angle M$ are right angles.

A ***rhombus*** is a parallelogram in which all sides are of equal length. Rhombus $QRST$ is a parallelogram in which $QR = RS = ST = TQ$.

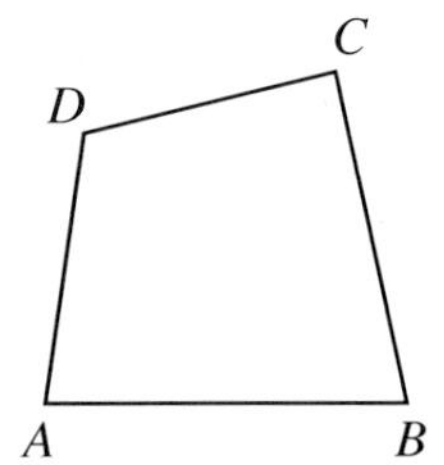

A ***square*** is a rectangle in which all sides are of equal length. Therefore, square $WXYZ$ is also a parallelogram in which $\angle W$, $\angle X$, $\angle Y$, and $\angle Z$ are right angles, and $WX = XY = YZ = ZW$.

If we draw a large quadrilateral like the one shown at the left and then measure each of its four angles, is the sum 360°? It should be. If we do the same with several other quadrilaterals of different shapes and sizes, is the sum of the four measures 360° in each case? It should be.

Relying on what you have just verified by experimentation, it seems reasonable to make the following statement:

- **The sum of the measures of the angles of a quadrilateral is 360°.**

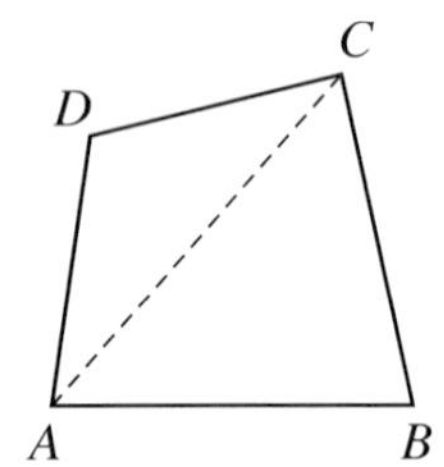

To prove informally that this statement is true, we draw diagonal $\overline{AC}$, whose endpoints are the vertices of the two opposite angles, $\angle A$ and $\angle C$. Then:

(1) Diagonal $\overline{AC}$ divides quadrilateral $ABCD$ into two triangles, $\triangle ABC$ and $\triangle ADC$.

(2) The sum of the measures of the angles of $\triangle ABC$ is 180°, and the sum of the measures of the angles of $\triangle ADC$ is 180°.

(3) The sum of the measures of all the angles of $\triangle ABC$ and $\triangle ADC$ together is 360°.

(4) Hence, $m\angle A + m\angle B + m\angle C + m\angle D = 360$.

The Family of Parallelograms

Listed below are some relationships that are true for the family of parallelograms that includes rectangles, rhombuses, and squares.

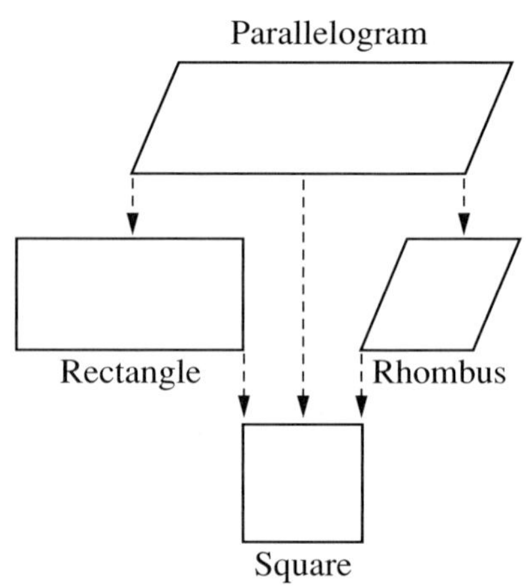

1. All rectangles, rhombuses, and squares are members of the family of parallelograms. Therefore, any property of the family of parallelograms must also be a property of rectangles, rhombuses, and squares.

2. A square is a member of the family of rectangles. Therefore, any property of the family of rectangles must also be a property of squares.

3. A square is a member of the family of rhombuses. Therefore, any property of the family of rhombuses must also be a property of squares.

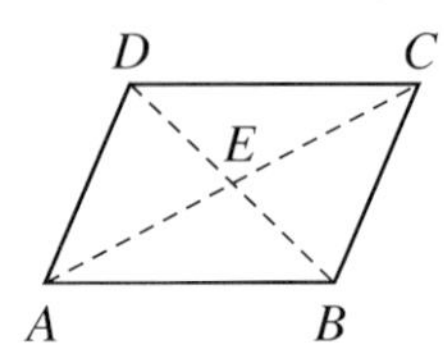

In parallelogram $ABCD$ at the left, opposite sides are parallel. Thus, $\overline{AB} \parallel \overline{DC}$ and $\overline{AD} \parallel \overline{BC}$. The following statements, which will be proved in a higher level course, are true for any parallelogram:

- **Opposite sides of a parallelogram are congruent.** Here, $\overline{AB} \cong \overline{DC}$ and $\overline{AD} \cong \overline{BC}$.

- **Opposite angles of a parallelogram are congruent.** Here, $\angle A \cong \angle C$ and $\angle B \cong \angle D$.
- **Consecutive angles of a parallelogram are supplementary.** Here, $m\angle A + m\angle B = 180$, $m\angle B + m\angle C = 180$, and so forth.
- **A diagonal divides a parallelogram into two congruent triangles.** Here, $\triangle ABC \cong \triangle CDA$.
- **The diagonals of a parallelogram bisect each other.** Here, $AE = EC$ and $DE = EB$.

Since rhombuses, rectangles, and squares are members of the family of parallelograms, these statements will be true for any rhombus, any rectangle, and any square. In addition,

- **The diagonals of a rectangle are congruent.**
- **The diagonals of a rhombus are perpendicular to each other.**
- **The diagonals of a rhombus bisect its angles.**

Informal Proofs for Statements About Angles in a Parallelogram

The following statements about parallelogram $ABCD$ can help to prove informally that *consecutive angles of a parallelogram are supplementary*.

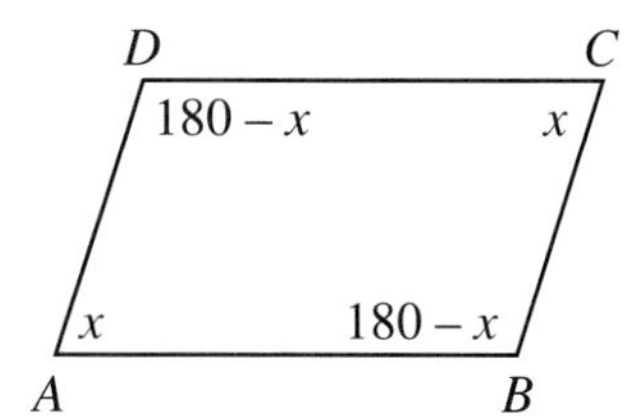

(1) $\overline{DC} \parallel \overline{AB}$ and $\overline{AD}$ is a transversal (opposite sides of a parallelogram are parallel, and a transversal intersects these lines).

(2) If $m\angle A = x$, then $m\angle D = 180 - x$ ($\angle A$ and $\angle D$ are interior angles on the same side of a transversal, and these angles have been shown to be supplementary).

(3) Similarly, looking at $\overline{AD} \parallel \overline{BC}$ and the transversal $\overline{AB}$, we can say: If $m\angle A = x$, then $m\angle B = 180 - x$ (the reason is the same as given above).

(4) Also, looking at $\overline{AB} \parallel \overline{DC}$ and the transversal $\overline{BC}$, we can say: If $m\angle B = 180 - x$, then $m\angle C = x$.

Therefore, we have proved that:

- **Consecutive angles of a parallelogram are supplementary.**

We can use this statement to prove informally that:

- **Opposite angles of a parallelogram are equal in measure.**

We reason as follows:

(1) In parallelogram $ABCD$, we have shown that $m\angle A = x$, $m\angle B = 180 - x$, $m\angle C = x$, and $m\angle D = 180 - x$.

(2) Thus, $m\angle A = m\angle C$ and $m\angle B = m\angle D$.

These statements are also true for any rhombus, any rectangle, and any square.

EXAMPLE

$ABCD$ is a parallelogram where $m\angle A = 2x + 50$ and $m\angle C = 3x + 40$.

a. Find the value of x.

b. Find the measure of each angle.

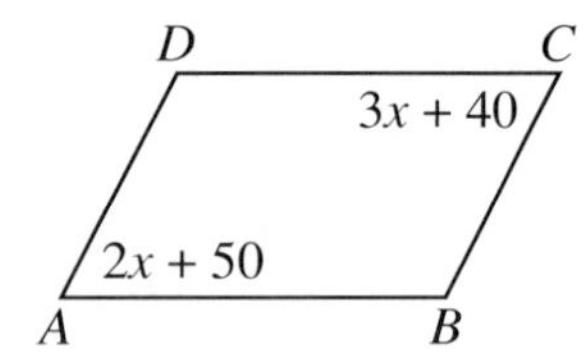

Solutions **a.** In $\square ABCD$, $m\angle C = m\angle A$ because the opposite angles of a parallelogram are equal in measure. Thus:

$$3x + 40 = 2x + 50$$

$$x = 10$$

Answer: $x = 10$

b. By substitution:

$$m\angle A = 2x + 50 = 2(10) + 50 = 70$$

$$m\angle C = 3x + 40 = 3(10) + 40 = 70$$

Since $m\angle B + m\angle A = 180$, $m\angle B + 70 = 180$, and $m\angle B = 110$.
Since $m\angle D = m\angle B$, $m\angle D = 110$.

Answer: $m\angle A = 70$, $m\angle B = 110$, $m\angle C = 70$, $m\angle D = 110$

EXERCISES WITH OPEN-RESPONSE PROBLEMS

In 1–6, in each case:

a. Copy the given statement. Is it true or false?
b. Write the converse of the given statement. Is it true or false?
c. Write the inverse of the given statement. Is it true or false?
d. Write the contrapositive of the given statement. Is it true or false?

1. If a polygon is a trapezoid, it is a quadrilateral.
2. If a polygon is a rectangle, it is a parallelogram.

3. If a polygon is a rhombus, it is a parallelogram.
4. If a polygon is a rhombus, it is a square.
5. If a polygon is a parallelogram, it is a square.
6. If two angles are opposite angles of a parallelogram, they are congruent.

In 7–10, the angle measures are represented in each quadrilateral. In each case:
a. Find the value of x.
b. Find the measure of each angle of the quadrilateral.

7.

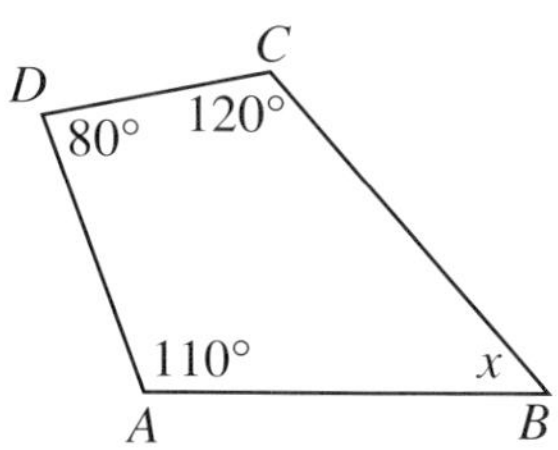

8.

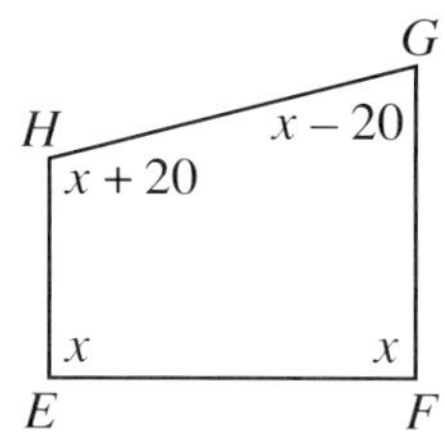

9.

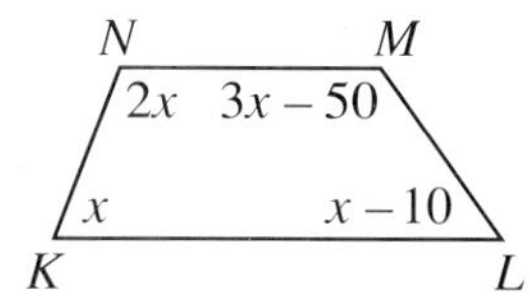

10.

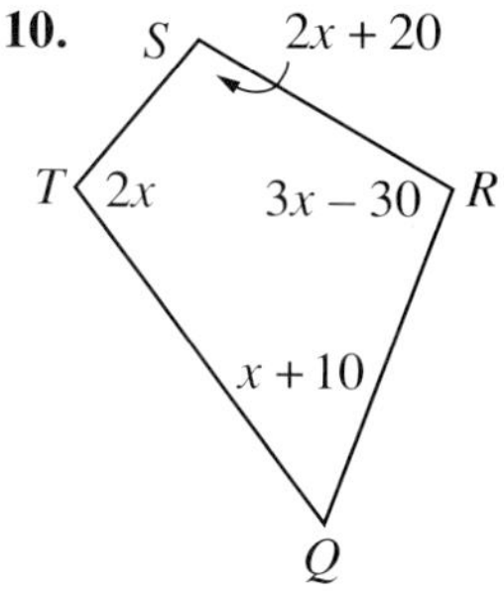

In 11 and 12, polygon $ABCD$ is a parallelogram.

11. $AB = 3x + 8$; $DC = x + 12$.
Find AB and DC.

12. $m\angle A = 5x - 40$; $m\angle C = 3x + 20$.
Find $m\angle A$, $m\angle B$, $m\angle C$, and $m\angle D$.

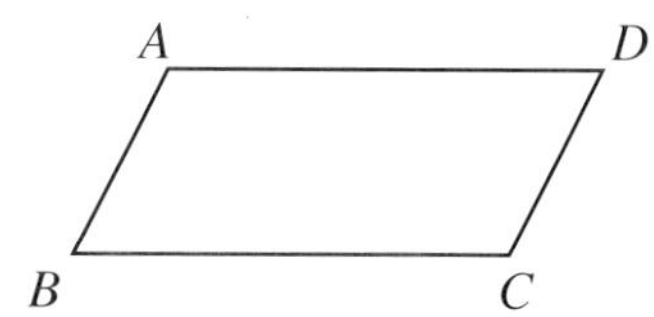

Ex. 11–12

In 13 and 14, polygon $ABCD$ is a rectangle.

13. $BC = 4x - 5$, $AD = 2x + 3$.
Find BC and AD.

14. $m\angle A = 5x - 10$. Find the value of x.

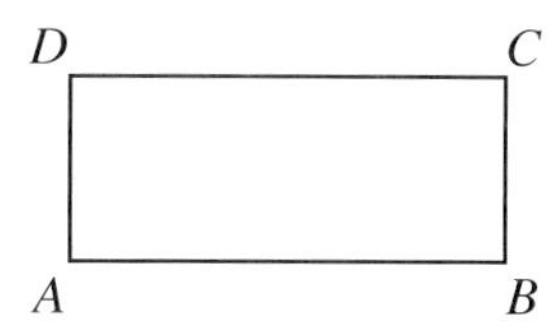

Ex. 13–14

15. $ABCD$ is a square. If $AB = 8x - 6$ and $BC = 5x + 12$, find the length of each side of the square.

16. In rhombus $KLMN$, $KL = 3x$, $LM = 2(x + 3)$. Find the length of each side of the rhombus.

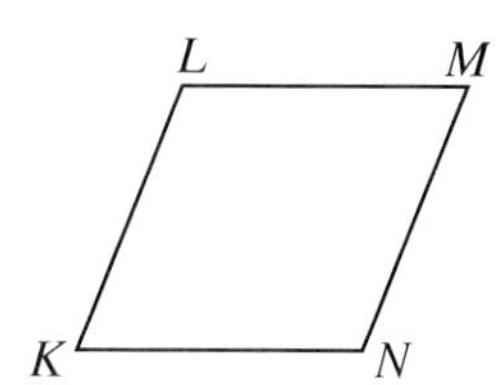

17. In the figure at the right, $\triangle MPT$ is an isosceles triangle in which $PM = TM$. $\overline{RS}$ is drawn parallel to $\overline{PT}$, forming an isosceles trapezoid in which $PR = TS$. (In an ***isosceles trapezoid*** exactly one pair of opposite sides is parallel, while the sides that are not parallel are congruent.)

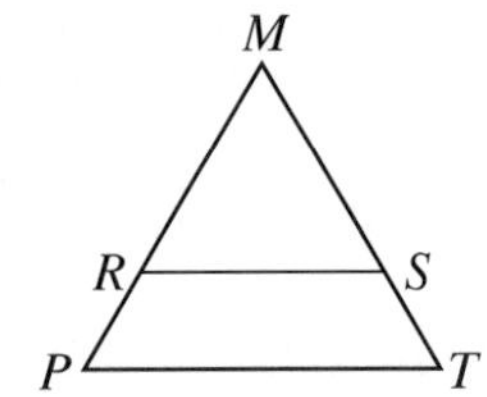

a. If $PR = x + 3$, $TS = 2x + 2$, $RS = x + 6$, and $PT = 8x + 3$, find the length of each side.
b. If $m\angle P = 60$, find the measures of all four angles in the isosceles trapezoid.

18. In the figure at the right, the angle measures of quadrilateral *ABCD* are 70°, 65°, 85°, and 140°. Four exterior angle are drawn.

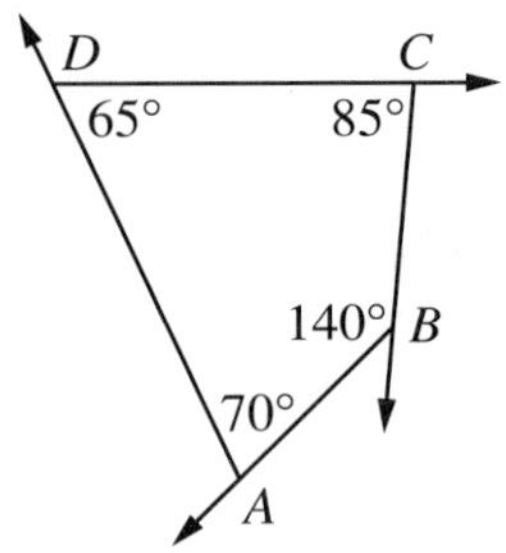

a. What is the sum of the measures of the interior angles of the quadrilateral?
b. What is the sum of the measures of the exterior angles of the quadrilateral?
c. Will the answer to part **b** be true for the sum of the measures of the exterior angles of any quadrilateral?

In 19–26, in each case: **a.** Graph the points and connect them with straight lines in order, forming a quadrilateral. **b.** Tell what kind of quadrilateral was drawn. **c.** Find the area of the quadrilateral.

19. $P(0, 0)$, $Q(5, 0)$, $R(5, 4)$, $S(0, 4)$

20. $C(8, -1)$, $A(9, 3)$, $L(4, 3)$, $F(3, -1)$

21. $H(-4, 0)$, $O(0, 0)$, $M(0, 4)$, $E(-4, 4)$

22. $F(5, 1)$, $A(5, 5)$, $R(0, 5)$, $M(-2, 1)$

23. $B(-2, -2)$, $A(2, -2)$, $R(2, 2)$, $N(-2, 2)$

24. $P(-3, 0)$, $O(0, 0)$, $N(2, 2)$, $D(-1, 2)$

25. $M(-1), -1)$, $I(4, -1)$, $L(4, 4)$, $K(-1, 4)$

26. $P(4, 3)$, $L(-2, 3)$, $O(0, 0)$, $W(3, 0)$

27. On graph paper, draw rectangle *ABCD*, where *A* is point (1, 1), *B* is on the same horizontal line as *A*, $AB = 6$, and the area of *ABCD* is 12.
a. How many rectangles are possible?
b. What are the possible coordinates of *B*?
c. What are the possible coordinates of *C*?
d. What are the possible coordinates of *D*?

28. On graph paper, draw parallelogram *EFGH*, where *E* is point $(-2, -1)$, *F* is on the same horizontal line as *E*, $EF = 7$, and the area of *EFGH* is 28.
a. What are the possible coordinates of *F*?
b. How many parallelograms are possible?
c. Write one possible pair of coordinates of *G*.
d. If point *G* has the coordinates given in answer to part **c**, what are the coordinates of *H*?

11-4 LINE REFLECTIONS AND LINE SYMMETRY

What Is a Transformation?

In the game of pool, 15 object balls numbered 1 through 15 are placed in a triangular framework called a rack (see Figure A). Let us suppose that the object balls are taken out of the rack, mixed up, and thrown back, as shown in Figure B. Most of the object balls have changed their positions, although a few, such as 2 and 11, remain in their original places.

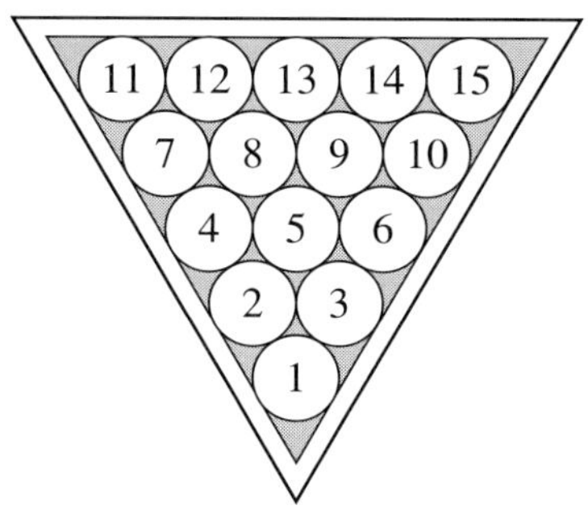

Figure A

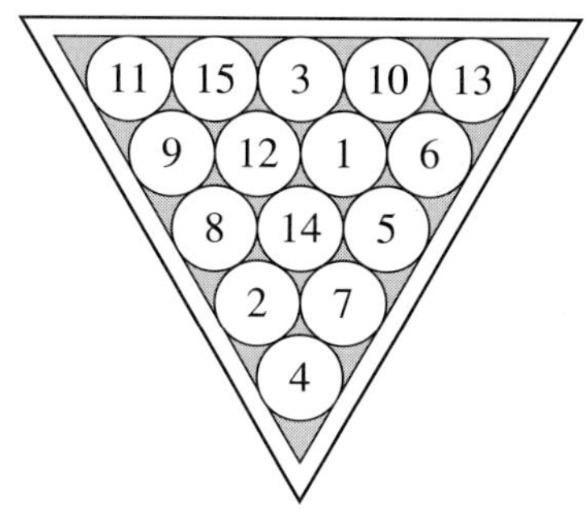

Figure B

The new rack of 15 object balls is merely a *change*, or a ***transformation***, of the 15 object balls in the original rack.

Imagine that each object ball is like a point, and that the rack containing the object balls is like a plane that contains an infinite number of points. In the same way that object balls in a rack change their positions, points, under a *transformation of the plane*, will move about and change their positions in the plane. At times, some of the points in the plane may remain fixed. After the transformation, or change, takes place, however, the plane must once again appear full and complete, without any missing points, just as the rack of 15 object balls appears full and complete in Figure B.

Comparing the positions of the object balls in the two racks, we find that 1 is replaced by 4, 2 is still 2, 3 is replaced by 7, and so on. The result is a ***one-to-one correspondence*** between the two sets, each of 15 object balls. In other words, each object ball is replaced by one and only one object ball until the rack is again complete.

We will extend this idea to points in a plane. An infinite number of transformations can take place in a plane. In this chapter we will study only a few special transformations.

Line Reflection

It is often possible to see the objects along the shore of a body of water reflected in the water. If a picture of such a scene is folded, the objects can be made to coincide with their images. Each point of the reflection is an image point of the corresponding point of the object. The line along which the picture is folded is the ***line of reflection***, and the correspondence between the object points and the image points is called a ***line reflection***. This common experience is used in mathematics to study congruent figures.

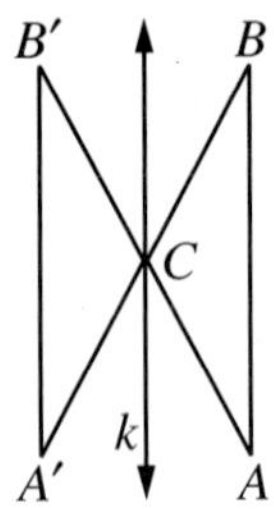

If the figure at the left were folded along line k, $\triangle ABC$ would coincide with $\triangle A'B'C$. Line k is the line of reflection, the image of A is A' (in symbols, $A \to A'$), and the image of B is B' ($B \to B'$). Point C is a fixed point because it is *on* the line of reflection. In other words, C is its own image ($C \to C$). Under a reflection in line k, then, the image of $\triangle ABC$ is $\triangle A'B'C$ ($\triangle ABC \to \triangle A'B'C$).

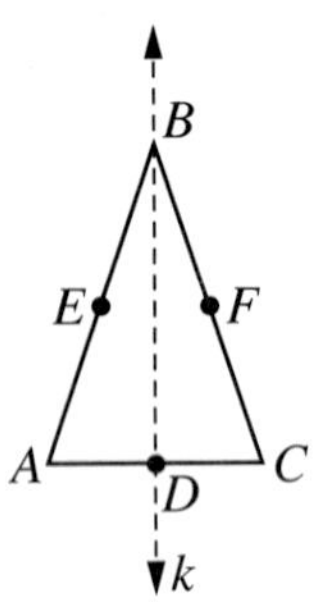

If we imagine that isosceles triangle ABC, shown at the left, is folded so that A falls on C, the line along which it fold, k, is a reflection line. Every point of the triangle has as its image a point of the triangle. Points B and D are fixed points because they are on the line of reflection.

Thus, under the line reflection in k:

1. All points of ΔABC are reflected so that $A \to C$, $C \to A$, $E \to F$, $F \to E$, $B \to B$, $D \to D$, and so on.
2. The sides of ΔABC are reflected; that is, $\overline{AB} \to \overline{CB}$, a statement verifying that the legs of an isosceles triangle are congruent. Also, $\overline{AC} \to \overline{CA}$, showing that the base is its own image.
3. The angles of ΔABC are reflected; that is, $\angle BAD \to \angle BCD$, a statement verifying that the base angles of an isosceles triangle are congruent. Also, $\angle ABC \to \angle CBA$, showing that the vertex angle is its own image.

Looking at isosceles triangle ABC and reflection line k, we can note some properties of a line reflection:

1. Distance is preserved (unchanged).

$$\overline{AB} \to \overline{CB} \quad \text{and} \quad AB = CB$$

$$\overline{AD} \to \overline{CD} \quad \text{and} \quad AD = CD$$

2. Angle measure is preserved.

$$\angle BAD \rightarrow \angle BCD \quad \text{and} \quad m\angle BAD = m\angle BCD$$

$$\angle BDA \rightarrow \angle BDC \quad \text{and} \quad m\angle BDA = m\angle BDC$$

3. The line of reflection is the perpendicular bisector of every segment joining a point to its image.
4. A figure is always congruent to its image.

Reflection in the y-axis In the figure, $\triangle ABC$ is reflected in the y-axis. Its image under the reflection is $\triangle A'B'C'$. From the figure, we see that:

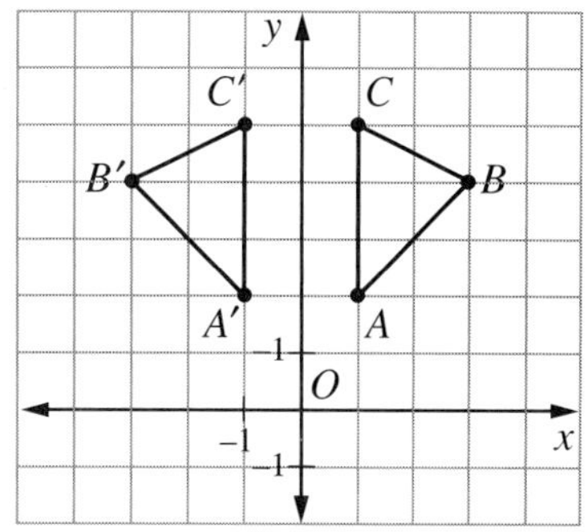

$$A(1, 2) \rightarrow A'(-1, 2)$$

$$B(3, 4) \rightarrow B'(-3, 4)$$

$$C(1, 5) \rightarrow C'(-1, 5)$$

For each point and its image under a reflection in the y-axis, the y-coordinate of the image is the same as the y-coordinate of the point; the x-coordinate of the image is the opposite of the x-coordinate of the point.

From these examples, we form a general rule:

- **Under a reflection in the y-axis, the image of $P(x, y)$ is $P'(-x, y)$.**

Reflection in the x-axis In the figure, $\triangle ABC$ is reflection in the x-axis. Its image under the reflection is $\triangle A'B'C'$. From the figure, we see that:

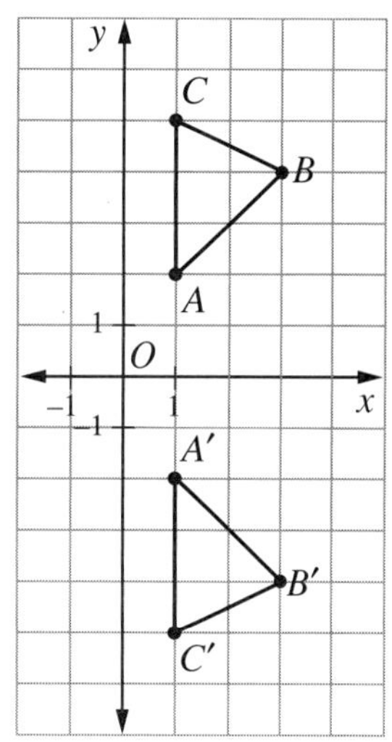

$$A(1, 2) \rightarrow A'(1,-2)$$

$$B(3, 4) \rightarrow B'(3,-4)$$

$$C(1, 5) \rightarrow C'(1,-5)$$

For each point and its image under a reflection in the x-axis, the x-coordinate of the image is the same as the x-coordinate of the point; the y-coordinate of the image is the opposite of the y-coordinate of the point.

From these examples, we form a general rule:

- **Under a reflection in the x-axis, the image of $P(x, y)$ is $P'(x, -y)$.**

EXAMPLES

1. **a.** On your paper, draw a segment and label the endpoints A and B.
b. Draw any line m.
c. Sketch the image of $\overline{AB}$ under a reflection in m.

Solutions (1) Draw $\overline{AB}$ and line m.

(2) Hold a ruler perpendicular to line m and touching point A. Measure the distance from A to line m. Find a point along the ruler that is the same distance from m as A but that is on the opposite side of m. Label this point A'.

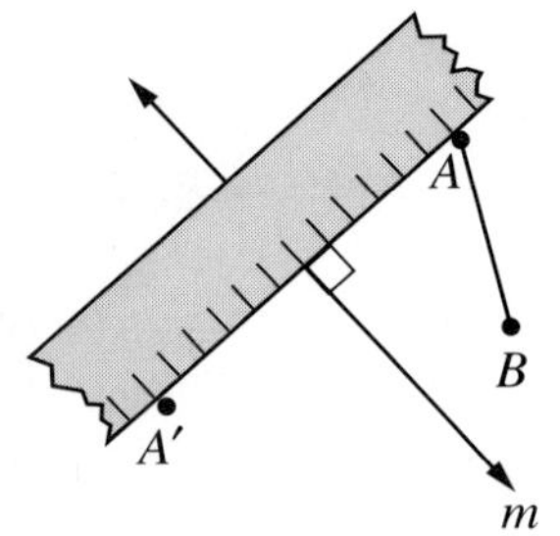

(3) Repeat step (2) for point B to locate B'.

(4) Draw $\overline{A'B'}$, the image of $\overline{AB}$.

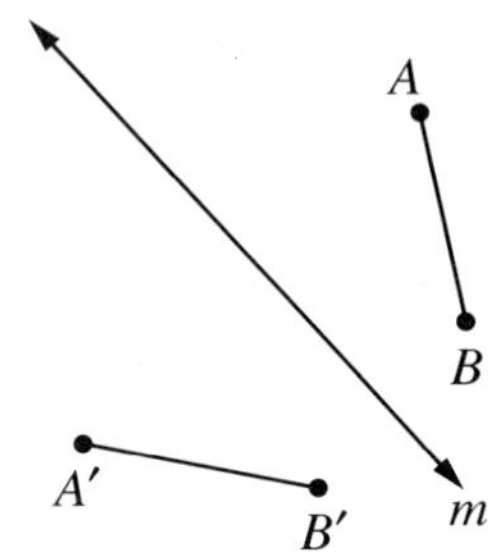

---

2. On graph paper:
a. Plot $A(3,-1)$.
b. Plot A', the image of A under a reflection in the y-axis, and write its coordinates.
c. Plot A'', the image of A under the reflection in the x-axis, and write its coordinates.

Solutions **a, b.**

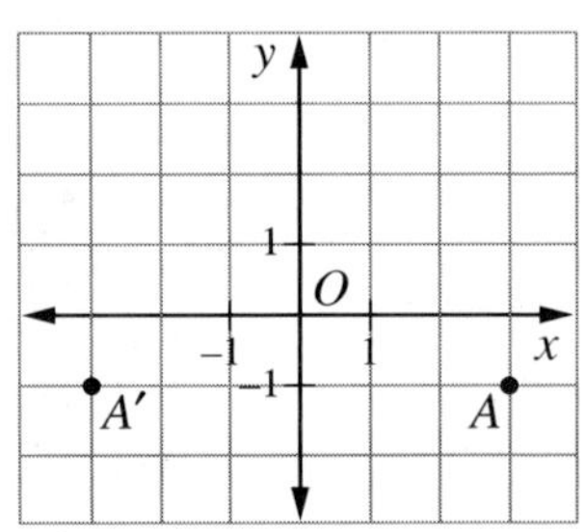

$A \to A'$
$(3,-1) \to (-3,-1)$

c.

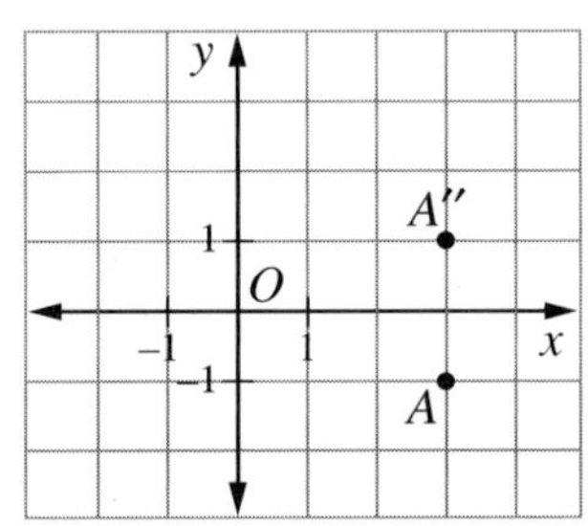

$A \to A''$
$(3,-1) \to (3, 1)$

EXERCISES

In 1–4, in each case: **a.** Copy the figure and line m on your paper. **b.** Using a ruler, sketch the image of the given figure under a reflection in m.

1.

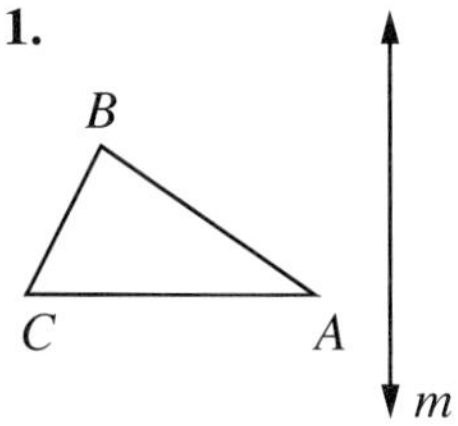

2.

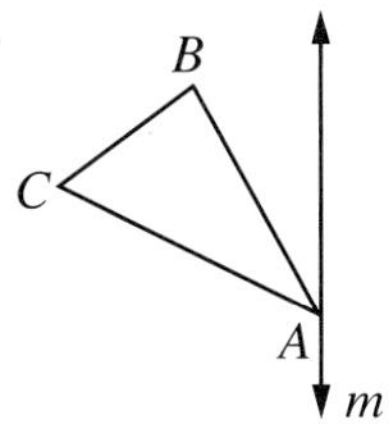

3.

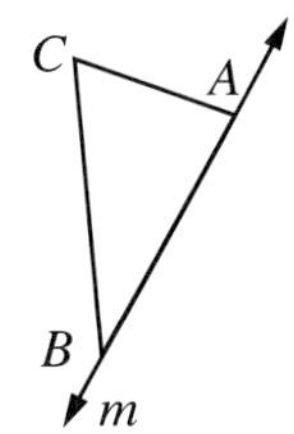

4.

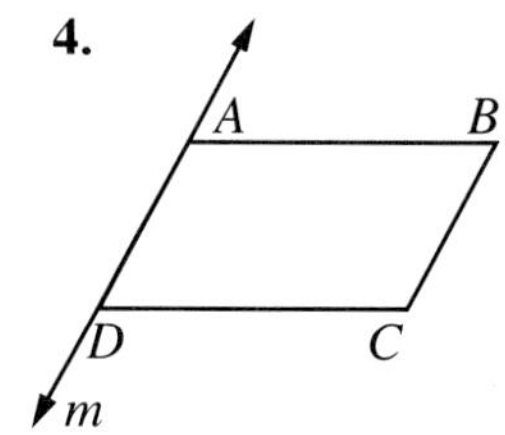

5. **a.** Draw square $ABCD$.
b. Draw m, a line of reflection for which the image of A is B.
c. Draw n, a line of reflection for which the image of A is C.
d. Draw p, a line of reflection for which the image of A is D.

Use graph paper for exercises 6–15.

In 6–10: **a.** Graph each point and its image under a reflection in the x-axis. **b.** Write the coordinates of the image point.

6. (2, 5) **7.** (1, 3) **8.** (−2, 3) **9.** (2, −4) **10.** (0, 2)

In 11–15: **a.** Graph each point and its image under a reflection in the y-axis. **b.** Write the coordinates of the image point.

11. (3, 5) **12.** (1, 4) **13.** (2, −3) **14.** (−2, 3) **15.** (−1, 0)

Line Symmetry

In nature, in art, and in industry, many forms have a pleasing, attractive appearance because of a balanced arrangement of their parts. We say that such forms have symmetry.

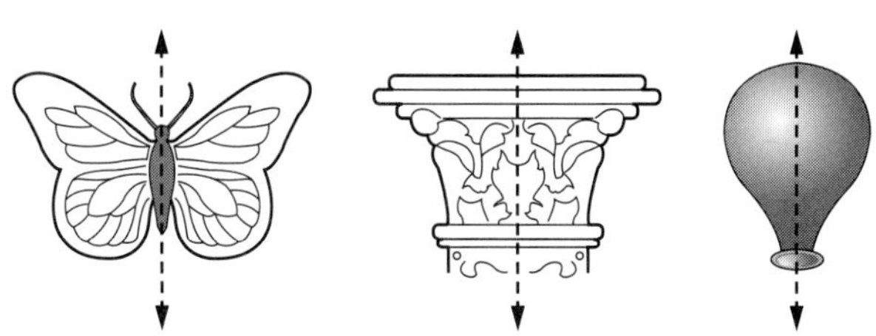

In each of the figures above, there is a line on which the figure could be folded so that the parts of the figure on opposite sides of the line would coincide.

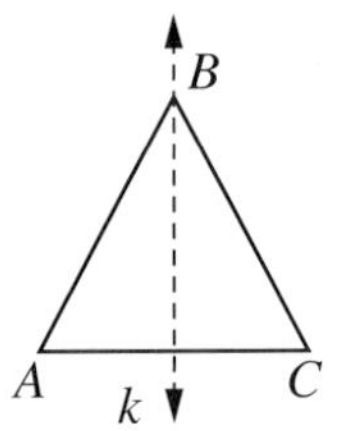

If we think of that line as a line of reflection, each point of the figure has as its image a point of the figure. This line of reflection is a line of symmetry, or ***axis of symmetry***, and the figure has ***line symmetry***.

An isosceles triangle has line symmetry. In the diagram on the left, the line of reflection, k, is an axis of symmetry and isosceles triangle ABC is symmetric with respect to the line through its vertex that is perpendicular to its base.

It is possible for a figure to have more than one axis of symmetry. In the rectangle at the right, line XY is an axis of symmetry and line VW is a second axis of symmetry.

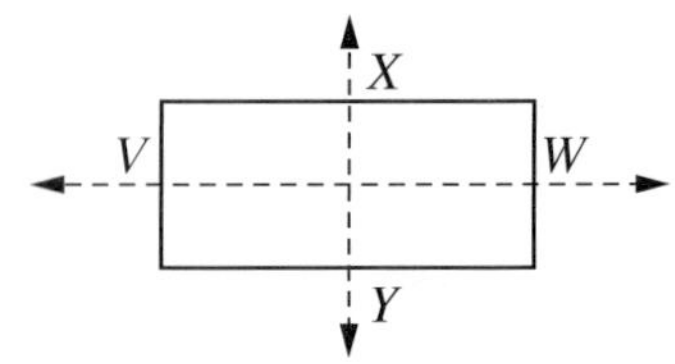

Lines of symmetry may be found for some letters and for some words, as shown at the right.

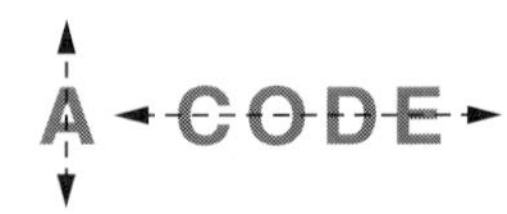

Not every figure, however, has line symmetry. If $\square ABCD$ at the right is reflected in diagonal $\overline{BD}$, the image of A is A' and the image of C is C'. Points A' and C', however, are not points of the original parallelogram. The image of $\square ABCD$ under a reflection in $\overline{BD}$ is $\square A'BC'D$. Therefore, $\square ABCD$ is not symmetric with respect to $\overline{BD}$.

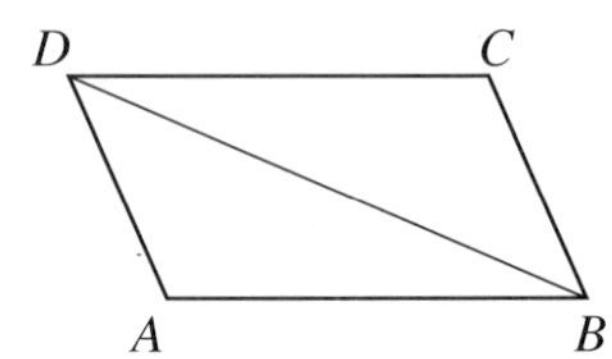

We have used diagonal $\overline{BD}$ as a line of reflection, but note that it is not a line of symmetry. In other words, there is no line along which the parallelogram can be folded so that points of the parallelogram on one side of the line will coincide with points of the parallelogram on the other.

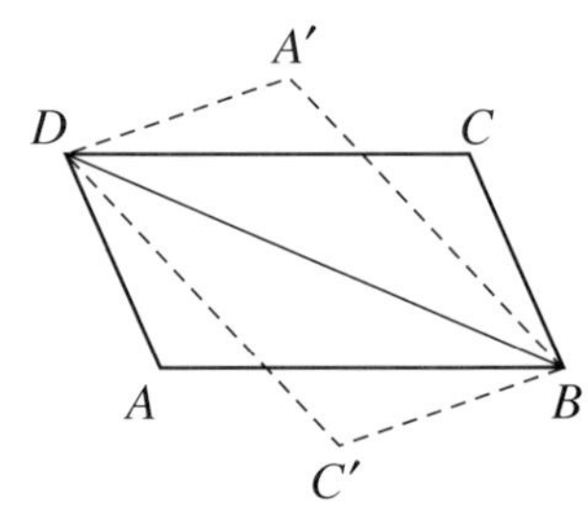

EXAMPLES

1. How many lines of symmetry does the letter **H** have?

Solution The horizontal line through the crossbar is a line of symmetry. The vertical line midway between the vertical segments is also a line of symmetry.

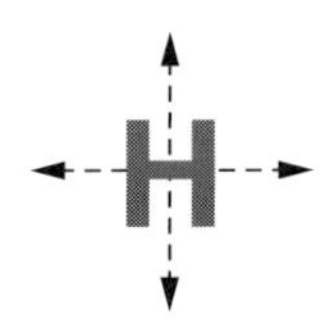

Answer: The letter **H** has two lines of symmetry.

2. Draw $\triangle ABC$ whose vertices are $A(-2,-1)$, $B(2,-1)$, $C(0, 5)$. What line is a line of symmetry for the triangle?

Solution

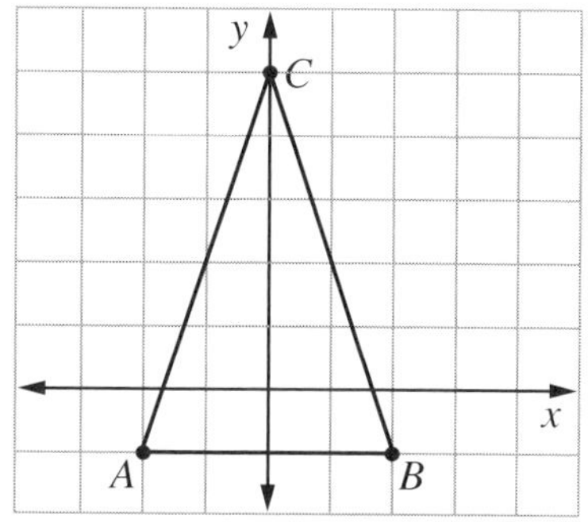

Answer: The y-axis is a line of symmetry for $\triangle ABC$.

EXERCISES WITH OPEN-RESPONSE PROBLEMS

1. **a.** Using the printed capital letters of the alphabet, write on your paper all letters that have line symmetry.
b. Indicate the lines of symmetry.

2. **a.** Copy each of the following "words."
b. Draw a line of symmetry, or indicate that the word does not have line symmetry by writing "None."

(*1*) MOM (*2*) DAD (*3*) SIS (*4*) OTTO
(*5*) BOOK (*6*) RADAR (*7*) un (*8*) NOON
(*9*) HIKE (*10*) SWI S (*11*) OHHO (*12*) CHOKED

In 3–14, for each geometric figure named: **a.** Sketch the figure. **b.** Tell the number of lines of symmetry, if any, that the figure has, and sketch them on your drawing.

3. rectangle
4. equilateral triangle
5. parallelogram
6. isosceles triangle
7. rhombus
8. regular hexagon
9. trapezoid
10. scalene triangle
11. circle
12. regular octagon
13. square
14. regular pentagon

15. **a.** Draw rectangle $PQRS$, whose vertices are $P(-5,-2)$, $Q(5,-2)$, $R(5, 2)$, and $S(-5, 2)$.
b. What two lines are lines of symmetry for the rectangle?

16. **a.** Draw rectangle $ABCD$, whose vertices are $A(2, 0)$, $B(2, 5)$, $C(4, 5)$, and $D(4, 0)$.
b. Draw the axes of the symmetry for the rectangle.

11-5 POINT REFLECTIONS AND POINT SYMMETRY

Another kind of reflection involves a point. In the figure at the right, $\triangle A'B'C'$ is the image of $\triangle ABC$ under a reflection in point P. If a line segment is drawn connecting any point to its image, then the point of reflection is the midpoint of that segment. In the figure:

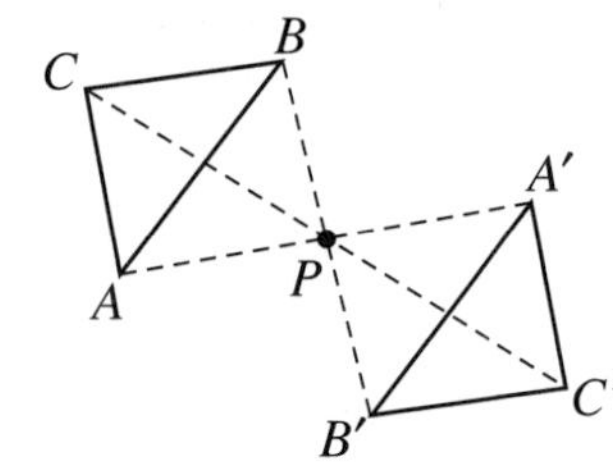

Point A' is on $\overleftrightarrow{AP}$, $AP = PA'$, and P is the midpoint of $\overline{AA'}$.

Point B' is on $\overleftrightarrow{BP}$, $BP = PB'$, and P is the midpoint of $\overline{BB'}$.

Point C' is on $\overleftrightarrow{CP}$,$CP = PC'$, and P is the midpoint of $\overline{CC'}$.

In parallelogram $ABCD$, shown at the right, diagonals $\overline{AC}$ and $\overline{BD}$ intersect at E. Point E is the midpoint of $\overline{AC}$ and of $\overline{BD}$. Therefore, under a reflection in point E, $A \rightarrow C$ and $C \rightarrow A$, $B \rightarrow D$ and $D \rightarrow B$. Similarly, $F \rightarrow G$ and $G \rightarrow F$, and every point of the parallelogram has its image on the parallelogram.

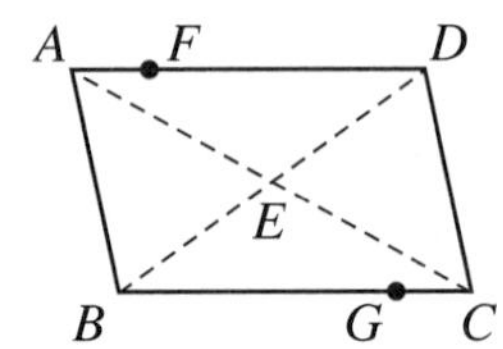

Properties of Point Reflections

Looking at parallelogram $ABCD$ and point of reflection E, we can note some properties of ***point reflection***:

1. Distance is preserved.

$$\overline{AB} \rightarrow \overline{CD} \quad \text{and} \quad AB = CD$$

$$\overline{AD} \rightarrow \overline{CB} \quad \text{and} \quad AD = CB$$

2. Angle measure is preserved.

$$\angle BAD \rightarrow \angle DCB \quad \text{and} \quad m\angle BAD = m\angle DCB$$

$$\angle ABC \rightarrow \angle CDA \quad \text{and} \quad m\angle ABC = m\angle CDA$$

3. The point of reflection is the midpoint of every segment formed by joining a point to its image.

$$AE = EC \text{ and } BE = ED$$

4. A figure is always congruent to its image.

Reflection in the Origin

The origin, $O(0, 0)$, is the most common point of reflection in the coordinate plane.

$$A(2, 1) \rightarrow A'(-2, -1)$$
$$B(1, 4) \rightarrow B'(-1, -4)$$
$$C(-2, 2) \rightarrow C'(2, -2)$$

The image of $\triangle ABC$ under a reflection in the origin is $\triangle A'B'C'$.

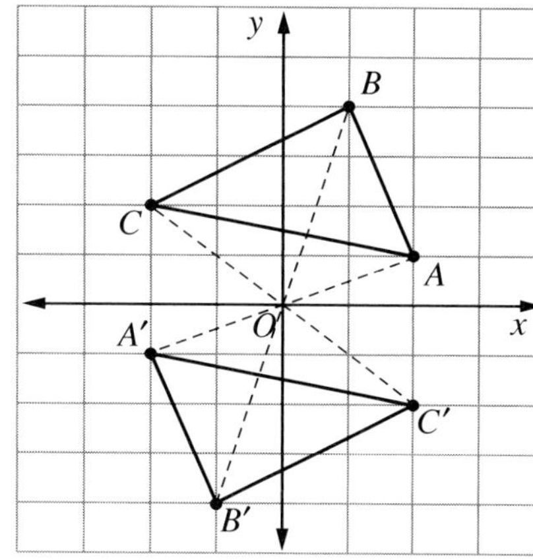

EXAMPLE

a. On your paper, draw any triangle ABC.

b. Sketch the image of $\triangle ABC$ under a reflection in point A.

Solutions (1) Draw a triangle and label it $\triangle ABC$.

(2) Hold a ruler on $\overline{AB}$ and measure the distance from B to A. Since A is the point of reflection, the image of B is on the line $\overleftrightarrow{AB}$. Locate the point along the ruler that is the same distance from A as B but that is on the opposite side of A. Label it point B'.

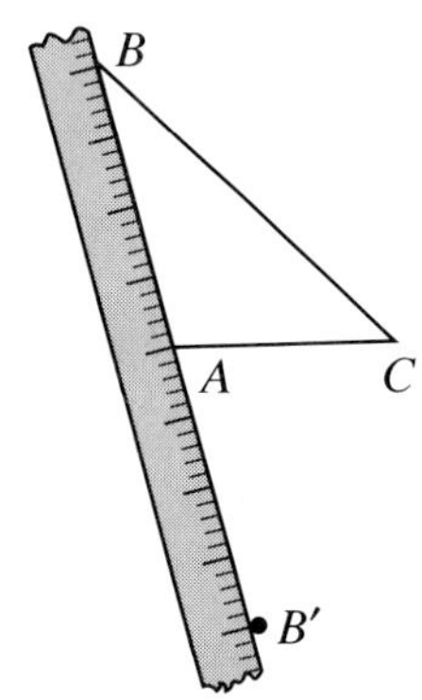

(3) Repeat step (2) for C by placing the ruler on $\overline{AC}$ to locate point C'.

(4) Draw $\triangle AB'C'$, the image of $\triangle ABC$.

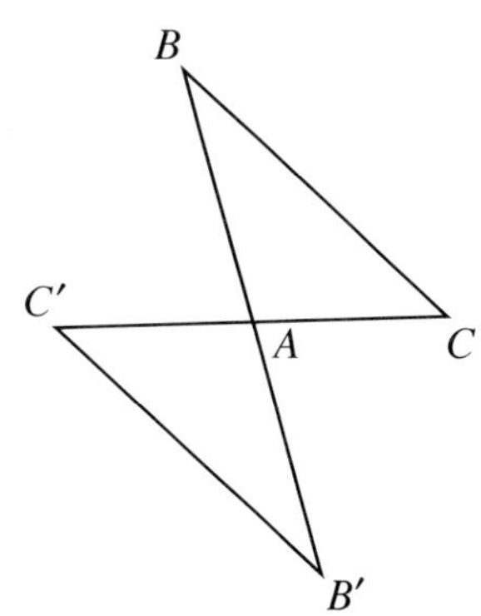

EXERCISES WITH OPEN-RESPONSE PROBLEMS

In 1–4, on your paper, copy each figure. Using a ruler, sketch the image of the figure under a reflection in point A.

1.

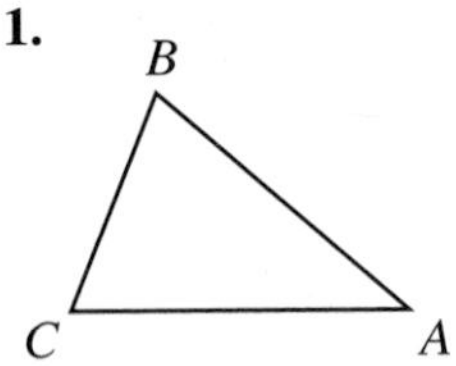

2.

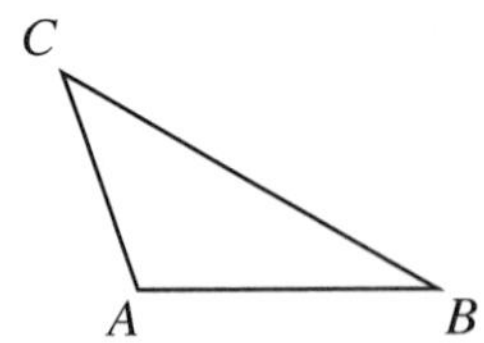

3.

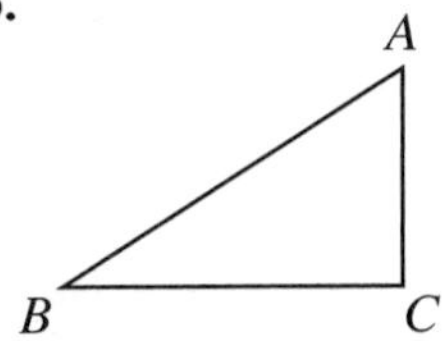

4.

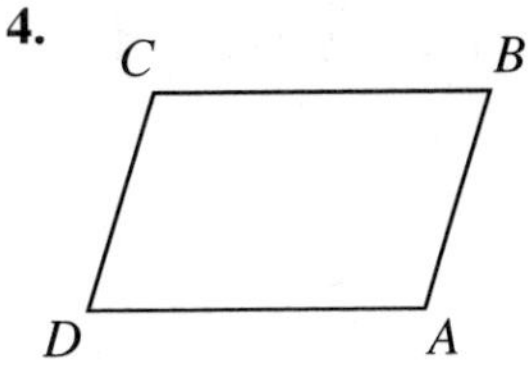

5. In the figure, $\triangle ABC \cong \triangle DBE$. Find the image of each of the following under a reflection in B:

a. A **b.** B **c.** C

d. D **e.** E **f.** $\overline{AC}$

g. $\overline{AB}$ **h.** $\overline{DE}$

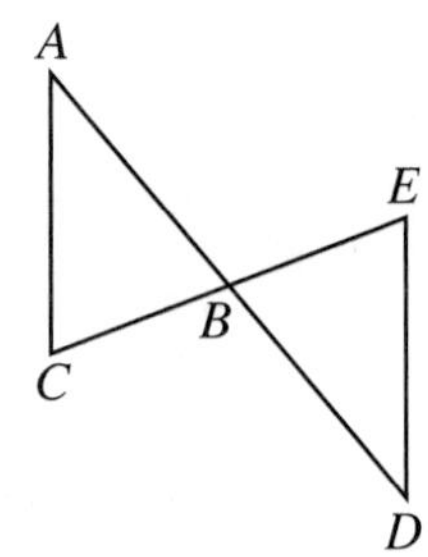

In 6–9, in each case: **a.** Graph the point and its reflection in the origin. **b.** State the coordinates of the image point.

6. $A(4, 3)$ **7.** $B(-3, 2)$ **8.** $C(-2,-1)$ **9.** $D(5,-4)$

Point Symmetry

In each of the figures shown above, the design is built around a central point. For every point in the figure, there is another point at the same distance from the center, so that the center is the midpoint of the segment joining the pair of points. Under a point reflection through the center, each point has as its image, another point of the figure. The figure has ***point symmetry***.

Parallelogram $ABCD$ at the right has point symmetry under a reflection in point E, the intersection of its diagonals.

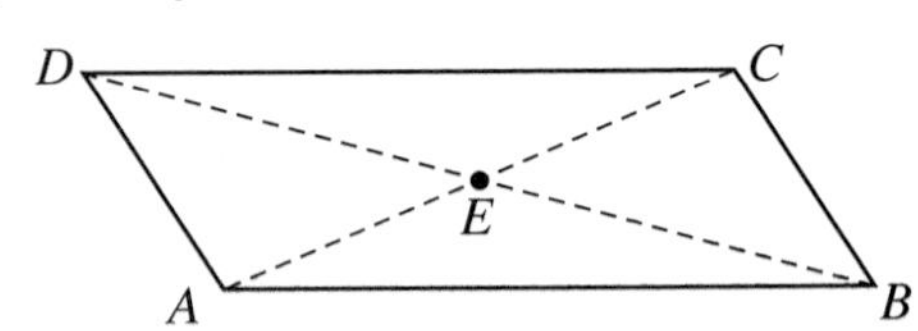

It is possible for a figure to have both line symmetry and point symmetry at the same time. In the square at the left, there are four lines of symmetry. Note that the point of symmetry lies at the intersection of these four lines.

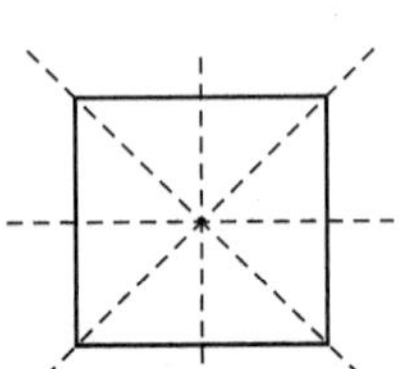

Points of symmetry may be found for some letters and for some words, as shown at the left.

EXERCISES WITH OPEN-RESPONSE PROBLEMS

1. **a.** Using the printed capital letters of the alphabet, write on your paper all of the letters that have point symmetry.
   b. Show the point of symmetry for each letter.

In 2–7, tell whether each figure has point symmetry.

2.  **3.** 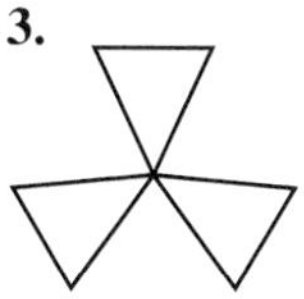 **4.** 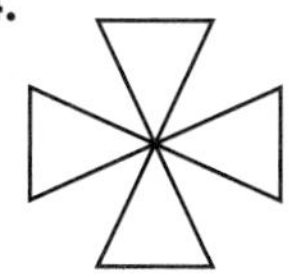 **5.** 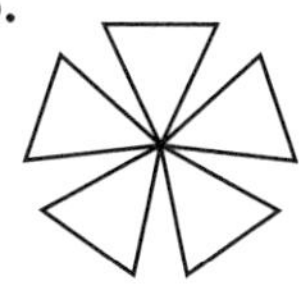 **6.** 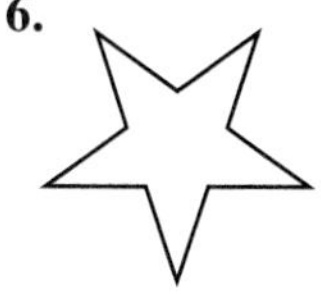 **7.** 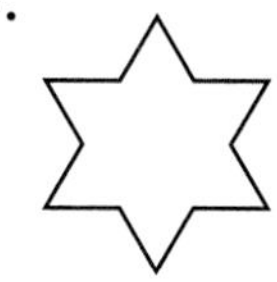

8. Which of the following has point symmetry?

(1) WOW (2) O (3) SIS (4) OTTO
(5) pop (6) pod (7) un (8) NOON
(9) HOHO (10) OHHO (11) SWI S (12) SOS

9. Draw a figure that has point symmetry but does not have line symmetry.
10. Draw a figure that has both point symmetry and line symmetry.
11. Which polygons have point symmetry?
(1) All polygons (2) All regular polygons
(3) Only squares (4) Regular polygons with an even number of sides

11-6 TRANSLATIONS

It is often useful or necessary to move objects from one place to another. If we move a desk from one place in the room to another, each leg moves the same distance in the same direction.

A ***translation*** moves every point in the plane the same distance in the same direction.

If $\triangle A'B'C'$ is the image of $\triangle ABC$ under a translation, $AA' = BB' = CC'$ and $\overline{AA'} \parallel \overline{BB'} \parallel \overline{CC'}$. The size and shape of the figure are unchanged, so that $\triangle ABC \cong \triangle A'B'C'$. Thus, as with reflections, a figure is congruent to its image under a translation.

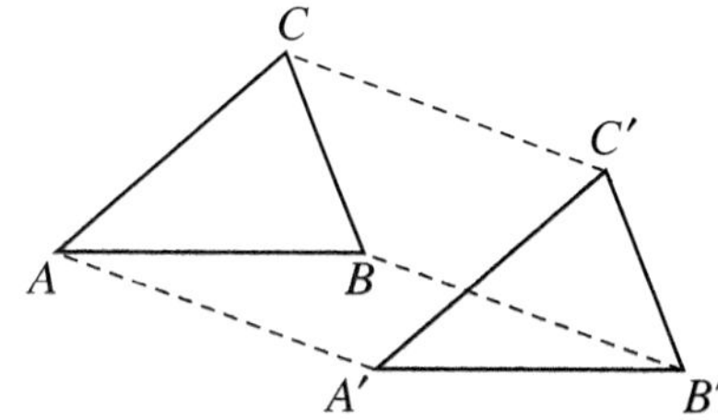

In the figure at the right, $\triangle ABC$ is translated by moving every point 4 units to the right and 5 units down. From the figure, we see that:

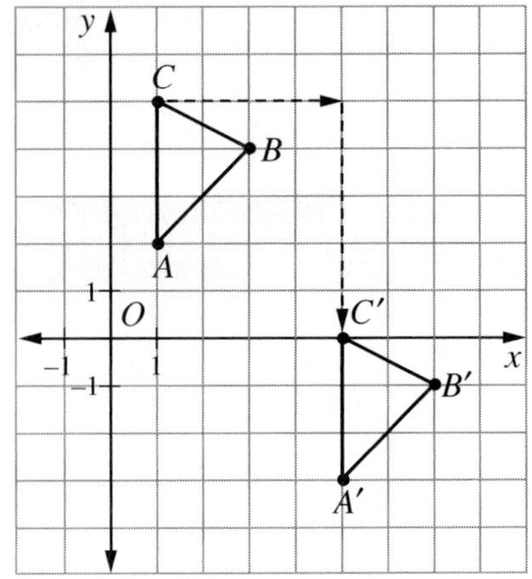

$$A(1, 2) \rightarrow A'(5, -3)$$

$$B(3, 4) \rightarrow B'(7, -1)$$

$$C(1, 5) \rightarrow C'(5, 0)$$

For each point and its image under a translation that moves every point 4 units to the right (+4) and 5 units down (−5), the x-coordinate of the image is 4 more than the x-coordinate of the point ($x \rightarrow x + 4$); the y-coordinate of the image is 5 less than the y-coordinate of the point ($y \rightarrow y - 5$).

From this example, we form a general rule:

- **Under a translation of *a* units in the horizontal direction and *b* units in the vertical direction, the image of $P(x, y)$ is $P'(x + a, y + b)$.**

Patterns used for decorative purposes such as wallpaper or borders on clothing often appear to have ***translational symmetry***. True translational symmetry would be possible, however, only if the pattern could repeat without end.

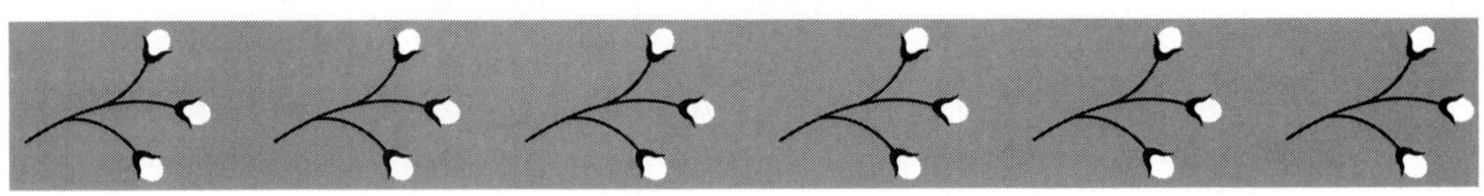

EXERCISES WITH OPEN-RESPONSE PROBLEMS

For 1 and 2, the diagram consists of nine congruent rectangles.

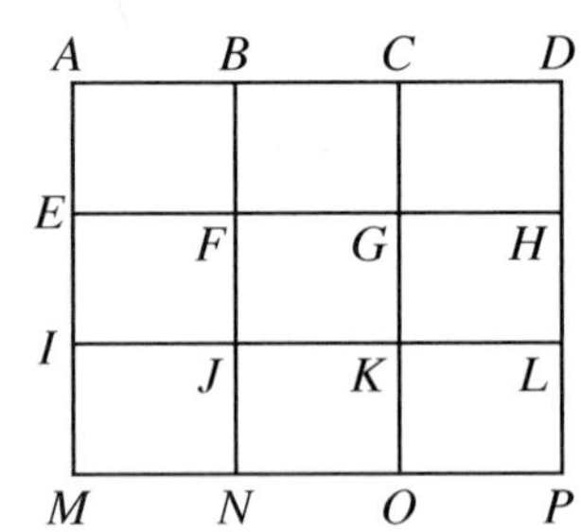

Ex. 1–2

1. Under a translation, the image of A is G. Find the image of each of the given points under the same translation.

a. J **b.** B **c.** I **d.** F **e.** E

2. Under a translation, the image of K is J. Find the image of each of the given points under the same translation.

a. J **b.** B **c.** O **d.** L **e.** G

3. In the diagram, $ADEH$ is a parallelogram. Points B, C, G, and F divide $\overline{AD}$ and $\overline{EH}$ into congruent segments. Under a translation, the image of A is G. Under the same translation, tell whether or not the image of each given point is a point of the diagram.
   a. G **b.** B **c.** C **d.** F

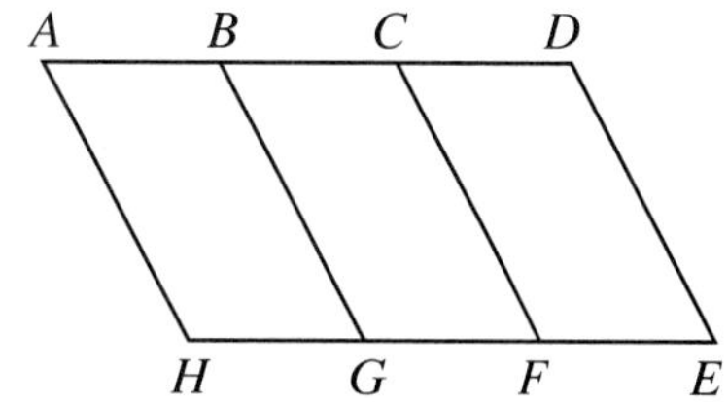

4. **a.** On graph paper, draw and label $\triangle ABC$, whose vertices have the coordinates $A(1, 2)$, $B(6, 3)$, and $C(4, 6)$.
   b. Under the translation $P(x, y) \rightarrow P'(x + 5, y - 3)$, every point moves 5 units to the right and 3 units down. For example, under this translation, the image of $A(1, 2)$ is $A'(1 + 5, 2 - 3)$ or $A'(6, -1)$. If, under this translation, the image of B is B' and the image of C is C', find the coordinates of B' and C'.
   c. On the same graph drawn in part **a**, draw and label $\triangle A'B'C'$.
5. **a.** On graph paper, draw and label $\triangle ABC$ if the coordinates of A are $(-2, -2)$, the coordinates of B are $(2, 0)$, and the coordinates of C are $(3, -3)$.
   b. On the same graph, draw and label $\triangle A'B'C'$, the image of $\triangle ABC$ under a translation whose rule is $(x, y) \rightarrow (x - 4, y + 7)$.
   c. Give the coordinates of the vertices of $\triangle A'B'C'$.
6. Which of the following is the rule of the translation in which every point moves 6 units to the right on a graph?
   (1) $(x, y) \rightarrow (x, y + 6)$ (2) $(x, y) \rightarrow (x + 6, y)$
   (3) $(x, y) \rightarrow (x + 6, y + 6)$ (4) $(x, y) \rightarrow (x - 6, y)$
7. In a translation, every point moves 4 units down. Write a rule for this translation by completing the sentence $(x, y) \rightarrow \ldots$.
8. The coordinates of $\triangle ABC$ are $A(2, 1)$, $B(4, 1)$, and $C(5, 5)$.
   a. On graph paper, draw and label $\triangle ABC$.
   b. Write a rule for the translation in which the image of A is $C(5, 5)$.
   c. Use the rule from part **b** to find the coordinates of B', the image of B, and C', the image of C, under this translation.
   d. On the graph drawn in part **a**, draw and label $\triangle CB'C'$, the image of $\triangle ABC$.
9. The coordinates of $\triangle ABC$ are $A(2, 1)$, $B(4, 1)$, and $C(5, 5)$.
   a. On graph paper, draw and label $\triangle ABC$.
   b. Under a transation, the image of C, C', is at $B(4, 1)$. Find the coordinates of A', the image of A, and of B', the image of B, under this same translation.
   c. On the graph drawn in part **a**, draw and label $\triangle A'B'C'$, the image of $\triangle ABC$.
   d. How many points, if any, are fixed points under this translation?

11-7 ROTATIONS

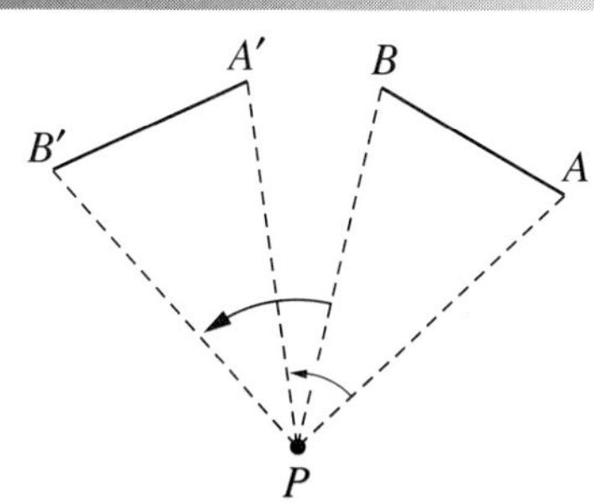

Think of what happens to all of the points of the steering wheel of a car as the wheel is turned. Except for the fixed point in the center, every point moves through a part of a circle, or arc, so that the position of each point is changed by a ***rotation*** of the same number of degrees.

In the figure, if A is rotated to A', then B is rotated the same number of degrees to B', and $m\angle APA' = m\angle BPB'$. Since P is the center of rotation, $PA = PA'$ and $PB = PB'$.

In general, a rotation preserves distance and angle measure. Under a rotation, a figure is congruent to its image. Unless otherwise stated, a rotation is in the counterclockwise direction.

H S N

Many letters, as well as designs in the shapes of wheels, stars, and polygons, have ***rotational symmetry***. Each figure shown at the right has rotational symmetry.

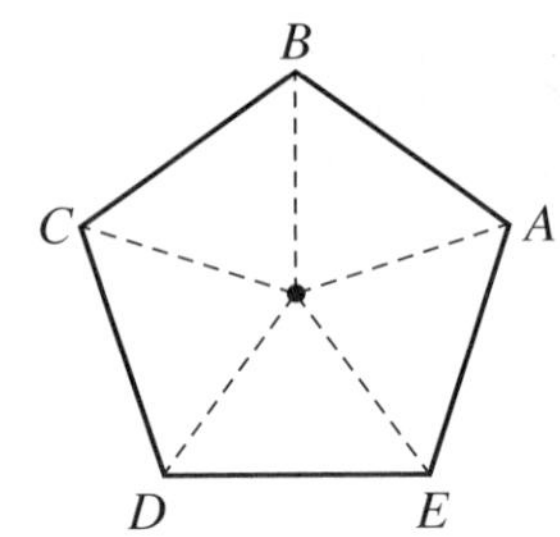

Any regular polygon has rotational symmetry. When regular pentagon $ABCDE$ is rotated $\frac{360^\circ}{5}$, or 72°, about its center, the image of every point of the figure is a point of the figure. Under this rotation, $A \to B$, $B \to C$, $C \to D$, $D \to E$, and $E \to A$.

The figure would also have rotational symmetry if rotated through a multiple of 72° (144°, 216°, or 288°). If it were rotated through 360°, every point would be its own image. Since this is true for every figure, we do not usually consider a 360° rotation as rotational symmetry.

EXAMPLE

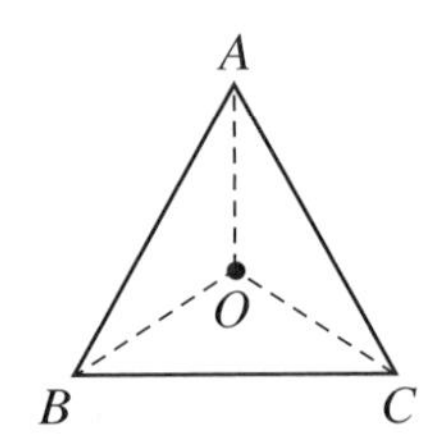

Point O is at the center of equilateral triangle ABC so that $OA = OB = OC$. Find the image of each of the following under a rotation of 120° about O:

a. A **b.** B **c.** C **d.** $\overline{AB}$
e. $\angle CAB$

Answers:

a. B **b.** C **c.** A **d.** $\overline{BC}$ **e.** $\angle ABC$

EXERCISES WITH OPEN-RESPONSE PROBLEMS

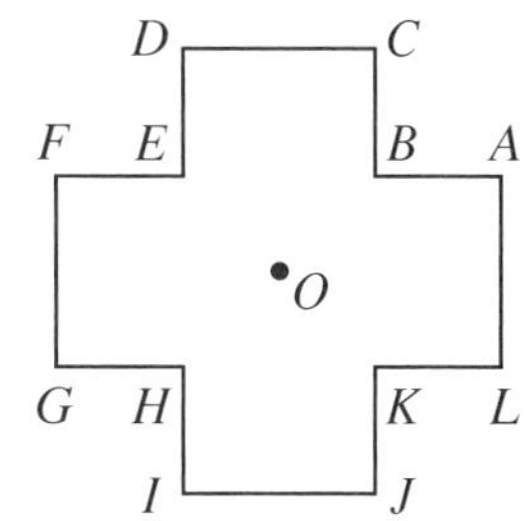

Ex. 1–2

1. What is the image of each of the given points under a rotation of 90° in the counterclockwise direction about O?
a. A **b.** B **c.** C **d.** G **e.** H **f.** J **g.** K

2. What is the image of each of the given points under a rotation of 90° in the clockwise direction about O?
a. A **b.** B **c.** C **d.** G **e.** H **f.** J **g.** K

In 3–8, for each geometric figure named: **a.** Sketch the figure. **b.** If the figure has rotational symmetry, mark the center of rotation with a dot. **c.** If the figure has rotational symmetry, give the measure of the smallest angle for which the symmetry exists.

3. Rectangle **4.** Parallelogram **5.** Rhombus
6. Trapezoid **7.** Regular hexagon **8.** Regular pentagon

11-8 COMPOSITIONS OF TRANSFORMATIONS

The combination of two transformations is called a ***composition of transformations*** when the first transformation produces an image, and the second transformation is performed on that image.

We start again with $\triangle ABC$, whose vertices are $A(1, 2)$, $B(5, 5)$, and $C(5, 2)$. For the composition of transformations, we will consider two line reflections. First, by reflecting $\triangle ABC$ in the y-axis, we form $\triangle A'B'C'$, or $\triangle$I. Then, by reflecting $\triangle A'B'C'$ in the x-axis, we form $\triangle A''B''C''$, or $\triangle$II. We observe how the vertices are related:

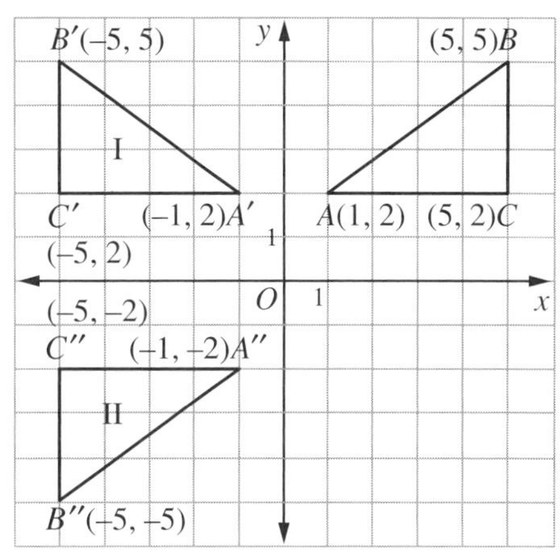

$$A(1, 2) \rightarrow A'(-1, 2) \rightarrow A''(-1, -2)$$

$$B(5, 5) \rightarrow B'(-5, 5) \rightarrow B''(-5, -5)$$

$$C(5, 2) \rightarrow C'(-5, 2) \rightarrow C''(-5, -2)$$

Now let us compare the original triangle, $\triangle ABC$, and the final image, $\triangle A''B''C''$, with the image of $\triangle ABC$ under a reflection in the origin. We observe:

- **The composition of a line reflection in the y-axis, followed by a line reflection in the x-axis, is equivalent to a single transformation, namely, a reflection through point O, the origin.**

We reflected the triangle first in the y-axis and then in the x-axis. If we had reflected the triangle first in the x-axis and then in the y-axis, would this composition also be equivalent to a reflection through point O, the origin? The answer is yes. However, not all compositions will act in the same way. In general, compositions of transformations are *not* commutative.

EXAMPLES

1. The coordinates of the endpoints of $\overline{RS}$ are $R(3, -2)$ and $S(5, -1)$.

a. Graph the line segment $\overline{RS}$.

b. Graph on the same set of axes the image $\overline{R'S'}$, a reflection of $\overline{RS}$ in the x-axis.

c. Graph on the same set of axes $\overline{R''S''}$, a reflection of $\overline{R'S'}$ in the origin.

d. Name the single transformation by which $\overline{RS} \rightarrow \overline{R''S''}$.

Solutions **a, b.** $R(3, -2) \rightarrow R'(3, 2)$
$S(5, -1) \rightarrow S'(5, 1)$

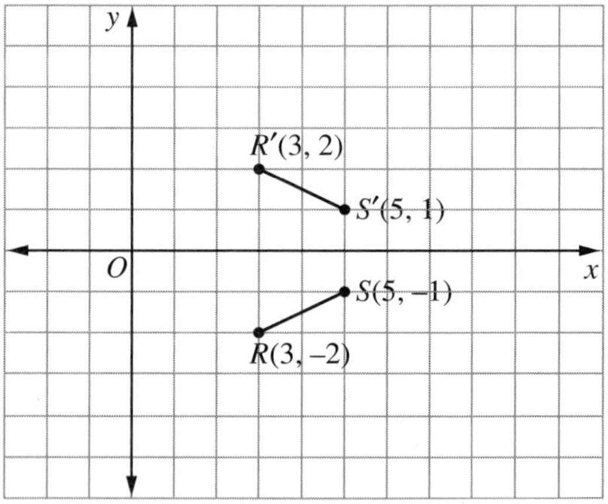

c. $R'(3, 2) \rightarrow R''(-3, -2)$
$S'(5, 1) \rightarrow S''(-5, -1)$

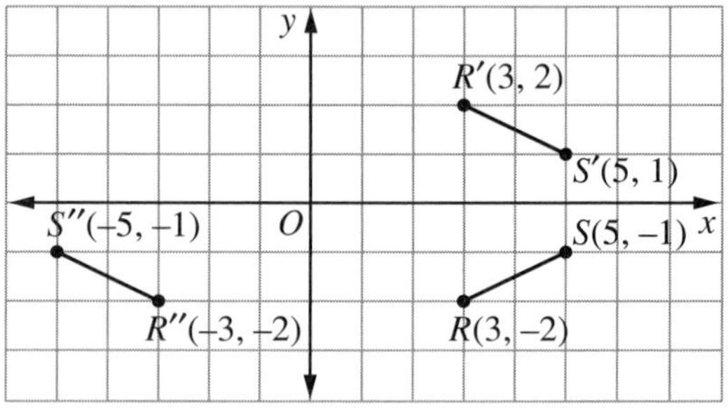

d. $\overline{RS} \rightarrow \overline{R''S''}$ is the same as a reflection of $\overline{RS}$ in the y-axis.
$R(3, -2) \rightarrow R''(-3, -2)$
$S(5, -1) \rightarrow S''(-5, -1)$

2. The coordinates of $\triangle ABC$ are $A(1, 3)$, $B(1, 1)$, and $C(5, 1)$. Is the image $\triangle A''B''C''$ the same if it is reflected in the origin, then rotated 90° clockwise about the origin as the image is if it is rotated 90° clockwise about the origin, then reflected 90° in the origin? Explain your answer.

Solution Under a reflection in the origin, then rotated 90° clockwise about the origin:

$A(1, 3) \rightarrow A'(-1, -3)$	$A'(-1, -3) \rightarrow A''(-3, 1)$
$B(1, 1) \rightarrow B'(-1, -1)$	$B'(-1, -1) \rightarrow B''(-1, 1)$
$C(5, 1) \rightarrow C'(-5, -1)$	$C'(-5, -1) \rightarrow C''(-1, 5)$

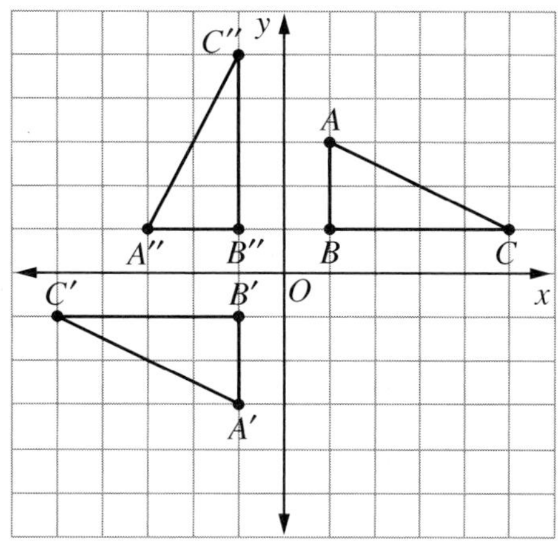

Rotated 90° clockwise about the origin, then under a reflection in the origin:

$A(1, 3) \rightarrow A'(3, -1)$	$A'(3, -1) \rightarrow A''(-3, 1)$
$B(1, 1) \rightarrow B'(1, -1)$	$B'(1, -1) \rightarrow B''(-1, 1)$
$C(5, 1) \rightarrow C'(1, -5)$	$C'(1, -5) \rightarrow C''(-1, 5)$

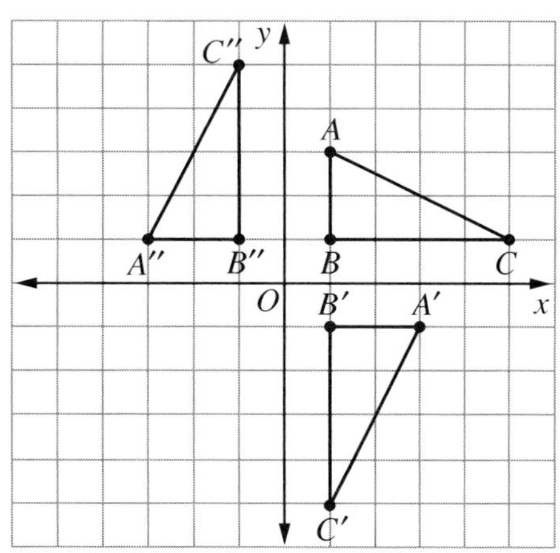

In each case $\triangle A''B''C''$ is the same.

EXERCISES WITH OPEN-RESPONSE PROBLEMS

1. On graph paper draw $\triangle ABC$ whose vertices are $A(-5, 2)$, $B(-3, 6)$, and $C(0, 0)$.

a. Using the same set of axes, graph $\triangle A'B'C'$, the image of $\triangle ABC$ under a reflection in the origin.

b. Using the same set of axes, graph $\triangle A''B''C''$, a reflection of $\triangle A'B'C'$ in the y-axis.

c. Name the single transformation by which $\triangle ABC \to \triangle A''B''C''$.

In 2 and 3, the vertices of $\triangle LMN$ are $L(1, 3)$, $M(1, 1)$, and $N(5, 1)$.

2. a. Draw $\triangle LMN$ and its image, $\triangle L'M'N'$, under a point reflection through the origin.

b. Using the same graph, reflect $\triangle L'M'N'$ in the x-axis to form its image, $\triangle L''M''N''$.

c. What single transformation is equivalent to the composition of a point reflection through the origin followed by a reflection in the x-axis?

3. a. Draw $\triangle LMN$ and its image $\triangle L'M'N'$ under a reflection in the x-axis.

b. Using the same graph, reflect $\triangle L'M'N'$ in the y-axis to form its image, $\triangle L''M''N''$.

c. What single transformation is equivalent to the composition of a reflection in the x-axis followed by a reflection in the y-axis?

4. Given $\triangle DEF$ and $\triangle D''E''F''$ as shown on the grid.

a. Describe one pair of transformations that will result in $\triangle DEF \to \triangle D''E''F''$.

b. Name a single transformation that will give the same result as the pair described in **a**.

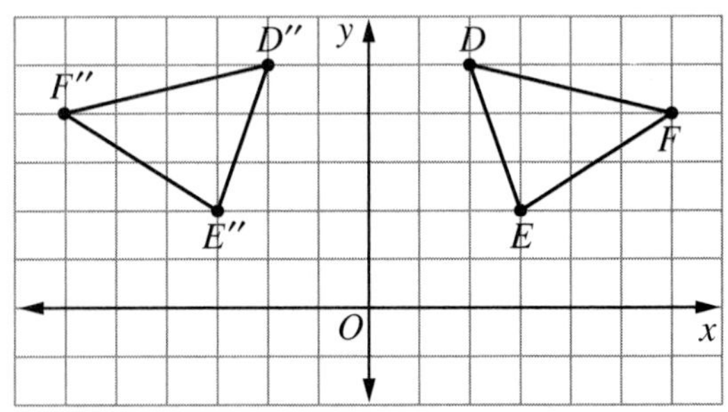

5. On graph paper draw parallelogram $RSTU$ with vertices $R(-8, -4)$, $S(-6, -2)$, $T(-3, -2)$, and $U(-5, -4)$.

a. Find the coordinates of the point of symmetry for parallelogram $RSTU$.

b. Find the images of $\overline{RS}$ and $\overline{RU}$ under this point reflection.

6. The vertices of parallelogram $ABCD$ are $A(1, 1)$, $B(3, 5)$, $C(9, 5)$, and $D(7, 1)$.

a. Find the coordinates of the point of symmetry for parallelogram $ABCD$.

b. Find the images of $\overline{AB}$ and $\overline{AD}$ under this point reflection.

7. The coordinates of the endpoints of $\overline{YZ}$ are $Y(3, 4)$ and $Z(1, 1)$. Is the image $\overline{Y''Z''}$ the same if it is reflected in the y-axis, then rotated 90° counterclockwise about the origin as the image is if it is rotated 90° counterclockwise about the origin then reflected in the y-axis? Explain your answer.

8. The coordinates of $\triangle JKL$ are $J(2, 4)$, $K(2, 2)$, and $L(6, 2)$. Is the image $\triangle J''K''L''$ the same if it is reflected in the origin, then rotated 180° counterclockwise about the origin as the image is if it is rotated 180° counterclockwise about the origin then reflected in the origin? Explain your answer.

CHAPTER SUMMARY

A ***simple closed curve*** starts and ends at the same point and does not cross itself. A ***polygon*** is a simple closed curve that consists of line segments. A ***regular polygon*** is a polygon in which all sides are congruent and all angles are congruent.

Two triangles are congruent if the corresponding parts in one of the following sets are congruent:

1. Two sides and the included angle.
2. Two angles and the included side.
3. Three sides.

The sum of the measures of the angles of a quadrilateral is 360°. The opposite angles of a parallelogram are congruent. The consecutive angles of a parallelogram are supplementary.

Among the transformations that can take place in a plane are the following:

1. ***Line reflection:*** the line of reflection is the perpendicular bisector of every segment joining a point and its image.
2. ***Point reflection:*** the point of reflection is the midpoint of every segment formed by joining a point and its image.
3. ***Translation:*** every point is moved the same distance in the same direction.
4. ***Rotation about point* P*:*** every point moves along a part of a circle by the same number of degrees.

A ***composition of transformations*** is a combination of two transformations where the second transformation is performed on the image of the first.

VOCABULARY

11-1 Geometric figure Plane geometric figure Simple closed curve Interior region Exterior region Boundary Polygon Side Vertex Triangle Quadrilateral Pentagon Hexagon Heptagon Octagon Nonagon Decagon Regular polygon

11-2 Congruent polygons Corresponding sides Corresponding angles

11-3 Quadrilateral Consecutive angles Opposite angles Trapezoid Isosceles trapezoid Parallelogram Rectangle Rhombus Square

11-4 Transformation One-to-one correspondence Line reflection Axis of symmetry Line symmetry

11-5 Point reflection Point symmetry

11-6 Translation Translational symmetry

11-7 Rotation Rotational symmetry

11-8 Composition of transformations

REVIEW EXERCISES WITH OPEN-RESPONSE PROBLEMS

1. In $\triangle ABC$, D is the midpoint of $\overline{AC}$, and $\overline{BD}$ intersects $\overline{AC}$ at right angles.

a. State a reason why $\triangle ABD \cong \triangle CBD$.

b. If $AB = 3x + 1$ and $CB = 5x - 7$, find x.

c. What is the image of A under a reflection in $\overleftrightarrow{BD}$?

d. What is the line of symmetry for $\triangle ABC$?

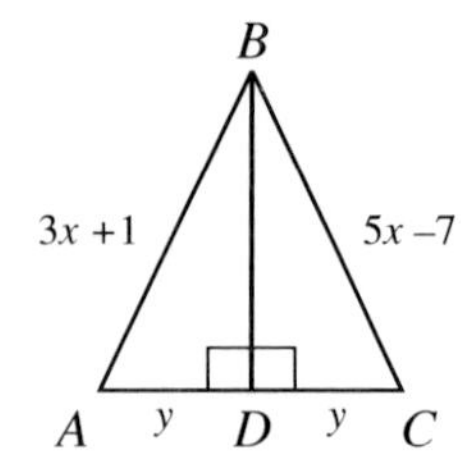

2. If $FLAT$ is a square where $FL = 9y - 2$ and $LA = 7y + 16$, find y.

3. In quadrilateral $ABCD$, $m\angle A = 100$, $m\angle B = 70$, $m\angle C = x$, and $m\angle D = x + 50$. Find $m\angle C$.

4. a. Graph points $A(1, 1)$, $B(5, 1)$, and $C(5, 4)$.

b. What are the coordinates of point D if $ABCD$ is a rectangle?

5. a. Graph points $P(-2,-4)$ and $Q(2,-4)$.

b. What are the coordinates of R and S if $PQRS$ is a square? (Two answers are possible.)

6. a. Graph points $S(3, 0)$, $T(0, 4)$, $A(-3, 0)$, and $R(0,-4)$, and draw the rhombus $STAR$.

b. Find the area of $STAR$ by adding the areas of the triangles into which the axes separate the rhombus.

7. a. Graph points $P(2, 0)$, $L(1, 1)$, $A(-1, 1)$, $N(-2, 0)$, $E(-1,-1)$, and $T(1,-1)$, and draw the hexagon $PLANET$.

b. Find the area of $PLANET$. (*Hint:* Use the x-axis to separate the hexagon into two parts.)

8. Two of the vertices of rectangle $ABCD$ are $(-3,-4)$ and $(3, 5)$. What are the coordinates of the other two vertices if the y-axis is a line of symmetry for the rectangle?

9. Which condition does *not* demonstrate that two triangles are congruent?

(1) a.s.a. ≅ a.s.a. (2) s.a.s. ≅ s.a.s.
(3) a.a.a. ≅ a.a.a. (4) s.s.s. ≅ s.s.s.

For 10–14, use square $ABCD$, where E, F, G, and H are midpoints of the sides of the square, and diagonals $\overline{AC}$ and $\overline{BD}$ meet at O.

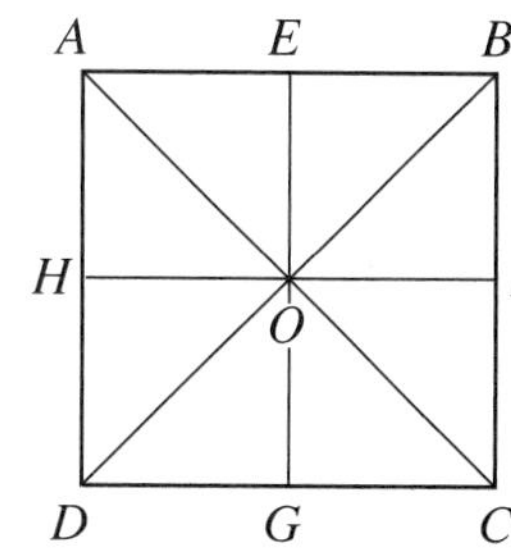

Ex. 10–14

10. What is the image of each of the following under a reflection in $\overleftrightarrow{EG}$?
a. A **b.** E **c.** $\overline{AO}$ **d.** $\angle HDO$

11. What is the image of each of the following under a reflection in O?
a. A **b.** E **c.** $\overline{AO}$ **d.** $\angle HDO$

12. What is the image of each of the following under a rotation of 90° about O?
a. A **b.** E **c.** $\overline{AO}$ **d.** $\angle HDO$

13. Under a certain translation, the image of A is O. What is the image of each of the following under the same translation?
a. H **b.** O **c.** $\overline{AH}$ **d.** $\angle OAH$

14. Under which transformation is the figure *not* symmetric?
(1) line reflection (2) point reflection
(3) translation (4) 90° rotation

15. Look at the handwritten word "chump" in the box. Does it have any type of symmetry?

16. Many business organizations have a symbol or trademark that identifies the company. Find in a newspaper or magazine, or draw, a symbol that has:
a. line symmetry **b.** point symmetry **c.** rotational symmetry
d. no symmetry

17. Which printed capital letters of the alphabet have point symmetry but *not* line symmetry?

18. Which triangle shows the image of $\triangle XYZ$ after a reflection in the x-axis followed by a reflection in the origin?
(1) I (2) II (3) III (4) IV

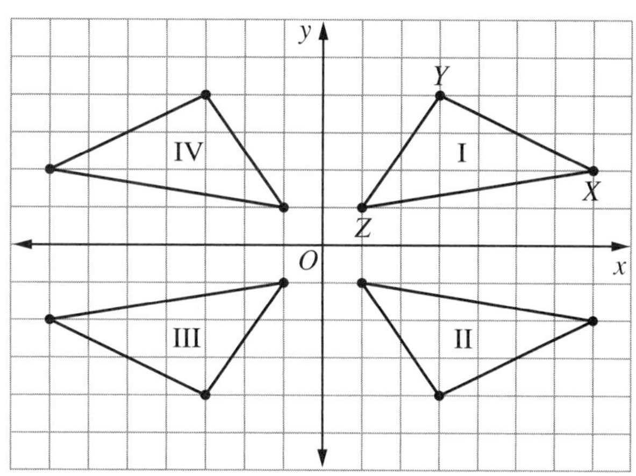

CUMULATIVE REVIEW WITH OPEN-RESPONSE PROBLEMS

1. Express the sum of $3x^2 - 2x + 5$ and $8 + 2x - x^2$ in simplest form.
2. Write the product $(2x - 5)(x - 3)$ as a polynomial.
3. Find the area of $\triangle ABC$ if the coordinates of the vertices are $A(0, -1)$, $B(6, -1)$, and $C(6, 4)$.
4. What is the quotient when $12x^4 - 18x^3 - 6x^2$ is divided by $6x^2$?
5. Graph the solution set of $3(x - 2) \leq 5x$.
6. Which of the following is the graph of $-2 \leq x < 3$

(1) 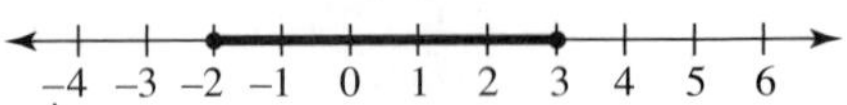

(2) 

(3) 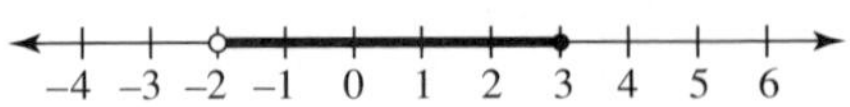

(4) 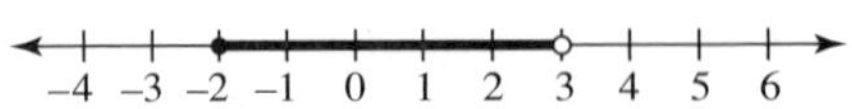

7. If $0.000072 = 7.2 \times 10^n$, what is the value of n?
8. For what value of n does $0.00045 = 4.5 \times 10^n$?
9. In each of the following, perform the indicated operation and tell whether the number is rational or irrational.

   a. $8 \times \pi$ **b.** $\pi \div \pi$ **c.** $\frac{\sqrt{2}}{2}$
10. Subtract $3x^2 - 5x - 2$ from $x^2 - 7x + 2$ and express the difference in simplest form.
11. In the diagram at the right, $\overline{AB} \parallel \overline{CD}$, $\overline{EF}$ intersects $\overline{AB}$ at G and $\overline{CD}$ at H. If $m\angle AGE$ is 32 degrees less than $m\angle CHF$, find the measure of each of the following:

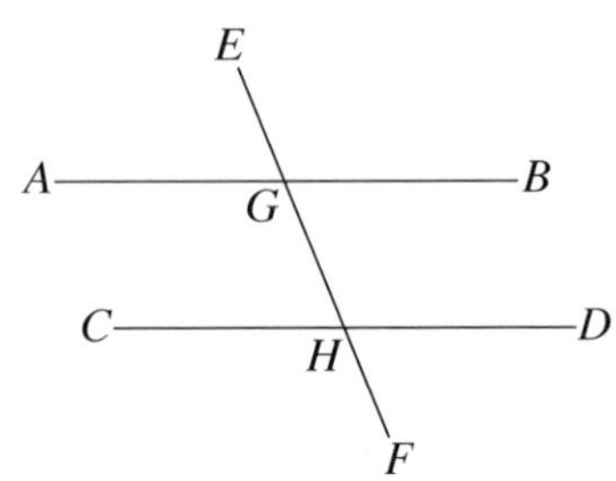

   a. $m\angle AGE$ **b.** $m\angle CHF$ **c.** $m\angle BGH$
   d. $m\angle BGE$ **e.** $m\angle AGH$ **f.** $m\angle DHG$
   g. $m\angle CHG$ **h.** $m\angle FHD$
12. At the beginning of the year, Latonya's salary increased by 3%. If her weekly salary is now \$254.41, what was her weekly salary last year?

Exploration

Make a portfolio of pictures (snapshots, postcards, pictures from magazines or newspapers) of buildings that illustrate geometric shapes. Identify the shapes or combination of shapes and any symmetry that you observe.

Chapter 12

Ratio and Proportion

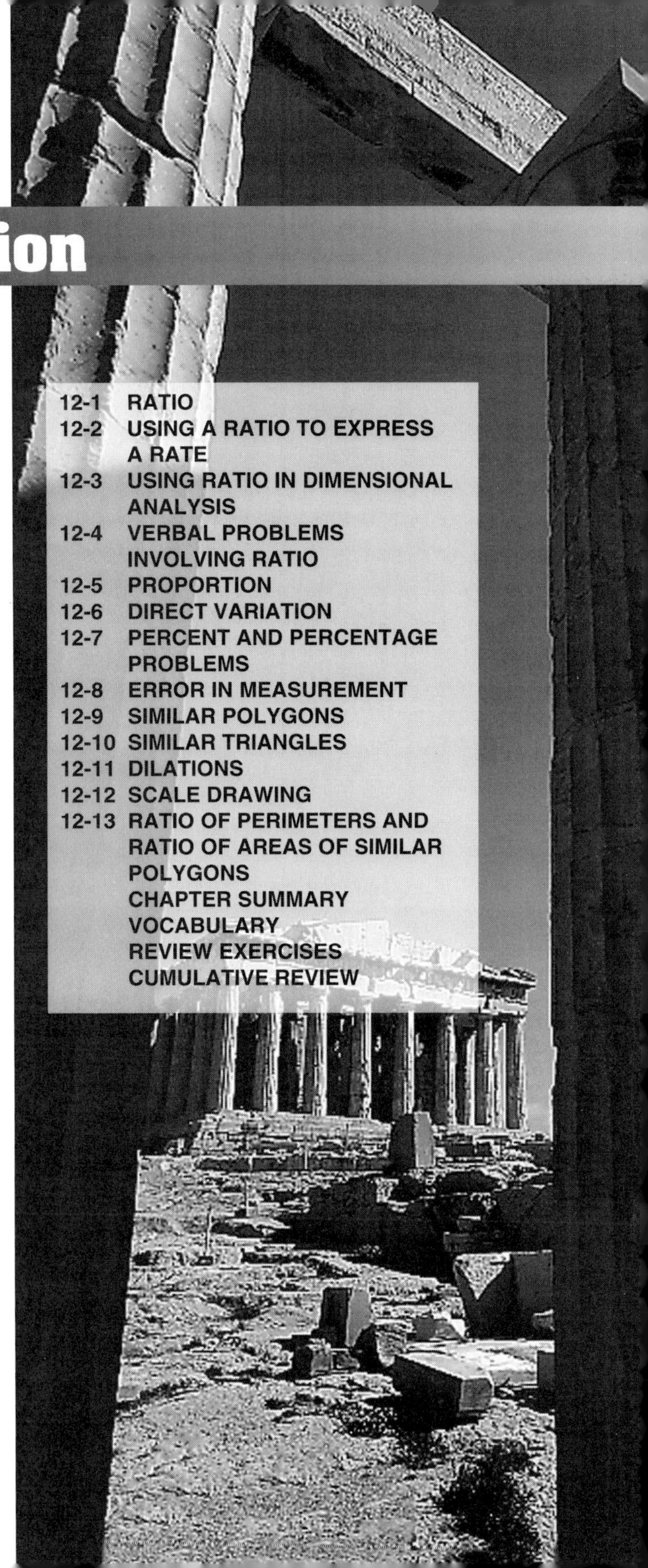

The ancient Greeks used a special ratio called the *golden section* in their art and architecture.

In a ***golden section***, a line segment is divided into two parts, a and b, so that the ratio (or comparison) of the longer part, a, to the shorter part, b, is equal to the ratio of the entire segment, $a + b$, to its longer part, a. In symbols, we write:

$$\frac{a}{b} = \frac{a + b}{a}$$

By solving the above equation, the Greeks found measures to form the ratio of the length to the width in the ***golden rectangle***, thought to be the most pleasing and artistic of all rectangles.

The Greeks used the golden rectangle to design the Parthenon, an ancient temple begun in the year 447 B.C., that still stands today. Every golden rectangle in the Parthenon, whether in the shape of the structure or in the spaces between its columns, reduces to the same ratio, $\frac{\text{length}}{\text{width}} = \frac{a}{b}$. Some ratios, such as $\frac{13}{8}$, $\frac{89}{55}$, and $\frac{233}{144}$, are close, but not equal, to the dimensions of the golden rectangle.

The actual ***golden ratio***, which is an irrational number, is the value $\frac{1 + \sqrt{5}}{2}$. We can use a calculator to find a rational approximation for the golden ratio. The calculator display is a decimal that is being compared to 1, just as a is compared to b, and length is compared to width. How close are the other ratios mentioned above to this rational approximation?

12-1 RATIO

A ***ratio***, which is a comparison of two numbers by division, is the quotient obtained when the first number is divided by the second, nonzero number. A ratio may be considered as an ordered pair of numbers, and finding the ratio of two numbers is a binary operation.

Since a ratio is the quotient of two numbers divided in a definite order, care must be taken to write each ratio in its intended order. For example, the ratio of 3 to 1 is written as

$$\frac{3}{1} \text{ (a fraction)} \qquad \text{or} \qquad 3:1 \text{ (using a colon)}$$

while the ratio of 1 to 3 is written as

$$\frac{1}{3} \text{ (a fraction)} \qquad \text{or} \qquad 1:3 \text{ (using a colon)}$$

In general, the ratio of a to b can be expressed as

$$\frac{a}{b} \qquad \text{or} \qquad a \div b \qquad \text{or} \qquad a:b$$

To find the ratio of two quantities, both quantities must be expressed in the same unit of measure before finding their quotient. For example, to compare the value of a nickel and a penny, we first convert the nickel to 5 pennies and then find the ratio, which is $\frac{5}{1}$ or 5:1. Therefore, a nickel is worth 5 times as much as a penny. The ratio has no unit of measure.

Equivalent Ratios

Since the ratio $\frac{5}{1}$ is a fraction, we can use the multiplication property of 1 to find many ***equivalent ratios***. For example:

$$\frac{5}{1} = \frac{5}{1} \cdot \frac{2}{2} = \frac{10}{2} \qquad \frac{5}{1} = \frac{5}{1} \cdot \frac{3}{3} = \frac{15}{3} \qquad \frac{5}{1} = \frac{5}{1} \cdot \frac{x}{x} = \frac{5x}{1x} \qquad (x \neq 0)$$

From the last example, we see that $5x$ and $1x$ represent two numbers whose ratio is 5:1.

In general, if a, b, and x are numbers ($b \neq 0$, $x \neq 0$), ax and bx represent two numbers whose ratio is $a:b$ because

$$\frac{a}{b} = \frac{a}{b} \cdot 1 = \frac{a}{b} \cdot \frac{x}{x} = \frac{ax}{bx}$$

Also, since a ratio such as $\frac{24}{16}$ is a fraction, we can divide the numerator and the denominator of the fraction by the same nonzero number to find equivalent ratios.

For example:

$$\frac{24}{16} = \frac{24 \div 2}{16 \div 2} = \frac{12}{8} \qquad \frac{24}{16} = \frac{24 \div 4}{16 \div 4} = \frac{6}{4} \qquad \frac{24}{16} = \frac{24 \div 8}{16 \div 8} = \frac{3}{2}$$

A ratio is expressed in ***simplest form*** when both terms of the ratio are whole numbers and when there is no whole number other than 1 that is a factor of both of these terms. Therefore, to express the ratio $\frac{24}{16}$ in simplest form, we divide both terms by 8, the greatest common factor of 24 and 16, thus obtaining the ratio $\frac{3}{2}$, as shown above.

Continued Ratio

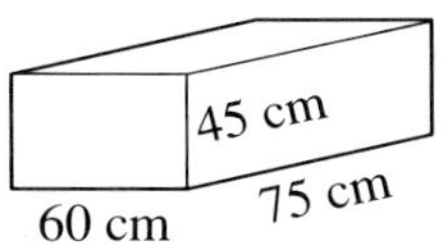

Comparisons can also be made for three or more quantities. For example, the length of a rectangular solid is 75 centimeters, the width is 60 centimeters, and the height is 45 centimeters. The ratio of the length to the width is 75:60, and the ratio of the width to the height is 60:45. We can write these two ratios in an abbreviated form as the continued ratio 75:60:45.

A ***continued ratio*** is a comparison of three or more quantities in a definite order. Here, the ratio of the measures of the length, width, and height (in that order) of the rectangular solid is 75:60:45 or, in simplest form, 5:4:3.

In general, the ratio of the numbers a, b, and c ($b, c \neq 0$) is $\boldsymbol{a:b:c}$.

EXAMPLES

1. An oil tank with a capacity of 200 gallons contains 50 gallons of oil.

a. Find the ratio of the number of gallons of oil in the tank to the capacity of the tank.

b. What part of the tank is full?

Solutions **a.** $\text{Ratio} = \frac{\text{number of gallons of oil in tank}}{\text{capacity of tank}} = \frac{50}{200} = \frac{1}{4}$ *Answer*

b. The tank is $\frac{1}{4}$ full. *Answer*

---

2. Compute the ratio of 6.4 ounces to 1 pound.

Solution First, express both quantities in the same unit of measure.
Since 1 pound (lb) = 16 ounces (oz),

$$\text{Ratio} = \frac{6.4 \text{ oz}}{1 \text{ lb}} = \frac{6.4 \text{ oz}}{16 \text{ oz}} = \frac{6.4}{16} = \frac{64}{160} = \frac{2}{5}$$

Calculator Solution On a calculator, divide 6.4 ounces by 16 ounces.

Enter: 6.4 [÷] 16 [=]

Display: [0.4]

Change the decimal in the display to a fraction: $0.4 = \frac{4}{10} = \frac{2}{5}$.

Or think of the display as a numerator of 0.4 compared to a denominator of 1:

$$0.4 = \frac{0.4}{1} = \frac{0.4}{1} \cdot \frac{10}{10} = \frac{4}{10} = \frac{2}{5}$$

Answer: The ratio is 2:5.

---

3. Express the ratio $1\frac{3}{4}$ to $1\frac{1}{2}$ in simplest form.

Solution $$\text{Ratio} = 1\frac{3}{4} : 1\frac{1}{2} = 1\frac{3}{4} \div 1\frac{1}{2} = \frac{7}{4} \div \frac{3}{2} = \frac{7}{4} \times \frac{2}{3} = \frac{14}{12} = \frac{7}{6}$$

Calculator Solution ***Enter:*** (1 + 3 ÷ 4) ÷ (1 + 1 ÷ 2) =

Display: 1.1666667

Recognize the repeating decimal in the display, and convert it to a fraction:

$$1.1666667 = 1.166666\ldots = 1\frac{1}{6} = \frac{7}{6}$$

Answer: The ratio in simplest form is $\frac{7}{6}$ or 7:6.

EXERCISES WITH OPEN-RESPONSE PROBLEMS

In 1–5, express each ratio: **(a)** as a fraction **(b)** using a colon.

1. 36 to 12 **2.** 48 to 24 **3.** 40 to 25 **4.** 2 to 3 **5.** 5 to 4

6. Express each ratio in simplest form.

a. $\frac{8}{32}$ **b.** $\frac{40}{5}$ **c.** $\frac{12}{28}$ **d.** $\frac{36}{27}$ **e.** $\frac{36}{24}$

f. 20:10 **g.** 15:45 **h.** 18:18 **i.** 48:20 **j.** 21:35

k. $3x:2x$ **l.** $1y:4y$ **m.** $3c:5c$ **n.** $7x:7y$ **o.** $12s:4s$

7. In each of the following, the larger number is how many times the smaller number?

a. 10, 5 **b.** 18, 6 **c.** 12, 8 **d.** 25, 10 **e.** 15, 25

8. If the ratio of two numbers is 10:1, how many times the smaller number is the larger number?

9. If the ratio of two numbers is 8:1, the smaller number is what fractional part of the larger number?

10. Tell whether each ratio is equal to $\frac{3}{2}$.

a. $\frac{30}{20}$ **b.** $\frac{9}{4}$ **c.** $\frac{8}{12}$ **d.** 9:6 **e.** $\frac{45}{30}$ **f.** 18:6

11. In each case, name the ratios that are equal.

a. $\frac{2}{3}, \frac{6}{9}, \frac{10}{30}, \frac{28}{36}, \frac{50}{75}$ **b.** 10:8, 20:16, 15:13, 4:5, 50:40

12. For each given ratio, find three pairs of numbers such that each pair has that ratio.
a. $\frac{1}{2}$ **b.** $\frac{1}{5}$ **c.** 3:1 **d.** 4:1 **e.** $\frac{3}{4}$ **f.** 2:3

13. Using a colon, express in simplest form the ratio of all pairs of:
a. equal numbers (not zero) **b.** nonzero numbers whose difference is 0

14. Express each ratio in simplest form.
a. $\frac{3}{4}$ to $\frac{1}{4}$ **b.** $1\frac{1}{8}$ to $\frac{3}{8}$ **c.** 1.2 to 2.4 **d.** .75 to .25 **e.** 6:0.25

15. Express each ratio in simplest form.
a. 80 m to 16 m **b.** 75 g to 100 g **c.** 36 cm to 72 cm
d. 54 g to 90 g **e.** 75 cm to 35 mm **f.** 32 cg to 80 mg
g. 36 mL to 3 L **h.** 150 m to 5 km **i.** 500 g to 2 kg

16. Express each ratio in simplest form.
a. $1\frac{1}{2}$ hr. to $\frac{1}{2}$ hr. **b.** 3 in. to $\frac{1}{2}$ in. **c.** 1 ft. to 1 in.
d. 1 yd. to 1 ft. **e.** $\frac{1}{3}$ yd. to 6 in. **f.** 12 oz. to 3 lb.
g. 1 hr. to 15 min. **h.** $6 to 50 cents **i.** 2 mi. to 880 yd.

17. A baseball team played 162 games and won 90.
a. What is the ratio of the number of games won to the number of games played?
b. For every nine games played, how many games were won?

18. A student did six of ten problems correctly.
a. What is the ratio of the number right to the number wrong?
b. For every two answers that were wrong, how many answers were right?

19. A cake recipe calls for $1\frac{1}{4}$ cups of milk to $1\frac{3}{4}$ cups of flour. What is the ratio of the number of cups of milk to the number of cups of flour in this recipe?

20. The perimeter of a rectangle is 30 feet, and the width is 5 feet. Find the ratio of the length of the rectangle to its width.

21. In a freshman class, there are b boys and g girls. Express the ratio of the number of boys to the total number of pupils.

22. The length of a rectangle is represented by $3x$ and its width by $2x$. Find the ratio of the width of the rectangle to its perimeter.

23. In each case, represent in terms of x two numbers whose ratio is:
a. 3 to 4 **b.** 5 to 3 **c.** 1 to 4 **d.** 1:2 **e.** 3:5

24. In each case, represent in terms of x three numbers that have the continued ratio:
a. 1 to 2 to 3 **b.** 3 to 4 to 5 **c.** 1:3:4 **d.** 2:3:5

25. The ages of three teachers are 48, 28, and 24 years. Find, in simplest form, the continued ratio of these ages from oldest to youngest.

26. In a rectangular solid, let ℓ = length, w = width, and h = height. Express, in simplest form, the continued ratio $\ell:w:h$ for each set of measures.
a. $\ell = 30$ m, $w = 25$ m, $h = 10$ m **b.** $\ell = 2$ yd., $w = 2$ ft., $h = 4$ ft.
c. ℓ $1\frac{1}{2}$ ft., $w = \frac{3}{4}$ ft., $h = 2$ ft. **d.** $\ell = 80$ cm, $w = 0.4$ m, $h = 40$ cm
e. $\ell = 1$ yd., $w = 1$ ft., $h = 1$ in. **f.** ℓ = twice width w, h = one-half width w

27. Taya and Jed collect coins. The ratio of the number of coins in their collections, in some order, is 4 to 3. If Taya has 60 coins in her collection, how many coins could Jed have?

12-2 USING A RATIO TO EXPRESS A RATE

A ratio is a comparison of two quantities, usually having the same units of measure. A ***rate***, like a ratio, is a comparison of two quantities, but these quantities may have *different units of measures*.

For example, if a plane flies 1,920 kilometers in 3 hours, its *rate* of speed is a ratio that compares the distance traveled to the time that the plane was in flight.

$$\text{Rate} = \frac{\text{distance}}{\text{time}} = \frac{1{,}920\text{ km}}{3\text{ h}} = \frac{640\text{ km}}{1\text{ h}} = 640\text{ km/h}$$ [This is read as 640 kilometers per hour.]

A rate is expressed in ***lowest terms*** when the numbers in its ratio are in simplest form as in $\frac{640}{1}$.

A ratio is written without units of measure, such as 640 : 1, but a rate is always written with the appropriate units of measure, such as 640 kilometers per hour.

EXAMPLES

1. Kareem scored 175 points in seven basketball games. Express, in lowest terms, the average rate of the number of points Kareem scored per game.

Solution

$$\text{Rate} = \frac{175\text{ points}}{7\text{ games}} = \frac{25}{1} = 25\text{ points per game}$$

Answer: Kareem scored points at an average rate of $\frac{25}{1}$ or 25 points per game.

---

2. There are 5 grams of salt in 100 cubic centimeters of a solution of salt and water. Express, in lowest terms, the ratio of the number of grams of salt to the number of cubic centimeters in the solution.

Solution

$$\frac{5\text{ g}}{100\text{ cm}^3} = \frac{1\text{ g}}{20\text{ cm}^3} = \frac{1}{20}\text{ g/cm}^3$$

Calculator Solution

Enter: 5 [÷] 100 [=]

Display: [0.05]

Answer: The solution contains $\frac{1}{20}$ gram or 0.05 gram of salt per cubic centimeter of solution.

EXERCISES WITH OPEN-RESPONSE PROBLEMS

In 1–6, express each rate in lowest terms.

1. The ratio of 36 apples to 18 people

2. The ratio of 48 patients to 6 nurses

3. The ratio of \$1.50 to 3 liters

4. The ratio of 96 cents to 16 grams

5. The ratio of 6.75 ounces to \$2.25

6. The ratio of 62 miles to 100 kilometers

7. If there are 250 tennis balls in 80 cans, how many tennis balls are in each can?

8. If an 11-ounce can of shaving cream costs 88 cents, what is the cost of each ounce of shaving cream in the can?

9. In each case, find the average rate of speed, expressed in miles per hour.

a. A vacationer traveled 230 miles in 4 hours.

b. A post office truck delivered mail on a 9-mile route in 2 hours.

c. A commuter drove 48 miles to work in $1\frac{1}{2}$ hours.

d. A race-car driver traveled 31 miles in 15 minutes.

10. In a supermarket, the regular size of Cleanright cleanser contains 14 ounces and costs 49 cents. The giant size of Cleanright cleanser, which contains 20 ounces, costs 66 cents.

a. Find, correct to the *nearest tenth* of a cent, the cost per ounce for the regular can.

b. Find, correct to the *nearest tenth* of a cent, the cost per ounce for the giant can.

c. Which is the better buy?

11. Johanna and Al use computers for word processing. Johanna can keyboard 920 words in 20 minutes, and Al can keyboard 1,290 words in 30 minutes. Who is faster at entering words on a keyboard?

12. Ronald runs 300 meters in 40 seconds. Carlos runs 200 meters in 30 seconds. Which boy is the faster runner for short races?

12-3 USING RATIO IN DIMENSIONAL ANALYSIS

Length, time, and weight are always expressed in terms of a unit of measure. In applications that use dimensions it is often necessary to change from one unit of measure to another by using ***dimensional analysis***. To do this, we multiply by a ratio whose terms are equal measures, so that, in effect, we are multiplying by the identity element 1. For example, since 12 inches = 1 foot,

$$\frac{1 \text{ ft}}{12 \text{ in.}} = \frac{12 \text{ in.}}{12 \text{ in.}} = 1 \qquad \text{and} \qquad \frac{1 \text{ ft}}{12 \text{ in.}} = \frac{1 \text{ ft}}{1 \text{ ft}} = 1$$

Notice how the ratios are chosen for the two examples that follow.

To change 30 inches to feet, multiply by $\frac{1\text{ foot}}{12\text{ inches}}$.

$$\begin{aligned} 30\text{ in.} &= 30\text{ in.} \cdot 1 \\ &= 30\text{ in.} \cdot \frac{1\text{ ft}}{12\text{ in.}} \\ &= \frac{30}{12}\text{ ft} \\ &= 2.5\text{ ft} \end{aligned}$$

Here, we cancel the unit of measure "inches" in the factors, while the unit of measure "feet" remains.

On the other hand, to change 6 feet to inches, multiply by $\frac{12\text{ inches}}{1\text{ foot}}$.

$$\begin{aligned} 6\text{ ft} &= 6\text{ ft} \cdot 1 \\ &= 6\text{ ft} \cdot \frac{12\text{ in.}}{1\text{ ft}} \\ &= 6(12)\text{ in.} \\ &= 72\text{ in.} \end{aligned}$$

Here, we cancel the unit of measure "feet" in the numerator and denominator of the factors so that only the unit of measure "inches" remains.

Often it is necessary to use more than one ratio to change to the required unit of measure. For example, three ratios are needed to change 24 ounces to kilograms.

1. First, we change ounces to pounds.

$$24\text{ oz} = 24\text{ oz} \cdot \frac{1\text{ lb}}{16\text{ oz}} = \frac{24}{16}\text{ lb} = 1.5\text{ lb}$$

2. Next, we change pounds to grams using the fact that 454 grams = 1 pound.

$$1.5\text{ lb} = 1.5\text{ lb} \cdot \frac{454\text{ g}}{1\text{ lb}} = 1.5(454)\text{ g} = 681\text{ g}$$

3. Finally, we change grams to kilograms.

$$681\text{ g} = 681\text{ g} \cdot \frac{1\text{ kg}}{1{,}000\text{ g}} = \frac{681}{1{,}000}\text{ kg} = 0.681\text{ kg}$$

We can write these three steps in one equation:

$$24\text{ oz} = 24\text{ oz} \cdot \frac{1\text{ lb}}{16\text{ oz}} \cdot \frac{454\text{ g}}{1\text{ lb}} \cdot \frac{1\text{ kg}}{1{,}000\text{ g}} = \frac{24(454)}{16(1{,}000)}\text{ kg}$$

We use a calculator to compute the value of $\frac{24\ (454)}{16\ (1{,}000)}$ kg.

Enter: (24 × 454) ÷ (16 × 1000) =

Display: 0.681

Therefore, 24 ounces is equal to 0.681 kilogram.

EXAMPLES

1. Change 60 miles per hour to feet per second.

Solution (1) Write 60 miles per hour as a fraction: $\frac{60 \text{ mi}}{1 \text{ hr}}$

(2) Change miles to feet. Multiply by a ratio with miles in the denominator to cancel miles in the numerator: $= \frac{60 \cancel{\text{mi}}}{1 \text{ hr}} \cdot \frac{5{,}280 \text{ ft}}{1 \cancel{\text{mi}}}$

(3) Change hours to minutes. Multiply by a ratio with hours in the numerator to cancel hours in the denominator: $= \frac{60(5{,}280) \text{ ft}}{1 \cancel{\text{hr}}} \cdot \frac{1 \cancel{\text{hr}}}{60 \text{ min}}$

(4) Change minutes to seconds. Multiply by a ratio with minutes in the numerator to cancel minutes in the denominator: $= \frac{60(5{,}280) \text{ ft}}{60 \cancel{\text{min}}} \cdot \frac{1 \cancel{\text{min}}}{60 \text{ sec}}$

(5) Compute and simplify: $= \frac{60(5{,}280) \text{ ft}}{60(60) \text{ sec}} = \frac{88 \text{ ft}}{1 \text{ sec}}$

Alternative Solution Write the ratios in one expression, and compute on a calculator.

$$\frac{60 \cancel{\text{mi}}}{1 \cancel{\text{hr}}} \cdot \frac{5{,}280 \text{ ft}}{1 \cancel{\text{mi}}} \cdot \frac{1 \cancel{\text{hr}}}{60 \cancel{\text{min}}} \cdot \frac{1 \cancel{\text{min}}}{60 \text{ sec}} = \frac{60(5{,}280) \text{ ft}}{60(60) \text{ sec}}$$

Enter: [( ] 60 [×] 5280 [ )] [÷] [( ] 60 [×] 60 [ )] [=]

Display: [88.]

Answer: 60 miles per hour = 88 feet per second

---

2. In France, apples cost 30 francs per kilogram. In the United States, apples cost $1.29 per pound. If the currency exchange rate is 6.90 francs per 1 dollar, in which country are the apples more expensive?

Solution Change francs per kilogram to dollars per pound.

$$\frac{30 \text{ francs}}{1 \text{ kg}} = \frac{30 \cancel{\text{francs}}}{1 \cancel{\text{kg}}} \cdot \frac{1 \text{ dollar}}{6.90 \cancel{\text{francs}}} \cdot \frac{1 \cancel{\text{kg}}}{1{,}000 \cancel{\text{g}}} \cdot \frac{454 \cancel{\text{g}}}{1 \text{ lb}} = \frac{30(454) \text{ dollars}}{6.90(1{,}000) \text{ lb}}$$

Enter: [( ] 30 [×] 454 [ )] [÷] [( ] 6.90 [×] 1000 [ )] [=]

Display: [1.973913]

Apples cost about \$1.97 per pound in France, and \$1.29 per pound in the United States.

Answer: Apples are more expensive in France.

EXERCISES WITH OPEN-RESPONSE PROBLEMS

In 1–8, make each indicated change.

1. 82 ft to yards
2. 2.3 mi to feet
3. 87 oz to pounds
4. 3 days to minutes
5. 4 hr to seconds
6. 34 cm to meters
7. 46 kg to grams
8. 1,000 kL to milliliters

In 9–15, change each quantity to the indicated measure, using the relationship given in parentheses.

9. 20 ft to centimeters (2.54 cm = 1 in)
10. 5,000 yd to kilometers (1 ft = 0.3048 m)
11. 12 m to feet (1 yd = 0.9144 m)
12. 300 fl oz to liters (1 L = 33.8 fl oz)
13. 2 mL to fluid ounces (1 fl oz = 29.6 mL)
14. 40 mi per hr to kilometers per hour (1 km = 0.623 mi)
15. 2 gal to liters (1 qt = 0.9467 L)

16. A highway sign in Canada gives the speed limit as 100 kilometers per hour. Tony is driving at 65 miles per hour. If 1 mile = 1.6 kilometers, by how much is Tony exceeding the speed limit? (Give your answer in kilometers per hour.)

17. While traveling in Germany, Rupert saw that fresh grapes cost 15 marks per kilogram. He recalls that in the United States he bought grapes for \$1.59 per pound. If the current exchange rate is 2.06 marks per dollar and 454 grams equals 1 pound, in which country are grapes more expensive?

18. Taylor had a painting for which she paid 1 million yen when she was traveling in Japan. At that time, the exchange rate was 1 dollar per 126 yen. A friend has offered her \$2,000 for the painting.

a. Is the price offered larger or smaller than the purchase price?

b. If she sells the painting, what will be her profit or loss, in dollars?

19. A cylindrical tank has a circular base whose radius r is 40.0 centimeters and whose length h is 2.07 meters.

a. Using the formula $V = \pi r^2 h$, compute the volume, V, in cubic centimeters, of the tank.

b. In the metric system, a volume of 1 cubic centimeter (cm^3) can hold 1 milliliter (mL) of liquid. Thus, 1 cm^3 = 1 mL. How many *liters* of home heating oil can the tank hold?

c. One liter is equivalent to 1.0563 quarts. Find, to the nearest *gallon*, the amount of heating oil the tank can hold.

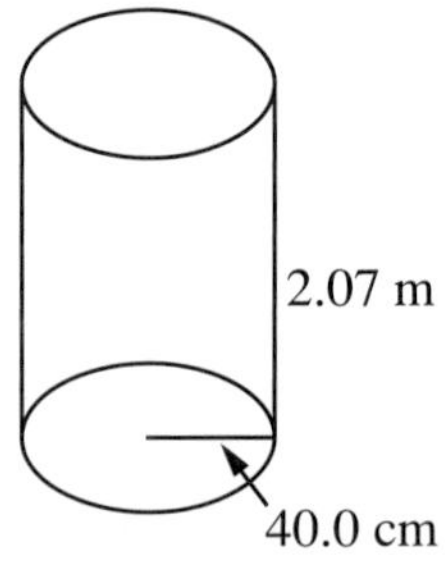

12-4 VERBAL PROBLEMS INVOLVING RATIO

Any pair of numbers in the ratio 3:5 can be found by multiplying 3 and 5 by the same non-zero number.

$3\,(3) = 9$	$3\,(7) = 21$	$3\,(0.3) = 0.9$	$3\,(x) = 3x$
$5\,(3) = 15$	$5\,(7) = 35$	$5\,(0.3) = 1.5$	$5\,(x) = 5x$
$9:15 = 3:5$	$21:35 = 3:5$	$0.9:1.5 = 3:5$	$3x:5x = 3:5$

Thus, for any non-zero number x, $3x:5x = 3:5$.

In general, when we know the ratio of two or more numbers, the terms of the ratio and a non-zero variable, x, can be used to express the numbers.

Any two numbers in the ratio $a:b$ can be written as ax and bx where x is a non-zero real number.

EXAMPLES

1. The perimeter of a triangle is 60 feet. If the sides are in the ratio 3:4:5, find the length of each side of the triangle.

Solution

Let $3x$ = length of first side,

$4x$ = length of second side,

and $5x$ = length of third side.

The perimeter of the triangle is 60 feet.

$$3x + 4x + 5x = 60$$
$$12x = 60$$
$$x = 5$$
$$3x = 15$$
$$4x = 20$$
$$5x = 25$$

Check:

$15:20:25 = 3:4:5$

$15 + 20 + 25 = 60$

Answer: The lengths of the sides are 15 feet, 20 feet, and 25 feet.

2. Two numbers have the ratio 2:3. The larger is 30 more than $\frac{1}{2}$ of the smaller. Find the numbers.

Solution Let $2x$ = smaller number,

and $3x$ = larger number.

The larger number is 30 more than $\frac{1}{2}$ of the smaller number.

$$3x = \frac{1}{2}(2x) + 30$$
$$3x = x + 30$$
$$3x - x = x + 30 - x$$
$$2x = 30$$
$$x = 15$$
$$2x = 30$$
$$3x = 45$$

Check: The ratio of 30 to 45 is 30:45 or 2:3. The larger number, 45, is 30 more than 15, which is $\frac{1}{2}$ of the smaller number.

Answer: The numbers are 30 and 45.

EXERCISES WITH OPEN-RESPONSE PROBLEMS

1. Two numbers are in the ratio 4:3. Their sum is 70. Find the numbers.
2. Find two numbers whose sum is 160 and that have the ratio 5:3.
3. Two numbers have the ratio 7:5. Their difference is 12. Find the numbers.
4. Find two numbers whose ratio is 4:1 and whose difference is 36.
5. A piece of wire 32 centimeters in length is divided into two parts that are in the ratio 3:5. Find the length of each part.
6. The sides of a triangle are in the ratio of 6:6:5. The perimeter of the triangle is 34 centimeters. Find the length of each side of the triangle.
7. The ratio of the number of boys in a school to the number of girls is 11 to 10. If there are 525 pupils in the school, how many of them are boys?
8. The perimeter of a triangle is 48 centimeters. The lengths of the sides are in the ratio 3:4:5. Find the length of each side.
9. The perimeter of a rectangle is 360 centimeters. If the ratio of its length to its width is 11:4, find the dimensions of the rectangle.
10. The ratio of the measures of two complementary angles is 2:3. Find the measure of each angle.
11. The ratio of the measures of two supplementary angles is 4:5. Find the measure of each angle.

12. In the figure, $\overleftrightarrow{AB}$ and $\overleftrightarrow{CD}$ are two parallel lines cut by a transversal. If the ratio of m∠3 to m∠6 is 1:2, find the measures of all eight angles.

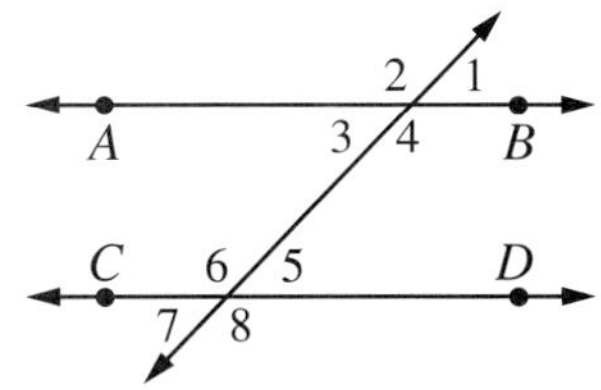

13. In $\triangle DEF$, the ratio of the measures of the three angles is 2:2:5.
    a. Find the measures of the three angles. **b.** What kind of triangle is $\triangle DEF$?
14. In isosceles triangle *ABC*, the ratio of the measures of vertex angle *A* and base angle *B* is 7:4. Find the measure of each angle of the triangle.
15. In a triangle, two sides have the same length. The ratio of each of these sides to the third side is 5:3. If the perimeter of the triangle is 65 inches, find the length of each side of the triangle.
16. The ratio of Carl's money to Donald's money is 7:3. If Carl gives Donald $20, the two then have equal amounts. Find the original amount that each one had.
17. Two numbers are in the ratio 3:7. The larger exceeds the smaller by 12. Find the numbers.
18. Two numbers are in the ratio 3:5. If 9 is added to their sum, the result is 41. Find the numbers.
19. In a basketball foul-shooting contest, the points made by Sam and Wilbur were in the ratio 7:9. Wilbur made 6 more points than Sam. Find the number of points made by each.
20. A chemist wishes to make $12\frac{1}{2}$ liters of an acid solution by using water and acid in the ratio 3:2. How many liters of each should she use?
21. In isosceles triangle *RST*, the ratio of the measures of vertex angle *T* and the exterior angle at *R* is 1:5. Find the measure of each interior angle of the triangle.

12-5 PROPORTION

A ***proportion*** is an equation that states that two ratios are equal. Since the ratio $\frac{4}{20}$ is equal to the ratio $\frac{1}{5}$, we may write the proportion

$$\frac{4}{20} = \frac{1}{5} \quad \text{or} \quad 4:20 = 1:5$$

Each of these proportions is read as "4 is to 20 as 1 is to 5." The general form of a proportion may be written as:

$$\frac{a}{b} = \frac{c}{d} \quad (b \neq 0, d \neq 0) \qquad \text{or} \qquad a:b = c:d$$

Each of these proportions is read as "*a* is to *b* as *c* is to *d*." There are four terms in this proportion, namely, *a*, *b*, *c*, and *d*. The first and fourth terms, *a* and *d*, are called the ***extremes*** of the proportion. The second and third terms, *b* and *c*, are the ***means***.

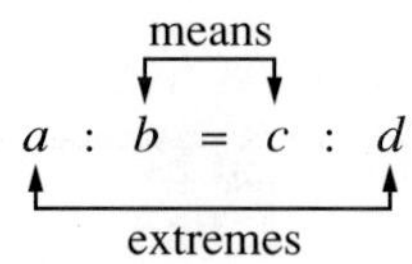

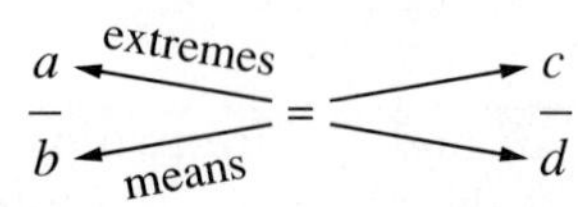

In the proportion $4:20 = 1:5$, the product of the means, 20(1), is equal to the product of the extremes, 4(5).

In the proportion $\frac{5}{15} = \frac{10}{30}$, the product of the means, 15(10), is equal to the product of the extremes, 5(30).

In any proportion $\frac{a}{b} = \frac{c}{d}$, we can show that the product of the means is equal to the product of the extremes, $ad = bc$.

Since $\frac{a}{b} = \frac{c}{d}$ is an equation, we can multiply both members by bd, the least common denominator of the fractions in the equation.

$$\frac{a}{b} = \frac{c}{d}$$

$$bd\left(\frac{a}{b}\right) = bd\left(\frac{c}{d}\right)$$

$$bd\left(a \cdot \frac{1}{b}\right) = bd\left(c \cdot \frac{1}{d}\right)$$

$$\left(b \cdot \frac{1}{b}\right)(a \cdot d) = \left(d \cdot \frac{1}{d}\right)(bc) \quad \text{(commutative and associative properties)}$$

$$1 \cdot (ad) = 1 \cdot (bc)$$

$$ad = bc$$

Therefore, we have shown that the following statement is always true:

- **In a proportion, the product of the means is equal to the product of the extremes.**

If a, b, c, and d are nonzero numbers and $\frac{a}{b} = \frac{c}{d}$, then $ad = bc$. There are three other proportions using a, b, c and d for which $ad = bc$.

$$\frac{a}{c} = \frac{b}{d} \qquad \frac{d}{b} = \frac{c}{a} \qquad \frac{d}{c} = \frac{b}{a}$$

For example, we know that $\frac{6}{4} = \frac{15}{10}$ is a proportion because $6(10) = 4(15)$. Therefore, each of the following is also a proportion.

$$\frac{6}{15} = \frac{4}{10} \qquad \frac{10}{4} = \frac{15}{6} \qquad \frac{10}{15} = \frac{4}{6}$$

EXAMPLES

1. Tell whether $\frac{4}{16} = \frac{5}{20}$ is a proportion.

Solution Four methods are shown here. The first two use paper and pencil; the last two make use of a calculator.

METHOD 1: Reduce each ratio to simplest form.

$$\frac{4}{16} = \frac{1}{4} \quad \text{and} \quad \frac{5}{20} = \frac{1}{4}$$

Since each ratio reduces to $\frac{1}{4}$, the ratios are equal and $\frac{4}{16} = \frac{5}{20}$ is a proportion.

METHOD 2: Show that the product of the means equals the product of the extremes.

$$\text{(second term)} \times \text{(third term)} = \text{(first term)} \times \text{(fourth term)}$$

$$16 \times 5 = 4 \times 20$$

$$80 = 80$$

Therefore, $\frac{4}{16} = \frac{5}{20}$ is a proportion.

METHOD 3: Use a calculator. Find the decimal form of each ratio.

Enter: 4 [÷] 16 [=]	***Enter:*** 5 [÷] 20 [=]
Display: [0.25]	***Display:*** [0.25]

Since the ratios are equal, $\frac{4}{16} = \frac{5}{20}$ is a proportion.

METHOD 4: Use a calculator. If the ratios are equal, then the first ratio minus the second ratio must equal 0. If the ratios are not equal, then their difference will not equal 0.

Enter: 4 [÷] 16 [−] 5 [÷] 20 [=]

Display: [0.]

Since the difference of the ratios is 0, the ratios are equal and $\frac{4}{16} = \frac{5}{20}$ is a proportion.

Answer: Yes, $\frac{4}{16} = \frac{5}{20}$ is a proportion.

2. Solve for q in the proportion $25:q = 5:2$.

Solution: Since $25:q = 5:2$ is a proportion, then the product of the means is equal to the product of the extremes. Therefore:

$$\begin{aligned} 5q &= 25(2) \\ &= 50 \\ q &= 10 \end{aligned}$$

Check: Reduce each ratio to simplest form.

$$\begin{aligned} 25:q &= 5:2 \\ 25:10 &\stackrel{?}{=} 5:2 \\ 5:2 &= 5:2 \text{ (True)} \end{aligned}$$

Answer: $q = 10$

---

3. Solve for x: $\dfrac{12}{x-2} = \dfrac{32}{x+8}$.

Solution:

$$\frac{12}{x-2} = \frac{32}{x+8}$$

$$\begin{aligned} 32(x-2) &= 12(x+8) \\ 32x - 64 &= 12x + 96 \\ 32x - 12x &= 96 + 64 \\ 20x &= 160 \\ x &= 8 \end{aligned}$$

(In a proportion, the product of the means is equal to the product of the extremes.)

Check:

$$\frac{12}{x-2} = \frac{32}{x+8}$$

$$\frac{12}{8-2} \stackrel{?}{=} \frac{32}{8+8}$$

$$\frac{12}{6} \stackrel{?}{=} \frac{32}{16}$$

$$2 = 2 \text{ (True)}$$

Answer: $x = 8$

---

4. The denominator of a fraction exceeds the numerator by 7. If 3 is subtracted from the numerator of the fraction and the denominator is unchanged, the value of the resulting fraction becomes $\frac{1}{3}$. Find the original fraction.

Solution Let x = numerator of original fraction,

$x + 7$ = denominator of original fraction,

and $\dfrac{x}{x+7}$ = original fraction.

Then $\dfrac{x-3}{x+7}$ = new fraction.

The value of the new fraction is $\frac{1}{3}$.

$$\begin{aligned} \frac{x-3}{x+7} &= \frac{1}{3} \\ 1(x+7) &= 3(x-3) \\ x+7 &= 3x-9 \\ 7+9 &= 3x-x \\ 16 &= 2x \\ 8 &= x \\ 15 &= x+7 \end{aligned}$$

Check: The original fraction was $\frac{8}{15}$.

The new fraction is $\frac{8-3}{15} = \frac{5}{15} = \frac{1}{3}$.

Answer: The original fraction was $\frac{8}{15}$.

EXERCISES WITH OPEN-RESPONSE PROBLEM

In 1–6, state, in each case, whether the given ratios may form a proportion.

1. $\frac{3}{4}, \frac{30}{40}$ **2.** $\frac{2}{3}, \frac{10}{5}$ **3.** $\frac{4}{5}, \frac{16}{25}$ **4.** $\frac{2}{5}, \frac{5}{2}$ **5.** $\frac{14}{18}, \frac{28}{36}$ **6.** $\frac{36}{30}, \frac{18}{15}$

In 7–14, find the missing term in each proportion.

7. $\frac{1}{2} = \frac{?}{8}$ **8.** $\frac{3}{5} = \frac{18}{?}$ **9.** $1:4 = 6:?$ **10.** $4:6 = ?:42$

11. $\frac{4}{?} = \frac{12}{60}$ **12.** $\frac{?}{9} = \frac{35}{63}$ **13.** $?:60 = 6:10$ **14.** $16:? = 12:9$

In 15–23, solve each equation. Check.

15. $\frac{x}{60} = \frac{3}{20}$ **16.** $\frac{5}{4} = \frac{x}{12}$ **17.** $\frac{30}{4x} = \frac{10}{24}$

18. $\frac{5}{15} = \frac{x}{x+8}$ **19.** $\frac{x}{12-x} = \frac{10}{30}$ **20.** $\frac{16}{8} = \frac{21-x}{x}$

21. $12:15 = x:45$ **22.** $\frac{5}{x+2} = \frac{4}{x}$ **23.** $\frac{3x+3}{3} = \frac{7x-1}{5}$

In 24–26, in each case solve for x in terms of the other variables.

24. $a:b = c:x$ **25.** $2r:s = x:3s$ **26.** $2x:m = 4r:s$

In 27–34, use a proportion to solve each problem.

27. The numerator of a fraction is 8 less than the denominator of the fraction. The value of the fraction is $\frac{3}{5}$. Find the fraction.

28. The denominator of a fraction exceeds twice the numerator of the fraction by 10. The value of the fraction is $\frac{5}{12}$. Find the fraction.

29. The denominator of a fraction is 30 more than the numerator of the fraction. If 10 is added to the numerator of the fraction and the denominator is unchanged, the value of the resulting fraction becomes $\frac{3}{5}$. Find the original fraction.

30. The numerator of a certain fraction is 3 times the denominator. If the numerator is decreased by 1 and the denominator is increased by 2, the value of the resulting fraction is $\frac{5}{2}$. Find the original fraction.

31. What number must be added to both the numerator and denominator of the fraction $\frac{7}{19}$ to make the resulting fraction equal to $\frac{3}{4}$?

32. The numerator of a fraction exceeds the denominator by 3. If 3 is added to the numerator and 3 is subtracted from the denominator, the resulting fraction is equal to $\frac{5}{2}$. Find the original fraction.

33. The numerator of a fraction is 7 less than the denominator. If 3 is added to the numerator and 9 is subtracted from the denominator, the new fraction is equal to $\frac{3}{2}$. Find the original fraction.

34. Slim Johnson was usually the best free-throw shooter on his basketball team. Early in the season, however, he had made only 9 of 20 shots. By the end of the season, he had made all the additional shots he had taken, thereby ending with a season record of 75%. How many additional shots had he taken?

12-6 DIRECT VARIATION

If the length of a side, s, of a square is 1 inch, then the perimeter, P, of the square is 4 inches. Also, if s is 2 inches, P is 8 inches; if s is 3 inches, P is 12 inches. These pairs of values are shown in the table at the right.

s	1	2	3
P	4	8	12

From the table, we observe that, as s varies, P also varies. Comparing each value of P to its corresponding value of s, we notice that all three sets of values result in the same ratio:

$$\frac{P}{s}: \quad \frac{4}{1} \qquad \frac{8}{2} = \frac{4}{1} \qquad \frac{12}{3} = \frac{4}{1}$$

If a relationship exists between two variables so that their ratio is a constant, such as $\frac{P}{s} = \frac{4}{1}$, that relationship between the variables is called a ***direct variation***.

In every direct variation, we say that one variable ***varies directly*** as the other, or that one variable is ***directly proportional*** to the other. The constant ratio is called a ***constant of variation***.

It is important to indicate the *order* in which the variables are being compared before stating the constant of variation. For example:

$$\text{In comparing } P \text{ to } s \text{, the constant of variation } = \frac{P}{s} = \frac{4}{1} = 4.$$

$$\text{However, in comparing } s \text{ to } P \text{, the constant of variation } = \frac{s}{P} = \frac{1}{4}.$$

Note that each proportion, $\frac{P}{s} = \frac{4}{1}$ and $\frac{s}{P} = \frac{1}{4}$, becomes $P = 4s$, the formula for the perimeter of a square.

In a direct variation, the values of both variables increase (or decrease) when we multiply (or divide) by the same factor, as shown below.

Here, when s is multiplied by a number, P is multiplied by that same number.

$\times$ by 2 (from 1 to 2; from 4 to 8)
$\times$ by $1\frac{1}{2}$ (from 2 to 3; from 8 to 12)

s	1	2	3
P	4	8	12

Here, when s is divided by a number, P is divided by that same number.

$\div$ by $1\frac{1}{2}$ (from 3 to 2; from 12 to 8)
$\div$ by 2 (from 2 to 1; from 8 to 4)

s	3	2	1
P	12	8	4

EXAMPLES

1. If x varies directly as y, and $x = 1.2$ when $y = 7.2$, find the constant of variation by comparing x to y.

Solution

$$\text{Constant of variation } = \frac{x}{y} = \frac{1.2}{7.2} = \frac{1}{6}$$

Answer: $\frac{1}{6}$

2. The table gives pairs of values for the variables x and y.

x	1	2	3	10	?
y	8	16	24	?	1,600

a. Show that one variable varies directly as the other.

b. Express the relationship between the variables as a formula.

c. Find the values missing in the table.

Solutions **a.** Compare the x-values to the y-values.

$$\frac{x}{y}: \quad \frac{1}{8} \qquad \frac{2}{16} = \frac{1}{8} \qquad \frac{3}{24} = \frac{1}{8}$$

Since all the given pairs of values result in a constant ratio, x and y vary directly.

b. Write a proportion that includes both variables and the constant of variation: $\frac{x}{y} = \frac{1}{8}$

Multiply, showing that the product of the means is equal to the product of the extremes, to obtain the formula: $y = 8x$

Answer: $y = 8x$

c. Substitute the known value in the proportion (or formula) that contains the constant ratio. Then solve the proportion (or equation) to find the unknown value.

When $x = 10$ (known), find y (unknown).

$$\frac{10}{y} = \frac{1}{8}$$

$$1y = (10)(8)$$

$$y = 80$$

When $y = 1{,}600$ (known), find x (unknown).

$$\frac{x}{1{,}600} = \frac{1}{8}$$

$$1{,}600 = 8x$$

$$200 = x$$

Answers: The missing values are $y = 80$ and $x = 200$.

---

3. There are about 90 calories in 20 grams of a cheese. Reggie ate 70 grams of this cheese. About how many calories were there in the cheese she ate if the number of calories varies directly as the weight of the cheese?

Solution Let x = number of calories in 70 grams of cheese.

$$\frac{x}{70} = \frac{90}{20} \begin{array}{l}\leftarrow \text{number of calories} \\ \leftarrow \text{number of grams of cheese}\end{array}$$

$20x = 90(70)$ (In a proportion, the product of the means is equal to the product of the extremes.)

$= 6{,}300$

$x = 315$

Check: $\frac{315}{70} \stackrel{?}{=} \frac{90}{20}$

$\frac{9}{2} = \frac{9}{2}$ (True)

Answer: There were about 315 calories in 70 grams of the cheese.

EXERCISES WITH OPEN-RESPONSE PROBLEM

In 1–9, in each case one value is given for each of two variables that *vary directly*. Find the constant of variation.

1. $x = 12, y = 3$ **2.** $d = 120, t = 3$ **3.** $y = 2, z = 18$

4. $P = 12.8, s = 3.2$ **5.** $t = 12, n = 8$ **6.** $i = 51, t = 6$

7. $s = 88, t = 110$ **8.** $A = 212, P = 200$ **9.** $r = 87, s = 58$

In 10–15, tell, in each case, whether one variable varies directly as the other. If it does, express the relation between the variables by means of a formula.

10.

p	3	6	9
s	1	2	3

11.

n	3	4	5
c	6	8	10

12.

x	4	5	6
y	6	8	10

13.

t	1	2	3
d	20	40	60

14.

x	2	3	4
y	−6	−9	−12

15.

x	1	2	3
y	1	4	9

In 16–18, in each case one variable varies directly as the other. Find the missing numbers, and write the formula that relates the variables.

16.

h	1	2	?
A	5	?	25

17.

h	4	8	?
S	6	?	15

18.

L	2	8	?
W	1	?	7

In 19–22, state whether the relation between the variables in each equation is a direct variation. In each case, give a reason for your answer.

19. $R + T = 80$ **20.** $15T = D$ **21.** $\frac{e}{i} = 20$ **22.** $bh = 36$

23. $C = 7N$ is a formula for the cost of any number of articles that sell for \$7 each.
- **a.** How do C and N vary?
- **b.** How will the cost of nine articles compare with the cost of three articles?
- **c.** If N is doubled, what change takes place in C?

24. $A = 12\ell$ is a formula for the area of any rectangle whose width is 12.
- **a.** Describe how A and ℓ vary.
- **b.** How will the area of a rectangle whose length is 8 inches compare with the area of a rectangle whose length is 4 inches?
- **c.** If ℓ is tripled, what change takes place in A?

25. The variable D varies directly as t. If $D = 520$ when $t = 13$, find D when $t = 9$.

26. Y varies directly as x. If $Y = 35$ when $x = -5$, find Y when $x = -20$.

27. A varies directly as h. $A = 48$ when $h = 4$. Find h when $A = 36$.

28. N varies directly as d. $N = 10$ when $d = 8$. Find N when $d = 12$.

In 29–46, the quantities vary directly. Solve algebraically.

29. If 3 pounds of apples cost \$0.89, what is the cost of 15 pounds of apples at the same rate?

30. If four tickets to a show cost \$17.60, what is the cost of seven such tickets?

31. If $\frac{1}{2}$ pound of meat sells for \$3.50, how much meat can be bought for \$8.75?

32. Willis scores an average of 7 foul shots in every 10 attempts. At the same rate, how many shots would he score in 200 attempts?

33. There are about 60 calories in 30 grams of canned salmon. About how many calories are there in a 210-gram can?

34. There are 81 calories in a slice of bread that weighs 30 grams. How many calories are there in a loaf of this bread that weighs 600 grams?

35. There are about 17 calories in three medium-size shelled peanuts. Joan ate 30 such peanuts. How many calories were there in the peanuts she ate?

36. A train traveled 90 miles in $1\frac{1}{2}$ hours. At the same rate, how long will the train take to travel 330 miles?

37. The weight of 20 meters of copper wire is 0.9 kilogram. Find the weight of 170 meters of the same wire.

38. A recipe calls for $1\frac{1}{2}$ cups of sugar for a 3-pound cake. How many cups of sugar should be used for a 5-pound cake?

39. In a certain concrete mixture, the ratio of cement to sand is 1:4. How many bags of cement would be used with 100 bags of sand?

40. A house that is assessed for \$12,000 pays \$960 in realty taxes. At the same rate what should be the realty tax on a house asssessed for \$16,500?

41. The scale on a map is given as 5 centimeters to 3.5 kilometers. How far apart are two towns if the distance between these two towns on the map is 8 centimeters?

42. David received \$8.75 in dividends on 25 shares of a stock. How much should Marie receive in dividends on 60 shares of the same stock?

43. A picture $3\frac{1}{4}$ inches long and $2\frac{1}{8}$ inches wide is to be enlarged so that its length will become $6\frac{1}{2}$ inches. What will be the width of the enlarged picture?

44. An 11-pound turkey costs \$9.79. At this rate, find:
a. the cost of a 14.4-pound turkey, rounded to the nearest cent
b. the cost of a 17.5-pound turkey, rounded to the nearest cent
c. the price per pound at which turkeys are sold
d. the largest size turkey, to the *nearest tenth* of a pound, that can be bought for \$20 or less.

45. If a man can buy p kilograms of candy for d dollars, represent the cost of n kilograms of this candy.

46. If a family consumes q liters of milk in d days, represent the amount of milk consumed in h days.

12-7 PERCENT AND PERCENTAGE PROBLEMS

Base, Rate, and Percent

Problems dealing with discounts, commissions, and taxes involve percents. A ***percent***, which is a *ratio* of a number to 100, is also called a *rate*. Here, the word *rate* is treated as a comparison of a quantity to the whole. For example, 8% (read as 8 percent) is the ratio of 8 to 100, or $\frac{8}{100}$.

A percent can be expressed as a fraction or as a decimal:

$$8\% \quad = \quad \frac{8}{100} \quad = \quad 0.08$$

If an item is taxed at a rate of 8%, then a \$50 pair of jeans will have an additional charge of \$4 tax. Here, three quantities are involved:

1. The ***base***, or the sum of money being taxed, which is \$50.
2. The ***rate***, or the rate of tax, which is 8% or 0.08 or $\frac{8}{100}$.
3. The ***percentage***, or the amount of tax being charged, which is \$4.

These three related terms may be written as a proportion or as a formula:

As a proportion:	*As a formula:*
$\frac{\textbf{Percentage}}{\textbf{Base}} = \textbf{rate}$	**Base × rate = percentage**
For example: $\frac{4}{50} = \frac{8}{100}$	For example: $50 \times \frac{8}{100} = 4$

Just as we have seen two ways to look at this problem involving sales tax, we will see more than one approach to every percentage problem. First, however, we need to look at a calculator.

Percent and the Calculator

To change a fraction to a decimal, we know that we divide the numerator of the fraction by its denominator. For example:

Enter: 7 [÷] 20 [=]
Display: [0.35]

This means $\frac{7}{20} = 0.35$.

To change a decimal to a percent, we multiply by 100. For example:

Enter: 0.35 [×] 100 [=]
Display: [35.]
This means 0.35 = 35%.

Therefore, to change a fraction to a percent, we combine both procedures: we divide the numerator of the fraction by its denominator, and then multiply by 100. For example:

Enter: 7 [÷] 20 [×] 100 [=]
Display: [35.]
This means $\frac{7}{20} = 35\%$.

Most calculators, however, have a ***percent key***, [%], which allows us to change a fraction to a percent in fewer entries.

On scientific calculators, percent is often shown as the *second function* of a key. In this case, a user must enter the [2nd] or [INV] or [SHIFT] key to access the percent key. Most calculators display the answer when the [%] key is pressed, but on some devices the [=] key must also be entered. (Check to see which method works on your calculator.)

METHOD 1. ***Enter:*** 7 [÷] 20 [%]
METHOD 2. ***Enter:*** 7 [÷] 20 [INV] [%]
METHOD 3. ***Enter:*** 7 [÷] 20 [%] [=]
METHOD 4. ***Enter:*** 7 [÷] 20 [INV] [%] [=]
Display: [35.]
This means $\frac{7}{20} = 35\%$.

Remember, when using a calculator, we must always use careful estimations and check our answers to see whether they make sense. Consider the following example.

A student scored 90% on a test of 50 items. If all answers are equal in value, how many questions did the student answer correctly?

In other words, to solve this problem: Find 90% of 50.

An incorrect solution on some calculators

Enter: 90 [%] [×] 50 [=]
Display: [4500.]

This answer makes no sense. To score 90%, the student must have answered 4,500 questions correctly, but there are only 50 questions.

A correct calculator solution

Enter: 50 [×] 90 [%]
Display: [45.]

This answer seems reasonable.

Check it: $\frac{45}{50} \stackrel{?}{=} \frac{90}{100} \rightarrow \frac{9}{10} = \frac{9}{10}$ (True)

The solutions shown above demonstrate that *order* can be important on some calculators when using the [%] key. (Try each set of keys shown above to determine which will give the correct answer on your calculator.)

In general, we use the [%] key as the last step.

PROCEDURE. To find the percentage of a number on a calculator:

1. First, enter the number or base.
2. Next, enter the multiplication key, [×].
3. Last, enter the rate of percent followed by the [%] key.

EXAMPLES

1. Find the amount of tax on a $60 radio when the rate of tax is 8%.

Solution METHOD 1. Use the proposition: $\frac{\text{Percentage}}{\text{Base}} = \text{rate}$, and let t = amount of tax.

Then, $\frac{\text{amount of tax}}{\text{sum to be taxed}} = \frac{8}{100}$ (This is a proportion because it consists of two equal ratios.)

$$\frac{t}{60} = \frac{8}{100}$$

$$100t = 480$$ (In a proportion, the product of the means is equal to the product of the extremes.)

$$t = 4.80$$

In terms of dollars and cents, the tax is $4.80.

METHOD 2. Use the formula: Base × rate = percentage; let t = amount of tax.

Change 8% to a fraction.

$$\text{Base} \times \text{rate} = \text{percentage}$$
$$60 \times 8\% = t$$
$$60 \times \frac{8}{100} = t$$
$$\frac{480}{100} = t$$
$$4.8 = t$$

or

Change 8% to a decimal.

$$\text{Base} \times \text{rate} = \text{percentage}$$
$$60 \times 8\% = t$$
$$60 \times 0.08 = t$$
$$4.80 = t$$

Whether the fraction or the decimal form of 8% is used, the tax is $4.80.

METHOD 3. Use a calculator and the formula: Base × rate = percentage.

The rate of tax, 8%, may be entered as a percent ⟶	*Enter:* 60 [×] 8 [%]
Or the rate of tax, 8%, may be entered as a fraction, $\frac{8}{100}$ ⟶	*Enter:* 60 [×] 8 [÷] 100 [=]
Or the rate of tax, 8%, may be entered as a decimal, 0.08 ⟶	*Enter:* 60 [×] 0.08 [=]
In each case, the percentage or tax shown in the display is \$4.80.	*Display:* [4.8]

Answer: The tax is \$4.80.

---

2. If 25% of a number is 80, find the number.

Solution The rate is 25%, which can be written as a fraction or a decimal. The percentage is given as 80, and the base is not known.

Let n = the number, or base.

METHOD 1. Use the proportion.

$$\frac{\text{Percentage}}{\text{Base}} = \text{rate}$$

$$\frac{80}{n} = \frac{25}{100}$$

$$25n = 8{,}000$$

$$n = 320$$

Check: Does 25% of 320 = 80?

$$0.25 \times 320 = 80$$

$$80.00 = 80 \text{ (True)}$$

Alternative Check: Reduce each ratio in the proportion.

$$\frac{80}{320} \stackrel{?}{=} \frac{25}{100}$$

$$\frac{1}{4} = \frac{1}{4} \text{ (True)}$$

METHOD 2. Use the formula. Substitute values for the rate and percentage, and solve the equation for the base, n.

$$\text{Base} \times \text{rate} = \text{percentage}$$

$$n \times 25\% = 80$$

Using fractions:

$$n \cdot \frac{25}{100} = 80$$

$$n \cdot \frac{\cancel{25}}{\cancel{100}} \cdot \frac{\cancel{100}}{\cancel{25}} = 80 \cdot \frac{100}{25}$$

$$n = \frac{8{,}000}{25}$$

$$= 320$$

Using decimals:

$$n \cdot 0.25 = 80$$

$$\frac{n\cancel{(0.25)}}{\cancel{0.25}} = \frac{80}{0.25}$$

$$n = \frac{80}{0.25}$$

$$= 320$$

Or, use a calculator to solve the equation. Compare each entry to the third step in the solution of the equation above.

Enter: 80 [×] 100 [÷] 25 [=]	***Enter:*** 80 [÷] 0.25 [=]
Display: 320.	***Display:*** 320.

Answer: The number is 320.

EXERCISES WITH OPEN-RESPONSE PROBLEMS

In 1–9, find each indicated percentage.

1. 2% of 36
2. 6% of 150
3. 15% of 48
4. 2.5% of 400
5. 60% of 56
6. 100% of 7.5
7. $12\frac{1}{2}$% of 128
8. $33\frac{1}{3}$% of 72
9. 150% of 18

In 10–17, find each number.

10. 20 is 10% of what number?
11. 64 is 80% of what number?
12. 8% of what number is 16?
13. 72 is 100% of what number?
14. 125% of what number is 45?
15. $37\frac{1}{2}$% of what number is 60?
16. $66\frac{2}{3}$% of what number is 54?
17. 3% of what number is 1.86?

In 18–25, find each percent.

18. 6 is what percent of 12?
19. 9 is what percent of 30?
20. What percent of 10 is 6?
21. What percent of 35 is 28?
22. 5 is what percent of 15?
23. 22 is what percent of 22?
24. 18 is what percent of 12?
25. 2 is what percent of 400?

26. A newspaper has 80 pages. If 20 of the 80 pages are devoted to advertising, what percent of the newspaper consists of advertising?

27. A test was passed by 90% of a class. If 27 students passed the test, how many students were in the class?

28. Marie bought a dress that was marked $24. The sales tax is 8%. **a.** Find the sales tax. **b.** Find the total amount Marie had to pay.

29. There were 120 planes on an airfield. If 75% of the planes took off for a flight, how many planes took off?

30. One year, the Ace Manufacturing Company made a profit of $480,000. This represented 6% of the volume of business for the year. What was the volume of business for the year?

31. The price of a new motorcycle that Mr. Klein bought was $5,430. Mr. Klein made a down payment of 15% of the price of the motorcycle and arranged to pay the rest in installments. How much was his down payment?

32. How much silver is in 75 kilograms of an alloy that is 8% silver?

33. In a factory, 54,650 parts were made. When they were tested, 4% were found to be defective. How many parts were good?

34. A baseball team won eight games, which was 50% of the total number of games it played. How many games did the team play?
35. Helen bought a coat at a "20% off" sale and saved \$24. What was the marked price of the coat?
36. A businessman is required to collect an 8% sales tax. One day, he collected \$280 in taxes. Find the total amount of sales he made that day.
37. A merchant sold a stereo speaker for \$150, which was 25% above its cost to her. Find the cost of the stereo speaker to the merchant.
38. Bill bought a wooden chess set at a sale. The original price was \$120; the sale price was \$90. By what percent was the original price reduced?
39. If the sales tax on \$150 is \$7.50, what is the percent of the sales tax?
40. Mr. Taylor took a 2% discount on a bill. He paid the balance with a check for \$76.44. What was the original amount of the bill?
41. After Mrs. Sims lost 15% of her investment, she had \$2,550 left. How much did she invest originally?
42. When a salesman sold a vacuum cleaner for \$110, he received a commission of \$8.80. What was the rate of commission?
43. At a sale, a camera was reduced \$8. This represented 10% of the original price. On the last day of the sale, the camera was sold for 75% of the original price. What was the final selling price of the camera?
44. The regular ticketed prices of four items at Grumbell's Clothier are as follows: coat, \$139.99; blouse, \$43.99; shoes, \$89.99; jeans, \$32.99.
    a. These four items were placed on sale at 20% off the regular price. Find, correct to the nearest cent, the sale prices of these four items.
    b. Describe two different ways to find the sale prices.
45. At Relli's Natural Goods, all items are being sold today at 30% off their regular prices. Customers must still pay an 8% tax, however, on these items. Edie, a good-natured owner, allows each customer to choose one of two plans at this sale:
    Plan 1: Deduct 30% of the cost of all items, then add 8% tax to the bill.
    Plan 2: Add 8% tax to the cost of all items, then deduct 30% of this total.
    Which plan, if either, saves the customer more money? Explain why.
46. In early March, Phil Kalb bought shares of stocks in two different companies.
    Stock *ABC* rose 10% in value in March, then decreased 10% in April.
    Stock *XYZ* fell 10% in value in March, then rose 10% in April.
    What percent of its original price is each of these stocks now worth?

12-8 ERROR IN MEASUREMENT

Even when great care is taken, error in measurement sometimes occurs. Although human error can be the cause of an inaccurate measurement, often the error is due to the limitations of the measuring equipment. For example, we can use a ruler to measure something to the nearest $\frac{1}{2}$ inch or—with a steady eye—to the nearest $\frac{1}{16}$ inch. But if we used this ruler to measure a pencil that was exactly

$6\frac{3}{70}$ inches long, we would only be able to say that it looked very close to 6 inches. The fraction is too small to measure, so our measurement has an error of $\frac{3}{70}$.

Margin of Error

Margin of error is a term that describes the maximum difference expected between a measurement and the true value. A margin of error is sometimes written with a plus-or-minus sign (±). A smaller margin of error implies that the measurement is more exact. For our ruler used above, the margin of error might be $\pm\frac{1}{32}$ inch.

EXAMPLES

1. The population of a town is 4,560 when rounded to the nearest 10 people. What is the range of numbers for the actual population? What is the margin of error?

Solution If there were 4,555 people, 4,555 would round up to 4,560:

$$4{,}560 - 4{,}555 = 5$$

If there were 4,564 people, 4,564 would round down to 4,560:

$$4{,}564 - 4{,}560 = 4$$

The range for the population is from 4,555 to 4,564. The maximum possible error is 5, so the margin of error is ±5.

---

2. A survey says that 59% of people prefer sneaker brand A to sneaker brand B, with a margin of error of 10%. What is the smallest possible percentage of people who prefer brand A to brand B?

Solution 59% − 10% = 49%

It is possible that only 49% of people actually prefer brand A sneakers.

Percentage Error

Knowing a margin of error is often not enough to tell how accurate a measurement is. A margin of error of 10 is large when compared with a measurement of 1 and small when compared with a measurement of 100,000,000. Therefore, the amount of error is often shown as a percentage of the true value.

PROCEDURE To find the percentage error in a measurement.

1. Record the measured reading from the measuring equipment.
2. Record the accepted value of the reading, taken from more reliable or trusted sources.
3. Find the difference between the measured reading and the accepted value.
4. Divide that difference by the accepted value.
5. Change the decimal value to a percent.

EXAMPLE

A 200-pound weight was placed on a bathroom scale. The scale dial displayed 210 pounds. If the scale is consistently off by the same percentage, how much does a man really weigh if his weight displayed on this scale is 168 pounds?

Solution Write the formula for percentage error.

$$\text{Error} = \left|\frac{\text{measured value} - \text{accepted value}}{\text{accepted value}}\right| \times 100\%$$

Substitute and evaluate.

$$= \left|\frac{210 - 200}{200}\right| \times 100\% = 5\%$$

Every measure on the bathroom scale is 5% too high.

$$\text{actual weight} \times 1.05 = \text{observed weight}$$

Substitute the observed weight.

$$\text{actual weight} \times 1.05 = 168$$

Solve.

$$\text{actual weight} = \frac{168}{1.05} = 160 \text{ pounds}$$

Answer: 160 pounds

Initial Error in Subsequent Calculations

When we substitute a measurement for a variable in an equation, we must be careful with our answer. The small error in our measurement may become a larger error in the answer. It is often useful to substitute the highest and lowest possible true values of the measurement in the equation to see how exact our answer truly is.

EXAMPLE

Shanayda has a small aquarium that she wants to fill with water. The dimensions of the aquarium, rounded to the nearest centimeter, are 16 centimeters, 10 centimeters, and 120 centimeters. What is the maximum number of liters of water that she can put in and still be sure that the aquarium will not overflow?
(1) 16 (2) 17 (3) 19 (4) 20

Solution This problem is asking for the smallest possible volume of the aquarium. If Shanayda pours in more than this smallest volume, the tank might overflow.

To find the smallest value, we need to start with the smallest possible true values for the measurements. Since each measurement is to the nearest centimeter, the true values can be at most 0.5 centimeter less than the measurements.

16 cm × 10 cm × 120 cm becomes 15.5 cm × 9.5 cm × 119.5 cm.

Use a calculator. 15.5 cm × 9.5 cm × 119.5 cm

17,596.375 cm^3 or 17,596.375 ml

Convert to liters. 17.596375 L

Shanayda can pour at most 17.596375 liters into the aquarium without an overflow. The answer choice of 17 is the closest she can come.

Answer: (2)

Error in Different Dimensions

In geometric calculations, error increases when dimension increases, such as when linear measure is used to find area or volume. Error decreases when dimension decreases, such as when area or volume is used to find length. To understand how this relationship works, study the table below.

Measurement and True Value	Percent Error of Measurement	Calculations	Percent Error of Calculation
Example 1			
Length of side of cube: measured: 9 inches true: 10 inches	$\left\lvert \frac{10-9}{10} \right\rvert = \frac{1}{10}$ **= 10%**	Volume of cube: $9 \times 9 \times 9 = 729$ $10 \times 10 \times 10 = 1{,}000$	$\left\lvert \frac{1{,}000-729}{1{,}000} \right\rvert = \frac{271}{1{,}000}$ = 0.271 = **27.1%** ERROR INCREASED
Example 2			
Volume of cube: measured: 900 cu in. true: 1,000 cu in.	$\left\lvert \frac{1{,}000-900}{1{,}000} \right\rvert = \frac{100}{1{,}000}$ **= 10%**	Length of side: $\sqrt[3]{900} = 9.655$ $\sqrt[3]{1{,}000} = 10$	$\left\lvert \frac{10-9.655}{10} \right\rvert = \frac{0.345}{10}$ = 0.0345 = **3.45%** ERROR DECREASED

Note: Since we are working with measurements, we can ignore the negative square roots.

EXERCISES WITH OPEN-RESPONSE PROBLEMS

1. In a physics experiment, the acceleration of gravity was found to be 9.5 meters per second squared. The accepted acceleration of gravity to this degree of accuracy is 9.8 meters per second squared. Find the approximate percentage error.
(1) 96.9% (2) 3.16% (3) 3.06% (4) 1.03%
2. A senator polled the people in his district to discover their opinion on two versions of a crime bill. The poll indicated that 45% preferred bill A, 35% preferred bill B, and 20% were undecided. The poll had a margin of error of $\pm 5\%$. Which of the following is a possibility for the true opinions of the people in the senator's district?
(1) 40% for bill A, 30% for bill B, and 30% undecided
(2) 40% for bill A, 40% for bill B, and 20% undecided
(3) 50% for bill A, 40% for bill B, and 10% undecided
(4) 50% for bill A, 25% for bill B, and 25% undecided
3. Douglas measured his backyard and found it to be 26 feet by 43 feet. He calculated the area and bought enough new sod to cover the yard. When he got home, he found that his measurements were each 2 feet too long. How much excess sod will he have?
(1) 142 square feet (2) 134 square feet (3) 86 square feet (4) 52 square feet
4. Tracy uses the measure of a side of a cube to calculate the volume. She finds the volume to be 216 cubic inches. Later, she realizes that the true volume of the cube is 343 cubic inches. To the nearest tenth of a percent, what was the error of her measure *for the length of the side*?
(1) 14.3% (2) 16.7% (3) 37.0% (4) 58.8%
5. An odometer measures the distance a vehicle travels. The odometer on Eric's car read 65,480 miles before he left Long Island. When he arrived at his cousin's house in Orange County, his odometer read 65,565 miles. The trip distance was actually 90 miles. On a trip from New York City to Albany, his odometer display changed from 66,450 miles to 66,620 miles. Assuming that the odometer is always incorrect by the same percentage, how far did he actually travel?
6. Harry was stopped on the New York State Thruway for driving at an excessive speed. The officer's radar recorder indicated a speed of 80 mph. Harry claimed that although he was driving too fast, his speedometer recorded a speed of only 70 miles per hour. If the radar has a margin of error of ± 3 mph, what is the minimum percentage error of Harry's speedometer?

Exercises 7–10: The temperature of Maddy's oven is always off by the same percentage error. She must set the oven temperature to 400° when she uses a recipe that calls for roasting at 375°. Finish the chart for Maddy so that she can easily convert from recipe temperature to displayed oven temperature. Round all temperatures to the *nearest 5 degrees*.

	Desired temperature	Setting on Maddy's Oven
7.	325	
8.	350	
	375	400
9.	400	
10.	425	

Exercises 11–12: A man weighed 150 pounds and his wife weighed 125 pounds on his doctor's accurate scale. Wearing identical clothes, the man later weighed himself on a scale in the shopping center. The shopping center scale displayed a weight of 153 pounds.

11. If the shopping center scale was off by a constant *amount,* what would his wife weigh on the shopping center scale?

12. If the shopping center scale was off by a constant *percentage error,* what would his wife weigh on the shopping center scale.

12-9 SIMILAR POLYGONS

Congruent polygons are figures that have the same shape and same size. Other geometric figures have the same shape but do *not* have the same size; examples are seen when a photographer enlarges or reduces a picture, and when an architect creates a small scale drawing or floor plan of a room.

Polygons that have the same shape but *not* the same size are called ***similar polygons***. The symbol for ***similar*** or ***is similar to*** is $\sim$.

Similar polygons are shown below. To study why they have the same shape, we start by pairing vertices: A with A', B with B', C with C', and D with D'.

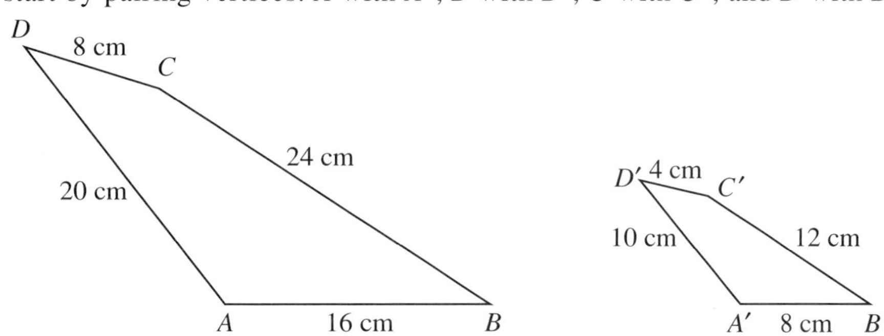

Similar polygons
$ABCD \sim A'B'C'D'$

By matching these vertices, we establish *corresponding parts*. In the diagram, there are four pairs of corresponding angles and four pairs of corresponding sides. Here, as in all similar figures, the corresponding angles are equal in measure or are congruent:

$$\angle A \cong \angle A' \qquad \angle B \cong \angle B' \qquad \angle C \cong \angle C' \qquad \angle D \cong \angle D'$$

The corresponding sides of these figures are *not* equal, but we note the ratio that is obtained when the lengths of these corresponding sides are compared:

$$\frac{AB}{A'B} = \frac{16}{8} = \frac{2}{1} \qquad \frac{BC}{B'C'} = \frac{24}{12} = \frac{2}{1} \qquad \frac{CD}{C'D'} = \frac{8}{4} = \frac{2}{1} \qquad \frac{DA}{D'A'} = \frac{20}{10} = \frac{2}{1}$$

Here, as in all similar figures, the *ratios* of the lengths of corresponding sides are equal. Because the ratios are equal, we say that the corresponding sides are in proportion. Thus:

$$\frac{AB}{A'B'} = \frac{BC}{B'C'} = \frac{CD}{C'D'} = \frac{DA}{D'A'}$$

Therefore, we say:

- **Two polygons are similar if their corresponding angles are congruent and their corresponding sides are in proportion.**

Conversely, we may also say:

- **If two polygons are similar, then their corresponding angles are congruent and their corresponding sides are in proportion.**

EXAMPLE

Rectangle $ABCD$ is similar to rectangle $A'B'C'D'$, $AB = 16$ centimeters, $AD = 12$ centimeters, and $A'B' = 8$ centimeters.

Find the length of $\overline{A'D'}$.

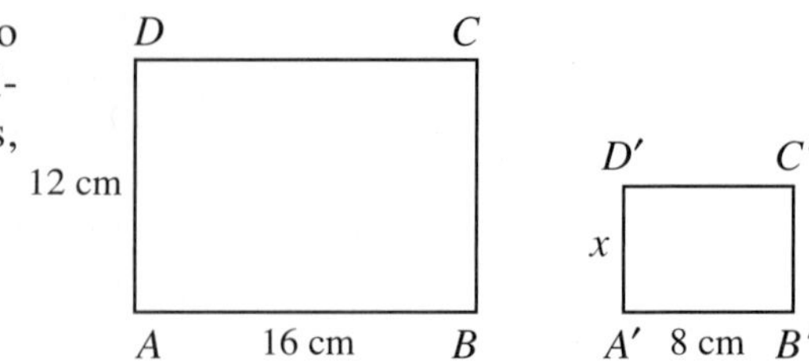

Solution In similar figures, the lengths of corresponding sides form a proportion.

Let x = length of $\overline{A'D'}$.

$$\frac{A'D'}{AD} = \frac{A'B'}{AB}$$

$$\frac{x}{12} = \frac{8}{16}$$

$$16(x) = 8(12)$$

$$16x = 96$$

$$x = 6$$

Check:

$$\frac{6}{12} \stackrel{?}{=} \frac{8}{16}$$

$$\frac{1}{2} = \frac{1}{2} \text{ (True)}$$

Answer: $A'D' = 6$ centimeters

EXERCISES WITH OPEN-RESPONSE PROBLEMS

1. Is square $ABCD$ similar to rectangle $EFGH$? Why?

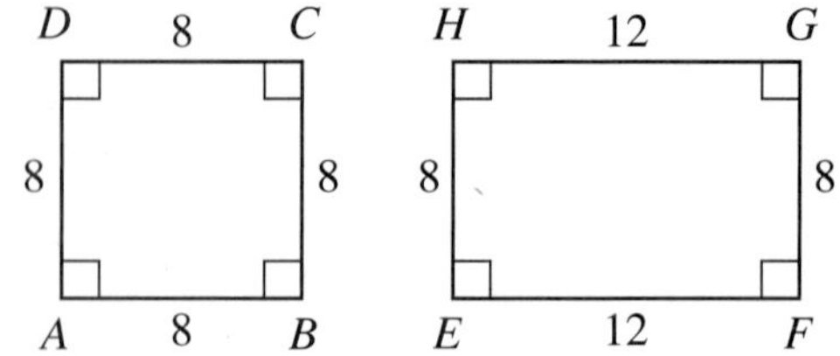

2. Is parallelogram $ABCD$ similar to rectangle $EFGH$? Why?

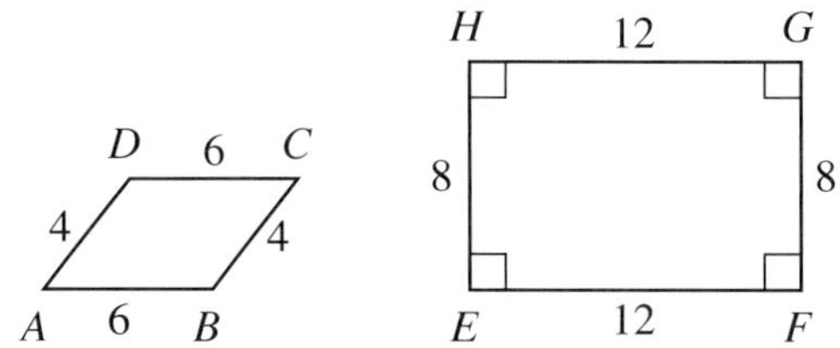

3. Is square $WXYZ$ similar to square $QRST$? Why?

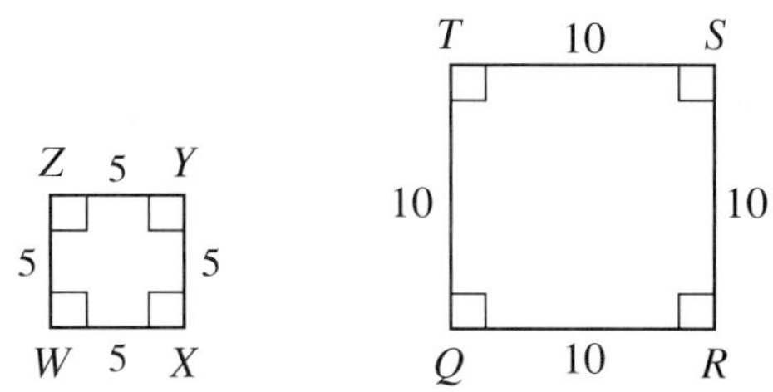

4. Are all squares similar? Why?

5. Are all rectangles similar? Why?

6. Are all parallelograms similar? Why?

7. Are all rhombuses similar? Why?

8. Using the dimensions shown on the similar rectangles below, find the number of centimeters in the length of $\overline{ZY}$.

9. A picture 12 centimeters long and 9 centimeters wide is to be enlarged so that its length will be 16 centimeters. How wide will the enlarged picture be?

10. Jenny wishes to enlarge a rectangular photograph that is $4\frac{1}{2}$ inches long and $2\frac{3}{4}$ inches wide so that it will be 9 inches long. How wide will the enlargement be?

In 11–14, in each case the two polygons are similar. Pairs of corresponding vertices are noted by using primes. Find the length of every side in each polygon.

11.

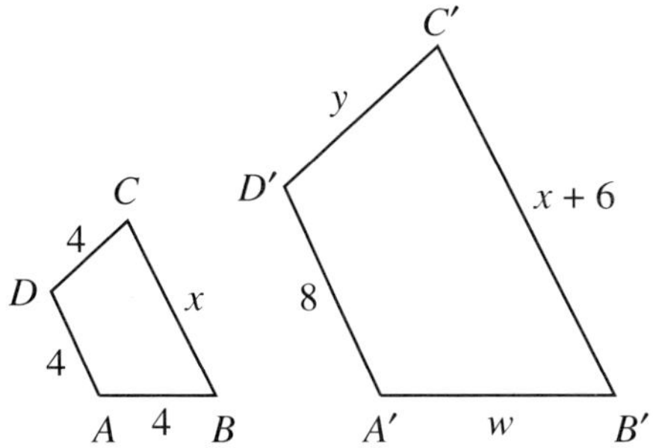

12.

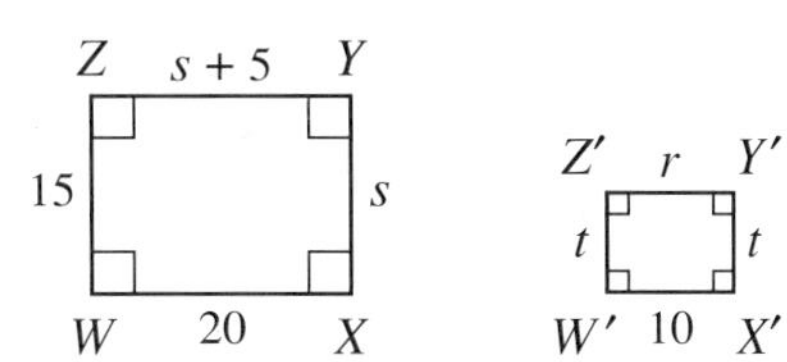

13.

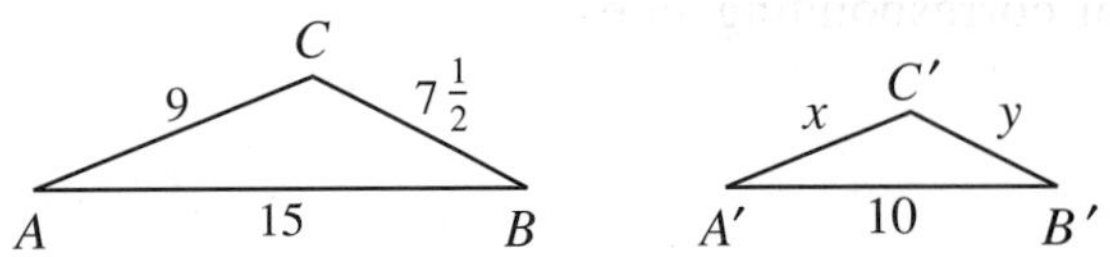

14.

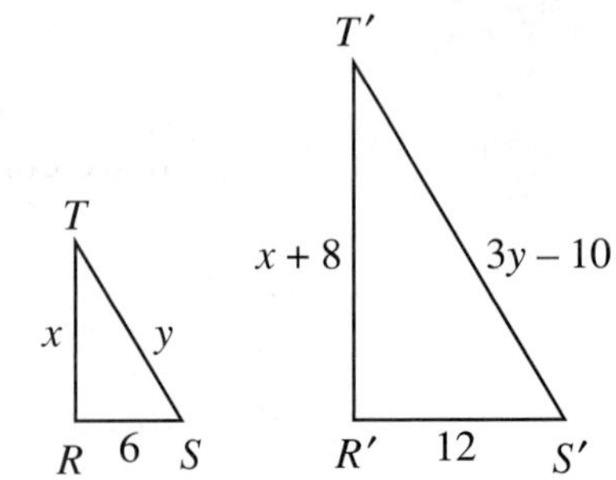

15. Tonette is deciding what size photographs she should order for her graduation package. In addition to a wallet size, standard photographs come in the following sizes: 3 inches by 5 inches, 5 inches by 7 inches, and 8 inches by 10 inches.

a. Which, if any, of these standard-size photographs are in proportion?

b. A standard wallet-size photograph measures 2 inches by 3 inches. In each of the following, find the missing measure so that the larger size photograph is in proportion to the 2-inch by 3-inch wallet size:

(*1*) 8″ by __	(*2*) 5″ by __	(*3*) 3″ by __
(*4*) __ by 10″	(*5*) __ by 7″	(*6*) __ by 5″

12-10 SIMILAR TRIANGLES

Two triangles are similar if and only if their corresponding angles are congruent and their corresponding sides are in proportion. There is, however, a shorter method to demonstrate that two triangles are similar:

1. Draw any triangle, such as $\triangle ABC$. Here, $m\angle A = 30$ and $m\angle B = 65$.

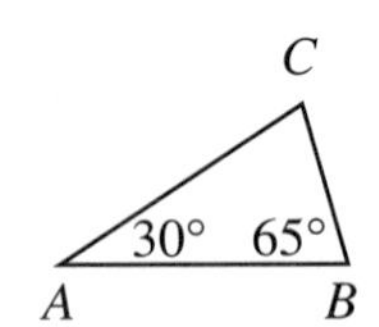

2. To draw a second triangle, first draw $\overline{A'B'}$ so that it is twice as large as side $\overline{AB}$.

3. Use a protractor to copy $\angle A$ at A' and $\angle B$ at B'. Here, a 30° angle is drawn at A' and a 65° angle is drawn at B'.

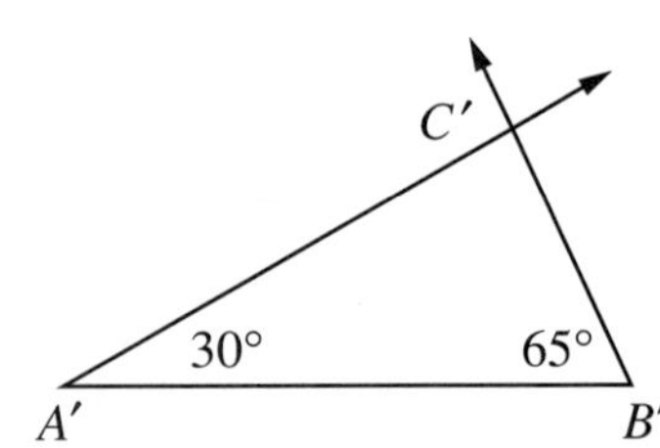

4. Let C' name the point of intersection of the two rays just drawn, namely, $\overrightarrow{A'C'}$ and $\overrightarrow{B'C'}$.

Triangle $A'B'C'$ appears to be similar to $\triangle ABC$. If we measure $\angle C$ and $\angle C'$, we discover that each angle contains 85°. Why is this so? The sum of the measures of the angles of every triangle is always 180°.

Therefore, if two angles of one triangle are equal in measure to two angles of a second triangle, the third or remaining angles in these triangles must also be equal in measure.

If we compare the lengths of corresponding sides in the larger triangle and in the smaller triangle, we discover that for each pair the lengths have a ratio of 2 : 1. For example, side $\overline{A'C'}$ is twice as long as side $\overline{AC}$.

Notice that we started with triangles in which only two pairs of corresponding angles were congruent. Now we have seen that all three corresponding angles must be congruent, and that the lengths of corresponding sides must have the same ratios and are therefore in proportion. Thus it seems reasonable to accept the truth of the following statements:

- **Two triangles are similar if two angles of one triangle are equal in measure to two corresponding angles of the other triangle [a.a. ≅ a.a.].**

or

- **Two triangles are similar if two angles of one triangle are congruent to two corresponding angles of the other triangle.**

Since two right triangles have a pair of congruent right angles, we can also say:

- **Two right triangles are similar if an acute angle in one triangle is equal in measure to an acute angle in the other triangle.**

EXAMPLES

1. $\triangle ABC$ and $\triangle DEF$ are similar.

a. Name three pairs of corresponding angles.

b. Name three pairs of corresponding sides.

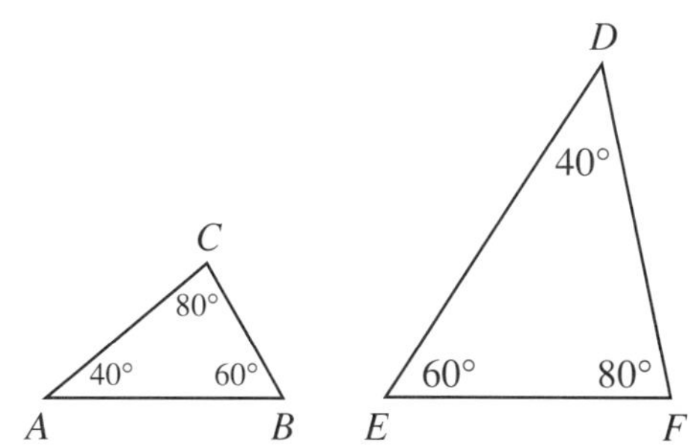

Solutions **a.** The corresponding angles are the pairs of angles whose measures are equal, that is, the angles that are congruent.

Since m$\angle A$ = 40 and m$\angle D$ = 40, $\angle D$ corresponds to $\angle A$.
Since m$\angle B$ = 60 and m$\angle E$ = 60, $\angle E$ corresponds to $\angle B$.
Since m$\angle C$ = 80 and m$\angle F$ = 80, $\angle F$ corresponds to $\angle C$.

Answer: **a.** $\angle D$ and $\angle A$; $\angle E$ and $\angle B$; $\angle F$ and $\angle C$

b. The sides that join the vertices of two pairs of congruent angles are corresponding sides.

Since $\angle A \cong \angle D$ and $\angle B \cong \angle E$, $\overline{AB}$ corresponds to $\overline{DE}$.

Since $\angle B \cong \angle E$ and $\angle C \cong \angle F$, $\overline{BC}$ corresponds to $\overline{EF}$.

Since $\angle C \cong \angle F$ and $\angle A \cong \angle D$, $\overline{CA}$ corresponds to $\overline{FD}$.

It is important to name corresponding sides by keeping the corresponding vertices in the *same order*. Note that we say $\overline{CA}$ corresponds to $\overline{FD}$. It would be confusing to say $\overline{CA}$ corresponds to $\overline{DF}$.

Answer: **b.** $\overline{DE}$ and $\overline{AB}$; $\overline{EF}$ and $\overline{BC}$; $\overline{FD}$ and $\overline{CA}$

---

2. The measures of the sides of a triangle are 2, 8, and 7 inches. If the longest side of a similar triangle measures 4 inches, find the length of the shortest side of that triangle.

Solution If $\triangle RST \sim \triangle DEF$, their corresponding sides are in proportion. The longest sides, $\overline{RS}$ and $\overline{DE}$, are corresponding sides; and the shortest sides, $\overline{RT}$ and $\overline{DF}$, are corresponding sides.

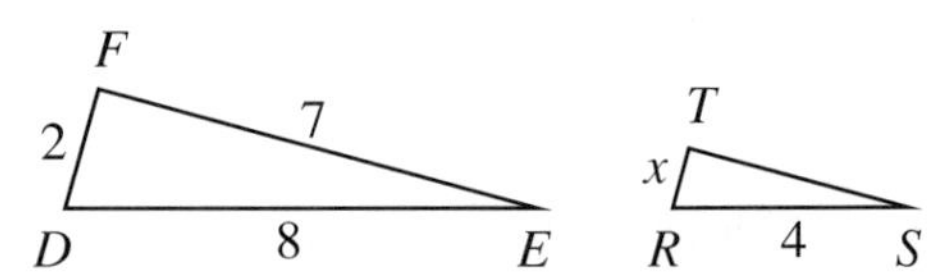

$$\frac{RT}{DF} = \frac{RS}{DE}$$

Let x = length RT. Then, since $DF = 2$, $RS = 4$, and $DE = 8$:

$$\frac{x}{2} = \frac{4}{8}$$
$$8(x) = 2(4)$$
$$8x = 8$$
$$x = 1$$

Answer: The length of the shortest side of $\triangle RST$ is 1 inch.

---

3. At the same time that a vertical flagpole casts a shadow 15 feet long, a vertical pole that is 6 feet high casts a shadow 5 feet long. Find the height of the flagpole.

Solution (1) Since the poles represented by $\overline{BC}$ and $\overline{EF}$ are vertical, $m\angle ABC = 90$ and $m\angle DEF = 90$.

(2) The sun's rays are parallel, and the ground acts as a transversal. Therefre, $m\angle A = m\angle D$ because they are corresponding angles of parallel lines cut by a transversal.

(3) Therefore, $\triangle DEF \sim \triangle ABC$ because two angles in one triangle are equal in measure to two corresponding angles in the other triangle.

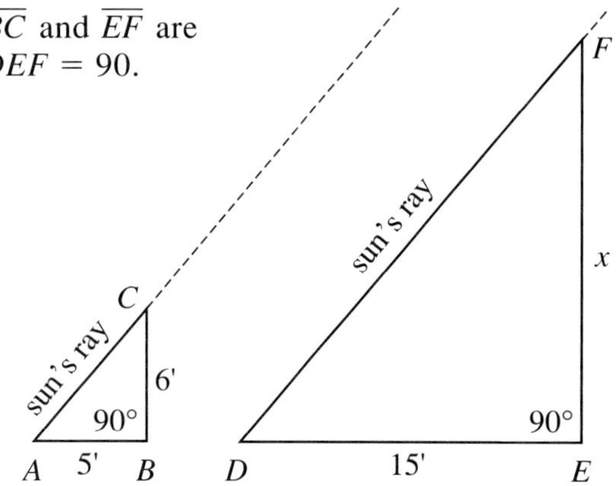

(4) The corresponding sides of $\triangle DEF$ and $\triangle ABC$ are in proportion. Therefore:

$$\frac{EF}{BC} = \frac{DE}{AB}$$

Let x = length EF. Then:

$$\frac{x}{6} = \frac{15}{5}$$

$$5(x) = 6(15)$$

$$5x = 90$$

$$x = 18$$

Answer: The height of the flagpole is 18 feet.

EXERCISES WITH OPEN-RESPONSE PROBLEMS

In 1–4, in each case the two triangles are similar. For each two triangles:
a. Name three pairs of corresponding angles.
b. Name three pairs of corresponding sides.

1.

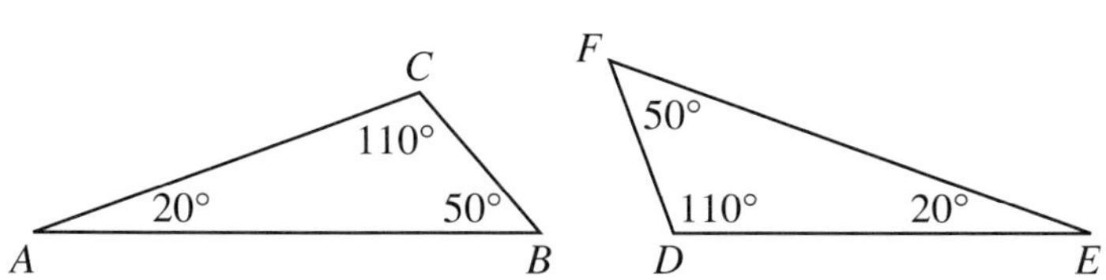

2.

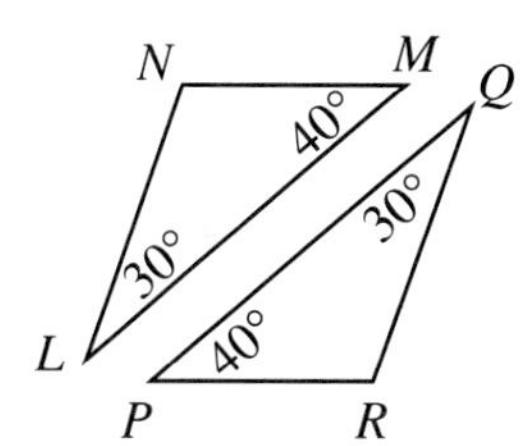

3.

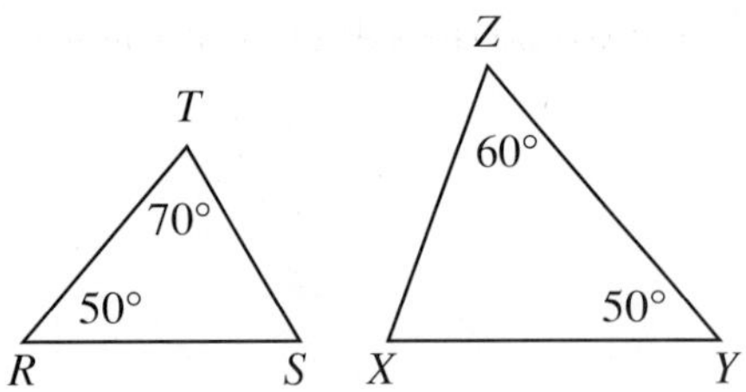

4.

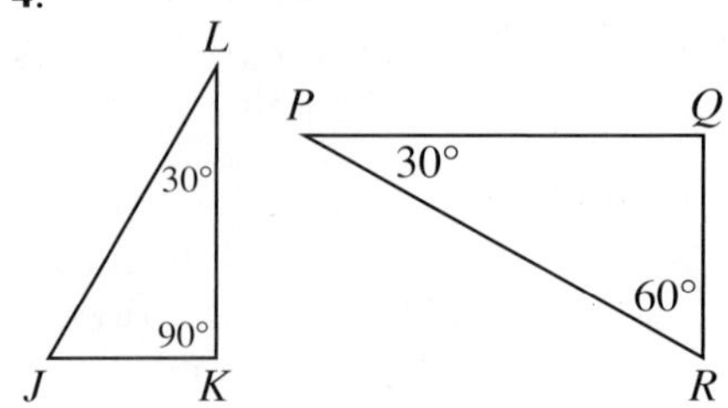

In 5 and 6, use a ruler and protractor to draw the triangles.

5. **a.** Draw $\triangle ABC$ in which $AB = 2$ inches, $m\angle A = 50$, and $m\angle B = 70$.
b. Draw $\triangle A'B'C'$ similar to $\triangle ABC$ so that $\angle A'$ corresponds to $\angle A$, $\angle B'$ corresponds to $\angle B$, and $A'B' = 3$ inches.

6. **a.** Draw right triangle RST in which $\angle S$ is a right angle, $m\angle R = 70$, and $RS = 3$ inches.
b. Draw $\triangle R'S'T' \sim \triangle RST$ so that $\overline{R'S'}$ corresponds to $\overline{RS}$ and $R'S' = 1\frac{1}{2}$ inches.

7. In $\triangle RST$, $m\angle R = 90$ and $m\angle S = 40$. In $\triangle XYZ$, $m\angle Y = 40$ and $m\angle Z = 50$.
a. Is $\triangle RST$ similar to $\triangle XYZ$? **b.** Why?

In 8 and 9, in each case select the triangles that are similar and tell why they are similar.

8.

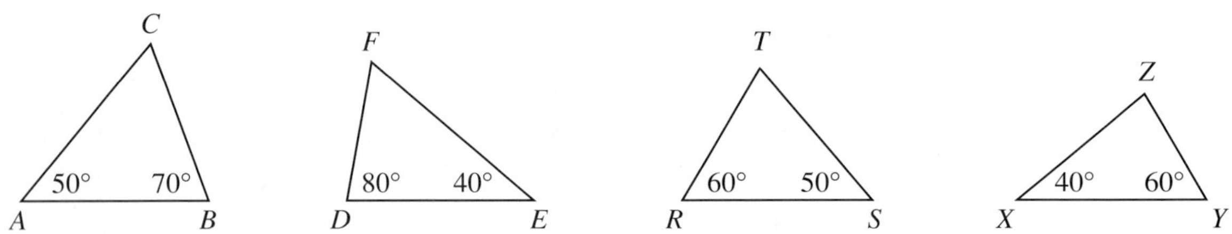

(1) (2) (3) (4)

9.

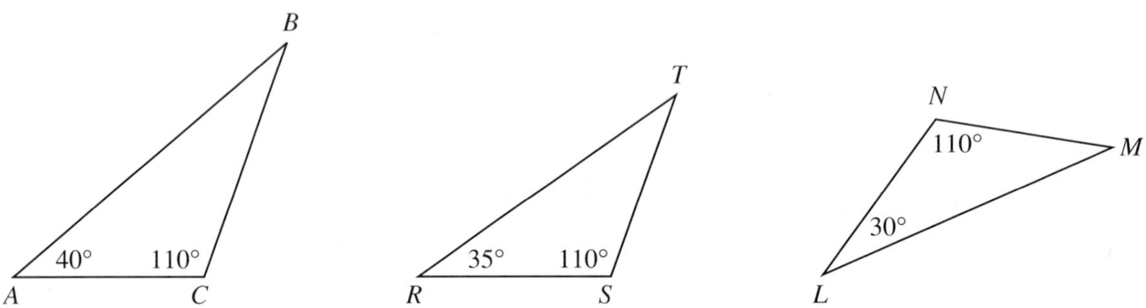

(1) (2) (3)

10. Triangle $ABC \sim \triangle RST$. Angle A corresponds to $\angle R$, $\angle B$ to $\angle S$, and $\angle C$ to $\angle T$. If $AB = 9$ meters, $AC = 6$ meters, and $RS = 3$ meters, find RT.

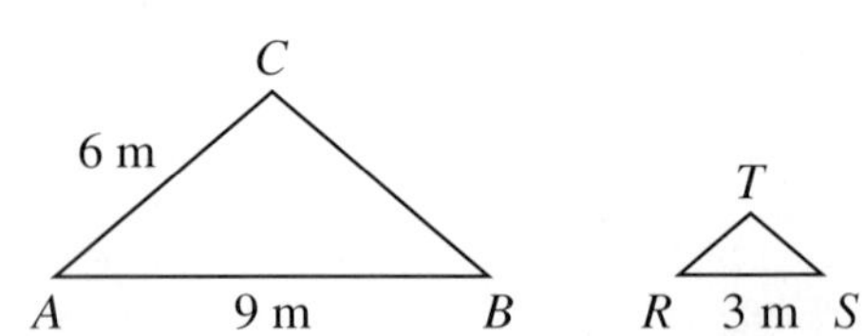

11. **a.** Is $\triangle ABC$ similar to $\triangle DEF$?
b. State the reason for the answer given in part **a**.
c. Find BC.

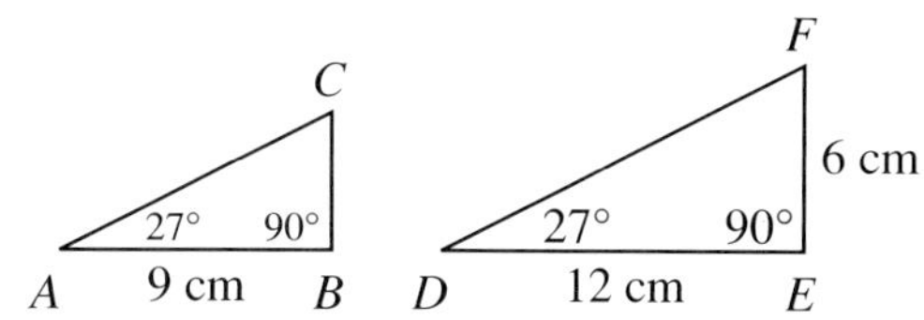

12.

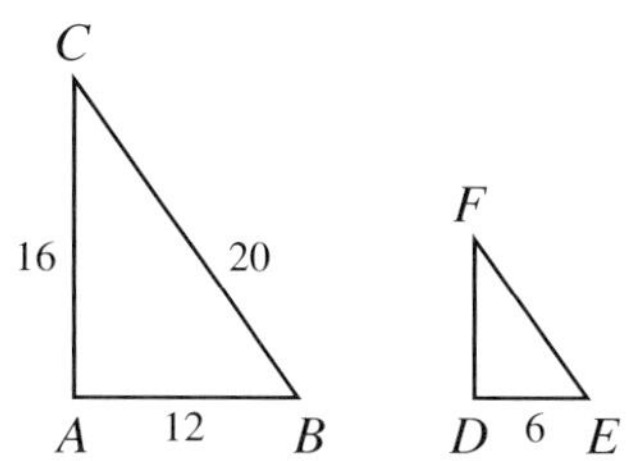

In the figure, $m\angle A = m\angle D$ and $m\angle C = m\angle F$. If $AB = 12$, $AC = 16$, $BC = 20$, and $DE = 6$, find DF and EF.

13.

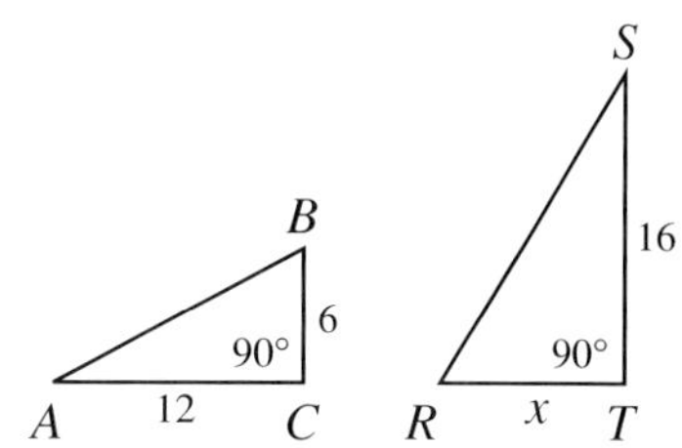

In right triangles ABC and RST, $m\angle A = m\angle S$.
a. Why is $\triangle ABC$ similar to $\triangle SRT$?
b. Find the value of x.

14. The lengths of the sides of a triangle are 24, 16, and 12. If the shortest side of a similar triangle measures 6, what is the length of the longest side of this triangle?

15. The lengths of the sides of a triangle are 36, 30, and 18. If the longest side of a similar triangle measures 9, what is the length of the shortest side of this triangle?

16. To find the height of a tree, a hiker measured its shadow and found it to be 12 feet. At the same time, a vertical rod 8 feet high cast a shadow of 6 feet.

a. Is $\triangle ABC$ similar to $\triangle DEF$?

b. Why?

c. Find the height of the tree.

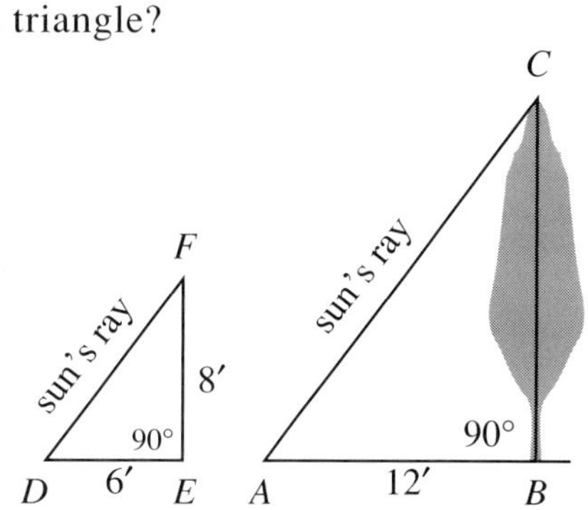

17. A certain tree casts a shadow 6 meters long. At the same time, a nearby boy 2 meters tall casts a shadow 4 meters long. Find the height of the tree.

18. A building casts a shadow 18 feet long. At the same time, a woman 5 feet tall casts a shadow 3 feet long. Find the number of feet in the height of the building.

19. $\overline{AB}$ represents the width of a river, and $\angle B$ and $\angle D$ are right angles. $\overline{AE}$ and $\overline{BD}$ are line segments. $BC = 16$ meters, $CD = 8$ meters, and $DE = 4$ meters.

a. Is $\triangle ABC$ similar to $\triangle EDC$? Why?

b. Find AB, the width of the river.

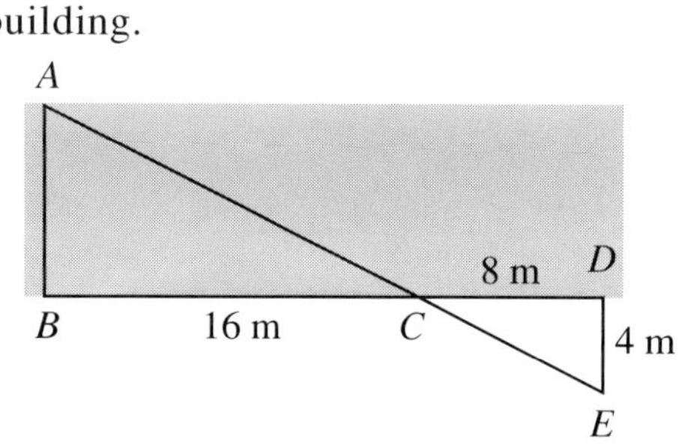

20. Using the figure in Exercise 19, find AB, the width of the river, if $BC = 240$ feet, $CD = 80$ feet, and $DE = 25$ feet.

21. In the figure, AD and CB are line segments, and $\overline{AB} \parallel \overline{DC}$.

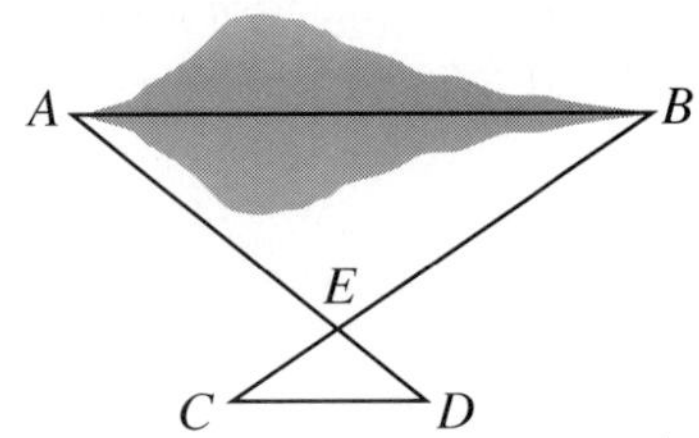

a. Why is $\triangle ABE \sim \triangle DCE$?

b. Find AB, the distance across the pond represented in the figure, if:

(*1*) $CE = 40$ m, $EB = 120$ m, and $CD = 50$ m

(*2*) $AE = 75$ m, $ED = 30$ m, and $CD = 36$ m

22. Calvin's height, to the *nearest inch*, is 5 feet 6 inches. At the same time that Calvin casts a 3-foot-long shadow, shadows are cast by a nearby house, a telephone pole, a tree, and his friend Barbara. Given the lengths of these shadows, find to the nearest inch:

a. the height of the house, whose shadow is 15 feet long

b. the height of the telephone pole, whose shadow is 6 feet 6 inches long

c. the height of the tree, whose shadow is 32 feet long

d. Barbara's height if her shadow is 2 feet 8 inches long

12-11 DILATIONS

In Chapter 11, we saw four types of geometric transformations: a line reflection, a point reflection, a translation, and a rotation. For each of these transformations, the original geometric figure and its image were congruent. In other words, length and angle measure remained unchanged under the transformations.

In another type of transformation, called a ***dilation***, angle measures remain unchanged but length is altered. Under a dilation, the image is larger or smaller than the original figure. In other words, the original figure and its image are *similar* to each other.

An example of a dilation occurs in projecting films or photographic images. As light passes through a frame of movie film or a photographic slide, it projects a larger, or dilated, image on a screen. In the diagram, $\overline{AB}$ represents the film and $\overline{A'B'}$ represents the image on the screen.

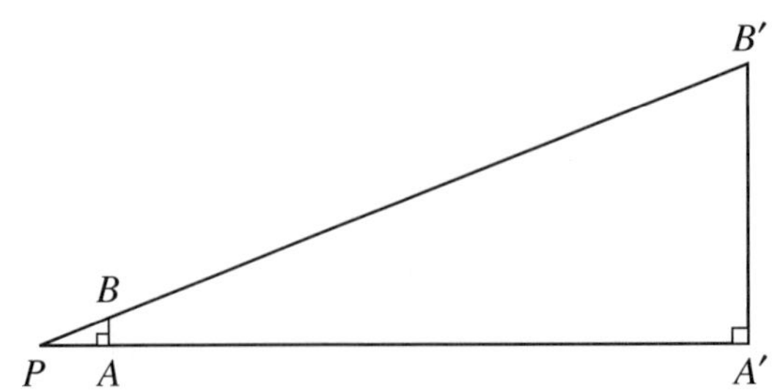

$\triangle PAB \sim \triangle PA'B'$

Think of $\triangle PAB$ and $\triangle PA'B'$ as the original triangle and its image, respectively. These figures have two pairs of congruent angles ($\angle PAB \cong \angle PA'B'$ and $\angle P \cong \angle P$).

Therefore, $\triangle PAB$ is similar to $\triangle PA'B'$, and their corresponding sides are in proportion.

In a dilation, there is one fixed point, called the ***center of dilation***. In our example, point P, the source of the light, is the center of dilation. Using this center P, we say that the image of A is A', the image of B is B', and the image of P is P. It follows that the image of $\overline{AB}$ is $\overline{A'B'}$, the image of $\overline{PA}$ is $\overline{PA'}$, and the image of $\overline{PB}$ is $\overline{PB'}$.

When we compare the length of one side of an image to the length of its corresponding side in the original, we form a ratio called the ***constant of dilation***. For example, if the height of the film, AB, is 32 millimeters and the height of its projection on the screen, $A'B'$, is 3.2 meters (or 3,200 mm), the constant of dilation is $\frac{A'B'}{AB} = \frac{3{,}200}{32} = 100$. Thus, a figure in the film is magnified 100 times on the screen.

EXAMPLE

In the diagram, $\angle ADE$ and $\angle ABC$ are right angles, $\triangle ADE \sim \triangle ABC$, $AB = 24$ centimeters, $BC = 18$ centimeters, and $AD = 16$ centimeters.

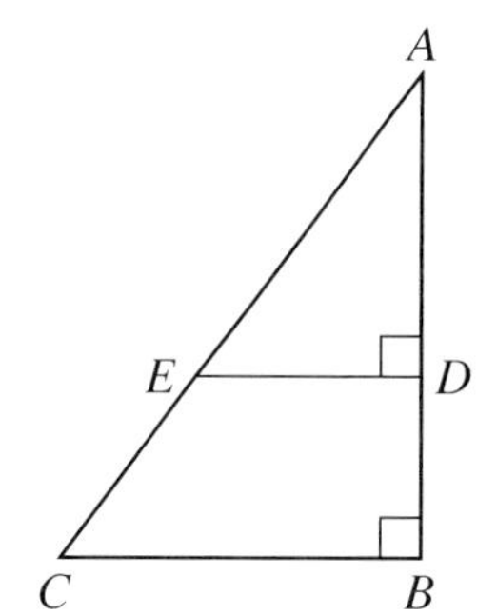

a. Find DE.
b. If B is the image of D under a dilation with center at A, what is the constant of dilation?
c. Find the area of $\triangle ABC$.
d. Find the area of $\triangle ADE$.
e. Find the area of trapezoid $BCED$.

Solutions **a.** Since $\triangle ADE \sim \triangle ABC$,

$$\frac{DE}{BC} = \frac{AD}{AB}$$

Let $x = DE$. Then:

$$\frac{x}{18} = \frac{16}{24}$$

$$24x = 16(18)$$

$$= 288$$

$$x = 12 \text{ cm}$$

b. Constant of dilation $= \frac{AB}{AD}$

$$= \frac{24}{16} = \frac{3}{2}$$

Check: The image of E is C, and the image of $\overline{DE}$ is $\overline{BC}$.

$$\frac{BC}{DE} = \frac{18}{12} = \frac{3}{2} = \text{constant of dilation}$$

c. Area of

$$\triangle ABC = \frac{1}{2}(\text{base})(\text{height})$$
$$= \frac{1}{2}(BC)(AB)$$
$$= \frac{1}{2}(18)(24)$$
$$= \frac{1}{2}(432) = 216$$

Area of $\triangle ABC = 216 \text{ cm}^2$

d. Area of

$$\triangle ADE = \frac{1}{2}(\text{base})(\text{height})$$
$$= \frac{1}{2}(DE)(AD)$$
$$= \frac{1}{2}(12)(16)$$
$$= \frac{1}{2}(192) = 96$$

Area of $\triangle ADE = 96 \text{ cm}^2$

e. Area of trapezoid $BCED$ = area of $\triangle ABC$ − area of $\triangle ADE$
= 216 − 96
= 120

An Alternative Solution Use the formula for the area of a trapezoid.

$$\text{The height } BD = AB - AD = 24 - 16 = 8$$

$$\text{Area of a trapezoid} = \frac{1}{2}h(b_1 + b_2)$$

$$\text{Area of trapezoid } BCED = \frac{1}{2}(BD)(BC + DE)$$
$$= \frac{1}{2}(8)(18 + 12)$$
$$= \frac{1}{2}(8)(30)$$
$$= 120 \text{ cm}^2$$

Answers: **a.** 12 cm **b.** $\frac{3}{2}$ **c.** 216 cm^2 **d.** 96 cm^2 **e.** 120 cm^2

EXERCISES WITH OPEN-RESPONSE PROBLEMS

In 1–3, under a dilation with center at O, the image of A is A' and the image of B is B'.

1. Find OA' and OB' when $OA = 12$, $OB = 18$, and the constant of dilation is:

a. 4 b. 1.5 c. $\frac{4}{3}$

2. Find the constant of dilation when $OA = 16$, $OB = 20$, and:

a. $OA' = 48$ b. $OB' = 24$ c. $OA' = 40$

3. Find AB if $A'B' = 20$ and the constant of dilation is $\frac{5}{4}$.

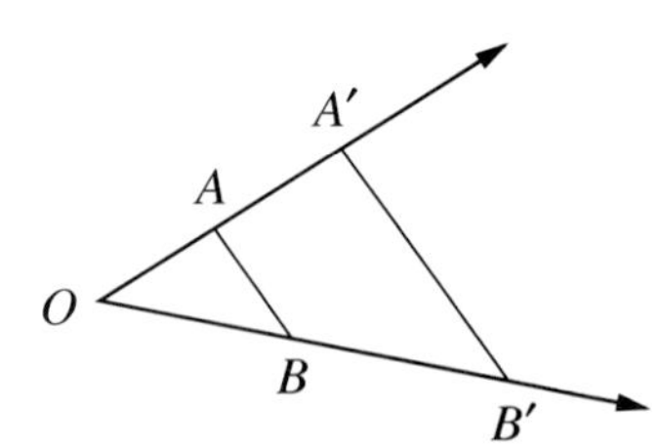

Ex. 1–3

For 4–9, use the accompanying figure showing $\triangle ABC$ and $\triangle AB'C'$.

4. Prove informally that $\triangle ABC \sim \triangle AB'C'$.

5. $BC = 3$ inches, $AC = 4$ inches, and $AC' = 8$ inches. Find $B'C'$.

6. $AC = 5$ meters, $BC = 2$ meters, and $B'C' = 8$ meters. Find AC'.

7. $BC = 10$ centimeters and $AC = 8$ centimeters. Find the ratio of $B'C'$ to AC'.

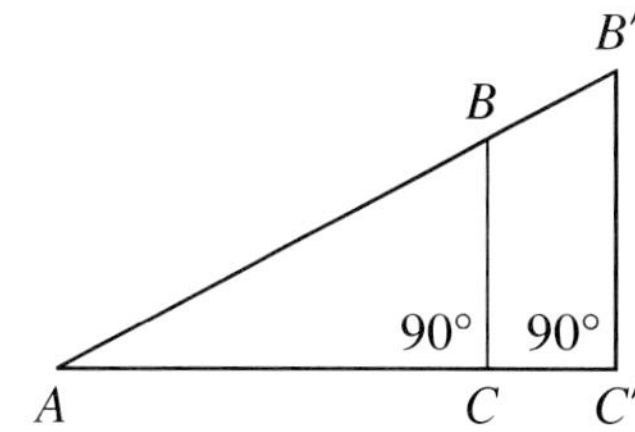

Ex. 4–9

8. $AC' = 48$ meters, $B'C' = 36$ meters, and $AC = 32$ meters.

a. Find BC.
b. Find the area of $\triangle AC'B'$.
c. Find the area of $\triangle ACB$.
d. Find the area of trapezoid $BCC'B'$.

9. Let B' be the image of B under a dilation with center at A. Let $AC = 12$, $BC = 9$, and $AB = 15$. The constant of dilation is $\frac{4}{3}$.

a. What is the image of C under this dilation?
b. Find $B'C'$, AC', and AB', using the constant of dilation.

For 10–12, use the accompanying figure, where the center of dilation is R, the image of S is M, the image of T is N, $RS = 8$, and $RT = 6$.

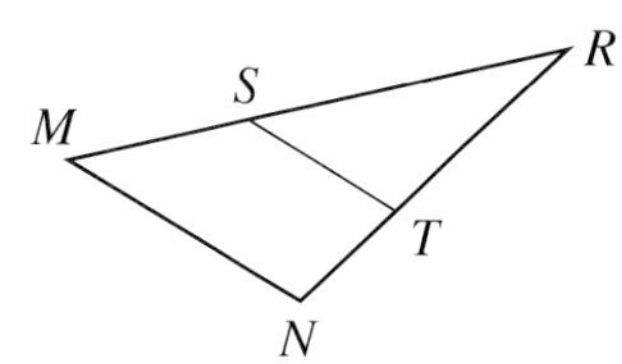

Ex. 10–12

10. If the constant of dilation is 2, find:
a. RM **b.** MS **c.** RN **d.** TN

11. Reanswer Exercise 10, using 3 as the constant of dilation.

12. Reanswer Exercise 10, using 1.25 as the constant of dilation.

13. Under a dilation with center O, the image of A is A', the image of B is B', $OA = 20$, and $OB = 30$.

a. When the constant of dilation is 0.6, find: (*1*) OA' (*2*) OB'
b. When the constant of dilation is $\frac{2}{5}$, find: (*1*) OA' (*2*) OB'
c. If $OA' = A'A$, what is the constant of dilation?

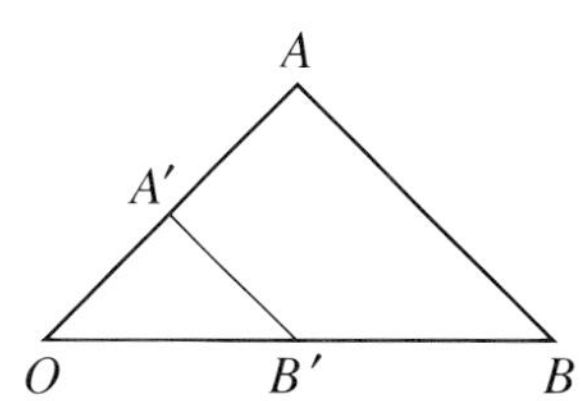

14. A teacher is using a projector to show slides. The light source is at L, the slide to be projected is at $\overline{AB}$, and the image is at $\overline{CD}$.

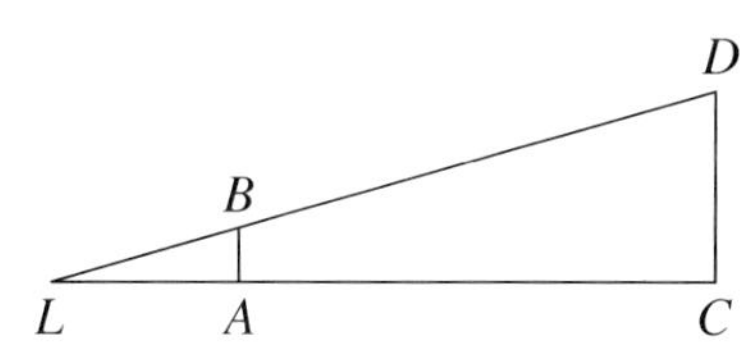

a. If $\overline{AB}$ and $\overline{CD}$ are each perpendicular to $\overline{LC}$, explain why $\triangle LAB \sim \triangle LCD$.
b. If the distance from L to A is 16 centimeters and the distance from L to C is 8 meters, what is the constant of dilation?
c. If the height of the slide, AB, is 32 millimeters, what is the height of the image, CD?
d. How far from $\overline{CD}$ should L be placed if the distance from L to A remains the same and the height of the image, CD, is to be 2.4 meters?

12-12 SCALE DRAWING

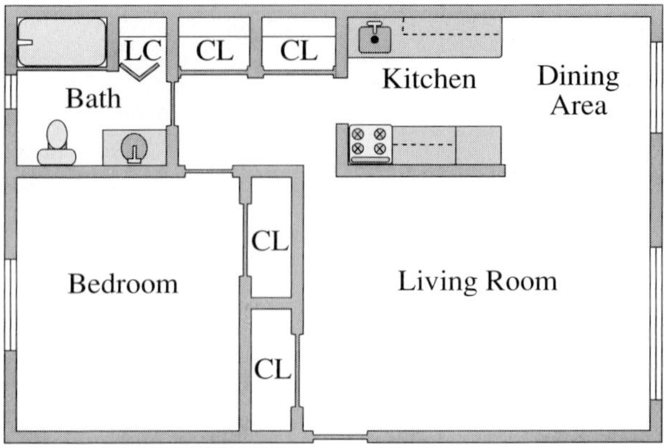

Before construction begins on any building, or before a skilled worker begins to make a tool or a part for a machine, careful plans must be drawn. To complete the required construction, a worker must have exact information about the project including size, shape, angle measure, and a multitude of other details. This information is usually presented in the form of a drawing such as the floor plan of the apartment shown at the right.

In order to show the plan of a building on paper, the size of the drawing must be smaller than the actual building. Or, in order to show the detail of a very small part for a machine, the size of the drawing may be larger than the actual part. A **scale drawing** is a sketch that is similar to the real object it represents. The lengths of the lines and the radii of the curved portions are proportional to the actual measurements and the angles are congruent to the actual angles. The scale of the drawing shows the relationship between the lengths of the lines in the drawing and the actual lengths of the object.

EXAMPLES

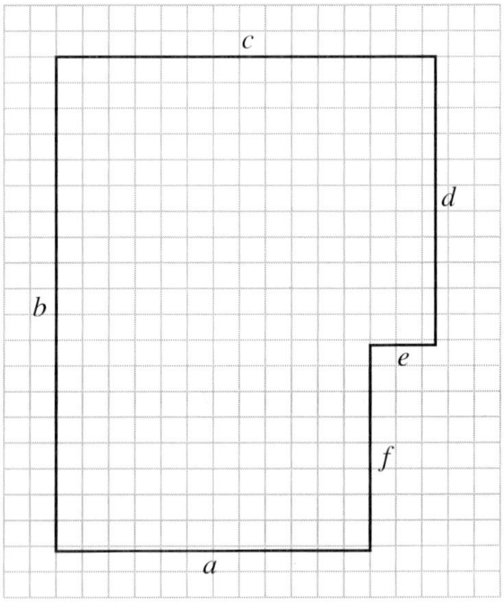

1. The sketch at the right shows the shape of a room whose dimensions are $a = 7.5$ feet, $b = 12$ feet, $c = 9.0$ feet, and $d = 7.0$ feet. A drawing is to be made with a scale of 1 inch = 2.5 feet.

a. Find the length required for each of the six edges in the scale drawing.

b. The length of each line in the drawing is what fractional part of the true length?

Solution **a.** Let a', b', c', d', e', and f' represent the corresponding lengths of the edges in the scale drawing. For each edge, write and solve the proportion

$$\frac{\text{length of edge in the drawing in inches}}{\text{true length in feet}} = \frac{1}{2.5}$$

Note that the numerator of each fraction is in inches and the denominator of each fraction is in feet.

$$\frac{a'}{7.5} = \frac{1}{2.5} \qquad \frac{b'}{12} = \frac{1}{2.5} \qquad \frac{c'}{9.0} = \frac{1}{2.5}$$

$$2.5a' = 7.5 \qquad 2.5b' = 12 \qquad 2.5c' = 9.0$$

$$a' = 3.0 \qquad b' = 4.8 \qquad c' = 3.6$$

$$\frac{d'}{7.0} = \frac{1}{2.5} \qquad a' + e' = c' \qquad d' + f' = b'$$

$$2.5d' = 7.0 \qquad 3.0 + e' = 3.6 \qquad 2.8 + f' = 4.8$$

$$d = 2.8 \qquad e' = 0.6 \qquad f' = 2.0$$

Answer: $a' = 3.0$ in., $b' = 4.8$ in., $c' = 3.6$ in., $d' = 2.8$ in., $e' = 0.6$ in., and $f' = 2.0$ in.

b. Multiply $\frac{1 \text{ in.}}{2.5 \text{ ft}}$ by $\frac{1 \text{ ft}}{12 \text{ in.}}$, a fraction equivalent to 1, to obtain a ratio that is independent of the measure. $\frac{1 \not{\text{in.}}}{2.5 \not{\text{ft}}} \times \frac{1 \not{\text{ft}}}{12 \not{\text{in.}}} = \frac{1}{30}$

Answer: Each length in the scale drawing is $\frac{1}{30}$ of the true length.

---

2. The scale on an architect's drawing is $\frac{1}{4}$ inch = 1 foot. What are the true dimensions of a room shown as 3 inches long and $2\frac{1}{2}$ inches wide?

Solution Use the proportion $\frac{\text{length of edge in the drawing in inches}}{\text{true length in feet}} = \frac{\frac{1}{4}}{1}$

Let ℓ = the true length of the room.
w = the true width of the room.

$$\frac{3}{\ell} = \frac{\frac{1}{4}}{1} \qquad \frac{2\frac{1}{2}}{w} = \frac{\frac{1}{4}}{1}$$

$$\frac{1}{4}\ell = 3 \qquad \frac{1}{4}w = 2\frac{1}{2}$$

$$\ell = 12 \qquad w = 10$$

Answer: The true dimensions of the room are 12 feet by 10 feet.

EXERCISES WITH OPEN-RESPONSE PROBLEMS

In 1-6, the scale of a drawing is given. In a drawing made to the given scale, the length of each line will be what fractional part of the true measure of the corresponding line?

1. 1 in. = 3 ft **2.** 1 in. = $2\frac{1}{4}$ ft **3.** $\frac{1}{4}$ in. = 1 ft

4. $\frac{1}{2}$ ft = 5 ft **5.** 2 in. = 7 ft 6 in. **6.** 3 in. = 8 ft 4 in.

In 7-10, use the given scale to find the true dimensions of each of the lengths given on the scale drawing shown at the right.

7. 1 in. = 2.5 ft **8.** 1 in. = 3 ft 3 in.

9. 2 in. = 1.5 ft **10.** $\frac{1}{4}$ in. = 1 ft

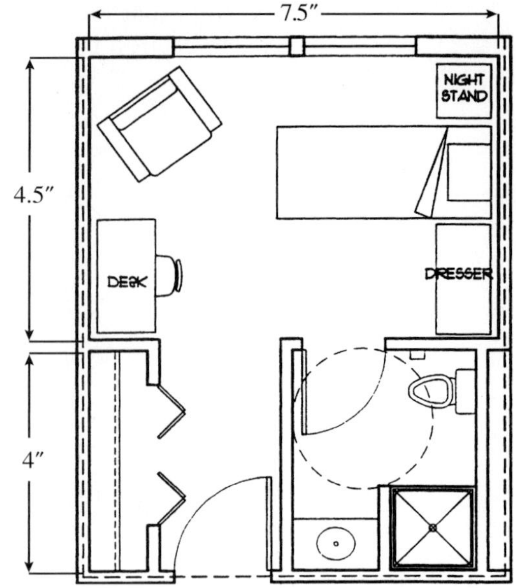

In 11-14, the true dimensions of a rectangular office are 18 feet by 24 feet. Find the dimensions needed for a drawing made to the given scale.

11. 1 in. = 4 ft **12.** 1 in. = 2.5 ft **13.** 2 in. = 9 ft **14.** $\frac{1}{2}$ in. = 3 ft

15. In a scale drawing, a wall 8.75 feet long is represented by a line 2.5 inches long. What was the scale used in the drawing?

16. A part for a machine is to be 0.75 centimeters long. A scale drawing of the part is 15 centimeters long.

a. One edge of the part is to be 0.50 centimeters long. What length should be used to show this edge on the drawing?

b. Another edge of the part is shown on the drawing by a line 5.0 centimeters long. What is the actual length of that edge of the part?

12-13 RATIO OF PERIMETERS AND RATIO OF AREAS OF SIMILAR POLYGONS

In the diagram, $\triangle ABC \sim \triangle A'B'C'$. Therefore, the corresponding angles are congruent and the corresponding sides are in proportion.

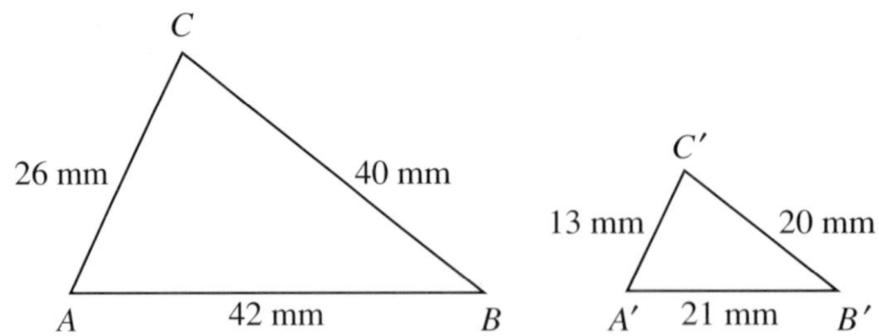

$$\frac{AB}{A'B'} = \frac{42}{21} = \frac{2}{1},$$

$$\frac{AC}{A'C'} = \frac{26}{13} = \frac{2}{1},$$

$$\frac{BC}{B'C'} = \frac{40}{20} = \frac{2}{1}$$

Ratio of Perimeters

The perimeter of $\triangle ABC = 42 + 26 + 40 = 108$ millimeters
The perimeter of $\triangle A'B'C' = 21 + 13 + 20 = 54$ millimeters

$$\frac{\text{Perimeter of } \triangle ABC}{\text{Perimeter of } \triangle A'B'C'} = \frac{108}{54} = \frac{2}{1}$$

The ***ratio of the perimeters*** of these similar triangles is the same as the ratio of the measures of the corresponding sides. If we looked at other numerical examples, we would reach the same conclusion.

Thus, we say:

- **If two triangles are similar, the ratio of the perimeters is equal to the ratio of the measures of the corresponding sides.**

This statement is also true for similar polygons of more than three sides, and we say:

- **If two polygons are similar, the ratio of the perimeters is equal to the ratio of the measures of the corresponding sides.**

For example, rectangles $PRST$ and $P'R'S'T'$ are similar.

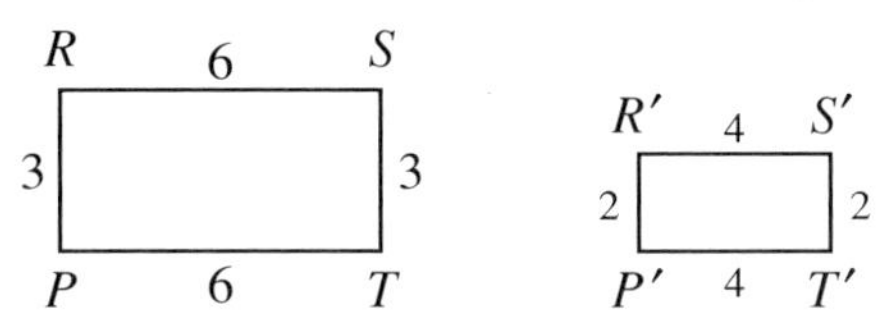

$$\frac{PR}{P'R'} = \frac{3}{2}, \quad \frac{RS}{R'S'} = \frac{6}{4} = \frac{3}{2}, \quad \frac{ST}{S'T'} = \frac{3}{2}, \quad \frac{TP}{T'P'} = \frac{6}{4} = \frac{3}{2}$$

$$\frac{\text{Perimeter of } PRST}{\text{Perimeter of } P'R'S'T'} = \frac{3 + 6 + 3 + 6}{2 + 4 + 2 + 4} = \frac{18}{12} = \frac{3}{2}$$

The ratio of the lengths of corresponding sides is $\frac{3}{2}$, and the ratio of the perimeters is also $\frac{3}{2}$.

Ratio of Areas

Let us now examine the ***ratio of the areas*** of the similar figures discussed in preceding sections. The ratio of the corresponding sides of similar triangles ABC and $A'B'C'$ is 2 : 1. The altitudes, $\overline{CD}$ and $\overline{C'D'}$, that we have drawn in these triangles are also in the same 2 : 1 ratio:

$$\frac{CD}{C'D'} = \frac{24}{12} = \frac{2}{1}$$

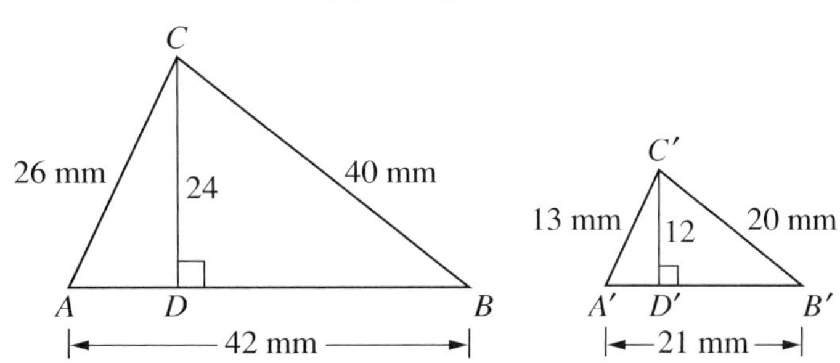

To compare areas, we use this formula:

$$\text{Area of a triangle} = \frac{1}{2}(\text{base} \times \text{altitude})$$

$$\text{Area of } \triangle ABC = \frac{1}{2}(42)(24) = 504$$

$$\text{Area of } \triangle A'B'C' = \frac{1}{2}(21)(12) = 126$$

$$\frac{\text{Area of } \triangle ABC}{\text{Area of } \triangle A'B'C'} = \frac{504}{126} = \frac{4}{1} = \left(\frac{2}{1}\right)^2$$

The ratio of the areas of these similar triangles is the *square* of the ratio of the measures of the corresponding sides. If we looked at other numerical examples, we would verify this conclusion.

Thus, we say:

- **If two triangles are similar, the ratio of the areas is the *square* of the ratio of the measures of the corresponding sides.**

Since this statement is also true for similar polygons of more than three sides, we say:

- **If two polygons are similar, the ratio of the areas is the *square* of the ratio of the measures of the corresponding sides.**

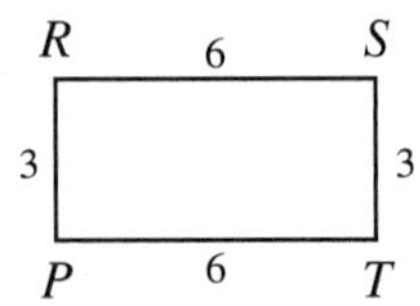

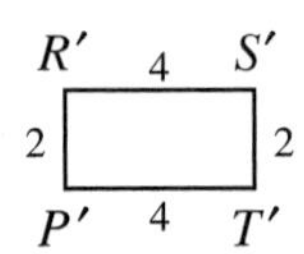

For example, $PRST$ and $P'R'S'T'$ are similar rectangles whose corresponding sides are in the ratio of 3:2.

To compare the areas of these rectangles, we use the formula

$$\text{Area of a rectangle} = \text{length} \times \text{width}$$

$$\frac{\text{Area of } PRST}{\text{Area of } P'R'S'T'} = \frac{(6)(3)}{(4)(2)} = \frac{18}{8} = \frac{9}{4} = \left(\frac{3}{2}\right)^2$$

Thus, the ratio of the lengths of corresponding sides is $\frac{3}{2}$, and the ratio of the areas is $\left(\frac{3}{2}\right)^2$ or $\frac{9}{4}$.

EXAMPLES

1. Right triangle ABC is similar to right triangle $A'B'C'$. As shown in the diagram, the measures in inches are $AB = 13$, $AC = 5$, $BC = 12$, and $B'C' = 8$.

a. Find the ratio of the measures of the corresponding sides.

b. Find $A'B'$ and $A'C'$.

c. Find the ratio of the perimeters of the triangles.

d. Find the ratio of their areas.

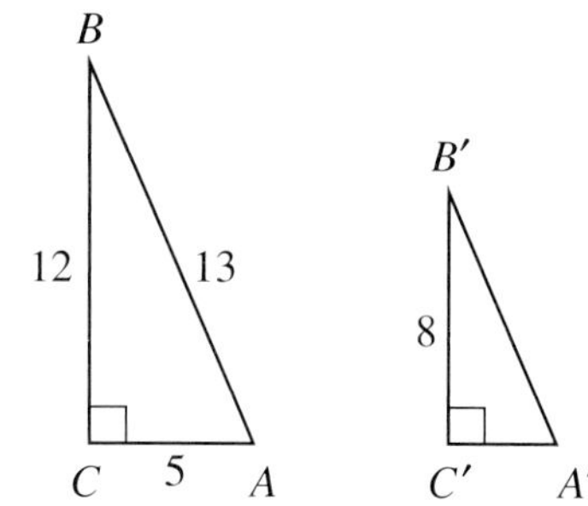

Solutions **a.** Since corresponding sides of similar triangles are in proportion, the ratio of the measures of each pair of sides of these triangles is the same as the ratio of the two given sides, $\frac{12}{8}$ or $\frac{3}{2}$.

b. The ratio of the measures of each pair of corresponding sides of these triangles is 3 : 2.

$$\frac{AB}{A'B'} = \frac{3}{2}$$
$$\frac{13}{A'B'} = \frac{3}{2}$$
$$3A'B' = 26$$
$$A'B' = \frac{26}{3} \text{ in.}$$

$$\frac{AC}{A'C'} = \frac{3}{2}$$
$$\frac{5}{A'C'} = \frac{3}{2}$$
$$3A'C' = 10$$
$$A'C' = \frac{10}{3} \text{ in.}$$

c. In similar triangles, the ratio of the perimeters is equal to the ratio of the measures of the corresponding sides.

The ratio of perimeters is $\frac{3}{2}$.

Instead, the ratio of the perimeters can be found by computing the perimeter of each triangle and finding the ratio:

$$\frac{\text{Perimeter of } \triangle ABC}{\text{Perimeter of } \triangle A'B'C'} = \frac{12 + 13 + 5}{8 + \frac{26}{3} + \frac{10}{3}} = \frac{30}{20} = \frac{3}{2}$$

d. In similar triangles, the ratio of the areas is the square of the ratio of the measures of the corresponding sides.

The ratio of the areas is $\left(\frac{3}{2}\right)^2 = \frac{9}{4}$.

Instead, the ratio can be found by computing and comparing the areas:

$$\frac{\text{Area of } \triangle ABC}{\text{Area of } \triangle A'B'C'} = \frac{\frac{1}{2}(5)(12)}{\frac{1}{2}\left(\frac{10}{3}\right)(8)} = \frac{\frac{60}{2}}{\frac{80}{6}} = \frac{\frac{60}{\cancel{2}}}{\frac{80}{\cancel{6}}} \cdot \frac{\overset{3}{\cancel{6}}}{\cancel{6}} = \frac{180}{80} = \frac{9}{4}$$

Answers: **a.** $\frac{3}{2}$ **b.** $A'B' = \frac{26}{3}$ inches, $A'C' = \frac{10}{3}$ inches **c.** $\frac{3}{2}$ **d.** $\frac{9}{4}$

---

2. Two triangles are similar. The sides of the smaller triangle have lengths of 4, 6, and 8 centimeters. The perimeter of the larger triangle is 27 centimeters. Find the length of the shortest side of the larger triangle.

Solution (1) In the smaller triangle, the length of the shortest side is 4 centimeters. Find the perimeter P:

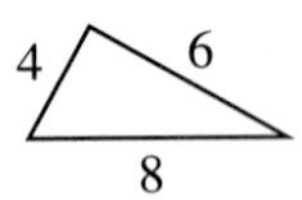

$P = 18$

$$P = 4 + 6 + 8 = 18$$

(2) In the larger triangle, the perimeter P' is 27 centimeters. Let x = length of shortest side.

(3) Since the triangles are similar, use the following proportion:

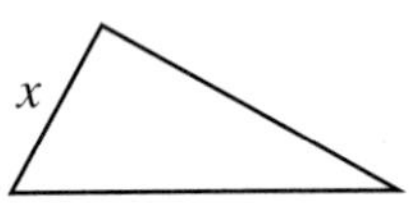

$P' = 27$

$$\frac{\text{Perimeter of smaller } \triangle}{\text{Perimeter of larger } \triangle} = \frac{\text{length of shortest side in smaller } \triangle}{\text{length of shortest side in larger } \triangle}$$

$$\frac{18}{27} = \frac{4}{x}$$

(4) Solve for x:

$$18x = 27(4)$$
$$= 108$$
$$x = 6$$

Answer: The length of the shortest side of the larger triangle is 6 centimeters.

---

3. Polygon $WXYZ \sim$ polygon $W'X'Y'Z'$.
If $WX = 3$ centimeters and $W'X' = 12$ centimeters, find:
a. the ratio of the perimeters.
b. the ratio of the areas.

Solutions **a.** $\frac{\text{Perimeter of polygon } WXYZ}{\text{Perimeter of polygon } W'X'Y'Z'} = \frac{WX}{W'X'} = \frac{3}{12} = \frac{1}{4}$

b. $\frac{\text{Area of polygon } WXYZ}{\text{Area of polygon } W'X'Y'Z'} = \left(\frac{WX}{W'X'}\right)^2 = \left(\frac{3}{12}\right)^2 = \frac{9}{144} = \frac{1}{16}$

Answers: **a.** $\frac{1}{4}$ **b.** $\frac{1}{16}$

---

4. The lengths of a pair of corresponding sides of two similar polygons are 4 inches and 6 inches. If the area of the first polygon is 20 square inches, find the area of the second polygon.

Solution Let x = area of second polygon.
Since the polygons are similar:

$$\frac{\text{Area of 1st polygon}}{\text{Area of 2nd polygon}} = \left(\frac{\text{length of side of 1st polygon}}{\text{length of side of 2nd polygon}}\right)^2$$

$$\frac{20}{x} = \left(\frac{4}{6}\right)^2$$

$$\frac{20}{x} = \frac{16}{36}$$

$$16x = 720$$

$$x = 45$$

Answer: The area of the second polygon is 45 square inches.

---

5. Two similar polygons have areas of 150 square centimeters and 54 square centimeters. Find the ratio of the lengths of two of their corresponding sides.

Solution (1) Find the ratio of their areas, in simplified form: $\frac{150}{54} = \frac{150 \div 6}{54 \div 6} = \frac{25}{9}$

(2) Since the ratio of the areas is the square of the ratio of the lengths of corresponding sides, find the *square roots* of both the numerator and the denominator: $\frac{\sqrt{25}}{\sqrt{9}} = \frac{5}{3}$

Calculator Solution To find the ratio of the lengths of corresponding sides of similar polygons, divide the *square root* of the area of the first polygon by the *square root* of the area of the second polygon.

Enter: 150 [√] [÷] 54 [√] [=]

Display: [1.66666667]

Write the decimal in the calculator display as a fraction. Recognize that 1.66666667 is a rational approximation for $1\frac{2}{3}$ or $\frac{5}{3}$.

Answer: The ratio of the lengths of two corresponding sides is 5:3.

EXERCISES WITH OPEN-RESPONSE PROBLEMS

In 1–6, in each case the ratio of the lengths of the sides of two similar polygons is given. **a.** What is the ratio of the perimeters? **b.** What is the ratio of the areas?

1. $\frac{3}{1}$ **2.** $\frac{4}{1}$ **3.** $\frac{1}{2}$ **4.** $\frac{1}{4}$ **5.** $\frac{6}{5}$ **6.** $\frac{10}{3}$

7. In the diagram $\triangle ABC \sim \triangle A'B'C'$. The measures in inches are $AB = 10$, $A'B' = 4$, $B'C' = 5$, and $C'A' = 6$.

a. Find the ratio of the lengths of the corresponding sides.

b. Find BC and CA.

c. Find the ratio of the perimeters.

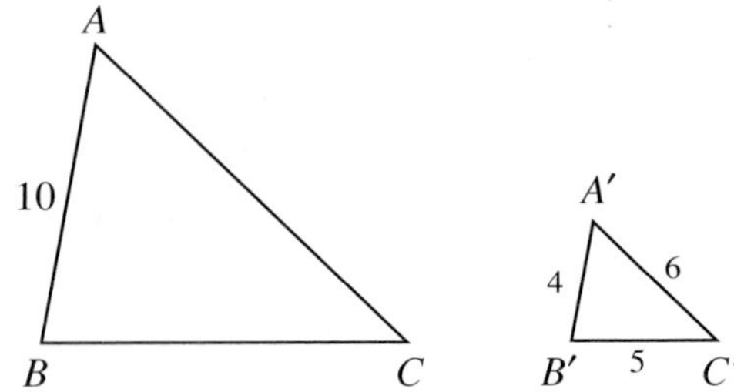

8. The lengths of the sides of a triangle are 8, 10, and 12. If the length of the shortest side of a similar triangle is 6, find the length of its longest side.

9. The sides of a triangle measure 7, 9, and 11. Find the perimeter of a similar triangle in which the shortest side has a length of 21.

10. The sides of a triangle measure 5, 12, and 13. Find the perimeter of a similar triangle in which the longest side measures 39.

11. Each side of a square has length 9 centimeters. If the length of each side is doubled, what is the ratio of the perimeter of the larger square thus formed to the perimeter of the smaller square?
(1) 36 : 1 (2) 18 : 1 (3) 9 : 1 (4) 2 : 1

12. The sides of a triangle have lengths 6, 8, and 10. What is the length of the shortest side of a similar triangle whose perimeter is 12?
(1) 6 (2) 8 (3) 3 (4) 4

13. Two triangles are similar. The sides of the smaller triangle have lengths of 6, 7, and 12. The perimeter of the larger triangle is 75. The length of the longest side of the larger triangle is:
(1) 18 (2) 2 (3) 36 (4) 4

14. The ratio of the perimeters of two similar polygons is 5 : 9. What is the ratio of the areas?

15. In the diagram $ABCD \sim A'B'C'D'$. The measures in inches are $AB = 8$, $BC = 12$, $CD = 4$, $DA = 10$, and $D'A' = 30$.

a. Find the ratio of the lengths of the corresponding sides.

b. Find $A'B'$, $B'C'$, and $C'D'$.

c. Find the ratio of the perimeters.

d. Find the ratio of the areas of these polygons.

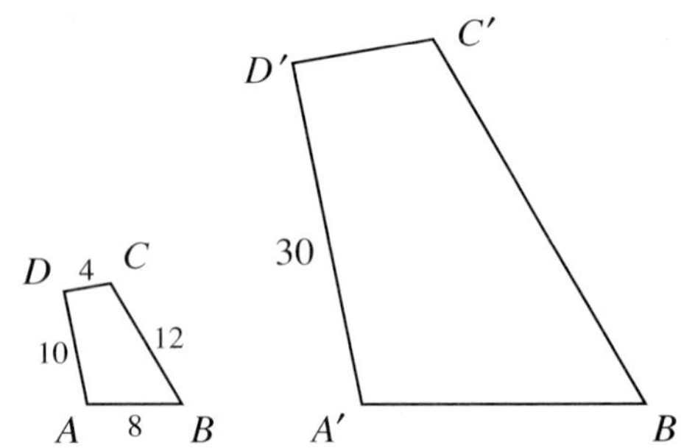

16. The lengths of a pair of corrresponding sides of two similar polygons are 3 centimeters and 11 centimeters. If the area of the first polygon is 18 square centimeters, find the area of the second polygon.

In 17–22, in each case the ratio of the areas of two similar polygons is given. Find the ratio of the lengths of two corresponding sides of the polygons.

17. $\frac{25}{1}$ **18.** $\frac{16}{9}$ **19.** $\frac{1}{4}$ **20.** $\frac{225}{49}$ **21.** $\frac{8}{50}$ **22.** $\frac{75}{108}$

23. The ratio of the perimeters of two similar polygons is 2 : 3. If the area of the smaller polygon is 6 square feet, what is the area of the larger polygon?

24. The length and width of a rectangle are each doubled, forming a new rectangle.
   a. What is the ratio of the perimeter of the new rectangle to the perimeter of the original rectangle?
   b. What is the ratio of the area of the new rectangle to the area of the original rectangle?

25. All regular polygons with the same number of sides are similar. The measure of the side of a regular pentagon is 12 centimeters, and the measure of the side of a larger regular pentagon is 15 centimeters.
   a. What is the ratio of the measures of the sides? **b.** What is the ratio of the areas?

26. The lengths of corresponding sides of two similar polygons are in the ratio 3 : 1. If the perimeter of the larger polygon exceeds the perimeter of the smaller polygon by 48 millimeters, find the perimeter of the smaller polygon.

27. The lengths of corresponding sides of two similar polygons are in the ratio 5 : 8. If the perimeter of the larger polygon is 10 inches less than twice the perimeter of the smaller polygon, find the perimeter of each polygon.

28. Two similar polygons have areas of 162 square centimeters and 32 square centimeters.
   a. Find the ratio of the areas in simplest form.
   b. Find the ratio of the lengths of two corresponding sides of the polygons.

29. If two similar parallelograms have areas of 720 square inches and 125 square inches, respectively, what is the ratio of two corresponding bases of these parallelograms?

30. The length of a large rectangle is twice the length of a smaller rectangle. The width of the larger rectangle is 3 times the width of the smaller.
   a. What is the ratio of the length of the larger rectangle to the length of the smaller rectangle?
   b. What is the ratio of the width of the larger rectangle to the width of the smaller one?
   c. Are the rectangles similar? Why?
   d. What is the ratio of the area of the larger rectangle to the area of the smaller one?

CHAPTER SUMMARY

A ***ratio***, which is a comparison of two numbers by division, is the quotient obtained when the first number is divided by a second, nonzero number. Quantities in a ratio are expressed in the same unit of measure before finding the quotient.

> ***Ratio of a to b:***
>
> $\frac{a}{b}$ or $a:b$

A ***rate*** is a comparison of two quantities that may have *different* units of measure, such as a rate of speed in miles per hour.

Dimensional analysis is used to change from one unit of measure to another. This is accomplished by multiplying a ratio whose terms are equal measures by another ratio that is equivalent to the identity element 1.

A ***proportion*** is an equation stating that two ratios are equal. Standard ways to write a proportion are shown at the right. The terms a and d are called the ***extremes***, and the terms b and c are the ***means***.

> ***Proportion:***
>
> $\frac{a}{b} = \frac{c}{d}$ or $a:b = c:d$

In a proportion, the product of the means is equal to the product of the extremes.

A ***direct variation*** is a relation between two variables such that their ratio is always the same value, called the ***constant of variation***. For example, the diameter of a circle is always twice the radius, so $\frac{d}{r} = 2$ shows a direct variation between d and r with a constant of variation 2.

A ***percent*** (%), which is a ratio of a number to 100, is also called a ***rate***. Here, the word *rate* is treated as a comparison of a quantity to a whole. In basic formulas, such as those used with discounts and taxes, the ***base*** and ***percentage*** are numbers, and the rate is a percent.

$$\frac{\textit{Percentage}}{\textit{Base}} = \textit{rate}$$

$$\textit{Base} \times \textit{rate} = \textit{percentage}$$

Margin of error describes the maximum difference expected between a measurement and the actual value. A margin of error is sometimes written with a plus-or-minus sign (±). The amount of error can also be expressed as a percentage of the true value.

Similar polygons have the same shape but not the same size. In other words, polygons are ***similar*** (~) if and only if their corresponding angles are congruent and corresponding sides are in proportion. Triangles are similar if and only if two angles of one triangle are equal in measure to two angles in the other triangle (a.a. ≅ a.a.).

Under a ***dilation***, which is a transformation of the plane, the original figure and its image are similar figures.

A ***scale drawing*** is a sketch that is similar to the real object it represents. The lengths of the lines and the radii of the curved portions are proportional to the actual measurements and the angles are congruent to the actual angles.

If two polygons are similar, and the ratio of their corresponding sides is $\frac{a}{b}$, then the ratio of their perimeters is $\frac{a}{b}$ and the ratio of their areas is $\left(\frac{a}{b}\right)^2$.

VOCABULARY

12-1 Ratio Equivalent ratios Simplest form Continued ratio

12-2 Rate Lowest terms

12-3 Dimensional analysis

12-5 Proportion Means Extremes

12-6	Direct variation	Directly proportional	Constant of variation		
12-7	Percent	Base	Rate	Percentage	Percent key [%]
12-8	Margin of error				
12-9	Similar polygons	Similar (is similar to)			
12-10	Similar triangles [a.a. ≅ a.a.]				
12-11	Dilation	Center of dilation	Constant of dilation		
12-12	Scale drawing				
12-13	Ratio of perimeters	Ratio of areas			

REVIEW EXERCISES WITH OPEN-RESPONSE PROBLEMS

In 1–4, express each ratio in simplest form.

1. 30:35 **2.** $8w$ to $12w$ **3.** $\frac{3}{8}$ to $\frac{5}{8}$ **4.** 75 millimeters to 15 centimeters

5. If 10 paper clips weigh 11 grams, what is the weight in grams of 150 paper clips?

6. Thelma can type 150 words in 3 minutes. At this rate, how many words can she type in 10 minutes?

7. What is the ratio of six nickels to four dimes?

In 8–10, in each case solve for x and check.

8. $\frac{8}{2x} = \frac{12}{9}$ **9.** $\frac{x}{x + 5} = \frac{1}{2}$ **10.** $\frac{4}{x} = \frac{6}{x + 3}$

11. The ratio of two numbers is 1:4, and the sum of these numbers is 40. Find the numbers.

In 12–15, in each case, select the numeral preceding the choice that makes the statement true.

12. In a class of 9 boys and 12 girls, the ratio of the number of girls to the number of students in the class is
(1) 3:4 (2) 4:3 (3) 4:7 (4) 7:4

13. The perimeter of a triangle is 45 centimeters, and the lengths of its sides are in the ratio 2:3:4. The length of the longest side is
(1) 5 cm (2) 10 cm (3) 20 cm (4) 30 cm

14. If $a:x = b:c$, then x equals
(1) $\frac{ac}{b}$ (2) $\frac{bc}{a}$ (3) $ac - b$ (4) $bc - a$

15. If snow is falling at the rate of 1.5 inches per hour, how many inches of snow will fall in x minutes?

(1) $1.5x$ (2) $\frac{40}{x}$ (3) $\frac{60}{x}$ (4) $\frac{x}{40}$

16. Seven percent of what number is 21?

17. What percent of 36 is 45?

18. The sales tax collected on each sale varies directly as the amount of the sale. What is the constant of variation if a sales tax of $0.63 is collected on a sale of $9.00?

19. When rounded to the nearest inch, the dimensions of a rectangular rug are 102 inches and 138 inches. What is the range for the area of the rug? What is the margin of error for the area?

20. In the diagram, $\triangle DEF \sim \triangle GHJ$. The measures in inches are $DE = 12$, $DF = 10$, $EF = 15$, and $GH = 9$.

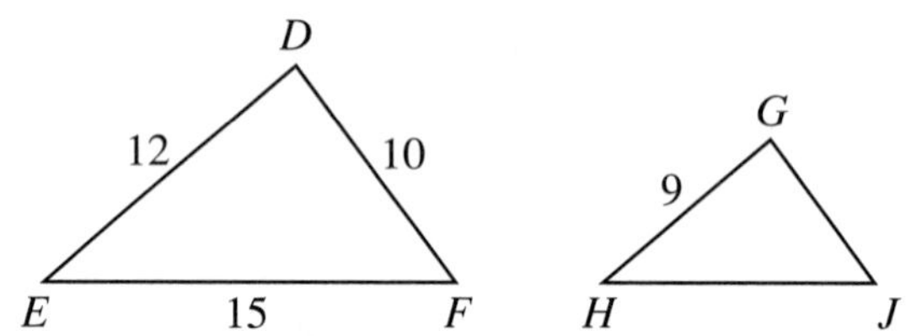

a. Find HJ and GJ.
b. What is the ratio of the lengths of the corresponding sides?
c. What is the ratio of the perimeters?
d. What is the ratio of the areas?

21. On a stormy February day, 28% of the students enrolled at Southside High School were absent. How many students are enrolled at Southside High School if 476 students were absent?

22. After a 5-inch-by-7-inch photograph is enlarged, its shorter side measures 15 inches. Find the length in inches of its longer side.

23. In the diagram, the two triangles are similar.

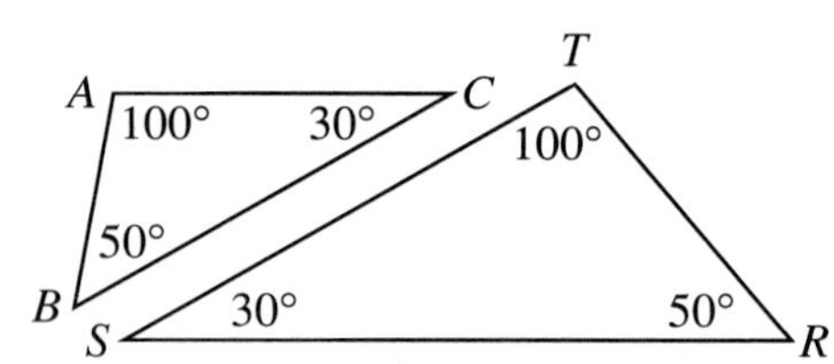

a. Name three pairs of corresponding angles.
b. Name three pairs of corresponding sides.
c. If $AB = 4$, $AC = 6$, and $ST = 9$, find RT.

24. The sides of a triangle measure 18, 20, and 24. If the shortest side of a similar triangle measures 12, find the length of its longest side.

25. In the diagram, $\triangle ABC \sim \triangle A'B'C'$. Find:
a. y **b.** z **c.** AC
d. $A'C'$ **e.** AB **f.** $A'B'$

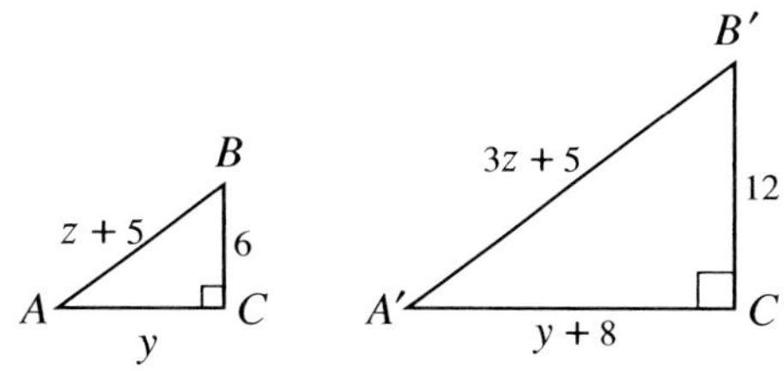

26. A student who is 5 feet tall casts an 8-foot shadow. At the same time, a tree casts a 40-foot shadow. How many feet tall is the tree?

27. In $\triangle ABC$, $m\angle A = 70$ and $m\angle B = 45$. In $\triangle DEF$, $m\angle D = 65$ and $m\angle E = 70$.
a. Is $\triangle ABC$ similar to $\triangle EFD$? Why?
b. If $AB = 5$, $DE = 12$, and $EF = 20$, find the length of $\overline{AC}$.

28. In the diagram, $\angle Y$ and $\angle ATR$ are right angles. The measures in inches are $YT = 4$, $AT = 3$, and $TR = 6$.

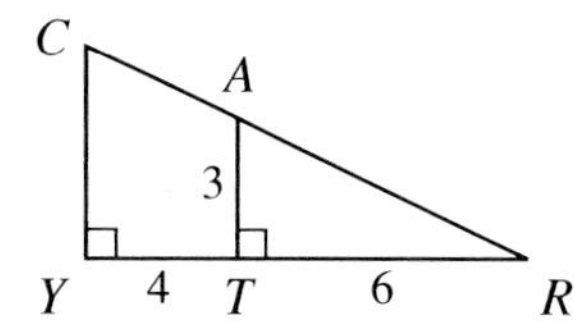

a. State a reason why $\triangle CRY \sim \triangle ART$.
b. Find CY.
c. Find the area of: (*1*) $\triangle ART$ (*2*) $\triangle CRY$ (*3*) trapezoid $CATY$
d. The diagram shows a dilation where R is the center of dilation, the image of T is Y, and the image of A is C. What is the constant of dilation?

29. A model of a cylindrical tank is made with a scale of 3 centimeters = 1.5 meters. The diameter of the model is 8 centimeters and the height is 12 centimeters.

a. Find the diameter and the height of the actual tank.
b. What is the ratio of the volume of the model to the volume of the real tank?

30. The ratio of the areas of two similar polygons is 49 : 4. What is the ratio of the lengths of two corresponding sides of these polygons?

31. If four carpenters can build four tables in 4 days, how long will it take one carpenter to build one table?

32. How many girls would have to leave a room in which there are 99 girls and 1 boy in order that 98% of the remaining persons would be girls?

CUMULATIVE REVIEW WITH OPEN-RESPONSE PROBLEMS

1. Find two consecutive integers whose squares differ by 25.

2. Solve and check: $4(2x - 1) = 5x + 5$

3. Parallelogram *ABCD* is separated into three rhombuses as shown in the diagram at the right. The length of $\overline{DA}$ is 20 centimeters.

a. Find the measure of $\overline{AB}$.

b. Find the ratio of the area of rhombus *AEFD* to the area of parallelogram *ABCD*.

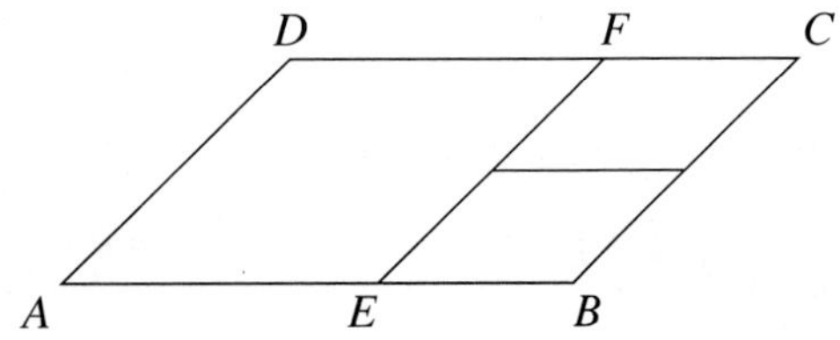

Exploration

Make a scale diagram of a room or an area of your home, your school, or some public building. Include the furnishings to scale. Or copy the diagram of a house from a newspaper or magazine. Draw an enlarged diagram and add furnishing to scale.

Chapter 13

Probability

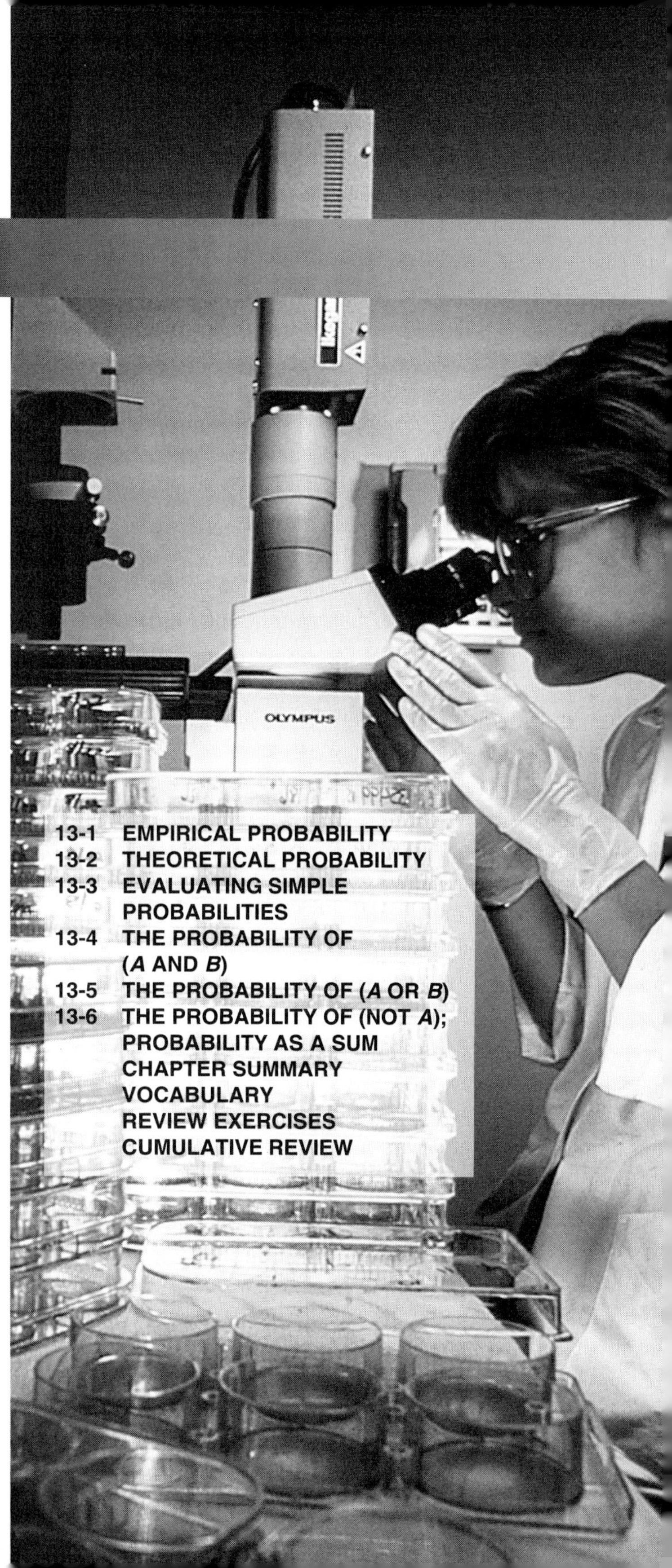

Mathematicians first studied probability by looking at situations involving games. Today, probability is used in a wide variety of fields. In medicine, it helps us to determine the chances of catching an infection or of controlling an epidemic, and the likelihood that a drug will be effective in curing a disease. In industry, probability tells us how long a manufactured product should last.

We can predict when more tellers are needed at bank windows, when and where traffic jams are likely to occur, and what kind of weather we may expect for the next few days. In biology, the study of genes inherited from one's parents and grandparents is a direct application of probability. While the list of applications is almost endless, all of them demand a strong knowledge of higher mathematics.

Probability is a branch of mathematics in which the chance of an event happening is assigned a numerical value that predicts how likely that event is to occur. Although this prediction tells us little about what may happen in an individual case, it can provide valuable information about what to expect in a large number of cases. For example, probability is used to set insurance rates.

Like the early mathematicians, we will begin a formal study of probability by looking at games and other rather simple applications.

13-1 EMPIRICAL PROBABILITY

A decision is sometimes reached by the toss of a coin: "Heads, we'll go to the movies; tails, we'll go bowling." When we toss a coin, we don't know whether the coin will land with the head facing upward or with the tail up. However, we believe that heads and tails have *equal chances* of happening whenever we toss a fair coin. We can describe this situation by saying that the ***probability*** of heads is $\frac{1}{2}$ and the ***probability*** of tails is $\frac{1}{2}$, symbolized as:

$$P(\text{heads}) = \frac{1}{2} \text{ or } P(\text{H}) = \frac{1}{2} \qquad P(\text{tails}) = \frac{1}{2} \text{ or } P(\text{T}) = \frac{1}{2}$$

Before we define probability, let us consider two more situations.

1. Suppose we toss a coin and it lands head up. If we were to toss the coin a second time, would the coin land tail up?

Is your answer "I don't know"? Good! We cannot say that the coin must now be tails because *we cannot predict the next result with certainty.*

2. Suppose we take an index card, and fold it down the center. If we then toss the card and let is fall, there are only three possible results. The card may land on its side, it may land on its edge, or it may form a tent when it lands. Can we say $P(\text{edge}) = \frac{1}{3}$, $P(\text{side}) = \frac{1}{3}$, and $P(\text{tent}) = \frac{1}{3}$?

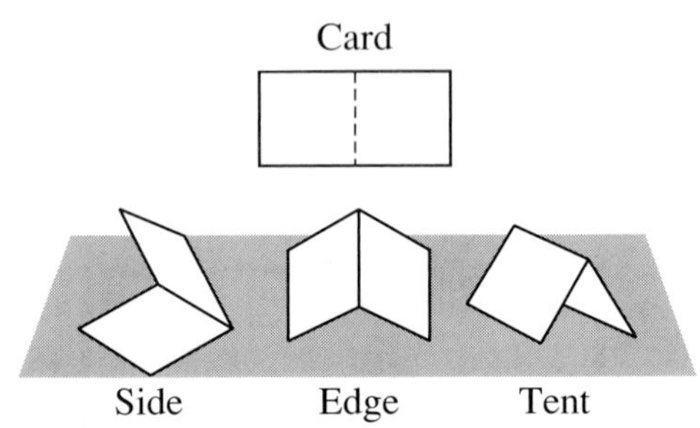

Again, your answer should be "I don't know." *We cannot assign a number as a probability until we have some evidence to support our claim.* In fact, if we were to gather evidence by tossing this card, we would find that the probabilities are *not* $\frac{1}{3}$, $\frac{1}{3}$, and $\frac{1}{3}$.

An Empirical Study

Let us go back to the problem of tossing a coin. While we cannot predict the result of one toss of a coin, we can still say that the probability of heads is $\frac{1}{2}$, based on observations made in an empirical study. In an ***empirical study***, we perform an experiment many times, keep records of the results, and then analyze these results.

For example, ten students decided to take turns tossing a coin. Each student completed 20 tosses and the number of heads was recorded as shown on the next page.

If we look at the results and try to think of the probability of heads as a fraction comparing the number of heads to the total number of tosses, only Maria, with 10 heads out of 20 tosses, had results where the probability was $\frac{10}{20}$, or $\frac{1}{2}$. This fraction is called the ***relative frequency***. Elizabeth had the lowest relative frequency of heads, $\frac{6}{20}$. Peter and Debbie tied for the highest relative frequency with $\frac{13}{20}$. This table does *not* mean that Maria had correct results while the other students were incorrect; the coins simply fell that way.

The students decided to combine their results, by expanding the chart, to see what happened when all 200 tosses of the coin were considered. As shown in columns 3 and 4 of the table below, the ***cumulative*** results are found by adding all the results up to a specific point. For example, in the second row, by adding the 8 heads that Albert tossed and the 13 heads that Peter tossed, we find that the cumulative number of heads is 21, the total number of heads tossed by Albert and Peter together. Similarly, when the 20 tosses that Albert made and the 20 tosses that Peter made are added, the cumulative number of tosses is 40.

	Number of Heads	Number of Tosses
Albert	8	20
Peter	13	20
Thomas	12	20
Maria	10	20
Elizabeth	6	20
Joanna	12	20
Kathy	11	20
Jeanne	7	20
Debbie	13	20
James	9	20

	(Col. 1)	(Col. 2)	(Col. 3)	(Col. 4)	(Col. 5)
	Number of Heads	**Number of Tosses**	**Cumulative Number of Heads**	**Cumulative Number of Tosses**	**Cumulative Relative Frequency**
Albert	8	20	8	20	8/20 = 0.400
Peter	13	20	21	40	21/40 = 0.525
Thomas	12	20	33	60	33/60 = 0.550
Maria	10	20	43	80	43/80 = 0.538
Elizabeth	6	20	49	100	49/100 = 0.490
Joanna	12	20	61	120	61/120 = 0.508
Kathy	11	20	72	140	72/140 = 0.514
Jeanne	7	20	79	160	79/160 = 0.494
Debbie	13	20	92	180	92/180 = 0.511
James	9	20	101	200	101/200 = 0.505

In column 5, the ***cumulative relative frequency*** is found by dividing the total number of heads at or above a row by the total number of tosses at or above that row. The cumulative relative frequency is shown as a fraction and then, for easy comparison, as a decimal. The decimal is given to the *nearest thousandth.*

While the relative frequency for individual students varied greatly, from $\frac{6}{20}$ for Elizabeth to $\frac{13}{20}$ for Peter and Debbie, the cumulative relative frequency was $\frac{101}{200}$, a number very close to $\frac{1}{2}$.

A ***graph*** of the results of columns 4 and 5 will tell us even more. In the graph, the horizontal axis is labeled "Number of tosses" to show the cumulative results of column 4; the vertical axis is labeled "Cumulative relative frequency of heads" to show the results of column 5.

CUMULATIVE RELATIVE FREQUENCY OF HEADS FOR 0 TO 200 TOSSES OF A COIN

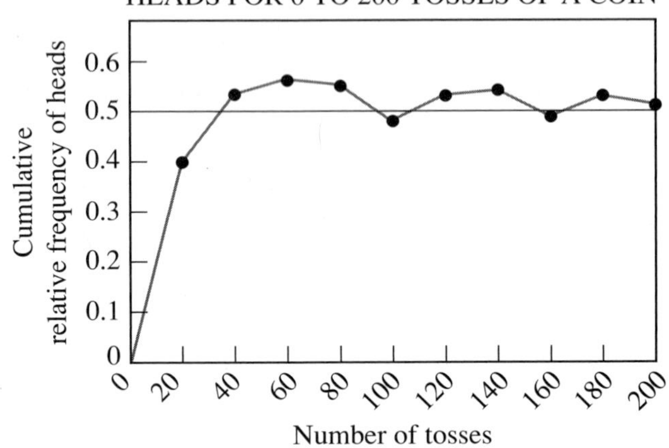

In the graph, we have plotted the points that represent the data in columns 4 and 5 of the preceding table, and we have connected these points to form a line graph. Notice how the line moves up and down around the relative frequency of 0.5, or $\frac{1}{2}$. The graph shows that the more times the coin is tossed, the closer the relative frequency comes to $\frac{1}{2}$. In other words, the line seems to *level out* at a relative frequency of $\frac{1}{2}$. We say that the cumulative relative frequency ***converges*** to the number $\frac{1}{2}$ and the coin will land heads up about one-half of the time.

Even though the cumulative relative frequency of $\frac{101}{200}$ is not exactly $\frac{1}{2}$, we sense that the line will approach the number $\frac{1}{2}$. When we use carefully collected evidence about tossing a fair coin to guess that the probability of heads is $\frac{1}{2}$, we have arrived at this conclusion *empirically*, that is, by experimentation and observation.

- **Empirical probability may be defined as the most accurate scientific estimate, based on a large number of trials, of the cumulative relative frequency of an event happening.**

Experiments in Probability

A single attempt at doing something, such as tossing a coin only once, is called a ***trial***. We perform ***experiments*** in probability by repeating the same trial many times. Experiments are aimed at finding the probabilities to be assigned to the occurrences of different events, such as heads or tails coming up on a coin. The objects used in an experiment may be classified into one of two categories:

1. ***Fair and unbiased objects*** have not been weighted or made unbalanced. An object is fair when the different possible results have *equal chances* of happening. Objects such as coins, cards, and spinners will always be treated in this book as fair objects, unless otherwise noted.
2. ***Biased objects*** have been tampered with or are weighted to give one result a *better chance* of happening than another. The folded index card described earlier in this section is a biased object because the probability of each of three results is not $\frac{1}{3}$. The card is weighted so that it will fall on its side more often than it will fall on its edge.

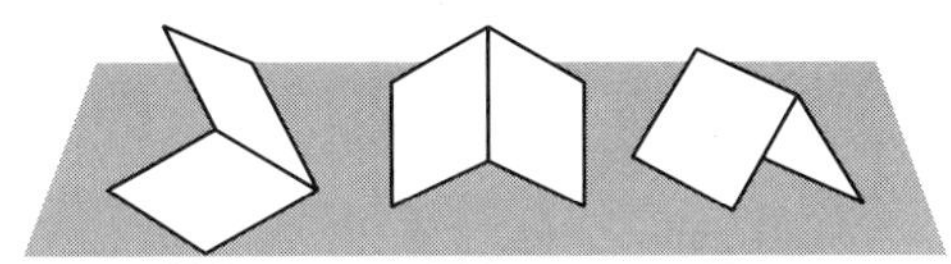

Side Edge Tent

EXAMPLES

You have seen how to determine empirical probability by the tables and graph previously shown in this section. Sometimes, however, it is possible to guess the probability that should be assigned to the result described before you start an experiment.

In 1–5, use common sense to guess the probability that might be assigned to each result. (The answers are given without comment here. You will learn how to determine these probabilities in the next section.)

1. A ***die*** is a six-sided solid object. Each side (or face) is a square. The sides are numbered 1, 2, 3, 4, 5, and 6. (The plural of *die* is *dice*.)

In rolling a fair die, find the probability of getting a 4, or $P(4)$.

Die

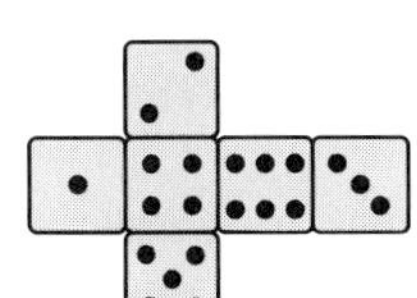

Faces of a die

Answer: $P(4) = \frac{1}{6}$

2. A ***standard deck of cards*** contains 52 cards. There are four suits: hearts, diamonds, spades, and clubs. Each suit contains 13 cards: 2, 3, 4, 5, 6, 7, 8, 9, 10, jack, queen, king, and ace. The diamonds and hearts are red: the spades and clubs are black.

In selecting a card from the deck without looking, find the probability of drawing: **a.** the 7 of diamonds **b.** a 7 **c.** a diamond

Answers: **a.** $P(7 \text{ of diamonds}) = \frac{1}{52}$

b. $P(7) = \frac{4}{52}$, or $\frac{1}{13}$

c. $P(\text{diamond}) = \frac{13}{52}$, or $\frac{1}{4}$

---

3. There are 10 ***digits*** in our numeral system: 0, 1, 2, 3, 4, 5, 6, 7, 8, and 9. In selecting a digit without looking, what is the probability it will be: **a.** the 8? **b.** an odd digit?

Answers: **a.** $P(8) = \frac{1}{10}$ **b.** $P(\text{odd}) = \frac{5}{10}$, or $\frac{1}{2}$

4. A jar contains eight marbles: three are white and the remaining five are blue. In selecting a marble without looking, what is the probability it will be blue? (All marbles are the same size.)

Answer: $P(\text{blue}) = \frac{5}{8}$

---

5. The English alphabet contains 26 letters. There are five ***vowels*** (A, E, I, O, and U); the other 21 letters are ***consonants***. If a person turns 26 tiles from a word game face down and each tile represents a different letter of the alphabet, what is the probability of turning over: **a.** the A? **b.** a vowel? **c.** a consonant?

Answers: **a.** $P(A) = \frac{1}{26}$ **b.** $P(\text{vowel}) = \frac{5}{26}$ **c.** $P(\text{consonant}) = \frac{21}{26}$

EXERCISES WITH OPEN-RESPONSE PROBLEMS

1. The figure at the right shows a disk, with an arrow that can be spun so that it has an equal chance of landing on one of four regions. The regions are equal in size and are numbered 1, 2, 3, and 4.

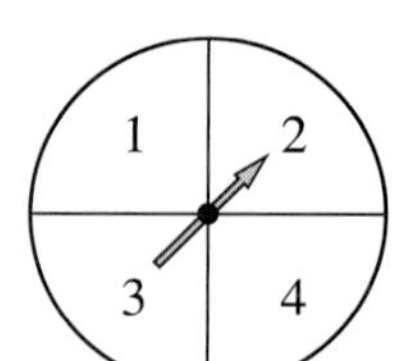

An experiment was conducted by five people to find the probability that the arrow will land on the 2. Each person spun the arrow 100 times. When the arrow landed on a line, the result did not count and the arrow was spun again.

a. Before doing the experiment, what probability would you assign to the arrow landing on the 2? (In symbols, $P(2) = \underline{\ ?\ }$.)

b. Copy and complete the table below to find the *cumulative* results of this experiment. In the last column, record the cumulative relative frequencies as fractions and as decimals to the *nearest thousandth*.

	Number of Times Arrow Landed on 2	**Number of Spins**	**Cumulative Number of Times Arrow Landed on 2**	**Cumulative Number of Spins**	**Cumulative Relative Frequency**
Barbara	29	100	29	100	$\frac{29}{100} = .290$
Tom	31	100	60	200	
Ann	19	100			
Eddie	23	100			
Cathy	24	100			

c. Did the experiment provide evidence that the probability you assigned in part **a** was correct?

In 2–7, in each case a *fair, unbiased object* is involved. These questions should be answered *without* conducting an experiment; take a guess.

2. A six-sided die is rolled; the sides are numbered 1, 2, 3, 4, 5, and 6. In rolling the die, find $P(5)$.

3. In drawing a card from a standard deck without looking, find $P(\text{any heart})$.

4. Each of 10 pieces of paper contains a different number from {0, 1, 2, 3, 4, 5, 6, 7, 8, 9}. The pieces of paper are folded and placed in a paper bag. In selecting a piece of paper without looking, find $P(7)$.

5. A jar contains five marbles, all the same size. Two marbles are black, and the other three are white. In selecting a marble without looking, find $P(\text{black})$.

6. Using the lettered tiles from a word game, a boy places 26 tiles, one tile for each letter of the alphabet, face down on a table. After mixing up the tiles, he picks one. What is the probability that the tile contains one of the letters in the word MATH?

7. A ***tetrahedron*** is a four-sided object. Each side, or face, is an equilateral triangle. The numerals 1, 2, 3, and 4 are used to number the different faces, as shown in the figure. A trial consists of rolling the tetrahedron and reading the number that is face down. Find $P(4)$.

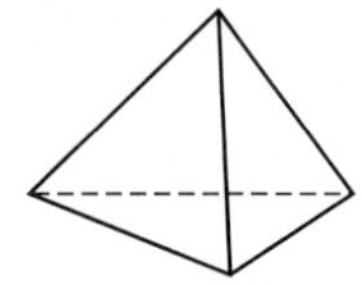

Tetrahedron

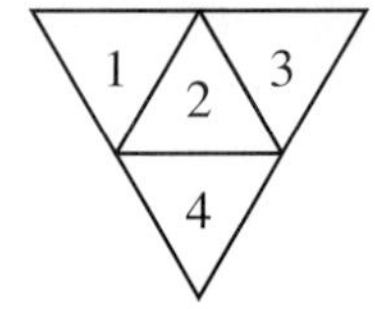

Faces of a tetrahedron

8. By yourself, or with some classmates, conduct any of the experiments described in Exercises 2–7 to verify that you have assigned the correct probability to the event or result described. A good experiment should contain at least 100 trials.

In 9–14, a *biased object* is described. A probability can be assigned to a result only by conducting an experiment to determine the cumulative relative frequency of the event. While you may wish to guess at the probability of the event before starting the experiment, conduct at least 100 trials to determine the best probability to be assigned.

9. An index card is folded in half and tossed. As described earlier in this section, the card may land in one of three positions: on its side, on its edge, or in the form of a tent. In tossing the folded card, find $P(\text{tent})$, the probability that the card will form a tent when it lands.

10. A paper cup is tossed. It can land in one of three positions: on its top, on its bottom, or on its side, as shown in the figure. In tossing the cup, find $P(\text{top})$, the probability that the cup will land on its top.

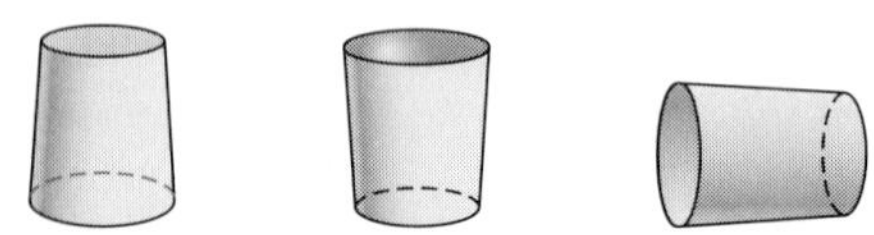

Top Bottom Side

11. A nickel and a quarter are glued or taped together so that the two faces seen are the head of the quarter and the tail of the nickel. (This is a very crude model of a weighted coin.) In tossing the coin, find $P(\text{head})$.

12. A paper cup in the shape of a cone is tossed. It can land in one of two positions: on its base or on its side, as shown in the figure. In tossing this cup, find $P(\text{side})$, the probability that the cup will land on its side.

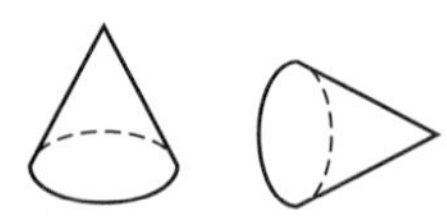

Base Side

13. A thumbtack is tossed. It may land either with the pin up or with the pin down, as shown in the figure. In tossing the thumbtack, find $P(\text{pin up})$.

Pin up Pin down

14. The first word on a page of a book is selected.
    a. What is the probability that the word contains at least one of the letters *a*, *e*, *i*, *o*, *u*, or *y*?
    b. What is the largest possible probability?
    c. What is the probability that the word does not contain at least one of the letters *a*, *e*, *i*, *o*, *u*, or *y*?
    d. What is the smallest possible probability?

13-2 THEORETICAL PROBABILITY

An empirical approach to probability is necessary whenever we deal with biased objects. However, common sense tells us that there is a simple way to define the probability of an event when we deal with fair, unbiased objects.

For example, let us suppose that Alma is playing a game in which each player must roll a die. To win, Alma must roll a number greater than 4. What is the probability that Alma will win on her next turn? Common sense tells us that:

1. The die has an equal chance of falling in any one of *six* ways:
   1, 2, 3, 4, 5, and 6.
2. There are *two* ways for Alma to win: rolling a 5 or a 6.
3. Therefore: $P(\text{Alma wins}) = \dfrac{\text{number of winning results}}{\text{number of possible results}} = \dfrac{2}{6} = \dfrac{1}{3}$

Terms and Definitions

Using the details of the preceding example, let us examine the correct terminology to be used.

An ***outcome*** is a result of some activity or experiment. In rolling a die, 1 is an outcome, 2 is an outcome, 3 is an outcome, and so on. There are six outcomes here.

A ***sample space*** is a set of all possible outcomes for the activity. When rolling a die, there are six possible outcomes in the sample space: 1, 2, 3, 4, 5, and 6. We also say that the ***outcome set*** is {1, 2, 3, 4, 5, 6,}.

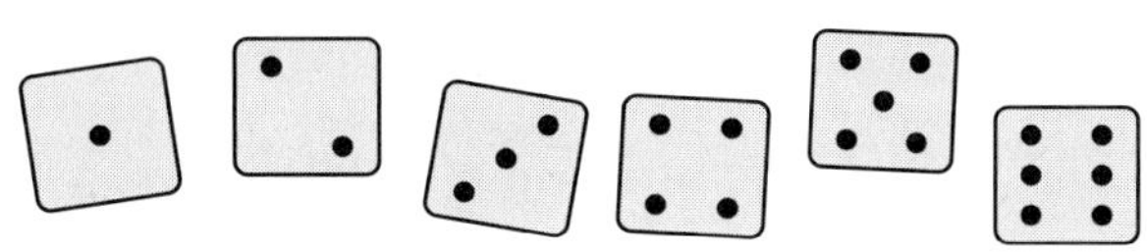

An ***event*** is a subset of the sample space. We use the term *event* in two ways. In ordinary conversation, it means a situation or happening. In the technical language of probability, it is the subset of the sample space that lists all of the outcomes for a given situation.

When we roll a die, we may define many different situations. Each of these is called an event.

1. For Alma, the event of rolling a number *greater than 4* contains only two outcomes: 5 and 6.
2. For Lee, a different event might be rolling a number *less than 5*. This event contains four outcomes: 1, 2, 3, and 4.

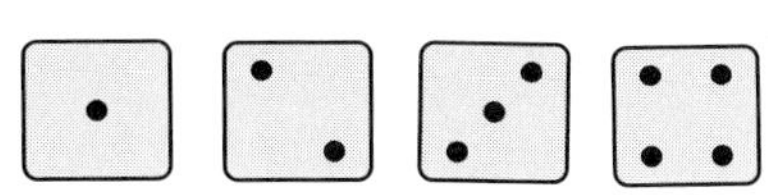

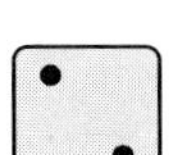

3. For Sandy, the event of rolling *a 2* contains only one outcome: 2. When there is only one outcome, we call this a ***singleton event***.

We can now define ***theoretical probability*** for fair, unbiased objects:

- **The theoretical probability of an event is the number of ways that the event can occur, divided by the total number of possibilities in the sample space.**

In symbolic form, we write:

$P(E) = \frac{n(E)}{n(S)}$, where	**$P(E)$** represents the probability of event E; **$n(E)$** represents the number of ways that event E can occur; **$n(S)$** represents the total number of possibilities, or the total number of possible outcomes in sample space S.

Let us now use this formula to write the probabilities of the three events described above.

1. For Alma, there are two ways to roll a number *greater than 4*, and there are six possible ways that the die may fall. We say:

E = the set of numbers greater than 4 = $\{5, 6\}$.

Since there are two outcomes in this event, $\boldsymbol{n(E) = 2}$.

S = the set of all possible outcomes = $\{1, 2, 3, 4, 5, 6\}$;

Since there are six outcomes in the sample space, $\boldsymbol{n(S) = 6}$.

Therefore:

$$P(E) = \frac{n(E)}{n(S)} = \frac{\text{number of ways to roll a number greater than 4}}{\text{total number of outcomes for rolling the die}} = \frac{2}{6}, \text{ or } \frac{1}{3}$$

2. For Lee, the probability of rolling a number *less than 5* is

$$P(E) = \frac{n(E)}{n(S)} = \frac{4}{6}, \text{ or } \frac{2}{3}$$

3. For Sandy, the probability of rolling *a 2* is

$$P(E) = \frac{n(E)}{n(S)} = \frac{1}{6}$$

Uniform Probability

A sample space is said to have ***uniform probability***, or to contain ***equally likely outcomes***, when each of the possible outcomes has an *equal chance* of occurring. In rolling a die, there are six possible outcomes in the sample space; each is equally likely to occur. Therefore:

$$P(1) = \frac{1}{6};\quad P(2) = \frac{1}{6};\quad P(3) = \frac{1}{6};\quad P(4) = \frac{1}{6};\quad P(5) = \frac{1}{6};\quad P(6) = \frac{1}{6}$$

We say that the die has uniform probability.

If, however, a die is weighted to make it biased, then one or more sides will have a probability greater than $\frac{1}{6}$, while one or more sides will have a probability less than $\frac{1}{6}$. A weighted die does not have uniform probability; the rule for theoretical probability does *not* apply to weighted objects.

Random Selection

When we select an object without looking, we are making a ***random selection***. Random selections are made when drawing a marble from a bag, when taking a card from a deck, or when picking a name out of a hat. In the same way, we may use the word *random* to describe the outcomes when tossing a coin or rolling a die; the outcomes happen without any special selection on our part.

PROCEDURE FOR FINDING THE SIMPLE PROBABILITY OF AN EVENT:

1. Count the total number of outcomes in the sample space: $n(S)$.
2. Count all the possible outcomes of the event E: $n(E)$.
3. Substitute these values in the formula for the probability of event E:

$$P(E) = \frac{n(E)}{n(S)}$$

Note: The probability of an event is usually written *as a fraction*. A standard calculator display, however, is in decimal form. Therefore, if you use a calculator when working with probability, it is important that you know fraction-decimal equivalents.

Recall that some common fractions have equivalent decimals that are terminating decimals.

$$\frac{1}{2} = 0.5 \qquad \frac{1}{4} = 0.25 \qquad \frac{1}{5} = 0.2 \qquad \frac{1}{8} = 0.125$$

Others have equivalent decimals that are repeating decimals.

$$\frac{1}{3} = 0.333\ldots \qquad \frac{2}{3} = 0.666\ldots \qquad \frac{1}{6} = 0.166\ldots \qquad \frac{1}{9} = 0.111\ldots$$

EXAMPLES

1. A standard deck of 52 cards is shuffled. Lillian draws a single card from the deck at random. What is the probability that the card is a jack?

Solution S = sample space of all possible outcomes, or 52 cards. Thus, $n(S) = 52$.

J = event of selecting a jack. There are four jacks in the deck: jack of hearts, of diamonds, of spades, and of clubs. Thus, $n(J) = 4$.

$$P(J) = \frac{n(J)}{n(S)} = \frac{\text{number of possible jacks}}{\text{total number of possible cards}} = \frac{4}{52}, \text{ or } \frac{1}{13} \quad \textit{Answer}$$

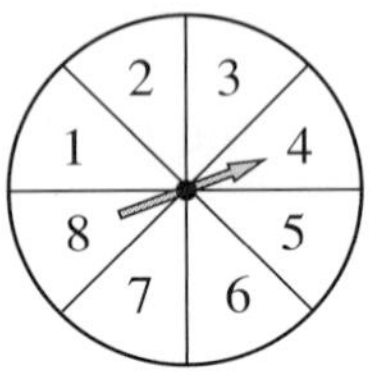

2. A spinner contains eight regions, numbered 1 through 8, as shown in the figure. The arrow has an equally likely chance of landing on any of the eight regions. If the arrow lands on a line, the result is not counted and the arrow is spun again.

Solutions and Answers:

a. How many possible outcomes are in the sample space S?

a. $n(S) = 8$

b. What is the probability that the arrow lands on the 4? Simply, what is $P(4)$?

b. Since there is only one way to land on the 4 out of eight possible numbers: $P(4) = \frac{1}{8}$.

c. List the set of possible outcomes for event O, in which the arrow lands on an odd number.

c. Event $O = \{1, 3, 5, 7\}$

d. Find the probability that the arrow lands on an odd number.

d. Since event $O = \{1, 3, 5, 7\}$, then $n(O) = 4$.

$$P(O) = \frac{n(O)}{n(S)} = \frac{4}{8}, \text{ or } \frac{1}{2}$$

EXERCISES WITH OPEN-RESPONSE PROBLEMS

1. A fair coin is tossed.

a. List the sample space.

b. What is $P(\text{head})$, the probability that a head will appear?

c. What is $P(\text{tail})$?

In 2–7, a fair die is tossed. For each question: **a.** List the possible outcomes for the event. **b.** State the probability of the event.

2. The number 3 appears.

3. An even number appears.

4. A number less than 3 appears.

5. An odd number appears.

6. A number greater than 3 appears.

7. A number greater than or equal to 3 appears.

In 8–13, a spinner is divided into five equal regions, numbered 1 through 5, as shown in the figure. An arrow is spun and lands in one of the regions. For each question: **a.** List the outcomes for the event. **b.** State the probability of the event.

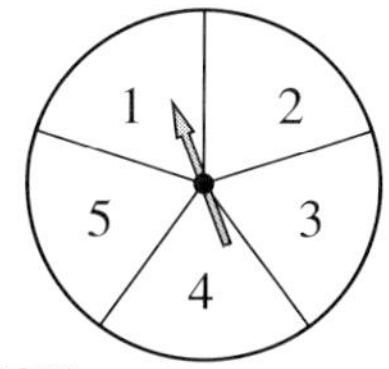

8. The number 3 appears.

9. An even number appears.

10. A number less than 3 appears.

11. An odd number appears.

12. A number greater than 3 appears.

13. A number greater than or equal to 3 appears.

14. A standard deck of 52 cards is shuffled, and one card is drawn. What is the probability that the card is:
a. the queen of hearts? **b.** a queen? **c.** a heart?
d. a red card? **e.** the 7 of clubs? **f.** a club?
g. an ace? **h.** a red 7? **i.** a black 10?
j. a picture card (king, queen, jack)?

15. A person does not know the answer to a test question and takes a guess. Find the probability that the answer is correct if the question is a: **a.** multiple-choice question with four choices. **b.** true-false question **c.** question where the choices given are "sometimes, always, or never."

16. A marble is drawn at random from a bag. Find the probability that the marble is green if the bag contains marbles whose colors are:
a. 3 blue, 2 green **b.** 4 blue, 1 green **c.** 5 red, 2 green, 3 blue
d. 6 blue, 4 green **e.** 3 green, 9 blue **f.** 5 red, 2 green, 9 blue

17. The digits of the number 1776 are written on disks and placed in a jar. What is the probability that the digit 7 will be chosen on a single random draw?

18. A class contains 16 boys and 14 girls. The teacher calls students at random to the chalkboard. What is the probability that the first person called is: **a.** a boy? **b.** a girl?

19. There are 840 tickets sold in a raffle. Jay bought five tickets, and Lynn bought four tickets. What is the probability that: **a.** Jay has the winning ticket? **b.** Lynn has the winning ticket?

20. A letter is chosen at random from a given word. Find the probability that the letter is a vowel if the word is: **a.** APPLE **b.** BANANA **c.** GEOMETRY **d.** MATHEMATICS

21. Explain why each of the following statements is *incorrect*.

a. Since there are 50 states, the probability that a baby born in the United States will be born in New Jersey is $\frac{1}{50}$.

b. Since there are 12 months in a year, the probability of being born in September is $\frac{1}{12}$.

c. A pin with a small head is tossed. Since the pin can fall either point up or point down, as shown at the right, the probability of falling point down is $\frac{1}{2}$.

d. Since there are 7 days in a week, the probability that a person attends religious services on a Wednesday is $\frac{1}{7}$.

22. The figures that follow are eight polygons: a square; a rectangle; a parallelogram (which is not a rectangle); a right triangle; an isosceles triangle (not containing a right angle); a trapezoid (not containing a right angle); an equilateral triangle; a regular hexagon.

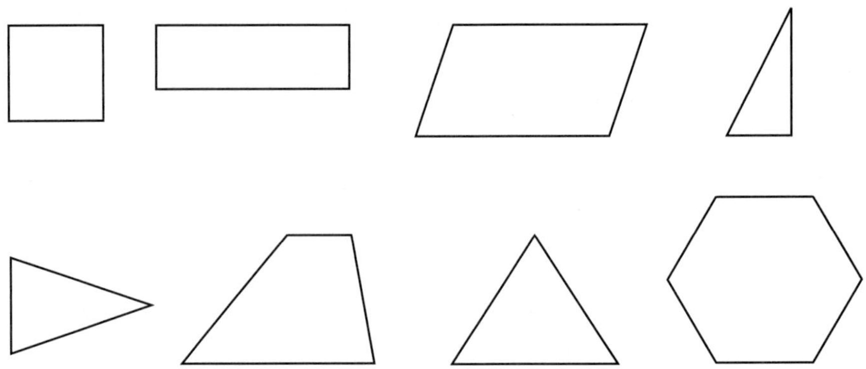

One of the figures is selected at random. What is the probability that this polygon:

a. contains a right angle?
b. is a quadrilateral?
c. is a triangle?
d. has at least one acute angle?
e. has all sides congruent?
f. has at least two sides congruent?
g. has less than five sides?
h. has an odd number of sides?
i. has four or more sides?
j. has at least two obtuse angles?

13-3 EVALUATING SIMPLE PROBABILITIES

We have called an event for which there is only one outcome a *singleton*. For example, when rolling a die, there is only one way to roll the singleton 3.

However, when rolling a die only once, more than one outcome is possible for some events to occur. For example:

1. The event of rolling an even number on a die = {2, 4, 6}.
2. The event of rolling a number less than 6 on a die = {1, 2, 3, 4, 5}.

The Impossible Case

On a single roll of a die, what is the probability that the number 7 will appear? We call this case an ***impossibility*** because there is no way in which this event can occur. In this example, event E = rolling a 7; $E = \{\ \}$ and $n(E) = 0$. The sample space S for rolling a die contains six possible outcomes, and $n(S) = 6$. Therefore:

$$P(E) = \frac{n(E)}{n(S)} = \frac{\text{number of ways to roll a 7}}{\text{total number of outcomes for the die}} = \frac{0}{6} = 0$$

In general, for any sample space S containing k possible outcomes, we say $n(S) = k$. For any impossible event E which cannot occur in any way, we say $n(E) = 0$. Thus, the probability of an impossible event is

$$\mathbf{P(E) = \frac{n(E)}{n(S)} = \frac{0}{k} = 0}$$

and we say:

- **The probability of an impossible event is 0.**

There are many other impossibilities where the probability must equal zero. For example, the probability of selecting the letter E from the word PROBABILITY is 0. Also, selecting a coin worth 9¢ from a bank containing a nickel, a dime, and a quarter is an impossible event.

The Certain Case

On a single roll of a die, what is the probability that a number less than 7 will appear? We call this case a ***certainty*** because every one of the possible outcomes in the sample space is also an outcome for this event. In this example, the event E = rolling a number less than 7, and $n(E) = 6$. The sample space S for rolling a die contains six possible outcomes, and $n(S) = 6$. Therefore:

$$P(E) = \frac{n(E)}{n(S)} = \frac{\text{number of ways to roll a number less than 7}}{\text{total number of outcomes for the die}} = \frac{6}{6} = 1$$

To describe an event E that is certain, we may use the sample space S itself, that is, $E = S$.

In general, for any sample space S containing k possible outcomes, we say $n(S) = k$. When the event E is certain, every possible outcome for the sample space is also an outcome for event E, or $n(E) = k$. Thus, the probability of a certainty is given as

$$\mathbf{P(E) = \frac{n(E)}{n(S)} = \frac{k}{k} = 1}$$

and we say:

- **The probability of an event that is certain to occur is 1.**

There are many other certainties where the probability must equal 1. Examples include the probability of selecting a consonant from the letters JFK or selecting a red sweater from a drawer containing only red sweaters.

The Probability of Any Event

The smallest possible probability is 0, for an impossible case; no probability can be less than 0. The largest possible probability is 1, for a certainty event; no probability can be greater than 1. Many other events, as seen earlier, however, have probabilities that fall between 0 and 1. Therefore, we conclude:

- **The probability of any event E must be equal to or greater than 0, and less than or equal to 1:**

$$0 \leq P(E) \leq 1$$

Subscripts in Sample Spaces

A sample space may sometimes contain two or more objects that are exactly alike. To distinguish one object from another, we make use of a device called a *subscript*. A ***subscript*** is a number, usually written in smaller size to the lower right of a term.

For example, a box contains six jellybeans: two red, three green, and one yellow. Using R, G, and Y to represent the colors red, green, and yellow, respectively, we can list this sample space in full, using subscripts:

$$\{R_1, R_2, G_1, G_2, G_3, Y_1\}$$

Since there is only one yellow jellybean, we could have listed the last outcome as Y instead of Y_1.

EXAMPLES

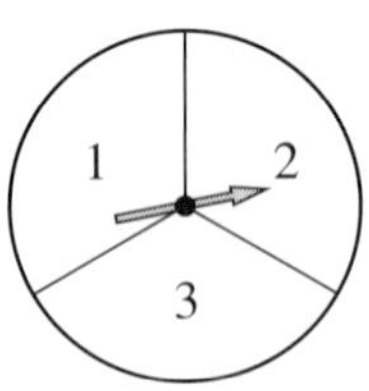

1. An arrow is spun once and lands on one of three equally likely regions, numbered 1, 2, and 3, as shown in the figure.

a. List the sample space for this experiment.

b. List all eight possible events for one spin of the arrow.

Solutions **a.** The sample space $S = \{1, 2, 3\}$. *Answer*

b. Since events are subsets of the sample space S, the eight possible events are the eight subsets of S:

$\{\ \}$ = the empty set, for *impossible* events.

$\{1, 2, 3\}$ = the set itself, for events with *certainty*.

$\{1\}, \{2\}, \{3\}$ = the *singleton* events.

$\{1, 2\}, \{1, 3\}, \{2, 3\}$ = the events with *two possible* outcomes. An event such as getting an odd number has two possible outcomes: 1 and 3.

Answer $\{\ \}, \{1\}, \{2\}, \{3\}, \{1, 2\}, \{1, 3\}, \{2, 3\}, \{1, 2, 3\}$

2. A bank contains a nickel, a dime, and a quarter. A person selects one of the coins. What is the probability that the coin is worth:

a. exactly 10¢? **b.** exactly 3¢? **c.** more than 3¢?

Solutions **a.** There is only one coin worth exactly 10¢: the dime. There are three coins in the bank. Thus:

$$P(\text{coin is worth } 10¢) = \frac{n(E)}{n(S)} = \frac{1}{3} \quad \textit{Answer}$$

b. There is no coin worth exactly 3¢. This is an impossible event. Thus:

$$P(\text{coin is worth } 3¢) = \frac{n(E)}{n(S)} = \frac{0}{3} = 0 \quad \textit{Answer}$$

c. Each of the three coins is worth more than 3¢. This is a certain event. Thus:

$$P(\text{coin is worth more than } 3¢) = \frac{n(E)}{n(S)} = \frac{3}{3} = 1 \quad \textit{Answer}$$

---

3. In the Sullivan family, there are two more girls than boys. At random, Mrs. Sullivan asks one of her children to go to the store. If she is equally likely to ask any one of her children, and the probability that she asks a girl is $\frac{2}{3}$, how many boys and how many girls are there in the Sullivan family?

Solution

$$\begin{aligned} \text{Let } x &= \text{number of boys.} \\ \text{Then, } x + 2 &= \text{number of girls,} \\ \text{and } 2x + 2 &= \text{number of children.} \end{aligned}$$

Then:

$$\begin{aligned} P(\text{girl}) &= \frac{\text{number of girls}}{\text{number of children}} \\ \frac{2}{3} &= \frac{x + 2}{2x + 2} \\ 2(2x + 2) &= 3(x + 2) \\ 4x + 4 &= 3x + 6 \\ x &= 2 \end{aligned}$$

$$\begin{aligned} \text{Then, } x + 2 &= 4 \\ \text{and } 2x + 2 &= 6 \end{aligned}$$

Check:

$$\begin{aligned} P(\text{girl}) &= \frac{\text{number of girls}}{\text{number of children}} \\ \frac{2}{3} &\stackrel{?}{=} \frac{4}{6} \\ \frac{2}{3} &= \frac{2}{3} \text{ (True)} \end{aligned}$$

Answer: There are two boys and four girls.

---

4. A letter is chosen at random from the word REED.
a. List the sample space using subscripts.
b. Find the probability of choosing an E.

Solutions **a.** The sample space is $\{R, E_1, E_2, D\}$. *Answer*

b. Since there are two ways to choose the letter E from the four letters:

$$P(E) = \frac{2}{4}, \text{ or } \frac{1}{2} \quad \textit{Answer}$$

EXERCISES WITH OPEN-RESPONSE PROBLEMS

1. A fair coin is tossed, and its sample space $S = \{H, T\}$.
a. List all four possible events for the toss of a fair coin.
b. Find the probability of each event named in part **a**.

In 2–9, a spinner is divided into seven equal regions, numbered 1 through 7 as shown in the figure. An arrow is spun to fall into one of the regions. For each question:

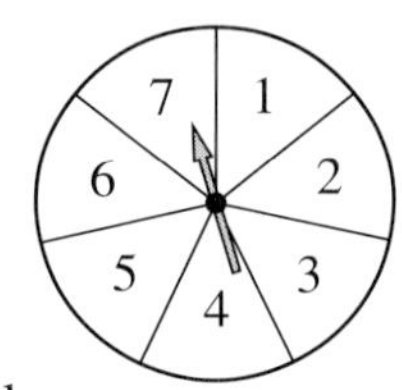

a. List the elements of the event, shown as a subset of $\{1, 2, 3, 4, 5, 6, 7\}$.
b. Find the probability that the arrow lands on the number described.

2. the number 5
3. an even number
4. a number less than 5
5. an odd number
6. a number greater than 5
7. a number greater than 1
8. a number greater than 7
9. a number less than 8

10. A marble is drawn at random from a bag. Find the probability that the marble is black if the bag contains marbles whose colors are:
a. 5 black, 2 green
b. 2 black, 1 green
c. 3 black, 4 green, 1 red
d. 9 black
e. 3 green, 4 red
f. 3 black

11. Ted has two quarters, three dimes, and one nickel in his pocket. He pulls out a coin at random. Find the probability that the coin is worth:
a. exactly 5¢
b. exactly 10¢
c. exactly 25¢
d. exactly 50¢
e. less than 25¢
f. less than 50¢
g. more than 25¢
h. more than 1¢
i. less than 1¢

12. A single fair die is rolled. Find the probability for each event.
a. The number 8 appears.
b. A whole number appears.
c. The number is less than 5.
d. The number is less than 1.
e. The number is less than 10.
f. The number is negative.

13. A standard deck of 52 cards is shuffled, and you pick a card at random. Find the probability that the card is:
a. a jack **b.** a club **c.** a star
d. a red club **e.** a card from the deck **f.** a black club
g. the jack of stars **h.** a 17 **i.** a red 17

14. A class contains 15 girls and 10 boys. The teacher calls on a student at random to answer a question. Express, *in decimal form*, the probability that the student called upon is:
a. a girl **b.** a boy **c.** a pupil in the class **d.** the teacher of the class

15. The last digit of a telephone number can be any of the following: 0, 1, 2, 3, 4, 5, 6, 7, 8, or 9. Express, *as a percent,* the probability that the last digit is: **a.** 7 **b.** odd **c.** more than 5
d. a whole number **e.** the letter R

16. A girl is holding five cards in her hand: the 3 of hearts, 3 of diamonds, 3 of clubs, 4 of diamonds, 7 of clubs. A player to her left takes one of these cards at random. (If the player takes the 3 of hearts, we describe the probability of this event as $\frac{1}{5}$.) Find the probability that the card selected from the five cards in the girl's hand is:
a. a 3 **b.** a diamond **c.** a 4
d. a black 4 **e.** a club **f.** the 4 of hearts
g. a 5 **h.** the 7 of clubs **i.** a red card
j. a number card **k.** a spade **l.** a number less than 8

17. The measures of the three interior angles of the triangle in the figure are given as 40°, 60°, and 80°. The measures of the exterior angles of the triangle are 140°, 120°, and 100°, respectively. One of the six angles is chosen at random. Find the probability that the angle is:

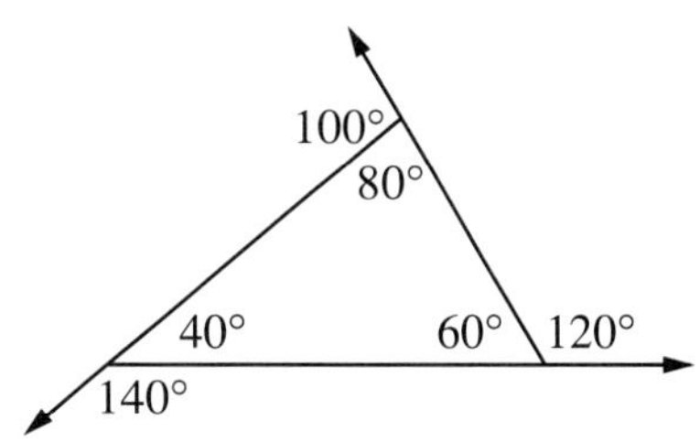

a. an interior angle **b.** a straight angle
c. a right angle **d.** an acute angle
e. a 60° angle **f.** an acute exterior angle
g. an angle whose measure is less than 180°

18. A sack contains 20 marbles. The probability of drawing a green marble is $\frac{2}{5}$. How many green marbles are in the sack?

19. There are three more boys than girls in the chess club. A member of the club is to be chosen at random to play in a tournament. Each member is equally likely to be chosen. If the probability that a girl is chosen is $\frac{3}{7}$, how many boys and how many girls are members of the club?

20. A box of candy contains caramels and nut clusters. There are six more caramels than nut clusters. If a piece of candy is to be chosen at random, the probability that it will be a caramel is $\frac{3}{5}$. How many caramels and how many nut clusters are in the box?

21. At a fair, each ride costs one, two, or four tickets. The number of rides that cost two tickets is three times the number of rides that cost one ticket Also, seven more rides cost four tickets than cost two tickets. Tycella, who has a book of tickets, goes on a ride at random. If the probability that the ride cost her four tickets is $\frac{4}{7}$, how many rides are there at the fair?

22. a. List three events for which the probability is 0.
b. List three events for which the probability is 1.

23. Explain why the following sentence is *incorrect*: "I did so well on the test that the probability of my passing is greater than 1."
24. Explain why the following sentence is *incorrect*: "The probability that we'll go swimming in Maine next December is less than 0."
25. Explain why this sentence is *correct*: "The probability that Tuesday comes after Monday is 100%."

In 26–30, a letter is chosen at random from a given word. For each question: **a.** List the elements of the event, using subscripts if needed. **b.** Find the probability of the event.

26. Selecting the letter E from the word EVENT
27. Selecting the letter S from the word MISSISSIPPI
28. Selecting a vowel from the word TRIANGLE
29. Selecting a vowel from the word RECEIVE
30. Selecting a consonant from the word SPRY

13-4 THE PROBABILITY OF (*A* AND *B*)

The connective ***and*** has been used in logic. As we will see, *and* is sometimes also used to describe events in probability.

Using Counting to Look at *P*(*A* and *B*)

For example, if a fair die is rolled, we can find simple probabilities. Let event A = rolling an even number. Then:

$$P(A) = \frac{n(A)}{n(S)} = \frac{3}{6}$$

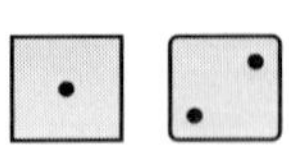

Let event B = rolling a number less than 3. Then:

$$P(B) = \frac{n(B)}{n(S)} = \frac{2}{6}$$

Now, what is the probability of obtaining a number on the die that is even **and** less than 3? We may think of this as ***event*** (***A* and *B***).

In logic, a sentence ***p and q***, written $p \wedge q$, is true only when p is true *and* q is true.

In probability, an outcome is in event (***A* and *B***) only when the outcome is in event A *and* the outcome is also in event B.

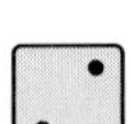

For the example above, the only outcome in the event (A and B) is 2, because 2 is even and 2 is less than 3. Since $n(A \text{ and } B) = 1$ and there are six outcomes on the die, or $n(S) = 6$:

$$P(A \text{ and } B) = \frac{n(A \text{ and } B)}{n(S)} = \frac{1}{6}$$

Let us consider another example in which a fair die is rolled. What is the probability of rolling a number that is both odd *and* a 4?

Event C = {odd numbers}

$$P(C) = \frac{n(C)}{n(S)} = \frac{3}{6}$$

Event D = {4}

$$P(D) = \frac{n(D)}{n(S)} = \frac{1}{6}$$

Then, event (C and D) = {die numbers that are odd *and* 4}.

Since there are *no* outcomes common to both event C and event D, $n(C \text{ and } D) = 0$. Therefore,

$$P(C \text{ and } D) = \frac{n(C \text{ and } D)}{n(S)} = \frac{0}{6} = 0.$$

Using Sets to Look at *P*(*A* and *B*)

We have seen that an outcome is in the event (A and B) only when the outcome is in event A *and* in event B. Let event A = rolling an even number on a die, and let event B = rolling a number less than 3. The following set diagram illustrates the event (A and B):

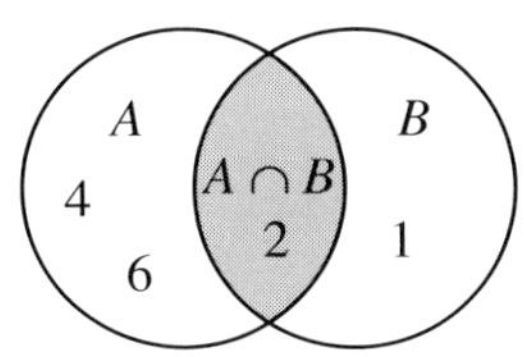

Event (A and B) is the set of numbers that are even *and* less than 3. Since event A is {2, 4, 6} and event B is {1, 2}, we can see that event (A and B) is {2}, the ***intersection*** of the two sets, or $A \cap B$. By counting the number of outcomes in this intersection, we say:

$$P(A \text{ and } B) = P(A \cap B) = \frac{n(A \cap B)}{n(S)} = \frac{1}{6}$$

In the diagram below, event (C and D) is the set of numbers that are odd and 4. Here, the intersection of event C and event D is empty, or $C \cap D = \{\ \}$. Thus:

$$P(C \text{ and } D) = P(C \cap D) = \frac{n(C \cap D)}{n(S)} = \frac{0}{6} = 0$$

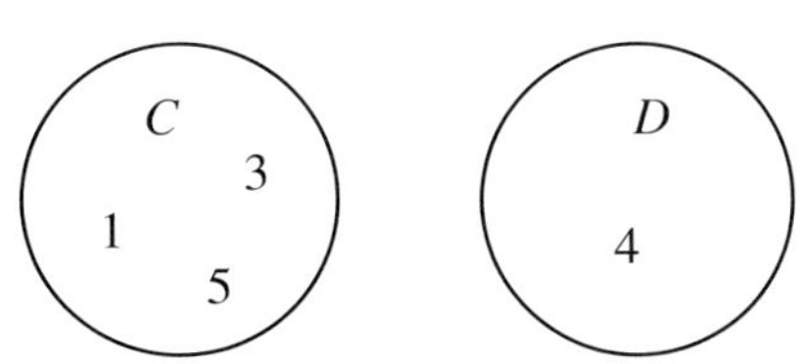

Conclusions

There is no simple rule or formula that works for all problems whereby the values of $P(A)$ and $P(B)$ can be used to find $P(A \text{ and } B)$. We must simply observe the intersection of the two sets and count the number of elements in that intersection, or we can count the number of outcomes that are common in both events.

KEEP IN MIND

Event (A and B) consists of the outcomes that are in event A *and* in event B. Event (A and B) may be regarded as the *intersection* of sets, namely, $\boldsymbol{A \cap B}$.

EXAMPLE

A fair die is rolled once. Find the probability of obtaining a number that is greater than 3 and less than 6.

Solution Event A = {numbers greater than 3} = {4, 5, 6}.
Event B = {numbers less than 6} = {1, 2, 3, 4, 5}.
Event (A and B) = {numbers greater than 3 *and* less than 6} = {4, 5}, or ($A \cap B$) = {4, 5}.

Therefore:

$$P(A \text{ and } B) = \frac{n(A \text{ and } B)}{n(S)} = \frac{2}{6} \quad \text{or} \quad \frac{n(A \cap B)}{n(S)} = \frac{2}{6} \ \textit{Answer}$$

EXERCISES WITH OPEN-RESPONSE PROBLEMS

1. A fair die is rolled once. The sides are numbered 1, 2, 3, 4, 5, and 6. Find the probability that the number rolled is:

a. greater than 2 and odd
b. less than 4 and even
c. greater than 2 and less than 4
d. less than 2 and even
e. less than 6 and odd
f. less than 4 and greater than 3

2. From a standard deck of cards, one card is drawn. Find the probability that the card is:

a. the king of hearts
b. a red king
c. a club king
d. a black jack
e. a diamond 10
f. a red club
g. the 2 of spades
h. a black 2
i. a red picture card

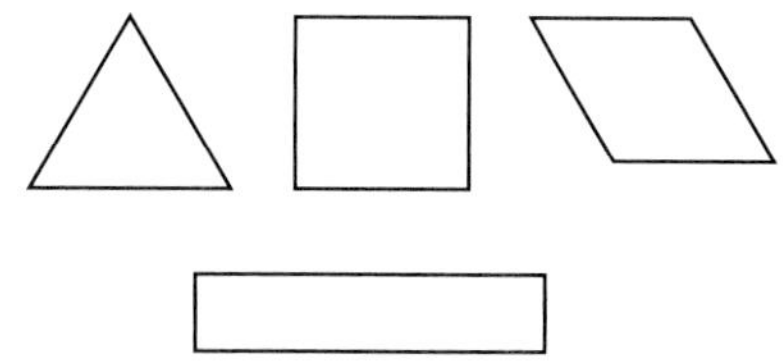

3. The set of polygons at the right consists of an equilateral triangle, a square, a rhombus, and a rectangle. One of the polygons is selected at random. Find the probability that the polygon contains:
  a. all sides congruent and all angles congruent
  b. all sides congruent and all right angles
  c. all sides congruent and two angles not congruent
  d. at least two congruent sides and at least two congruent angles
  e. at least three congruent sides and at least two congruent angles

4. At a St. Patrick's Day party, some of the boys and girls take turns singing songs. Of the five boys: Patrick and Terence are teenagers; younger boys include Brendan, Drew, and Kevin. Of the seven girls: Heather and Claudia are teenagers; younger girls include Maureen, Elizabeth, Gwen, Caitlin, and Kelly. Find the probability that the first song is sung by:
  a. a girl
  b. a boy
  c. a teenager
  d. someone under 13 years old
  e. a boy under 13
  f. a girl whose initial is C
  g. a teenage girl
  h. a girl under 13
  i. a boy whose initial is C
  j. a teenage boy

5. In a class of 30 students, 23 take science, 28 take math, and all take either science or math.
  a. How many students take both science *and* math?
  b. A student from the class is selected at random. Find:
    (*1*) *P*(takes science) (*2*) *P*(takes math) (*3*) *P*(takes science *and* math)

13-5 THE PROBABILITY OF (*A* OR *B*)

The connective ***or*** has been used in logic. As we will see, *or* is sometimes also used to describe events in probability.

Using Counting to Look at *P*(*A* or *B*)

For example, if a fair die is rolled, we can find simple probabilities. Let event A = rolling an even number. Then:

$$P(A) = \frac{n(A)}{n(S)} = \frac{3}{6}$$

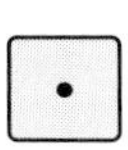

Let event C = rolling a number less than 2. Then:

$$P(C) = \frac{n(C)}{n(S)} = \frac{1}{6}$$

Now, what is the probability of obtaining a number on the die that is even *or* less than 2? We may think of this as ***event*** (***A* or *C***).

In logic, a sentence ***p or q***, written $\boldsymbol{p} \vee \boldsymbol{q}$, is true when p is true, or when q is true, or when both p and q are true.

In probability, an outcome is in event (**A** *or* **C**) when the outcome is in event A, or the outcome is in event C, or the outcome is in both event A and event C.

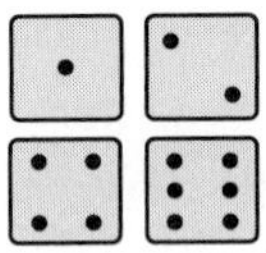

For the example above, there are four outcomes in event (A or C): 1, 2, 4, and 6. Each of these numbers is even, or it is less than 2. Since $n(A \text{ or } C) = 4$, and there are six outcomes on the die, or $n(S) = 6$:

$$P(A \text{ or } C) = \frac{n(A \text{ or } C)}{n(S)} = \frac{4}{6}$$

Observe that $P(A) = \frac{3}{6}$, $P(C) = \frac{1}{6}$, and $P(A \text{ or } C) = \frac{4}{6}$. In this case, it appears that $P(A) + P(C) = P(A \text{ or } C)$. Will this simple addition rule hold true for all problems? Before you say "yes," consider the next example, in which a fair die is rolled.

Event A = {even numbers}

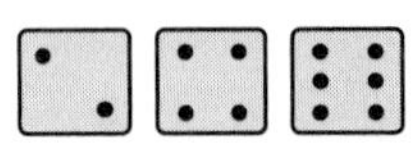

$$P(A) = \frac{n(A)}{n(S)} = \frac{3}{6}$$

Event B = {numbers less than 3}

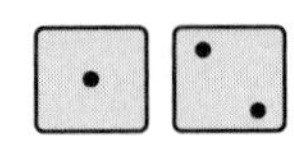

$$P(B) = \frac{n(B)}{n(S)} = \frac{2}{6}$$

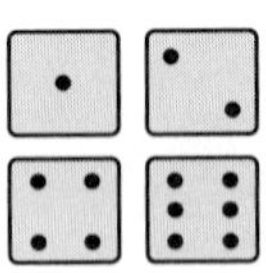

Then, event (A or B) = {numbers that are even or less than 3}. There are still only four outcomes in this new event: 1, 2, 4, and 6. Each of these numbers is even, or it is less than 3. Therefore:

$$P(A \text{ or } B) = \frac{n(A \text{ or } B)}{n(S)} = \frac{4}{6}$$

Here $P(A) = \frac{3}{6}$, $P(B) = \frac{2}{6}$, and $P(A \text{ or } B) = \frac{4}{6}$. In this case, the simple rule of addition does not work: $P(A) + P(B) \neq P(A \text{ or } B)$. What made this example different from $P(A \text{ or } C)$, shown previously?

A Rule for the Probability of (*A* or *B*)

Probability is based on counting the outcomes in a given event. For the event (A or B) in our example, we observe that the outcome 2 is found in event A *and* in event B. Therefore, we may describe the outcome 2 as the event (A and B).

We realize that the simple addition rule does not work for the event (A or B) because we have counted the outcome 2 twice: first in event A, then again in event B. Since we counted this outcome twice, we must take it away, or subtract it, *once*:

Thus, the rule becomes:	$n(A)$	$+$	$n(B)$	$-$	$n(A \text{ and } B)$	$=$	$n(A \text{ or } B)$
For this example:	3	$+$	2	$-$	1	$=$	4

Dividing each term by $n(S)$, we get an equivalent equation:

$$\frac{n(A)}{n(S)} + \frac{n(B)}{n(S)} - \frac{n(A \text{ and } B)}{n(S)} = \frac{n(A \text{ or } B)}{n(S)}$$

For this example:

$$\frac{3}{6} + \frac{2}{6} - \frac{1}{6} = \frac{4}{6}$$

Since $P(A \text{ or } B) = \dfrac{n(A \text{ or } B)}{n(S)}$, we can write a general rule:

$$\boldsymbol{P(A \text{ or } B) = P(A) + P(B) - P(A \text{ and } B)}$$

Using Sets to Look at *P*(*A* or *B*)

A set diagram can help us to understand the event (*A* or *B*) just discussed. That is, event (*A* or *B*) = {numbers that are even *or* less than 3}.

Since event (*A* or *B*) = {1, 2, 4, 6}, we recognize that event (*A* or *B*) is the ***union*** of the two sets, or $A \cup B$, as shown by the shaded region in the diagram.

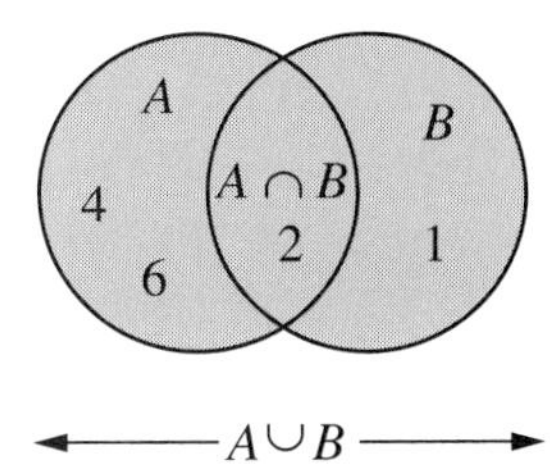

We recall that event (*A* and *B*) = {2}, which is the intersection of the two sets, or $A \cap B$. We can again see how the counting procedure works:

1. Count the number of elements in one set.
2. Add the number of elements from the second set.
3. Subtract the number of elements in their intersection since they were counted twice.

The result is the number of elements in the union of the two sets, or $n(A \cup B)$.

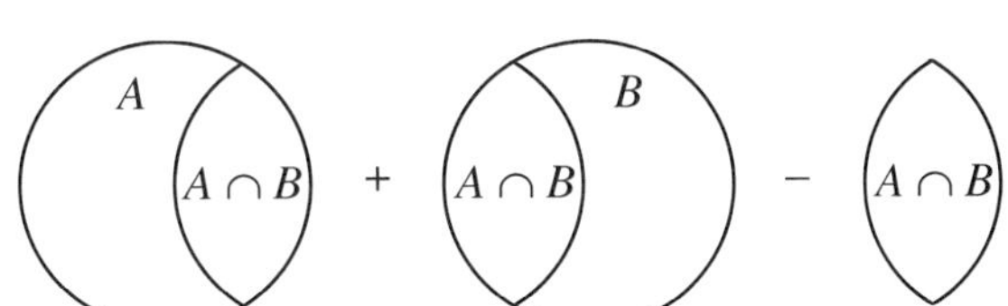

$$n(A) \quad + \quad n(B) \quad - \quad n(A \cap B)$$

In set terminology, the rule for probability becomes:

$$\boldsymbol{P(A \text{ or } B) = P(A \cup B) = P(A) + P(B) - P(A \cap B)}$$

Disjoint Sets

Again, let event (A or C) = {numbers that are even or less than 2}.

Here, event $A = \{2, 4, 6\}$ and event $C = \{1\}$, shown below by ***disjoint sets***. Since A and C are disjoint, there are no elements in their intersection, or $(A \cap C) = \{\ \}$.

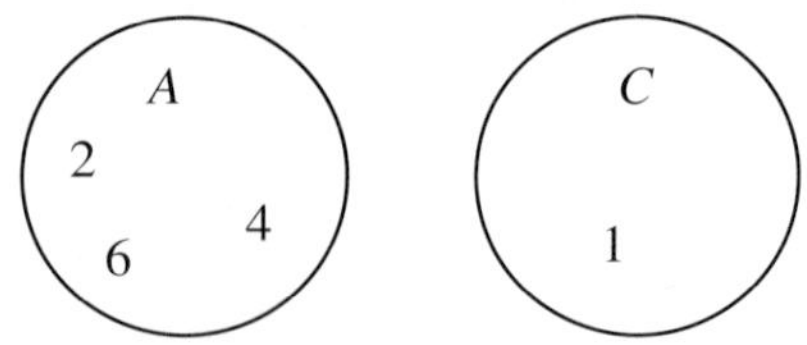

The probability of an empty set is 0. Thus:

$$P(A \cup C) = P(A) + P(C) - P(A \cap C)$$

$$P(A \cup C) = P(A) + P(C) - 0 = P(A) + P(C)$$

This result tells us that the simple addition rule first observed will always be true for events that do not intersect, or that have no outcomes in common.

For disjoint events:

$$\mathbf{P(A \text{ or } C) = P(A \cup C) = P(A) + P(C)}$$

KEEP IN MIND

Event (A or B) consists of all outcomes that are in event A, or in event B, or in both event A and event B. Event (A or B) may be regarded as the *union* of sets, namely, $\boldsymbol{A \cup B}$.

EXAMPLES

1. A standard deck of 52 cards is shuffled, and one card is drawn at random. Find the probability that the card is:
a. a king or an ace **b.** red or an ace

Solutions **a.** There are four kings in the deck, so $P(\text{king}) = \frac{4}{52}$.

There are four aces in the deck, so $P(\text{ace}) = \frac{4}{52}$.

Kings and aces are disjoint events, having no outcomes in common, so

$$P(\text{king or ace}) = P(\text{king}) + P(\text{ace}) = \frac{4}{52} + \frac{4}{52} = \frac{8}{52}$$

Answer: $P(\text{king or ace}) = \frac{8}{52}$

b. There are 26 red cards in the deck, so $P(\text{red}) = \frac{26}{52}$.

There are four aces in the deck, so $P(\text{ace}) = \frac{4}{52}$.

Two cards in the deck are red aces, so $P(\text{red and ace}) = \frac{2}{52}$. Then:

$$P(\text{red or ace}) = P(\text{red}) + P(\text{ace}) - P(\text{red and ace})$$
$$= \frac{26}{52} + \frac{4}{52} - \frac{2}{52}$$

Answer: $\frac{28}{52}$, or $\frac{7}{13}$

Alternative Solution By a counting procedure, there are 26 red cards and two more aces not already counted (the ace of spades, the ace of clubs). Therefore, there are 26 + 2, or 28, cards in this event. Then:

$$P(\text{red or ace}) = \frac{28}{52}, \text{ or } \frac{7}{13}.$$

Answer: $P(\text{red or ace}) = \frac{7}{13}$

---

2. There are two events, A and B. Given that $P(A) = .3$, $P(B) = .5$, and $P(A \cap B) = .1$, find $P(A \cup B)$.

Solution

$$P(A \cup B) = P(A) + P(B) - P(A \cap B)$$
$$= .3 + .5 - .1$$
$$= .8 - .1 = .7$$

Answer: $P(A \cup B) = .7$

EXERCISES WITH OPEN-RESPONSE PROBLEMS

1. A spinner consists of five regions, as shown in the figure, equally likely to occur when an arrow is spun. For a single spin of the arrow, find the probability that the number obtained is:

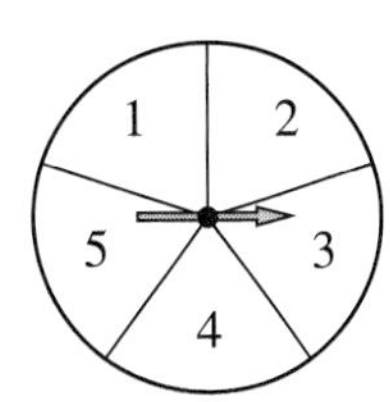

a. 4 **b.** 3 or 4
c. odd **d.** odd or 2
e. less than 4 **f.** 4 or less
g. 2 or 3 or 4 **h.** odd or 3

2. A fair die is rolled once. The sides are numbered 1, 2, 3, 4, 5, and 6. Find the probability that the number rolled is:
a. 4 **b.** 3 or 4 **c.** odd **d.** odd or 2
e. less than 4 **f.** 4 or less **g.** 2 or 3 or 4 **h.** odd or 3
i. less than 2 or more than 5 **j.** less than 5 or more than 2

3. From a standard deck of cards, one card is drawn. Find the probability that the card will be:
a. a queen or an ace **b.** a queen or a 7 **c.** a heart or a spade
d. a queen or a spade **e.** a queen or a red card **f.** a jack or a queen or a king
g. a 7 or a diamond **h.** a club or a red card **i.** an ace or a picture card

4. A bank contains two quarters, six dimes, three nickels, and five pennies. A coin is drawn at random. Find the probability that the coin is:
a. a quarter **b.** a quarter or a dime **c.** a dime or a nickel
d. worth 10 cents **e.** worth more than 10 cents **f.** worth 10 cents or less
g. worth 1 cent or more **h.** worth more than 1 cent **i.** a quarter, a nickel, or a penny

In 5–10, in each case choose the numeral preceding the word or expression that best completes the statement or answers the question.

5. If a single card is drawn from a standard deck, what is the probability that it is a 4 or a 9?
(1) $\frac{2}{52}$ (2) $\frac{8}{52}$ (3) $\frac{13}{52}$ (4) $\frac{26}{52}$

6. If a single card is drawn from a standard deck, what is the probability that it is a 4 or a diamond?
(1) $\frac{8}{52}$ (2) $\frac{16}{52}$ (3) $\frac{17}{52}$ (4) $\frac{26}{52}$

7. If $P(A) = .2$, $P(B) = .5$, and $P(A \cap B) = .1$, then $P(A \cup B) =$
(1) .6 (2) .7 (3) .8 (4) .9

8. If $P(A) = \frac{1}{3}$, $P(B) = \frac{1}{2}$, and $P(A \text{ and } B) = \frac{1}{6}$, then $P(A \text{ or } B) =$
(1) $\frac{2}{5}$ (2) $\frac{2}{3}$ (3) $\frac{5}{6}$ (4) 1

9. If $P(A) = \frac{1}{4}$, $P(B) = \frac{1}{2}$, and $P(A \cap B) = \frac{1}{8}$, then $P(A \cup B) =$
(1) $\frac{1}{8}$ (2) $\frac{5}{8}$ (3) $\frac{3}{4}$ (4) $\frac{7}{8}$

10. If $P(A) = .3$, $P(B) = .35$, and $(A \cap B) = \varnothing$, then $P(A \text{ or } B) =$
(1) .05 (2) .38 (3) .65 (4) 0

11. In a sophomore class of 340 students, some study Spanish, some study French, some study both languages, and some study neither language. If $P(\text{Spanish}) = .7$, $P(\text{French}) = .4$, and $P(\text{Spanish } \textit{and} \text{ French}) = .25$, find: **a.** the probability that a sophomore studies Spanish *or* French **b.** the number of sophomores who study one or more of these languages

12. Linda and Aaron recorded a compact disc together. Each sang some solos, and the two sang some duets. Aaron recorded twice as many duets as solos, and Linda recorded six more solos than duets. If a CD player selects one of these songs at random, the probability that it will select a duet is $\frac{1}{4}$. Find the number of: **a.** solos by Aaron **b.** solos by Linda **c.** duets **d.** songs recorded on the disc

13-6 THE PROBABILITY OF (NOT *A*); PROBABILITY AS A SUM

The Probability of (not *A*)

In rolling a fair die, we will let A represent rolling the number 4. We know that $P(A) = P(4) = \frac{1}{6}$ since there is only one outcome for this event.

Since it is *certain* that we roll a 4 or do not roll a 4, we can say:

$$P(4) + P(\text{not getting } 4) = 1$$

Then, by subtracting,

$$P(\text{not getting } 4) = 1 - P(4)$$
$$= 1 - \frac{1}{6} = \frac{5}{6}$$

Using sets to look at* P(*not* A)** The event (not A) is seen as the ***complement of set A, namely, $\bar{A}$. Therefore, we may also say:

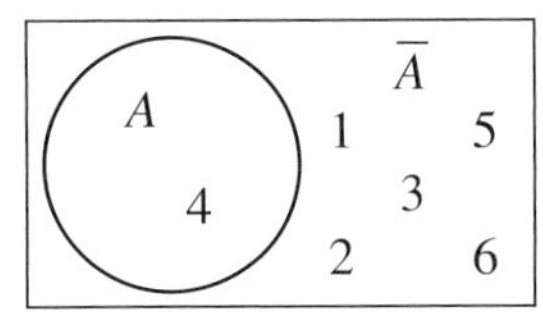

$$P(\text{not } A) = P(\bar{A}) = 1 - P(A)$$
$$= 1 - \frac{1}{6} = \frac{5}{6}$$

***A rule for the probability of (not* A)** In general, if $P(A)$ is the probability that some given result will occur, and $P(\text{not } A)$ is the probability that the given result will not occur, then:

$$\mathbf{P(A) + P(\text{not } A) = 1 \text{ or } P(A) = 1 - P(\text{not } A) \text{ or } P(\text{not } A) = 1 - P(A)}$$

Probability as a Sum

When sets are disjoint, we have seen that the probability of a union can be found by the rule $P(A \cup B) = P(A) + P(B)$. Since the possible outcomes that are singletons represent disjoint sets, we can say:

- **The probability of any event is equal to the sum of the probabilities of the singleton outcomes in the event.**

For example, when we draw a card from a standard deck, there are 52 singleton outcomes, each with a probability of $\frac{1}{52}$. Since all singleton events are disjoint, we can say:

$$P(\text{king}) = P(\text{king of hearts}) + P(\text{king of diamonds}) + P(\text{king of spades}) + P(\text{king of clubs})$$
$$= \frac{1}{52} + \frac{1}{52} + \frac{1}{52} + \frac{1}{52}$$

$$P(\text{king}) = \frac{4}{52}, \text{ or } \frac{1}{13}$$

We also say:

- **The sum of the probabilities of all possible singleton outcomes for any sample space must always equal 1.**

For example, in tossing a coin,

$$P(S) = P(\text{head}) + P(\text{tail}) = \frac{1}{2} + \frac{1}{2} = 1.$$

Also, in rolling a die,

$$\begin{aligned} P(S) &= P(1) + P(2) + P(3) + P(4) + P(5) + P(6) \\ &= \frac{1}{6} + \frac{1}{6} + \frac{1}{6} + \frac{1}{6} + \frac{1}{6} + \frac{1}{6} = 1 \end{aligned}$$

KEEP IN MIND

Event (not A) consists of outcomes from the sample space that are not in event A. Event (not A) may be regarded as the *complement of set* A, namely, $\bar{A}$.

EXAMPLES

1. A fair die is tossed. Find the probability of not rolling a number less than 5.

Solution On a die, the numbers less than 5 are 1, 2, 3, and 4. Since $P(\text{less than } 5) = \frac{4}{6}$, we can say:

$$\begin{aligned} P(\text{not less than } 5) &= 1 - P(\text{less than } 5) \\ &= 1 - \frac{4}{6} = \frac{2}{6}, \text{ or } \frac{1}{3} \ \textit{Answer} \end{aligned}$$

---

2. A letter is drawn at random from the word ERROR.

a. Find the probability of drawing each of the different letters in the word.

b. Demonstrate that the sum of these probabilities is 1.

Solutions **a.** $P(\text{E}) = \frac{1}{5}$; $P(\text{R}) = \frac{3}{5}$; $P(\text{O}) = \frac{1}{5}$ *Answer*

b. $P(\text{E}) + P(\text{R}) + P(\text{O}) = \frac{1}{5} + \frac{3}{5} + \frac{1}{5} = \frac{5}{5} = 1$ *Answer*

EXERCISES WITH OPEN-RESPONSE PROBLEMS

1. A fair die is rolled once. Find the probability that the number rolled is:
a. 3 **b.** not 3 **c.** even **d.** not even
e. less than 3 **f.** not less than 3 **g.** odd or even **h.** not odd or even

2. The weather bureau predicted a 30% chance of rain. Express *in fractional form*:
a. the probability that it will rain **b.** the probability that it will not rain

3. From a standard deck of cards, one card is drawn. Find the probability that the card is:
a. a club **b.** not a club **c.** a picture card **d.** not a picture card
e. not an 8 **f.** not a red 6 **g.** not the queen of spades

4. A bank contains three quarters, four dimes, and five nickels. A coin is drawn at random.
a. Find the probability of drawing: (*1*) a quarter; (*2*) a dime; (*3*) a nickel.
b. Demonstrate that the sum of the three probabilities given as answers in part **a** is 1.

5. One letter is selected at random from the word PICNICKING.
a. Find the probability of drawing each of the different letters in the word.
b. Demonstrate that the sum of these probabilities is 1.

6. If the probability of an event happening is $\frac{1}{7}$, what is the probability of that event not happening?

7. If the probability of an event happening is .093, what is the probability of that event not happening?

8. The square dart board shown at the right, whose side measures 30 inches, has at its center a shaded square region whose side measures 10 inches. If darts directed at the board are equally likely to land anywhere on the board, what is the probability that a dart does not land in the shaded region?

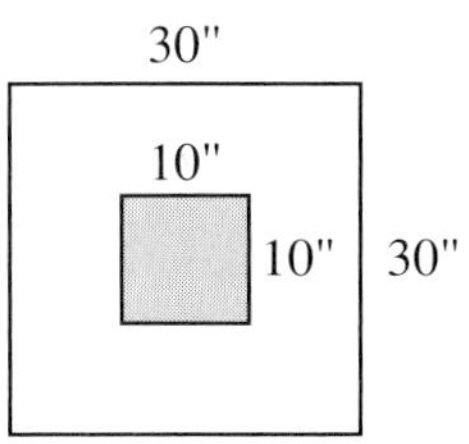

General Exercises

9. A jar contains seven marbles, all the same size: three are red and four are green. If a marble is chosen at random, find the probability that it is:
a. red **b.** green **c.** not red **d.** red or green **e.** red and green

10. A box contains 3 times as many black marbles as green, all the same size. If a marble is drawn at random, find the probability that it is:
a. black **b.** green **c.** not black **d.** black or green **e.** not green

11. The mail contained two letters, three bills, and five ads. Mr. Jacobsen picked up the first piece of mail without looking at it. Express, *in decimal form*, the probability that this piece of mail is:
a. a letter **b.** a bill **c.** an ad **d.** a letter or an ad
e. a bill or an ad **f.** not a bill **g.** not an ad **h.** a bill and an ad

12. A letter is chosen at random from the word PROBABILITY. Find the probability that the letter chosen is:
a. A **b.** B **c.** C **d.** A or B **e.** A or I
f. a vowel **g.** not a vowel **h.** A or B or L **i.** A or not A

13. A single card is drawn at random from a well-shuffled deck of 52 cards. Find the probability that the card is:

a. a 6 **b.** a club **c.** the 6 of clubs **d.** a 6 or a club
e. not a club **f.** not a 6 **g.** a 6 or a 7 **h.** not the 6 of clubs
i. a 6 and a 7 **j.** a black 6 **k.** a 6 or a black card

14. A telephone dial contains the ten digits: 0, 1, 2, 3, 4, 5, 6, 7, 8, 9. Mabel is dialing a friend. Find the probability that the last digit in her friend's telephone number is:

a. 6 **b.** 6 or more **c.** less than 6 **d.** 6 or odd
e. 6 or less **f.** not 6 **g.** 6 and odd **h.** not more than 6
i. less than 2 and more than 6 **j.** less than 2 or more than 6
k. less than 6 and more than 2 **l.** less than 6 or more than 2

CHAPTER SUMMARY

Probability is a branch of mathematics in which the chance of an event happening is predicted before it occurs. In ***empirical probability***, this prediction is based on an experiment with a large number of trials. In ***theoretical probability***, which applies only to fair and unbiased objects or situations, a formula is used:

$$P(E) = \frac{n(E)}{n(S)}$$

This formula states that the probability of an ***event*** E is equal to the number of ***outcomes*** in event E divided by the total number of possible outcomes in the ***sample space*** S.

The probability of an ***impossible*** event is 0, and the probability of a ***certain*** event is 1. All other probabilities are greater than 0 but less than 1, leading to this statement:

$$0 \leq P(E) \leq 1$$

Sets are often used to explain situations in probability. The ***event*** (**A and B**) is similar to the intersection of sets, $A \cap B$. ***Event*** (**A or B**) is similar to the union of sets, $A \cup B$. ***Event*** (**not A**) is similar to the complement of a set, $\overline{A}$. In general,

$$P(A \text{ or } B) = P(A) + P(B) - P(A \text{ and } B)$$

or, when using sets,

$$P(A \cup B) = P(A) + P(B) - P(A \cap B)$$

If A and B are disjoint sets, then

$$P(A \text{ or } B) = P(A) + P(B)$$

Also,

$$P(\text{not } A) = P(\overline{A}) = 1 - P(A)$$

VOCABULARY

13-1 Probability Empirical study Empirical probability Relative frequency Cumulative Cumulative relative frequency Trial Experiment Fair and unbiased objects Biased objects Die Standard deck of cards Digits Vowels Consonants

13-2 Outcome Sample space (S) Event (E) Singleton event

Theoretical probability: $P(E) = \frac{n(E)}{n(S)}$ Uniform probability

Equally likely outcomes Random selection

13-3 Impossibility Certainty Subscript

13-4 $P(A \text{ and } B)$ Intersection of sets

13-5 $P(A \text{ or } B)$ Union of sets Disjoint sets

13-6 $P(\text{not } A)$ Complement of a set Sum of probabilities

REVIEW EXERCISES WITH OPEN-RESPONSE PROBLEMS

1. A letter is chosen at random from the word "GYMNASIUM." Find the probability that the letter is a vowel.

2. A six-sided die is rolled; the sides are numbered 1, 2, 3, 4, 5, and 6. In rolling the die, find $P(4)$.

3. The probability that Glennon will get a hit the next time at bat is 35%. What is the probability that Glennon will not get a hit?

4. In right triangle ABC, one of the three angles is chosen at random. Find the probability that the angle is:

a. acute **b.** right **c.** obtuse **d.** reflex

5. A jar contains two red, three white, and four blue marbles. If a marble is drawn at random, find the probability that the marble is:

a. red **b.** not red **c.** red or blue **d.** green

6. A single card is drawn from a standard deck of 52 playing cards. Find the probability that the card is:

a. red **b.** a 4 **c.** a red 4 **d.** a heart
e. a 4 or a heart **f.** red or a 4 **g.** not a club **h.** red and a 3

7. How many different two-digit numbers can be formed using the digits of 1492 if each digit appears only once in a number?

8. A spinner is divided into six equal regions, numbered 1-6 as shown in the figure. An arrow is spun to fall into one of the equal regions. Find the probability that the arrow lands on

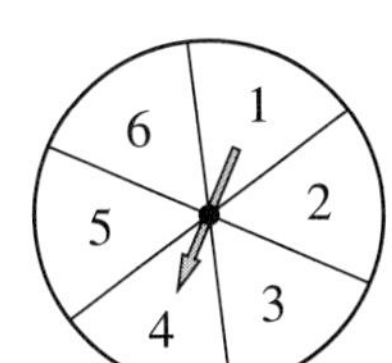

a. 5 **b.** an even number
c. an odd number **d.** a number greater than 3

9. A bag contains 15 marbles. The probability of drawing a blue marble is $\frac{2}{5}$. How many marbles in the bag are not blue?

10. A letter is chosen at random from the word "ANTHROPOLOGY." Find the probability that the letter chosen is

a. a vowel **b.** not a vowel **c.** M **d.** A or Y
e. R or O **f.** O or not O **g.** A or R or T **h.** S or P
i. a letter between G and M in the alphabet

11. A box of small tiles contains 4 times as many white tiles as green, all the same size and shape. If one of the tiles is taken from the box at random, what is the probability that it is:

a. white **b.** green **c.** not white **d.** white or green **e.** not green

In 12–13, in each case, select the numeral preceding the correct answer.

12. If $P(A) = .4$, $P(B) = .3$, and $P(A \cap B) = .2$, then $P(A \cup B) =$
(1) .9 (2) .7 (3) .5 (4) .12

13. If $P(E) = \frac{n}{5}$ for event E, which *cannot* be a value for n?
(1) 1 (2) 0 (3) 5 (4) 15

14. About how many times can you expect to turn up a head in 50 consecutive tosses of a fair coin?

15. A jar contains red and white marbles. The number of white marbles is 5 more than twice the number of red marbles. If the probability of drawing a red marble at random is $\frac{2}{7}$, how many red marbles are in the jar?

16. From a class of girls and boys, the probability that one student chosen at random will be a girl is $\frac{1}{3}$. If four boys leave the class, the probability that a student chosen at random will be a girl is $\frac{2}{5}$. How many boys and girls are there in the class before the four boys leave?

CUMULATIVE REVIEW WITH OPEN-RESPONSE PROBLEMS

1. The measures of the sides of a triangle are 6 centimeters, 10 centimeters, and 12 centimeters. The measure of a side of a similar triangle is 20 centimeters.

a. What are the possible measures of the sides of the larger triangle?
b. If one of the triangles that are similar to the original is chosen at random, what is the probability that the measure of each side in centimeters is an integer?

Exploration

Rachael had a box of disks, all the same size and shape. She removed 20 disks from the box, marked them, and then returned them to the box. After mixing the marked disks with the others in the box, she removed a handful of disks and recorded the total number of disks and the number of marked disks. Then she returned the disks to the box, mixed them, and removed another handful. She repeated this last step eight times. The chart below shows her results.

Disks	12	10	15	11	8	10	7	12	9	13
Marked Disks	3	4	5	4	3	2	2	5	3	5

a. Use the data to estimate the number of disks in the box.
b. Repeat Rachael's experiment using a box containing an unknown number of disks. Compare your estimate with the actual number of disks in the box.
c. Explain how this procedure could be used to estimate the number of fish in a pond.

Chapter 14

The Counting Principle, Permutations, and Combinations

Regular license plates in some states are made up of three letters followed by three numbers. You would think that this method along with special ''vanity'' plates would provide enough different license plates for all the cars registered in the state. In some cases, the Department of Motor Vehicles runs out of arrangements of letters and numbers and has to make up a new system for creating license plates. One solution may be to increase the number of letters or numbers to four of one or the other or both.

Many telephone companies have been having a similar problem. With the increased number of cellular phones, fax machines, and computers, the telephone companies run out of new numbers. All the possible different arrangements of seven-digit numbers is often exhausted within an area code. In order to create more new numbers, a telephone company will split an area code into two area codes. Then the same seven-digit telephone number can be linked to two different area codes. In this way, the number of available seven-digit numbers in the original location will be doubled.

In this chapter, you will learn how to count how many different license plates can be made from three letters and three numbers and how many seven-digit telephone numbers can be created in one area code.

14-1 THE COUNTING PRINCIPLE AND SAMPLE SPACES

So far, we have looked at simple problems involving a *single* activity, such as rolling one die or choosing one card. More realistic problems occur when there are *two or more* activities, such as rolling two dice or dealing a hand of five cards. Before studying the probability of events based on two or more activities, let us study an easy way to count the number of elements or "outcomes" in a sample space for two or more activities. For example:

A store offers five flavors of ice cream: vanilla, chocolate, strawberry, peach, and raspberry. A sundae can be made with either a hot fudge topping or a marshmallow topping. If a sundae consists of one flavor of ice cream and one topping, how many different sundaes are possible?

We will use initial letters to represent the five flavors of ice cream (V, C, S, P, R) and the two toppings (F, M). We can show the number of elements in the sample space in three ways:

1. ***Tree Diagram.*** The *tree diagram* at the right first branches out to show five flavors of ice cream. For each of these flavors the tree again branches out to show the two toppings. In all, there are 10 paths or branches to follow, each with one flavor of ice cream and one topping. These 10 branches show that the sample space consists of 10 possible outcomes—in this case, sundaes.

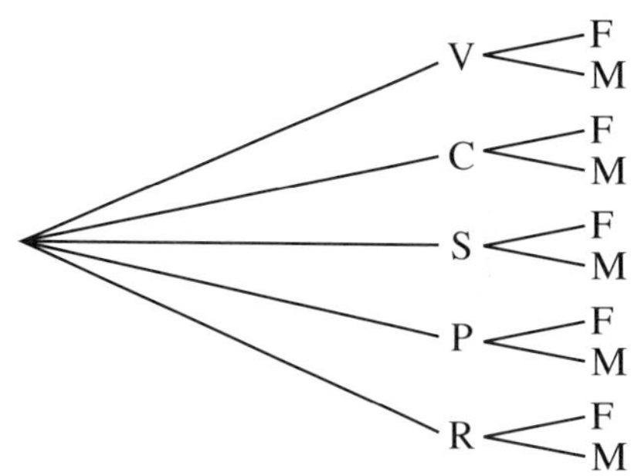

Tree diagram

2. ***List of Ordered Pairs.*** It is usual to order a sundae by telling the clerk the flavor of ice cream and the type of topping. This suggests a *listing of ordered pairs*. The first component of the ordered pair is the ice-cream flavor, and the second component is the type of topping. The set of pairs (ice cream, topping) is shown above.

$$\left\{\begin{matrix}(V, F), (C, F), (S, F), (P, F), (R, F), \\ (V, M), (C, M), (S, M), (P, M), (R, M)\end{matrix}\right\}$$

List of ordered pairs

These 10 ordered pairs show that the sample space consists of 10 possible sundaes.

3. ***Graph of Ordered Pairs.*** Instead of listing pairs, we may construct a *graph of the ordered pairs*. At the right, the five flavors of ice cream appear on the horizontal scale or line, and the two toppings are on the vertical line.

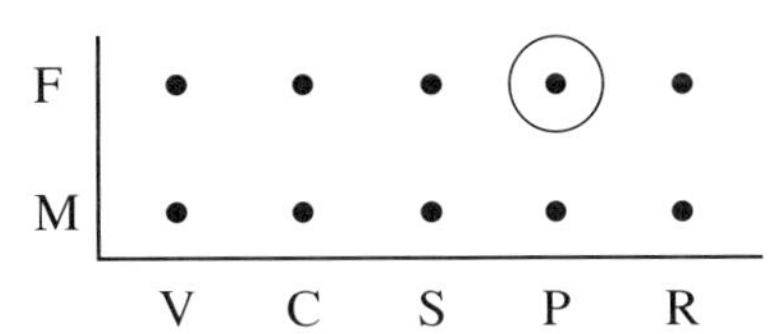

Graph of ordered pairs

Each point in the graph represents an ordered pair. For example, the point circled shows the ordered pair (P, F), that is, (peach ice cream, fudge topping). This graph of 10 points, or 10 ordered pairs, shows that the sample space consists of 10 possible sundaes.

Whether we use a tree diagram, a list of ordered pairs, or a graph of ordered pairs, we recognize that the sample space consists of 10 sundaes. The number of elements in the sample space can be found by multiplication:

$$\underbrace{\begin{matrix}\text{number of}\\ \text{ice-cream flavors}\end{matrix}}_{5} \times \underbrace{\begin{matrix}\text{number of}\\ \text{toppings}\end{matrix}}_{2} = \underbrace{\begin{matrix}\text{number of}\\ \text{possible sundaes}\end{matrix}}_{10}$$

Suppose the store offered 30 flavors of ice cream and seven possible toppings. To find the number of elements in the sample space, we multiply: $30 \times 7 = 210$ possible sundaes. This simple multiplication procedure is known as the ***counting principle***, because it enables us to count the number of elements in a sample space.

- **The Counting Principle: If one activity can occur in any of *m* ways and, following this, a second activity can occur in any of *n* ways, then both activities can occur in the order given in $m \cdot n$ ways.**

We can extend this rule to include three or more activities by extending the multiplication process. We can also display three or more activities by extending the branches on a tree diagram, or by listing ordered elements such as ordered triples and ordered quadruples.

For example, a coin is tossed three times in succession.

On the first toss, the coin may fall in either of two ways: a head or a tail.

On the second toss, the coin may also fall in either of two ways.

On the third toss, the coin may still fall in either of two ways.

By the counting principle, the sample space contains 2 • 2 • 2, or 8, possible outcomes.

By letting H represent a head and T represent a tail, we can illustrate the sample space by a tree diagram or by a list of ordered triples:

Three Tosses of a Coin

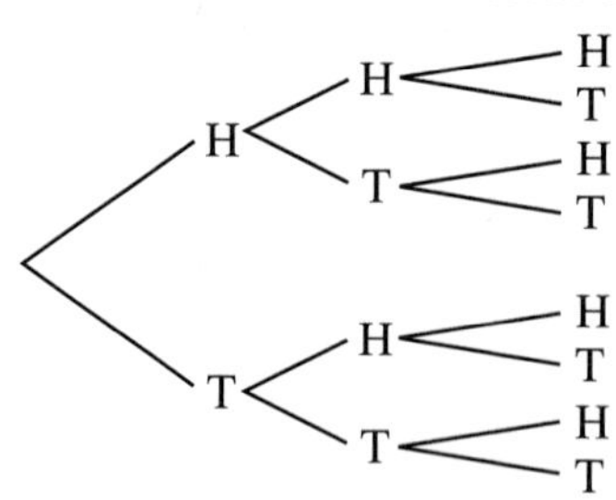

Tree diagram

(H, H, H)
(H, H, T)
(H, T, H)
(H, T, T)

(T, H, H)
(T, H, T)
(T, T, H)
(T, T, T)

List of ordered triples

We did *not* attempt to draw a graph of this sample space because we would need a horizontal scale, a vertical scale, and a third scale, making the graph three-dimensional. Although such a graph can be drawn, it is too difficult at this time. We can conclude that:

1. *Tree diagrams*, or *lists of ordered elements*, are effective ways to indicate any compound event of two or more activities.

2. *Graphs should be limited to ordered pairs*, or to events consisting of exactly two activities.

EXAMPLES

1. The school cafeteria offers four types of salads, three types of beverages, and five types of desserts. If a lunch consists of one salad, one beverage, and one dessert, how many possible lunches can be chosen?

Solution By the counting principle, we multiply the number of possibilities for each choice:

$$4 \cdot 3 \cdot 5 = 12 \cdot 5 = 60 \text{ possible lunches} \quad \textit{Answer}$$

---

2. There are 12 staircases going from the first floor to the second in our school. Roger goes up one staircase and then goes down a different staircase. How many possible ways can this event occur?

Solution Roger can choose any of 12 staircases going up. Since he goes down a different staircase, he now has 11 choices left for going down. By applying the counting principle, we see that there are $12 \cdot 11 = 132$ ways. *Answer*

EXERCISES WITH OPEN-RESPONSE PROBLEMS

1. Tell how many possible outfits consisting of one shirt and one pair of pants Terry can choose if he owns: **a.** 5 shirts, 2 pairs of pants **b.** 10 shirts, 4 pairs of pants **c.** 6 shirts, an equal number of pairs of pants

2. There are 10 doors into the school and eight staircases from the first floor to the second. How many possible ways are there for a student to go from outside the school to a classroom on the second floor?

3. A tennis club has 15 members: eight women and seven men. How many different teams may be formed consisting of one woman and one man on each team?

4. A dinner menu lists two soups, seven meats, and three desserts. How many different meals consisting of one soup, one meat, and one dessert are possible?

5. There are three ways to go from town A to town B. There are four ways to go from town B to town C. How many different ways are there to go from town A to town C, passing through town B?

6. The school cafeteria offers the menu shown.

Main Course	Dessert	Drink
Pizza	Ice cream	Milk
Frankfurter	Cookies	Juice
Ham sandwich	Jello	
Tuna sandwich	Apple pie	
Jelly sandwich		

a. How many meals consisting of one main course, one dessert, and one drink can be selected from this menu?
b. Joe hates ham and jelly. How many meals (again, one main course, one dessert, and one drink) can Joe select, not having ham and not having jelly?
c. If the pizza, frankfurters, ice cream, and cookies have been sold out, how many different meals can JoAnn select from the remaining menu?

7. A quarter and a penny are tossed simultaneously. Each coin may fall heads or tails. The tree diagram at the right shows the sample space involved.

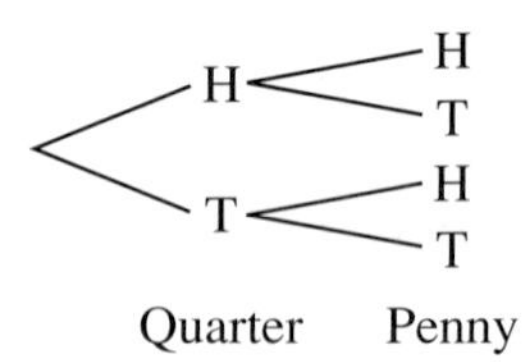

a. List the sample space as a set of ordered pairs.
b. Use the counting principle to demonstrate that there are four outcomes in the same space.
c. In how many outcomes do the coins both fall heads up?
d. In how many outcomes do the coins land showing one head and one tail?

8. A teacher gives a quiz consisting of two questions. Each question has as its answer either true (T) or false (F).
a. Using T and F, draw a tree diagram to show all possible ways the questions can be answered.
b. List this sample space as a set of ordered pairs.
c. State the relationship that exists between this sample space and the p and q columns in a table of truth values for two statements developed in the study of logic.

9. A quiz consists of three true-false questions.
a. How many possible ways are there to answer the questions on this test?
b. Draw a tree diagram to show the sample space.
c. List the ordered triples to show this sample space.

10. Elizabeth has a number of possible routes to school. She can take either A Street or B Street, and then turn onto Avenue X, Avenue Y, or Avenue Z.
a. Draw a tree diagram to show all the possible routes.
b. List the ordered pairs to show this sample space.

11. A test consists of multiple-choice questions. Each question has four choices. Tell how many possible ways there are to answer the questions on the test if the test consists of:
a. 1 question **b.** 3 questions **c.** n questions

12. Options on a bicycle include two types of handlebars, two types of seats, and a choice of 15 colors. The bike may also be ordered in ten-speeds, in three-speeds, or standard. How many possible versions of a bicycle can a customer choose from, if he selects a specific type of handlebars, type of seat, color, and speed?

13. Two dice are rolled simultaneously. Each die may land with one of six numbers face up.
a. Use the counting principle to determine the number of outcomes in this sample space.
b. Display the sample space by constructing a graph of the set of ordered pairs.

14. A state issues license plates consisting of letters and numbers. There are 26 letters, and the letters may be repeated on a plate; there are 10 digits, and the digits may be repeated. Tell how many possible license plates the state may issue when a license consists of:
a. 2 letters, followed by 3 numbers **b.** 2 numbers, followed by 3 letters
c. 4 numbers, followed by 2 letters
(*Note:* The license 1-ID is actually 0001-ID.)

15. An ice-cream company offers 31 different flavors. Hilda orders a double-scoop cone. In how many different ways can the clerk put the ice cream in the cone if: **a.** Hilda wants two different flavors? **b.** Hilda wants the same flavor on both scoops? **c.** Hilda cannot make up her mind and tells the clerk, "Surprise me"?

14-2 PROBABILITIES AND THE COUNTING PRINCIPLE; PREDICTING OUTCOMES

Independent Events

The probability of rolling 1 on the toss of a die is $\frac{1}{6}$. What is the probability of rolling a pair of 1's when two dice are tossed?

When we roll two dice, the number obtained on one die is completely, absolutely, without question, *independent* of the result obtained on the second die.

When the result of one activity in no way influences the result of a second activity, the result of these activities are called ***independent events***. In cases where two events are independent, we may extend the counting principle to find the probability that both independent events occur at the same time. For example:

$$P(1 \text{ on first die}) = \frac{1}{6}$$

$$P(1 \text{ on second die}) = \frac{1}{6}$$

$$P(1 \text{ on both dice}) = P(1 \text{ on first}) \bullet P(1 \text{ on second}) = \frac{1}{6} \bullet \frac{1}{6} = \frac{1}{36}$$

An event that consists of two or more independent events is called a ***compound event***. Activities that result in compound events include rolling two dice, tossing three coins, and spinning an arrow several times in succession.

We can extend the counting principle to help us find the probability of any compound event consisting of two or more independent events.

- **The Counting Principle for Probability: When E and F are independent events, and when the probability of event E is m ($0 \leq m \leq 1$) and the probability of event F is n ($0 \leq n \leq 1$), the probability of the compound event in which E and F occur jointly is the product $m \cdot n$.**

Note 1: The product $m \bullet n$ is within the range of values for a probability, namely, $0 \leq m \bullet n \leq 1$.

Note 2: Not all events are independent, and this simple product rule cannot be used to find the probability of every compound event.

Predicting Outcomes

It is often useful to use past experience to predict what to expect in the future. For example, we decide how much bread we will buy for a week based on how much we have used in the past. We decide which roads to take to avoid traffic, based on the traffic we encountered on these roads in the past. The rates set by insurance companies are based on life expectancies. The branch of biology known as genetics has established certain probability ratios that are useful to plant and animal breeders. Prediction based on past experiences is used in Example 2, which follows.

EXAMPLES

1. Mr. Gillen may take any of three buses, A or B or C, to get to the same train station. He may then take the 6th Avenue train or the 8th Avenue train to get to work. The buses and trains arrive at random and are equally likely to arrive. What is the probability that Mr. Gillen takes the B bus and the 6th Avenue train to get to work?

Solution

$$P(B \text{ bus}) = \frac{1}{3} \text{ and } P(\text{6th Ave. train}) = \frac{1}{2}$$

Since the train taken is independent of the bus taken:

$$P(B \text{ bus and 6th Ave. train}) = P(B \text{ bus}) \bullet P(\text{6th Ave. train})$$

$$= \frac{1}{3} \bullet \frac{1}{2} = \frac{1}{6}$$

Answer: The probability of taking the B bus and the 6th Avenue train is $\frac{1}{6}$.

2. The owner of a garden nursery knows that the color red will occur in one of about 15 seedlings of a certain plant. If the nursery owner wants to have about 200 red seedlings of this plant to sell, how many seedlings of this variety should he plant?

How to Proceed:	*Solution:*
(1) Write the proportion:	$P(\text{red}) = \dfrac{\text{number of red}}{\text{number of planted}}$
(2) Let x = total number of seedlings needed to get 200 red plants:	$\dfrac{1}{15} = \dfrac{200}{x}$
(3) Solve for x:	$x = 3{,}000$

Answer: To have about 200 red seedlings, the nursery owner must plant 3,000 seedlings.

EXERCISES WITH OPEN-RESPONSE PROBLEMS

(All the events described in Exercises 1–16 are independent events.)

1. A fair coin and a six-sided die are tossed simultaneously. What is the probability of obtaining:
a. a head on the coin? **b.** a 4 on the die?
c. a head on the coin and a 4 on the die jointly?

2. A fair coin and a six-sided die are tossed simultaneously. What is the probability of obtaining jointly:
a. a head and a 3? **b.** a head and an even number?
c. a tail and a number less than 5? **d.** a tail and a number greater than 4?

3. Two fair coins are tossed. What is the probability that both land heads up?

4. When I enter school, 3 of 4 times I use the main door. When I leave school, I use the main door only 1 of 3 times. On any given day, what is the probability that I both enter and leave by the main door?

5. In our school cafeteria the menu rotates so that $P(\text{hamburger}) = \frac{1}{4}$, $P(\text{apple pie}) = \frac{2}{3}$, and $P(\text{soup}) = \frac{4}{5}$. On any given day, what is the probability that the cafeteria offers hamburger, apple pie, and soup on the same menu?

6. A quiz consists of true-false questions only. Harry has not studied, and he guesses every answer. Find the probability that he will guess correctly to get a perfect score if the test consists of:
a. 1 question **b.** 4 questions **c.** n questions

7. The probability of the Tigers beating the Cougars is $\frac{2}{3}$. The probability of the Tigers beating the Mustangs is $\frac{1}{4}$. If the Tigers play one game with the Cougars and one game with the Mustangs, find the probability that the Tigers:
a. win both games **b.** lose both games

8. Three fair coins are tossed. **a.** Find $P(\text{H, H, H})$. **b.** Find $P(\text{T, T, T})$.

9. A fair spinner contains equal regions, numbered 1 through 8. If the arrow is spun twice, find the probability that it: **a.** lands on 7 both times **b.** does not land on 7 either time

10. As shown in the diagram, 6th Avenue runs north and south. The lights are not timed to accommodate traffic traveling along 6th Avenue; they are independent of one another. At each of the intersections shown, $P(\text{red light}) = .7$ and $P(\text{green light}) = .3$ for cars traveling along 6th Avenue.

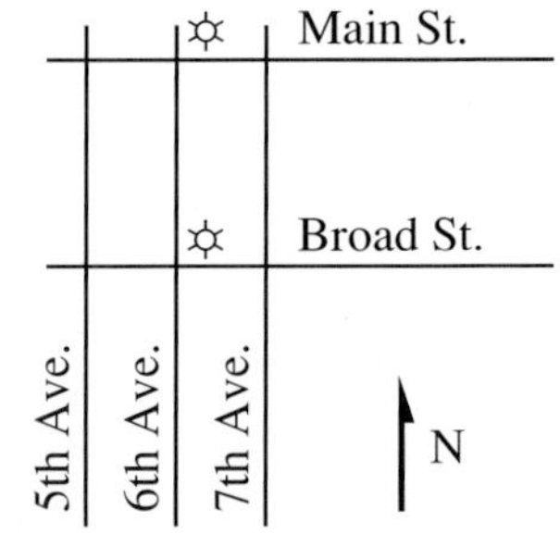

Find the probability that a car traveling north on 6th Avenue will be faced with each set of given conditions at the two traffic lights shown.

a. Both lights are red.
b. The first light is red, and the second is green.
c. Both lights are green.
d. The first light is green, and the second is red.
e. One of the two lights is red.
f. Both lights are the same color.

11. A manufacturer of radios knows that the probability of a defect in any of his products is $\frac{1}{400}$. If 10,000 radios are manufactured in January, how many are likely to be defective?

12. Past records from the weather bureau indicate that rain has fallen on 2 of every 7 days in August on Cape Cod. If Joan goes to Cape Cod for 2 weeks in August, how many days will it probably rain if the records hold true?

13. A fair coin is tossed 50 times and lands head up each time. What is the probability that it will land head up on the next toss? Explain your answer.

14. Dr. Velez knows that she will recommend a flu shot for 20% of her patients, and that an additional 15% will elect to take a flu shot. If she expects to work with 500 patients during this flu season, how many flu shots will she probably administer?

15. About how many times can you expect to turn up a head in 100 consecutive tosses of a coin?

16. About how many times can you expect to roll a 5 in 100 consecutive rolls of a single fair die?

17. A nationwide fast-food chain has a promotion, distributing to customers 2,000,000 coupons for the prizes shown in the following display.

1 Grand Prize:	$25,000 cash
2 Second-Place Prizes:	New car
100 Third-Place Prizes:	New TV
Fourth-Place Prizes:	Free meal
Consolation Prizes:	25¢ off any purchase

a. Find the probability of winning: (*1*) the grand prize (*2*) a new car (*3*) a new TV

b. If the probability of winning a free meal is $\frac{1}{400}$, how many coupons are marked as fourth-place prizes?

c. The fast-food chain announces that every customer has a 1 in 2 chance of winning a prize. How many coupons are marked "25¢ off any purchase"?

14-3 PROBABILITIES WITH TWO OR MORE ACTIVITIES

The counting principle, when applied to independent events, can be used to find the probability of a compound event.

For example, a family moves in next door. We have heard that our new neighbors include three children, but we do not know how many are boys and how many are girls. The counting principle tells us that the sample space consists of $2 \cdot 2 \cdot 2$, or 8, possibilities for the sex of the three children in this family.

Letting G represent a girl and B represent a boy, we can illustrate the sample space by a tree diagram or by a set of ordered triples, both shown below.

Three Children in a Family

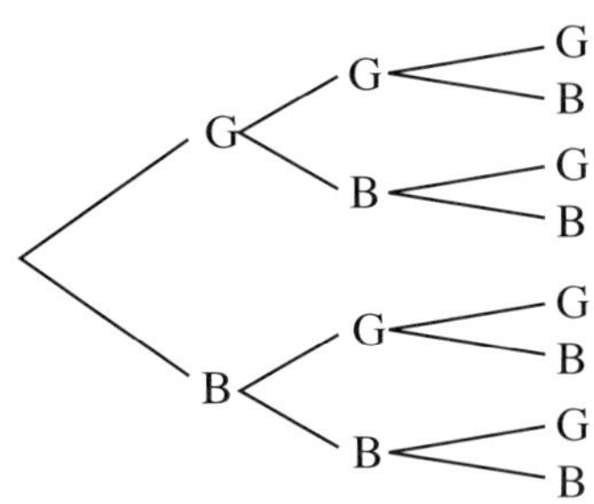

Tree diagram

(G, G, G)
(G, G, B)
(G, B, G)
(G, B, B)
(B, G, G)
(B, G, B)
(B, B, G)
(B, B, B)

List of ordered triples

We have learned that, to find the probability of any event, two values are needed:

1. The number of ways in which event E can occur, or $n(E)$.

2. The total number of possible outcomes in sample space S, or $n(S)$.

The sample space for three children in a family contains eight ordered triples, or $n(S) = 8$. Let us examine two probabilities for this sample space

1. *What is the probability that the family contains two girls and one boy?* By examining the eight outcomes in the sample space, we see that this event can happen in three possible ways: (G, G, B), (G, B, G), (B, G, G).
Therefore, $n(E) = 3$, and

$$P(\text{2 girls and 1 boy}) = \frac{n(E)}{n(S)} = \frac{3}{8}$$

2. *What is the probability that the family contains at least one boy?* When we say "at least one," there may be one boy or two boys or three boys. By examining the sample space of eight outcomes, we see that there are seven possible ways for this event to happen: (G, G, B), (G, B, G), (G, B, B), (B, G, G), (B, G, B), (B, B, G), (B, B, B).

Therefore, $n(E) = 7$ and

$$P(\text{at least one boy}) = \frac{n(E)}{n(S)} = \frac{7}{8}$$

Note: Since "at least one boy" has the same meaning as "not all girls," it is also correct to use the following alternative approach to the problem:

$$P(\text{at least one boy}) = P(\text{not all girls}) = 1 - P(\text{all girls}) = 1 - \frac{1}{8} = \frac{7}{8}$$

PROCEDURE FOR FINDING PROBABILITIES OF COMPOUND EVENTS:

1. List the sample space by constructing a tree diagram or a set of ordered elements.
2. Count the number of outcomes in the sample space, $n(S)$.
3. For the event E being described, count the number of outcomes from the sample space that are elements of the event, $n(E)$.
4. Substitute the numbers for $n(E)$ and $n(S)$ in the rule for the probability of an event E, namely, $P(E) = \frac{n(E)}{n(S)}$.

EXAMPLES

1. A fair coin is tossed 2 times in succession.
   a. List the sample space by using: (*1*) a tree diagram (*2*) a set of ordered pairs (*3*) a graph of ordered pairs
   b. Find the probability of each event: (*1*) Event A = the coin is head up both times. (*2*) Event B = 1 head and 1 tail are tossed.

Solutions **a.** (*1*) Tree diagram (*2*) Set of ordered pairs (*3*) Graph of ordered pairs

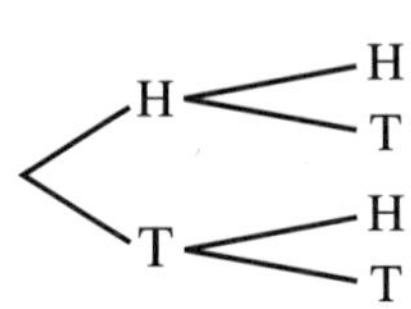

$$\left\{\begin{matrix}(\text{H, H}), (\text{H, T}),\\ (\text{T, H}), (\text{T, T})\end{matrix}\right\}$$

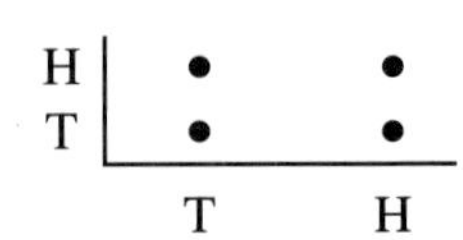

b. (*1*) Event A consists of one pair, a head on both tosses: (H, H).

Since $n(A) = 1$, and $n(S) = 4$:

$$P(A) = \frac{n(A)}{n(S)} = \frac{1}{4} \quad \textit{Answer}$$

(2) Event B consists of two pairs with one head and one tail: (H, T), (T, H).

Since $n(B) = 2$, and $n(S) = 4$:

$$P(B) = \frac{n(B)}{n(S)} = \frac{2}{4} \quad \textit{Answer}$$

---

2. In an experiment, the first step is to pick one number from the set {1, 2, 3}. The second step is to pick one number from the set {3, 5}.

a. Draw a tree diagram or list the sample space of all possible pairs that are outcomes.

b. Determine the probability that:
(1) both numbers are the same
(2) the sum of the numbers is even
(3) the first number is larger than the second

Solutions **a.** Tree diagram

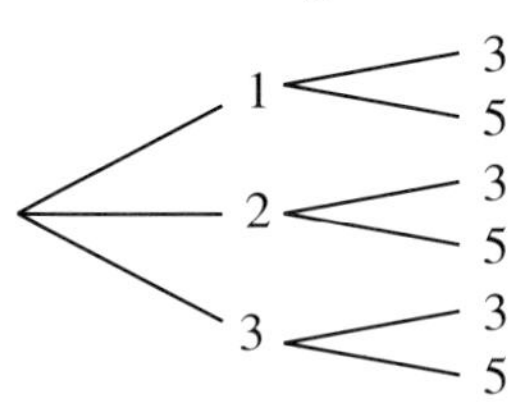

List of ordered pairs

$$\left\{\begin{matrix}(1, 3), (2, 3), (3, 3), \\ (1, 5), (2, 5), (3, 5)\end{matrix}\right\}$$

b. (1) Event A consists of only one pair with both numbers the same: (3, 3).

Since $n(A) = 1$, and $n(S) = 6$:

$$P(A) = \frac{n(A)}{n(S)} = \frac{1}{6} \quad \textit{Answer}$$

(2) Event B consists of four pairs with an even sum: (1, 3), (1, 5), (3, 3), (3, 5).

Since $n(B) = 4$, and $n(S) = 6$:

$$P(B) = \frac{n(B)}{n(S)} = \frac{4}{6} \quad \textit{Answer}$$

(3) Event C is the empty set { }, because there is no pair in which the first number is larger than the second.
Since $n(C) = 0$:

$$P(C) = \frac{n(C)}{n(S)} = \frac{0}{6} = 0 \quad \textit{Answer}$$

---

3. Two standard dice are rolled. Find the probability that the sum of the numbers on the dice is 8.

Solution The sample space consists of $6 \cdot 6$, or 36, outcomes, as shown by the graph of ordered pairs at the right.

For this event E, there are five ordered pairs in which the sum of the numbers on the dice is 8. These five pairs, encircled on the graph, are (2, 6), (3, 5), (4, 4), (5, 3), (6, 2).

Since $n(S) = 36$, and $n(E) = 5$:

$$P(E) = \frac{n(E)}{n(S)} = \frac{5}{36} \quad \textit{Answer}$$

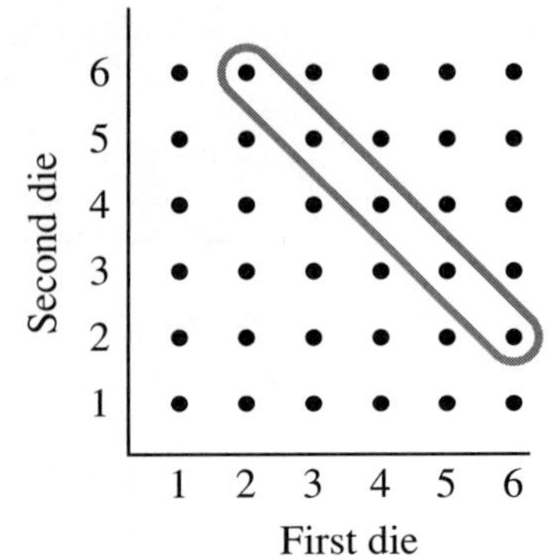

EXERCISES WITH OPEN-RESPONSE PROBLEMS

1. Two fair coins are tossed simultaneously.
a. Draw a tree diagram, or list the sample space of all possible pairs of outcomes.
b. Find P(both coins are tails).
c. Find P(no tails). **d.** Find P(at least one coin is a head).

2. In a family of two children, determine the following probabilities:
a. both are boys **b.** both are girls
c. one is a boy and one is a girl **d.** both are of the same sex
e. there is at least one girl **f.** the younger is a boy
g. the older is a girl and the younger is a boy.

3. In a family of three children, determine the probability that:
a. all are boys **b.** all are girls
c. all are of the same sex **d.** exactly two are boys
e. the youngest is a girl **f.** there is at least one boy
g. the oldest and the youngest are both girls

4. Draw a tree diagram or list the ordered elements of the sample space to indicate the possible sexes of four children in a family. Then reanswer the questions in Exercise 3a–g for the sample space of *four* children.

5. Three fair coins are tossed simultaneously.
a. Indicate the sample space as a tree diagram or as a set of ordered triples.
b. Find P(all are tails). **c.** Find P(there are exactly 2 tails).
d. Find P(there are at least 2 tails).

6. Two standard dice are rolled. To describe the sample space: **a.** Draw a tree diagram. **b.** List the ordered pairs.

7. Two standard dice are rolled. When the numbers on the dice are added, the smallest possible sum is 2, from the pair (1, 1). The largest possible sum is 12, from the pair (6, 6). Find the probability of rolling two dice to get a sum of:
a. 2 **b.** 3 **c.** 4 **d.** 5 **e.** 6 **f.** 7 **g.** 8 **h.** 9 **i.** 10 **j.** 11 **k.** 12

8. What is the sum of all the probabilities obtained in Exercise 7?

9. In a certain game, darts are thrown at two boards so that *each board* will contain exactly *one dart*. If any darts miss or land on a line, they are not counted and the person tries again.

As shown in the diagram, the first board contains three equally likely regions numbered {1, 2, 3}, and the second board contains four equally likely regions lettered {M, A, T, H}.

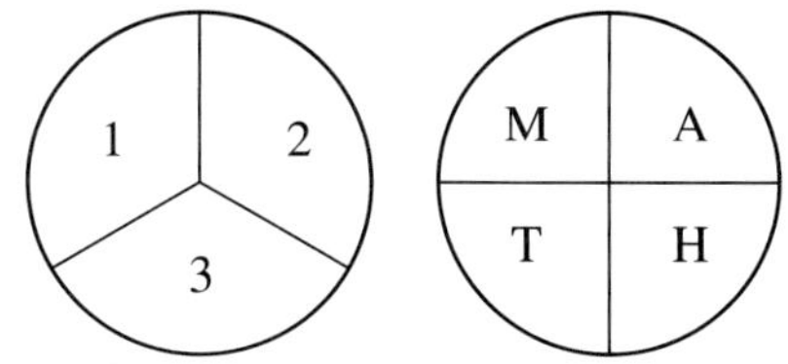

a. Draw a tree diagram or list the sample space of all possible pairs that are outcomes for placing one dart on each board.
b. Find $P(2, \text{M})$, the probability of obtaining the result (2, M).
c. Find $P(3, \text{T})$.
d. Find $P(\text{odd number, H})$.
e. Find $P(\text{even number, vowel})$.
f. Find $P(5, \text{vowel})$.
g. Find $P(\text{odd number, consonant})$.

10. One standard die is rolled twice. Verify that the sample space is the same as when two standard dice are rolled once.

11. One standard die is rolled twice. Find the probability of getting, on two rolls of the die:
a. the pair (3, 3) **b.** the pair (5, 2) **c.** the pair (7, 1)
d. a pair of identical even numbers **e.** a pair of identical odd numbers
f. a pair whose sum is even **g.** a pair whose sum is odd
h. a sum less than 5 **i.** a sum less than 15

In 12–14, select in each case the choice that best completes the statement.

12. When a coin and a standard die are tossed simultaneously, the number of outcomes in the sample space is:
(1) 8 (2) 2 (3) 12 (4) 36

13. A spinner shows three regions, numbered {1, 2, 3}, all equally likely to occur. When the arrow is spun twice, the number of pairs in the outcome set is:
(1) 6 (2) 2 (3) 3 (4) 9

14. Two coins and a standard die are tossed simultaneously. The number of outcomes in the sample space is:
(1) 10 (2) 24 (3) 3 (4) 8

14-4 PERMUTATIONS

Mrs. Hendrix, a teacher, has announced that she will call on three students of her class, Al, Betty, and Chris, to give oral reports today. How many possible ways are there for Mrs. Hendrix to choose the order in which these students will give their reports?

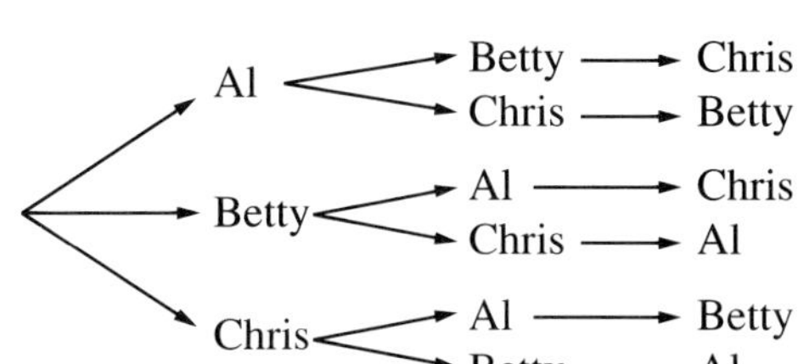

A tree diagram shows that there are six possible orders or arrangements. For example, Al, Betty, Chris is one possible arrangement; Al, Chris, Betty is another. Each of these arrangements is called a permutation. A ***permutation*** is an arrangement of objects or things in some specific order. (In discussing permutations, the words *objects* or *things* are used in a mathematical sense to include all elements in question, whether they are people, numbers, or inanimate objects.)

The six possible permutations in this case may also be shown as a set of ordered triples. Here, we let A represent Al, B represent Betty, and C represent Chris:

$$\{(A, B, C), (A, C, B), (B, A, C), (B, C, A), (C, A, B), (C, B, A)\}$$

Let us see, from another point of view, why there are six possible orders. We know that any one of three students can be called on to give the first report. Once the first report is given, the teacher may call on any one of the two remaining students. After the second report is given, the teacher must call on the one remaining student. Using the counting principle, we see that there are $3 \cdot 2 \cdot 1$, or 6, possible orders.

Consider another situation. A chef is preparing a recipe with 10 ingredients. He puts all of one ingredient in a bowl, followed by all of another ingredient, and so on. How many possible orders are there for placing the 10 ingredients in a bowl, using the stated procedure? Using the counting principle, we have:

$10 \cdot 9 \cdot 8 \cdot 7 \cdot 6 \cdot 5 \cdot 4 \cdot 3 \cdot 2 \cdot 1$, or 3,628,800, possible ways

Factorials

If there are more than 3 million possible ways of placing 10 ingredients in a bowl, can you imagine in how many ways 300 people who want to buy tickets for a football game can be arranged in a straight line? Using the counting principle, we have these factors: $300 \cdot 299 \cdot 298 \cdot 297 \cdot \cdots \cdot 3 \cdot 2 \cdot 1$. To deal with such an example, we make use of a ***factorial symbol*!**. We represent the product of these 300 numbers by the symbol 300!, read as "three hundred factorial" or "factorial 300."

In general, for any natural number n, we define n factorial or factorial n as follows:

$$\mathbf{n! = n(n - 1)(n - 2)(n - 3) \cdot \cdots \cdot 3 \cdot 2 \cdot 1}$$

Note that 1! is the natural number 1.

The factorial key, [x!] Many calculators have a key to evaluate factorials. On most scientific calculators, the factorial value is accessed by first pressing the [2nd] or [INV] or [SHIFT] key. For example, to evaluate 5!, most calculators accept one of these three methods:

Enter:

METHOD 1: 5 [2nd] [x!]

METHOD 2: 5 [INV] [x!]

METHOD 3: 5 [SHIFT] [x!]

Display: 120.

Of course, whether a calculator does or does not have a ***factorial key***, we can always use repeated multiplication to evaluate a factorial:

Enter: 5 [×] 4 [×] 3 [×] 2 [×] 1 [=]

Display: [120.]

All of the methods shown above indicate that $5! = 5 \cdot 4 \cdot 3 \cdot 2 \cdot 1 = 120$.

Calculator displays for factorials When using a calculator, we must keep in mind that factorial numbers are usually very large. If the number of digits in a factorial exceeds the number of places in the display, the scientific calculator will shift from standard decimal notation to scientific notation. Let us evaluate 13! on a scientific calculator:

On a calculator that displays only 8 digits

Enter: 13 [2nd] [x!]

Display: [6.2270208 9]

The answer is in scientific notation: $13! = 6.2270208 \times 10^9$.

On a calculator that displays 10 digits

Enter: 13 [2nd] [x!]

Display: [6227020800.]

The answer is in decimal notation: $13! = 6{,}227{,}020{,}800$.

Representing Permutations

We have said that permutations are arrangements of objects in different orders. For example, the number of different ways for four people to board a bus can be shown as 4!, or $4 \cdot 3 \cdot 2 \cdot 1$, or 24. There are 24 permutations, that is, 24 different arrangements, of these four people, in which all four of them get on the bus.

We may also represent this number of permutations by the symbol ${}_4P_4$. The symbol ${}_4P_4$ is read as: "the permutation of four objects taken four at a time," or as: "the permutation of four things taken four at a time." Here, the letter P represents the word *permutation*.

The small $_4$ written to the lower left of P tells us that four objects are available to be used in an arrangement, as four people waiting for a bus.

The small $_4$ written to the lower right of P tells us how many of these objects are to be used in each arrangement, as in all four people getting on the bus.

Thus, ${}_4P_4 = 4! = 4 \cdot 3 \cdot 2 \cdot 1 = 24$.

Similarly, ${}_5P_5 = 5! = 5 \cdot 4 \cdot 3 \cdot 2 \cdot 1 = 120$.

In the next section, we will study examples where not all the objects are used in the arrangement. We will also examine a calculator key used with permutations. For now, we make the following observation:

- **For any natural number n, the permutation of n objects taken n at a time can be represented as:**

$$ {}_nP_n = n! = n \cdot (n-1) \cdot (n-2) \cdot \cdots \cdot 3 \cdot 2 \cdot 1 $$

EXAMPLES

1. Compute the value of each expression.

a. $6!$ **b.** ${}_2P_2$ **c.** $\frac{7!}{3!}$

Solutions **a.** $6! = 6 \cdot 5 \cdot 4 \cdot 3 \cdot 2 \cdot 1 = 720$

Calculator Solution

Enter: 6 [2nd] [x!]

Display: [720.]

b. ${}_2P_2 = 2! = 2 \cdot 1 = 2$

c. $\frac{7!}{3!} = \frac{7 \cdot 6 \cdot 5 \cdot 4 \cdot \cancel{3 \cdot 2 \cdot 1}}{\cancel{3 \cdot 2 \cdot 1}} = 840$

Calculator Solution

Enter: 7 [2nd] [x!] [÷] 3 [2nd] [x!] [=]

Display: [840.]

Answers **a.** 120 **b.** 2 **c.** 840

---

2. Let any arrangement of letters be called a word even if it has no meaning. Consider the letters in {N, O, W}:

a. How many three-letter words can be formed if each letter is used only once in the word?

b. List the words.

Solutions **a.** Because we are arranging letters in different orders, this is a permutation. Thus, ${}_3P_3 = 3! = 3 \cdot 2 \cdot 1 = 6$ possible words. *Answer*

b. NOW, NWO, ONW, OWN, WNO, WON. *Answer*

---

3. Paul wishes to call Virginia, but he has forgotten her unlisted telephone number. He knows that the exchange is 555, and that the last four digits are 1, 4, 7, and 9, but he cannot remember their order. What is the maximum number of telephone calls that Paul may have to make in order to dial the correct number?

Solution The telephone number is 555-____. Since the last four digits will be an arrangement of 1, 4, 7, and 9, this is a permutation of four numbers, taken four at a time.

Thus, ${}_4P_4 = 4! = 4 \cdot 3 \cdot 2 \cdot 1 = 24$ possible orders.

Answer: The maximum number of calls that Paul may have to make is 24.

EXERCISES WITH OPEN-RESPONSE PROBLEMS

1. Compute the value of each expression.

a. $2!$ **b.** $4!$ **c.** $6!$ **d.** $7!$
e. $3! + 2!$ **f.** $(3 + 2)!$ **g.** ${}_3P_3$ **h.** ${}_8P_8$
i. ${}_5P_5$ **j.** $\frac{8!}{5!}$ **k.** $\frac{15!}{15!}$ **l.** $({}_3P_3) \cdot ({}_4P_4)$

2. Using the letters E, M, I, T: **a.** How many words of four letters can be found if each letter is used only once in the word? **b.** List these words.

3. In how many different ways can five students be arranged in a row?

4. How many three-letter arrangements of the letters X, Y and Z can be made if each letter is used only once in each arrangement?

5. How many different four-digit numbers can be made using the digits 2, 4, 6, and 8 if each digit appears only once in each number?

6. In a game of cards, Gary held exactly one club, one diamond, one heart, and one spade. In how many different ways can Gary arrange these four cards in his hand?

7. There are nine players on a baseball team. The manager must establish a batting order for the players at each game. The pitcher will bat last. How many different batting orders are possible for the eight remaining players on the team?

In 8–10: **a.** Write each answer in factorial form. **b.** Write each answer, after using a scientific calculator, in scientific notation.

8. In how many different ways may 60 people line up to buy tickets at a theater?

9. We learn the alphabet in a certain order, starting with A, B, C, and ending with Z. How many possible orders are there for listing the letters of the English alphabet?

10. In how many different ways can a librarian put 35 different novels on a shelf, with one book following another?

14-5 MORE ABOUT PERMUTATIONS

At times, we deal with situations involving permutations in which we are given n objects, but we use fewer than n objects in each arrangement. For example, Mr. Brown has announced that he will call students from the first row to explain homework problems at the board. The students in the first row are George, Helene, Jay, Karla, and Lou. If there are only two homework problems, and each problem is to be explained by a different student, in how many ways may Mr. Brown select students to go to the board?

We know that the first problem can be assigned to any of five students. Once this problem is explained, the second problem can be assigned to any of the four remaining students. By the counting principle, there are $5 \cdot 4$, or 20, possible selections, as shown in the following table:

Homework Problem	First	Second
Students to choose from	5	4

Another example involves the eight basketball players on a team. In how many ways can three of them be seated on a bench?

By the counting principle, the number of possible seating arrangements is $8 \cdot 7 \cdot 6 = 336$. Observe that the first factor, 8, in the table below is the number of players on the team; once the first player is chosen, the next factor is 7, and so on. Each factor is 1 less than the preceding factor since each time a player is seated one fewer player is available for the next seat.

Seat	First	Second	Third
Players to choose from	8	7	6

The number of factors, 3, is the number of players to be seated or the number of seats available.

Using the language of permutations, we say that the number of permutations of eight different objects taken three at a time is 336. In symbols:

$$_8P_3 = 8 \cdot 7 \cdot 6 = 336$$

The Symbols for Permutations

In general, if we have a set of n different objects, and we make arrangements of r objects from this set, we represent the number of arrangements by the symbol $_nP_r$. The subscript r, representing the number of factors being used, must be less than or equal to n, the total number of objects in the set. Thus:

- **For numbers n and r, where $r \leq n$, the permutation of n objects, taken r at a time, is found by the formula:**

$$_nP_r = \underbrace{n(n-1)(n-2)\cdots}_{r \text{ factors}}$$

Permutations and the Calculator

There are many ways to use a calculator to evaluate a permutation. In the three solutions presented here, we will evaluate the permutation $_8P_3$, which we know is equal to 336.

$$_8P_3 = \underbrace{8 \cdot 7 \cdot 6}_{3 \text{ factors}} = 336$$

Solution 1: Using the Multiplication Key, [×]

A permutation is simply a multiplication of factors. In ${}_8P_3$, the first factor, 8, is multiplied by (8 − 1), or 7, and then by (8 − 2), or 6. Here, exactly three factors have been multiplied.

Enter: 8 [×] 7 [×] 6 [=]

Display: [336.]

This approach can be used for any permutation. In the general permutation ${}_nP_r$, where $r \le n$, the first factor n is multiplied by $(n - 1)$, and then by $(n - 2)$, and so on, until exactly r factors have been multiplied.

Solution 2: Using the Factorial Key, [x!]

When evaluating ${}_8P_3$, it appears as if we start to evaluate 8! but then we stop after multiplying only three factors. There is a way to get the product 8 • 7 • 6 using factorials. As shown here, we divide 8! by 5! to get 8 • 7 • 6, or 336. Also, notice that 8 − 3 = 5 demonstrates a subtraction procedure that helps us find the factorial to use in the denominator:

$$ {}_8P_3 = \frac{8!}{(8-3)!} = \frac{8!}{5!} = \frac{8 \cdot 7 \cdot 6 \cdot \cancel{5 \cdot 4 \cdot 3 \cdot 2 \cdot 1}}{\cancel{5 \cdot 4 \cdot 3 \cdot 2 \cdot 1}} = 8 \cdot 7 \cdot 6 = 336 $$

We now use this approach on a calculator.

Enter: 8 [2nd] [x!] [÷] [(] 8 [−] 3 [)] [2nd] [x!] [=]

Display: [336.]

This approach shows us that there are two formulas that can be used for the general permutation ${}_nP_r$, where $r \le n$:

$$ {}_nP_r = \underbrace{n(n-1)(n-2)\cdots}_{r \text{ factors}} \quad \text{or} \quad {}_nP_r = \frac{n!}{(n-r)!} $$

On a calculator, then, to evaluate any permutation ${}_nP_r$, where $r \le n$, we divide $n!$ by $(n - r)!$

Solution 3: Using the Permutation Key, [nPr]

On some scientific calculators, there is a ***permutation key,*** [nPr], which is usually accessed by first pressing the [2nd] or [INV] or [SHIFT] key. To evaluate ${}_8P_3$ using this key, we first enter 8, then press the [2nd] and [nPr] keys, enter 3, and finally press the [=] key.

Enter: 8 [2nd] [nPr] 3 [=]

Display: [336.]

If this key is on a calculator, it can be used to evaluate any permutation ${}_nP_r$, where $r \le n$. Usually, we first enter n; then press the [2nd] and [nPr] keys; then enter r; and press finally the [=] key. However, not all scientific calculators that have a **nPr** key use this same order for entering n and r; it may be necessary to experiment or to study the calculator manual to find the correct sequence for a particular calculator.

EXAMPLES

1. Evaluate ${}_6P_2$.

Solution This is a permutation of six objects, taken two at a time.

Use the formula

$${}_nP_r = \underbrace{n(n-1)\cdots}_{r \text{ factors}}$$

$${}_6P_2 = \underbrace{6 \cdot 5}_{2 \text{ factors}} = 30$$

or

Use the formula

$${}_nP_r = \frac{n!}{(n-r)!}$$

$${}_6P_2 = \frac{6!}{(6-2)!} = \frac{6!}{4!} = \frac{6 \cdot 5 \cdot \cancel{4 \cdot 3 \cdot 2 \cdot 1}}{\cancel{4 \cdot 3 \cdot 2 \cdot 1}} = 30$$

Calculator Solutions

Enter:

METHOD 1: 6 **×** 5 **=**

METHOD 2: 6 **2nd** **x!** **÷** **(** 6 **−** 2 **)** **2nd** **x!** **=**

METHOD 3: 6 **2nd** **nPr** 2 **=**

Display: 30.

Answer: ${}_6P_2 = 30$

---

2. There are 12 horses in a race. Winning horses are those that cross the finish line in first, second, and third place, commonly called win, place, and show. How many possible winning orders are there for a race with 12 horses?

Solution This is a permutation of 12 objects, taken three at a time, since there are three winning positions in a face. Thus:

$${}_{12}P_3 = \underbrace{12 \cdot 11 \cdot 10}_{3 \text{ factors}} = 1{,}320$$

Answers: There are 1,320 possible winning orders.

---

3. How many three-letter words can be formed from the letters L, O, G, I, C if each letter is used only once in a word?

Solution Forming three-letter words from a set of five letters is a permutation of five, taken three at a time. Thus:

$$_5P_3 = \underbrace{5 \cdot 4 \cdot 3}_{\text{3 factors}} = 60 \quad \text{or} \quad _5P_3 = \frac{5!}{(5-3)!} = \frac{5!}{2!} = \frac{5 \cdot 4 \cdot 3 \cdot \cancel{2 \cdot 1}}{\cancel{2 \cdot 1}} = 60$$

Calculator Solutions

Enter:

METHOD 1: 5 [×] 4 [×] 3 [=]

METHOD 2: 5 [2nd] [x!] [÷] [(] 5 [−] 3 [)] [2nd] [x!] [=]

METHOD 3: 5 [2nd] [nPr] 3 [=]

Display: 60.

Answer: 60 words

---

4. A lottery ticket contains a four-digit number. How many possible four-digit numbers are there when:

a. a digit may appear only once in the number?
b. a digit may appear more than once in the number?

Solutions **a.** If a digit appears only once in a four-digit number, this is a permutation of 10 digits, taken four at a time. Thus:

$$_{10}P_4 = 10 \cdot 9 \cdot 8 \cdot 7 = 5{,}040$$

b. If a digit may appear more than once, we can choose any of 10 digits for the first position, then any of 10 digits for the second position, and so forth. By the counting principle:

$$10 \cdot 10 \cdot 10 \cdot 10 = 10{,}000$$

Answers: **a.** 5,040 **b.** 10,000

EXERCISES WITH OPEN-RESPONSE PROBLEMS

1. Evaluate each expression.

a. $_6P_3$ **b.** $_{10}P_2$ **c.** $_{25}P_2$ **d.** $_4P_3$ **e.** $_{20}P_2$ **f.** $_{11}P_4$

g. $_{22}P_3$ **h.** $_{10}P_4$ **i.** $_7P_6$ **j.** $_{101}P_3$ **k.** $_8P_5$ **l.** $_6P_6$

2. In each case, how many three-letter words can be formed from the given letters, if each letter is used only once in a word?
   a. LION b. TIGER c. MONKEY d. LEOPARD e. MAN

3. There are 30 students in a class. Every day, the teacher calls on different students to write homework problems on the board, with each problem done by only one student. In how many ways can the teacher call students to the board if the homework consists of:
   a. only 1 problem? b. 2 problems? c. 3 problems?

4. Tell how many possible winning orders there are for a horse race where three horses finish in winning positions and there are:
   a. 7 horses b. 9 horses c. 11 horses d. n horses

5. How many different ways are there to label the three vertices of the scalene triangle shown at the right, using no letter more than once, when:
   a. we use the letters R, S, T?
   b. we use all the letters of the English alphabet?

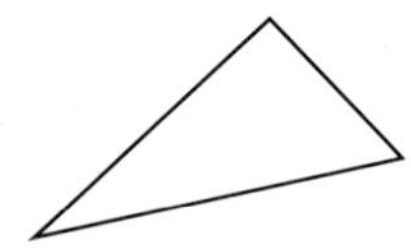

6. A class of 31 students elects four people to office, namely, a president, vice president, secretary, and treasurer. In how many possible ways can four people be elected from this class?

7. How many possible ways are there to write two initials, using the letters of the English alphabet, if: a. an initial may appear only once in each pair? b. the same initial may be used twice?

In 8–12, use a calculator to evaluate each permutation.

8. ${}_{27}P_4$ 9. ${}_{98}P_4$ 10. ${}_{217}P_3$ 11. ${}_{10}P_{10}$ 12. ${}_{36}P_5$

13. Which of these two values, if either, is larger: ${}_9P_9$ or ${}_9P_8$? Explain your answer.

14. Which expression has the greatest value?
   (1) ${}_{60}P_5$ (2) ${}_{45}P_6$ (3) ${}_{24}P_7$ (4) ${}_{19}P_7$

14-6 PROBABILITY WITHOUT REPLACEMENT; PROBABILITY WITH REPLACEMENT

Without Replacement

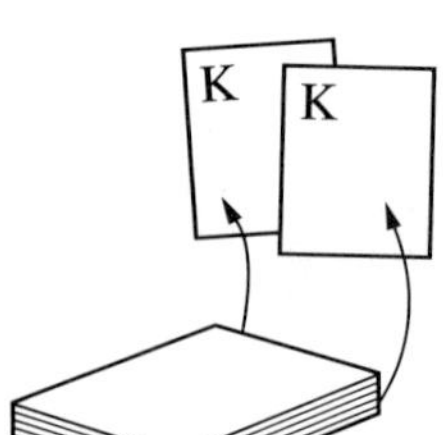

Two cards are drawn at random from an ordinary pack of 52 cards. In this situation, a single card is drawn from a deck of 52 cards, and then a second card is drawn from the remaining 51 cards in the deck. What is the probability that both cards drawn are kings?

On the first draw, there are four kings in the deck of 52 cards. Thus, $P(\text{first king}) = \frac{4}{52}$. If a second card is drawn without replacing the first king selected, there are now only three kings in the 51 cards remaining. Therefore, $P(\text{second king}) = \frac{3}{51}$.

By the counting principle:

$$
\begin{aligned}
P(\text{both kings}) &= P(\text{first king}) \bullet P(\text{second king}) \\
&= \frac{4}{52} \bullet \frac{3}{51} \\
&= \frac{1}{13} \bullet \frac{1}{17} \\
&= \frac{1}{221}
\end{aligned}
$$

This is called a problem *without replacement* because the first king drawn was not placed back into the deck. Typical problems without replacement include spending coins from one's pocket, eating jellybeans from a jar, and choosing students to give reports.

With Replacement

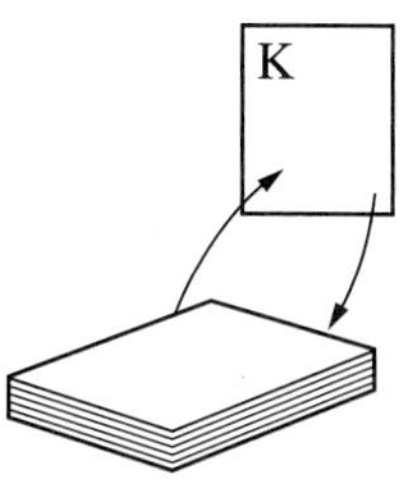

A card is drawn at random from an ordinary deck, the card is placed back into the deck, and a second card is then drawn and replaced. In this situation, it is clear the deck contains 52 cards each time that a card is drawn. What is the probability that each time the card drawn is a king?

On the first draw, there are four kings in the deck of 52 cards. Thus, $P(\text{first king}) = \frac{4}{52}$. If the first king drawn is now placed back into the deck, then, on the second draw, there are again four kings in the deck of 52 cards. Thus, $P(\text{second king}) = \frac{4}{52}$.

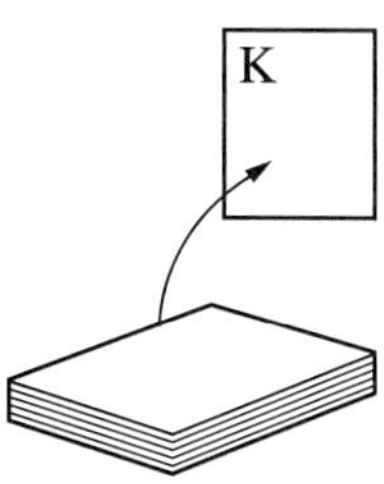

By the counting principle:

$$
\begin{aligned}
P(\text{both kings}) &= P(\text{first king}) \bullet P(\text{second king}) \\
&= \frac{4}{52} \bullet \frac{4}{52} \\
&= \frac{1}{13} \bullet \frac{1}{13} \\
&= \frac{1}{169}
\end{aligned}
$$

This is called a problem *with replacement* because the first card drawn was placed back into the deck. In this case, since the card drawn is replaced, the number of cards in the deck remains constant.

Rolling two dice is similar to drawing two cards with replacement because the number of faces on each die remains constant, as did the number of cards in the deck. Typical problems with replacement include rolling dice, tossing coins (each coin always has two sides), and spinning arrows.

KEEP IN MIND

1. If the problem does not specifically mention *with replacement* or *without replacement*, ask yourself: "Is this problem with or without replacement?"
2. For many compound events, the probability can be determined most easily by using the counting principle.
3. Every probability problem can always be solved by:
   - counting the number of elements in the same space, $n(S)$;
   - counting the number of outcomes in the event, $n(E)$;
   - and substituting these numbers in the probability formula,

$$P(E) = \frac{n(E)}{n(S)}.$$

EXAMPLES

1. A fair die is thrown three times. What is the probability that a 5 comes up each time?

Solution This is a problem with replacement.

On the first toss, there is one way to obtain a 5 so $P(5) = \frac{1}{6}$.

On the second toss, there is one way to obtain a 5 so $P(5) = \frac{1}{6}$.

On the third toss, there is one way to obtain a 5 so $P(5) = \frac{1}{6}$.

By the counting principle:

$$\begin{aligned} P(\text{rolling 5 each time}) &= P(\text{5 on first}) \bullet P(\text{5 on second}) \bullet P(\text{5 on third}) \\ &= \frac{1}{6} \bullet \frac{1}{6} \bullet \frac{1}{6} \\ &= \frac{1}{216} \quad \textit{Answer} \end{aligned}$$

---

2. If two cards are drawn from an ordinary deck without replacement, what is the probability that the cards form a pair?

Solution On the first draw, any card at all may be chosen, so:

$$P(\text{any card}) = \frac{52}{52}$$

There are now 51 cards left in the deck. Of these 51, there are three that match the first card taken, to form a pair, so:

$$P(\text{second card forms a pair}) = \frac{3}{51}$$

Then:

$$\begin{aligned} P(\text{pair}) &= P(\text{any card}) \bullet P(\text{second card forms a pair}) \\ &= \frac{52}{52} \bullet \frac{3}{51} \\ &= \frac{1}{1} \bullet \frac{1}{17} = \frac{1}{17} \quad \textit{Answer} \end{aligned}$$

3. A jar contains four white marbles and two blue marbles, all the same size. A marble is drawn at random and not replaced. A second marble is then drawn from the jar. Find the probability that: **a.** both marbles are white **b.** both marbles are blue **c.** both marbles are the same color

Solutions **a.** On the first draw:

$$P(\text{white}) = \frac{4}{6}$$

Since the white marble drawn is not replaced, five marbles, of which three are white, are left in the jar. Thus, on the second draw:

$$P(\text{white}) = \frac{3}{5}$$

Then:

$$P(\text{both white}) = \frac{4}{6} \bullet \frac{3}{5} = \frac{12}{30}, \text{ or } \frac{2}{5} \quad \textit{Answer}$$

b. When we start with a full jar of six marbles, on the first draw:

$$P(\text{blue}) = \frac{2}{6}$$

Since the blue marble drawn is not replaced, five marbles, of which only one is blue, are left in the jar. Thus, on the second draw:

$$P(\text{blue}) = \frac{1}{5}$$

Then:

$$P(\text{both blue}) = \frac{2}{6} \bullet \frac{1}{5} = \frac{2}{30}, \text{ or } \frac{1}{15} \quad \textit{Answer}$$

c. If both marbles are the same color, both are white or both are blue. These are disjoint events, so,

$$P(A \text{ or } B) = P(A) + P(B)$$

Therefore:

$$\begin{aligned} P(\text{both white or both blue}) &= P(\text{both white}) + P(\text{both blue}) \\ &= \frac{4}{6} \bullet \frac{3}{5} + \frac{2}{6} \bullet \frac{1}{5} \\ &= \frac{12}{30} + \frac{2}{30} \\ &= \frac{14}{30}, \text{ or } \frac{7}{15} \quad \textit{Answer} \end{aligned}$$

---

4. Fred has two quarters and one nickel in his pocket. The pocket has a hole in it, and a coin drops out. Fred picks up the coin and puts it back into his pocket. A few minutes later, a coin drops out of his pocket again.

a. Draw a tree diagram or list the sample space for all possible pairs that are outcomes to describe the coins that fell.

b. What is the probability that the same coin fell out of Fred's pocket both times?

c. What is the probability that the two coins that fell have a total value of 30 cents?

d. What is the probability that a quarter fell out at least once?

Solutions **a.** Because there are two quarters, use subscripts. The three coins are $\{Q_1, Q_2, N\}$, where Q represents a quarter and N represents a nickel. This is a problem with replacement.

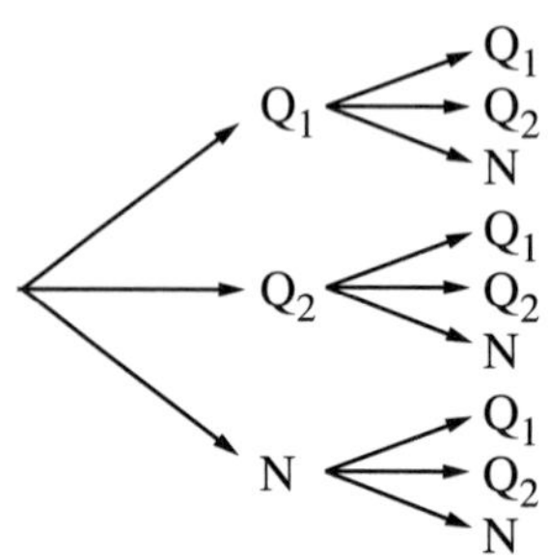

or

$$\left\{\begin{matrix} (Q_1, Q_1), (Q_1, Q_2), (Q_1, N), \\ (Q_2, Q_1), (Q_2, Q_2), (Q_2, N), \\ (N, Q_1), (N, Q_2), (N, N) \end{matrix}\right\}$$

b. Of the nine outcomes, three name the same coin both times: (Q_1, Q_1); (Q_2, Q_2); (N, N), so:

$$P(\text{same coin}) = \frac{3}{9}, \text{ or } \frac{1}{3} \quad \textit{Answer}$$

c. Of the nine outcomes, four consist of a quarter and a nickel, which total 30 cents: (Q_1, N); (Q_2, N); (N, Q_1); (N, Q_2), so:

$$P(\text{coins total 30 cents}) = \frac{4}{9} \quad \textit{Answer}$$

d. Of the nine outcomes, eight contain one or more quarters, the equivalent of at least one quarter. The only outcome not counted is (N, N), so:

$$P(\text{at least one quarter}) = \frac{8}{9} \quad \textit{Answer}$$

EXERCISES WITH OPEN-RESPONSE PROBLEMS

1. A jar contains two red and five yellow marbles. If one marble is drawn at random, what is the probability that it is: **a.** red? **b.** yellow?

2. A jar contains two red and five yellow marbles. A marble is drawn at random and then replaced. A second draw is made at random. Find the probability that:
a. both marbles are red
b. both marbles are yellow
c. both marbles are the same color
d. the marbles are different in color

3. A jar contains two red and five yellow marbles. A marble is drawn at random. Then without replacement, a second draw is made at random. Find the probability that:
a. both marbles are red
b. both marbles are yellow
c. both marbles are the same color
d. the marbles are different in color

4. In an experiment, the arrow is spun twice on a wheel containing four equally likely regions, numbered 1 through 4, as shown in the figure.

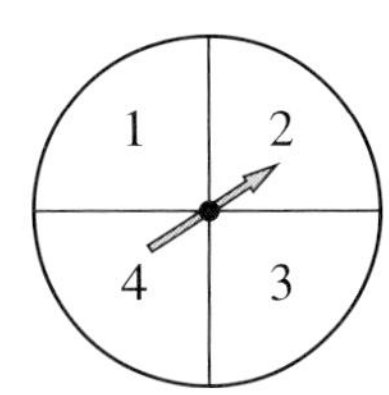

a. Indicate the sample space by drawing a tree diagram or writing a set of ordered pairs.
b. Find the probability of spinning the digits 2 and 3 in that order.
c. Find the probability that the same digit is spun both times.
d. What is the probability that the first digit spun is larger than the second?

5. Sal has a bag of hard candies: three are lemon and two are grape. He ate two of the candies while waiting for a bus, selecting them at random one after another.
a. Using subscripts, draw a tree diagram or list the sample space of all possible outcomes showing which candies were eaten.
b. Find the probability that:
(*1*) both candies were lemon
(*2*) neither candy was lemon
(*3*) the candies were the same flavor
(*4*) at least one candy was lemon

6. Carol has five children: three girls and two boys. One of her children was late for lunch. Later that day, one of her children was late for supper.

a. Indicate the sample space by a tree diagram or list of ordered pairs showing which children were late.

b. If each child was equally likely to be late, find the probability that:

(*1*) both children who were late were girls (*2*) both children who were late were boys

(*3*) the same child was late both times

(*4*) at least one of the children who was late was a boy

7. Several players start playing a game with a full deck of 52 cards. Each player draws two cards at random, one at a time. Find the probability that:

a. Flo draws two jacks **b.** Frances draws two hearts

c. Jerry draws two red cards **d.** Mary draws two picture cards

e. Ann does not draw a pair **f.** Stephen draws two black kings

g. Carrie draws a 5 and a 10 in that order **h.** Bill draws a heart and a club in that order

8. Saverio had four coins: a half dollar, a quarter, a dime, and a nickel. He chose one of the coins and put it in a bank. Later he chose another coin and also put that in the bank.

a. Indicate the sample space of coins saved.

b. If each coin was equally likely to have been saved, find the probability that the coins saved:

(*1*) were worth a total of 35¢ (*2*) added to an even amount

(*3*) included the half dollar (*4*) were worth a total of less than 30¢

9. Farmer Brown must wake up before sunrise to start his chores. Dressing in the dark, he reaches into a drawer and pulls out two loose socks. There are eight white socks and six red socks in the drawer.

a. Find the probability that both socks are: (*1*) white (*2*) red (*3*) the same color

b. Find the minimum number of socks Farmer Brown must pull out of the drawer to guarantee that he will get a matching pair.

10. Tillie had three quarters and four dimes in her purse. She took out a coin at random, but it slipped from her hand and fell back into her purse. She reached in and again picked out a coin at random. Find the probability that:

a. both coins were quarters **b.** both coins were dimes

c. the coins were a dime and a quarter, in any order

d. the same coin was picked both times

e. the coins picked totaled less than 40¢

f. the coins picked totaled exactly 30¢

g. at least one of the coins picked was a quarter

11. A jar contained nine orange disks and three blue disks. A girl chose one at random and then, without replacing it, chose another. Letting O represent orange and B represent blue:

a. Find the probability of each of the following outcomes:

(*1*) (O, O) (*2*) (O, B) (*3*) (B, O) (*4*) (B, B)

b. Now, find for the disks chosen, the probability that:

(*1*) neither was orange (*2*) only one was blue

(*3*) at least one was orange (*4*) they were the same color

(*5*) at most one was orange (*6*) they were the same disk

14-7 COMBINATIONS

Comparing Permutations and Combinations

Before we discuss combinations, let us start with a problem we know how to solve.

PROBLEM 1. Ann, Barbara, Carol, and Dave are the only members of a school club. In how many different ways can they elect a president and a treasurer for the club?

Any one of 4 students can be elected as president. After this happens, any one of the 3 remaining students can be elected as treasurer. Thus, there are $4 \cdot 3$, or 12, possible arrangements. Using initials to represent the students involved, we can write these 12 arrangements:

$$\begin{array}{llll} (A, B), & (B, A), & (B, C), & (C, B), \\ (A, C), & (C, A), & (B, D), & (D, B), \\ (A, D), & (D, A), & (C, D), & (D, C). \end{array}$$

This is the number of permutations of 4 things, taken 2 at a time. We could simply have said:

$$_4P_2 = 4 \cdot 3 = 12$$

Answer: 12 arrangements

Now let us consider two problems of a different type involving the members of the same school club.

PROBLEM 2. Ann, Barbara, Carol, and Dave are the only members of a school club. In how many ways can they choose 2 people to represent the club at student council meetings?

Looking carefully at the list of 12 arrangements given in answer to the first problem, we can see that, while (A, B) and (B, A) are two different slates for president and treasurer, sending Ann and Barbara to the student council is exactly the same as sending Barbara and Ann. For this problem, let us match up answers that are duplicates:

$$\begin{array}{ll} (A, B) \leftrightarrow (B, A) & (B, C) \leftrightarrow (C, B) \\ (A, C) \leftrightarrow (C, A) & (B, D) \leftrightarrow (D, B) \\ (A, D) \leftrightarrow (D, A) & (C, D) \leftrightarrow (D, C) \end{array}$$

Although order is important in listing slates of officers in the first problem, there is no reason to order the elements in this one. In fact, if we think of 2 representatives to the student council as a *set* of 2 students, we can list all possible sets of 2 representatives as follows:

{A, B} {B, C}
{A, C} {B, D}
{A, D} {C, D}

From this list we can find the number of ***combinations*** of 4 things, taken 2 at a time, written in symbols as ${}_4C_2$. The answer to this question is found by dividing the number of permutations of 4 things, taken 2 at a time, by 2! orders. Thus:

$$ {}_4C_2 = \frac{{}_4P_2}{2!} = \frac{4 \cdot 3}{2 \cdot 1} = \frac{12}{2} = 6 $$

Answer: 6 combinations

PROBLEM 3. Ann, Barbara, Carol, and Dave are the only members of a school club. In how many ways can they select a 3-person committee to work on the club's next project?

Is *order* important to this answer? If 3 officers were to be elected, such as a president, a treasurer, and a secretary, then order would be important and a permutation would be found. However, a committee is merely a set of people. In listing the elements of a set, order is not important; thus, a combination is used. Let us compare the permutations and combinations that follow.

Permutations						*Combinations*
(A, B, C)	(A, C, B)	(B, A, C)	(B, C, A)	(C, A, B)	(C, B, A)	{A, B, C}
(A, B, D)	(A, D, B)	(B, A, D)	(B, D, A)	(D, A, B)	(D, B, A)	{A, B, D}
(A, C, D)	(A, D, C)	(C, A, D)	(C, D, A)	(D, A, C)	(D, C, A)	{A, C, D}
(B, C, D)	(B, D, C)	(C, B, D)	(C, D, B)	(D, B, C)	(D, C, B)	{B, C, D}

While there are 24 permutations, written as ordered triples, there are only 4 combinations, written as sets. For example, in the first row of permutations, there are 3!, or $3 \cdot 2 \cdot 1$, or 6, ordered triples containing A, B, and C. However, there is only one set that contains A, B, and C, namely, the set {A, B, C}.

Therefore, the number of ways to select a 3-person committee from a group of 4 people is the number of *combinations* of 4 things, taken 3 at a time, written in symbols as ${}_4C_3$. The answer to this problem is found by dividing the number of permutations of 4 things, taken 3 at a time, by 3! orders. Thus:

$$ {}_4C_3 = \frac{{}_4P_3}{3!} = \frac{4 \cdot 3 \cdot 2}{3 \cdot 2 \cdot 1} = \frac{4 \cdot \cancel{3} \cdot \cancel{2}}{\cancel{3} \cdot \cancel{2} \cdot 1} = 4 $$

Answer: 4 combinations

- **In general, for counting numbers n and r, where $r \leq n$, the number of combinations of n things, taken r at a time, is found by using the formula**

$$_nC_r = \frac{_nP_r}{r!}$$

On most scientific calculators, there is a ***combination key***, [nCr], which is usually accessed by first pressing the [2nd] or [INV] or [SHIFT] key. For example, let us evaluate $_4C_3$.

Enter: 4 [2nd] [nCr] 3 [=]

Display: [4.]

Not all scientific calculators use the order of keystrokes shown here. The sequence to use with the [nCr] key on a particular calculator will be the same as the sequence that is used with the [nPr] key.

Note. The notation $\binom{n}{r}$ also represents the number of combinations of n things, taken r at a time. Thus:

$$\binom{n}{r} = {_nC_r} \quad \text{or} \quad \binom{n}{r} = \frac{_nP_r}{r!}$$

Some Relationships Involving Combinations

1. Given a group of 5 people, how many different 5-person committees can be formed? Common sense tells us that there is only 1 such committee, namely, the committee consisting of all 5 people. Using combinations, we see that

$$_5C_5 = \frac{_5P_5}{5!} = \frac{5 \cdot 4 \cdot 3 \cdot 2 \cdot 1}{5 \cdot 4 \cdot 3 \cdot 2 \cdot 1} = 1$$

Also,

$$_3C_3 = \frac{_3P_3}{3!} = \frac{3 \cdot 2 \cdot 1}{3 \cdot 2 \cdot 1} = 1 \quad \text{and} \quad _4C_4 = \frac{_4P_4}{4!} = \frac{4 \cdot 3 \cdot 2 \cdot 1}{4 \cdot 3 \cdot 2 \cdot 1} = 1.$$

Observation 1: **For any counting number n, $_nC_n = 1$.**

2. Given a group of 5 people, in how many different ways can we select a committee consisting of no people, or 0 people? Common sense tells us that there is only 1 way to select no one. Thus, using combinations, $_5C_0 = 1$. Let us agree to the following generalization:

Observation 2: **For any counting number n, $_nC_0 = 1$.**

3. In how many ways can we select a committee of 2 people from a group of 7 people? Since a committee is a combination,

$$_7C_2 = \frac{_7P_2}{2!} = \frac{7 \cdot 6}{2 \cdot 1} = \frac{7 \cdot \overset{3}{\cancel{6}}}{\cancel{2} \cdot 1} = 21$$

Now, given a group of 7 people, in how many ways can 5 people *not* be appointed to the committee? Since each set of people not appointed is a combination:

$$_7C_5 = \frac{_7P_5}{5!} = \frac{7 \cdot 6 \cdot 5 \cdot 4 \cdot 3}{5 \cdot 4 \cdot 3 \cdot 2 \cdot 1} = \frac{7 \cdot \overset{3}{\cancel{6}} \cdot \cancel{5} \cdot \cancel{4} \cdot \cancel{3}}{\cancel{5} \cdot \cancel{4} \cdot \cancel{3} \cdot \cancel{2} \cdot 1} = 21$$

Notice that $_7C_2 = {_7C_5}$. In other words, starting with a group of 7 people, the number of sets of 2 people that can be selected is equal to the number of sets of 5 people that can be *not* selected. In the same way it can be shown that $_7C_3 = {_7C_4}$, that $_7C_6 = {_7C_1}$, and that $_7C_7 = {_7C_0}$.

In general, starting with n objects, the number of ways to choose r objects for a combination is equal to the number of ways to *not* choose $(n - r)$ objects for the combination.

Observation 3: **For whole numbers *n* and *r*, where $r \leq n$,**

$$_nC_r = {_nC_{n-r}}$$

KEEP IN MIND

Permutations	*Combinations*
$_nP_r = \underbrace{n(n-1)(n-2)\ldots}_{r \text{ factors}}$	$_nC_r = \frac{_nP_r}{r!}$
1. *Order is important.* Think of ordered elements such as ordered pairs and ordered triples. 2. An *arrangement* indicates a permutation.	1. *Order is not important.* Think of sets. 2. A *committee*, or a *selection of a group*, indicates a combination.

EXAMPLES

1. Evaluate: $_{10}C_3$.

Solution This is the number of combinations of 10 things, taken 3 at a time. Thus:

$$_{10}C_3 = \frac{_{10}P_3}{3!} = \frac{10 \cdot 9 \cdot 8}{3 \cdot 2 \cdot 1} = \frac{10 \cdot \overset{3}{\cancel{9}} \cdot \overset{4}{\cancel{8}}}{\cancel{3} \cdot \cancel{2} \cdot 1} = 120$$

Calculator Solution

Enter: 10 [2nd] [nCr] 3 [=]

Display: [120.]

Answer: 120

---

2. Evaluate: $\binom{25}{23}$.

Solution (1) This is an alternative form for the number of combinations of 25 things, taken 23 at a time: $\binom{25}{23} = {}_{25}C_{23}$.

(2) Since ${}_nC_r = {}_nC_{n-r}$, we write: ${}_{25}C_{23} = {}_{25}C_2$.

(3) Using ${}_{25}C_2$, perform the easier computation. Thus:

$$\binom{25}{23} = {}_{25}C_{23} = {}_{25}C_2 = \frac{{}_{25}P_2}{2!} = \frac{25 \cdot \overset{12}{\cancel{24}}}{\cancel{2} \cdot 1} = 300$$

Calculator Solution

Enter:

METHOD 1: 25 [2nd] [nCr] 23 [=]

METHOD 2: 25 [2nd] [nCr] 2 [=]

Display: [300.]

Answer: 300

---

3. There are 10 teachers in the science department. How many 4-person committees can be formed in the department if Mrs. Martens and Dr. Blumenthal, 2 of the teachers, must be on each committee?

Solution Since Mrs. Martens and Dr. Blumenthal must be on each committee, the problem becomes one of filling 2 positions on a committee from the remaining 8 teachers.

$${}_8C_2 = \frac{{}_8P_2}{2!} = \frac{8 \cdot 7}{2 \cdot 1} = \frac{\overset{4}{\cancel{8}} \cdot 7}{\cancel{2} \cdot 1} = 28$$

Answer: 28 committees

---

4. There are 6 points in a plane, no 3 of which are collinear. How many straight lines can be drawn using pairs of these 6 points?

Solution Whether joining points A and B, or points B and A, only 1 line exists, namely, $\overleftrightarrow{AB}$. Since order is not important here, this is a combination of 6 points, taken 2 at a time.

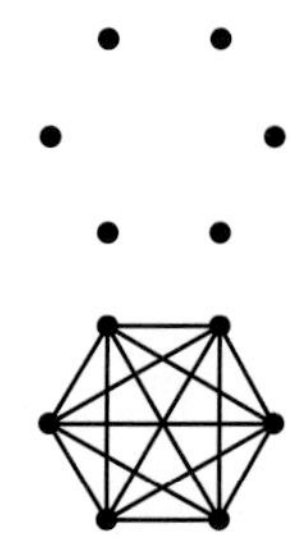

$$_6C_2 = \frac{_6P_2}{2!} = \frac{6 \cdot 5}{2 \cdot 1} = \frac{\overset{3}{\cancel{6}} \cdot 5}{\cancel{2} \cdot 1} = 15$$

Answer: 15 lines

---

5. Kathy Dwyer is a teacher at a local high school. In her class, there are 10 boys and 20 girls. Find the number of ways in which Ms. Dwyer can select a team of 3 students from the class to work on a group project if the team consists of:
a. any 3 students **b.** 1 boy and 2 girls **c.** 3 girls **d.** at least 2 girls

Solutions **a.** The class contains 10 boys and 20 girls, for a total of 30 students. Since order is not important on a team, this is a combination of 30 students, taken 3 at a time.

$$_{30}C_3 = \frac{_{30}P_3}{3!} = \frac{30 \cdot 29 \cdot 28}{3 \cdot 2 \cdot 1} = \frac{\overset{10}{\cancel{30}} \cdot 29 \cdot \overset{14}{\cancel{28}}}{\cancel{3} \cdot \cancel{2} \cdot 1} = 4{,}060$$

b. This is a compound event. To find the number of ways to select 1 boy out of 10 boys for a team, use $_{10}C_1$. To find the number of ways to select 2 girls out of 20 for the team, use $_{20}C_2$. Then, by the counting principle, multiply the results.

$$_{10}C_1 \cdot {_{20}C_2} = \frac{10}{1} \cdot \frac{20 \cdot 19}{2 \cdot 1} = \frac{10}{1} \cdot \frac{\overset{10}{\cancel{20}} \cdot 19}{\cancel{2} \cdot 1}$$
$$= 10 \cdot 190 = 1{,}900$$

c. This is another compound event, in which 0 boy out of 10 boys is selected. Recall that $_{10}C_0 = 1$. Thus:

$$_{10}C_0 \cdot {_{20}C_3} = 1 \cdot \frac{20 \cdot 19 \cdot 18}{3 \cdot 2 \cdot 1} = 1 \cdot \frac{20 \cdot 19 \cdot \overset{3}{\cancel{18}}}{\cancel{3} \cdot \cancel{2} \cdot 1}$$
$$= 1 \cdot 1{,}140 = 1{,}140$$

d. A team of at least 2 girls can consist of exactly 2 girls (see part **b**) or exactly 3 girls (see part **c**). Since these events are disjoint, add the solutions to parts **b** and **c**:

$$1{,}900 + 1{,}140 = 3{,}040 \text{ teams}$$

Answers: **a.** 4,060 teams **b.** 1,900 teams **c.** 1,140 teams **d.** 3,040 teams

EXERCISES WITH OPEN-RESPONSE PROBLEMS

In 1–12, evaluate each expression.

1. ${}_{15}C_2$ **2.** ${}_{12}C_3$ **3.** ${}_{10}C_4$ **4.** ${}_{25}C_1$ **5.** ${}_{13}C_0$ **6.** ${}_{14}C_{14}$

7. ${}_{9}C_8$ **8.** ${}_{20}C_{18}$ **9.** $\binom{7}{3}$ **10.** $\binom{9}{4}$ **11.** $\binom{17}{17}$ **12.** $\binom{8}{0}$

13. Find the number of combinations of 6 things, taken 3 at a time.
14. How many different committees of 3 people can be chosen from a group of 9 people?
15. How many different subsets of exactly 7 elements can be formed from the set {0, 1, 2, 3, 4, 5, 6, 7, 8, 9}?
16. For each given number of noncollinear points in a plane, how many straight lines can be drawn?
a. 3 **b.** 4 **c.** 5 **d.** 7
e. 8 **f.** 10 **g.** n

17. A coach selects players for a team. If, while making this first selection, the coach pays no attention to the positions that individuals will play, how many teams may be formed?
a. Of 14 candidates, Coach Richko needs 5 for a basketball team.
b. Of 16 candidates, Coach Jones needs 11 for a football team.
c. Of 13 candidates, Coach Greves needs 9 for a baseball team.

18. A disc jockey has 25 recordings at hand, but has time to play only 22 on the air. How many groups of 22 recordings can be selected?
19. There are 14 teachers in a mathematics department.
a. How many 4-person committees can be formed in the department?
b. How many 4-person committees can be formed if Mr. McDonough, 1 of the 14, must be on the committee?
c. How many 4-person committees can be formed if Mr. Goldstein and Mrs. Friedel, 2 of the 14, must be on the committee?

20. There are 12 Republicans and 10 Democrats on a senate committee. From this group, a 3-person subcommittee is to be formed. Find the number of 3-person subcommittees that consist of:
a. any members of the senate committee **b.** Democrats only
c. 1 Republican and 2 Democrats **d.** at least 2 Democrats
e. John Clark, who is a Democrat, and any 2 Republicans

21. Sue Bartling loves to read mystery books and car-repair manuals. On a visit to the library, Sue finds 9 new mystery books and 3 car-repair manuals. She borrows 4 of these books. Find the number of different sets of 4 books Sue can borrow if:
a. all are mystery books **b.** exactly 2 are mystery books
c. only 1 is a mystery book **d.** all are car-repair manuals

22. Mr. Spartalis owns 4 suits, 8 shirts, and 10 ties. He plans to travel and packs 2 suits, 4 shirts, and 3 ties. How many different sets of clothing could Mr. Spartalis have chosen?

23. Cards are drawn at random from a 52-card deck. Find the number of different 5-card poker hands possible consisting of:

a. any 5 cards from the deck
b. 3 aces and 2 kings
c. 4 queens and any other card
d. 5 spades
e. 2 aces and 3 picture cards
f. 5 jacks

24. Consider the following formulas:

$$(1)\ {}_nC_r = \frac{{}_nP_r}{r!}$$

$$(2)\ {}_nC_r = \frac{n!}{(n-r)!r!}$$

$$(3)\ {}_nC_r = \frac{n(n-1)(n-2)\ldots(n-r+1)}{r(r-1)(r-2)\ldots(1)}$$

a. Evaluate ${}_8C_3$ using each of the 3 formulas.
b. Evaluate ${}_{11}C_7$ using each of the 3 formulas.
c. Can all 3 formulas be used to find the combination of n things, taken r at a time?

25. How many committees consisting of 7 people or more can be formed from a group of 10 people?

26. There are 12 roses growing in Heather's garden. How many different bunches, each consisting of more than 7 roses, can Heather cut to bring into the house?

14-8 PROBABILITY AND COMBINATIONS

The probability of an event E is defined as $P(E) = \frac{n(E)}{n(S)}$. We have seen that $n(E)$ and $n(S)$ are sometimes found by counting permutations. In the examples that follow, we will see that $n(E)$ and $n(S)$ are sometimes found by counting combinations.

As explained in Section 14-7, if a probability question is asked in which order is not important, $n(E)$ and $n(S)$ should be found by using combinations.

EXAMPLES

1. Two cards are drawn at random from a standard deck of 52 cards, without replacement. What is the probability that both cards drawn are fives?

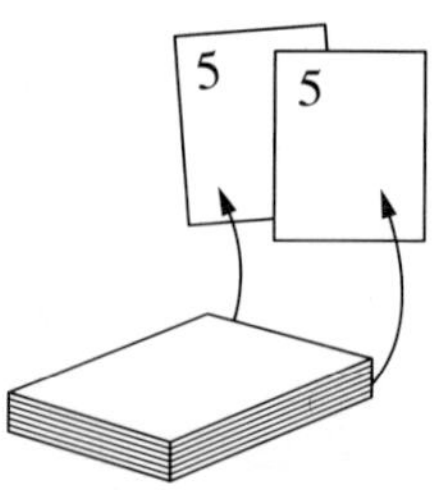

Solutions METHOD 1: *Counting Principle*

(1) On the first draw, there are 4 fives in a deck of 52 cards, so

$P(\text{first } 5) = \frac{4}{52}$.

(2) Assume a five is drawn. Then, on the second draw, there are 3 fives in the 51 cards remaining, so

$P(\text{second } 5) = \frac{3}{51}$.

(3) Apply the counting principle of probability.

$$P(\text{both } 5) = P(\text{first } 5) \cdot P(\text{second } 5) = \frac{4}{52} \cdot \frac{3}{51} = \frac{1}{13} \cdot \frac{1}{17} = \frac{1}{221}$$

METHOD 2: *Permutations*

(1) For the sample space S, treat the number of ways to draw 2 cards from a deck of 52 as a permutation. Then $n(S) = {}_{52}P_2$.

(2) For event E, treat the number of ways to draw 2 cards out of 4 fives in the deck as a permutation. Then $n(E) = {}_4P_2$.

(3) Apply the rule of probability.

(4) Simplify the fraction.

If event E = both cards are fives:

$$P(E) = \frac{n(E)}{n(S)} = \frac{{}_4P_2}{{}_{52}P_2} = \frac{4 \cdot 3}{52 \cdot 51} = \frac{1}{221}$$

Answer: $P(\text{both } 5\text{'s}) = \frac{1}{221}$

METHOD 3: *Combinations*

(1) For sample space S, each set of 2 cards from a deck of 52 is a combination. Thus, $n(S) = {}_{52}C_2$.

(2) For event E, each set of 2 cards from any of 4 fives in the deck is a combination. Thus, $n(E) = {}_4C_2$.

(3) Apply the rule of probability.

(4) Simplify the fraction.

If event E = both cards are fives,

$$P(E) = \frac{n(E)}{n(S)} = \frac{{}_4C_2}{{}_{52}C_2} = \frac{\frac{4 \cdot 3}{2 \cdot 1}}{\frac{52 \cdot 51}{2 \cdot 1}} = \frac{6}{1{,}326} = \frac{1}{221}$$

Answer: $P(\text{both } 5\text{'s}) = \frac{1}{221}$

Calculator Solutions The expressions found for Method 1 (the counting principle) and Method 2 (permutations) can be entered on a calculator. The calculator display, however, is a rational approximation in which the decimal is *close to*, but not equal to, the exact answer, $\frac{1}{221}$.

Enter:

METHOD 1: 4 [÷] 52 [×] 3 [÷] 51 [=]

METHOD 2: 4 [2nd] [nPr] 2 [÷] 52 [2nd] [nPr] 2 [=]

Display: [0.004524887]

Note. On certain calculators, a special feature converts a decimal to a fraction and vice versa, allowing a fractional answer to appear in the display. If a specific calculator lacks this feature, then only a pencil-and-paper approach can be used to find the exact answer, $\frac{1}{221}$.

Some probability questions, such as the one just asked, may be solved by using either permutations or combinations. This is not necessarily true, however, for all probability questions. It is wise to follow this basic rule:

- **In general, use permutations where order is important, and combinations where order is not important.**

---

2. In a school organization, there are 4 sophomores and 5 juniors. A committee of 4 people is to be selected from this group. What is the probability that 2 sophomores and 2 juniors will be on the committee?

Solution (1) The school organization has a total of 9 people (4 sophomores and 5 juniors). For sample space S, each 4-person committee from a group of 9 people is a combination.

$$n(S) = {}_9C_4 = \frac{9 \cdot 8 \cdot 7 \cdot 6}{4 \cdot 3 \cdot 2 \cdot 1} = \frac{9 \cdot \overset{2}{\cancel{8}} \cdot 7 \cdot \cancel{6}}{\cancel{4} \cdot \cancel{3} \cdot \cancel{2} \cdot 1} = 126$$

(2) The number of ways to choose 2 of 4 sophomores for a committee is ${}_4C_2$. The number of ways to choose 2 of 5 juniors for a committee is ${}_5C_2$. Since event E is a compound event consisting of two independent events, apply the counting principle.

$$n(E) = {}_4C_2 \cdot {}_5C_2 = \frac{4 \cdot 3}{2 \cdot 1} \cdot \frac{5 \cdot 4}{2 \cdot 1} = \frac{\overset{2}{\cancel{4}} \cdot 3}{\cancel{2} \cdot 1} \cdot \frac{5 \cdot \overset{2}{\cancel{4}}}{\cancel{2} \cdot 1} = 6 \cdot 10 = 60$$

(3) Apply the rule of probability and simplify the fraction.

$$P(E) = \frac{n(E)}{n(S)} = \frac{{}_4C_2 \cdot {}_5C_2}{{}_9C_4} = \frac{6 \cdot 10}{126} = \frac{60}{126} = \frac{10}{21}$$

Answer: $P(\text{2 sophomores, 2 juniors}) = \frac{10}{21}$

3. An urn contains 4 white marbles and 5 blue marbles, all of equal size. Three marbles are drawn at random with no replacement. What is the probability that *at least* 2 marbles drawn are blue?

Solution (1) For sample space S, 3 marbles are drawn from a total of 9.

$$n(S) = {}_9C_3 = \frac{\overset{3}{\cancel{9}} \cdot \overset{4}{\cancel{8}} \cdot 7}{\cancel{3} \cdot \cancel{2} \cdot 1} = 84$$

(2) In event E, if *at least* 2 marbles are blue, then either exactly 2 are blue or exactly 3 are blue.

$$\text{Let } A = \{\text{combinations of 2 blue and 1 white}\},$$

$$\text{and } B = \{\text{combinations of 3 blue and 0 white}\}.$$

Then:

$$n(A) = {}_5C_2 \cdot {}_4C_1 = \frac{5 \cdot \overset{2}{\cancel{4}}}{\cancel{2} \cdot 1} \cdot \frac{4}{1} = 10 \cdot 4 = 40$$

$$n(B) = {}_5C_3 \cdot {}_4C_0 = \frac{5 \cdot \overset{2}{\cancel{4}} \cdot \cancel{3}}{\cancel{3} \cdot \cancel{2} \cdot 1} \cdot 1 = 10 \cdot 1 = 10$$

(3) Since A and B are disjoint sets:

$$n(E) = n(A) + n(B) = 40 + 10 = 50$$

(4) Apply the rule of probability.

$$P(E) = \frac{n(E)}{n(S)} = \frac{50}{84} = \frac{25}{42}$$

Answer: $P(\text{at least 2 blue}) = \frac{25}{42}$

EXERCISES WITH OPEN-RESPONSE PROBLEMS

In 1–4, evaluate each expression.

1. $\frac{{}_4C_2}{{}_{11}C_2}$ **2.** $\frac{{}_7C_3}{{}_{12}C_3}$ **3.** $\frac{{}_4C_3 \cdot {}_6C_2}{{}_{10}C_5}$ **4.** $\frac{{}_{12}C_0 \cdot {}_5C_2}{{}_{17}C_2}$

5. **a.** Let $n(S)$ = number of subsets of 2 elements that can be formed from the set $\{1, 2, 3, 4, 5\}$. Find $n(S)$.

b. Let $n(E)$ = number of subsets of 2 elements that can be formed from the set $\{1, 2, 3, 4, 5\}$ where both elements are odd digits. Find $n(E)$.

c. Using parts **a** and **b**, find the probability that a 2-element subset of $\{1, 2, 3, 4, 5\}$ contains 2 odd digits.

6. **a.** In how many ways can 2 cards be drawn from a 52-card deck?
b. In how many ways can 2 hearts be drawn from a 52-card deck?
c. If 2 cards are drawn from a 52-card deck, what is the probability that both cards are hearts?

7. Two cards are drawn from a 52-card deck without replacement. Find the probability that both cards are:
a. aces **b.** red **c.** spades **d.** picture cards

8. Three cards are drawn from a 52-card deck without replacement. Find the probability that the 3 cards are:
a. black **b.** diamonds **c.** kings **d.** 3 of a kind

9. The English Department consists of 5 men and 7 women.
a. How many different 3-person committees can be chosen from the 12 teachers in the English department?
b. Find the probability that the 3-person committee includes:
(*1*) 3 men (*2*) 3 women (*3*) 2 men and 1 woman
(*4*) 1 man and 2 women (*5*) at least 2 women

10. Find the probability that a 4-person team consists of 2 girls and 2 boys if the team is chosen at random from:
a. 3 girls and 3 boys **b.** 5 girls and 2 boys
c. 2 girls and 6 boys **d.** 3 girls and 1 boy

11. In a contest, 2 students out of 5 are to be chosen for a final playoff. All students have an equal chance of winning. The students include 4 girls (Liz, JoAnn, Claudia, Rosemary), and 1 boy (Drew).
a. In how many ways can the 2 finalists be chosen?
b. Find the probability that the finalists are:
(*1*) 2 girls (*2*) a girl and a boy (*3*) 2 boys (*4*) Claudia and Drew
c. Find the probability that one of the finalists is Liz.

12. On an examination, a student is to select any 4 questions out of 7 to answer. All questions are of equal difficulty.
a. How many different sets of 4 questions can be selected?
b. How many sets of 4 questions will include question 1?
c. If a student answers any 4 questions out of 7 on the test, what is the probability that he selected question 1?

13. An urn contains 5 red marbles and 3 white marbles, all of equal size. If 3 marbles are drawn at random with no replacement, find the probability of drawing:
a. 2 red and 1 white **b.** 1 red and 2 white
c. 3 red **d.** 3 white **e.** at least 2 red
f. 3 of the same color **g.** at most 1 red

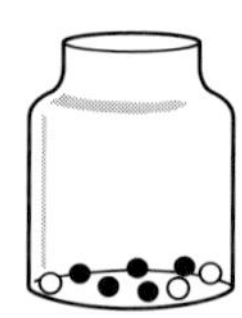

14. A box of filled donuts contains 6 strawberry, 5 raspberry, and 1 grape. Renee selects 2 donuts, not knowing what the filling is. Find the probability that Renee selected:
a. 2 strawberry **b.** 2 raspberry **c.** 2 grape
d. 1 strawberry and 1 raspberry
e. 2 with the same filling

15. Baby Stephen has a box containing 4 red blocks, 3 yellow blocks, and 2 orange blocks. He picks 3 blocks at random. Find the probability that Stephen selected:

a. 3 red blocks **b.** 3 yellow blocks **c.** 3 orange blocks
d. 2 red blocks and 1 of another color **e.** at least 2 red blocks
f. 2 yellow blocks and 1 of another color **g.** no red blocks
h. 3 blocks of the same color **i.** 1 block of each color

14-9 A MIXTURE OF PROBABILITY PROBLEMS

In this section, a variety of probability questions are presented. In some cases, permutations should be used. In other cases, combinations should be used. When answering these questions, the following should be kept in mind:

1. If the question asks "How many?" or "In how many ways?", the answer will be a whole number.
2. If the question asks "What is the probability?", the answer x will be a value between 0 and 1 inclusive, that is, $0 \leq x \leq 1$.
3. If order is important, use permutations.
4. If order is not important, use combinations.

EXAMPLES

In 1–4, Ms. Fenstermacher must select 4 students to represent the class in a spelling bee. Her best students include 3 girls (Callie, Daretta, and Kim) and 4 boys (Bandu, Carlos, Sanjit, and Uwe).

1. Ms. Fenstermacher decides to select the 4 students from the 7 best by drawing names from a hat. How many different groups of 4 are possible?

Solution In choosing a group of 4 students out of 7, order is not important. Therefore, use combinations.

$${}_7C_4 = \frac{{}_7P_4}{4!} = \frac{7 \cdot 6 \cdot 5 \cdot 4}{4 \cdot 3 \cdot 2 \cdot 1} = \frac{7 \cdot \cancel{6} \cdot 5 \cdot \cancel{4}}{\cancel{4} \cdot \cancel{3} \cdot \cancel{2} \cdot 1} = 35$$

Answer: 35 groups

---

2. What is the probability that the 4 students selected for the spelling bee will consist of 2 girls and 2 boys?

Solution (1) The answer to Example 1 shows 35 possible groups, or $n(S) = 35$.

(2) Since order is not important in choosing groups, use combinations.

The number of ways to choose 2 girls out of 3 is

$${}_3C_2 = \frac{3 \cdot \cancel{2}}{\cancel{2} \cdot 1} = 3$$

The number of ways to choose 2 boys out of 4 is

$${}_4C_2 = \frac{\overset{2}{\cancel{4}} \cdot 3}{\cancel{2} \cdot 1} = 6$$

Event E is the compound event of choosing 2 girls and 2 boys, or $n(E) = {}_3C_2 \cdot {}_4C_2 = 3 \cdot 6 = 18$.

(3) Thus, $P(E) = \frac{{}_3C_2 \cdot {}_4C_2}{{}_7C_4} = \frac{3 \cdot 6}{35} = \frac{18}{35}$

Answer: $P(\text{2 girls, 2 boys}) = \frac{18}{35}$

---

3. The students chosen are Callie, Kim, Carlos, and Sanjit. In how many ways can these 4 students be called upon in the spelling bee?

Solution The number of ways in which 4 students can be called upon in a spelling bee means that someone is first, someone else is second, and so on. Since order is important, use permutations.

$${}_4P_4 = 4! = 4 \cdot 3 \cdot 2 \cdot 1 = 24$$

---

4. For the group consisting of Callie, Kim, Carlos, and Sanjit, what is the probability that the first 2 students called upon will be girls?

Solution This question may be answered using various methods, two of which are shown below.

METHOD 1: *Counting Principle*

There are 2 girls out of 4 students who may be called first. Once a girl is called upon, only 1 girl remains in the 3 students not yet called upon. Apply the counting principle of probability.

$$P(\text{first 2 are girls}) = P(\text{girl first}) \cdot P(\text{girl second})$$

$$= \frac{2}{4} \cdot \frac{1}{3} = \frac{2}{12} = \frac{1}{6}$$

METHOD 2: *Permutations*

Let $n(S)$ equal the number of ways to call 2 of the 4 students in order, and let $n(E)$ equal the number of ways to call 2 of the 2 girls in order.

$$P(E) = \frac{n(E)}{n(S)} = \frac{{}_2P_2}{{}_4P_2} = \frac{2 \cdot 1}{4 \cdot 3} = \frac{2}{12} = \frac{1}{6}$$

Answer: $P(\text{first 2 are girls}) = \frac{1}{6}$

EXERCISES WITH OPEN-RESPONSE PROBLEMS

1. The Art Club consists of 4 girls (Jennifer, Joanna, Gloria, Teresa) and 2 boys (Joe and Arthur).
  a. In how many ways can the club elect a president and a treasurer?
  b. Find the probability that the 2 officers elected are both girls.
  c. How many 2-person teams can be selected to work on a project?
  d. Find the probability that a 2-person team consists of:
    (*1*) 2 girls (*2*) 2 boys (*3*) 1 girl and 1 boy
    (*4*) 2 people, each having J as a first initial

2. A committee of 4 is to be chosen from 4 men and 3 women.
  a. How many different 4-member committees are possible?
  b. How many 4-member committees contain 3 men and 1 woman?
  c. What is the probability that a 4-member committee will contain exactly 1 woman?
  d. What is the probability that a man will be on the committee?

3. A committee of 6 people is to be chosen from 9 available people.
  a. How many 6-person committees can be chosen?
  b. The committee, when chosen, has 4 students and 2 teachers. Find the probability that a 3-person subcommittee from this group includes:
    (*1*) students only (*2*) exactly 1 teacher (*3*) at least 2 students

4. A box of chocolate-covered candies contains 7 caramels and 3 creams all exactly the same in appearance. Jim selects 4 pieces of candy.
  a. Find the number of selections possible of 4 pieces of candy that include:
    (*1*) 4 caramels (*2*) 1 caramel and 3 creams (*3*) 2 caramels and 2 creams (*4*) any 4 pieces
  b. Find the probability that Jim's selection included:
    (*1*) 4 caramels (*2*) 1 caramel and 3 creams (*3*) 2 caramels and 2 creams (*4*) no caramels

5. Two cards are drawn at random from a 52-card deck without replacement. Find the probability of drawing:
  a. the ace of clubs and jack of clubs in either order
  b. an ace and a jack in that order **c.** 2 jacks **d.** an ace and a jack in either order
  e. 2 clubs **f.** a red ace and a black jack in either order

6. Mrs. Carberry has 4 quarters and 3 nickels in her purse. If she takes 3 coins out of her purse without looking at them, find the probability that the 3 coins are worth:
  a. exactly 75 cents **b.** exactly 15 cents **c.** exactly 35 cents
  d. exactly 55 cents **e.** more than 10 cents **f.** less than 40 cents

7. A 3-digit numeral is formed by selecting digits at random from {2, 4, 6, 7} without repetition. Find the probability that the number formed:
  a. is less than 700 **b.** is greater than 600
  c. contains only even digits **d.** is an even number

8. **a.** There are 10 runners on the track team. If 4 runners are needed for a relay race, how many different relay teams are possible?
  b. Once the relay team is chosen, in how many different orders can the 4 runners run the race?
  c. If Nicolette is on the relay team, what is the probability that she will lead off the race?

9. Lou Grant is an editor at a newspaper employing 10 reporters and 3 photographers.
a. If Lou selects 2 reporters and 1 photographer to cover a story, from how many possible 3-person teams can he choose?
b. If Lou hands out 1 assignment per reporter, in how many ways can he assign the first 3 stories to his 10 reporters?
c. If Lou plans to give the first story to Rossi, a reporter, in how many ways can he now assign the first 3 stories?
d. If 3 out of 10 reporters are chosen at random to cover a story, what is the probability that Rossi is on this team?

10. Chris, Willie, Tim, Matt, Juan, Bob, and Steffen audition for roles in the school play.
a. If 2 male roles in the play are those of the hero and the clown, in how many ways can the director select 2 of the 7 boys for these roles?
b. Chris and Willie got the 2 leading male roles.
(*1*) In how many ways can the director select a group of 3 of the remaining 5 boys to work in a crowd scene?
(*2*) How many of these groups of 3 will include Tim?
(*3*) Find the probability that Tim is in the crowd scene.

11. There are 8 candidates for 3 seats in the student government. The candidates include 3 boys (Alberto, Peter, Thomas) and 5 girls (Elizabeth, Maria, Joanna, Rosa, Danielle). If all candidates have an equal chance of winning, find the probability that the winners include:
a. 3 boys **b.** 3 girls **c.** 1 boy and 2 girls **d.** at least 2 girls
e. Maria, Peter, and anyone else **f.** Danielle and any other two candidates

12. A gumball machine contains 6 lemon-, 4 lime-, 3 cherry-, and 2 orange-flavored gumballs. Five coins are put into the machine, and 5 gumballs are obtained.
a. How many different sets of 5 gumballs are possible?
b. How many of these will contain 2 lemon and 3 lime gumballs?
c. Find the probability that the 5 gumballs dispensed by the machine include:
(*1*) 2 lemon and 3 lime (*2*) 3 cherry and 2 orange (*3*) 2 lemon, 2 lime, and 1 orange
(*4*) lemon only (*5*) lime only (*6*) no lemon

13. The letters in the word HOLIDAY are rearranged at random.
a. How many 7-letter words can be formed?
b. Find the probability that the first 2 letters are vowels.
c. Find the probability of no vowels in the first 3 letters.

14. At a bus stop, 5 people enter a bus that has only 3 empty seats.
a. In how many different ways can 3 of the 5 people occupy these empty seats?
b. If Mrs. Costa is 1 of the 5 people, what is the probability that she will *not* get a seat?
c. If Ann and Bill are 2 of the 5 people, what is the probability that they both will get seats?

CHAPTER SUMMARY

The ***counting principle*** states that, if one activity occurs in m ways and a second, following activity occurs in n ways, then both activities can occur in the order given in $m \bullet n$ ways. This principle can be extended to the probability of ***independent events***, that is, events in which the result of one activity in no way influences the result of a second activity: If E and F are independent events, where $P(E) = m$ and $P(F) = n$, then

$$P(E \text{ and } F) = m \cdot n$$

A ***permutation*** is an arrangement of objects in some specific order. The permutation of n objects taken n at a time, ${}_nP_n$, is equal to ***factorial n***:

$${}_nP_n = n! = n \bullet (n - 1) \bullet (n - 2) \bullet \cdots \bullet 3 \bullet 2 \bullet 1$$

The permutation of n objects taken r at a time, where $r \leq n$, is equal to

$${}_nP_r = \underbrace{n \bullet (n - 1) \bullet (n - 2) \bullet \cdots}_{r \text{ factors}} \qquad \text{or} \qquad {}_nP_r = \frac{n!}{(n - r)!}$$

A ***combination*** is a set of objects in which order is *not* important, as in a committee. The formula for the combination of n objects taken r at a time $(r \leq n)$ is

$${}_nC_r = \frac{{}_nP_r}{r!}$$

By definition: ${}_nC_n = 1$; ${}_nC_0 = 1$; and ${}_nC_r = {}_nC_{n-r}$.

Some calculators have a ***factorial key*** **x!** and/or a ***permutation key*** **nPr** and a ***combination key*** **nCr** to help in the computation of permutations and combinations.

Compound events, which involve two or more activities, sometimes include *replacement* of outcomes in the sample space and sometimes describe situations *without replacement*.

VOCABULARY

14-1 Tree diagram List of ordered pairs Graph of ordered pairs Counting principle

14-2 Independent events Compound event Counting Principle for Probability

14-4 Permutation Factorial Factorial key, **x!**

14-5 Permutation key, **nPr**

14-6 Without replacement With replacement

14-7 Combination Combination key, **nCr**

REVIEW EXERCISES WITH OPEN-RESPONSE PROBLEMS

1. If three fair coins are tossed simultaneously, what is the total number of outcomes in the sample space?

2. There are five staircases going from the first to the second floor in school. Find the total number of ways that Andrea can go from the first floor to the second and then return to the first.

3. If two fair six-sided dice are tossed, and their sum is noted, find:
a. $P(\text{sum is } 8)$ **b.** $P(\text{sum is greater than } 9)$
c. $P(\text{sum is odd})$ **d.** $P(\text{sum is } 12 \text{ or less})$

4. Assume that $P(\text{male}) = P(\text{female})$. In a family of three children, what is the probability that all three children are of the same sex?

5. Compute the value of each expression:
a. $5!$ **b.** $\frac{10!}{8!}$ **c.** ${}_4P_4$ **d.** ${}_{30}P_2$ **e.** ${}_7P_3$ **f.** ${}_{24}P_6$ **g.** $9!$
h. ${}_5P_5$ **i.** ${}_{12}P_2$ **j.** ${}_5C_5$ **k.** ${}_{12}C_3$ **l.** ${}_{40}C_{38}$

6. From a list of 10 books, Gwen selects 4 to read over the summer. In how many ways can Gwen make a selection of 4 books?

7. Mrs. Moskowitz, the librarian, checks out the 4 books Gwen has chosen. In how many different orders can the librarian stamp these 4 books?

8. A coach must choose a team of 5 from 8 candidates. How many different teams can be chosen?

9. How many 3-letter words can be formed from the letters in QUESTION if each letter is used only once in a word?

10. A SYZYGY is a nearly straight-line configuration of three celestial bodies such as the Earth, Moon, and Sun during an eclipse.
a. How many different 6-letter arrangements can be made using the letters in the word SYZYGY?
b. Find the probability that the first letter in an arrangement of SYZYGY is (*1*) G (*2*) Y (*3*) a vowel

In 11 and 12, select in each case the choice that best answers the question or completes the statement.

11. How many different arrangements of four letters can be made from the word HELP if each letter is used only once in each arrangement?
(1) 1 (2) 256 (3) 24 (4) 4

12. How many different four-digit numbers can be formed from the digits 1, 2, 3, and 4 if each digit is used only once?
(1) 24 (2) 16 (3) 12 (4) 4

13. Two darts were thrown at a pentagon-shaped dartboard like that at the right until each dart held in one of the five equally likely numerical regions.

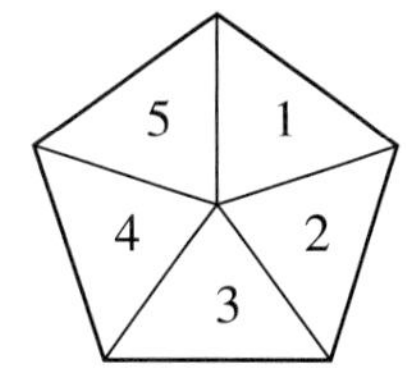

a. Draw a tree diagram or write a sample space to show the numbered regions for the two darts.
b. Find the probability that both darts landed in the same region.
c. Find the probability that both darts landed in odd-numbered regions.
d. Find the probability that at least one dart landed in the region marked 5.
e. If the score is the sum of the numbers of the regions hit, find the probability of getting an odd-numbered score.

14. A bank contains a quarter, a dime, and a nickel. The bank is shaken until one coin falls out, and then is shaken again until a second coin falls out. All coins are equally likely to fall out.
a. Draw a tree diagram or list the sample space to show the pairs of the first and second coins that fall out.
b. Find the probability that the *sum* of the values of the coins is:
(*1*) exactly 30¢ (*2*) exactly 20¢ (*3*) 30¢ or less
(*4*) less than 40¢ (*5*) equal to an odd number of cents
c. Find the probability that the first coin is a quarter.
d. Find the probability that neither coin is a dime.

15. **a.** How many 3-digit numerals can be formed using the digits 2, 3, 4, 5, 6, 7, and 8 if repetition is not allowed?
b. What is the probability that such a 3-digit numeral is greater than 400?

16. **a.** If a 4-member committee is formed from 3 girls and 6 boys in a club, how many committees can be formed?
b. Find the probability that the 4-member committee consists of:
(*1*) 2 girls and 2 boys (*2*) 1 girl and 3 boys (*3*) 4 boys
c. What is the probability that a boy is on the committee?

17. A committee of 5 is to be chosen from 4 men and 3 women.
a. How many different 5-person committees are possible?
b. Find the probability that the committee includes:
(*1*) 2 men and 3 women (*2*) 3 men and 2 women
(*3*) at least 2 women (*4*) all women
c. In how many ways can this 5-person committee select a chairperson and a secretary?
d. If Hilda and Joan are on the committee, what is the probability that one is selected as chairperson and the other as secretary?

CUMULATIVE REVIEW WITH OPEN-RESPONSE PROBLEMS

1. Temperature in degrees Celsius (C) and degrees Fahrenheit (F) are related by the formula $F = \frac{9}{5}C + 32$.
   a. Solve the formula for C.
   b. To the nearest degree, what is the temperature in degrees Celsius if the temperature in degrees Fahrenheit is 77.

2. Which number is irrational?

   $\sqrt{\frac{1}{16}}$ $\sqrt{21}$ $\sqrt{169}$

   Explain why the number you chose is irrational.

3. Solve for x: $3(x + 1) = 1.4 - x$

4. The mean (average) weight of three packages is 72 pounds. One of the packages weighs 32 pounds. The other two packages have the same weight. Find the weights of each of these two packages.

5. The model for the background of a company logo is shown in the diagram. It consists of three circles with the same center. The radii of the circles have lengths of 2 inches, 4 inches, and 6 inches.
   a. What is the area of the shaded region to the *nearest tenth of a square inch?*
   b. To the *nearest percent*, what percent of the logo is shaded?

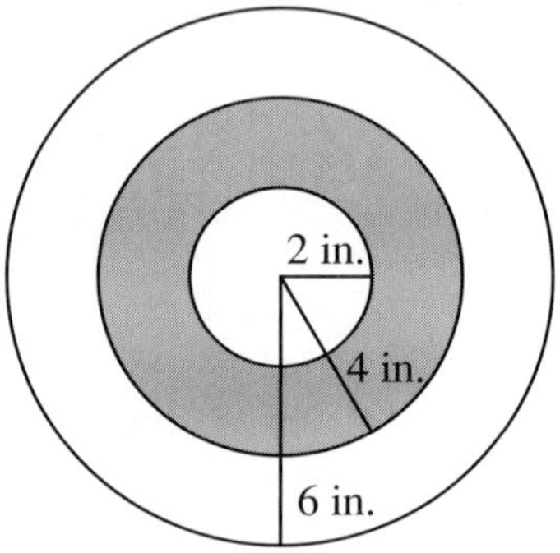

6. a. The image of $A(x, y)$ under a reflection in the x-axis is B, and the image of B under a reflection in the y-axis is C. What are the coordinates of C?
   b. The image of $A(x, y)$ under a reflection in the y-axis is D, and the image of D under a reflection in the x-axis is E. What are the coordinates of E?
   c. Under what transformation is E the image of A?

Exploration

From a standard deck of 52 cards, what is the probability of drawing a spade and a queen in that order?

Chapter 15

Statistics

Every 4 years, each major political party in the United States holds a convention to select the party's nominee for president of the United States. Before these conventions are held, each candidate assembles a staff whose job is to plan a successful campaign. This plan relies heavily on statistics: on the collection and organization of data, on the results of opinion polls, and on information about the factors that influence the way people vote. At the same time, newspaper reporters and television commentators assemble other data to keep the public informed on the progress of the candidates.

Election campaigns are just one example of the use of statistics to organize data in a way that enables us to use available information to evaluate the current situation and to plan for the future.

15-1 COLLECTING DATA

In our daily lives, we often deal with problems that involve many related items of numerical information called ***data***. For example, in the daily newspaper we can find data dealing with sports, with business, with politics, even with the weather.

Statistics is the study of numerical data. There are three typical steps in a statistical study:

Step 1. The collection of data.

Step 2. The organization of these data into tables, charts, and graphs.

Step 3. The drawing of conclusions from an analysis of these data.

When these three steps, which describe and summarize the formation and use of a set of data, are included in a statistical study, the study is often called ***descriptive statistics***. You will study these steps in this first course. In some cases, a fourth step, in which the analyzed data are used to predict trends and future events, is added. You will not study this type of statistics in this course.

Data can be collected in a number of ways, including the following:

1. A written *questionnaire* or list of questions that a person can answer by checking one of several categories or supplying written responses.
2. An *interview*, either in person or by telephone, in which answers are given verbally and responses are recorded by the person asking the questions.
3. A *log* or a diary, such as a hospital chart or an hourly recording of the outdoor temperature, in which a person records information on a regular basis.

Sampling

A statistical study may be useful in situations such as the following:

1. A doctor wants to know how effective a new medicine will be in curing a disease.
2. A quality-control team wants to know the expected life span of flashlight batteries made by its company.
3. A company advertising on television wants to know the most frequently watched TV shows so that its ads will be seen by the greatest number of people.

When conducting a statistical study, it is not always possible to obtain information about every person, object, or situation to which the study applies. Unlike a ***census***, in which every person is counted, some statistical studies use only a ***sample***, or portion, of the items being investigated.

To find effective medicines, pharmaceutical companies usually conduct tests in which a sample, or small group, of the patients having the disease under study receive the medicine. If the manufacturer of flashlight batteries tested the life span of every battery made, he would soon have a warehouse filled with dead batteries. He tests only a sample of the batteries to determine their average life span. An advertiser cannot contact every person owning a TV set to see which shows are being watched. Instead, the advertiser studies TV ratings released by a firm that conducts polls based on a small sample of TV viewers.

For any statistical study, whether based on a census or a sample, to be useful, data must be collected carefully and correctly.

Techniques of Sampling

We must be careful when choosing samples:

1. **The sample must be fair, to reflect the entire population being studied.** To know what an apple pie tastes like, it is not necessary to eat the entire pie. Eating a sample, such as a piece of the apple pie, would be a fair way of knowing how the pie tastes. However, eating only the crust or only the apples would be an unfair sample that would not tell us what the entire pie tastes like.

2. **The sample must contain a reasonable number of the items being tested or counted.**
   If a medicine is generally effective, it must work for many people. The sample tested cannot include only one or two patients. Similarly, the manufacturer of flashlight batteries cannot make claims based on testing five or 10 batteries. A better sample might include 100 batteries.

3. **Patterns of sampling or random selection should be employed in a study.** The manufacturer of flashlight batteries might test every 1,000th battery to come off the assembly line. Instead, he might select the batteries to be tested at random.

These techniques will help to make the sample, or the small group, *representative* of the entire population of items being studied. From the study of the small group, reasonable conclusions can be drawn about the entire group.

EXAMPLE

To determine which television programs are the most popular in a large city, a poll is conducted by selecting people at random at a street corner and interviewing them. Outside of which location would the interviewer be most likely to find a fair sample?

(1) a ball park (2) a concert hall (3) a supermarket

Solution People outside a ball park may be going to a game or purchasing tickets for a game in the future; this sample may be biased in favor of sports programs. Similarly, those outside a concert hall may favor musical or cultural programs. The best (that is, the fairest) sample or cross section of people for the three choices given would probably be found outside a supermarket.

Answer: (3)

EXERCISES WITH OPEN-RESPONSE PROBLEMS

In 1–8, in each case a sample of students is to be selected and the height of each student is to be measured to determine the average height of a student in high school. For each sample:

a. Tell whether the sample is fair or unfair.
b. If the sample is unfair, explain why.

1. The basketball team **2.** The senior class **3.** All 14-year-old students **4.** All girls

5. Every tenth person selected from an alphabetical list of all students

6. Every fifth person selected from an alphabetical list of all boys

7. The first three students who report to the nurse on Monday

8. The first three students who enter each homeroom on Tuesday

In 9–14, in each case the Student Organization wishes to interview a sample of students to determine the general interests of the student body. Two questions will be asked: "Do you want more pep rallies for sports events? Do you want more dances?" For each location, tell whether the Student Organization would find a fair sample at that place.

9. The gym, after a game

10. The library

11. The lunchroom

12. The cheerleaders' meeting

13. The next meeting of the Junior Prom committee

14. A homeroom section chosen at random

15. A statistical study is useful when reliable data are collected. At times, however, people may exaggerate or lie when answering a question. Of the six questions that follow, find the *three* questions that will most probably produce the largest number of *unreliable* answers.
a. What is your height? **b.** What is your weight?
c. What is your age? **d.** In which state do you live?
e. What is your income? **f.** How many people are in your family?

16. List the three steps necessary to conduct a statistical study.

15-2 ORGANIZING DATA

Data are often collected in an unorganized and random manner. For example, a teacher recorded the number of days each of 25 students in her class was absent last month. These absences were as follows:

0, 3, 1, 0, 4, 2, 1, 3, 5, 0, 2, 0, 0
0, 4, 0, 1, 1, 2, 1, 0, 7, 3, 1, 0.

How many students were absent fewer than 2 days? What was the number of days on which the largest number of students were absent? How many students were absent more than 5 days? To answer questions such as these, we find it helpful to organize the data into a table.

Preparing a Table

In the left column of the accompanying table, we list the data values (or the number of absences) in order. We start with the largest number, 7, at the top and go down to the smallest number, 0.

For each data value, we place a ***tally*** mark, 𝍷, in the row for that number. For example, the first data value in the teacher's list is 0, so we place a tally in the 0 row; the second value is 3, so we place a tally in the 3 row. We follow this procedure until a tally for each data value is recorded in the proper row. To simplify counting, we write every fifth tally as a diagonal mark passing through the first four tallies: 𝍸.

Absences	Tally
7	𝍷
6	
5	𝍷
4	𝍷𝍷
3	𝍷𝍷𝍷
2	𝍷𝍷𝍷
1	𝍸 𝍷
0	𝍸 𝍷𝍷𝍷𝍷

Once the data have been organized, we can count the number of tally marks in each row and add a column for the ***frequency***, that is, the number of times that a value occurs in the set of data. When there are no tally marks in a row, as for the row showing six absences, the frequency is 0. The sum of all of the frequencies is called the ***total frequency***. In this case, the total frequency is 25. (It is always wise to check the total frequency to be sure that no data value was overlooked in tallying.)

From the table it is now easy to see that 15 students were absent fewer than 2 days, that 0 was the number of days on which the largest number of students (nine) were absent, and that one student was absent more than 5 days.

Absences	Tally	Frequency
7	𝍷	1
6		0
5	𝍷	1
4	𝍷𝍷	2
3	𝍷𝍷𝍷	3
2	𝍷𝍷𝍷	3
1	𝍸 𝍷	6
0	𝍸 𝍷𝍷𝍷𝍷	9
Total frequency		25

Grouped Data

A teacher marked a set of 32 test papers. The grades or scores earned by the students were as follows:

90, 85, 74, 86, 65, 62, 100, 95, 77, 82, 50, 83, 77, 93, 73, 72,
98, 66, 45, 100, 50, 89, 78, 70, 75, 95, 80, 78, 83, 81, 72, 75.

Because of the large number of different scores, it is convenient to organize these data into ***groups*** or ***intervals***, which must be equal in size. Here we will use six intervals: 41–50, 51–60, 61–70, 71–80, 81–90, 91–100. Each interval has a length of 10, found by subtracting the starting point of an interval from the starting point of the next higher interval.

For each test score, we now place a tally mark in the row for the interval that includes that score. For example, the first two scores in the list above are 90 and 85, so we place two tally marks in the interval 81–90. The next score is 74, so we place a tally mark in the interval 71–80. When all of the scores have been tallied, we write the frequency for each interval.

This table, containing a set of intervals and the corresponding frequency for each interval, is an example of ***grouped data***.

Interval	Tally	Frequency
91–100	𝍸 |	6
81–90	𝍸 |||	8
71–80	𝍸 𝍸 |	11
61–70	||||	4
51–60		0
41–50	|||	3

Rules for grouping data When unorganized data are grouped into intervals, we must follow certain rules in setting up the intervals:

1. The intervals must cover the complete range of values. The ***range*** is the difference between the highest and lowest values.

2. The intervals must be equal in size.

3. The number of intervals should be between 5 and 15. The use of too many or too few intervals does not make for effective grouping of data. We usually use a large number of intervals, for example, 15, only when we have a large set of data, such as hundreds of scores.

4. Every score to be tallied, from the highest to the lowest, must fall into one and only one interval. Thus, the intervals should not overlap. When an interval ends with a counting number, the following interval begins with the next counting number.

These rules tell us that there are many ways to set up tables, all of them correct, for the same set of data. For example, here is another correct way to group the 32 unorganized test scores given at the beginning of this section. Note that the length of the interval here is 8.

Interval	Tally	Frequency
93–100	𝍸 |	6
85–92	||||	4
77–84	𝍸 ||||	9
69–76	𝍸 ||	7
61–68	|||	3
53–60		0
45–52	|||	3

Constructing a Stem-and-Leaf Diagram

Another method of displaying data is called a ***stem-and-leaf diagram***. The stem-and-leaf diagram groups the data without losing the individual data values.

A group of 30 students were asked to record the length of time, in minutes, spent on math homework yesterday. They reported the following data:

38, 15, 22, 20, 25, 44, 5, 40, 38, 22, 20, 35, 20, 0, 36,
27, 37, 26, 33, 25, 17, 45, 22, 30, 18, 48, 12, 10, 24, 27.

To construct a stem-and-leaf diagram for the lengths of time given, we begin by choosing part of the data values to be the stem. Since every score is a one- or two-digit number, we will choose the tens digit as a convenient stem. For the one-digit numbers, 0 and 5, the stem is 0; for the other data values, the stem is 1, 2, 3, or 4. Then the units digit will be the leaves. We construct the diagram as follows:

Step 1. List the stems, starting with 4, under one another with a vertical line to the right.

Stem	Leaves
4	
3	
2	
1	
0	

Step 2. Enter each score by writing its leaf (the units digit) to the right of the vertical line, following the appropriate stem (its tens value). For example, enter 38 by writing 8 to the right of the vertical line, after stem 3.

Stem	Leaves
4	
3	8
2	
1	
0	

Step 3. Add the other scores to the diagram until all are entered.

Stem	Leaves
4	4 0 5 8
3	8 8 5 6 7 3 0
2	2 0 5 2 0 0 7 6 5 2 4 7
1	5 7 8 2 0
0	5 0

Step 4. Arrange the leaves in order after each stem.

Stem	Leaves
4	0 4 5 8
3	0 3 5 6 7 8 8
2	0 0 0 2 2 2 4 5 5 6 7 7
1	0 2 5 7 8
0	0 5

EXAMPLES

1. The following data consist of the weights, in kilograms, of a group of 30 students:

70, 43, 48, 72, 53, 81, 76, 54, 58, 64, 51, 53, 75, 62, 84, 67, 72, 80, 88, 65, 60, 43, 53, 42, 57, 61, 55, 75, 82, 71.

a. Copy and complete the table to group these data.

b. Based on the grouped data, which interval contains the greatest number of students?

c. How many students weigh less than 70 kilograms?

Interval	Tally	Frequency (number)
80–89		
70–79		
60–69		
50–59		
40–49		

Solutions **a.**

Interval	Tally	Frequency (number)
80–89	卌	5
70–79	卌 ||	7
60–69	卌 |	6
50–59	卌 |||	8
40–49	||||	4

b. The interval 50–59 contains the greatest number of students, 8. *Answer*

c. The three lowest intervals, namely 40–49, 50–59, and 60–69, show weights less than 70 kilograms. Add the frequencies in these three intervals: $4 + 8 + 6 = 18$. *Answer*

---

2. Draw a stem-and-leaf diagram for the data in Example 1.

Solution Let the tens digit be the stem and the units digit the leaf.

(1) Enter the data values in the order given:

Stem	Leaves
8	1 4 0 8 2
7	0 2 6 5 2 5 1
6	4 2 7 5 0 1
5	3 4 8 1 3 3 7 5
4	3 8 3 2

(2) Arrange the leaves in order after each stem:

Stem	Leaves
8	0 1 2 4 8
7	0 1 2 2 5 5 6
6	0 1 2 4 5 7
5	1 3 3 3 4 5 7 8
4	2 3 3 8

EXERCISES WITH OPEN-RESPONSE PROBLEMS

1. **a.** Copy and complete the table to group the data, which are the heights, in centimeters, of 36 students:

162, 173, 178, 181, 155, 162, 168, 147, 180,
171, 168, 183, 157, 158, 180, 164, 160, 171,
183, 174, 166, 175, 169, 180, 149, 170, 150,
158, 162, 175, 171, 163, 158, 163, 164, 177.

Interval	Tally	Frequency (number)
180–189		
170–179		
160–169		
150–159		
140–149		

b. Use the grouped data to answer the following questions:

(*1*) How many students are less than 160 centimeters in height?
(*2*) How many students are 160 centimeters or more in height?
(*3*) Which interval contains the greatest number of students?
(*4*) Which interval contains the least number of students?

2. **a.** Copy and complete the table to group the data, which gives the life span, in hours, of 50 flashlight batteries:

73, 81, 92, 80, 108, 76, 84, 102, 58, 72,
82, 100, 70, 72, 95, 105, 75, 84, 101, 62,
63, 104, 97, 85, 106, 72, 57, 85, 82, 90,
54, 75, 80, 52, 87, 91, 85, 103, 78, 79,
91, 70, 88, 73, 67, 101, 96, 84, 53, 86.

Interval	Tally	Frequency (number)
50–59		
60–69		
70–79		
80–89		
90–99		
100–109		

b. Use the grouped data to answer the following questions:

(*1*) How many flashlight batteries lasted for 80 or more hours?
(*2*) How many flashlight batteries lasted fewer than 80 hours?
(*3*) Which interval contains the greatest number of batteries?
(*4*) Which interval contains the least number of batteries?

3. The following data consists of the hours spent each week watching television, as reported by a group of 38 teenagers:

13, 20, 17, 36, 25, 21, 9, 32, 20, 17, 12, 19, 5, 8, 11, 28, 25, 18,
19, 22, 4, 6, 0, 10, 16, 3, 27, 31, 15, 18, 20, 17, 3, 6, 19, 25, 4, 7.

a. Construct a table to group these data, using intervals of 0–4; 5–9; 10–14; 15–19; 20–24; 25–29; 30–34; 35–39.

b. Construct a table to group these data, using intervals of 0–7; 8–15; 16–23; 24–31; 32–39.

4. The following data shows test scores for 30 students:

90, 83, 87, 71, 62, 46, 67, 72, 75, 100, 93, 81, 74, 75, 82,
83, 83, 84, 92, 58, 95, 98, 81, 88, 72, 59, 95, 50, 73, 93.

a. Copy and complete the table, using these intervals of length 10.

Interval	Frequency
91–100	
81–90	
71–80	
61–70	
51–60	
41–50	

b. Copy and complete the table, using these intervals of length 12.

Interval	Frequency
89–100	
77–88	
65–76	
53–64	
41–52	

c. For the grouped data in part **a**, which interval contains the greatest number of students?

d. For the grouped data in part **b**, which interval contains the greatest number of students?

e. Do the answers for parts **c** and **d** indicate the same general region of test scores, such as "scores in the eighties"? Explain your answer.

5. For the ungrouped data from Exercise 3, tell why each of the following sets of intervals is not correct for grouping the data.

a.

Interval
25–38
13–24
0–12

b.

Interval
30–39
20–29
10–19
5–9
0–4

c.

Interval
32–40
24–32
16–24
8–16
0–8

d.

Interval
33–40
25–32
17–24
9–16
1–8

6. Display the data from Exercise 3 in a stem-and-leaf diagram.

7. Display the data from Exercise 1 in a stem-and-leaf diagram. Use the first two digits of the numbers as the stems.

15-3 THE HISTOGRAM

Section 15-2 explained how to organize data by grouping them into intervals of equal length. After the data have been organized, a graph can be used to visualize the intervals and their frequencies.

The table at the right shows the distribution of test scores for 32 students in a class. The data have been organized into six intervals, each having a length of ten.

Test Scores (intervals)	Frequency (number of scores)
91–100	6
81–90	8
71–80	11
61–70	4
51–60	0
41–50	3

We can use a *histogram* to display the data graphically. A ***histogram*** is a vertical bar graph in which each interval is represented by the width of its bar and the frequency of the interval is represented by the height of its bar. The bars are placed next to each other to show that, as one interval ends, the next interval begins.

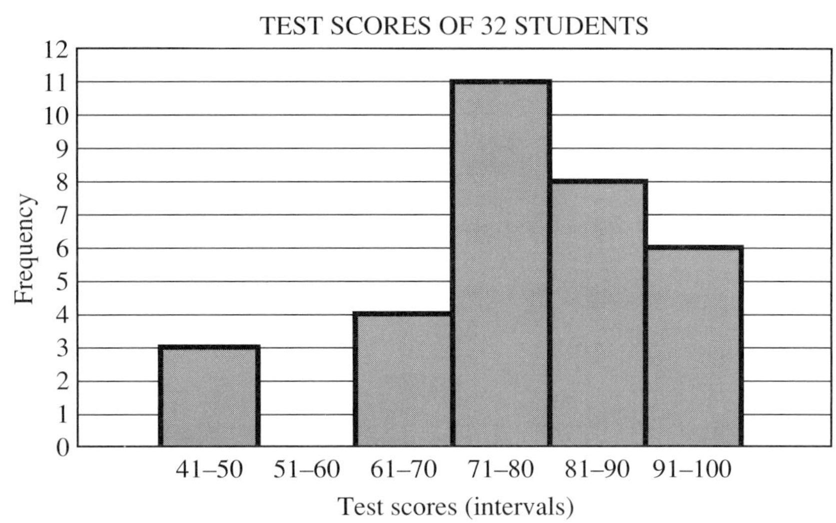

In the above histogram, the intervals are listed on the horizontal axis in the order of increasing scores, and the frequency scale is shown on the vertical axis. The first bar shows that 3 students had test scores in the interval 41–50. Since no student scored in the interval 51–60, there is no bar for this interval. Then, 4 students scored between 61 and 70; 11 had test scores between 71 and 80; 8 had scores between 81 and 90; and 6 scored between 91 and 100.

Except for an interval having a frequency of 0, as for 51–60 in this example, there are no gaps between the bars drawn in a histogram.

Since the histogram displays the frequency, or number, of scores in each interval, we sometimes call this graph a ***frequency histogram***.

EXAMPLES

1. The table at the right represents the number of miles per gallon of gasoline obtained by 40 drivers of compact cars in a large city. Construct a frequency histogram based on the data.

Interval	Frequency
16–19	5
20–23	11
24–27	8
28–31	5
32–35	7
36–39	3
40–43	1

Solution (1) Draw and label a vertical scale to show frequencies. The scale starts at 0 and increases to include the highest frequency in any one interval (here, it is 11).

(2) Draw and label intervals of equal length on a horizontal scale. Label the horizontal scale, telling what the numbers represent.

(3) Draw the bars vertically, leaving no gaps between the intervals.

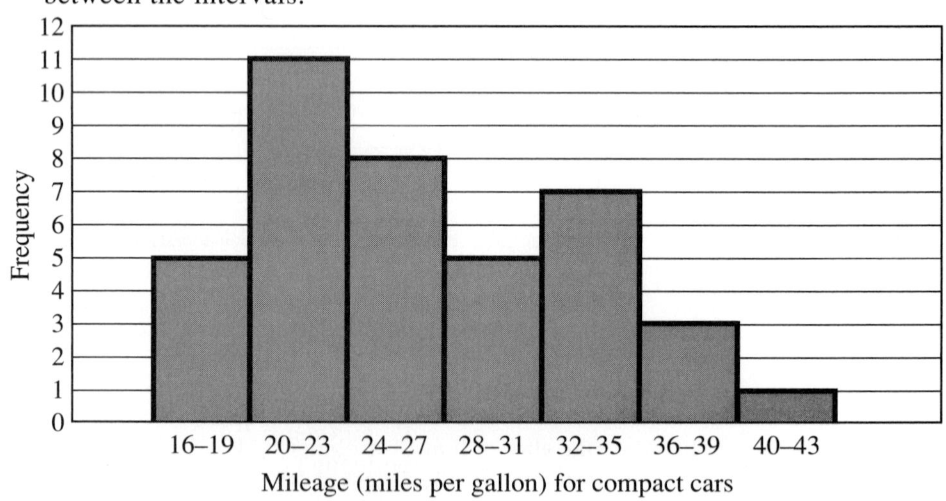

2. Use the histogram constructed in Example 1 to answer the following questions:

a. In what interval is the greatest frequency found?

Answer: 20–23

b. What is the number (or frequency) of cars reporting mileages between 28 and 31 miles per gallon?

Answer: 5

c. For what interval are the fewest cars reported?

Answer: 40–43

d. How many of the cars reported mileages greater than 31 miles per gallon?

Solution Add the frequencies for the three highest intervals. There are 7 in (32–35), 3 in (36–39), and 1 in (40–43), so $7 + 3 + 1 = 11$. *Answer*

e. What percent of the cars reported mileages from 24 to 27 miles per gallon?

Solution The interval 24–27 has a frequency of 8. The total frequency for this survey is 40. Then, $\frac{8}{40} = \frac{1}{5} = 20\%$. *Answer*

EXERCISES WITH OPEN-RESPONSE PROBLEMS

In 1–3, in each case construct a frequency histogram for the grouped data.

1.

Interval	Frequency
91–100	5
81–90	9
71–80	7
61–70	2
51–60	4

2.

Interval	Frequency
30–34	5
25–29	10
20–24	10
15–19	12
10–14	0
5–9	2

3.

Interval	Frequency
1–3	24
4–6	30
7–9	28
10–12	41
13–15	19
16–18	8

4. For the table of grouped scores given in Exercise 3, answer the following questions:

a. What is the total frequency, or the total number of data values, in the table?

b. What interval contains the greatest frequency?

c. The number of scores reported for the interval 4–6 is what percent of the total number of scores?

d. How many scores from 10 through 18 were reported?

5. Towering Ted McGurn is the star of the school's basketball team. The number of points scored by Ted in his last 20 games are as follows:

36, 32, 28, 30, 33, 36, 24, 33, 29, 30,
30, 25, 34, 36, 34, 31, 36, 29, 30, 34.

a. Copy and complete the table to find the number (or frequency) in each interval.

b. Construct a frequency histogram based on the data found in part **a**.

c. Which interval contains the greatest frequency?

d. In how many games did Ted score 32 or more points?

e. In what percent of these 20 games did Ted score fewer than 26 points?

Interval	Tally	Frequency
35–37		
32–34		
29–31		
26–28		
23–25		

6. Thirty students on the track team were timed in the 200-meter dash. Each student's time was recorded to the *nearest tenth* of a second. The times were as follows:

29.3, 31.2, 28.5, 37.6, 30.9, 26.0, 32.4, 31.8, 36.6, 35.0,
38.0, 37.0, 22.8, 35.2, 35.8, 37.7, 38.1, 34.0, 34.1, 28.8,
29.6, 26.9, 36.9, 39.6, 29.9, 30.0, 36.0, 36.1, 38.2, 37.8.

a. Copy and complete the table to find the number (or frequency) in each interval.

b. Construct a frequency histogram for the given data.

c. From the data, determine the number of students who ran the 200-meter dash in under 29 seconds.

d. If a student on the track team is chosen at random, what is the probability that he or she ran the 200-meter dash in fewer than 29 seconds?

Interval	Tally	Frequency
37.0–40.9		
33.0–36.9		
29.0–32.9		
25.0–28.9		
21.0–24.9		

15-4 THE MEAN, THE MEDIAN, AND THE MODE

In a statistical study, after we have collected the data, organized them, and presented them graphically, we then analyze the data and summarize our findings. To do this, we often look for a representative, or typical, score.

Averages in Arithmetic

In your previous study of arithmetic, you learned how to find the average of two or more numbers. For example, to find the average of 17, 25, and 30:

Step 1. Add these three numbers: $17 + 25 + 30 = 72$.

Step 2. Divide this sum by 3 since there are three numbers: $73 \div 3 = 24$.

The average of the three numbers is 24.

Averages in Statistics

The word ***average*** has many different meanings. For example, there is an *average* of test scores, a batting *average*, the *average* television viewer, an *average* intelligence, and the *average* size of a family. These averages are *not* necessarily found by the same rule or procedure. Because of this confusion, in statistics we speak of ***measures of central tendency***. These measures are numbers that usually fall somewhere in the center of a set of organized data.

We will discuss three measures of central tendency: the *mean*, the *median*, and the *mode*.

The Mean

In statistics, the arithmetic average previously studied is called the ***mean*** of a set of numbers. It is also called the *arithmetic mean* or the *numerical average*. The mean is found in the same way as the arithmetic average is found.

PROCEDURE. To find the mean of a set of N numbers, add the numbers and divide the sum by N.

For example, if Ralph's grades on five tests in science during this marking period are 93, 80, 86, 72, and 94, he can find the mean of his test grades as follows:

Step 1. Add the five data values: $93 + 80 + 86 + 72 + 94 = 425$.

Step 2. Divide this sum by 5, the number of tests: $425 \div 5 = 85$.

The mean (arithmetic average) is 85.

Let us consider another example. In a car wash, there are seven employees whose ages are 17, 19, 20, 17, 46, 17, and 18. What is the mean of the ages of these employees?

Here, we add the seven ages to get a sum of 154. Then, $154 \div 7 = 22$. While the mean age of 22 is the correct answer, this measure does *not* truly represent the data. Only one person is older than 22, while six people are under 22. For this reason, we will look at another measure of central tendency that will eliminate the extreme case (the employee aged 46) that is *distorting* the data.

The Median

The ***median*** is the middle value for a set of data arranged in numerical order. For example, the median of the ages 17, 19, 20, 17, 46, 17, and 18 for the car-wash employees can be found in the following manner:

Step 1. Arrange the ages in numerical order: 17, 17, 17, 18, 19, 20, 46.

Step 2. Find the middle number: 17, 17, 17, 18, 19, 20, 46.

The median is 18 because there are three ages less than 18, and three ages greater than 18. The median, 18, is a better indication of the typical age of the employees than the mean, 22, because there are so many younger people working at the car wash.

Now, let us suppose that one of the car-wash employees has a birthday, and her age changes from 17 to 18. What is now the median age?

Step 1. Arrange the ages in numerical order: 17, 17, 18, 18, 19, 20, 46.

Step 2. Find the middle number: 17, 17, 18, 18, 19, 20, 46.

The median, or middle value, is still 18. We can no longer say that there are three ages less than 18 because one of the three youngest employees is now 18. We can say, however, that:

1. the median is 18 because there are three ages less than or equal to 18 and three ages greater than or equal to 18; or
2. the median is 18 because, when the data values are arranged in order, there are three values below this median, or middle number, and three values above it.

Recently, the car wash hired a new employee whose age is 21. The data now include eight ages, an even number, so there is no middle value. What is now the median age?

Step 1. Arrange the ages in numerical order: 17, 17, 18, 18, 19, 20, 21, 46.

Step 2. There is no single middle number. Find the *two* middle numbers: 17, 17, 18, 18, 19, 20, 21, 46.

Step 3. Find the mean (arithmetic average) of the two middle numbers: $\frac{18 + 19}{2} = 18\frac{1}{2}$

The median is now $18\frac{1}{2}$. There are four ages less than this center value of $18\frac{1}{2}$ and four ages greater than $18\frac{1}{2}$.

PROCEDURE. To find the median of a set of N numbers:

1. Arrange the numbers in numerical order.
2. If N is odd, find the middle number. This number is the median.
3. If N is even, find the mean (arithmetic average) of the two middle numbers. This average is the median.

The Mode

The ***mode*** is the data value that appears most often in a given set of data. It is usually best to arrange the data in numerical order before finding the mode.

Let us consider some examples of finding the mode:

1. The ages of employees in a car wash are 17, 17, 17, 18, 19, 20, 46. The mode, which is the number appearing most often, is 17.
2. The number of hours spent by each of the six students in reading a book are 6, 6, 8, 11, 14, 21. The mode, or number appearing most frequently, is 6. In this case, however, the mode is not a useful measure of central tendency. A better indication is given by the mean or the median.
3. The number of photographs printed from each of Renee's last six rolls of film are 8, 8, 9, 11, 11, and 12. Since 8 appears twice and 11 appears twice, we say that there are two modes: 8 and 11. We do not take the average of these two numbers since the mode tells us where most of the scores appear; we simply report both numbers. When *two modes* appear within a set of data, we say that the data are ***bimodal***.
4. The number of people living in each house on Meryl's street are 2, 2, 3, 3, 4, 5, 5, 6, 8. These data have *three* modes: 2, 3, and 5.
5. Ralph's test scores in science are 72, 80, 86, 93, and 94. Here, every number appears the same number of times, once. Since *no* number appears more often than the others, we define such data as ***having no mode***.

PROCEDURE. To find the mode for a set of data, be aware that:

1. If one number appears most often in the data, that number is the mode.
2. If two or more numbers appear more often than all other data values, and these numbers appear with the same frequency, then each of these numbers is a mode.
3. If each number in the set of data appears with the same frequency, there is no mode.

KEEP IN MIND

Three measures of central tendency are:

1. The *mean*, or *mean average*, found by adding N data values and then dividing the sum by N.

2. The *median*, or *middle* score, found only when data are arranged in numerical order.
3. The *mode*, or the value that appears *most often*.

EXAMPLES

1. The weights, in pounds, of five players on the basketball team are 168, 174, 181, 195, and 182. Find the average weight of a player on this team.

Solution The word *average*, by itself, indicates the *mean*. Therefore:
(1) Add the five weights: $168 + 174 + 181 + 195 + 182 = 900$.
(2) Divide the sum by 5, the number of players: $900 \div 5 = 180$.

Calculator Solution

Enter: [( ] 168 [+] 174 [+] 181 [+] 195 [+] 182 [ )] [÷] 5 [=]

Display: 180.

Answer: 180 lb

---

2. Renaldo has marks of 75, 82, and 90 on three mathematics tests. What mark must he obtain on the next test to have an average of exactly 85 for the four math tests?

Solution The word *average*, by itself, indicates the *mean*.
Let x = Renaldo's mark on the fourth test. Then:

The sum of the four test marks divided by 4 is 85.

$$\frac{75 + 82 + 90 + x}{4} = 85$$

$$\frac{247 + x}{4} = 85$$

$$247 + x = 340$$

$$x = 93$$

Check:

$$\frac{75 + 82 + 90 + 93}{4} \stackrel{?}{=} 85$$

$$\frac{340}{4} \stackrel{?}{=} 85$$

$$85 = 85 \quad \text{(True)}$$

Answer: Renaldo must obtain a mark of 93 on his fourth math test.

---

3. Find the median for each distribution.
a. 3, 2, 5, 5, 1 **b.** 9, 8, 8, 7, 4, 3, 3, 2, 0, 0

Solutions **a.** Arrange the data in numerical order: 1, 2, 3, 5, 5.
The median is the middle value. 1, 2, 3, 5, 5

Answer: median = 3

b. Since there is an even number of values, there are two middle values. Find the mean (average) of these two middle values:

9, 8, 8, 7, 4, 3, 3, 2, 0, 0

$$\frac{4+3}{2} = \frac{7}{2} = 3\frac{1}{2}$$

Answer: median = $3\frac{1}{2}$ or 3.5

---

4. Find the mode for each distribution.
a. 2, 9, 3, 7, 3 **b.** 3, 4, 5, 4, 3, 7, 2 **c.** 1, 2, 3, 4, 5, 6, 7

Solutions **a.** Arrange the data in numerical order: 2, 3, 3, 7, 9.
The mode, or most frequent value, is 3.

Answer: mode = 3

b. Arrange the data in numerical order: 2, 3, 3, 4, 4, 5, 7.
Both 3 and 4 appear twice. There are two modes.

Answer: modes = 3 and 4

c. Every value occurs the same number of times in the data set 1, 2, 3, 4, 5, 6, 7.

Answer: There is no mode.

EXERCISES WITH OPEN-RESPONSE PROBLEMS

1. Sid received grades of 92, 84, and 70 on three tests. Find his test average.
2. Sarah's grades were 80 on two of her tests and 90 on each of three other tests. Find her test average.
3. Louise received a grade of x on two of her tests and of y on each of three other tests. Represent her average for all the tests in terms of x and y.
4. Andy has grades of 84, 65, and 76 on three social studies tests. What grade must he obtain on the next test to have an average of exactly 80 for the four tests?
5. Rosemary has grades of 90, 90, 92, and 78 on four English tests. What grade must she obtain on the next test so that her average for the five tests will be 90?

6. The first three test scores are shown below for each of four students. A fourth test will be given, and averages taken for all four tests. Each student hopes to maintain an average of 85. Find the score needed by each student on the fourth test to have an 85 average, or explain why such an average is not possible.
   a. Pat: 78, 80, 100 **b.** Bernice: 79, 80, 81 **c.** Helen: 90, 92, 95 **d.** Al: 65, 80, 80

7. The average weight of Sue, Pam, and Nancy is 55 kilograms.
   a. What is the total weight of the three girls?
   b. Agnes weighs 60 kilograms. What is the average weight of the four girls: Sue, Pam, Nancy, and Agnes?

8. For the first 6 days of a week, the average rainfall in Chicago was 1.2 inches. On the last day of the week, 1.9 inches of rain fell. What was the average rainfall for the week?

9. The smallest of three consecutive integers is 32. What is the mean of the three integers?

10. The average of three consecutive even integers is 20. Find the integers.

11. The mean of three numbers is 31. The second is 1 more than twice the first. The third is 4 less than 3 times the first. Find the numbers.

12. If the heights, in centimeters, of a group of students are 180, 180, 173, 170, and 167, what is the mean height of these students?

13. Find the median for each set of data.
   a. 3, 4, 7, 8, 12 **b.** 2, 9, 10, 10, 12 **c.** 3.2, 4, 4.1, 5, 5
   d. 3, 4, 7, 8, 12, 13 **e.** 2, 9, 10, 10 **f.** 3.2, 4, 4.1, 5

14. Find the median for each set of data after placing the values into numerical order.
   a. 1, 2, 5, 3, 4 **b.** 2, 9, 2, 9, 7 **c.** 3, 8, 12, 7, 1, 0, 4
   d. 80, 83, 97, 79, 25 **e.** 3.2, 8.7, 1.4 **f.** 2, 0.2, 2.2, 0.02, 2.02
   g. 21, 24, 23, 22, 20, 24, 23, 21, 22, 23 **h.** 5, 7, 9, 3, 8, 7, 5, 6

15. What is the median age of a family whose members are 42, 38, 14, 13, 10, and 8 years old?

16. What is the median age of a class in which 14 students are 14 years old and 16 students are 15 years old?

17. In a charity collection, ten people gave amounts of \$1, \$2, \$1, \$1, \$3, \$1, \$2, \$1, \$1, and \$1.50. What was the median donation?

18. The test scores for an examination were 62, 67, 67, 70, 90, 93, and 98. What is the median test score?

19. What is the median for the digits 1, 2, 3, . . . , 9?

20. What is the median for the counting numbers from 1 through 100?

21. Find the mode for each distribution.
   a. 2, 2, 3, 4, 8 **b.** 2, 2, 3, 8, 8 **c.** 2, 2, 8, 8, 8 **d.** 2, 3, 4, 7, 8
   e. 2, 2, 3, 8, 8, 9, 9 **f.** 1, 2, 1, 2, 1, 2, 1 **g.** 1, 2, 3, 2, 1, 2, 3, 2, 1 **h.** 3, 19, 21, 75, 0, 6
   i. 3, 2, 7, 6, 2, 7, 3, 1, 4, 2, 7, 5 **j.** 19, 21, 18, 23, 19, 22, 18, 19, 20

22. A set of data consists of six numbers: 7, 8, 8, 9, 9, and x. Find the mode for these six numbers when:
   a. $x = 9$ **b.** $x = 8$ **c.** $x = 7$ **d.** $x = 6$

23. A set of data consists of the values 2, 4, 5, x, 5, 4. Find a possible value of x such that:
a. there is no mode because all scores appear an equal number of times **b.** there is only one mode **c.** there are two modes.

24. For each set of data, find: (1) the mean (2) the median (3) the mode
a. 7, 3, 5, 11, 9 **b.** 22, 38, 18, 14, 22, 30
c. $5\frac{1}{2}, 2\frac{1}{4}, 7, 5\frac{3}{4}, 4\frac{1}{2}$ **d.** 1, 0.01, 1.1, 0.12, 1, 1.03

25. For the set of data 5, 5, 6, 7, 7, which statement is true?
(1) mean = mode (2) median = mode (3) mean = median (4) mean < median

26. For the set of data 8, 8, 9, 10, 15, which statement is true?
(1) mean < median (2) mean > mode (3) median < mode (4) mean = median

27. When the data consists of 3, 4, 5, 4, 3, 4, 5, which statement is true?
(1) mean > median (2) mean > mode (3) median < mode (4) mean = median

28. For which set of data is there no mode?
(1) 2, 1, 3, 1, 2 (2) 1, 2, 3, 3, 3 (3) 1, 2, 4, 3, 5 (4) 2, 2, 3, 3, 3

29. For which set of data is there more than one mode?
(1) 8, 7, 7, 8, 7 (2) 8, 7, 4, 5, 6 (3) 8, 7, 5, 7, 6, 5 (4) 1, 2, 2, 3, 3, 3

30. For which set of data does the median equal the mode?
(1) 3, 3, 4, 5, 6 (2) 3, 3, 4, 5 (3) 3, 3, 4 (4) 3, 4

31. For which set of data will the mean, median, and mode all be equal?
(1) 1, 2, 5, 5, 7 (2) 1, 2, 5, 5, 8, 9 (3) 1, 1, 1, 2, 5 (4) 1, 1, 2

32. The weekly salaries of six employees in a small firm are \$340, \$345, \$345, \$350, \$350, and \$520.
a. For these six salaries, find: (*1*) the mean (*2*) the median (*3*) the mode
b. If negotiations for new salaries are in session and you represent management, which measure of central tendency will you use as the average salary? Explain your answer.
c. If negotiations are in session and you represent the labor union, which measure of central tendency will you use as an average salary? Explain your answer.

33. In a certain school district, bus service is provided for students living more than $1\frac{1}{2}$ miles from school. The distances from school to home for ten students are 0, $\frac{1}{2}$, $\frac{1}{2}$, 1, 1, 1, 1, $1\frac{1}{2}$, $3\frac{1}{2}$, and 10 miles.
a. For these data, find: (*1*) the mean (*2*) the median (*3*) the mode
b. How many of the ten students are entitled to bus service?
c. Explain why the mean is not a good measure of central tendency to describe the average distance between home and school for these students.

34. Last month, a carpenter used 12 boxes of nails each of which contained nails of only one size. The sizes marked on the boxes were $\frac{3''}{4}, \frac{3''}{4}, \frac{3''}{4}, \frac{3''}{4}, \frac{3''}{4}, \frac{3''}{4}, \frac{3''}{4}, \frac{3''}{4}$, 1″, 1″, 2″, and 2″.
a. For these data, find: (*1*) the mean (*2*) the median (*3*) the mode
b. Describe the average-size nail used by the carpenter, using at least one of these measures of central tendency. Explain your answer.

15-5 MEASURES OF CENTRAL TENDENCY AND GROUPED DATA

Intervals of Length 1

In a statistical study, when the range is small, we can use intervals of length 1 to group the data. For example, a class of 25 students reported the number of books read during the first half of the school year. The data were as follows:

5, 3, 5, 3, 1, 8, 2, 4, 2, 6,
3, 8, 8, 5, 3, 4, 5, 8, 5, 3,
3, 5, 6, 2, 3.

These data, for which the values range from 1 to 8, can be organized into a table such as the one shown at the right, with each value representing an interval. Since 25 students were included in this study, the total frequency, N, is 25, as shown below the "Frequency" column. We can use this table, with intervals of length 1, to find the mode, median, and mean for this data.

Interval	Frequency
8	4
7	0
6	2
5	6
4	2
3	7
2	3
1	1

$N = 25$

Mode Since the greatest frequency, 7, appears for interval 3, the mode for these data is 3.

In general:

- **For a set of grouped data, the mode is the value of the interval that contains the greatest frequency.**

Median We have learned that the median for a set of data in numerical order is the middle value.

1, 2, 2, 2, 3, 3, 3, 3, 3, 3, 3, 4, 4, 5, 5, 5, 5, 5, 5, 6, 6, 8, 8, 8, 8

For these 25 numbers, the median, or middle value, is 4. There are 12 numbers larger than or equal to 4, and 12 numbers smaller than or equal to 4.

When the data are grouped in the table that follows, a simple counting procedure can be used to find the median. Since the total frequency is 25, the median will be the 13th number. There are 12 numbers above and 12 below the median.

Interval	Frequency
8	4
7	0
6	2
5	6
4	2
3	7
2	3
1	1

When we add the frequencies of the first four intervals, starting at the top, we find that these intervals include data for:

$$4 + 0 + 2 + 6 = 12 \text{ students.}$$

Therefore, the next lower interval must include the median, the value for the 13th student. This interval contains the data value 4.

When we add the frequencies of the first three intervals, starting at the bottom, we find that these intervals include data for:

$$1 + 3 + 7 = 11 \text{ students.}$$

The next higher interval contains two scores, one for the 12th student and one that is the median, or the value for the 13th student. Again this is the interval that contains the data value 4.

In general:

- **For a set of grouped data, the median is the value of the interval that contains the middle data value.**

Mean By adding the four 8's in the ungrouped data, we see that four students reading eight books each result in a total of 32 books: $8 + 8 + 8 + 8 = 32$. We can arrive at this same number by using the grouped intervals in the table; we multiply the interval 8 by the frequency 4; thus, $(4)(8) = 32$. Applying this multiplication shortcut to each line of the table, we obtain the third column of the following table:

Interval	Frequency	(Interval) • (Frequency)
8	4	8 • 4 = 32
7	0	7 • 0 = 0
6	2	6 • 2 = 12
5	6	5 • 6 = 30
4	2	4 • 2 = 8
3	7	3 • 7 = 21
2	3	2 • 3 = 6
1	1	1 • 1 = 1
	$N = 25$	Total = 110

The **total** (110) represents **the sum of all 25 items of data**. We can check this by adding the 25 scores in the unorganized data.

Finally, to find the mean, we divide the total number, 110, by the number of items, 25: thus, $110 \div 25 = 4.4$, which is the mean.

PROCEDURE. To find the mean for N values in a table of grouped data when the length of the interval is 1:

1. For each interval, multiply the interval value by its corresponding frequency.
2. Find the sum of these products.
3. Divide this sum by the total frequency, N.

Intervals Other Than Length 1

There are definite mathematical procedures to find the mean, median, and mode for grouped data with intervals other than length 1, but we will not study them at this time. Instead, we will simply identify the intervals that contain some of these measures of central tendency.

For example, a small industrial plant surveyed 50 workers to find the number of miles each person commuted to work. The commuting distances were reported, to the nearest mile, as follows:

0, 0, 1, 1, 2, 2, 2, 3, 3, 4, 4, 4, 5, 5, 6, 6, 6, 7, 7, 7, 9,
10, 10, 10, 10, 10, 10, 10, 10, 12, 12, 14, 15, 17, 17,
18, 22, 23, 25, 28, 30, 32, 32, 33, 34, 34, 36, 37, 37, 52.

These data were organized into a table with intervals of length 10, as follows:

Interval (commuting distance)	Frequency (number of workers)
50–59	1
40–49	0
30–39	9
20–29	4
10–19	15
0–9	21

$N = 50$

Modal interval In the table, interval 0–9 contains the greatest frequency, 21. We say that interval 0–9 is the ***group mode***, or ***modal interval***, because this group of numbers has the greatest frequency.

The modal interval is *not* the same as the mode. The modal interval is a group of numbers; the mode is usually a single number. For this example, the original data (before being placed into the table) show that the number appearing most often is 10. Hence, the mode is 10. The modal interval, which is 0–9, tells us that, of the six intervals in the table, the most frequently occurring commuting distance is 0 to 9 miles. The modal interval for this grouping is 0–9.

Both the mode and the modal interval depend on the concept of greatest frequency. For the mode, we look for a single number that has the greatest frequency. For the modal interval, we look for the interval that has the greatest frequency.

Interval containing the median To find the interval containing the median, we follow the procedure described earlier in this section. For 50 numbers, the median, or middle number, will be at a point where 25 numbers are above the median and 25 are below it.

Interval	Frequency
50–59	1
40–49	0
30–39	9
20–29	4
10–19	15
0–9	21

$N = 50$

Counting the frequencies in the table from the uppermost interval and moving downward, we add $1 + 0 + 9 + 4 = 14$. Since there are 15 numbers in the next lower interval, and $14 + 15 = 29$, we see that 25 numbers will be reached somewhere in that interval, 10–19.

Counting from the bottom interval and moving up, we have 21 numbers in the first interval. Since there are 15 numbers in the next higher interval, and $21 + 15 = 36$, we see that 25 numbers will be reached somewhere in that interval, 10–19. This is the same result that we obtained when we moved downward. The interval containing the median for this grouping is 10–19.

In this course, we will not deal with problems in which the median is not found in any interval.

As a final note, there is no simple procedure to identify the interval that contains the mean when data are grouped using intervals other than length 1. This problem will also be studied in higher level courses.

When data are grouped using intervals of length other than 1, there is no simple procedure to identify the interval containing the mean. However, the mean can be approximated by assuming that the data are equally distributed throughout each interval. The mean is then found by using the midpoint of each interval as the value of each entry in the interval. This problem is studied in higher level courses.

EXAMPLE

In the table, the data indicate the heights, in inches, of 17 basketball players. For these data find:

a. the mode **b.** the median **c.** the mean

Height (inches)	Frequency (number)
77	2
76	0
75	5
74	3
73	4
72	2
71	1

Solutions **a.** The greatest frequency, 5, occurs for the interval where heights are 75 inches. The mode, or height appearing most often, is 75.

Answer: mode = 75

Height	Frequency
77	2
76	0
75	5
74	3
73	4
72	2
71	1

b. For 17 players, the median is the middle number (the 9th number), so there are 8 heights greater than or equal to the median, and 8 heights less than or equal to the median.

Counting the frequencies going down, we have:

$2 + 0 + 5 = 7$ and $2 + 0 + 5 + 3 = 10$.

Thus, the mid height must lie in interval 74.

Counting the frequencies going up, we have:

$1 + 2 + 4 = 7$ and $1 + 2 + 4 + 3 = 10$.

Thus, the mid height must lie in interval 74.

Answer: median = 74

c. (*1*) Multiply each height given as an interval by the frequency for that interval.

(*2*) Find the total of these products: 1,258.

(*3*) Divide this total, 1,258, by the total frequency ($N = 17$) to obtain the mean, 74.

Answer: mean = 74

$$
\begin{aligned}
77 \cdot 2 &= 154 \\
76 \cdot 0 &= 0 \\
75 \cdot 5 &= 375 \\
74 \cdot 3 &= 222 \\
73 \cdot 4 &= 292 \\
72 \cdot 2 &= 144 \\
71 \cdot 1 &= \underline{71} \\
\text{Total} &= 1{,}258
\end{aligned}
$$

$$
\begin{array}{r}
74 \\
17\overline{)1258} \\
\underline{119} \\
68 \\
\underline{68}
\end{array}
$$

Calculator Solution The memory capability of the scientific calculator can be used to perform the calculation in part **c** of this example. Your calculator may have some of the following memory keys:

[STO ▸] [SUM] [RCL] [EXC] [M+] [M−] [MR] [MC] [MRC]

The [STO ▸] key is used to store individual data values, and the [SUM] or [M+] key to add to the number already in memory. If your calculator has no [STO ▸] key, use [SUM] or [M+], but be certain that the memory is cleared before you begin. The memory can be cleared by pressing [MC] once, by pressing [MRC] twice, or by entering 0 into the display and pressing [EXC]. In the last case the number in the display and the number in the memory will be exchanged so that 0 is in the memory.

Some calculators have more than one memory, and you may need to enter a letter or a number after you press a memory key to indicate in which memory the number is to be stored.

METHOD 1

Enter: 77 [×] 2 [=] [STO ▸]
Display: 154.

Enter: 75 [×] 5 [=] [SUM]
Display: 375.

Enter: 74 [×] 3 [=] [SUM]
Display: 222.

Enter: 73 [×] 4 [=] [SUM]
Display: 292.

Enter: 72 [×] 2 [=] [SUM]
Display: 144.

Enter: 71 [×] 1 [=] [SUM]
Display: 71.

Enter: [RCL] [÷] 17 [=]
Display: 74.

METHOD 2

Enter: [MC] 77 [×] 2 [=] [SUM]
Display: 154.

Enter: 75 [×] 5 [=] [SUM]
Display: 375.

Enter: 74 [×] 3 [=] [SUM]
Display: 222.

Enter: 73 [×] 4 [=] [SUM]
Display: 292.

Enter: 72 [×] 2 [=] [SUM]
Display: 144.

Enter: 71 [×] 1 [=] [SUM]
Display: 71.

Enter: [RCL] [÷] 17 [=]
Display: 74.

The following method does not use the memory:

Enter: [(] 77 [×] 2 [+] 75 [×] 5 [+] 74 [×] 3 [+] 73 [×] 4 [+] 72 [×] 2 [+] 71 [×] 1 [)] [÷] 17 [=]
Display: 74.

EXERCISES WITH OPEN-RESPONSE PROBLEMS

In 1–3, the data are grouped in each table in intervals of length 1. Find:
a. the total frequency **b.** the mean **c.** the median **d.** the mode

1.

Interval	Frequency
10	1
9	2
8	3
7	2
6	4
5	3

2.

Interval	Frequency
15	3
16	2
17	4
18	1
19	5
20	6

3.

Interval	Frequency
25	4
24	0
23	3
22	2
21	4
20	5
19	2

4. On a test consisting of 20 questions, 15 students received the following scores:

17, 14, 16, 18, 17, 19, 15, 15,
16, 13, 17, 12, 18, 16, 17.

a. On your answer paper, copy and complete the table shown at the right.

b. Find the median score.

c. Find the mode.

d. Find the mean.

Grade	Frequency
20	
19	
18	
17	
16	
15	
14	
13	
12	

Interval (minutes)	Number of People (frequency)
6	12
5	20
4	36
3	20
2	12

5. A questionnaire was distributed to 100 people. The table at the left shows the time taken, in minutes, to complete the questionnaire.

a. For these data, find:
(*1*) the mean (*2*) the median (*3*) the mode

b. How are the three measures found in part **a** related for these data?

Size of Suit (interval)	Number Sold (frequency)
48	1
46	1
44	3
42	5
40	3
38	8
36	2
34	2

6. A store owner kept a tally of the sizes of suits purchased in her store, as shown in the table at the right.

a. For this set of data find:
(*1*) the total frequency (*2*) the mean
(*3*) the median (*4*) the mode

b. Which measure of central tendency should the store owner use to describe the average suit sold?

In 7–9, data are grouped in each table in intervals other than length 1. Find: **a.** the total frequency **b.** the interval that contains the median **c.** the modal interval

7.

Interval	Frequency
55–64	3
45–54	8
35–44	7
25–34	6
15–24	2

8.

Interval	Frequency
4–9	12
10–15	13
16–21	9
22–27	12
28–33	15
34–39	10

9.

Interval	Frequency
126–150	4
101–125	6
76–100	6
51–75	3
26–50	7
1–25	2

10. Test scores for a class of 20 students are as follows:

93, 84, 97, 98, 100, 78, 86, 100, 85, 92,
72, 55, 91, 90, 75, 94, 83, 60, 81, 95.

a. On your answer paper, copy and complete the table shown at the right.

b. Find the modal interval.

c. Find the interval that contains the median.

Test Scores	Frequency
91–100	
81–90	
71–80	
61–70	
51–60	

11. The following data consist of the weights, in pounds, of 35 adults:

176, 154, 161, 125, 138, 142, 108, 115, 187, 158, 168,
162, 135, 120, 134, 190, 195, 117, 142, 133, 138, 151,
150, 168, 172, 115, 148, 112, 123, 137, 186, 171, 166,
166, 179.

a. Copy and complete the table shown at the right.

b. Construct a frequency histogram based on the grouped data.

c. In what interval is the median for these grouped data?

d. What is the modal interval?

Interval	Frequency
180–199	
160–179	
140–159	
120–139	
100–119	

15-6 QUARTILES, PERCENTILES, AND CUMULATIVE FREQUENCY

Quartiles

When the values in a set of data are listed in numerical order, the median separates the set into two equal parts. The numbers that separate the set into four equal parts are called ***quartiles***.

The heights, in inches, of 20 students are shown in the following list. The median, which is the average of the 10th and 11th data values, is shown here enclosed in a box.

Lower half | Upper half

53, 60, 61, 63, 64, 65, 65, 65, 65, 66, 66, 67, 67, 68, 69, 70, 70, 71, 71, 73

↑
66
Median

Ten heights are listed in the lower half, 53–66. The middle value for these 10 heights, which is the average of the 5th and 6th values from the lower end, or 64.5, separates the lower half into two equal parts.

Ten heights are also listed in the upper half, 66–73. The middle value for these 10 heights, which is the average of the 5th and 6th values from the upper end, or 69.5, separates the upper half into two equal parts.

The 20 data values are now separated into four equal parts, or quarters.

53, 60, 61, 63, 64, | 64.5 | 65, 65, 65, 65, 66, | 66 | 66, 67, 67, 68, 69, | 69.5 | 70, 70, 71, 71, 73

First quartile: 64.5 | Median, Second quartile: 66 | Third quartile: 69.5

The numbers that separate the data into four equal parts are the quartiles. For this set of data:

1. Since one quarter of the heights are less than or equal to 64.5 inches, 64.5 is the ***lower quartile***, or ***first quartile***.
2. Since two quarters of the heights are less than or equal to 66 inches, 66 is the ***second quartile***. The second quartile is always the same as the median.
3. Since three quarters of the heights are less than or equal to 69.5 inches, 69.5 is the ***upper quartile***, or ***third quartile***.

A ***whisker-box plot*** is a diagram that uses the quartile values to display information about a set of data. To draw a whisker-box plot, we first draw a scale that includes numbers from the smallest to the highest value of a set of data. For example, for the set of heights of the 20 students, the scale should include the numbers from 53 to 73. The second step is to place a dot above the numbers on the scale that are the lowest values, the first quartile, the median, the third quartile, and the highest value, as shown below.

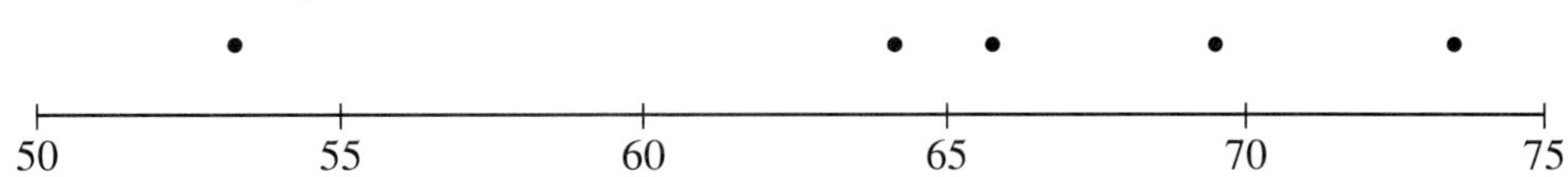

Next, we draw a box between the dots that represent the lower and upper quartiles, and we draw a vertical line in the box through the point that represents the median.

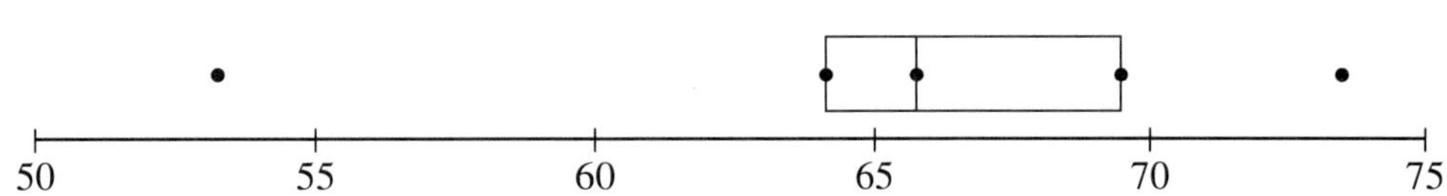

Finally, we add the whiskers by drawing a line segment joining the dots that represent the lowest data value and the lower quartile, and a second line segment joining the dots that represent the highest data value and the upper quartile.

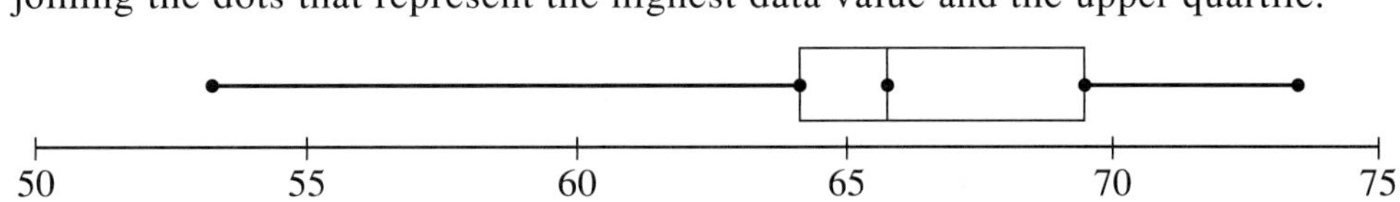

The box indicates the ranges of the middle half of the set of data. The long whisker at the left shows us that the data are more scattered at the lower than at the higher end.

PROCEDURE. To find the quartile values for a set of data:

1. Arrange the data in ascending order, left to right.
2. Find the median for the set of data. The median is the second quartile value.
3. Find the middle value for the lower half of the data. This number is the first, or lower, quartile value.
4. Find the middle value for the upper half of the data. This number is the third, or upper, quartile value.

EXAMPLE

Find the quartile values for the following set of data:

8, 5, 12, 9, 6, 2, 14, 7, 10, 17, 11, 8, 14, 5

Solution (1) Arrange the data in numerical order:
2, 5, 5, 6, 7, 8, 8, 9, 10, 11, 12, 14, 14, 17.

(2) Find the median. Since there are 14 data values in the set, the median is the average of the 7th and 8th values.

$$\text{Median} = \frac{8 + 9}{2} = 8.5$$

Therefore, 8.5 is the second quartile.

(3) Find the first quartile. There are seven values less than 8.5. The middle value is the 4th value from the lower end of the set of data, 6. Therefore, 6 is the first, or lower, quartile.

(4) Find the third quartile. There are seven values greater than 8.5. The middle value is the 4th value from the upper end of the set of data, 12. Therefore, 12 is the third, or upper, quartile.

Answer: The first quartile is 6, the second quartile is 8.5, and the third quartile is 12.

Note: The quartiles 6, 8.5, and 12 separate the data values into four equal parts:

2, 5, 5, [6], 7, 8, 8, ↑[8.5] 9, 10, 11, [12], 14, 14, 17

Even though the number of data values, 14, in the original set is not divisible by 4, the values that remain after the quartile values have been selected can be divided into groups with three numbers in each group.

Percentiles

A ***percentile*** is a number that tells us what percent of the total number of data values lie at or below a given measure.

Let us consider again the set of data values representing the heights of 20 students. In this set of 20 heights, to find, for example, the percentile rank of 65, we separate the data into the values that are less than or equal to 65 and those that are greater than or equal to 65, so that the four 65's in the set are divided equally between the two groups:

53, 60, 61, 63, 64, 65, 65, 65, 65, 66, 66, 67, 67, 68, 69, 70, 70, 71, 71, 73
↑

Half of 4, or 2, of the 65's are in the lower group and half in the upper group. Since there are seven data values in the lower group, we find what percent 7 is of 20, the total number of values:

$$\frac{7}{20} = 0.35 = 35\%$$

Therefore, 65 is at the 35th percentile (35%ile).

To find the percentile rank of 69, we separate the data into the values that are less than or equal to 69 and those that are greater or equal to 69:

53, 60, 61, 63, 64, 65, 65, 65, 65, 66, 66, 67, 67, 68, 69, 70, 70, 71, 71, 73
↑

Because 69 occurs only once, we will include it as half of a data value in the lower group and half of a data value in the upper group. Therefore there are $14\frac{1}{2}$ or 14.5 data values in the lower group.

$$\frac{14.5}{20} = 0.725 = 72.5\%$$

Because percentiles are usually written to the nearest whole percent, we say that 69 is at the 73rd percentile (73%ile).

EXAMPLE

Find the percentile rank of 87 in the following set of 30 marks:

56, 65, 65, 67, 72, 73, 75, 77, 77, 78, 78, 78, 80, 80, 80,
82, 83, 85, 85, 85, 86, 87, 87, 87, 88, 90, 92, 93, 95, 98.

Solution (1) Find the sum of the number of marks less than 87 and one-half of the number of 87's:

Number of marks less than 87 =	21
Half of the number of 87's (0.5 × 3) =	1.5
	22.5

(2) Divide the sum by the total number of marks:

$$\frac{22.5}{30} = 0.75$$

(3) Change the decimal value to a percent: 0.75 = 75%.

Answer: A mark of 87 is at the 75th percentile.

Note that 87 is also the upper quartile mark.

Cumulative Frequency

In a school, a final examination was given to all 240 students taking biology. The test grades of these students were then grouped into a table, as shown at the right. At the same time, a histogram of the results was constructed, as shown below.

Interval (test scores)	Frequency (number)
91–100	45
81–90	60
71–80	75
61–70	40
51–60	20

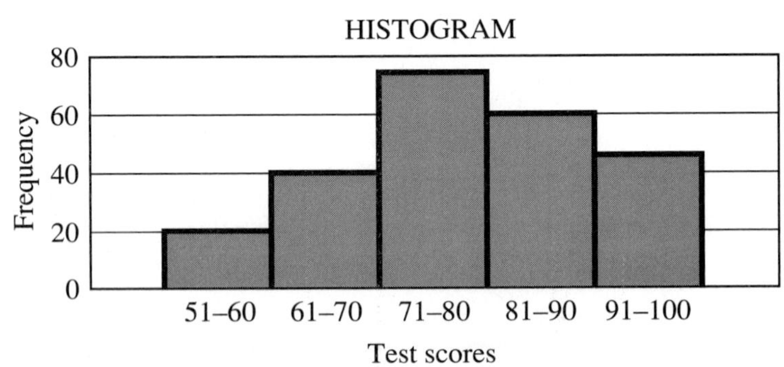

From the table and the histogram, we can see that 20 students scored in the interval 51–60, 40 students scored in the interval 61–70, and so forth. We can use these data to construct a new type of histogram that will answer this question: "How many students scored *below* a certain grade?"

By answering the following questions, we will gather some information before constructing the new histogram:

1. How many students scored 60 or less on the test?
From the lowest interval, 51–60, we know that 20 students scored 60 or less. *Answer:* 20

2. How many students scored 70 or less on the test?
By adding the frequencies for the two lowest intervals, 51–60 and 61–70, we see that 20 + 40, or 60, students scored 70 or less. *Answer:* 60

3. How many students scored 80 or less on the test?
By adding the frequencies for the three lowest intervals, 51–60, 61–70, and 71–80, we see that 20 + 40 + 75, or 135, students scored 80 or less. *Answer:* 135

4. How many students scored 90 or less on the test?
Here, we add the frequencies in the four lowest intervals.
Thus, 20 + 40 + 75 + 60, or 195, students scored 90 or less. *Answer:* 195

5. How many students scored 100 or less on the test?
By adding the five lowest frequencies, 20 + 40 + 75 + 60 + 45, we see that 240 students scored 100 or less. This result makes sense because 240 students took the test and all of them scored 100 or less. *Answer:* 240

Constructing a cumulative frequency histogram The answers to the five questions we have just asked were found by adding or *accumulating* the frequencies for the intervals in the grouped data to find the ***cumulative frequency***. The histogram that displays these accumulated figures is called a ***cumulative frequency histogram***.

Interval	Frequency	Cumulative Frequency
91–100	45	240
81–90	60	195
71–80	75	135
61–70	40	60
51–60	20	20

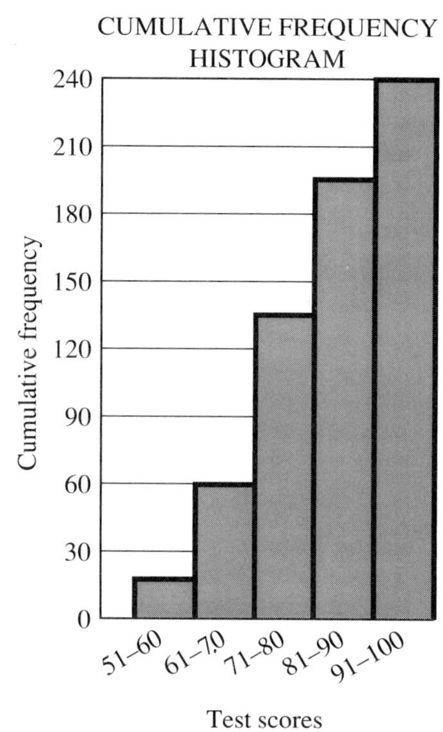

To find the cumulative frequency for each interval, we add the frequency for that interval to the frequencies for the intervals with lower values.

To draw a cumulative frequency histogram, we use the cumulative frequencies to determine the heights of the bars.

Using a percent scale on the cumulative frequency histogram For our example of the 240 biology students and their scores, the frequency scale for the cumulative frequency histogram goes from 0 to 240 (the total frequency for these data). We can replace the scale above of the cumulative frequency histogram with a different one that expresses the cumulative frequency in percents. Since 240 students represent 100% of the students taking biology, we write 100% to correspond to a cumulative frequency of 240. Similarly, since 0 students represent 0% of the students taking biology, we write 0% to correspond to a cumulative frequency of 0.

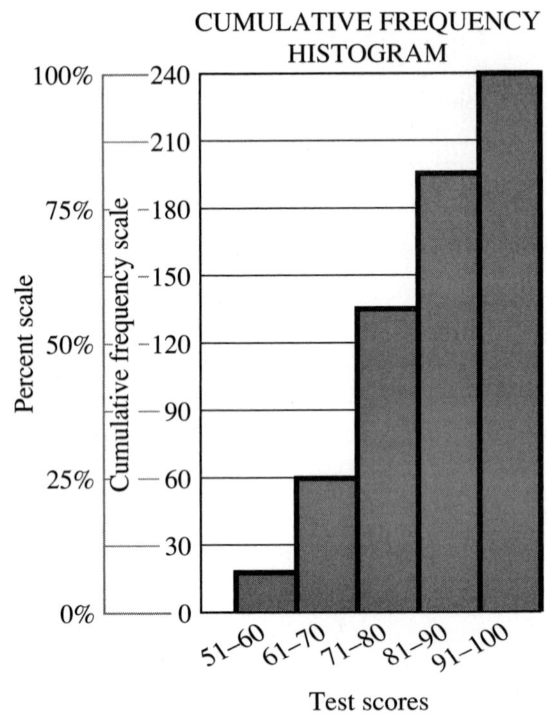

If we divide the percent scale into four equal parts, we can label the three added divisions as 25%, 50%, and 75%. Thus the graph relates each cumulative frequency to a percent of the total number of biology students. For example, 120 students (half of the total number) corresponds to 50%.

Let us use the percent scale to answer the question, "What percent of the students scored 70 or below on the test?" The height of each bar represents both the number of students and the percent of the students who had scores at or below the largest number in the interval represented by that bar. Since 25%, or a quarter, of the scores were 70 or below, we say that 70 is an approximate value for the lower quartile, or 25%ile.

From the histogram, we can see that about 56% of the students had scores at or below a number in the 71–80 interval. Thus, the second quartile, the median, and the 56%ile are in the 71–80 interval.

The histogram also shows that about 81% of the students had scores at or below a number in the 81–90 interval. For these data, the upper quartile and the 81%ile are between 81 and 90.

From the cumulative frequency histogram, we can also conveniently read the approximate percentiles for the scores that are the end values of the intervals. For example, to find the percentile for a score of 60, the right-end score of the first interval, we draw a vertical line segment beginning at 60 on the score scale and ending at the height of the first interval, as shown by the heavy broken line in the histogram below. From this point, we draw a horizontal line to the percent scale. The fact that the horizontal line crosses the percent scale at about one-third the distance between 0% and 25% tells us that approximately 8% of the students scored 60 or below 60. Thus, the 8th percentile is a good estimate for a score of 60.

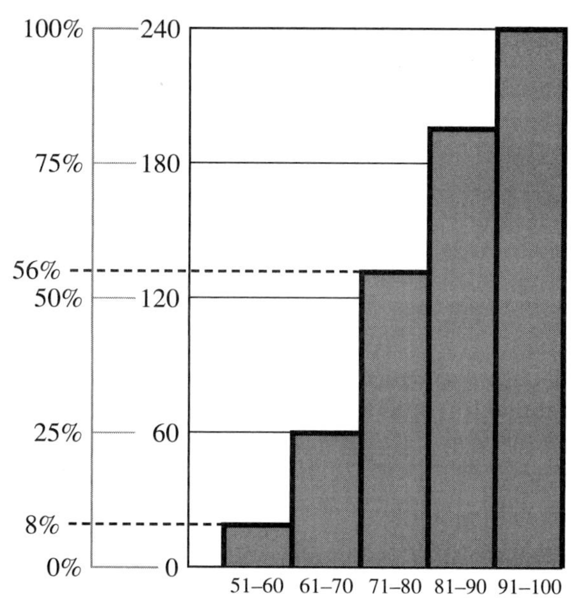

Similarly, we can read from the graph that 80 on the score scale corresponds to approximately 56% on the percent scale. Therefore, we know that approximately 56% of the students scored 80 or less than 80. Thus, the 56th percentile is a good estimate for a score of 80.

EXAMPLE

For the table of values at the right, in what interval is the lower quartile found?

Interval	Frequency
41–50	4
31–40	3
21–30	6
11–20	7
1–10	4

Solution The total frequency = 4 + 3 + 6 + 7 + 4 = 24.

The median divides the data into two equal groups with 12 data values in each group. The lower quartile is the middle value of the lower group of 12 data values, that is, the average of the 6th and 7th values.

There are fewer than six data values in the 1–10 interval and more than seven values in the two intervals 1–10 and 11–20. Therefore, both the 6th and 7th values lie in the 11–20 interval, and their average also lies in that interval.

Answer: The lower quartile is in the 11–20 interval.

EXERCISES WITH OPEN-RESPONSE PROBLEMS

In 1–4, for each set of data, find: **a.** the median **b.** the lower quartile **c.** the upper quartile

1. 12, 17, 20, 21, 25, 27, 29, 30, 32, 33, 33, 37, 40, 42, 44

2. 67, 70, 72, 77, 78, 78, 80, 84, 86, 88, 90, 92

3. 0, 0, 1, 1, 1, 1, 2, 2, 2, 2, 3, 3, 3, 3, 5, 7, 9, 9

4. 3.6, 4.0, 4.2, 4.3, 4.5, 4.8, 4.9, 5.0

5. In the table at the right, data are given for the heights, in inches, of 22 basketball players.

a. Copy and complete the table.

b. Draw a cumulative frequency histogram.

c. Find the height that is the lower quartile.

d. Find the height that is the upper quartile.

Height (inches)	Frequency	Cumulative Frequency
77	2	
76	2	
75	7	
74	5	
73	3	
72	2	
71	1	

In 6–8, data are grouped into tables. For each set of data:
a. Construct a cumulative frequency histogram.
b. Find the interval in which the lower quartile lies.
c. Find the interval in which the median lies.
d. Find the interval in which the upper quartile lies.

6.

Interval	Frequency
41–50	8
31–40	5
21–30	2
11–20	5
1–10	4

7.

Interval	Frequency
25–29	3
20–24	1
15–19	3
10–14	9
5–9	4

8.

Interval	Frequency
1–4	6
5–8	3
9–12	7
13–16	2
17–20	2

9. For the data shown at the right:

a. Construct a histogram.

b. Construct a cumulative frequency histogram.

c. In what interval is the median?

d. The value 10 occurs twice in the data. What is the percentile rank of 10?

Interval	Frequency
21–25	5
16–20	4
11–15	6
6–10	3
1–5	2

Interval	Frequency
33–37	4
28–32	3
23–27	7
18–22	12
13–17	8
8–12	5
3–7	1

10. For the data given in the table at the left:

a. Construct a cumulative frequency histogram.

b. In what interval is the median?

c. In what interval is the upper quartile?

d. What percent of scores are 17 or less?

e. In what interval is the 25th percentile?

11. A group of 400 students were asked to state the number of minutes that each spends watching television in 1 day. The cumulative frequency histogram shown below summarizes the responses as percents.

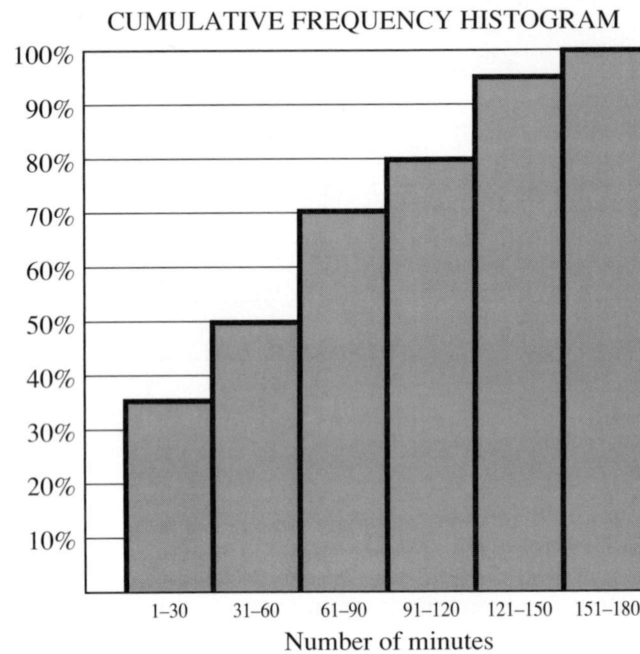

a. What percent of the students questioned watch television for 90 minutes or less each day?
b. How many of the students watch television for 90 minutes or less each day?
c. In what interval is the upper quartile?
d. In what interval is the lower quartile?
e. If one of these students is picked at random, what is the probability that he or she watches 30 minutes or less of television each day?

12. Cecilia's average for 4 years is 86. Her average is the upper quartile for her class of 250 students. At least how many students in her class have averages that are less than or equal to Cecilia's?

13. The lower quartile for a group of data was 40. These data consisted of the heights, in inches, of 680 children. At most, how many of these children measured more than 40 inches?

In 14–15, select, in each case, the numeral preceding the correct answer.

14. On a standardized test, Sally scored at the 80th percentile. This means that
(1) Sally answered 80 questions correctly.
(2) Sally answered 80% of the questions correctly.
(3) Of the students who took the test, about 80% had the same score as Sally.
(4) Of the students who took the test, about 80% had scores that were less than or equal to Sally's score.

15. For a set of data consisting of test scores, the 50th percentile is 87. Which of the following could be *false*?
(1) 50% of the scores are 87.
(2) 50% of the scores are 87 or less.
(3) Half of the scores are at least 87.
(4) The median is 87.

15-7 CIRCLE GRAPHS

A **circle graph** shows how percents or parts of a whole are distributed. The whole circle represents 100% of the data. The part is shown by a **sector** of the circle that is formed by two radii and an arc. The angle between the two radii is called the **central angle** of the sector.

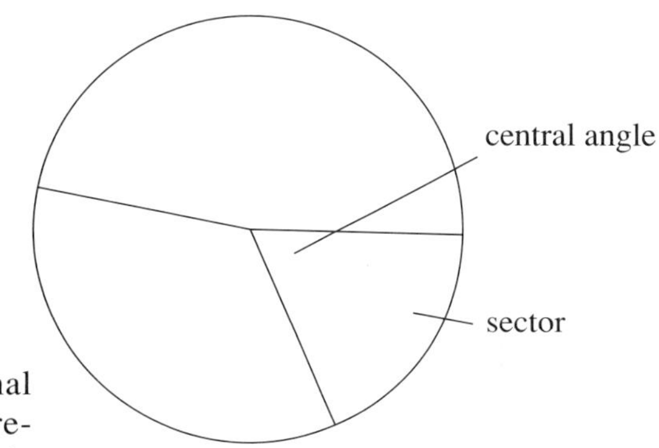

The area of a section is proportional to the fraction or percentage it represents. The central angle of the sector is equal to the given percentage of 360° (the whole circle). For example, a sector representing 25% of the data would have a central angle equal to $25\% \times 360° = 0.25 \times 360° = 90°$; a sector representing 12% would have a central angle equal to 12% of $360° = 0.12 \times 360° = 43.2°$.

PROCEDURE

1. Find the number that represents the whole if it is not given.
2. Find what part of the whole each category represents. It is usually most convenient to express each part as a decimal.
3. Calculate the central angle for each sector: multiply 360° by each part of the whole found in Step 2.
4. Draw a circle. Use a protractor to construct the central angle for each sector. Label each sector.

EXAMPLES

1. In a survey, 400 people named their favorite TV sport: baseball, 115; football, 129; basketball, 99; tennis, 57. Draw a circle graph representing the results of the survey.

Solution The total number, 400, is given.

baseball: $\frac{115}{400} = 0.2875$	$0.2875 \times 360° = 103.5°$
football: $\frac{129}{400} = 0.3225$	$0.3225 \times 360° = 116.1°$
basketball: $\frac{99}{400} = 0.2475$	$0.2475 \times 360° = 89.1°$
tennis: $\frac{57}{400} = 0.1425$	$0.1425 \times 360° = 51.3°$

Check: $103.5° + 116.1° + 89.1° + 51.3° = 360°$
Note that the sum may not always be exact due to rounding.

FAVORITE SPORTS

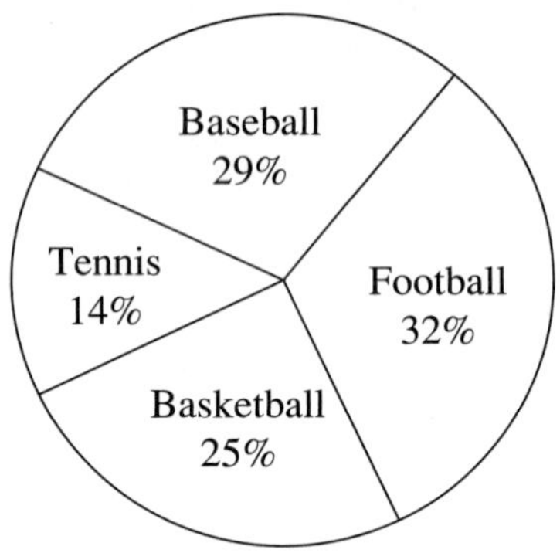

---

2. This circle graph shows votes in a student council election. If 1,200 students voted, how many voted for each candidate?

ELECTION RESULTS

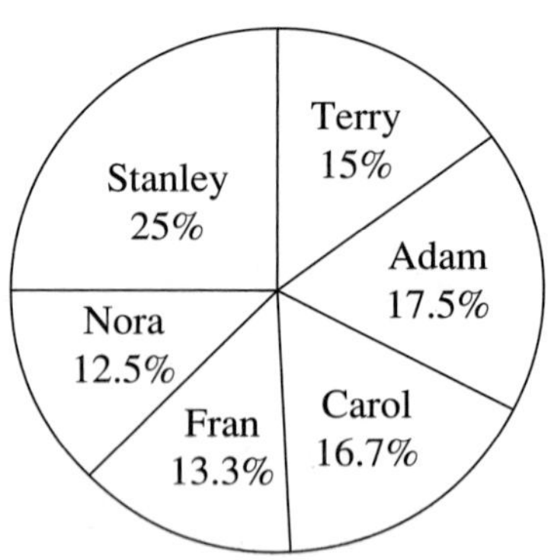

Solution Use your calculator.

Adam:	17.5% of 1,200 = 210
Carol:	16.7% of 1,200 = 200.4 (round to 200)
Fran:	13.3% of 1,200 = 159.6 (round to 160)
Nora:	12.5% of 1,200 = 150
Stanley:	25% of 1,200 = 300
Terry:	15% of 1,200 = 180

Check: 210 + 200 + 160 + 150 + 300 + 180 = 1,200

EXERCISES WITH OPEN-RESPONSE PROBLEMS

1. A class of 24 students was surveyed about their favorite toppings on pizza. The results are shown in the table below.

Sausage	Bacon	Pepperoni	Mushrooms	Peppers
9	4	3	6	2

In a circle graph showing the results of this survey, what would be the measure of the central angle, in degrees, for each sector of the graph?

2. A class of 30 students was surveyed to find each person's favorite subject. The results are shown in the table below.

Subject	Number of Students
English	8
Social Studies	9
Mathematics	6
Science	6
Foreign Language	1

In a circle graph showing the results of this survey, what is the measure of the central angle, in degrees, for each sector?

3. The data in the table at the right is to be displayed in a circle graph.
   a. Find the degree measure of the central angle for each percentage given.
   b. Construct a circle graph for the given data.

Favorite Lunch Food	Percentage of Students
Pizza	42
Hamburgers	34
Hot Dogs	12
Yogurt	8
Cottage cheese	4

4. The following circle graph represents 34 people's favorite ice cream flavor. How many people preferred each flavor?

FAVORITE ICE CREAM FLAVORS

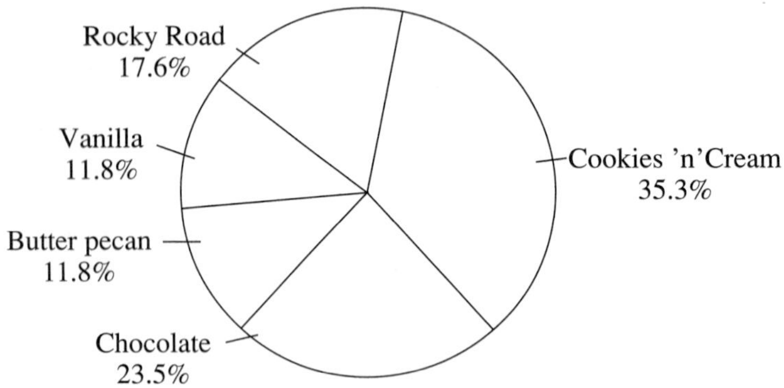

5. The table shows favorite types of television programs for a group of 50 students. **a.** Find the percent of the group that prefer each type of program. **b.** Create a circle graph to display the data.

Program Type	Number of Students
News	7
Sports	11
Drama	8
Talk	5
Comedy	14
Science	3
Cooking	2

6. Inez has totaled her expenses for last year and represented the results in this circle graph. If her total expenses were \$18,000, find the amount spent in each category.

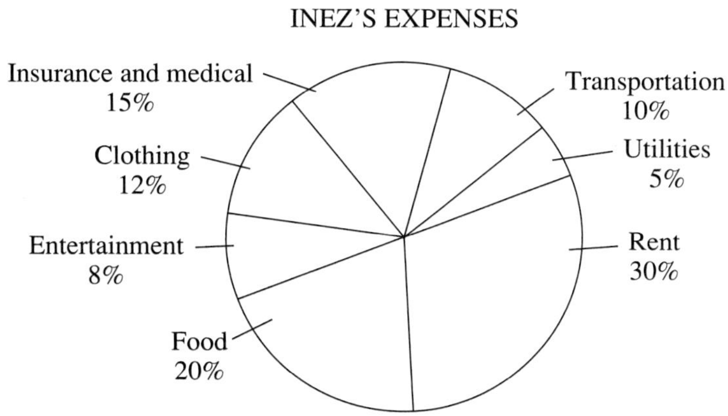

CHAPTER SUMMARY

Statistics is the study of numerical data. In a statistical study, data are collected, organized into tables and graphs, and analyzed to draw conclusions.

Tables and ***stem-and-leaf diagrams*** are used to organize data. A table should have between five and 15 intervals that include all data values, are of equal size, and do not overlap.

A ***histogram*** is a bar graph in which the height of a bar represents the frequency of the data values represented by that bar. A ***cumulative frequency histogram*** is a bar graph in which the height of the bar represents the total frequency of the data values that are less than or equal to the upper endpoint of that bar.

The mean, median, and mode are three measures of ***central tendency***. The ***mean*** is the sum of the data values divided by the total frequency. The ***median*** is the middle value when the data values are placed in numerical order. The ***mode*** is the data value that has the largest frequency.

Quartile values separate the data into four equal parts. A ***whisker-box plot*** displays a set of data values using the lowest value, the first quartile, the median, the third quartile, and the largest value as significant measures. The ***percentile*** rank tells what percent of the data values lie at or below a given measure.

A ***circle graph*** displays a set of data as percents or parts of a whole. The circle is divided into parts or ***sectors*** to represent each data value. The ***central angle*** of each sector in the graph is determined by the percentage of the whole of each data value.

VOCABULARY

15-1 Data Statistics Descriptive statistics Sample Census

15-2 Tally Frequency Grouped data Range Stem-and-leaf diagram

15-3 Histogram Frequency histogram

15-4 Measures of central tendency Average Mean Median Mode Bimodal

15-5 Group mode Modal interval

15-6 Quartiles Lower or first quartile Second quartile Upper or third quartile Whisker-box plot Percentile Cumulative frequency Cumulative frequency histogram

15-7 Circle graph Sector Central angle

REVIEW EXERCISES WITH OPEN-RESPONSE PROBLEMS

1. The weights, in kilograms, of five adults are 53, 72, 68, 70, and 72. Find:
a. the mean **b.** the median **c.** the mode

2. Steve's test scores are 82, 94, and 91. What grade must Steve earn on a fourth test so that the mean of his four scores will be exactly 90?

3. Express, in terms of y, the mean of $3y - 2$ and $7y + 18$.

4. For a certain day, temperature readings were 72°, 75°, 79°, 83°, 83°, and 88°. For these data, find: **a.** the mean **b.** the median **c.** the mode

In 5–7, select, in each case, the numeral preceding the correct answer.

5. Which set of data has more than one mode?
(1) 3, 4, 3, 4, 3, 5 (2) 1, 3, 5, 7, 1, 2, 4
(3) 9, 3, 2, 8, 3, 3 (4) 9, 3, 2, 3, 8, 2, 7

6. For which set of data are the mean, median, and mode all equal?
(1) 2, 2, 5 (2) 2, 5, 5 (3) 1, 3, 3, 5 (4) 1, 1, 5, 5

7. For the set of data 3, 2, 7, 1, 2, which statement is true?
(1) median = mean (2) median = mode
(3) median $>$ mean (4) median $>$ mode

8. Paul worked the following numbers of hours each week over a 20-week period:

15, 3, 7, 6, 2, 14, 9, 25, 8, 12,
8, 8, 15, 0, 8, 12, 28, 10, 14, 10.

a. Copy and complete the table at the right.

b. Draw a frequency histogram of the data.

c. In what interval does the median lie?

d. Which interval contains the lower quartile?

Interval	Frequency
24–29	
18–23	
12–17	
6–11	
0–5	

9. The table at the right shows the scores of 25 test papers.

a. Find the mean score.
b. Find the median score.
c. Find the mode.
d. Copy and complete the table.
e. Draw a cumulative frequency histogram.
f. Find the percentile rank of 90.
g. What is the probability that a paper chosen at random has a score of 80?

Score	Frequency	Cumulative Frequency
60	1	
70	9	
80	8	
90	2	
100	5	

10. The electoral votes cast for the winning presidential candidate in elections from 1900 and 1994 are as follows:

292, 336, 321, 435, 277, 404, 382, 444, 472, 523, 449, 432,
303, 442, 457, 303, 486, 301, 520, 297, 489, 525, 426, 370.

a. Organize the data in a stem-and-leaf diagram. (Use the first digit as the stems, and the last two digits as the leaves.)

b. Find the median number of electoral votes.

c. Find the first- and third-quartile values.

d. Draw a whisker-box plot to display the data.

11. The ages of 21 high school students are shown in the table at the right.

a. What is the median age?

b. What is the percentile rank of age 15?

c. When the ages of these 21 students are combined with the ages of 20 additional students, the median age remains unchanged. What is the smallest possible number of students under 16 in the second group?

Age	Frequency
18	1
17	4
16	2
15	7
14	2
13	5

12. In a recent survey, 1,200 people were asked whether they liked cats or dogs better. A circle graph was constructed to show the results. The central angles for two of the three sectors are shown in the diagram. How many people liked cats better?

DOGS AND CATS

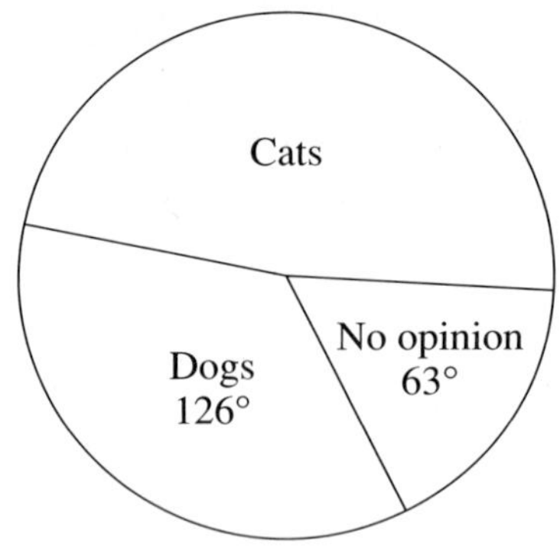

CUMULATIVE REVIEW WITH OPEN-RESPONSE PROBLEMS

1. Solve and check: $x - 4 = 3 - 0.4x$
2. Write in simplest form the sum of $3x - 5x^2$ and $x^2 - 2x + 7$.
3. In a bridge club, there are three more women than men. How many persons are members of the club if the probability is 0.6 that a member of the club chosen at random is a woman?
4. Find the measures of the angles of an isosceles triangle if the measure of the vertex angle is 20 degrees less than the sum of the measures of the two base angles.
5. Huy worked on an assignment for four days. Each day he worked half as long as he worked the day before and spent 3.75 hours on the assignment.
   a. How long did Huy work on the assignment each day?
   b. Find the mean number of hours that Huy worked each day.

Exploration

Often, when the weather forecast is given, we hear phrases such as "The probability of rain tomorrow is 40%." How is this prediction determined? Use an encyclopedia or books on meteorology or interview a meteorologist.

Chapter 16

Graphing Linear Functions and Relations

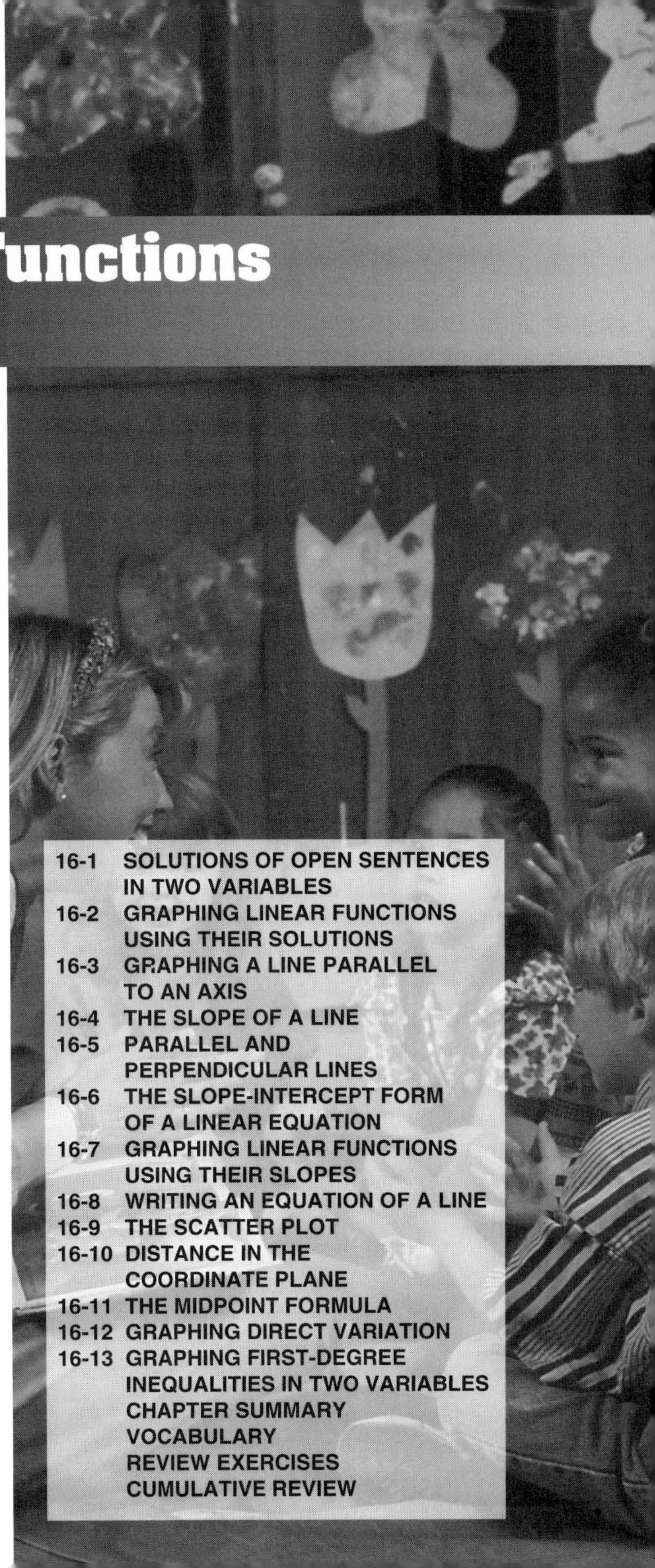

The Tiny Tot Day Care Center charges \$100 a week for children who stay at the center between 8:00 A.M. and 5:00 P.M. If a child is not picked up by 5:00 P.M., the center charges an additional \$4.00 per hour or any fraction of an hour.

Mr. Shubin often has to work late and is unable to pick up his daughter on time. For Mr. Shubin, the weekly cost of day care is a *function* of time; that is, his total cost depends on the time he arrives at the center.

This function for the weekly cost of day care can be expressed in terms of an equation in two variables: $y = 100 + 4x$. The variable x represents the total time, in hours, after 5:00 P.M. that a child remains at the center, and the variable y represents the cost, in dollars, of day care for the week.

16-1 SOLUTIONS OF OPEN SENTENCES IN TWO VARIABLES

Charita wants to enclose a rectangular garden. How much fencing will she need if the width of the garden is to be 5 feet? The answer to this question depends on the length of the garden. Therefore, we say that the amount of fencing, y, is a ***function*** of the length of the garden, x. The formula for the perimeter of a rectangle can be used to express y in terms of x:

$$P = 2\ell + 2w$$

$$\begin{aligned} y &= 2x + 2(5) \\ &= 2x + 10 \end{aligned}$$

Some possible values for the amount of fencing can also be shown in a table. As the length of the garden changes, the amount of fencing changes, as the following table indicates.

x	$2x + 10$	y
1	$2(1) + 10$	12
3	$2(3) + 10$	16
4.5	$2(4.5) + 10$	19
7.2	$2(7.2) + 10$	24.4
9.1	$2(9.1) + 10$	28.2

When we use a calculator:

Enter: 2 [×] 1 [+] 10 [=]

Display: [12.]

We repeat the same sequence, replacing 1 by each value of x in turn.

Each solution to $y = 2x + 10$ is a pair of numbers. In writing each pair, we place the value of x first and the value of y second. For example, when $x = 1$, $y = 12$. Thus, (1, 12) is a solution. When $x = 3$, $y = 16$. Therefore, (3, 16) is another solution. It is not possible to list all ordered pairs in the solution set of the equation because the solution set is infinite. However, it is possible to determine whether a given ordered pair is a member of the solution set. Replace x with the first element of the pair (the abscissa) and y with the second element of the pair (the ordinate). If the result is a true statement, the ordered pair is a solution of the equation.

(1.5, 13) is a solution of $y = 2x + 10$ because $13 = 2(1.5) + 10$ is true.

(4, 14) is *not* a solution of $y = 2x + 10$ because $14 = 2(4) + 10$ is false.

Note: Since in the equation $y = 2x + 10$, x represents the length of a garden, only positive numbers are acceptable replacements for x. Therefore, although $(-2, 6)$ is a solution of $y = 2x + 10$, it is not possible for -2 to be the length of a garden.

EXAMPLES

1. Find five members of the solution set of the sentence $3x + y = 7$.

How to Proceed: / *Solution:*

(1) Transform the equation into an equivalent equation with y alone as one member:

$$\begin{array}{rcl} 3x + y &=& 7 \\ \underline{-3x\quad} && \underline{-3x\quad} \\ y &=& -3x + 7 \end{array}$$

(2) Choose any five values for x. Since no replacement set is given, any real numbers can be used:

(3) For each selected value of x, determine y:

x	$-3x + 7$	y
-2	$-3(-2) + 7$	13
0	$-3(0) + 7$	7
$\frac{1}{3}$	$-3\left(\frac{1}{3}\right) + 7$	6
3	$-3(3) + 7$	-2
5	$-3(5) + 7$	-8

Answer: $(-2,13)$, $(0,7)$, $\left(\frac{1}{3}, 6\right)$, $(3,-2)$, $(5,-8)$

Note that many other solutions are also possible.

---

2. Determine whether each given ordered pair is a solution of the inequality $y - 2x \geq 4$.

a. $(4, 0)$ **b.** $(-1, 2)$ **c.** $(0, 6)$

Solutions

a.
$$\begin{aligned} y - 2x &\geq 4 \\ 0 - 2(4) &\overset{?}{\geq} 4 \\ 0 - 8 &\overset{?}{\geq} 4 \end{aligned}$$
(False)

b.
$$\begin{aligned} y - 2x &\geq 4 \\ 2 - 2(-1) &\overset{?}{\geq} 4 \\ 2 + 2 &\overset{?}{\geq} 4 \end{aligned}$$
(True)

c.
$$\begin{aligned} y - 2x &\geq 4 \\ 6 - 2(0) &\overset{?}{\geq} 4 \\ 6 - 0 &\overset{?}{\geq} 4 \end{aligned}$$
(True)

Answers: **a.** Not a solution **b.** A solution **c.** A solution

---

3. The cost of renting a car for 1 day is \$34.00 plus \$0.25 per mile. Let x represent the number of miles the car was driven, and let y represent the rental cost, in dollars, for a day.

a. Write an equation for the rental cost of the car in terms of the number of miles driven.

b. Find the missing member of each of the following ordered pairs, which are elements of the solution set of the equation written in part **a**, and explain the meaning of the pair.

(*1*) (155, ?) (*2*) (?, 39)

Solutions **a.** Rental cost is \$34.00 plus \$0.25 times the number of miles.

$$y = 34.00 + 0.25x$$

b. (*1*) Since (155, ?) = (x, y), let $x = 155$. Then:

$$\begin{aligned} y &= 34 + 0.25x \\ &= 34 + 0.25(155) \\ &= 34 + 38.75 \\ &= 72.75 \end{aligned}$$

When $x = 155$ miles, the rental cost $y = \$72.75$.

(2) Since (?, 39) = (x, y), let $y = 39$. Then:

$$\begin{aligned} y &= 34 + 0.25x \\ 39 &= 34 + 0.25x \\ -34 &\quad -34 \\ 5 &= 0.25x \\ \frac{5}{0.25} &= \frac{0.25x}{0.25} \\ 20 &= x \end{aligned}$$

When the rental cost $y = \$39$, the car was driven $x = 20$ miles.

Answers: **a.** $y = 34.00 + 0.25x$

b. (*1*) (155, 72.75) The car was driven 155 miles, and the cost was \$72.75.

(2) (20, 39) The car was driven 20 miles, and the cost was \$39.00.

EXERCISES WITH OPEN-RESPONSE PROBLEMS

In 1–5, find the missing member in each ordered pair if the second member of the pair is twice the first member.

1. (3, ?) **2.** (0, ?) **3.** (−2, ?) **4.** (?, 11) **5.** (?, −8)

In 6–10, find the missing member in each ordered pair if the first member of the pair is 4 more than the second member.

6. (?, 5) **7.** $\left(?, \frac{1}{2}\right)$ **8.** (?, 0) **9.** $\left(9\frac{1}{4}, ?\right)$ **10.** (−8, ?)

In 11–25, state whether each given ordered pair of numbers is a solution of the sentence.

11. $y = 5x$; (3, 15) **12.** $y = 4x$; (16, 4) **13.** $y = 3x + 1$; (7, 22)

14. $3x - 2y = 0$, (3, 2) **15.** $y > 4x$; (2, 10) **16.** $y < 2x + 3$; (0, 2)

17. $3y > 2x + 1$; (4, 3) **18.** $2x + 3y \leq 9$; (0, 3) **19.** $x + y = 8$; (4, 5)

20. $4x + 3y = 2$; $\left(\frac{1}{4}, \frac{1}{3}\right)$ **21.** $3x = y + 4$; (−7,−1) **22.** $x - 2y = 15$; (1,−7)

23. $y > 6x$; (−1,−2) **24.** $3x < 4y$; (5, 2) **25.** $5x - 2y \leq 19$; (3,−2)

26. Which ordered pair is a solution of $x + y = 6$?
(1) (8,−2) (2) (2,−6) (3) (7,13) (4) (3, 9)

27. Which ordered pair is a solution of $y > 4x$?
(1) (5, 3) (2) (7, 2) (3) (1,−1) (4) $\left(-2, -\frac{1}{4}\right)$

28. Which ordered pair is *not* a solution of $y < 2x + 1$?
(1) (1, 8) (2) (5, 2) (3) (3,−1) (4) (0,−4)

In 29–32, in each case: **a.** Write an equation that expresses the relationship between x and y. **b.** Find two ordered pairs in the solution set of the equation that you wrote in part **a.**

29. The cost, y, of renting a bicycle is $5.00 plus $2.50 times the number of hours, x, that the bicycle is used. Assume that x can be a fractional part of an hour.

30. In an isosceles triangle whose perimeter is 54 centimeters, the base is x centimeters and the measure of each leg is y centimeters.

31. Jules has $265 at the beginning of the month. What amount, y, does he have left after x days if he spends $8.50 a day?

32. At the beginning of the day, the water in the swimming pool was 2 feet deep. Throughout the day, water was added at a rate of 0.4 of a foot per hour. What was the depth, y, of the water in the pool after being filled for x hours?

33. (1) At the Riverside Amusement Park, rides are paid for with tokens purchased at a central booth. Some rides require two tokens, and others one token. Tomas bought 10 tokens and spent them all on x two-token rides and y one-token rides. On how many two-token and how many one-token rides did Tomas go?

(2) Minnie used 10 feet of fence to enclose three sides of a rectangular pen whose fourth side was the side of a garage. She used x feet of fence for each side perpendicular to the garage and y feet of fence for the side parallel to the garage. What were the dimensions of the pen that Minnie built?

a. Write an equation that can be used to solve both problems.

b. Write the six solutions for problem (1).

c. Which of the six solutions for problem (1) are *not* solutions for problem (2)?

d. Write two solutions for problem (2) that are *not* solutions for problem (1).

e. There are only six solutions for problem (1) but infinitely many solutions for problem (2). Explain why.

16-2 GRAPHING LINEAR FUNCTIONS USING THEIR SOLUTIONS

Think of two numbers that add up to 6. If x and y represent these numbers, then $x + y = 6$ is an equation showing that their sum is 6. If the replacement set for both x and y is the set of real numbers, we can find infinitely many solutions for the equation $x + y = 6$. Some of the solutions are shown in the following table.

x	8	7	6	5	4.5	4	3	2.2	2	0	−2
y	−2	−1	0	1	1.5	2	3	3.8	4	6	8

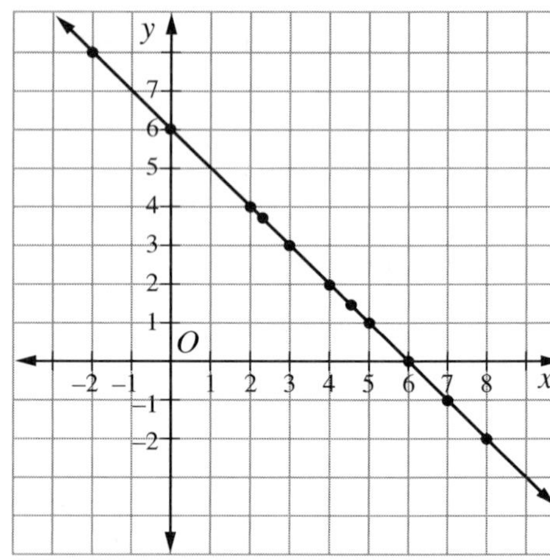

Each ordered pair of numbers, such as $(8,-2)$, locates a point in the coordinate plane. When all the points that have the coordinates associated with the number pairs in the table are located, the points appear to lie on a straight line, as shown at the right.

In fact, if the replacement set for both x and y is the set of real numbers, the following is true:

> All of the points whose coordinates are solutions of $x + y = 6$ lie on this same straight line, and all of the points whose coordinates are not solutions of $x + y = 6$ do *not* lie on this line.

This line, which is the set of all the points and only the points whose coordinates make the equation $x + y = 6$ true, is called the ***graph*** of $x + y = 6$.

A first-degree equation in two variables can be written in the form

$$Ax + By + C = 0$$

where A, B, and C are real numbers, with A and B not both 0. For example, $x + y = 6$ can be written as $1x + 1y - 6 = 0$, with $A = 1$, $B = 1$, and $C = -6$. We call an equation of the form $Ax + By + C = 0$ a ***linear equation*** since its graph is a straight line that contains all points whose coordinates make the equation true. The replacement set for both variables is the set of real numbers unless otherwise indicated.

When we graph a linear equation, we can determine the line by plotting two points whose coordinates satisfy that equation. However, we always plot a third point as a check on the first two. If the third point lies on the line determined by the first two points, we have probably made no error.

PROCEDURE. To graph a linear equation by means of its solutions:

1. Transform the equation into an equivalent equation that is solved for y in terms of x.
2. Find three solutions of the equation by choosing values for x and computing corresponding values for y.
3. In the coordinate plane, graph the ordered pairs of numbers found in Step 2.
4. Draw the line that passes through the points graphed in Step 3.

EXAMPLES

1. Does point $(2,-3)$ lie on the graph of $x - 2y = -4$?

Solution If point $(2,-3)$ is to lie on the graph of $x - 2y = -4$, it must be a solution of $x - 2y = -4$.

$$x - 2y = -4$$

$$(2) - 2(-3) \stackrel{?}{=} -4$$

$$2 + 6 \stackrel{?}{=} -4$$

$$8 \neq -4$$

Since $8 = -4$ is not true, point $(2,-3)$ does not lie on line $x - 2y = -4$.

Answer: No

---

2. What must be the value of d if $(d, 4)$ lies on line $3x + y = 10$?

Solution The coordinate $(d, 4)$ must satisfy $3x + y = 10$.

Let $x = d$,
and $y = 4$.
Then:

$$3d + 4 = 10$$

$$3d = 6$$

$$d = 2$$

Answer: The value of d is 2.

---

3. a. Write the following verbal sentence as an equation: The sum of twice the abscissa of a point and the ordinate of that point is 4.

b. Graph the equation written in part **a**.

Solution **a.** Let x = the abscissa of the point, and y = the ordinate of the point. Then:

$$2x + y = 4 \quad \textit{Answer}$$

b. *How to Proceed:* / *Solution:*

(1) Transform the equation into an equivalent equation that has y alone as one member.

$$2x + y = 4$$
$$y = -2x + 4$$

(2) Determine three solutions of the equation by choosing values for x and computing the corresponding values for y.

x	$-2x + 4$	y
0	$-2(0) + 4$	4
1	$-2(1) + 4$	2
2	$-2(2) + 4$	0

(3) Plot the points that are associated with the three solutions.

(4) Draw a line through the points that were plotted. Label the line with its equation.

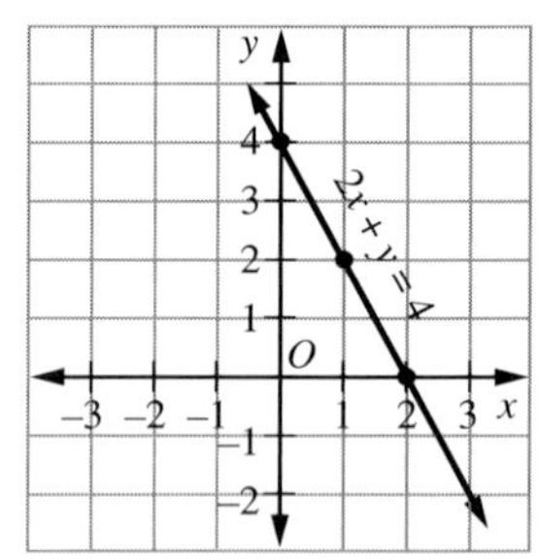

EXERCISES WITH OPEN-RESPONSE PROBLEMS

1. For each part, state whether the pair of values for x and y satisfies the equation $2x - y = 6$.
a. $x = 4, y = 2$ **b.** $x = 0, y = 6$ **c.** $x = 4, y = -2$

For 2–5, state in each case whether the point whose coordinates are given is on the graph of the given equation.

2. $x + y = 7$; $(4, 3)$

3. $2y + x = 7$; $(1, 3)$

4. $3x - 2y = 8$; $(2, 1)$

5. $2y = 3x - 5$; $(-1, -4)$

In 6–9, find in each case the number that can replace k so that the resulting ordered number pair will be on the graph of the given equation.

6. $x + 2y = 5$; $(k, 2)$

7. $3x + 2y = 22$; $(k, 5)$

8. $x + 3y = 10$; $(13, k)$

9. $x - y = 0$; (k, k)

In 10–13, find in each case a value that can replace k so that the graph of the resulting equation will pass through the point whose coordinates are given.

10. $x + y = k$; $(2, 5)$

11. $x - y = k$; $(5, -3)$

12. $5y - 2x = k$; $(-2, -1)$

13. $x + y = k$; $(0, 0)$

In 14–19, solve each equation for y in terms of x.

14. $3x + y = -1$

15. $4x - y = 6$

16. $2y = 6x$

17. $12x = \frac{3}{2}y$

18. $4x + 2y = 8$

19. $6x - 3y = 5$

In 20–22: **a.** Find the missing values of the variable needed to complete each table. **b.** Plot the points described by the pairs of values in each completed table; then, draw a line through the points.

20. $y = 4x$

x	y
0	?
1	?
2	?

21. $y = 3x + 1$

x	y
−1	?
0	?
1	?

22. $x + 2y = 3$

x	y
−1	?
2	?
5	?

In 23–46, graph each equation.

23. $y = 2x$

24. $y = 5x$

25. $y = -3x$

26. $y = -x$

27. $x = 2y - 3$

28. $x = \frac{1}{2}y + 1$

29. $y = x + 3$

30. $y = 2x - 1$

31. $y = 3x + 1$

32. $y = -2x + 4$

33. $x + y = 8$

34. $x - y = 5$

35. $y - x = 0$

36. $3x + y = 12$

37. $x - 2y = 0$

38. $y - 3x = -5$

39. $2x - y = 6$

40. $3x - y = -6$

41. $x + 3y = 12$

42. $2x + 3y = 6$

43. $3x - 2y = -4$

44. $x - 3y = 9$

45. $2x = y - 4$

46. $4x = 3y$

47. **a.** Through points $(0, -2)$ and $(4, 0)$, draw a straight line.
b. Write the coordinates of two other points on the line drawn in part **a**.

48. **a.** Through points $(-2, 3)$ and $(1, -3)$, draw a straight line.
b. Does point $(0, -1)$ lie on the line drawn in part **a**?

In 49–53: **a.** Write each verbal sentence as an equation. **b.** Graph the equation.

49. The ordinate of a point is twice the abscissa.

50. The ordinate of a point is 2 more than the abscissa.

51. The sum of the ordinate and the abscissa of a point is 6.

52. The difference of the ordinate and the abscissa of a point is 1.

53. Twice the ordinate of a point decreased by 3 times the abscissa is 6.

16-3 GRAPHING A LINE PARALLEL TO AN AXIS

Lines Parallel to the *x*-Axis

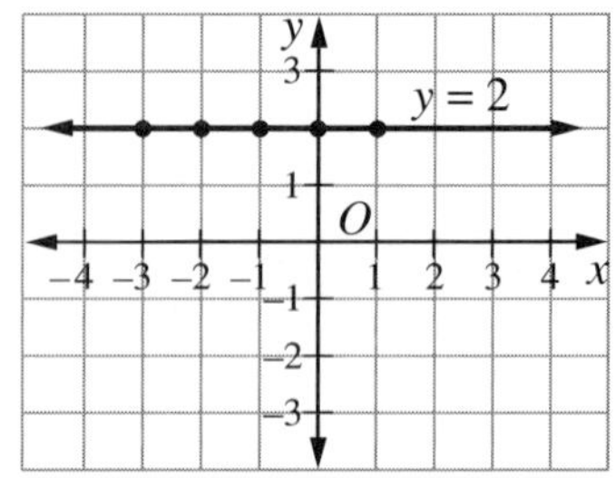

An equation such as $y = 2$ can be graphed in the coordinate plane. Any pair of values whose y-coordinate is 2, no matter what the x-coordinate is, makes the equation $y = 2$ true. Therefore, $(-3, 2)$, $(-2, 2)$, $(-1, 2)$, $(0, 2)$, $(1, 2)$, or any other pair $(a, 2)$ for all values of a, are points on the graph of $y = 2$. As shown at the right, the graph of $y = 2$ is a horizontal line parallel to the x-axis and 2 units above it.

The ***y*-intercept** of a line is the y-coordinate of the point at which the line intersects the y-axis. From the graph, we see that the y-intercept of the line $y = 2$ is 2. In general:

- **An equation of the line parallel to the x-axis with y-intercept b is $y = b$.**

Lines Parallel to the *y*-Axis

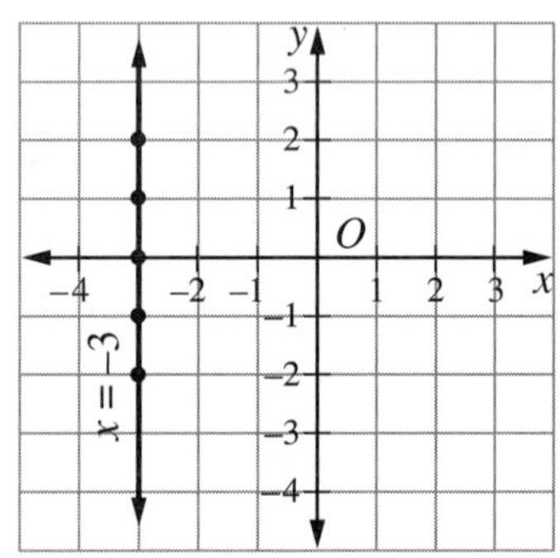

An equation such as $x = -3$ can be graphed in the coordinate plane. Any pair of values whose x-coordinate is -3, no matter what the y-coordinate is, makes the equation $x = -3$ true. Therefore, $(-3, -2)$, $(-3, -1)$, $(-3, 0)$, $(-3, 1)$, $(-3, 2)$, or any other pair $(-3, b)$ for all values of b, are points on the graph of $x = -3$. As shown at the right, the graph of $x = -3$ is a vertical line parallel to the y-axis and 3 units to the left of it.

The ***x*-intercept** of a line is the x-coordinate of the point at which the line intersects the x-axis. From the graph, we see that the x-intercept of the line $x = -3$ is -3. In general:

- **An equation of the line parallel to the y-axis with x-intercept a is $x = a$.**

EXERCISES

In 1–15, draw the graph of each equation.

1. $x = 6$ **2.** $x = 4$ **3.** $x = 0$ **4.** $x = -3$ **5.** $x = -5$

6. $y = 4$ **7.** $y = 5$ **8.** $y = 0$ **9.** $y = -4$ **10.** $y = -7$

11. $x = \frac{1}{2}$ **12.** $y = 1\frac{1}{2}$ **13.** $y = -2\frac{1}{2}$ **14.** $x = -\frac{3}{2}$ **15.** $y = 3.5$

16. Write an equation of the line that is parallel to the x-axis and whose y-intercept is:
a. 1 **b.** 5 **c.** -4 **d.** -8 **e.** -2.5

17. Write an equation of the line that is parallel to the y-axis and whose x-intercept is:
a. 3 **b.** 10 **c.** $4\frac{1}{2}$ **d.** -6 **e.** -10

18. Which statement is true about the graph of the equation $y = 6$?
(1) It is parallel to the y-axis. (2) It is parallel to the x-axis.
(3) It goes through the origin. (4) It has an x-intercept.

19. Which statement is true about the graph of the equation $x = 5$?
(1) It goes through the origin. (2) It is parallel to the x-axis.
(3) It is parallel to the y-axis. (4) It has a y-intercept.

20. Which statement is true about the graph of the equation $y = x$?
(1) It is parallel to the x-axis. (2) It is parallel to the y-axis.
(3) It goes through $(2, -2)$. (4) It goes through the origin.

21. The cost of admission to an amusement park on Family Night is $25.00 for a family of any size.
a. What is the cost of admission for the Gauger family of six persons?
b. Write an equation for the cost of admission, y, for a family of x persons.

16-4 THE SLOPE OF A LINE

Meaning of the Slope of a Line

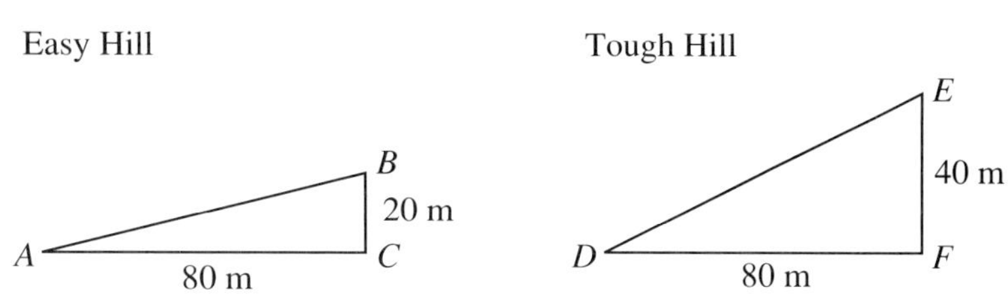

It is more difficult to hike up Tough Hill, shown above, than to hike up Easy Hill. Tough Hill rises 40 meters vertically over a horizontal distance of 80 meters, whereas Easy Hill rises only 20 meters vertically over the same horizontal distance of 80 meters. Therefore, Tough Hill is steeper than Easy Hill. To compare the steepness of roads $\overline{AB}$ and $\overline{DE}$, which lead up the two hills, we compare their ***slopes***.

The slope of road $\overline{AB}$ is the ratio of the change in vertical distance, CB, to the change in horizontal distance, AC:

$$\text{slope of road } \overline{AB} = \frac{\text{change in vertical distance, } CB}{\text{change in horizontal distance, } AC} = \frac{20 \text{ m}}{80 \text{ m}} = \frac{1}{4}$$

Also:

$$\text{slope of road } \overline{DE} = \frac{\text{change in vertical distance, } FE}{\text{change in horizontal distance, } DF} = \frac{40 \text{ m}}{80 \text{ m}} = \frac{1}{2}$$

Finding the Slope of a Line

To find the slope of the line determined by the two points (2, 3) and (5, 8), as shown at the right, we write the ratio of the difference in y-values to the difference in x-values, as follows:

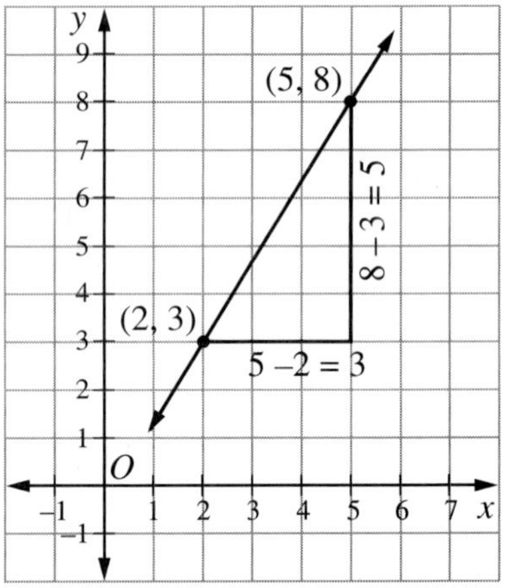

$$\text{slope} = \frac{\text{difference in } y\text{-values}}{\text{difference in } x\text{-values}}$$

$$= \frac{8-3}{5-2} = \frac{5}{3}$$

Suppose we change the order of the points in performing the computation. We then have:

$$\text{slope} = \frac{\text{difference in } y\text{-values}}{\text{difference in } x\text{-values}} = \frac{3-8}{2-5} = \frac{-5}{-3} = \frac{5}{3}$$

The result of both computations is the same. When we compute the slope of a line that is determined by two points, it does not matter which point is considered as the first point and which the second.

Also, when we find the slope of a line using two points on the line, it does not matter which two points on the line we use because all segments of a line have the same slope as the line.

PROCEDURE. To find the slope of a line:

1. Select any two points on the line.

2. Find the vertical change, that is, the change in y-values, in going from the point on the left to the point on the right.

3. Find the horizontal change, that is, the change in x-values, in going from the point on the left to the point on the right.

4. Write the ratio of the vertical change to the horizontal change.

For example, in the graph:

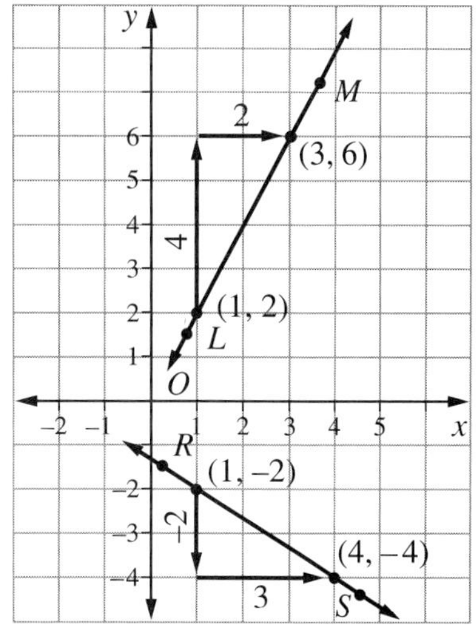

$$\text{slope of } \overleftrightarrow{LM} = \frac{\text{vertical change}}{\text{horizontal change}}$$

$$= \frac{4}{2} = \frac{2}{1}, \text{ or } 2$$

$$\text{slope of } \overleftrightarrow{RS} = \frac{\text{vertical change}}{\text{horizontal change}}$$

$$= \frac{-2}{3} = -\frac{2}{3}$$

In general, as shown at the right, the slope, m, of a line that passes through any two points $P_1(x_1, y_1)$ and $P_2(x_2, y_2)$, where $x_1 \neq x_2$, is the ratio of the difference of the y-values of these points to the difference of the corresponding x-values. Thus:

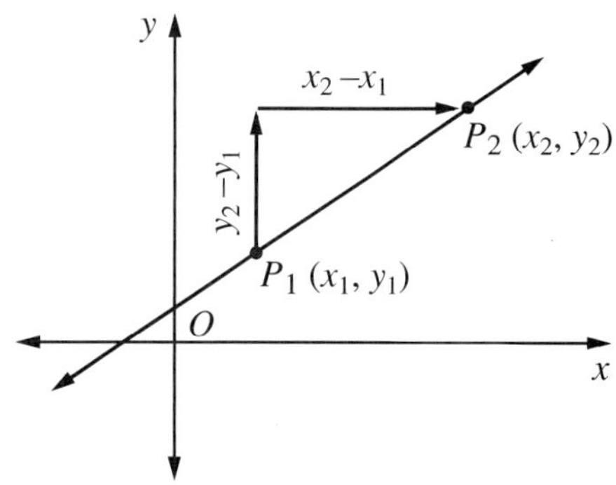

$$\text{slope of a line} = \frac{\text{difference in } y\text{-values}}{\text{difference in } x\text{-values}}$$

Therefore,

$$\text{slope of } \overleftrightarrow{P_1P_2} = m = \frac{y_2 - y_1}{x_2 - x_1}$$

The difference in x-values, $x_2 - x_1$, can be represented by Δx, read as "delta x." Similarly, the difference in y-values, $y_2 - y_1$, can be represented by Δy, read as "delta y." Therefore, we write:

$$\textbf{Slope of a line} = m = \frac{\Delta y}{\Delta x}$$

Positive Slopes

Examining $\overleftrightarrow{AB}$ from left to right and observing the path of a point from C to D, for example, we see that the line is rising. As the x-values increase, the y-values also increase. Between point C and point D, the change in y is 1, and the change in x is 2. Since both Δy and Δx are positive, the slope of $\overleftrightarrow{AB}$ must be positive. Thus:

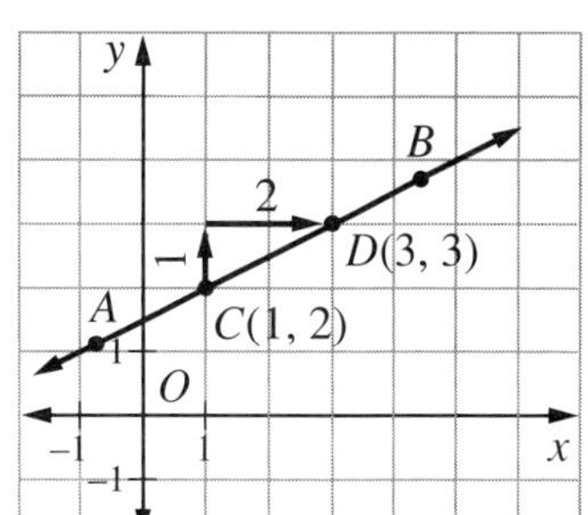

$$\text{slope} = m = \frac{\Delta y}{\Delta x} = \frac{1}{2}$$

This example illustrates:

- **PRINCIPLE 1. As a point moves from left to right along a line that is *rising*, y increases as x increases and the slope of the line is *positive*.**

Negative Slopes

Now, examining $\overleftrightarrow{EF}$ from left to right and observing the path of a point from C to D, we see that the line is falling. As the x-values increase, the y-values decrease. Between point C and point D, the change in y is -2, and the change in x is 3. Since Δy is negative and Δx is positive, the slope of $\overleftrightarrow{EF}$ must be negative. Thus:

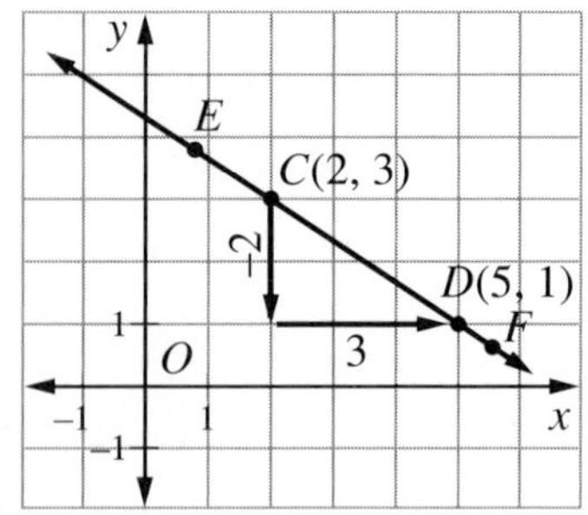

$$\text{slope} = m = \frac{\Delta y}{\Delta x} = \frac{-2}{3} = -\frac{2}{3}$$

This example illustrates:

- **PRINCIPLE 2. As a point moves from left to right along a line that is *falling*, y decreases as x increases and the slope of the line is *negative*.**

Zero Slope

On the graph, $\overleftrightarrow{GH}$ is parallel to the x-axis. We consider a point moving along $\overleftrightarrow{GH}$ from left to right, for example, from C to D. As the x-values increase, the y-values are unchanged. Between point C and point D, the change in y is 0, and the change in x is 3. Since Δy is 0 and Δx is 3, the slope of $\overleftrightarrow{GH}$ must be 0. Thus:

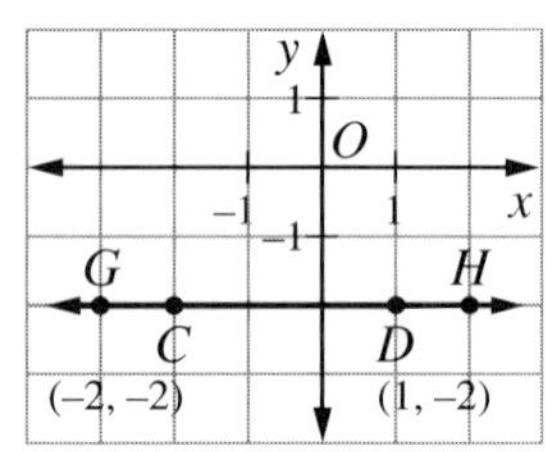

$$\text{slope} = m = \frac{\Delta y}{\Delta x} = \frac{0}{3} = 0$$

This example illustrates:

- **PRINCIPLE 3. If a line is parallel to the x-axis, its slope is 0.**

Note: The slope of the x-axis itself is also 0.

No Slope

On the graph, $\overleftrightarrow{LM}$ is parallel to the y-axis. We consider a point moving upward along $\overleftrightarrow{LM}$, for example, from C to D. The x-values are unchanged, but the y-values increase. Between point C and point D, the change in y is 3, and the change in x is 0. Since the slope of $\overleftrightarrow{LM} = \frac{\Delta y}{\Delta x} = \frac{3}{0}$, and a number cannot be divided by 0, $\overleftrightarrow{LM}$ has no defined slope.

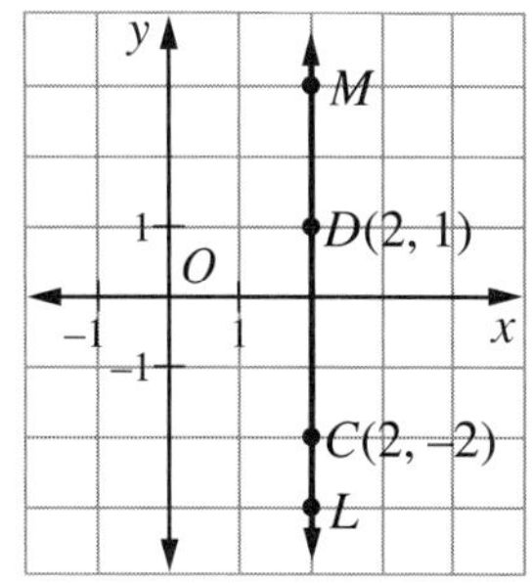

This example illustrates:

- **PRINCIPLE 4. If a line is parallel to the y-axis, it has no defined slope.**

Note: The y-axis itself has no defined slope.

A vertical line and a horizontal line are perpendicular to each other.

EXAMPLES

1. Find the slope of the line that is determined by points $(-2, 4)$ and $(4, 2)$.

Solution Plot points $(-2, 4)$ and $(4, 2)$. Let point $(-2, 4)$ be $P_1(x_1, y_1)$, and let point $(4, 2)$ be $P_2(x_2, y_2)$. Then, $x_1 = -2$, $y_1 = 4$ and $x_2 = 4$, $y_2 = 2$.

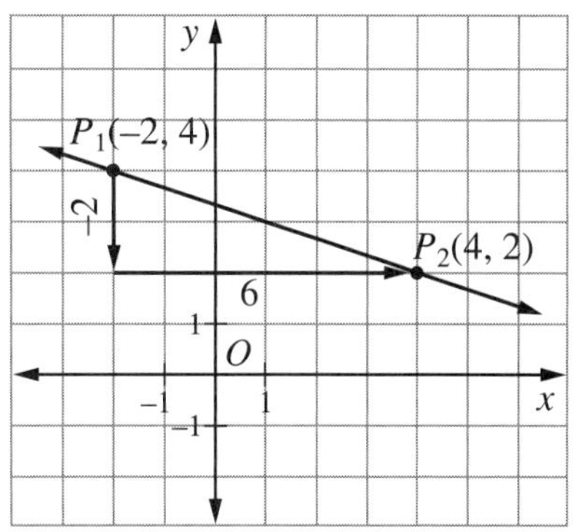

$$\text{slope of } \overleftrightarrow{P_1P_2} = \frac{\Delta y}{\Delta x} = \frac{y_2 - y_1}{x_2 - x_1}$$

$$= \frac{(2) - (4)}{(4) - (-2)}$$

$$= \frac{2 - 4}{4 + 2}$$

$$= \frac{-2}{6}$$

$$= -\frac{1}{3} \quad \textit{Answer}$$

2. Through point $(2, -1)$, draw the line whose slope is $\frac{3}{2}$.

How to Proceed: *Solution:*

(1) Graph point $A(2, -1)$:

(2) Note that, since slope $= \frac{\Delta y}{\Delta x} = \frac{3}{2}$, when y changes 3, then x changes 2. Start at point $A(2, -1)$ and move 3 units upward and 2 units to the right to locate point B:

(3) Start at B and repeat these movements to locate point C:

(4) Draw a line that passes through points A, B, and C:

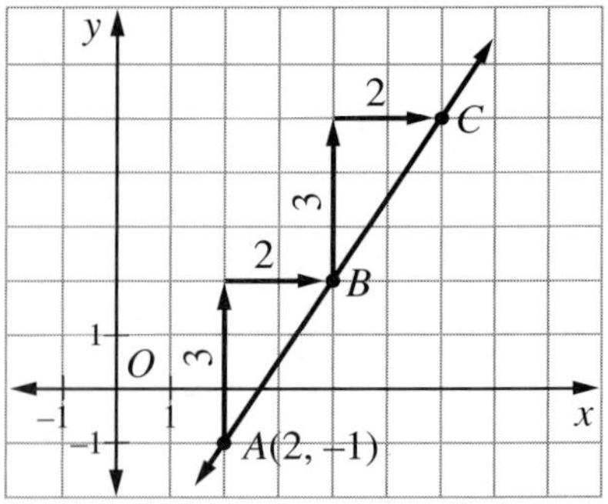

KEEP IN MIND

A fundamental property of a straight line is that its slope is constant. Therefore, any two points on a line may be used to compute the slope of the line.

EXERCISES WITH OPEN-RESPONSE PROBLEMS

In 1–6: **a.** Tell whether each line has a positive slope, a negative slope, a slope of zero, or no slope. **b.** Find the slope of each line. If the line has no slope, indicate that fact.

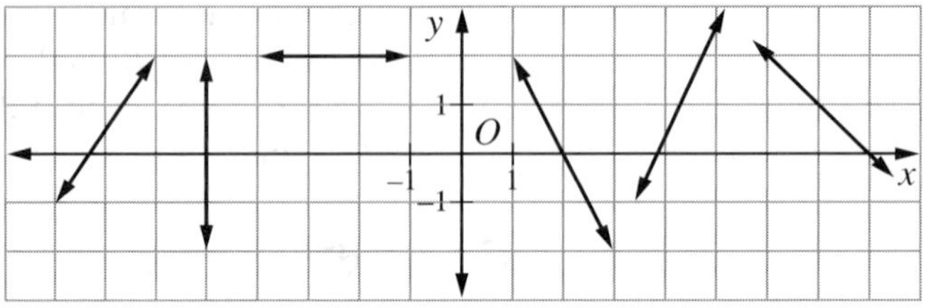

1. **2.** **3.** **4.** **5.** **6.**

In 7–15, in each case: **a.** Plot both points, and draw the line that they determine. **b.** Find the slope of this line.

7. (0, 0) and (4, 4) **8.** (0, 0) and (4, 8) **9.** (0, 0) and (3, −6)

10. (1, 5) and (3, 9) **11.** (7, 3) and (1, −1) **12.** (−2, 4) and (0, 2)

13. (5, −2) and (7, −8) **14.** (4, 2) and (8, 2) **15.** (−1, 3) and (2, 3)

In 16–27, in each case, draw a line with the given slope, m, through the given point.

16. (0, 0); $m = 2$ **17.** (1, 3); $m = 3$ **18.** (2, −5); $m = 4$

19. (−4, 5); $m = \frac{2}{3}$ **20.** (3, 1); $m = 0$ **21.** (−3, −4); $m = -2$

22. $(1, -5)$; $m = -1$ **23.** $(2, 4)$; $m = -\frac{3}{2}$ **24.** $(-2, 3)$; $m = -\frac{1}{3}$

25. $(-1, 0)$; $m = -\frac{5}{4}$ **26.** $(0, 2)$; $m = -\frac{2}{3}$ **27.** $(-2, 0)$; $m = \frac{1}{2}$

28. Points $A(2, 4)$, $B(8, 4)$, and $C(5, 1)$ are the vertices of $\triangle ABC$. Find the slope of each side of $\triangle ABC$.

29. Points $A(3, -2)$, $B(9, -2)$, $C(7, 4)$, and $D(1, 4)$ are the vertices of a quadrilateral.

a. Graph the points and draw quadrilateral $ABCD$.

b. What type of quadrilateral does $ABCD$ appear to be?

c. Compute the slope of $\overline{BC}$ and the slope of $\overline{AD}$.

d. What is true of the slope of $\overline{BC}$ and the slope of $\overline{AD}$?

e. If two segments such as $\overline{AD}$ and $\overline{BC}$, or two lines such as $\overleftrightarrow{AD}$ and $\overleftrightarrow{BC}$, are parallel, what appears to be true of their slopes?

f. Since $\overline{AB}$ and $\overline{CD}$ are parallel, what might be true of their slopes?

g. Compute the slope of $\overline{AB}$ and the slope of $\overline{DC}$.

h. Is the slope of $\overline{AB}$ equal to the slope of $\overline{DC}$?

16-5 PARALLEL AND PERPENDICULAR LINES

Parallel Lines

Two lines in the same plane are either intersecting lines or parallel lines. Recall that parallel lines are lines in the same plane that have no points in common or that have all points in common.

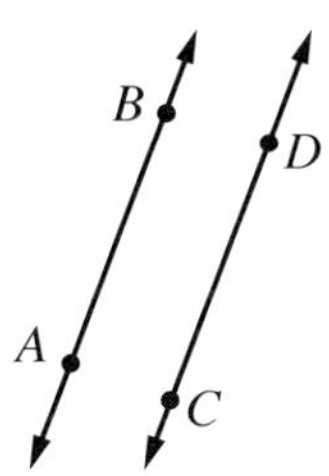

We indicate that $\overleftrightarrow{AB}$ is parallel to $\overleftrightarrow{CD}$ by writing $\overleftrightarrow{AB} \parallel \overleftrightarrow{CD}$. These parallel lines, extended indefinitely, will never intersect and therefore have no points in common. However, using the fact that lines are parallel if they have all points in common, we may also write $\overleftrightarrow{AB} \parallel \overleftrightarrow{AB}$ and $\overleftrightarrow{CD} \parallel \overleftrightarrow{CD}$. The word *lines* means straight lines that extend indefinitely in both directions. Segments and rays are also said to be parallel if the lines that contain them are parallel.

The Slopes of Parallel Lines

In the figure, $\overleftrightarrow{AB} \parallel \overleftrightarrow{CD}$. Use the coordinates given in the diagram to write:

$$\text{Slope of } \overleftrightarrow{AB} = \frac{y_2 - y_1}{x_2 - x_1} = \frac{6 - 2}{4 - 2} = \frac{4}{2} = \frac{2}{1}$$

$$\text{Slope of } \overleftrightarrow{CD} = \frac{y_2 - y_1}{x_2 - x_1} = \frac{6 - 4}{7 - 6} = \frac{2}{1}$$

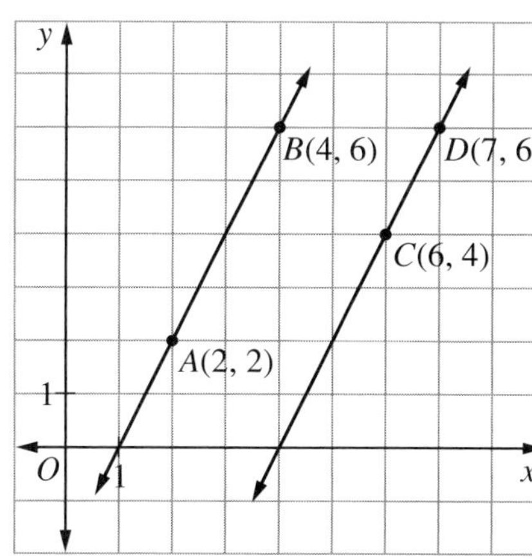

This example illustrates the following statement:

- **Two nonvertical lines are parallel if and only if their slopes are equal.**

In other words:

When the slope of $\overleftrightarrow{AB} = m_1$, and the slope of $\overleftrightarrow{CD} = m_2$, $\overleftrightarrow{AB}$ is *parallel* to $\overleftrightarrow{CD}$ if and only if:

$$\boldsymbol{m_1 = m_2}$$

Perpendicular Lines

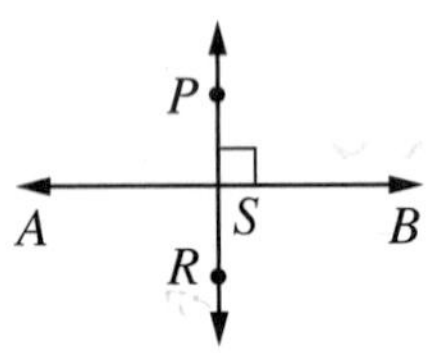

Recall that perpendicular lines are two lines that intersect to form right angles. For example, in the diagram $\overleftrightarrow{AB}$ and $\overleftrightarrow{PR}$ intersect at S to form four right angles: $\angle BSP$, $\angle PSA$, $\angle ASR$, and $\angle RSB$. Therefore, $\overleftrightarrow{PR} \perp \overleftrightarrow{AB}$.

The Slopes of Perpendicular Lines

In the figure, $\overleftrightarrow{AB} \perp \overleftrightarrow{BC}$. Use the coordinates of A, B, and C to write:

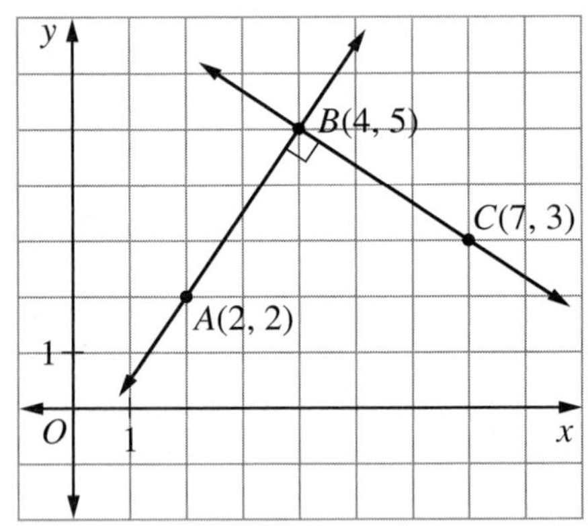

$$\text{Slope of } \overleftrightarrow{AB} = \frac{y_2 - y_1}{x_2 - x_1} = \frac{5 - 2}{4 - 2} = \frac{3}{2}$$

$$\text{Slope of } \overleftrightarrow{BC} = \frac{y_2 - y_1}{x_2 - x_1} = \frac{3 - 5}{7 - 4} = \frac{-2}{3} = -\frac{2}{3}$$

In this example, where $\overleftrightarrow{AB} \perp \overleftrightarrow{BC}$, the product of the slopes of the lines is -1, that is, $\left(\frac{3}{2}\right)\left(-\frac{2}{3}\right) = -1$.

One number is the ***negative reciprocal*** of another if the product of the two numbers is -1.

In general terms, if $a \neq 0$ and $b \neq 0$, we may say that $-\frac{b}{a}$ is the negative reciprocal of $\frac{a}{b}$ since $\left(\frac{a}{b}\right)\left(-\frac{b}{a}\right) = -1$.

Thus, to find the negative reciprocal of a fraction, we simply invert the fraction and change its sign.

- **Two nonvertical lines are perpendicular if and only if the slope of one line is the negative reciprocal of the slope of the other line.**

In other words:

When the slope of $\overleftrightarrow{AB} = m_1$, and the slope of $\overleftrightarrow{CD} = m_2$, $\overleftrightarrow{AB}$ is *perpendicular* to $\overleftrightarrow{CD}$ if and only if:

$$m_1 = -\frac{1}{m_2} \quad \text{or} \quad m_2 = -\frac{1}{m_1} \quad \text{or} \quad m_1 \cdot m_2 = -1$$

EXAMPLES

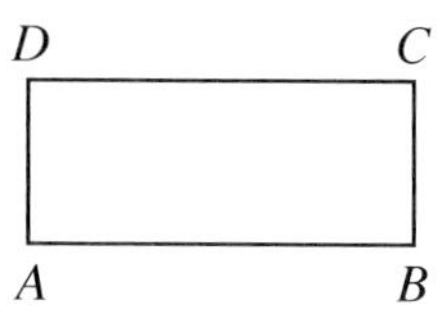

1. A rectangle is a four-sided polygon with two pairs of parallel sides and four right angles. In the diagram, *ABCD* is a rectangle.

a. Name two pairs of parallel lines.

b. Name four pairs of perpendicular lines.

Answers: **a.** $\overleftrightarrow{AB} \parallel \overleftrightarrow{CD}$, $\overleftrightarrow{AD} \parallel \overleftrightarrow{BC}$

b. $\overleftrightarrow{AB} \perp \overleftrightarrow{BC}$, $\overleftrightarrow{BC} \perp \overleftrightarrow{CD}$, $\overleftrightarrow{CD} \perp \overleftrightarrow{DA}$, $\overleftrightarrow{DA} \perp \overleftrightarrow{AB}$

---

2. Show by means of slopes that quadrilateral *ABCD* with vertices *A*(2, 1), *B*(4, −2), *C*(4, 1), and *D*(2, 4) has two pairs of parallel sides.

Solution Find the slope of each side of the quadrilateral.

Slope of $\overline{AB} = \frac{-2-1}{4-2} = \frac{-3}{2} = -\frac{3}{2}$

$\overline{BC}$ is a vertical line and has no slope.

Slope of $\overline{CD} = \frac{4-1}{2-4} = \frac{3}{-2} = -\frac{3}{2}$

$\overline{DA}$ is a vertical line and has no slope.

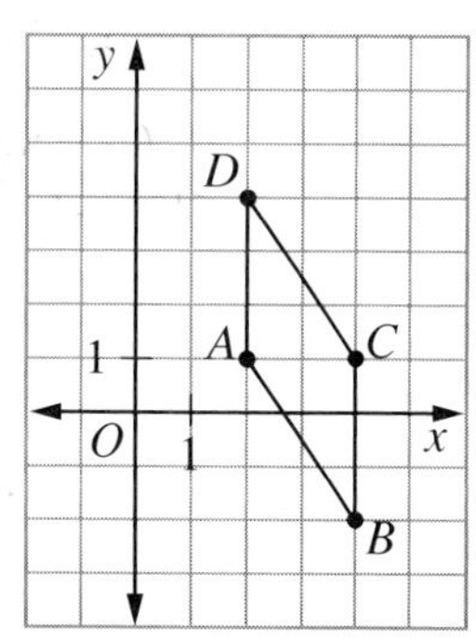

Answer: $\overline{AB}$ and $\overline{CD}$ have the same slope and are parallel.
$\overline{BC}$ and $\overline{DA}$ are vertical lines and are parallel.

---

3. Show, by means of slopes, that the triangle whose vertices are $A(0, 2)$, $B(2, 3)$, and $C(1, 5)$ is a right triangle.

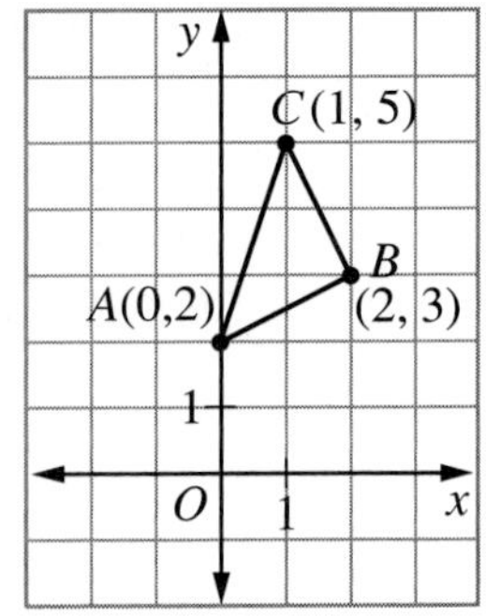

Solution Show that two sides of the triangle are perpendicular by finding two slopes that are negative reciprocals.

Slope of $\overline{AB} = \frac{3-2}{2-0} = \frac{1}{2}$

Slope of $\overline{BC} = \frac{5-3}{1-2} = \frac{2}{-1} = -2$

The slope of $\overline{AB}$ is the negative reciprocal of the slope of $\overline{BC}$ because $\left(\frac{1}{2}\right)(-2) = -1$.

Therefore, $\overline{AB} \perp \overline{BC}$ and $\angle ABC$ is a right angle.

Answer: Triangle ABC is a right triangle because it has a right angle.

EXERCISES WITH OPEN-RESPONSE PROBLEMS

In 1–6, in each case the coordinates of A, B, C, and D are given. Use slope to determine whether or not $\overline{AB} \parallel \overline{CD}$.

1. $A(1, 5), B(3, 9), C(2, 2), D(4, 6)$
2. $A(1, 1), B(3, 7), C(2, 0), D(5, 9)$
3. $A(1,-5), B(3, 3), C(2, 6), D(4,-2)$
4. $A(1, 5), B(3, 6), C(2, 2), D(4, 6)$
5. $A(-1, 5), B(1, 1), C(1, 2), D(3,-2)$
6. $A(1, 3), B(1, 9), C(2, 2), D(2, 6)$

In 7–12, in each case the coordinates of A, B, C, and D are given. Use slope to determine whether or not $\overline{AB} \perp \overline{CD}$.

7. $A(1, 5), B(3, 9), C(2, 3), D(8, 0)$
8. $A(1, 1), B(3, 3), C(5, 0), D(0, 5)$
9. $A(1,-5), B(1, 3), C(2, 6), D(-4, 6)$
10. $A(1, 5), B(3, 4), C(2, 2), D(3, 0)$
11. $A(-1, 1), B(3, 0), C(1, 0), D(-1, 8)$
12. $A(1, 7), B(4, 9), C(4, 0), D(0, 6)$

13. A parallelogram is a quadrilateral with two pairs of parallel sides. Show that $ABCD$ is a parallelogram if the coordinates of the vertices are $A(-2, 3)$, $B(2, 7)$, $C(8, 5)$, and $D(4, 1)$.

14. A rectangle is a quadrilateral with four right angles. Show that $EFGH$ is a rectangle if the coordinates of the vertices are $E(-3,-5)$, $F(3,-1)$, $G(1, 2)$, and $H(-5,-2)$.

16-6 THE SLOPE-INTERCEPT FORM OF A LINEAR EQUATION

When drawing the graph of an equation such as $2x - y = 1$, we found it convenient to solve for y in terms of x:

$$2x - y = 1$$

$$-y = -2x + 1$$

$$\frac{-y}{-1} = \frac{-2x}{-1} + \frac{1}{-1}$$

$$y = 2x - 1$$

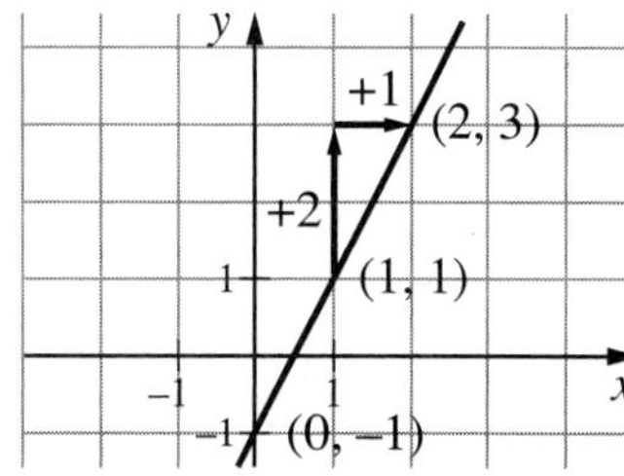

There are two numbers in the equation: the coefficient of x, 2, and the constant term, -1. Each of these numbers gives us important information about the graph shown above.

1. The coefficient of x is the slope of the line determined by

$$\text{slope} = \frac{\text{change in } y}{\text{change in } x} = \frac{2}{1} = 2$$

2. The constant term is the y-intercept.

We recall that the y-intercept of the graph is the y-coordinate of the point at which the graph intersects the y-axis. For every point on the y-axis, $x = 0$. When $x = 0$,

$$\begin{aligned} y &= 2(0) - 1 \\ &= 0 - 1 \\ &= -1 \end{aligned}$$

The equation we have been studying can be compared to a general equation of the same type:

Equation: $y = 2x - 1$	*Equation*: $y = mx + b$
slope $= \frac{2}{1} = 2$	*slope* $= m$
y-intercept $= -1$	*y-intercept* $= b$

The following statement, a conditional, is true for the general equation:

- **If the equation of a line is written in the form $y = mx + b$, then the slope of the line is m and the y-intercept is b.**

The converse of this statement is also true:

- **If the slope of a line is m and its y-intercept is b, then the equation of the line is $y = mx + b$.**

An equation of the form $y = mx + b$ is called the ***slope-intercept form of a linear equation***.

EXAMPLES

1. Find the slope and the y-intercept of the line that is the graph of the equation $4x + 2y = 10$.

How to Proceed:	*Solution:*
Transform the equation into an equivalent equation of the form $y = mx + b$ by solving for y in terms of x:	$4x + 2y = 10$ $2y = -4x + 10$ $y = -2x + 5$
Then m (the coefficient of x) is the slope:	slope $= -2$
and b (the constant term) is the y-intercept:	y-intercept $= 5$

Answer: Slope $= -2$, y-intercept $= 5$

---

2. Write an equation of a line whose slope is $\frac{1}{2}$ and whose y-intercept is -4.

How to Proceed:	*Solution:*
(1) Write the equation $y = mx + b$:	$y = mx + b$
(2) Replace m by the numerical value of the slope:	$y = \frac{1}{2}x + b$
(3) Replace b by the numerical value of the y-intercept:	$y = \frac{1}{2}x + (-4)$ $= \frac{1}{2}x - 4$

Answer: $y = \frac{1}{2}x - 4$

EXERCISES WITH OPEN-RESPONSE PROBLEMS

In 1–16, in each case, find the slope and the y-intercept of the line that is the graph of the equation.

1. $y = 3x + 1$ **2.** $y = x - 3$ **3.** $y = 2x$ **4.** $y = x$

5. $y = \frac{1}{2}x + 5$ **6.** $y = -2x + 3$ **7.** $y = -3x$ **8.** $y = -2$

9. $y = -\frac{2}{3}x + 4$ **10.** $y - 3x = 7$ **11.** $2x + y = 5$ **12.** $3y = 6x + 9$

13. $2y = 5x - 4$ **14.** $\frac{1}{2}x + \frac{3}{4} = \frac{1}{3}y$ **15.** $4x - 3y = 0$ **16.** $2x = 5y - 10$

In 17–24, in each case write an equation of the line whose slope and y-intercept are, respectively:

17. 2 and 7 **18.** -1 and -3 **19.** 0 and -5 **20.** -3 and 0

21. $\frac{2}{3}$ and 1 **22.** $\frac{1}{2}$ and 0 **23.** $-\frac{1}{3}$ and 2 **24.** $-\frac{3}{2}$ and 0

25. Write equations for three lines each of which has a slope of 2.

26. Write equations for three lines each of which has a y-intercept of -4.

27. What do the graphs of the equations $y = 4x$, $y = 4x + 2$, and $y = 4x - 2$ all have in common?

28. How are the graphs of $y = mx + b$ affected when m is always replaced by the same number and b is replaced by different numbers?

29. What do the lines that are the graphs of the equations $y = 2x + 1$, $y = 3x + 1$, and $y = -4x + 1$ all have in common?

30. How are the graphs of $y = mx + b$ affected when b is always replaced by the same number and m is replaced by different numbers?

In 31–34, state in each case whether the lines are parallel.

31. $y = 3x + 2$, $y = 3x - 5$ **32.** $y = -2x - 6$, $y = 2x + 6$

33. $y = 4x - 8$, $y - 4x = 3$ **34.** $y = 2x$, $2y - 4x = 9$

35. Which of the following statements is true of the graph of the equation $y = -3x$?
(1) It is parallel to the x-axis. (2) It is parallel to the y-axis.
(3) Its slope is -3. (4) It does not have a y-intercept.

36. Which of the following statements is true of the graph of the equation $y = 8$?
(1) It is parallel to the x-axis. (2) It is parallel to the y-axis.
(3) It has no slope. (4) It goes through the origin.

16-7 GRAPHING LINEAR FUNCTIONS USING THEIR SLOPES

The slope and the y-intercept of a line can be used to draw the graph of a linear function.

EXAMPLE

Draw the graph of $2x + 3y = 9$ using the slope-intercept method.

How to Proceed: | *Solution:*

(1) Transform the equation into the form $y = mx + b$:

$$2x + 3y = 9$$
$$3y = -2x + 9$$
$$y = \frac{-2}{3}x + 3$$

(2) Find the slope of the line (the coefficient of x):

$$\text{slope} = \frac{-2}{3}$$

(3) Find the y-intercept of the line (the constant):

$$y\text{-intercept} = 3$$

(4) On the y-axis, graph point A, whose ordinate is the y-intercept:

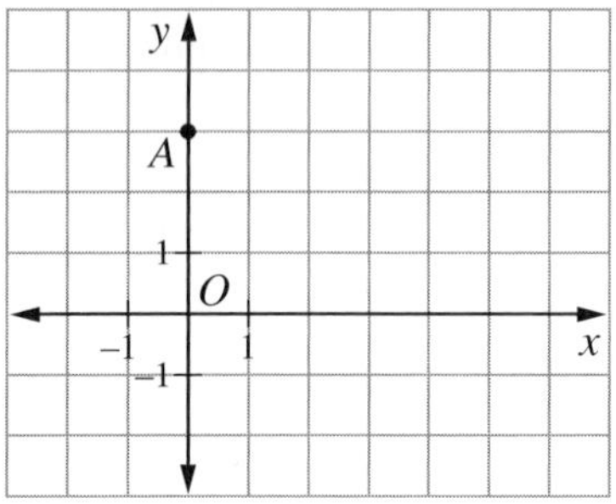

(5) Use the slope to find two more points on the line. Since slope $= \frac{\Delta y}{\Delta x} = \frac{-2}{3}$, when y changes by -2, x changes by 3. Therefore, start at point A and move 2 units down and 3 units to the right to locate point B. Then start at point B and repeat the procedure to locate point C:

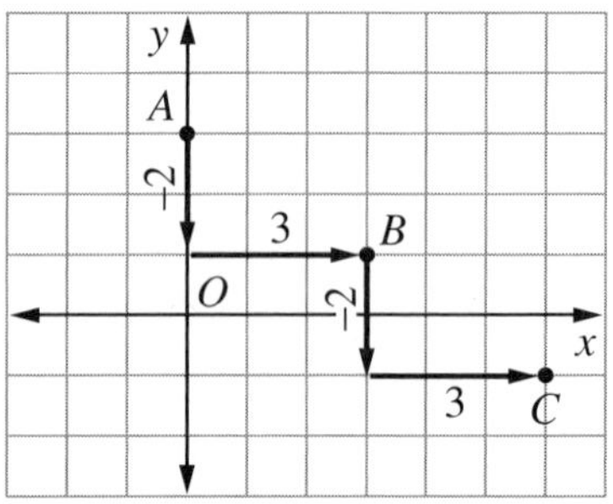

(6) Draw the line that passes through the three points:
This line is the graph of $2x + 3y = 9$.

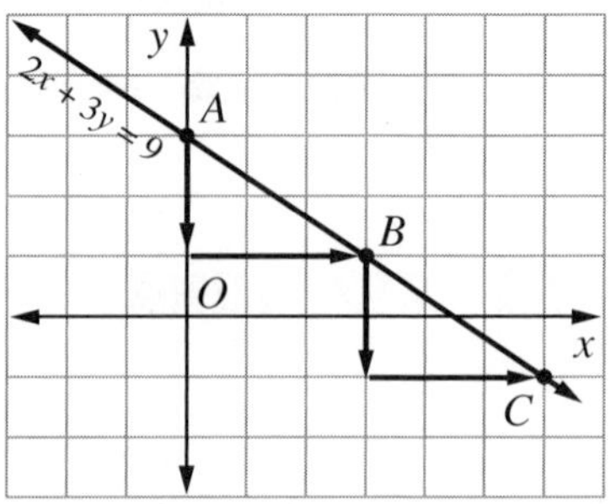

Check: To check a graph, select two or more points on the line drawn and substitute their x- and y-values in the original equation. For example, check this graph using points (4.5, 0), (3, 1), and (6, −1).

$2x + 3y = 9$	$2x + 3y = 9$	$2x + 3y = 9$
$2(4.5) + 3(0) = 9$	$2(3) + 3(1) = 9$	$2(6) + 3(-1) = 9$
$9 + 0 = 9$	$6 + 3 = 9$	$12 - 3 = 9$
$9 = 9$	$9 = 9$	$9 = 9$

EXERCISES WITH OPEN-RESPONSE PROBLEMS

In 1–24, graph each equation using the slope-intercept method.

1. $y = 2x + 3$ **2.** $y = 2x - 5$ **3.** $y = 2x$ **4.** $y = x - 2$

5. $y = 2x - 2$ **6.** $y = 3x - 2$ **7.** $y = 3x$ **8.** $y = 5x$

9. $y = -2x$ **10.** $y = \frac{2}{3}x + 2$ **11.** $y = \frac{1}{2}x - 1$ **12.** $y = \frac{3}{2}x$

13. $y = \frac{1}{3}x$ **14.** $y = -\frac{4}{3}x + 5$ **15.** $y = -\frac{3}{4}x$ **16.** $y - 2x = 8$

17. $3x + y = 4$ **18.** $2y = 4x + 6$ **19.** $3y = 4x + 9$ **20.** $4x - y = 3$

21. $3x + 4y = 12$ **22.** $2x = 3y + 6$ **23.** $4x + 3y = 0$ **24.** $2x - 3y - 6 = 0$

25. a. Draw the line through (−2,−3) whose slope is 2.
b. What appears to be the y-intercept of this line?
c. Use the slope of the line and the answer to part **b** to write an equation of the line.
d. Do the coordinates of point (−2,−3) satisfy the equation written in part **c**? Perform a check, and answer yes or no.

26. a. Draw the line through (3, 5) whose slope is $\frac{2}{3}$.
b. What appears to be the y-intercept of this line?
c. Use the slope of the line and the answer to part **b** to write an equation of the line.
d. Do the coordinates of point (3, 5) satisfy the equation written in part **c**? Perform a check and answer yes or no.

27. a. Is (1, 1) a point on the graph of $3x - 2y = 1$?
b. What is the slope of $3x - 2y = 1$?
c. Draw the graph of $3x - 2y = 1$ using point (1, 1) and the slope of the line.
d. Why is it easier to use point (1, 1) rather than the y-intercept to draw the graph?

16-8 WRITING AN EQUATION OF A LINE

We have learned to draw a line given this information:

- the coordinates of two points on the line, or
- the coordinates of one point and the slope of the line.

With this same information, we can also write an equation of a line.

EXAMPLES

1. Write an equation of the line that has a slope of 4 and that passes through point (3, 5).

How to Proceed:	*Solution:*
(1) In the equation of a line, $y = mx + b$, replace m by the given slope, 4:	$y = mx + b$ $y = 4x + b$
(2) Be aware that, since the given point, (3, 5), is on the line, its coordinates satisfy the equation $y = 4x + b$. Replace x by 3 and y by 5:	$(5) = 4(3) + b$
(3) Solve the resulting equation to find the value of b, the y-intercept:	$5 = 12 + b$ $-7 = b$
(4) In $y = 4x + b$, replace b by -7:	$y = 4x - 7$

Answer: $y = 4x - 7$

---

2. Write an equation of the line that passes through points (2, 5) and (4, 11).

How to Proceed:	*Solution:*
(1) Find the slope of the line that passes through the two given points, (2, 5) and (4, 11):	Let P_1 be (2, 5). $[x_1 = 2, y_1 = 5]$, and P_2 be (4, 11). $[x_2 = 4, y_2 = 11]$ $m = \frac{y_2 - y_1}{x_2 - x_1}$ $= \frac{11 - 5}{4 - 2} = \frac{6}{2} = 3$
(2) In $y = mx + b$, replace m by the slope, 3:	$y = mx + b$ $= 3x + b$

(3) Select one point that is on the line, for example, (2, 5). Its coordinates must satisfy the equation $y = 3x + b$. Replace x by 2 and y by 5:	$(5) = 3(2) + b$
(4) Solve the resulting equation to find the value of b, the y-intercept:	$5 = 6 + b$ $-1 = b$
(5) In $y = 3x + b$, replace b by -1:	$y = 3x - 1$
(6) Check whether the coordinates of the second point, (4, 11), satisfy the equation $y = 3x - 1$:	*Check:* $11 \stackrel{?}{=} 3(4) - 1$ $11 = 11$ (True)

Answer: $y = 3x - 1$

EXERCISES WITH OPEN-RESPONSE PROBLEMS

In 1–6, in each case write an equation of the line that has the given slope, m, and that passes through the given point.

1. $m = 2$; $(1, 4)$ **2.** $m = 2$; $(-3, 4)$ **3.** $m = -3$; $(-2, -1)$

4. $m = \frac{1}{2}$; $(4, 2)$ **5.** $m = \frac{-3}{4}$; $(0, 0)$ **6.** $m = \frac{-5}{3}$; $(-3, 0)$

In 7–12, in each case write an equation of the line that passes through the given points.

7. $(1, 4)$; $(3, 8)$ **8.** $(3, 1)$; $(9, 7)$ **9.** $(1, 2)$; $(10, 14)$

10. $(0, -1)$; $(6, 8)$ **11.** $(-2, -5)$; $(-1, -2)$ **12.** $(0, 0)$; $(-3, 5)$

13. Write an equation of the line that is:

a. parallel to the line $y = 2x - 4$, and has a y-intercept of 7.

b. parallel to the line $y - 3x = 6$, and has a y-intercept of -2.

c. parallel to the line $2x + 3y = 12$, and that passes through the origin.

d. perpendicular to the line $y = 3x + 2$, and has a y-intercept of 2.

e. perpendicular to the line $3y + 4x = 18$, and that passes through the origin.

14. Write an equation of the line that is:

a. parallel to the line $y = 4x + 1$, and that passes through point $(2, 3)$.

b. parallel to the line $2y - 6x = 9$, and that passes through point $(-2, 1)$.

c. parallel to the line $y = 4x + 3$, and that has the same y-intercept as the line $y = 5x - 3$.

d. parallel to the line $y = -\frac{1}{2}x$, and that has the same y-intercept as the line $2y = 7x + 6$.

e. perpendicular to the line $2y - 4x = 8$, and that passes through point $(6, -6)$.

f. perpendicular to the line $2y - 5x = 15$, and has an x-intercept of 3.

15. A triangle is determined by the three points $A(3, 5)$, $B(6, 4)$, and $C(1, -1)$. Write the equation of each line:

a. $\overleftrightarrow{AB}$ **b.** $\overleftrightarrow{BC}$ **c.** $\overleftrightarrow{CA}$

16. Latonya sends her film to be developed to a mail-order house that charges a fee per roll plus a fixed amount for postage and handling; that is, the amount is the same no matter how many rolls of film are developed. Last month Latonya paid $7.00 to have two rolls of film developed, and this month she paid $13.00 to have five rolls developed.

a. Write two ordered pairs such that the abscissa, x, is the number of rolls of film developed and the ordinate, y, is the total cost of developing the film.

b. Write an equation for the total cost, y, for x rolls of film.

c. What is the replacement set for x? **d.** What is the replacement set for y?

17. Every repair bill at Chickie's Service Spot includes a fixed charge for an estimate of the repairs plus an hourly fee for labor. Jack paid $123 for a TV repair, including 3 hours of labor. John paid $65 for a VCR repair, including 1 hour of labor.

a. Write two ordered pairs (x,y), where x represents the number of hours of labor and y is the total cost for repairs.

b. Write an equation in the form $y = mx + b$ that expresses the value of y, the total cost of repairs, in terms of x, the number of hours of labor.

c. What is the fixed charge for an estimate at Chickie's Service Spot?

d. What is the hourly fee for labor at Chickie's?

e. How are the answers to parts **c** and **d** related to the equation written in part **b**?

16-9 THE SCATTER PLOT

Statistics are often used to study the relationship between two different sets of values. For example, a dietician may want to study the relationship between the number of calories from fat in a person's diet and the level of cholesterol in that person's blood, or a merchant may want to study the relationship between the amount spent on advertising and gross sales. Although these examples involving ***two-valued statistics*** require complex statistical methods, we can investigate similar but simpler problems by looking at graphs.

The chart below shows the recorded heights in inches and weights in pounds for the last 24 persons who enrolled in a health club.

Height	Weight	Height	Weight	Height	Weight	Height	Weight
69	160	75	180	66	145	71	165
67	160	76	155	66	130	66	155
63	135	70	175	68	160	67	140
73	185	73	170	68	140	78	210
71	215	68	190	72	170	72	160
79	225	74	190	77	195	69	145

Two-valued statistics can be displayed on a graph called a **scatter plot** that relates two sets of data. The data, heights and weights, from the chart on page 614 are represented in the accompanying scatter plot.

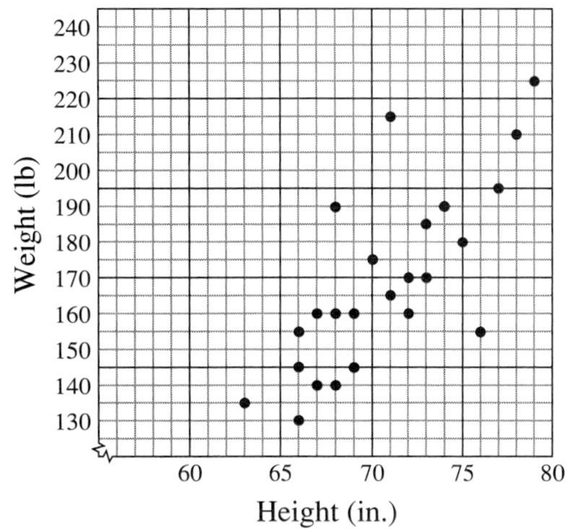

The graph shows that, in general, as heights increase, weights increase. Note that you want to draw a line that approximates the relationship shown in the graph. Begin by finding the mean of the heights and the mean of the weights, using a calculator to perform the computations.

For the heights:

Enter: (69 + 75 + 66 + 71 + 67 + 76
+ 66 + 66 + 63 + 70 + 68 + 67
+ 73 + 73 + 68 + 78 + 71 + 68
+ 72 + 72 + 79 + 74 + 77 + 69)
÷ 24 =

Display: 70.75

For the weights:

Enter: (160 + 180 + 145 + 165 + 160
+ 155 + 130 + 155 + 135 + 175
+ 160 + 140 + 185 + 170 + 140
+ 210 + 215 + 190 + 170 + 160
+ 225 + 190 + 195 + 145) ÷ 24 =

Display: 168.9583333

The mean of the heights is 70.75, and the mean of the weights is, to the nearest integer, 169. Locate point (70.75, 169) on the graph and, through this point, draw a straight line that best represents the data.

It is possible to draw many different straight lines that are close to the given data. Choose a line such that approximately half of the data values are above the line and half are below the line, as shown in the accompanying graph. The line we have drawn includes point (66, 145). This line is often called the **line of best fit** or a **trend line** for the data displayed.

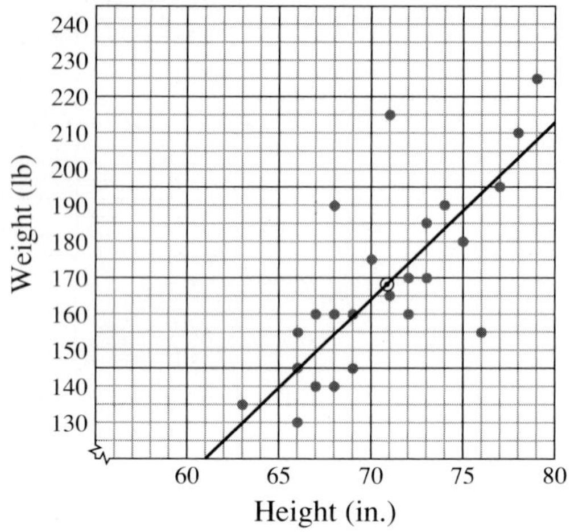

You can use points (70.75, 169) and (66, 145) to write an equation of the line that we drew. First approximate the slope of the line.

$$\text{slope} = \frac{169 - 145}{70.75 - 66} = \frac{24}{4.75} \approx 5.0526316$$

To simplify the equation, round the slope to 5. Then, using point (70.75, 169) and slope 5, apply the point-slope form of a line to find an equation.

$$(y - 169) = 5(x - 70.75)$$
$$y - 169 = 5x - 353.75$$
$$y = 5x - 184.75$$

Once again, round values to write a simplified equation:

$$y = 5x - 185$$

This is only one of many possible equations that can be used to approximate a person's weight, y, in pounds when given the person's height, x, in inches.

Observations for these data sets:

1. To see the value of the equation $y = 5x - 185$, find estimated weights using the actual heights given by the 24 people. In two cases, estimated weights are equal to actual weights; in 11 cases, estimated weights are higher than actual weights; and in 11 cases, estimated weights are lower than actual weights. The average difference between estimated and actual weights for this group is 0.2 pound, a very small error.

2. Predictions can be made based on the equation. For example, a person whose height is 64 inches should have a weight close to

$$y = 5x - 185 = 5(64) - 185 = 135 \text{ pounds.}$$

- **In general, when given a relation that involves two sets of data:**
  1. **In some cases, one or more linear equations can be found to approximate the relationship between the data sets.**
  2. **In other cases, the graphed data points are so scattered that it is not possible to write a linear equation to approximate the given relation.**

EXAMPLE

The accompanying table shows the number of calories and the number of grams of carbohydrates in a $\frac{1}{2}$-cup serving of ten different canned or frozen vegetables.

Carbohydrates	9	23	4	5	19	8	12	7	13	17
Calories	45	100	20	25	110	35	50	30	70	80

a. Draw a graph of the data. Let the horizontal axis represent grams of carbohydrates, and the vertical axis represent number of calories.
b. Find the mean number of grams of carbohydrates in a serving of vegetables.
c. Find the mean number of calories in a serving of vegetables.
d. On the graph, draw a line that approximates the data in the table.
e. Write an equation of the line that you drew in part **d**.

Solutions **a.**

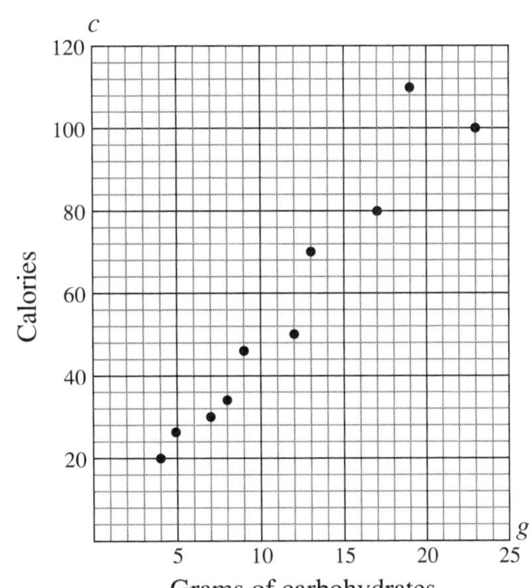

Carbohydrates	9	23	4	5	19	8	12	7	13	17
Calories	45	100	20	25	110	35	50	30	70	80

b. Use a calculator to compute the mean.

Enter: [( ] 9 [+] 23 [+] 4 [+] 5 [+] 19 [+] 8 [+] 12 [+] 7 [+] 13 [+] 17 [ )] [÷] 10 [=]

Display: 11.7

Answer: The mean number of grams of carbohydrates is 11.7.

c. Use a calculator to compute the mean.

Enter: [( ] 45 [+] 100 [+] 20 [+] 25 [+] 110 [+] 35 [+] 50 [+] 30 [+] 70 [+] 80 [ )] [÷] 10 [=]

Display: 56.5

Answer: The mean number of calories is 56.5.

d. Use the means, (11.7, 56.5), as the coordinates of one point. Since it is reasonable that 0 gram of carbohydrates would have 0 calorie, use (0, 0) as a second point. When you draw this line, approximately half of the points are above the line and half are below it.

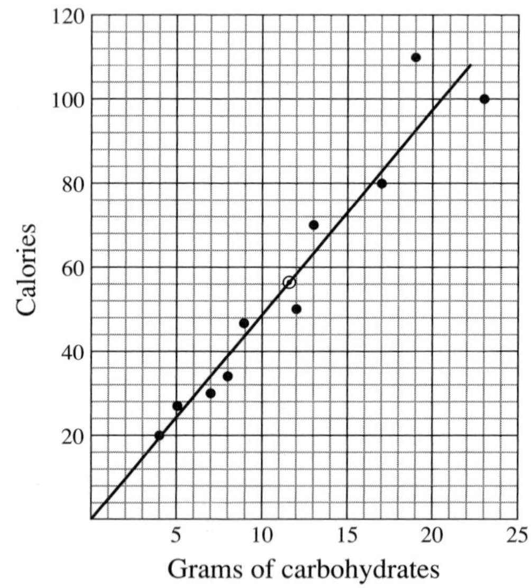

e. Use points (11.7, 56.5) and (0, 0).

Find the slope: $\text{slope} = \frac{56.5 - 0}{11.7 - 0} \approx 4.83$

Find an equation:

$$y - 56.5 = 4.83(x - 11.7)$$
$$y - 56.5 = 4.83x - 56.511$$
$$y = 4.83x - 0.011$$

Simplify numerical values: $y = 4.8x$

Answer: $y = 4.8x$ is an acceptable equation to approximate the line drawn in part **d**.

EXERCISES WITH OPEN-RESPONSE PROBLEMS

1. When Gina bought a new car, she decided to keep a record of how much gas she uses. Each time she puts gas in the car, she records the number of gallons of gas purchased and the number of miles driven since the last fill-up. Her record for the first 2 months is as follows:

Gallons of gas	10	12	9	6	11	10	8	12	10	7
Miles driven	324	375	290	190	345	336	250	375	330	225

a. Draw a graph of the data. Let the horizontal axis represent the number of gallons of gas, and the vertical axis represent the number of miles driven.
b. Find the mean number of gallons of gasoline per fill-up.
c. Find the mean number of miles driven between fill-ups.
d. On the graph, draw a line that approximates the data in the table.
e. Write an equation of the line that you drew in part **d**.

2. According to the curator of zoology at the Rochester Museum and Science Center, if you count the number of chirps produced by a tree cricket in 15 seconds and add 40, you will have a close approximation of the actual air temperature in degrees Fahrenheit. To test this statement, Alexa recorded, on several summer evenings, the temperature in degrees Fahrenheit and the number of chirps of a tree cricket in 1 minute. She obtained the following results:

Chirps/min.	150	170	140	125	108	118	145	68	95	110
Temperature	80	85	76	71	65	70	76	56	63	67

a. Write an equation that states the relationship between the number of chirps per minute of the tree cricket, C, to the temperature in degrees Fahrenheit, F.
b. Draw a graph of the data given in the table. Record the number of chirps per minute on the horizontal axis, and the temperature on the vertical axis.
c. Draw the graph of the equation that you wrote in part **a** on the same set of axes that you used for part **b**.
d. Can the data that Alexa recorded be represented by the equation that you wrote for part **a**? Explain your answer.

3. Gemma made a record of the cost and length of each of the 14 long-distance telephone calls that she made in the past month. Her record is given below.

Minutes	3.2	1.0	19.4	0.8	4.3	34.8	2.9
Cost	\$0.35	\$0.11	\$2.12	\$0.09	\$0.47	\$3.78	\$0.22
Minutes	2.5	7.1	10.9	5.8	1.5	1.4	7.6
Cost	\$0.28	\$0.79	\$1.21	\$0.65	\$0.20	\$0.19	\$0.89

a. Draw a graph of the data. Let the horizontal axis represent the number of minutes, and the vertical axis represent the cost of the call.
b. Find the mean number of minutes per call.
c. Find the mean cost of the calls.
d. On the graph, draw a line that approximates the data in the table.
e. Write an equation of the line that you drew in part **d**.

4. Using the data and graph from Exercise 3:
a. Find the median number of minutes per call.
b. Find the median cost of the calls.
c. Using the answers to parts **a** and **b** as the coordinates of a point, draw a line that approximates the data.
d. Write an equation of the line that you drew in part **c**.

5. Marissa believes that names influence the way people, animals, and plants perform. She kept the following list of the numbers of blooms produced last year by 10 of her hybrid tea roses.

Name	Number of Blooms	Name	Number of Blooms
Brandy	6	Queen Elizabeth	18
Chicago Peace	17	Talisman	7
JFK	18	Medallion	5
Mirandy	14	Honor	10
Peace	21	Fragrant Cloud	5

a. Graph the data, comparing the number of letters in the name of the rose to the number of blooms that the rose produced last year.
b. Marissa believes that the more letters in the name of the rose, the more blooms it will produce. Do you agree with Marissa? Explain your answer on the basis of the graph that you drew.

16-10 DISTANCE IN THE COORDINATE PLANE

The Distance Formula

Recall that the distance between any two points on a number line is the difference between the coordinates assigned to these points. For example, the distance between G and C, written as GC, is 4 units, found by subtracting the coordinates of these points: $GC = 5 - 1 = 4$. The distance CG must equal the distance GC. Therefore, we make use of absolute value in finding distance so that the difference of the coordinates will be independent of the order in which they are subtracted. Thus, we write:

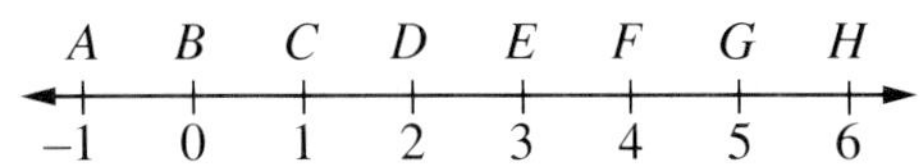

$$GC = |5 - 1| = |4| = 4 \qquad \text{and} \qquad CG = |1 - 5| = |-4| = 4$$

We can now state the following definition:

- ***Definition:*** The ***distance between any two points on the real number line*** is the absolute value of the difference of the coordinates of the two points.

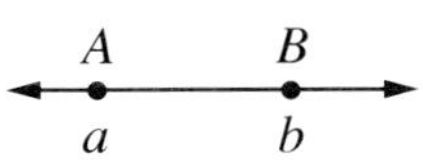

Hence, on the real number line, the distance between point A, whose coordinate is a, and point B, whose coordinate is b, can be represented by

$$AB = |a - b| \qquad \text{or} \qquad AB = |b - a|$$

In the coordinate plane, when two points are on the same horizontal line, they have the same y-coordinate. For example, points $A(2, 1)$ and $B(6, 1)$ have the same ordinate, 1. The distance AB between A and B, two points with the same ordinate, is the absolute value of the difference between their x-coordinates.

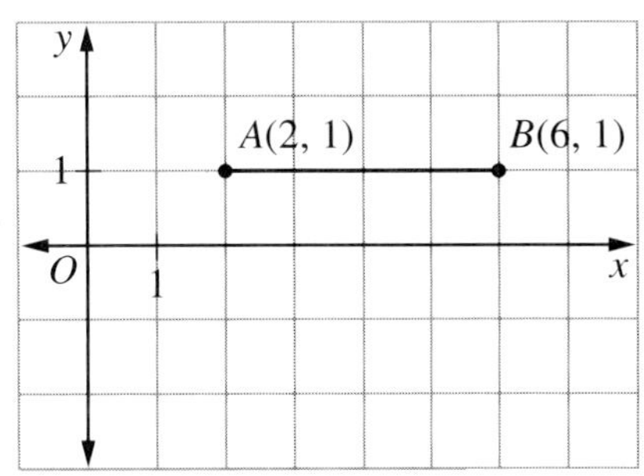

$$AB = |6 - 2| = |4| = 4$$

and

$$AB = |2 - 6| = |-4| = 4$$

When two points are on the same vertical line, they have the same x-coordinate. For example, points $C(2, 4)$ and $D(2, -1)$ have the same abscissa, 2. The distance CD between C and D, two points with the same abscissa, is the absolute value of the difference between their y-coordinates.

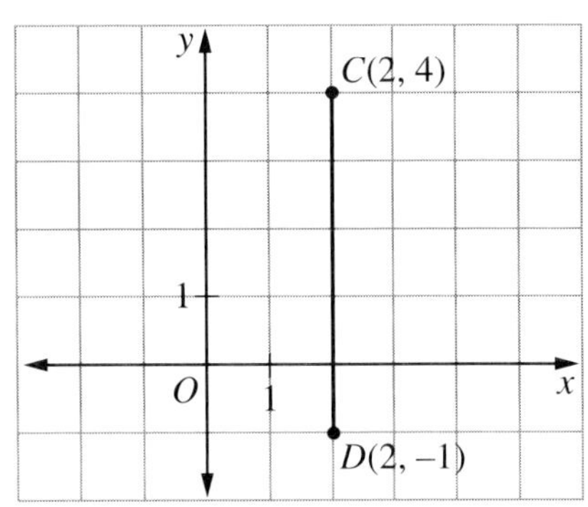

$$CD = |4 - (-1)| = |4 + 1| = |5| = 5$$

and $\quad CD = |-1 - 4| = |-5| = 5$

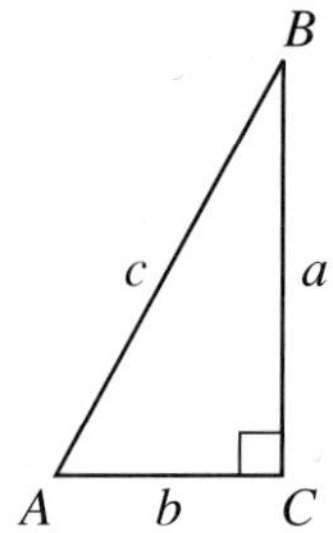

The method that we used in these special cases can help us to find the distance between any two points. First, recall the Pythagorean Theorem from earlier studies. This theorem, which can be proved, applies only to right triangles. For example, in the accompanying diagram, $\triangle ABC$ is a right triangle in which the right angle is $\angle C$, the longest side, or the hypotenuse, is $\overline{AB}$, and the legs are $\overline{BC}$ and $\overline{AC}$. Let $BC = a$, $AC = b$, and $AB = c$. The theorem is stated here, first in words and then in symbols.

- ***Pythagorean Theorem.*** If a triangle is a right triangle, then the square of the length of the longest side (the hypotenuse) is equal to the sum of the squares of the lengths of the other two sides (the legs).

In symbols: $(AB)^2 = (BC)^2 + (AC)^2$

or $c^2 = a^2 + b^2$

Now, to find the distance d from $A(2, 3)$ to $B(5, 7)$, two points that have different abscissas and different ordinates, first draw a horizontal line through A and a vertical line through B. These two lines intersect at point $C(5, 3)$. The distance from $A(2, 3)$ to $C(5, 3)$ is the absolute value of the difference of their abscissas:

$$AC = |5 - 2| = 3$$

Also, the distance from $B(5, 7)$ to $C(5, 3)$ is the absolute value of the difference of their ordinates:

$$BC = |7 - 3| = 4$$

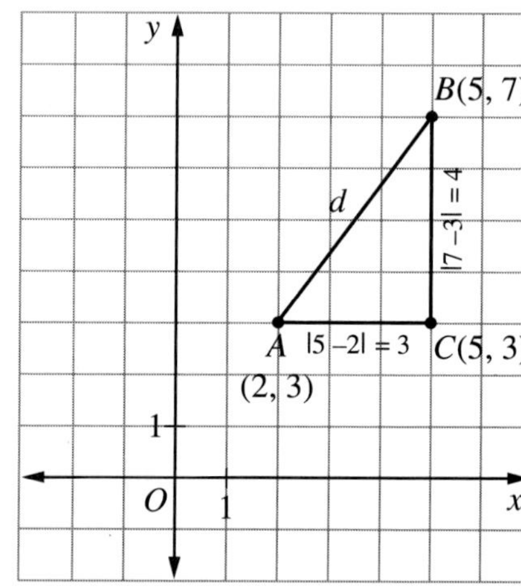

Since $\triangle ABC$ is a right triangle whose hypotenuse is $\overline{AB}$, we can use the Pythagorean Theorem to find AB.

$$\begin{aligned} c^2 &= a^2 + b^2 \\ (AB)^2 &= (BC)^2 + (AC)^2 \\ &= 4^2 + 3^2 \\ &= 16 + 9 \\ &= 25 \\ AB &= 5 \end{aligned}$$

Repeat the steps that you used to find AB to derive a formula for the distance d between any two points $R(x_1, y_1)$ and $S(x_2, y_2)$. It is possible to use the Pythagorean Theorem to derive the distance formula only because the axes that were chosen for the coordinate system are perpendicular to each other, and the scales on the axes are the same.

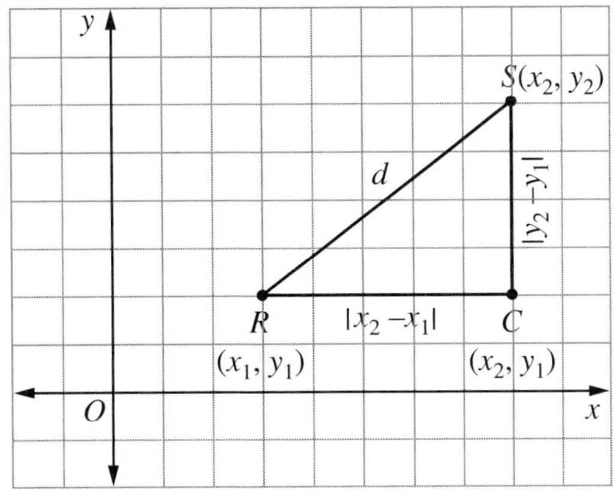

1. Form right triangle RCS by drawing a vertical line through S and a horizontal line through R with the two lines intersecting at C.

2. The coordinates of C are (x_2, y_1).

3. Therefore, $RC = |x_2 - x_1|$ and $CS = |y_2 - y_1|$. Let $RS = d$.

4.
$$\begin{aligned}(RS)^2 &= (RC)^2 + (CS)^2 \\ d^2 &= |x_2 - x_1|^2 + |y_2 - y_1|^2 \\ &= (x_2 - x_1)^2 + (y_2 - y_1)^2\end{aligned}$$

5. Take the positive square root of each member of the equation:
$d = \sqrt{(x_2 - x_1)^2 + (y_2 - y_1)^2}$

We can now state the ***distance formula*** in general terms:

- The distance d between any two points (x_1, y_1) and (x_2, y_2) in the coordinate plane is given by the formula

$$d = \sqrt{(x_2 - x_1)^2 + (y_2 - y_1)^2}$$

EXAMPLES

1. Find the distance between $A(-4, -2)$ and $B(8, 3)$.

Solution Use the distance formula.
Let $A = (x_1, y_1)$. Then $x_1 = -4$, $y_1 = -2$.
Let $B = (x_2, y_2)$. Then $x_2 = 8$, $y_2 = 3$.

$$\begin{aligned} d &= \sqrt{(x_2 - x_1)^2 + (y_2 - y_1)^2} \\ AB &= \sqrt{(8 - (-4))^2 + (3 - (-2))^2} \\ &= \sqrt{(8 + 4)^2 + (3 + 2)^2} \\ &= \sqrt{12^2 + 5^2} \\ &= \sqrt{144 + 25} \\ &= \sqrt{169} \\ &= 13 \end{aligned}$$

Answer: $AB = 13$

Calculator Solution The calculation can be done on a scientific calculator by using the following sequence of keys:

Enter: [( ] [( ] 8 [−] 4 [+/−] [ )] [x^2] [+] [( ]
3 [−] 2 [+/−] [ )] [x^2] [ )] [$\sqrt{x}$] [=]

Display: 13.

---

2. Express the length of $\overline{CD}$ in radical form if the coordinates of C are $(-4, 2)$ and the coordinates of D are $(-3, 5)$.

Solution

Let $C = (x_1, y_1)$. Then $x_1 = -4$, $y_1 = 2$.

Let $D = (x_2, y_2)$. Then $x_2 = -3$, $y_2 = 5$.

$$\begin{aligned} CD &= \sqrt{(x_2 - x_1)^2 + (y_2 - y_1)^2} \\ &= \sqrt{[-3 - (-4)]^2 + [5 - 2]^2} \\ &= \sqrt{(-3 + 4)^2 + (5 - 2)^2} \\ &= \sqrt{(1)^2 + (3)^2} = \sqrt{1 + 9} = \sqrt{10} \end{aligned}$$

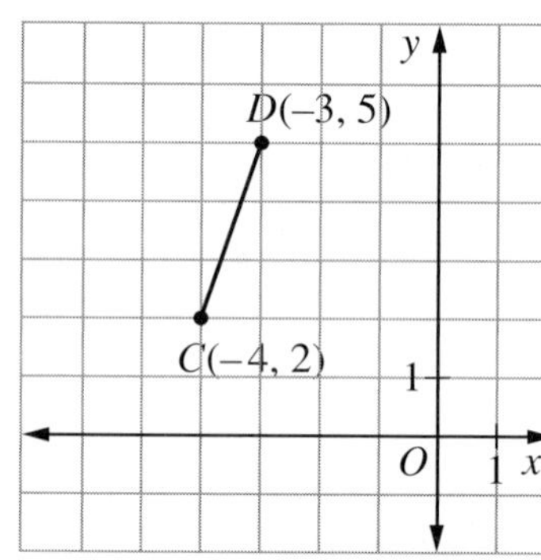

Answer: $CD = \sqrt{10}$.

Note. The only acceptable answer for this question is $\sqrt{10}$. Recall that $\sqrt{10}$ is an irrational number whose exact value is shown by the radical. Any solution found by using a calculator will be a *rational approximation* of $\sqrt{10}$.

Proofs Involving Specific Cases

By combining the distance formula with facts learned earlier about geometric figures, it is possible to determine specific relationships for given figures. Let us consider this example, in which specific coordinates are given.

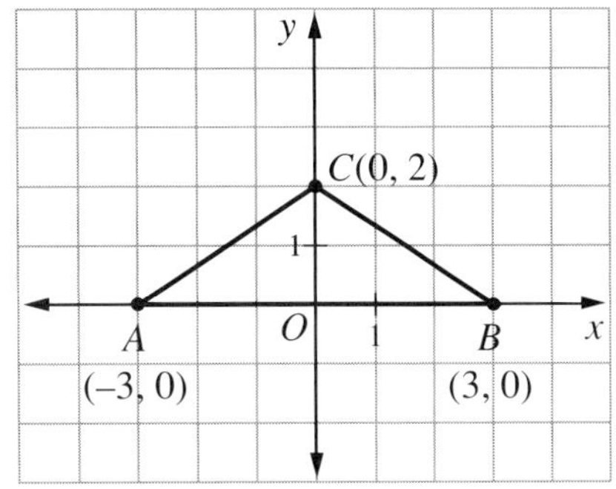

Given: $\triangle ABC$ with vertices $A(-3, 0)$, $B(3, 0)$, and $C(0, 2)$.

Show: $\triangle ABC$ is isosceles.

Plan: Find AB, BC, and CA. If two of these distances are equal, then $\triangle ABC$ is isosceles.

Solution

(1) $AB = |x_2 - x_1| = |3 - (-3)| = |3 + 3| = |6| = 6$

(2) $BC = \sqrt{(x_2 - x_1)^2 + (y_2 - y_1)^2} = \sqrt{(3 - 0)^2 + (0 - 2)^2}$
$= \sqrt{(3)^2 + (-2)^2} = \sqrt{9 + 4} = \sqrt{13}$

(3) $CA = \sqrt{(x_2 - x_1)^2 + (y_2 - y_1)^2} = \sqrt{[0 - (-3)]^2 + [2 - 0]^2}$
$= \sqrt{(0 + 3)^2 + (2 - 0)^2}$
$= \sqrt{(3)^2 + (2)^2} = \sqrt{9 + 4} = \sqrt{13}$

(4) Since $BC = CA$, $\triangle ABC$ is isosceles.

Note. It is also possible to use the general distance formula in step (1) to find distance AB:

$$AB = \sqrt{(x_2 - x_1)^2 + (y_2 - y_1)^2} = \sqrt{[3 - (-3)]^2 + [0 - 0]^2}$$
$$= \sqrt{(3 + 3)^2 + (0)^2} = \sqrt{(6)^2 + (0)^2} = \sqrt{36 + 0} = \sqrt{36} = 6$$

EXAMPLES

1. *Given:* The quadrilateral with vertices $A(2, 2)$, $B(5, -2)$, $C(9, 1)$, and $D(6, 5)$.

Show: $ABCD$ is a rhombus.

Plan: Show that $ABCD$ is a quadrilateral with four sides of equal length.

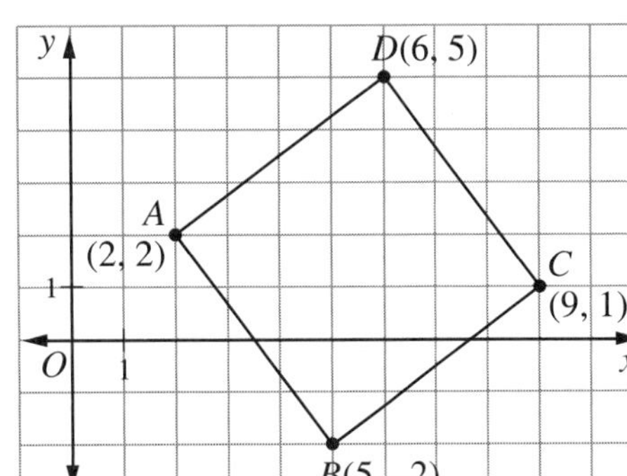

Solution

(1) $d = \sqrt{(x_2 - x_1)^2 + (y_2 - y_1)^2}$

(2) $AB = \sqrt{(5 - 2)^2 + (-2 - 2)^2} = \sqrt{(3)^2 + (-4)^2}$
$= \sqrt{9 + 16} = \sqrt{25} = 5$

(3) $BC = \sqrt{[9 - 5]^2 + [1 - (-2)]^2} = \sqrt{(4)^2 + (1 + 2)^2}$
$= \sqrt{(4)^2 + (3)^2} = \sqrt{16 + 9} = \sqrt{25} = 5$

(4) $CD = \sqrt{(9 - 6)^2 + (1 - 5)^2} = \sqrt{(3)^2 + (-4)^2}$
$= \sqrt{9 + 16} = \sqrt{25} = 5$

(5) $DA = \sqrt{(6 - 2)^2 + (5 - 2)^2} = \sqrt{(4)^2 + (3)^2}$
$= \sqrt{16 + 9}$
$= \sqrt{25} = 5$

(6) $AB = BC = CD = DA$.

(7) Since *ABCD* is an equilateral quadrilateral, it is a rhombus.

---

2. *Given:* The triangle with vertices $A(4, -1)$, $B(5, 6)$, and $C(1, 3)$.

Show: $\triangle ABC$ is an isosceles right triangle.

Plan: To show that $\triangle ABC$ is isosceles, show that two sides are equal in length. To show that $\triangle ABC$ is a right triangle, test the Pythagorean Theorem.

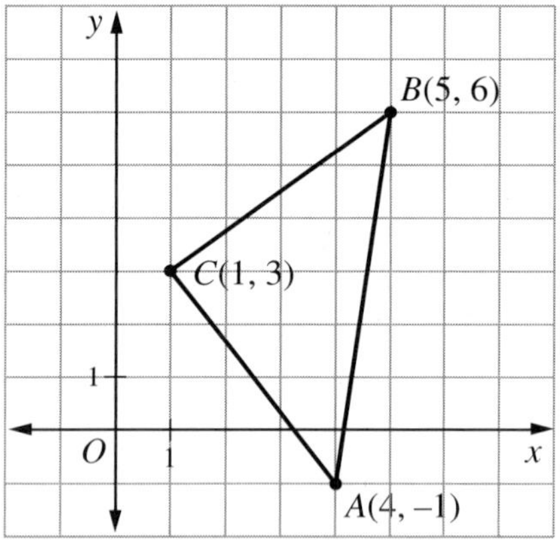

Solution

(1) $d = \sqrt{(x_2 - x_1)^2 + (y_2 - y_1)^2}$

(2) $AB = \sqrt{[5 - 4]^2 + [6 - (-1)]^2} = \sqrt{(5 - 4)^2 + (6 + 1)^2}$
$= \sqrt{(1)^2 + (7)^2} = \sqrt{1 + 49} = \sqrt{50}$

(3) $BC = \sqrt{(5 - 1)^2 + (6 - 3)^2} = \sqrt{(4)^2 + (3)^2}$
$= \sqrt{16 + 9} = \sqrt{25} = 5$

(4) $CA = \sqrt{(4 - 1)^2 + (-1 - 3)^2} = \sqrt{(3)^2 + (-4)^2}$
$= \sqrt{9 + 16} = \sqrt{25} = 5$

(5) $BC = CA$. Therefore, $\triangle ABC$ is an isosceles triangle.

(6) The longest side of $\triangle ABC$ is $\overline{AB}$. Use the Pythagorean Theorem to determine whether the triangle is a right triangle.

$$\begin{aligned} (AB)^2 &\stackrel{?}{=} (BC)^2 + (CA)^2 \\ (\sqrt{50})^2 &\stackrel{?}{=} (5)^2 + (5)^2 \\ 50 &\stackrel{?}{=} 25 + 25 \\ 50 &= 50 \end{aligned}$$

Therefore, since $(AB)^2 = (BC)^2 + (CA)^2$, $\triangle ABC$ is a right triangle.

(7) From steps (5) and (6), $\triangle ABC$ is an isosceles right triangle.

EXERCISES WITH OPEN-RESPONSE PROBLEMS

Throughout this section, answers that are not whole numbers should be left in radical form.

In 1–12, find the distance between each pair of points.

1. (2, 7) and (12, 7)
2. (−3, 4) and (5, 4)
3. (−1, −2) and (−1, 4)
4. (0, 0) and (4, 3)
5. (−6, −8) and (0, 0)
6. (1, 4) and (4, 8)
7. (4, 2) and (−2, 10)
8. (6, 4) and (3, 6)
9. (−5, 0) and (−9, 6)
10. (0, 0) and (−2, 5)
11. $(0, c)$ and $(b, 0)$
12. (0, 0) and (a, b)

In 13–18, find in each case the lengths of the sides of a triangle whose vertices are:

13. (0, 0), (8, 0), (4, 3)
14. (1, 5), (5, 5), (5, 1)
15. (3, 6), (−1, 3), (5, −5)
16. (6, −3), (0, 4), (8, −1)
17. (−1, 7), (0, 0), (8, 4)
18. (−4, 2), (−1, 6), (5, 4)

19. Find the length of the shortest side of the triangle whose vertices are $R(-2, -1)$, $S(1, 3)$, and $T(1, 10)$.

In 20–23, show that each triangle with the given vertices is isosceles.

20. (2, 3), (5, 7), (1, 4)
21. (1, 0), (5, 0), (3, 4)
22. (7, −1), (2, −2), (3, 3)
23. (4, −7), (−3, −4), (7, 0)

24. Points $A(-1, 2)$ and $B(-1, -3)$ have the same abscissa.
  a. Find AB by using the absolute value of the difference of the ordinates.
  b. Find AB by using the distance formula.
  c. Can the distance formula be used to find the distance between two points that have the same abscissa or two points that have the same ordinate? Explain your answer.
  d. Does $\sqrt{(x_2 - x_1)^2}$ always equal $|x_2 - x_1|$? Explain your answer.

25. Plot the following points: $A(5, 3)$, $B(-5, 3)$, $C(-5, -3)$, $D(5, -3)$
Connect these points in the order given. What kind of quadrilateral is $ABCD$? Prove your conclusion informally.

26. Show, in each case, that the quadrilateral whose vertices are given is a parallelogram:
a. (1, 2), (2, 5), (5, 7), (4, 4) **b.** (−1, 1), (−3, 4), (1, 5), (3, 2)

27. Show, in each case, that the quadrilateral whose vertices are given is a rhombus:
a. (1, 1), (5, 3), (7, 7), (3, 5) **b.** (−3, 2), (−2, 6), (2, 7), (1, 3)

28. **a.** Show that the quadrilateral whose vertices are (1, 4), (4, 9), (−1, 12), (−4, 7) is equilateral.
b. Show that the diagonals of this quadrilateral are equal in length.
c. What type of quadrilateral have you shown this to be?

29. The vertices of trapezoid $ABCD$ are $A(1,-4)$, $B(10,-4)$, $C(9, 2)$, and $D(2, 2)$.
a. Show that $ABCD$ is an isosceles trapezoid.
b. Show that the length of diagonal $\overline{AC}$ equals the length of diagonal $\overline{BD}$.

30. Plot the following points: $A(-2, 0)$, $B(0, -3)$, $C(2, 0)$. Connect the points in the order given. What kind of triangle is $\triangle ABC$? Give an informal proof of your conclusion.

31. The vertices of $\triangle RST$ are (1, 1), (9, 4), and (1, 7), respectively.
a. Show that $\triangle RST$ is isosceles. **b.** Show that $\triangle RST$ is *not* a right triangle.

32. Show that the line segments joining points (−1, 3), (9, 3), and (4, 8) form an isosceles right triangle.

33. Points (1, 1), (7, 1), (7, 4), and (1, 4) are the vertices of a rectangle. Show that the diagonals are equal in length.

34. In each case, draw the quadrilateral that has the indicated points as its vertices. What type of quadrilateral is formed? Prove your conclusion informally.
a. (1, 1), (4, 1), (4,−2), (1,−2) **b.** (2, 2), (3, 4), (6, 4), (5, 2)
c. (−1,−1), (−1,−3), (3,−3), (3,−1) **d.** (6, 0), (9, 4), (6, 8), (3, 4)

16-11 THE MIDPOINT FORMULA IN COORDINATE GEOMETRY

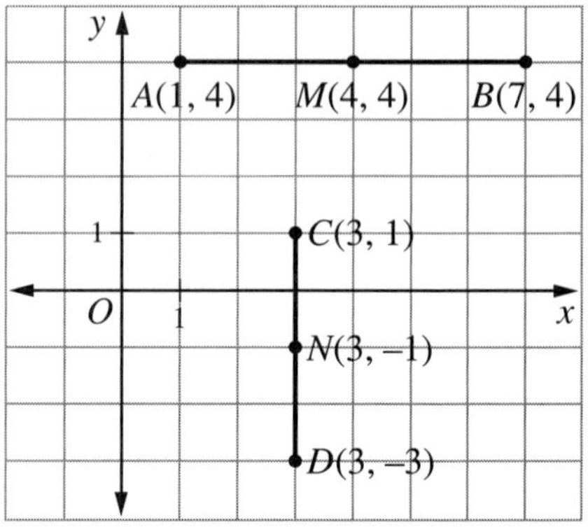

In the figure, $A(1, 4)$ and $B(7, 4)$ determine a horizontal segment, $\overline{AB}$, whose midpoint M can be found by counting units. Since $AB = 6$ units, then $AM = 3$ units, $MB = 3$ units, and the coordinates of midpoint M are (4, 4).

Notice that the binary operation of averaging can be used on the coordinates of A and B to determine the coordinates of midpoint M. Here, the ordinate of point M is the same as the ordinate of points A and B:

$$\text{Abscissa of } M\text{: } \frac{1+7}{2} = \frac{8}{2} = 4\text{; ordinate of } M = \frac{4+4}{2} = \frac{8}{2} = 4.$$

Similarly, $C(3, 1)$ and $D(3,-3)$ determine a vertical segment, $\overline{CD}$, whose midpoint N can be found by counting units. Since $CD = 4$ units, then $CN = 2$ units, $ND = 2$ units, and the coordinates of midpoint N are $(3,-1)$. Again notice that the binary operation of averaging can be used on the coordinates of C and D to determine the coordinates of midpoint N. Here, the abscissa of point N is the same as the abscissa of points C and D:

$$\text{Abscissa of } N\text{: } \frac{3+3}{2} = \frac{6}{2} = 3\text{; ordinate of } N\text{: } \frac{1+(-3)}{2} = \frac{-2}{2} = -1.$$

These examples suggest the following postulates.

- **POSTULATE.** The abscissa of the midpoint of a line segment parallel to the x-axis is the average of the abscissas of the endpoints.
- **POSTULATE.** The ordinate of the midpoint of a line segment parallel to the y-axis is the average of the ordinates of the endpoints.

- **PRINCIPLE 5. If a line segment joins the midpoints of two sides of a triangle, the segment is parallel to the third side and its length is one-half the length of the third side.**

These postulates can be used with Principle 5 to demonstrate the following principle.

- **PRINCIPLE 6. Each coordinate of the midpoint of a line segment is equal to the average of the corresponding coordinates of the endpoints of the line segment.**

Given: $\overline{AB}$ with coordinates $A(x_1, y_1)$ and $B(x_2, y_2)$. The midpoint of $\overline{AB}$ is point M.

Show: The coordinates of midpoint M are $\left(\frac{x_1 + x_2}{2}, \frac{y_1 + y_2}{2}\right)$.

Plan: Draw right triangle ABC with hypotenuse $\overline{AB}$ and legs parallel to the x-axis and y-axis. Find the coordinates of the midpoint of each leg. From these coordinates, obtain the coordinates of midpoint M.

Statements

1. Through $A(x_1, y_1)$, draw a line parallel to the x-axis.
   Through $B(x_2, y_2)$, draw a line parallel to the y-axis.
   Let C be the point at which the lines intersect.
2. The coordinates of C are (x_2, y_1).

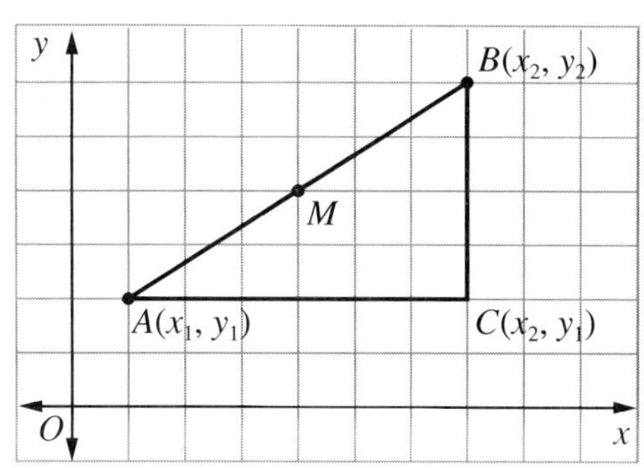

3. Let D be the midpoint of $\overline{BC}$. Then the coordinates of D are $\left(x_2, \frac{y_1 + y_2}{2}\right)$.
4. Let E be the midpoint of $\overline{AC}$. Then the coordinates of E are $\left(\frac{x_1 + x_2}{2}, y_1\right)$.
5. $\overline{ME} \parallel \overline{BC} \parallel$ y-axis. M and E have the same x-coordinate.
   $\overline{MD} \parallel \overline{AC} \parallel$ x-axis. M and D have the same y-coordinate.
6. Therefore, the coordinates of M are $\left(\frac{x_1 + x_2}{2}, \frac{y_1 + y_2}{2}\right)$.

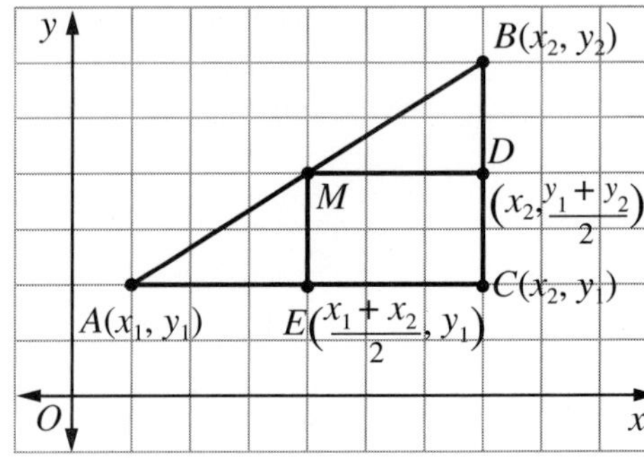

Using general coordinates, we now write the formula for the midpoint:

- **If the coordinates of the endpoints of $\overline{AB}$ are $A(x_1, y_1)$ and $B(x_2, y_2)$ and if $M(x_m, y_m)$ is the midpoint of $\overline{AB}$, then:**

$$x_m = \frac{x_1 + x_2}{2} \quad \text{and} \quad y_m = \frac{y_1 + y_2}{2}$$

EXAMPLES

1. Find the coordinates of the midpoint of the segment whose endpoints are $(-2, 3)$ and $(4, -3)$.

Solution Let $(x_1, y_1) = (-2, 3)$.
Then $x_1 = -2$ and $y_1 = 3$.

Let $(x_2, y_2) = (4, -3)$.
Then $x_2 = 4$ and $y_2 = -3$.

Let $(x_m, y_m) =$ coordinates of midpoint M.

$$x_m = \frac{x_1 + x_2}{2} = \frac{-2 + 4}{2} = \frac{2}{2} = 1$$

$$y_m = \frac{y_1 + y_2}{2} = \frac{3 + (-3)}{2} = \frac{0}{2} = 0$$

$(x_m, y_m) = (1, 0)$

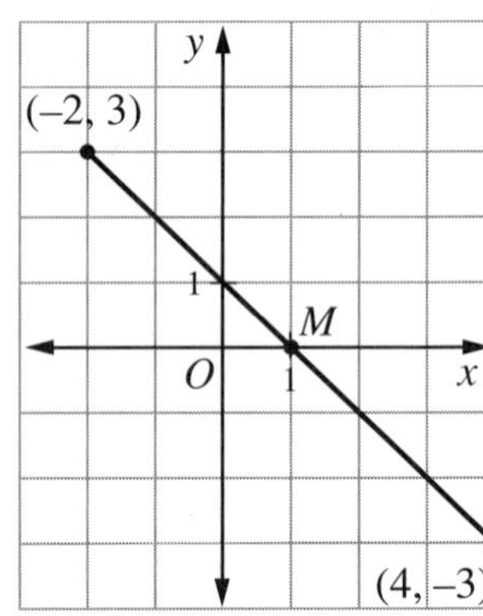

Answer: The coordinates of the midpoint are $(1, 0)$.

---

2. Midpoint M of $\overline{AB}$ has coordinates $(6, 5)$. If the coordinates of A are $(4, 1)$, what are the coordinates of B?

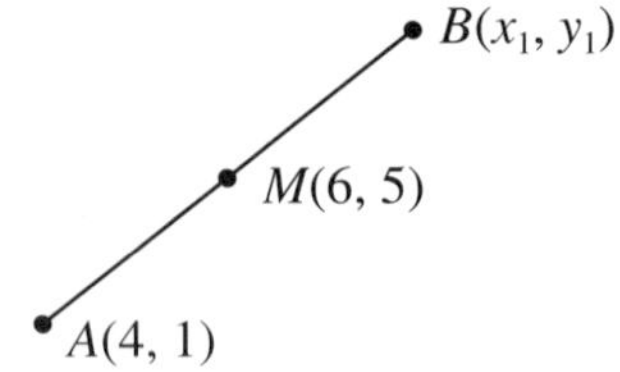

Solution (1) Draw $\overline{AB}$ with midpoint M.
Label $A(4, 1)$ and $M(6, 5)$.

(2) Since the coordinates of B are unknown, label $B(x_1, y_1)$.
Then $A(4, 1)$ is (x_2, y_2), so, $x_2 = 4$ and $y_2 = 1$.
Also, $M(6, 5)$ is (x_m, y_m), so, $x_m = 6$ and $y_m = 5$.

(3) Write and solve two equations, one involving abscissas and one involving ordinates:

$x_m = \frac{x_1 + x_2}{2}$	$y_m = \frac{y_1 + y_2}{2}$
$6 = \frac{x_1 + 4}{2}$	$5 = \frac{y_1 + 1}{2}$
$12 = x_1 + 4$	$10 = y_1 + 1$
$8 = x_1$	$9 = y_1$

Therefore, $B(x_1, y_1) = (8, 9)$.

Check: Is the midpoint of the segment with endpoints at (4, 1) and (8, 9) equal to (6, 5)?

$$\left(\frac{4 + 8}{2}, \frac{1 + 9}{2}\right) \stackrel{?}{=} (6, 5)$$

$$\left(\frac{12}{2}, \frac{10}{2}\right) \stackrel{?}{=} (6, 5)$$

$$(6, 5) = (6, 5) \text{ (True)}$$

(4) *Check* the answer, as shown.

Answer: The coordinates of B are (8, 9).

---

3. Given the quadrilateral whose vertices are $A(-2, 2)$, $B(1, 4)$, $C(2, 8)$, and $D(-1, 6)$.

a. Find the coordinates of the midpoint of diagonal $\overline{AC}$.
b. Find the coordinates of the midpoint of diagonal $\overline{BD}$.
c. Show that $ABCD$ is a parallelogram.

Plan: Show that both diagonals have the same midpoint.

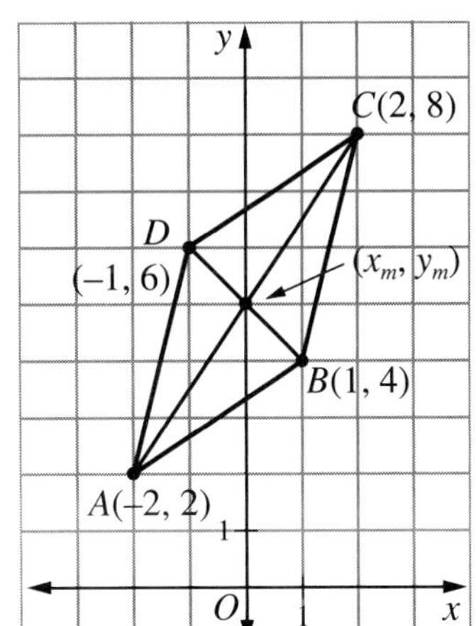

Solutions **a.** (1) Let $A(-2, 2)$ be (x_1, y_1), and let $C(2, 8)$ be (x_2, y_2). Thus, $x_1 = -2$, $y_1 = 2$, $x_2 = 2$, $y_2 = 8$.

(2) Let the midpoint of $\overline{AC}$ be (x_m, y_m). By the midpoint formula:

$$(x_m, y_m) = \left(\frac{x_1 + x_2}{2}, \frac{y_1 + y_2}{2}\right)$$

$$= \left(\frac{-2 + 2}{2}, \frac{2 + 8}{2}\right)$$

$$= \left(\frac{0}{2}, \frac{10}{2}\right) = (0, 5)$$

Answer: The midpoint of diagonal $\overline{AC}$ is (0, 5).

b. (1) Let $D(-1, 6)$ be (x_1, y_1) and let $B(1, 4)$ be (x_2, y_2). Thus, $x_1 = -1$, $y_1 = 6$, $x_2 = 1$, $y_2 = 4$.

(2) Let the midpoint of $\overline{BD}$ be (x_m, y_m). By the midpoint formula:

$$(x_m, y_m) = \left(\frac{x_1 + x_2}{2}, \frac{y_1 + y_2}{2}\right)$$
$$= \left(\frac{-1 + 1}{2}, \frac{6 + 4}{2}\right)$$
$$= \left(\frac{0}{2}, \frac{10}{2}\right) = (0, 5)$$

Answer: The midpoint of diagonal $\overline{BD}$ is (0, 5).

c. (1) Since point $E(0, 5)$ is the midpoint of each diagonal, the diagonals bisect each other at point E.

(2) Since $ABCD$ is a quadrilateral whose diagonals bisect each other, it is a parallelogram.

EXERCISES WITH OPEN-RESPONSE PROBLEMS

1. Find the coordinates of the midpoint of the line segment that joins each of the following pairs of points:

a. (6, 8), (4, 10) **b.** (2, 6), (8, 4) **c.** (2, 11), (7, 6)
d. (5, −6), (−1, 9) **e.** (−4, −8), (−6, −5) **f.** (12, 11), (−1, −2)
g. (2, 0), (5, 9) **h.** (−2, 7), (0, −5) **i.** (0, 0), (8, 10)
j. (2, −7), (−2, 7) **k.** $(5c, 2c)$, $(c, 8c)$ **l.** $(m, 0)$, $(0, n)$

2. In each case, find the abscissa of the midpoint of the line segment whose endpoints are:

a. (4, 8), (10, 12) **b.** (6, −6), (12, −4) **c.** (−7, −3), (−5, −7)
d. (0, −5), (−7, 0) **e.** $(-4c, 2d)$, $(8c, 6d)$ **f.** $(a, 0)$, $(0, b)$

3. In each case, find the ordinate of the midpoint of the line segment whose endpoints are:

a. (6, 10), (8, 6) **b.** (8, −8), (14, −4) **c.** (6, −4), (−9, 1)
d. (−7, 6), (0, 0) **e.** $(a, 2b)$, $(3a, 4b)$ **f.** $(c, 0)$, $(0, d)$

4. In a circle, A and B are endpoints of a diameter, and P is the center of the circle. In each case, find the coordinates of P, given the following coordinates for A and B:

a. (3, 4), (7, 8) **b.** (−5, −2), (3, 7) **c.** (0, 0), (10, 4)
d. (8, 5), (−4, 5) **e.** (1, 8), (1, −14) **f.** $(a, 0)$, $(0, b)$

5. In $\triangle ABC$, the midpoints of sides $\overline{AB}$, $\overline{BC}$, and $\overline{CA}$ are points D, E, and F, respectively. In each case, find the coordinates of D, E, and F, given the following coordinates of the vertices:

a. $A(2, 8)$, $B(4, 12)$, $C(6, 4)$ **b.** $A(-5, 2)$, $B(7, 4)$, $C(3, -6)$
c. $A(8, 0)$, $B(0, 5)$, $C(4, 6)$ **d.** $A(-2, 0)$, $B(0, -8)$, $C(0, 0)$

6. In each case, determine the coordinates of the midpoints of the sides of a quadrilateral with vertices:

a. $A(0, 0)$, $B(10, 0)$, $C(7, 5)$, $D(3, 5)$ **b.** $P(-3, 3)$, $Q(11, 3)$, $R(7, 7)$, $S(1, 7)$

7. M is the midpoint of $\overline{CD}$. The coordinates of point C are (8, 4) and of point M are (8, 10). Find the coordinates of point D.

8. M is the midpoint of $\overline{AB}$. In each case, given the coordinates of points A and M, find the coordinates of point B.

a. $A(4, 3)$, $M(4, 9)$ **b.** $A(4, 8)$, $M(6, 10)$ **c.** $A(2, 6)$, $M(0, 3)$
d. $A(6, 10)$, $M(7, -2)$ **e.** $A(5, -1)$, $M(-1, 1)$ **f.** $A(a, 0)$, $M(2a, b)$

9. Given points $A(-4, 6)$, $B(2, 7)$, and $C(r, s)$. If B is the midpoint of $\overline{AC}$, find the value of r and the value of s.

10. In parallelogram $ABCD$, the coordinates of A are (6, 7), and the coodinates of C are (12, 3). What are the coordinates of the point of intersection of the diagonals?

11. Points $A(2, 3)$, $B(7, 5)$, $C(8, 8)$, and $D(3, 6)$ are the vertices of a quadrilateral.

a. Show that the diagonals of quadrilateral $ABCD$ bisect each other.
b. Show that $ABCD$ is a parallelogram.

12. Show that points $A(1, -3)$, $B(5, -2)$, $C(8, 2)$, and $D(4, 1)$ are the vertices of a parallelogram.

13. Show that the diagonals of the quadrilateral whose vertices are $A(3, 5)$, $B(6, 4)$, $C(7, 8)$, and $D(4, 12)$ do *not* bisect each other.

14. Points $A(4, 0)$, $B(14, 0)$, and $C(8, 6)$ are the vertices of $\triangle ABC$. Show that the length of the line segment that joins the midpoints of $\overline{CA}$ and $\overline{CB}$ is equal to one-half of AB.

15. Points $A(2, 2)$, $B(6, -6)$, $C(8, 2)$, and $D(4, 4)$ are the vertices of polygon $ABCD$.

a. Show that $\overline{LM}$, the line segment that joins the midpoints of $\overline{AD}$ and $\overline{DC}$, is congruent and parallel to $\overline{PN}$, the line segment that joins the midpoints of $\overline{BA}$ and $\overline{BC}$.
b. If $\overline{MN}$ and $\overline{LP}$ are also drawn, what type of quadrilateral is $LMNP$?

In 16–22, answers that are not whole numbers should be left in radical form.

16. Given quadrilateral $ABCD$, whose vertices are $A(0, 0)$, $B(6, 8)$, $C(16, 8)$, and $D(10, 0)$.

a. Using graph paper, draw quadrilateral $ABCD$.
b. If R is the midpoint of $\overline{AB}$, S the midpoint of $\overline{BC}$, and T the midpoint of $\overline{AD}$, find: (*1*) RS (*2*) ST (*3*) RT
c. Show that RST is a right triangle.

17. The vertices of quadrilateral $ABCD$ are $A(4, 0)$, $B(13, 3)$, $C(12, 6)$, and $D(3, 3)$.

a. Find the coordinates of the midpoint of $\overline{AC}$.
b. Find the coordinates of the midpoint of $\overline{BD}$. **c.** Show that $ABCD$ is a parallelogram.
d. Find the lengths of diagonals $\overline{AC}$ and $\overline{BD}$. **e.** Show that $ABCD$ is a rectangle.

18. The vertices of a triangle are $R(7, 1)$, $S(2, 1)$, and $T(4, 7)$. Find the length of the median from R to $\overline{ST}$.

19. In $\triangle ABC$, the vertices are $A(4, 8)$, $B(-2, 6)$, and $C(0, -4)$. Find the lengths of medians $\overline{AM}$, $\overline{BN}$, and $\overline{CP}$ in this triangle.

20. The vertices of $\triangle ABC$ are $A(0, 1)$, $B(6, 1)$, and $C(3, 5)$.

a. Show that $\triangle ABC$ is isosceles. **b.** Find the length of the altitude from C to $\overline{AB}$.

21. The vertices of $\triangle RST$ are $R(11, -1)$, $S(13, 10)$, and $T(3, 5)$.

a. Show that $\triangle RST$ is isosceles. **b.** Find the length of the altitude from S to $\overline{RT}$.

22. The vertices of $\triangle ABC$ are $A(3, 1)$, $B(9, 9)$, and $C(9, 1)$.

a. Show that M, the midpoint of $\overline{AB}$, is equidistant from A, B, and C.
b. Show that $\triangle ABC$ is a right triangle.

16-12 GRAPHING DIRECT VARIATION

Recall that, when two variables represent quantities that are *directly proportional* or that *vary directly*, the ratio of the two variables is a constant. For example, when lemonade is made from a frozen concentrate, the amount of lemonade, y, that is obtained varies directly as the amount of concentrate, x, that is used. If we can make 3 cups of lemonade by using 1 cup of concentrate, the relationship can be expressed as

$$\frac{y}{x} = \frac{3}{1} \quad \text{or} \quad y = 3x$$

The *constant of variation* is 3.

The equation $y = 3x$ can be represented by a line in the coordinate plane, as shown in the graph below.

x	$3x$	y
0	3(0)	0
1	3(1)	3
2	3(2)	6
3	3(3)	9

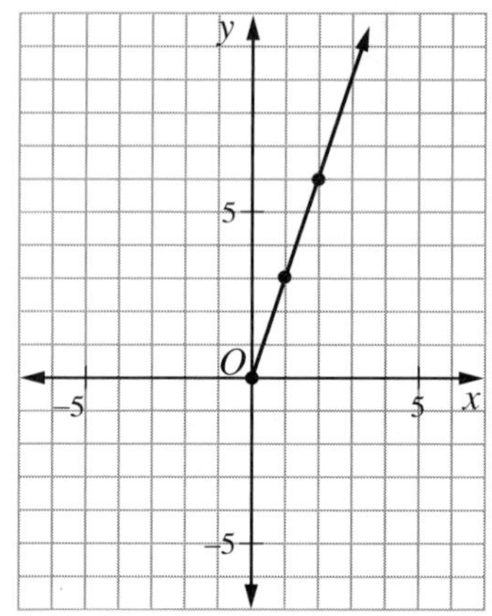

Here, the replacement set for x and y is the set of positive numbers and 0. Thus, the graph includes only points in the first quadrant.

In our example, if 0 cup of frozen concentrate is used, 0 cup of lemonade can be made. Thus, the ordered pair (0,0) is a member of the solution set of $y = 3x$.

Using any two points from the table above, for example, (0,0) and (1,3), we can write:

$$\frac{\Delta y}{\Delta x} = \frac{3 - 0}{1 - 0} = \frac{3}{1} = 3$$

Thus, we see that the slope of the line is also the constant of variation.

The unit of measure for the lemonade and for the frozen concentrate is the same, namely, cups. However, no unit of measure is associated with the ratio which, in this case, is $\frac{3}{1}$ or 3.

$$\frac{\text{lemonade}}{\text{concentrate}} = \frac{3 \text{ cups}}{1 \text{ cup}} = \frac{3}{1} \text{ or } \frac{6 \text{ cups}}{2 \text{ cups}} = \frac{3}{1} \text{ or } \frac{7.5 \text{ cups}}{2.5 \text{ cups}} = \frac{3}{1}$$

There are many applications of direct variation in business and science, and it is important to recognize how the choice of unit can affect the constant of variation. For example, if a machine is used to pack boxes of cereal in cartons, the rate at which the machine works can be expressed in cartons per minute or in cartons per second. If the machine fills a carton every 6 seconds, it will fill 10 cartons in 1 minute. The rate can be expressed as:

$$\frac{1 \text{ carton}}{6 \text{ seconds}} = \frac{1}{6} \text{ carton/second or } \frac{10 \text{ cartons}}{1 \text{ minute}} = 10 \text{ cartons/minute}$$

As shown below, each of these rates can be represented by a graph, where the rate, or constant of variation, is the slope of the line.

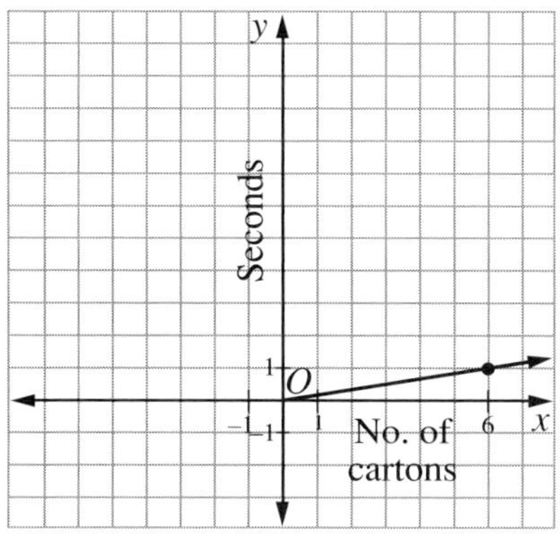

Cartons per second

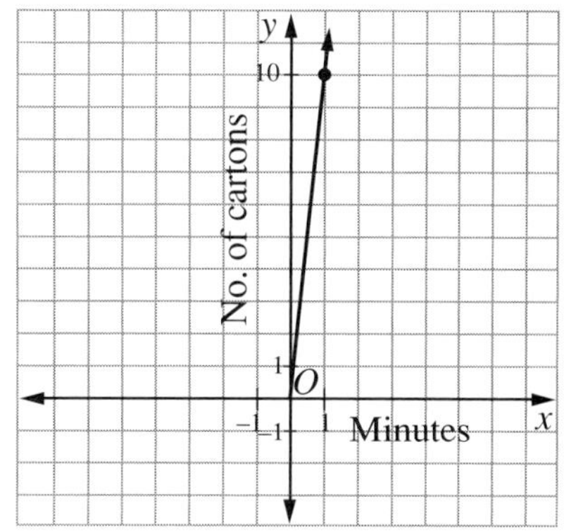

Cartons per minute

The legend of the graph, that is, the units in which the rate is expressed, must be clearly stated if the graph is to be meaningful.

EXAMPLE

The amount of flour needed to make a white sauce varies directly as the amount of milk used. To make a white sauce, a chef used 2 cups of flour and 8 cups of milk. Write an equation and draw the graph of the relationship.

Solution Let x = number of cups of milk, and y = number of cups of flour.

Then:

$$\frac{y}{x} = \frac{2}{8}$$

$$8y = 2x$$

$$y = \frac{2}{8}x$$

$$= \frac{1}{4}x$$

x	$\frac{1}{4}x$	y
2	$\frac{1}{4}(2)$	$\frac{1}{2}$
4	$\frac{1}{4}(4)$	1
6	$\frac{1}{4}(6)$	$\frac{3}{2}$

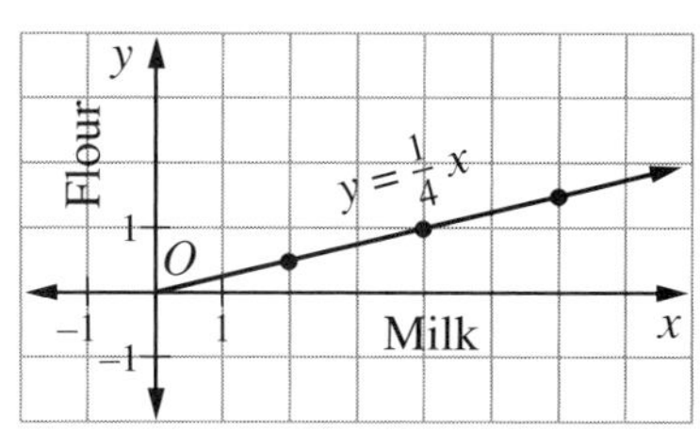

EXERCISES WITH OPEN-RESPONSE PROBLEMS

In 1–9, y varies directly as x. In each case: **a.** What is the constant of variation? **b.** Write an equation for y in terms of x. **c.** Using an appropriate scale, draw the graph of the equation written in part **b**. **d.** What is the slope of the line drawn in part **c**?

1. The perimeter of a square (y) is 12 centimeters when the length of a side of the square (x) is 3 centimeters.
2. Jeanne can type 90 words (y) in 2 minutes (x).
3. A printer can type 160 characters (y) in 10 seconds (x).
4. A cake recipe uses 2 cups of flour (y) to $1\frac{1}{2}$ cups of sugar (x).
5. The length of a photograph (y) is 12 centimeters when the length of the negative from which it is developed (x) is 1.2 centimeters.
6. There are 20 slices (y) in 12 ounces of bread (x).
7. Twelve slices of cheese (y) weigh 8 ounces (x).
8. Willie averages 3 hits (y) for every 12 times at bat (x).
9. There are about 20 calories (y) in three crackers (x).
10. If a car travels at a constant rate of speed, the distance that it travels varies directly as time. If a car travels 75 miles in 2.5 hours, it will travel 110 feet in 2.5 seconds.
    Find the constant of variation **a.** in miles per hour. **b.** in feet per second.

16-13 GRAPHING FIRST-DEGREE INEQUALITIES IN TWO VARIABLES

When a line is graphed in the coordinate plane, the line is a ***plane divider*** because it separates the plane into two regions called ***half-planes***. One of these regions is a half-plane on one side of the line; the other is a half-plane on the other side of the line.

Let us consider, for example, the horizontal line $y = 3$ as a plane divider. As shown in the graph below, the line $y = 3$ and the two half-planes that it forms determine three sets of points:

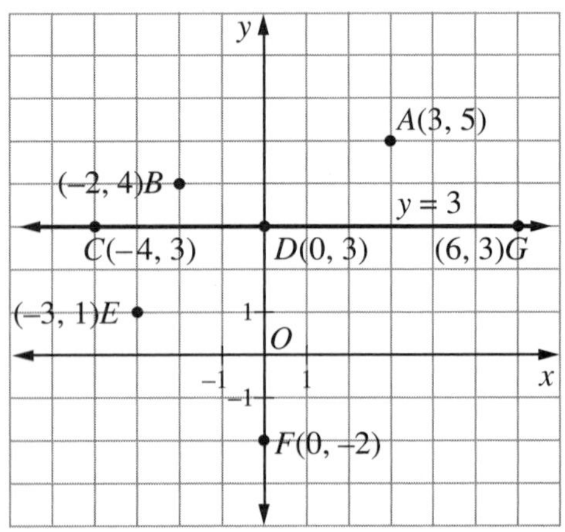

1. The half-plane above the line $y = 3$ is the set of all points whose y-coordinates are greater than 3, that is, $y > 3$. For example, at point A, $y = 5$; at point B, $y = 4$.
2. The line $y = 3$ is the set of all points whose y-coordinates are equal to 3. For example, $y = 3$ at each point $C(-4, 3)$, $D(0, 3)$, and $G(6, 3)$.
3. The half-plane below the line $y = 3$ is the set of all points whose y-coordinates are less than 3, that is, $y < 3$. For example, at point E, $y = 1$; at point F, $y = -2$.

Together, the three sets of points form the entire plane.

To graph an inequality in the coordinate plane, we proceed as follows:

1. On the plane, represent the plane divider, for example, $y = 3$, by a *dashed line* to show that this divider does not belong to the graph of the half-plane.

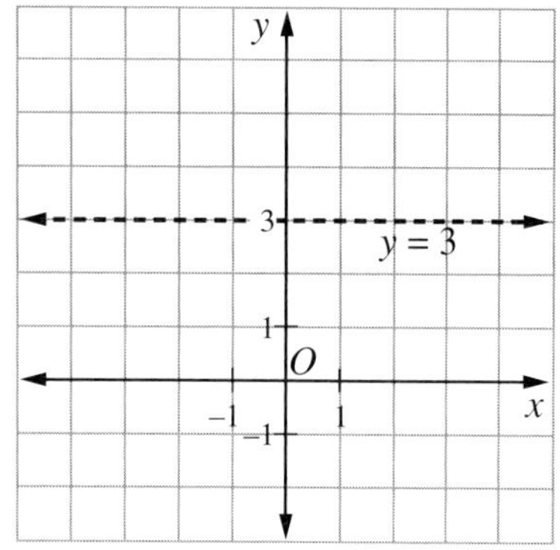

2. *Shade the region* of the half-plane whose points satisfy the inequality. To graph $y > 3$, shade the region *above* the plane divider.

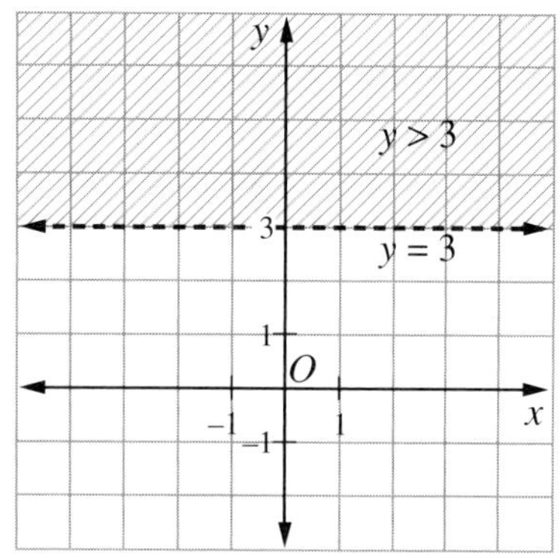

Graph of $y > 3$

To graph $y < 3$, shade the region *below* the plane divider.

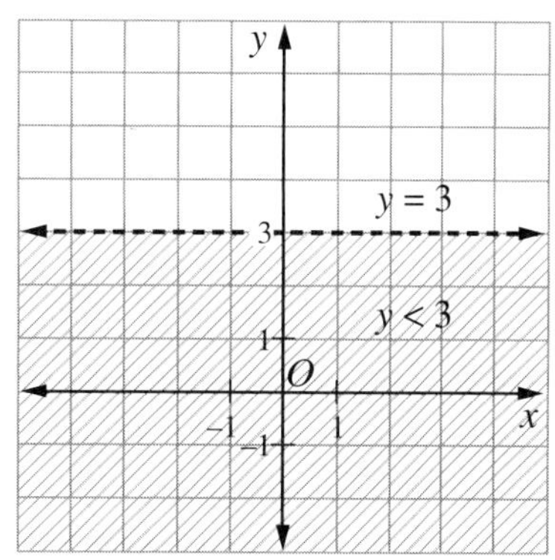

Graph of $y < 3$

Let us consider another example, where the plane divider is not a horizontal line. To graph the inequality $y > 2x$ or $y < 2x$, we use a dashed line to indicate that the line $y = 2x$ is not a part of the graph. This dashed line acts as a boundary line for the half-plane being graphed.

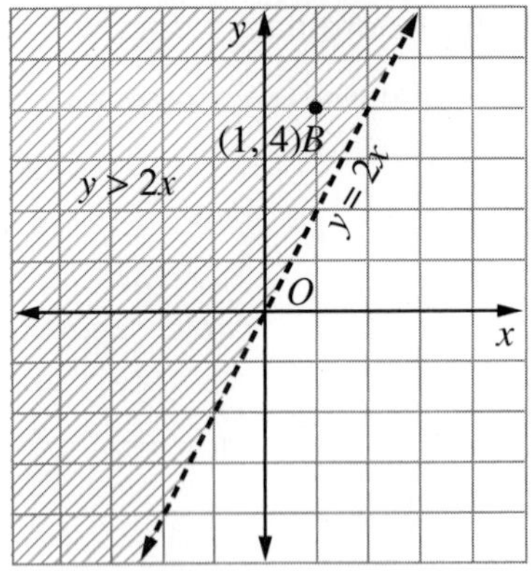

Graph of $y > 2x$

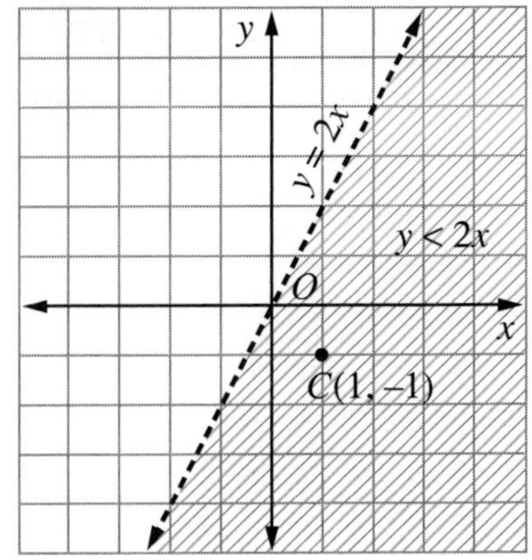

Graph of $y < 2x$

The graph of $y > 2x$ is the shaded half-plane above the line $y = 2x$. It is the set of all points in which the y-coordinate is *greater than* twice the x-coordinate.

The graph of $y < 2x$ is the shaded half-plane below the line $y = 2x$. It is the set of all points in which the y-coordinate is *less than* twice the x-coordinate.

A condition involving both inequality and equality, such as $y \geq 2x$, means $y > 2x$ or $y = 2x$.

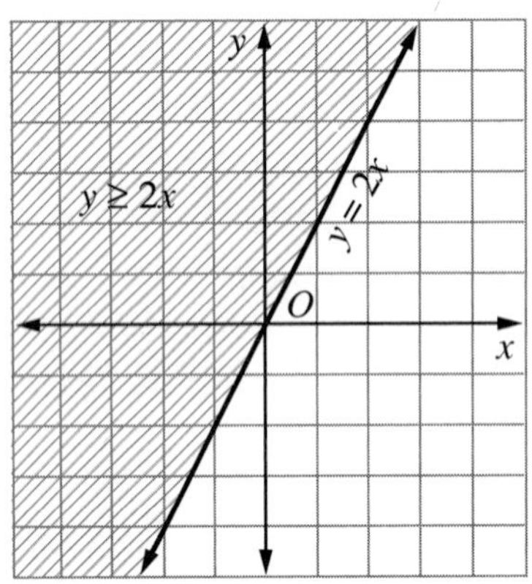

Graph of $y \geq 2x$

The graph of $y \geq 2x$ is the *union* of the two disjoint sets of points. To indicate that $y = 2x$ is part of the graph of $y \geq 2x$, we draw the graph of $y = 2x$ as a solid line. We shade the region above $y = 2x$ to indicate that $y > 2x$ is part of the graph of $y \geq 2x$.

In general:

- **When the equation of a line is written in the form $y = mx + b$, the half-plane above the line is the graph of $y > mx + b$ and the half-plane below the line is the graph of $y < mx + b$.**

To check whether the correct half-plane has been chosen as the graph of a linear inequality, we select any point in that half-plane. If the selected point satisfies the inequality, every point in that half-plane satisfies the inequality. On the other hand, if the point chosen does not satisfy the inequality, then the other half-plane is the graph of the inequality.

EXAMPLES

1. Graph the inequality $y - 2x \geq 2$.

How to Proceed: | *Solution:*

(1) Transform the sentence into one having y as the left member:

$$y - 2x \geq 2$$
$$y \geq 2x + 2$$

(2) Make a table of values to graph the plane divider, $y = 2x + 2$, and draw it as a dashed line.

x	$2x + 2$	y
-1	$-2 + 2$	0
0	$0 + 2$	2
1	$2 + 2$	4

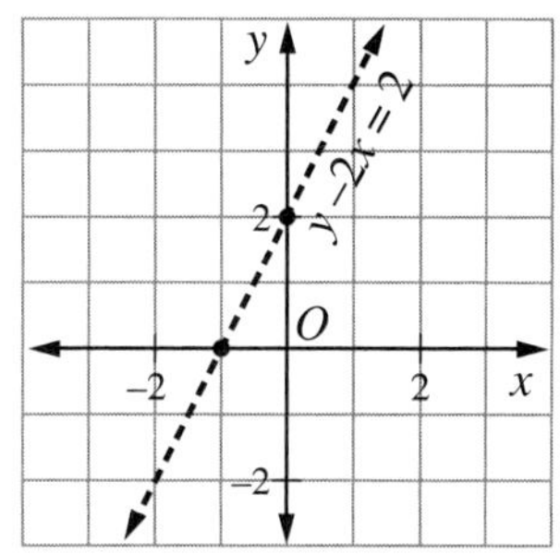

(3) Shade the half-plane above the line: This region and the line are the required graph: the half-plane is the graph of $y - 2x > 2$, and the line is the graph of $y - 2x = 2$. Note that the line is now drawn solid to show that it is part of the graph.

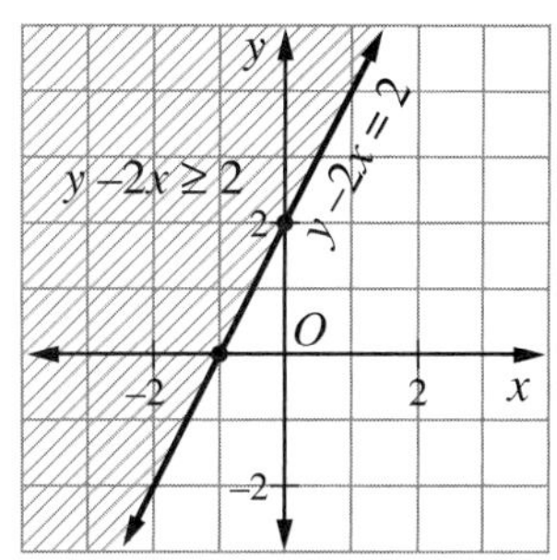

(4) Check the solution. Choose any point in the half-plane selected as the solution to see whether it satisfies the original inequality, $y - 2x \geq 2$:

Select point (0, 5), which is in the shaded region.

$$y - 2x \geq 2$$
$$(5) - 2(0) \geq 2 \ (?)$$
$$5 \geq 2 \ \text{(True)}$$

The above graph is the graph of $y - 2x \geq 2$.

2. Graph each of the following sentences in the coordinate plane:

a. $x > 1$ **b.** $x \leq 1$ **c.** $y \geq 1$ **d.** $y < 1$

Solutions

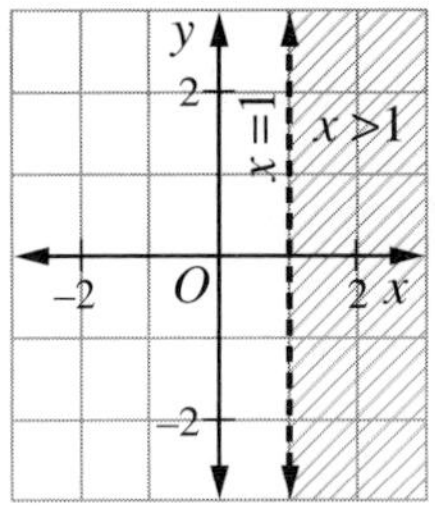

a. $x > 1$

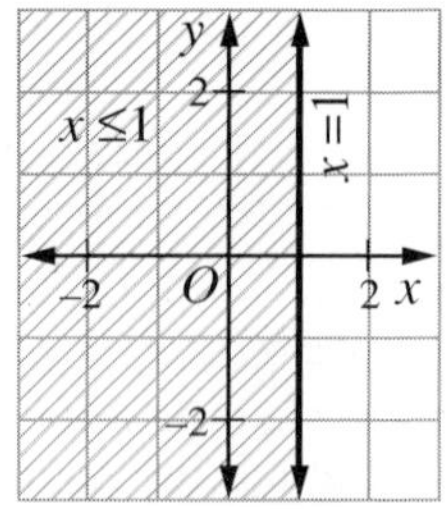

b. $x \leq 1$

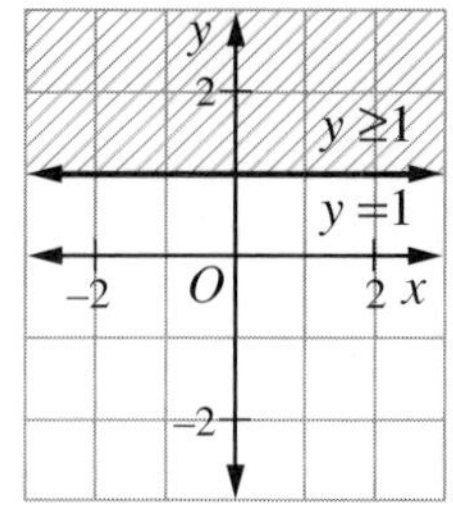

c. $y \geq 1$

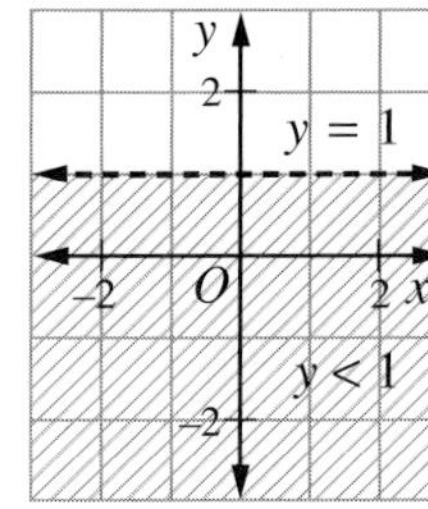

d. $y < 1$

EXERCISES

In 1–6, transform each sentence into one whose left member is y.

1. $y - 2x > 0$ **2.** $5x > 2y$ **3.** $y - x \geq 3$

4. $2x + y \leq 0$ **5.** $3x - y \geq 4$ **6.** $4y - 3x \leq 12$

In 7–30, graph each sentence in the coordinate plane.

7. $x > 4$ **8.** $x \leq -2$ **9.** $y > 5$ **10.** $y \leq -3$

11. $x \geq 6$ **12.** $y \leq 0$ **13.** $y > 4x$ **14.** $y \leq 3x$

15. $y < x - 2$ **16.** $y \geq \frac{1}{2}x + 3$ **17.** $x + y < 4$ **18.** $x + y \geq 4$

19. $x + y \leq -3$ **20.** $y - x \geq 5$ **21.** $x - y \leq -1$ **22.** $x - 2y \leq 4$

23. $2x + y - 4 \leq 0$ **24.** $y - x + 6 > 0$ **25.** $2y - 6x > 0$ **26.** $3x + 4y \leq 0$

27. $2x - 3y \geq 6$ **28.** $\frac{1}{2}y > \frac{1}{3}x + 1$ **29.** $9 - x \geq 3y$ **30.** $x < 2y + 7$

In 31–33: **a.** Write each verbal sentence as an open sentence. **b.** Graph each open sentence in the coordinate plane.

31. The ordinate of a point is equal to or greater than 3 more than the abscissa.

32. The sum of the abscissa and the ordinate of a point is less than or equal to 5.

33. The ordinate of a point decreased by 3 times the abscissa is greater than or equal to 2.

CHAPTER SUMMARY

The solutions of equations or inequalities in two variables are ordered pairs of numbers. The set of points whose coordinates make an equation or inequality true is the graph of that equation or inequality. The ***graph*** of a ***linear function*** of the form $Ax + By + C = 0$ is a straight line.

A linear function can be written in the ***slope-intercept form*** $y = mx + b$, where m is the slope and b is the y-intercept of the line that is the graph of the function. A line parallel to the x-axis has a slope of 0, and a line parallel to the y-axis has no slope. The ***slope*** of a line is the ratio of the change in the vertical direction to the change in the horizontal direction. If (x_1, y_1) and (x_2, y_2) are two points on a line, the slope of the line is $m = \frac{y_2 - y_1}{x_2 - x_1}$.

Parallel lines have equal slopes. If two ***nonvertical lines*** are perpendicular, the slope of one is the negative reciprocal of the slope of the other.

The equation of a line can be written using the coordinates of two points on the line or the coordinates of one point and the slope of the line.

In two-valued statistics, a relationship between two different sets of data is studied and can be graphed in a ***scatter plot***. In some, but not all, cases, one or more straight lines may be found and a linear equation may be written to approximate the relationship between data sets. This line is often called the ***line of best fit*** or a ***trend line***.

In a coordinate plane, the ***distance formula*** is used to find the measure of any line segment:

$$AB = d = \sqrt{(x_2 - x_1)^2 + (y_2 - y_1)^2}$$

If $A(x_1, y_1)$ and $B(x_2, y_2)$ are two points in the coordinate plane, then the coordinates of the ***midpoint*** of $\overline{AB}$ are $\left(\frac{x_1 + x_2}{2}, \frac{y_1 + y_2}{2}\right)$.

If y varies directly as x, the ratio of y to x is a constant. ***Direct variation*** can be represented by a line through the origin whose slope is the ***constant of variation***.

The graph of $y = mx + b$ separates the plane into two ***half-planes***. The half-plane above the graph of $y = mx + b$ is the graph of $y > mx + b$, and the half-plane below the graph of $y = mx + b$ is the graph of $y < mx + b$.

VOCABULARY

16-1 Function

16-2 Graph Linear equation

16-3 y-intercept x-intercept

16-4 Slope

16-5 Parallel lines Perpendicular lines Negative reciprocal

16-6	Slope-intercept form of a linear equation
16-9	Two-valued statistics Scatter plot Line of best fit
16-10	Pythagorean Theorem Distance formula
16-11	Midpoint formula
16-12	Directly proportional Vary directly Constant of variation
16-13	Plane divider Half-plane

REVIEW EXERCISES WITH OPEN-RESPONSE PROBLEMS

1. What is the slope of the graph of $y = -2x + 5$?

2. Solve the following equation for y in terms of x: $3x - 2y = 12$.

3. Write an equation of the line whose slope is -1 and whose y-intercept is 7.

4. What is the slope of the line that passes through points (4, 5) and (6, 1)?

In 5–10, graph each equation or inequality.

5. $y = -x + 2$ **6.** $y = 3$ **7.** $y = \frac{2}{3}x$

8. $x + 2y = 8$ **9.** $y - x > 2$ **10.** $2x - y \geq 4$

In 11–15, refer to the coordinate graph to answer each question.

11. What is the slope of line k?

12. What is the y-intercept of line k?

13. What is the equation of line m?

14. Write an equation of the line that is parallel to line k and passes through the origin.

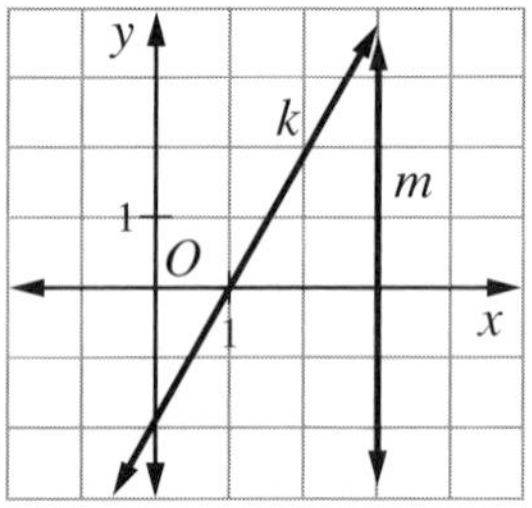

15. Write an equation of the line that is perpendicular to line k and passes through the point (0, 2).

16. If point $(d, 3)$ lies on the graph of $3x - y = 9$, find the value of d.

17. The Tiny Tot Day Care Center has changed its rates. It now charges $150 a week for children who stay at the center between 8:00 A.M. and 5:00 P.M. If a child is not picked up by 5:00 P.M., the center charges an additional $2.00 per hour or any fractional part of an hour.

a. Write an equation for the cost of day care for a week, y, in terms of the number of hours, x, that a child stays beyond 5:00 P.M.

b. What is the charge for 1 week for a child who was picked up at the following times: Monday at 5:00, Tuesday at 5:10, Wednesday at 6:00, Thursday at 5:00, and Friday at 5:25?

18. Aurora buys oranges every week. The accompanying table indicates the weights and the cost of her last 10 purchases of oranges.

Weight (lb)	2.2	1.2	3.6	4.5	1.0	2.5	1.8	5.0	3.5	1.7
Cost ($)	1.22	0.60	1.04	1.58	0.50	0.89	0.95	1.88	1.46	0.70

a. Draw a graph of the data. Let the horizontal axis represent the weights of the oranges, and the vertical axis represent the costs.
b. Find the mean number of pounds of oranges purchased.
c. Find the mean weekly cost of the oranges.
d. On the graph, draw a line that approximates the data in the table.
e. Write an equation of the line drawn in part **d.**

19. Each time Raphael put gasoline into his car, he recorded the number of gallons of gas he needed to fill the tank and the number of miles driven since the last fillup. The chart below shows his record for one month.

Gallons of Gasoline	12	8.5	10	4.5	13
Number of Miles	370	260	310	140	420

a. Plot points on a graph to represent the information in the chart. Let the horizontal axis represent the number of gallons of gasoline and the vertical axis the number of miles driven.
b. Find the mean number of gallons of gasoline per fillup.
c. Find the mean number of miles driven between fillups.
d. Locate a point on the graph that represents the mean number of gallons of gasoline per fillup.
e. Draw, through the point that you plotted in **d**, a line that best represents the information in the chart.
f. Write an equation for the line that you drew in **e**.
g. Raphael drove 200 miles since his last fillup. How many gallons of gasoline should he expect to need to fill the tank?

In 20–26, in each case, select the numeral preceding the correct answer.

20. Which point does *not* lie on the graph of $3x - y = 9$?
(1) $(1, -6)$ (2) $(2, 3)$ (3) $(3, 0)$ (4) $(0, -9)$

21. Which ordered pair is in the solution set of $y < 2x - 4$?
(1) $(0, -5)$ (2) $(2, 0)$ (3) $(3, 3)$ (4) $(0, 2)$

22. Which equation has a graph parallel to the graph of $y = 5x - 2$?
(1) $y = -5x$ (2) $y = 5x + 3$ (3) $y = -2x$ (4) $y = 2x - 5$

23. The graph of $2x + y = 8$ intersects the x-axis at point
(1) $(0, 8)$ (2) $(8, 0)$ (3) $(0, 4)$ (4) $(4, 0)$

24. Which equation represents a line perpendicular to the line $y = 4x - 3$?
(1) $y = 4x + 2$ (2) $y = -\frac{1}{4}x - 2$ (3) $y = \frac{1}{4}x + 1$ (4) $y = -4x + 5$

25. What is the slope of the graph of the equation $y = 4$?
(1) 1 (2) 0 (3) -4 (4) 4

26. In which ordered pair is the abscissa 3 more than the ordinate?
(1) (1, 4) (2) (1, 3) (3) (3, 1) (4) (4, 1)

27. **a.** Plot points $A(5,-2)$, $B(3, 3)$, $C(-3, 3)$, and $D(-5,-2)$.
b. Draw polygon $ABCD$.
c. What kind of polygon is $ABCD$?
d. Find DA and BC.
e. Find the length of the altitude from C to $\overline{DA}$.
f. Find the area of $ABCD$.
g. Write an equation for the line of symmetry of $ABCD$.

28. The vertices of right triangle ABC are $A(1, 2)$, $B(4, 2)$, and $C(4, 4)$. Find the length of the hypotenuse of $\triangle ABC$.

29. What is the midpoint of the segment whose endpoints are $(-2, 7)$ and $(-6, -11)$?

30. Midpoint M of $\overline{AB}$ has coordinates (5, 4). If the coordinates of A are (2, 8), what are the coordinates of B?

31. The vertices of parallelogram $ABCD$ are $A(-3, -1)$, $B(3, -1)$, $C(7, 5)$, and $D(1, 5)$. Find the coordinates of the intersection of diagonals $\overline{AC}$ and $\overline{BD}$.

32. Three vertices of rectangle $ABCD$ are $A(-2, 2)$, $B(7, -2)$, $C(7, 6)$. Find the coordinates of vertex D.

33. Use coordinate geometry to show that the diagonals of a rectangle are equal. (*Hint:* Draw a rectangle in the coordinate plane with vertices on points of the grid.)

CUMULATIVE REVIEW WITH OPEN-RESPONSE PROBLEMS

1. Solve then check:
a. $\dfrac{x}{x-3} = \dfrac{9}{8}$ **b.** $4(x - 2) + 12 = 19 - x$

2. Write the inverse of "If today is not Monday, then I play tennis."

Exploration

Work with one or two classmates to take a survey of foot lengths and heights, in inches, of 25 students. Record the measurements in a chart.

a. Plot points on a graph to represent the information in the chart. Let the horizontal axis represent foot lengths and the vertical axis represent heights.
b. Find the mean number of inches for the foot lengths.
c. Find the mean number of inches for the heights.
d. Using the answers to parts **b** and **c** as the coordinates of a point, draw a line that approximates the data.
e. Write an equation of the line that you drew in part **d**.
f. Based on the equation, make predictions of foot lengths for three heights that are not points on the graph.

Chapter 17

Systems of Linear Open Sentences in Two Variables

Architects often added an outdoor stairway to a building as a design feature or as an approach to an entrance above ground level. But stairways are an obstacle to handicapped persons, and most buildings are now approached by means of ramps in addition to or in place of stairways.

In designing a ramp, the architect must keep the slant or slope gradual enough to easily accommodate a wheelchair. If the slant is too gradual, however, the ramp may become inconveniently long or may even require turns to fit it into the available space. The architect will also want to include design features that harmonize with the rest of the building and its surroundings.

Solving a problem such as the design of a ramp often involves determining the solution of several equations or inequalities at the same time.

17-1 USING A GRAPH TO SOLVE A SYSTEM OF LINEAR EQUATIONS

Consistent Equations

The perimeter of a rectangle is 10 feet. When we let x represent the width of the rectangle and y represent the length, the equation $2x + 2y = 10$ expresses the perimeter of the rectangle. This equation, which can be simplified to $x + y = 5$, has infinitely many solutions.

If we also know that the length of the rectangle is 1 foot more than the width, the dimensions of the rectangle can be represented by the equation $y = x + 1$. We want both of the equations, $x + y = 5$ and $y = x + 1$, to be true for the same pair of numbers. The two equations are called a ***system of simultaneous equations***.

Let x = width of rectangle,
and y = length of rectangle.
Then:
Perimeter:

$$2x + 2y = 10$$

$$x + y = 5$$

Measures of sides:

$$y = x + 1$$

System of simultaneous equations:

$$x + y = 5$$

$$y = x + 1$$

The solution of a system of simultaneous equations in two variables is an ordered pair of numbers that satisfies both equations.

The graphs of $x + y = 5$ and $y = x + 1$, drawn in a coordinate plane using the same set of axes, are shown at the right. The possible solutions of $x + y = 5$ are all coordinates of the points on the line $x + y = 5$. The possible solutions of $y = x + 1$ are all coordinates of the points on the line $y = x + 1$. The coordinates of the point of intersection, (2, 3), are a solution of both equations. The ordered pair (2, 3) is a solution of the system of simultaneous equations.

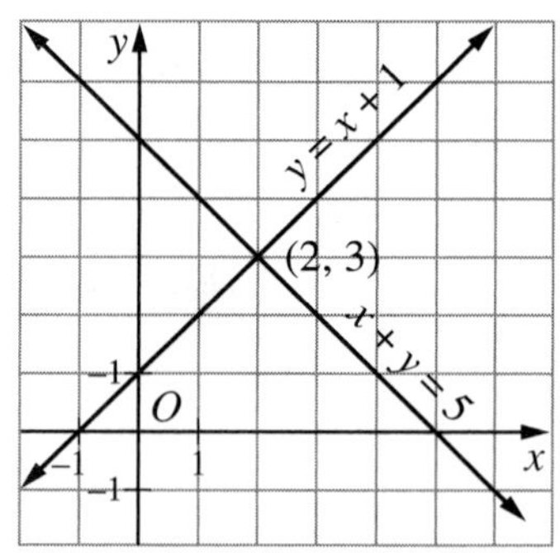

Check: Substitute (2, 3) in both equations:

$$x + y = 5 \qquad\qquad y = x + 1$$

$$2 + 3 \stackrel{?}{=} 5 \qquad\qquad 3 \stackrel{?}{=} 2 + 1$$

$$5 = 5 \text{ (True)} \qquad\qquad 3 = 3 \text{ (True)}$$

Since two straight lines can intersect in no more than one point, there is no other ordered pair that is a solution of this system. Therefore, $x = 2$ and $y = 3$; or (2, 3) is the solution of the system of equations. The width of the rectangle is 2 feet, and the length of the rectangle is 3 feet.

When two lines are graphed in the same coordinate plane on the same set of axes, one and only one of the following three possibilities can occur. The pair of lines will:

1. intersect in one point and have one ordered number pair in common;
2. be parallel and have no ordered number pairs in common;
3. coincide, that is, be the same line with an infinite number of ordered number pairs in common.

If a system of linear equations such as $x + y = 5$ and $y = x + 1$ has one common solution, it is called a ***system of consistent equations***.

Inconsistent Equations

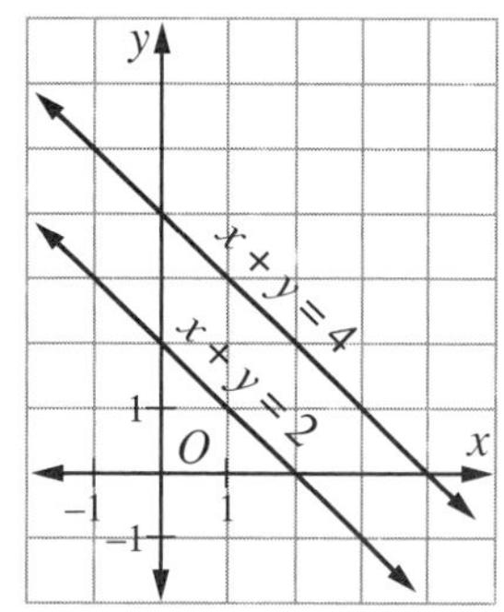

Sometimes, as shown at the right, when two linear equations are graphed in a coordinate plane using the same set of axes, the lines are parallel and fail to intersect, as in the case of $x + y = 2$ and $x + y = 4$. There is no common solution for the system of equations $x + y = 2$ and $x + y = 4$. It is obvious that there can be no ordered number pair (x, y) such that the sum of those numbers, $x + y$, is both 2 and 4. Since the solution set of the system has no members, it is the empty set.

If a system of linear equations such as $x + y = 2$ and $x + y = 4$ has no common solution, it is called a ***system of inconsistent equations***. The graphs of two inconsistent linear equations are lines that have equal slopes or lines that have no slopes. Such lines will be parallel.

Dependent Equations

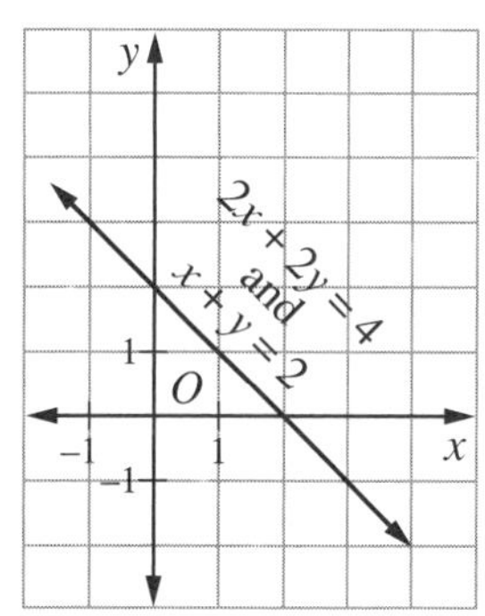

Sometimes, as shown at the right, when two linear equations are graphed in a coordinate plane using the same set of axes, the graphs turn out to be the same line; that is, they coincide. This happens in the case of the equations $x + y = 2$ and $2x + 2y = 4$. Every one of the infinite number of solutions of $x + y = 2$ is also a solution of $2x + 2y = 4$. Thus, we see that $2x + 2y = 4$ and $x + y = 2$ are equivalent equations with identical solutions. We note that, when both sides of the equation $2x + 2y = 4$ are divided by 2, the result is $x + y = 2$.

If a system of two linear equations, for example, $x + y = 2$ and $2x + 2y = 4$, is such that every solution of one of the equations is also a solution of the other, it is called a ***system of dependent equations***. The graphs of two dependent linear equations are the same line.

PROCEDURE. To solve a pair of linear equations graphically:

1. Graph one equation in a coordinate plane.
2. Graph the other equation using the same set of coordinate axes.
3. Find the common solution, that is, the ordered numbers pair associated with the point of intersection of the two graphs.
   (*Exceptions:* The graphs of inconsistent equations do not intersect; their solution is the empty set. The graphs of dependent equations are the same line; their solution is the set of coordinates of all points on that line).
4. Check the solution by verifying that the ordered pair satisfies both equations.

EXAMPLE

Solve graphically and check:

$$2x + y = 8$$
$$y - x = 2$$

Solution (1) Select three values of x and substitute in the equation $y = -2x + 8$ to find the corresponding values of y. Then graph $2x + y = 8$, or $y = -2x + 8$.

x	$-2x + 8$	y
1	$-2(1) + 8$	6
3	$-2(3) + 8$	2
4	$-2(4) + 8$	0

(2) Graph $y - x = 2$, or $y = x + 2$.

x	$x + 2$	y
0	$0 + 2$	2
1	$1 + 2$	3
2	$2 + 2$	4

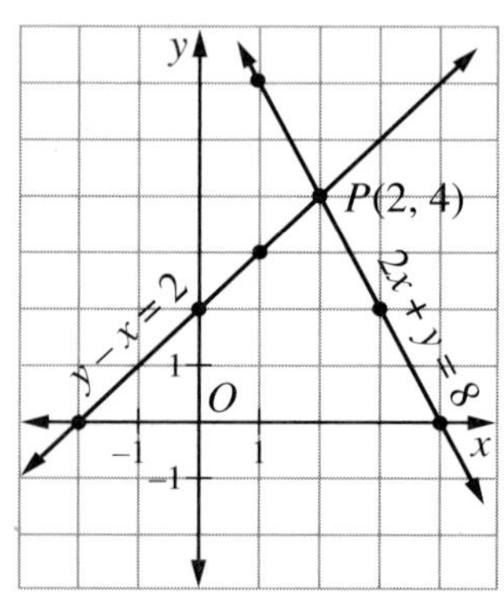

(3) Read the coordinates of the point of intersection: $P(2, 4)$.

(4) *Check:* Substitute (2, 4) in both equations:

$$2x + y = 8 \qquad y - x = 2$$
$$2(2) + 4 \stackrel{?}{=} 8 \qquad 4 - 2 \stackrel{?}{=} 2$$
$$8 = 8 \text{ (True)} \qquad 2 = 2 \text{ (True)}$$

Answer: (2, 4), or $x = 2$, $y = 4$, or the solution set $\{(2, 4)\}$

EXERCISES WITH OPEN-RESPONSE PROBLEMS

In 1–30, solve each system of equations graphically, and check.

1. $y = 3x - 3$; $y = 2x$

2. $y = 2x + 5$; $y = x + 4$

3. $y = -2x + 3$; $y = \frac{1}{2}x + 3$

4. $x + y = 7$; $x - y = 1$

5. $x + y = 4$; $x - y = 0$

6. $x + y = -4$; $x - y = 6$

7. $x + y = 1$; $x + 3y = 9$

8. $y = 2x + 1$; $x + 2y = 7$

9. $x - y = 4$; $x + 2y = 10$

10. $x + 2y = 17$; $y = 2x + 1$

11. $y = 3x$; $2x + y = 10$

12. $x + 3y = 9$; $x = 3$

13. $y - x = -2$; $x - 2y = 4$

14. $3x + y = 6$; $y = 3$

15. $4x - y = 9$; $2x + y = 12$

16. $y = 2x + 4$; $x = y - 5$

17. $x + y = 3$; $2x - y = -9$

18. $y - 3x = 12$; $y = -3$

19. $2x - y = -1$; $x = y + 1$

20. $3x + y = -9$; $x + 3y = -11$

21. $x = 3$; $y = 4$

22. $y = \frac{1}{3}x - 3$; $2x - y = 8$

23. $3x + y = 13$; $x + 6y = -7$

24. $x = 0$; $y = -5$

25. $2x = y + 9$; $6x + 3y = 15$

26. $5x - 3y = 9$; $5y = 13 - x$

27. $x = 0$; $y = 0$

28. $x + y + 2 = 0$; $x = y - 8$

29. $y + 2x + 6 = 0$; $y = 2x$

30. $7x - 4y + 7 = 0$; $3x - 5y + 3 = 0$

In 31–36, in each case: **a.** Graph both equations. **b.** State whether the system is consistent, inconsistent, or dependent.

31. $x + y = 1$; $x + y = 3$

32. $x + y = 5$; $2x + 2y = 10$

33. $y = 2x + 1$; $y = 3x + 3$

34. $2x - y = 1$; $2y = 4x - 2$

35. $y - 3x = 2$; $y = 3x - 2$

36. $x + 4y = 6$; $x = 2$

37. **a.** Are there any ordered number pairs that satisfy both the equations $2x + y = 7$ and $2x = 5 - y$? **b.** Explain your answer.

38. **a.** Are there any ordered number pairs that satisfy the equation $y - x = 4$ but do *not* satisfy the equation $2y = 8 + 2x$? **b.** Explain your answer.

In 39–42, in each case: **a.** Write a system of two first-degree equations involving the two variables x and y that represent the conditions stated in the problem. **b.** Solve the system graphically.

39. The sum of two numbers is 8. Their difference is 2. Find the numbers.

40. The sum of two numbers is 3. The larger number is 5 more than the smaller number. Find the numbers.

41. The perimeter of a rectangle is 12 meters. Its length is twice its width. Find the dimensions of the rectangle.

42. The perimeter of a rectangle is 14 centimeters. Its length is 3 centimeters more than its width. Find the length and the width.

43. **a.** The U-Drive-It car rental agency rents cars for \$50 a day with free unlimited mileage. Write an equation to show the cost of renting a car from U-Drive-It for 1 day, y, if the car is driven for x miles.

b. The Safe Travel car rental agency rents cars for \$30 a day plus \$0.20 a mile. Write an equation to show the cost of renting a car from Safe Travel for 1 day, y, if the car is driven for x miles.

c. Draw, on the same set of axes, the graphs of the equations written in parts **a** and **b**.

d. If Greg will drive the car he rents for 200 miles, which agency offers the less expensive car?

e. If Sarah will drive the car she rents for 50 miles, which agency offers the less expensive car?

f. If Philip finds that the price for both agencies will be the same, how far is he planning to drive the car?

17-2 USING ADDITION TO SOLVE A SYSTEM OF LINEAR EQUATIONS

In the preceding section, graphs were used to find solutions of systems of simultaneous equations. Since most of the solutions were integers, the values of x and y were easily read from the graphs. However, it is not always possible to read values accurately from a graph.

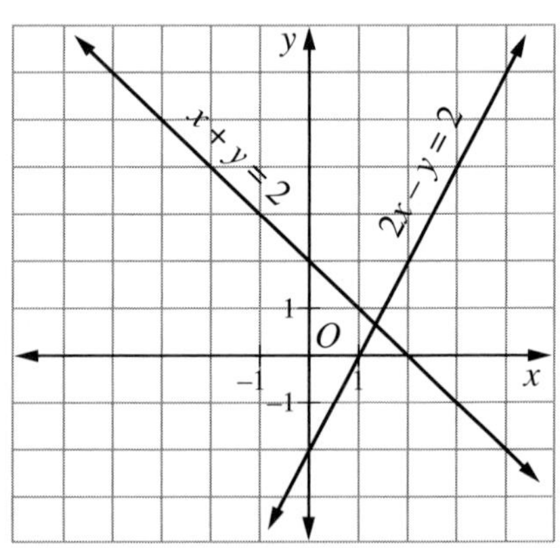

For example, the graphs of the system of equations $2x - y = 2$ and $x + y = 2$ are shown at the right. The solution of this system of equations is not a pair of integers. We could approximate the solution and then determine whether our approximation was correct by checking. However, there are other, more direct methods of solution.

Algebraic methods can be used to solve a system of linear equations in two variables. Solutions by these methods often take less time and lead to more accurate results than the graphic method used in Section 17-1.

Systems of equations that have the same solution set are called ***equivalent systems***. For example, the following two systems are equivalent systems because they have the same solution set, $\left\{\left(\frac{4}{3}, \frac{2}{3}\right)\right\}$:

System A	System B
$x + y = 2$	$x = \frac{4}{3}$
$2x - y = 2$	$y = \frac{2}{3}$

To solve a system of linear equations such as system A, whose solution set is not obvious, we transform it into an equivalent system such as system B, whose solution set is obvious. To do this, we make use of the properties of equality.

System A can be solved by the ***addition method*** as follows:

(1) Since the coefficients of y in the two equations are additive inverses, add the equations to obtain one equation in one variable:

$$\begin{aligned} 2x - y &= 2 \\ x + y &= 2 \\ \hline 3x \quad &= 4 \end{aligned}$$

(2) Solve the resulting equation for x:

$$x = \frac{4}{3}$$

(3) Replace x by its value in either of the given equations:

$$x + y = 2$$

$$\frac{4}{3} + y = 2$$

(4) Solve the resulting equation for y:

$$\begin{aligned} -\frac{4}{3} \quad &= -\frac{4}{3} \\ \hline y &= \frac{2}{3} \end{aligned}$$

(5) *Check:* Substitute $\frac{4}{3}$ for x and $\frac{2}{3}$ for y in each of the given equations, and show that these values make the given equations true:

$$\begin{aligned} 2x - y &= 2 \\ 2\left(\frac{4}{3}\right) - \frac{2}{3} &\stackrel{?}{=} 2 \\ \frac{8}{3} - \frac{2}{3} &\stackrel{?}{=} 2 \\ \frac{6}{3} &\stackrel{?}{=} 2 \\ 2 &= 2 \text{ (True)} \end{aligned} \qquad \begin{aligned} x + y &= 2 \\ \frac{4}{3} + \frac{2}{3} &\stackrel{?}{=} 2 \\ \frac{6}{3} &\stackrel{?}{=} 2 \\ 2 &= 2 \text{ (True)} \end{aligned}$$

EXAMPLES

Solve the system of equations and check.

a. $x + 3y = 13$
$x + y = 5$

How to Proceed: | *Solution:*

(1) Since the coefficients of the variable x are the same in both equations, write an equation equivalent to equation B by multiplying both sides of equation B by -1. Now, since the coefficients of x are additive inverses, add the two equations so that the resulting equation involves one variable, y:

$$\begin{aligned} x + 3y &= 13 \quad [A] \\ x + y &= 5 \quad [B] \\ x + 3y &= 13 \\ -x - y &= -5 \\ \hline 2y &= 8 \end{aligned}$$

(2) Solve the resulting equation for variable y:

$$y = 4$$

(3) Replace y by its value in either of the given equations:

$$\begin{aligned} x + y &= 5 \quad [B] \\ x + 4 &= 5 \end{aligned}$$

(4) Solve the resulting equation for the remaining variable, x:

$$x = 1$$

(5) *Check:* Substitute 1 for x and 4 for y in each of the given equations to verify that the resulting sentences are true:

$$x\ 3y = 13 \qquad\qquad x + y = 5$$

$$2 + 3(4) \stackrel{?}{=} 13 \qquad\qquad 1 + 4 \stackrel{?}{=} 5$$

$$13 = 13 \text{ (True)} \qquad\qquad 5 = 5 \text{ (True)}$$

Answer: Since $x = 1$ and $y = 4$, the solution is $(1, 4)$, or the solution set is $\{(1, 4)\}$.

b. $5a + b = 13$
$4a - 3b = 18$

How to Proceed: | *Solution:*

(1) Multiply both members of equation $[A]$ by 3 to obtain an equivalent equation, $[A']$: The coefficient of b in $[A']$ is the additive inverse of the coefficient of b in $[B]$:

$$\begin{aligned} 5a + b &= 13 \quad [A] \\ 4a - 3b &= 18 \quad [B] \\ 3[A]\colon\ 15a + 3b &= 39 \quad [A'] \\ 4a - 3b &= 18 \quad [B] \\ \hline \end{aligned}$$

(2) Add the corresponding members of equations $[B]$ and $[A']$ to eliminate variable b:

$$19a = 57$$

(3) Solve the resulting equation for variable a:

$$a = 3$$

(4) Replace a by its value in either of the given equations:

$$\begin{aligned} 5a + b &= 13 \quad [A] \\ 5(3) + b &= 13 \end{aligned}$$

(5) Solve the resulting equation for the remaining variable, b:

$$\begin{aligned} 15 + b &= 13 \\ b &= -2 \end{aligned}$$

(6) *Check:* Substitute 3 for a and -2 for b in the *original* equations:

$$\begin{aligned} 5a + b &= 13 \\ 5(3) + (-2) &\stackrel{?}{=} 13 \\ 15 + (-2) &\stackrel{?}{=} 13 \\ 13 &= 13 \end{aligned} \qquad \begin{aligned} 4a - 3b &= 18 \\ 4(3) - 3(-2) &\stackrel{?}{=} 18 \\ 12 + 6 &\stackrel{?}{=} 18 \\ 18 &= 18 \text{ (True)} \end{aligned}$$

Answer: Since $a = 3$ and $b = -2$, $(a, b) = (3, -2)$, or the solution set is $\{(3, -2)\}$.

c. $7x = 5 - 2y$
$3y = 16 - 2x$

How to Proceed:	*Solution:*
(1) Transform each of the given equations [A] and [B] into equivalent equations [A'] and [B'] in which the terms containing the variables appear on one side and the constant appears on the other side:	$7x = 5 - 2y$ [A] $3y = 16 - 2x$ [B] $7x + 2y = 5$ [A'] $2x + 3y = 16$ [B']
(2) To eliminate y, multiply both sides of equation [A''] by 3, and multiply both sides of equation [B''] by -2. This results in the equivalent equations [A'] and [B'], in which the coefficients of y are additive inverses:	$3(7x + 2y = 5)$ $-2(2x + 3y = 15)$ $21x + 6y = 15$ [A''] $\underline{-4x + 6y = -32}$ [B'']
(3) Add the corresponding members of equations [A'] and [B'] to eliminate variable y:	$17x = -17$
(4) Solve the resulting equation for variable x:	$x = -1$
(5) Replace x by its value in any equation containing both variables:	$3y = 16 - 2x$ [B] $= 16 - 2(-1)$
(6) Solve the resulting equation for the remaining variable, y:	$3y = 16 + 2$ $3y = 18$ $y = 6$

(7) *Check:* Substitute -1 for x and 6 for y in each of the original equations to verify the answer. The check is left to you.

Answer: Since $x = -1$ and $y = 6$, the solution is $(-1, 6)$, or the solution set is $\{(-1, 6)\}$.

EXERCISES

In 1 and 2, state, in each case, which of the given ordered pairs is the solution of the system of equations.

1. $x + y = 5$; $x - y = -1$ (2, 3) (2, −3) (3, −2) (3, 2)

2. $6x + 2y = 14$; $3x + 2y = 8$ (1, 2) (1, 4) (2, 1) (4, −2)

In 3–47, solve each system of equations by using addition to eliminate one of the variables. Check.

3. $x + y = 12$
$x - y = 4$

4. $a + b = 13$
$a - b = 5$

5. $r + s = -6$
$r - s = -10$

6. $3x + y = 16$
$2x + y = 11$

7. $c - 2d = 14$
$c + 3d = 9$

8. $x + y = 10$
$x - y = 0$

9. $a - 4b = -8$
$a - 2b = 0$

10. $x + 2y = 8$
$x - 2y = 4$

11. $8a + 5b = 9$
$2a - 5b = -4$

12. $4x + 5y = 23$
$4x - y = 5$

13. $-2m + 4n = 13$
$6m + 4n = 9$

14. $3a - b = 3$
$a + 3b = 11$

15. $3r + s = 6$
$r + 3s = 10$

16. $4x - y = 10$
$2x + 3y = 12$

17. $5m + 3n = 14$
$2m + n = 6$

18. $2c - d = -1$
$c + 3d = 17$

19. $2m + n = 12$
$m + 2n = 9$

20. $r - 3s = -11$
$3r + s = 17$

21. $a + 3b = 4$
$2a - b = 1$

22. $3x + 4y = 26$
$x - 3y = 0$

23. $5x + 8y = 1$
$3x + 4y = -1$

24. $x - y = -1$
$3x - 2y = 3$

25. $5a - 2b = 3$
$2a - b = 0$

26. $5x - 2y = 20$
$2x + 3y = 27$

27. $2x - y = 26$
$3x - 2y = 42$

28. $2x + 3y = 6$
$3x + 5y = 15$

29. $5r - 2s = 8$
$3r - 7s = -1$

30. $3x + 7y = -2$
$2x + 3y = -3$

31. $4x + 3y = -1$
$5x + 4y = 1$

32. $4a - 6b = 15$
$6a - 4b = 10$

33. $5y + x = -8$
$x = 7$

34. $x + y = 4$
$y = x$

35. $2x + y = 17$
$5x = 25 + y$

36. $5r + 3s = 30$
$2r = 12 - 3s$

37. $6r = s$
$5r = 2s - 14$

38. $3a - 7 = 7b$
$4a = 3b + 22$

39. $3x - 4y = 2$
$x = 2(7 - y)$

40. $3x + 5(y + 2) = 1$
$8y = -3x$

41. $\frac{1}{3}x + \frac{1}{4}y = 10$
$\frac{1}{3}x - \frac{1}{2}y = 4$

42. $\frac{1}{2}a + \frac{1}{3}b = 8$
$\frac{3}{2}a - \frac{4}{3}b = -4$

43. $c - 2d = 1$
$\frac{2}{3}c + 5d = 26$

44. $2a = 3b$
$\frac{2}{3}a - \frac{1}{2}b = 2$

45. $0.04x + 0.06y = 26$
$x + y = 500$

46. $0.03x + 0.05y = 17$
$x + y = 400$

47. $0.03x = 0.06y + 9$
$x + y = 600$

17-3 USING SUBSTITUTION TO SOLVE A SYSTEM OF LINEAR EQUATIONS

Another algebraic method, called the ***substitution method***, can be used to eliminate one of the variables when solving a system of equations. When we use this method, we apply the substitution principle to transform one of the equations of the system into an equivalent equation that involves only one variable.

EXAMPLE

Solve the system of equations and check.

a. $4x + 3y = 27$
$y = 2x - 1$

How to Proceed:	*Solution:*	
(1) Since in equation [B], both y and $2x - 1$ name the same number, eliminate y in equation [A] by replacing it with $2x - 1$:	$4x + 3y = 27$	[A]
	$y = 2x - 1$	[B]
	$4x + 3(y) = 27$	[A]
	$4x + 3(2x - 1) = 27$	
	$4x + 6x - 3 = 27$	
	$10x - 3 = 27$	
	$10x = 30$	
(2) Solve the resulting equation for x:	$x = 3$	
(3) Replace x with its value in any equation involving both variables:	$y = 2x - 1$	[B]
	$= 2(3) - 1$	
(4) Solve the resulting equation for y:	$y = 6 - 1$	
	$= 5$	

(5) *Check:* Substitute 3 for x and 5 for y in each of the given equations to verify that the resulting sentences are true.

$$\begin{aligned} 4x + 3y &= 27 \\ 4(3) + 3(5) &\stackrel{?}{=} 27 \\ 12 + 15 &\stackrel{?}{=} 27 \\ 27 &= 27 \text{ (True)} \end{aligned} \qquad \begin{aligned} y &= 2x - 1 \\ 5 &\stackrel{?}{=} 2(3) - 1 \\ 5 &\stackrel{?}{=} 6 - 1 \\ 5 &= 5 \text{ (True)} \end{aligned}$$

Answer: Since $x = 3$ and $y = 5$, the solution is (3, 5), or the solution set is $\{(3, 5)\}$.

b. $3x - 4y = 26$
$x + 2y = 2$

How to Proceed:	*Solution:*	
(1) Transform one of the equations into an equivalent equation in which one of the variables is expressed in terms of the other. In equation $[B]$, solve for x in terms of y, to obtain equation $[B']$:	$3x - 4y = 26$	$[A]$
	$x + 2y = 2$	$[B]$
	$x = 2 - 2y$	$[B']$
(2) Eliminate x in equation $[A]$ by replacing it with $2 - 2y$, the expression for x in equation $[B']$:	$3(x) - 4y = 26$	$[A]$
	$3(2 - 2y) - 4y = 26$	
	$6 - 6y - 4y = 26$	
	$6 - 10y = 26$	
	$-10y = 20$	
(3) Solve the resulting equation for y:	$y = -2$	
(4) Replace y by its value in any equation, A, B, or B', involving both variables:	$x = 2 - 2y$	$[B']$
	$= 2 - 2(-2)$	
(5) Solve the resulting equation for x:	$x = 2 + 4 = 6$	

(6) *Check:* Substitute 6 for x and -2 for y in each of the given equations to verify that the resulting sentences are true. This check is left to you.

Answer: Since $x = 6$ and $y = -2$, the solution is $(6,-2)$, or the solution set is $\{(6,-2)\}$.

EXERCISES

In 1–20, solve each system of equations by using substitution to eliminate one of the variables. Check.

1. $y = x$
$x + y = 14$

2. $x = y$
$5x - 4y = -2$

3. $y = 2x$
$x + y = 21$

4. $x = 4y$
$2x + 3y = 22$

5. $a = -2b$
$5a - 3b = 13$

6. $r = -3s$
$3r + 4s = -10$

7. $y = x + 1$
$x + y = 9$

8. $x = y - 2$
$x + y = 18$

9. $y = x + 3$
$3x + 2y = 26$

10. $y = 2x + 1$
$x + y = 7$

11. $a = 3b + 1$
$5b - 2a = 1$

12. $a + b = 11$
$3a - 2b = 8$

13. $3m - 2n = 11$
$m + 2n = 9$

14. $a - 2b = -2$
$2a - b = 5$

15. $7x - 3y = 23$
$x + 2y = 13$

16. $4b - 8c = 24$
$3b + 2c = 2$

17. $4d - 3h = 25$
$3d - 12h = 9$

18. $2x = 3y$
$4x - 3y = 12$

19. $4y = -3x$
$5x + 8y = 4$

20. $2x + 3y = 7$
$4x - 5y = 25$

In 21–36, solve each system of equations by using an algebraic method that seems convenient. Check.

21. $y = 3x$
$y - x = 18$

22. $s + r = 0$
$r - s = 6$

23. $3a - b = 13$
$2a + 3b = 16$

24. $m + 2n = 14$
$3n + m = 18$

25. $x = y$
$4x - 5y = -2$

26. $y = x - 2$
$3x - y = 16$

27. $-2c = d$
$6c + 5d = -12$

28. $3x + 8y = 16$
$5x + 10y = 25$

29. $y = 3x$
$\frac{1}{3}x + \frac{1}{2}y = 11$

30. $a - \frac{2}{3}b = 4$
$\frac{3}{5}a + b = 15$

31. $3(y - 6) = 2x$
$3x + 5y = 11$

32. $x + y = 300$
$0.1x + 0.3y = 78$

33. $3d = 13 - 2c$
$\frac{3c + d}{2} = 8$

34. $3x = 4y$
$\frac{3x + 8}{5} = \frac{3y - 1}{2}$

35. $\frac{a}{3} + \frac{a + b}{6} = 3$
$\frac{b}{3} - \frac{a - b}{2} = 6$

36. $a + 3(b - 1) = 0$
$2(a - 1) + 2b = 16$

17-4 USING SYSTEMS OF EQUATIONS TO SOLVE VERBAL PROBLEMS

You have previously learned how to solve word problems by using one variable. Frequently, however, a problem can be solved more easily by using two variables rather than one variable.

For example, we can use two variables to solve the following problem:

The sum of two numbers is 8.6. Three times the larger number decreased by twice the smaller is 6.3. What are the numbers?

First, we will represent each number by a different variable.
Let x = the larger number and y = the smaller number.
Now use the conditions of the problem to write two equations:

The sum of the numbers is 8.6	$x + y = 8.6$
Three times the larger decreased by twice the smaller is 6.3.	$3x - 2y = 6.3$

Solve this system of equation to find the numbers,

$$\begin{array}{rl}
x + y = 8.6 & \text{[A]} \\
3x - 2y = 6.3 & \text{[B]} \\
\underline{2x + 2y = 17.2} & \text{[A]} \times 2 \\
5x \quad = 23.5 & \\
x = 4.7 & \\
x + y = 8.6 & \\
4.7 + y = 8.6 & \\
y = 3.9 &
\end{array}$$

The two numbers are 4.7 and 3.9.

PROCEDURE. To solve a word problem by using a system of two equations involving two variables:

1. Use two different variables to represent the different unknown quantities in the problem.
2. Translate two relationships in the problem into a system of two equations.
3. Solve the system of equations to determine the answer(s) to the problem.
4. Check the answer(s) in the original word problem.

EXAMPLES

1. The owner of a men's clothing store bought six belts and eight hats for \$140. A week later, at the same prices, he bought nine belts and six hats for \$132. Find the price of a belt and the price of a hat.

Solution

Let b = the price in dollars, of a belt,

and h = the price, in dollars, of a hat.

6 belts and 8 hats cost \$140.

$$\underbrace{6b}\quad \underbrace{+\ 8h}\quad \underbrace{=\ 140} \qquad [A]$$

9 belts and 6 hats cost \$132.

$$\underbrace{9b}\quad \underbrace{+\ 6h}\quad \underbrace{=\ 132} \qquad [B]$$

(1) To eliminate h, multiply both members of equation $[B]$ by 4 and both members of equation $[A]$ by -3. Then, add the equations and solve for b.

$$\begin{aligned} 4(9b + 6h = 132) &\rightarrow \quad 36b + 24h = 528 \\ -3(6b + 8h = 140) &\rightarrow \quad \underline{-18b - 24h = -420} \\ &\qquad\quad 18b \qquad\;\; = 108 \\ &\qquad\qquad\qquad b = 6 \end{aligned}$$

(2) In equation $[A]$, substitute 6 for b and solve for h.

$$\begin{aligned} 6b + 8h &= 140 \\ 36 + 8h &= 140 \\ 8h &= 104 \\ h &= 13 \end{aligned}$$

Check: 6 belts and 8 hats cost 6(\$6) + 8(\$13) = \$36 + \$104 = \$140.
9 belts and 6 hats cost 9(\$6) + 6(\$13) = \$54 + \$78 = \$132.

Answer: A belt costs \$6; a hat costs \$13.

2. When Angelo cashed a check for \$170, the bank teller gave him 12 bills, some \$20 bills and the rest \$10 bills. How many bills of each denomination did Angelo receive?

Solution (1) Represent the unknowns using two variables.
Let x = number of \$10 bills,
and y = number of \$20 bills.

In this problem, part of the information is in terms of the *number* of bills (the bank teller gave Angelo 12 bills) and part of the information is in terms of the *value* of the bills (the check was for \$170).

(2) Write one equation using the *number* of bills: $x + y = 12 \quad [A]$

Write a second equation using the *value* of the bills. The value of x \$10 bills is $10x$, the value of y \$20 bills is $20y$, and the total value is \$170: $10x + 20y = 170 \ [B]$

(3) Solve the system of equations:

$$\begin{aligned} -10\,[A]:\ -10(x + y = 12) &\rightarrow -10x - 10y = -120 \\ [B]:\ 10x + 20y = 170 &\rightarrow \underline{\ \ 10x + 20y = \ \ \ 170} \\ &\qquad\qquad\quad 10y = \ \ \ \ 50 \\ &\qquad\qquad\qquad y = 5 \end{aligned}$$

Substitute 5 for y in equation $[A]$:

$$\begin{aligned} x + y &= 12 \\ x + 5 &= 12 \\ x &= 7 \end{aligned}$$

(4) *Check* the number of bills:
$x = 7$ \$10 bills
$y = 5$ \$20 bills
$7 + 5 = 12$ bills (True)

Check the value of the bills:
7 \$10 bills are worth \$70.
5 \$20 bills are worth \$100.
Total value = \$70 + \$100 = \$170 (True)

Answer: Angelo received seven \$10 bills and five \$20 bills.

EXERCISES WITH OPEN-RESPONSE PROBLEMS

In 1–29, solve each problem algebraically, using two variables.

1. The sum of two numbers is 36. Their difference is 24. Find the numbers.
2. The sum of two numbers is 74. The larger number is 3 more than the smaller number. Find the numbers.
3. The sum of two numbers is 104. The larger number is 1 less than twice the smaller number. Find the numbers.
4. The difference between two numbers is 34. The larger exceeds 3 times the smaller by 4. Find the numbers.
5. If 5 times the smaller of two numbers is subtracted from twice the larger, the result is 16. If the larger is increased by 3 times the smaller, the result is 63. Find the numbers.
6. One number is 15 more than another. The sum of twice the larger and 3 times the smaller is 182. Find the numbers.
7. The perimeter of a rectangle is 50 centimeters. The length is 9 centimeters more than the width. Find the length and the width of the rectangle.
8. A rectangle has a perimeter of 38 feet. The length is 1 foot less than 3 times the width. Find the dimensions of the rectangle.
9. Two angles are supplementary. The larger angle measures 120° more than the smaller. Find the degree measure of each angle.
10. Two angles are supplementary. The larger angle measures 15° less than twice the smaller. Find the degree measure of each angle.
11. Two angles are complementary. The measure of the larger angle is 30° more than the measure of the smaller angle. Find the degree measure of each angle.
12. The larger of two complementary angles measures 6° less than twice the smaller angle. Find the degree measure of each angle.
13. In an isosceles triangle, each base angle measures 30° more than the vertex angle. Find the degree measures of the three angles of the triangle.
14. At a quick-lunch counter, three pretzels and one cup of soda cost \$2.75. Two pretzels and one cup of soda cost \$2.00. Find the cost of a pretzel and the cost of a cup of soda.
15. On one day, four gardeners and four helpers earned \$360. On another day, working the same number of hours and at the same rate of pay, five gardeners and six helpers earned \$480. How much does a gardener and how much does a helper earn each day?
16. A baseball manager bought four bats and nine balls for \$76.50. On another day, she bought three bats and one dozen balls at the same prices and paid \$81.00. How much did she pay for each bat and each ball?

17. Mrs. Black bought 2 pounds of veal and 3 pounds of pork, for which she paid \$20.00. Mr. Cook, paying the same prices, paid \$11.25 for 1 pound of veal and 2 pounds of pork. Find the price of a pound of veal and the price of a pound of pork.

18. One day, Mrs. Rubero paid \$18.70 for 4 kilograms of brown rice and 3 kilograms of white rice. Next day, Mrs. Rubero paid \$13.30 for 3 kilograms of brown rice and 2 kilograms of white rice. If the prices were the same on each day, find the price per kilogram for each type of rice.

19. Tickets for a high school dance cost \$1.00 each if purchased in advance of the dance, but \$1.50 each if bought at the door. If 100 tickets were sold and \$120 was collected, how many tickets were sold in advance and how many were sold at the door?

20. A dealer sold 200 tennis racquets. Some were sold at \$18 each, and the rest at \$33 each. The total receipts from these sales were \$4,800. How many racquets did the dealer sell at \$18 each?

21. Mrs. Rinaldo changed a \$100 bill in a bank. She received \$20 bills and \$10 bills. The number of \$20 bills was 2 more than the number of \$10 bills. How many bills of each kind did she receive?

22. Linda spent \$3.60 for stamps to mail packages. Some were 30¢ stamps and the rest were 20¢ stamps. The number of 20¢ stamps was 2 less than the number of 30¢ stamps. How many stamps of each kind did Linda buy?

23. A dealer has some hard candy worth \$2.00 a pound and some worth \$3.00 a pound. He wishes to make a mixture of 80 pounds that he can sell for \$2.20 a pound. How many pounds of each kind should he use?

24. At the Savemore Supermarket, 3 pounds of squash and 2 pounds of eggplant cost \$2.85. The cost of 4 pounds of squash and 5 pounds of eggplant is \$5.41. What is the cost of 1 pound of squash, and what is the cost of 1 pound of eggplant?

25. One year, Roger Jackson and his wife Wilma together earned \$47,000. If Roger earned \$4,000 more than Wilma earned that year, how much did each earn?

26. The sum of two numbers is 900. When 4% of the larger is added to 7% of the smaller, the sum is 48. Find the numbers.

27. Mrs. Moto invested \$1,400, part at 5% and part at 8%. Her total annual income from both investments was \$100. Find the amount she invested at each rate.

28. Mr. Stein invested a sum of money in bonds yielding 4% a year and another sum in bonds yielding 6% a year. In all, he invested \$4,000. If his total annual income from the two investments was \$188, how much did he invest at each rate?

29. Mr. May invested \$21,000, part at 8% and the rest at 6%. If the annual incomes from both investments were equal, find the amount invested at each rate.

17-5 GRAPHING THE SOLUTION SET OF A SYSTEM OF INEQUALITIES

To find the solution set of a system of inequalities, we must find the ordered pairs that satisfy the open sentences of the system. We do this by a graphic procedure that is similar to the method used in finding the solution set of a system of equations.

EXAMPLES

1. Graph the solution set of this system:

$$x > 2$$
$$y < -2$$

Solution (1) Graph $x > 2$ by first graphing the plane divider $x = 2$, which, in the figure at the right, is represented by the dashed line labeled ℓ. The half-plane to the right of this line is the graph of the solution set of $x > 2$.

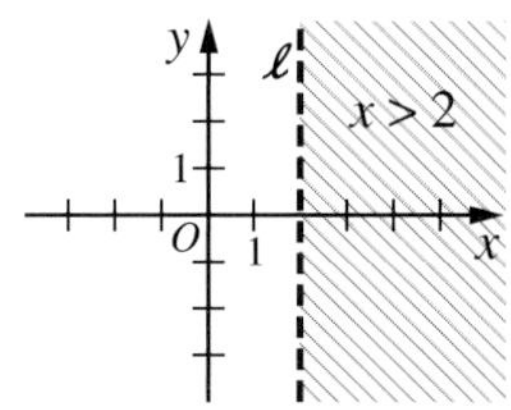

(2) Using the same set of axes, graph $y < -2$ by first graphing the plane divider $y = -2$, which, in the figure at the right, is represented by the dashed line labeled m. The half-plane below this line is the graph of the solution set of $y < -2$.

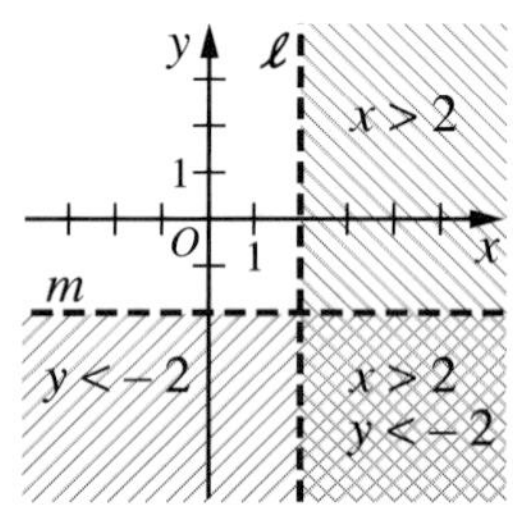

The solution set of the system $x > 2$ and $y < -2$ consists of the intersection of the solution sets of $x > 2$ and $y < -2$.

Therefore, the crosshatched region in the lower figure, which is the intersection of the graphs made in steps (1) and (2), is the graph of the solution set of the system $x > 2$ and $y < -2$.

All points in this region, and no others, satisfy both inequalities of the system. For example, point $(4,-3)$, which lies in the region, satisfies the system because its x-value satisfies one of the given inequalities, $4 > 2$, and its y-value satisfies the other inequality, $-3 < -2$.

2. Graph the solution set of $3 < x < 5$ in a coordinate plane.

Solution The inequality $3 < x < 5$ means $3 < x$ and $x < 5$. This may be written as $x > 3$ and $x < 5$.

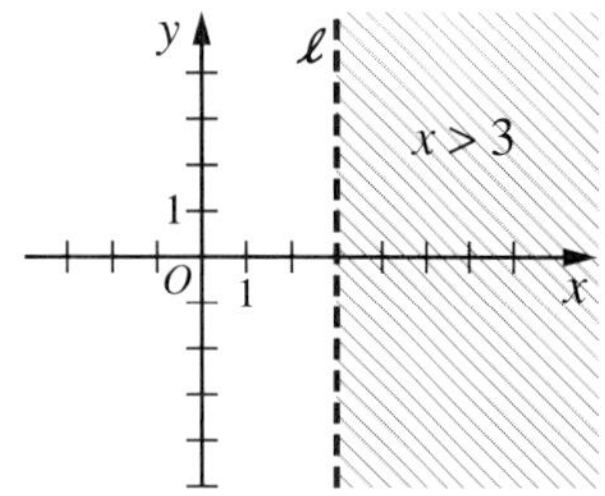

(1) Graph $x > 3$ by first graphing the plane divider $x = 3$, which, in the figure at the right, is represented by the dashed line labeled ℓ. The half-plane to the right of line $x = 3$ is the graph of the solution set of $x > 3$.

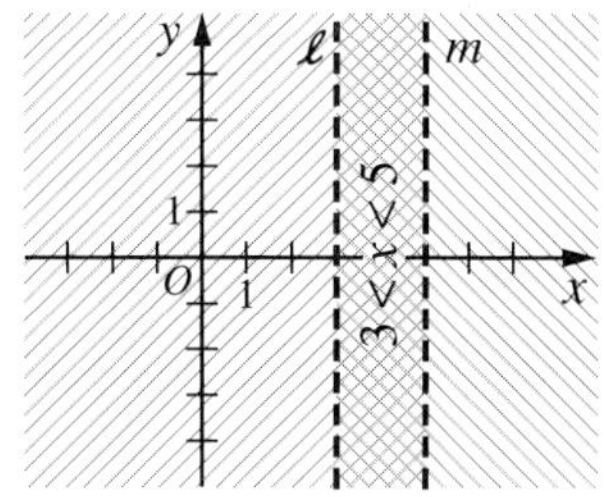

(2) Using the same set of axes, graph $x < 5$ by first graphing the plane divider $x = 5$, which, in the figure at the right, is represented by the dashed line labeled m. The half-plane to the left of line $x = 5$ is the graph of the solution set of $x < 5$.

The crosshatched region, which is the intersection of the graphs made in steps (1) and (2), is the graph of the solution set of $x > 3$ and $x < 5$, or $3 < x < 5$.

All points in this region, and no others, satisfy $3 < x < 5$. For example, point (4, 3), which lies in the region and whose x-value is 4, satisfies $3 < x < 5$ because $3 < 4 < 5$ is a true statement.

---

3. Graph the following system of inequalities and label the solution set R:

$$x + y \geq 4$$

$$y \leq 2x - 3$$

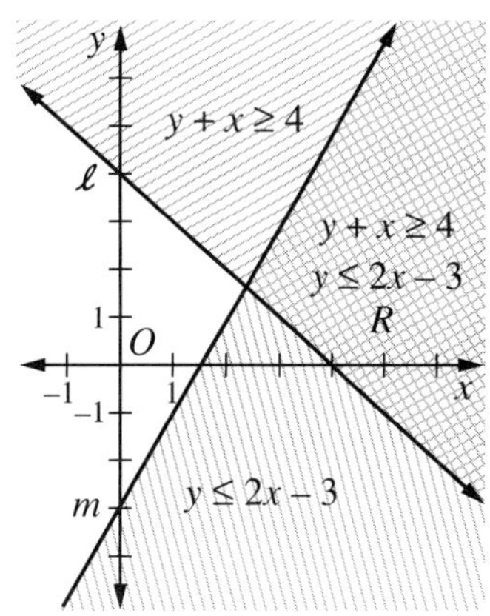

Solution (1) Graph $x + y \geq 4$ by first graphing the plane divider $x + y = 4$, which, in the figure at the right, is represented by the solid line labeled ℓ. The line $x + y = 4$ and the half-plane above this line together form the graph of the solution set of $x + y \geq 4$.

(2) Using the same set of axes, graph $y \leq 2x - 3$ by first graphing the plane divider $y = 2x - 3$, which, in the figure at the right, is represented by the solid line labeled m. The line $y = 2x - 3$ and the half-plane below this line together form the graph of the solution set of $y \leq 2x - 3$.

The crosshatched region labeled R, the intersection of the graphs made in steps (1) and (2), is the solution set of the system $x + y \geq 4$ and $y \leq 2x - 3$. Any point in the region R, such as (5, 2), will satisfy $x + y \geq 4$ because $5 + 2 \geq 4$, that is, $7 \geq 4$, is true, and will at the same time satisfy $y \leq 2x - 3$ because $2 \leq 2(5) - 3$, or $2 \leq 7$, is true.

EXERCISES

In 1–24, graph each system of inequalities and label the solution set S.

1. $x \geq 1$ $y > -2$	**2.** $x < 2$ $y \geq 3$	**3.** $x > 0$ $y > 0$	**4.** $x < 0$ $y > 0$
5. $y \geq x$ $x < 2$	**6.** $y \leq x$ $x \geq -1$	**7.** $y \geq 1$ $y < x - 1$	**8.** $y \leq 5$ $y < x + 3$
9. $y > x$ $y < 2x + 3$	**10.** $y \geq 2x$ $y > x + 3$	**11.** $y \leq 2x + 3$ $y \geq -x$	**12.** $y - x \geq 5$ $y - 2x \leq 7$
13. $y > x - 3$ $y > -x + 5$	**14.** $y \geq -2x + 4$ $y < x - 5$	**15.** $y < -x + 7$ $y \geq 2x + 1$	**16.** $y < x - 1$ $x + y \geq 2$
17. $x + y \leq 8$ $y > x - 4$	**18.** $y + 3x \geq 6$ $y < 2x - 4$	**19.** $x + y > 3$ $x - y < 6$	**20.** $x - y \leq -2$ $x + y \geq 2$
21. $2x + y \leq 6$ $x + y - 2 > 0$	**22.** $2x + 3y \geq 6$ $x + y - 4 \leq 0$	**23.** $y \geq x$ $x = 0$	**24.** $x + y \leq 3$ $y - 2x = 0$

In 25–33, in each case, graph the solution set in a coordinate plane.

25. $1 < x < 4$	**26.** $-5 \leq x \leq -1$	**27.** $-4 \leq x < 0$
28. $2 < y \leq 6$	**29.** $-2 \leq y \leq 3$	**30.** $0 < y \leq 4$
31. $(x > 2) \wedge (x \leq 7)$	**32.** $(x < 1) \vee (x \geq 4)$	**33.** $(y \geq -1) \wedge (y < 5)$

CHAPTER SUMMARY

A system of two linear equations in two variables may be:

1. ***Consistent***; its solution is one ordered pair of numbers, resulting in coincident lines.
2. ***Inconsistent***; its solution is the empty set. (The graphs of the equations are parallel with no points in common.)
3. ***Dependent***; its solution is an infinite set of number pairs. (The graphs of the equations are the same line.)

The solution of a consistent linear system in two variables may be found ***graphically*** by determining the coordinates of the point of intersection of the graphs or ***algebraically*** by using ***addition*** or ***substitution***. Systems of linear equations can be used to solve verbal problems.

The solution set of a system of inequalities can be shown on a graph as the ***intersection*** of the solution sets of the inequalities.

VOCABULARY

17-1 System of simultaneous equations Consistent equations Inconsistent equations Dependent equations

17-2 Equivalent systems Addition method

17-3 Substitution method

REVIEW EXERCISES WITH OPEN-RESPONSE PROBLEMS

1. Solve the following system of equations for x:

$$5x - 2y = 22$$
$$x + 2y = 2$$

In 2–4, solve each system of equations *graphically* and check.

2. $x + y = 6$
$y = 2x - 6$

3. $y = -x$
$2x + y = 3$

4. $2y = x + 4$
$x - y + 4 = 0$

In 5–7, solve each system of equations by using *addition* to eliminate one of the variables. Check.

5. $2x + y = 10$
$x + y = 3$

6. $x + 4y = 1$
$5x - 6y = -8$

7. $3c + d = 0$
$c - 4d = 52$

In 8–10, solve each system of equations by using *substitution* to eliminate one of the variables. Check.

8. $x + 2y = 7$
$x = y - 8$

9. $3r + 2s = 20$
$r = -2s$

10. $x + y = 7$
$2x + 3y = 21$

In 11–16, solve each system of equations by using an algebraic method that seems convenient. Check.

11. $x + y = 0$
$3x + 2y = 5$

12. $5a + 3b = 17$
$4a - 5b = 21$

13. $t + u = 12$
$t = \frac{1}{3}u$

14. $3a = 4b$
$4a - 5b = 2$

15. $x + y = 1{,}000$
$0.06x = 0.04y$

16. $10t + u = 24$
$t + u = \frac{1}{7}(10u + t)$

In 17–19, solve each problem algebraically, using two variables.

17. The sum of two numbers is 7. Their difference is 18. Find the numbers.

18. At a store, three notebooks and two pencils cost \$2.80. At the same prices, two notebooks and five pencils cost \$2.60. Find the cost of one notebook and one pencil.

19. Two angles are complementary. The larger angle measures 15° less than twice the smaller angle. Find the degree measure of each angle.

20. **a.** Solve this system of equations algebraically:

$$x - y = 3$$

$$x + 3y = 9$$

b. On a set of coordinate axes, graph the system of equations given in part **a**.

In 21–23, graph each system of inequalities and label the solution set A.

21. $y > 2x - 3$
$y \leq 5 - x$

22. $y \leq \frac{1}{2}x$
$x \geq -4$

23. $2x + y < 4$
$x - y < -2$

CUMULATIVE REVIEW WITH OPEN-RESPONSE PROBLEMS

1. Draw the graph of the solution set of $2x - 7 \leq 1$.

2. For what value of x is the following statement false?

If $3(x + 7) = 9$, then $x + 1 = 5$

3. The measure of the vertex angle of an isosceles triangle is three times the measure of each of the base angles. Find the measure of each angle of the triangle.

4. At the beginning of the year, Latonya's salary increased by 3%. If her weekly salary is now \$254.41, what was her weekly salary last year?

Exploration

Jason and his brother Robby live 1.25 miles from school. Jason rides his bicycle to school and Robby walks. The graph below shows their relative positions on their way to school one morning.

Use the information provided by the graph to describe Robby's and Jason's journeys to school. Who left first? Who arrived at school first? If each boy took the same route to school, when did Jason pass Robby? How fast did Robby walk? How fast did Jason ride his bicycle?

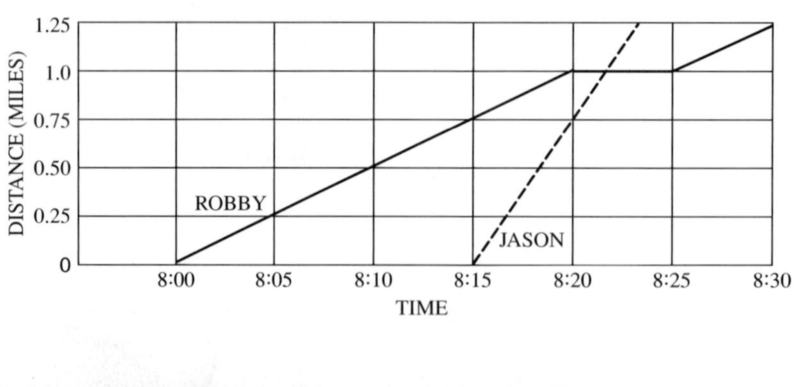

Chapter 18

Special Products and Factors

The owners of a fruit farm intend to extend their orchards by planting 100 new apple trees. The trees are to be planted in rows, with the same number of trees in each row. In order to decide how the trees are to be planted, the owners will use whole numbers that are factors of 100 to determine the possible arrangements:

1 row of 100 trees
2 rows of 50 trees each
4 rows of 25 trees each
5 rows of 20 trees each
10 rows of 10 trees each
20 rows of 5 trees each
25 rows of 4 trees each
50 rows of 2 trees each
100 rows of 1 tree each

From this list of possibilities, the arrangement that best fits the dimensions of the land to be planted can be chosen.

From earliest times, the study of factors and prime numbers has fascinated mathematicians, leading to the discovery of many important principles. In this chapter you will extend what you know about the factors of whole numbers, study the products of special binomials, and learn to write polynomials in factored form.

18-1 FACTORS AND FACTORING

When two numbers are multiplied, the result is called their ***product***. The numbers that are multiplied are ***factors*** of the product. Since $3(5) = 15$, the numbers 3 and 5 are factors of 15.

Factors of a product can be found by using division. If the remainder is 0, then the divisor and the quotient are factors of the dividend. For example, $35 \div 5 = 7$. Thus, $35 = 5(7)$, and 5 and 7 are factors of 35.

Factoring a number is the process of finding those numbers whose product is the given number. Usually, when we factor, we are finding the factors of an integer and we find only those factors that are integers. We call this *factoring over the set of integers.*

Every positive integer that is the product of two positive integers is also the product of the opposites of those integers.

$$+21 = (+3)(+7) \qquad +21 = (-3)(-7)$$

Every negative integer that is the product of a positive integer and a negative integer is also the product of the opposites of those integers.

$$-21 = (+3)(-7) \qquad -21 = (-3)(+7)$$

Usually, when we factor a positive integer, we write only the positive integral factors.

Two factors of any number are 1 and the number itself. To find other integral factors, if they exist, we use division, as stated above. We let the number being factored be the dividend, and we divide this number in turn by the whole numbers 2, 3, 4, and so on. If the quotient is an integer, then both the divisor and the quotient are factors of the dividend.

For example, to find the factors of 126, we will divide using a calculator. Pairs of factors of 126 are listed at the right.

		Pairs of Factors of 126
We know that 1 and 126 are factors.		1 • 126
Enter: 126 ÷ 2 =	***Display:*** 63.	2 • 63
Enter: 126 ÷ 3 =	***Display:*** 42.	3 • 42
Enter: 126 ÷ 4 =	***Display:*** 31.5	—
Enter: 126 ÷ 5 =	***Display:*** 25.2	—
Enter: 126 ÷ 6 =	***Display:*** 21.	6 • 21
Enter: 126 ÷ 7 =	***Display:*** 18.	7 • 18
Enter: 126 ÷ 8 =	***Display:*** 15.75	—
Enter: 126 ÷ 9 =	***Display:*** 14.	9 • 14
Enter: 126 ÷ 10 =	***Display:*** 12.6	—
Enter: 126 ÷ 11 =	***Display:*** 11.454545	—
Enter: 126 ÷ 12 =	***Display:*** 10.5	—

When the quotient is smaller than the divisor (here, $10.5 < 12$), we have found all possible positive integral factors.

The factors of 126 are 1, 2, 3, 6, 7, 9, 14, 18, 21, 42, 63, and 126.

Recall that a ***prime number*** is an integer greater than 1 that has no positive integral factors other than itself and 1. The first seven prime numbers are 2, 3, 5, 7, 11, 13, 17. Integers greater than 1 that are not prime are called ***composite numbers***.

In general, a positive integer greater than 1 is a prime or can be expressed as the product of prime factors. Although the factors may be written in any order, there is one and only one combination of prime factors whose product is a given composite number. As shown at the right, a prime factor may occur in the product more than once.

$$21 = 3 \cdot 7$$

$$20 = 2 \cdot 2 \cdot 5 \text{ or } 2^2 \cdot 5$$

To express a positive integer, for example 280, as the product of primes, we start with any pair of positive integers, say 28 and 10, whose product is the given number. Then, we factor these factors and continue to factor the factors until all are primes. Finally, we rearrange these factors in numerical order, as shown at the right.

$$280 = 28 \cdot 10$$

$$280 = 2 \cdot 14 \cdot 2 \cdot 5$$

$$280 = 2 \cdot 2 \cdot 7 \cdot 2 \cdot 5$$

$$280 = 2 \cdot 2 \cdot 2 \cdot 5 \cdot 7$$

or

$$280 = 2^3 \cdot 5 \cdot 7$$

Expressing each of two integers as the product of prime factors makes it possible to discover the greatest integer that is a factor of both of them. We call this factor the ***greatest common factor*** of these integers.

Let us find the greatest common factor of 180 and 54.

$$\begin{array}{rcllllll}
180 & = & 2 \cdot 2 \cdot & 3 \cdot & 3 \cdot 5 & \text{or } 2^2 \cdot 3^2 \cdot 5 & \\
 & & \downarrow & \downarrow & \downarrow & & \\
54 & = & 2 \cdot 3 \cdot & 3 \cdot & 3 & \text{or } 2 \cdot 3^3 & \\
 & & \downarrow & \downarrow & \downarrow & & \\
\text{Greatest common factor} & = & 2 \cdot & 3 \cdot & 3 & \text{or } 2 \cdot 3^2 & \text{or } 18
\end{array}$$

We see that the greatest number of times that 2 appears as a factor of both 180 and 54 is once; the greatest number of times that 3 appears as a factor of both 180 and 54 is twice. Therefore, the greatest common factor of 180 and 54 is $2 \cdot 3 \cdot 3$, or $2 \cdot 3^2$, or 18.

The greatest common factor of two or more monomials is the product of the greatest common factor of their numerical coefficients and the highest power of every variable that is a factor of each monomial.

For example, let us find the greatest common factor of $24a^3b^2$ and $18a^2b$.

$$
\begin{array}{rcccccccccccc}
24a^3b^2 & = & 2 \cdot & 2 \cdot 2 \cdot & 3 \cdot & & a \cdot & a \cdot & a \cdot & b & \cdot\, b & \\
 & & \downarrow & & \downarrow & & \downarrow & \downarrow & & \downarrow & & \\
18a^2b & = & 2 \cdot & & 3 \cdot & 3 \cdot & a \cdot & a \cdot & & b & & \\
 & & \downarrow & & \downarrow & & \downarrow & \downarrow & & \downarrow & & \\
\text{Greatest common factor} & = & 2 \cdot & & 3 \cdot & & a \cdot & a \cdot & & b & = & 6a^2b
\end{array}
$$

The greatest common factor of $24a^3b^2$ and $18a^2b$ is $6a^2b$.
When we are expressing an algebraic factor, such as $6a^2b$, we will agree that:

- **Numerical coefficients need not be factored.** (6 need not be written as $2 \cdot 3$.)
- **Powers of variables need not be represented as the product of several equal factors.** (a^2b need not be written as $a \cdot a \cdot b$.)

EXAMPLES

1. Write all positive integral factors of 72.

Solution

We know that 1 and 72 are factors.

		Pairs of Factors of 72
		$1 \cdot 72$
Enter: 72 [÷] 2 [=]	***Display:*** 36.	$2 \cdot 36$
Enter: 72 [÷] 3 [=]	***Display:*** 24.	$3 \cdot 24$
Enter: 72 [÷] 4 [=]	***Display:*** 18.	$4 \cdot 18$
Enter: 72 [÷] 5 [=]	***Display:*** 14.4	
Enter: 72 [÷] 6 [=]	***Display:*** 12.	$6 \cdot 12$
Enter: 72 [÷] 7 [=]	***Display:*** 10.285714	
Enter: 72 [÷] 8 [=]	***Display:*** 9.	$8 \cdot 9$

It is not necessary to divide by 9, since 9 has just been listed as a factor.

Answer: The factors of 72 are 1, 2, 3, 4, 6, 8, 9, 12, 18, 24, 36, and 72.

2. Express 700 as a product of prime factors.

Solution

$$700 = 2 \cdot 350$$
$$700 = 2 \cdot 2 \cdot 175$$
$$700 = 2 \cdot 2 \cdot 5 \cdot 35$$
$$700 = 2 \cdot 2 \cdot 5 \cdot 5 \cdot 7 \text{ or } 2^2 \cdot 5^2 \cdot 7$$

Answer: $2^2 \cdot 5^2 \cdot 7$

---

3. Find the greatest common factor of the monomials $60r^2s^4$ and $36rs^2t$.

Solution

$$60r^2s^4 = 2 \cdot 2 \cdot 3 \cdot 5 \cdot r \cdot r \cdot s \cdot s \cdot s \cdot s$$
$$36rs^2t = 2 \cdot 2 \cdot 3 \cdot 3 \cdot r \cdot s \cdot s \cdot t$$
$$\text{Greatest common factor} = 2 \cdot 2 \cdot 3 \cdot r \cdot s \cdot s$$

Answer: $12rs^2$

EXERCISES

In 1–10, tell whether each integer is a prime.

1. 5 **2.** 8 **3.** 13 **4.** 18 **5.** 73

6. 36 **7.** 41 **8.** 49 **9.** 57 **10.** 1

In 11–20, express each integer as a product of prime numbers.

11. 35 **12.** 18 **13.** 144 **14.** 77 **15.** 128

16. 400 **17.** 202 **18.** 129 **19.** 590 **20.** 316

In 21–26, write all the positive integral factors of each given number.

21. 26 **22.** 50 **23.** 36 **24.** 88 **25.** 100 **26.** 242

27. The product of two integers is 144. Find the second factor if the first factor is:

a. 2 **b.** 8 **c.** -18 **d.** 36 **e.** -48

28. The product of two monomials is $36x^3y^4$. Find the second factor if the first factor is:

a. $3x^2y^3$ **b.** $6x^3y^2$ **c.** $12xy^2$ **d.** $-9x^3y$ **e.** $18x^3y^2$

In 29–36, find, in each case, the greatest common factor of the given integers.

29. 10; 15 **30.** 12; -28 **31.** 14; 35 **32.** 18; -24; 36

33. 75; 50 **34.** 72; 108 **35.** -144; 200 **36.** 96; 156; 175

In 37–48, find, in each case, the greatest common factor of the given monomials.

37. $4x$; $4y$ **38.** 6; $12a$ **39.** $4r$; $6r^2$ **40.** $8xy$, $6xz$

41. $10x^2$; $15xy^2$ **42.** $7c^3d^3$; $-14c^2d$ **43.** $36xy^2z$; $-27xy^2z^2$ **44.** $50m^3n^2$; $75m^3n$

45. $24ab^2c^3$; $18ac^2$ **46.** $14a^2b$; $13ab$ **47.** $36xyz$; $25xyz$ **48.** $2ab^2c$; $3x^2yz$

18-2 COMMON MONOMIAL FACTORS

To *factor a polynomial* over the set of integers means to express the given polynomial as the product of polynomials whose coefficients are integers. For example, since $2(x + y) = 2x + 2y$, the polynomial $2x + 2y$ can be written in factored form as $2(x + y)$. The monomial 2 is a factor of each term of the polynomial $2x + 2y$. Therefore, 2 is called a ***common monomial factor*** of the polynomial $2x + 2y$.

To factor a polynomial, we look first for the ***greatest common monomial factor***, that is, the greatest monomial that is a factor of each term of the polynomial. For example:

1. Factor $4rs + 8st$. There are many common factors such as 2, 4, $2s$, and $4s$. The greatest common monomial factor is $4s$. We divide $4rs + 8st$ by $4s$ to obtain the quotient $r + 2t$, which is the second factor. Therefore, the polynomial $4rs + 8st = 4s(r + 2t)$.
2. Factor $3x + 4y$. We notice that 1 is the only common factor. The second factor is $3x + 4y$. We say that $3x + 4y$ is a ***prime polynomial***. A polynomial with integers as coefficients is a prime polynomial if its only factors are 1 and the polynomial itself.

PROCEDURE. To factor a polynomial whose terms have a common monomial factor:
1. Find the greatest monomial that is a factor of each term of the polynomial.
2. Divide the polynomial by the monomial factor. The quotient is the other factor.
3. Express the polynomial as the indicated product of the two factors.

We can check by multiplying the factors to obtain the original polynomial.

EXAMPLE

Write in factored form: $6c^3d - 12c^2d^2 + 3cd$

Solution (1) $3cd$ is the greatest common factor of $6c^3d$, $12c^2d^2$, and $3cd$.

(2) To find the other factor, divide $6c^3d - 12c^2d^2 + 3cd$ by $3cd$.

$$(6c^3d - 12c^2d^2 + 3cd) \div 3cd = \frac{6c^3d}{3cd} - \frac{12c^2d^2}{3cd} + \frac{3cd}{3cd}$$
$$= 2c^2 - 4cd + 1$$

Answer: $6c^3d - 12c^2d^2 + 3cd = 3cd(2c^2 - 4cd + 1)$

EXERCISES

In 1–52, write each expression in factored form.

1. $2a + 2b$ **2.** $5c + 5d$ **3.** $8m + 8n$ **4.** $3x - 3y$
5. $7l - 7n$ **6.** $6R - 6r$ **7.** $bx + by$ **8.** $sr - st$
9. $xc - xd$ **10.** $4x + 8y$ **11.** $3m - 6n$ **12.** $12t - 6r$
13. $15c - 10d$ **14.** $12x - 18y$ **15.** $18c - 27d$ **16.** $8x + 16$
17. $6x - 18$ **18.** $8x - 12$ **19.** $7y - 7$ **20.** $8 - 4y$
21. $6 - 18c$ **22.** $y^2 - 3y$ **23.** $2x^2 + 5x$ **24.** $3x^2 - 6x$
25. $32x + x^2$ **26.** $rs^2 - 2r$ **27.** $ax - 5ab$ **28.** $3y^4 + 3y^2$
29. $10x - 15x^3$ **30.** $2x - 4x^3$ **31.** $p + prt$ **32.** $s - sr$
33. $hb + hc$ **34.** $\pi r^2 + \pi R^2$ **35.** $\pi r^2 + \pi r\ell$ **36.** $\pi r^2 + 2\pi rh$
37. $4x^2 + 4y^2$ **38.** $3a^2 - 9$ **39.** $5x^2 + 5$ **40.** $12y^2 - 4y$
41. $3ab^2 - 6a^2b$ **42.** $10xy - 15x^2y^2$ **43.** $21r^3s^2 - 14r^2s$
44. $2x^2 + 8x + 4$ **45.** $3x^2 - 6x - 30$ **46.** $ay - 4aw - 12a$
47. $c^3 - c^2 + 2c$ **48.** $2ma + 4mb + 2mc$ **49.** $9ab^2 - 6ab - 3a$
50. $15x^3y^3z^3 - 5xyz$ **51.** $8a^4b^2c^3 + 12a^2b^2c^2$ **52.** $28m^4n^3 - 70m^2n^4$

53. The perimeter of a rectangle is represented by $2\ell + 2w$. Express the perimeter as a product of two factors.

In 54–57, each expression represents the area of a rectangle. Write this expression as the product of two factors.

54. $5x + 5y$ **55.** $18x + 6$ **56.** $x^2 + 2x$ **57.** $4x^3 + 6x^2$

18-3 THE SQUARE OF A MONOMIAL

To ***square a monomial*** means to multiply the monomial by itself. For example:

$$(3x)^2 = (3x)(3x) = (3)(3)(x)(x) = (3)^2(x)^2 \text{ or } 9x^2$$

$$(5y^2)^2 = (5y^2)(5y^2) = (5)(5)(y^2)(y^2) = (5)^2(y^2)^2 \text{ or } 25y^4$$

$$(-6b^4)^2 = (-6b^4)(-6b^4) = (-6)(-6)(b^4)(b^4) = (-6)^2(b^4)^2 \text{ or } 36b^8$$

$$(4c^2d^3)^2 = (4c^2d^3)(4c^2d^3) = (4)(4)(c^2)(c^2)(d^3(d^3) = (4)^2(c^2)^2(d^3)^2 \text{ or } 16c^4d^6$$

PROCEDURE. To square a monomial:
1. Find the square of the numerical coefficient.
2. Find the square of each literal factor by multiplying its exponent by 2.
3. The square of the monomial is the product of the expressions found in steps 1 and 2.

When a monomial is a square, its numerical coefficient is a square and the exponent of each variable is an even number. This statement holds true for each of the results shown above.

EXAMPLE

Square each monomial mentally.

		Think:	*Write:*
a.	$(4s^3)^2$	$= (4)^2 \cdot (a^3)^2$	$= 16a^6$
b.	$\left(\frac{2}{5}ab\right)^2$	$= \left(\frac{2}{5}\right)^2 \cdot (a)^2(b)^2$	$= \frac{4}{25}a^2b^2$
c.	$(-7xy^2)^2$	$= (-7)^2 \cdot (x)^2 \cdot (y^2)^2$	$= 49x^2y^4$
d.	$(0.3y^2)^2$	$= (0.3)^2 \cdot (y^2)^2$	$= 0.09y^4$

EXERCISES

In 1–20, square each monomial mentally.

1. $(a^2)^2$ **2.** $(b^3)^2$ **3.** $(-d^5)^2$ **4.** $(rs)^2$ **5.** $(m^2n^2)^2$

6. $(-x^3y^2)^2$ **7.** $(3x^2)^2$ **8.** $(-5y^4)^2$ **9.** $(9ab)^2$ **10.** $(10x^2y^2)^2$

11. $(-12cd^3)^2$ **12.** $\left(\frac{3}{4}a\right)^2$ **13.** $\left(\frac{5}{7}xy\right)^2$ **14.** $\left(-\frac{7}{8}a^2b^2\right)^2$ **15.** $\left(\frac{x}{6}\right)^2$

16. $\left(-\frac{4x^2}{5}\right)^2$ **17.** $(0.8x)^2$ **18.** $(0.5y^2)^2$ **19.** $(0.1xy)^2$ **20.** $(-0.6a^2b)^2$

21. In each case, represent the area of a square whose side is represented by:

a. $4x$ **b.** $10y$ **c.** $\frac{2}{3}x$ **d.** $1.5x$ **e.** $3x^2$ **f.** $4x^2y^2$

18-4 MULTIPLYING THE SUM AND THE DIFFERENCE OF TWO TERMS

When two binomials are multiplied, the product contains *four terms*. This fact is demonstrated in the diagram at the right and is also shown by using the distributive property.

	a	b
c	ac	bc
d	ad	bd

$$(a + b)(c + d) = a(c + d) + b(c + d)$$
$$= ac + ad + bc + bd$$

If two of the four terms of the product are similar, the similar terms can be combined so that the product is a trinomial. For example:

$$(x + 2)(x + 3) = x(x + 3) + 2(x + 3)$$
$$= x^2 + 3x + 2x + 6 = x^2 + 5x + 6$$

There is, however, a special case in which the sum of two terms is multiplied by the difference of the *same* two terms. In this case, the sum of the two middle terms of the product is 0, and the product is a binomial, as in the following examples:

$$(a + 4)(a - 4) = a(a - 4) + 4(a - 4)$$
$$= a^2 - 4a + 4a - 16$$
$$= a^2 - 16$$

$$(3x^2 + 5y)(3x^2 - 5y) = 3x^2(3x^2 - 5y) + 5y(3x^2 - 5y)$$
$$= 9x^4 - 15x^2y + 15x^2y - 25y^2$$
$$= 9x^4 - 25y^2$$

These examples illustrate the following procedure, which enables us to find the products mentally:

PROCEDURE. To multiply the sum of two terms by the difference of the same two terms:

1. Square the first term.

2. From this result, subtract the square of the second term.

$$(a + b)(a - b) = a^2 - b^2$$

EXAMPLES

Find each product mentally.

	Think:	*Write:*
a. $(y + 7)(y - 7)$	$= (y)^2 - (7)^2$	$= y^2 - 49$
b. $(3a + 4b)(3a - 4b)$	$= (3a)^2 - (4b)^2$	$= 9a^2 - 16b^2$

EXERCISES

In 1–22, find each product mentally.

1. $(x + 8)(x - 8)$
2. $(y + 10)(y - 10)$
3. $(m - 4)(m + 4)$
4. $(n - 9)(n + 9)$
5. $(10 + a)(10 - a)$
6. $(12 - b)(12 + b)$
7. $(c + d)(c - d)$
8. $(r - s)(r + s)$
9. $(3x + 1)(3x - 1)$
10. $(5c + 4)(5c - 4)$
11. $(8x + 3y)(8x - 3y)$
12. $(5r - 7s)(5r + 7s)$
13. $(x^2 + 8)(x^2 - 8)$
14. $(3 - 5y^2)(3 + 5y^2)$
15. $\left(a + \frac{1}{2}\right)\left(a - \frac{1}{2}\right)$
16. $(r + 0.5)(r - 0.5)$
17. $(0.3 + m)(0.3 - m)$
18. $(ab + 8)(ab - 8)$
19. $(r^3 - 2s^4)(r^3 + 2s^4)$
20. $(a + 5)(a - 5)(a^2 + 25)$
21. $(x - 3)(x + 3)(x^2 + 9)$
22. $(a + b)(a - b)(a^2 + b^2)$

In 23–26, express the area of each rectangle whose length ℓ and width w are given.

23. $\ell = x + 7, w = x - 7$
24. $\ell = 2x + 3, w = 2x - 3$
25. $\ell = c + d, w = c - d$
26. $\ell = 2a + 3b, w = 2a - 3b$

18-5 FACTORING THE DIFFERENCE OF TWO SQUARES

An expression of the form $a^2 - b^2$ is called a ***difference of two squares***. Factoring an expression that is the difference of two squares is the reverse of multiplying the sum of two terms by the difference of the same two terms. Since the product $(a + b)(a - b)$ is $a^2 - b^2$, the factors of $a^2 - b^2$ are $(a + b)$ and $(a - b)$. Therefore:

$$\mathbf{a^2 - b^2 = (a + b)(a - b)}$$

PROCEDURE. To factor a binomial that is a difference of two squares:
Express each of its terms as the square of a monomial; then apply the rule $a^2 - b^2 = (a + b)(a - b)$.

Remember that a monomial is a square if and only if its numerical coefficient is a square and the exponent of each of its variables is an even number.

EXAMPLES

1. Factor each polynomial mentally.

	Think:	*Write:*
a. $r^2 - 1$	$= (r)^2 - (1)^2$	$= (r + 1)(r - 1)$
b. $25x^2 - \frac{1}{49}y^2$	$= (5x)^2 - \left(\frac{1}{7}y\right)^2$	$= \left(5x + \frac{1}{7}y\right)\left(5x - \frac{1}{7}y\right)$
c. $0.04 - c^6d^4$	$= (0.2)^2 - (c^3d^2)^2$	$= (0.2 + c^3d^2)(0.2 - c^3d^2)$

---

2. Express $x^2 - 100$ as the product of two binomials.

Solution Since $x^2 - 100$ is a difference of two squares, the factors of $x^2 - 100$ are $(x + 10)$ and $(x - 10)$.

Answer: $(x + 10)(x - 10)$

EXERCISES WITH OPEN-RESPONSE PROBLEMS

In 1–9: If possible, express each binomial as the difference of the squares of monomials, as in the first step of the preceding examples; if not possible, tell why.

1. $y^2 - 64$ **2.** $4r^2 - b^2$ **3.** $r^2 + s^2$

4. $t^2 - 7$ **5.** $9n^2 - 16m^2$ **6.** $c^2 - 0.09d^2$

7. $p^2 - \frac{9}{25}q^2$ **8.** $16a^4 - 25b^6$ **9.** $-9 + m^2$

In 10–45, factor each binomial.

10. $a^2 - 4$ **11.** $b^2 - 25$ **12.** $c^2 - 100$ **13.** $r^2 - 16$

14. $s^2 - 49$ **15.** $t^2 - 81$ **16.** $9 - x^2$ **17.** $144 - c^2$

18. $121 - m^2$ **19.** $16a^2 - b^2$ **20.** $25m^2 - n^2$ **21.** $d^2 - 4c^2$

22. $r^4 - 9$ **23.** $x^4 - 64$ **24.** $25 - s^4$ **25.** $100x^2 - 81y^2$

26. $64e^2 - 9f^2$ **27.** $r^2s^2 - 144$ **28.** $w^2 - \frac{1}{64}$ **29.** $s^2 - \frac{1}{100}$

30. $\frac{1}{81} - t^2$ **31.** $49x^2 - \frac{1}{9}$ **32.** $\frac{4}{25} - \frac{49d^2}{81}$ **33.** $\frac{1}{9}r^2 - \frac{64s^2}{121}$

34. $x^2 - 0.64$ **35.** $y^2 - 1.44$ **36.** $0.04 - 49r^2$ **37.** $0.16m^2 - 9$

38. $81n^2 - 0.01$ **39.** $0.81x^2 - y^2$ **40.** $64a^2b^2 - c^2d^2$ **41.** $25r^2s^2 - 9t^2u^2$

42. $81m^2n^2 - 49x^2y^2$ **43.** $49m^4 - 64n^4$ **44.** $25x^6 - 121y^{10}$ **45.** $x^4y^8 - 144a^6b^{10}$

In 46–50, each given polynomial represents the area of a rectangle. Express the area as the product of two binomials.

46. $x^2 - 4$ **47.** $y^2 - 9$ **48.** $t^2 - 49$ **49.** $t^2 - 64$ **50.** $4x^2 - y^2$

In 51–53, express the area of each shaded region as: **a.** the difference of the areas shown and **b.** the product of two binomials.

51.

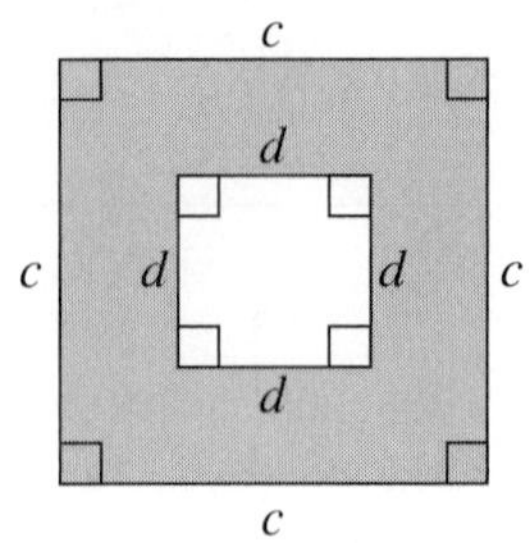

52.

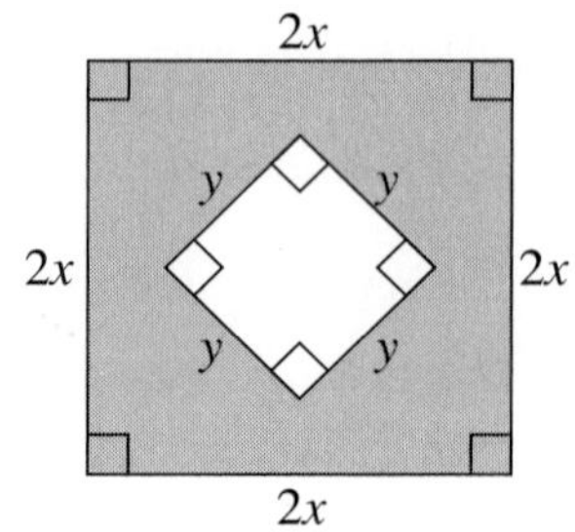

53.

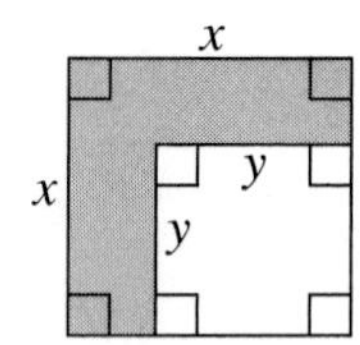

In 54 and 55, express the area of each shaded region as the product of two binomials.

54.

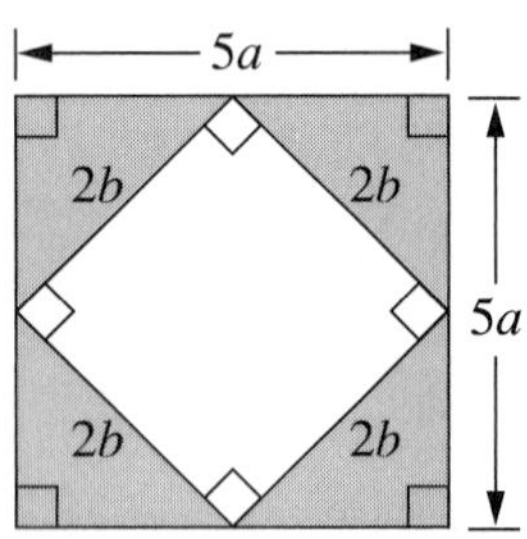

55.

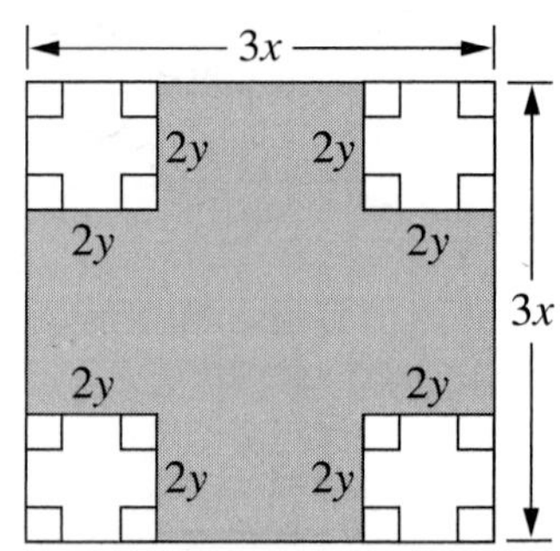

18-6 MULTIPLYING BINOMIALS

We have used the distributive property to multiply two binomials of the form $ax + b$ and $cx + d$. The pattern of the multiplication in the following example shows us how to find the product of two binomials mentally.

$$\begin{aligned}(2x - 3)(4x + 5) &= 2x(4x + 5) - 3(4x + 5)\\ &= 2x(4x) + 2x(5) - 3(4x) - 3(5)\\ &= 8x^2 + 10x - 12x - 15\\ &= 8x^2 - 2x - 15\end{aligned}$$

Examine the trinomial $8x^2 - 2x - 15$ and note the following:

1. The first term of the trinomial is the product of the first terms of the binomials:

$$(2x)(4x) = 8x^2$$

2. The middle term of the trinomial is the sum of two terms, the product of the outer terms, and the product of the inner terms of the binomials:

$$\underbrace{(2x - \overbrace{3)(4x}^{-12x} + 5)}_{+10x} \qquad \textit{Think:} \quad (-12x) + (+10x) = -2x$$

3. The last term of the trinomial is the product of the last terms of the binomials:

$$(-3)(+5) = -15$$

PROCEDURE. To find the product of two binomials of the form $ax + b$ and $cx + d$:
1. Multiply the first terms of the binomials.
2. Multiply the first term of each binomial by the last term of the other binomial (the outer terms and the inner terms), and add these products.
3. Multiply the last terms of the binomials.
4. Combine the results obtained in steps 1, 2, and 3.

EXAMPLES

Multiply.

a. $(x - 5)(x - 7)$

Solution

$$\underbrace{(x - \overbrace{5)(x}^{-5x} - 7)}_{-7x}$$

Think:
1. $(x)(x) = x^2$
2. $(-5x) + (-7x) = -12x$
3. $(-5)(-7) = +35$

Write: $(x - 5)(x - 7) = x^2 - 12x + 35$ *Answer*

b. $(3y - 8)(4y + 3)$

Solution

$$\overset{-32y}{(3y - 8)(4y + 3)}_{+9y}$$

Think:

1. $(3y)(4y) = 12y^2$
2. $(-32y) + (+9y) = -23y$
3. $(-8)(+3) = -24$

Write: $(3y - 8)(4y + 3) = 12y^2 - 23y - 24$ *Answer*

EXERCISES

In 1–38, perform each indicated operation mentally.

1. $(x + 5)(x + 3)$ **2.** $(y + 9)(y + 2)$ **3.** $(6 + d)(3 + d)$ **4.** $(x - 10)(x - 5)$

5. $(y - 1)(y - 9)$ **6.** $(8 - c)(3 - c)$ **7.** $(x + 7)(x - 2)$ **8.** $(y + 11)(y - 4)$

9. $(m - 15)(m + 2)$ **10.** $(n - 20)(n + 3)$ **11.** $(5 - t)(9 + t)$ **12.** $(2x + 1)(x + 1)$

13. $(3x + 2)(x + 5)$ **14.** $(c - 5)(3c - 1)$ **15.** $(m - 6)(3m + 2)$ **16.** $(y + 8)^2$

17. $(a - 4)^2$ **18.** $(y + 5)^2$ **19.** $(1 - t)^2$ **20.** $(2x + 1)^2$

21. $(3x - 2)^2$ **22.** $(7x + 3)(2x - 1)$ **23.** $(2y + 3)(3y + 2)$ **24.** $(5k - 3)(2k - 5)$

25. $(2y + 3)(2y + 3)$ **26.** $(3x + 4)^2$ **27.** $(2x - 5)^2$ **28.** $(3t - 2)(4t + 7)$

29. $(5y - 4)(5y - 4)$ **30.** $(2t + 3)(5t + 1)$ **31.** $(2a - 1)(2a - 3)$ **32.** $(5x + 7)(3x - 4)$

33. $(2c - 3d)(5c - 2d)$ **34.** $(4a - 3b)(3a + b)$ **35.** $(5a + 7b)(5a - 7b)$

36. $(5a + 7b)(5a + 7b)$ **37.** $(5a + 7b)(7a + 5b)$ **38.** $(5a + 7b)(7a - 5b)$

39. Represent the area of a rectangle whose length and width are:
a. $(x + 5)$ and $(x + 4)$ **b.** $(2x + 3)$ and $(x - 1)$

40. Represent the area of a square each of whose sides is:
a. $(x + 6)$ **b.** $(x - 2)$ **c.** $(2x + 1)$ **d.** $(3x - 2)$

18-7 FACTORING TRINOMIALS

We have learned that $(x + 3)(x + 5) = x^2 + 8x + 15$. Therefore, factors of $x^2 + 8x + 15$ are $(x + 3)$ and $(x + 5)$. Factoring a trinomial of the form $ax^2 + bx + c$ is the reverse of multiplying binomials of the form $dx + e$ and $fx + g$. When we factor a trinomial of this form, we list the possible pairs of factors, using combinations of factors of the first and last terms, and test the pairs, one by one, until we find the correct middle term.

For example, let us factor $x^2 + 7x + 10$.

1. The product of the first terms of the binomials must be x^2. Therefore, for each first term, we use x. We write:

$$x^2 + 7x + 10 = (x \quad)(x \quad)$$

2. Since the product of the last terms of the binomials must be $+10$, these last terms must be either both positive or both negative. The pairs of integers whose products is $+10$ are

$(+1)(+10)$ $\qquad$ $(+5)(+2)$ $\qquad$ $(-1)(-10)$ $\qquad$ $(-2)(-5)$

3. From the products obtained in steps 1 and 2, we see that the possible pairs of factors are

$$(x + 10)(x + 1) \qquad (x - 10)(x - 1)$$
$$(x + 5)(x + 2) \qquad (x - 5)(x - 2)$$

4. Now, we test each pair of factors. For example:

$(x + 10)(x + 1)$ is not correct because the middle term, $(+10x) + (+1x)$, is $+11x$, not $+7x$.

$+10x$
$(x + 10)(x + 1)$
$+1x$

$(x + 5)(x + 2)$ is correct because the middle term, $(+5x) + (+2x)$, is $+7x$.

$+5x$
$(x + 5)(x + 2)$
$+2x$

None of the remaining pairs of factors is correct because each would result in a negative middle term.

5. The factors of $x^2 + 7x + 10$ are $(x + 5)(x + 2)$.

Observe that, in this trinomial, the first and last terms are both positive: x^2 and $+10$. Since the middle term of the trinomial is also *positive*, the last terms of both binomial factors must be *positive* ($+5$ and $+2$).

In this example, we chose to use x times x as the factors of x^2 when selecting the first terms of the binomial factors. If we had chosen $-x$ times $-x$ as the factors of x^2, the factors of $x^2 + 7x + 10$ would have been written as $(-x - 5)(-x - 2)$. Every trinomial that has two binomial factors also has the opposites of these binomials as factors. Usually, however, we write the pair of factors whose first terms have positive coefficients as the factors of a trinomial.

PROCEDURE. To factor a trinomial of the form $ax^2 + bx + c$, find two binomials that have the following characteristics:

1. The product of the first terms of the binomials is equal to the first term in the trinomial (ax^2).
2. The product of the last terms of the binomials is equal to the last term of the trinomial (c).
3. When the first term of one binomial is multiplied by the last term of the other binomial and the sum of these products is found, the result is equal to the middle term of the trinomial (bx).

EXAMPLES

Factor.

a. $y^2 - 8y + 12$

Solution (1) The product of the first terms of the binomials must be y^2. Therefore, for each first term, we use y. We write:

$$y^2 - 8y + 12 = (y \quad)(y \quad)$$

(2) Since the product of the last terms of the binomials must be $+12$, these last terms must be either both positive or both negative. The pairs of integers whose product is $+12$ are:

$(+1)(+12)$	$(+2)(+6)$	$(+3)(+4)$
$(-1)(-12)$	$(-2)(-6)$	$(-3)(-4)$

(3) The possible factors are:

$(y + 1)(y + 12)$	$(y + 2)(y + 6)$	$(y + 3)(y + 4)$
$(y - 1)(y - 12)$	$(y - 2)(y - 6)$	$(y - 3)(y - 4)$

(4) When we find the middle term in each of the trinomial products, we see that only the factors $(y - 6)\,(y - 2)$ yield a middle term of $-8y$.

$-6y$
$(y - 6)(y - 2)$
$-2y$

Answer: $y^2 - 8y + 12 = (y - 6)(y - 2)$

When the first and last terms are both positive (y^2 and $+ 12$) and the middle term of the trinomial is *negative*, the last terms of both binomial factors must be *negative* (-6 and -2).

b. $c^2 + 5c - 6$

Solution (1) The product of the first terms of the binomials must be c^2. Therefore, for each first term, we use c. We write:

$$c^2 + 5c - 6 = (c \quad)(c \quad)$$

(2) Since the product of the last terms of the binomials must be -6, one of these last terms must be positive, the other negative. The pairs of integers whose product is -6 are:

$$(+1)(-6) \qquad (+3)(-2)$$
$$(-1)(+6) \qquad (-3)(+2)$$

(3) The possible factors are:

$$(c + 1)(c - 6) \qquad (c + 3)(c - 2)$$
$$(c - 1)(c + 6) \qquad (c - 3)(c + 2)$$

(4) When we find the middle term of each of the trinomial products, we see that only the factors $(c - 1)(c + 6)$ yield a middle term of $+5c$.

$-1c$
$(c - 1)(c + 6)$
$+6c$

Answer: $c^2 + 5c - 6 = (c - 1)(c + 6)$

c. $2x^2 - 7x - 15$

Solution (1) Since the product of the first terms of the binomials must be $2x^2$, we use as one of these terms $2x$, and as the other, x. We write:

$$2x^2 - 7x - 15 = (2x \quad)(x \quad)$$

(2) Since the product of the last terms of the binomials must be -15, one of these last terms must be positive, the other negative. The pairs of integers whose product is -15 are:

$$(+1)(-15) \qquad (+3)(-5)$$
$$(-1)(+15) \qquad (-3)(+5)$$

(3) These four pairs of integers will form eight pairs of binomial factors since the way in which the integers are combined with the first terms will produce different pairs of factors: $(2x + 1)(x - 15)$ is not the same product as $(2x - 15)(x + 1)$. The possible pairs of factors are:

$$(2x + 1)(x - 15) \quad (2x + 3)(x - 5) \quad (2x - 1)(x + 15) \quad (2x - 3)(x + 5)$$
$$(2x + 15)(x - 1) \quad (2x + 5)(x - 3) \quad (2x - 15)(x + 1) \quad (2x - 5)(x + 3)$$

(4) When we find the middle term of each of the trinomial products, we find that only the factors $(2x + 3)(x - 5)$ yield a middle term of $-7x$.

$+3x$
$(2x + 3)(x - 5)$
$-10x$

Answer: $2x^2 - 7x - 15 = (2x + 3)(x - 5)$

In factoring a trinomial of the form $ax^2 + bx + c$, when a is a positive integer ($a > 0$):

1. The coefficients of the first terms of the binomial factors are usually written as positive integers.
2. If the last term, c, is positive, the last terms of the binomial factors must be either both positive or both negative.
3. If the last term, c, is negative, one of the last terms of the binomial factors must be positive and the other negative.

EXERCISES

In 1–45, factor each trinomial:

1. $a^2 + 3a + 2$
2. $c^2 + 6c + 5$
3. $x^2 + 8x + 7$
4. $r^2 + 12r + 11$
5. $m^2 + 5m + 4$
6. $y^2 + 12y + 35$
7. $x^2 + 11x + 24$
8. $a^2 + 11a + 18$
9. $16 + 17c + c^2$
10. $x^2 + 2x + 1$
11. $z^2 + 10z + 25$
12. $a^2 - 8a + 7$
13. $a^2 - 6a + 5$
14. $x^2 - 5x + 6$
15. $x^2 - 11x + 10$
16. $y^2 - 6y + 8$
17. $15 - 8y + y^2$
18. $x^2 - 10x + 24$
19. $c^2 - 14c + 40$
20. $x^2 - 16x + 48$
21. $x^2 - 14x + 49$
22. $x^2 - x - 2$
23. $x^2 - 6x - 7$
24. $y^2 + 4y - 5$
25. $z^2 - 12z - 13$
26. $c^2 - 2c - 15$
27. $c^2 + 2c - 35$
28. $x^2 - 7x - 18$
29. $z^2 + 9z - 36$
30. $x^2 - 13x - 48$
31. $x^2 - 16x + 64$
32. $2x^2 + 5x + 2$
33. $2x^2 + 7x + 6$
34. $3x^2 + 10x + 8$
35. $16x^2 + 8x + 1$
36. $2x^2 + x - 3$
37. $3x^2 + 2x - 5$
38. $2x^2 + x - 6$
39. $4x^2 - 12x + 5$
40. $10a^2 - 9a + 2$
41. $18y^2 - 23y - 6$
42. $x^2 + 3xy + 2y^2$
43. $r^2 - 3rs - 10s^2$
44. $3a^2 - 7ab + 2b^2$
45. $4x^2 - 5xy - 6y^2$

In 46–48, express each polynomial as the product of two binomial factors.

46. $x^2 + 9x + 18$
47. $x^2 - 9x + 14$
48. $y^2 - 5y - 24$

In 49–51, each trinomial represents the area of a rectangle. In each case, find the binomials that could be expressions for the dimensions of the rectangle.

49. $x^2 + 8x + 7$
50. $x^2 + 9x + 20$
51. $3x^2 + 14x + 15$

In 52–54, each trinomial represents the area of a square. In each case, find the binomial that could be an expression for the measure of each side of the square.

52. $x^2 + 10x + 25$
53. $81x^2 + 18x + 1$
54. $4x^2 + 12x + 9$

18-8 FACTORING A POLYNOMIAL COMPLETELY

Some polynomials, such as $x^2 + 4$ and $x^2 + x + 1$, cannot be factored into other polynomials with integral coefficients. We say that these polynomials are *prime over the set of integers.*

Factoring a polynomial completely means finding the *prime factors* of the polynomial over a designated set of numbers. In this book, whenever we factor a polynomial, we will continue the process of factoring until all factors other than monomial factors are prime factors over the set of integers.

PROCEDURE. To factor a polynomial completely:

1. Look for the greatest common factor. If there is one, factor the given polynomial. Then examine each factor.
2. If any of these factors is a trinomial, see whether it can be factored. If so, find its binomial factors.
3. If any of these factors is a binomial, see whether it is a difference of two squares. If so, factor it as such.
4. Write the answer as the product of all the factors. Make certain that in the answer all factors other than monomial factors are prime factors.

EXAMPLES

Factor.

a. $by^2 - 4b$

How to Proceed:	*Solution:*
(1) Find the greatest common factor:	$by^2 - 4b = b(y^2 - 4)$
(2) Factor the difference of two squares:	$by^2 - 4b = b(y + 2)(y - 2)$ *Answer*

b. $3x^2 - 6x - 24$

How to Proceed:	*Solution:*
(1) Find the greatest common factor:	$3x^2 - 6x - 24 = 3(x^2 - 2x - 8)$
(2) Factor the trinomial:	$3x^2 - 6x - 24 = 3(x - 4)(x + 2)$ *Answer*

c. $x^4 - 16$

How to Proceed:	*Solution:*
(1) Find the greatest common factor if there is one.	
(2) Since there is no G.C.F., factor $x^4 - 16$ as the difference of two squares:	$x^4 - 16 = (x^2 + 4)(x^2 - 4)$
(3) Factor $x^2 - 4$ as the difference of two squares:	$x^4 - 16 = (x^2 + 4)(x + 2)(x - 2)$ *Answer*

EXERCISES

In 1–46, factor each polynomial completely.

1. $2a^2 - 2b^2$ **2.** $6x^2 - 6y^2$ **3.** $4x^2 - 4$ **4.** $ax^2 - ay^2$

5. $cm^2 - cn^2$ **6.** $st^2 - s$ **7.** $2x^2 - 18$ **8.** $2x^2 - 32$

9. $3x^2 - 27y^2$ **10.** $18m^2 - 8$ **11.** $12a^2 - 27b^2$ **12.** $63c^2 - 7$

13. $x^3 - 4x$ **14.** $y^3 - 25y$ **15.** $z^3 - z$ **16.** $4a^3 - ab^2$

17. $4c^3 - 49c$ **18.** $9db^2 - d$ **19.** $4a^2 - 36$ **20.** $x^4 - 1$

21. $y^4 - 81$ **22.** $\pi R^2 - \pi r^2$ **23.** $\pi c^2 - \pi d^2$ **24.** $100x^2 - 36y^2$

25. $ax^2 + 3ax + 2a$ **26.** $3x^2 + 6x + 3$ **27.** $4r^2 - 4r - 48$ **28.** $x^3 + 7x^2 + 10x$

29. $4x^2 - 6x - 4$ **30.** $a^2y + 10ay + 25y$ **31.** $d^3 - 8d^2 + 16d$

32. $2ax^2 - 2ax - 12a$ **33.** $abx^2 - ab$ **34.** $z^6 - z^2$

35. $16x^2 - x^2y^4$ **36.** $x^4 + x^2 - 2$ **37.** $a^4 - 10a^2 + 9$

38. $y^4 - 13y^2 + 36$ **39.** $2x^2 + 12x + 8$ **40.** $5x^4 + 10x^2 + 5$

41. $2a^2b + 7ab + 3b$ **42.** $16x^2 - 16x + 4$ **43.** $25x^2 + 100xy + 100y^2$

44. $18m^2 + 24m + 8$ **45.** $12a^2 - 5ab - 2b^2$ **46.** $10a^3 + 20a^2 + 10a$

CHAPTER SUMMARY

Factoring a number or a polynomial is the process of finding those numbers or polynomials whose product is the given number or polynomial. A ***prime number or polynomial*** has only two factors, 1 and itself. A ***composite number*** is an integer greater than 1 that is not prime. A composite number has more than two factors.

To factor a polynomial completely:

1. Factor out the greatest common monomial factor if there is one.
2. Write any factor of the form $a^2 - b^2$ as $(a + b)(a - b)$.

3. Write any factor of the form $ax^2 + bx + c$ as the product of two binomial factors if possible.
4. Write the given polynomial as the product of these factors.

VOCABULARY

18-1 Factor   Factoring a number   Prime number   Composite number   Greatest common factor

18-2 Common monomial factor   Greatest common monomial factor   Prime polynomial

18-3 Square of a monomial

18-5 Difference of two squares

18-8 Factoring a polynomial completely

REVIEW EXERCISES WITH OPEN-RESPONSE PROBLEMS

1. Express 250 as a product of prime numbers.
2. What is the greatest common factor of $8ax$ and $4ay$?
3. Find the greatest common factor of $16a^3bc^2$ and $24a^2bc^4$.

In 4–6, square each monomial

4. $(3g^3)^2$
5. $(-4x^4)^2$
6. $(0.2c^2y)^2$

In 7–12, find each product.

7. $(x - 5)(x + 9)$
8. $(y - 8)(y - 6)$
9. $(ab + 4)(ab - 4)$
10. $(3d + 1)(d - 2)$
11. $(2w + 1)^2$
12. $(2x + 3c)(x + 4c)$

In 13–27, in each case factor completely.

13. $6x + 27b$
14. $3y^2 + 10y$
15. $m^2 - 81$
16. $x^2 - 16h^2$
17. $x^2 - 4x - 5$
18. $y^2 - 9y + 14$
19. $64b^2 - 9$
20. $121 - k^2$
21. $x^2 - 8x + 16$
22. $a^2 - 7a - 30$
23. $x^2 - 16x + 60$
24. $16y^2 - 16$
25. $x^2 + 6bx - 16b^2$
26. $2x^2 - 9xy - 5y^2$
27. $3x^3 - 6x^2 - 24x$
28. Express the product $(k + 15)(k - 15)$ as a binomial.
29. Express as a binomial: $4ez^2(4e - z)$.
30. If the length and width of a rectangle are represented by $2x - 3$ and $3x - 2$, respectively, express the area of the rectangle as a trinomial.
31. Find the trinomial that represents the area of a square if the measure of a side is $8m + 1$.

32. If $9x^2 + 30x + 25$ represents the square of a number, find the binomial that represents the number.

33. Factor completely: $60a^2 + 37a - 6$.

34. A group of people wanted to form committees each of which had the same number of persons; everyone in the group was to serve on a committee. When the group tried to make 2, 3, 4, 5, or 6 committees, there was always one extra person. However, they were able to make more than 6 but fewer than 12 committees of equal size.
a. What was the smallest possible number of persons in the group?
b. How many persons were on each committee that was formed?

CUMULATIVE REVIEW WITH OPEN-RESPONSE PROBLEMS

1. The cost of a hamburger and a cola is \$2.70. The cost of a hamburger and fries is \$2.90. The cost of a hamburger, a cola, and fries is \$3.70. What is the cost of a hamburger?

2. **a.** Write an equation of a line whose slope is -2 and whose y-intercept is 5.
b. Sketch the graph of the equation written in **a**.
c. Is $(-1, 3)$ on the graph of the equation written in **a**?
d. If $(k, 1)$ is a point on the graph of the equation written in **a**, what is the value of k?

3. Let x be a number chosen from the positive integers less than 31. After x is chosen and not replaced, let y be a second number chosen at random from the numbers that remain.
a. What is the probability that $\sqrt{x}$ is rational?
b. What is the probability that $\sqrt{x} < 5$?
c. What is the probability that $\sqrt{x} + \sqrt{y}$ is rational?
d. What is the probability that $\sqrt{x + y}$ is rational?

4. Of the four polynomials given below, which is different from the others? Explain why it is different.

$x^2 - 9 \qquad x^2 - 2x + 1 \qquad x^2 - 2x - 1 \qquad x^3 + 5x^2 + 6x$

Exploration

Explain how the expression $(a + b)(a - b) = a^2 - b^2$ can be used to find the product of two consecutive even or two consecutive odd integers.

Chapter 19

Algebraic Fractions, and Equations and Inequalities Involving Fractions

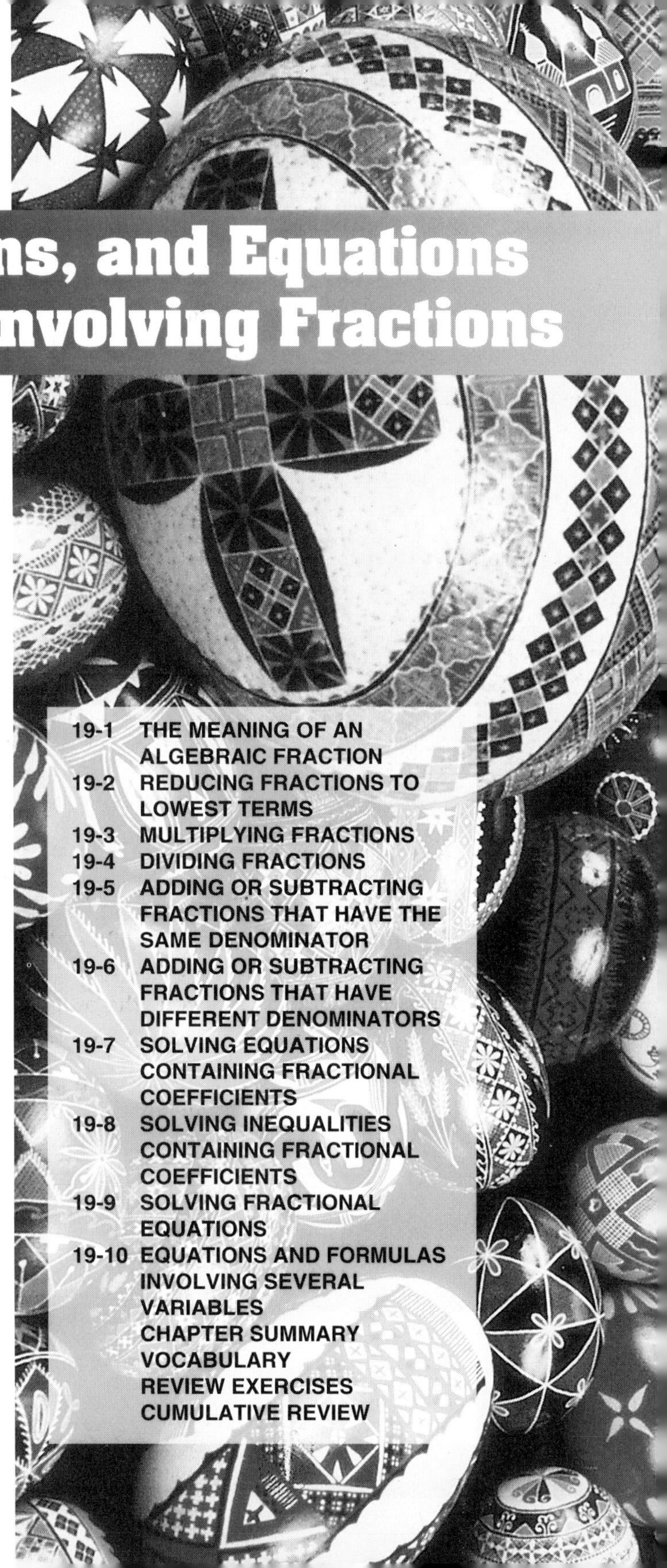

Although people today are making greater use of calculators, computers, and the metric system, leading to increased use of decimal fractions, ***common fractions*** still surround us.

We use common fractions in everyday measures: $\frac{1}{4}$-inch nail, $2\,\frac{1}{2}$-yard gain in football, $\frac{1}{2}$ pint of cream, $1\frac{1}{3}$ cups of flour. We buy $\frac{1}{2}$ dozen eggs, not 0.5 dozen eggs. We describe 15 minutes as $\frac{1}{4}$ hour, not 0.25 hour. Items are sold at a third $\left(\frac{1}{3}\right)$ off, or at a fraction of the original price.

Fractions are also used when sharing. For example, Andrea designed some beautiful Ukrainian eggs this year. She gave one-fifth of the eggs to her grandparents. Then she gave one-fourth of the eggs she had left to her parents. Next, she presented her aunt with one-third of the eggs that remained. Finally, she gave one-half of the eggs she had left to her brother, and she kept six eggs. Can you use some problem-solving skills to discover how many Ukrainian eggs Andrea designed?

In this chapter, you will learn operations with algebraic fractions and methods to solve equations and inequalities that involve fractions.

19-1 THE MEANING OF AN ALGEBRAIC FRACTION

A *fraction* is a quotient of any number divided by any nonzero number. For example, the arithmetic fraction $\frac{3}{4}$ indicates the quotient of 3 and 4.

An ***algebraic fraction*** is a quotient of two algebraic expressions. An algebraic fraction that is the quotient of two polynomials is called a ***rational expression***. Here are some examples of algebraic fractions:

$$\frac{2}{5} \quad \frac{x}{2} \quad \frac{2}{x} \quad \frac{a}{b} \quad \frac{4c}{3d} \quad \frac{x+5}{x-2} \quad \frac{x^2+4x+3}{x+1}$$

The fraction $\frac{a}{b}$ means that the number represented by a, the numerator, is to be divided by the number represented by b, the denominator. Since division by 0 is not possible, the value of the denominator, b, cannot be 0. An algebraic fraction is defined or has meaning only for values of the variables for which the denominator is not 0.

EXAMPLE

Find the value of x for which $\frac{12}{x-9}$ has no meaning.

Solution $\frac{12}{x-9}$ is not defined when the denominator, $x - 9$, is equal to 0.

Let $x - 9 = 0$. Then $x = 9$.

Answer: 9

EXERCISES

In 1–10, find, in each case, the value of the variable for which the fraction is not defined.

1. $\frac{2}{x}$ **2.** $\frac{-5}{6x}$ **3.** $\frac{12}{y^2}$ **4.** $\frac{1}{x-5}$ **5.** $\frac{x}{x-8}$

6. $\frac{7}{2-x}$ **7.** $\frac{y+5}{y+2}$ **8.** $\frac{10}{2x-1}$ **9.** $\frac{2y+3}{4y+2}$ **10.** $\frac{1}{x^2-4}$

In 11–15, represent the answer to each problem as a fraction.

11. What is the cost of one piece of candy if five pieces cost c cents?

12. What is the cost of 1 meter of lumber if p meters cost 98 cents?

13. If a piece of lumber $10x + 20$ centimeters in length is cut into y pieces of equal length, what is the length of each of the pieces?

14. What fractional part of an hour is m minutes?

15. If the perimeter of a square is $4x + 2y$, what is the length of each side of the square?

19-2 REDUCING FRACTIONS TO LOWEST TERMS

A fraction is said to be ***reduced to lowest terms*** when its numerator and denominator have no common factor other than 1 or -1.

Each of the fractions $\frac{5}{10}$ and $\frac{a}{2a}$ can be expressed in lowest terms as $\frac{1}{2}$.

The arithmetic fraction $\frac{5}{10}$ is reduced to lowest terms when both its numerator and denominator are divided by 5:

$$\frac{5}{10} = \frac{5 \div 5}{10 \div 5} = \frac{1}{2}$$

In the same way, the algebraic fraction $\frac{a}{2a}$ is reduced to lowest terms when both its numerator and denominator are divided by a, where $a \neq 0$:

$$\frac{a}{2a} = \frac{1a \div a}{2a \div a} = \frac{1}{2}$$

Fractions that are equal in value are called ***equivalent fractions***. Thus, $\frac{5}{10}$ and $\frac{1}{2}$ are equivalent fractions, and both are equivalent to $\frac{a}{2a}$, where $a \neq 0$.

The examples shown above illustrate the ***division property of a fraction***: If the numerator and the denominator of a fraction are divided by the same nonzero number, the resulting fraction is equal to the original fraction.

In general, for any numbers x, y, and a, where $y \neq 0$ and $a \neq 0$:

$$\boldsymbol{\frac{ax}{ay} = \frac{ax \div a}{ay \div a} = \frac{x}{y}}$$

When a fraction is reduced to lowest terms, we list the values of the variables that must be excluded so that the original fraction is equivalent to the reduced form and also has meaning. For example:

$$\frac{4x}{5x} = \frac{4x \div x}{5x \div x} = \frac{4}{5} \text{ (where } x \neq 0\text{)}, \qquad \frac{cy}{dy} = \frac{cy \div y}{dy \div y} = \frac{c}{d} \text{ (where } y \neq 0, d \neq 0\text{)}$$

When reducing a fraction, the division of the numerator and the denominator by a common factor may be indicated by a ***cancellation***.

Here, we use cancellation to divide the numerator and the denominator by 3:

$$\frac{3(x+5)}{18} = \frac{\overset{1}{\cancel{3}}(x+5)}{\underset{6}{\cancel{18}}} = \frac{x+5}{6}$$

Here, we use cancellation to divide the numerator and the denominator by $(a - 3)$:

$$\frac{a^2 - 9}{3a - 9} = \frac{\overset{1}{\cancel{(a-3)}}(a+3)}{3\underset{1}{\cancel{(a-3)}}} = \frac{a+3}{3}$$

(where $a \neq 3$)

By reexamining one of the examples just seen, we can show that the *multiplication property of 1* is used whenever a fraction is reduced:

$$\frac{3(x+5)}{18} = \frac{3 \cdot (x+5)}{3 \cdot 6} = \frac{3}{3} \cdot \frac{(x+5)}{6} = \mathbf{1} \cdot \frac{(x+5)}{6} = \frac{x+5}{6}$$

PROCEDURE. To reduce a fraction to its lowest terms:

Method 1

1. Factor completely both the numerator and the denominator.
2. Determine the greatest common factor of the numerator and the denominator.
3. Express the given fraction as the procuct of two fractions, one of which has as its numerator and its denominator the greatest common factor determined in step 2.
4. Use the multiplication property of 1.

Method 2

1. Factor both the numerator and the denominator.
2. Divide both the numerator and the denominator by their greatest common factor.

EXAMPLES

1. Reduce $\frac{15x^2}{35x^4}$ to lowest terms.

Solution

METHOD 1

$$\frac{15x^2}{35x^4} = \frac{3}{7x^2} \cdot \frac{5x^2}{5x^2}$$

$$= \frac{3}{7x^2} \cdot 1$$

$$= \frac{3}{7x^2}$$

Answer: $\frac{3}{7x^2}$ $(x \neq 0)$

METHOD 2

$$\frac{15x^2}{35x^4} = \frac{3 \cdot 5x^2}{7x^2 \cdot 5x^2}$$

$$= \frac{3 \cdot \overset{1}{\cancel{5x^2}}}{7x^2 \cdot \underset{1}{\cancel{5x^2}}} = \frac{3}{7x^2}$$

Answer: $\frac{3}{7x^2}$ $(x \neq 0)$

2. Express $\frac{2x^2 - 6x}{10x}$ as an equivalent fraction in lowest terms.

Solution

METHOD 1

$$\frac{2x^2 - 6x}{10x} = \frac{2x(x-3)}{2x \cdot 5}$$
$$= \frac{(x-3)}{5} \cdot \frac{2x}{2x}$$
$$= \frac{(x-3)}{5} \cdot 1$$
$$= \frac{x-3}{5}$$

Answer: $\frac{x-3}{5}$ $(x \neq 0)$

METHOD 2

$$\frac{2x^2 - 6x}{10x} = \frac{2x(x-3)}{10x}$$
$$= \frac{\overset{1}{\cancel{2x}}(x-3)}{\underset{5}{\cancel{10x}}}$$
$$= \frac{x-3}{5}$$

Answer: $\frac{x-3}{5}$ $(x \neq 0)$

---

3. Reduce each fraction to lowest terms.

a. $\frac{x^2 - 16}{x^2 - 5x + 4}$

Solution:

$$\frac{x^2 - 16}{x^2 - 5x + 4} = \frac{(x+4)(x-4)}{(x-1)(x-4)}$$
$$= \frac{(x+4)\overset{1}{\cancel{(x-4)}}}{(x-1)\underset{1}{\cancel{(x-4)}}}$$
$$= \frac{x+4}{x-1}$$

Answer: $\frac{x+4}{x-1}$ $(x \neq 1, x \neq 4)$

b. $\frac{2 - x}{4x - 8}$

Solution:

$$\frac{2 - x}{4x - 8} = \frac{-x + 2}{4x - 8}$$
$$= \frac{-x + 2}{4x - 8}$$
$$= \frac{-1(x-2)}{4(x-2)}$$
$$= \frac{-1\overset{1}{\cancel{(x-2)}}}{4\underset{1}{\cancel{(x-2)}}} = -\frac{1}{4}$$

Answer: $-\frac{1}{4}$ $(x \neq 2)$

EXERCISES WITH OPEN-RESPONSE PROBLEMS

In 1–52, reduce each fraction to lowest terms.

1. $\frac{4}{12}$ **2.** $\frac{27}{36}$ **3.** $\frac{24c}{36d}$ **4.** $\frac{9r}{10r}$

5. $\frac{ab}{cb}$ **6.** $\frac{3ay^2}{6by^2}$ **7.** $\frac{5xy}{9xy}$ **8.** $\frac{2abc}{4abc}$

9. $\frac{15x^2}{5x}$ 10. $\frac{5x^2}{25x^4}$ 11. $\frac{27a}{36a^2}$ 12. $\frac{8xy^2}{24x^2y}$

13. $\frac{+12a^2b}{-8ac}$ 14. $\frac{-20x^2y^2}{-90xy^2}$ 15. $\frac{-32a^3b^3}{+48a^3b^3}$ 16. $\frac{+5xy}{+45x^2y^2}$

17. $\frac{3x + 6}{4}$ 18. $\frac{8y - 12}{6}$ 19. $\frac{5x - 35}{5x}$ 20. $\frac{8m^2 + 40m}{8m}$

21. $\frac{2ax + 2bx}{6x^2}$ 22. $\frac{5a^2 - 10a}{5a^2}$ 23. $\frac{12ab - 3b^2}{3ab}$ 24. $\frac{6x^2y + 9xy^2}{12xy}$

25. $\frac{18b^2 + 30b}{9b^3}$ 26. $\frac{4x}{4x + 8}$ 27. $\frac{7d}{7d + 14}$ 28. $\frac{5y}{5y + 5x}$

29. $\frac{2a^2}{6a^2 - 2ab}$ 30. $\frac{14}{7r - 21s}$ 31. $\frac{12a + 12b}{3a + 3b}$ 32. $\frac{x^2 - 9}{3x + 9}$

33. $\frac{x^2 - 1}{5x - 5}$ 34. $\frac{1 - x}{x - 1}$ 35. $\frac{3 - b}{b^2 - 9}$ 36. $\frac{2s - 2r}{s^2 - r^2}$

37. $\frac{16 - a^2}{2a - 8}$ 38. $\frac{x^2 - y^2}{3y - 3x}$ 39. $\frac{2b(3 - b)}{b^2 - 9}$ 40. $\frac{r^2 - r - 6}{3r - 9}$

41. $\frac{x^2 + 7x + 12}{x^2 - 16}$ 42. $\frac{x^2 + x - 2}{x^2 + 4x + 4}$ 43. $\frac{3y - 3}{y^2 - 2y + 1}$ 44. $\frac{x^2 - 3x}{x^2 - 4x + 3}$

45. $\frac{x^2 - 25}{x^2 - 2x - 15}$ 46. $\frac{a^2 - a - 6}{a^2 - 9}$ 47. $\frac{a^2 - 6a}{a^2 - 7a + 6}$ 48. $\frac{2x^2 - 50}{x^2 + 8x + 15}$

49. $\frac{r^2 - 4r - 5}{r^2 - 2r - 15}$ 50. $\frac{48 + 8x - x^2}{x^2 + x - 12}$ 51. $\frac{2x^2 - 7x + 3}{(x - 3)^2}$ 52. $\frac{x^2 - 7xy + 12y^2}{x^2 + xy - 20y^2}$

53. **a.** Use substitution to find the *numerical* value of $\frac{x^2 - 5x}{x - 5}$, and then reduce each *numerical fraction* to lowest terms when:
(*1*) $x = 7$ (*2*) $x = 10$ (*3*) $x = 20$ (*4*) $x = 2$ (*5*) $x = -4$

b. What pattern, if any, do you observe for the answers to part **a**?

c. Can substitution be used to reduce $\frac{x^2 - 5x}{x - 5}$ when $x = 5$? Explain your answer.

d. Reduce the *algebraic fraction* $\frac{x^2 - 5x}{x - 5}$ to lowest terms.

e. Using the answer to part **d**, find the value of $\frac{x^2 - 5x}{x - 5}$, reduced to lowest terms, when $x = 38{,}756$.

19-3 MULTIPLYING FRACTIONS

The product of two fractions is a fraction with the following properties:

1. The numerator is the product of the numerators of the given fractions.
2. The denominator is the product of the denominators of the given fractions.

In general, for any numbers a, b, x, and y, when $b \neq 0$ and $y \neq 0$:

$$\frac{a}{b} \cdot \frac{x}{y} = \frac{ax}{by}$$

We can find the product of $\frac{7}{27}$ and $\frac{9}{4}$ in lowest terms by using either of two methods:

METHOD 1	METHOD 2
$\frac{7}{27} \cdot \frac{9}{4} = \frac{7 \cdot 9}{27 \cdot 4} = \frac{63}{108} = \frac{7 \cdot \overset{1}{\cancel{9}}}{12 \cdot \underset{1}{\cancel{9}}} = \frac{7}{12}$	$\frac{7}{27} \cdot \frac{9}{4} = \frac{7 \cdot \overset{1}{\cancel{9}}}{\underset{3}{\cancel{27}} \cdot 4} = \frac{7}{12}$

Notice that Method 2 requires less computation than Method 1 since the reduced form of the product was obtained by dividing the numerator and the denominator by a common factor *before* the product was found. This method may be called the ***cancellation method***.

The properties that apply to the multiplication of arithmetic fractions also apply to the multiplication of algebraic fractions.

To multiply $\frac{5x^2}{7y}$ by $\frac{14y^2}{15x^3}$ and express the product in lowest terms, we may use either of the two methods shown above. In this example, $x \neq 0$ and $y \neq 0$.

METHOD 1

$$\frac{5x^2}{7y} \cdot \frac{14y^2}{15x^3} = \frac{5x^2 \cdot 14y^2}{7y \cdot 15x^3} = \frac{70x^2y^2}{105x^3y} = \frac{2y}{3x} \cdot \frac{35x^2y}{35x^2y} = \frac{2y}{3x} \cdot 1 = \frac{2y}{3x}$$

METHOD 2 (the cancellation method)

$$\frac{5x^2}{7y} \cdot \frac{14y^2}{15x^3} = \frac{\overset{1}{\cancel{5x^2}}}{\underset{1}{\cancel{7y}}} \cdot \frac{\overset{2y}{\cancel{14y^2}}}{\underset{3x}{\cancel{15x^3}}} = \frac{2y}{3x}$$

PROCEDURE. To multiply fractions:

Method 1

1. Multiply the numerators of the given fractions.
2. Multiply the denominators of the given fractions.
3. Reduce the resulting fraction, if possible, to lowest terms.

Method 2

1. Use cancellation to divide any numerator and any denominator by all common factors.
2. Multiply the remaining numerators, and multiply the remaining denominators, to write the product in reduced form.

For problems in which the numerator, the denominator, or both are polynomials that are not monomials, it is helpful to factor the numerator and the denominator before applying the cancellation method. For example:

$$\frac{3x + 15}{4y} \bullet \frac{2}{3} = \frac{\overset{1}{\cancel{3}}(x + 5)}{\underset{2}{\cancel{4}}y} \bullet \frac{\overset{1}{\cancel{2}}}{\underset{1}{\cancel{3}}} = \frac{x + 5}{2y} \quad \text{(where } y \neq 0\text{)}$$

Here is another example:

$$\frac{x^2 - 4}{2x} \bullet \frac{2}{x^2 - 3x + 2} = \frac{\overset{1}{\cancel{(x - 2)}}(x + 2)}{\underset{1}{\cancel{2}}x} \bullet \frac{\overset{1}{\cancel{2}}}{(x - 1)\underset{1}{\cancel{(x - 2)}}} = \frac{x + 2}{x(x - 1)}$$

(where $x \neq 0, 1, 2$)

EXAMPLES

1. Multiply, and express the product in reduced form: $\frac{5a^3}{9bx} \bullet \frac{6bx}{a^2}$.

How to Proceed:	*Solution:*
(1) Divide the numerators and the denominators by the common factors $3bx$ and a^2:	$\frac{5a^3}{9bx} \bullet \frac{6bx}{a^2} = \frac{\overset{5a}{\cancel{5a^3}}}{\underset{3}{\cancel{9bx}}} \bullet \frac{\overset{2}{\cancel{6bx}}}{\underset{1}{\cancel{a^2}}}$
(2) Multiply the remaining numerators, and then multiply the remaining denominators.	$= \frac{10a}{3}$

Answer: $\frac{10a}{3}$ $(a \neq 0, b \neq 0, x \neq 0)$

---

2. Multiply, and simplify the product: $\frac{x^2 - 5x + 6}{3x} \bullet \frac{2}{4x - 12}$.

Solution

$$\frac{x^2 - 5x + 6}{3x} \bullet \frac{2}{4x - 12} = \frac{\overset{1}{\cancel{(x - 3)}}(x - 2)}{3x} \bullet \frac{\overset{1}{\cancel{2}}}{\underset{2}{\cancel{4}}\underset{(1)}{\cancel{(x - 3)}}} = \frac{x - 2}{6x}$$

Answer: $\frac{x - 2}{6x}$ $(x \neq 0, 3)$

EXERCISES

In 1–39, find each product in lowest terms.

1. $\frac{8}{12} \cdot \frac{30}{36}$
2. $36 \cdot \frac{5}{9}$
3. $\frac{1}{2} \cdot 20x$
4. $\frac{5}{d} \cdot d^2$
5. $\frac{x^2}{36} \cdot 20$
6. $mn \cdot \frac{8}{m^2n^2}$
7. $\frac{24x}{35y} \cdot \frac{14y}{8x}$
8. $\frac{12x}{5y} \cdot \frac{15y^2}{36x^2}$
9. $\frac{m^2}{8} \cdot \frac{32}{3m}$
10. $\frac{6r^2}{5s^2} \cdot \frac{10rs}{6r^3}$
11. $\frac{30m^2}{18n} \cdot \frac{6n}{5m}$
12. $\frac{24a^3b^2}{7c^3} \cdot \frac{21c^2}{12ab}$
13. $\frac{7}{8} \cdot \frac{2x+4}{21}$
14. $\frac{3a+9}{15a} \cdot \frac{a^3}{18}$
15. $\frac{5x-5y}{x^2y} \cdot \frac{xy^2}{25}$
16. $\frac{12a-4}{b} \cdot \frac{b^3}{12}$
17. $\frac{ab-a}{b^2} \cdot \frac{b^3-b^2}{a}$
18. $\frac{x^2-1}{x^2} \cdot \frac{3x^2-3x}{15}$
19. $\frac{2r}{r-1} \cdot \frac{r-1}{10}$
20. $\frac{7s}{s+2} \cdot \frac{2s+4}{21}$
21. $\frac{8x}{2x+6} \cdot \frac{x+3}{x^2}$
22. $\frac{1}{x^2-1} \cdot \frac{2x+2}{6}$
23. $\frac{a^2-9}{3} \cdot \frac{12}{2a-6}$
24. $\frac{x^2-x-2}{3} \cdot \frac{21}{x^2-4}$
25. $\frac{a(a-b)^2}{4b} \cdot \frac{4b}{a(a^2-b^2)}$
26. $\frac{(a-2)^2}{4b} \cdot \frac{16b^3}{4-a^2}$
27. $\frac{a^2-7a-8}{2a+2} \cdot \frac{5}{a-8}$
28. $\frac{x^2+6x+5}{9y^2} \cdot \frac{3y}{x+1}$
29. $\frac{y^2-2y-3}{2c^3} \cdot \frac{4c^2}{2y+2}$
30. $\frac{4a-6}{4a+8} \cdot \frac{6a+12}{5a-15}$
31. $\frac{x^2-25}{4x^2-9} \cdot \frac{2x+3}{x-5}$
32. $\frac{4x+8}{6x+18} \cdot \frac{5x+15}{x^2-4}$
33. $\frac{y^2-81}{(y+9)^2} \cdot \frac{10y+90}{5y-45}$
34. $\frac{8x}{2x^2-8} \cdot \frac{8x+16}{32x^2}$
35. $\frac{2-x}{2x} \cdot \frac{3x}{3x-6}$
36. $\frac{x^2-3x+2}{2x^2-2} \cdot \frac{2x}{x-2}$
37. $\frac{b^2+81}{b^2-81} \cdot \frac{81-b^2}{81+b^2}$
38. $\frac{d^2-25}{4-d^2} \cdot \frac{5d^2-20}{d+5}$
39. $\frac{a^2+12a+36}{a^2-36} \cdot \frac{36-a^2}{36+a^2}$
40. What is the value of $\frac{x^2-4}{6x+12} \cdot \frac{4x-12}{x^2-5x+6}$ when $x = 65{,}908$?

19-4 DIVIDING FRACTIONS

We know that the operation of division may be defined by means of the multiplicative inverse, the reciprocal. A quotient can be expressed as the product of the dividend and the reciprocal of the divisor. Thus:

$$8 \div 5 = \frac{8}{1} \cdot \frac{1}{5} = \frac{8 \cdot 1}{1 \cdot 5} = \frac{8}{5} \quad \text{and} \quad \frac{8}{7} \div \frac{5}{3} = \frac{8}{7} \cdot \frac{3}{5} = \frac{8 \cdot 3}{7 \cdot 5} = \frac{24}{35}$$

We use the same rule to divide algebraic fractions. In general, for any numbers a, b, c, and d, when $b \neq 0$, $c \neq 0$, and $d \neq 0$:

$$\frac{a}{b} \div \frac{c}{d} = \frac{a}{b} \cdot \frac{d}{c} = \frac{ad}{bc}$$

PROCEDURE. To divide by a fraction, multiply the dividend by the reciprocal of the divisor.

EXAMPLES

Divide.

a. $\frac{16c^3}{21d^2} \div \frac{24c^4}{14d^3}$

Solution Multiply the dividend by the reciprocal of the divisor:

$$\frac{16c^3}{21d^2} \div \frac{24c^4}{14d^3} = \frac{\overset{2}{\cancel{16c^3}}}{\underset{3}{\cancel{21d^2}}} \cdot \frac{\overset{2d}{\cancel{14d^3}}}{\underset{3c}{\cancel{24c^4}}} = \frac{4d}{9c}$$

Answer: $\frac{4d}{9c}$ $(c \neq 0,\ d \neq 0)$

b. $\frac{8x + 24}{x^2 - 25} \div \frac{4x}{x^2 + 8x + 15}$

How to Proceed:	*Solution:*
(1) Multiply the dividend by the reciprocal of the divisor:	$\frac{8x + 24}{x^2 - 25} \div \frac{4x}{x^2 + 8x + 15}$ $= \frac{8x + 24}{x^2 - 25} \cdot \frac{x^2 + 8x + 15}{4x}$
(2) Factor the numerators and denominators, and divide by the common factors:	$= \frac{\overset{2}{\cancel{8}}(x + 3)}{\underset{1}{\cancel{(x + 5)}}(x - 5)} \cdot \frac{\overset{1}{\cancel{(x + 5)}}(x + 3)}{\underset{1}{\cancel{4}}x}$
(3) Multiply the remaining numerators and then the remaining denominators:	$= \frac{2(x + 3)^2}{x(x - 5)}$

Answer: $\frac{2(x + 3)^2}{x(x - 5)}$ $(x \neq 0, 5, -5, -3)$

Note. If $x = 5$ or -5 or -3, the fractions being divided will not be defined. If $x = 0$, the reciprocal of the divisor will not be defined.

EXERCISES

In 1–25, in each case divide and express the quotient in lowest terms.

1. $\frac{7}{10} \div \frac{21}{5}$

2. $\frac{12}{35} \div \frac{4}{7}$

3. $8 \div \frac{1}{2}$

4. $\frac{x}{9} \div \frac{x}{3}$

5. $\frac{3x}{5y} \div \frac{21x}{2y}$

6. $\frac{7ab^2}{10cd} \div \frac{14b^3}{5c^2d^2}$

7. $\frac{xy^2}{x^2y} \div \frac{x}{y^3}$

8. $\frac{6a^2b^2}{8c} \div 3ab$

9. $\frac{4x + 4}{9} \div \frac{3}{8x}$

10. $\frac{3y^2 + 9y}{18} \div \frac{5y^2}{27}$

11. $\frac{a^3 - a}{b} \div \frac{a^3}{4b^3}$

12. $\frac{x^2 - 1}{5} \div \frac{x - 1}{10}$

13. $\frac{x^2 - 5x + 4}{2x} \div \frac{2x - 2}{8x^2}$

14. $\frac{4a^2 - 9}{10} \div \frac{10a + 15}{25}$

15. $\frac{b^2 - b - 6}{2b} \div \frac{b^2 - 4}{b^2}$

16. $\frac{a^2 - ab}{4a} \div (a^2 - b^2)$

17. $\frac{12y - 6}{8} \div (2y^2 - 3y + 1)$

18. $\frac{(x - 2)^2}{4x^2 - 16} \div \frac{21x}{3x + 6}$

19. $\frac{x^2 - 2xy - 8y^2}{x^2 - 16y^2} \div \frac{5x + 10y}{3x + 12y}$

20. $\frac{x^2 - 4x + 4}{3x - 6} \div (2 - x)$

21. $(9 - y^2) \div \frac{y^2 + 8y + 15}{2y + 10}$

22. $\frac{x - 1}{x + 1} \bullet \frac{2x + 2}{x + 2} \div \frac{4x - 4}{x + 2}$

23. $\frac{x + y}{x^2 + y^2} \bullet \frac{x}{x - y} \div \frac{(x + y)^2}{x^4 - y^4}$

24. $\frac{2a + 6}{a^2 - 9} \div \frac{3 + a}{3 - a} \bullet \frac{a + 3}{4}$

25. $\frac{(a + b)^2}{a^2 - b^2} \div \frac{a + b}{b^2 - a^2} \bullet \frac{a - b}{(a - b)^2}$

26. For what value(s) of a is $\frac{a^2 - 2a + 1}{a^2} \div \frac{a^2 - 1}{a}$ undefined?

27. Find the value of $\frac{y^2 - 6y + 9}{y^2 - 9} \div \frac{10y - 30}{y^2 + 3y}$ when $y = 70$.

28. If $x \div y = a$ and $y \div z = \frac{1}{a}$, what is the value of $x \div z$?

19-5 ADDING OR SUBTRACTING FRACTIONS THAT HAVE THE SAME DENOMINATOR

We know that the sum (or difference) of two arithmetic fractions that have the same denominator is another fraction whose numerator is the sum (or difference) of the numerators and whose denominator is the common denominator of the given fractions. We use the same rule to add algebraic fractions that have the same nonzero denominator. Thus:

Arithmetic fractions	*Algebraic fractions*
$\frac{5}{7} + \frac{1}{7} = \frac{5 + 1}{7} = \frac{6}{7}$	$\frac{a}{x} + \frac{b}{x} = \frac{a + b}{x}$
$\frac{5}{7} - \frac{1}{7} = \frac{5 - 1}{7} = \frac{4}{7}$	$\frac{a}{x} - \frac{b}{x} = \frac{a - b}{x}$

PROCEDURE. To add (or subtract) fractions that have the same denominator:

1. Write a fraction whose numerator is the sum (or difference) of the numerators and whose denominator is the common denominator of the given fractions.
2. Reduce the resulting fraction to lowest terms.

EXAMPLES

Add or subtract as indicated. Reduce each answer to lowest terms. ($x \neq 0$).

a. $\frac{5}{4x} + \frac{9}{4x} - \frac{8}{4x}$

Solution:

$$\frac{5}{4x} + \frac{9}{4x} - \frac{8}{4x}$$
$$= \frac{5 + 9 - 8}{4x}$$
$$= \frac{6}{4x}$$
$$= \frac{3}{2x} \quad \textit{Answer}$$

b. $\frac{4x + 7}{6x} - \frac{2x - 4}{6x}$

Solution:

$$\frac{4x + 7}{6x} - \frac{2x - 4}{6x}$$
$$= \frac{(4x + 7) - (2x - 4)}{6x}$$
$$= \frac{4x + 7 - 2x + 4}{6x}$$
$$= \frac{2x + 11}{6x} \quad \textit{Answer}$$

Note: In Example b, since the fraction bar is a symbol of grouping, we enclose numerators that have more than one term in parentheses. In this way, we can see all the signs that need to be changed for the subtraction.

EXERCISES

In 1–38, in each case add or subtract (combine) the fractions as indicated. Reduce each answer to lowest terms. In all cases, assume that the fractions are defined.

1. $\frac{1}{8} + \frac{4}{8}$
2. $\frac{9}{15} - \frac{6}{15}$
3. $\frac{2}{x} + \frac{3}{x}$
4. $\frac{11}{4c} + \frac{5}{4c} - \frac{6}{4c}$
5. $\frac{3x}{4} + \frac{2x}{4}$
6. $\frac{12y}{5} - \frac{4y}{5}$
7. $\frac{2c}{5} - \frac{3d}{5}$
8. $\frac{x}{2} - \frac{y}{2} + \frac{z}{2}$
9. $\frac{x}{a} + \frac{y}{a}$
10. $\frac{5r}{t} - \frac{2s}{t}$
11. $\frac{9}{8x} + \frac{6}{8x}$
12. $\frac{8}{9y} + \frac{4}{9y} - \frac{3}{9y}$
13. $\frac{6a}{4x} + \frac{5a}{4x}$
14. $\frac{11b}{3y} - \frac{4b}{3y}$
15. $\frac{19c}{12d} + \frac{9c}{12d}$
16. $\frac{6}{10c} + \frac{9}{10c} - \frac{3}{10c}$

17. $\frac{2x + 1}{2} + \frac{3x + 6}{2}$

18. $\frac{4x + 12}{16x} + \frac{8x + 4}{16x}$

19. $\frac{5x - 4}{3} - \frac{2x + 1}{3}$

20. $\frac{12a - 15}{12a} - \frac{9a - 6}{12a}$

21. $\frac{5}{x + 2} + \frac{3}{x + 2}$

22. $\frac{2}{a - b} - \frac{1}{a - b}$

23. $\frac{r}{y - 2} + \frac{x}{y - 2}$

24. $\frac{x}{x + 1} + \frac{1}{x + 1}$

25. $\frac{2x}{x + 3} + \frac{6}{x + 3}$

26. $\frac{y}{y^2 - 4} - \frac{2}{y^2 - 4}$

27. $\frac{4x + 1}{3x + 2} + \frac{6x - 3}{3x + 2}$

28. $\frac{3c - 7}{2c - 3} + \frac{c + 9}{2c - 3}$

29. $\frac{6y - 4}{4y + 3} + \frac{7 - 2y}{4y + 3}$

30. $\frac{9d + 6}{2d + 1} - \frac{7d + 5}{2d + 1}$

31. $\frac{8x - 4}{2x + 6} - \frac{4x - 6}{2x + 6}$

32. $\frac{6x - 5}{x^2 - 1} - \frac{5x - 6}{x^2 - 1}$

33. $\frac{a^2 + 3ab}{a + b} + \frac{b^2 - ab}{a + b}$

34. $\frac{x^2 - 2xy}{x - 2y} - \frac{xy - 2y^2}{x - 2y}$

35. $\frac{8x - 8}{6x - 5} - \frac{2x - 7}{6x - 5} + \frac{6x - 9}{6x - 5}$

36. $\frac{a + 4b}{a^2 - b^2} + \frac{4a - 7b}{a^2 - b^2} - \frac{3a - b}{a^2 - b^2}$

37. $\frac{r^2 + 4r}{r^2 - r - 6} + \frac{8 - r^2}{r^2 - r - 6}$

38. $\frac{4m^2 + 7m}{2m^2 + 5m + 2} - \frac{1 + 7m}{2m^2 + 5m + 2}$

In 39–42, copy and complete the table, showing the results of adding, subtracting, multiplying, and dividing the expressions that represent a and b.

	a	b	$a + b$	$a - b$	$a \cdot b$	$a \div b$ or $\frac{a}{b}$
39.	$\frac{12}{y}$	$\frac{3}{y}$				
40.	$\frac{3x}{8}$	$\frac{x}{8}$				
41.	$\frac{r}{t}$	$\frac{p}{t}$				
42.	$\frac{7k}{2x}$	$\frac{5k}{2x}$				

19-6 ADDING OR SUBTRACTING FRACTIONS THAT HAVE DIFFERENT DENOMINATORS

Finding Equivalent Fractions

Section 19-2 explained how to reduce both numerical and algebraic fractions to lowest terms by using the division property of a fraction. Now, let us think backwards. Equivalent fractions can also be found by using multiplication, specifically the multiplication property of 1.

Arithmetic fractions	*Algebraic fractions*
$\frac{3}{5} = \frac{3}{5} \bullet 1 = \frac{3}{5} \bullet \frac{2}{2}$	$\frac{a}{b} = \frac{a}{b} \bullet 1 = \frac{a}{b} \bullet \frac{x}{x}$
$= \frac{6}{10}$	$= \frac{ax}{bx} \quad (b \neq 0, x \neq 0)$

These examples illustrate the ***multiplication property of a fraction***: If the numerator and the denominator of a fraction are multiplied by the same nonzero number, the resulting fraction is equivalent to the original fraction.

Addition and Subtraction of Fractions

In arithmetic, in order to add (or subtract) fractions that have *different* denominators, we change these fractions to equivalent fractions that have the *same* denominator, called the ***common denominator***. Then we add (or subtract) the equivalent fractions.

For example, to add $\frac{3}{4}$ and $\frac{1}{6}$, we use any common denominator that has 4 and 6 as factors.

METHOD 1: One common denominator is the product of the denominator. Here, the common denominator is 4 • 6, or 24.

$$\frac{3}{4} + \frac{1}{6} = \frac{3}{4} \bullet \frac{6}{6} + \frac{1}{6} \bullet \frac{4}{4}$$
$$= \frac{18}{24} + \frac{4}{24}$$
$$= \frac{22}{24}$$
$$= \frac{11}{12} \quad \textit{Answer}$$

METHOD 2: To simplify work, we use the ***lowest common denominator*** (L.C.D.) of 4 and 6, which is 12.

$$\frac{3}{4} + \frac{1}{6} = \frac{3}{4} \bullet \frac{3}{3} + \frac{1}{6} \bullet \frac{2}{2}$$
$$= \frac{9}{12} + \frac{2}{12}$$
$$= \frac{11}{12} \quad \textit{Answer}$$

To find the L.C.D. of two fractions, we factor the denominators of the fractions completely. The L.C.D. is the product of all of the factors of the first denominator times the factors of the second denominator that are not factors of the first.

$$4 = 2 \bullet 2$$
$$6 = 2 \quad \bullet 3$$
$$\text{L.C.D.} = 2 \bullet 2 \bullet 3 = 12$$

Then, to change each fraction to an equivalent form that has the L.C.D. as the denominator, we multiply by $\frac{x}{x}$, where x is the number by which the original denominator must be multiplied to obtain the L.C.D.

$\frac{3}{4}\left(\frac{?}{}\right) = \frac{}{12}$	$\frac{1}{6}\left(\frac{?}{}\right) = \frac{}{12}$
$\frac{3}{4}\left(\frac{3}{3}\right) = \frac{9}{12}$	$\frac{1}{6}\left(\frac{2}{2}\right) = \frac{2}{12}$

Algebraic fractions are added in the same manner as arithmetic fractions, as shown in the examples that follow.

PROCEDURE. To add (or subtract) fractions that have different denominators:

1. Choose a common denominator for the fractions. The common denominator is a multiple of each of the denominators of the given fractions. The L.C.D. is the smallest common denominator.
2. Change each fraction to an equivalent fraction having the chosen common denominator.
3. Write a fraction whose numerator is the sum (or difference) of the numerators of the new fractions and whose denominator is the common denominator.
4. Reduce the resulting fraction to lowest terms.

EXAMPLES

1. Add: $\frac{5}{a^2b} + \frac{2}{ab^2}$.

Solution:

$$a^2b = a \cdot a \cdot b$$
$$ab^2 = a \quad \cdot b \cdot b$$

$$\text{L.C.D.} = a \cdot a \cdot b \cdot b = a^2b^2$$

$$\frac{5}{a^2b} + \frac{2}{ab^2} = \frac{5}{a^2b} \cdot \frac{b}{b} + \frac{2}{ab^2} \cdot \frac{a}{a}$$
$$= \frac{5b}{a^2b^2} + \frac{2a}{a^2b^2}$$
$$= \frac{5b + 2a}{a^2b^2}$$

Answer: $\frac{5b + 2a}{a^2b^2}$ $(a \neq 0, b \neq 0)$

2. Subtract: $\frac{2x + 5}{3} - \frac{x - 2}{4}$.

Solution:

$$\text{L.C.D.} = 3 \cdot 4 = 12$$

$$\frac{2x + 5}{3} - \frac{x - 2}{4} = \frac{4}{4} \cdot \frac{(2x + 5)}{3} - \frac{3}{3} \cdot \frac{(x - 2)}{4}$$
$$= \frac{8x + 20}{12} - \frac{3x - 6}{12}$$
$$= \frac{(8x + 20) - (3x - 6)}{12}$$
$$= \frac{8x + 20 - 3x + 6}{12}$$
$$= \frac{5x + 26}{12}$$

Answer: $\frac{5x + 26}{12}$

3. Express as a fraction in simplest form: $y + 1 - \frac{1}{y-1}$.

Solution:

L.C.D. = $y - 1$

$$y + 1 - \frac{1}{y-1}$$

$$= \frac{y+1}{1} - \frac{1}{y-1}$$

$$= \frac{(y+1)(y-1)}{1(y-1)} - \frac{1}{y-1}$$

$$= \frac{y^2-1}{y-1} - \frac{1}{y-1}$$

$$= \frac{y^2-1-1}{y-1}$$

$$= \frac{y^2-2}{y-1}$$

Answer: $\frac{y^2-2}{y-1}$ $(y \neq 0)$

4. Subtract: $\frac{5x}{x^2-4} - \frac{3}{x-2}$.

Solution:

$x^2 - 4 = (x-2)(x+2)$

$x - 2 = 1 \cdot (x-2)$

L.C.D. = $(x-2)(x+2)$

$$\frac{5x}{x^2-4} - \frac{3}{x-2}$$

$$= \frac{5x}{(x-2)(x+2)} - \frac{3}{(x-2)}$$

$$= \frac{5x}{(x-2)(x+2)} - \frac{3(x+2)}{(x-2)(x+2)}$$

$$= \frac{5x-(3x+6)}{(x-2)(x+2)}$$

$$= \frac{5x-3x-6}{(x-2)(x+2)}$$

$$= \frac{2x-6}{(x-2)(x+2)} \quad \text{or} \quad \frac{2x-6}{x^2-4}$$

Answer: $\frac{2x-6}{x^2-4}$ $(x \neq 2, -2)$

EXERCISES

In 1–11, in each case, find the lowest common denominator for two fractions whose denominators are given.

1. 2; 3 **2.** 6; 5 **3.** x; $4x$ **4.** $12r$; 8 **5.** xy; yz

6. $12x^2$; $15y^2$ **7.** $5x$; $15(x + y)$ **8.** $4(a + b)$; $12a$

9. $4(y + z)$; $12(y + z)$ **10.** $(x^2 - 9)$; $(x + 3)$ **11.** $2x - 1$; $4x^2 - 1$

In 12–47, in each case, add or subtract (combine) the fractions as indicated. Reduce each answer to lowest terms.

12. $\frac{5}{3} + \frac{3}{2}$ **13.** $\frac{9}{5} - \frac{2}{3}$ **14.** $\frac{7}{4} + \frac{10}{3}$ **15.** $\frac{4}{10} - \frac{7}{100}$ **16.** $\frac{5}{6} + \frac{1}{12}$

17. $\frac{7}{8} - \frac{1}{4}$ **18.** $\frac{1}{3} - \frac{1}{6}$ **19.** $\frac{5}{4} + \frac{3}{2} - \frac{1}{3}$ **20.** $\frac{x}{3} + \frac{x}{2}$ **21.** $\frac{d}{3} - \frac{d}{5}$

22. $\frac{5x}{6} - \frac{2x}{3}$ **23.** $\frac{y}{6} + \frac{y}{5} - \frac{y}{2}$ **24.** $\frac{ab}{5} + \frac{ab}{4}$ **25.** $\frac{8x}{5} - \frac{3x}{4} + \frac{7x}{10}$ **26.** $\frac{5a}{6} - \frac{3a}{4}$

27. $\frac{a}{7} + \frac{b}{14}$ **28.** $\frac{9}{4x} + \frac{3}{2x}$ **29.** $\frac{1}{2x} - \frac{1}{x} + \frac{3}{8x}$ **30.** $\frac{9a}{8b} - \frac{3a}{4b}$ **31.** $\frac{1}{a} + \frac{1}{b}$

32. $\frac{2}{a^2} - \frac{5}{b}$ **33.** $\frac{1}{xy} + \frac{1}{yz}$ **34.** $\frac{5}{rs} + \frac{9}{st}$ **35.** $\frac{x}{3ab} - \frac{y}{2bc}$

36. $\frac{9}{ab} + \frac{2}{bc} - \frac{3}{ac}$ **37.** $\frac{1}{x^2} + \frac{3}{xy} - \frac{5}{y^2}$ **38.** $\frac{a - 3}{3} + \frac{a + 1}{6}$ **39.** $\frac{x + 7}{3} - \frac{2x - 3}{5}$

40. $\frac{3y - 4}{5} - \frac{y - 2}{4}$ **41.** $\frac{a - b}{4} - \frac{a + b}{6}$ **42.** $\frac{x + 5}{2x} + \frac{2x - 1}{4x}$ **43.** $\frac{d + 6}{d} + \frac{d - 3}{4d}$

44. $\frac{b - 3}{5b} - \frac{b + 2}{10b}$ **45.** $\frac{3b + 1}{5b} - \frac{4b - 3}{4b}$ **46.** $\frac{y - 4}{4y^2} + \frac{3y - 5}{3y}$ **47.** $\frac{3c - 7}{2c} - \frac{3c - 3}{6c^2}$

In 48–50, represent the perimeter of each polygon in simplest form.

48. A triangle whose sides are represented by $\frac{x}{2}$, $\frac{3x}{5}$, and $\frac{7x}{10}$.

49. A rectangle whose length is represented by $\frac{x + 3}{4}$, and whose width is represented by $\frac{x - 4}{3}$.

50. An isosceles triangle whose equal legs are each represented by $\frac{2x - 3}{7}$, and whose base is represented by $\frac{6x - 18}{21}$.

In 51 and 52, find, in each case, a representation in simplest form for the indicated length.

51. If the perimeter of a triangle is represented by $\frac{17x}{24}$, and two of the sides are represented by $\frac{3x}{8}$ and $\frac{2x - 5}{12}$, respectively, find a representation for the third side.

52. If the perimeter of a rectangle is represented by $\frac{14x}{15}$, and each length is represented by $\frac{x + 2}{3}$, find a representation for each width.

In 53–71, write each expression as a fraction in lowest terms.

53. $5\frac{2}{3}$ **54.** $9\frac{3}{4}$ **55.** $5 + \frac{1}{x}$ **56.** $9 - \frac{7}{s}$ **57.** $m + \frac{1}{m}$

58. $d - \frac{7}{5d}$ **59.** $\frac{a}{b} + c$ **60.** $3 + \frac{5}{x + 1}$ **61.** $6 - \frac{4}{x - y}$ **62.** $7 + \frac{2a}{b + c}$

63. $t + \frac{1}{t + 1}$ **64.** $s - \frac{1}{s - 1}$ **65.** $5 - \frac{2x}{x + y}$ **66.** $\frac{4}{y - 2} + 4$ **67.** $8 + \frac{c + 2}{c - 3}$

68. $7 - \frac{x + y}{x - y}$ **69.** $a + 1 + \frac{1}{a + 1}$ **70.** $x - 5 - \frac{x}{x + 3}$ **71.** $\frac{2x - 1}{x + 2} + 2x - 3$

In 72–98, in each case combine the fractions as indicated, and reduce the answer to lowest terms.

72. $\frac{5}{x - 3} + \frac{7}{2x - 6}$ **73.** $\frac{9}{y + 1} - \frac{3}{4y + 4}$ **74.** $\frac{2}{3a - 1} + \frac{7}{15a - 5}$

75. $\frac{10}{3x - 6} + \frac{3}{2x - 4}$ **76.** $\frac{11x}{8x - 8} - \frac{3x}{4x - 4}$ **77.** $\frac{3}{2x - 3y} + \frac{5}{3y - 2x}$

78. $\frac{2a}{4a - 8b} + \frac{3b}{3a - 6b}$

79. $\frac{3x - 2}{2x + 2} + \frac{4x - 1}{3x + 3}$

80. $\frac{5x + 2}{6x - 3} - \frac{3x - 5}{8x - 4}$

81. $\frac{1}{x - 5} + \frac{1}{x + 5}$

82. $\frac{9}{y + 4} - \frac{6}{y - 4}$

83. $\frac{7}{a + 3} + \frac{4}{2 - a}$

84. $\frac{7}{x - 2} + \frac{3}{x}$

85. $\frac{9}{c + 8} - \frac{2}{c}$

86. $\frac{2a + b}{a - b} + \frac{a}{b}$

87. $\frac{5}{y^2 - 9} + \frac{3}{y - 3}$

88. $\frac{6}{y^2 - 16} - \frac{5}{y + 4}$

89. $\frac{9}{a^2 - b^2} + \frac{3}{b - a}$

90. $\frac{3y}{y^2 - 4} - \frac{4}{2y - 4}$

91. $\frac{x}{x^2 - 36} - \frac{4}{3x + 18}$

92. $\frac{9}{a^2 - ab} + \frac{3}{ab - b^2}$

93. $\frac{1}{y - 3} + \frac{2}{y + 4} + \frac{2}{3}$

94. $\frac{1}{(x + 2)^3} - \frac{1}{(x + 2)^2} + \frac{1}{x + 2}$

95. $\frac{7a}{(a - 1)(a + 3)} + \frac{2a - 5}{(a + 3)(a + 2)}$

96. $\frac{5}{r^2 - 4} - \frac{3}{r^2 + 3r - 10}$

97. $\frac{x + 2y}{3x + 12y} - \frac{6x - y}{x^2 + 3xy - 4y^2}$

98. $\frac{2a + 7}{a^2 - 2a - 15} - \frac{3a - 4}{a^2 - 7a + 10}$

99. Find the value of $\frac{4x - 1}{2x - 5} + \frac{9}{5 - 2x}$: **a.** when $x = 375$. **b.** when $x = -243$

19-7 SOLVING EQUATIONS CONTAINING FRACTIONAL COEFFICIENTS

Here are examples of equations that contain fractional coefficients:

$$\frac{1}{2}x = 10 \quad or \quad \frac{x}{2} = 10, \qquad \frac{1}{3}x + 60 = \frac{5}{6}x \quad or \quad \frac{x}{3} + 60 = \frac{5x}{6}$$

Each of these equations can be solved by transforming it into an equivalent equation that does not contain fractional coefficients. This can be done by multiplying both sides of the equation by a common denominator for all the fractions present in the equation. We usually multiply by the lowest common denominator, the L.C.D.

Note that the equation $0.5x = 10$ can also be written as $\frac{1}{2}x = 10$, illustrating the fact that in an equation a coefficient that is a decimal fraction can be replaced with a coefficient that is a common fraction.

PROCEDURE. To solve an equation that contains fractional coefficients:
1. Find the L.C.D. of all coefficients.
2. Multiply both members of the equation by the L.C.D.
3. Solve the resulting equation using the usual methods.
4. Check in the original equation.

EXAMPLES

1. Solve and check: $\frac{x}{3} + \frac{x}{5} = 8$

How to Proceed:	*Solution:*	*Check:*
(1) Write the equation:	$\frac{x}{3} + \frac{x}{5} = 8$	$\frac{x}{3} + \frac{x}{5} = 8$
(2) Find the L.C.D.:	L.C.D. $= 3 \cdot 5 = 15$	$\frac{15}{3} + \frac{15}{5} \stackrel{?}{=} 8$
(3) Multiply both members of the equation by the L.C.D.:	$15\left(\frac{x}{3} + \frac{x}{5}\right) = 15(8)$	$5 + 3 \stackrel{?}{=} 8$ $8 = 8$ ✔
(4) Use the distributive property:	$15\left(\frac{x}{3}\right) + 15\left(\frac{x}{5}\right) = 15(8)$	
(5) Simplify:	$5x + 3x = 120$	
(6) Solve for x:	$8x = 120$ $x = 15$ *Answer*	

2. Solve:

a. $\frac{3x}{4} = 20 + \frac{x}{4}$

Solution:

$$\frac{3x}{4} = 20 + \frac{x}{4}$$

L.C.D. $= 4$

$$4\left(\frac{3x}{4}\right) = 4\left(20 + \frac{x}{4}\right)$$

$$4\left(\frac{3x}{4}\right) = 4(20) + 4\left(\frac{x}{4}\right)$$

$$3x = 80 + x$$

$$2x = 80$$

$$x = 40 \quad \textit{Answer}$$

b. $\frac{2x + 7}{6} - \frac{2x - 9}{10} = 3$

Solution:

$$\frac{2x + 7}{6} - \frac{2x - 9}{10} = 3$$

L.C.D. $= 30$

$$30\left(\frac{2x + 7}{6} - \frac{2x - 9}{10}\right) = 30(3)$$

$$30\left(\frac{2x + 7}{6}\right) - 30\left(\frac{2x - 9}{10}\right) = 30(3)$$

$$5(2x + 7) - 3(2x - 9) = 90$$

$$10x + 35 - 6x + 27 = 90$$

$$4x + 62 = 90$$

$$4x = 28$$

$$x = 7 \quad \textit{Answer}$$

In Examples **2 a** and **b**, the check is left to you.

3. A woman purchased stock in the AAA Company over 3 months. In the first month, she purchased one-half of her present number of shares. In the second month, she bought two-fifths of her present number of shares. In the third month, she purchased 14 more shares. How many shares of AAA stock did the woman purchase?

Solution

Let x = total number of shares of stock purchased.

Then $\frac{1}{2}x$ = number of shares purchased in month 1,

and $\frac{2}{5}x$ = number of shares purchased in month 2.

The sum of the shares purchased over 3 months is the total number of shares.

$$\textit{Month 1} + \textit{month 2} + \textit{month 3} = \textit{Total}$$

$$\frac{1}{2}x + \frac{2}{5}x + 14 = x$$

$$10\left(\frac{1}{2}x + \frac{2}{5}x + 14\right) = 10(x)$$

$$5x + 4x + 140 = 10x$$

$$9x + 140 = 10x$$

$$140 = x$$

Check: month 1 $= \frac{1}{2}(140) = 70$, month 2 $= \frac{2}{5}(140) = 56$, month 3 $= 14$

$70 + 56 + 14 = 140$ (True)

Answer: 140 shares

---

4. In a child's coin bank, there is a collection of nickels, dimes, and quarters that amounts to \$3.20. There are 3 times as many quarters as nickels, and 5 more dimes than nickels. How many coins of each kind are there?

Solution

Let x = the number of nickels.
Then $3x$ = the number of quarters,
and $x + 5$ = the number of dimes.

Also, $0.05x$ = the value of the nickels,
$0.25(3x)$ = the value of the quarters,
and $0.10(x + 5)$ = the value of the dimes.

Write the equation for the value of the coins. To simplify the equation, which contains coefficients that are decimal fractions with denominators of 100, multiply each side of the equation by 100.

The total value of the coins is \$3.20.

$$0.05x + 0.25(3x) + 0.10(x + 5) = 3.20$$
$$100[0.05x + 0.25(3x) + 0.10(x + 5)] = 100[3.20]$$
$$5x + 25(3x) + 10(x + 5) = 320$$
$$5x + 75x + 10x + 50 = 320$$
$$90x + 50 = 320$$
$$90x = 270$$
$$x = 3 \quad \text{number of nickels}$$
$$\text{then } 3x = 3(3) = 9 \quad \text{number of quarters}$$
$$\text{and } x + 5 = 3 + 5 = 8 \quad \text{number of dimes}$$

Check: The value of 3 nickels = 0.05(3) = 0.15
The value of 9 quarters = 0.25(9) = 2.25
The value of 8 dimes = 0.10(8) = 0.80
\$3.20

Answer: There are 3 nickels, 9 quarters, and 8 dimes.

EXERCISES WITH OPEN-RESPONSE PROBLEMS

In 1–50, solve each equation and check.

1. $\frac{x}{7} = 3$ **2.** $\frac{1}{6}t = 18$ **3.** $\frac{3x}{5} = 15$ **4.** $\frac{5}{7}n = 35$ **5.** $\frac{x + 8}{4} = 6$

6. $\frac{m - 2}{9} = 3$ **7.** $0.8 = 8y$ **8.** $2z = 0.08$ **9.** $6t = 18.6$ **10.** $0.4r = 16$

11. $1.3t = 0.39$ **12.** $0.25a = 6$ **13.** $\frac{2r + 6}{5} = -4$ **14.** $\frac{5y - 30}{7} = 0$ **15.** $\frac{5x}{2} = \frac{15}{4}$

16. $\frac{m - 5}{35} = \frac{5}{7}$ **17.** $\frac{2x + 1}{3} = \frac{6x - 9}{5}$ **18.** $\frac{3y + 1}{4} = \frac{44 - y}{5}$

19. $\frac{x}{5} + \frac{x}{3} = \frac{8}{15}$ **20.** $10 = \frac{x}{3} + \frac{x}{7}$ **21.** $\frac{r}{3} - \frac{r}{6} = 2$

22. $\frac{3t}{5} - \frac{t}{5} = 3$ **23.** $1 = \frac{7r}{8} - \frac{3r}{8}$ **24.** $\frac{3t}{4} - 6 = \frac{t}{12}$

25. $\frac{a}{2} + \frac{a}{3} + \frac{a}{4} = 26$ **26.** $\frac{7y}{12} - \frac{1}{4} = 2y - \frac{5}{3}$ **27.** $\frac{y + 2}{4} - \frac{y - 3}{3} = \frac{1}{2}$

28. $\frac{t - 3}{6} - \frac{t - 25}{5} = 4$ **29.** $\frac{3m + 1}{4} = 2 - \frac{3 - 2m}{6}$ **30.** $0.7x - 0.4 = 1$

31. $0.03y - 1.2 = 8.7$ **32.** $0.4x + 0.08 = 4.24$ **33.** $0.5x - 0.3x = 8$

34. $2c + 0.5c = 50$ **35.** $0.08y - 0.9 = 0.02y$ **36.** $1.7x = 30 + 0.2x$

37. $1.5y - 1.69 = 0.2y$

38. $0.08c = 1.5 + 0.07c$

39. $0.8m + 2.6 = 0.2m + 9.8$

40. $0.05x - 0.25 = 0.02x + 0.44$

41. $0.13x - 1.4 = 0.08x + 7.6$

42. $0.06y + 40 - 0.03y = 70$

43. $0.02(x + 5) = 8$

44. $0.05(x - 8) = 0.07x$

45. $0.4(x - 9) = 0.3(x + 4)$

46. $0.06(x - 5) = 0.04(x + 8)$

47. $0.04x + 0.03(2{,}000 - x) = 75$

48. $0.02x + 0.4(1{,}500 - x) = 48$

49. $0.05x + 10 = 0.06(x + 50)$

50. $0.08x = 0.03(x + 200) - 4$

51. The sum of one-half of a number and one-third of that number is 25. Find the number.

52. The difference between one-fifth of a positive number and one-tenth of that number is 10. Find the number.

53. If one-half of a number is increased by 20, the result is 35. Find the number.

54. If two-thirds of a number is decreased by 30, the result is 10. Find the number.

55. If the sum of two consecutive integers is divided by 3, the quotient is 9. Find the integers.

56. If the sum of two consecutive odd integers is divided by 4, the quotient is 10. Find the integers.

57. In an isosceles triangle, each of the congruent sides is two-thirds of the base. The perimeter of the triangle is 42. Find the length of each side of the triangle.

58. The larger of two numbers is 12 less than 5 times the smaller. If the smaller number is equal to one-third of the larger number, find the numbers.

59. The larger of two numbers exceeds the smaller by 14. If the smaller number is equal to three-fifths of the larger, find the numbers.

60. Separate 90 into two parts such that one part is one-half of the other part.

61. Separate 150 into two parts such that one part is two-thirds of the other part.

62. Four vegetable plots of unequal lengths and of equal widths are arranged as shown. The length of the third plot is one-fourth the length of the second plot. The length of the fourth plot is one-half the length of the second plot. The length of the first plot is 10 feet more than the length of the fourth plot. If the total length of the four plots is 100 feet, find the length of each plot.

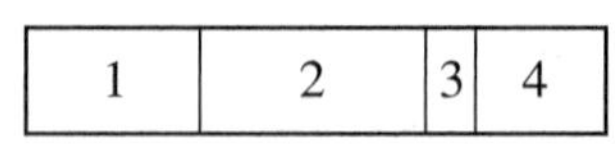

63. Sam is now one-sixth as old as his father. In 4 years, Sam will be one-fourth as old as his father will be then. Find the ages of Sam and his father now.

64. Robert is one-half as old as his father. Twelve years ago, he was one-third as old as his father was then. Find their present ages.

65. A coach finds that, of the students who try out for track, 65% qualify for the team and 90% of those who qualify remain on the team throughout the season. What is the smallest number of students who must try out for track in order to have 30 on the team at the end of the season?

66. A bus that runs once daily between the villages of Alpaca and Down makes only two stops between, at Billow and at Comfort. Today, the bus left Alpaca with some passengers. At Billow, one-half of the passengers got off, and six new ones got on. At Comfort, again one-half of the passengers got off, and, this time, five new ones got on. At Down, the last 13 passengers on the bus got off. How many passengers were aboard when the bus left Alpaca?

67. Sally spent half of her money on a present for her mother, and she spent half of what remained on a treat for herself. If Sally had $2.00 left, how much money did she have originally?

68. Bob planted some lettuce seedlings in his garden. A few days later, after one-third of these seedlings had been eaten by a rabbit, Bob planted 15 new lettuce seedlings. A week later, again one-third of the seedlings had been eaten, leaving 22 seedlings unharmed. How many lettuce seedlings had Bob planted originally?

69. May has 3 times as many dimes as nickels. In all, she has $1.40. How many coins of each type does she have?

70. Mr. Jantzen bought some cans of soup at 39¢ per can, and some packages of frozen vegetables at 59¢ per package. He bought twice as many packages of vegetables as cans of soup. If the total bill was $9.42, how many cans of soup did he buy?

71. Roger has $2.30 in dimes and nickels. There are 5 more dimes than nickels. Find the number of each kind of coin that he has.

72. Bess has $2.80 in quarters and dimes. The number of dimes is 7 less than the number of quarters. Find the number of each kind of coin that she has.

73. A movie theater sold student tickets for $3.25 and full-price tickets for $5. On Saturday, the theater sold 16 more full-price tickets than student tickets. If the total sales on Saturday were $740, how many of each kind of ticket were sold?

74. Is it possible to have $4.50 in dimes and quarters, and have twice as many quarters as dimes? Explain.

75. Is it possible to have $6.00 in nickels, dimes, and quarters, and have the same number of each kind of coin? Explain.

76. Mr. Symms invested a sum of money in 7% bonds. He invested $400 more than this sum in 8% bonds. If the total annual interest from these two investments is $257, how much did he invest at each rate?

77. Mr. Charles borrowed a sum of money at a 10% interest rate. He borrowed a second sum, which was $1,500 less than the first sum, at an 11% interest rate. If the total annual interest he has to pay on these two loans is $202.50, how much did he borrow at each rate?

19-8 SOLVING INEQUALITIES CONTAINING FRACTIONAL COEFFICIENTS

In our modern world, many problem-solving situations involve inequalities. A potential buyer may offer *at most* one amount for a house, while the seller will accept *no less than* another amount. The chart at the right helps us to translate words into algebraic symbols.

Inequalities that contain fractional coefficients are handled in much the same way as equations that contain fractional coefficients.

Words	Symbols
a is greater than b.	$a > b$
a is less than b.	$a < b$
a is at least b. a is no less than b.	$a \geq b$
a is at most b. a is no greater than b.	$a \leq b$

PROCEDURE. To solve an inequality that contains fractional coefficients:
1. Find the L.C.D., a positive number.
2. Multiply both sides of the inequality by the L.C.D.
3. Solve the resulting inequality using the usual methods.

EXAMPLES

1. Solve the inequality, and graph the solution set on a number line:

a. $\frac{x}{3} - \frac{x}{6} > 2$

Solution:

$$\frac{x}{3} - \frac{x}{6} > 2$$

$$6\left(\frac{x}{3} - \frac{x}{6}\right) > 6(2)$$

$$6\left(\frac{x}{3}\right) - 6\left(\frac{x}{6}\right) > 12$$

$$2x - x > 12$$

$$x > 12$$

Since no domain was given, use the domain of real numbers for the variable.

Answer: $x > 12$

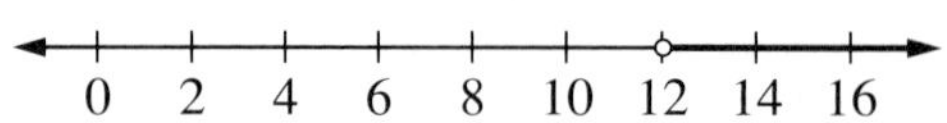

b. $\frac{3y}{2} + \frac{8 - 4y}{7} \leq 3$

Solution:

$$\frac{3y}{2} + \frac{8 - 4y}{7} \leq 3$$

$$14\left(\frac{3y}{2} + \frac{8 - 4y}{7}\right) \leq 14(3)$$

$$14\left(\frac{3y}{2}\right) + 14\left(\frac{8 - 4y}{7}\right) \leq 42$$

$$21y + 16 - 8y \leq 42$$

$$13y \leq 26$$

$$y \leq 2$$

Use the domain of real numbers for the variable.

Answer: $y \leq 2$

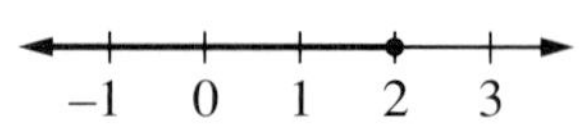

---

2. The smaller of two integers is one-third of the larger, and their sum is greater than 100. Find the smallest possible integers.

Solution

Let $x =$ the larger integer.

Then $\frac{1}{3}x =$ the smaller integer.

$$\underbrace{\textit{Their sum}}_{x + \frac{1}{3}x} \underbrace{\textit{is greater than}}_{>} \underbrace{\textit{100.}}_{100}$$

$$3\left(x + \frac{1}{3}x\right) > 3(100)$$

$$3x + x > 300$$

$$4x > 300$$

$$x > 75$$

$$\text{Then } \frac{1}{3}x > 25$$

The smallest integer greater than 25 is 26. This is one-third of 78.

Check: 78 and 26 $26 \stackrel{?}{=} \frac{1}{3}(78)$ | $78 + 26 \stackrel{?}{>} 100$

$26 = 26$ (True) | $104 > 100$ (True)

Answer: The smallest possible integers are 26 and 78.

EXERCISES WITH OPEN-RESPONSE PROBLEMS

In 1–21, solve each inequality, and graph the solution set on a number line.

1. $\frac{1}{4}x - \frac{1}{5}x > \frac{9}{20}$ **2.** $y - \frac{2}{3}y < 5$ **3.** $\frac{5}{6}c > \frac{1}{3}c + 3$ **4.** $\frac{x}{4} - \frac{x}{8} \leq \frac{5}{8}$

5. $\frac{y}{6} \geq \frac{y}{12} + 1$ **6.** $\frac{y}{9} - \frac{y}{4} > \frac{5}{36}$ **7.** $\frac{t}{10} \leq 4 + \frac{t}{5}$ **8.** $1 + \frac{2x}{3} \geq \frac{x}{2}$

9. $2.5x - 1.6x > 4$ **10.** $2y + 3 \geq 0.2y$ **11.** $\frac{3x - 1}{7} > 5$ **12.** $\frac{5y - 30}{7} \leq 0$

13. $2d + \frac{1}{4} < \frac{7d}{12} + \frac{5}{3}$ **14.** $\frac{4c}{3} - \frac{7}{9} \geq \frac{c}{2} + \frac{7}{6}$ **15.** $\frac{2m}{3} \geq \frac{7 - m}{4} + 1$

16. $\frac{3x - 30}{6} < \frac{x}{3} - 2$ **17.** $\frac{6x - 3}{2} > \frac{37}{10} + \frac{x + 2}{5}$ **18.** $\frac{2y - 3}{3} + \frac{y + 1}{2} < 10$

19. $\frac{2r - 3}{5} - \frac{r - 3}{3} \leq 2$ **20.** $\frac{3t - 4}{3} \geq \frac{2t + 4}{6} + \frac{5t - 1}{9}$ **21.** $\frac{2 - a}{2} \leq \frac{a + 2}{5} - \frac{2a + 3}{6}$

22. If one-third of an integer is increased by 7, the result is at most 13. Find the largest possible integer.

23. If two-fifths of an integer is decreased by 11, the result is at least 4. Find the smallest possible integer.

24. The sum of one-fifth of an integer and one-tenth of that integer is less than 40. Find the greatest possible integer.

25. The difference between three-fourths of a positive integer and one-half of that integer is greater than 28. Find the smallest possible integer.

26. The smaller of two integers is two-fifths of the larger, and their sum is less than 40. Find the largest possible integers.

27. The smaller of two positive integers is five-sixths of the larger, and their difference is greater than 3. Find the smallest possible integers.

28. Talk and Tell Answering Service offers customers two monthly options.

Option 1: Measured Service	Option 2: Unmeasured Service
base rate = \$15	unlimited calls = \$20
each call = 10¢	

Find the least number of calls for which unmeasured service is cheaper than measured service.

29. Paul spent one-half of his pocket money for a book, and then spent one-half of what remained for a record. If he had less than \$3 left, what was the greatest amount of money he could have had originally?

30. Mary bought some cans of vegetables at 89¢ per can, and some cans of soup at 99¢ per can. If she bought twice as many cans of vegetables as cans of soup, and paid at least \$10, what is the least number of cans of vegetables she could have bought?

31. A coin bank contains nickels, dimes, and quarters. The number of dimes is 7 more than the number of nickels, and the number of quarters is twice the number of dimes. If the total value of the coins is no greater than \$7.20, what is the greatest possible number of nickels in the bank?

32. Rhoda is two-thirds as old as her sister Alice. Five years from now, the sum of their ages will be less than 60. What is the largest possible integral value for each sister's present age?

33. Bill is $1\frac{1}{4}$ times as old as his cousin Mary. Four years ago, the difference between their ages was greater than 3. What is the smallest possible integral value for each cousin's present age?

34. Mr. Drew invested a sum of money at $7\frac{1}{2}$ %. He invested a second sum, which was \$200 less than the first, at 7%. If the total annual interest from these two investments is at least \$160, what is the smallest amount he could have invested at $7\frac{1}{2}$ %?

35. Mr. Lehtimaki wanted to sell his house. He advertised an asking price but knew that he would accept, as a minimum, nine-tenths of the asking price. Mrs. Paatela offered to buy the house, but her maximum offer was seven-eighths of the asking price. If the difference between the seller's lowest acceptance price and the buyer's maximum offer was at least \$3,000, find:
a. the minimum asking price for the house
b. the minimum amount Mr. Lehtimaki, the seller, would accept
c. the maximum amount offered by Mrs. Paatela, the buyer

36. An express train started from the depot with some passengers. At the first stop, one third of the passengers got off, and three new passengers got on. At the second stop, again one-third of the passengers got off, and this time six new ones got on. At the next and last stop, the remaining passengers, fewer than 45, left the train. What was the greatest possible number of passengers on the train when it started?

19-9 SOLVING FRACTIONAL EQUATIONS

An equation is called a ***fractional equation*** when a variable appears in the *denominator* of one, or more than one, of its terms. For example,

$$\frac{1}{3} + \frac{1}{x} = \frac{1}{2} \quad \text{and} \quad \frac{2}{3d} + \frac{1}{3} = \frac{11}{6d} - \frac{1}{4}$$

are fractional equations. To simplify such an equation, we can clear it of fractions by multiplying both sides by the lowest common denominator (L.C.D.) of all denominators in the equation. Then, we solve the simpler equation.

As is true of all algebraic fractions, a fractional equation has meaning only when values of the variable do not lead to a denominator of 0.

KEEP IN MIND

When both members of an equation are multiplied by a variable expression that may represent 0, the resulting equation may not be equivalent to the given equation. Each solution, therefore, must be checked in the given equation.

EXAMPLES

Solve and check.

a. $\frac{1}{3} + \frac{1}{x} = \frac{1}{2}$

Solution Multiply both members of the equation by the L.C.D., $6x$.

$$\frac{1}{3} + \frac{1}{x} = \frac{1}{2}$$

$$6x\left(\frac{1}{3} + \frac{1}{x}\right) = 6x\left(\frac{1}{2}\right)$$

$$6x\left(\frac{1}{3}\right) + 6x\left(\frac{1}{x}\right) = 6x\left(\frac{1}{2}\right)$$

$$2x + 6 = 3x$$

$$6 = x$$

Check:

$$\frac{1}{3} + \frac{1}{x} = \frac{1}{2}$$

$$\frac{1}{3} + \frac{1}{6} \stackrel{?}{=} \frac{1}{2}$$

$$\frac{2}{6} + \frac{1}{6} \stackrel{?}{=} \frac{1}{2}$$

$$\frac{3}{6} \stackrel{?}{=} \frac{1}{2}$$

$$\frac{1}{2} = \frac{1}{2} \text{ (True)}$$

Answer: $x = 6$

b. $\frac{5x + 10}{x + 2} = 7$

Solution Multiply both sides of the equation by the L.C.D., $x + 2$.

$$\cancel{(x + 2)}\left(\frac{5x + 10}{\cancel{x + 2}}\right) = (x + 2)(7)$$

$$5x + 10 = 7x + 14$$

$$10 = 2x + 14$$

$$-4 = 2x$$

$$-2 = x$$

Check:

$$\frac{5x + 10}{x + 2} = 7$$

$$\frac{5(-2) + 10}{-2 + 2} \stackrel{?}{=} 7$$

$$\frac{-10 + 10}{-2 + 2} \stackrel{?}{=} 7$$

$$\frac{0}{0} = 7 \text{ (False)}$$

The only possible value of x is a value for which the equation has no meaning because it leads to a denominator of 0. Therefore, there is no solution for this equation.

Answer: The solution set is $\varnothing$ or $\{\ \}$.

EXERCISES WITH OPEN-RESPONSE PROBLEMS

In 1–24, solve each equation, and check.

1. $\frac{10}{x} = 5$ **2.** $\frac{15}{y} = 3$ **3.** $\frac{6}{x} = 12$ **4.** $\frac{8}{b} = -2$

5. $\frac{3}{2x} = \frac{1}{2}$ **6.** $\frac{15}{4x} = \frac{1}{8}$ **7.** $\frac{7}{3y} = -\frac{1}{3}$ **8.** $\frac{4}{5y} = -\frac{1}{10}$

9. $\frac{10}{x} + \frac{8}{x} = 9$ **10.** $\frac{15}{y} - \frac{3}{y} = 4$ **11.** $\frac{7}{c} + \frac{1}{c} = 16$ **12.** $\frac{9}{2x} = \frac{7}{2x} + 2$

13. $\frac{30}{x} = 7 + \frac{18}{2x}$ **14.** $\frac{y - 2}{2y} = \frac{3}{8}$ **15.** $\frac{5}{c} + 6 = \frac{17}{c}$ **16.** $\frac{x - 5}{x} + \frac{3}{x} = \frac{2}{3}$

17. $\frac{y + 9}{2y} + 3 = \frac{15}{y}$ **18.** $\frac{1}{3a} + \frac{5}{12} = \frac{2}{a}$ **19.** $\frac{b - 6}{b} - \frac{1}{6} = \frac{4}{b}$ **20.** $\frac{7}{8} - \frac{x - 3}{x} = \frac{1}{4}$

21. $\frac{5 + x}{2x} - 1 = \frac{x + 1}{x}$ **22.** $\frac{2 + x}{6x} = \frac{3}{5x} + \frac{1}{30}$ **23.** $\frac{a}{a + 2} - \frac{a - 2}{a} = \frac{1}{a}$ **24.** $\frac{x + 1}{2x} + \frac{2x + 1}{3x} = 1$

In 25–28, explain why each fractional equation has no solution.

25. $\frac{6x}{x} = 3$ **26.** $\frac{4a + 4}{a + 1} = 5$ **27.** $\frac{2}{x} = 4 + \frac{2}{x}$ **28.** $\frac{x}{x - 1} + 2 = \frac{1}{x - 1}$

In 29–46, solve each equation, and check.

29. $\frac{6}{3x - 1} = \frac{3}{4}$ **30.** $\frac{2}{3x - 4} = \frac{1}{4}$ **31.** $\frac{5x}{x + 1} = 4$ **32.** $\frac{3}{5 - 3a} = \frac{1}{2}$

33. $\frac{4z}{7 + 5z} = \frac{1}{3}$ **34.** $\frac{1 - r}{1 + r} = \frac{2}{3}$ **35.** $\frac{3}{y} = \frac{2}{5 - y}$ **36.** $\frac{5}{a} = \frac{7}{a - 4}$

37. $\frac{2}{m} = \frac{5}{3m - 1}$ **38.** $\frac{2}{a - 4} = \frac{5}{a - 1}$ **39.** $\frac{12}{2 - x} = \frac{15}{7 + x}$ **40.** $\frac{12y}{8y + 5} = \frac{2}{3}$

41. $\frac{1}{2} = \frac{6a + 3}{24a - 2}$ **42.** $\frac{10}{20 + b} = \frac{30}{8 - b}$ **43.** $\frac{9}{x - 1} = 2 + \frac{1}{x - 1}$

44. $\frac{3}{8x - 1} = 1 - \frac{2}{8x - 1}$ **45.** $\frac{y}{y + 1} - \frac{1}{y} = 1$ **46.** $\frac{x}{x^2 - 9} = \frac{1}{x + 3}$

47. If 24 is divided by a number, the result is 6. Find the number.

48. If 10 is divided by a number, the result is 30. Find the number.

49. The sum of 20 divided by a number, and 7 divided by the same number, is 9. Find the number.

50. If 3 times a number is increased by one-third of that number, the result is 280. Find the number.

51. When the reciprocal of a number is decreased by 2, the result is 5. Find the number.

52. The numerator of a fraction is 8 less than the denominator of the fraction. The value of the fraction is $\frac{3}{5}$. Find the fraction.

53. The numerator and denominator of a fraction are in the ratio 3:4. When the numerator is decreased by 4 and the denominator is increased by 2, the value of the new fraction, in simplest form, is $\frac{1}{2}$. Find the original fraction.

19-10 EQUATIONS AND FORMULAS INVOLVING SEVERAL VARIABLES

Sometimes we need to solve an equation involving several variables, such as x, a, and b, for one of those variables, say x. In this case we express x in terms of the other variables.

To determine what steps should be used in the solution, we use a simpler related equation in which all of the variables except the one for which we are solving are replaced by constants. Then we solve the given equation using the same procedure that was used for the simpler equation.

EXAMPLE

Solve for x and check: $\frac{3x}{2a} + b = 3b$.

Solution:

Step 1: Write and solve a simpler equation by letting $a = 1$ and $b = 1$:

$$\frac{3x}{2} + 1 = 3$$
$$2\left(\frac{3x}{2} + 1\right) = 2(3)$$
$$2\left(\frac{3x}{2}\right) + 2(1) = 2(3)$$
$$3x + 2 = 6$$
$$-2 = -2$$
$$3x = 4$$
$$x = \frac{4}{3}$$

Step 2: Now solve the given equation, using the same process as in step 1.

$$\frac{3x}{2a} + b = 3b$$
$$2a\left(\frac{3x}{2a} + b\right) = 2a(3b)$$
$$2a\left(\frac{3x}{2a}\right) + 2a(b) = 2a(3b)$$
$$3x + 2ab = 6ab$$
$$-2ab = -2ab$$
$$3x = 4ab$$
$$x = \frac{4ab}{3}$$

Check the given equation:

$$\frac{3x}{2a} + b = 3b$$
$$\frac{\not{3}\left(\frac{4ab}{\not{3}}\right)}{2a} + b \stackrel{?}{=} 3b$$
$$\frac{4ab}{2a} + b \stackrel{?}{=} 3b$$
$$\frac{\overset{2}{\not{4}}a b}{\underset{1}{\not{2}\not{a}}} + b \stackrel{?}{=} 3b$$
$$2b + b \stackrel{?}{=} 3b$$
$$3b = 3b$$

(True)

Answer: $x = \frac{4ab}{3}$

EXERCISES

In 1–12, solve each equation for x, and check.

1. $\frac{x}{5} = t$ **2.** $\frac{x}{c} = d$ **3.** $\frac{x}{3a} = b$ **4.** $\frac{x}{3} = \frac{b}{4}$

5. $\frac{x}{a} - \frac{b}{3} = 0$ **6.** $\frac{x}{3} + b = 4b$ **7.** $\frac{r}{x} = t$ **8.** $\frac{t}{x} - k = 0$

9. $\frac{x - 4b}{5} = 8b$ **10.** $\frac{a + b}{x} = c$ **11.** $\frac{mx}{r} + d = 2d$ **12.** $\frac{x}{3a} + \frac{x}{5a} = 8$

In 13–18, solve each formula for the indicated variable.

13. $C = \frac{360}{n}$ for n **14.** $R = \frac{E}{I}$ for I **15.** $v = \frac{s}{t}$ for t

16. $F = \frac{mv^2}{gr}$ for m **17.** $V = \frac{L}{RA}$ for R **18.** $F = 32 + \frac{9}{5}C$ for C

19. If $S = \frac{1}{2}at^2$, express a in terms of S and t.

20. If $A = p + prt$, express t in terms of A, r, and p.

In 21 and 22, find, in each case, the value of the variable indicated when the other variables have the stated values.

21. If $\frac{D}{R} = T$, find the value of R when $D = 120$ and $T = 4$.

22. If $A = \frac{1}{2}h(b + c)$, find the value of h when $A = 48$, $b = 12$, and $c = 4$.

23. If $x = \frac{by}{c}$ and $y = \frac{c^2}{a}$ and $b = \frac{a}{c}$, where none of the variables equals 0, which of the following expressions is true?

(1) $x = 1$ (2) $x = a$ (3) $x = b$ (4) $x = c$

CHAPTER SUMMARY

An ***algebraic fraction*** is the quotient of two algebraic expressions. If the algebraic expressions are polynomials, the fraction is called a ***rational expression***. An algebraic fraction is defined only if values of the variables do not produce a denominator of 0.

Fractions that are equal in value are called ***equivalent fractions***. A fraction is reduced to lowest terms when an equivalent fraction is found such that its numerator and denominator have no common factor other than 1 or -1.

Operations with algebraic fractions follow the same rules as operations with arithmetic fractions.

- *Multiplication:* $\frac{a}{x} \cdot \frac{b}{y} = \frac{ab}{xy}$ $(x \neq 0, y \neq 0)$
- *Division:* $\frac{a}{x} \div \frac{b}{y} = \frac{a}{x} \cdot \frac{y}{b} = \frac{ay}{bx}$ $(x \neq 0, y \neq 0, b \neq 0)$
- *Addition/subtraction with the same denominator:*
$$\frac{a}{c} + \frac{b}{c} = \frac{a+b}{c} \quad (c \neq 0), \qquad \frac{x}{z} - \frac{y}{z} = \frac{x-y}{z} \quad (z \neq 0)$$
- *Addition/subtraction with different denominators* (first, obtain the common denominator):
$$\frac{a}{b} + \frac{c}{d} = \frac{a}{b}\left(\frac{d}{d}\right) + \left(\frac{b}{b}\right)\frac{c}{d} = \frac{ad}{bd} + \frac{bc}{bd} = \frac{ad+bc}{bd} \quad (b \neq 0, d \neq 0)$$

A ***fractional equation*** is an equation in which a variable appears in the denominator of one or more terms. To simplify a fractional equation, or any equation or inequality containing fractional coefficients, multiply both sides by the *lowest common denominator* (L.C.D.) to eliminate the fractions. Then solve the simpler equation or inequality.

VOCABULARY

19-1 Algebraic fraction Rational expression

19-2 Reduced to lowest terms Equivalent fractions Division property of a fraction Cancellation

19-6 Multiplication property of a fraction Common denominator Lowest common denominator

19-9 Fractional equation

REVIEW EXERCISES WITH OPEN-RESPONSE PROBLEMS

1. What fractional part of 1 centimeter is x millimeters?

2. For what value of y is the fraction $\frac{y-1}{y-4}$ undefined?

In 3–6, reduce each fraction to lowest terms.

3. $\frac{8bg}{12bg}$ **4.** $\frac{14d}{7d^2}$ **5.** $\frac{5x^2 - 60}{5}$ **6.** $\frac{8y^2 - 12y}{8y}$

In 7–15, in each case, perform the indicated operation, and express the answer in lowest terms.

7. $\frac{3x^2}{4} \bullet \frac{8}{9x}$ **8.** $\frac{2x - 2}{3y} \bullet \frac{3xy}{2x}$ **9.** $6c^2 \div \frac{c}{2}$

10. $\frac{3a}{7b} \div \frac{18a}{35}$ **11.** $\frac{5m}{6} - \frac{m}{6}$ **12.** $\frac{9}{k} - \frac{3}{k} + \frac{4}{k}$

13. $\frac{ax}{3} + \frac{ax}{4}$ **14.** $\frac{5}{xy} - \frac{2}{yz}$ **15.** $\frac{4x + 5}{3x} + \frac{2x + 1}{3x}$

16. If the sides of a triangle are represented by $\frac{b}{2}$, $\frac{5b}{6}$, and $\frac{2b}{3}$, express the perimeter of the triangle in simplest form.

17. If $a = 2$, $b = 3$, and $c = 4$, what is the sum of $\frac{b}{a}$ and $\frac{a}{c}$?

In 18–23, solve each equation, and check.

18. $\frac{k}{20} = \frac{3}{4}$ **19.** $\frac{x - 3}{10} = \frac{4}{5}$ **20.** $\frac{y}{2} - \frac{y}{6} = 4$

21. $\frac{6}{m} = \frac{20}{m} - 2$ **22.** $\frac{2t}{5} - \frac{t - 2}{10} = 2$ **23.** $\frac{3}{x} + 4 = \frac{3}{x}$

In 24–26, solve each equation for r in terms of the other variables.

24. $\frac{S}{h} = 2\pi r$ **25.** $\frac{a}{r} - n = 0$ **26.** $\frac{c}{2r} = \pi$

In 27–32, in each case perform the indicated operation and express the answer in lowest terms.

27. $\dfrac{x^2 - 5x}{x^2} \bullet \dfrac{x}{2x - 10}$

28. $\dfrac{2a}{a + b} + \dfrac{2b}{a + b}$

29. $\dfrac{y + 7}{5} - \dfrac{y + 3}{4}$

30. $\dfrac{x^2 - 25}{12} \div \dfrac{x^2 - 10x + 25}{4}$

31. $\dfrac{c - 3}{12} + \dfrac{c + 3}{8}$

32. $\dfrac{3a - 9a^2}{a} \div (1 - 9a^2)$

33. Mr. Vroman deposited a sum of money in the bank. After a few years, he found that the interest equaled one-fourth of his original deposit and he had a total sum, deposit plus interest, of $2,400 in the bank. What was the original deposit?

34. One-third of the result obtained by adding 5 to a certain number is equal to one-half of the result obtained when 5 is subtracted from the number. Find the number.

35. Of the total number of points scored by the winning team in a basketball game, one-fifth were scored in the first quarter, one-sixth were scored in the second quarter, one-third were scored in the third quarter, and 27 were scored in the fourth quarter. How many points did the winning team score?

CUMULATIVE REVIEW WITH OPEN-RESPONSE PROBLEMS

1. The cost of two cups of coffee and a bagel is $1.75. The cost of four cups of coffee and three bagels is $4.25. What is the cost of a cup of coffee and the cost of a bagel?
2. Factor completely: $12x^3 - 27x$

Exploration

Some rational numbers can be written as terminating decimals and others as infinitely repeating decimals.

Write each of the following fractions as a decimal:

$$\frac{1}{2}, \frac{1}{4}, \frac{1}{5}, \frac{1}{8}, \frac{1}{10}, \frac{1}{16}, \frac{1}{20}, \frac{1}{25}, \frac{1}{50}, \frac{1}{100}$$

What do you observe about the decimals? Write each denominator in factored form. What do you observe?

Write each of the following fractions as a decimal:

$$\frac{1}{3}, \frac{1}{6}, \frac{1}{9}, \frac{1}{11}, \frac{1}{12}, \frac{1}{15}, \frac{1}{18}, \frac{1}{22}, \frac{1}{24}, \frac{1}{30}$$

What do you observe about the decimals? Write each denominator in factored form. What do you observe?

Write a statement about terminating the infinitely repeating decimals based on your observations.

Chapter 20

Operations with Radicals

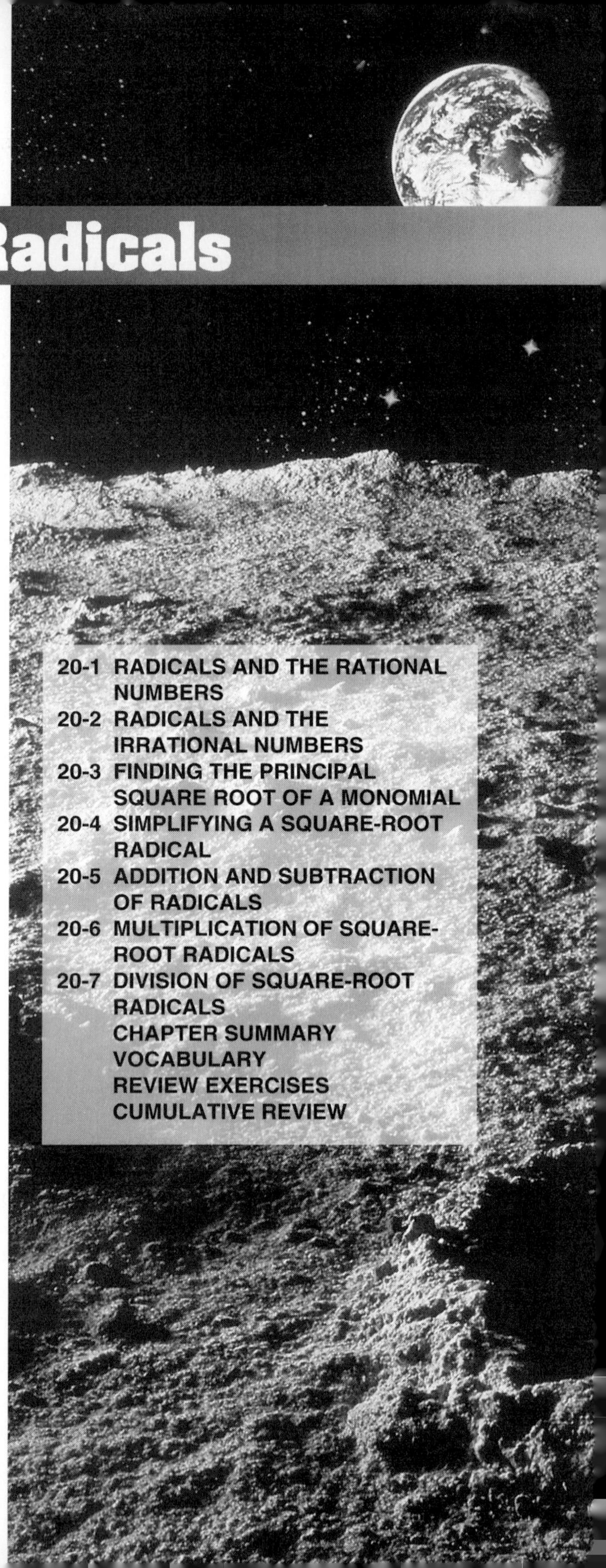

Whenever we send a satellite into space, or we send men and women to the moon, technicians at earthbound space centers monitor activities closely. They continually make small corrections to help the spacecraft stay on course.

The distance from the earth to the moon varies, from 221,460 miles to 252,700 miles, and both the earth and the moon are constantly rotating in space. A tiny error can send the craft thousands of miles off course. Why do such errors occur?

Space centers rely heavily on sophisticated computers, but computers and calculators alike work with *approximations* of numbers, not necessarily with *exact* values.

We have learned that irrational numbers, such as $\sqrt{2}$ and $\sqrt{5}$, are shown on a calculator as decimal approximations of their true values. All irrational numbers, which are examples of radicals, are nonrepeating decimals that never end. How can we work with them?

In this chapter, we will learn techniques to compute with radicals and to find exact answers. We will also look at methods for working with radicals on a calculator to understand how to minimize errors when using these devices.

20-1 RADICALS AND THE RATIONAL NUMBERS

In Chapter 1, we briefly studied the inverse operations of *squaring* a number and finding the *square root* of a number. Recall the following:

To ***square*** a number means to multiply the number by itself	To find the ***square root*** of a number means to find the value that, when multiplied by itself, is equal to the given number.
To square 8, we write: $8^2 = 8 \cdot 8 = 64$	To express the square root of 64, we write: $\sqrt{64} = 8$
On a calculator: ***Enter:*** 8 [x^2] ***Display:*** [64.]	On a calculator: ***Enter:*** 64 [$\sqrt{x}$] ***Display:*** [8.]

The symbol $\sqrt{\ }$ is called the ***radical sign***, and the quantity under the radical sign is the ***radicand***. For example, in $\sqrt{64}$, which we read as "the square root of 64," the radicand is 64.

A ***radical***, which is any term containing both a radical sign and a radicand, is a root of a quantity. For example, $\sqrt{64}$ is a radical.

- **In general, the square root of b is x (written as $\sqrt{b} = x$) if and only if $x \geq 0$ and $x^2 = b$.**

Some radicals, such as $\sqrt{4}$ and $\sqrt{9}$, are rational numbers; others, such as $\sqrt{2}$ and $\sqrt{3}$, are irrational numbers. We begin this study of radicals by examining radicals that are rational numbers.

Perfect Squares

Any number that is the square of a rational number is called a ***perfect square***. For example, perfect squares include numbers such as 9, 0, 1.96, and $\frac{4}{49}$.

$$3 \cdot 3 = 9 \qquad 0 \cdot 0 = 0 \qquad (1.4)(1.4) = 1.96 \qquad \frac{2}{7} \cdot \frac{2}{7} = \frac{4}{49}$$

Then, by applying the inverse operation, we know that:

- **The square root of every perfect square is a *rational* number.**

$$\sqrt{9} = 3 \qquad \sqrt{0} = 0 \qquad \sqrt{1.96} = 1.4 \qquad \sqrt{\frac{4}{49}} = \frac{2}{7}$$

Radicals That Are Square Roots

Certain generalizations can be made for *all* radicals that are square roots, whether they are rational numbers or irrational numbers:

1. Since the square root of 36 is a number whose square is 36, we can write the statement $(\sqrt{36})^2 = 36$. We notice that $(\sqrt{36})^2 = (6)^2 = 36$.

   - **In general, for every real nonnegative number n: $(\sqrt{n})^2 = n$.**

2. Since $(+6)(+6) = 36$ and $(-6)(-6) = 36$, both $+6$ and -6 are square roots of 36. This example illustrates the following statement:

   - **Every positive number has two square roots: one root is a positive number called the *principal square root*, and the other root is a negative number. These two roots have the same absolute value.**

To indicate the positive or principal square root only, place a radical sign over the number:	$\sqrt{25} = 5$	$\sqrt{0.49} = 0.7$
To indicate the negative square root only, place a negative sign before the radical:	$-\sqrt{25} = -5$	$-\sqrt{0.49} = -0.7$
To indicate both square roots, place both a positive and a negative sign before the radical:	$\pm\sqrt{25} = \pm 5$	$\pm\sqrt{0.49} = \pm 0.7$

3. Every real number, when squared, is either positive or 0. Therefore:

   - **The square root of a negative number does not exist in the set of real numbers.**

   For example, $\sqrt{-25}$ does not exist in the set of real numbers because there is no real number that, when multiplied by itself, equals -25.

Calculators and Square Roots

The key strokes required to find a square root of a number can vary depending on the type of calculator being used, and we need to check the calculator manual or try the following methods to see which is correct for a specific calculator. On a calculator, the ***square-root key*** is typically shown as [√x].

The key strokes required to find the principal square root of 25:

Enter:
METHOD 1: 25 [√x]
METHOD 2: [√x] 25 [=]

Display: 5.

The key strokes required to find the negative square root of 25:

Enter:
METHOD 1: 25 [√x] [+/–]
METHOD 2: [√x] 25 [=] [+/–]

Display: –5.

(In this book, we will use *Method 1*, in which the radicand is entered first.)

Cube Roots and Other Roots

A ***cube root*** of a number is one of the three equal factors of the number. For example, 2 is a cube root of 8 because $2 \cdot 2 \cdot 2 = 8$, or $2^3 = 8$. A cube root of 8 is written as $\sqrt[3]{8}$.

- **Finding a cube root of a number is the inverse operation of cubing a number. In general, the cube root of b is x (written as $\sqrt[3]{b} = x$) if and only if $x^3 = b$.**

We have said that $\sqrt{-25}$ does not exist in the set of real numbers. However, $\sqrt[3]{-8}$ does exist in the set of real numbers. Since $(-2)^3 = (-2)(-2)(-2) = -8$, then $\sqrt[3]{-8} = -2$.

In the set of real numbers, every number has one cube root. The cube root of a positive number is positive, and the cube root of a negative number is negative.

In the symbol $\sqrt[n]{b}$, n, an integer that indicates the root to be taken, is called the ***index*** of the radical. Here are two examples:

In $\sqrt[3]{8}$, read as "the cube root of 8," the index is 3.

In $\sqrt[4]{16}$, read as "the fourth root of 16," the index is 4. Since $2^4 = 16$, 2 is one of the four equal factors of 16, and $\sqrt[4]{16} = 2$.

When no index appears, the index is understood to be 2. Thus, $\sqrt{25} = \sqrt[2]{25} = 5$.

Calculators and Roots

Some scientific calculators have a ***root key***, [$\sqrt[x]{\ }$], which is usually accessed by pressing the [2nd] or [INV] or [SHIFT] key. The key strokes needed to find the root of a number will vary according to the calculator being used. If a calculator has a root key, we need to check the manual or experiment to determine the correct sequence needed to find a root. Two common methods of using the root key are presented below.

In *Method 1*, we first enter the radicand (followed by the sign-change key if the radicand is negative). Then we press the [2nd] and [$\sqrt[x]{\ }$] keys, enter the index, and finally press the [=] key.

In *Method 2*, we first enter the index. Then we press the [2nd] and [$\sqrt[x]{\ }$] keys, enter the radicand (followed by the sign-change key if the radicand is negative), and finally press the [=] key.

To show that $\sqrt[3]{64} = 4$:

Enter:

METHOD 1:
64 [2nd] [$\sqrt[x]{\ }$] 3 [=]

METHOD 2:
3 [2nd] [$\sqrt[x]{\ }$] 64 [=]

Display: [4.]

To show that $\sqrt[3]{-64} = -4$:

Enter:

METHOD 1:
64 [+/–] [2nd] [$\sqrt[x]{\ }$] 3 [=]

METHOD 2:
3 [2nd] [$\sqrt[x]{\ }$] 64 [+/–] [=]

Display: [-4.]

(In this book, we will use *Method 1*, in which the radicand is entered first.)

To evaluate the root of a fraction on a calculator, enclose the radicand in parentheses. For example, to evaluate $\sqrt[3]{\frac{8}{125}}$, which equals $\frac{2}{5}$, on a calculator, we proceed as follows:

Enter: [( ] 8 [÷] 125 [ )] [2nd] [$\sqrt[x]{\ }$] 3 [=]

Display: [0.4]

Since the square root of a number has an unwritten index of 2, the root key can also be used to find square roots when an index of 2 is entered. For example, to show that $\sqrt{64} = 8$, we can think of this expression as $\sqrt[2]{64}$.

Enter: 64 [2nd] [$\sqrt[x]{\ }$] 2 [=]

Display: [8.]

EXAMPLES

1. Find the principal square root of 361.

Solution Since $19 \cdot 19 = 361$, then $\sqrt{361} = 19$.

Calculator Solutions

Enter: 361 [$\sqrt{x}$]

Display: [19.]

Enter: 361 [2nd] [$\sqrt[x]{\ }$] 2 [=]

Display: [19.]

Answer: 19

---

2. Find the value of $-\sqrt{0.0016}$.

Solution:

Since $(0.04)(0.04) = 0.0016$, then $-\sqrt{0.0016} = -0.04$.

Calculator Solution:

Enter: 0.0016 [$\sqrt{x}$] [+/−]

Display: [−0.04]

Answer: -0.04

---

3. Is the quantity $(\sqrt{13})^2$ a rational or an irrational number? Explain your answer.

Solution Since $(\sqrt{n})^2 = n$ for $n \geq 0$, then $(\sqrt{13})^2 = 13$.

Answer: The quantity $(\sqrt{13})^2$ is a rational number since its square root, 13, can be written as the quotient of two integers, $\frac{13}{1}$, where the denominator is not 0.

---

4. Solve for x: $x^2 = 36$.

Solution If $x^2 = a$, then $x = \pm\sqrt{a}$ when a is a positive number.

$$\begin{aligned} x^2 &= 36 \\ x &= \pm\sqrt{36} \\ &= \pm 6 \end{aligned}$$

Check: $x^2 = 36$ $\quad$ $(+6)^2 \stackrel{?}{=} 36$ $\quad$ $36 = 36$ (True)

$x^2 = 36$ $\quad$ $(-6)^2 \stackrel{?}{=} 36$ $\quad$ $36 = 36$ (True)

Answer: $x = +6$ or $x = -6$; the solution set is $\{+6, -6\}$.

EXERCISES WITH OPEN-RESPONSE PROBLEMS

In 1–5, state the index and the radicand of each radical.

1. $\sqrt{36}$ **2.** $\sqrt[3]{125}$ **3.** $\sqrt[4]{81}$ **4.** $\sqrt[5]{32}$ **5.** $\sqrt[n]{1}$

In 6–15, find the principal square root of each number.

6. 81 **7.** 1 **8.** 121 **9.** 225 **10.** 900
11. $\frac{1}{9}$ **12.** $\frac{4}{25}$ **13.** 0.49 **14.** 1.44 **15.** 0.04

In 16–40, express each radical as a rational number with the appropriate signs.

16. $\sqrt{16}$ **17.** $\sqrt{81}$ **18.** $\sqrt{121}$ **19.** $-\sqrt{64}$ **20.** $-\sqrt{144}$
21. $\sqrt{0}$ **22.** $\pm\sqrt{100}$ **23.** $\pm\sqrt{169}$ **24.** $\sqrt{400}$ **25.** $-\sqrt{625}$
26. $\sqrt{\frac{1}{4}}$ **27.** $-\sqrt{\frac{9}{16}}$ **28.** $\pm\sqrt{\frac{25}{81}}$ **29.** $\sqrt{\frac{49}{100}}$ **30.** $\pm\sqrt{\frac{144}{169}}$
31. $\sqrt{0.64}$ **32.** $-\sqrt{1.44}$ **33.** $\pm\sqrt{0.09}$ **34.** $-\sqrt{0.01}$ **35.** $\pm\sqrt{0.0004}$
36. $\sqrt[3]{1}$ **37.** $\sqrt[4]{81}$ **38.** $\sqrt[5]{32}$ **39.** $\sqrt[3]{-8}$ **40.** $-\sqrt[3]{-125}$

In 41–60, evaluate each radical by using a calculator.

41. $\sqrt{10.24}$ **42.** $\sqrt{46.24}$ **43.** $\sqrt[3]{2.197}$ **44.** $\sqrt[3]{-3375}$ **45.** $\sqrt[4]{4096}$
46. $\sqrt[5]{-1024}$ **47.** $-\sqrt[3]{-1000}$ **48.** $\sqrt[9]{512}$ **49.** $\sqrt{1.5625}$ **50.** $\sqrt[4]{0.1296}$
51. $\sqrt{8.41}$ **52.** $\sqrt[3]{0.125}$ **53.** $-\sqrt{32.49}$ **54.** $\sqrt[3]{-0.125}$ **55.** $\sqrt{5184}$
56. $\sqrt[4]{70.7281}$ **57.** $\sqrt{0.0576}$ **58.** $\sqrt[5]{-0.32768}$ **59.** $\pm\sqrt{5.76}$ **60.** $\sqrt[4]{1874161}$

In 61–74, find the value of each expression in simplest form.

61. $\sqrt{(8)^2}$ **62.** $\sqrt{\left(\frac{1}{2}\right)^2}$ **63.** $\sqrt{(0.7)^2}$ **64.** $\sqrt{\left(\frac{9}{3}\right)^2}$
65. $\sqrt{\left(\frac{9}{5}\right)^2}$ **66.** $(\sqrt{4})^2$ **67.** $(\sqrt{36})^2$ **68.** $(\sqrt{11})^2$

69. $(\sqrt{39})^2$ **70.** $(\sqrt{97})(\sqrt{97})$ **71.** $\sqrt{36} + \sqrt{49}$ **72.** $\sqrt{100} - \sqrt{25}$

73. $(\sqrt{17})^2 + (\sqrt{7})(\sqrt{7})$ **74.** $\sqrt{(-9)^2} - (\sqrt{83})^2$

75. In each part, compare the given number with its square, using one of the symbols: $>$, $<$, $=$.

a. $\frac{1}{2}$ **b.** $\frac{3}{4}$ **c.** 1 **d.** $\frac{3}{2}$ **e.** 4 **f.** 100

76. Compare the positive real number n with its square n^2 when:

a. $n < 1$ **b.** $n = 1$ **c.** $n > 1$

77. In each part, compare the principal square root of the given number with the number.

a. $\frac{1}{9}$ **b.** $\frac{4}{25}$ **c.** 1 **d.** $\frac{49}{25}$ **e.** 4 **f.** 9

78. Compare $\sqrt{m}$, when m is a positive real number, with the number m, when:

a. $m < 1$ **b.** $m = 1$ **c.** $m > 1$

In 79–90, solve each equation for the variable when the replacement set is the set of real numbers.

79. $x^2 = 4$ **80.** $y^2 = 100$ **81.** $z^2 = \frac{4}{81}$ **82.** $x^2 = 0.49$

83. $x^2 - 16 = 0$ **84.** $y^2 - 36 = 0$ **85.** $2x^2 = 50$ **86.** $3y^2 - 27 = 0$

87. $x^3 = 8$ **88.** $y^3 = 1$ **89.** $y^4 = 81$ **90.** $z^5 = 32$

In 91–94, find, in each case: **a.** the length of each side of a square that has the given area **b.** the perimeter of the square.

91. 36 square feet **92.** 196 square yards

93. 121 square centimeters **94.** 225 square meters

95. Express in terms of x the perimeter of a square whose area is represented by x^2 $(x > 0)$.

96. Write each of the integers from 101 to 110 as the sum of the smallest possible number of perfect squares.

20-2 RADICALS AND THE IRRATIONAL NUMBERS

We have learned that $\sqrt{n}$ is a rational number when n is a perfect square. What type of number is $\sqrt{n}$ when n is *not* a perfect square? As an example, let us examine $\sqrt{5}$ using a calculator.

Enter: 5 [$\sqrt{x}$]

Display: [2.2360679]

Can we state that $\sqrt{5} = 2.2360679$? Or is 2.2360679 a rational approximation of $\sqrt{5}$? To answer this question, we will clear the calculator and find the square of 2.2360679.

Enter: 2.2360679 [x^2]

Display: [4.9999996]

We know that if $\sqrt{b} = x$, then $x^2 = b$. The calculator displays shown above demonstrate that $\sqrt{5} \neq 2.2360679$ because $(2.2360679)^2 \neq 5$. The rational number 2.2360679 is an approximate value for $\sqrt{5}$.

We recall from Chapter 1 that a number such as $\sqrt{5}$ is called an *irrational number*. Irrational numbers cannot be expressed in the form $\frac{a}{b}$, where a and b are integers and $b \neq 0$; furthermore, irrational numbers cannot be expressed as terminating or repeating decimals. The example above illustrates the truth of the following statement:

- **If n is any positive number that is *not* a perfect square, then $\sqrt{n}$ is an irrational number.**

Radicals and Estimation

The radical $\sqrt{5}$ represents the *exact value* of the irrational number whose square is 5. The calculator display for $\sqrt{5}$ is a ***rational approximation*** of the irrational number; it is a number that is close to, but not equal to, $\sqrt{5}$. There are other values correctly rounded from the calculator display that are also approximations of $\sqrt{5}$.

Rational Approximations of $\sqrt{5}$

Exact Value		
$\sqrt{5}$	2.2360679	calculator display, to seven decimal places
	2.236068	rounded to six decimal places (nearest millionth)
	2.23607	rounded to five decimal places (nearest hundred-thousandth)
	2.2361	rounded to four decimal places (nearest ten-thousandth)
	2.236	rounded to three decimal places (nearest thousandth)
	2.24	rounded to two decimal places (nearest hundredth)

Each rational approximation of $\sqrt{5}$, as seen above, indicates that $\sqrt{5}$ is greater than 2 but less than 3. This fact can be further demonstrated by placing the square of $\sqrt{5}$, which is 5, between the squares of two consecutive integers, one less than 5 and one greater than 5, and then finding the square root of each number.

$$\text{Since } 4 < 5 < 9, \text{ then } \sqrt{4} < \sqrt{5} < \sqrt{9}, \text{ or } 2 < \sqrt{5} < 3.$$

In the same way, to get a quick estimate of any square-root radical, we place its square *between* the squares of two consecutive integers. Then we take the square root of each term to show between which two consecutive integers the radical lies. For example, to estimate $\sqrt{73}$:

$$\text{Since } 64 < 73 < 81, \text{ then } \sqrt{64} < \sqrt{73} < \sqrt{81}, \text{ or } 8 < \sqrt{73} < 9.$$

Basic Rules for Radicals That Are Irrational Numbers

There are general rules to follow when working with radicals, especially those that are irrational numbers:

1. **If the degree of accuracy is not specified in a question, it is best to give the *exact* answer in radical form.**

   In other words, if the answer involves a radical, leave the answer in radical form. For example, the sum of 2 and $\sqrt{5}$ is written as $2 + \sqrt{5}$, an *exact* value.

2. **If the degree of accuracy is not specified in a question and a rational approximation is given in place of an answer that includes a radical, the approximation should be correct to *three or more* significant digits.**

   For example, an exact answer is $2 + \sqrt{5}$. By using a calculator, a student discovers that $2 + \sqrt{5}$ is approximately $2 + 2.2360679 = 4.2360679$. *Acceptable answers* would include the calculator display and correctly rounded approximations of the display to three or more digits:

4.2360679	(eight digits)	4.2361	(five digits)
4.236068	(seven digits)	4.236	(four digits)
4.23607	(six digits)	4.24	(three digits).

   Unacceptable answers would include values that are not rounded correctly, as well as values with fewer than three significant digits such as 4.2 (two digits) and 4 (one digit).

3. **When the solution to a problem involving radicals has two or more steps, no radical should be rounded until the very last step.**
   For example, let us find the value of 3 times $\sqrt{5}$, rounded to the *nearest hundredth*.

   An Incorrect Solution

   $$\begin{aligned} 3(\sqrt{5}) &= 3(2.2360679) \\ &= 3(2.24) \text{ rounded too early} \\ &= 6.72 \text{ an incorrect value} \end{aligned}$$

   A Correct Solution

   $$\begin{aligned} 3(\sqrt{5}) &= 3(2.2360679) \\ &= 6.7082037 \\ &= 6.71 \text{ rounded on the last step} \end{aligned}$$

The correct answer is 6.71.

EXAMPLES

1. Between which two consecutive integers is $-\sqrt{42}$?

How to Proceed:	*Solution:*
(1) Place 42 between the squares of consecutive integers:	$36 < 42 < 49$
(2) Take the square root of each number:	$\sqrt{36} < \sqrt{42} < \sqrt{49}$
(3) Simplify terms:	$6 < \sqrt{42} < 7$

(4) Multiply each term of the inequality by -1. Recall that, when an inequality is multiplied by a negative number, the order of the inequality is reversed.

$$-6 > -\sqrt{42} > -7$$

or

$$-7 < -\sqrt{42} < -6$$

Answer: $-\sqrt{42}$ is between -7 and -6, or $-7 < -\sqrt{42} < -6$.

---

2. State whether each of the following expressions is a rational or an irrational number.

a. $\sqrt{56}$

Solution 56 is a positive integer, but it is not a perfect square. There is no rational number that, when squared, equals 56. Therefore, $\sqrt{56}$ is irrational.

Calculator Solution

Step (1):

Enter: 56 [$\sqrt{x}$]

Display: [7.4833147]

To show that the calculator displays a rational approximation, first clear the display. Then show that the square of 7.4833147 does not equal 56.

Step (2):

Enter: 7.4833147 [x^2]

Display: [55.999998]

Since $(7.4833147)^2 \neq 56$, then $\sqrt{56} \neq 7.4833147$. Therefore, $\sqrt{56}$ is irrational.

Answer: $\sqrt{56}$ is an irrational number.

b. $\sqrt{8.0656}$

Calculator Solution

Step (1):

Enter: 8.0656 [$\sqrt{x}$]

Display: [2.84]

It appears that there is a rational square root of 8.0656. To demonstrate that this is true, clear the display and square 2.84.

Step (2):

Enter: 2.84 [x^2]

Display: [8.0656]

Since $(2.84)^2 = 8.0656$, then $\sqrt{8.0656} = 2.84$. Therefore, $\sqrt{8.0656}$ is rational.

Answer: $\sqrt{8.0656}$ is a rational number.

(*Other Hints:* If n is a positive rational number written as a terminating decimal, then n^2 has twice as many decimal places as n. For example, the square of 2.84 has four decimal places. Also, since the last digit of 2.84 is 4, note that the last digit of 2.84^2 must be 6 because $4^2 = 16$.)

---

3. Approximate the value of the expression $\sqrt{9^2 + 11} - 4$:
a. to the *nearest hundredth* **b.** to the *nearest thousandth*

Calculator Solution

Enter: METHOD 1: 9 [x^2] [+] 11 [=] [$\sqrt{x}$] [−] 4 [=]

METHOD 2: [(] 9 [x^2] [+] 11 [)] [$\sqrt{x}$] [−] 4 [=]

Display: 5.5916630

a. To round to the *nearest hundredth*, look at the thousandths place:
5.5916630
↑
Since this digit (1) is *less than 5*, *drop* this digit and all digits to the right. Thus, 5.59 remains.

b. To round to the *nearest thousandth*, look at the ten-thousandths place:
5.5916630
↑
Since this digit (6) is *greater than 5*, drop this digit and all digits to the right but *add 1* to the thousandths place.
Thus, 5.591 + 0.001 = 5.592.

Answers: **a.** 5.59 **b.** 5.592

EXERCISES WITH OPEN-RESPONSE PROBLEMS

In 1–10, between which consecutive integers is each given number?

1. $\sqrt{8}$ **2.** $\sqrt{13}$ **3.** $\sqrt{40}$ **4.** $-\sqrt{2}$ **5.** $-\sqrt{14}$

6. $\sqrt{52}$ **7.** $\sqrt{73}$ **8.** $-\sqrt{125}$ **9.** $\sqrt{143}$ **10.** $-\sqrt{150}$

In 11–16, in each case, order the given numbers, starting with the smallest.

11. $2, \sqrt{3}, -1$ **12.** $4, \sqrt{17}, 3$ **13.** $-\sqrt{15}, -3, -4$

14. $0, \sqrt{7}, -\sqrt{7}$ **15.** $5, \sqrt{21}, \sqrt{30}$ **16.** $-\sqrt{11}, -\sqrt{23}, -\sqrt{19}$

In 17–31, state whether each number is rational or irrational.

17. $\sqrt{25}$ **18.** $\sqrt{40}$ **19.** $-\sqrt{36}$ **20.** $-\sqrt{54}$ **21.** $-\sqrt{150}$

22. $\sqrt{400}$ **23.** $\sqrt{\frac{1}{2}}$ **24.** $-\sqrt{\frac{4}{9}}$ **25.** $\sqrt{0.36}$ **26.** $\sqrt{0.1}$

27. $\sqrt{1156}$ **28.** $\sqrt{951}$ **29.** $\sqrt{6.1504}$ **30.** $\sqrt{2672.89}$ **31.** $\sqrt{5.8044}$

In 32–46, for each irrational number, write a rational approximation: **a.** as shown on a calculator display **b.** rounded to four significant digits

32. $\sqrt{2}$ **33.** $\sqrt{3}$ **34.** $\sqrt{21}$ **35.** $\sqrt{39}$ **36.** $\sqrt{80}$

37. $\sqrt{90}$ **38.** $\sqrt{108}$ **39.** $\sqrt{23.5}$ **40.** $\sqrt{88.2}$ **41.** $-\sqrt{115.2}$

42. $\sqrt{28.56}$ **43.** $\sqrt{67.25}$ **44.** $\sqrt{4389}$ **45.** $\sqrt{123.7}$ **46.** $\sqrt{134.53}$

In 47–52, in each case, find to the *nearest tenth* of a centimeter the length of a side of a square whose area is the given measure.

47. 8 square centimeters **48.** 29 square centimeters **49.** 96 square centimeters
50. 140 square centimeters **51.** 200 square centimeters **52.** 288 square centimeters

In 53–56, in each case, find c if $c = \sqrt{a^2 + b^2}$ and a and b have the given values.

53. $a = 6, b = 8$ **54.** $a = 5, b = 12$ **55.** $a = 15, b = 20$ **56.** $a = 15, b = 36$

In 57–64, use a calculator to approximate each expression: **a.** to the *nearest hundredth* **b.** to the *nearest thousandth*.

57. $\sqrt{7} + 7$ **58.** $\sqrt{2} + \sqrt{6}$ **59.** $\sqrt{8} + \sqrt{8}$ **60.** $\sqrt{50} + 17$
61. $2 + \sqrt{3}$ **62.** $19 - \sqrt{3}$ **63.** $\sqrt{130} - 4$ **64.** $\sqrt{27} + 4.0038$

In 65–68, find the perimeter of each figure, rounded to the *nearest hundredth*.

65.

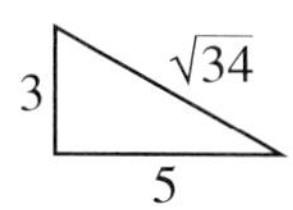

66.

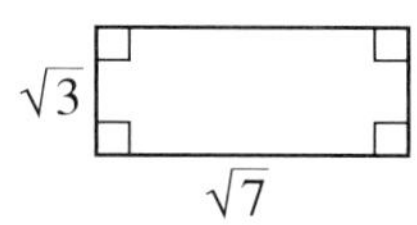

67.

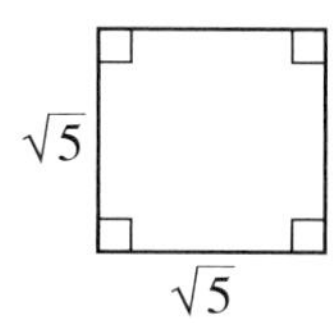

68.

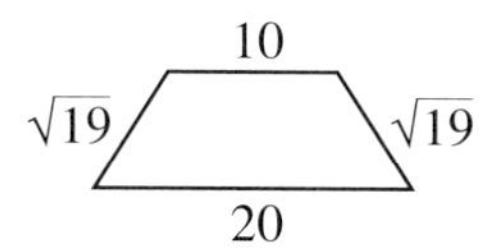

69. Both $\sqrt{58}$ and $\sqrt{58.01}$ are irrational numbers. Find a rational number n such that $\sqrt{58} < n < \sqrt{58.01}$.

70. **a.** Use a calculator to evaluate $\sqrt{999999}$.
b. When you use a calculator to square your answer to part **a**, is the result 999,999?
c. Is $\sqrt{999999}$ rational or irrational? Explain your answer.

20-3 FINDING THE PRINCIPAL SQUARE ROOT OF A MONOMIAL

Just as we can find the principal square root of a number that is a perfect square, we can find the principal square roots of variables and monomials that represent perfect squares.

Since $(6)(6) = 36$, then $\sqrt{36} = 6$.

Since $(x)(x) = x^2$, then $\sqrt{x^2} = x \ (x \geq 0)$.

Since $(a^2)(a^2) = a^4$, then $\sqrt{a^4} = a^2$

Since $(6a^2)(6a^2) = 36a^4$, then $\sqrt{36a^4} = 6a^2$.

In the last case, where the square root contains both numerical and variable factors, we can determine the square root by finding the square roots of its factors and multiplying:

$$\sqrt{36a^4} = \sqrt{36} \bullet \sqrt{a^4} = 6 \bullet a^2 = 6a^2$$

PROCEDURE. To find the square root of a monomial that has two or more factors, write the indicated product of the square roots of its factors.

Note: In our work, we limit the domain of all variables under a radical sign to *nonnegative* numbers.

EXAMPLE

Find the principal square root of each monomial. Assume that all variables represent positive numbers.

a. $\sqrt{25y^2}$ **b.** $\sqrt{1.44a^6b^2}$ **c.** $\sqrt{1369m^{10}}$

Solutions

a. $\sqrt{25y^2} = (\sqrt{25})(\sqrt{y^2}) = (5)(y) =$

b. $\sqrt{1.44a^6b^2} = (\sqrt{1.44})(\sqrt{a^6})(\sqrt{b^2}) = (1.2)(a^3)(b) =$

c. $\sqrt{1369m^{10}} = (\sqrt{1369})(\sqrt{m^{10}}) = (37)(m^5) =$

Answers:
a. $5y$
b. $1.2a^3b$
c. $37m^5$

EXERCISES WITH OPEN-RESPONSE PROBLEMS

In 1–20, find the principal square root of each monomial. Assume that all variables represent positive numbers.

1. $\sqrt{4a^2}$ **2.** $\sqrt{16d^2}$ **3.** $\sqrt{49z^2}$ **4.** $\sqrt{\frac{16}{25}r^2}$ **5.** $\sqrt{0.81w^2}$

6. $\sqrt{9c^2}$ **7.** $\sqrt{36y^4}$ **8.** $\sqrt{c^2d^2}$ **9.** $\sqrt{x^4y^2}$ **10.** $\sqrt{r^8s^6}$

11. $\sqrt{4x^2y^2}$ **12.** $\sqrt{36a^6b^4}$ **13.** $\sqrt{144a^4b^2}$ **14.** $\sqrt{169x^4y^2}$ **15.** $\sqrt{0.36m^2}$

16. $\sqrt{0.49a^2b^2}$ **17.** $\sqrt{289a^{10}}$ **18.** $\sqrt{484k^{12}}$ **19.** $\sqrt{70.56b^2x^{10}}$ **20.** $\sqrt{10.89y^{16}}$

In 21–24, where all variables represent positive numbers:
a. Represent each side of the square whose area is given.
b. Represent the perimeter of that square.

21. $49c^2$ **22.** $64x^2$ **23.** $100x^2y^2$ **24.** $144a^2b^2$

20-4 SIMPLIFYING A SQUARE-ROOT RADICAL

If a square-root radical is not a perfect square, such as $\sqrt{50}$, there is a way to simplify the radical. To understand this procedure, let us first consider some radicals that are perfect squares. We know that $\sqrt{36} = 6$ and that $\sqrt{400} = 20$.

Since $\sqrt{4 \cdot 9} = \sqrt{36} = 6$	Since $\sqrt{16 \cdot 25} = \sqrt{400} = 20$
and $\sqrt{4} \cdot \sqrt{9} = 2 \cdot 3 = 6$,	and $\sqrt{16} \cdot \sqrt{25} = 4 \cdot 5 = 20$,
then $\sqrt{4 \cdot 9} = \sqrt{4} \cdot \sqrt{9}$.	then $\sqrt{16 \cdot 25} = \sqrt{16} \cdot \sqrt{25}$.

These examples illustrate the following property of square-root radicals:

- **The square root of a product of nonnegative numbers is equal to the product of the square roots of these numbers.**

In general, if *a* and *b* are nonnegative numbers:

$$\sqrt{a \cdot b} = \sqrt{a} \cdot \sqrt{b} \quad \text{and} \quad \sqrt{a} \cdot \sqrt{b} = \sqrt{a \cdot b}$$

Now we will apply this rule to $\sqrt{50}$, a square-root radical that is not a perfect square.

1. Express 50 as the product of 25 and 2, where 25 is the *greatest perfect-square* factor of 50: $\sqrt{50} = \sqrt{25 \cdot 2}$
2. Write the product of the two square roots: $= \sqrt{25} \cdot \sqrt{2}$
3. Replace $\sqrt{25}$ with 5 to obtain the simplified expression $5\sqrt{2}$: $= 5\sqrt{2}$

To demonstrate that $\sqrt{50} = 5\sqrt{2}$, evaluate each expression with a calculator.

Evaluate $\sqrt{50}$:	Evaluate $5\sqrt{2}$:
Enter: 50 [$\sqrt{x}$]	***Enter:*** 5 [×] 2 [$\sqrt{x}$] [=]
Display: 7.0710678	***Display:*** 7.0710678

The expression $5\sqrt{2}$ is called the *simplest form* of $\sqrt{50}$. ***The simplest form of a square-root radical*** is one in which the radicand has no perfect-square factor other than 1, and there is no fraction under the radical sign.

PROCEDURE. To simplify the square root of a product:
1. Find, if possible, two factors of the radicand, one of which is the largest perfect-square factor.
2. Express the square root of the product as the product of the square roots of the factors.
3. Find the square root of the factor that is a perfect square.

EXAMPLE

Simplify each expression. Assume that $y > 0$.

a. $\sqrt{18}$ **b.** $4\sqrt{50}$ **c.** $\frac{1}{2}\sqrt{48}$ **d.** $\sqrt{4y^3}$

Solutions

a. $\sqrt{18} = \sqrt{9 \cdot 2} = \sqrt{9} \cdot \sqrt{2} = 3\sqrt{2}$

b. $4\sqrt{50} = 4\sqrt{25 \cdot 2} = 4\sqrt{25} \cdot \sqrt{2} = 4 \cdot 5\sqrt{2} = 20\sqrt{2}$

c. $\frac{1}{2}\sqrt{48} = \frac{1}{2}\sqrt{16 \cdot 3} = \frac{1}{2}\sqrt{16} \cdot \sqrt{3} = \frac{1}{2} \cdot 4\sqrt{3} = 2\sqrt{3}$

d. $\sqrt{4y^2} = \sqrt{4y^2 \cdot y} = \sqrt{4y^2} \cdot \sqrt{y} = 2y\sqrt{y}$

Answers:
a. $3\sqrt{2}$
b. $20\sqrt{2}$
c. $2\sqrt{3}$
d. $2y\sqrt{y}$

EXERCISES WITH OPEN-RESPONSE PROBLEMS

In 1–35, simplify each expression. Assume that all variables represent positive numbers.

1. $\sqrt{8}$ **2.** $\sqrt{12}$ **3.** $\sqrt{20}$ **4.** $\sqrt{28}$ **5.** $\sqrt{40}$

6. $\sqrt{27}$ **7.** $\sqrt{54}$ **8.** $\sqrt{63}$ **9.** $\sqrt{90}$ **10.** $\sqrt{98}$

11. $\sqrt{99}$ **12.** $\sqrt{108}$ **13.** $\sqrt{162}$ **14.** $\sqrt{175}$ **15.** $\sqrt{300}$

16. $3\sqrt{8}$ **17.** $4\sqrt{12}$ **18.** $2\sqrt{20}$ **19.** $4\sqrt{90}$ **20.** $2\sqrt{45}$

21. $3\sqrt{200}$ **22.** $\frac{1}{2}\sqrt{72}$ **23.** $\frac{1}{4}\sqrt{48}$ **24.** $\frac{3}{4}\sqrt{96}$ **25.** $\frac{2}{3}\sqrt{63}$

26. $5\sqrt{24}$ **27.** $2\sqrt{80}$ **28.** $7\sqrt{45}$ **29.** $8\sqrt{9x}$ **30.** $\sqrt{3x^3}$

31. $\sqrt{49x^5}$ **32.** $\sqrt{36r^2s}$ **33.** $\sqrt{128x^6}$ **34.** $\sqrt{243xy^2}$ **35.** $\sqrt{720n^7}$

36. A student simplified the expression $\sqrt{192}$ by writing $\sqrt{192} = \sqrt{16 \cdot 12} = 4\sqrt{12}$

a. Explain why $4\sqrt{12}$ is *not* the simplest form of $\sqrt{192}$.
b. Find the simplest form of $\sqrt{192}$.
c. Find, to the *nearest ten thousandth*, the value of:
(*1*) $\sqrt{192}$ (*2*) $4\sqrt{12}$ (*3*) your answer to part **b**

37. The expression $\sqrt{48}$ is equivalent to
(1) $2\sqrt{3}$ (2) $4\sqrt{12}$ (3) $4\sqrt{3}$ (4) $16\sqrt{3}$

38. The expression $4\sqrt{2}$ is equivalent to
(1) $\sqrt{8}$ (2) $\sqrt{42}$ (3) $\sqrt{32}$ (4) $\sqrt{64}$

39. The expression $3\sqrt{18}$ is equivalent to
(1) $\sqrt{54}$ (2) $3\sqrt{2}$ (3) $9\sqrt{2}$ (4) $3\sqrt{6}$

40. The expression $3\sqrt{3}$ is equivalent to
(1) $\sqrt{9}$ (2) $\sqrt{6}$ (3) $\sqrt{12}$ (4) $\sqrt{27}$

In 41–44, for each expression:

a. Use a calculator to find a rational approximation of the expression.
b. Write the original expression in simplest radical form.
c. Use a calculator to find a rational approximation of the answer to part **b**.
d. State whether the approximations in parts **a** and **c** are equal.

41. $\sqrt{300}$ **42.** $\sqrt{180}$ **43.** $2\sqrt{288}$ **44.** $\frac{1}{3}\sqrt{252}$

45. a. Does $\sqrt{9 + 16} = \sqrt{9} + \sqrt{16}$? Explain your answer.
b. Is finding a square root always distributive over addition?

46. a. Does $\sqrt{25 - 9} = \sqrt{25} - \sqrt{9}$? Explain your answer.
b. Is finding a square root always distributive over subtraction?

20-5 ADDITION AND SUBTRACTION OF RADICALS

Radicals are exact values. Computations sometimes involve many radicals, as in the following example:

$$\sqrt{12} + \sqrt{75} + 3\sqrt{3}$$

Rather than use approximations obtained with a calculator to find this sum, it may be important to express the answer as an *exact* value in radical form. To learn how to perform computations with radicals, we need to recall some algebraic concepts previously learned.

Adding and Subtracting Like Radicals

We have learned how to add like terms in algebra:

$$2x + 5x + 3x = 10x$$

If we replace the variable x by an irrational number, $\sqrt{3}$, it follows that this statement must be true:

$$2\sqrt{3} + 5\sqrt{3} + 3\sqrt{3} = 10\sqrt{3}$$

Like radicals, such as $2\sqrt{3}$ and $5\sqrt{3}$, are radicals that have the same index and the same radicand. In the same way, $4\sqrt[3]{7}$ and $9\sqrt[3]{7}$ are like radicals.

To demonstrate that the sum of like radicals is written as a single term, we can also apply a concept used to add like terms, namely, the distributive property:

$$2\sqrt{3} + 5\sqrt{3} + 3\sqrt{3} = (2 + 5 + 3)\sqrt{3} = 10\sqrt{3}$$

Similarly,

$$6\sqrt{7} - \sqrt{7} = 6\sqrt{7} - 1\sqrt{7} = (6 - 1)\sqrt{7} = 5\sqrt{7}$$

PROCEDURE. To add or subtract like radicals:
1. Add or subtract the coefficient of the radicals.
2. Multiply the sum or difference obtained by the common radical.

Adding and Subtracting Unlike Radicals

Unlike radicals are radicals that have different radicands or different indexes, or are different in both. For example:

$3\sqrt{5}$ and $3\sqrt{2}$ are unlike radicals because their radicands are different.

$\sqrt[3]{2}$ and $\sqrt{2}$ are unlike radicals because their indexes are different.

$9\sqrt{10}$ and $\sqrt[4]{3}$ are unlike radicals because their radicands and indexes are different.

The sum of unlike radicals cannot always be expressed as a single term.

The sum of $\sqrt{5}$ and $\sqrt{2}$ is

$$\sqrt{5} + \sqrt{2}$$

The difference of $5\sqrt{7}$ and $\sqrt{11}$ is

$$5\sqrt{7} - \sqrt{11}$$

However, it is sometimes possible to transform unlike radicals into equivalent like radicals. These like radicals can then be combined into a single term. Let us return to the problem posed at the start of this section:

$$\begin{aligned}\sqrt{12} + \sqrt{75} + 3\sqrt{3} &= \sqrt{4 \cdot 3} + \sqrt{25 \cdot 3} + 3\sqrt{3}\\ &= \sqrt{4} \cdot \sqrt{3} + \sqrt{25} \cdot \sqrt{3} + 3\sqrt{3}\\ &= 2\sqrt{3} + 5\sqrt{3} + 3\sqrt{3}\\ &= 10\sqrt{3}\end{aligned}$$

How does this problem compare to the one above in which like radicals were added?

PROCEDURE. To combine unlike radicals:
1. Simplify each radical, if possible.
2. Combine like radicals by using the distributive property.
3. Indicate the sum or difference of the unlike radicals.

EXAMPLES

1. Combine: $8\sqrt{5} + \sqrt{5} - 2\sqrt{5}$

Solution $8\sqrt{5} + \sqrt{5} - 2\sqrt{5} = (8 + 1 - 2)\sqrt{5} = 7\sqrt{5}$ *Answer*

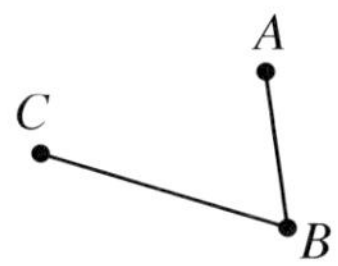

2. A scientist carefully measured the distance from A to B, and from B to C, as shown in the diagram, to determine the total length, in centimeters, of very fine wire that she needed for an experiment. Using formulas, she discovered that $AB = 5\sqrt{3}$ centimeters and $BC = 4\sqrt{12}$ centimeters.

a. Find, in simplest radical form, the total number of centimeters of wire needed.

b. Approximate your answer to part **a** to the *nearest thousandth* of a centimeter.

Solution **a.**

$$\begin{aligned} 5\sqrt{3} + 4\sqrt{12} &= 5\sqrt{3} + 4\sqrt{4 \cdot 3} \\ &= 5\sqrt{3} + 4\sqrt{4} \cdot \sqrt{3} \\ &= 5\sqrt{3} + 4 \cdot 2 \cdot \sqrt{3} \\ &= 5\sqrt{3} + 8\sqrt{3} \\ &= (5 + 8)\sqrt{3} \\ &= 13\sqrt{3} \end{aligned}$$

b. Use a calculator.

Enter: 13 [×] 3 [$\sqrt{x}$] [=]

Display: 22.5166605

↑

Rounded to the *nearest thousandth:*

$$\begin{array}{r} 22.516 \\ +\ \ .001 \\ \hline 22.517 \end{array}$$

Answer: **a.** $13\sqrt{3}$ centimeters or **b.** 22.517 centimeters

EXERCISES WITH OPEN-RESPONSE PROBLEMS

In 1–28, in each case, combine the radicals. Assume that all variables represent positive numbers

1. $8\sqrt{2} + 7\sqrt{2}$ **2.** $8\sqrt{5} + \sqrt{5}$ **3.** $5\sqrt{3} + 2\sqrt{3} + 8\sqrt{3}$

4. $14\sqrt{6} - 2\sqrt{6}$ **5.** $7\sqrt{2} - \sqrt{2}$ **6.** $4\sqrt{3} + 2\sqrt{3} - 6\sqrt{3}$

7. $5\sqrt{3} + \sqrt{3} - 2\sqrt{3}$ **8.** $4\sqrt{7} - \sqrt{7} - 5\sqrt{7}$

9. $3\sqrt{5} + 6\sqrt{2} - 3\sqrt{2} + \sqrt{5}$ **10.** $9\sqrt{x} + 3\sqrt{x}$

11. $15\sqrt{y} - 7\sqrt{y}$ **12.** $\sqrt{2} + \sqrt{50}$ **13.** $\sqrt{27} + \sqrt{75}$

14. $\sqrt{80} - \sqrt{5}$ **15.** $\sqrt{72} - \sqrt{50}$ **16.** $\sqrt{12} - \sqrt{48} + \sqrt{3}$

17. $3\sqrt{32} - 6\sqrt{8}$ **18.** $5\sqrt{27} - \sqrt{108} + 2\sqrt{75}$ **19.** $3\sqrt{8} - \sqrt{2}$

20. $5\sqrt{8} - 3\sqrt{18} + \sqrt{3}$ **21.** $3\sqrt{50} - 5\sqrt{18}$ **22.** $\sqrt{98} - 4\sqrt{8} + 3\sqrt{128}$

23. $\frac{2}{3}\sqrt{18} - \sqrt{72}$ **24.** $\sqrt{7a} + \sqrt{28a}$ **25.** $\sqrt{81x} + \sqrt{25x}$

26. $\sqrt{100b} - \sqrt{64b} + \sqrt{9b}$ **27.** $3\sqrt{3x} - \sqrt{12x}$ **28.** $\sqrt{3a^2} + \sqrt{12a^2}$

29. Express in simplest radical form the sum of:

a. $3\sqrt{2}$ and $\sqrt{98}$ **b.** $5\sqrt{3}$ and $\sqrt{27}$ **c.** $2\sqrt{12}$ and $4\sqrt{75}$

30. Express in simplest radical form the difference between:

a. $4\sqrt{2}$ and $\sqrt{18}$ **b.** $\sqrt{75}$ and $\sqrt{12}$ **c.** $4\sqrt{48}$ and $8\sqrt{12}$

In 31–33, in each case, select the numeral preceding the correct choice.

31. The difference $5\sqrt{2} - \sqrt{32}$ is equivalent to
(1) $\sqrt{2}$ (2) $9\sqrt{2}$ (3) $-11\sqrt{2}$ (4) $18\sqrt{2}$

32. The sum $3\sqrt{8} + 6\sqrt{2}$ is equivalent to
(1) $9\sqrt{10}$ (2) $\sqrt{72}$ (3) $18\sqrt{10}$ (4) $12\sqrt{2}$

33. The sum of $\sqrt{12}$ and $\sqrt{27}$ is equivalent to
(1) $\sqrt{39}$ (2) $5\sqrt{6}$ (3) $13\sqrt{3}$ (4) $5\sqrt{3}$

In 34 and 35: **a.** Express the perimeter of the figure in simplest radical form. **b.** Using a calculator, approximate the expression obtained in part **a** to the *nearest thousandth*.

34.

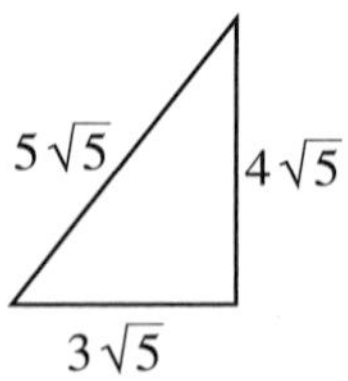

35.

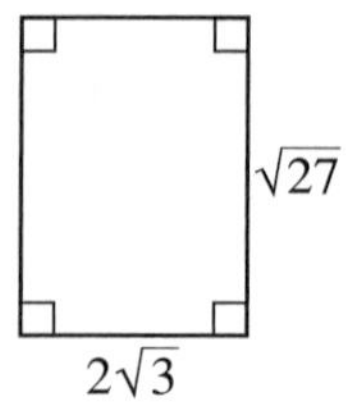

20-6 MULTIPLICATION OF SQUARE-ROOT RADICALS

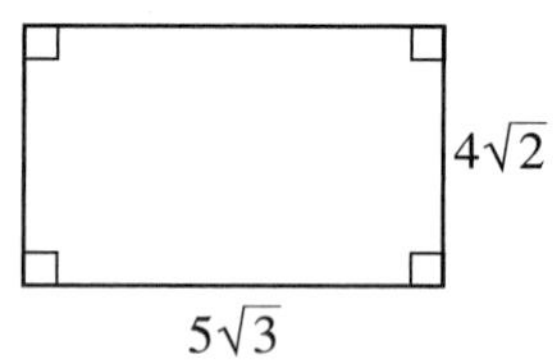

To find the area of the rectangle pictured at the right, we multiply $5\sqrt{3}$ by $4\sqrt{2}$.

We have learned that $\sqrt{a} \cdot \sqrt{b} = \sqrt{ab}$ when a and b are nonnegative numbers.

To multiply $4\sqrt{2}$ by $5\sqrt{3}$, we use the commutative and associative laws of multiplication as follows:

$$(4\sqrt{2})(5\sqrt{3}) = (4)(5)(\sqrt{2})(\sqrt{3}) = (4 \cdot 5)(\sqrt{2 \cdot 3}) = 20\sqrt{6}$$

In general, if a and b are nonnegative numbers:

$$x\sqrt{a} \cdot y\sqrt{b} = xy\sqrt{ab}$$

PROCEDURE. To multiply two monomial square roots:
1. Multiply the coefficients to find the coefficient of the product.
2. Multiply the radicands to find the radicand of the product.
3. If possible, simplify the result.

EXAMPLES

1. **a.** Multiply $3\sqrt{6} \cdot 5\sqrt{2}$, and write the product in simplest radical form.
   b. Check the work performed in part **a** using a calculator.

Solutions **a.**

$$\begin{aligned} 3\sqrt{6} \cdot 5\sqrt{2} &= 3 \cdot 5\sqrt{6 \cdot 2} \\ &= 15\sqrt{12} \\ &= 15\sqrt{4} \cdot \sqrt{3} \\ &= 15 \cdot 2 \cdot \sqrt{3} \\ &= 30\sqrt{3} \textit{ Answer} \end{aligned}$$

b. To check, evaluate $3\sqrt{6} \cdot 5\sqrt{2}$.

Enter: 3 [×] 6 [√x] [×] 5 [×] 2 [√x] [=]

Display: [51.961524]

Clear the display. Then evaluate $30\sqrt{3}$.

Enter: 30 [×] 3 [√x] [=]

Display: [51.961524]

Therefore, $3\sqrt{6} \cdot 5\sqrt{2} = 30\sqrt{3}$ appears to be true.

---

2. Find the value of $(2\sqrt{3})^2$.

Solution

$$\begin{aligned} (2\sqrt{3})^2 &= 2\sqrt{3} \cdot 2\sqrt{3} \\ &= 2 \cdot 2\sqrt{3 \cdot 3} = 4\sqrt{9} = 4 \cdot 3 = 12 \textit{ Answer} \end{aligned}$$

---

3. Find the indicated product: $\sqrt{3x} \cdot \sqrt{6x}$ $(x > 0)$.

Solution

$$\begin{aligned} \sqrt{3x} \cdot \sqrt{6x} &= \sqrt{3x \cdot 6x} \\ &= \sqrt{18x^2} \\ &= \sqrt{9x^2 \cdot 2} = \sqrt{9x^2} \cdot \sqrt{2} = 3x\sqrt{2} \textit{ Answer} \end{aligned}$$

EXERCISES WITH OPEN-RESPONSE PROBLEMS

In 1–24: In each case multiply, or raise to the power, as indicated. Then simplify the result. Assume that all variables represent positive numbers.

1. $\sqrt{3} \cdot \sqrt{3}$ **2.** $\sqrt{7} \cdot \sqrt{7}$ **3.** $\sqrt{a} \cdot \sqrt{a}$ **4.** $\sqrt{2x} \cdot \sqrt{2x}$

5. $\sqrt{12} \cdot \sqrt{3}$ **6.** $2\sqrt{18} \cdot 3\sqrt{8}$ **7.** $\sqrt{14} \cdot \sqrt{2}$ **8.** $\sqrt{60} \cdot \sqrt{5}$

9. $3\sqrt{6} \cdot \sqrt{3}$ **10.** $5\sqrt{8} \cdot 7\sqrt{3}$ **11.** $\frac{2}{3}\sqrt{24} \cdot 9\sqrt{3}$ **12.** $5\sqrt{6} \cdot \frac{2}{3}\sqrt{15}$

13. $(-4\sqrt{a})(3\sqrt{a})$ **14.** $\left(-\frac{1}{2}\sqrt{y}\right)(-6\sqrt{y})$ **15.** $(\sqrt{2})^2$

16. $(\sqrt{y})^2$ **17.** $(\sqrt{t})^2$ **18.** $(3\sqrt{6})^2$

19. $\sqrt{25x} \cdot \sqrt{4x}$ **20.** $\sqrt{27a} \cdot \sqrt{3a}$ **21.** $\sqrt{15x} \cdot \sqrt{3x}$

22. $\sqrt{9a} \cdot \sqrt{ab}$ **23.** $(\sqrt{5x})^2$ **24.** $(2\sqrt{t})^2$

In 25–30: **a.** Perform each indicated operation. **b.** State whether the product is an irrational number or a rational number.

25. $(5\sqrt{12})(4\sqrt{3})$ **26.** $(3\sqrt{2})(2\sqrt{32})$ **27.** $(4\sqrt{6})(9\sqrt{3})$

28. $(8\sqrt{5})\left(\frac{1}{2}\sqrt{10}\right)$ **29.** $\left(\frac{2}{3}\sqrt{5}\right)^2$ **30.** $\left(\frac{1}{6}\sqrt{8}\right)\left(\frac{1}{2}\sqrt{18}\right)$

In 31–34, in each case, find the area of the square in which the length of each side is the given number.

31. $\sqrt{2}$ **32.** $2\sqrt{3}$ **33.** $6\sqrt{2}$ **34.** $5\sqrt{3}$

In 35 and 36: **a.** Express the area of the figure in simplest radical form. **b.** Check the work performed in part **a** by using a calculator.

35.

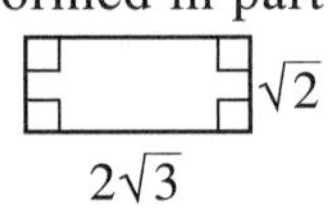

36.

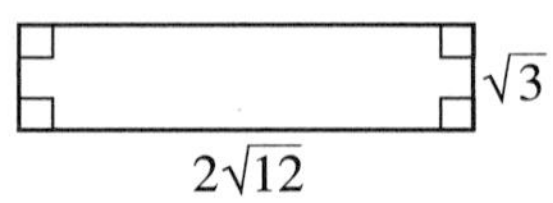

37. Two square-root radicals that are *irrational numbers* are multiplied.
a. Give two examples where the product of these radicals is also an irrational number.
b. Give two examples where the product of these radicals is a rational number.

20-7 DIVISION OF SQUARE-ROOT RADICALS

$$\text{Since } \sqrt{\frac{4}{9}} = \frac{2}{3} \text{ and } \frac{\sqrt{4}}{\sqrt{9}} = \frac{2}{3}, \text{ then } \sqrt{\frac{4}{9}} = \frac{\sqrt{4}}{\sqrt{9}}.$$

$$\text{Since } \sqrt{\frac{16}{25}} = \frac{4}{5} \text{ and } \frac{\sqrt{16}}{\sqrt{25}} = \frac{4}{5}, \text{ then } \sqrt{\frac{16}{25}} = \frac{\sqrt{16}}{\sqrt{25}}.$$

These examples illustrate the following property of square-root radicals:

- **The square root of a fraction that is the quotient of a nonnegative number and a positive number is equal to the square root of the numerator divided by the square root of the denominator.**

In general, if a is a nonnegative and b is positive:

$$\sqrt{\frac{a}{b}} = \frac{\sqrt{a}}{\sqrt{b}} \quad \textbf{and} \quad \frac{\sqrt{a}}{\sqrt{b}} = \sqrt{\frac{a}{b}}$$

We use this principle to divide $\sqrt{72}$ by $\sqrt{8}$. In this example, notice that the quotient of two irrational numbers is a *rational* number.

$$\frac{\sqrt{72}}{\sqrt{8}} = \sqrt{\frac{72}{8}} = \sqrt{9} = 3$$

We can divide other radical terms by using the property of fractions:

$$\frac{ac}{bd} = \frac{a}{b} \bullet \frac{c}{d}$$

Note, in the following example, that the quotient of two irrational numbers is also irrational:

$$\frac{6\sqrt{10}}{3\sqrt{2}} = \frac{6}{3} \bullet \frac{\sqrt{10}}{\sqrt{2}} = \frac{6}{3} \bullet \sqrt{\frac{10}{2}} = 2\sqrt{5}$$

In general, if a is nonnegative, b is positive, and $y \neq 0$:

$$\frac{x\sqrt{a}}{y\sqrt{b}} = \frac{x}{y}\sqrt{\frac{a}{b}}$$

PROCEDURE. To divide two monomial square roots:

1. Divide the coefficients to find the coefficient of the quotient.
2. Divide the radicands to find the radicand of the quotient.
3. If possible, simplify the result.

EXAMPLE

a. Divide $8\sqrt{48}$ by $4\sqrt{2}$, and simplify the quotient.

b. Check the work performed in part **a** using a calculator.

Solutions **a.** $8\sqrt{48} \div 4\sqrt{2} = \frac{8}{4}\sqrt{\frac{48}{2}} = 2\sqrt{24} = 2\sqrt{4 \bullet 6} = 2 \bullet \sqrt{4} \bullet \sqrt{6}$

$= 2 \bullet 2 \bullet \sqrt{6} = 4\sqrt{6}$ *Answer*

b. To check, evaluate $8\sqrt{48} \div 4\sqrt{2}$. Note the use of parentheses to group terms in the denominator.

Enter: 8 [×] 48 [√x] [÷] [(] 4 [×] 2 [√x] [)] [=]

Display: [9.797959]

Clear the display. Then evaluate $4\sqrt{6}$.

Enter: 4 [×] 6 [√x] [=]

Display: [9.797959]

Since the displays are equal, it appears that the work in part **a** is correct.

EXERCISES

In 1–16, divide. Then simplify the quotient.

1. $\sqrt{72} \div \sqrt{2}$ **2.** $\sqrt{75} \div \sqrt{3}$ **3.** $\sqrt{70} \div \sqrt{10}$ **4.** $\sqrt{14} \div \sqrt{2}$

5. $8\sqrt{48} \div 2\sqrt{3}$ **6.** $\sqrt{24} \div \sqrt{2}$ **7.** $\sqrt{150} \div \sqrt{3}$ **8.** $21\sqrt{40} \div \sqrt{5}$

9. $9\sqrt{6} \div 3\sqrt{6}$ **10.** $7\sqrt{3} \div 3\sqrt{3}$ **11.** $2\sqrt{2} \div 8\sqrt{2}$ **12.** $\sqrt{9y} \div \sqrt{y}$

13. $\frac{12\sqrt{20}}{3\sqrt{5}}$ **14.** $\frac{20\sqrt{50}}{4\sqrt{2}}$ **15.** $\frac{25\sqrt{24}}{5\sqrt{2}}$ **16.** $\frac{3\sqrt{54}}{6\sqrt{3}}$

In 17–22, state whether each quotient is a rational number or an irrational number.

17. $\frac{\sqrt{7}}{5}$ **18.** $\frac{\sqrt{50}}{\sqrt{2}}$ **19.** $\frac{\sqrt{18}}{\sqrt{3}}$ **20.** $\frac{\sqrt{49}}{\sqrt{7}}$ **21.** $\frac{\sqrt{9}}{\sqrt{16}}$ **22.** $\frac{\sqrt{18}}{\sqrt{25}}$

In 23–28, simplify each expression.

23. $\sqrt{\frac{36}{49}}$ **24.** $\sqrt{\frac{3}{4}}$ **25.** $4\sqrt{\frac{5}{16}}$ **26.** $\sqrt{\frac{8}{49}}$ **27.** $10\sqrt{\frac{8}{25}}$ **28.** $\frac{2}{3}\sqrt{\frac{144}{64}}$

In 29–32, in each case the area A of a parallelogram and the measure of its base b are given. **a.** Find the height h of the parallelogram, expressed in simplest form. **b.** Check the work performed in part **a** using a calculator.

29. $A = 7\sqrt{12}$, $b = 7\sqrt{3}$ **30.** $A = \sqrt{640}$, $b = \sqrt{32}$

31. $A = 8\sqrt{45}$, $b = 2\sqrt{15}$ **32.** $A = 2\sqrt{98}$, $b = \sqrt{32}$

33. Each of five real numbers: $\sqrt{\frac{1}{4}}$ $\sqrt{3}$ $\sqrt{0.4}$ $\sqrt{0.64}$ $\sqrt{\frac{2}{9}}$ is written on a separate piece of paper. The five pieces of paper are placed in a bag, and one piece is chosen at random. Find the probability that the number on the chosen paper is:

a. a rational number **b.** an irrational number **c.** equal to $\frac{1}{2}$

d. equal to 0.2 **e.** equal to 0.8 or $\frac{1}{2}$ **f.** a real number

CHAPTER SUMMARY

A ***radical***, which is the root of a quantity, is written in its general form as $\sqrt[n]{b}$. A radical consists of a ***radicand***, b, placed under a ***radical sign***, $\sqrt{}$, with an ***index***, n.

Finding the square root of a quantity is the inverse operation of squaring. A ***square-root radical*** has an index of 2, which is generally not written. Thus, $\sqrt[2]{49} = \sqrt{49} = 7$ because $7^2 = 49$. In general, for nonnegative numbers b and x, $\sqrt{b} = x$ if and only if $x^2 = b$.

If k is a positive number that is a perfect square, then $\sqrt{k}$ is a rational number. If k is positive but not a perfect square, then $\sqrt{k}$ is an irrational number.

Every positive number has two square roots: a positive root called the ***principal square root***, and a negative root. These roots have the same absolute value:

Principal Square Root	*Negative Square Root*	*Both Roots*
$\sqrt{x^2} = \lvert x \rvert$	$-\sqrt{x^2} = -\lvert x \rvert$	$\pm\sqrt{x^2} = \pm x$

Finding the ***cube root*** of a number is the inverse operation of cubing. Thus, $\sqrt[3]{64} = 4$ because $4^3 = 4 \cdot 4 \cdot 4 = 64$. In general, $\sqrt[3]{b} = x$ if and only if $x^3 = b$.

Like radicals have the same radicand and the same index, as in $2\sqrt{7}$ and $3\sqrt{7}$. ***Unlike radicals*** can differ in their radicands $(\sqrt{7}$ and $\sqrt{2})$, in their indexes $(\sqrt{7}$ and $\sqrt[3]{7})$, or in both $(\sqrt[4]{7}$ and $\sqrt[3]{6})$.

A square root of a positive integer is ***simplified*** by factoring out its greatest perfect square. The simplified radical then contains no perfect square factor in its radicand other than 1. When a radical is irrational, the radical expresses its *exact* value. Calculators display only ***rational approximations*** of radicals that are irrational numbers.

Operations with radicals include:

1. Addition and subtraction: Combine like radicals by adding or subtracting their coefficients, and then multiplying this result by their common radical. Unlike radicals, unless transformed to equivalent like radicals, cannot be combined into a single term.

2. Multiplication and division: For all radicals whose denominators are not equal to 0, multiply (or divide) coefficients, multiply (or divide) radicands, and simplify. The general rules for these operations are as follows:

$$x\sqrt{a} \cdot y\sqrt{b} = xy\sqrt{ab} \qquad \frac{x\sqrt{a}}{y\sqrt{b}} = \frac{x}{y}\sqrt{\frac{a}{b}}$$

VOCABULARY

20-1 Square Square root Radical sign ($\sqrt{\ }$) Radicand Radical Perfect square Principal square root Square-root key $\boxed{\sqrt{x}}$ Cube root Index Root key $\boxed{\sqrt[x]{\ }}$

20-2 Rational approximation

20-4 Simplest form of a square-root radical

20-5 Like radicals Unlike radicals

REVIEW EXERCISES WITH OPEN-RESPONSE PROBLEMS

1. Write the principal square root of 1,225.

2. Order the following numbers from smallest to largest: $\sqrt{18}$, $2\sqrt{2}$, 3.

In 3–6, write each given number without a radical.

3. $\sqrt{\frac{9}{25}}$ **4.** $-\sqrt{49}$ **5.** $\sqrt[3]{-27}$ **6.** $\pm\sqrt{1.21}$

In 7–10, in each case, solve for the variable, using the set of real numbers as the replacement set.

7. $y^2 = 81$ **8.** $m^2 = 0.09$ **9.** $3x^2 = 300$ **10.** $k^2 - 144 = 0$

11. a. Use calculator to evaluate $\sqrt{315.4176}$.

b. Is $\sqrt{315.4176}$ a rational or an irrational number? Explain your answer.

In 12–15, find each indicated root. Assume that all variables represent positive numbers.

12. $\sqrt{400x^2}$ **13.** $\sqrt{4y^4}$ **14.** $\sqrt{c^{10}d^2}$ **15.** $\sqrt{0.01m^{16}}$

In 16–19, simplify each expression. Assume that $b > 0$.

16. $\sqrt{180}$ **17.** $3\sqrt{18}$ **18.** $\frac{1}{2}\sqrt{28}$ **19.** $\sqrt{48b}$

In 20–23, in each case, combine the radicals.

20. $\sqrt{18} + \sqrt{8} - \sqrt{32}$ **21.** $3\sqrt{20} - 2\sqrt{45}$

22. $2\sqrt{50} - \sqrt{98} + \frac{1}{2}\sqrt{72}$ **23.** $\sqrt{75} - 3\sqrt{12}$

In 24–29, in each case, multiply or divide, and then simplify.

24. $8\sqrt{2} \cdot 2\sqrt{2}$ **25.** $(3\sqrt{5})^2$ **26.** $2\sqrt{7} \cdot \sqrt{70}$

27. $\sqrt{98} \div \sqrt{2}$ **28.** $\frac{16\sqrt{21}}{2\sqrt{7}}$ **29.** $\frac{5\sqrt{162}}{9\sqrt{50}}$

In 30 and 31, in each case, select the numeral preceding the correct choice.

30. The expression $\sqrt{108} - \sqrt{3}$ is equivalent to

(1) $\sqrt{105}$ (2) $35\sqrt{3}$ (3) $5\sqrt{3}$ (4) 6

31. The sum of $9\sqrt{2}$ and $\sqrt{32}$ is

(1) $9\sqrt{34}$ (2) $13\sqrt{2}$ (3) $10\sqrt{34}$ (4) 15

In 32–34, for each irrational number given, write a rational approximation: **a.** as shown on a calculator display **b.** rounded to four significant digits

32. $\sqrt{194}$ **33.** $\sqrt[3]{16}$ **34.** $-\sqrt{0.7}$

35. The area of a square is 28 square meters.
a. Find, to the *nearest thousandth* of a meter, the length of one side of the square.
b. Find, to the *nearest thousandth* of a meter, the perimeter of the square.
c. Explain why the answer to part **b** is *not* equal to 4 times the answer to part **a**.

CUMULATIVE REVIEW WITH OPEN-RESPONSE PROBLEMS

1. Write as a polynomial in simplest form: $(2x + 3)(x - 7)$

2. What is the product of $-3.5x^2$ and $-6.2x^3$.

Exploration

(1) On a sheet of graph paper, draw the positive ray of the real number line. Draw a square, using the interval from 0 to 1 as one side of the square. Draw the diagonal of this square from 0 to its opposite vertex. The length of the diagonal is $\sqrt{2}$. Why? Place the point of a pair of compasses at 0 and the pencil at the opposite vertex of the square so that the measure of the opening of the pair of compasses is $\sqrt{2}$. Keep the point of the compasses at 0 and use the pencil to mark the endpoint of a segment of length $\sqrt{2}$ on the number line.

(2) Using the same number line, draw a rectangle whose dimensions are $\sqrt{2}$ by 1, using the interval on the number line from 0 to $\sqrt{2}$ as one side. Draw the diagonal of this rectangle from 0 to the opposite vertex. The length of the diagonal is $\sqrt{3}$. Place the point of a pair of compasses at 0 and the pencil at the opposite vertex of the rectangle so that the measure of the opening of the pair of compasses is $\sqrt{3}$. Keep the point of the compasses at 0 and use the pencil to mark the endpoint of a segment of length $\sqrt{3}$ on the number line.

(3) Repeat step (2), drawing a rectangle whose dimensions are $\sqrt{3}$ by 1 to locate $\sqrt{4}$ on the number line.

(4) Explain how these steps can be used to locate $\sqrt{n}$ for any positive integer n.

Chapter 21
Quadratic Equations

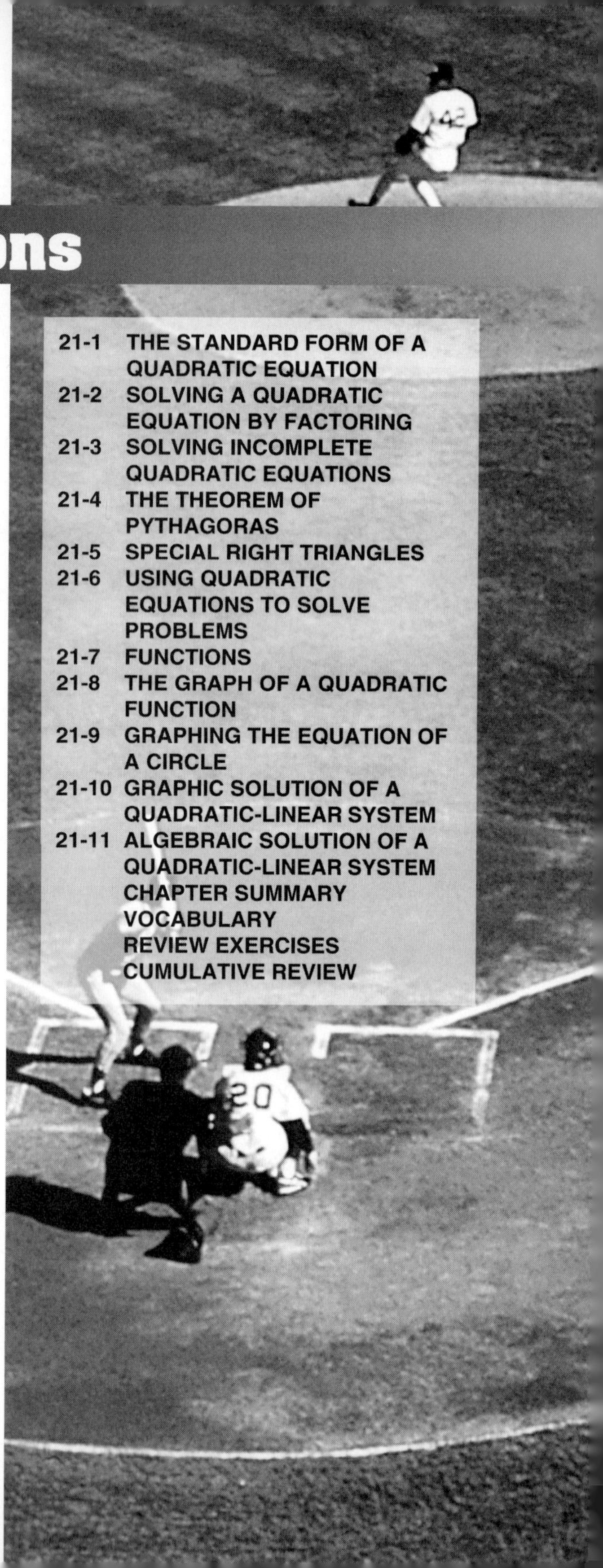

When a baseball is hit, the maximum height to which it rises is determined by the force with which the ball was hit and the angle at which it was hit. This height can be found by using an equation, as can the height of the ball at any moment and the distance between the batter and the point where the ball hits the ground.

An equation that expresses y, the height of the baseball above the ground, in terms of x, its horizontal distance from the point at which it was hit, can be written in the form

$$y = ax^2 + bx + c$$

An equation of this form is called a *quadratic function* or a *second-degree polynomial function*. In this chapter, we will study graphs of quadratic functions and some methods used to solve quadratic equations.

21-1 THE STANDARD FORM OF A QUADRATIC EQUATION

The equation $x^2 - 3x - 10 = 0$ is a polynomial equation in one variable. This equation is of degree two, or second degree, because the greatest exponent of the variable x is 2. The equation is in ***standard form*** because all terms are collected, in descending order of exponents, in one member, and the other member is 0.

A ***polynomial equation*** of ***degree two*** is also called a ***quadratic equation***. In general, the standard form of a quadratic equation in one variable is

$$\mathbf{ax^2 + bx + c = 0}$$

where a, b, and c are real numbers and $a \neq 0$.

EXAMPLE

Transform the equation $x(x - 4) = 5$ into an equivalent quadratic equation in standard form.

Solution

$$x(x - 4) = 5$$
$$x^2 - 4x = 5$$
$$x^2 - 4x - 5 = 0 \text{ Answer}$$

EXERCISES WITH OPEN-RESPONSE PROBLEM

In 1–9, transform each equation into an equivalent quadratic equation in standard form, and state the values of a, b, and c.

1. $x^2 + 9x = 10$ **2.** $2x^2 + 7x = 3x$ **3.** $x^2 = 3x - 8$

4. $4x + 3 = x^2$ **5.** $3x^2 = 27x$ **6.** $x(x - 3) = 10$

7. $x^2 = 5(x + 4)$ **8.** $\frac{x^2}{2} + 3 = \frac{x}{4}$ **9.** $x^2 = 6 - \frac{x}{2}$

10. Explain why $x^2 + 2(3x + 1) = x(x - 7)$ *cannot* be transformed into an equivalent quadratic equation in standard form.

21-2 SOLVING A QUADRATIC EQUATION BY FACTORING

The following examples:

$$5 \times 0 = 0 \qquad (-2) \times 0 = 0 \qquad \tfrac{1}{2} \times 0 = 0$$
$$0 \times 7 = 0 \qquad 0 \times (-3) = 0 \qquad 0 \times 0 = 0$$

illustrate that, whenever the product of two or more factors is 0, at least one of the factors is 0.

In general:

- **If a and b are real numbers, then:**

$$ab = 0 \quad \text{if and only if } a = 0 \text{ or } b = 0.$$

This principle is used to solve quadratic equations. For example, to solve the quadratic equation $x^2 - 3x + 2 = 0$, we can write it as $(x - 2)(x - 1) = 0$. The factors $(x - 2)$ and $(x - 1)$ represent real numbers whose product is 0. The equation will be true if either one of the factors is 0, that is, if $(x - 2) = 0$ or if $(x - 1) = 0$.

$$x^2 - 3x + 2 = 0$$

$$(x - 2)(x - 1) = 0$$

$$\begin{array}{r|r} x - 2 = 0 & x - 1 = 0 \\ +2 = +2 & +1 = +1 \\ \hline x = 2 & x = 1 \end{array}$$

A check will show that both 2 and 1 are values of x for which the equation is true.

Check for $x = 2$:	*Check for* $x = 1$:
$x^2 - 3x + 2 = 0$	$x^2 - 3x + 2 = 0$
$(2)^2 - 3(2) + 2 \stackrel{?}{=} 0$	$(1)^2 - 3(1) + 2 \stackrel{?}{=} 0$
$4 - 6 + 2 \stackrel{?}{=} 0$	$1 - 3 + 2 \stackrel{?}{=} 0$
$0 = 0$ (True)	$0 = 0$ (True)

Since both 2 and 1 satisfy the equation $x^2 - 3x + 2 = 0$, the solution set of this equation is $\{2, 1\}$. The roots of the equation are 2 and 1.

PROCEDURE. To solve a quadratic equation by factoring:

1. If necessary, transform the equation into standard form.
2. Factor the quadratic expression.
3. Set each factor containing the variable equal to 0.
4. Solve each of the resulting equations.
5. Check by substituting each value of the variable in the original equation.

EXAMPLES

1. Solve and check: $x^2 - 7x = -10$.

How to Proceed:	*Solution:*
	$x^2 - 7x = -10$
(1) Write the equation in standard form:	$x^2 - 7x + 10 = 0$
(2) Factor the quadratic expression:	$(x - 2)(x - 5) = 0$
(3) Let each factor = 0:	$x - 2 = 0 \mid x - 5 = 0$
(4) Solve each equation:	$x = 2 \mid x = 5$

(5) Check both values in the original equation.

Check for $x = 2$:	*Check for* $x = 5$:
$x^2 - 7x = -10$	$x^2 - 7x = -10$
$(2)^2 - 7(2) \stackrel{?}{=} -10$	$(5)^2 - 7(5) \stackrel{?}{=} -10$
$4 - 14 \stackrel{?}{=} -10$	$25 - 35 \stackrel{?}{=} -10$
$-10 = -10$ (True)	$-10 = -10$ (True)

Answer: $x = 2$ or $x = 5$; the solution set is $\{2, 5\}$.

---

2. List the members of the solution set of $2x^2 = 3x$.

How to Proceed:	*Solution:*
(1) Write the equation in standard form:	$2x^2 = 3x$
(2) Factor the quadratic expression:	$2x^2 - 3x = 0$
(3) Let each factor = 0:	$x(2x - 3) = 0$
(4) Solve each equation:	$x = 0 \mid 2x - 3 = 0$
	$2x = 3$
The check is left to you.	$x = \frac{3}{2}$

Answer: The solution set is $\left\{0, \frac{3}{2}\right\}$.

Note: We never divide both sides of an equation by an expression containing a variable. If we had divided $2x^2 = 3x$ by x, we would have obtained the equation $2x = 3$, whose solution is $\frac{3}{2}$, but would have lost the solution $x = 0$.

---

3. Find the solution set of the equation $x(x - 8) = 2x - 25$.

How to Proceed:	*Solution:*
	$x(x - 8) = 2x - 25$
(1) Use the distributive property:	$x^2 - 8x = 2x - 25$
(2) Write the equation in standard form:	$x^2 - 10x + 25 = 0$
(3) Factor the quadratic expression:	$(x - 5)(x - 5) = 0$
(4) Let each factor = 0:	$x - 5 = 0 \mid x - 5 = 0$
(5) Solve each equation:	$x = 5 \mid x = 5$

Answer: $x = 5$; the solution set is $\{5\}$.

Every quadratic equation has two roots, but the two roots are not always different numbers. Sometimes the roots are the same number, as in Example 3 where both roots are 5. Such a root, called a ***double root***, is written only once in the solution set.

---

4. The height h of a ball thrown into the air with an initial vertical velocity of 24 feet per second from a height of 6 feet above the ground is given by the equation

$$h = -16t^2 + 24t + 6$$

where t is the time, in seconds, that the ball has been in the air. After how many seconds is the ball at a height of 14 feet?

Solution In the equation, let $h = 14$ and solve for t.

$$h = -16t^2 + 24t + 6$$

$$14 = -16t^2 + 24t + 6$$

$$0 = -16t^2 + 24t - 8$$

$$\frac{0}{-8} = \frac{-16t^2 + 24t - 8}{-8}$$

$$0 = 2t^2 - 3t + 1$$

$$0 = (2t - 1)(t - 1)$$

$2t - 1 = 0$	$t - 1 = 0$
$2t = 1$	$t = 1$
$t = \frac{1}{2}$	

Check: Use a calculator to show that the height is 14 feet when $t = \frac{1}{2}$ and when $t = 1$.

Enter: 16 [+/–] [×] [(] 1 [÷] 2 [)] [x^2] [+] 24 [×] 1 [÷] 2 [+] 6 [=]

Display: 14.

Enter: 16 [+/–] [×] 1 [x^2] [+] 24 [×] 1 [+] 6 [=]

Display: 14.

Answer: The ball is at a height of 14 feet after $\frac{1}{2}$ second as it rises and after 1 second as it falls.

EXERCISES WITH OPEN-RESPONSE PROBLEMS

In 1–48, solve each equation and check.

1. $x^2 - 3x + 2 = 0$
2. $z^2 - 5z + 4 = 0$
3. $x^2 - 8x + 16 = 0$
4. $r^2 - 12r + 35 = 0$
5. $c^2 + 6c + 5 = 0$
6. $m^2 + 10m + 9 = 0$
7. $x^2 + 2x + 1 = 0$
8. $y^2 + 11y + 24 = 0$
9. $x^2 - 4x - 5 = 0$
10. $x^2 - 5x - 6 = 0$
11. $x^2 + x - 6 = 0$
12. $x^2 + 2x - 15 = 0$
13. $r^2 - r - 72 = 0$
14. $x^2 - x - 12 = 0$
15. $x^2 - 49 = 0$
16. $z^2 - 4 = 0$
17. $m^2 - 64 = 0$
18. $3x^2 - 12 = 0$
19. $d^2 - 2d = 0$
20. $s^2 - s = 0$
21. $x^2 + 3x = 0$
22. $z^2 + 8z = 0$
23. $2x^2 - 5x + 2 = 0$
24. $3x^2 - 10x + 3 = 0$
25. $3x^2 - 8x + 4 = 0$
26. $5x^2 + 11x + 2 = 0$
27. $x^2 - x = 6$
28. $y^2 - 3y = 28$
29. $c^2 - 8c = -15$
30. $2m^2 + 7m = -6$
31. $r^2 = 4$
32. $x^2 = 121$
33. $y^2 = 6y$
34. $s^2 = -4s$
35. $y^2 = 8y + 20$
36. $2x^2 - x = 15$
37. $x^2 = 9x - 20$
38. $30 + x = x^2$
39. $x^2 + 3x - 4 = 50$
40. $2x^2 + 7 = 5 - 5x$
41. $\frac{1}{3}x^2 + \frac{4}{3}x + 1 = 0$
42. $\frac{1}{2}x^2 - \frac{7}{6}x = 1$
43. $x(x - 2) = 35$
44. $y(y - 3) = 4$
45. $x(x + 3) = 40$
46. $\frac{x + 2}{2} = \frac{12}{x}$
47. $\frac{y + 3}{3} = \frac{6}{y}$
48. $\frac{x}{3} = \frac{12}{x}$

49. The height h of a ball thrown into the air with an initial vertical velocity of 48 feet per second from a height of 5 feet above the ground is given by the equation

$$h = -16t^2 + 48t + 5$$

where t is the time in seconds that the ball has been in the air. After how many seconds is the ball at a height of 37 feet?

50. A batter hit a baseball at a height of 4 feet with a force that gave the ball an initial vertical velocity of 64 feet per second. The equation

$$h = -16t^2 + 64t + 4$$

gives the height h of the baseball t seconds after the ball was hit. If the ball was caught at a height of 4 feet, how long after the batter hit the ball was the ball caught?

21-3 SOLVING INCOMPLETE QUADRATIC EQUATIONS

A quadratic equation that has no first-degree term is called an ***incomplete quadratic equation*** or a ***pure quadratic equation***. When $b = 0$, the general form of a quadratic equation is

$$\mathbf{ax^2 + c = 0}$$

where $a \neq 0$.

For example, $x^2 - 25 = 0$ is an incomplete quadratic equation in standard form. This equation can be solved by factoring the left member.

$$x^2 - 25 = 0$$

$$(x - 5)(x + 5) = 0$$

$$x - 5 = 0 \quad \Big| \quad x + 5 = 0$$

$$x = 5 \quad \Big| \quad x = -5$$

Another method that we may use to solve $x^2 - 25 = 0$ applies the following principle:

- **Every positive real number has two real square roots, one of which is the opposite of the other.**

Solution

$$x^2 - 25 = 0$$

$$x^2 = 25$$

$$x = \sqrt{25} \quad \Big| \quad x = -\sqrt{25}$$

$$x = 5 \quad \Big| \quad x = -5$$

Check for $x = 5$:	*Check for* $x = -5$:
$x^2 - 25 = 0$	$x^2 - 25 = 0$
$(5)^2 - 25 \stackrel{?}{=} 0$	$(-5)^2 - 25 \stackrel{?}{=} 0$
$25 - 25 \stackrel{?}{=} 0$	$25 - 25 \stackrel{?}{=} 0$
$0 = 0$ (True)	$0 = 0$ (True)

For the equation $x^2 - 25 = 0$, $x = 5$ or $x = -5$, a solution that may be written as $x = \pm 5$.

The solution set is $\{5, -5\}$. Since both 5 and -5 satisfy the equation $x^2 - 25 = 0$, the solution may be written as $\{5, -5\}$ or as $x = \pm 5$.

PROCEDURE. To solve an incomplete quadratic equation of the form $ax^2 + c = 0$:

1. Transform the equation into the form $x^2 = n$, where n is a nonnegative real number. If n is a negative number, the equation has no solution in the set of real numbers.
2. Take the square root of each side of the equation. There are two square roots of the nonnegative number n, $+\sqrt{n}$ and $-\sqrt{n}$.
3. Check both roots, $+\sqrt{n}$ and $-\sqrt{n}$, in the original equation.

EXAMPLES

1. Find the solution set: $7y^2 = 3y^2 + 36$.

Solution

$$7y^2 = 3y^2 + 36$$
$$7y^2 - 3y^2 = 36$$
$$4y^2 = 36$$
$$y^2 = 9$$
$$y = \sqrt{9} = 3 \mid y = -\sqrt{9} = -3$$

Check:

$y = -3$

$$7y^2 = 3y^2 + 36$$
$$7(-3)^2 = 3(-3)^2 + 36$$
$$7(9) = 3(9) + 36$$
$$63 = 27 + 36$$
$$63 = 63 \quad \text{(True)}$$

The check for $y = 3$ is left to you.

Answer: The solution set is $\{3, -3\}$. (The roots are rational numbers.)

---

2. Solve: $4x^2 - 14 = 2x^2$.

Solution

$$4x^2 - 14 = 2x^2$$
$$4x^2 - 2x^2 - 14 = 0$$
$$2x^2 - 14 = 0$$
$$2x^2 = 14$$
$$x^2 = 7$$
$$x = \sqrt{7} \mid x = -\sqrt{7}$$

Check:

$x = \sqrt{7}$

$$4x^2 - 14 = 2x^2$$
$$4(\sqrt{7})^2 - 14 = 2(\sqrt{7})^2$$
$$4(7) - 14 = 2(7)$$
$$28 - 14 = 14$$
$$14 = 14 \quad \text{(True)}$$

The check for $x = -\sqrt{7}$ is left to you.

Answer: $x = \pm\sqrt{7}$ (The roots are irrational numbers.)

---

3. The areas of two similar triangles are in the ratio of 4 : 1. The length of a side of the smaller triangle is 5 inches. Find the length of the corresponding side in the larger triangle.

Solution Recall that the ratio of the areas of two similar polygons is the square of the ratio of the measures of the sides.

$$\frac{A}{A'} = \frac{(s)^2}{(s')^2}$$

$$\frac{4}{1} = \frac{(s)^2}{(5)^2}$$

$$\frac{4}{1} = \frac{s^2}{25}$$

$$s^2 = 100$$

$$s = \pm\sqrt{100}$$

$$= \pm 10$$

Check:

$$\frac{A}{A'} = \frac{(s)^2}{(s')^2}$$

$$\frac{4}{1} \stackrel{?}{=} \frac{(10)^2}{(5)^2}$$

$$\frac{4}{1} \stackrel{?}{=} \frac{100}{25}$$

$$4 = 4 \text{ (True)}$$

There are two values of s that make the equation true, but the negative value cannot be a solution to the problem because the length of a side cannot be a negative number.

Answer: The length of the corresponding side in the larger triangle is 10 inches.

EXERCISES WITH OPEN-RESPONSE PROBLEMS

In 1–15, solve each equation and check.

1. $x^2 = 4$

2. $a^2 - 36 = 0$

3. $\frac{1}{2}x^2 = 50$

4. $5y^2 = 45$

5. $3k^2 = 147$

6. $2x^2 - 32 = 0$

7. $r^2 - 11 = 70$

8. $4x^2 - 46 = 210$

9. $2x^2 - 11 = 39$

10. $2x^2 + 3x^2 = 720$

11. $6x^2 - 4x^2 = 98$

12. $4y^2 - 2.5 = y^2 + 4.25$

13. $\frac{y^2}{3} = 12$

14. $\frac{x}{9} = \frac{144}{x}$

15. $\frac{4x}{25} = \frac{9}{x}$

In 16–24, in each case, solve for x in simplest radical form.

16. $x^2 = 10$ **17.** $x^2 = 20$ **18.** $3x^2 = 6$

19. $2x^2 - 16 = 0$ **20.** $x^2 + 25 = 100$ **21.** $x^2 - 4 = 14$

22. $8x^2 - 6x^2 = 126$ **23.** $3x^2 - 28 = 2x^2 + 33$ **24.** $\frac{2x}{9} = \frac{6}{x}$

In 25–27, find, in each case, the positive value of x, correct to the *nearest thousandth.*

25. $x^2 = 24$ **26.** $4x^2 - 160 = 0$ **27.** $7x^2 = x^2 + 198$

In 28–33, in each case, solve for x in terms of the other variable(s).

28. $x^2 = b^2$ **29.** $x^2 = 25a^2$ **30.** $9x^2 = r^2$

31. $4x^2 - a^2 = 0$ **32.** $x^2 + a^2 = c^2$ **33.** $x^2 + b^2 = c^2$

In 34–39, in each case, solve for the indicated variable in terms of the other variable(s).

34. Solve for s: $s^2 = A$. **35.** Solve for r: $A = \pi r^2$. **36.** Solve for r: $S = 4\pi r^2$.

37. Solve for r: $V = \pi r^2 h$. **38.** Solve for t: $s = \frac{1}{2}gt^2$. **39.** Solve for v: $F = \frac{mv^2}{gr}$.

40. How many feet of fencing are needed to enclose a square garden that has an area of 36 square feet?

41. The length of a rectangular flower bed is 3 times the width. The area of the bed is 108 square meters. What are the dimensions of the bed?

42. The height of a triangular metal plate is 6 times the measure of the base. The area of the plate is 120 square inches. In simplest radical form, what is the measure of the base?

43. The areas of two similar polygons are in the ratio 36:1. The length of a side of the smaller polygon is 2 centimeters. Find the length of the corresponding side of the larger polygon.

44. Two similar triangles have areas of 40 and 32. The length of a side of the smaller triangle is 8. Find the length of the corresponding side of the larger triangle.

21-4 THE THEOREM OF PYTHAGORAS

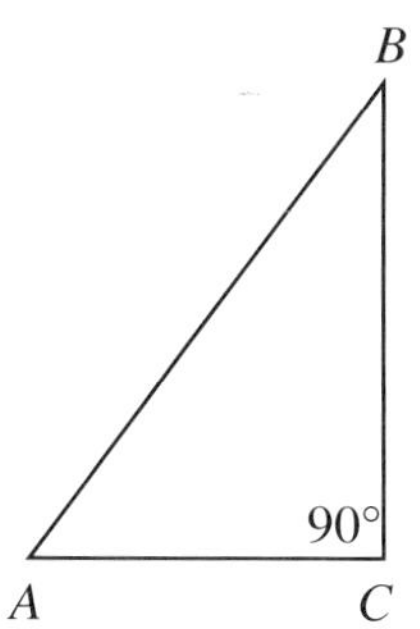

The figure at the left represents a *right triangle*. Recall that such a triangle contains one and only one right angle. In right triangle ABC, side $\overline{AB}$, which is opposite the right angle, is called the *hypotenuse*. The hypotenuse is the longest side of the triangle. The other two sides of the triangle, $\overline{BC}$ and $\overline{AC}$, form the right angle. They are called the *legs* of the right triangle.

More than 2,000 years ago, the Greek mathematician Pythagoras demonstrated the following property of the right triangle, which is called the ***Pythagorean Theorem***:

- **In a right triangle, the square of the length of the hypotenuse is equal to the sum of the squares of the lengths of the other two sides.**

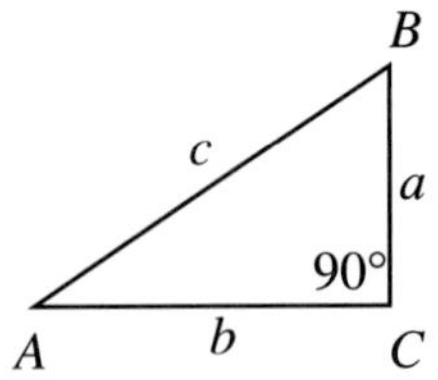

If we represent the length of the hypotenuse of right triangle ABC by c and the lengths of the other two sides by a and b, the Theorem of Pythagoras may be written as the following formula:

$$c^2 = a^2 + b^2$$

Geometrically, this relation means that, if squares are built on the hypotenuse and on each leg of a right triangle, the area of the square on the hypotenuse is equal to the sum of the areas of the squares on the legs.

For example, $\triangle ABC$ at the right represents a right triangle in which the length of side $\overline{AC} = 4$ centimeters, the length of side $\overline{BC} = 3$ centimeters, and the length of the hypotenuse $\overline{AB} = 5$ centimeters.

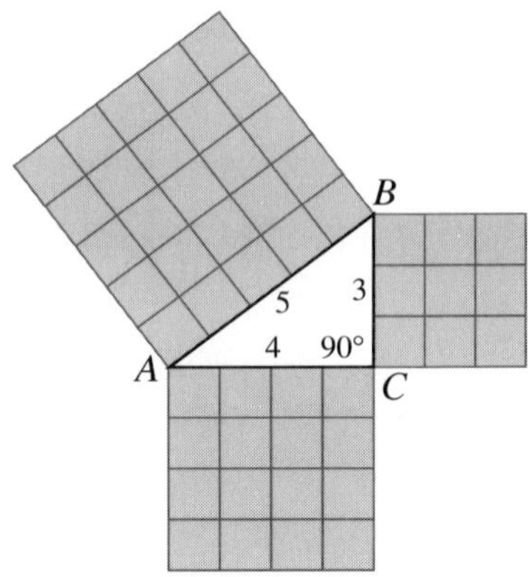

The area of the square on side $\overline{BC} = 9$ square centimeters.

The area of the square on side $\overline{AC} = 16$ square centimeters.

The area of the square on hypotenuse $\overline{AB} = 25$ square centimeters.

Here we see that

$$25 \text{ cm}^2 = 9 \text{ cm}^2 + 16 \text{ cm}^2$$

$$(5 \text{ cm})^2 = (3 \text{ cm})^2 + (4 \text{ cm})^2$$

Or simply: $$5^2 = 3^2 + 4^2$$

Another Approach to the Pythagorean Theorem

Similar triangles can be used to demonstrate the truth of the Pythagorean Theorem. Recall that two triangles are similar if two angles of one are congruent to two angles of the other, and also that the measures of the corresponding sides of similar triangles are in proportion.

Triangle ABC at the right is a right triangle with $\angle ACB$ the right angle.

Altitude $\overline{CD}$ has been drawn to side $\overline{AB}$, forming two right angles at D: $\angle ADC$ and $\angle BDC$.

Let $BC = a$, $AC = b$, $AB = c$, $AD = x$, and $DB = c - x$.

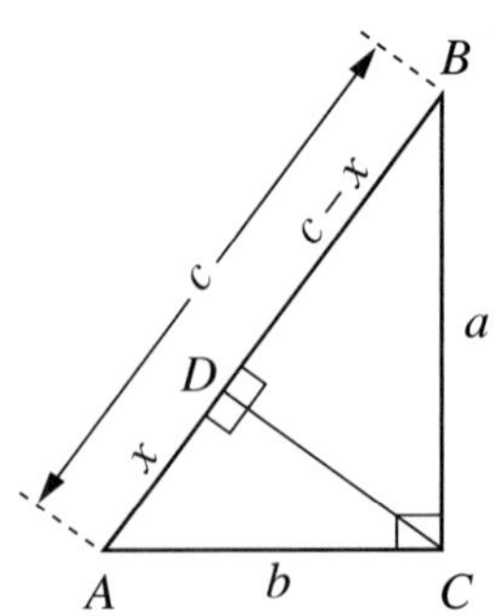

There are now three right triangles in the diagram: $\triangle ABC$ and the newly formed $\triangle ACD$ and $\triangle CBD$. Triangles ACD and CBD are each similar to $\triangle ABC$.

$\triangle ACD \sim \triangle ABC$ because $\angle ADC \cong \angle ACB$ and $\angle A \cong \angle A$. Corresponding sides are in proportion:

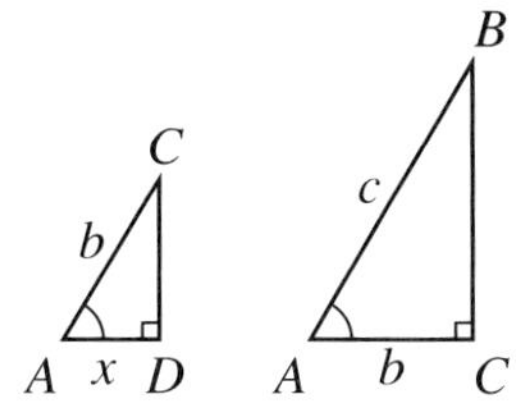

$$\frac{AC}{AD} = \frac{AB}{AC}$$

$$\frac{b}{x} = \frac{c}{b}$$

$$b^2 = cx$$

$\triangle CBD \sim \triangle ABC$ because $\angle CDB \cong \angle ACB$ and $\angle B \cong \angle B$. Corresponding sides are in proportion:

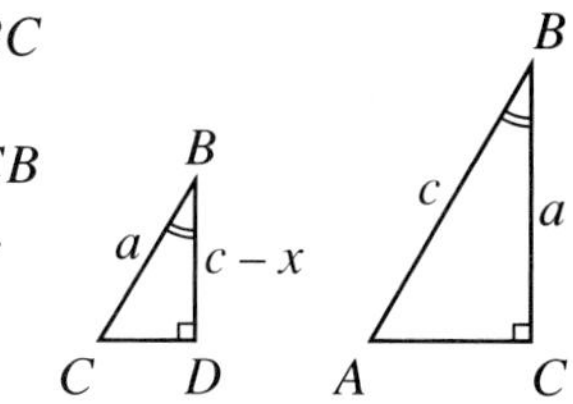

$$\frac{CB}{BD} = \frac{AB}{BC}$$

$$\frac{a}{c - x} = \frac{c}{a}$$

$$a^2 = c^2 - cx$$

Now, we add these two equations:

$$a^2 + b^2 = c^2 - cx + cx$$

$$a^2 + b^2 = c^2$$

Logical Statements of the Pythagorean Theorem

Three logical statements can be made for any right triangle where c represents the length of the hypotenuse, and a and b represent the lengths of the other two sides. All of the following statements are true.

1. The *conditional* form of the Pythagorean Theorem:
   If a triangle is a right triangle, then the square of the length of the hypotenuse is equal to the sum of the squares of the lengths of the other two sides. (If a triangle is a right triangle, then $c^2 = a^2 + b^2$.)

2. The *converse* of the Pythagorean Theorem:
   If the square of the length of the longest side of a triangle is equal to the sum of the squares of the lengths of the other two sides, the triangle is a right triangle. (If $c^2 = a^2 + b^2$ in a triangle, then the triangle is a right triangle.)

3. The *biconditional* form of the Pythagorean Theorem:
   A triangle is a right triangle if and only if $c^2 = a^2 + b^2$.

EXAMPLES

1. A ladder is placed 5 feet from the foot of a wall. The top of the ladder reaches a point 12 feet above the ground. Find the length of the ladder.

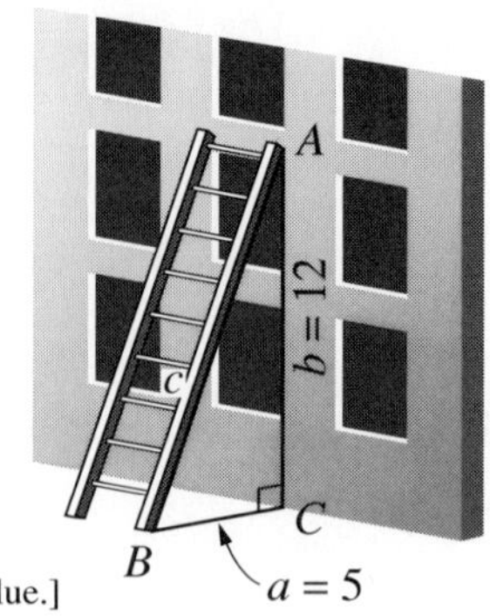

Solution Let the length of the hypotenuse = c, the length of side $a = 5$, the length of side $b = 12$.

$$\begin{aligned} c^2 &= a^2 + b^2 \text{ [Theorem of Pythagoras]} \\ &= 5^2 + 12^2 \\ &= 25 + 144 \\ &= 169 \\ c &= \sqrt{169} = 13 \\ \text{or } c &= -\sqrt{169} = -13 \text{ [Reject the negative value.]} \end{aligned}$$

Calculator Solution

(1) Enter $a^2 + b^2$ to find c^2. ***Enter:*** 5 [x^2] [+] 12 [x^2] [=] ***Display:*** 169.

(2) Use the square-root key to find c, the principal square root. ***Enter:*** 169 [$\sqrt{x}$] ***Display:*** 13.

These steps can be combined in a single entry:

Enter: [(] 5 [x^2] [+] 12 [x^2] [)] [$\sqrt{x}$] [=]

Display: 13.

Answer: The length of the ladder is 13 feet.

---

2. The hypotenuse of a right triangle is 20 centimeters long and one leg is 16 centimeters long.

a. Find the length of the other leg.

b. Find the area of the triangle.

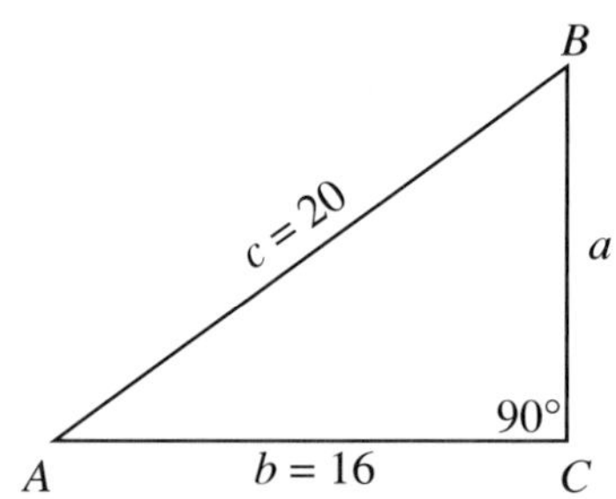

Solutions **a.** Let the length of the unknown leg = a; the length of the hypotenuse $c = 20$, the length of side $b = 16$.

$$\begin{aligned} c^2 &= a^2 + b^2 \text{ [Theorem of Pythagoras]} \\ 20^2 &= a^2 + 16^2 \\ 400 &= a^2 + 256 \\ 144 &= a^2 \\ 12 &= a \\ \text{or } -12 &= a \text{ [Reject the negative value.]} \end{aligned}$$

Answer: The length of the other leg is 12 centimeters

b. Area of $\triangle ABC = \frac{1}{2}bh$

$$= \frac{1}{2}(16)(12)$$
$$= 96$$

Answer: The area of $\triangle ABC$ is 96 square centimeters.

---

3. Express in simplest radical form the length of a side of a square whose diagonal is 10.

Solution In square $QRST$, $\angle QRS$ is a right angle. Therefore, $\triangle QRS$ is a right triangle. Let x = the length of each side of the square.

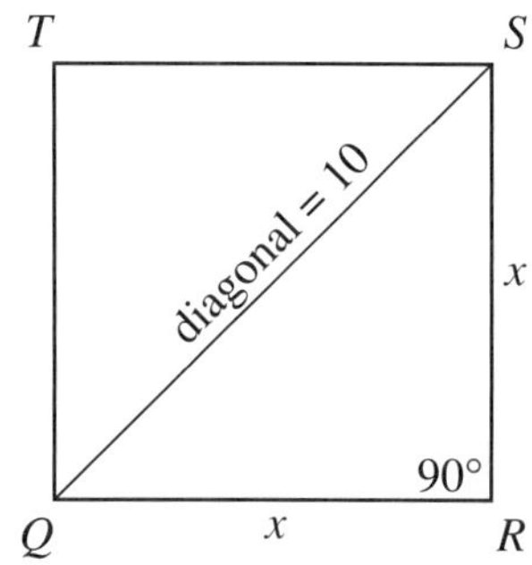

$$a^2 + b^2 = c^2$$
$$x^2 + x^2 = 10^2$$
$$2x^2 = 100$$
$$x^2 = 50$$
$$x = \pm\sqrt{50}$$
$$= \pm\sqrt{25 \cdot 2} \text{ [Reject the negative value.]}$$
$$= 5\sqrt{2}$$

Answer: The length of a side of the square is $5\sqrt{2}$.

---

4. Is a triangle whose sides measure 8 centimeters, 7 centimeters, and 4 centimeters a right triangle?

Solution If the triangle is a right triangle, the longest side, whose measure is 8, must be the hypotenuse. Then:

$$c^2 = a^2 + b^2$$
$$8^2 \stackrel{?}{=} 7^2 + 4^2$$
$$64 \stackrel{?}{=} 49 + 16$$
$$64 \neq 65$$

Answer: The triangle is not a right triangle.

Pythagorean Triples

Let us find the length of the hypotenuse of a right triangle whose legs measure 3 and 4.

$$\begin{aligned} c^2 &= a^2 + b^2 \\ &= (3)^2 + (4)^2 \\ &= 9 + 16 \\ &= 25 \\ c &= 5 \end{aligned}$$

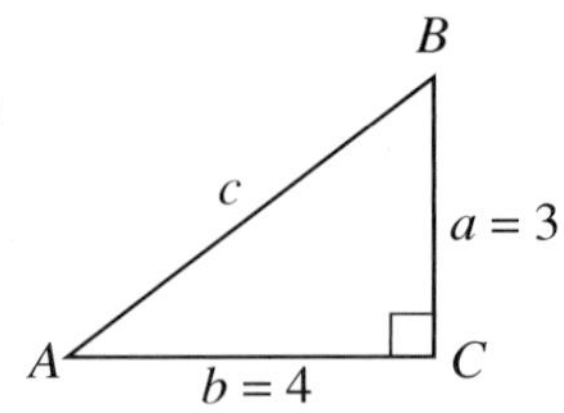

The length of the hypotenuse is 5.

When three integers are so related that the sum of the squares of two of them is equal to the square of the third, the set of three integers is called a ***Pythagorean triple***. For example, (3, 4, 5) is a Pythagorean triple because the numbers 3, 4, and 5 satisfy the relation $a^2 + b^2 = c^2$.

In general, if we multiply each number of a Pythagorean triple by some positive number x, then the new triple created is also a Pythagorean triple because it will satisfy the relation $a^2 + b^2 = c^2$. For example:

If (3, 4, 5) is a Pythagorean triple, then $(3x, 4x, 5x)$ is also a Pythagorean triple for a similar triangle where the ratio of corresponding sides of the second triangle to corresponding sides of the first triangle is $x:1$.

Let $x = 2$. Then $(3x, 4x, 5x) = (6, 8, 10)$ and $6^2 + 8^2 = 10^2$.
Let $x = 3$. Then $(3x, 4x, 5x) = (9, 12, 15)$, and $9^2 + 12^2 = 15^2$.
Let $x = 10$. Then $(3x, 4x, 5x) = (30, 40, 50)$, and $30^2 + 40^2 = 50^2$.

Thus, if the lengths of the legs of a right triangle are 30 and 40, we may recognize quickly that these are multiples of the first two numbers of the (3, 4, 5) Pythagorean triple. Thus, $x = 10$ and the length of the hypotenuse is $5 \cdot 10 = 50$.

Here are other examples of Pythagorean triples that occur frequently:
(5, 12, 13) or, in general, $(5x, 12x, 13x)$, where x is a positive integer.
(8, 15, 17) or, in general, $(8x, 15x, 17x)$, where x is a positive integer.

With a knowledge of these Pythagorean triples, we can solve some right triangle problems without writing lengthy algebraic solutions.

EXERCISES WITH OPEN-RESPONSE PROBLEMS

In 1–9, c represents the length of the hypotenuse of a right triangle and a and b represent the lengths of the legs. For each right triangle, find the length of the side whose measure is not given.

1. $a = 3, b = 4$
2. $a = 8, b = 15$
3. $c = 10, a = 6$
4. $c = 13, a = 12$
5. $c = 17, b = 15$
6. $c = 25, b = 20$
7. $a = \sqrt{2}, b = \sqrt{2}$
8. $a = 4, b = 4\sqrt{3}$
9. $a = 5\sqrt{3}, c = 10$

In 10–15, c represents the length of the hypotenuse of a right triangle and a and b represent the lengths of the legs. For each right triangle:

a. Express the length of the third side in simplest radical form.
b. Express the length of the third side to the *nearest hundredth*.

10. $a = 2, b = 3$ **11.** $a = 3, b = 3$ **12.** $a = 4, c = 8$

13. $a = 7, b = 1$ **14.** $b = \sqrt{3}, c = \sqrt{15}$ **15.** $a = 4\sqrt{2}, c = 8$

In 16–19, find x in each case and express irrational results in simplest radical form.

16.

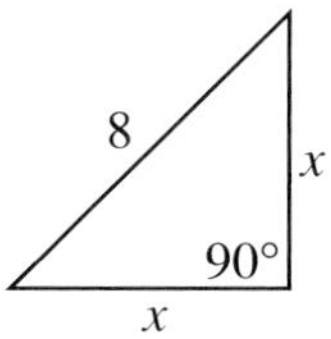

17.

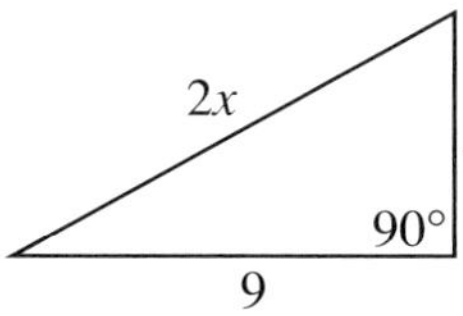

18.

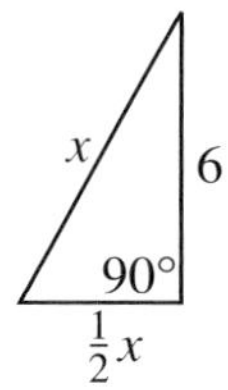

19.

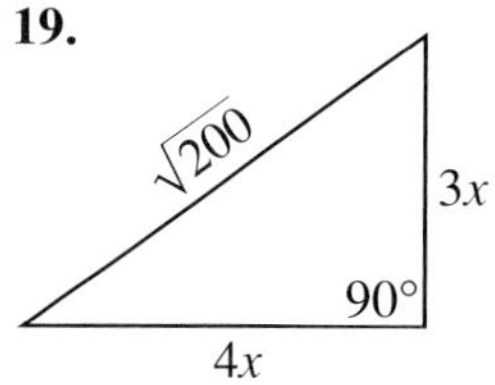

20. A ladder 39 feet long leans against a building and reaches the ledge of a window. If the foot of the ladder is 15 feet from the foot of the building, how high is the window ledge above the ground?

21. Mr. Rizzo placed a ladder so that it reached a window 15 feet above the ground when the foot of the ladder was 5 feet from the wall. Find the length of the ladder to the *nearest tenth* of a foot.

22. Miss Murray traveled 24 kilometers north and then 10 kilometers east. How far was she from her starting point?

23. One day, Ronnie left his home at A and reached his school at C by walking along $\overline{AB}$ and $\overline{BC}$, the sides of a rectangular open field that was muddy. When he was ready to return home, the field was dry and Ronnie decided to take a shortcut by walking diagonally across the field along $\overline{CA}$. How much shorter was the trip home than the trip to school?

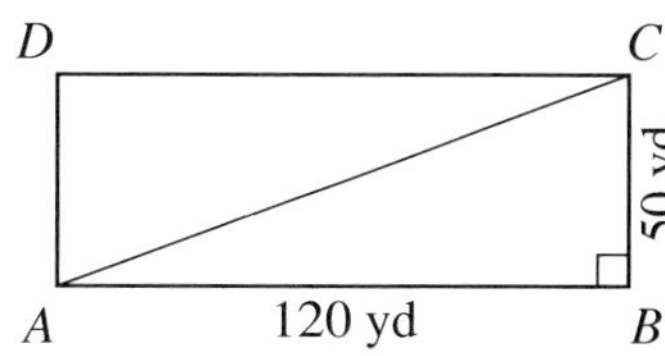

In 24–27, find, in each case, the diagonal of a rectangle whose sides have the given measurements.

24. 7 inches and 24 inches **25.** 9 centimeters and 40 centimeters

26. 28 feet and 45 feet **27.** 11 meters and 60 meters

28. The diagonal of a rectangle measures 65 centimeters. One side is 33 centimeters long.
a. Find the length of the other side.
b. Find the area of the rectangle.

29. Find the area of a rectangle in which the diagonal is 8.9 meters and one side is 8 meters.

30. Approximate, to the *nearest inch*, the measure of the base of a rectangle whose diagonal measures 25 inches and whose altitude measures 18 inches.

In 31–35, in each case approximate, to the *nearest tenth* of a meter, the length of a diagonal of a square whose side has the given measurement.

31. 2 meters **32.** 4 meters **33.** 5 meters **34.** 6 meters **35.** 7 meters

36. A baseball diamond has the shape of a square 90 feet on each side. Approximate, to the *nearest tenth* of a foot, the distance from home plate to second base.

37. In the figure, ABC is an isosceles triangle in which sides $\overline{AC}$ and $\overline{CB}$ are of equal length, that is, $AC = CB$. $\overline{CD}$, which is drawn so that $\angle CDB$ is a right angle, is the altitude to the base of the triangle. Altitude $\overline{CD}$ divides the base $\overline{AB}$ into two segments of equal measure, $AD = DB$. Each of sides $\overline{AC}$ and $\overline{CB}$ measures 26 centimeters, and base $\overline{AB}$ measures 20 centimeters.

a. Find the length of the altitude drawn to the base.
b. Find the area of $\triangle ABC$.

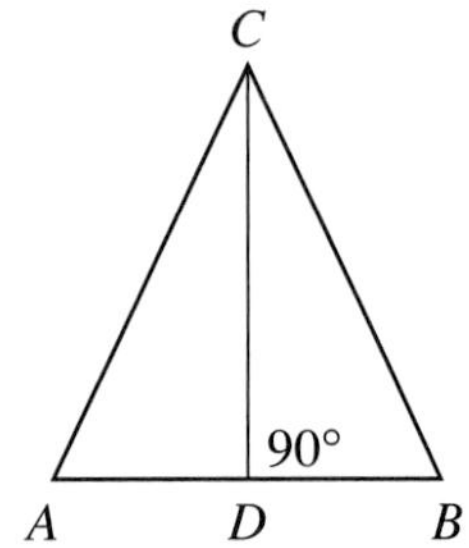

38. In an isosceles triangle, each of the equal sides measures 29 inches and the altitude drawn to the base measures 21 inches.

a. Find the length of the base. **b.** Find the area of the triangle.

In 39–44, find, in each case, the length of the altitude of an equilateral triangle when each of its sides has the given measure. (Express each answer in simplest radical form.)

39. 4 centimeters **40.** 6 centimeters **41.** 8 centimeters
42. 10 centimeters **43.** 5 centimeters **44.** 15 centimeters

In 45–48, each set of measures represents the lengths of the sides of a triangle. In each case, is the triangle a right triangle?

45. 6 yards, 8 yards, 10 yards **46.** 7 meters, 4 meters, 5 meters
47. 16 centimeters, 63 centimeters, 65 centimeters **48.** 10 feet, 15 feet, 20 feet

49. **a.** On graph paper, draw the triangle whose vertices are $A(0, 5)$, $B(4, 5)$, $C(4, 9)$.
b. Find the length of each side of $\triangle ABC$.
c. Find the area of $\triangle ABC$.

50. **a.** On graph paper, draw the rectangle whose vertices are $A(0, 0)$, $B(9, 0)$, $C(9, 7)$, $D(0, 7)$.
b. Find the measure of each side of the rectangle.
c. Use the length and width of the rectangle to find the measure of the diagonal in radical form.
d. Express the measure of the diagonal to the *nearest tenth*.

51. The length of a side of a square is represented by s, and the length of a diagonal by d. Show that $d = s\sqrt{2}$.

In 52 and 53, select the numeral that best answers the question or completes the sentence.

52. The set of integers $\{3, 4, 5\}$ is a Pythagorean triple. Another such set is
(1) $\{2, 4, 5\}$ (2) $\{6, 8, 12\}$ (3) $\{6, 10, 14\}$ (4) $\{5, 12, 13\}$

53. The lengths of three sides of a triangle are given. Which triangle is a right triangle?
(1) 2, 4, 6 (2) 7, 8, 12 (3) 12, 16, 20 (4) 18, 36, 39

21-5 SPECIAL RIGHT TRIANGLES

The 30–60-Degree Right Triangle

In a special right triangle, called the ***30–60-degree right triangle***, we can find the lengths of all three sides when given the length of only one side. We will develop the relationship of these sides by starting with an equilateral triangle.

(1) In equilateral triangle ABC, we know that the measure of each angle is 60°.

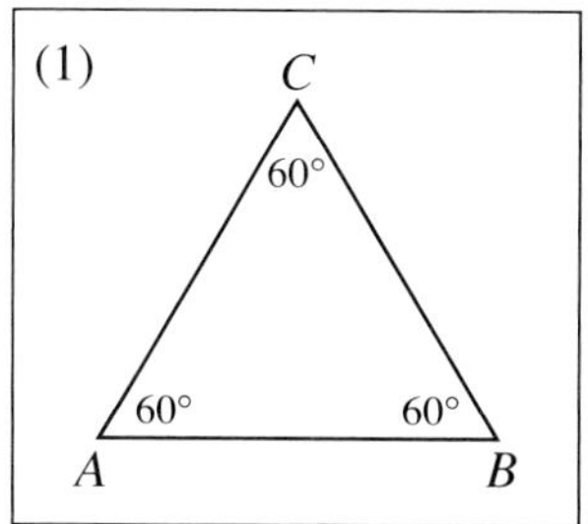

(2) We draw altitude $\overline{CD}$ to side $\overline{AB}$, forming two right angles at D. Since $\overline{AC} \cong \overline{BC}$, and $\overline{CD} \cong \overline{CD}$, it follows that $\triangle ADC \cong \triangle BDC$ by hy. leg ≅ hy. leg.

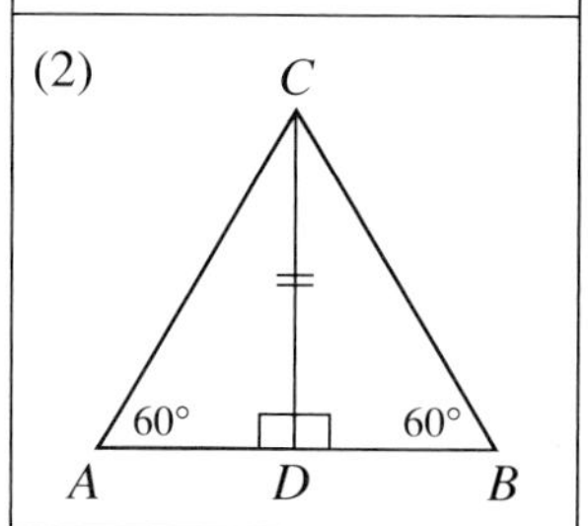

(3) Since $\triangle ADC \cong \triangle BDC$, their corresponding parts are congruent. Thus:

a. $\angle ACD \cong \angle BCD$, and the measure of each of these angles is 30°.

b. $\overline{AD} \cong \overline{BD}$. If we let $AD = x$, and $BD = x$, it follows that each side of equilateral triangle ABC has a length of $2x$.

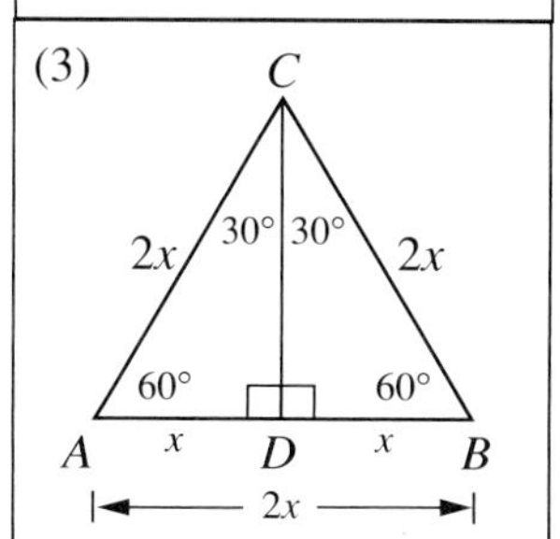

(4) In right triangle ACD, $\overline{AC}$ is the hypotenuse. Since $AD = x$ and $AC = 2x$, we apply the Pythagorean Theorem to find the length of leg $\overline{CD}$:

$$\begin{aligned}(CD)^2 + (AD)^2 &= (AC)^2 \\ (CD)^2 + (x)^2 &= (2x)^2 \\ (CD)^2 + x^2 &= 4x^2 \\ (CD)^2 &= 3x^2 \\ CD &= \sqrt{3x^2} = \sqrt{x^2} \cdot \sqrt{3} \\ &= x\sqrt{3}\end{aligned}$$

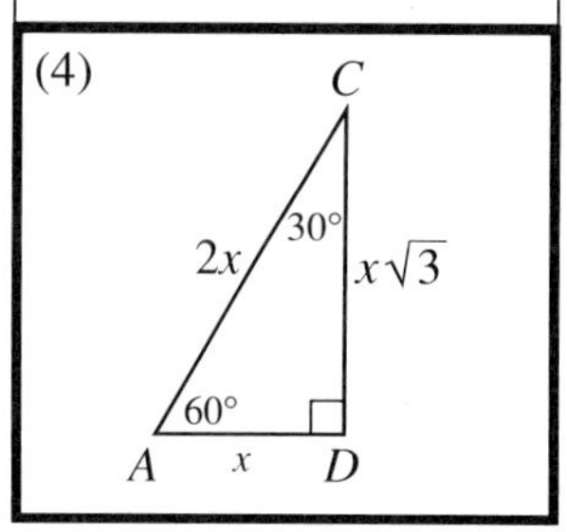

The final diagram, (4), shows the 30–60-degree right triangle and the following relationships:

1. If the side opposite the 30° angle has a length of x, then the side opposite the 60° angle has a length of $x\sqrt{3}$, and the side opposite the 90° angle has a length of $2x$.
2. The length of the hypotenuse is *twice* the length of the side opposite the 30° angle. Thus, the side opposite the 30° angle is *half* the length of the hypotenuse.
3. The length of the side opposite the 60° angle is equal to the length of the side opposite the 30° angle multiplied by $\sqrt{3}$.
4. The ratio of the lengths of the sides is $1:\sqrt{3}:2$.

EXAMPLES

1. Find the length of the segments marked x and y.

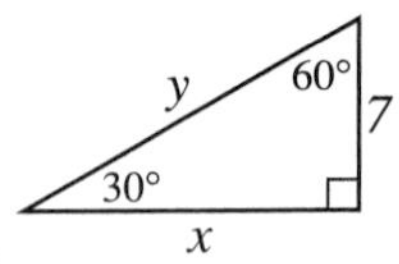

Solution Let k = length of side opposite 30° angle, or $k = 7$.

Then $k\sqrt{3}$ = length of side opposite 60° angle, or $k\sqrt{3} = 7\sqrt{3}$.

and $2k$ = length of side opposite 90° angle, or $2k = 2(7) = 14$.

Answer: $x = 7\sqrt{3}$; $y = 14$.

---

2. Express in radical form the length of an altitude of an equilateral triangle whose side measures 12 centimeters.

Solution

METHOD 1

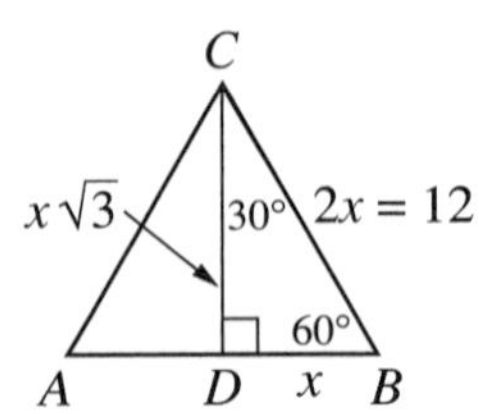

Draw altitude $\overline{CD}$, creating a 30–60-degree right triangle.

Label the lengths of the sides as in the figure:

x (opposite 30° angle)
$2x$ (opposite 90° angle)
$x\sqrt{3}$ (opposite 60° angle)

Since $CB = 2x = 12$,

then $DB = x = 6$,

and $CD = x\sqrt{3} = 6\sqrt{3}$.

METHOD 2

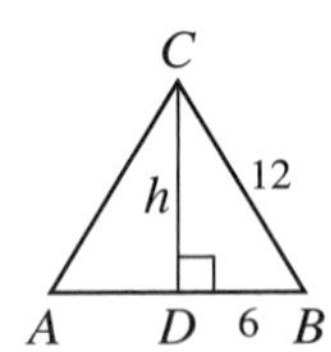

Draw altitude $\overline{CD}$, which bisects base $\overline{AB}$. Thus, $CB = 12$ and $DB = 6$.
Let h = length of altitude $\overline{CD}$.

For right triangle CDB, the Pythagorean Theorem states:

$$(h)^2 + (6)^2 = (12)^2$$
$$h^2 + 36 = 144$$
$$h^2 = 108$$
$$h = \sqrt{108} = \sqrt{36} \cdot \sqrt{3}$$
$$= 6\sqrt{3}$$

Answer: The altitude measures $6\sqrt{3}$ cm.

---

3. In right triangle ABC, $m\angle A = 30$, $m\angle B = 60$, and $AC = 6$ inches. Find BC to the *nearest tenth of an inch.*

Solution

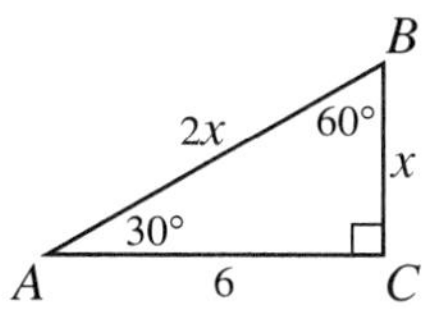

METHOD 1

Since the length of the hypotenuse is twice the length of the leg opposite the 30° angle:

Let x = length of $\overline{BC}$,

and $2x$ = length of $\overline{AB}$.

$$(2x)^2 = (x)^2 + (6)^2$$
$$4x^2 = x^2 + 36$$
$$3x^2 = 36$$
$$x^2 = 12$$
$$x = \sqrt{12} = \sqrt{4} \cdot \sqrt{3}$$
$$= 2\sqrt{3}$$

METHOD 2

Label the sides:

$BC = x$ (opposite 30° angle)

$AB = 2x$ (opposite 90° angle)

$AC = x\sqrt{3}$ (opposite 60° angle)

Since $AC = x\sqrt{3} = 6$:

$$\frac{x\sqrt{3}}{\sqrt{3}} = \frac{6}{\sqrt{3}}$$
$$x = \frac{6}{\sqrt{3}} \cdot \frac{\sqrt{3}}{\sqrt{3}}$$
$$= \frac{6\sqrt{3}}{3} = 2\sqrt{3}$$

Evaluate $x = 2\sqrt{3}$ using a calculator.

Enter: 2 [×] 3 [√x] [=]

Display: 3.46410161

Since it is not necessary to simplify the radical when an approximate value is required, we could have evaluated $\sqrt{12}$ or $\frac{6}{\sqrt{3}}$.

Enter: 12 [√x]

Display: 3.46410161

Enter: 6 [÷] 3 [√x] [=]

Display: 3.46410161

In all cases, round the number displayed to the nearest tenth, 3.5.

Answer: BC = 3.5 in., to the *nearest tenth of an inch.*

EXERCISES WITH OPEN-RESPONSE PROBLEMS

Note. In all exercises, irrational answers should be left in radical form unless otherwise indicated.

In 1–5, find the lengths of the segments marked x and y.

1.

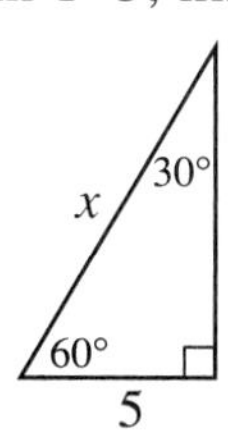

2.

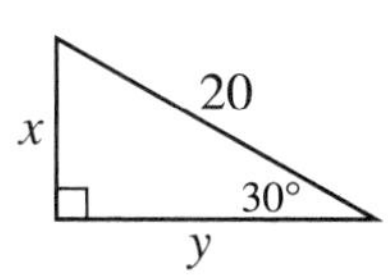

3.

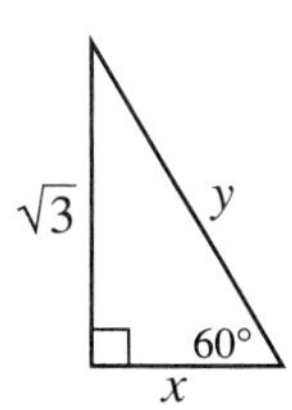

4.

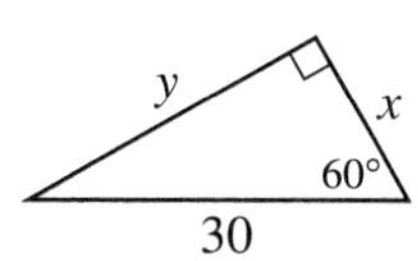

5.

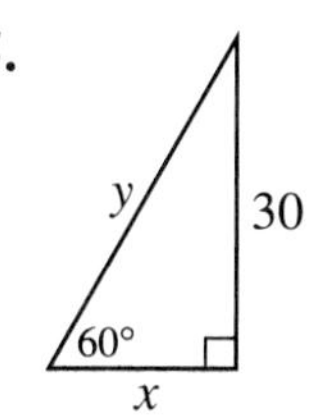

6. Express in radical form the length of the altitude of an equilateral triangle whose side has a unit length of:

a. 10 **b.** 16 **c.** 4 **d.** 7 **e.** 15 **f.** s

In 7 and 8, select the *numeral* preceding the choice that best completes the statement.

7. If one side of an equilateral triangle has a length of 6, the length of its altitude is

(1) 6 (2) $3\sqrt{3}$ (3) 3 (4) $6\sqrt{3}$

8. If the altitude of an equilateral triangle measures $7\sqrt{3}$, then one side of the triangle measures

(1) 7 (2) 14 (3) 21 (4) $14\sqrt{3}$

9. In inches, the lengths of two sides of a triangle are 10 and 14 and the angle included between these sides measures 30°. Find the length of the altitude on the side whose length is 14 inches.

10. In a right triangle, the measure of one acute angle is double the measure of the other acute angle. If the length of the shorter leg of the triangle is 3 centimeters, find the length of the hypotenuse.

11. If one angle of a right triangle measures 60° and the length of the hypotenuse is 8 centimeters, find the length of the side opposite the 60° angle, correct to the *nearest tenth of a centimeter*.

12. In isosceles triangle ABC, vertex angle C measures 120° and the length of $\overline{AC}$ is 8 inches. **a.** Find the length of the altitude on $\overline{AB}$. **b.** Find AB to the *nearest tenth of an inch*.

13. In a rhombus that contains a 60° angle, the length of each side is 10 millimeters. Find the length of: **a.** the shorter diagonal **b.** the longer diagonal in radical form.

14. In $\triangle ABC$, $\angle B$ is the right angle, $\overline{DC} \cong \overline{DA}$, $m\angle CDB = 60$, and $DB = 5$ yards. Find the following measures:

a. $m\angle DCB$ **b.** $m\angle CDA$

c. $m\angle DCA$ **d.** $m\angle CAD$

e. $m\angle ACB$ **f.** DC

g. CB **h.** DA

i. AC **j.** AB

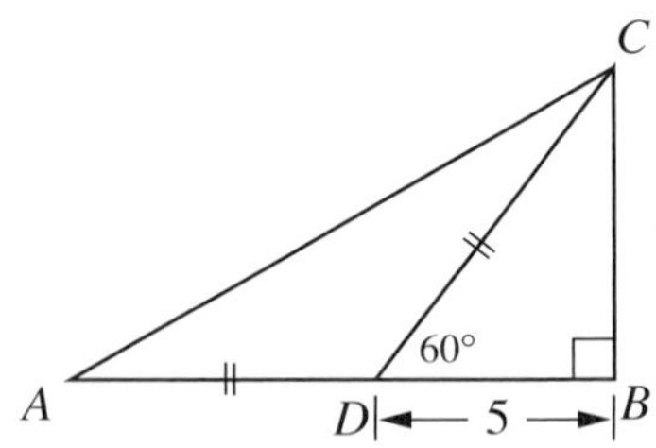

In 15–19, $\triangle ABC$ is a right triangle with altitude $\overline{CD}$ drawn to hypotenuse $\overline{AB}$, $m\angle B = 60$, and $m\angle A = 30$.

15. If $DB = 2$ feet, find CB, AB, AD, CD, and AC.

16. If $CB = 6$ feet, find AB, DB, AD, CD, and AC.

17. If $AB = 4$, find CB, DB, AD, CD, and AC.

18. If $AC = 8\sqrt{3}$, find CB, AB, DB, AD, and CD.

19. If $CD = 6\sqrt{3}$, find DB, BC, AB, AD, and AC.

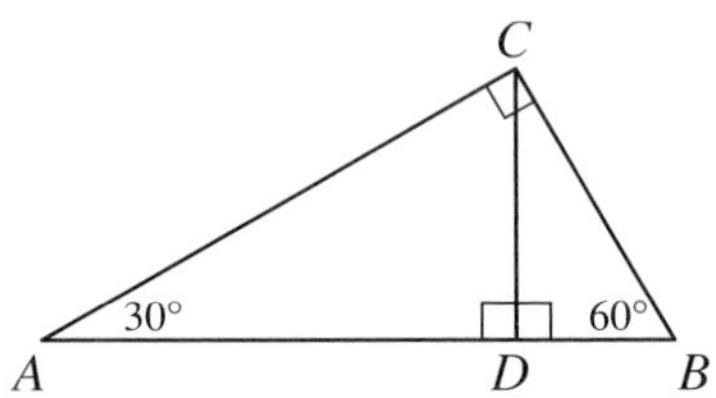

Ex. 15–19

The Isosceles Right Triangle

In another special right triangle, called the ***45–45-degree right triangle*** or the ***isosceles right triangle***, we can find the lengths of all three sides when given the length of only one side. We can develop the relationships of these sides by starting with a square.

(1) In square $ABCD$, we know that every angle has a measure of 90°, and all four sides are congruent.

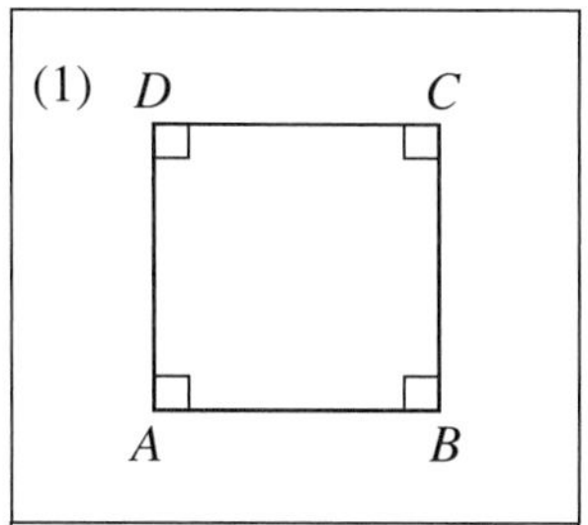

(2) We draw diagonal $\overline{AC}$. Since $\angle ABC$ and $\angle ADC$ are right angles, $\overline{AB} \cong \overline{AD}$, and $\overline{AC} \cong \overline{AC}$, it follows that $\triangle ABC \cong \triangle ADC$ by hy. leg $\cong$ hy. leg.

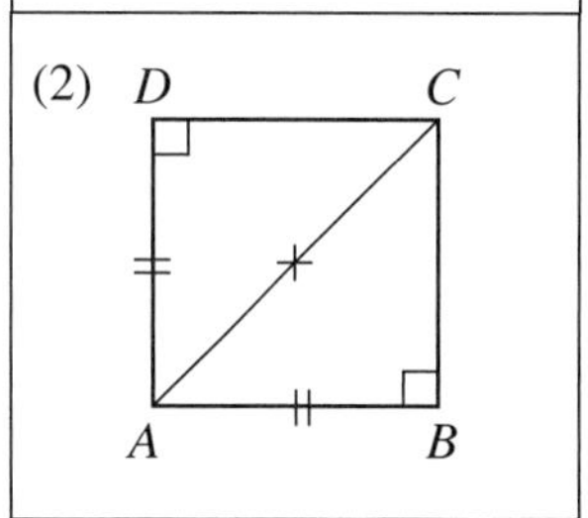

(3) *a.* Since $\triangle ABC \cong \triangle ADC$, their corresponding angles are congruent. Thus, $\angle CAB \cong \angle CAD$, and $\angle ACB \cong \angle ACD$. Each of these angles measures 45°.

b. Since all four sides of the square are congruent, let each side have a length of x.

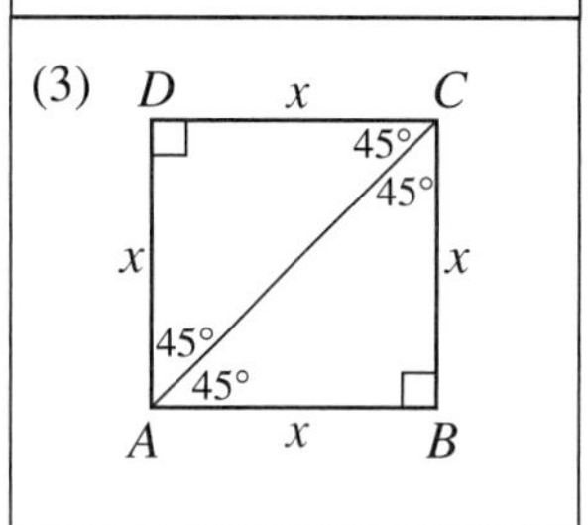

(4) In right triangle ABC, $\overline{AC}$ is the hypotenuse. Since $AB = x$ and $BC = x$, we apply the Pythagorean Theorem to find the length of hypotenuse $\overline{AC}$:

$$\begin{aligned}(AC)^2 &= (AB)^2 + (BC)^2 \\ &= x^2 \quad + x^2 \\ &= 2x^2 \\ AC &= \sqrt{2x^2} = \sqrt{x^2} \cdot \sqrt{2} \\ &= x\sqrt{2}\end{aligned}$$

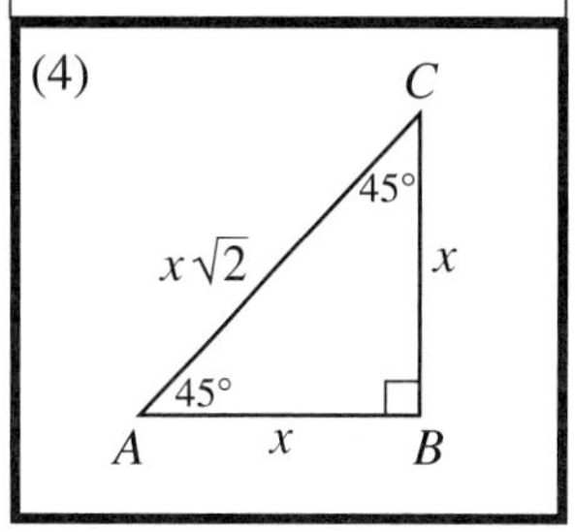

The final diagram, (4), shows the isosceles right triangle and the following relationships:

1. If each side opposite a 45° angle has a length of x, then the side opposite the 90° angle has a length of $x\sqrt{2}$.
2. In an isosceles right triangle, the length of the hypotenuse is equal to the length of a leg multiplied by $\sqrt{2}$.
3. The ratio of the lengths of the sides is $1:1:\sqrt{2}$.

EXAMPLE

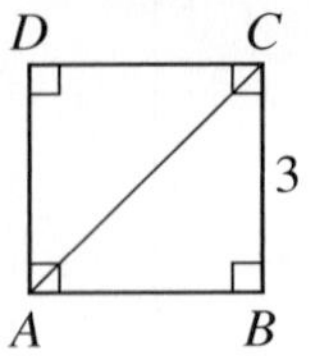

In square $ABCD$, the length of a side is 3 centimeters and $\overline{AC}$ is a diagonal.

a. Find, in radical form, the number of centimeters in the length of $\overline{AC}$.

b. Find AC, to the *nearest tenth of a centimeter.*

Solutions *METHOD 1*

a. In the isosceles right triangle formed by the square's diagonal, label the length of each leg x and the length of the hypotenuse $x\sqrt{2}$.

Since $CB = AB = x = 3$:

$$AC = x\sqrt{2} = 3\sqrt{2}$$

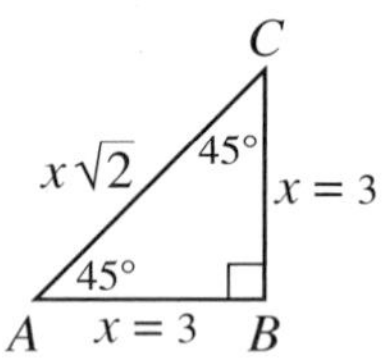

b. ***Enter:*** 3 [×] 2 [$\sqrt{x}$] [=]

Display: 4.24264068

Round to *tenths:* 4.2

METHOD 2

a. In square $ABCD$, $AB = 3$ and $BC = 3$.

Let h = length of $\overline{AC}$.

By the Pythagorean Theorem:

$$\begin{aligned} h^2 &= (3)^2 + (3)^2 \\ &= 9 \quad + 9 \\ &= 18 \\ h &= \sqrt{18} = \sqrt{9} \cdot \sqrt{2} \\ &= 3\sqrt{2} \end{aligned}$$

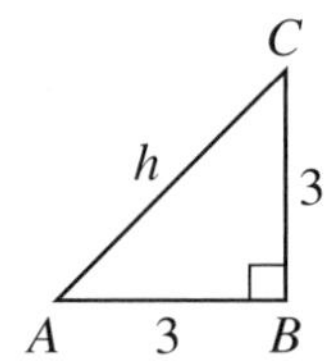

b. ***Enter:*** 18 [$\sqrt{x}$]

or

Enter: 3 [×] 2 [$\sqrt{x}$] [=]

Display: 4.24264068

Round to *tenths:* 4.2

Answer: **a.** $AC = 3\sqrt{2}$ cm

b. $AC = 4.2$ cm, to the *nearest tenth of a centimeter.*

EXERCISES WITH OPEN-RESPONSE PROBLEMS

Note. In all exercises, irrational answers should be left in radical form unless otherwise indicated.

In 1–5, find, in each case, the lengths of the segments marked x and y.

1.

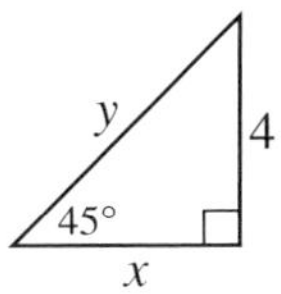

2.

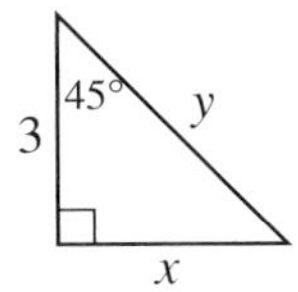

3.

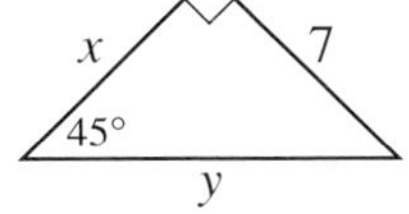

4.

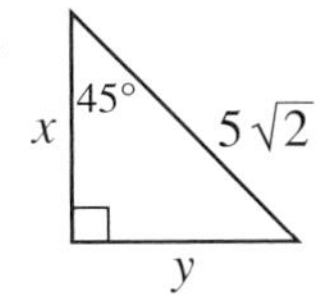

5.

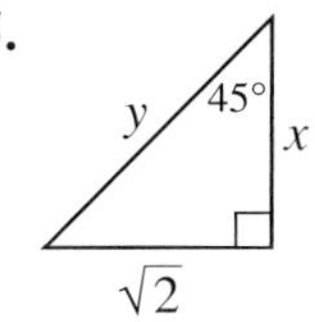

6. Find the length of a side of a square whose diagonal has a measure of:
a. $2\sqrt{2}$ **b.** $6\sqrt{2}$ **c.** $8\sqrt{2}$ **d.** $\sqrt{2}$ **e.** 4

7. Find the length of a diagonal of a square if the length of one of its sides is:
a. 6 **b.** 2 **c.** 15 **d.** $3\sqrt{2}$ **e.** s

8. If the diagonal of a square has a length of $4\sqrt{2}$, the perimeter of the square is
(1) 8 (2) 16 (3) $16\sqrt{2}$ (4) 32

9. The length of a leg of an isosceles right triangle is 20 centimeters. Find the length of the hypotenuse: **a.** in radical form **b.** to the *nearest tenth of a centimeter*

10. The lengths of the bases of an isosceles trapezoid are 7 inches and 15 inches. Each leg makes an angle of 45° with the longer base. Find the length of the altitude of the trapezoid.

11. In square $PQRS$, diagonals $\overline{PR}$ and $\overline{QS}$ intersect at M. If $PM = 3$, find the length of:
a. a diagonal of the square **b.** a side of the square

Miscellaneous Problems with Both Special Right Triangles

In 12 and 13, $\triangle ABC$ is isosceles with base angles at A and B.

12. Find the length of side $\overline{AC}$ if altitude $\overline{CD}$ has a length of 6 and each base angle has a degree measure of:
a. 30 **b.** 45 **c.** 60

13. Find the length of altitude $\overline{CD}$ if $AC = 18$ and each base angle has a degree measure of:
a. 30 **b.** 45 **c.** 60

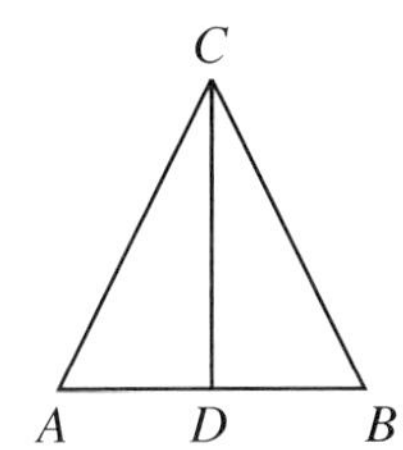

Ex. 12–13

14. In $\triangle ABC$, D is a point on $\overline{AC}$, $m\angle BDC = 45$, $m\angle BAC = 30$, and $BC = 10$.
Find DC, DB, AB, and AC.

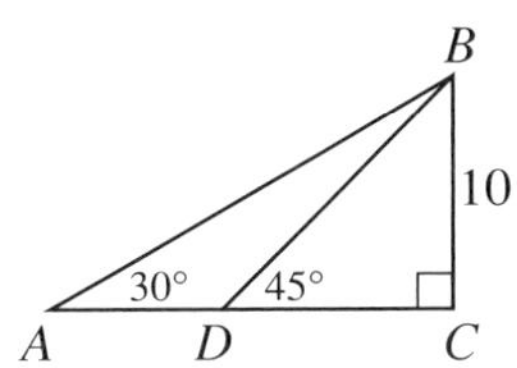

Ex. 14

15. In $\triangle RST$, P is a point on $\overline{RT}$, $m\angle SPT = 60$, $m\angle SRT = 45$, and $PT = \sqrt{3}$.
Find PS, ST, RT, and RS.

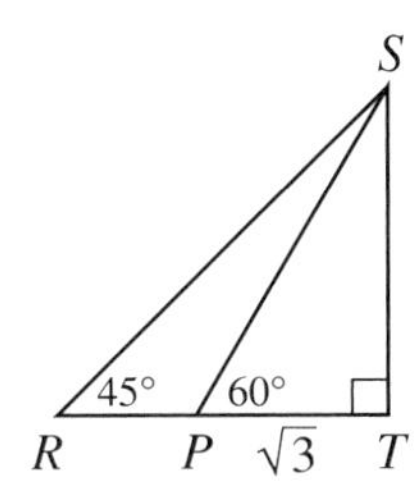

Ex. 15

16. In quadrilateral $ABCD$, the length of diagonal $\overline{AC}$ is 12, $m\angle BAC = 30$, $m\angle BCA = 60$, $m\angle ACD = 45$, and $m\angle CAD = 45$.

a. Find the *exact* length of each side of $ABCD$.

b. Express the length of each side of $ABCD$ to the *nearest hundredth*.

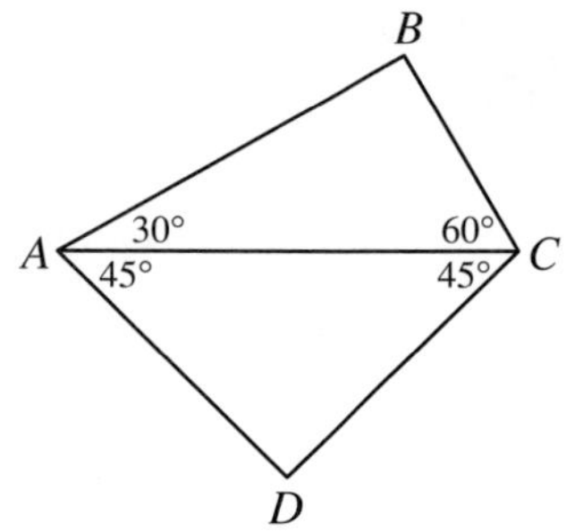

21-6 USING QUADRATIC EQUATIONS TO SOLVE PROBLEMS

The use of an algebraic strategy to solve a problem may require the solution of a quadratic equation. Often, one of the two roots of the quadratic equation is positive and the other negative. Sometimes both the positive and the negative roots are solutions to the problem. At times, however, as in problems where the variable represents a length, only the positive root is an acceptable solution.

Many quadratic equations can be solved by factoring. Of course, as shown in the Alternative Solution for Example 1, the quadratic formula can be used to solve any quadratic equation.

EXAMPLES

1. The square of a number decreased by 4 times the number equals 21. Find the number.

Solution Solve a quadratic equation by *factoring*.
Let $x =$ the number.

The square of a number decreased by 4 times the number equals 21.

$$x^2 - 4x = 21$$

$$x^2 - 4x - 21 = 0$$

$$(x - 7)(x + 3) = 0$$

$$x - 7 = 0 \quad \Big| \quad x + 3 = 0$$

$$x = 7 \quad \Big| \quad x = -3$$

Check both values in the original equation.

Check for $x = 7$:

$$x^2 - 4x = 21$$
$$(7)^2 - 4(7) \stackrel{?}{=} 21$$
$$49 - 28 \stackrel{?}{=} 21$$
$$21 = 21 \text{ (True)}$$

Check for $x = -3$:

$$x^2 - 4x = 21$$
$$(-3)^2 - 4(-3) \stackrel{?}{=} 21$$
$$9 + 12 \stackrel{?}{=} 21$$
$$21 = 21 \text{ (True)}$$

Answer: The number is 7 or -3.

---

2. The product of two consecutive, positive, even integers is 80. Find the integers.

Solution

Let $x =$ the first positive, even integer.
Then $x + 2 =$ the next consecutive, positive, even integer.

The product of two consecutive, positive, even integers is 80.

$$x(x + 2) = 80$$
$$x^2 + 2x = 80$$
$$x^2 + 2x - 80 = 0$$
$$(x - 8)(x + 10) = 0$$

$$x - 8 = 0 \quad\Big|\quad x + 10 = 0$$
$$x = 8 \quad\Big|\quad x = -10$$ [Reject. The required integer is positive.]
$$x + 2 = 10$$

Check: The product of the two consecutive, positive, even integers 8 and 10 is (8)(10), or 80.

Answer: The integers are 8 and 10.

---

3. The base of a parallelogram measures 7 centimeters more than its altitude. If the area of the parallelogram is 30 square centimeters, find the measure of its base and the measure of its altitude.

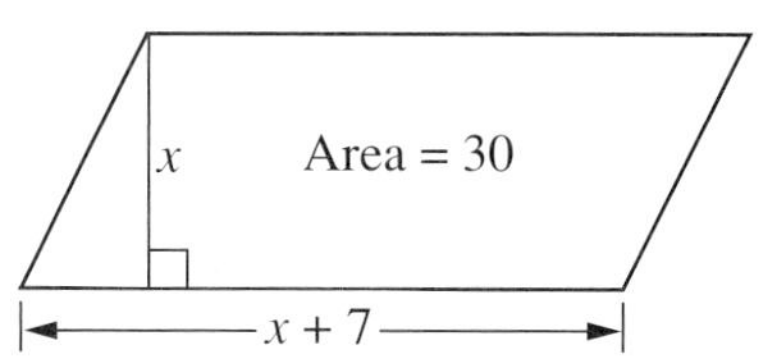

Solution

Let $x =$ the number of centimeters in the altitude.
Then, $x + 7 =$ the number of centimeters in the base.

The area of the parallelogram is 30.

$$\text{Base} \cdot \text{Altitude} = \text{Area}$$
$$(x + 7)x = 30$$
$$x^2 + 7x = 30$$
$$x^2 + 7x - 30 = 0$$
$$(x - 3)(x + 10) = 0$$

$$x - 3 = 0 \quad \Big| \quad x + 10 = 0$$
$$x = 3 \quad \Big| \quad x = -10 \text{ [Reject. The altitude cannot be negative.]}$$
$$x + 7 = 10 \quad \Big|$$

Check: When the base of a parallelogram measures 10 and its altitude measures 3, the area is 10 • 3, or 30. Also, the measure of the base, 10, is 7 more than the measure of the altitude, 3.

Answer: The altitude measures 3 centimeters; the base measures 10 centimeters.

EXERCISES WITH OPEN-RESPONSE PROBLEMS

In 1–24, in each case, use a quadratic equation to solve.

1. The square of a number added to 3 times the number is 28. Find the number.
2. The square of a certain number minus 9 times the number is 36. Find the number.
3. The length of a rectangular garden is 4 meters more than the width. The area of the garden is 60 square meters. Find the dimensions of the garden.
4. The altitude of a parallelogram measures 11 centimeters less than its base. The area of the parallelogram is 80 square centimeters. Find the measure of its base and the measure of its altitude.
5. Find two positive numbers in the ratio 2 : 3 whose product is 2,400.
6. The ratio of the measures of the length and width of a rectangle is 5 : 7. The area of the rectangle is 1,715 square feet. Find the dimensions of the rectangle.
7. The length of each of a pair of parallel sides of a square is increased by 2 meters, and the length of each of the other two sides of the square is decreased by 2 meters. The area of the rectangle formed is 32 square meters. Find the measure of one side of the original square.
8. The larger of two numbers is 5 more than the smaller. The product of the numbers is 204. Find the numbers.
9. The product of two consecutive integers is 56. Find the integers.
10. The product of two consecutive odd integers is 143. Find the integers.
11. Find two consecutive positive integers such that the square of the smaller increased by 4 times the larger is 64.

12. Find three consecutive integers such that the product of the second integer and the third integer is 20.

13. Joan's rectangular garden is 6 meters long and 4 meters wide. She plans to double the area of the garden by increasing each side by the same amount. Find the number of meters by which each dimension must be increased.

14. The altitude of a triangle measures 5 centimeters less than its base. The area of the triangle is 42 square centimeters. Find the lengths of the base and the altitude.

15. One leg of a right triangle is 1 centimeter longer than the other leg. The hypotenuse measures 5 centimeters. Find the measure of each leg of the triangle.

16. The hypotenuse of a right triangle is 1 inch longer than one leg and 8 inches longer than the other. Find the length of each side of the triangle.

17. The ratio of the lengths of the legs of a right triangle is 7:24. The length of the hypotenuse is 50. Find the length of each leg of the triangle.

18. The hypotenuse of a right triangle is 3 meters longer than twice the length of one of the legs. The measure of the other leg is 12 meters. Find the measure of the hypotenuse of the triangle.

19. The sum of two numbers is 10. The sum of their squares is 52. Find the numbers.

20. Find two consecutive integers such that the sum of their squares is 61.

21. The perimeter of a rectangle is 20 inches, and its area is 16 square inches. Find the dimensions of the rectangle.

22. The perimeter of a rectangle is 54 centimeters, and its area is 140 square centimeters. Find the dimensions of the rectangle.

23. The length and width of a rectangle are in the ratio 2:3. The area of the rectangle is 300 square feet.

a. Find the length and width of the rectangle in simplest radical form.

b. Find the length and width of the rectangle to the *nearest hundredth* of a foot.

24. The measures of the base and altitude of a triangle are in the ratio 4:5. The area of the triangle is 120 square meters.

a. Find the measures of the base and altitude of the triangle in simplest radical form.

b. Find the measures of the base and altitude to the *nearest hundredth* of a meter.

25. The length of a rectangle is 4 inches more than the width. The area of the rectangle is 4 square inches.

a. Write an equation that can be used to find the length and width of the rectangle.

b. Use the quadratic formula to find the length and width in radical form.

c. Express the length and width to the *nearest hundredth* of an inch.

21-7 FUNCTIONS

A relation is any set of ordered pairs. Certain relations, however, behave in a special way, leading to an important concept in mathematics called a *function*.

- **DEFINITION.** A ***function*** is a relation in which each element of the domain corresponds to one and only one element in the range.

Relation A: Squaring a Number
Every element in the domain is matched to its square by the rule $x \to x^2$ or $y = x^2$.

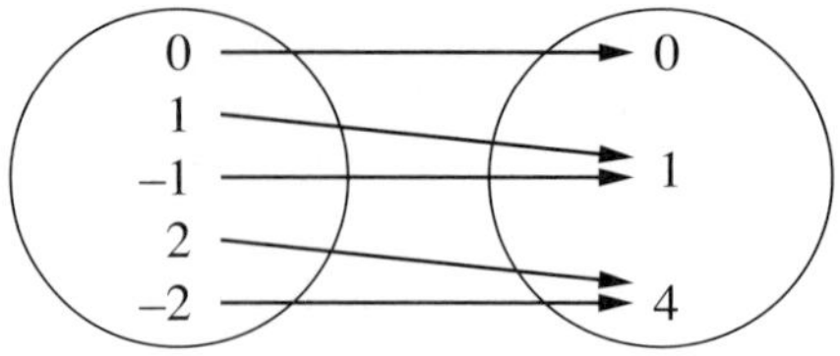

$x \to x^2$ or $y = x^2$
A Function

Since every number has *one and only one* square, this relation is a *function.*

Relation B: Finding a Square Root
Every element in the domain is matched to its square root(s) by the rule $x \to \pm\sqrt{x}$ or $y = \pm\sqrt{x}$.

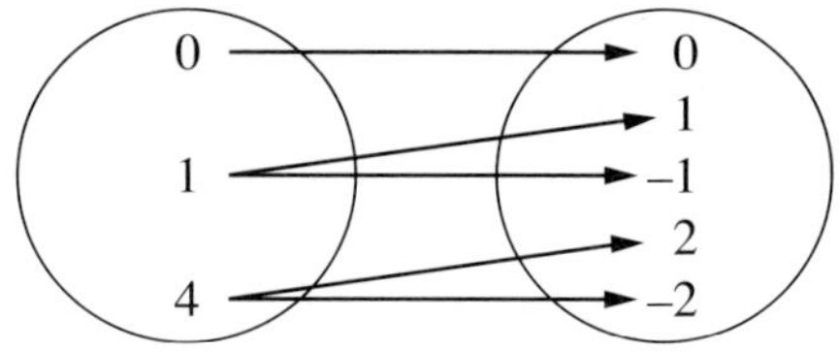

$x \to \pm\sqrt{x}$ or $y = \pm\sqrt{x}$
Not a Function

Since every nonzero number has *two* square roots, this relation is *not* a function.

Another interesting discovery can be made by comparing the set of ordered pairs for these two relations. Relation *A* (Squaring) is a function: Relation *B* (Finding a Square Root) is not a function:

Different first elements

(0, 0), (1, 1), (−1, 1), (2, 4), (−2, 4)
(0, 0), (1, 1), (1, −1), (4, 2), (4, −2)

Same first element   Same first element

If two or more ordered pairs have the same first element, then some element of the domain corresponds to more than one element in the range. This fact allows us to form an alternative definition of function.

- **DEFINITION.** A ***function*** is a relation in which no two ordered pairs have the same first element.

In this section, we will study coordinate graphs of relations and functions. In a coordinate graph, the domain is represented by values on the horizontal or x-axis, and the range is read from values on the vertical or y-axis.

When a relation is graphed on the coordinate plane, a method is available to determine rather quickly whether the relation *is* or *is not* a function. This method is called the *vertical-line test for a function.*

- ***Vertical-Line Test for a Function.*** **If each vertical line drawn through the graph of a relation intersects the graph at one and only one point, then the relation is a function.**

Let us apply this test to the relations studied above.

Relation A (Squaring a Number): $x \to x^2$ or $y = x^2$

In Figure 1 on the next page, the domain = {0, 1, −1, 2, −2}, and the graph contains exactly five points. A vertical line passing through any of these points will intersect one and only one point in the relation.

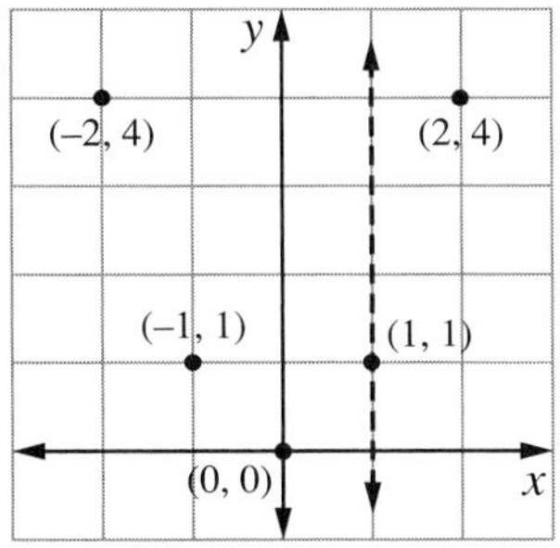

Figure 1

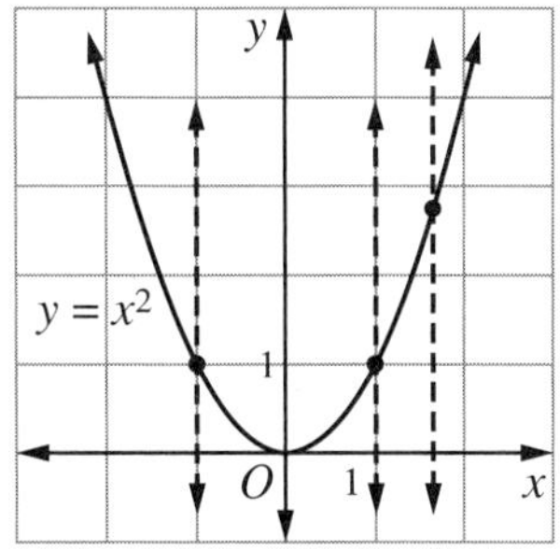

Figure 2

In Figure 2, the relation $y = x^2$ is graphed for the set of all real numbers, or domain = {real numbers}. The vertical-line test demonstrates geometrically that every real number has one and only one square.

We say that $y = x^2$ is a function of x or $y = x^2$ is a function. This means that $\{(x, y) \mid y = x^2\}$ or the set of ordered pairs (x, x^2) is a function.

Relation B (Finding a Square Root): $x \rightarrow \pm\sqrt{x}$ *or* $y = \pm\sqrt{x}$

In Figure 3, the domain = {0, 1, 4}, and the graph contains exactly five points. A vertical line passing through (1, 1) also passes through (1, −1), indicating that this relation is *not* a function.

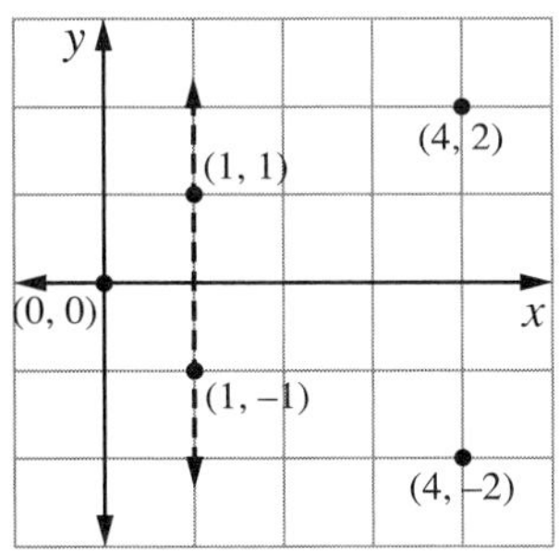

Figure 3

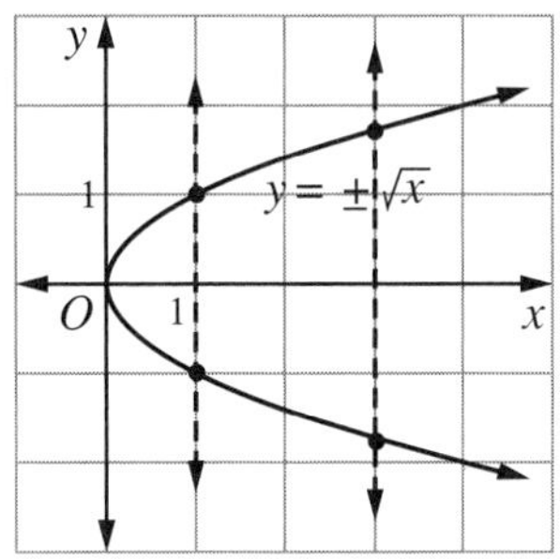

Figure 4

By squaring both members of the equation $y = \pm\sqrt{x}$, we form the equation $y^2 = x$. The graph of $y^2 = x$ is shown in Figure 4, where the domain or set of x values = {positive real numbers and zero}. The vertical-line test demonstrates geometrically that

$y = \pm\sqrt{x}$ is *not* a function of x or $y = \pm\sqrt{x}$ is *not* a function.

KEEP IN MIND

A ***relation*** is any set of ordered pairs.

A ***function*** is a relation in which each element of the domain corresponds to one and only one element in the range, *or* a relation in which no two ordered pairs have the same first element.

Note: For relation B, we have stated that $y = \pm\sqrt{x}$ is *not* a function of x. However, if we write $\{(y, x) \mid y = \pm\sqrt{x}\}$, then we can say x is a function of y. This can be demonstrated geometrically by interchanging the x- and y-axes. However, we will make the following agreements unless stated otherwise:

1. When the variables x and y are used, the ordered pairs are written as (x, y) and the horizontal axis is the x-axis.
2. When testing for a function, we will test to see whether the relation is a function of x, or of the variable represented on the horizontal axis.
3. If a relation is not a function of the variable represented on the horizontal axis, we will simply state that the relation is not a function.

EXAMPLES

In 1–3, the domain = {2, 4, 6}. State whether the indicated relation is or is not a function.

1.

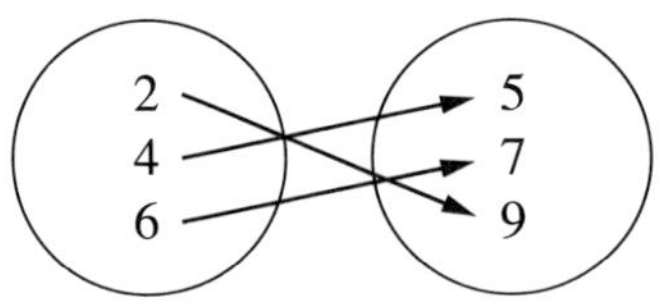

Solution: Every element in the domain corresponds to one and only one element in the range, {5, 7, 9}.

Answer: The relation is a function.

2.

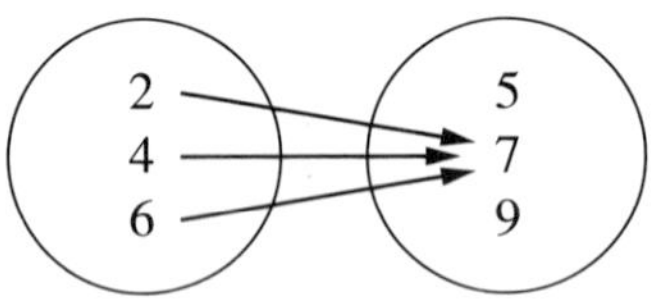

Solution: Every element in the domain corresponds to one and only one element, 7, in the range.

Answer: The relation is a function.

3.

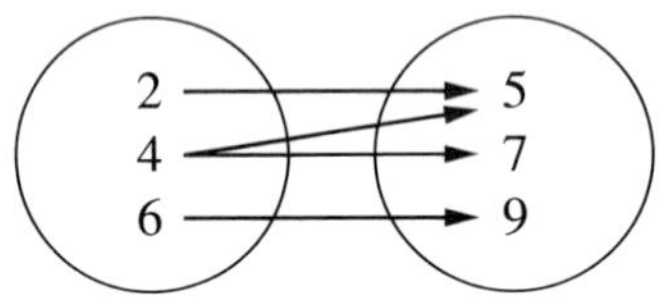

Solution: Element 4 in the domain corresponds to 5 and 7, that is, to *more than one* element, in the range.

Answer: The relation is *not* a function.

---

4. Which relation, if any, is a function?

$C = \{(0, 6), (0, 9), (1, 4), (2, 7)\}$ $\quad$ $D = \{(0, 6), (1, 6), (2, 6), (3, 6)\}$

Solution Since relation C contains two ordered pairs with the same first element, namely, (0, 6) and (0, 9), C is not a function.

In relation D, however, no two ordered pairs have the same first element.

Answer: Relation D is a function.

---

5. Which graph represents a relation that is a function?

(1)
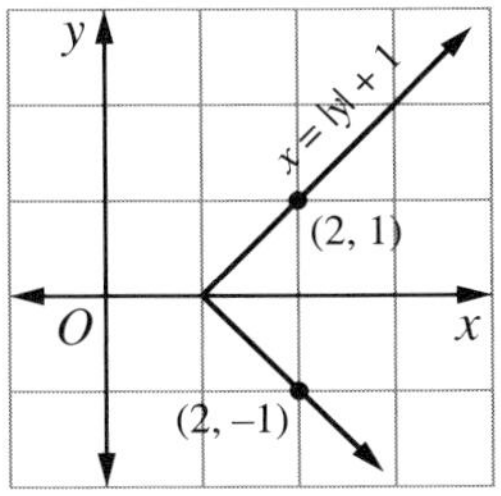

(2)
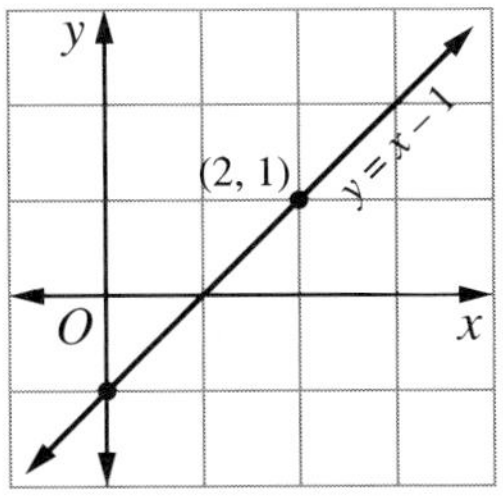

(3)
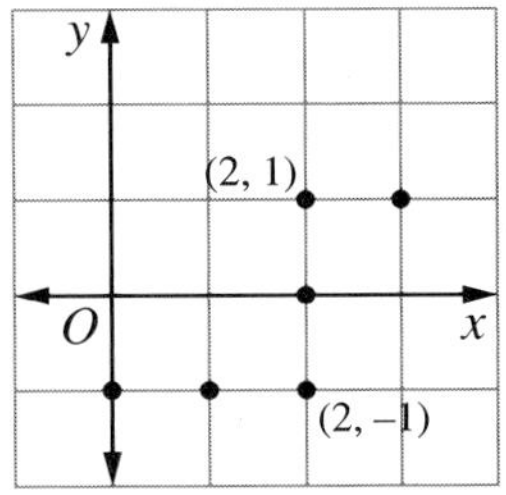

(4)
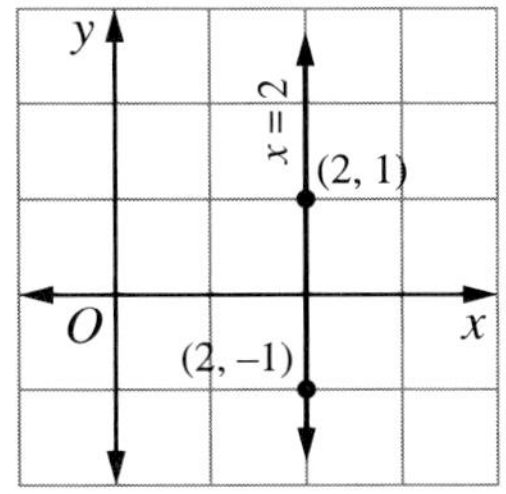

Solution A vertical line drawn through graphs (1), (3), and (4) at $x = 2$ indicates that $(2, -1)$ and $(2, 1)$ are on each graph. These relations are not functions because each graph contains more than one point for some elements of the domain. Only graph (2) passes the vertical-line test.

Answer: Graph (2) represents a relation that is a function.

EXERCISES

In 1–6, the domain in each relation is $\{1, 2, 3\}$. **a.** State whether the relation is or is not a function. **b.** If the relation is not a function, explain why.

1.
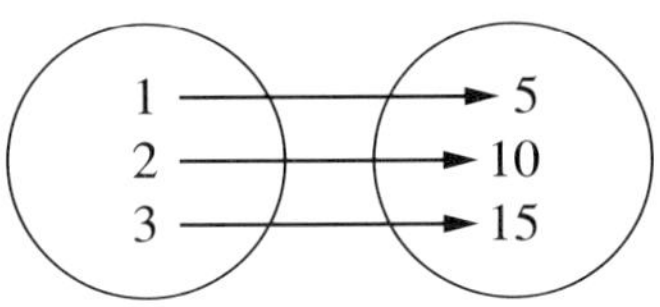

2.
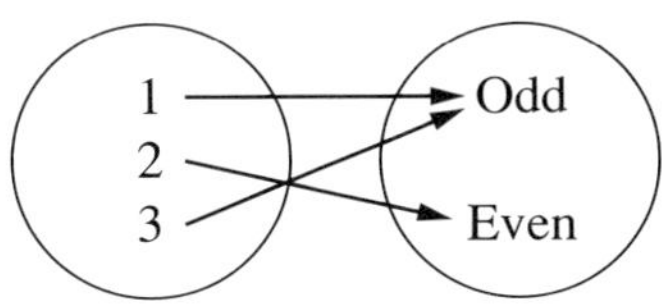

3.
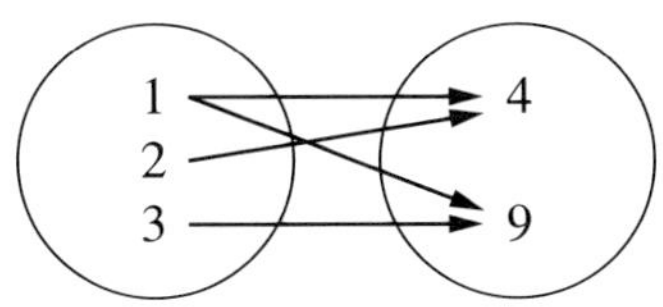

4.
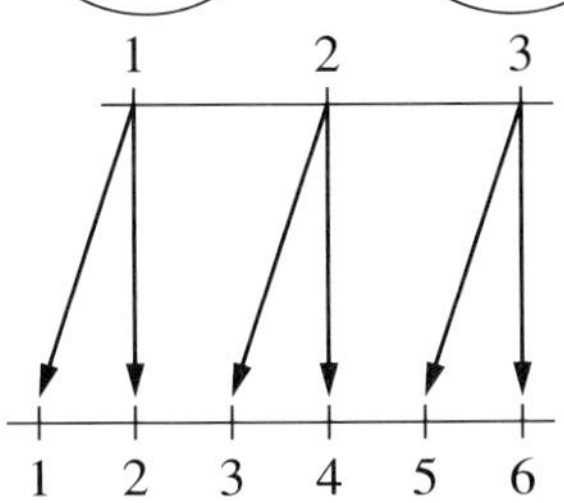

5.
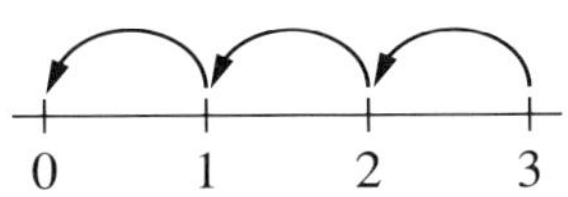

6.
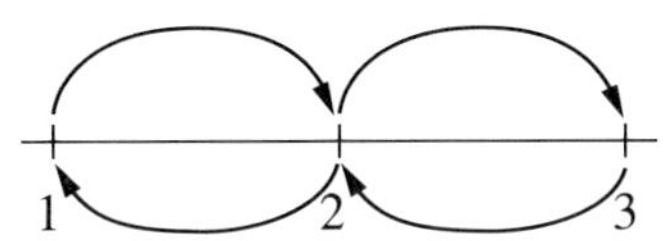

In 7–10, for each relation: **a.** State the domain. **b.** State the range. **c.** Tell whether the relation is or is not a function.

7. $\{(3, 7), (4, 9), (5, 11), (6, 13)\}$

8. $\{(4, 3), (3, 2), (2, 4), (3, 4)\}$

9. $\{(7, 1), (7, 2), (7, 3), (7, 4), (7, 5)\}$

10. $\{(1, 7), (2, 7), (3, 7), (4, 7), (5, 7)\}$

11. Relation $A = \{(6, 3), (8, 2), (10, 1), (x, 0)\}$. Name all possible values that x may *not* represent if relation A is a function.

12. Relation B is indicated by the rule $x \rightarrow y$, where $y < x$. The domain and the range of the relation are elements of the set $\{-2, -1, 0, 1, 2\}$.

a. Write all ordered pairs that belong to relation B.

b. Is relation B a function? Explain your answer.

13. A function F is given by the rule $x \to x^2 - 2$. If the domain of this function is $\{-5, \sqrt{3}, \sqrt{5}, 3\}$, list all ordered pairs in function F.

14. The domain of a function K is $\{-3, 0, \sqrt{2}, \sqrt{7}, 4\}$, and its rule is $y = 2x + 1$.
a. Write all ordered pairs in function K. **b.** Write the range of function K.

In 15–22, determine whether or not each graph represents a function. (*Hint.* Use the vertical-line test.)

15.

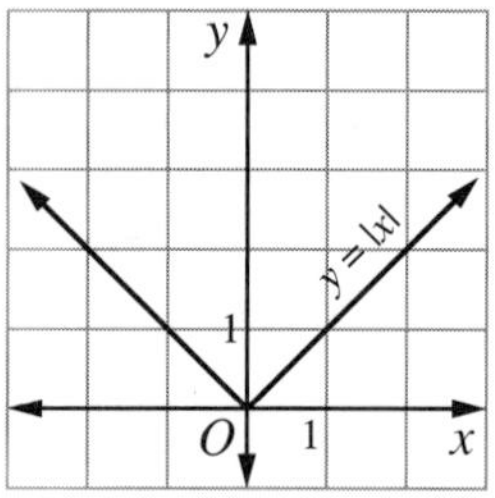

16.

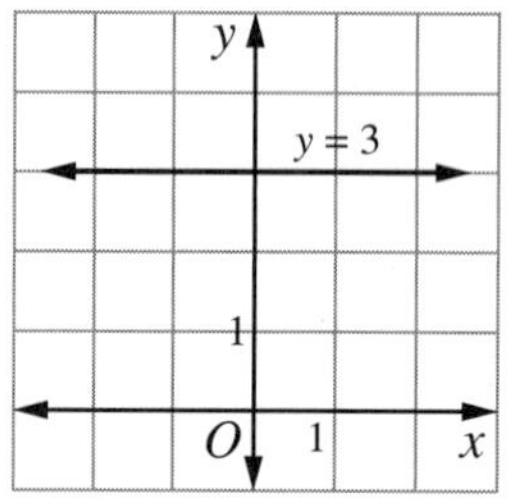

17.

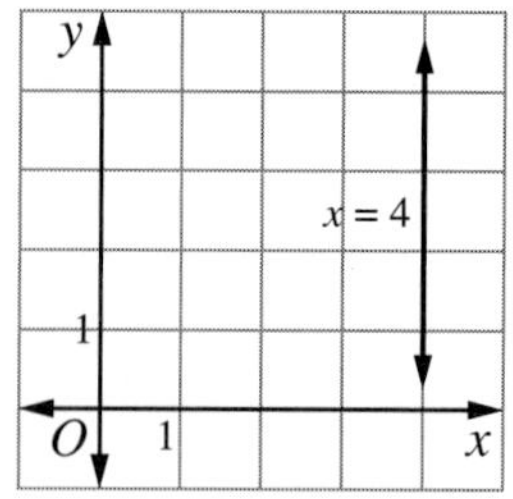

18.

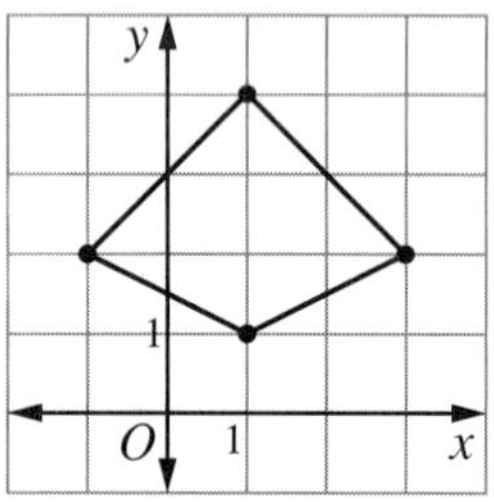

19.

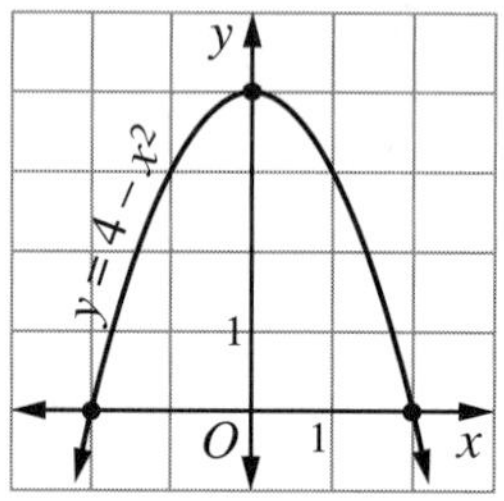

20.

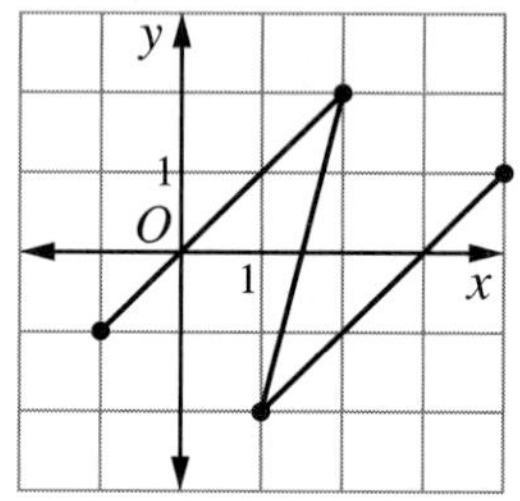

21.

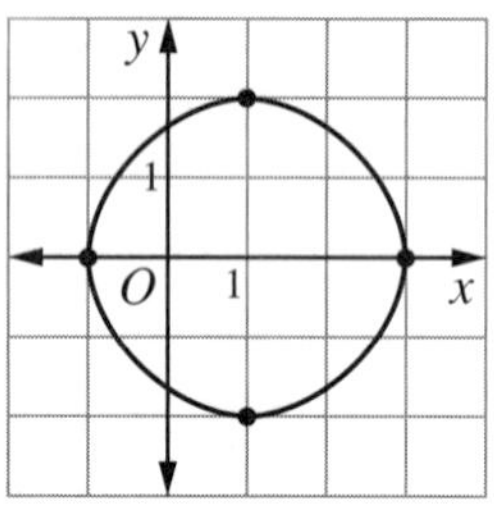

22.

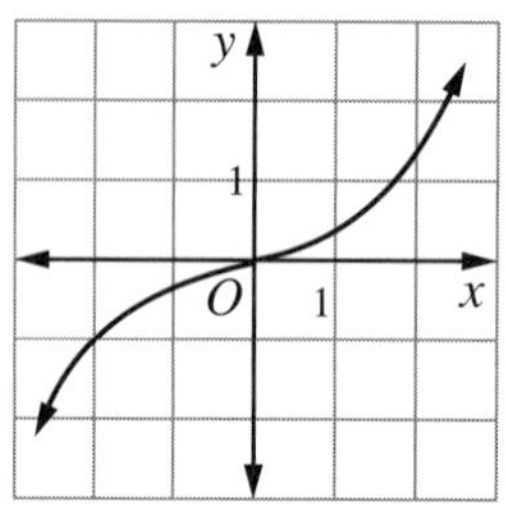

23. For the relation given in Exercise 15, state: **a.** the domain **b.** the range

24. For the relation given in Exercise 21, state: **a.** the domain **b.** the range

In 25–32, in each case, let $D \to R$ represent a relation such that elements of the domain, D, correspond to elements of the range, R. **a.** Is the relation a function? **b.** Explain your answer.

25. Person → birthday
26. Birthday → person
27. Houses → zip codes
28. States → zip codes
29. Homes → people
30. People → homes
31. Telephone numbers → people
32. Cost of item in store → sales tax

33. Describe three functions in our world that are similar to the functions in Exercises 25–32.

21-8 THE GRAPH OF A QUADRATIC FUNCTION

The Graph of a Parabola

The graph of a first-degree polynomial function (a linear function) is a straight line. For example, the graph of $y = 2x - 5$ is a straight line. Two or three points are sufficient to draw this graph.

The graph of a ***second-degree polynomial function*** (a quadratic function), however, is not a straight line, and a larger number of points are needed to draw its graph.

EXAMPLES

1. Graph the quadratic function $y = 2x^2$ for values of x from $x = -3$ to $x = 3$ inclusive, that is, for $-3 \leq x \leq 3$.

Solution (1) Make a table using integral values of x from -3 to 3.

x	$2x^2$	y
-3	$2(-3)^2$	18
-2	$2(-2)^2$	8
-1	$2(-1)^2$	2
0	$2(0)^2$	0
1	$2(1)^2$	2
2	$2(2)^2$	8
3	$2(3)^2$	18

(2) Plot the points associated with each ordered pair (x, y).

(3) Draw a smooth curve through the points. Notice that the graph of $y = 2x^2$ is a curve. This curve is called a *parabola*.

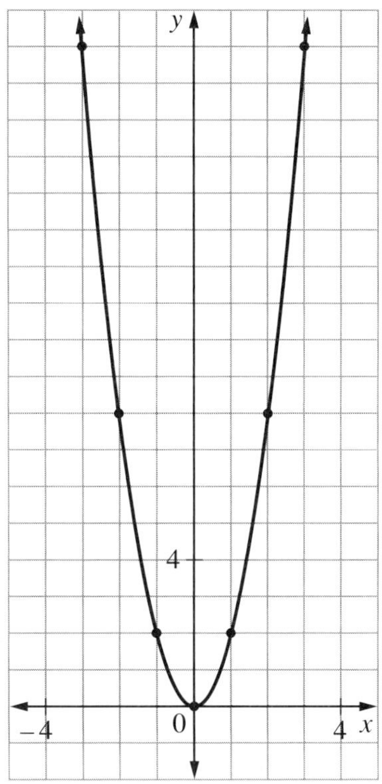

- **The graph of a quadratic function of the form $y = ax^2 + bx + c$, where a, b, and c are real numbers and $a \neq 0$, is a *parabola*.**

Using the following graph of the parabola $y = x^2 - 4x$, we observe:

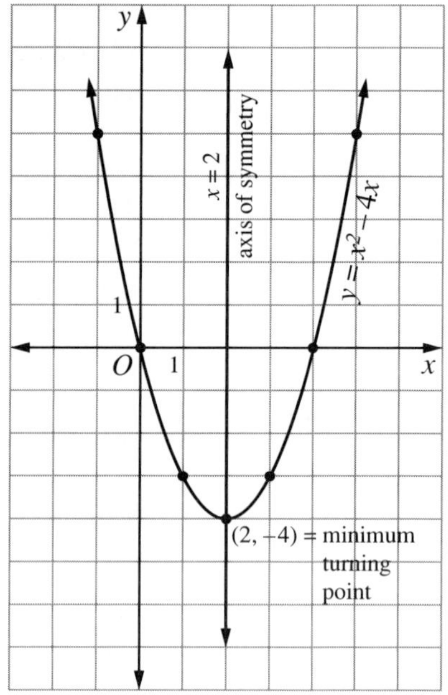

1. The parabola is symmetric with respect to a line of reflection, or an ***axis of symmetry***, whose equation is $x = 2$. Just as in a mirror image, every point on the parabola to the left of $x = 2$ matches a point on the parabola to the right of $x = 2$, and vice versa.
2. The parabola $y = x^2 - 4x$ *opens upward* and contains a ***minimum point***, $(2, -4)$, located on its axis of symmetry. This point is called the minimum because it contains the smallest value of y for the graph. Notice that, as x increases from -1 to 5, y decreases until the minimum point is reached; then the graph turns, as y starts to increase. For this reason, $(2, -4)$ is also called the ***turning point*** of this parabola.
3. By applying the vertical-line test to this graph, we see that every element of the domain has one and only one element in the range. In other words, for every x, there is one and only one y. Therefore, the parabola $y = x^2 - 4x$ is the graph of a function. The domain of this function is the set of all real numbers, and the range is the set of real numbers greater than or equal to -4.

In symbols: Domain $= \{x | x \in \text{real numbers}\}$

Range $= \{y | y \geq -4 \text{ and } y \in \text{real numbers}\}$

---

2. Graph the quadratic equation $y = -x^2 + 2x + 5$, using all integral values of x from $x = -2$ to $x = 4$ inclusive.

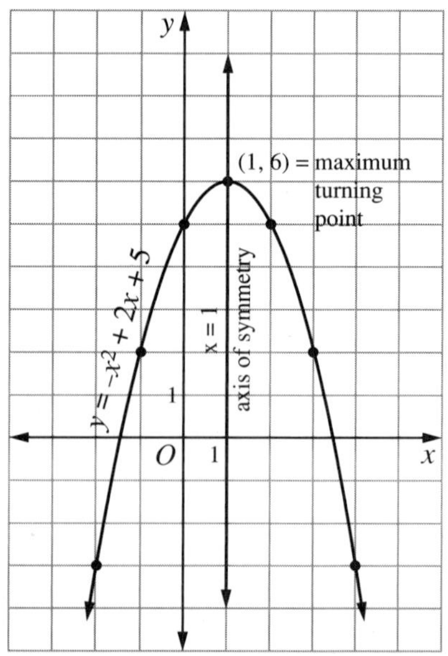

Solution (1) Develop a table of values for $-2 \leq x \leq 4$.

x	$-x^2 + 2x + 5$	y
−2	$-(-2)^2 + 2(-2) + 5$	−3
−1	$-(-1)^2 + 2(-1) + 5$	2
0	$-(0)^2 + 2(0) + 5$	5
1	$-(1)^2 + 2(1) + 5$	6
2	$-(2)^2 + 2(2) + 5$	5
3	$-(3)^2 + 2(3) + 5$	2
4	$-(4)^2 + 2(4) + 5$	−3

(2) Plot the point associated with each ordered pair (x, y): $(-2, -3)$, $(-1, 2)$, $(0, 5)$, etc.

(3) Draw the parabola through these points.

Using the graph of $y = -x^2 + 2x + 5$, we observe:

1. The equation of the ***axis of symmetry*** is $x = 1$.
2. This parabola *opens downward* and contains a ***maximum point*** (1, 6) located on its axis of symmetry. As x increases from -2 to 4, y increases until the maximum point is reached, then turns, and y starts to decrease. For this reason, (1, 6) is also called the ***turning point*** of this parabola.
3. The parabola $y = -x^2 + 2x + 5$ is the graph of a function whose domain is the set of all real numbers and whose range is the set of real numbers less than or equal to 6.

In symbols: Domain $= \{x \mid x \in \text{real numbers}\}$

Range $= \{y \mid y \leq 6 \text{ and } y \in \text{real numbers}\}$

Rules for Graphing the Parabola

For any equation of the form $y = ax^2 + bx + c$, where a, b, and c are real numbers and $a \neq 0$, certain rules can be helpful in graphing the equation.

1. If a is *positive*, the parabola *opens upward* and contains a *minimum point*, as in the graph of $y = x^2 - 4x$ shown on page 782. Here, $a = 1$, a positive number.
2. If a is *negative*, the parabola *opens downward* and contains a *maximum point*, as in the graph of $y = -x^2 + 2x + 5$ shown on page 782. Here, $a = -1$, a negative number.
3. An equation of the *axis of symmetry* is $x = \frac{-b}{2a}$.

For example, in $y = x^2 - 4x$,
$a = 1$ and $b = -4$.
Equation of axis of symmetry:

$$x = \frac{-b}{2a}$$

$$x = \frac{-(-4)}{2(1)}, \text{ or } x = \frac{4}{2}$$

$$x = 2$$

For example, in $y = -x^2 + 2x + 5$,
$a = -1$ and $b = 2$.
Equation of axis of symmetry:

$$x = \frac{-b}{2a}$$

$$x = \frac{-(2)}{2(-1)}, \text{ or } x = \frac{-2}{-2}$$

$$x = 1$$

4. The *turning point* of the parabola is always found on the axis of symmetry. Therefore, the x-coordinate of the turning point equals $\frac{-b}{2a}$, and the y-coordinate of the turning point is found by substituting $\frac{-b}{2a}$ for x into the equation $y = ax^2 + bx + c$.

Since the axis of symmetry for the parabola $y = x^2 - 4x$ is given by the equation $x = 2$, the x-coordinate of the turning point is 2. Then, by substituting 2 for x in $y = x^2 - 4x$, we see that $y = (2)^2 - 4(2) = 4 - 8 = -4$. The turning point is $(2, -4)$, as shown on the graph.

KEEP IN MIND

For the graph of the parabola $y = ax^2 + bx + c$:

1. The axis of symmetry is a vertical line whose equation is $x = \frac{-b}{2a}$.
2. If a is positive, the parabola opens upward. If a is negative, the parabola opens downward.
3. Every parabola $y = ax^2 + bx + c$ is a function whose domain is the set of real numbers. If a is positive, the range is the set of real numbers greater than or equal to the y value of the minimum point. If a is negative, the range is the set of real numbers less than or equal to the y value of the maximum point.

EXAMPLES

1. Write an equation of the axis of symmetry of the graph of $y = 3x^2 + 12x - 2$.

Solution (1) The axis of symmetry for the parabola $y = 3x^2 + 12x - 2$ is a vertical line whose equation is $x = \frac{-b}{2a}$.

(2) Comparing $y = 3x^2 + 12x - 2$ to $y = ax^2 + bx + c$, observe that $a = 3$ and $b = 12$.

(3) Thus, $x = \frac{-b}{2a}$ becomes $x = \frac{-(12)}{2(3)}$, or $x = \frac{-12}{6}$, or $x = -2$.

Answer: $x = -2$

---

2. Which is an equation of the graph shown?

(1) $y = x^2 - 4x + 4$

(2) $y = x^2 + 4x + 4$

(3) $y = -x^2 - 4x + 4$

(4) $y = -x^2 + 4x + 4$

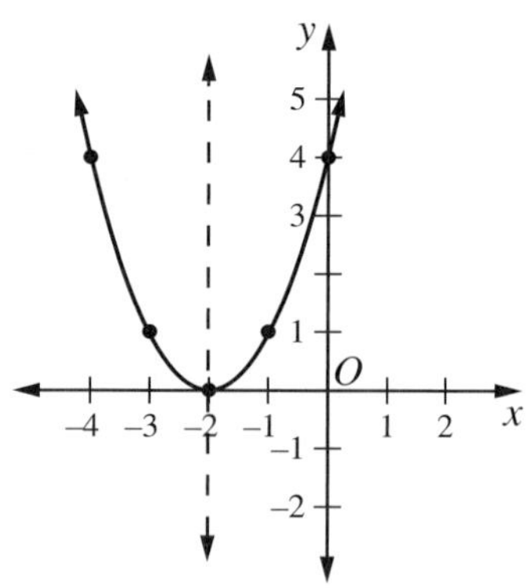

Solution Since the parabola opens upward, the leading coefficient, a, must be positive. Therefore, eliminate choices (3) and (4).

Find the equation of the axis of symmetry, $x = \frac{-b}{2a}$, for choices (1) and (2). In (1), $x = \frac{-(-4)}{2(1)}$, or $x = 2$. In (2), $x = \frac{-(4)}{2(1)}$, or $x = -2$. According to the graph, choice (2) is correct.

Answer: (2)

Note. If you attempt to find the answer by substituting coordinates of points on the graph, namely, $(-4, 4)$, $(-3, 1)$, $(-2, 0)$, $(-1, 1)$, or $(0, 4)$, be careful! Point $(0, 4)$ is a solution to all four equations given as choices. If you use this method, it is recommended that you check *three or more* coordinates before claiming that the equation is correct.

EXERCISES WITH OPEN-RESPONSE PROBLEMS

In 1–20, graph each quadratic equation. Use the integral values for x indicated in parentheses to prepare the necessary table of values.

1. $y = x^2$ $(-3 \le x \le 3)$

2. $y = 3x^2$ $(-2 \le x \le 2)$

3. $4x^2 = y$ $(-2 \le x \le 2)$

4. $5x^2 = y$ $(-2 \le x \le 2)$

5. $y = -x^2$ $(-3 \le x \le 3)$

6. $y = -2x^2$ $(-2 \le x \le 2)$

7. $y = \frac{1}{2}x^2$ $(-4 \le x \le 4)$

8. $-\frac{1}{2}x^2 = y$ $(-4 \le x \le 4)$

9. $y = x^2 + 1$ $(-3 \le x \le 3)$

10. $x^2 - 1 = y$ $(-3 \le x \le 3)$

11. $y = x^2 - 4$ $(-3 \le x \le 3)$

12. $-x^2 + 4 = y$ $(-3 \le x \le 3)$

13. $y = x^2 - 2x$ $(-2 \le x \le 4)$

14. $x^2 + 2x = y$ $(-4 \le x \le 2)$

15. $y = -x^2 + 2x$ $(-2 \le x \le 4)$

16. $y = x^2 - 6x + 8$ $(0 \le x \le 6)$

In 17–28: **a.** Write an equation of the axis of symmetry for the given parabola. **b.** Tell whether the turning point of the parabola is a maximum point or a minimum point. **c.** Find the coordinates of the turning point.

17. $y = x^2 - 10x$

18. $y = x^2 + 14x - 1$

19. $y = -x^2 + 12x + 4$

20. $y = 2x^2 + 3$

21. $y = -2x^2 + 8x - 1$

22. $y = 3x^2 - 18x$

23. $y = x^2 - x$

24. $y = -x^2 + 8x - 6$

25. $y = -x^2 + 5x$

26. $y = \frac{1}{2}x^2 + 3x - \frac{1}{2}$

27. $y = \frac{1}{2}x^2 - 5x$

28. $y = 3 - x^2$

In 29–36, select the numeral preceding the choice that best answers the question or completes the sentence.

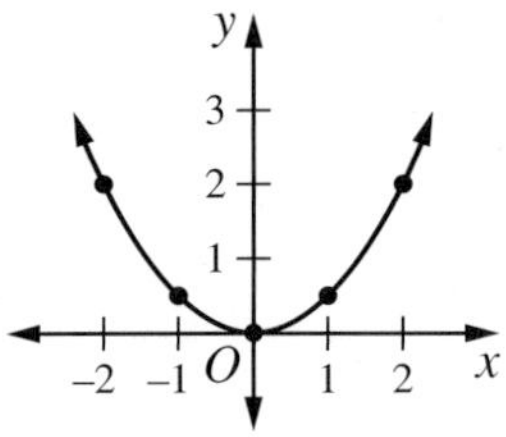

Ex. 29

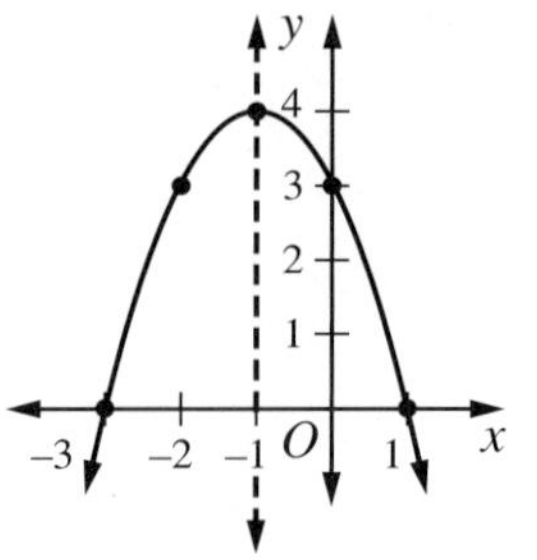

Ex. 30

29. Which is an equation of the accompanying graph?

(1) $y = 2x^2$ (2) $y = -2x^2$ (3) $y = \frac{1}{2}x^2$ (4) $y = -\frac{1}{2}x^2$

30. Which is an equation of the accompanying graph?

(1) $y = x^2 + 2x - 3$ (2) $y = -x^2 - 2x + 3$
(3) $y = -x^2 + 2x + 3$ (4) $y = x^2 - 2x + 3$

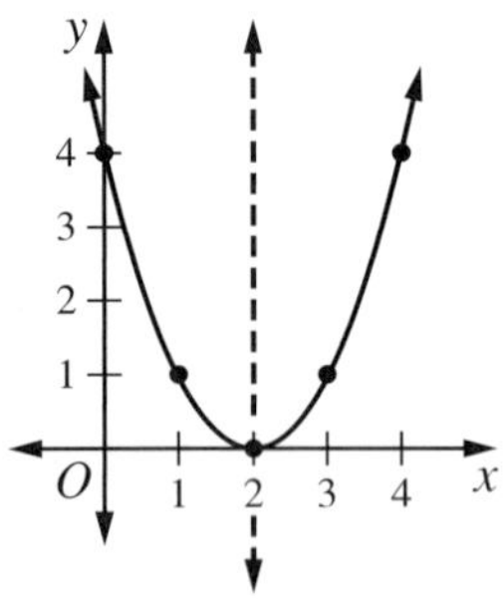

Ex. 31

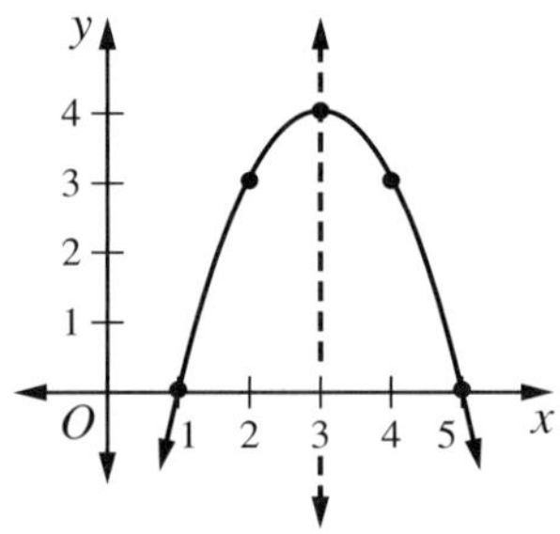

Ex. 32

31. Which is an equation of the accompanying graph?

(1) $y = x^2 - 4x + 4$ (2) $y = x^2 + 4x + 4$
(3) $y = -x^2 + 4x + 4$ (4) $y = -x^2 - 4x + 4$

32. Which is an equation of the accompanying graph?

(1) $y = x^2 - 6x + 5$ (2) $y = x^2 - 6x - 5$
(3) $y = -x^2 + 6x + 5$ (4) $y = -x^2 + 6x - 5$

33. Which is an equation of the axis of symmetry for the graph of $y = x^2 - 6x + 15$?

(1) $y = 3$ (2) $x = 3$ (3) $y = -3$ (4) $x = -3$

34. Which are the coordinates of the turning point of the parabola $y = x^2 + 10x$?

(1) $(-5, 25)$ (2) $(-5, -25)$ (3) $(5, 75)$ (4) $(5, -25)$

35. If $x = 2$ is an equation of the axis of symmetry of the parabola $y = ax^2 - 8x + 3$, then a equals

(1) -2 (2) 2 (3) -4 (4) 4

36. If $x = -3$ is an equation of the axis of symmetry of the parabola $y = x^2 + bx - 9$, the value of b is
(1) -6 (2) 6 (3) 3 (4) -3

37. **a.** Draw a graph of the equation $y = -x^2 + 10x - 21$, using all integral values of x from $x = 2$ to $x = 8$ inclusive.
b. Write an equation for the axis of symmetry. **c.** Write the coordinates of the turning point.

38. For the equation $y = x^2 - 2x - 5$:
a. Write an equation for the axis of symmetry.
b. Write the coordinates of the turning point. **c.** Draw a graph.

39. A batter hit a baseball at a height 3 feet off the ground, with an initial vertical velocity of 64 feet per second. Let x represent the time in seconds, and y represent the height of the baseball. The height of the ball can be determined over a limited period of time by using the equation $y = -16x^2 + 64x + 3$.
a. Make a table, using integral values of x from 0 to 4 to find values of y.
b. Graph the equation. Let one horizontal unit = 1 second, and one vertical unit = 10 feet. (See the suggested graph at the right.)
c. If the ball was caught after 4 seconds, what was its height when it was caught?
d. From the table and the graph, determine
(*1*) the maximum height reached by the baseball
(*2*) the time required for the ball to reach this height

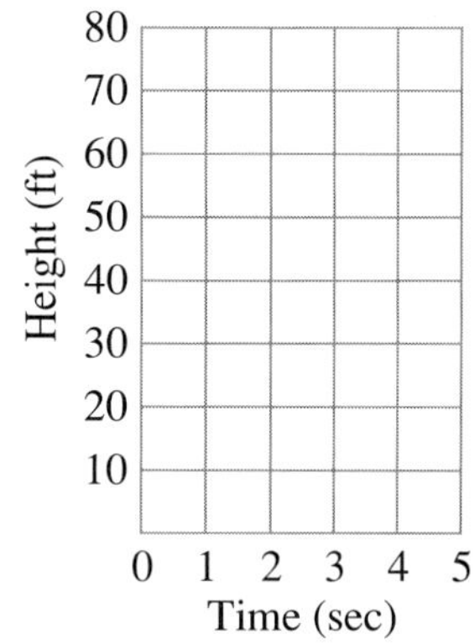

21-9 GRAPHING THE EQUATION OF A CIRCLE

We have seen that the graph of $y = ax^2 + bx + c$, a quadratic equation in which only one variable is raised to the second power, is a parabola.

Now we will consider a quadratic equation in which both variables are raised to the second power: $ax^2 + by^2 = c$.

To obtain the graph of a circle, we will use the following procedure.

PROCEDURE. To obtain a graph of the equation $ax^2 + by^2 = c$:

1. Determine the y-intercepts and the x-intercepts.
2. Solve the equation for y in terms of x.
3. Based on the findings of steps 1 and 2, develop a table of values.
4. Plot the points and join them.

EXAMPLE

Sketch the graph of $x^2 + y^2 = 9$.

Solution (1) Determine the intercepts.

To obtain the y-intercepts, let $x = 0$.

$$x^2 + y^2 = 9$$
$$0 + y^2 = 9$$
$$y^2 = 9$$
$$y = \pm 3$$

To obtain the x-intercepts, let $y = 0$.

$$x^2 + y^2 = 9$$
$$x^2 + 0 = 9$$
$$x^2 = 9$$
$$x = \pm 3$$

(2) Solve the equation for y in terms of x.

$$x^2 + y^2 = 9$$
$$y^2 = 9 - x^2$$
$$y = \pm\sqrt{9 - x^2}$$

(3) The x-intercepts are ± 3. When $x > +3$ and when $x < -3$, y has no real value because the radicand is negative.

Develop a table of values by taking $x = 0$, $x = \pm 1$, $x = \pm 2$, and $x = \pm 3$.

Evaluate each radical by using a calculator, and round to the *nearest tenth.*

x	$\pm\sqrt{9 - x^2}$	y
0	$\pm\sqrt{9 - 0} = \pm\sqrt{9}$	± 3
± 1	$\pm\sqrt{9 - 1} = \pm\sqrt{8}$	± 2.8
± 2	$\pm\sqrt{9 - 4} = \pm\sqrt{5}$	± 2.2
± 3	$\pm\sqrt{9 - 9} = \pm\sqrt{0}$	0

(4) Plot the points. Join them with a smooth curve to obtain a circle whose center is at the origin and whose radius measures 3 units.

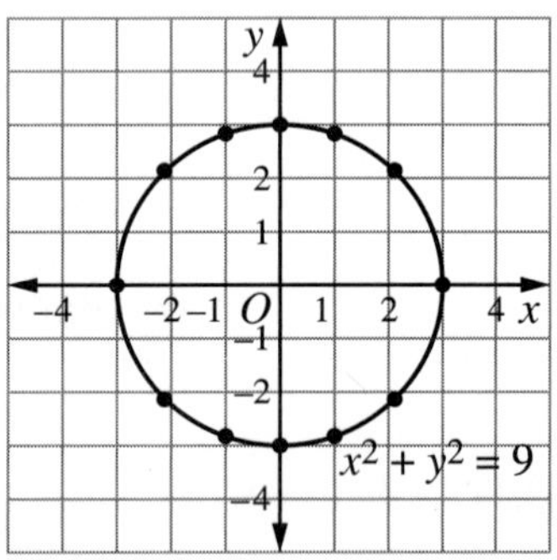

- **In general, when $a = b$ and a, b, c have the same sign, the graph of the equation $ax^2 + by^2 = c$ is a *circle*.**

When $a = b \neq 1$, the equation can be reduced. For example, $2x^2 + 2y^2 = 32$ is equivalent to $x^2 + y^2 = 16$. Following the procedure for obtaining a graph, we would find that this graph is a circle whose center is at the origin and whose radius measures 4 units. Thus, we say:

- **The graph of the equation $x^2 + y^2 = r^2$ is a circle whose center is at the origin and whose radius measures r units.**

Note that, for this graph of a circle, the origin is a point of symmetry and the axes are lines of symmetry. The circle has an infinite number of lines of symmetry. A vertical line passes through two points on the circle. Therefore, a circle does *not* represent a function.

EXERCISES

In 1-3, in each case, write the coordinates of the y-intercepts and x-intercepts.

1. $5x^2 + 5y^2 = 45$ **2.** $x^2 + y^2 = 7$ **3.** $4x^2 + 4y^2 = 36$

In 4-6, graph each equation.

4. $x^2 + y^2 = 25$ **5.** $3x^2 + 3y^2 = 27$ **6.** $2x^2 + 2y^2 = 18$

In 7-10, in each case, write an equation of the circle with the given length of the radius and the given point as the center.

7. $r = 2$, (0, 0) **8.** $r = 5$, (0, 0) **9.** $r = 1$, (0, 0) **10.** $r = \sqrt{5}$, (0, 0)

In 11 and 12, select the numeral preceding the choice that best answers the question.

11. Which equation has a circle as its graph?
(1) $3x^2 = 15 + 3y^2$ (2) $3x^2 = 15 - 3y^2$ (3) $3x^2 = 15 - y^2$ (4) $3x^2 = 15 + y^2$

12. Which equation is *not* the graph of a circle?
(1) $x^2 + y^2 = 10$ (2) $2x^2 + 2y^2 = 20$ (3) $-3x^2 - 3y^2 = -30$ (4) $2x^2 + 3y^2 = 6$

13. Which is an equation of the circle shown in the accompanying diagram?
(1) $2x^2 + 2y^2 = 4$ (2) $x^2 + y^2 = 4$
(3) $x^2 - y^2 = 2$ (4) $2x^2 - 2y^2 = 4$

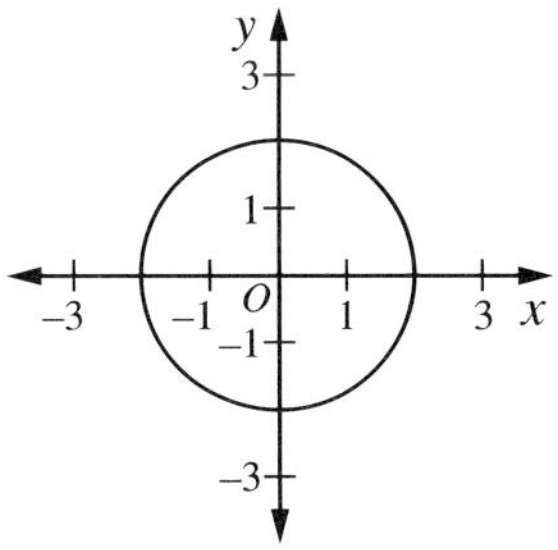

21-10 GRAPHIC SOLUTION OF A QUADRATIC-LINEAR SYSTEM

In Chapter 17, we learned how to solve a *system of linear equations* by graphing. For example, given the system of linear equations $\begin{Bmatrix} y = -\frac{1}{2}x + 4 \\ y = 2x - 1 \end{Bmatrix}$, we graph the equations on the same set of axes, as shown. Since the lines intersect at (2, 3), the values $x = 2$ and $y = 3$ will satisfy both equations of this system.

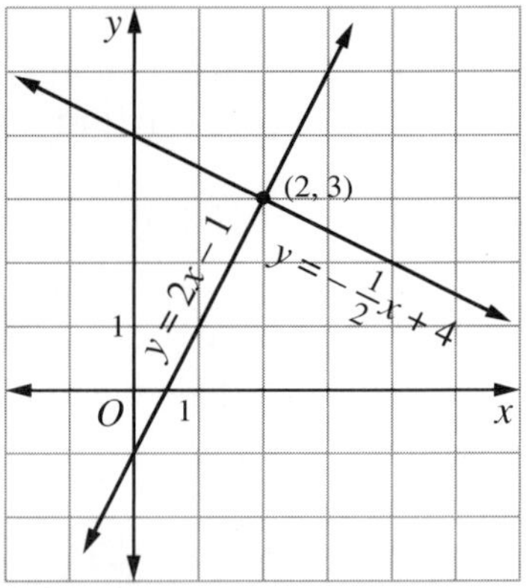

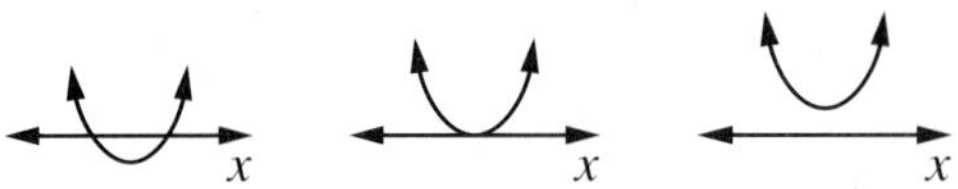

A ***quadratic-linear system*** consists of a quadratic equation and a linear equation. We have studied graphs of such systems in preceding sections of this chapter. For example, we learned that a parabola (the quadratic equation) may intersect the x-axis (the linear equation) at two, one, or no points.

The graphs of quadratic equations may include parabolas and circles. As shown in the examples that follow, *any line* may be used, not just the x-axis.

EXAMPLES

1. a. Draw the graph of the equation $y = x^2 - 6x + 6$.

b. On the same axes, draw the graph of the equation $y = x - 4$.

c. Determine the coordinates of the points of intersection of the graphs in parts **a** and **b**, and check these coordinates in both equations.

Solutions **a.** The equation $y = x^2 - 6x + 6$ is a parabola. An equation of the axis of symmetry is $x = \frac{-b}{2a}$.

Thus, for $y = x^2 - 6x + 6$, the axis of symmetry is $x = \frac{-(-6)}{2(1)}$, or $x = 3$.

Choose some values of x smaller than 3 and some values of x larger than 3 to develop a table of values.

x	$x^2 - 6x + 6$	y
0	$(0)^2 - 6(0) + 6$	6
1	$(1)^2 - 6(1) + 6$	1
2	$(2)^2 - 6(2) + 6$	−2
3	$(3)^2 - 6(3) + 6$	−3
4	$(4)^2 - 6(4) + 6$	−2
5	$(5)^2 - 6(5) + 6$	1
6	$(6)^2 - 6(6) + 6$	6

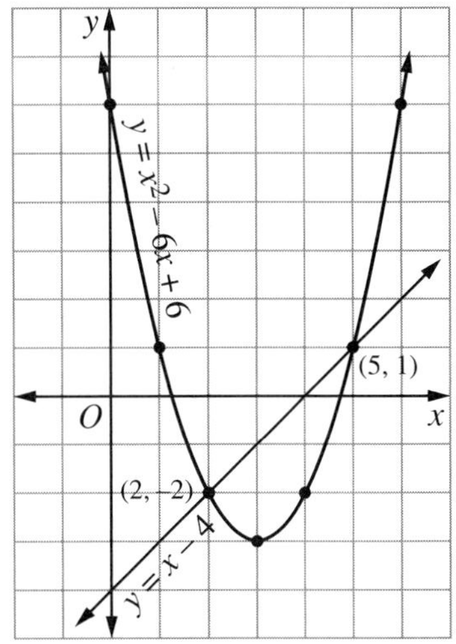

Plot the points.
Draw and label the parabola.

b. The linear equation $y = x - 4$ may be graphed by various methods.

METHOD 1: Table of Values

x	$x - 4$	y
0	$0 - 4$	−4
1	$1 - 4$	−3
2	$2 - 4$	−2

Develop a table of values to find the coordinates of three points. Plot the points. Draw and label the straight line, extended to cross the parabola at all possible points of intersection.

METHOD 2: Slope-Intercept

$$\text{In } y = \frac{1}{1}x - 4 \begin{cases} \text{slope} = \dfrac{\Delta y}{\Delta x} = \dfrac{1}{1} \\ y\text{-intercept} = -4 \end{cases}$$

Plot point $(0,-4)$ on the y-axis. From $(0,-4)$, use the slope to plot other points, showing that y increases by 1 as x increases by 1. Draw and label the straight line, extended to cross the parabola at two points.

The graph of equation $y = x - 4$ is shown on the graph for part **a**.

c. The graphs from parts **a** and **b** intersect at two points:

$$(x, y) = (2, -2) \qquad (x, y) = (5, 1)$$

Each ordered pair must be checked in each of the two original equations. Thus, four checks are needed.

Check for x = 2 and y = −2	
$y = x^2 - 6x + 6$	$y = x - 4$
$(-2) \stackrel{?}{=} (2)^2 - 6(2) + 6$	$(-2) \stackrel{?}{=} (2) - 4$
$-2 \stackrel{?}{=} 4 - 12 + 6$	$-2 = -2$ (True)
$-2 = -2$ (True)	

Check for x = 5 and y = 1	
$y = x^2 - 6x + 6$	$y = x - 4$
$(1) \stackrel{?}{=} (5)^2 - 6(5) + 6$	$(1) \stackrel{?}{=} (5) - 4$
$1 \stackrel{?}{=} 25 - 30 + 6$	$1 = 1$ (True)
$1 = 1$ (True)	

Answers: **a** and **b.** See the graph.
c. $\{(2, -2), (5, 1)\}$

---

2. Solve the system of equations graphically: $x^2 + y^2 = 9$
$y = 3 - x$

Solution (1) The graph of $x^2 + y^2 = 9$ is a circle whose center is at the origin with a radius of 3. Mark the x- and y-intercepts and graph the circle or use a compass.

(2) The graph of $y = 3 - x$ is a line. Select values of x and develop a table of values.

x	$3 - x$	y
0	$3 - 0$	3
3	$3 - 3$	0
4	$3 - 4$	-1

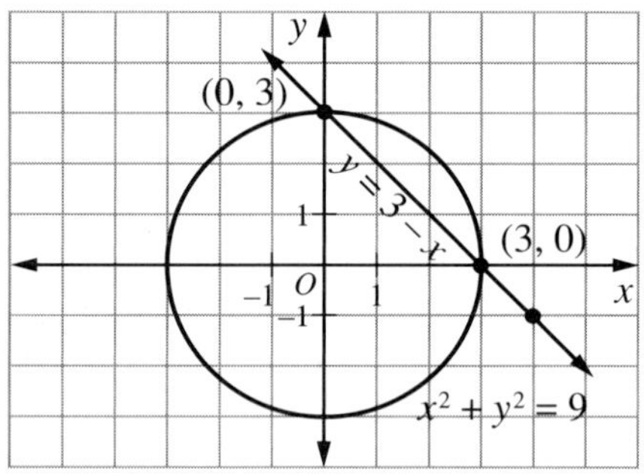

Plot the points and draw the line.

(3) The graphs intersect at (0, 3) and (3, 0). Each ordered pair is a solution of the system. The check is left to the student.

Answer: $\{(0, 3), (3, 0)\}$.

EXERCISES WITH OPEN-RESPONSE PROBLEMS

1. Name the maximum number of points of intersection for:
a. two lines **b.** a line and a parabola **c.** a line and a circle

2. What does it mean if the two graphs of a system do not intersect?

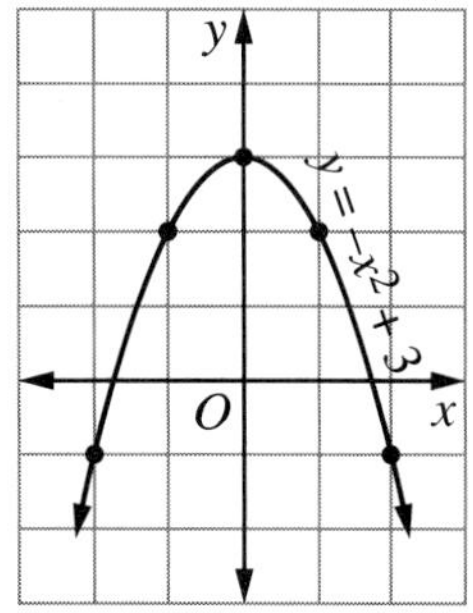

In 3–8, the graph of $y = -x^2 + 3$ is shown. Find the solution set for each system of equations by inspecting the graph and naming the coordinates of the points of intersection.

3. $y = -x^2 + 3$
$y = 2$

4. $y = -x^2 + 3$
$y = -1$

5. $y = -x^2 + 3$
$y = 3$

6. $y = -x^2 + 3$
$y = 5$

7. $y = -x^2 + 3$
$x = 1$

8. $y = -x^2 + 3$
$x = -2$

9. a. Draw the graph of $y = x^2 - 4x + 5$, using integral values for $0 \leq x \leq 5$.
b. On the same set of axes, draw the graph of $y = -x + 5$.
c. Determine the coordinates of the points of intersection of the graphs in parts **a** and **b**, and check these coordinates in both equations.

In 10–18, solve each quadratic-linear system graphically, showing the graphs of both equations on the same axes for the interval $-2 \leq x \leq 2$. Check the solutions in both equations.

10. $y = x^2$
$y = x + 2$

11. $y = -x^2 + 4$
$y = x + 2$

12. $y = x^2 - 2$
$y = 2x - 2$

13. $y = 2 - x^2$
$y = x + 3$

14. $y = 2x^2$
$y = -2x + 4$

15. $y = 2x^2 - 4$
$y = 2x$

16. $y = x^2 - 3$
$y = -x - 1$

17. $y = 6 - 2x^2$
$2x + y = 2$

18. $y = \frac{1}{2}x^2 + 1$
$x + y = 1$

In 19–27, in each case: **a.** Graph the given quadratic equation over an appropriate interval.
b. On the same set of axes, graph the linear equation.
c. Determine the coordinates of the points of intersection of the graphs in parts **a** and **b**, and check these coordinates in both equations.

19. $y = x^2 - 2x - 4$
$y = x$

20. $y = x^2 + 2x + 1$
$y = 2x + 5$

21. $y = 4x - x^2$
$y = x - 4$

22. $y = x^2 - 8x + 15$
$y = -x + 5$

23. $y = -x^2 + 6x - 5$
$y = 3$

24. $y = x^2 + 4x + 1$
$y = 2x + 1$

25. $y = x^2 + x - 4$
$y = 2x - 2$

26. $y = x^2 - 5x + 6$
$y = -2x + 6$

27. $y = -x^2 + 3x + 2$
$y = x - 1$

In 28–33, solve each system of equations graphically, and check.

28. $x^2 + y^2 = 25$
$y = x - 1$

29. $x^2 + y^2 = 16$
$x - y = 4$

30. $x^2 + y^2 = 4$
$x + y = 2$

31. $x^2 + y^2 = 9$
$y = 3$

32. $x^2 + y^2 = 36$
$x = -6$

33. $2x^2 + 2y^2 = 50$
$x = 6$

21-11 ALGEBRAIC SOLUTION OF A QUADRATIC-LINEAR SYSTEM

If the graphs of a system of equations do not meet at integer values, it is not possible to read the exact values for x and y from the graph itself. To find these values for x and y, we must use an algebraic solution.

Systems of Linear Equations

In Chapter 17, we learned two algebraic methods to solve systems of linear equations. Let us review these methods with linear systems before applying them to quadratic-linear systems.

As seen in the graph at the right, at the point of intersection of the two lines, x is some value between 0 and 1, and y appears to equal 2. Let us solve this system of equations algebraically.

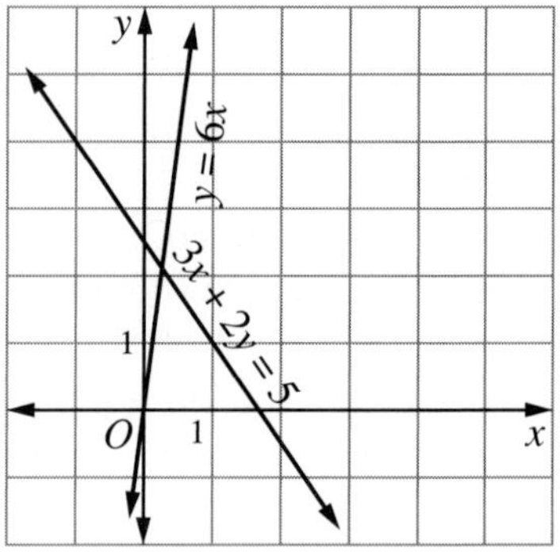

METHOD 1: Using Substitution

Solve the system of equations and check: $3x + 2y = 5$
$y = 6x$

How to Proceed:	*Solution:*
	$3x + 2y = 5$
(1) Since y and $6x$ are equal values in the second equation, replace y by $6x$ in the first equation:	$3x + 2(6x) = 5$
(2) Solve the resulting equation for x:	$3x + 12x = 5$
	$15x = 5$
	$x = \frac{5}{15}$ or $\frac{1}{3}$
(3) Replace x with its value in any equation containing both variables, and solve for y:	$y = 6x$
	$= 6\left(\frac{1}{3}\right)$
	$= 2$

(4) *Check* the values of x and y in *both original* equations:

$3x + 2y = 5$	$y = 6x$
$3\left(\frac{1}{3}\right) + 2(2) \stackrel{?}{=} 5$	$(2) \stackrel{?}{=} 6\left(\frac{1}{3}\right)$
$1 + 4 \stackrel{?}{=} 5$	$2 = 2$ (True)
$5 = 5$ (True)	

Answer: $x = \frac{1}{3}$ and $y = 2$, or $(x, y) = \left(\frac{1}{3}, 2\right)$

METHOD 2: Using Addition

Solve the system of equations and check: $3x + 2y = 5$
$y = 6x$

How to Proceed:	*Solution:*
(1) Transform the equations to align corresponding terms. Here, variables are on one side and constants on the other:	$3x + 2y = 5 \rightarrow \quad 3x + 2y = 5$ $y = 6x \rightarrow -6x + \quad y = 0$
(2) Eliminate one variable. To eliminate x, multiply both sides of the first equation by 2:	$2(3x + 2y = 5) \rightarrow \quad 6x + 4y = 10$ $-6x + \quad y = 0 \rightarrow \underline{-6x + \quad y = 0}$
(3) Add like terms in both equations (equals added to equals are equal), and solve for y:	$5y = 10$ $y = 2$
(4) Replace y by its value in any equation containing both variables, and solve for x:	$y = 6x$ $2 = 6x$ $\frac{1}{3} = x$

(5) *Check* in both original equations (shown in Method 1).

Answer: $x = \frac{1}{3}$ and $y = 2$, or $(x, y) = \left(\frac{1}{3}, 2\right)$

Quadratic-Linear Systems

The graphs of $y = x^2 - 4x + 3$ and $y = \frac{1}{2}x + 1$ intersect at two points. One solution to this quadratic-linear system can be read from the graph, namely, $(x, y) = (4, 3)$. However, the second solution for this system does not occur at a point where x and y are integers. Although the graph shows that x is some value between 0 and 1, and y is some value between 1 and 2, we are not able to identify the exact values of x and y graphically.

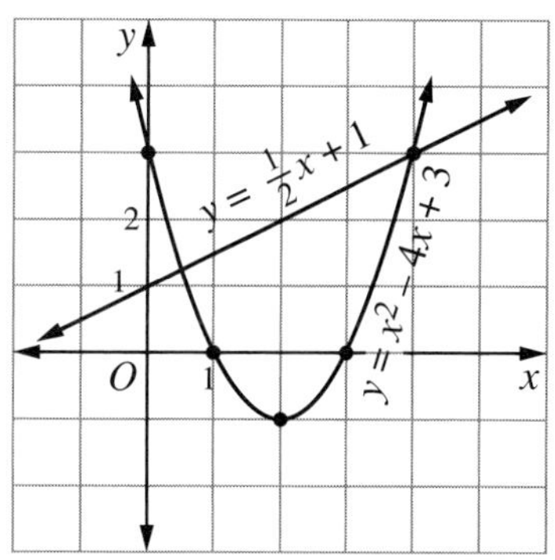

To find the exact values of the coordinates of both points of intersection, we use an algebraic technique. The approach generally used to solve a quadratic-linear system is the *substitution method*. The solution of the system graphed above is shown in Example 1.

EXAMPLES

1. Solve and check: $y = x^2 - 4x + 3$
$y = \frac{1}{2}x + 1$

How to Proceed:	*Solution:*
(1) Since y is expressed in terms of x in the linear equation, substitute the expression $\frac{1}{2}x + 1$ for y in the quadratic equation, forming one equation with one unknown:	$y = x^2 - 4x + 3$ $\frac{1}{2}x + 1 = x^2 - 4x + 3$
(2) To eliminate fractions as coefficients, multiply by the least common denominator, 2:	$2\left(\frac{1}{2}x + 1\right) = 2(x^2 - 4x + 3)$ $x + 2 = 2x^2 - 8x + 6$
(3) Transform the quadratic equation to standard form:	$0 = 2x^2 - 9x + 4$
(4) Solve the quadratic equation by any method. Here, factor to find two values for x:	$0 = (2x - 1)(x - 4)$ $2x - 1 = 0 \;\mid\; x - 4 = 0$ $2x = 1 \;\mid\; x = 4$ $x = \frac{1}{2}$
(5) Substitute each value of x in the linear equation of the system to find the corresponding value of y:	In $y = \frac{1}{2}x + 1$, substitute $\frac{1}{2}$ and 4 for x. If $x = \frac{1}{2}$, $y = \frac{1}{2}\left(\frac{1}{2}\right) + 1 = \frac{1}{4} + 1 = \frac{5}{4}$ If $x = 4$, $y = \frac{1}{2}(4) + 1 = 2 + 1 = 3$
(6) Write the solution as coordinates:	$(x, y) = \left(\frac{1}{2}, \frac{5}{4}\right) \;\mid\; (x, y) = (4, 3)$

(7) Check each ordered pair in each original equation for a total of four checks.

Check for $x = \frac{1}{2}$ *and* $y = \frac{5}{4}$

$y = x^2 - 4x + 3$	$y = \frac{1}{2}x + 1$
$\left(\frac{5}{4}\right) \stackrel{?}{=} \left(\frac{1}{2}\right)^2 - 4\left(\frac{1}{2}\right) + 3$	$\left(\frac{5}{4}\right) \stackrel{?}{=} \frac{1}{2}\left(\frac{1}{2}\right) + 1$
$\frac{5}{4} \stackrel{?}{=} \frac{1}{4} - 2 + 3$	$\frac{5}{4} \stackrel{?}{=} \frac{1}{4} + 1$
$\frac{5}{4} = \frac{5}{4}$ (True)	$\frac{5}{4} = \frac{5}{4}$ (True)

Check for $x = 4$ *and* $y = 3$

$y = x^2 - 4x + 3$	$y = \frac{1}{2}x + 1$
$(3) \stackrel{?}{=} (4)^2 - 4(4) + 3$	$(3) \stackrel{?}{=} \frac{1}{2}(4) + 1$
$3 \stackrel{?}{=} 16 - 16 + 3$	$3 \stackrel{?}{=} 2 + 1$
$3 = 3$ (True)	$3 = 3$ (True)

Answer: $\left\{\left(\frac{1}{2}, \frac{5}{4}\right), (4, 3)\right\}$

---

2. Solve and check: $x^2 + y^2 = 25$
$2x - y = 5$

How to Proceed:	*Solution:*
(1) In the linear equation, solve for one variable in terms of the other. Here, solve for y in terms of x:	$2x - y = 5$ $-y = -2x + 5$ $y = 2x - 5$
(2) Substitute this expression for y in the quadratic equation:	$x^2 + y^2 = 25$ $x^2 + (2x - 5)^2 = 25$
(3) Simplify. Then transform the resulting quadratic equation to standard form:	$x^2 + 4x^2 - 20x + 25 = 25$ $5x^2 - 20x + 25 = 25$ $5x^2 - 20x = 0$
(4) Solve the quadratic equation. Here, factor to find two values for x:	$5x(x - 4) = 0$ $5x = 0 \quad \vert \quad x - 4 = 0$ $x = 0 \quad \vert \quad x = 4$
(5) Substitute each value of x in $y = 2x - 5$, a form of the linear equation of the system, to find the corresponding value of y:	In $y = 2x - 5$, substitute 0 and 4 for x. If $x = 0$, $y = 2(0) - 5 = 0 - 5 = -5$ If $x = 4$, $y = 2(4) - 5 = 8 - 5 = 3$
(6) Write the solution as coordinates:	$(x, y) = (0, -5)$ $(x, y) = (4, 3)$

(7) Four checks are required. These are left to the student.

Answer: $\{(0, -5), (4, 3)\}$

EXERCISES WITH OPEN-RESPONSE PROBLEMS

In 1–6, solve each system of equations by using an algebraic method that seems convenient. Check.

1. $y = 3x$
$y - x = 18$

2. $3a - b = 13$
$2a + 3b = 16$

3. $-2c = d$
$6c + 5d = -12$

4. $a - \frac{2}{3}b = 4$
$\frac{3}{5}a + b = 15$

5. $3(y - 6) = 2x$
$3x + 5y = 11$

6. $x + y = 300$
$0.25x + 0.75y = 195$

In 7–36, solve each system of equations algebraically, and check all solutions.

7. $y = x^2 - 2x$
$y = x$

8. $y = x^2 + 5$
$y = x + 5$

9. $y = x^2 - 4x + 3$
$y = x - 1$

10. $y = x^2 + 2x + 1$
$y = x + 3$

11. $x^2 + y = 9$
$y = x + 9$

12. $x^2 + y = 2$
$y = -x$

13. $x^2 + 2y = 5$
$y = x + 1$

14. $x^2 - y = 5$
$y = 3x - 1$

15. $y = x^2 - 4x + 9$
$y - 1 = 2x$

16. $x^2 - 2y = 11$
$y = x - 4$

17. $y = x^2 - x$
$x + y = 4$

18. $y = 2x^2 + 2$
$y = 5x$

19. $y = 2x^2 - 6x + 5$
$y = x + 2$

20. $y = 2x^2 + 2x + 3$
$y - x = 3$

21. $y = 3x^2 - 8x + 5$
$x + y = 3$

22. $y = x^2 + 4x$
$y = \frac{1}{2}x + 2$

23. $y = x^2 + 3x + 1$
$y = \frac{1}{3}x + 2$

24. $y = x^2 - 6x + 8$
$y = -\frac{1}{2}x + 2$

25. $y = x^2 - 4x + 4$
$3y = x$

26. $y = x^2 - 2x - 1$
$x + 2y = 0$

27. $y = x^2 + 4x + 5$
$2x + 3y = 0$

28. $xy = 6$
$y = x - 1$

29. $xy = 9$
$x = y$

30. $xy = -1$
$x + y = 0$

31. $x^2 + y^2 = 25$
$x = 2y - 5$

32. $x^2 + y^2 = 100$
$y = x + 2$

33. $x^2 + y^2 = 50$
$x = y$

34. $x^2 + y^2 = 40$
$y = 3x$

35. $x^2 + y^2 = 26$
$x = 5y$

36. $x^2 + y^2 = 2$
$y = x - 2$

37. The graphs of $y = x^2 - x$ and $y = x - 1$ are shown at the right.

a. Name the coordinates of the point of intersection.

b. Solve the system algebraically:
$$y = x^2 - x$$
$$y = x - 1$$

c. Explain why the solution set in part **b** consists of only one ordered pair of values for x and y.

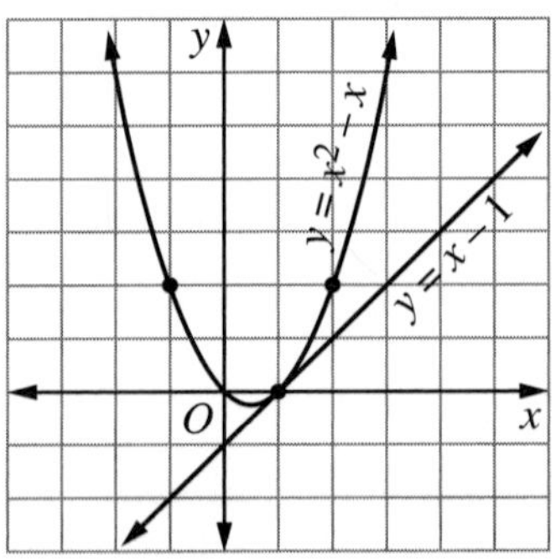

38. The graphs of $y = x^2 - x$ and $y = x - 2$ are shown at the right. Explain why the solution set for the indicated system of equations is the empty set.

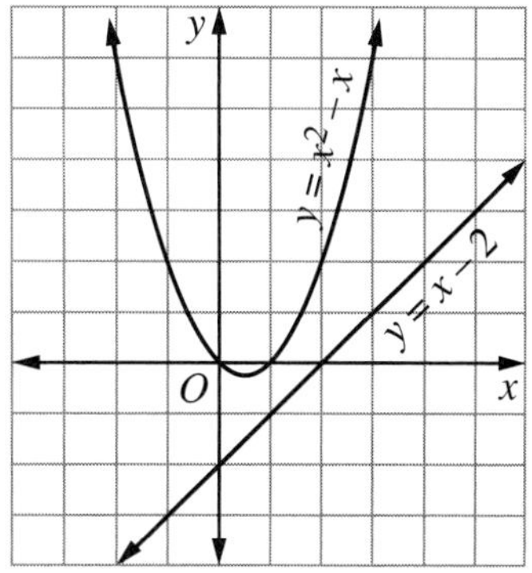

39. The doorway to a store is shaped in the form of an arch whose equation is $h = -\frac{3}{4}x^2 + 6x$, where x represents a horizontal distance, in feet, from a given point, and h is the height, in feet, of the door at that point. The coordinates (0, 0) are assigned to one point at the base of the doorway.

a. How wide is the doorway at its base?

b. What is the maximum height of the doorway?

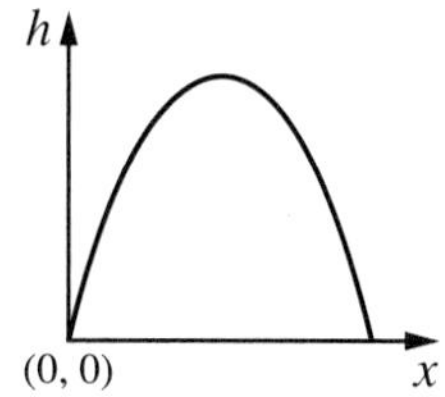

40. A batter hit a baseball with an initial velocity of 80 feet per second. The ball was 3 feet off the ground when it was hit. Let t represent the time, in seconds, and h represent the height, in feet, of the ball above the ground. For a limited time period, this height can be determined by the equation $h = -16t^2 + 80t + 3$.

a. Make a table of integral values of t from 0 to 5 to find values of h.

b. Graph the equation for this interval. In the set of axes, let 1 horizontal unit = 0.25 second, and let 1 vertical unit = 10 feet.

c. If the ball was caught at a height of 3 feet, for how long did it travel before being caught?

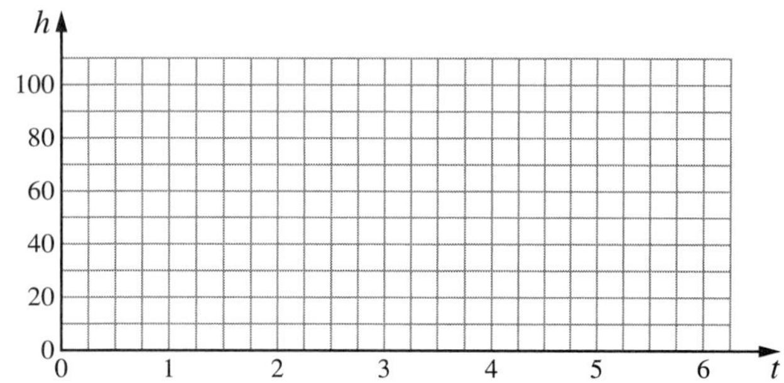

d. Find the maximum height reached by the ball.

e. How long did it take for the ball to reach its maximum height?

f. Find the times, to the *nearest tenth* of a second, when the ball was at a height of 50 feet. (*Hint:* Solve a quadratic-linear system in which the linear equation is $h = 50$.)

CHAPTER SUMMARY

The standard form of a ***quadratic equation*** in one variable is

$$ax^2 + bx + c = 0$$

where a, b, and c are real numbers and $a \neq 0$. A quadratic equation with rational roots can be solved by factoring the quadratic expression over the set of integers and setting each factor equal to 0.

An ***incomplete quadratic equation*** is one in which b is 0. An incomplete quadratic equation can be solved by writing an equivalent equation of the form $x^2 = n$, where n is a nonnegative real number, and taking the square root of each side of the equation.

In any right triangle, the square of the length of the hypotenuse is equal to the sum of the squares of the lengths of the other two sides. This relationship is called the ***Theorem of Pythagoras***. When c is the length of the longest side of a triangle and a and b are the lengths of the other two sides, the triangle is a right triangle if and only if $c^2 = a^2 + b^2$.

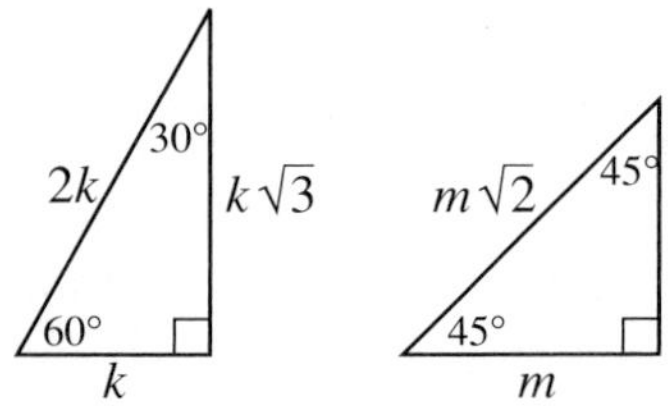

The lengths of the sides of a ***30–60-degree right triangle*** are in the ratio $1:\sqrt{3}:2$. The lengths of the sides of an ***isosceles right triangle*** (a 45–45-degree right triangle) are in the ratio $1:1:\sqrt{2}$.

A ***relation*** is any set of ordered pairs. The ***domain*** of the relation is the set of all first elements of the ordered pairs, and the ***range*** is the set of all second elements of these pairs.

A ***function*** is a relation in which each element of the domain corresponds to one and only one element in the range, or a relation in which no two ordered pairs have the same first element.

The graph of a quadratic function of the form

$$y = ax^2 + bx + c$$

where a, b, and c are real numbers and $a \neq 0$, is a ***parabola***.

Every second-degree equation of the form $y = ax^2 + bx + c$, where $a \neq 0$, is a function whose domain is the set of real numbers.

The graph of the second-degree equation $ax^2 + by^2 = c$ is a ***circle*** when a, b, and c have the same signs and $a = b$.

A ***system of equations*** consists of two or more equations. The simultaneous solution of a system of equations may be found by graphing the equations on the same set of axes or by using an algebraic method.

A system may consist of linear equations only or of both quadratic and linear equations. When the system consists of a parabola ($y = ax^2 + bx + c$) and the equation of the x-axis ($y = 0$), the x-coordinates of the points at which the parabola intersects the x-axis are the roots of the quadratic equation $ax^2 + bx + c = 0$.

VOCABULARY

21-1 Polynomial equation of degree two Quadratic equation Standard form of a quadratic equation

21-2 Double root

21-3 Incomplete quadratic equation Pure quadratic equation

21-4 Pythagorean Theorem Pythagorean triple

21-5 30–60-degree right triangle 45–45-degree right triangle Isosceles right triangle

21-7 Relation Function Vertical-line test for a function

21-8 Second-degree polynomial function Parabola Axis of symmetry Minimum point Turning point Maximum point

21-10 Quadratic-linear system

REVIEW EXERCISES WITH OPEN-RESPONSE PROBLEMS

In 1–6, solve each equation and check all roots.

1. $x^2 - 11x + 18 = 0$ **2.** $y^2 - 36 = 0$ **3.** $x^2 - 2x = 15$

4. $2k^2 + 5k + 3 = 3$ **5.** $w(w + 7) = 18$ **6.** $\frac{t-1}{3} = \frac{4}{t}$

7. What is the positive root of $m^2 = 16 - 6m$?

8. Solve for x in simplest radical form: $2x^2 - 36 = 0$.

In 9–11, in each case a and b represent the lengths of the legs of a right triangle, and c represents the length of the hypotenuse. Find the missing length.

9. $a = 5, b = 12$ **10.** $c = 41, b = 40$ **11.** $a = 2\sqrt{5}, b = 4$

12. The legs of a right triangle measure 2 and 6. Find, in radical form, the length of the hypotenuse of the triangle.

13. The diagonal path in a rectangular garden measures 34 meters. If the width of the garden is 16 meters, find the number of meters in the length of the garden.

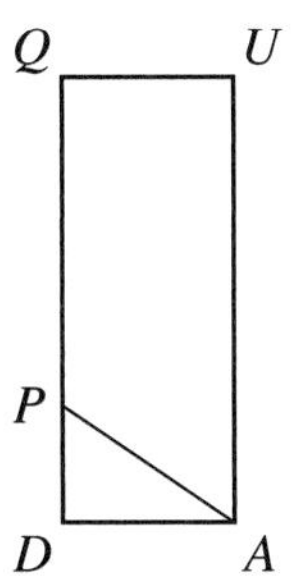

14. In rectangle $QUAD$, P is a point on $\overline{QD}$, $PD = 3$, $PA = 5$, and QP equals the sum of PD and PA.
a. Find QP. **b.** Find AD. **c.** Find AU.
d. Find the area of $\triangle PAD$.
e. Find the area of trapezoid $QUAP$.

In 15–21, select the numeral preceding the choice that best completes the statement or answers the question.

15. The diagonals of a rhombus are 18 and 24. The length of a side of the rhombus is (1) 15 (2) $15\sqrt{2}$ (3) $15\sqrt{3}$ (4) 30

16. The length of the altitude of an equilateral triangle whose side has a length of 10 is (1) $5\sqrt{2}$ (2) $5\sqrt{3}$ (3) $10\sqrt{2}$ (4) $10\sqrt{3}$

17. The length of the diagonal of a square is $8\sqrt{2}$. The length of one side of the square is (1) 16 (2) 8 (3) $16\sqrt{2}$ (4) 4

18. Which is an equation of the accompanying graph?
(1) $y = -x^2 + 2x + 3$ (2) $y = -x^2 - 2x + 3$
(3) $y = x^2 + 2x + 3$ (4) $y = x^2 - 2x + 3$

19. The coordinates of the turning point in the graph at the right are
(1) (1, 4) (2) (−1, 4)
(3) (4, −1) (4) (−3, 1)

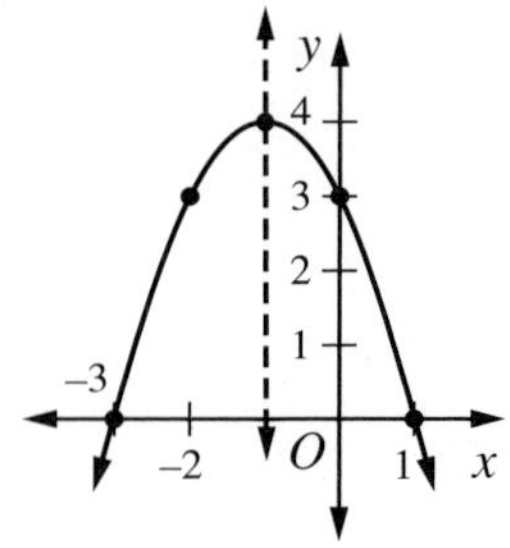

Ex. 18–19

20. An equation for the axis of symmetry of the graph of the parabola $y = 2x^2 - 8x + 10$ is
(1) $x = 2$ (2) $x = -2$ (3) $x = 4$ (4) $y = 4$

21. Which equation is a function?
(1) $x^2 + y^2 = 36$ (2) $x^2 + 4y^2 = 36$ (3) $x^2 - y^2 = 36$ (4) $x^2 + y = 36$

22. The square of a positive number is 6 more than 5 times the number. Find the number.

23. A rectangle has an area of 70. If the sides of the rectangle are represented by x and $3x - 1$, find the lengths of these sides.

24. The sum of the squares of two consecutive positive integers is 85. Find the integers.

25. Graph the quadratic equation $y = x^2 + 2x - 3$, using integral values of x from $x = -4$ to $x = 2$, inclusive.

26. **a.** Draw a graph of $y = -x^2 + 4x - 1$ for an appropriate interval.
b. On the same set of axes, draw the graph of $x + y = 3$.
c. Determine the coordinates of the points of intersection of the graphs in parts **a** and **b**, and check.

27. At an apple orchard, people can pick their own apples. They are given a choice of paying either \$0.50 per pound for the apples or a single price of \$6.00 for all the apples that can fit in a bag.
a. Let x = number of pounds of apples, and y = cost of the apples. Write two equations: one for the cost of apples by the pound and one for the bag price.
b. Graph the two equations on one set of coordinate axes.
c. When is the per pound price the better bargain? When is the bag price the better bargain? Explain your answer.

28. The relation {(5, 6), (7, 8), (x, 9), (10, x)} is a function. Name all possible values that x *cannot* equal.

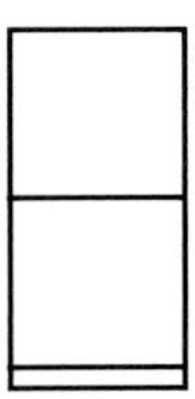

29. Two square pieces are cut from a rectangular piece of carpet as shown in the diagram. The area of the original piece is 136 square feet, and the width of the small rectangle that is left is 1 foot. Find the dimensions of the original piece of carpet.

30. **a.** On graph paper, draw $\triangle ABC$ with vertices $A(-1,-3)$, $B(4, 9)$, and $C(4,-3)$.
b. What kind of triangle is $\triangle ABC$?
c. Find AC and BC.
d. Find AB, using your answers to part **c**.

CUMULATIVE REVIEW WITH OPEN-RESPONSE PROBLEMS

1. Let b and c be two different constants chosen from the set $\{2, 3, -4\}$.
a. How many different equations of the form $x^2 + bx + c = 0$ are possible?
b. How many of these equations can be solved by factoring over the set of integers?
c. What is the probability that, if b and c are chosen at random from the given set, the equation $x^2 + bx + c = 0$ can be solved by factoring?

2. Huy worked on an assignment for 4 days. Each day he worked half as long as he had worked the day before. The total time he spent on the assignment was $3\frac{3}{4}$ hours.
a. How long did Huy spend on the assignment each day?
b. Find the mean number of hours that Huy worked each day.

3. Evaluate $3x^0 + 5x^{-2} + 6x^2$ when $x = 5$.

4. What is the truth value of the statement "If x is an integer, then x is a rational number" when $x = -\frac{3}{4}$?

5. Write the sum $\frac{x+1}{2} + \frac{1-x}{6}$ as a fraction in lowest terms.

6. Find two consecutive integers the sum of whose squares is 421.

Exploration

Write the square of each integer from 2 to 20. Write the prime factorization of each of these squares. What do you observe about the prime factorization of these numbers?

If the prime factorization of n is $n = a^3 \bullet b^2 \bullet c^4$, is n a perfect square? Is $\sqrt{n}$ rational or irrational?

Chapter 22

Trigonometry of the Right Triangle

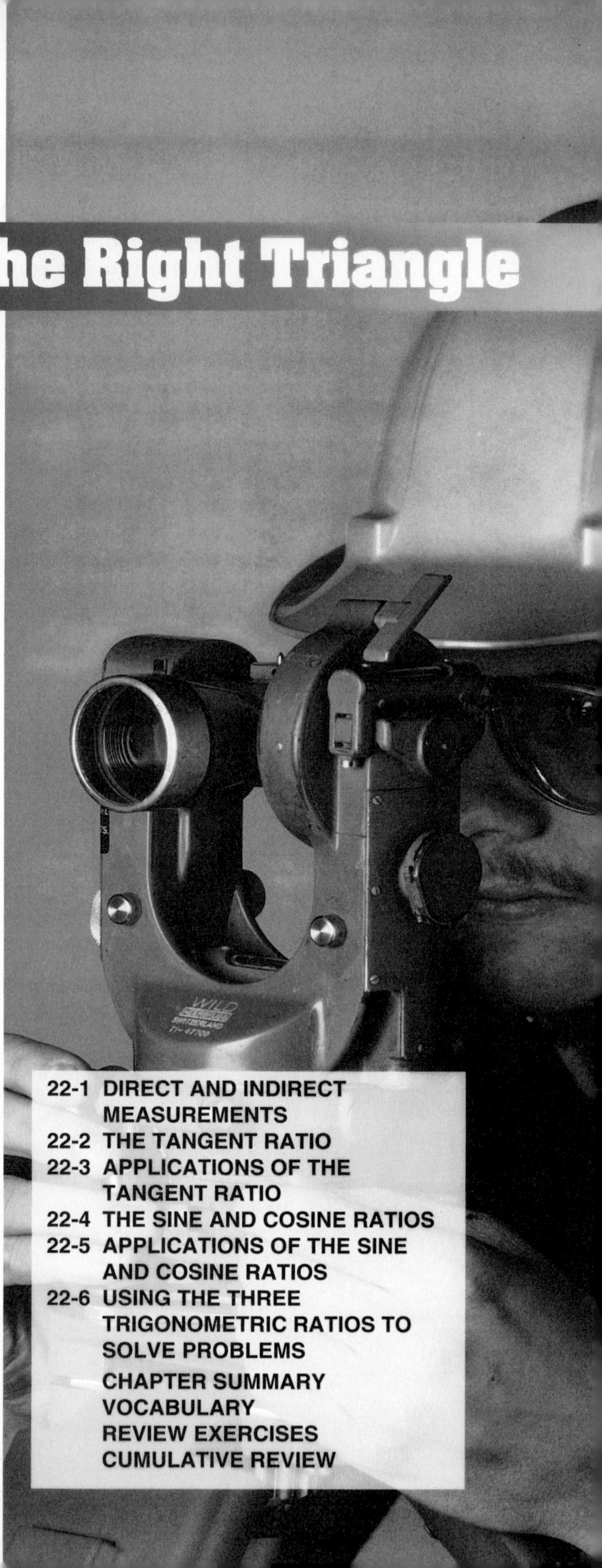

The accurate measurement of land has been a critical challenge throughout the history of civilization. Today's land measurement problems are not unlike those George Washington might have solved by using measurements made with a transit, but the modern surveyor has available a total work-station including EDM (electronic distance measuring) and a theodolite for angle measurement. Although modern instruments can perform many measurements and calculations, the surveyor needs to understand the principles of indirect measurement and trigonometry to correctly interpret and apply these results.

In this chapter, we will begin the study of a branch of mathematics called *trigonometry*. The word *trigonometry* is Greek in origin and means "*measurement of triangles*." Although the trigonometric functions have applications beyond the study of triangles, in this chapter we will limit the applications to the study of similar right triangles.

22-1 DIRECT AND INDIRECT MEASUREMENTS

The solution of many problems requires the measurement of line segments and angles. When we use a ruler or tape measure to determine the length of a segment, or a protractor to find the measure of an angle, we are taking a ***direct measurement*** of the segment or the angle.

In many situations, however, it is inconvenient or impossible to make a measurement directly. For example, it is difficult to make the direct measurements needed to answer the following questions:

What is the height of a 100-year-old oak tree?
What is the width of a river?
What is the distance to the sun?

We can answer these questions by using methods that involve ***indirect measurement***. Starting with some known lengths of segments or angle measures, we apply a formula or a mathematical relationship to *indirectly* find the measurement in question. In geometry, when we worked with similar triangles and the Pythagorean Theorem, we used indirect measurement.

One application of ***trigonometry*** relates the ratio of the lengths of two sides of a right triangle to an acute angle of the triangle. By combining a knowledge of trigonometry with that of similar triangles, we can measure line segments and angles indirectly. Engineers, surveyors, physicists, and astronomers frequently use these trigonometric methods in their work.

22-2 THE TANGENT RATIO

Naming Sides

In a right triangle, the *hypotenuse*, which is the longest side, is opposite the right angle. The other two sides in a right triangle are called the *legs*. However, in trigonometry of the right triangle, we call these legs the ***opposite side*** and the ***adjacent side*** to describe the relationship of these sides to one of the acute angles in the triangle.

Notice that $\triangle ABC$ is the same right triangle in both figures below, but the position names we apply to the legs change with respect to the angles.

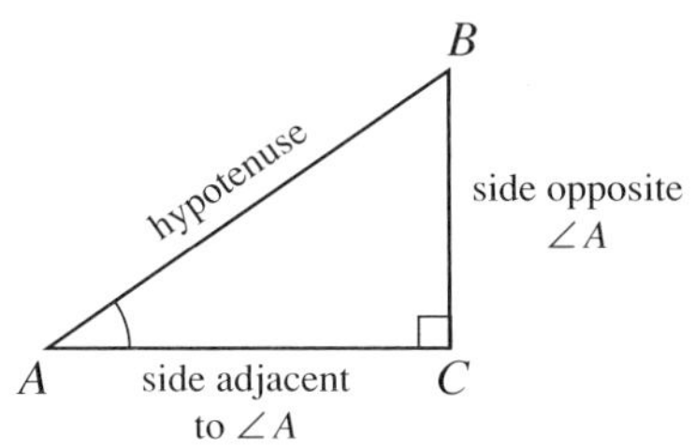

In $\triangle ABC$: $\overline{BC}$ is *opposite* $\angle A$;
$\overline{AC}$ is *adjacent* to $\angle A$.

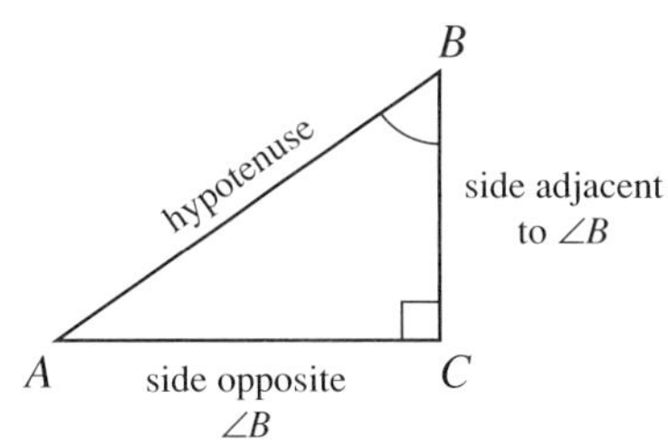

In $\triangle ABC$: $\overline{AC}$ is *opposite* $\angle B$;
$\overline{BC}$ is *adjacent* to $\angle B$.

Similar Triangles

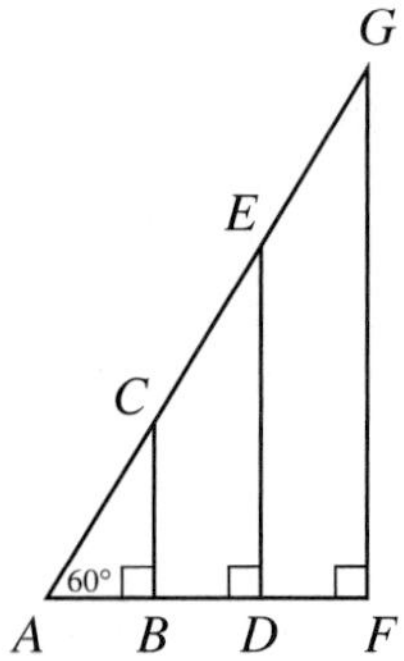

Three right triangles are drawn to coincide at vertex A. Since each triangle contains a right angle as well as $\angle A$, we know that all three triangles are *similar* by a.a. $\cong$ a.a.:

$$\triangle ABC \sim \triangle ADE \sim \triangle AFG$$

Also, since corresponding sides of similar triangles are in proportion, we can say:

$$\frac{CB}{BA} = \frac{ED}{DA} = \frac{GF}{FA}$$

Since the measure of $\angle A$ is 60°, each of these similar triangles is a 30–60-degree right triangle. In Chapter 21, we learned that, when the side opposite the 30° angle measures x, then the side opposite the 60° angle measures $x\sqrt{3}$. In the triangles shown here, the side opposite the 30° angle is the side *adjacent* to the 60° angle, or the side *adjacent* to $\angle A$.

The triangles seen in the preceding figure are now separated, and two lengths are indicated for each triangle.

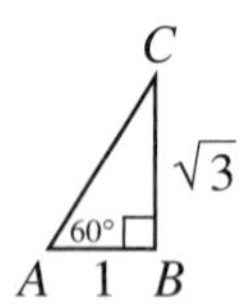

In $\triangle ABC$:
$\overline{CB}$ is opposite $\angle A$;
$\overline{BA}$ is adjacent to $\angle A$.

$$\frac{CB}{BA} = \frac{\sqrt{3}}{1}$$

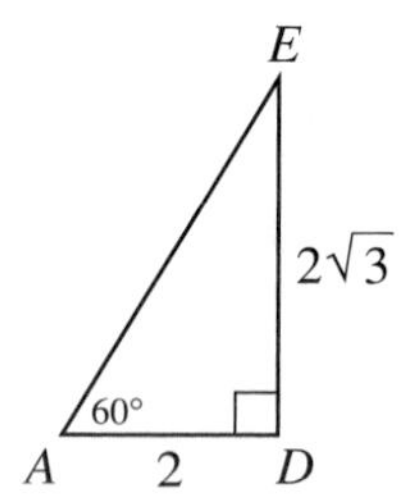

In $\triangle ADE$:
$\overline{ED}$ is opposite $\angle A$;
$\overline{DA}$ is adjacent to $\angle A$

$$\frac{ED}{DA} = \frac{2\sqrt{3}}{2} = \frac{\sqrt{3}}{1}$$

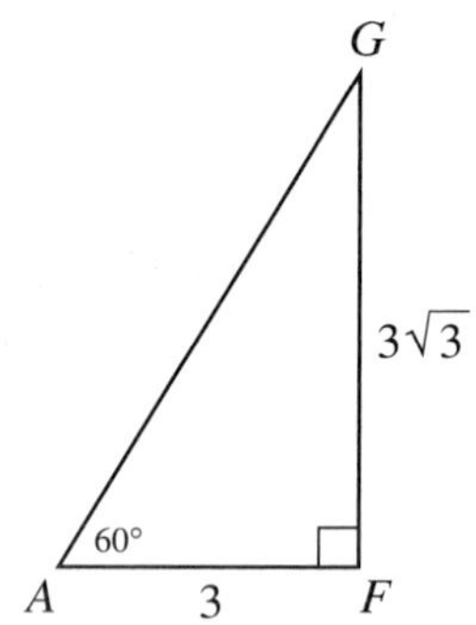

In $\triangle AFG$:
$\overline{GF}$ is opposite $\angle A$;
$\overline{FA}$ is adjacent to $\angle A$.

$$\frac{GF}{FA} = \frac{3\sqrt{3}}{3} = \frac{\sqrt{3}}{1}$$

In all three triangles, the angle referred to in naming the sides is a 60° angle. We see that the ratio of the lengths of the opposite and adjacent sides is $\frac{\sqrt{3}}{1}$, a constant value. Thus, the ratio is

$$\frac{\text{length of side } \textit{opposite} \text{ 60° angle}}{\text{length of side } \textit{adjacent} \text{ to 60° angle}} = \frac{\sqrt{3}}{1} = \sqrt{3}$$

This ratio will be constant for a 60° angle in any right triangle, no matter how large or how small the sides of the triangles are.

What we have shown to be true for right triangles with a 60° angle will also be true for a right triangle with any other acute angle. The ratio of the length of the opposite side to the length of the adjacent side is constant for all right triangles with congruent acute angles. This ratio is called the *tangent of the angle.*

- **Definition.** The ***tangent of an acute angle of a right triangle*** is the ratio of the length of the side opposite the acute angle to the length of the side adjacent to the acute angle.

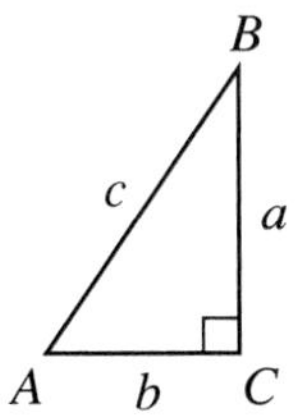

For right triangle ABC, with $m\angle C = 90$, the definition of the *tangent of* $\angle A$ is as follows:

$$\text{tangent } A = \frac{\text{length of side opposite } \angle A}{\text{length of side adjacent to } \angle A} = \frac{BC}{AC} = \frac{a}{b}$$

By using "tan A" as an abbreviation for tangent A, "opp" as an abbreviation for the length of the leg opposite $\angle A$, and "adj" as an abbreviation for the length of the leg adjacent to $\angle A$, we can shorten the way in which we write the relationship given above as follows:

$$\boxed{\tan A = \frac{\text{opp}}{\text{adj}} = \frac{BC}{AC} = \frac{a}{b}}$$

Finding Tangent Ratios on a Calculator

When the length of the side adjacent to the 60° angle in a 30–60-degree right triangle is x, the length of the side opposite the 60° angle is $x\sqrt{3}$, and we know that

$$\tan 60° = \frac{x\sqrt{3}}{x} = \sqrt{3}$$

But how can we find the constant value of the tangent ratio when the right triangle has an angle of 40° or 76°? Since we want to work with the value of this ratio for any right triangle, no matter what the measures of the acute angles may be, mathematicians have compiled tables of the tangent values for angles with measures from 0° to 90°. Also, a scientific calculator has the ability to display the value of this ratio for any angle. We will use a calculator to determine these values.

When a scientific calculator is turned on, it will usually be operating in degrees, as indicated by the letters DEG in the display. Most scientific calculators have a key labeled **DRG**. This key changes the system of angle measure used by the calculator. As **DRG** is pressed, the letters in the display will change from DEG to RAD to GRAD and then back to DEG. In this book we will always be working in degrees.

Case 1. Given an angle measure, find the tangent ratio.

We saw that tan 60° is equal to $\sqrt{3}$. The calculator will display this value as an approximate decimal. To find tan 60° on some calculators, 60 is entered first and then the key labeled **TAN** is pressed. On other calculators, 60 is entered after **TAN** is pressed.

Try the following sequences of keys to determine which is correct for your calculator. Be sure that the calculator is in degree mode before you start.

METHOD 1: ***Enter:*** 60 **TAN**

METHOD 2: ***Enter:*** **TAN** 60 **=**

Display: 1.7320508

The value given in the calculator display is the rational approximation of $\sqrt{3}$, the value of tan 60° that we found using ratios.

The algebraic form of this process is shown at the right. Given an angle measure, the tangent ratio is found (stated here to the *nearest ten-thousandth*).

$$\tan 60^\circ = x$$
$$\tan 60^\circ = 1.7321$$

Case 2. Given a tangent ratio, find the angle measure.

The value of the tangent ratio is different for each different angle measure from 0° to 90°. Therefore, if we know the value of the tangent ratio, we can find the measure of the acute angle that has this tangent ratio. The calculator key used to do this is labeled **TAN^{-1}** and is usually accessed by first pressing **2nd**, **INV**, or **SHIFT**. We can think of the meaning of $\tan^{-1}$ as "the angle whose tangent is." Therefore, if $\tan A = 0.9004$, then A is the angle whose tangent is 0.9004.

To find the measure of $\angle A$ from the calculator, we use one of the following sequences of keys.

METHOD 1: ***Enter:*** 0.9004 **2nd** **TAN^{-1}**

METHOD 2: ***Enter:*** **2nd** **TAN^{-1}** 0.9004 **=**

Display: 41.999872

The measure of $\angle A$ to the *nearest degree* is 42°.

The algebraic form of this process is shown at the right. Given a tangent ratio, the angle measure is found (stated here to the *nearest degree*).

$$\tan x = 0.9004$$
$$\tan 42^\circ = 0.9004$$

EXAMPLE

In $\triangle ABC$, $\angle C$ is a right angle, $BC = 3$, $AC = 4$, and $AB = 5$.

a. Find: *(1)* $\tan A$ *(2)* $\tan B$ *(3)* $m\angle A$ to the *nearest degree* *(4)* $m\angle B$ to the *nearest degree*

b. Show that the acute angles of the triangle are complementary.

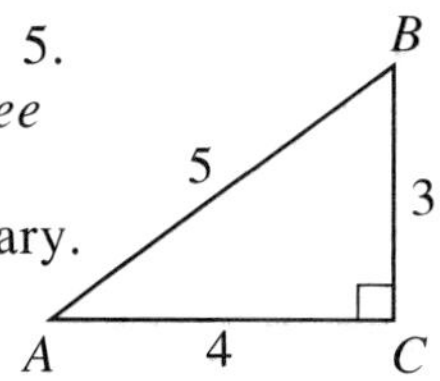

Solutions **a.** (*1*) $\tan A = \frac{\text{opp}}{\text{adj}} = \frac{BC}{AC} = \frac{3}{4}$ *Answer*

(*2*) $\tan B = \frac{\text{opp}}{\text{adj}} = \frac{AC}{BC} = \frac{4}{3}$ *Answer*

Use a calculator to find the measures of $\angle A$ and $\angle B$.

(*3*) ***Enter:*** (3 ÷ 4) 2nd TAN⁻¹

Display: 36.869898

To the nearest degree, $m\angle A = 37$ *Answer*

(*4*) ***Enter:*** (4 ÷ 3) 2nd TAN⁻¹

Display: 53.130102

To the nearest degree, $m\angle A = 53$ *Answer*

b. $m\angle A + m\angle B = 37 + 53 = 90$. Therefore, the acute angles of $\triangle ABC$ are complementary. *Answer*

EXERCISES WITH OPEN-RESPONSE PROBLEMS

In 1–4, find: **a.** $\tan A$ **b.** $\tan B$

1.

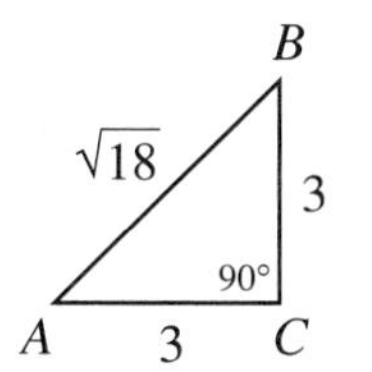

2.

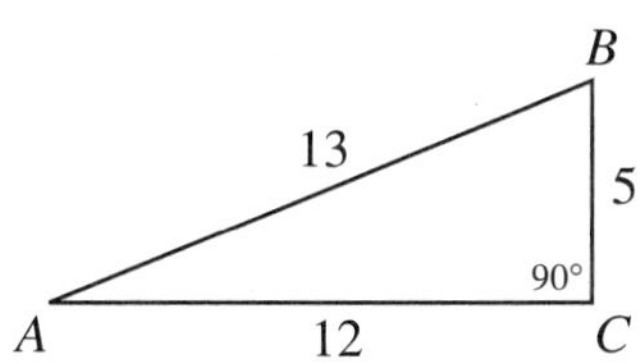

3.

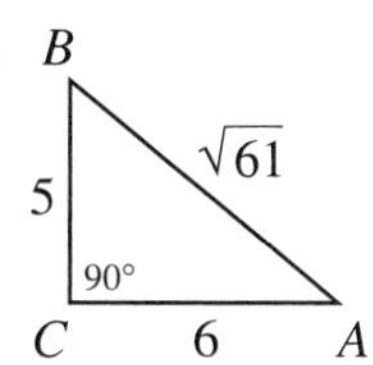

4.

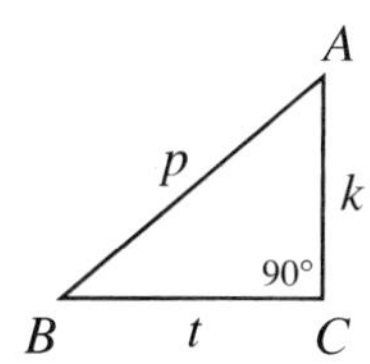

5. In $\triangle ABC$, $m\angle C = 90$, $AC = 6$, and $AB = 10$. Find $\tan A$.

6. In $\triangle RST$, $m\angle T = 90$, $RS = 13$, and $ST = 12$. Find $\tan S$.

In 7–14, use a calculator to find each of the following to the *nearest ten-thousandth*:

7. $\tan 10°$ **8.** $\tan 30°$ **9.** $\tan 70°$ **10.** $\tan 45°$

11. $\tan 1°$ **12.** $\tan 89°$ **13.** $\tan 36°$ **14.** $\tan 60°$

In 15–26, in each of the following, use a calculator to find the measure of $\angle A$ to the *nearest degree*:

15. $\tan A = 0.0875$	**16.** $\tan A = 0.3640$	**17.** $\tan A = 0.5543$	**18.** $\tan A = 1.0000$
19. $\tan A = 2.0503$	**20.** $\tan A = 3.0777$	**21.** $\tan A = 0.3754$	**22.** $\tan A = 0.7654$
23. $\tan A = 1.8000$	**24.** $\tan A = 0.3500$	**25.** $\tan A = 0.1450$	**26.** $\tan A = 2.9850$

27. Does the tangent of an angle increase or decrease as the degree measure of the angle varies from 1° to 89°?

28. **a.** Use a calculator to find tan 20° and tan 40°.

b. Is the tangent of the angle doubled when the measure of the angle is doubled?

29. In $\triangle ABC$, $m\angle C = 90$, $AC = 6$, and $BC = 6$.

a. Find $\tan A$. **b.** Find the measure of $\angle A$.

30. In $\triangle ABC$, $m\angle C = 90$, $BC = 4$, and $AC = 9$.

a. Find $\tan A$. **b.** Find the measure of $\angle A$ to the *nearest degree*.

c. Find $\tan B$. **d.** Find the measure of $\angle B$ to the *nearest degree*.

31. In $\triangle ABC$, $\angle C$ is a right angle, $m\angle A = 45$, $AC = 4$, $BC = 4$, and $AB = 4\sqrt{2}$.

a. Using the given lengths, write the ratio for $\tan A$.

b. Use a calculator to find tan 45°.

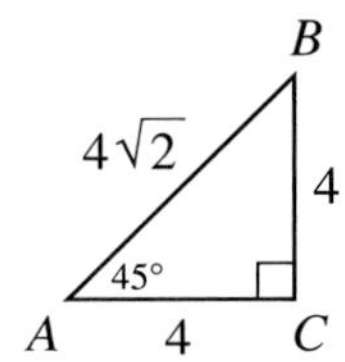

32. In $\triangle RST$, $\angle T$ is a right angle and r, s, and t are lengths of sides. Using these lengths:

a. Write the ratio for $\tan R$.

b. Write the ratio for $\tan S$.

c. Use parts **a** and **b** to find the numerical value of the product $(\tan R)(\tan S)$.

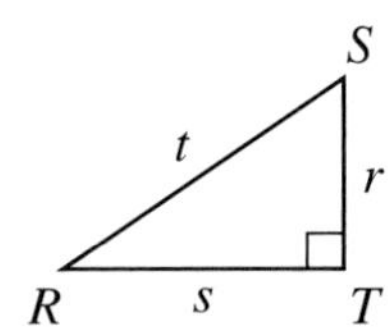

22-3 APPLICATIONS OF THE TANGENT RATIO

The tangent ratio is often used to make indirect measurements when it is possible to find the measures of a leg and an acute angle of a right triangle.

Angle of Elevation and Angle of Depression

When a telescope or some similar instrument is used to sight the top of a telephone pole, the instrument is *elevated* (tilted upward) from a horizontal position. Here, $\overleftrightarrow{OT}$ is the *line of sight* and $\overleftrightarrow{OA}$ is the horizontal line. The ***angle of elevation*** is the angle determined by the rays that are parts of the horizontal line and the line of sight when looking upward. Here, $\angle TOA$ is the *angle of elevation*.

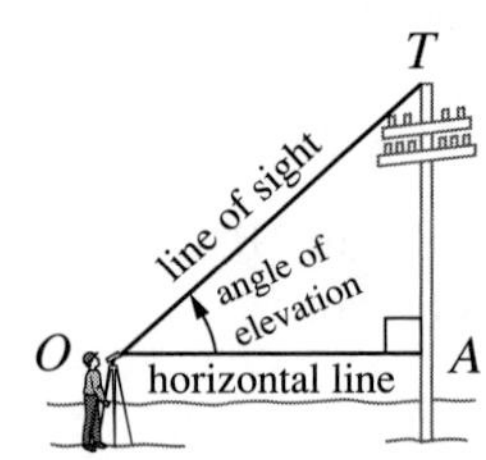

When an instrument is used to sight a boat from a cliff, the instrument is *depressed* (tilted downward) from a horizontal position. Here, $\overleftrightarrow{OB}$ is the *line of sight* and $\overleftrightarrow{OH}$ is the horizontal line. The ***angle of depression*** is the angle determined by the rays that are parts of the horizontal line and of the line of sight when looking downward. Here, $\angle HOB$ is the *angle of depression.*

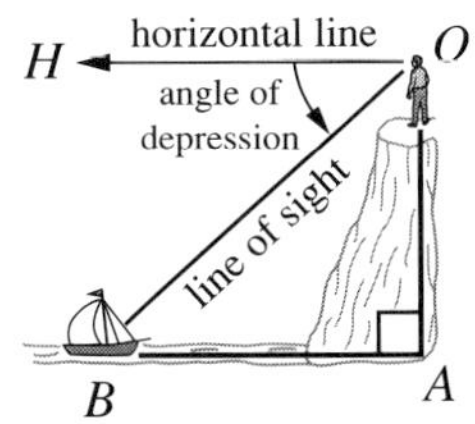

Note that, if $\overleftrightarrow{BA}$ is a horizontal line and $\overleftrightarrow{BO}$ is the line of sight from the boat to the top of the cliff, $\angle ABO$ is called the angle of elevation from the boat to the top of the cliff. Since $\overleftrightarrow{HO} \parallel \overleftrightarrow{BA}$ and $\overleftrightarrow{OB}$ is a transversal, alternate interior angles are congruent, namely, $\angle HOB \cong \angle ABO$. Thus, the angle of elevation measured from B to O is congruent to the angle of depression measured from O to B.

Using the Tangent Ratio to Solve Problems

PROCEDURE To solve a problem by using the tangent ratio:

1. For the given problem, make a diagram that includes a right triangle. Label the known measures of the sides and angles. Identify the unknown quantity by a variable.
2. If for the right triangle either (1) the lengths of two legs or (2) the length of one leg and the measure of one acute angle are known, write a formula for the tangent of an acute angle.
3. Substitute known values in the formula and solve the resulting equation for the unknown value.

EXAMPLES

1. Find to the *nearest degree* the measure of the angle of elevation of the sun when a vertical pole 6 meters high casts a shadow 8 meters long.

Solution The angle of elevation of the sun is the same as $\angle A$, the angle of elevation to the top of the pole from A. Since the segments mentioned in the problem are the legs of a right triangle opposite and adjacent to $\angle A$, use the tangent ratio.

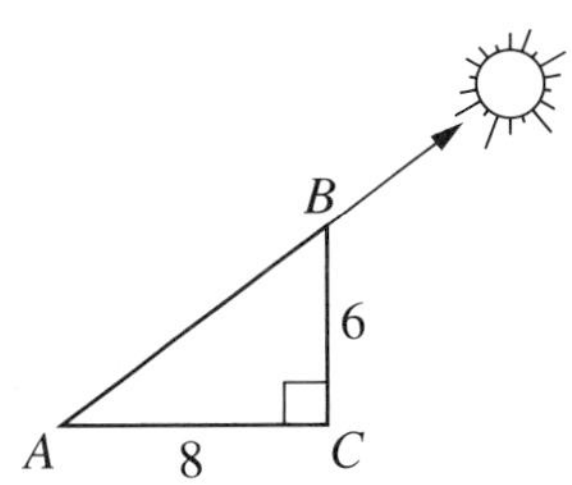

$$\tan A = \frac{\text{opp}}{\text{adj}} = \frac{BC}{AC} = \frac{6}{8}$$

Enter: (6 ÷ 8) 2nd TAN^{-1}

Display: 36.8698976

Answer: To the *nearest degree*, the angle of elevation of the sun is 37°.

2. At a point on the ground 40 meters from the foot of a tree, the measure of the angle of elevation of the top of the tree is 42°. Find the height of the tree to the *nearest meter*.

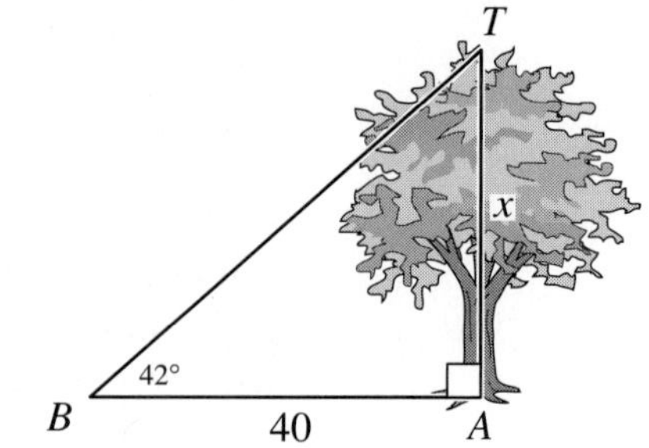

Solution Let T be the top of the tree, A be the foot of the tree, and B be the point on the ground 40 meters from A. Draw $\triangle ABT$, and label the diagram: $m\angle B = 42$, $AB = 40$.

Let x = height of tree (AT). The height of the tree is the length of the perpendicular from the top of the tree to the ground. Since the problem involves the measure of an acute angle and the measures of the legs of a right triangle, use the tangent ratio:

$$\tan B = \frac{\text{opp}}{\text{adj}}$$

$$\tan B = \frac{AT}{BA}$$

Substitute the given values: $$\tan 42° = \frac{x}{40}$$

Solve for x: $$x = 40 \tan 42°$$

Use a calculator for the computation:

Enter: 40 [×] 42 [TAN] [=]
Display: [36.0161617]

Answer: To the *nearest meter*, the height of the tree is 36 m.

---

3. From the top of a lighthouse 165 feet above sea level, the measure of the angle of depression of a boat at sea is 35°. Find to the *nearest foot* the distance from the boat to the foot of the lighthouse.

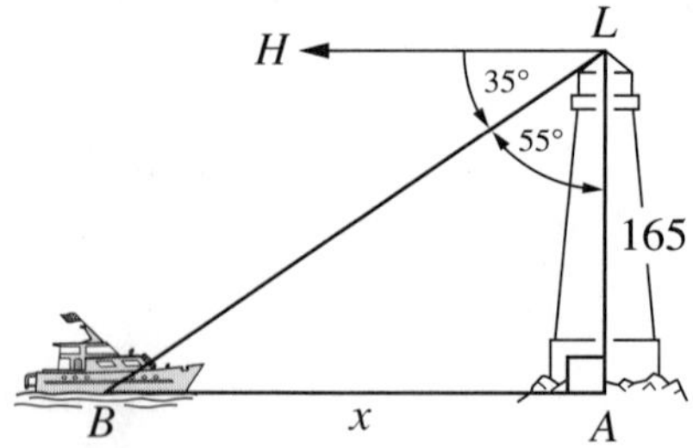

Solution Let L be the top of the lighthouse, LA be the length of the perpendicular from L to sea level, and B be the position of the boat. Draw right triangle ABL, and draw $\overleftrightarrow{HL}$, the horizontal line through L.

Since $\angle HLB$ is the angle of depression, $m\angle HLB = 35$, $m\angle LBA = 35$, and $m\angle BLA = 90 - 35 = 55$.

Let x = distance from boat to lighthouse (BA).

METHOD 1

Using $\angle BLA$, $\overline{BA}$ is the opposite side and $\overline{LA}$ is the adjacent side. Form the tangent ratio:

$$\tan \angle BLA = \frac{BA}{LA}$$

$$\tan 55° = \frac{x}{165}$$

$$x = 165 \tan 55°$$

Enter: 165 [×] 55 [TAN] [=]

Display: [235.644421]

METHOD 2

Using $\angle LBA$, $\overline{LA}$ is the opposite side and $\overline{BA}$ is the adjacent side. Form the tangent ratio.

$$\tan \angle LBA = \frac{LA}{BA}$$

$$\tan 35° = \frac{165}{x}$$

$$x \tan 35° = 165$$

$$x = \frac{165}{\tan 35°}$$

Enter: 165 [÷] 35 [TAN] [=]

Display: [235.644421]

Answer: To the *nearest foot*, the boat is 236 ft. from the foot of the lighthouse.

EXERCISES WITH OPEN-RESPONSE PROBLEMS

In 1–9, in each given triangle, find the length of the side marked x to the *nearest foot* or the number of degrees contained in the angle marked x to the *nearest degree*.

1.

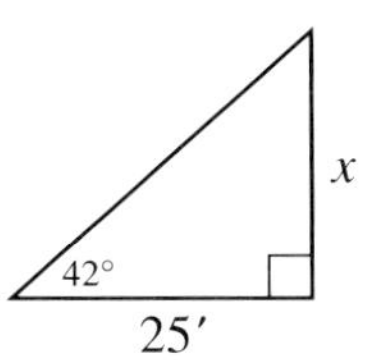

2.

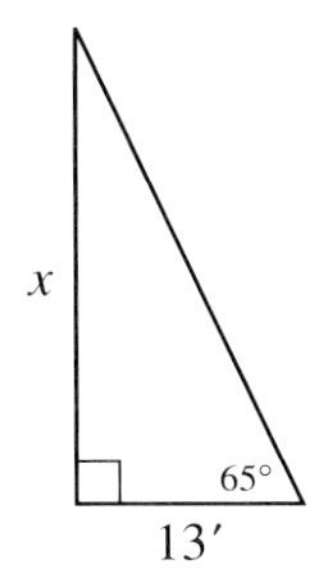

3.

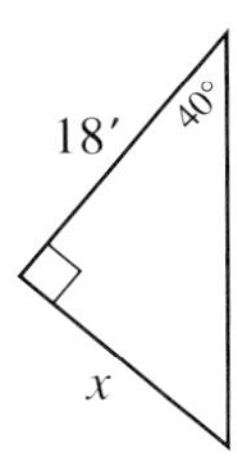

4.

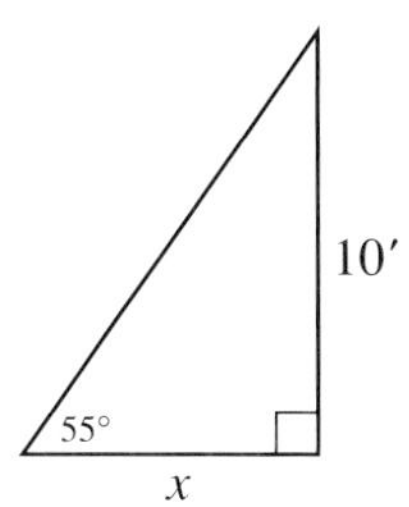

5.

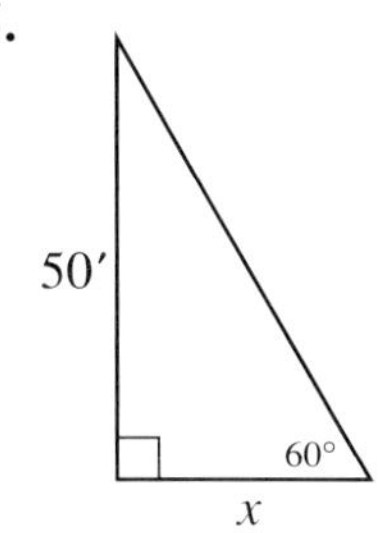

6.

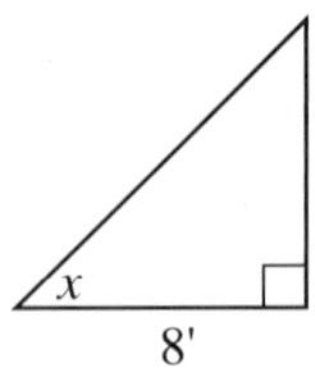

7.

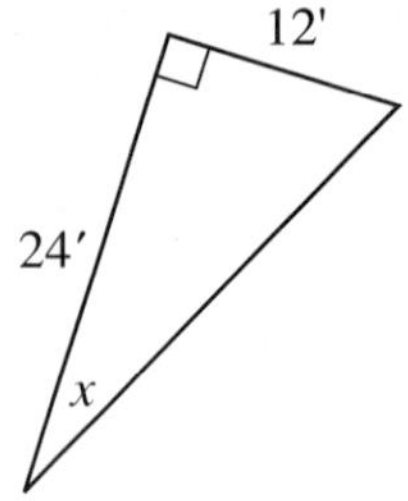

8.

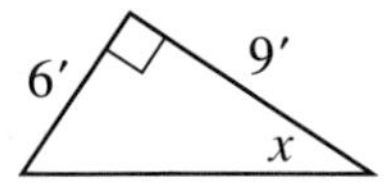

9.

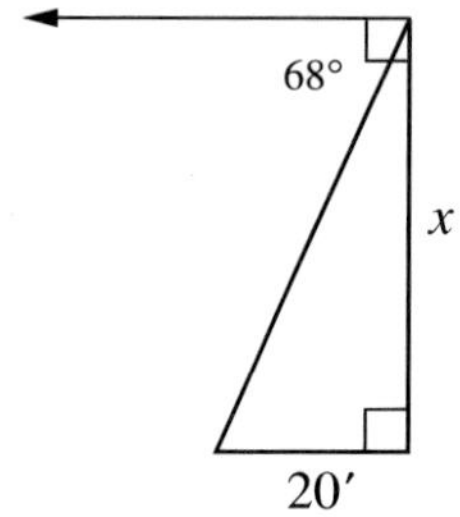

10. At a point on the ground 50 meters from the foot of a tree, the measure of the angle of elevation of the top of the tree is 48°. Find the height of the tree to the *nearest meter*.

11. A ladder is leaning against a wall. The foot of the ladder is 6.5 feet from the wall. The ladder makes an angle of 74° with the level ground. How high on the wall does the ladder reach? Round the answer to the *nearest tenth of a foot*.

12. A boy visiting New York City views the Empire State Building from a point, A, on the ground, that is 940 feet from the foot, C, of the building. The angle of elevation of the top, B, of the building as seen by the boy has a measure of 53°. Find the height of the building to the *nearest foot*.

13. Find to the *nearest meter* the height of a building if its shadow is 18 meters long when the angle of elevation of the sun has a measure of 38°.

14. From the top of a lighthouse 50 meters high, the angle of depression of a boat out at sea has a measure of 15°. Find to the *nearest meter* the distance from the boat to the foot of the lighthouse, which is at sea level.

15. From the top of a school 60 feet high, the measure of the angle of depression to the road in front of the school is 38°. Find to the *nearest foot* the distance from the road to the school.

16. Find to the *nearest degree* the measure of the angle of elevation of the sun when a student 170 centimeters tall casts a shadow 170 centimeters long.

17. Find to the *nearest degree* the measure of the angle of elevation of the sun when a woman 150 centimeters tall casts a shadow 40 centimeters long.

18. A ladder leans against a building. The top of the ladder reaches a point on the building that is 18 feet above the ground. The foot of the ladder is 7 feet from the building. Find to the *nearest degree* the measure of the angle that the ladder makes with the level ground.

19. In rhombus $ABCD$, diagonals $\overline{AC}$ and $\overline{BD}$ meet at M. If $BD = 14$ and $AC = 20$, find to the *nearest degree*:
a. $m\angle BCM$ **b.** $m\angle MBC$ **c.** $m\angle ABC$ **d.** $m\angle BCD$

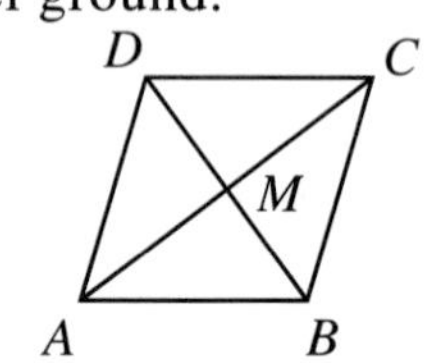

22-4 THE SINE AND COSINE RATIOS

Since the tangent is the ratio of the lengths of the two legs of a right triangle, it is not directly useful in solving problems in which the hypotenuse is involved. In trigonometry of the right triangle, two ratios that involve the hypotenuse are called the *sine of an angle* and the *cosine of an angle*.

As in our discussion of the tangent ratio, we recognize that the figure at the right shows three similar triangles because each triangle contains $\angle A$ and a right angle (a.a. $\cong$ a.a.). Thus, $\triangle ABC \sim \triangle AB'C' \sim \triangle AB''C''$.

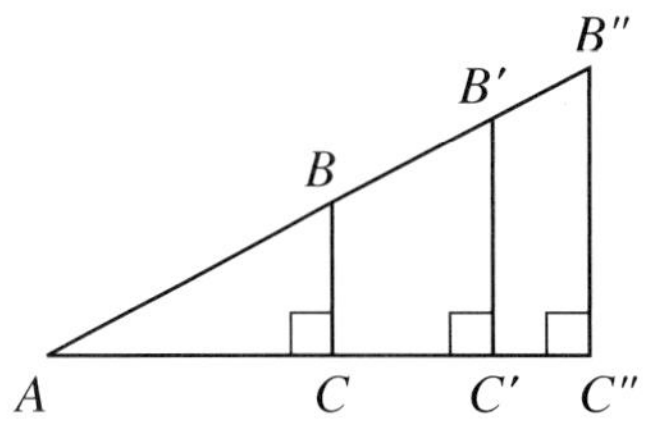

The Sine Ratio

From the figure, we see that

$$\frac{BC}{AB} = \frac{B'C'}{AB'} = \frac{B''C''}{AB''} = \text{a constant.}$$

This ratio is called the *sine of* $\angle A$.

- **Definition.** The ***sine of an acute angle of a right triangle*** is the ratio of the length of the side opposite the acute angle to the length of the hypotenuse.

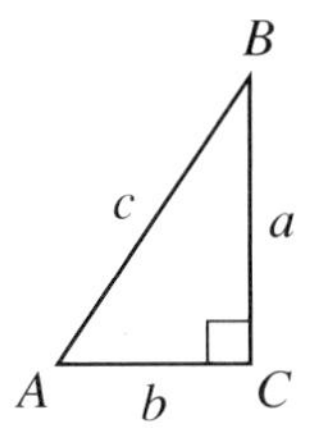

In right triangle ABC, with $m\angle C = 90$, the definition of the *sine of* $\angle A$ is:

$$\text{sine } A = \frac{\text{length of side opposite } \angle A}{\text{length of hypotenuse}} = \frac{BC}{AB} = \frac{a}{c}$$

By using "sin A" as an abbreviation for sine A, "opp" as an abbreviation for the length of the leg opposite $\angle A$, and "hyp" as an abbreviation for the length of the hypotenuse, we can shorten the way in which we write the definition of sine A as follows:

$$\boxed{\sin A = \frac{\text{opp}}{\text{hyp}} = \frac{BC}{AB} = \frac{a}{c}}$$

The Cosine Ratio

From the preceding figure, which shows $\triangle ABC \sim \triangle AB'C' \sim \triangle AB''C''$, we see that

$$\frac{AC}{AB} = \frac{AC'}{AB'} = \frac{AC''}{AB''} = \text{a constant.}$$

This ratio is called the *cosine of* $\angle A$.

- **Definition.** The ***cosine of an acute angle of a right triangle*** is the ratio of the length of the side adjacent to the acute angle to the length of the hypotenuse.

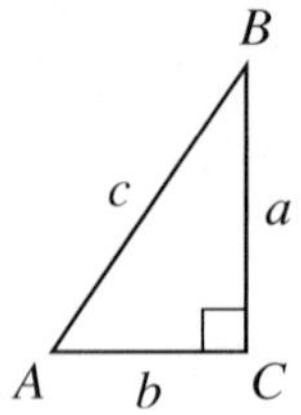

In right triangle ABC, with $m\angle C = 90$, the definition of the *cosine of* $\angle A$ is:

$$\text{cosine } A = \frac{\text{length of side adjacent to } \angle A}{\text{length of hypotenuse}} = \frac{AC}{AB} = \frac{b}{c}$$

By using "cos A" as an abbreviation for cosine A, "adj" as an abbreviation for the length of the leg adjacent to $\angle A$, and "hyp" as an abbreviation for the length of the hypotenuse, we can shorten the way in which we write the definition of cosine A as follows:

$$\cos A = \frac{\text{adj}}{\text{hyp}} = \frac{AC}{AB} = \frac{b}{c}$$

Finding Sine and Cosine Ratios on a Calculator

Case 1. Given an angle measure, find the sine or cosine ratio.

On a scientific calculator, we use the keys labeled SIN and COS to display the values of the sine and cosine of an angle. The sequence of keys that a calculator requires for tangent will be the same as the sequence for sine or cosine.

For example, to find sin 50° and cos 50°, we use one of the following:

METHOD 1: ***Enter:***	50 SIN	***Enter:*** 50 COS
METHOD 2: ***Enter:***	SIN 50 =	***Enter:*** COS 50 =
Display:	0.76604444	***Display:*** 0.64278761

These calculator solutions are shown in algebraic form as follows:

$$\sin 50° = x$$
$$\sin 50° = 0.76604444$$

$$\cos 50° = x$$
$$\cos 50° = 0.64278761$$

Case 2. Given a sine or cosine ratio, find the angle measure.

A scientific calculator will also find the measure of $\angle A$ when sin A or cos A is given. To do this, we use the keys labeled SIN^{-1} and COS^{-1}. These are usually the second functions of SIN and COS and are usually accessed by first pressing 2nd, INV, or SHIFT. We can think of the meaning of $\sin^{-1}$ as "the angle whose sine is." Therefore, if $\sin A = 0.2588$, then A is the angle whose sine is 0.2588.

To find the measure of $\angle A$ from the calculator, we use one of the following sequences of keys:

METHOD 1: ***Enter:*** 0.2588 [2nd] [SIN⁻¹]
METHOD 2: ***Enter:*** [2nd] [SIN⁻¹] 0.2588 [=]
Display: [14.9988703]

The measure of $\angle A$ to the *nearest degree* is 15°.

The algebraic form of this process is shown at the right, and the angle measure is written to the *nearest degree*. A similar form is used for cosine ratios.

$$\sin x = 0.2588$$
$$\sin 15° = 0.2588$$

EXAMPLE

In $\triangle ABC$, $\angle C$ is a right angle, $BC = 7$, $AC = 24$, and $AB = 25$. Find:
a. $\sin A$ **b.** $\cos A$ **c.** $\sin B$ **d.** $\cos B$
e. $m\angle B$ to the *nearest degree*.

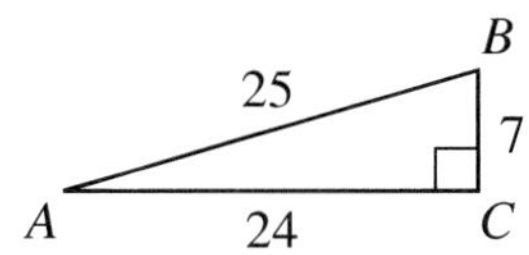

Solutions **a.** $\sin A = \frac{\text{opp}}{\text{hyp}} = \frac{BC}{AB} = \frac{7}{25}$ *Answer*

b. $\cos A = \frac{\text{adj}}{\text{hyp}} = \frac{AC}{AB} = \frac{24}{25}$ *Answer*

c. $\sin B = \frac{\text{opp}}{\text{hyp}} = \frac{AC}{AB} = \frac{24}{25}$ *Answer*

d. $\cos B = \frac{\text{adj}}{\text{hyp}} = \frac{BC}{AB} = \frac{7}{25}$ *Answer*

e. Use a calculator. Start with the ratio in part **c** and use [SIN⁻¹]; or start with the ratio in part **d** and use [COS⁻¹].

METHOD 1

$$\sin B = \frac{24}{25}$$

Enter: [(] 24 [÷] 25 [)] [2nd] [SIN⁻¹] [=]
Display: [73.7397952]

METHOD 2

$$\cos B = \frac{7}{25}$$

Enter: [(] 7 [÷] 25 [)] [2nd] [COS⁻¹] [=]
Display: [73.7397952]

Answer: $m\angle B = 74$ to the *nearest degree*.

EXERCISES WITH OPEN-RESPONSE PROBLEMS

In 1–4, find: **a.** $\sin A$ **b.** $\cos A$ **c.** $\sin B$ **d.** $\cos B$

1. **2.** **3.** **4.**

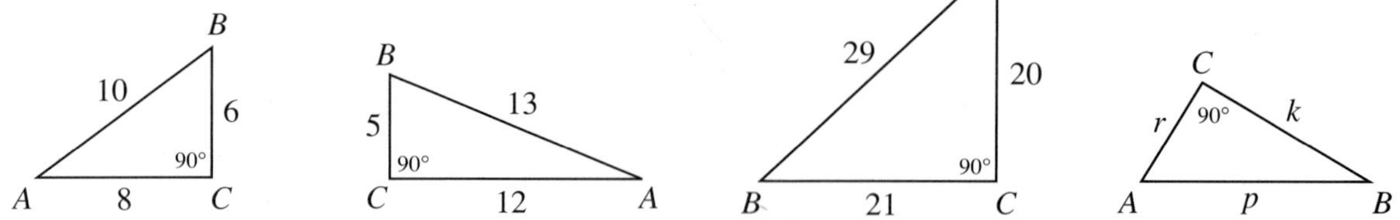

5. In $\triangle ABC$, $m\angle C = 90$, $AC = 4$, and $BC = 3$. Find $\sin A$.
6. In $\triangle RST$, $m\angle S = 90$, $RS = 5$, and $ST = 12$. Find $\cos T$.

In 7–18, for each of the following, use a calculator to find the trigonometric function value to the *nearest ten-thousandth.*

7. $\sin 18°$ **8.** $\sin 42°$ **9.** $\sin 58°$ **10.** $\sin 76°$
11. $\sin 1°$ **12.** $\sin 89°$ **13.** $\cos 21°$ **14.** $\cos 35°$
15. $\cos 40°$ **16.** $\cos 59°$ **17.** $\cos 74°$ **18.** $\cos 88°$

In 19–36, for each of the following, use a calculator to find the measure of $\angle A$ to the *nearest degree.*

19. $\sin A = 0.1908$ **20.** $\sin A = 0.8387$ **21.** $\sin A = 0.3420$
22. $\cos A = 0.9397$ **23.** $\cos A = 0.0698$ **24.** $\cos A = 0.8910$
25. $\sin A = 0.8910$ **26.** $\sin A = 0.9986$ **27.** $\cos A = 0.9986$
28. $\sin A = 0.1900$ **29.** $\cos A = 0.9750$ **30.** $\sin A = 0.8740$
31. $\cos A = 0.8545$ **32.** $\sin A = 0.5800$ **33.** $\cos A = 0.5934$
34. $\cos A = 0.2968$ **35.** $\sin A = 0.1275$ **36.** $\cos A = 0.8695$

37. As an angle increases in measure from 1° to 89°:
  a. Does the sine of the angle increase or decrease?
  b. Does the cosine of the angle increase or decrease?

38. **a.** Use a calculator to find $\sin 25°$ and $\sin 50°$.
  b. If the measure of an angle is doubled, is the sine of the angle also doubled?

39. **a.** Use a calculator to find $\cos 25°$ and $\cos 50°$.
  b. If the measure of an angle is doubled, is the cosine of the angle also doubled?

40. In $\triangle ABC$, $m\angle C = 90$, $BC = 20$, and $BA = 40$.
  a. Find $\sin A$ to the *nearest ten-thousandth.*
  b. Find the measure of $\angle A$.

41. In $\triangle ABC$, $m\angle C = 90$, $AC = 40$, and $AB = 80$.
a. Find cos A to the *nearest ten-thousandth*.
b. Find the measure of $\angle A$.

42. In $\triangle ABC$, $\angle C$ is a right angle, $AC = 8$, $BC = 15$, and $AB = 17$. Find:
a. sin A **b.** cos A **c.** sin B **d.** cos B
e. the measure of $\angle A$ to the *nearest degree*
f. the measure of $\angle B$ to the *nearest degree*

43. In $\triangle RST$, $m\angle T = 90$, $ST = 11$, $RT = 60$, and $RS = 61$. Find:
a. sin R **b.** cos R **c.** sin S **d.** cos S
e. the measure of $\angle R$ to the *nearest degree*
f. the measure of $\angle S$ to the *nearest degree*

44. In right triangle ABC, $\angle C$ is the right angle, $BC = 1$, $AC = \sqrt{3}$, and $AB = 2$.
a. Using the given lengths, write the ratios for sin A and cos A.
b. Use a calculator to find sin 30° and cos 30°.
c. What differences, if any, exist between the answers to parts **a** and **b**?

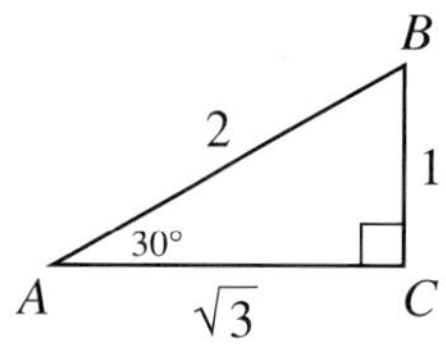

In 45 and 46, complete each sentence by filling in a degree measure that makes the sentence true.

45. **a.** sin 70° = cos ____
b. sin 23° = cos ____
c. sin 38° = cos ____
d. sin x° = cos ____

46. **a.** cos 50° = sin ____
b. cos 17° = sin ____
c. cos 82° = sin ____
d. cos x° = sin ____

47. In $\triangle ABC$, where C is a right angle, show that sin A = cos B.
48. In $\triangle ABC$, if $m\angle C = 90$, and sin A = cos A, find $m\angle A$.

22-5 APPLICATIONS OF THE SINE AND COSINE RATIOS

Since the sine and cosine ratios each have the length of the hypotenuse of a right triangle as the second term of the ratio, we can use these ratios to solve problems in the following cases:

1. We know the length of one leg and the measure of one acute angle and want to find the length of the hypotenuse.
2. We know the length of the hypotenuse and the measure of one acute angle and want to find the length of a leg.
3. We know the lengths of the hypotenuse and one leg and want to find the measure of an acute angle.

EXAMPLES

1. While flying a kite, Betty lets out 300 feet of string, which makes an angle of 38° with the ground. To the *nearest foot*, what is the height of the kite above the ground? (Assume that the string is stretched so that it is straight.)

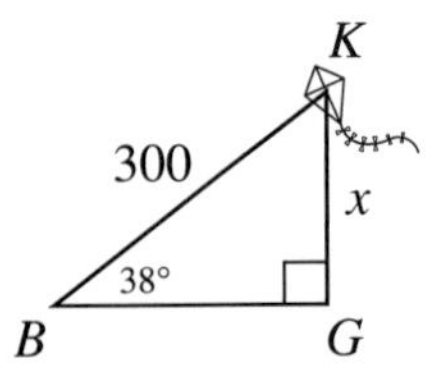

Solution Let K be the position of the kite in the air, B be the point on the ground at which Betty is holding the end of the string, and G be the point on the ground directly below the kite, as shown in the diagram. The height of the kite is the length of the perpendicular from the ground to the kite. Therefore, $m\angle G = 90$, $m\angle B = 38$, and the length of the string $BK = 300$ feet.

Let $x = KG$, the height of the kite. We know the length of the hypotenuse and the measure of one acute angle and want to find the length of a leg, KG.

In *Method 1* below, since leg KG is opposite $\angle B$, we can use $\sin B = \frac{\text{opp}}{\text{hyp}}$.

In *Method 2* below, since leg KG is adjacent to $\angle K$, we can use $\cos K = \frac{\text{adj}}{\text{hyp}}$ with $m\angle K = 90 - 38 = 52$.

	METHOD 1	METHOD 2
1. Write the ratio to be used in terms of $\triangle BKG$:	$\sin B = \frac{KG}{BK}$	$\cos K = \frac{KG}{BK}$
2. Substitute the given values:	$\sin 38° = \frac{x}{300}$	$\cos 52° = \frac{x}{300}$
3. Solve for x:	$x = 300 \sin 38°$	$x = 300 \cos 52°$
4. Compute using a calculator:		

METHOD 1: ***Enter:*** 300 [×] 38 [SIN] [=]

METHOD 2: ***Enter:*** 300 [×] 52 [COS] [=]

METHODS 1 and 2: ***Display:*** 184.69844

Answer: The height of the kite is 185 ft. to the *nearest foot*.

---

2. A wire reaches from the top of a pole to a stake in the ground 3.5 meters from the foot of the pole. The wire makes an angle of 65° with the ground. Find to the *nearest meter* the length of the wire.

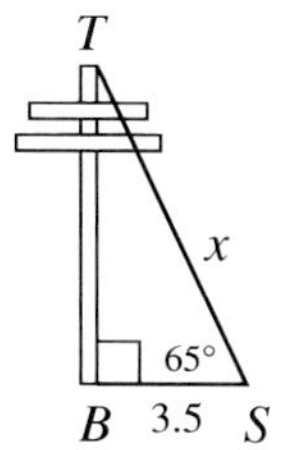

Solution In $\triangle BTS$, $\angle B$ is a right angle, $BS = 3.5$, $m\angle S = 65$, and $m\angle T = 90 - 65 = 25$.

Let $x = ST$, the length of the wire.

Since we know the length of one leg and the measure of one acute angle and want to find the length of the hypotenuse, we can use either the sine or the cosine ratio.

METHOD 1

$$\cos S = \frac{\text{adj}}{\text{hyp}}$$
$$= \frac{BS}{ST}$$
$$\cos 65° = \frac{3.5}{x}$$
$$x \cos 65° = 3.5$$
$$x = \frac{3.5}{\cos 65°}$$

Enter: 3.5 [÷] 65 [COS] [=]

Display: [8.2817055]

METHOD 2

$$\sin T = \frac{\text{opp}}{\text{hyp}}$$
$$= \frac{BS}{ST}$$
$$\sin 25° = \frac{3.5}{x}$$
$$x \sin 25° = 3.5$$
$$x = \frac{3.5}{\sin 25°}$$

Enter: 3.5 [÷] 25 [SIN] [=]

Display: [8.2817055]

Answer: The wire is 8 m long to the *nearest meter*.

---

3. A ladder 25 feet long leans against a building and reaches a point 23.5 feet above the ground. Find to the *nearest degree* the angle that the ladder makes with the ground.

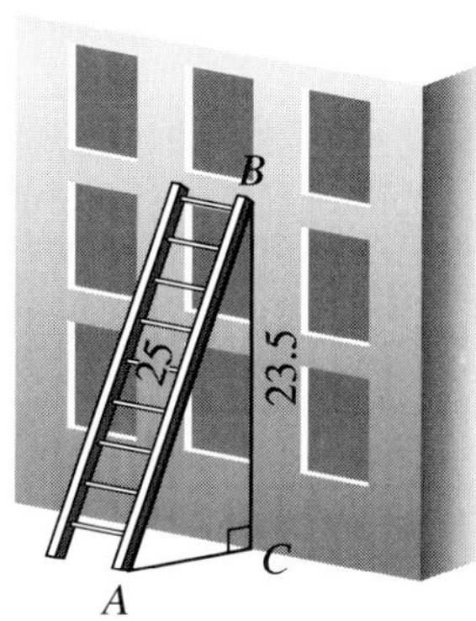

Solution In right triangle ABC, hypotenuse $\overline{AB}$ has a length of 25 and $BC = 23.5$. Since the problem involves $\angle A$, $\overline{BC}$ (its opposite side), and $\overline{AB}$ (the hypotenuse), the sine ratio is used.

$$\sin A = \frac{\text{opp}}{\text{hyp}}$$
$$= \frac{23.5}{25}$$

Enter: [(] 23.5 [÷] 25 [)] [2nd] [SIN^{-1}]

Display: [70.051556]

Answer: The angle has a measure of 70° to the *nearest degree*.

EXERCISES WITH OPEN-RESPONSE PROBLEMS

In 1–8: In each given triangle, find to the *nearest centimeter* the length of the side marked x.

1.

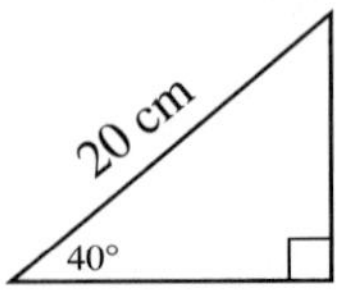

2.

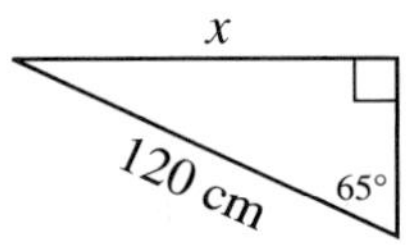

3.

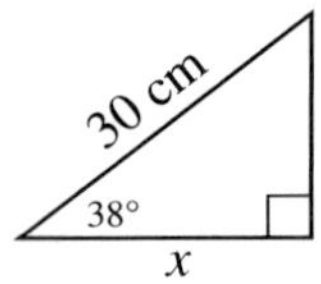

4.

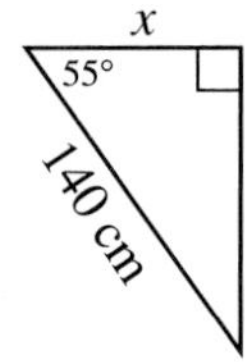

5.

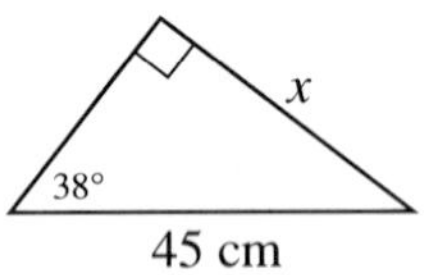

6.

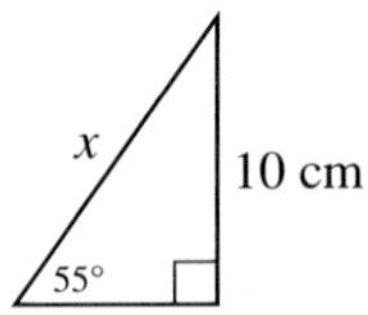

7.

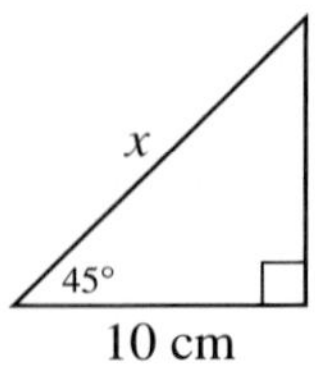

8.

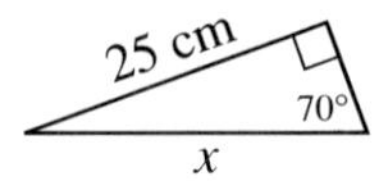

In 9–12: In each given triangle, find to the *nearest degree* the measure of the angle marked x.

9.

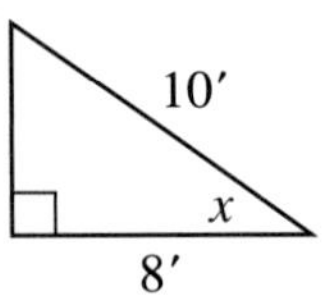

10.

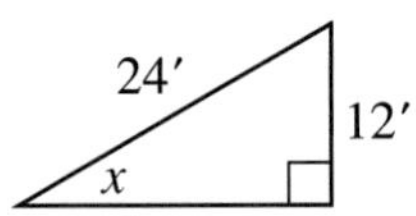

11.

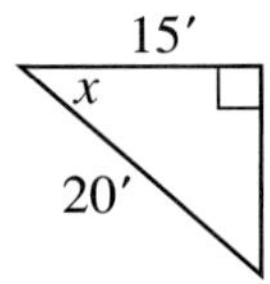

12.

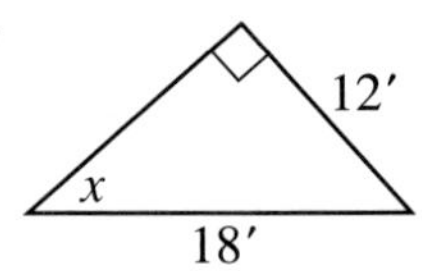

13. A wooden beam 6 meters long leans against a wall and makes an angle of 71° with the ground. Find to the *nearest tenth of a meter* how high up the wall the beam reaches.

14. A boy flying a kite lets out 400 feet of string, which makes an angle of 52° with the ground. Assuming that the string is stretched taut, find to the *nearest foot* how high the kite is above the ground.

15. A ladder that leans against a building makes an angle of 75° with the ground and reaches a point on the building 9.7 meters above the ground. Find to the *nearest meter* the length of the ladder.

16. From an airplane that is flying at an altitude of 3,000 feet, the angle of depression of an airport ground signal measures 27°. Find to the *nearest hundred feet* the distance between the airplane and the airport signal.

17. A 20-foot pole that is leaning against a wall reaches a point that is 18 feet above the ground. Find to the *nearest degree* the measure of the angle that the pole makes with the ground.

18. To reach the top of a hill that is 1 kilometer high, one must travel 8 kilometers up a straight road that leads to the top. Find to the *nearest degree* the measure of the angle that the road makes with the horizontal.

19. A 20-foot ladder leans against a building and makes an angle of 72° with the ground. Find to the *nearest foot* the distance between the foot of the ladder and the building.

20. A wire 2.4 meters in length is attached from the top of a post to a stake in the ground. The measure of the angle that the wire makes with the ground is 35°. Find to the *nearest tenth of a meter* the distance from the stake to the foot of the post.

21. An airplane rises at an angle of 14° with the ground. Find to the *nearest 10 feet* the distance the airplane has flown when it has covered a horizontal distance of 1,500 feet.

22. Henry is flying a kite. The kite string makes an angle of 43° with the ground. If Henry is standing 100 meters from a point on the ground directly below the kite, find to the *nearest meter* the length of the kite string, which is stretched taut.

23. The top of a 40-foot ladder touches a point on the wall that is 36 feet above the ground. Find to the *nearest degree* the measure of the angle that the ladder makes with the wall.

24. In a park, a slide 9 meters long is built over a horizontal distance of 6 meters along the ground. Find to the *nearest degree* the measure of the angle that the slide makes with the horizontal.

22-6 USING THE THREE TRIGONOMETRIC RATIOS TO SOLVE PROBLEMS

When the conditions of a problem can be modeled by a right triangle for which the measures of one side and an acute angle or of two sides are known, the trigonometric ratios can be used to find the measure of another side or of an acute angle.

PROCEDURE. To solve a problem by using trigonometric ratios:

1. Draw the right triangle described in the problem.
2. Label the sides and angles with the given values.
3. Assign a variable to represent the measure to be determined.
4. Select the appropriate trigonometric ratio.
5. Substitute in the trigonometric ratio, and solve the resulting equation.

Given $\triangle ABC$ with $m\angle C = 90$:

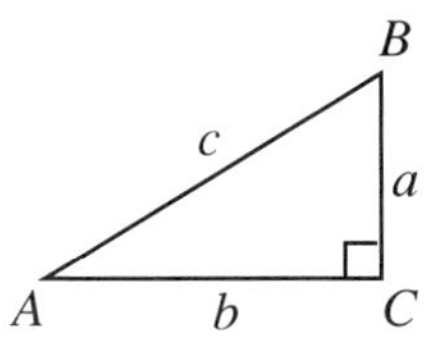

$$\sin A = \frac{\text{opp}}{\text{hyp}} = \frac{BC}{AB} = \frac{a}{c} \qquad \sin B = \frac{\text{opp}}{\text{hyp}} = \frac{AC}{AB} = \frac{b}{c}$$

$$\cos A = \frac{\text{adj}}{\text{hyp}} = \frac{AC}{AB} = \frac{b}{c} \qquad \cos B = \frac{\text{adj}}{\text{hyp}} = \frac{BC}{AB} = \frac{a}{c}$$

$$\tan A = \frac{\text{opp}}{\text{adj}} = \frac{BC}{AC} = \frac{a}{b} \qquad \tan B = \frac{\text{opp}}{\text{adj}} = \frac{AC}{BC} = \frac{b}{a}$$

EXAMPLE

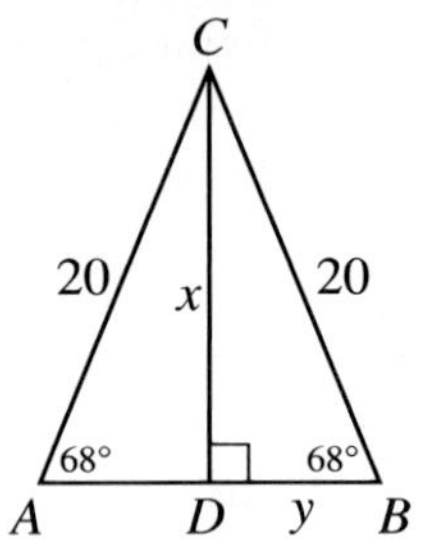

Given: In isosceles triangle ABC, $AC = CB = 20$ and $m\angle A = m\angle B = 68$. $\overline{CD}$ is an altitude.

Find: **a.** Length of altitude $\overline{CD}$ to the *nearest tenth.*
b. Length of $\overline{AB}$ to the *nearest tenth.*

Solutions **a.** In right $\triangle BDC$, $\sin B = \frac{CD}{CB}$

Let $x = CD$.

$$\sin 68° = \frac{x}{20}$$

$$x = 20 \sin 68°$$
$$= 18.543677$$
$$= 18.5$$

Answer: $CD = 18.5$ to the *nearest tenth.*

b. Since the altitude drawn to the base of an isosceles triangle bisects the base, $AB = 2DB$. Therefore, find DB in $\triangle BDC$ and double it to find AB.

In right $\triangle BDC$, $\cos B = \frac{DB}{CB}$

Let $y = DB$.

$$\cos 68° = \frac{y}{20}$$

$$y = 20 \cos 68° = 7.492132$$

$$AB = 2y = 2(7.492132) = 14.984264$$

Answer: $AB = 15.0$ to the *nearest tenth.*

EXERCISES WITH OPEN-RESPONSE PROBLEMS

In 1–8: In each given right triangle, find to the *nearest foot* the length of the side marked x; or find to the *nearest degree* the measure of the angle marked x.

1.

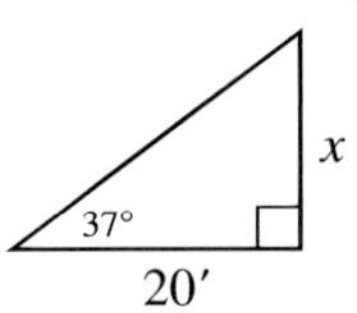

2.

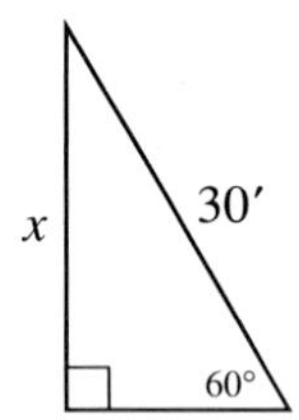

3.

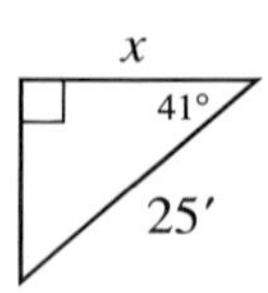

4.

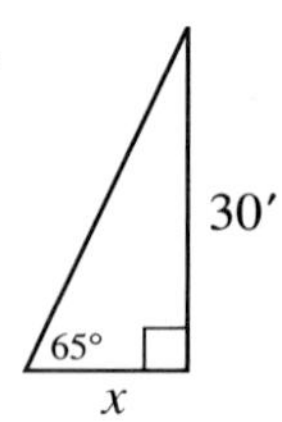

5.

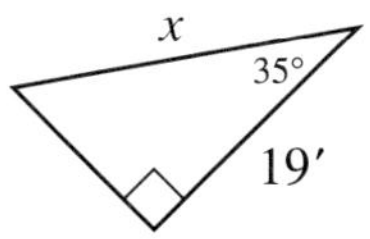

6.

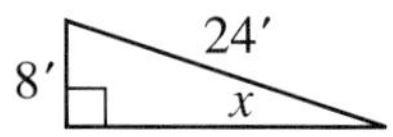

7.

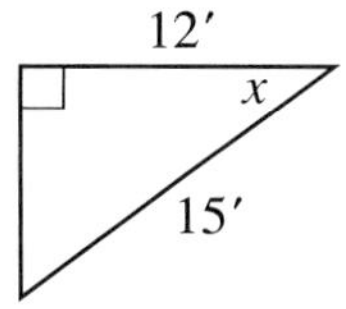

8.

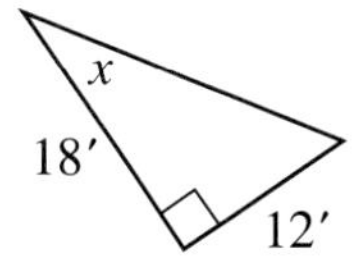

9. In $\triangle ABC$, $m\angle A = 42$, $AB = 14$, and $\overline{BD}$ is the altitude to $\overline{AC}$. Find BD to the *nearest tenth*.

10. In $\triangle ABC$, $\overline{AC} \cong \overline{BC}$, $m\angle A = 50$, and $AB = 30$. Find to the *nearest tenth* the length of the altitude from vertex C.

11. The legs of a right triangle measure 3 and 4. Find to the *nearest degree* the measure of the smallest angle of this triangle.

12. The length of hypotenuse $\overline{AB}$ of right triangle ABC is twice the length of leg $\overline{BC}$. Find the number of degrees in $\angle B$.

13. The longer side of a rectangle measures 10, and a diagonal makes an angle of 27° with this side. Find to the *nearest integer* the length of the shorter side.

14. In rectangle $ABCD$, diagonal $\overline{AC}$ measures 11 centimeters and side $\overline{AB}$ measures 7 centimeters. Find to the *nearest degree* the measure of $\angle CAB$.

15. Find to the *nearest meter* the height of a church spire that casts a shadow of 50 meters when the angle of elevation of the sun measures 68°.

16. From the top of a lighthouse 190 feet high, the angle of depression of a boat out at sea measures 34°. Find to the *nearest foot* the distance from the boat to the foot of the lighthouse.

17. A straight road to the top of a hill is 2,500 meters long and makes an angle of 12° with the horizontal. Find to the *nearest hundred meters* the height of the hill.

18. A wire attached to the top of a pole reaches a stake in the ground 20 feet from the foot of the pole and makes an angle of 58° with the ground. Find to the *nearest foot* the length of the wire.

19. An airplane climbs at an angle of 11° with the ground. Find to the *nearest hundred feet* the distance the airplane has traveled when it has attained an altitude of 400 feet.

20. Find to the *nearest degree* the measure of the angle of elevation of the sun if a child 90 centimeters high casts a shadow 180 centimeters long.

21. In right triangle ABC, $\overline{CD}$ is the altitude to hypotenuse $\overline{AB}$, $AB = 25$ meters, and $AC = 20$ meters. Find lengths AD, DB, and CD and, to the *nearest degree*, the measure of $\angle B$.

22. The lengths of the diagonals of a rhombus are 10 centimeters and 24 centimeters.

a. Find the perimeter of the rhombus. **b.** Find to the *nearest degree* the measure of the angle that the longer diagonal makes with a side of the rhombus.

23. The altitude on the hypotenuse of a right triangle divides the hypotenuse into segments whose measures are 9 inches and 4 inches. Find to the *nearest degree* the measure of the smaller acute angle of the original triangle.

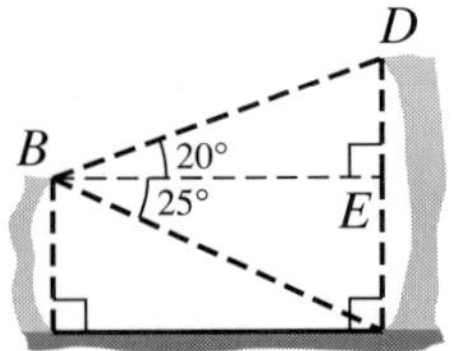

Ex. 24

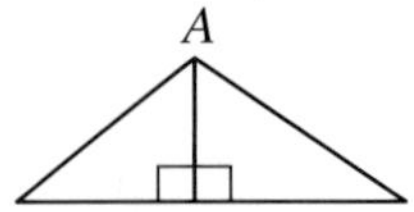

Ex. 25

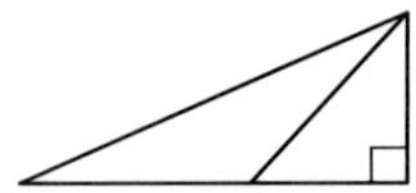

Ex. 26

24. $\overline{AB}$ and $\overline{CD}$ represent cliffs on opposite sides of a river 120 meters wide. From B, the angle of elevation of D measures 20° and the angle of depression of C measures 25°. Find to the *nearest meter*: **a.** the height of the cliff represented by $\overline{AB}$ **b.** the height of the cliff represented by $\overline{CD}$

25. In $\triangle ABC$, $AB = 30$ feet, $m\angle B = 42$, $m\angle C = 36$, and $\overline{AD}$ is an altitude. **a.** Find to the *nearest foot* the length of $\overline{AD}$. **b.** Using the result of part **a**, find to the *nearest foot* the length of $\overline{DC}$.

26. Points A, B, and D are on level ground. CD represents the height of a building, $AD = 85$ feet, and $m\angle D = 90$. At A, the angle of elevation of the top of the building, $\angle CAD$, measures 49°. At B, the angle of elevation of the top of the building, $\angle CBD$, measures 26°. **a.** Find CD to the *nearest foot*. **b.** Find BD to the *nearest foot*.

27. Angle D in quadrilateral $ABCD$ is a right angle, and diagonal $\overline{AC}$ is perpendicular to $\overline{BC}$. $BC = 20$, $m\angle B = 35$, and $m\angle DAC = 65$.
a. Find AC to the *nearest integer*.
b. Using the result of part **a**, find DC to the *nearest integer*.

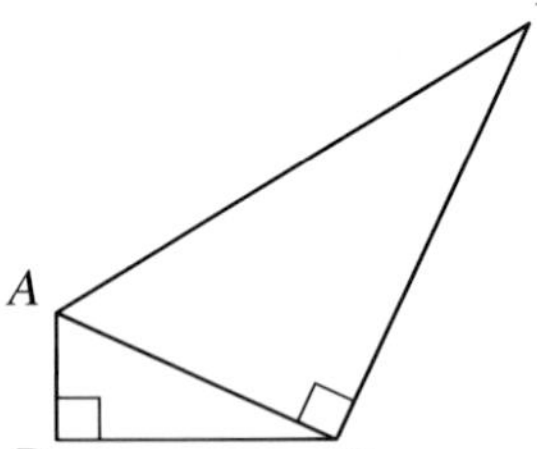

28. The diagonals of a rectangle each measure 22 and intersect at an angle whose measure is 110°. Find to the *nearest integer* the length and width of the rectangle.

29. In rhombus $ABCD$, the measure of diagonal $\overline{AC}$ is 80 and $m\angle BAC = 42$.
a. Find to the *nearest integer* the length of diagonal $\overline{BD}$.
b. Find to the *nearest integer* the length of a side of the rhombus.

30. In right triangle ABC, the length of hypotenuse $\overline{AB}$ is 100 and $m\angle A = 18$.
a. Find AC and BC to the *nearest integer*.
b. Show that the results of part **a** are approximately correct by using the relationship $(AB)^2 = (AC)^2 + (BC)^2$.

CHAPTER SUMMARY

The trigonometric functions associate the measure of each acute angle A with a number that is the ratio of the measures of two sides of a right triangle. The three most commonly used trigonometric functions are ***sine***, ***cosine***, and ***tangent***. In the application of trigonometry to the right triangle, these ratios are defined as follows:

$$\sin A = \frac{\text{opp}}{\text{hyp}} \qquad \cos A = \frac{\text{adj}}{\text{hyp}} \qquad \tan A = \frac{\text{opp}}{\text{adj}}$$

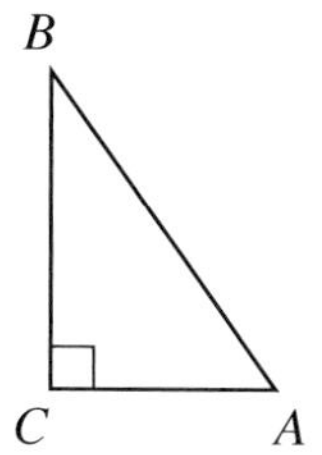

In right triangle ABC, with hypotenuse $\overline{AB}$:
$\overline{BC}$ is opposite $\angle A$ and adjacent to $\angle B$;
$\overline{AC}$ is opposite $\angle B$ and adjacent to $\angle A$.

$$\sin A = \frac{BC}{AB} \qquad \cos B = \frac{BC}{AB}$$

$$\cos A = \frac{AC}{AB} \qquad \sin B = \frac{AC}{AB}$$

$$\tan A = \frac{BC}{AC} \qquad \tan B = \frac{AC}{BC}$$

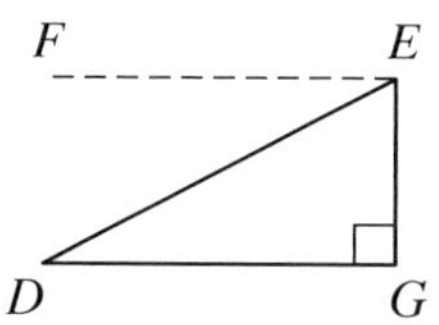

An ***angle of elevation***, $\angle GDE$ in the diagram, is an angle between a horizontal line and a line of sight to an object at a higher elevation. An ***angle of depression***, $\angle FED$ in the diagram, is an angle between a horizontal line and a line of sight to an object at a lower elevation.

VOCABULARY

22-1 Direct measurement Indirect measurement Trigonometry

22-2 Opposite side Adjacent side Tangent of an acute angle of a right triangle

22-3 Angle of elevation Angle of depression

22-4 Sine of an acute angle of a right triangle Cosine of an acute angle of a right triangle

REVIEW EXERCISES WITH OPEN-RESPONSE PROBLEMS

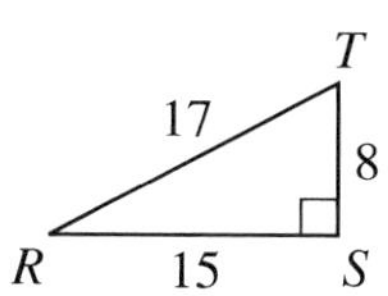

In 1–6, refer to $\triangle RST$ and express the value of each ratio as a fraction.

1. $\sin R$ **2.** $\tan T$ **3.** $\sin T$
4. $\cos R$ **5.** $\cos T$ **6.** $\tan R$

In 7–10: In each given triangle, find to the *nearest centimeter* the length of the side marked x.

7.

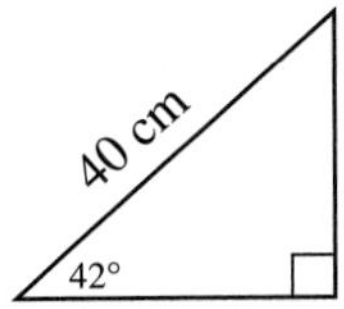

8.

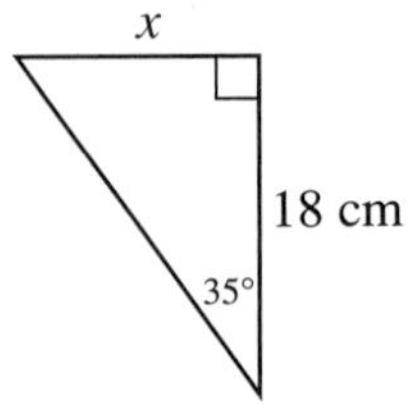

9.

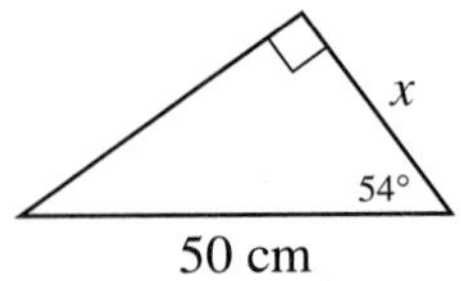

10.

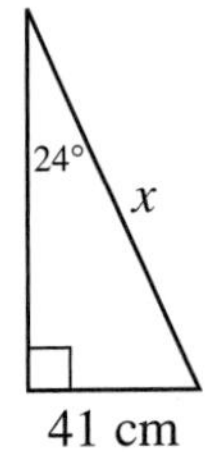

11. If $\cos A = \sin 30°$, and $0° \leq A \leq 90°$, what is the measure of $\angle A$?

12. In right triangle ACB, if $m\angle C = 90$, $m\angle A = 66$, and $AC = 100$, find BC to the *nearest integer.*

13. In right triangle ABC, $m\angle C = 90$, $m\angle B = 28$, and $BC = 30$ feet. Find AB to the *nearest foot.*

14. In $\triangle ABC$, $m\angle C = 90$, $\tan A = 0.7$, and $AC = 40$. Find BC.

15. In $\triangle ABC$, $m\angle C = 90$, $AB = 30$, and $BC = 15$. What is the measure, in degrees, of $\angle A$?

16. In $\triangle ABC$, $m\angle C = 90$, $BC = 5$, and $AC = 9$. Find to the *nearest degree* the measure of $\angle A$.

17. Find to the *nearest meter* the height of a building if its shadow is 42 meters long when the angle of elevation of the sun measures 42°.

18. A 5-foot wire attached to the top of a tent pole reaches a stake in the ground 3 feet from the foot of the pole. Find to the *nearest degree* the measure of the angle made by the wire with the ground.

19. While flying a kite, Doris let out 400 feet of string. Assuming that the string is stretched taut and makes an angle of 48° with the ground, find to the *nearest foot* how high the kite is.

20. In rectangle $ABCD$, diagonal $\overline{AC}$ is drawn. If $m\angle BAC = 62$ and $BC = 20$, find to the *nearest integer:* **a.** the length of $\overline{AB}$ **b.** the length of $\overline{AC}$

21. Jill made the following entry on her calculator:

Enter: 1.5 [2nd] [SIN^{-1}]

Explain why the calculator displayed an error message.

22. The vertices of a rectangle are $A(-1,-1)$, $B(-1, 7)$, $C(8, 7)$, and $D(8,-1)$.

a. Find the coordinates of E on $\overline{AB}$ such that the area of $\triangle EBC$ is one-fourth the area of $ABCD$.

b. Find $m\angle BCE$ to the *nearest degree.*

23. Orlo sights the top of a tree through a transit that is 5 feet above the ground. The horizontal distance between the transit and the tree is 20 feet. If the angle of elevation of the top of the tree is 50°, find the height of the tree to the *nearest foot.*

CUMULATIVE REVIEW WITH OPEN-RESPONSE PROBLEMS

1. The measure of each of the base angles of an isosceles triangle is 15 degrees more than twice the measure of the vertex angle. Find the measure of each angle of the triangle.

2. Find the length of a leg of an isosceles right triangle if the length of the hypotenuse is $\sqrt{90}$.

3. The vertices of quadrilateral $ABCD$ are $A(-2, 4)$, $B(3, 5)$, $C(4, -2)$, and $D(-3, -1)$.

a. Show that the diagonals are perpendicular.

b. Show that $ABCD$ has two pairs of congruent sides.

c. Show that $ABCD$ is not a parallelogram.

4. The length of an altitude of an equilateral triangle is 12 centimeters.

a. Find the length of a side of the triangle to the *nearest centimeter*.

b. Find the area of the triangle to the *nearest square centimeter*.

5. Prove that the midpoint of the hypotenuse of a right triangle is equidistant from the three vertices. (*Hint:* Graph the triangle on a coordinate grid.)

Exploration

Write a formula for the area of a regular polygon in terms of the measure of a side, s, and the number of sides, n.

Chapter 23

Locus

In the 1700's and 1800's, elaborate gardens were planted on the grounds of palaces and mansions in England and France. Many of these gardens contained mazes built of hedges, usually seven feet or more in height. People would walk through the maze, get disoriented and lost, and then attempt to find a path leading to a way out.

The accompanying photograph shows an aerial view of one such maze. It was said that if a person placed his or her right hand on one of the walls of hedges and continued to walk, the path would lead to an exit. Is that true of the maze in this photograph?

Today we deal with different mazes. This is usually the case when we attempt to trace a path using a street map, a road map, or a diagram of bus and train routes. In mathematics, a ***locus*** is nothing more than a *path*. This chapter contains an introduction to the study of locus.

23-1 THE MEANING OF LOCUS

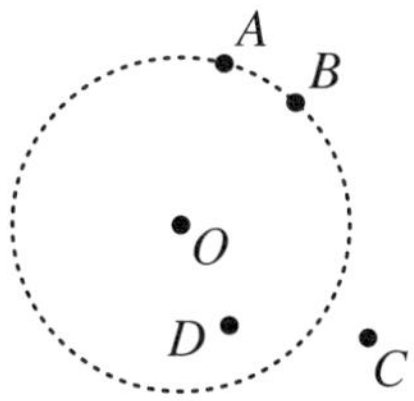

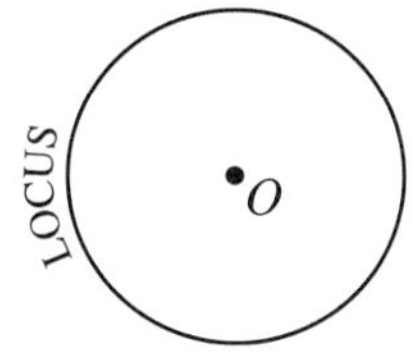

If a point is moving so that it is always satisfying a given condition, its path is called a *locus*, which is the Latin word for "place." The plural of *locus* is *loci* (pronounced "LOW sigh").

For example, we can think of a point moving in a plane so that it is always 1 centimeter from a given point O in the plane. If $OA = 1$ centimeter and $OB = 1$ centimeter, the point will move through positions A and B. Since $OC \neq 1$ centimeter and $OD \neq 1$ centimeter, however, the point will *not* move through positions C and D.

The path, or locus, is a circle whose center is O and whose radius is 1 centimeter in length. This circle is shown as a smooth or continuous curve, indicating that the locus consists of an infinite set of points.

- **Definition.** A ***locus*** is the set of all points, and only those points, that satisfy a given condition or set of conditions.

This definition of locus contains two important phrases: *all points* and *only those points*. The example given above helps us to understand what these phrases mean. These meanings are related to each other in a logical sense as follows:

1. *All points* ($p \rightarrow q$): If a point is on the locus, then the point satisfies the given conditions. Here, points on the circle, such as A and B, satisfy the given conditions.
2. *Only those points* ($\sim p \rightarrow \sim q$): If a point is not on the locus, then the point does not satisfy the given conditions. Here, points that are *not* on the circle, such as C and D, do *not* satisfy the given conditions.

We recall that the statement ($\sim p \rightarrow \sim q$) is the inverse of the statement ($p \rightarrow q$). Later in this chapter, when we prove that a locus is correct, it will be necessary to prove that both statements are true: the conditional and its inverse. For now, let us simply try to discover a locus.

23-2 DISCOVERING A LOCUS

PROCEDURE. To discover a probable locus:

1. Make a diagram that contains the fixed lines or points that are given.
2. Decide what condition must be satisfied, and locate one point that meets the given condition.
3. Then locate several other points that satisfy the given condition. These points should be sufficiently close together to develop the shape or the nature of the locus.
4. Use the points to draw a line or smooth curve that appears to be the locus.
5. Describe in words the geometric figure that appears to be the locus.

Note. In this chapter, we will assume that all given points, segments, rays, lines, and circles lie in the same plane and the desired locus lies in that plane also.

EXAMPLES

1. What is the probable locus of points equidistant from the endpoints of a given line segment?

Solution Apply steps 1–5 of the Procedure above.

Figure 1

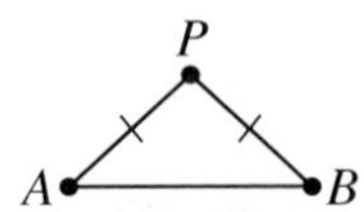

Figure 2

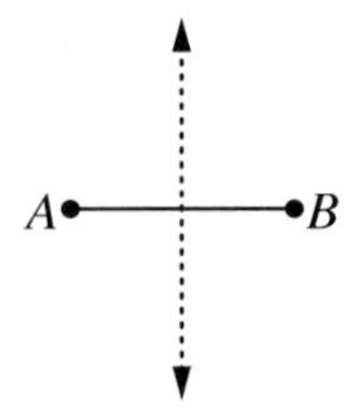

Figure 3

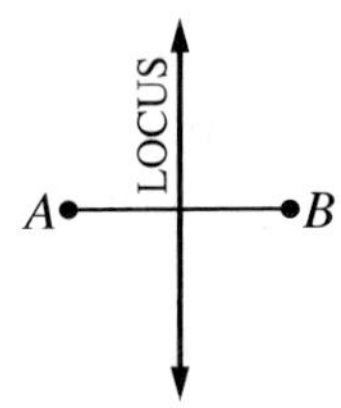

Figure 4

(1) Make a diagram: $\overline{AB}$ is the given line segment (Figure 1).
(2) Decide the condition to be satisfied: P is to be equidistant from A and B, or $PA = PB$ (Figure 2).
(3) Locate several points equidistant from A and B (Figure 3).
(4) Through these points, draw the straight line that appears to be the locus (Figure 4).
(5) Describe the locus in words. (See *Answer*.)

Answer: The probable locus is a straight line that is the perpendicular bisector of the given line segment.

---

2. What is the probable locus of points in the interior of an angle equidistant from the rays that form the sides of the given angle?

Solution

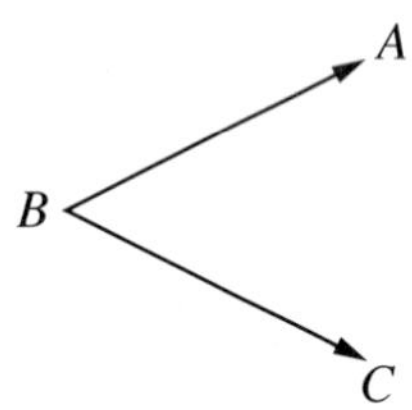

Figure 1

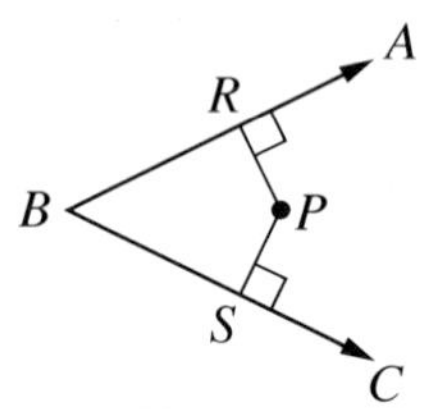

Figure 2

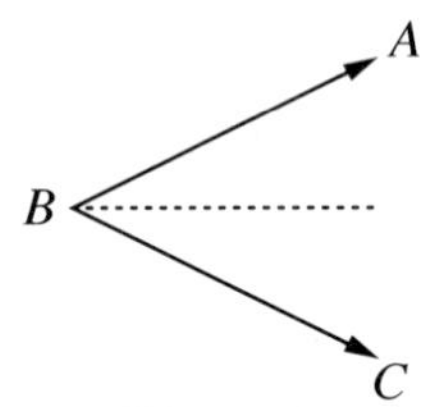

Figure 3

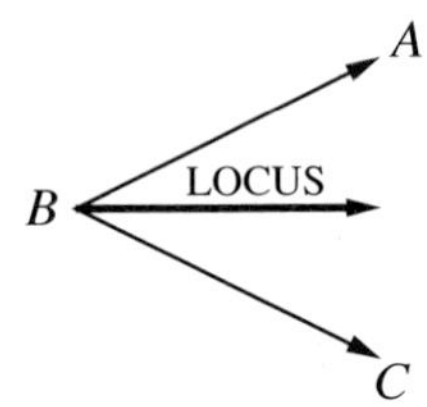

Figure 4

(1) Draw a diagram: $\angle ABC$ is the given angle (Figure 1).
(2) Decide the condition to be satisfied: P is to be equidistant from $\overrightarrow{BA}$ and $\overrightarrow{BC}$, the rays that are the sides of $\angle ABC$. Thus, for any point P, $PR = PS$, where $\overline{PR} \perp \overrightarrow{BA}$ and $\overline{PS} \perp \overrightarrow{BC}$ (Figure 2).
(3) Locate several points equidistant from $\overrightarrow{BA}$ and $\overrightarrow{BC}$ (Figure 3).
(4) Through these points, draw the ray that appears to be the locus (Figure 4).
(5) Describe the locus in words. (See *Answer*.)

Answer: The probable locus is a ray that is the bisector of $\angle ABC$.

EXERCISES WITH OPEN-RESPONSE PROBLEMS

In each of the following, use the procedure illustrated in the preceding examples to discover the probable locus.

1. What is the locus of the outer end of the hour hand of a clock during a 12-hour period?
2. What is the locus of the center of a train wheel that is moving along a straight, level track?
3. What is the locus of a car that is being driven down a street equidistant from the two opposite parallel curbs?
4. A dog is tied to a stake by a rope 6 meters long. Discover the boundary of the surface over which he may move.
5. A boy walks through an open field that is bounded on two sides by straight, intersecting roads. He walks so that he is always equidistant from the two intersecting roads. Determine his path.
6. There are two stationary floats on a lake. A girl swims so that she is always equidistant from both floats. Determine her path.
7. A dime is rolled along a horizontal line so that the dime always touches the line. What is the locus of the center of the dime?
8. What is the locus of a point on a steering wheel of a car that makes a complete revolution? Note that the center of the wheel remains in a fixed position.
9. What is the locus of points that are 10 centimeters from a given point?
10. What is the locus of points equidistant from two points, A and B, that are 8 meters apart?
11. What is the locus of points equidistant from two parallel lines 8 meters apart?
12. What is the locus of points 4 inches away from a given line, $\overleftrightarrow{AB}$?
13. What is the locus of points 3 inches from each of two parallel lines that are 6 inches apart?
14. What is the locus of points that are equidistant from two opposite sides of a square?
15. What is the locus of points that are equidistant from the vertices of two opposite angles of a square?
16. What is the locus of the center of a penny that is rolling around a quarter if the edges of the two coins are always touching each other?
17. What is the locus of points in the interior of a circle whose radius measures 3 inches if the points are 2 inches from the circle?
18. What is the locus of points in the exterior of a circle whose radius measures 3 inches if the points are 2 inches from the circle?
19. What is the locus of points 2 inches from a circle whose radius measures 3 inches?
20. Circle O has a radius of length r, and it is given that $r > m$.
    a. What is the locus of points in the exterior of circle O and at a distance m from the circle?
    b. What is the locus of points in the interior of circle O and at a distance m from the circle?
    c. What is the locus of points at a distance m from circle O?

21. ***Concentric circles*** have the same center. What is the locus of points equidistant from two concentric circles whose radii measure 10 centimeters and 18 centimeters?

22. A series of isosceles triangles are drawn, each of which has a fixed segment, $\overline{AB}$, as its base. What is the locus of the vertices of the vertex angles of all such isosceles triangles?

23. Triangle ABC is drawn with a fixed base, $\overline{AB}$, and an altitude whose measure is 3 feet. What is the locus of points that can indicate vertex C in all such triangles?

23-3 FIVE FUNDAMENTAL LOCI

There are five fundamental loci, each based on a different set of conditions. In each of the following, a condition is stated at the left, and the locus that fits the condition is then both drawn and described in words at the right.

Condition	Locus
Given A and B. Find points equidistant from these two fixed points. 	The locus of points is the perpendicular bisector, $\overleftrightarrow{CD}$, of the segment, $\overline{AB}$, joining the two points. 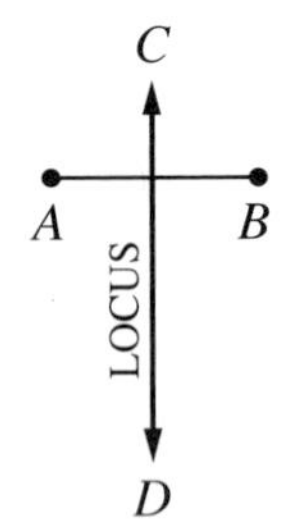
Given $\overleftrightarrow{AB}$ intersecting $\overleftrightarrow{CD}$. Find points equidistant from these two intersecting lines. 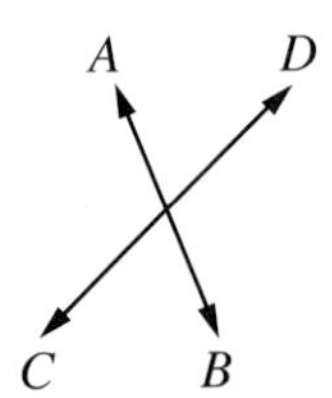	The locus of points is a pair of lines, $\overleftrightarrow{RS}$ and $\overleftrightarrow{LM}$, that bisect the angles formed by the intersecting lines. 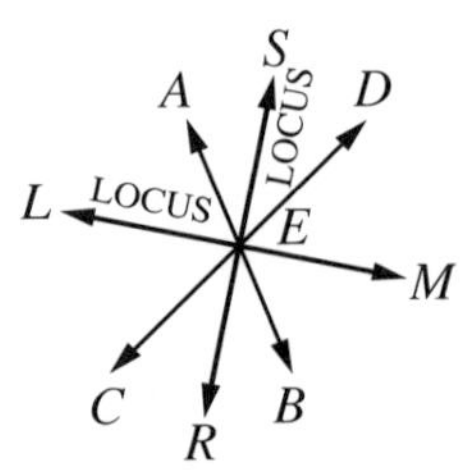

Condition	Locus
Given $\overleftrightarrow{AB} \parallel \overleftrightarrow{CD}$. Find points equidistant from these two parallel lines. 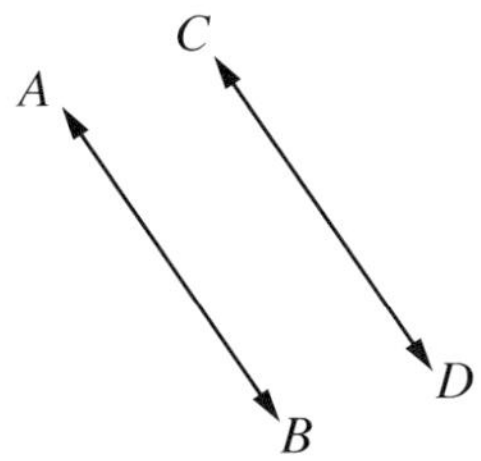	The locus of points is a third line, $\overleftrightarrow{EF}$, parallel to the two lines and midway between them. 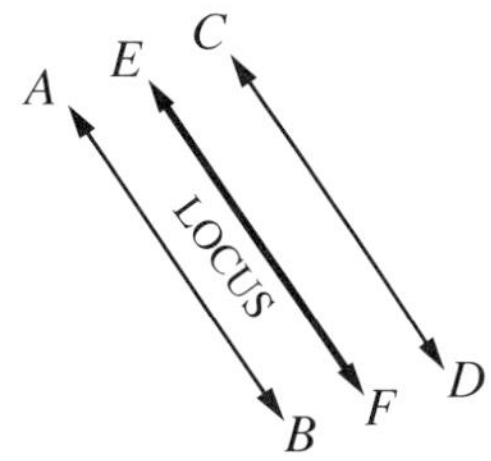
Given $\overleftrightarrow{AB}$ and a distance d. Find points that are at distance d from the line. 	The locus of points is a pair of lines, $\overleftrightarrow{CD}$ and $\overleftrightarrow{EF}$, each parallel to the given line, $\overleftrightarrow{AB}$, and at distance d from the line. 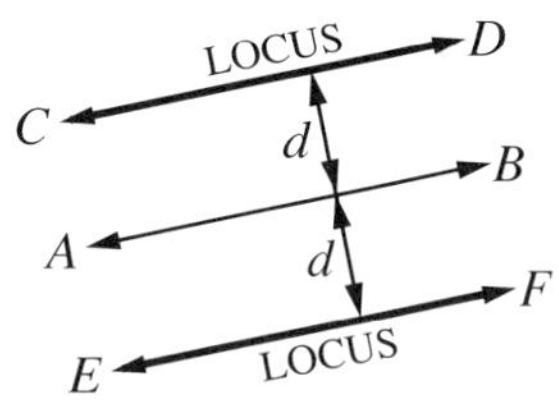
Given point A and distance d. Find points that are at distance d from fixed point A. 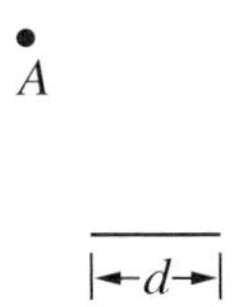	The locus of points is a circle whose center is point A and the length of whose radius is distance d. 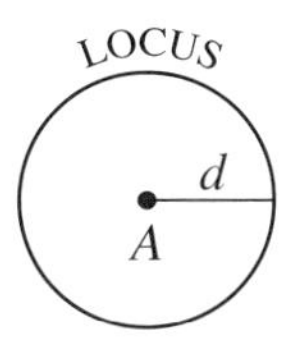

The fifth fundamental locus is generally treated as a definition rather than a statement to be proved. Thus:

- **Definition.** The locus of points at a given distance from a given point is a ***circle*** whose center is the given point and the length of whose radius is the given distance.

EXERCISES WITH OPEN-RESPONSE PROBLEMS

In 1–3, identify each required locus.

1. The locus of points equidistant from the endpoints of a segment is the _____ of the segment.
2. The locus of points in the interior of an angle that are equidistant from the sides of the angle is the _____ of that angle.
3. The locus of the midpoints of all radii of a given circle is _____.

4. If two lines intersect, on how many lines do points equidistant from the intersecting lines lie?
5. If two lines are parallel, on how many lines do points equidistant from the parallel lines lie?
6. Circle O has a radius whose length is r. What is the locus of points outside circle O and at a distance k from the circle?

In 7–14, select the numeral preceding the choice that best completes the statement or answers the question.

7. The locus of points at a given distance from a line is (1) a circle (2) one line (3) two parallel lines (4) two intersecting lines
8. The locus of points equidistant from two fixed points is (1) one circle (2) one line (3) two circles (4) two lines
9. The locus of points equidistant from the four vertices of a given rectangle is (1) one line (2) two lines (3) one point (4) a circle
10. A series of circles with congruent radii pass through a given point. The locus of the centers of all such circles is (1) a point (2) a circle (3) a line (4) a quadrilateral
11. $ABCD$ is a quadrilateral. The locus of points equidistant from $\overline{AB}$ and $\overline{AD}$ must include point C if $ABCD$ is a (1) trapezoid (2) rectangle (3) parallelogram (4) rhombus
12. What is the locus of points in the interior of square $ABCD$ and equidistant from sides $\overline{AB}$ and $\overline{BC}$? (1) point D (2) diagonal $\overline{BD}$ (3) diagonal $\overline{AC}$ (4) sides $\overline{AD}$ and $\overline{DC}$
13. What is the locus of points on or in the interior of square $ABCD$ and equidistant from vertices B and D? (1) $\overline{AC}$ (2) $\overline{BD}$ (3) $\overline{AD}$ (4) $\overline{DC}$
14. If a side of square $ABCD$ has length 10 centimeters, which points are 10 centimeters from vertex A? (1) B, C (2) C, D (3) B, D (4) B, C, D

23-4 LOCUS AND COORDINATE GEOMETRY

A locus may also be described in terms of coordinate geometry by naming specific points or by writing one or more equations.

A locus is the set of all points, and only those points, that satisfy a given condition or set of conditions. When the coordinates of all points on a locus determined by a given set of conditions satisfy an equation, and when all points *not* on the locus do *not* satisfy the equation, such an equation is called the ***equation of the locus***.

In the remaining sections of this chapter, we will study equations related to loci. We begin by examining the equations of loci that are straight lines.

Case 1. Every point whose abscissa is 3 lies on a line parallel to the y-axis and 3 units to the right of that axis. Every point whose abscissa is *not* 3, such as $P(-1, 3)$, does *not* lie on that line. Therefore, the locus of points whose abscissas are 3 is the line whose equation is $x = 3$.

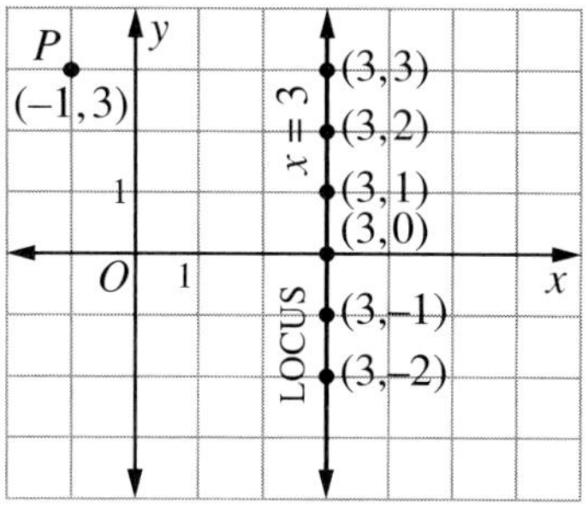

Using this example, we now generalize:

- **The locus of points whose abscissas are the constant a is the line that is parallel to the y-axis and whose equation is $x = a$.**

Case 2. Every point whose ordinate is -2 lies on a line parallel to the x-axis and 2 units below that axis. Every point whose ordinate is *not* -2, such as $P(-2, 3)$, does *not* lie on that line. Therefore, the locus of points whose ordinates are -2 is the line whose equation is $y = -2$.

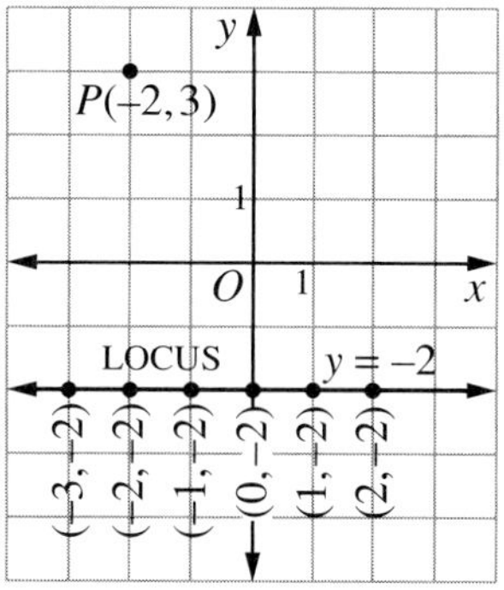

Using this example, we now generalize:

- **The locus of points whose ordinates are the constant b is the line that is parallel to the x-axis and whose equation is $y = b$.**

Case 3. Every point on the line shown has its ordinate equal to the sum of 3 times its abscissa and -1. Every point *not* on the line, such as $P(-3, 2)$, does *not* satisfy these conditions. Therefore, the locus of points described is a line whose slope is 3 and whose y-intercept is -1. The equation of this line is $y = 3x - 1$.

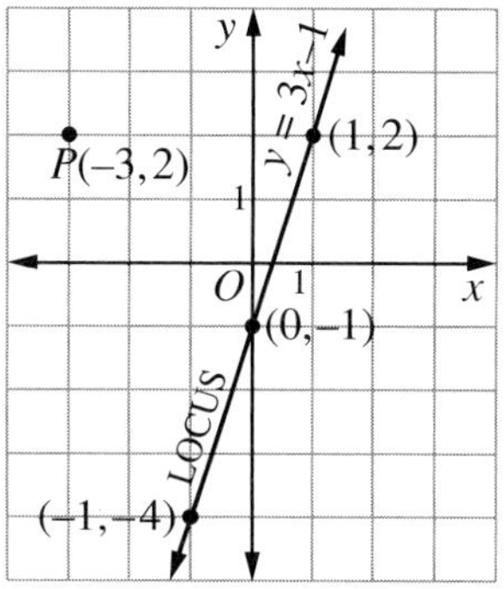

Using this example, we now generalize:

- **The locus of points in which the ordinate of each point is the sum of m times its abscissa and the constant b is a line whose slope is m, whose y-intercept is b, and whose equation is $y = mx + b$.**

EXAMPLE

Write an equation of the locus of points in which the ordinate of each point is 3 more than 4 times the abscissa of that point.

Solution Replace *ordinate* by y, and *abscissa* by x.
Thus, y is 3 more than 4 times x, or $y = 4x + 3$.

Answer: $y = 4x + 3$.

EXERCISES WITH OPEN-RESPONSE PROBLEMS

In 1–16, in each case:
a. Write an equation of the locus of points that satisfy the given conditions.
b. Graph the locus of points described.

Points in which the:

1. ordinates are 8.
2. abscissas are -5.
3. abscissas are 0.
4. ordinates are -4.
5. ordinate is equal to its abscissa.
6. ordinate is 2 times its abscissa.
7. ordinate is 2 more than its abscissa.
8. ordinate is 5 less than its abscissa.
9. ordinate is $\frac{3}{4}$ of its abscissa.
10. ordinate is 3 less than twice its abscissa.
11. abscissa and ordinate have a sum of 8.
12. ordinate is 4 more than 3 times its abscissa.
13. coordinates have a sum of -3.
14. abscissa decreased by its ordinate is 4.
15. abscissa is 4 less than one-third its ordinate.
16. ordinate is 3 more than one-half its abscissa.

17. Write an equation of the locus of points:
a. 2 units from the x-axis and above it.
b. 5 units from the y-axis and to the right of it.
c. 4 units from the y-axis and to the left of it.
d. 3 units from the x-axis and below it.
e. equidistant from the x-axis and y-axis and whose coordinates have the same sign.
f. equidistant from the x-axis and y-axis and whose coordinates have opposite signs.

18. Write an equation of the line passing through point (3, 4) and perpendicular to the x-axis.
19. Write an equation of the line passing through point $(-1, 3)$ and parallel to the y-axis.
20. Write an equation of the line passing through point (5, 2) and perpendicular to the y-axis.
21. Write an equation of the line passing through point $(-2, -4)$ and parallel to the x-axis.

In 22–27, in each case, determine whether point $(8, -2)$ is on the locus whose equation is given.

22. $x + y = 6$
23. $y = 10 - x$
24. $x = 8$
25. $x - y = 10$
26. $y = \frac{1}{4}x$
27. $x + 4y = 0$

In 28–33, in each case, find the value of k such that the graph of the given equation will pass through point (2, 3).

28. $x + y = k$ **29.** $2x - y = k$ **30.** $y = 3x + k$
31. $y = k$ **32.** $x = 2y + k$ **33.** $3x - 2y = k$

In 34–38, select the numeral preceding the choice that best completes the statement.

34. A point is on the locus whose equation is $x - y = 3$. If the ordinate of the point is 2, its abscissa is (1) 1 (2) -1 (3) 5 (4) -5

35. A point is on the locus whose equation is $3x - y = 12$. If the abscissa of the point is 3, its ordinate is (1) 5 (2) -3 (3) 3 (4) 4

36. An equation that represents the locus of points 9 units to the left of the y-axis is (1) $x = 9$ (2) $x = -9$ (3) $y = 9$ (4) $y = -9$

37. For $k > 0$, an equation that represents the locus of points k units below the x-axis is (1) $x = k$ (2) $x = -k$ (3) $y = k$ (4) $y = -k$

38. The equation $y = -x$ describes the locus of points equidistant from the x-axis and the y-axis in quadrants (1) I and IV (2) II and III (3) I and III (4) II and IV

23-5 LOCUS, COORDINATES, FIXED POINTS, AND THE CIRCLE

In this section, we will use coordinate geometry to find the equations of loci related to fixed points.

One Fixed Point: The Circle

Case 1. A circle whose center is the origin.

The origin $O(0, 0)$ is a fixed point. The locus of points 5 units from point O is a circle with its center at O and a radius whose length is 5. Let $P(x, y)$ be any point on the circle. Since the *distance* between the fixed point O and any point P on the circle is 5, let us apply these values to the *distance formula*.

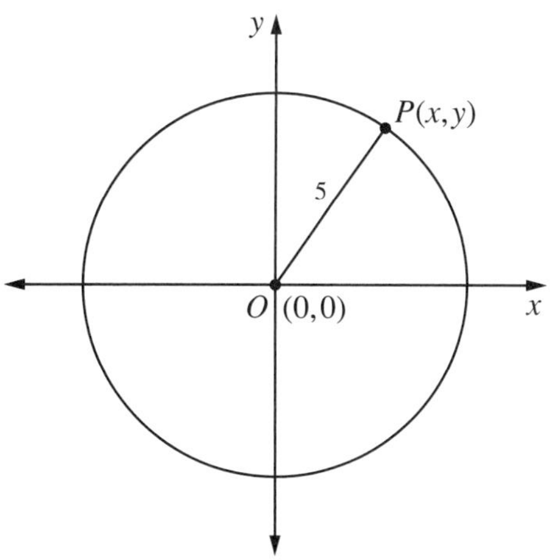

The coordinates of P are (x, y).
Let $(x, y) = (x_1, y_1)$.

The coordinates of O are (0, 0).
Let $(0, 0) = (x_2, y_2)$.
Let distance $d = 5$.

The distance formula is: $\sqrt{(x_1 - x_2)^2 + (y_1 - y_2)^2} = d$

By substitution:

$$\sqrt{(x - 0)^2 + (y - 0)^2} = 5$$

$$\sqrt{x^2 + y^2} = 5$$

Squaring both sides gives: $(\sqrt{x^2 + y^2})^2 = (5)^2$

The equation of the circle is: $x^2 + y^2 = 25$

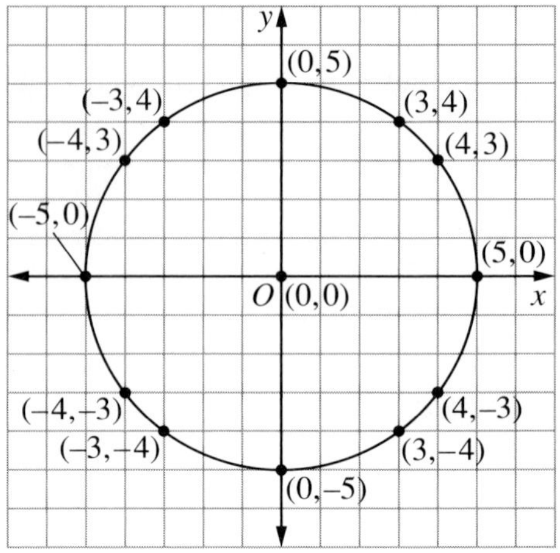

The locus of points 5 units from the origin $O(0, 0)$, that is, the circle whose equation is $x^2 + y^2 = 25$, is drawn on graph paper. We note that the circle passes through the axes at (5, 0), (0, 5), (−5, 0), and (0,−5). The coordinates of each of these points, when substituted into the equation $x^2 + y^2 = 25$, will produce a true statement.

The circle also passes through the points in which one of the x- or y-coordinates is +3 or −3, and the other coordinate is +4 or −4.

Now let us draw a radius of length 5 from the origin to one of these points on the circle, for example (3, 4). When a line is drawn from this point perpendicular to the x-axis, a 3-4-5 right triangle is formed. We recall that the distance formula was derived from the Pythagorean Theorem; thus, $3^2 + 4^2 = 5^2$. In the same way, for any point $P(x, y)$ that is on the circle but not on one of the axes, a right triangle can be formed where $x^2 + y^2 = 5^2$, or $x^2 + y^2 = 25$.

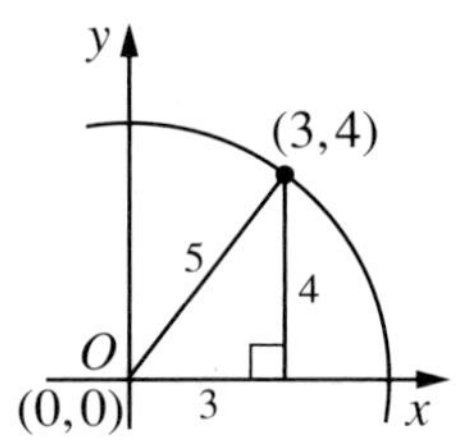

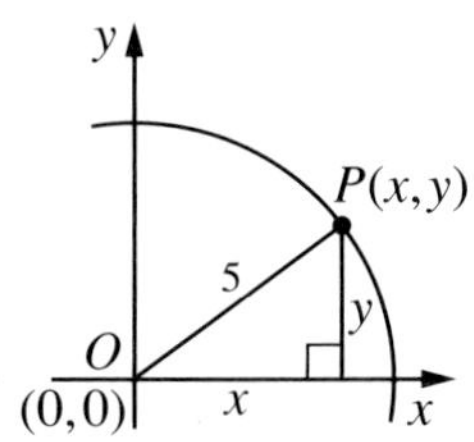

The Equation of the Locus

In general, the locus of points at a given distance r from the origin (0, 0) is a circle whose center is (0, 0) and whose radius has a length of r. Thus:

- **PRINCIPLE 7. The equation of a circle whose center is the origin (0, 0) and whose radius has a length of r is the equation $x^2 + y^2 = r^2$.**

Circle

Center = (0, 0)
Length of radius = r

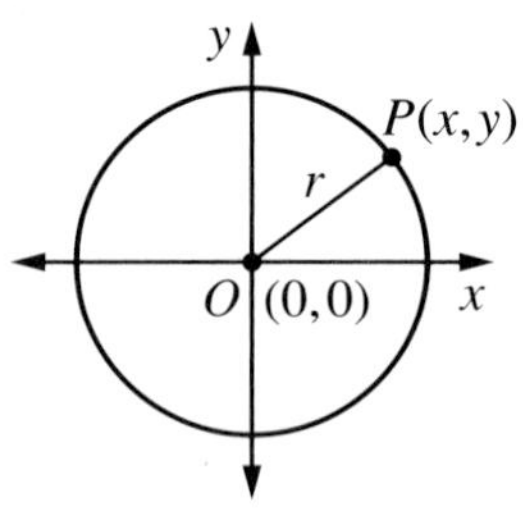

Equation: $x^2 + y^2 = r^2$

This can be proved using the distance formula with the fixed point $O(0, 0)$, the point $P(x, y)$ on the locus, and the distance r.

Case 2. A circle whose center is not *the origin.*

Point $A(4, 2)$ is a fixed point. The locus of points 3 units from point A is a circle with a center at A and a radius of length 3. We let $P(x, y)$ be any point on the circle, and again make use of the distance formula to find the equation of this circle. Here, distance $d = 3$.

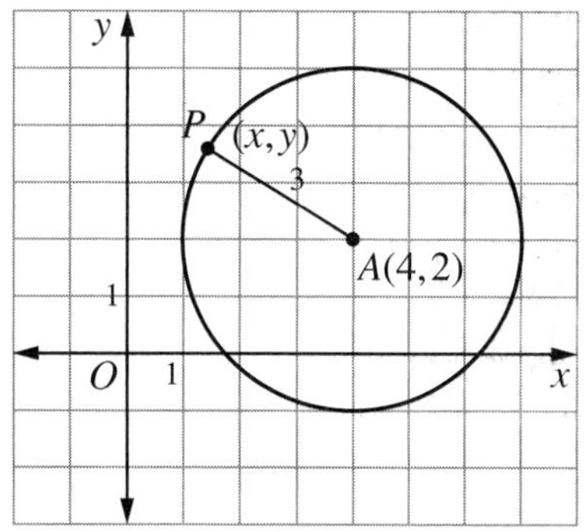

The coordinates of P are (x, y).
Let $(x, y) = (x_1, y_1)$.

The coordinates of A are $(4, 2)$.
Let $(4, 2) = (x_2, y_2)$.

The distance formula is: $\sqrt{(x_1 - x_2)^2 + (y_1 - y_2)^2} = d$

By substitution: $\sqrt{(x - 4)^2 + (y - 2)^2} = 3$

Squaring both sides gives: $\left[\sqrt{(x - 4)^2 + (y - 2)^2}\right]^2 = (3)^2$

The equation of the circle is: $(x - 4)^2 + (y - 2)^2 = 9$

In general, the locus of points at a given distance r from any fixed point, $A(a, b)$, is a circle whose center is (a, b) and whose radius has length r. Thus:

- **PRINCIPLE 8. The equation of a circle whose center is fixed point (a, b) and whose radius has a length of r is the equation $(x - a)^2 + (y - b)^2 = r^2$.**

This can be proved using the distance formula with the fixed point $A(a, b)$, the point $P(x, y)$ on the locus, and the distance r.

Circle

Center $= (a, b)$
Length of radius $= r$

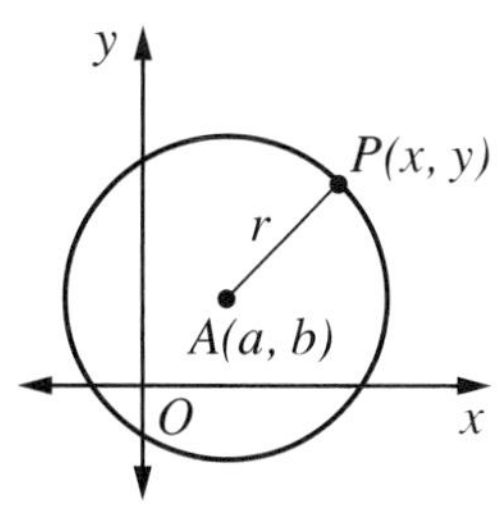

Equation:
$(x - a)^2 + (y - b)^2 = r^2$

Two Fixed Points: The Perpendicular Bisector

PROCEDURE. To find an equation of the locus of points equidistant from two fixed points A and B:

1. Draw the segment, $\overline{AB}$, and locate its midpoint, M.
2. Through M, draw a line, $\overleftrightarrow{CD}$, perpendicular to $\overline{AB}$. The equation of $\overleftrightarrow{CD}$ (the perpendicular bisector of $\overline{AB}$) is the equation of the locus of points equidistant from A and B.

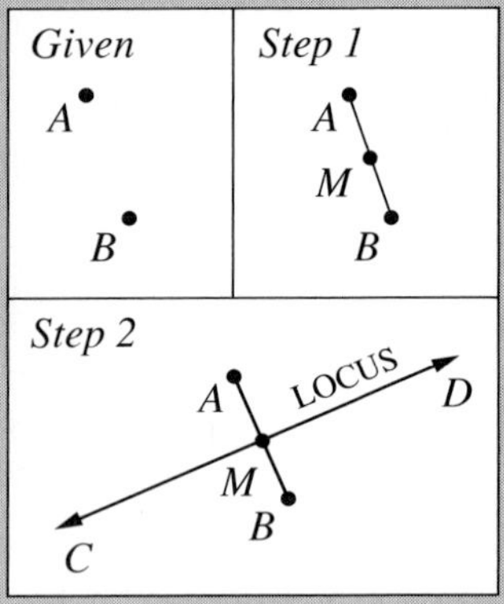

In Cases 1 and 2 below, we will examine a locus parallel to a coordinate axis. In Case 3, the locus is not parallel to either the x- or the y-axis.

Case 1. $A(2, 2)$ and $B(6, 2)$ are two fixed points. The midpoint of $\overline{AB}$ is $(4, 2)$. Since $\overline{AB}$ is parallel to the x-axis, then $\overleftrightarrow{CD}$ (the perpendicular bisector of $\overline{AB}$) is parallel to the y-axis, passing through $(4, 2)$.

The equation of $\overleftrightarrow{CD}$ is $x = 4$.

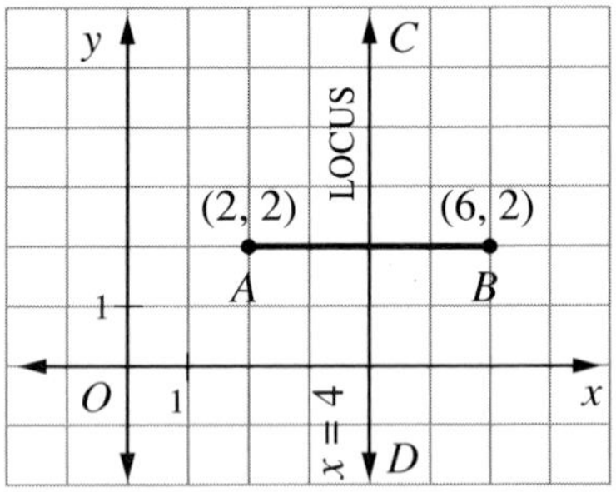

Case 2. $A(3, 1)$ and $B(3, -5)$ are two fixed points. The midpoint of $\overline{AB}$ is $(3, -2)$. Since $\overline{AB}$ is parallel to the y-axis, then $\overleftrightarrow{CD}$ is parallel to the x-axis, passing through $(3, -2)$.

The equation of $\overleftrightarrow{CD}$ is $y = -2$.

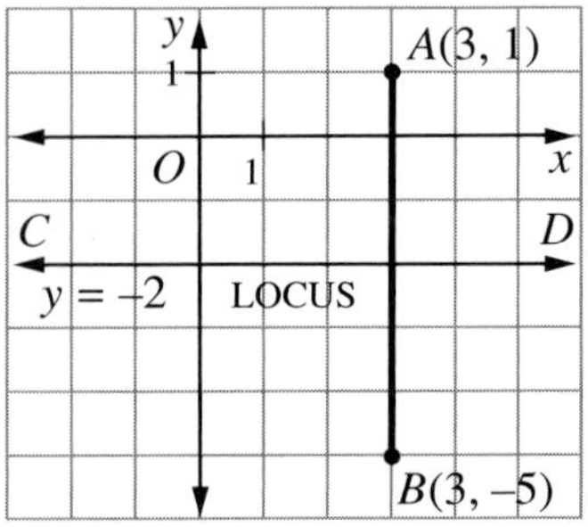

Case 3. $A(3, 6)$ and $B(5, 2)$ are two fixed points. Since $\overline{AB}$ is not parallel to either axis, we must find the slope of $\overline{AB}$ as well as its midpoint.

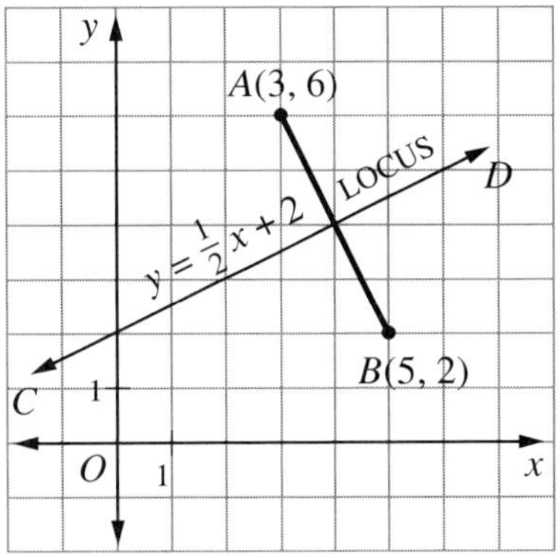

The coordinates of A are $(3, 6)$.
Let $(3, 6) = (x_1, y_1)$.

The coordinates of B are $(5, 2)$.
Let $(5, 2) = (x_2, y_2)$.

$$\text{Midpoint of } \overline{AB} = \left(\frac{x_1 + x_2}{2}, \frac{y_1 + y_2}{2}\right)$$
$$= \left(\frac{3 + 5}{2}, \frac{6 + 2}{2}\right) = (4, 4)$$

Since $\overleftrightarrow{CD} \perp \overline{AB}$, the slope of $\overleftrightarrow{CD}$ is the negative reciprocal of the slope of $\overline{AB}$.

$$\text{Slope of } \overline{AB} = \frac{y_2 - y_1}{x_2 - x_1} = \frac{2 - 6}{5 - 3} = \frac{-4}{2} = \frac{-2}{1}.$$

Therefore, the slope of $\overleftrightarrow{CD}$ is $+\frac{1}{2}$.

$\overleftrightarrow{CD}$ is a line passing through point $(4, 4)$ and having a slope of $\frac{1}{2}$. The equation of $\overleftrightarrow{CD}$ is found by the point-slope formula:

$$(y - y_1) = m(x - x_1)$$
$$(y - 4) = \frac{1}{2}(x - 4)$$
$$y - 4 = \frac{1}{2}x - 2$$
$$y = \frac{1}{2}x + 2$$

The equation of $\overleftrightarrow{CD}$ is $y = \frac{1}{2}x + 2$.

Note. If $\overleftrightarrow{CD}$ is carefully drawn through $(4, 4)$ with a slope of $\frac{1}{2}$, it is possible to observe that the line has a y-intercept of $+2$. Thus, the equation $y = \frac{1}{2}x + 2$ can be confirmed from the graph. It is recommended, however, that the equation of the locus be found by using formulas, *not* by reading directly from the graph.

EXAMPLES

1. If R and T are two distinct points on a plane, describe fully the locus of points equidistant from R and T.

Answer: The locus of points equidistant from points R and T is the perpendicular bisector of $\overline{RT}$.

---

2. If the coordinates of A are (0, 4), describe fully the locus of points 6 units from A.

Answer: The locus of points 6 units from A is a circle whose center is $A(0, 4)$ and whose radius measures 6 units.

---

3. A circle has this equation: $(x - 3)^2 + (y + 7)^2 = 64$. Find the coordinates of the center of the circle and the length of the radius.

Solution (1) Write the general form of a circle with center (a, b) and radius r: $(x - a)^2 + (y - b)^2 = r^2$

(2) Rewrite the given equation in general form:

$$(x - 3)^2 + (y + 7)^2 = 64$$

$$(x - 3)^2 + (y - [-7])^2 = 8^2$$

(3) Compare the two equations: $a = 3, b = -7, r = 8$

Answer: The center is at $(3, -7)$, and the radius measures 8 units.

---

4. a. Write an equation of the locus of points whose distance from the origin is $\sqrt{13}$.

b. Determine whether point $(-2, 3)$ is on the locus.

Solutions **a.** The locus of points whose distance from the origin is r is a circle whose center is $O(0, 0)$, whose radius has length r, and whose equation is $x^2 + y^2 = r^2$. Since $r = \sqrt{13}$ in this circle, the equation of the locus is:

$$x^2 + y^2 = (\sqrt{13})^2$$

$$x^2 + y^2 = 13$$

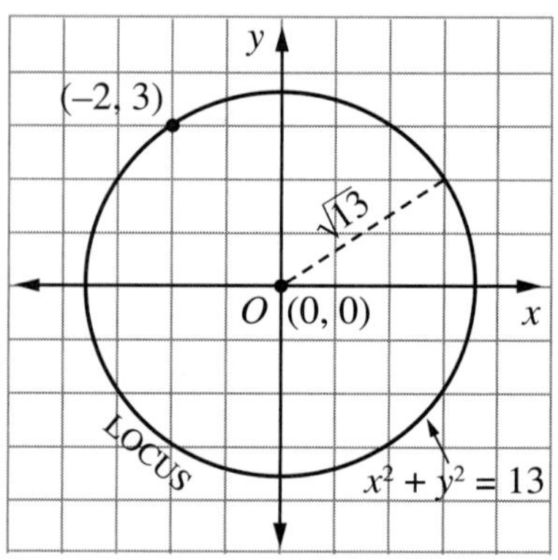

b. If point $(-2, 3)$ is on the locus, then $x = -2$ and $y = 3$ must satisfy the equation.

$$x^2 + y^2 = 13$$
$$(-2)^2 + (3)^2 \stackrel{?}{=} 13$$
$$4 + 9 \stackrel{?}{=} 13$$
$$13 = 13 \quad \text{(True)}$$

Answers: **a.** $x^2 + y^2 = 13$
b. Point $(-2, 3)$ is on the locus.

EXERCISES WITH OPEN-RESPONSE PROBLEMS

In 1–4, in each case, find: **a.** the coordinates of the center of the circle **b.** the length of the radius

1. Circle O **2.** Circle A **3.** Circle B **4.** Circle C

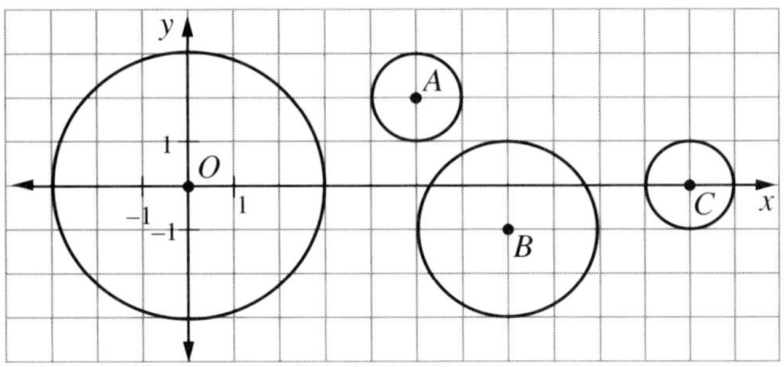

Ex. 1–4

In 5–10, in each case, describe fully the locus of points for the given equation.

5. $x^2 + y^2 = 36$ **6.** $x^2 + y^2 = 100$ **7.** $x^2 + y^2 = 5$
8. $x^2 + y^2 = 17$ **9.** $2x^2 + 2y^2 = 50$ **10.** $3x^2 + 3y^2 = 3$

11. Write an equation of the locus of points whose distance from the origin is
a. 2 **b.** 8 **c.** $\sqrt{3}$ **d.** $\sqrt{7}$ **e.** $\frac{1}{2}$

12. **a.** Write an equation of the locus of points whose distance from the origin is 13.
b. Tell whether or not each given point is on the locus.
(*1*) (0, 13) (*2*) (5, −12) (*3*) (8, 10) (*4*) (−13, 0)
(*5*) (7, 11) (*6*) (14, −1) (*7*) (−12, −5) (*8*) $(11, 4\sqrt{3})$

13. **a.** Write an equation of the locus of points whose distance from the origin is 5.
b. Without constructing the circle, determine the coordinates of every point on the locus whose abscissa is 3.

In 14–19, for each given equation of the circle, find: **a.** the coordinates of the center **b.** the length of the radius.

14. $(x - 10)^2 + (y - 7)^2 = 14^2$
15. $(x - 3)^2 + (y - 12)^2 = 49$
16. $(x + 6)^2 + (y - 1)^2 = 144$
17. $(x - 5)^2 + (y + 8)^2 = 20$
18. $(x - 2)^2 + y^2 = 25$
19. $x^2 + (y + 6)^2 = 1$

In 20–25, in each case, write an equation of the circle whose center and length of radius are given.

20. center (3, 4), radius = 4
21. center (5, 8), radius = 15
22. center (0, 9), radius = 20
23. center (2, −3), radius = 6
24. center (−2, −8), radius = 1
25. center (−6, 0), radius = $\sqrt{10}$

In 26–31, in each case: **a.** On graph paper, draw the locus of points equidistant from the two given points. **b.** Write the equation of the locus drawn.

26. (0, 2) and (0, 10)
27. (3, 0) and (9, 0)
28. (2, 6) and (2, 12)
29. (9, 3) and (15, 3)
30. (−3, 1) and (7, 1)
31. (−2, −4) and (−2, 8)

32. Write the coordinates of the point on the x-axis that is equidistant from points (8, 5) and (12, 5).

33. Write the coordinates of the point on the y-axis that is equidistant from points (4, 3) and (4, −9).

34. The equation $x = 3$ represents the locus of points equidistant from which of the following pairs of points?

(1) (0, 2) and (0, 4) (2) (0, 3) and (3, 0) (3) (10, 0) and (−4, 0) (4) (2, 0) and (6, 0)

In 35–38, in each case:
a. On a graph, draw the locus of points equidistant from the two given points.
b. Which of the following is the equation of the locus drawn?

(1) $y = x$ (2) $y = x + 2$ (3) $y = -x$ (4) $y = -x + 2$

35. (0, 2) and (2, 0)
36. (0, 2) and (−2, 0)
37. (0, 0) and (2, 2)
38. (0, 0) and (−2, 2)

In 39–44, in each case:
a. On graph paper, plot the given points A and B.
b. Find the coordinates of the midpoint of $\overline{AB}$.
c. Find the slope of $\overline{AB}$.
d. If $\overleftrightarrow{CD}$ is the locus of points equidistant from points A and B, what is the slope of $\overleftrightarrow{CD}$?
e. On graph paper, draw $\overleftrightarrow{CD}$.
Write an equation of $\overleftrightarrow{CD}$.

$B(-1, 4)$
40. $A(-2, 1)$, $B(2, 3)$
41. $A(4, 0)$, $B(0, 4)$
$B(3, -2)$
43. $A(-2, 0)$, $B(4, 2)$
44. $A(1, 4)$, $B(3, 0)$

45. The coordinates of the vertices of $\triangle ABC$ are $A(-1,-3)$, $B(7,-3)$, and $C(2, 4)$.
 a. Find the coordinates of midpoint M of side $\overline{AB}$.
 b. Write an equation of the locus of points equidistant from A and B.
 c. Is point C equidistant from A and B? Explain why.
 d. Describe fully the locus of points 3 units from point A.

46. The coordinates of the vertices of $\triangle ABC$ are $A(1, 7)$, $B(1, -1)$, and $C(6, 3)$.
 a. Find the coordinates of midpoint M of side $\overline{AB}$.
 b. Write an equation of the locus of points equidistant from A and B.
 c. Is $\triangle ABC$ isosceles? Explain your answer.
 d. Write an equation of the locus of points 4 units from point B.
 e. Is M a point on the locus described in part **d**? Explain why.

47. Write an equation of the locus of points equidistant from the circles whose equations are $x^2 + y^2 = 4$ and $x^2 + y^2 = 64$.

23-6 LOCUS, COORDINATES, AND FIXED LINES

One Fixed Line

Given one fixed line, the locus of points at a given distance d from the line is a pair of lines, each parallel to the given line. To find an equation of the locus, it is necessary to find an equation for each of the two parallel lines.

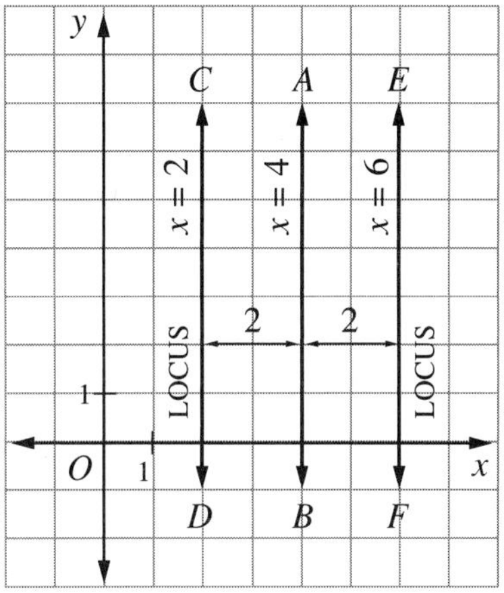

As an example, let $\overleftrightarrow{AB}$ be a fixed line whose equation is $x = 4$. The locus of points at a given distance of 2 units from $\overleftrightarrow{AB}$ is a pair of lines, each parallel to $\overleftrightarrow{AB}$. By counting 2 units to the left of $\overleftrightarrow{AB}$ and 2 units to the right of $\overleftrightarrow{AB}$, we can locate the lines, $\overleftrightarrow{CD}$ and $\overleftrightarrow{EF}$.

The description of the locus is a sentence that includes an equation for each of these two parallel lines: $x = 2$ or $x = 6$. This sentence can be written in words or in logic symbols.

Locus: $x = 2$ or $x = 6$

or $(x = 2) \lor (x = 6)$

Two Fixed Parallel Lines

Given two fixed parallel lines, the locus of points equidistant from the two lines is a third line, parallel to the other two and midway between them. The equation of the locus is the equation of this third parallel line.

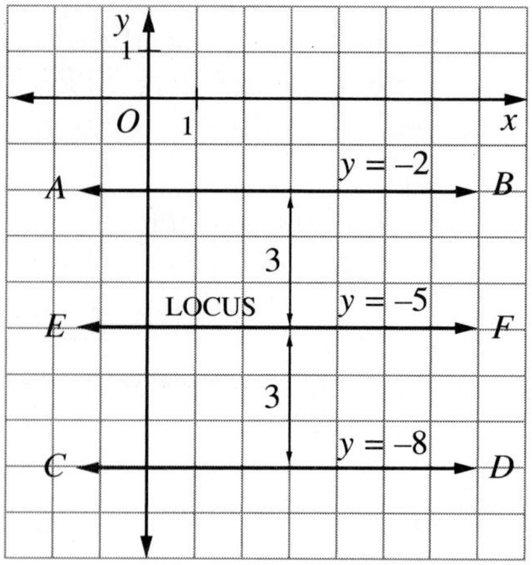

As an example, let $\overleftrightarrow{AB}$ and $\overleftrightarrow{CD}$ be fixed parallel lines whose equations are $y = -2$ and $y = -8$, respectively. The locus of points equidistant from $\overleftrightarrow{AB}$ and $\overleftrightarrow{CD}$ is the line, $\overleftrightarrow{EF}$, which is midway between them and parallel to them. Since the distance between these two parallel lines is 6 units, we count 3 units down from $\overleftrightarrow{AB}$ or 3 units up from $\overleftrightarrow{CD}$ to locate $\overleftrightarrow{EF}$, whose equation is $y = -5$.

Locus: $y = -5$

Two Fixed Intersecting Lines

Given two fixed lines that intersect, the locus of points equidistant from the two lines is a pair of lines that bisect the angles formed by the given intersecting lines. The description of this locus includes the equations of two lines.

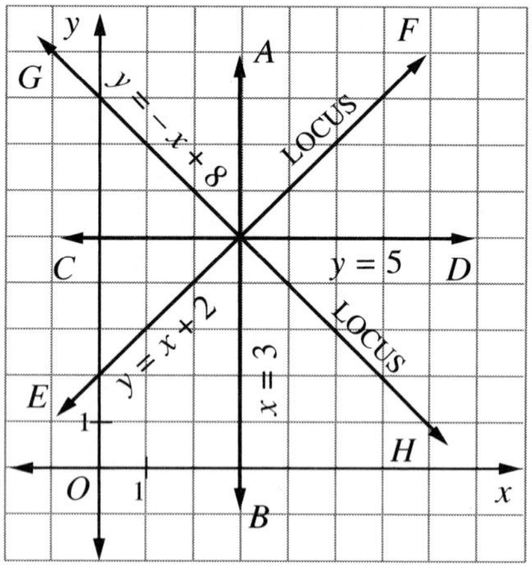

As an example, let $\overleftrightarrow{AB}$ and $\overleftrightarrow{CD}$ be fixed intersecting lines whose equations are $x = 3$ and $y = 5$, respectively. By drawing the angle bisectors through the point of intersection, (3, 5), we can locate the lines, $\overleftrightarrow{EF}$ and $\overleftrightarrow{GH}$. The description of the locus is a pair of equations for these two lines, found by determining their slopes and y-intercepts.

Locus: $y = x + 2$ or $y = -x + 8$

EXAMPLE

What is the equation of the locus of points equidistant from two lines whose equations are $y = \frac{1}{2}x + 6$ and $y = \frac{1}{2}x + 2$?

Solution The lines whose equations are $y = \frac{1}{2}x + 6$ and $y = \frac{1}{2}x + 2$ are parallel lines, each with a slope of $\frac{1}{2}$. The locus of points equidistant from the two lines is a third line, midway between them and parallel to them.

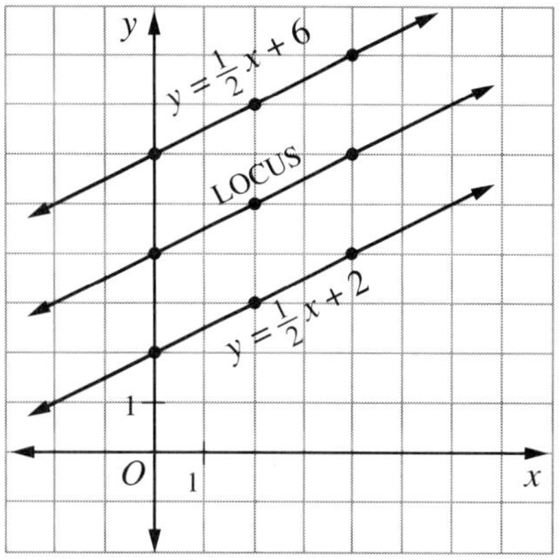

The equation of the locus must have a slope of $\frac{1}{2}$ and must intercept the y-axis midway between 6 and 2, that is, at $\frac{6 + 2}{2} = \frac{8}{2} = 4$.

Answer: The equation of the locus is $y = \frac{1}{2}x + 4$.

EXERCISES WITH OPEN-RESPONSE PROBLEMS

In all of the following exercises, it is helpful to use graph paper to draw the fixed line or lines, and then draw the locus described. In this way, an equation of the locus can be read from the graph.

In 1–4, in each case, write the equations of the locus based on the given conditions.

1. 3 units from line $x = 5$

2. 2 units from line $y = 7$

3. 5 units from line $y = 1$

4. 4 units from line $x = -4$

In 5–10, the equations of two fixed lines are given. In each case, write an equation of the locus of points equidistant from the two fixed lines.

5. $x = 4$ and $x = 10$

6. $y = 3$ and $y = -8$

7. $y = -3$ and $y = -13$

8. $x = 2$ and $x = 9$

9. $y = x$ and $y = x + 2$

10. $y = -\frac{1}{2}x + 7$ and $y = -\frac{1}{2}x + 1$

In 11–14, select the numeral preceding the choice that best answers the question.

11. Which is an equation of the locus of points 5 units from the y-axis?
(1) $y = 5$ (2) $x = 5$ or $x = -5$ (3) $y = 5$ or $y = -5$ (4) $x^2 + y^2 = 25$

12. Which is an equation of the locus of points equidistant from two lines whose equations are $y = 3x + 5$ and $y = 3x - 1$?
(1) $y = 3x$ (2) $y = 3x + 2$ (3) $y = 3x + 4$ (4) $y = 3$

13. Which is an equation of the locus of points equidistant from the x-axis and the y-axis?
(1) $x = y$ (2) $x = 0$ or $y = 0$ (3) $x = y$ or $x = -y$ (4) $x^2 + y^2 = r^2$

14. Which is an equation of the locus of points equidistant from two intersecting lines whose equations are $y = x + 3$ and $y = -x + 3$?
(1) $x = 0$ or $y = 3$ (2) $x = 3$ or $y = 0$ (3) $x = 0$ or $y = 6$ (4) $x = 6$ or $y = 0$

15. a. On graph paper, draw the intersecting lines whose equations are $y = 2x$ and $y = -2x + 4$
b. Draw the locus of points equidistant from these two intersecting lines.
c. Write an equation of the locus described in part **b.**

16. **a.** On graph paper, draw the intersecting lines whose equations are $y = \frac{1}{2}x + 1$ and $y = -\frac{1}{2}x + 5$.
b. Draw the locus of points equidistant from these intersecting lines.
c. Write an equation of the locus described in part **b**.

17. The vertices of $\triangle ABC$ are $A(5, 2)$, $B(10, 2)$ and $C(7, 8)$. Imagine that $\triangle ABK$ is equal in area to $\triangle ABC$, and both triangles have $\overline{AB}$ as a base. Write an equation of the locus of points that could be point K in $\triangle ABK$.

23-7 INTERSECTIONS OF LOCI

Given a condition, a locus of points is determined. Given a second condition, a second locus of points is determined. The points at which the two loci intersect, and only those points, satisfy both conditions.

PROCEDURE. To find points that satisfy two conditions:
1. Draw the locus of points that satisfy the first condition.
2. Draw the locus of points that satisfy the second condition.
3. Locate the points of intersection of these loci. These points must be the required points because they satisfy both conditions.

The intersection of loci may result in a single point, in two points, or in more than two points, depending on the given conditions. In cases where the loci do not intersect, there is no point that satisfies both of the given conditions.

Since we have studied loci with and without coordinates, we make the following observations:

1. Given *general* conditions (see Example 1), the loci are sketched on blank paper.
2. Given conditions involving *numbers* (see Example 2), the loci can be sketched on blank paper or on graph paper.
3. Given conditions involving *coordinates* (see Example 3), the loci are sketched on graph paper.

EXAMPLES

1. What is the number of points equidistant from two parallel lines and also equidistant from two points on one of the lines?

Solution (1) Sketch two parallel lines, ℓ and m. Place two points, A and B, on one of the lines.

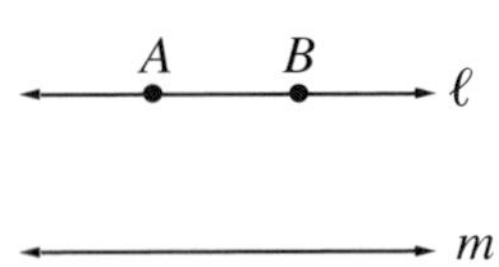

(2) The locus of points equidistant from parallel lines ℓ and m is a third line, k, parallel to ℓ and m, and midway between them. Draw line k.

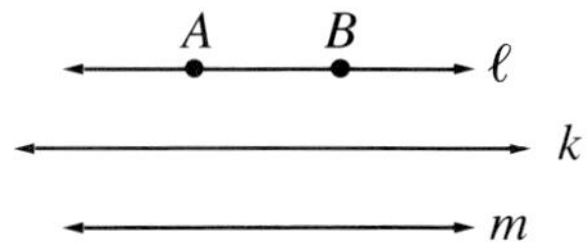

(3) The locus of points equidistant from the two points A and B is $\overleftrightarrow{CD}$, which is the $\perp$ bisector of $\overline{AB}$. Draw $\overleftrightarrow{CD}$. Extend the line to show all possible points of intersection.

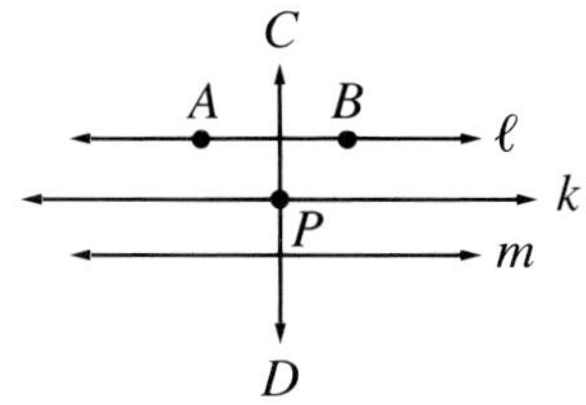

(4) Since $\overleftrightarrow{CD}$ intersects line k at *only one point*, P, this is the only point that satisfies both conditions. Point P is *both* equidistant from parallel lines ℓ and m *and also* equidistant from the two points A and B.

Answer: 1 point

---

2. How many points in a plane are 2 units from a given line and 3 units from a given point on the line?

Solution (1) Draw a line, m, and place one point, A, on the line to represent the given conditions.

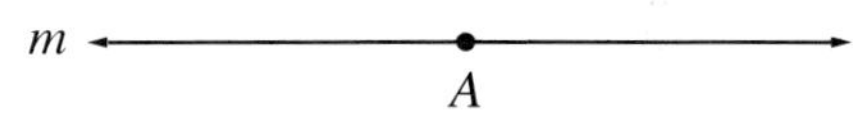

(2) The locus of points 2 units from given line m is a pair of parallel lines, k and t, each 2 units from line m. Draw lines k and t, and show that each is 2 units from line m.

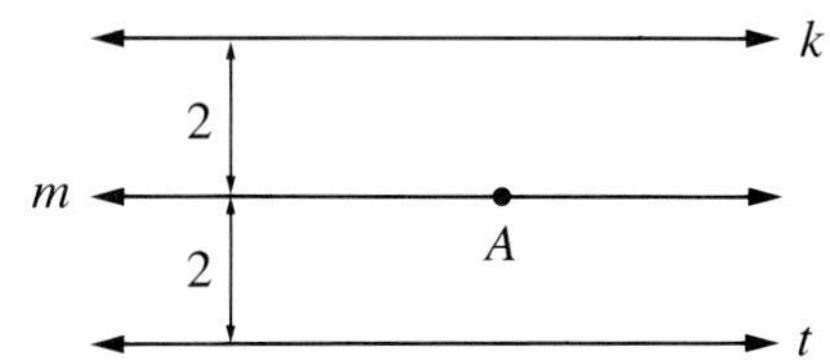

(3) The locus of points 3 units from point A is a circle whose center is A and whose radius has length 3. Draw circle A (use compasses for best results).

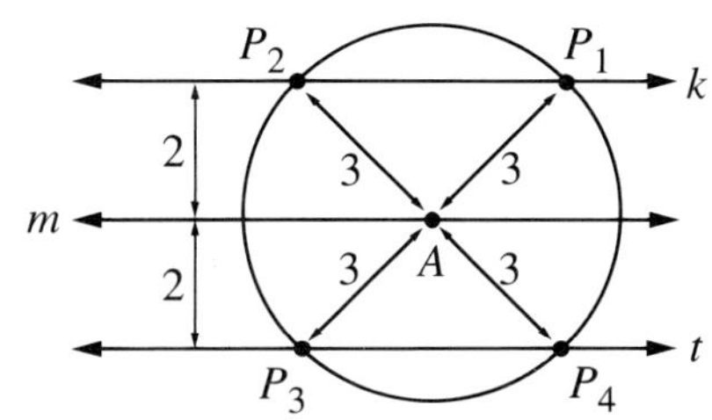

(4) The circle intersects the locus of the first condition at four points: P_1, P_2, P_3, and P_4. Since each of these four points is 3 units from point A and also 2 units from line m, there are four points that satisfy both conditions.

Answer: 4 points

Note. If you use graph paper, you will find it easier to draw parallel lines and count given units.

---

3. **a.** Write an equation of the locus for all points 3 units from the y-axis.
b. Describe fully the locus of points r units from point $A(1, 3)$.
c. Find the number of points that simultaneously satisfy the conditions stated in parts **a** and **b** when: (1) $r = 1$ (2) $r = 4$.
d. Write the coordinates of all points that simultaneously satisfy the conditions stated in parts **a** and **b** when $r = 2$.

Solutions **a.** The locus of points 3 units from the y-axis consists of two lines, each parallel to the y-axis, lines $x = 3$ and $x = -3$.
b. The locus of points is a circle with center $A(1, 3)$ and radius r.

c. (1) When the circle has a radius of length $r = 1$, there are no points in the intersection of the loci.

Graph, part **c** (1): $r = 1$.

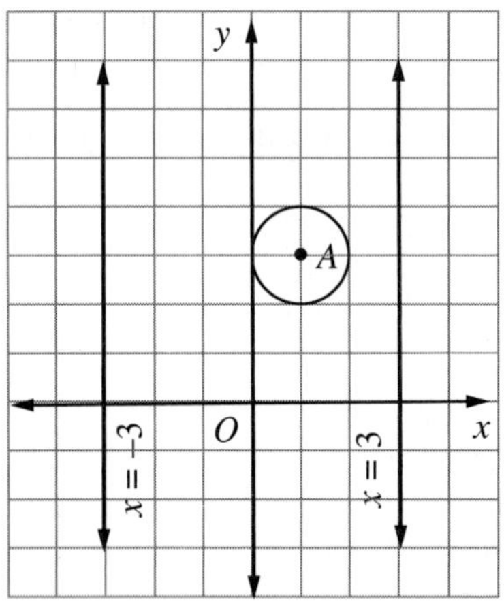

(2) When the circle has a radius of length $r = 4$, there are three points (P_1, P_2, P_3) in the intersection of the loci.

Graph, part **c** (2): $r = 4$.

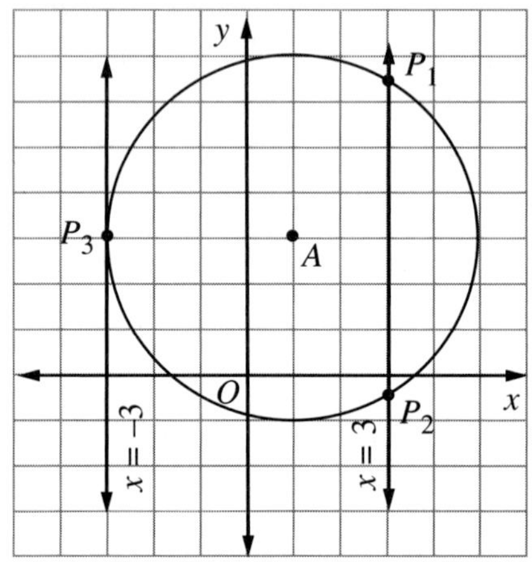

d. When the circle has a radius of length $r = 2$, there is one point, P_1, in the intersection of the loci. The coordinates of P_1 are $(3, 3)$.

Graph, part **d**: $r = 2$

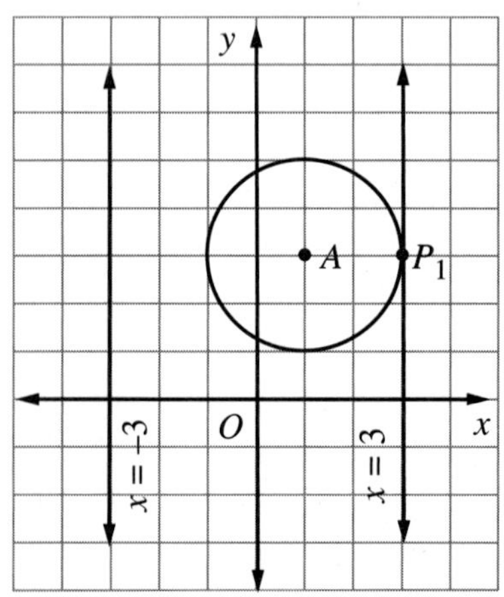

Answers:

a. $x = 3$ or $x = -3$, or $x = \pm 3$
b. a circle with center $A(1, 3)$ and radius r
c. (1) no points (2) 3 points
d. $(3, 3)$

EXERCISES WITH OPEN-RESPONSE PROBLEMS

1. What is the number of points in a plane at a given distance from a given line and also equidistant from two points on the given line?
2. How many points are there in a plane that are 4 centimeters from a given line and also 5 centimeters from a given point on that line?
3. How many points are there in a plane that are 5 centimeters from a given line and also 5 centimeters from a given point on that line?
4. Two points, A and B, are 6 inches apart. How many points are there that are equidistant from both A and B and also 5 inches from A?
5. How many points are there that are equidistant from two given points, A and B, and also 2 inches from the line passing through A and B?
6. $\overleftrightarrow{LM}$ and $\overleftrightarrow{RS}$ are two parallel lines 10 millimeters apart, and A is a point on $\overleftrightarrow{LM}$. How many points are there that are equidistant from $\overleftrightarrow{LM}$ and $\overleftrightarrow{RS}$ and 7 millimeters from A?
7. Two lines, $\overleftrightarrow{AB}$ and $\overleftrightarrow{CD}$, intersect at E. How many points are 2 units from E and also equidistant from $\overleftrightarrow{AB}$ and $\overleftrightarrow{CD}$?
8. Point P is 1 unit from a line, $\overleftrightarrow{AB}$. How many points in the plane are 2 units from $\overleftrightarrow{AB}$ and also 4 units from P?
9. Given $\angle ABC$, what is the number of points in the interior of the angle that are equidistant from the sides of the angle, and also equidistant from points A and B?

In 10–14, use the treasure map, which shows that tree A is 8 meters north of a house and tree B is 18 meters from tree A. Copy the map; then state, based on each set of conditions, the number of points where the treasure may be buried.

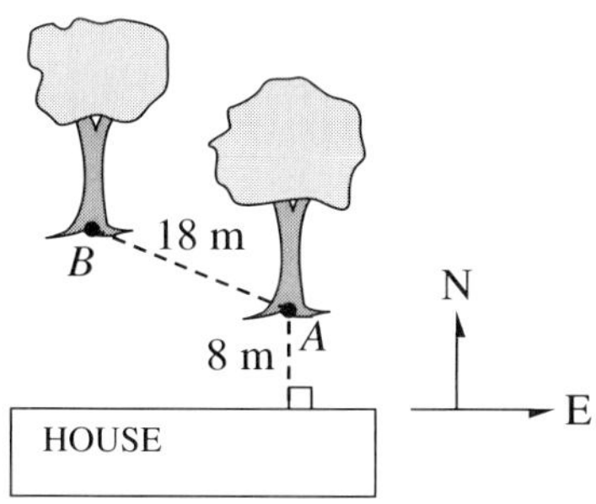

10. The treasure is 6 meters from A and 12 meters from B.
11. The treasure is 10 meters north of the house and 3 meters from A.
12. The treasure is 10 meters from A and 15 meters from B.
13. The treasure is equidistant from A and the line representing the north wall of the house, and also equidistant from A and B.
14. The treasure is equidistant from A and B and also 10 meters from B.

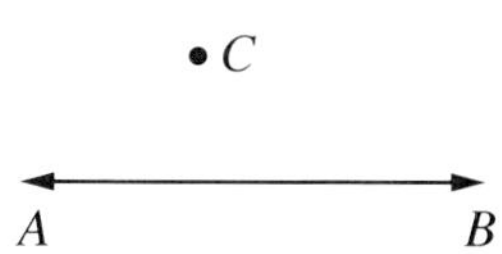

15. Given a line, $\overleftrightarrow{AB}$, and a point, C, that is 4 units from $\overleftrightarrow{AB}$.
    a. Sketch the locus of points equidistant from C and $\overleftrightarrow{AB}$.
    b. Sketch the locus of points 4 units from $\overleftrightarrow{AB}$.
    c. How many points simultaneously satisfy the conditions described in parts **a** and **b**?

16. Two points, A and B, are 7 inches apart. How many points are there that are 12 inches from A and also 4 inches from B?

In 17–26, select the numeral preceding the choice that best completes the statement or answers the question.

17. The number of points equidistant from two given parallel lines and at the same time equidistant from two given points on one of these lines is (1) 1 (2) 2 (3) 3 (4) 4

18. The number of points that are at a given distance from a given line and also equidistant from two given points on the line is (1) 1 (2) 2 (3) 3 (4) 4

19. The number of points in a plane 1 centimeter from a given line and 2 centimeters from a given point on the line is (1) 1 (2) 2 (3) 0 (4) 4

20. The number of points in a plane 2 centimeters from a given line and 1 centimeter from a given point on the line is (1) 1 (2) 2 (3) 0 (4) 4

21. Point C is 2 units from a line, $\overleftrightarrow{AB}$. How many points in $\overleftrightarrow{AB}$ are 3 units from point C?
(1) 1 (2) 2 (3) 0 (4) 4

22. $\overline{AB}$ is 1 centimeter long. How many points in the plane are 2 centimeters from both A and B?
(1) 1 (2) 2 (3) 0 (4) 4

23. Parallel lines k and t are 6 millimeters apart, and A is a point on line t. The number of points equidistant from k and t and also 3 millimeters from point A is (1) 1 (2) 2 (3) 0 (4) 4

24. $\overleftrightarrow{AB}$ and $\overleftrightarrow{CD}$ are parallel and are 6 inches apart. Point P is on $\overleftrightarrow{AB}$. The number of points equidistant from lines $\overleftrightarrow{AB}$ and $\overleftrightarrow{CD}$ and also 5 inches from point P is
(1) 1 (2) 2 (3) 0 (4) 4

25. Point P is 7 units from a given line. The number of points that are 3 units from the line and also 10 units from point P is (1) 1 (2) 2 (3) 3 (4) 4

26. In $\triangle ABC$, the point equidistant from the three vertices, A, B, and C, is the intersection of
(1) the angle bisectors
(2) the perpendicular bisectors of the sides
(3) the medians
(4) none of the above

Locus and Coordinates

27. Write an equation of the locus of points equidistant from points $(0, -3)$ and $(0, 7)$.

28. **a.** Draw the locus of points equidistant from points (4, 1) and (4, 5), and write an equation for this locus.
b. Draw the locus of points equidistant from points (3, 2) and (9, 2), and write an equation for this locus.
c. Find the number of points that satisfy both conditions stated in parts **a** and **b**. Give the coordinates of each point found.

29. **a.** Represent graphically the locus of points (*1*) 3 units from line $x = 1$ (2) 4 units from line $y = -2$
b. Write the equations for the loci represented in part **a**.
c. Find the coordinates of the points of intersection of these loci.

30. **a.** Represent graphically the locus of points: (*1*) 8 units from the y-axis (*2*) 10 units from the origin
b. Write equations for the loci represented in part **a**.
c. Find the coordinates of the points of intersection of these loci.

31. The vertices of $\triangle RST$ are $R(2, 3)$, $S(6, 3)$, and $T(3, 10)$.
a. Draw the graph and write an equation of the locus of points equidistant from R and S.
b. Draw a line, $\overleftrightarrow{TW}$, that passes through T and is parallel to $\overleftrightarrow{RS}$, and write an equation of $\overleftrightarrow{TW}$.
c. Write the coordinates of the points of intersection of the lines drawn in parts **a** and **b**.

32. **a.** Draw the locus of points equidistant from the circles whose equations are $x^2 + y^2 = 4$ and $x^2 + y^2 = 36$. Write an equation of the locus.
b. Draw the locus of points 4 units from the x-axis. Write an equation of the locus.
c. Find the number of points that satisfy both conditions stated in parts **a** and **b**. Write the coordinates of each of the points found.

33. **a.** Using graph paper, draw the triangle whose vertices are $A(1, 2)$, $B(7, 2)$, and $C(5, 6)$.
b. Draw the locus of points equidistant from A and B.
c. Write an equation of the locus drawn in answer to part **b**.
d. Write an equation of the line passing through A and C.
e. Find the coordinates of point P if P is a point on $\overleftrightarrow{AC}$ that is equidistant from points A and B.

34. The vertices of $\triangle ABC$ are $A(0, 0)$, $B(6, 0)$, and $C(0, 8)$.
a. Write an equation of the perpendicular bisector of $\overline{AB}$.
b. Write an equation of the perpendicular bisector of $\overline{AC}$.
c. Write the coordinates of the intersection of these loci.
d. Write the coordinates of the midpoint of $\overline{BC}$.
e. Will the perpendicular bisectors of the sides of $\triangle ABC$ pass through the same point?

35. **a.** Write an equation of the locus of points 2 units from the x-axis.
b. Describe fully the locus of points at a distance d from $P(2, 6)$.
c. Find the number of points that simultaneously satisfy the conditions in parts **a** and **b** for each of the following values of d:
(*1*) $d = 2$ (*2*) $d = 4$ (*3*) $d = 6$ (*4*) $d = 8$ (*5*) $d = 10$

36. **a.** Write the equation of the locus of points 5 units from point $(4, 0)$.
b. Describe fully the locus of points at distance d from the x-axis.
c. Find the number of points that simultaneously satisfy the conditions in parts **a** and **b** when:
(*1*) $d = 1$ (*2*) $d = 5$ (*3*) $d = 10$
d. Find the coordinates for all points 5 units from point $(4, 0)$ and also 5 units from the x-axis.
e. Find the coordinates for all points 5 units from point $(4, 0)$ and also 3 units from the x-axis.

37. The vertices of $\triangle RST$ are $R(1, 2)$, $S(9, 2)$, and $T(9, 8)$.
a. Write an equation of the locus of points equidistant from R and S.
b. Write an equation of the locus of points equidistant from S and T.
c. Find the coordinates of all points that simultaneously satisfy the conditions in parts **a** and **b**.
d. Write an equation of the locus of points 5 units from T.
e. Find the coordinates of all points that are 5 units from T and also intersect one side of $\triangle RST$.

CHAPTER SUMMARY

A ***locus*** is the set of all points, and only those points, that satisfy a given set of conditions. Locus (plural, *loci*) may be studied with or without coordinates.

A locus may be both drawn and described fully in words. When coordinates are given as part of the condition, the locus may also be described by an equation.

The conditions for fundamental loci typically include fixed points, fixed lines, and given distances, as seen in the following loci.

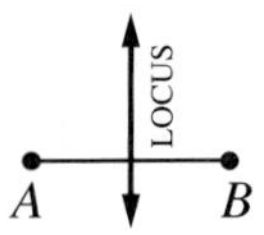

Locus 1

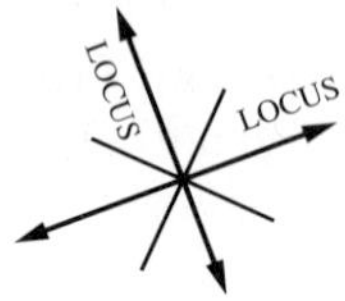

Locus 2

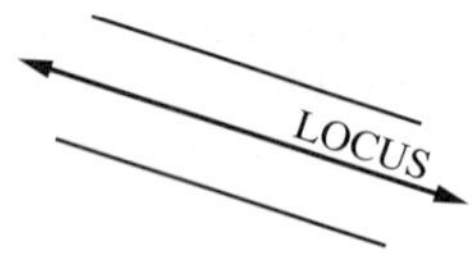

Locus 3

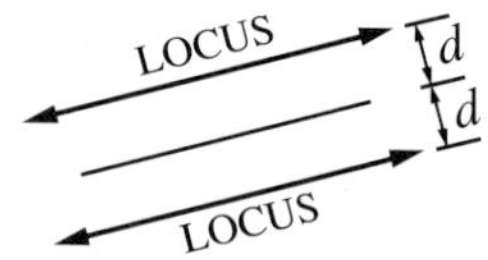

Locus 4

1. The locus of points equidistant from two fixed points, A and B, is the perpendicular bisector of the segment $\overline{AB}$.
2. The locus of points equidistant from two intersecting lines is a pair of lines that bisect the angles formed by the intersecting lines.
3. The locus of points equidistant from two parallel lines is a third line, parallel to both lines and midway between them.
4. The locus of points at a given distance d from a fixed line is a pair of lines, each parallel to the fixed line.

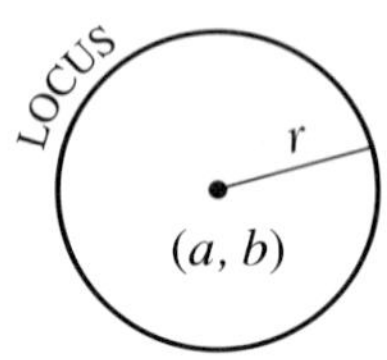

Locus 5

5. The locus of points at a given distance r from a fixed point is a circle whose center is the point and whose radius has length r.

   The equation of a ***circle*** whose center is point (a, b) and whose radius has length r is

   $$(x - a)^2 + (y - b)^2 = r^2$$

   When the center of this circle is the origin, its equation is $x^2 + y^2 = r^2$.

VOCABULARY

23-1 Locus Circle

23-2 Concentric circles

REVIEW EXERCISES WITH OPEN-RESPONSE PROBLEMS

In 1–5, describe fully each of the following loci. Points that are:

1. equidistant from two parallel lines.
2. equidistant from two points.

3. equidistant from two intersecting lines.
4. at a given distance d from line $x = 0$.
5. at a given distance d from point (3, 1).

In 6–13, write an equation for each locus.

6. Points in which the abscissa is 3 less than the ordinate.
7. Points on a line that is parallel to the x-axis and passes through point $(7, -1)$.
8. Points 8 units from the y-axis.
9. Points 8 units from the origin.
10. Points equidistant from $A(3, 2)$ and $B(3, 6)$.
11. Points equidistant from $A(3, 2)$ and $D(-9, 2)$.
12. Points equidistant from the two lines $x = 7$ and $x = 13$.
13. Points 3 units from fixed point (2, 6).

In 14–17, for each circle whose equation is given, find:
a. the coordinates of the center **b.** the length of the radius

14. $x^2 + y^2 = 16$
15. $3x^2 + 3y^2 = 108$
16. $(x - 5)^2 + y^2 = 100$
17. $(x + 1)^2 + (y - 9)^2 = 10$

In 18–25, select the numeral preceding the choice that best completes the statement or answers the question.

18. An equation that represents the locus of all points 12 units to the right of the y-axis is
(1) $y = 12$ (2) $x = 12$ (3) $y = -12$ (4) $x = -12$

19. Which of the following are the coordinates of a point on the circle whose equation is $x^2 + y^2 = 17^2$?
(1) (1, 16) (2) $(-8, 15)$ (3) $(10, -7)$ (4) $(12, -12)$

20. An equation of the locus of points equidistant from points (6, 8) and $(-2, 8)$ is
(1) $x = 2$ (2) $x = 4$ (3) $y = 2$ (4) $y = 8$

21. The number of points equidistant from both the x-axis and the y-axis and also 3 units from the origin is
(1) 1 (2) 2 (3) 3 (4) 4

22. Parallel lines p and q are 6 centimeters apart, and point A is on line p. The number of points equidistant from p and q and also 3 centimeters from point A is
(1) 1 (2) 2 (3) 3 (4) 4

23. How many points are 3 units from $\overleftrightarrow{AB}$ and 4 units from B?
(1) 1 (2) 2 (3) 3 (4) 4

24. The number of points equidistant from two parallel lines and also equidistant from two points on one of the lines is
(1) 1 (2) 2 (3) 0 (4) 4

25. If the length of $\overline{CD}$ is 20 millimeters, how many points are simultaneously 15 millimeters from C and 10 millimeters from D?
(1) 1 (2) 2 (3) 0 (4) 4

26. Given $\triangle ABC$ with vertices $A(-1, 1)$, $B(5, 1)$, and $C(2, 3)$.
a. Draw the locus of points equidistant from A and B.
b. Write an equation of the locus described in part **a**.
c. Draw the locus of points 4 units from point C.
d. Write an equation of the locus described in part **c**.
e. Write the coordinates of the points that simultaneously satisfy the loci described in parts **a** and **c**.

27. **a.** On graph paper, draw the locus of points 3 units from the line whose equation is $x = 1$.
b. Write an equation of the locus described in part **a**.
c. Describe fully the locus of points at a distance d from (2, 7).
d. Write an equation of the locus described in part **c**.
e. How many points simultaneously satisfy the conditions in parts **a** and **b** when: *(1)* $d = 1$ *(2)* $d = 3$ *(3)* $d = 5$
f. Find the coordinates of all points that are 3 units from line $x = 1$ and also 2 units from (2, 7).

28. **a.** Write an equation of the locus of points 13 units from the origin.
b. Write an equation of the locus of points equidistant from $x = 0$ and $x = 10$.
c. Solve algebraically the system of equations written in answer to parts **a** and **b** to find the coordinates of points 13 units from the origin and equidistant from lines $x = 0$ and $x = 10$.

29. **a.** Write an equation of the locus of points 10 units from the origin.
b. Write an equation of the locus of points whose ordinate is 2 units more than its abscissa.
c. Solve algebraically the system of equations written in answer to parts **a** and **b** to find the coordinates of each point 10 units from the origin whose ordinate is 2 more than its abscissa.

CUMULATIVE REVIEW WITH OPEN-RESPONSE PROBLEMS

1. A hexagon has angles whose measures are 100° and 140°. The measures of the other four angles are in the ratio 2:3:3:4. Find the measures of the angles of the hexagon.

2. Write an equation of the line that intersects the y-axis at (0, 1) and whose slope is $-\frac{2}{3}$.

3. In the figure, $\overline{DE} \parallel \overline{AB}$, $AD = 2$ meters, $CD = 6$ meters and AB is 3 meters longer than DE. Find DE and AB.

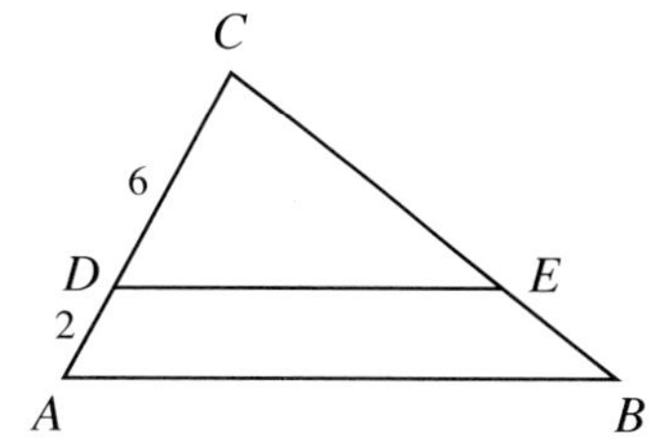

4. Find the roots of $2x^2 - 5x + 1 = 0$ **a.** in simplest radical form **b.** to the *nearest hundredth.*

Exploration

Create a treasure map like the one shown on page 853 for exercises 10-14. Write out instructions for finding the treasure using loci. Exchange maps with a classmate and try to find the treasure using the instructions given. After a reasonable time, discuss with your classmate how easy or difficult it was to find the treasure. If you were unable to find the treasure, discuss with your classmate how the instructions could be improved to make it easier to complete the task.

Chapter 24

Constructions

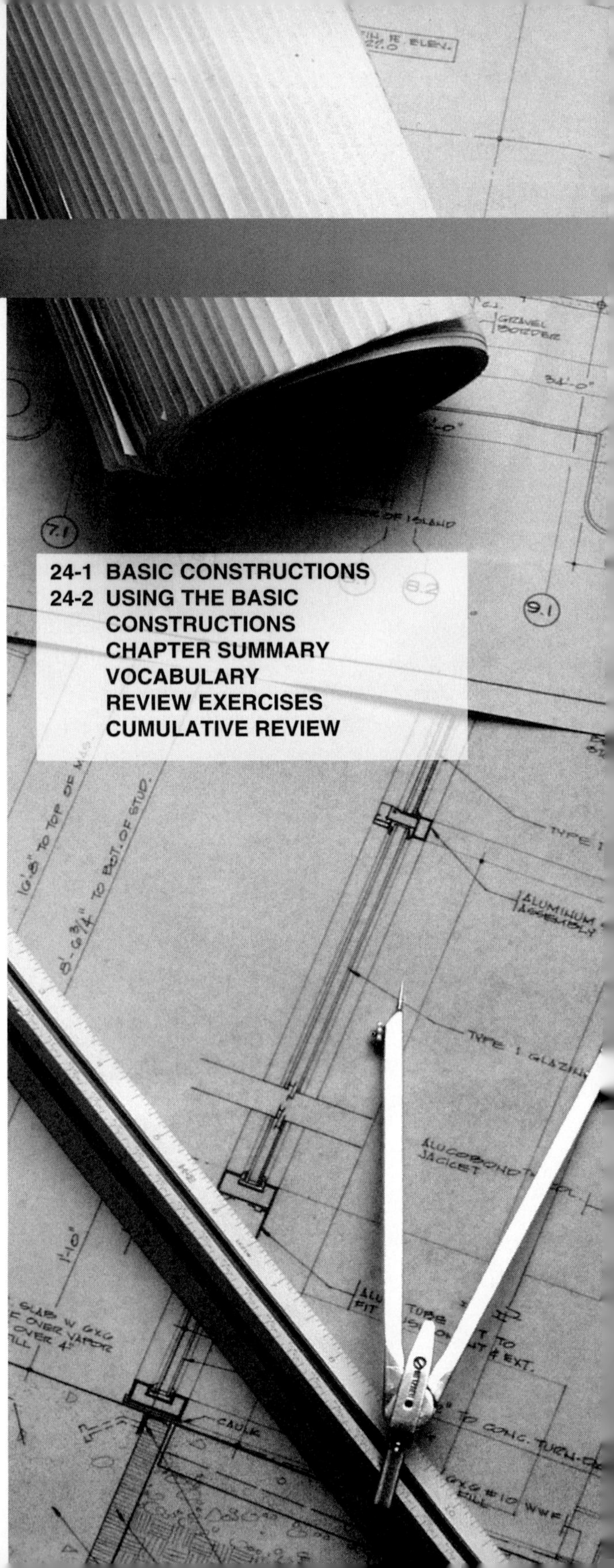

When blueprints are *drawn*, a draftsperson may use rulers to measure lengths, protractors to measure angles, and parallel rulers to draw parallel lines. Today, other devices such as drafting machines and computer-aided design programs are also used.

When you *draw* geometric figures in your work, you are also permitted to use tools such as a ruler and protractor. However, when you *construct* a geometric figure, these devices are *not* allowed.

To ***construct*** a geometric figure means that a specific design is accurately made by using only two instruments:

1. A pair of ***compasses***, which is a V-shaped device used to locate points at a given distance from a fixed point, and to describe circles and circular arcs.
2. A ***straightedge***, which is an unmarked ruler or a flat bar used to draw a straight line by connecting two given points.

This chapter presents important constructions in geometry, methods to perform these constructions, and plans for proving that the constructions are valid.

24-1 BASIC CONSTRUCTIONS

The six constructions presented in this section are the basic procedures. In all of these constructions, a pair of compasses is used. The point of the compasses is fixed as the center of a circle, while the pencil point draws an arc, a part of a circle. If the distance between the points of the pair of compasses is not changed, all arcs drawn are arcs of congruent circles and all line segments from a center to a point on an arc are radii of congruent circles.

To prove that a construction has accomplished what was required, we use the following postulate.

- **POSTULATE. Radii of congruent circles are congruent.**

Construction 1

To construct a line segment congruent (equal in length) to a given line segment.

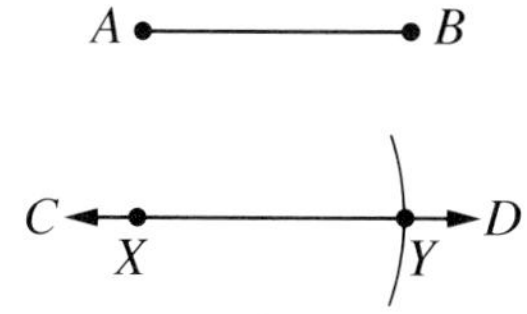

Given: $\overline{AB}$.

Construct: A line segment, $\overline{XY}$, congruent to $\overline{AB}$.

How to Proceed:

1. With a straightedge, draw any line, $\overleftrightarrow{CD}$, and mark a point, X, on it.
2. On $\overline{AB}$, place the compasses so that the point is at A and the pencil point is at B.
3. Keeping the setting on your compasses, place the point at X and draw an arc intersecting $\overleftrightarrow{CD}$ at Y.

Conclusion: $\overline{XY} \cong \overline{AB}$.

Plan for Proof: The construction makes $\overline{XY} \cong \overline{AB}$ because $\overline{XY}$ and $\overline{AB}$ are radii of congruent circles.

Construction 2

To bisect a given line segment.
[To construct the perpendicular bisector of a given line segment.]

Given: $\overline{AB}$.

Construct: A bisector of $\overline{AB}$.

How to Proceed:

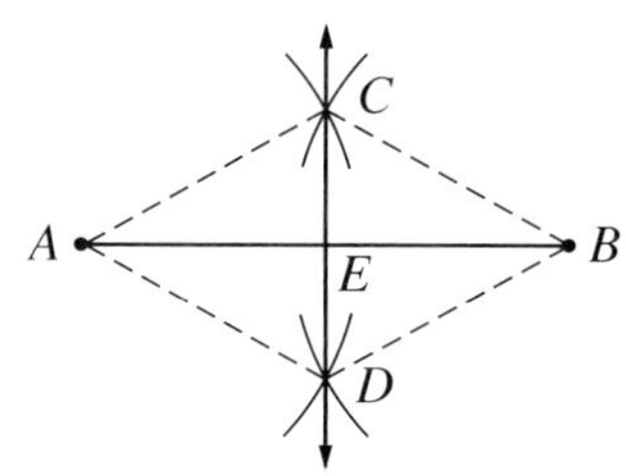

1. Open the compasses so that the distance between the point and the pencil point (this distance will be called the radius) is more than half the length of $\overline{AB}$.
2. Using point A as a center, draw one arc above $\overline{AB}$ and one arc below $\overline{AB}$.
3. Using the same radius and point B as a center, draw another pair of arcs, one

above $\overline{AB}$ and the other below $\overline{AB}$, that intersect the first pair of arcs at points C and D.

4. Use a straightedge to draw $\overleftrightarrow{CD}$ intersecting $\overline{AB}$ at E.

Conclusion: The construction makes $\overleftrightarrow{CD}$ the perpendicular bisector of $\overline{AB}$. Therefore, $\overline{AE} \cong \overline{EB}$.

Plan for Proof: If $\overline{CA}$, $\overline{CB}$, $\overline{DA}$, and $\overline{DB}$ are drawn, the construction makes $\overline{CA} \cong \overline{CB}$ and $\overline{DA} \cong \overline{DB}$ (radii of congruent circles are congruent). Therefore, $\overleftrightarrow{CD}$ is the perpendicular bisector of $\overline{AB}$ because two points, each equidistant from the endpoints of a line segment, determine the perpendicular bisector of the line segment.

Construction 3

To construct an angle congruent to a given angle.

Given: $\angle ABC$ and point D.

Construct: At point D, an angle congruent to $\angle ABC$.

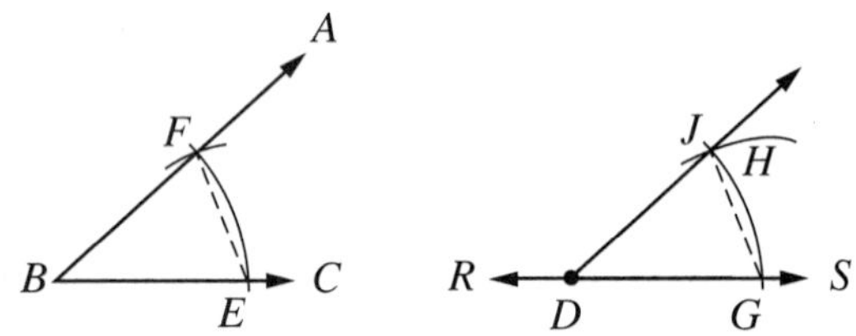

How to Proceed:

1. Through point D, draw any line, $\overleftrightarrow{RS}$.
2. With B as a center and any convenient radius, draw an arc that intersects $\overrightarrow{BC}$ at E and $\overrightarrow{BA}$ at F.
3. With D as a center and using the same radius as in step 2, draw an arc that intersects $\overrightarrow{DS}$ at G. Label this arc $\overparen{GJ}$.
4. With the compasses, measure distance EF. With G as a center and a radius whose length is EF, draw an arc that intersects $\overparen{GJ}$ at H.
5. Draw $\overrightarrow{DH}$.

Conclusion: $\angle HDS \cong \angle ABC$.

Plan for Proof: If $\overline{GH}$ is drawn, the construction makes $\overline{DH} \cong \overline{BF}$, $\overline{DG} \cong \overline{BE}$, $\overline{GH} \cong \overline{EF}$. Therefore, $\triangle DHG \cong \triangle BFE$ by s.s.s. $\cong$ s.s.s., and $\angle HDS \cong \angle ABC$.

Construction 4

To bisect a given angle.
[To construct the bisector of a given angle.]

Given: $\angle ABC$.

Construct: The bisector of $\angle ABC$.

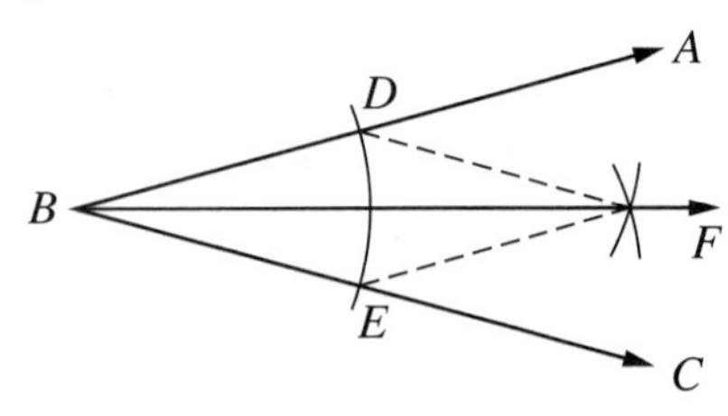

How to Proceed:

1. With B as a center and any convenient radius, draw an arc that intersects $\overrightarrow{BA}$ at D and $\overrightarrow{BC}$ at E.
2. With D and E as centers and with equal radii of sufficient length, draw arcs that intersect at F.
3. Draw $\overrightarrow{BF}$.

Conclusion: Since $\angle ABF \cong \angle CBF$, then $\overrightarrow{BF}$ bisects $\angle ABC$.

Plan for Proof: If $\overline{FD}$ and $\overline{FE}$ are drawn, then the construction makes $\overline{BD} \cong \overline{BE}$ and $\overline{DF} \cong \overline{EF}$. Also, $\overline{BF} \cong \overline{BF}$. Thus $\triangle BDF \cong \triangle BEF$ by s.s.s. $\cong$ s.s.s., and, $\angle ABF \cong \angle CBF$.

Construction 5

To construct a line perpendicular to a given line through a given point on the line.

Given: Point P is on $\overleftrightarrow{AB}$.

Construct: A line through point P perpendicular to $\overleftrightarrow{AB}$.

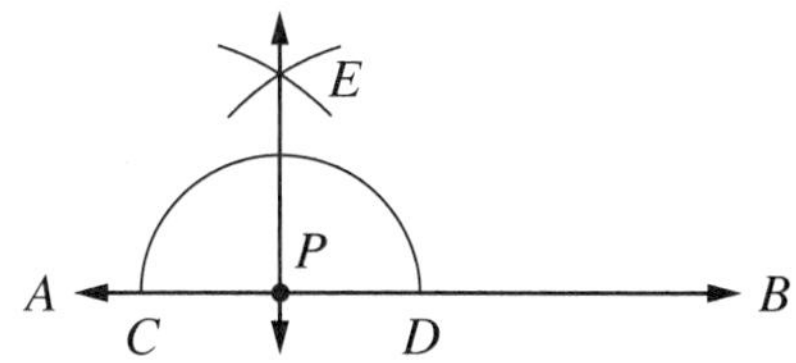

How to Proceed:

1. With P as a center and any convenient radius, draw an arc that intersects $\overleftrightarrow{AB}$ at points C and D.
2. With C and D as centers and with a radius greater in length than the one used before, draw arcs that intersect at E.
3. Draw $\overleftrightarrow{EP}$.

Conclusion: $\overleftrightarrow{EP}$ is perpendicular to $\overleftrightarrow{AB}$ at P.

Plan for Proof: The construction bisects straight angle APB. Angles APE and BPE are right angles. Therefore, $\overleftrightarrow{EP}$ is perpendicular to $\overleftrightarrow{AB}$ because two lines that intersect at right angles are perpendicular to each other.

Construction 6

To construct a line perpendicular to a given line through a given point outside the line.

Given: Point P is outside $\overleftrightarrow{AB}$.

Construct: A line through point P perpendicular to $\overleftrightarrow{AB}$.

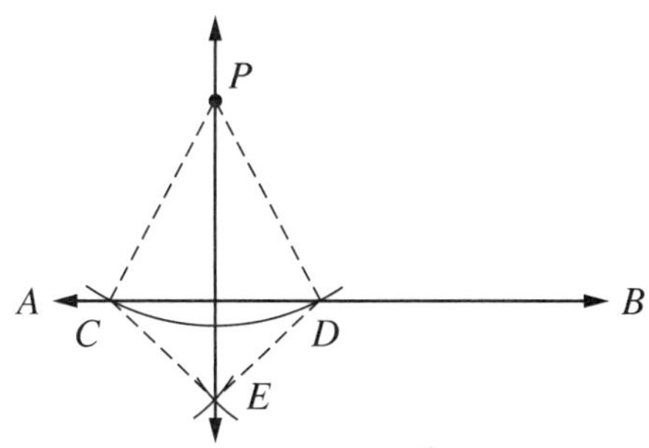

How to Proceed:

1. With P as a center and any convenient radius, draw an arc that intersects $\overleftrightarrow{AB}$ at C and D.
2. With C and D as centers and with a radius greater in length than $\frac{1}{2}(CD)$, draw arcs that intersect at E.
3. Draw $\overleftrightarrow{EP}$.

Conclusion: $\overleftrightarrow{EP}$ is perpendicular to $\overleftrightarrow{AB}$.

Plan for Proof: If $\overline{PC}$, $\overline{PD}$, $\overline{EC}$, and $\overline{ED}$ are drawn, the construction makes $\overline{PC} \cong \overline{PD}$ and $\overline{EC} \cong \overline{ED}$. Therefore, $\overleftrightarrow{PE}$ is the perpendicular bisector of $\overline{CD}$ because two points each equidistant from the endpoints of a line segment determine the perpendicular bisector of the line segment. Hence, $\overleftrightarrow{PE}$ is perpendicular to $\overleftrightarrow{AB}$.

EXERCISES WITH OPEN-RESPONSE PROBLEMS

In 1–6, draw each given figure and do the required constructions. (Where needed, enlarge the given figures for your convenience.)

1. *Given:* Line segment AB.

Constructions:
a. Copy $\overline{AB}$.
b. Construct a line segment whose measure is twice AB.
c. Construct the perpendicular bisector of $\overline{AB}$.
d. Construct a line segment whose measure is $2\frac{1}{2}(AB)$.

2. *Given:* Acute angle A.

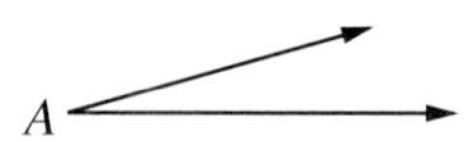

Constructions:
a. Copy $\angle A$.
b. Construct an angle whose measure is $2(m\angle A)$.
c. Construct the bisector of $\angle A$.
d. Construct an angle whose measure is $1\frac{1}{2}(m\angle A)$.

3. *Given:* Line segments a and b.

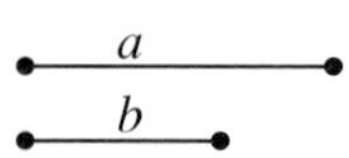

Constructions:
a. Construct an isosceles triangle with base a and leg b.
b. Construct an equilateral triangle with side a.
c. Construct a line segment whose measure is $a + b$.
d. Construct a line segment whose measure is $a - b$.

4. *Given:* Angle whose measure is 60° and whose vertex is P.

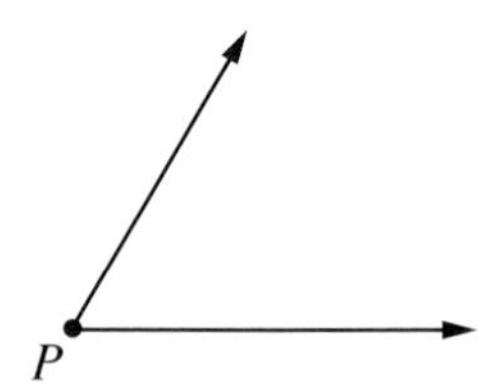

Constructions:
a. Construct an angle whose measure is 30° and whose vertex is P.
b. Construct an angle whose measure is 15° and whose vertex is P.
c. Construct an angle whose measure is 75° and whose vertex is P.

5. *Given:* Point P not on line ℓ.

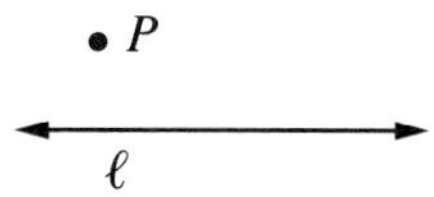

Constructions:

a. Construct a line through P that is perpendicular to ℓ.
b. Construct an angle whose measure is 45°.
c. Construct an angle whose measure is 135°.

6. *Given:* Acute $\triangle ABC$.

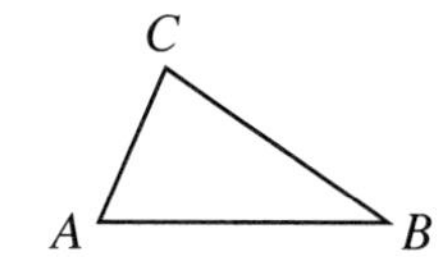

Constructions:

a. Construct by s.s.s. a triangle that is congruent to $\triangle ABC$.
b. Construct by s.a.s. a triangle that is congruent to $\triangle ABC$.
c. Construct by a.s.a. a triangle that is congruent to $\triangle ABC$.

7. Draw any two points, A and P. Construct A', the image of A under a reflection in P.

8. Draw any line, $\overleftrightarrow{PQ}$, and any point, A, that is not on $\overleftrightarrow{PQ}$. Construct A', the image of A under a reflection in $\overleftrightarrow{PQ}$.

9. Draw any two points, A and P.

a. Construct A', the image of A under a rotation of 90° about P in the counterclockwise direction.
b. Construct A'', the image of A under a rotation of 45° about P in the clockwise direction.

10. Prove Construction 1.
11. Prove Construction 2.
12. Prove Construction 3.
13. Prove Construction 4.
14. Prove Construction 5.
15. Prove Construction 6.

24-2 USING THE BASIC CONSTRUCTIONS

Each of the constructions that follows uses one or more of the basic constructions. In each case, some definition or previously shown principle is used to develop the method of construction.

Construction 7

To construct an altitude of a given triangle.

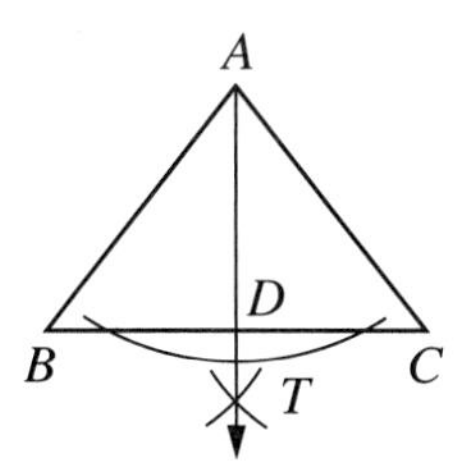

Acute Triangle

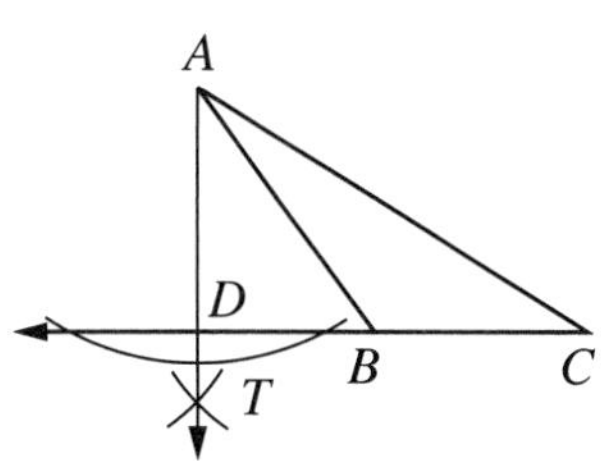

Obtuse Triangle

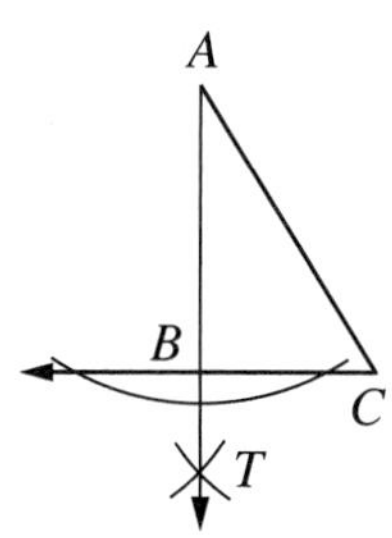

Right Triangle

Given: $\triangle ABC$.

Construct: The altitude from vertex A to side $\overline{CB}$.

How to Proceed:

Through point A, construct $\overrightarrow{AT}$, a ray perpendicular to $\overline{CB}$. Extend $\overleftrightarrow{CB}$ through an endpoint of $\overline{CB}$ if necessary.

Conclusion:

In acute triangle ABC, $\overrightarrow{AT}$ intersects $\overline{CB}$ at D. $\overline{AD}$ is the altitude from vertex A to side $\overline{CB}$.

In obtuse triangle ABC, $\overrightarrow{AT}$ intersects $\overleftrightarrow{CB}$, the line containing $\overline{CB}$, at D. $\overline{AD}$ is the altitude from vertex A to side $\overline{CB}$.

In right triangle ABC, $\overrightarrow{AT}$ intersects $\overline{CB}$ at B. $\overline{AB}$ is the altitude from vertex A to side $\overline{CB}$.

Plan for Proof: For each type of triangle ABC, $\overrightarrow{AT}$ is perpendicular to $\overline{CB}$. Therefore, $\overline{AD}$ is the altitude from vertex A to side $\overline{CB}$ because an altitude of a triangle is the line segment from a vertex perpendicular to the opposite side (or the line containing that side).

Note. If the three altitudes of a triangle are constructed, they will all intersect in the same point. In the case of the acute triangle, the three altitudes will intersect in the interior of the triangle; in the case of the obtuse triangle, in the exterior of the triangle; and in the case of the right triangle, at the vertex of the right angle. Hence, the three altitudes of a triangle are *concurrent*.

Construction 8

To construct a median of a given triangle.

Given: $\triangle ABC$.

Construct: The median from vertex A to side $\overline{BC}$.

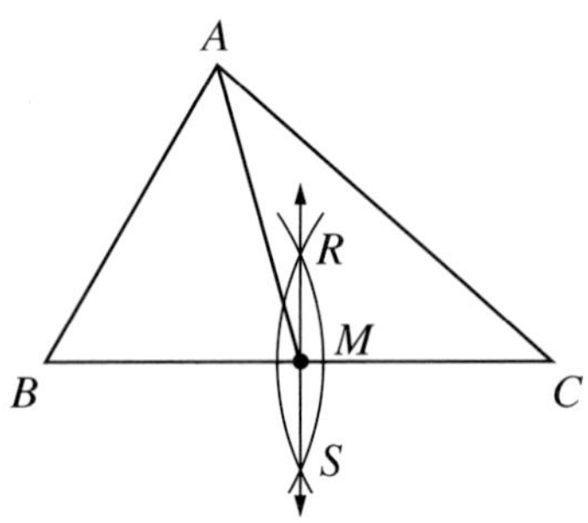

How to Proceed:

1. Construct $\overleftrightarrow{RS}$, the perpendicular bisector of $\overline{BC}$.
2. $\overleftrightarrow{RS}$ intersects $\overline{BC}$ at M.
3. Draw $\overline{AM}$.

Conclusion: $\overline{AM}$ is the median from vertex A to side $\overline{BC}$.

Plan for Proof: Since $\overleftrightarrow{RS}$ is the perpendicular bisector of $\overline{BC}$, point M is the midpoint of $\overline{BC}$. Therefore, $\overline{AM}$ is the median from vertex A to side $\overline{BC}$ because a median of a triangle is the line segment that joins a vertex and the midpoint of the opposite side.

Note. If the three medians of a triangle are constructed, they will all intersect in the same point in the interior of the triangle. Thus, the three medians of a triangle are *concurrent*.

Construction 9

To construct the bisector of an angle of a given triangle.

Given: $\triangle ABC$.

Construct: The bisector of $\angle A$.

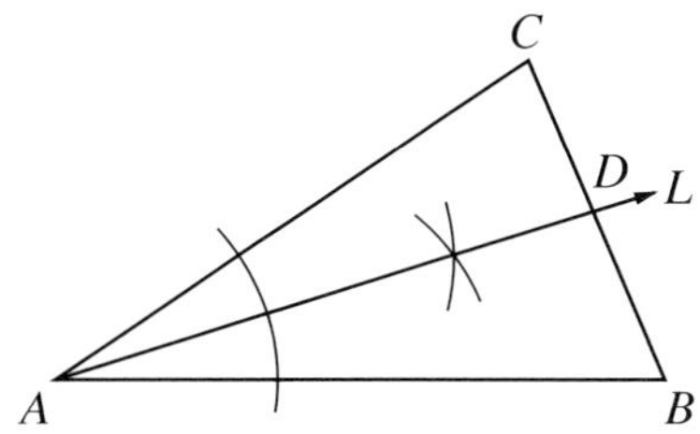

How to Proceed:

1. Construct $\overrightarrow{AL}$, the bisector of $\angle A$.
2. Label as D the point where $\overrightarrow{AL}$ intersects $\overline{BC}$, the side opposite $\angle A$.

Conclusion: $\overline{AD}$ is the bisector of $\angle A$ in $\triangle ABC$.

Plan for Proof: Since $\overrightarrow{AL}$ bisects $\angle A$, $\overline{AD}$ is the bisector of $\angle A$ in $\triangle ABC$ because a bisector of an angle of a triangle is the line segment, drawn from a vertex, that bisects that angle and that terminates in the opposite side.

Note. If the bisectors of the three angles of a triangle are constructed, they will all intersect in the same point in the interior of the triangle. Hence, the three angle bisectors of a triangle are *concurrent*.

Construction 10

To construct a line parallel to a given line through a given external point.

Given: $\overleftrightarrow{AB}$ and external point P.

Construct: Through point P, a line parallel to $\overleftrightarrow{AB}$.

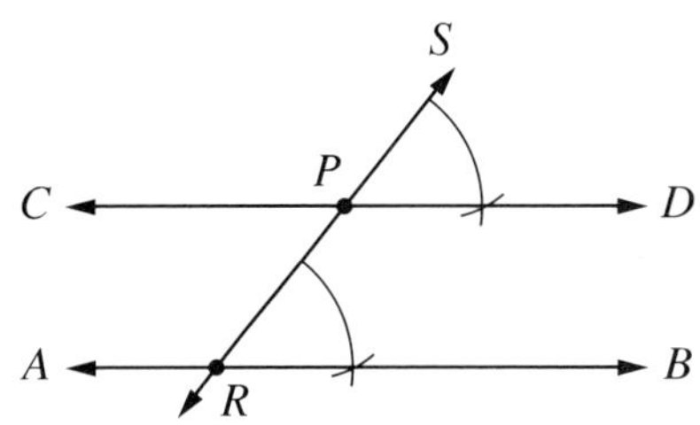

How to Proceed:

1. Through P, draw any line, intersecting $\overleftrightarrow{AB}$ at R. Let S be any point on the ray opposite $\overrightarrow{PR}$.
2. At P, construct $\angle SPD \cong \angle PRB$, to make a pair of congruent corresponding angles.

Conclusion: $\overleftrightarrow{PD}$ passes through P and is parallel to $\overleftrightarrow{AB}$.

Plan for Proof: The construction makes $\angle SPD \cong \angle PRB$. Therefore, $\overleftrightarrow{AB} \parallel \overleftrightarrow{PD}$ because, if two lines are cut by a transversal making a pair of corresponding angles congruent, the lines are parallel.

Construction 11

To construct an angle of 60° with a given point as vertex, or to construct an equilateral triangle.

Given: Point A.

Construct: An angle containing 60° whose vertex is A.

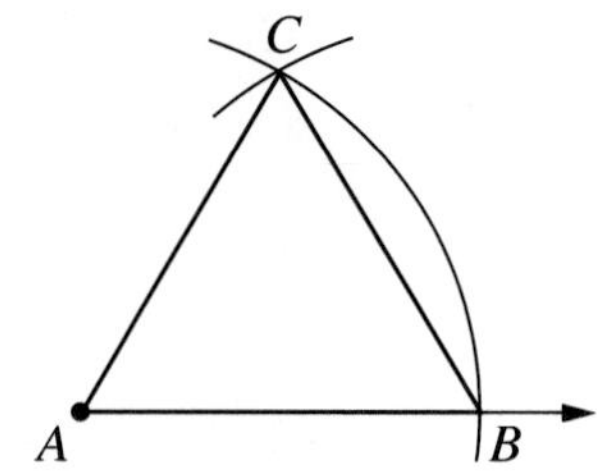

How to Proceed:

1. Draw a ray with endpoint A, and any point B.
2. Using A as the center and a radius whose length is equal to AB, draw an arc.
3. Using B as the center and a radius whose length is equal to AB, draw another arc that intersects the first arc at C.
4. Draw $\overline{CA}$ and $\overline{CB}$, forming equilateral triangle ABC.

Conclusion: $\angle CAB$ contains 60°.

Plan for Proof: The construction makes $\triangle ABC$ an equilateral triangle. Since an equilateral triangle is equiangular, $m\angle CAB = \frac{1}{3}(180)$, or $m\angle CAB = 60$.

Construction 12

To construct a triangle similar to a given triangle on a given line segment as a base.

Given: $\triangle ABC$ and $\overline{RS}$.

Construct: A triangle similar to $\triangle ABC$ on segment $\overline{RS}$, which corresponds to side $\overline{AB}$ of $\triangle ABC$.

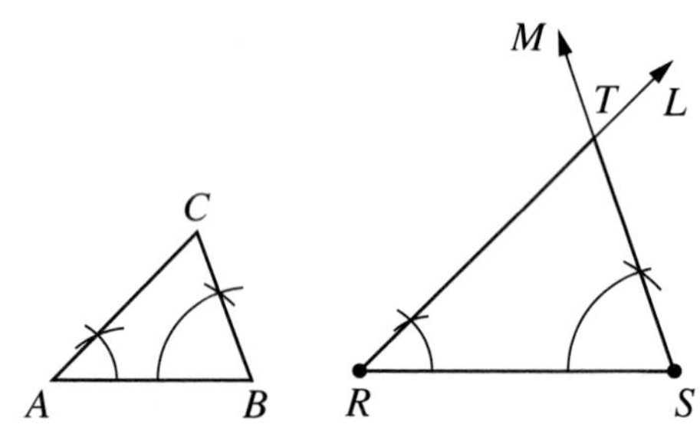

How to Proceed:

1. At R, construct $\angle SRL$ congruent to $\angle BAC$.
2. With $\overrightarrow{SR}$ as one ray, construct $\angle RSM$ congruent to $\angle ABC$.
3. Represent by T the point of intersection of $\overrightarrow{RL}$ and $\overrightarrow{SM}$.

Conclusion: The required triangle is $\triangle RST$.

Plan for Proof: The construction makes $\angle R \cong \angle A$ and $\angle S \cong \angle B$. Therefore, $\triangle RST \sim \triangle ABC$ because two triangles are similar if two angles of one triangle are congruent to two angles of the other triangle.

Construction 13

To divide a given line segment into any number of congruent parts.

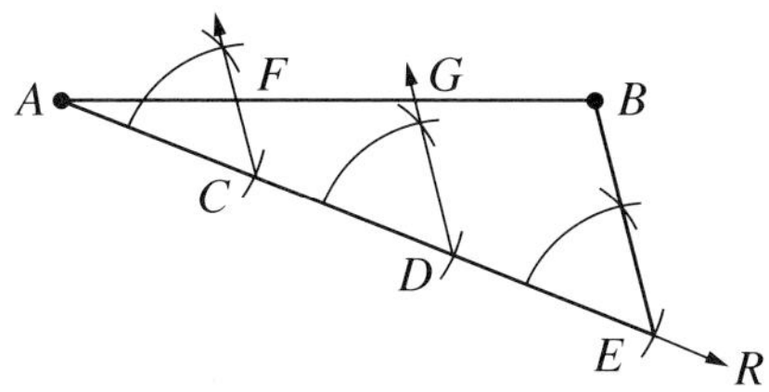

Given: $\overline{AB}$.

Construct: Congruent segments on $\overline{AB}$ such that $\overline{AB}$ is divided into any number of congruent parts (for example, three congruent parts).

How to Proceed:

1. Draw $\overrightarrow{AR}$, making any convenient angle, BAR.
2. On $\overrightarrow{AR}$, start at A and lay off any convenient length, AC, the required number of times, so that $AC = CD = DE$. Draw $\overline{EB}$.
3. Through D, construct $\overrightarrow{DG} \parallel \overline{EB}$. Through C, construct $\overrightarrow{CF} \parallel \overline{EB}$.

Conclusion: $\overline{AF}$, $\overline{FG}$, and $\overline{GB}$ are the three congruent parts of $\overline{AB}$.

Note: This construction also determines proportional segments. For example, $AF:FB = 1:2$.

Plan for Proof: The construction makes $\overline{AC} \cong \overline{CD} \cong \overline{DE}$ and also makes $\overrightarrow{CF} \parallel \overrightarrow{DG} \parallel \overline{EB}$. Therefore, $\overline{AF} \cong \overline{FG} \cong \overline{GB}$ because, if three or more parallel lines cut off segments of equal length on one transversal, they cut off segments of equal length on any transversal.

Construction 14

To construct a circle circumscribed about a given triangle.

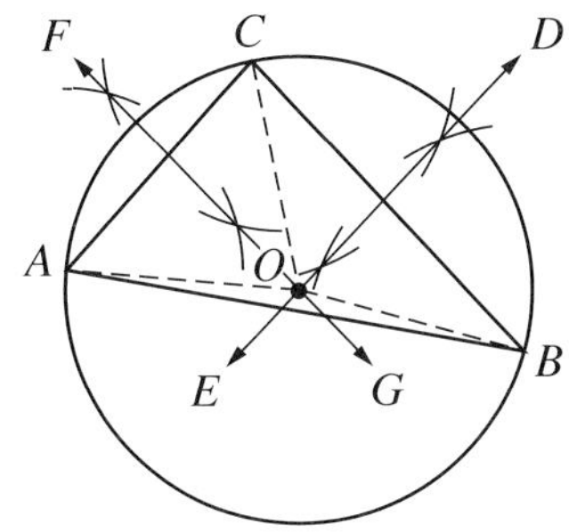

Given: $\triangle ABC$.

Construct: A circle circumscribed about $\triangle ABC$.

How to Proceed:

1. Construct $\overleftrightarrow{DE}$ and $\overleftrightarrow{FG}$, the perpendicular bisectors of $\overline{BC}$ and $\overline{AC}$.
2. Using as the center point O, the point of intersection of the perpendicular bisectors, construct a circle with radius of length OA.

Conclusion: Circle O is the required circle.

Plan for Proof: Draw $\overline{OA}$, $\overline{OC}$, and $\overline{OB}$. Since $\overleftrightarrow{FG}$ is the perpendicular bisector of $\overline{AC}$, $OA = OC$ because any point on the perpendicular bisector of a line segment is equidistant from the endpoints of the segment. Similarly, since $\overleftrightarrow{DE}$ is the perpendicular bisector of $\overline{BC}$, $OC = OB$. Hence, $OA = OB = OC$. Therefore, a circle with O as its center and a radius whose length is OA passes through A, B, and C; that is, circle O circumscribes $\triangle ABC$.

EXERCISES WITH OPEN-RESPONSE PROBLEMS

In 1–6, draw each given figure and do the required constructions. (Where needed, draw separate figures for parts **a**, **b**, etc., and enlarge the given figures for your convenience.)

1. *Given:* Obtuse triangle ABC.

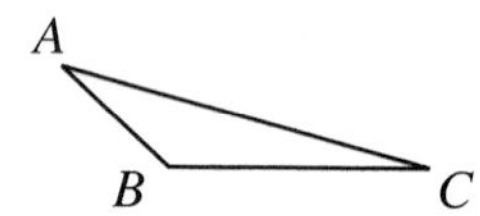

Constructions:

a. Construct the median to $\overline{AC}$.
b. Construct the bisector of $\angle A$.
c. Construct the altitude to $\overline{AC}$.
d. Construct the altitude to $\overline{BC}$.
e. Construct the perpendicular bisector of $\overline{BC}$.

2. *Given:* Parallel lines ℓ and m.

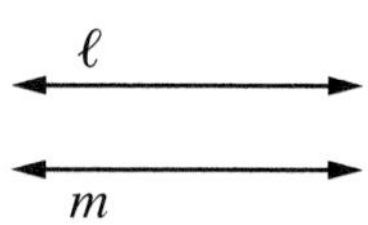

Constructions:

a. Construct a line that is parallel to ℓ and to m, and is above ℓ.
b. Construct a line that is parallel to ℓ and to m, and is below m.
c. Construct a line that is parallel to ℓ and to m, and is anywhere between ℓ and m.
d. Construct a line that is parallel to ℓ and to m, and is exactly midway between ℓ and m.

3. *Given:* Line segments a and b and $\angle A$.

Constructions:

a. Construct a rectangle with base a and altitude b.
b. Construct a square with side a.
c. Construct a parallelogram with consecutive sides a and b and an angle congruent to $\angle A$.
d. Construct a rhombus with side a and an angle congruent to $\angle A$.

4. *Given:* $\triangle ABC$ and $\overline{PQ}$.

Construct: $\triangle PQR$ so that $\frac{AB}{PQ} = \frac{BC}{QR} = \frac{CA}{RP}$.

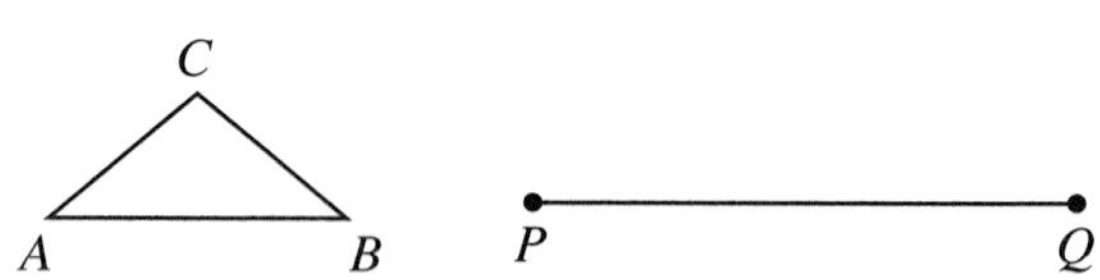

5. *Given:* $\overline{AB}$.

Constructions:

a. Divide $\overline{AB}$ into four congruent parts, using Construction 13.

b. Divide $\overline{AB}$ into four congruent parts, using the construction of a perpendicular bisector.

c. Explain when the method used in part **a** *must* be used to divide a given line segment into any number of congruent parts and when the method involving the perpendicular bisector can be used.

6. *Given:* Right triangle ABC.

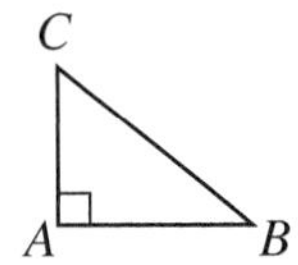

Construct:

a. Circumscribe a circle about $\triangle ABC$.

b. Describe the position of hypotenuse $\overline{BC}$ with respect to the circumscribed circle.

7. Draw $\triangle ABC$ and point A'. Construct $\triangle A'B'C'$, the image of $\triangle ABC$ under a translation.

8. Draw $\triangle ABC$ and point P.

a. Construct $\triangle A'B'C'$, the image of $\triangle ABC$ under a dilation with center at P and constant of dilation 2.

b. Construct $\triangle A''B''C''$, the image of $\triangle ABC$ under a dilation with center at P and constant of dilation $\frac{1}{2}$.

CHAPTER SUMMARY

To ***construct*** a geometric figure means that a specific design is accurately made by using only two instruments:

1. A pair of ***compasses***, which is a device used to locate points at a given distance from a fixed point and to describe circles.
2. A ***straightedge***, which is an unmarked ruler used to draw straight lines by connecting two given points.

Basic geometric constructions include (1) copying a segment; (2) copying an angle; (3) bisecting a segment; (4) bisecting an angle; and (5) constructing a line perpendicular to a given line through a point on the given line, as well as (6) through a point outside the given line.

These basic constructions, combined with definitions and previously proved theorems, allow us to construct more elaborate geometric figures and discover many more geometric relations.

VOCABULARY

Construct (A pair of) compasses Straightedge

REVIEW EXERCISES WITH OPEN-RESPONSE PROBLEMS

In 1–6, draw each given figure and do the required constructions. (Where needed, enlarge the given figures for convenience.)

1. *Given:* $\overline{AB}$.

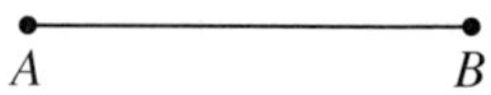

Constructions:

a. Construct a 30°–60° right triangle with hypotenuse $\overline{AB}$.
b. Construct an isosceles right triangle with hypotenuse $\overline{AB}$.
c. Construct an isosceles right triangle with leg $\overline{AB}$.

2. *Given:* $\overline{AB}$.

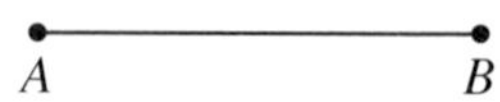

Constructions:

a. Construct $\overline{A'B'}$ so that $AB:A'B' = 1:4$.
b. Construct $\overline{A'B'}$ so that $AB:A'B' = 2:5$.
c. Construct $\overline{A'B'}$ so that $AB:A'B' = 3:4$.
d. Construct P on $\overline{AB}$ so that $AP:PB = 1:3$.
e. Construct P on $\overline{AB}$ so that $AP:PB = 1:4$.
f. Construct P on $\overline{AB}$ so that $AP:PB = 2:3$.

3. *Given:* $\overline{PR}$ and $\overline{PS}$.

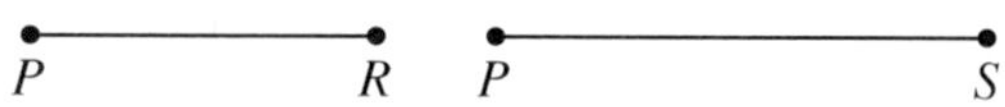

Constructions:

a. Construct a right triangle with legs $\overline{PR}$ and $\overline{PS}$.
b. Construct a right triangle with leg $\overline{PR}$ and hypotenuse $\overline{PS}$.

4. *Given:* $\overline{AB}$ and $\overline{CD}$.

Constructions:

a. Construct an isosceles triangle with base $\overline{AB}$ and with the sum of the measures of the two legs equal to CD.
b. Construct an isosceles triangle with the measure of the altitude equal to AB and the sum of the measures of the two legs equal to CD.
c. Construct an equilateral triangle with the perimeter equal to CD.
d. Construct an isosceles triangle with the perimeter equal to CD and the measure of the base equal to one-half the measure of a leg.

5. *Given:* $\overline{AB}$ and $\overline{AC}$.

Constructions:

a. Construct a parallelogram with adjacent sides $\overline{AB}$ and $\overline{AC}$ and $m\angle A = 60$.
b. Construct a parallelogram with adjacent sides $\overline{AB}$ and $\overline{AC}$ and $m\angle A = 30$.
c. Construct a parallelogram with adjacent sides $\overline{AB}$ and $\overline{AC}$ and $m\angle A = 120$.

6. *Given:* $\overline{PQ}$ and $\angle P$.

Constructions:

a. Construct a rhombus with side $\overline{PQ}$ and one angle congruent to $\angle P$.

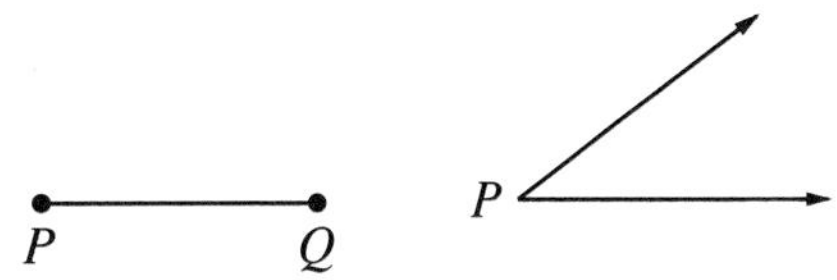

b. Construct a rhombus with side $\overline{PQ}$ and one angle equal in measure to $\frac{1}{2}(m\angle P)$.

CUMULATIVE REVIEW WITH OPEN-RESPONSE PROBLEMS

1. The coordinates of A are $(-1, 1)$, the coordinates of B are $(1, 0)$, and the coordinates of C are $(1, 5)$.
a. What is the slope of $\overleftrightarrow{AB}$?
b. What is the slope of $\overleftrightarrow{AC}$?
c. What is the slope of $\overleftrightarrow{BC}$?
d. Is $\triangle ABC$ a right triangle? Explain your answer.

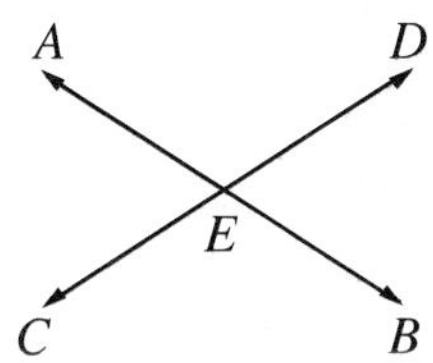

2. In the diagram, $\overleftrightarrow{AB}$ intersects $\overleftrightarrow{CD}$ at E. If $m\angle AEC = 7x - 35$ and $m\angle BED = 4x + 7$, find $m\angle AEC$ and $m\angle BEC$.

3. Triangle DEF is congruent to triangle RST. If DF is represented by $4x - 5$, RS is represented by $5x + 2$, and RT is represented by $2x + 9$, find each of the following:

a. x **b.** DF **c.** RT **d.** DE

4. If $p \vee q$ is false, what is the truth value of each of the following?

a. p **b.** q **c.** $\sim p \wedge \sim q$ **d.** $p \vee \sim q$ **e.** $\sim p \rightarrow q$

5. List the following in order, starting with the smallest: 9^{150}, 5^{200}, 20^{100}.

6. *Given:* If I am hungry and I am not busy then I stop for lunch.
I am hungry.
I do not stop for lunch.

Let H represent "I am hungry."
Let B represent "I am busy."
Let L represent "I stop for lunch."

a. Write each of the given statements in terms of H, B, and L.
b. What must the truth values of H, B, and L be for the first statement to be false?

7. The map and graph describe a car trip from Albany to Skaneatales along Route 90, city streets in Syracuse, and Route 20.

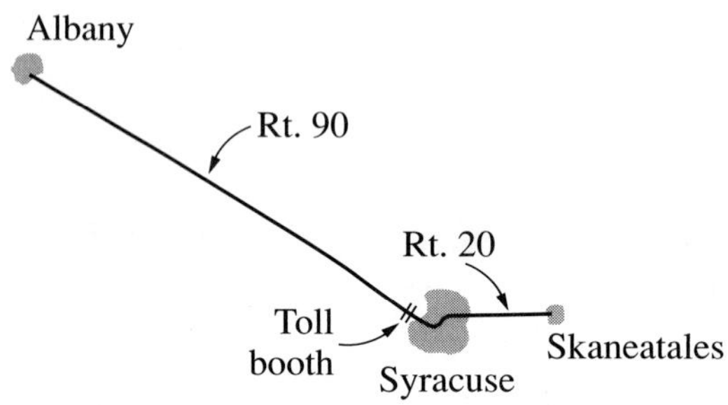

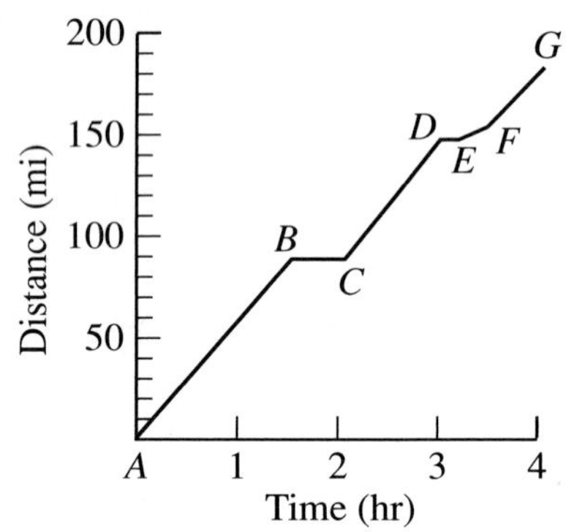

a. Describe the trip, making use of both the map and the graph to explain the progress of the journey.
b. Draw a graph that shows the speed at which the car traveled during the trip. Let the horizontal axis represent time in hours and the vertical axis represent speed in miles per hour.

8. Solve the following system of equations **a.** graphically **b.** algebraically.

$$y = x^2 - 4x - 1$$
$$3x + y = 1$$

9. The length of a rectangle is 12 centimeters longer than the width. The area of the rectangle is 100 square centimeters.
a. Find the width of the rectangle in simplest radical form.
b. Express the width of the rectangle to the *nearest hundredth* of a centimeter.
c. Find the length of the rectangle to the *nearest hundredth* of a centimeter.

10. Use compasses and a straightedge, or use a computer program, to do this exercise.

a. Draw any triangle and label the vertices A, B, and C.
b. Construct k, the $\perp$ bisector of $\overline{AB}$.
c. Construct ℓ, the $\perp$ bisector of $\overline{BC}$.
d. Let point F be the intersection of lines k and ℓ. Explain why F is equidistant from A and B, and also equidistant from B and C.
e. Construct m, the $\perp$ bisector of $\overline{AC}$. Explain why F must be a point on line m.
f. Using $FA = FB = FC$, construct a circle whose center is F in such a way that the circle contains vertices A, B, and C of $\triangle ABC$.

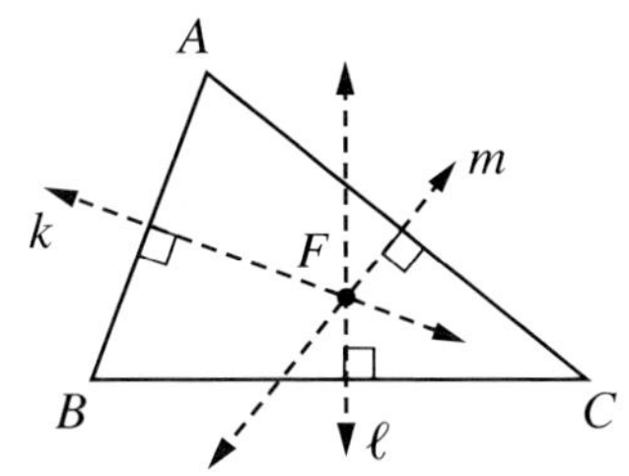

11. The vertices of quadrilateral $ABCD$ are $A(2, 0)$, $B(4, 6)$, $C(-8, 0)$ and $D(-2, -2)$.

a. Show that $ABCD$ is an isosceles trapezoid.
b. Show that $E(1, -3)$ is on $\overleftrightarrow{AB}$ and on $\overleftrightarrow{CD}$.
c. Show that triangle BCE is a right triangle.

Exploration

Draw any triangle. Construct lines that separate the triangle into triangles, each of which has a line of symmetry. (*Hint:* Every isosceles triangle has a line of symmetry.)

Chapter Opening Photo Credits:

Chapter 1	Westlight: *Crowd at sport game*
Chapter 2	Comstock Inc: *Teacher/Student classroom*
Chapter 3	Image Bank Inc: *Basketball Players*
Chapter 4	The Stock Market Photo Agency: *Lightning over city*
Chapter 5	Comstock Inc: *Stock Market Paper*
Chapter 6	The Stock Market Photo Agency: *Retail Store*
Chapter 7	The Stock Market Photo Agency: *Relay Race, passing the Baton*
Chapter 8	*Quilt that uses the Log Cabin pattern (no credit)*
Chapter 9	Comstock Inc: *Man in Office*
Chapter 10	FPG: *Carpenter building a deck*
Chapter 11	Image Bank: *Cutting Cloth using a Pattern*
Chapter 12	The Stock Market Photo Agency: *Parthenon, Athens, Greece*
Chapter 13	The Stock Market Photo Agency: *Microscope*
Chapter 14	The Stock Market Photo Agency: *Alan Schein*
Chapter 15	The Stock Market Photo Agency: *Convention*
Chapter 16	Comstock Inc: *Day School*
Chapter 17	Photo Researchers: *Wheel Chair*
Chapter 18	Image Bank: *Orchard*
Chapter 19	The Stock Market Photo Agency: *Ukranian Easter Eggs*
Chapter 20	Image Bank: *Earth viewed from the Moon*
Chapter 21	Image Bank: *Baseball Game*
Chapter 22	The Stock Market Photo Agency: *Surveying Equipment*
Chapter 23	Image Bank: *Garden Maze Pattern*
Chapter 24	The Stock Market Photo Agency: *Compasses and Ruler*

Index

N

O

P

Q